Synthesis of Best-Seller

Synthesis of Best-Seller Drugs

Ruben Vardanyan
Victor Hruby
Department of Chemistry and Biochemistry
University of Arizona

ELSEVIER

AMSTERDAM • BOSTON • HEIDELBERG • LONDON
NEW YORK • OXFORD • PARIS • SAN DIEGO
SAN FRANCISCO • SINGAPORE • SYDNEY • TOKYO

Academic Press is an imprint of Elsevier

AP

Academic Press is an imprint of Elsevier
125 London Wall, London EC2Y 5AS, UK
525 B Street, Suite 1800, San Diego, CA 92101-4495, USA
50 Hampshire Street, 5th Floor, Cambridge, MA 02139, USA
The Boulevard, Langford Lane, Kidlington, Oxford OX5 1GB, UK

ISBN: 978-0-12-411492-0

British Library Cataloguing in Publication Data

A catalogue record for this book is available from the British Library

Library of Congress Cataloging-in-Publication Data

A catalog record for this book is available from the Library of Congress

For information on all Academic Press publications
visit our website at http://store.elsevier.com/

Dedication

Dedicated to my daughters, Anna, Marina, and Irena, established personalities and professionals whom I infinitely love and am very proud of.

Ruben Vardanyan

Dedicated to my three sons, Timothy, Stephen, and Patrick, for encouraging my passion for science.

Victor Hruby

Contents

Preface

This book represents our effort to give a panoramic view of the most popular medications on the pharmaceutical market, accenting the reader's attention on the uses of these medications and schemes of synthesis of the best-selling pharmaceutical drugs in 2010s.

Although there are numerous books on the medicinal chemistry and drug design, this book is much different than the other books.

Here we present the synthesis of various groups of drugs, classifying them in accordance with the order and manner in which they are traditionally presented in gold-standard pharmacological curriculums, focusing on recently developed synthetic methods for the preparation of the most important best-selling drugs.

This book covers the basic, essential pharmacological classes of drugs and is separated into 38 chapters. It includes 208 schemes of synthesis, 463 figures, and 4597 references.

The basic philosophy of this book is to help understand where we are and what we are doing in the area of drug creation.

Recent years have seen major advances in the understanding of how to improve the chances of discovering all categories of drugs: essential, new-first-in-class, best-in-class, and yet-to-be-developed.

The approaches and technologies split into empirical and in silico efforts at many different levels, and modern paradigms of drug design include sophisticated titles such as computational chemistry and molecular modeling, combinatorial synthesis and high-throughput screening, target-based molecular modeling strategies, structure-based drug-design approaches, fragment-based drug design, diversity-oriented synthesis, and others.

But is there any design approach that one can claim to be perfect?

Most of the drugs now in use were discovered by chance through painstaking trial and error rather than by design.

Formal approaches cannot replace a scientist's intuition that is based on special knowledge and experience. The way to enhance the productivity in Big Pharma is not only by improving and "perfecting" techniques such as combinatorial chemistry or high-throughput screening, cutting-edge tools, and software programs for design, but also by giving scientists the freedom to use their imagination to explore the whole drug universe rather than focusing only on the details of some narrow area of a single class of drugs.

The tools themselves are no substitute for first-rate scientific minds—a fool with a tool is still a fool.

The journey of a new drug from the scientist's idea to the pharmacy shelf is a filled with victories and defeats, detours and delays. We fully agree with the words of Sir James Whyte Black, winner of the 1988 Nobel Prize in Physiology or Medicine: "The most fruitful basis for the discovery of a new drug is to start with an old drug."

Finally, we think that this book could become a resource for both newcomers to the field of medicinal chemistry, organic chemistry, and pharmacology, and for experienced researchers and scientists in academia and industry wanting to know about the existing remedies for different diseases and about synthetic routes implemented for the synthesis of best-selling pharmaceutical drugs in 2010s.

We earnestly hope that the time we spent writing this book has resulted in the kind of information that will interest those who work or plan to begin work in the area of medicinal drug design and synthesis.

Ruben Vardanyan
Victor Hruby

Chapter 1

General Anesthetics

The use of inhaled ether for surgical anesthesia was first demonstrated in 1846. Since then, the development of new safe anesthetics has contributed greatly to the advancement of surgery and other invasive procedures.

General anesthesia is the state of controlled, reversible unconsciousness and loss of protective reflexes, controlled level of nervous system suppression to allow adequate surgical access, obstetric, and diagnostic procedures to be completed painlessly. The patient receives medications for amnesia, analgesia, muscle paralysis, and sedation. Anesthesia includes the following components: analgesia (absence of pain), amnesia (absence of memory), suppression of reflexes such as bradycardia, laryngospasm, and loss of skeletal muscle tonicity.

In medical practice, general anesthesia is a complex procedure involving: preanesthetic assessment, administration of general anaesthetic drugs, cardiorespiratory monitoring, analgesia, airway management, and fluid management.

In a typical clinical procedure, the patient is premedicated with a sedative intended to relieve anxiety and facilitate the induction of anesthesia itself. For this purpose it is accepted to use tranquilizers, such as diazepam, lorazepam, or midazolam, or a central nervous system depressant-barbiturate such as thiopental, methohexital, or the hypnotic agent propofol. Sedation is followed by intravenous injection of an opioid analgesic such as morphine, fentanyl, alfentanil, or ketamine, which has a wide range of effects in humans, including analgesia and anesthesia. In addition, a nondepolarizing curare-like derivative like vecuronium or d-tubocurarine, or a depolarizing drug such as succinylcholine, is administered to induce muscle paralysis. After connection to artificial respiration, general anesthesia is maintained by a mixture of oxygen and nitrous oxide, often in combination with a volatile agent such as halothane, enflurane, or isoflurane. At the conclusion of the surgery, muscle relaxation is reversed by neostigmine or other anticholinesterase, and normal breathing is restored.

The ideal general anesthesia must include the aforementioned characteristics, as well as have a wide therapeutic index and no significant side effects.

The underlying neurocellular mechanisms by which the state of general anesthesia is achieved are only just beginning to be understood.

Components of general anesthesia formally are divided into two groups: noninhalation, (barbiturates, ketamine, and etomidate), and inhalation (halothane, enflurane, isoflurane, desflurane and nitrous oxide). Since the publication of our first book [1] in 2006 where the synthesis of all above mentioned

Synthesis of Best-Seller Drugs. http://dx.doi.org/10.1016/B978-0-12-411492-0.00001-8

drugs are described, no novel entities that address fundamentally new general anesthesia approaches have entered the clinic. That is the reason we limit discussion to the structural formulas of anesthetic agents in current clinical use.

1.1 NONINHALATION COMPONENTS FOR GENERAL ANESTHESIA

Tranquilizers

Benzodiazepines—diazepam (**1.1.1**), lorazepam (**1.1.2**), and midazolam (**1.1.3**)—which have anxiolytic, sedative, and anticonvulsant effects, and cause amnesia and muscle relaxation, are frequently used to relieve patient's anxiety during anesthesia. (Fig. 1.1.)

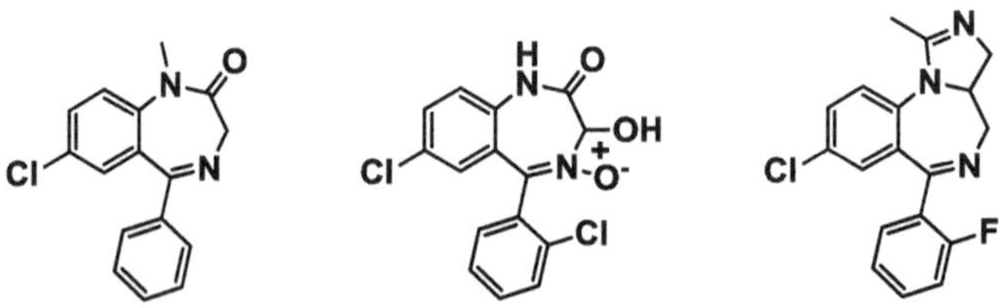

Diazepam 1.1.1 **Lorazepam 1.1.2** **Midazolam 1.1.3**

FIG. 1.1 **Benzodiazepines used during anesthesia.**

Central Nervous System Depressants (Barbiturates and Hypnotic Agents)

Two barbiturates primarily used in surgical practice are thiopental (**1.1.4**) and methohexital (**1.1.5**). Barbiturates are hypnotics, and at therapeutic doses have a very weak analgesic and muscle relaxant effect. Intravenous injection of a therapeutic dose of propofol (**1.1.6**) produces hypnosis rapidly with minimal excitation, usually within 40 seconds from the start of an injection. Etomidate (**1.1.7**) classified as a sedative hypnotic drug because of the quick loss of consciousness upon intravenous administration. It has an anticonvulsant activity and does not display analgesic characteristics. Duration of its action depends on the administered dose. (Fig. 1.2.)

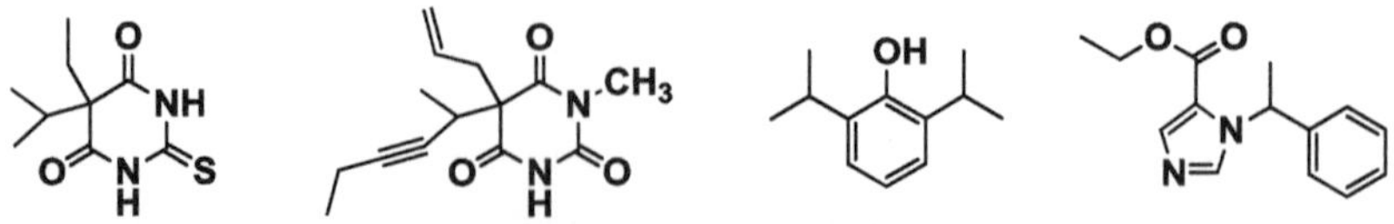

Thiopental 1.1.4 **Methohexital 1.1.5** **Propofol 1.1.6** **Etomidate 1.1.7**

FIG. 1.2 **Barbiturates and hypnotic agents used during anesthesia.**

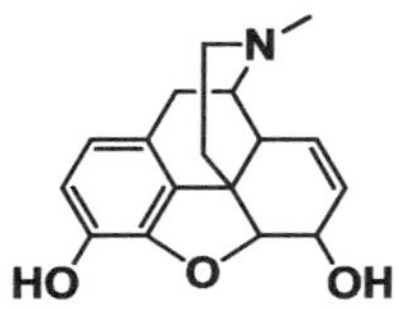

Morphine
1.1.8

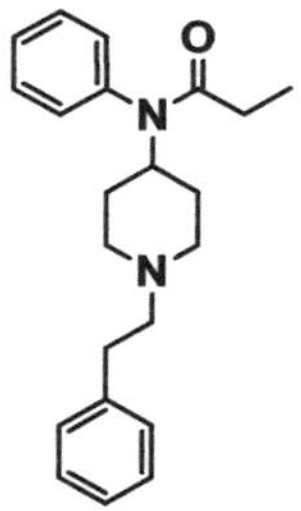

Fentanyl
1.1.9

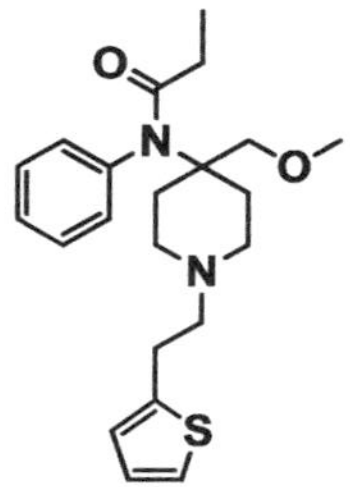

Sufentanil
1.1.10

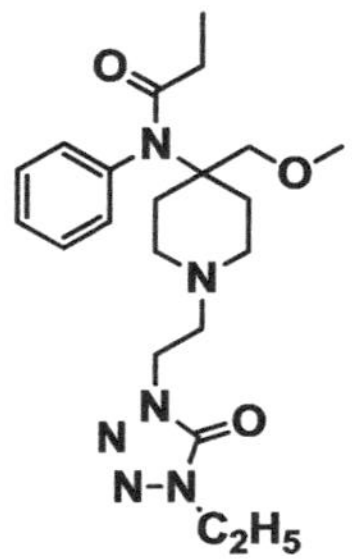

Alfentanil
1.1.11

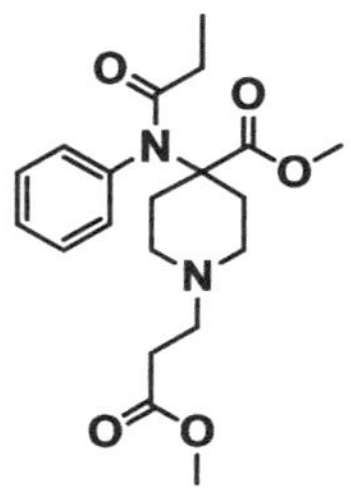

Remifentanil
1.1.12

FIG. 1.3 Opioid analgesics used during anesthesia.

Opioid Analgesics

Opioid analgesics, in particular morphine (**1.1.8**), fentanyl (**1.1.9**), alfentanil (**1.1.10**), and sufentanil (**1.1.11**), are widely used in the practice of anesthesiology as adjuncts. Recently, remifentanil (**1.1.12**) became popular for the maintenance of anesthesia due to its short-acting nature. (Fig. 1.3.)

Ketamine

Ketamine (**1.1.13**) is a specific drug for noninhalation narcosis which is used in brief surgical procedures.

It causes a condition known as dissociative anesthesia, which ensures amnesia and analgesia, and preserves normal respiration and muscle tonicity in the patient. Ketamine is practically void of muscle relaxant capabilities. (Fig. 1.4.)

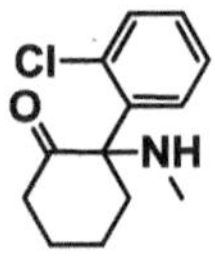

Ketamine 1.1.13

FIG. 1.4 **Structure of ketamine.**

Muscle Relaxants

Nondepolarizing neuromuscular-blocking agents as d-tubocurarine (**1.1.14**) and vecuronium (**1.1.15**), or a depolarizing drug such as succinylcholine (**1.1.16**), are used in surgery as muscle relaxants. Currently, atracurium (**1.1.17**) and rocuronium (**1.1.18**) are considered safer alternatives. (Fig. 1.5.)

Reversible Acetylcholinesterase Inhibitors

Neostigmine (**1.1.19**) is a parasympathomimetic that acts as a reversible acetylcholinesterase inhibitor. It is used at the end of an operation to reverse the effects of nondepolarizing muscle relaxants. (Fig. 1.6.)

1.2 INHALATION ANESTHETICS

The fluorinated inhalation anaesthetics used in current practice are halothane (**1.2.1**), sevoflurane (**1.2.2**), desflurane (**1.2.3**), isoflurane (**1.2.4**), nitrous oxide (**1.2.5**) (Fig. 1.7.), and xenon, which produce dose-dependent central nervous system, cardiovascular, and respiratory depressant effects. Despite proof that inhaled anesthetics act on multiple molecular targets, it is hypothesized that their mechanism of action could be better explained by physical phenomena

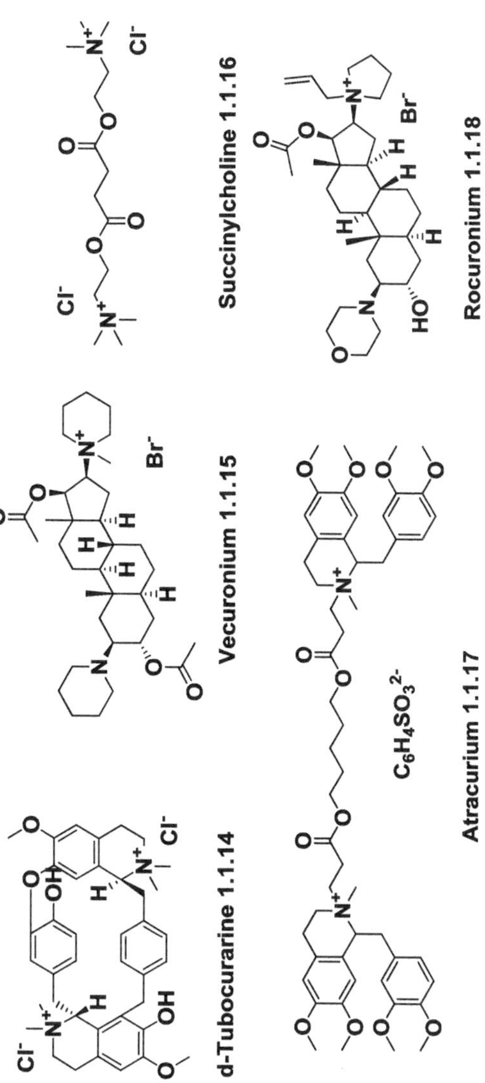

FIG. 1.5 Muscle relaxants used during anesthesia.

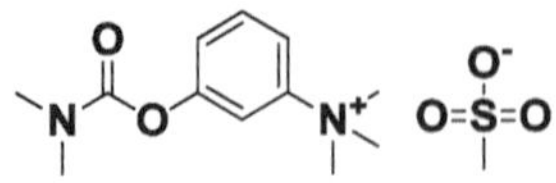

Neostigmine 1.1.19

FIG. 1.6 Structure of neostigmine.

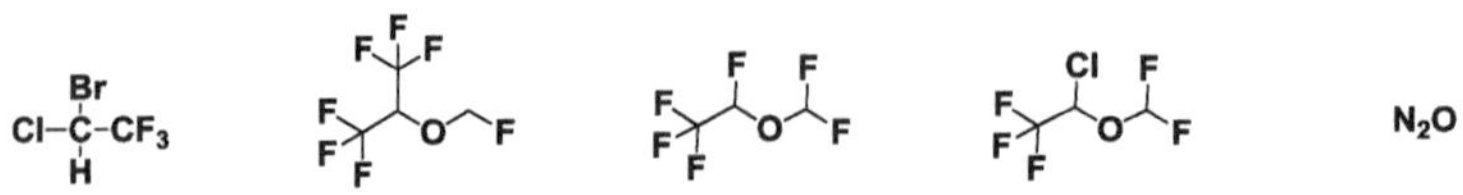

Halothane 1.2.1 Sevoflurane 1.2.2 Desflurane 1.2.3 Isoflurane 1.2.4 Nitrous oxide 1.2.5

FIG. 1.7 Inhalation anaesthetics used during anesthesia.

than by chemical interaction. "Anesthetics have been used for 160 years, and how they work is one of the great mysteries of neuroscience," says the famous contemporary anesthesiologists James Sonner about current inhalation anesthetics. The inhalation anesthetics are implicated in a variety of adverse viscerotoxic reactions. In general, they have been proven to produce very few nonpredicted toxicities. Hepatitis caused by halothane now seems to be the only major problem in this regard with these drugs in current practice.

"Older" inhalation anesthetics, such as diethyl ether, chloroform, chloroethane, trichloroethane, cyclopropane, methoxyflurane, and some others, are not used in medicine anymore.

General anesthetic pharmacology is unique because so many types of molecules possess this activity, including, haloalkanes, ethers, barbiturates, quaternary ammonium salts, steroids, simple gases, and other organic compounds. General anesthetics produce a widespread neurodepression in the central nervous system by enhancing inhibitory neurotransmission and reducing excitatory neurotransmission. However, the action mechanisms of general anesthetics are not completely understood. The last two decades have seen tremendous advances in the understanding of the mechanisms underlying general anesthesia. Theories about the mechanisms of anesthesia are presented in *Foye's Principles of Medicinal Chemistry* [2] and several reviews [3-11].

Of the described drugs, none is included in the list of Top 200 Drugs by sales for the 2010s.

REFERENCES

1. Vardanyan, R. S.; Hruby, V. J. *Synthesis of Essential Drugs;* Elsevier, 2006.
2. Maher, T. M. General anesthetics. In *Foye's Principles of Medicinal Chemistry,* 6th ed.; Lemke, T. L., Williams, D. A., Eds. Lippincott Williams & Wilkins, 2007; pp 490–503.
3. Uwe, R.; Bernd, A. Molecular and neuronal substrates for general anesthetics. *Nat. Rev. Neurosci.* **2004,** *5* (9), 709–720.
4. Yong, S. Molecular mechanisms of general anesthesia. *Korean J. Anesthesiol.* **2010,** *59* (1), 3–8.

5. Alkire, M. T.; Hudetz, A. G.; Tononi, G. Consciousness and anesthesia. *Science* **2008,** *322* (5903), 876–880.
6. Franks, N. P. General anaesthesia: from molecular targets to neuronal pathways of sleep and arousal. *Nat. Rev. Neurosci.* **2008,** *9,* 370–386.
7. Forman, S. A. Molecular approaches to improving general anesthetics. *Anesthesiol. Clin.* **2010,** *28* (4), 761–771.
8. Lambert, D. G. Mechanisms of action of general anaesthetic drugs. *Anaesth. Intens. Care Med.* **2011,** *12* (4), 141–143.
9. Diao, S.; Ni, J.; Shi, X.; Liu, P.; Xia, W. Mechanisms of action of general anesthetics. *Front. Biosci. (Landmark Ed.)* **2014,** *19* (5), 747–757.
10. Vlisides, P.; Xie, Z. Neurotoxicity of general anesthetics: an update. *Curr. Pharm. Des.* **2012,** *18* (38), 6232–6240.
11. Forman, S. A. Molecular approaches to improving general anesthetics. *Anesthesiol. Clin.* **2010,** *28* (4), 761–771.

Chapter 2

Local Anesthetics

Local anesthetics are drugs that causes reversible local anesthesia and a loss of nociception. When they are used on specific nerve pathways, effects such as analgesia (loss of pain sensation) and paralysis (loss of muscle power) can be achieved.

Local anesthetics are medications used for the purpose of temporary and reversible elimination of painful feelings in specific areas of the body by blocking nerve transmission, to cause local loss of feeling and preventing muscle activity in the process without affecting consciousness.

Local anesthetics are used in many different ways and in various situations requiring local pain relief, beginning with simple procedures such as removing a small piece of the outer layer of damaged skin, to complicated operations such as organ transplants. Local anesthetics are widely used in clinical use for pain relief in situations ranging from dental and ophthalmological procedures to gynecological interventions.

Local anesthetics are used for pain relief, soreness, itching, and irritation associated with disturbance of the integrity of the skin and mucous membranes (cuts, bites, wounds, rashes, allergic conditions, fungal infections, skin sores, and cracking).

Significant improvements in anesthesia technique permit surgeons to do many procedures that previously required sedation, and to perform surgical intervention without sedation, just under pure local anesthesia, thus increasing patient safety and convenience.

Local anesthetics can be classified according to the principal means of their clinical use, as well as by how they fit into specific chemical classes of compounds.

From the medical point of view, local anesthetics can be differentiated in the following manner:

- Topical anesthesia
  Local use of drugs of this kind on the mucous membranes of the nose, mouth, larynx, tracheobronchial tree, eyes, urinary tract, and gastrointestinal tract causes superficial anesthesia.
- Infiltration anesthesia
  The direct introduction of local anesthetic into the skin or deeper tissue for surgical intervention is called infiltration anesthesia.
- Block or regional anesthesia
  The introduction of local anesthetic into an individual nerve or group of nerves during minor surgical interventions with the purpose of blocking feeling and

Synthesis of Best-Seller Drugs. http://dx.doi.org/10.1016/B978-0-12-411492-0.00002-X

motor action is frequently called block or regional anesthesia. This method is often used during surgical intervention in the shoulder, arm, neck, or leg.

- Spinal anesthesia
  Spinal anesthesia is the introduction of local anesthetics directly into the spinal fluid, which causes a sympathetic blockage or loss of feeling, as well as muscle relaxation resulting from the interaction of anesthetic with every spinal nerve tract. This method is used during major surgical interventions.
- Epidural anesthesia
  This term is understood to be an introduction of local anesthetic into the spinal cord membrane of the intervertebral space. It is used during obstetrical and gynecological interventions that do not require a fast development of anesthesia.

The first synthetic local anesthetic drug, procaine, appeared in clinical practice in 1905. It was followed by thousands of compounds with analogous properties. However, only 10 to 12 of are still used in practice and anesthesia is still waiting for an ideal agent. In 1947, lidocaine was introduced followed by bupivacaine in 1963.

From the chemical point of view, general anesthetics can be classified as esters of *p*-aminobenzoic acid and dialkylaminoalkanols, or as anilides of N,N-dialkyl substituted α-amino acids. (Fig. 2.1.)

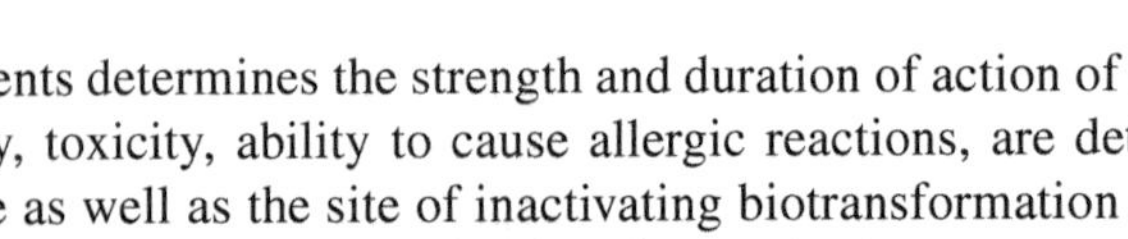

FIG. 2.1 Generalized chemical structures of general anesthetics.

Substituents determines the strength and duration of action of these drugs. The stability, toxicity, ability to cause allergic reactions, are determined by the structure as well as the site of inactivating biotransformation of the drug: either by enzymatic hydrolysis in the plasma (aminoester anesthetics), or decomposition in the liver (aminoamide anesthetics).

It is interesting that a number of antihistamine, anticholinergic, and adrenergic drugs with related chemical structures, also exhibit local anesthetic properties. It is believed that they interact with internal axoplasmic membranes, reducing the ion flow, particularly, the flow of sodium ions inside nerve cells. Because the binding process of anesthetics to ion channels is reversible, when the implemented drug is metabolized, the nerve cell function is completely restored. The mechanism of benzocaine action differs slightly from that mentioned above. It presumably acts by diffusing across the phospholipid membrane and then stretching it out. This deforms the sodium channels, which in turn, and in a unique manner, lowers sodium conduction.

An analogous mechanism of stretching (changing the fluidity) of the membrane was also suggested for explanation of the action mechanism of other local anesthetics.

Many excellent reviews on therapeutic considerations for using local anesthetic drugs their actions, mechanisms, pharmacokinetics, clinical indications, and side effects published in literature [1-17].

The synthesis of practically all of below mentioned drugs could be summarized generally as it is shown on (Fig. 2.2.), and is described in details in our previous book [18].

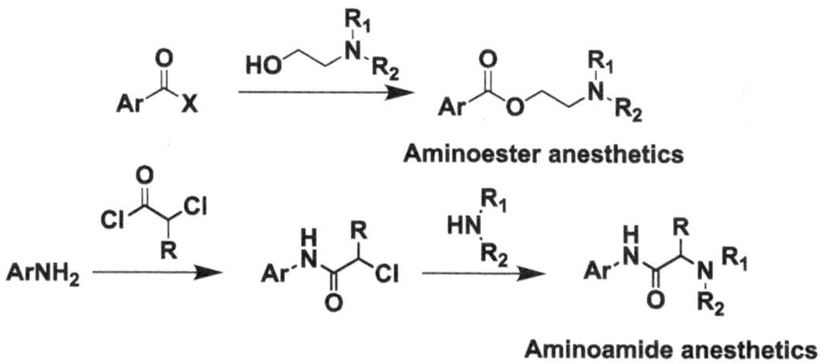

FIG. 2.2 Generalized approach for the synthesis of general anesthetics.

2.1 AMINOESTER ANESTHETICS

Benzocaine (**2.1.1**), procaine (novocaine) (**2.1.2**), chloroprocaine (**2.1.3**), tetracaine (**2.1.4**), dimethocaine (**2.1.5**), proparacaine (**2.1.6**), and cyclomethycaine (**2.1.7**) are representative of aminoester anesthetics. (Fig. 2.3.)

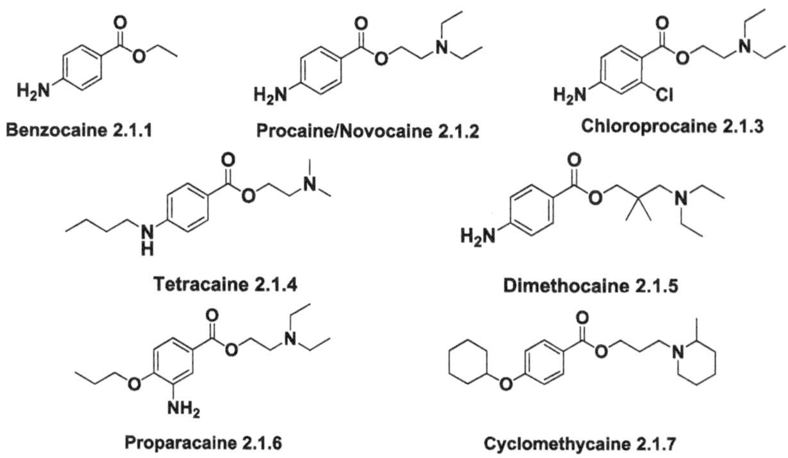

FIG. 2.3 Aminoester anesthetics.

2.2 AMINOAMIDE ANESTHETICS

Lidocaine (**2.2.1**), prilocaine (**2.2.2**), etidocaine (**2.2.3**), mepivacaine (**2.2.4**), bupivacaine (**2.2.5**), levobupivacaine (**2.2.6**), ropivacaine (**2.2.7**), and articaine (**2.2.8**) are representatives of aminoamide anesthetics. Later, an S-(-)-enantiomer of bupivacaine, named levobupivacaine (**2.2.6**), and the closely related ropivacaine (**2.2.7**) were introduced clinically. The latest development in local anesthetics is articaine (**2.2.8**) [19-23]. (Fig. 2.4.)

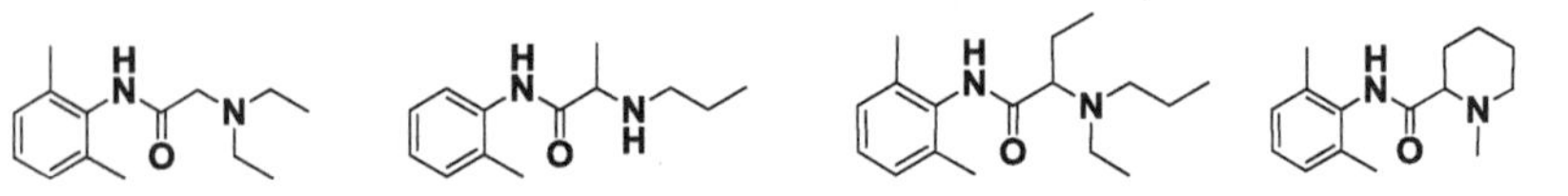

Lidocaine 2.2.1 **Prilocaine 2.2.2** **Etidocaine 2.2.3** **Mepivacaine 2.2.4**

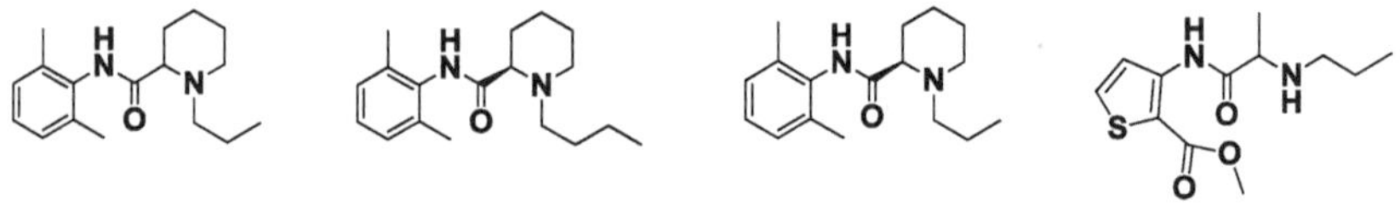

Bupivacaine 2.2.5 **Levobupivacaine 2.2.6** **Ropivacaine 2.2.7** **Articaine 2.2.8**

FIG. 2.4 Aminoamide anesthetics.

Of the described drugs, none is included in the list of Top 200 Drugs by sales for the 2010s.

REFERENCES

1. Lu, M. C. Inhibitors of nerve conduction: local anesthetics. In *Foye's Principles of Medicinal Chemistry*, 6th ed.; Lemke, T. L., Williams, D. A., Eds.; Lippincott Williams & Wilkins, 2007; pp 462–479.
2. Wiles, M. D.; Nathanson, M. H. Local anaesthetics and adjuvants-future developments. *Anaesthesia* **2010,** *65* (Suppl. 1), 22–37.
3. Yanagidate, F.; Strichartz, G. R. Local anesthetics. In *Handbook of Experimental Pharmacology*; Kauser, K., Zeither, A-M., Eds.; Vol. 177; Springer, 2007; pp 95–127(Analgesia).
4. Howe, J. P.; Fee, J. P. H. Local anaesthetics. In *Pharmacology for Anaesthesiologists;* Fee, J. P. H., Bovill, J. G., Eds.; CRC Press, 2005; pp 79–92.
5. Rippel, R. Local anesthetics. *Pharmaceuticals (Basel)* **2000,** *2,* 539–553.
6. Wang, G. K. Local anesthetics. In *Neural Mechanisms of Anesthesia;* Antognini, J. F., Carstens, E., Douglas, E. R., Eds.; 2003; pp 427–440.
7. Congedo, E.; Sgreccia, M.; De Cosmo, G. New drugs for epidural analgesia. *Curr. Drug Targets* **2009,** *10* (8), 696–706.
8. Ruetsch, Y. A.; Boni, T.; Borgeat, A. From cocaine to ropivacaine: the history of local anesthetic drugs. *Curr. Top. Med. Chem.* **2001,** *1* (3), 175–182.
9. Josee, J. G.; Walczak, J.-S.; Beaulieu, P. Recent advances in the pharmacological management of pain. *Drugs* **2007,** *67* (15), 2121–2133.
10. Josee, J. G.; Walczak, J.-S.; Beaulieu, P. Anesthetic agents for advanced regional anesthesia: a north American perspective. *Drugs* **2005,** *65* (6), 745–759.

11. Scholz, A. Mechanisms of (local) anaesthetics on voltage-gated sodium and other ion channels. *Br. J. Anaesth.* **2002,** *89* (1), 52–61.
12. Whiteside, J. B.; Wildsmith, J. A. W. Developments in local anesthetic drugs. *Br. J. Anaesth.* **2001,** *87* (1), 27–35.
13. Colvin, L. A.; McClure, J. H. Local anesthetics: structure-activity relationships and their role in pain treatment. *Pain Rev.* **1997,** *4* (1), 59–77.
14. Rees, D. C.; Hill, D. R. Drugs in anesthetic practice. *Ann. Rep. Med. Chem.* **1996,** *31,* 41–50.
15. Ruetsch, Y. A.; Boni, T.; Borgeat, A. From cocaine to ropivacaine: the history of local anesthetic drugs. *Curr. Top. Med. Chem.* **2001,** *1* (3), 175–182.
16. Buckenmaier, C. C., III; Bleckner, L. L. Anesthetic agents for advanced regional anesthesia: a North American perspective. *Drugs* **2005,** *65* (6), 745–759.
17. Sabrine, A.; Lyons, G. New local anesthetic analgesics. In *Pain: Current Understanding, Emerging Therapies and Novel Approaches to Drug Discovery;* Bountra, C., Munglani, R., Schmidt, W. K., Eds.; Marcel Dekker, 2003; pp 795–801.
18. Vardanyan, R. S.; Hruby, V. J. *Synthesis of Essential Drugs;* Elsevier, 2006.
19. Yapp, K. E.; Hopcraft, M. S.; Parashos, P. Articaine: a review of the literature. *Br. Dent. J.* **2011,** *210* (7), 323–329.
20. Ruschig, H.; Schorr, M.; Muschaweck, R.; Rippel, R. 3-Aminoacylamino thiophenes, US 3855243 (1974).
21. Kadushkin, A. V.; Trofimkin, Yu. I.; Granik, V. G. Method for the manufacture of methyl 4-methyl-3-[2-(1-propylamino]thiophene-2- carboxylate hydrochloride which does not require vacuum distillation purification, RU 2184730 (2002).
22. Zheng, A.; Zhang, Y.; Hu, Y.; Zhang, J.; Yang, Z.; Li, L. Method for preparation of Articaine hydrochloride, CN 102060840 (2011).
23. Ma, O.; Xuejuan, X.; Li, B.; Sun, P. Method for preparing of articaine hydrochloride CN 102321067 (2012).

Chapter 3

Analgesics

Pain is a symptom of many human diseases and different analgesics are routinely used to relieve the pain. Ninety percent of diseases are associated with pain. Pain is a very important protective phenomenon that accompanies many pathological conditions. However, in fulfilling its function of signaling, it can, upon excessive intensity, aggravate the course of the primary disease, and can in some cases, such as severe trauma, facilitate the development of shock.

Pain that resolves quickly is called *acute*; long-lasting pain is called *chronic*. Pain can be classified as nociceptive, neuropathic, or psychogenic/idiopathic. The mechanical-, thermal-, chemical-, or inflammation-induced pain in a localized area that is transmitted by nociceptors at the site of tissue damage is described as *nociceptive* pain, whereas pain caused by damage or disease affecting any part of the nervous system, even in the absence of visible injury or inflammation, is called *neuropathic* pain. Pain caused by mental, emotional, or behavioral factors is described as *psychogenic* pain. Pain is an extremely complex process involving multiple interrelated central and peripheral pathways. The possible multiple mechanisms for producing analgesia are very complex and at least 10–15 neuromodulators of the pain response have been found in different studies. In general, analgesia can be obtained by decreasing excitation or increasing inhibition in the nervous system.

The problem of alleviating painful sensations is as old as mankind itself. It can probably be said with a fair degree of confidence that the isolation of morphine, the oldest of the known pain-relieving drugs, from opium poppy in the 19th century served as the initiation for the intensive development of the chemistry, pharmacology, and pharmacy.

Traditional pain therapies rely on drugs long-known to have analgesic properties such as nonsteroidal antiinflammatory drugs (NSAIDs), cyclooxygenase (COX) inhibitors, opioid analgesics, and analgesic adjuvants, which include several classes of compounds and which have been introduced to pain therapy from the arsenal of drugs used for treatment of other medical conditions, but which have been found, in some instances, in the clinical setting, to be effective in pain control. They are antidepressants, anticonvulsants, voltage-gated ion-channel blockers as local anesthetics, antiarrhythmics, α_2-adrenergic agonists, *N*-methyl-D-aspartate (NMDA) antagonists, cannabinoids, and some others. Corticosteroids are used in the treatment of pain associated with arthritis, neuropathic pain cancer bone metastases. Pain can be described as mild to moderate and moderate to severe.

Synthesis of Best-Seller Drugs. http://dx.doi.org/10.1016/B978-0-12-411492-0.00003-1

Current methods of classifying pain are diverse [1]. Acute pain now is divided into injury, postoperative, and flare pain; chronic pain is subdivided into nociceptive (osteoarthritis, rheumatoid), neuropathic-central (poststroke, multiple sclerosis, spinal cord injury, migraine, HIV related), neuropathic-peripheral (postherpetic neuralgia, diabetic neuropathy); visceral (internal organ, pancreatitis, inflammatory bowel syndrome); mixed (lower back pain, cancer, fibromyalgia) [2].

3.1 NONSTEROIDAL ANTIINFLAMMATORY DRUGS

COX inhibitors or NSAIDs, are generally effective for mild or moderate pain. COX inhibitors are compounds that block COX enzymes, which are produced in response to inflammation and reduce pain and inflammation from many medical conditions. They inhibit two different enzymes called COX-1 and COX-2 are enzymes responsible for the production of prostaglandins.

COX-1 is responsible for the synthesis of prostaglandin and thromboxane in many types of cells; COX-2 is inducible enzyme and produces prostaglandins for inflammatory response and in some events in the central nervous system. Traditional NSAIDs–nonselective COX inhibitors can produce significant adverse effects on the stomach and gastrointestinal tract, including ulcers and bleeding, can cause liver and kidney problems. For most patients with moderate to severe pain, NSAIDs do not produce complete pain relief. Synthesis of each of the NSAIDs and COX inhibitors is described in the literature, particularly in our previous book [3], and are described below.

p-Aminophenol Derivatives

Acetaminophen (**3.1.1**) (Fig. 3.1.);

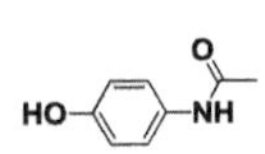

3.1.1. Acetaminophen

FIG. 3.1 Structure of acetaminophen.

Salicylic Acid Derivatives

Aspirin (**3.1.2**), diflunisal (**3.1.3**), salsalate (**3.1.4**), Asacol (**3.1.5**). (Fig. 3.2.)

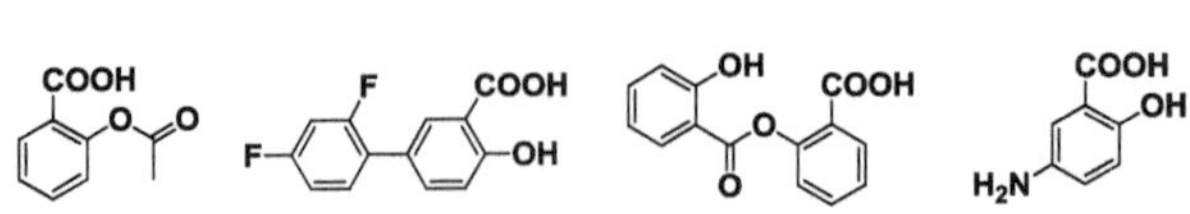

3.1.2 Aspirin 3.1.3 Diflunisal 3.1.4 Salsalate 3.1.5 Asacol

FIG. 3.2 Salicylic acid derivatives.

Asacol–Mesalazine

Asacol (**3.1.8**) is the single NSAID included in the on the list of Top 200 Drugs by sales for the 2010s [4]. Several synthetic schemes have been proposed for its synthesis. One scheme (Scheme 3.1) involves catalytic hydrogenation of 3-nitrobenzoic acid (**3.1.6**) in acidic solution. The process is believed to involve the formation of an intermediate product, 3-hydroxylaminobenzoic acid (**3.1.7**), which is rearranged to form expected 5-amino-2-hydroxybenzoic acid mesalazine (**3.1.8**) [5]. Nitration of salicylic acid (**3.1.9**) and consequent hydrogenation of the nitro group of (**3.1.10**) with H_2 in situ, obtained from the Zn/HCl system, also gives mesalazine (**3.1.8**) [6]. Another method involves diazo coupling reaction with salicylic acid and further reduction of formed azo compounds (**3.1.11**) to mesalazine (**3.1.8**) [7,8].

Mesalazine blocks production of prostaglandins and leukotrienes and has other beneficial effects on the inflammatory cascade and belongs to NSAIDs. In medicinal practice it is widely used on inflammatory bowel disease, which includes Crohn disease and ulcerative colitis. Inflammatory bowel disease is a relapsing and remitting condition characterized by chronic inflammation at various sites in the gastrointestinal tract, which results in diarrhea and abdominal pain.

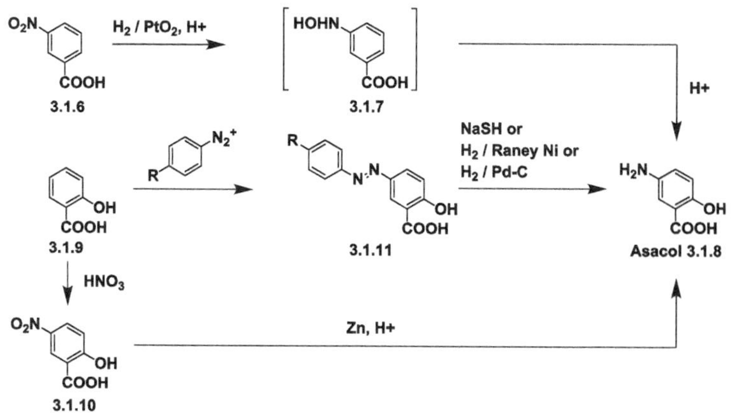

SCHEME 3.1 Synthesis of Asacol.

Anthranilic Acid Derivatives

Mefenamic acid (**3.1.12**), flufenamic acid (**3.1.13**), meclofenamic acid (**3.1.14**), tolfenamic acid (**3.1.15**), niflumic acid (**3.1.16**) (Fig. 3.3.).

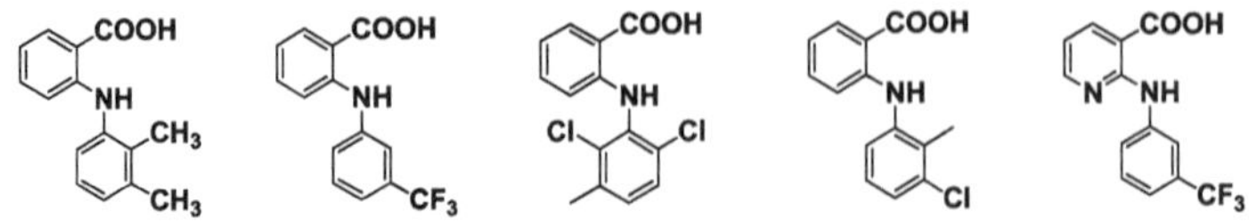

Mefenamic acid 3.1.12 | Flufenamic acid 3.1.13 | Meclofenamic acid 3.1.14 | Tolfenamic acid 3.1.15 | Niflumic acid 3.1.16

FIG. 3.3 Anthranilic acid derivatives.

Arylpropionic Acid Derivatives

Ibuprofen (**3.1.17**), ketoprofen (**3.1.18**), naproxen (**3.1.19**), fenoprofen (**3.1.20**), flurbiprofen (**3.1.21**), loxoprofen (**3.1.22**), dexketoprofen (**3.1.23**) (Fig. 3.4.).

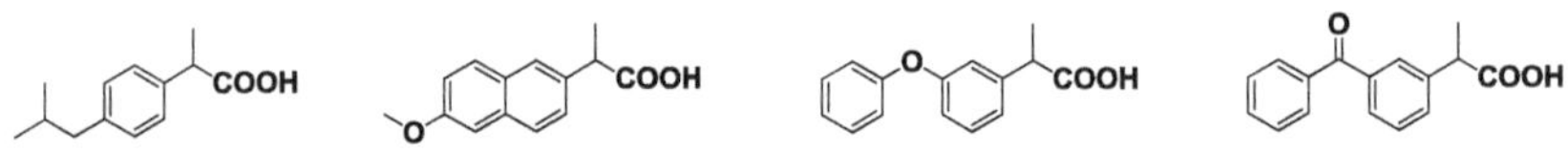

Ibuprofen 3.1.17 | Naproxen 3.1.18 | Fenoprofen 3.1.19 | Ketoprofen 3.1.20

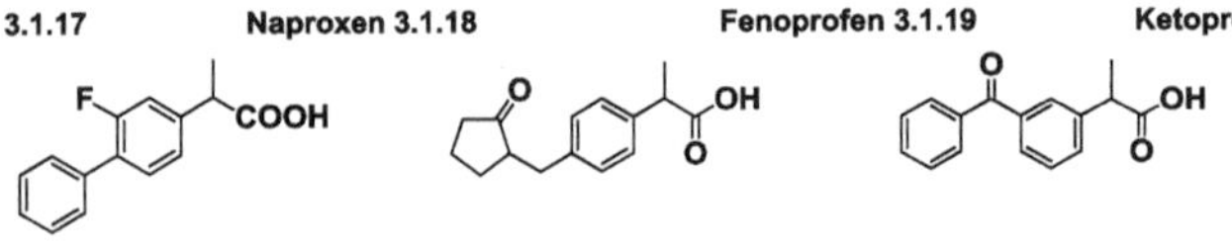

Flurbiprofen 3.1.21 | Loxoprofen 3.1.22 | Dexketoprofen 3.1.23

FIG. 3.4 Arylpropionic acid derivatives.

Arylacetic Acid Derivatives

Diclofenac (**3.1.24**), fenclofenac (**3.1.25**) (Fig. 3.5.).

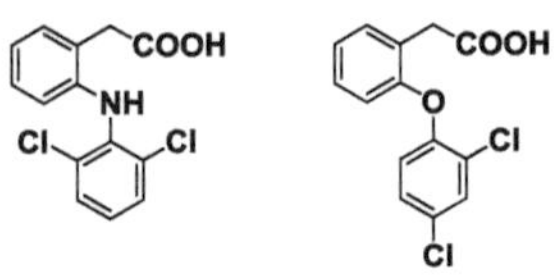

Diclofenac 3.1.24 | Fenclofenac 3.1.25

FIG. 3.5 Arylacetic acid derivatives.

Indolyl-/Indoleacetic Acid Derivatives

Indomethacin (**3.1.26**), zomepirac (**3.1.27**), ketorolac (**3.1.28**), etodolac (**3.1.29**), sulindac (**3.1.30**) (Fig. 3.6.).

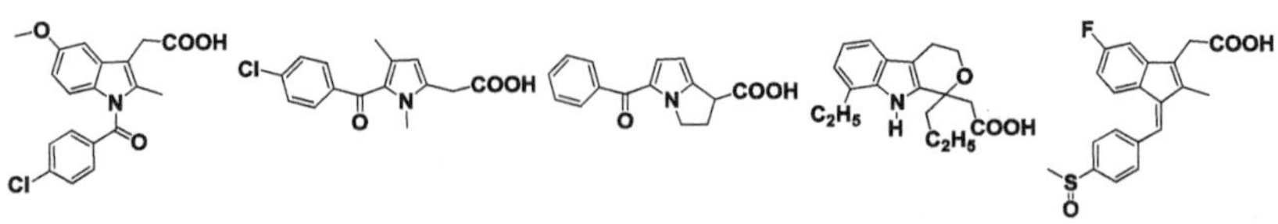

Indometacin 3.1.26 | Zomepirac 3.1.27 | Ketorolac 3.1.28 | Etodolac 3.1.29 | Sulindac 3.1.30

FIG. 3.6 Indolyl/indoleacetic acid derivatives.

Oxicames

Piroxicam (**3.1.31**), isoxicam (**3.1.32**), meloxicam (**3.1.33**), tenoxicam (**3.1.34**), lornoxicam (**3.1.35**), droxicam (**3.1.36**). Most oxicams are unselective inhibitors of the COX enzymes. Meloxicam has a slight preference for COX-2 (Fig. 3.7.).

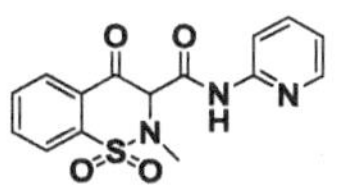
Piroxicam 3.1.31

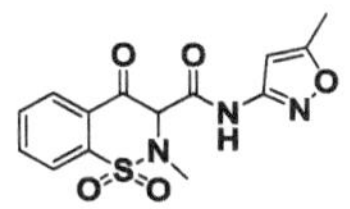
Isoxicam 3.1.32

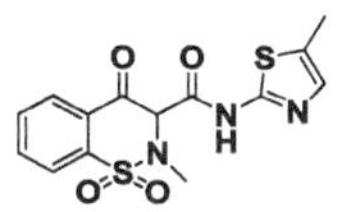
Meloxicam 3.1.33

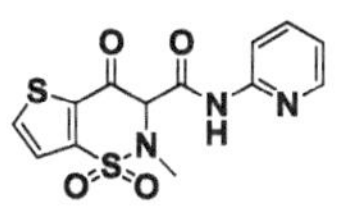
Tenoxicam 3.1.34

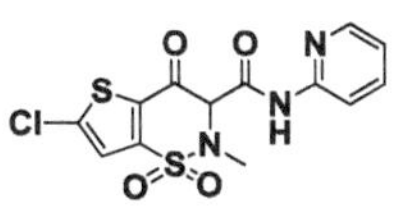
Lornoxicam 3.1.35

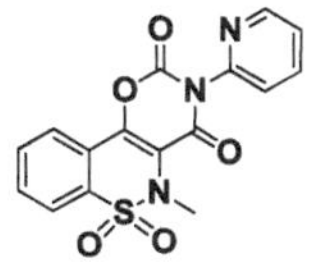
Droxicam 3.1.36

FIG. 3.7 **Oxicames.**

Pyrazolones

Phenylbutazone (**3.1.37**), sulfinpyrazone (**3.1.38**), metamizole sodium (**3.1.39**). (Fig. 3.8.). In many countries, the pyrazolone derivatives, such as dipyrone, antipyrine, aminopyrine, and propyphenazone, are widely used analgesics. Metamizole sodium was generally considered a safe and effective NSAID that was especially potent when it came to reducing fever until the 1970s. The United States has suspended the marketing authorization of metamizole sodium because of a few reports of agranulocytosis, but it is still available without a prescription in many countries.

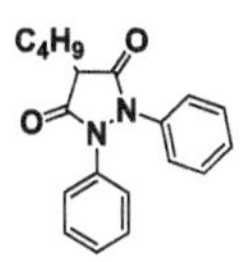
Phenylbutazone 3.1.37

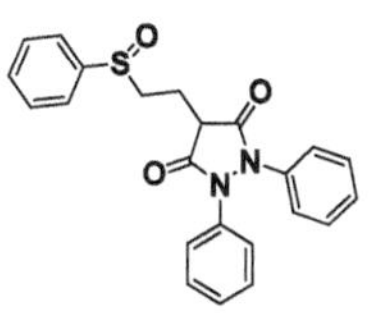
Sulfinpyrazone 3.1.38

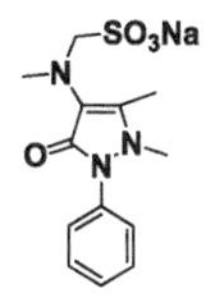
Metamizole sodium 3.1.39

FIG. 3.8 **Pyrazolones.**

Most NSAIDs act as nonselective inhibitors of COX, inhibiting both the COX-1 and COX-2 isoenzymes. The two main adverse reactions associated with NSAIDs relate to gastrointestinal and renal effects. In theory, the COX-2 selectivity inhibition could relieve pain while minimizing the gastrointestinal adverse effects with significantly lower incidence of gastrointestinal ulceration than traditional NSAIDs. This hypothesis gave birth to a new path in the pain sciences—creation of selective COX-2 inhibitors.

Selective COX-2 Inhibitors

Compounds like celecoxib (**3.1.40**), rofecoxib (**3.1.41**), valdecoxib (**3.1.42**), parecoxib (**3.1.43**) (Fig. 3.9.) recently appeared on the market. But all of them, except celecoxib, have been withdrawn from the market because of the increased risk of cardiovascular events (heart attack and stroke), serious skin reactions. It is easy to see structural similarities between pyrazolone derivatives and selective COX-2 inhibitors. In general both of them are bis-diaryl-substituted (Y-shaped) five-membered heterocycle derivatives.

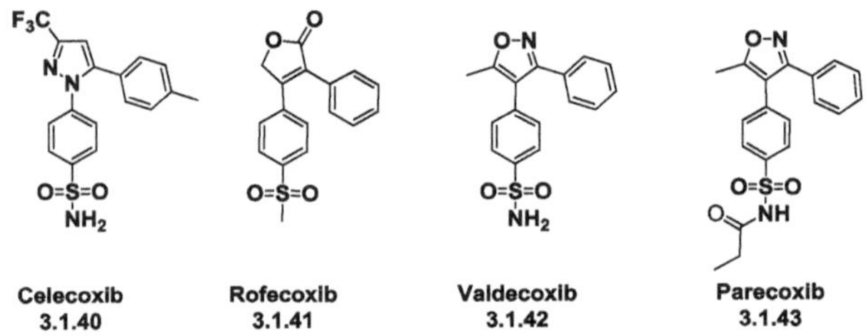

FIG. 3.9 Selective COX-2 inhibitors.

Celecoxib–Celebrex

Celecoxib (**3.1.40**) is a selective COX-2 inhibitor included on the list of Top 200 Drugs by sales for the 2010s. It was synthesized by a reaction of 4-sulfamoylaphenylhydrazine (**3.1.47**) with 4,4,4-trifluoro-1-(p-tolyl)butane-1,3-dione (**3.1.46**) obtained via interaction of 4-methyl-acetophenone (**3.1.44**) with *N*-(trifluoroacetyl)imidazole (**3.1.45**) in the presence of sodium *bis*(trimethylsilyl) amide [9]. Celecoxib is used in the treatment of osteoarthritis, rheumatoid arthritis, acute pain, and menstrual symptoms (Scheme 3.2.).

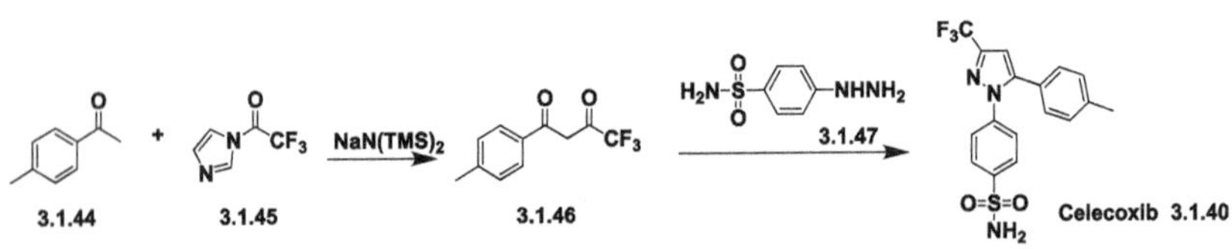

SCHEME 3.2 Synthesis of celecoxib.

3.2 OPIOID ANALGESICS

Morphine and other opioid analgesics [10-13] are considered the most effective analgesics for acute and chronic pain relief. Opioid analgesics include natural alkaloids of opium (morphine, codeine), their analogues (hydrocodone and hydromorphone, oxycodone, and oxymorphone), derivatives of morphinane (levorphanol) and a number of synthetic compounds: derivatives of phenylpiperidine (meperidine, promedol), 4-anilidopiperidines (fentanyl, sufentanil, alfentanil), and derivatives of diphenylheptane (methadone, propoxyphene) are effective pain relievers and continue to be the mainstays of acute and chronic pain management. However, their drawback includes adverse events such as physical and psychological dependence and addiction, psychotropic effects, abuse and addiction, nausea and vomiting, drowsiness, miosis, and constipation. Opioids in medicinal practice are classifiable as opium alkaloids, semisynthetic opium alkaloid derivatives; opium alkaloid analogues with simplified or complicated derivatives of 4,5-epoxyorphinan nucleus such as morphinans, decahydroisoquinolines, benzomorphans, 4-phenylpiperidines, and 4-anylidopiperidines. Synthesis of all of them is described in our previous book [3]. Opioids bind to specific μ-, δ-, and κ-opioid receptors in the nervous system and other tissues. Another receptor of clinical importance is the opioid-receptor-like receptor 1 (ORL1/nociceptin receptor), which is involved in pain responses as well as having a major role in the development of tolerance to μ-opioid agonists used as analgesics. Opioids are subdivided into three large subgroups according to their action on opioid receptors: agonists, mixed agonists–antagonists, and antagonists. Opioid agonists have an affinity for opioid receptors, imitating the activity of endogenous opioid analgesics. Mixed agonists–antagonists can be semisynthetic derivatives of morphine or peptide analogues of endogenous opioids that display agonistic activity in some opioid receptors and antagonistic activity in others. Opioid antagonists bind to opioid receptors but do not activate them. These compounds are not used for analgesia. Their therapeutic value is in relieving side effects that result from either absolute of relative overdoses or intolerance of drugs by patients, and also in treating cases of opioid dependency [14]. It seems that the main classes of opioid ligands have been developed as a result of simplification of morphine molecule-simplifications of 4,5-epoxyorphinane moiety (Fig. 3.10.).

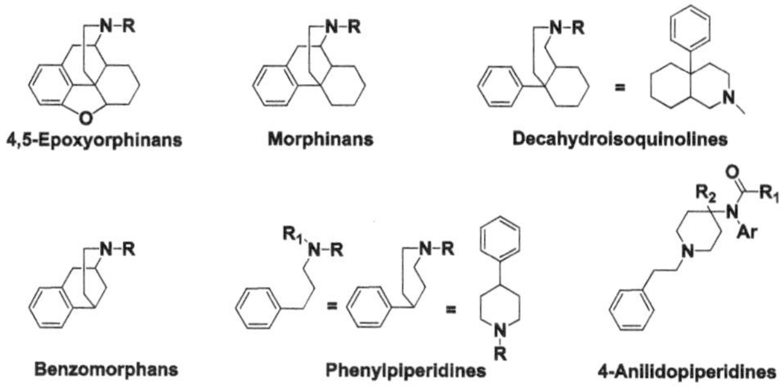

FIG. 3.10 **Simplified derivatives of 4,5-epoxyorphinan core: morphinans, decahydroisoquinolines, benzomorphans, 4-phenylpiperidines, and 4-anylidopiperidines.**

4,5-Epoxyorphinan Opium Alkaloids

In clinical medicine, morphine (**3.2.1**) is primarily used to treat both acute and chronic severe pain and regarded as the gold standard of analgesics. Codeine (**3.2.2**) is used to treat mild to moderate pain and to relieve cough, as well as to treat diarrhea and irritable bowel syndrome (Fig. 3.11.).

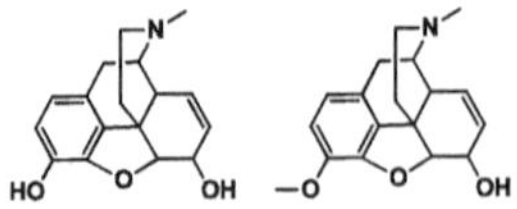

FIG. 3.11 **Structures of morphine and codeine.**

Morphine–Kadian–MS Contin

Morphine (**3.2.1**) (see Fig. 3.11.) is the oldest known analgesic used to relieve pain. Kadian and MS Contin are brand names of time-released formulation of morphine sulphate, usually taken for chronic pain included on the list of Top 200 Drugs by sales for the 2010s. The drug is typically prescribed to cancer patients and persons with severe damaging traumas.

It's source is opium, the dried, milky sap of unripe opium poppy bulbs whose analgesic properties have been known for more than 3000 years. The method used to obtain the opium is to extract from the plant the morphine, thebaine, and other alkaloids (~50) and separate each from the other. However, different schemes for the synthesis of morphine have been proposed [15-32].

Semisynthetic 4,5-Epoxyorphinan Derivatives (Agonists)

This row of clinically used compounds is represented by codeine (**3.2.2**), hydrocodone (**3.2.3**), hydromorphone (**3.2.4**), oxycodone (**3.2.5**), oxymorphone (**3.2.6**), and buprenorphine (**3.2.7**). Thebaine (**3.2.8**) is not used therapeutically, but it is widely used in industry for the synthesis of the above-listed variety of derivatives and many others (Fig. 3.12.). Hydromorphone is a very potent centrally acting analgesic. It is used to relieve moderate to severe pain and severe, painful dry coughing. Hydrocodone is used to treat moderate to severe pain and as an antitussive. The particular niche in which hydrocodone is most commonly used is as an intermediate centrally acting analgesic. Oxymorphone is indicated for the relief of moderate to severe pain and as a preoperative medication to maintain anesthesia. Oxycodone is effective for managing many types of pain from moderate to moderately severe, acute or chronic.

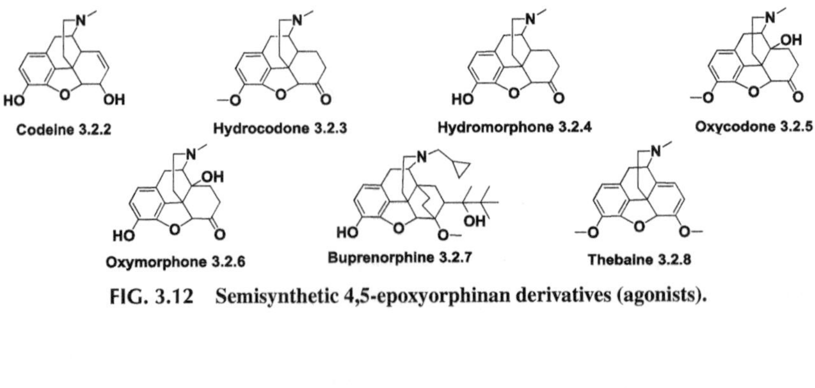

FIG. 3.12 **Semisynthetic 4,5-epoxyorphinan derivatives (agonists).**

Hydrocodone (**3.2.3**), oxycodone (**3.2.5**), oxymorphone (**3.2.6**), and buprenorphine (**3.2.7**) are opioids that are included on the list of Top 200 Drugs by sales for the 2010s [4].

Hydrocodone–Tussionex

Hydrocodone (**3.2.3**) is chemically related to morphine and codeine. It is synthesized by the isomerization of codeine (**3.2.2**) (internal reduction-oxidation [redox] reaction, or 1-3-hydrogen shift of the carbinol hydrogen to olefinic position) using a palladium or platinum catalyst (Scheme 3.3.) [33].

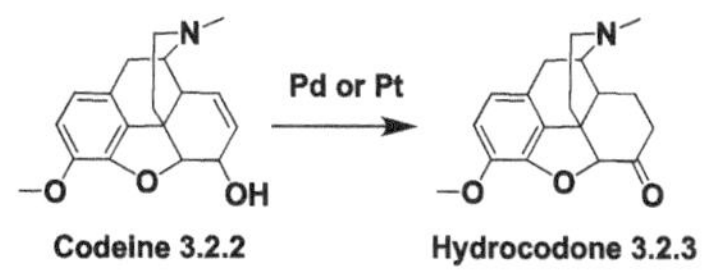

SCHEME 3.3 **Synthesis of hydrocodone.**

This drug is also synthesized by the hydrogenation of codeinone [34] and by oxidation of dihydrocodeine [35]. Hydrocodone is an orally active opioid analgesic and antitussive.

Oxycodone–OxyContin

Oxycodone also is included on the list of Top 200 Drugs by sales for the 2010s. It is synthesized from thebaine (**3.2.4**), which transforms into intermediate 14-hydroxycodeinone (**3.2.9**) during oxidation with hydrogen peroxide in formic acid. After the selective hydrogenation of the double bond, the desired oxycodone (**3.2.5**) has been synthesized (Scheme 3.4.).

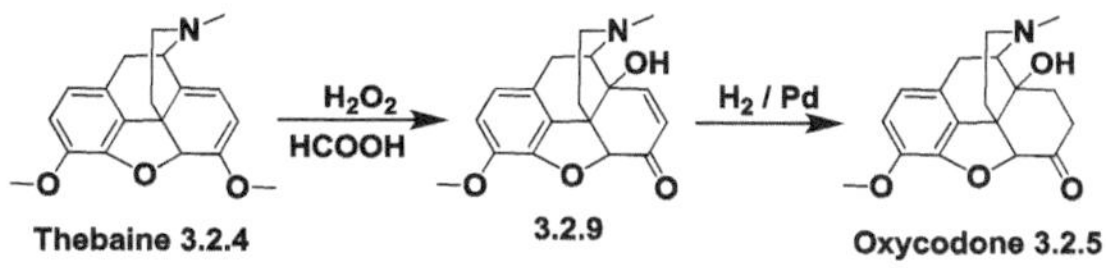

SCHEME 3.4 **Synthesis of oxycodone.**

Unlike hydrocodone, it is used as an analgesic in combination with other drugs such as aspirin or acetaminophen. Oxycodone is similar to morphine in terms of durational efficacy, and is intended for oral use. Side effects are analogous to those of morphine. It is intended for relieving moderate to severe pain in surgical and gynecological interventions and for postoperational pain.

Oxymorphone–Opana

Oxymorphone (**3.2.5**), Included on the list of Top 200 Drugs by sales for the 2010s, is a product of demethylation of oxycodone (**3.2.5**) by hydrogen bromide [36,37] (Scheme 3.5.).

SCHEME 3.5 Synthesis of oxymorphone.

Oxymorphone is approximately 10 times more active than morphine. Euphoric effects, as well as vomiting, are significantly stronger than those produced by morphine. Oxymorphone also displays poor antitussive activity. It is intended for relieving moderate to severe pain in surgical and gynecological interventions and for postoperational pain. Side effects are analogous to those of morphine.

Another semisynthetic 4,5-epoxyorphinan derivative is buprenorphine, which is also included on the list of Top 200 Drugs by sales for the 2010s [4].

Buprenorphine–Suboxone

Buprenorphine (**3.2.7**), which is included on the list of Top 200 Drugs by sales for the 2010s, has an unique pharmacological profile. It has an extremely high binding affinity at the μ, δ, and κ receptors, as well as at the ORL-1/nociceptin receptor. It is approximately 20 to 40 times more potent than morphine as a partial agonist at the μ receptor. It works as a competitive antagonist at the κ receptor and as a partial agonist at the μ and ORL-1/nociceptin receptors. In addition, it is indicated for the treatment of moderate to severe chronic pain and used in the management of opioid dependence and for detoxification.

Synthesis of buprenorphine (Scheme 3.6.) is an eight-step process started from thebaine [38,39]. A Diels-Alder reaction of thebaine (**3.2.4**) with methyl vinyl ketone gives a 6,14 *endo*-etheno bridge in the 4,5-epoxyorphinan system (**3.2.10**) and an acetyl substituent at the 7α position. Hydrogenation of generated double bond (**3.2.11**) followed by a Grignard reaction with t-butylmagnesium chloride creates a compound (**3.2.12**). It undergoes a von Braun *N*-demethylation with cyanogen bromide, producing the corresponding *N*-nitrile (**3.2.13**). N-nitrile hydrolyzed to the secondary amine (**3.2.14**), which was acylated with cyclopropyl methyl carbonyl chloride to (**3.2.15**) followed by lithium aluminum hydride reduction of the carbonyl group (**3.2.16**). Phenolic *O*-demethylation furnishes the requested buprenorphine (**3.2.8**).

Semisynthetic 4,5-Epoxymorphinan Derivatives (Antagonists)

Opioid antagonists (Fig. 3.13.) are agents that displace opioid molecules from their receptors, and block opioids from attaching to and activating those receptors. Such qualities are very important, to reverse toxic effects of opioid overmedication or overdose.

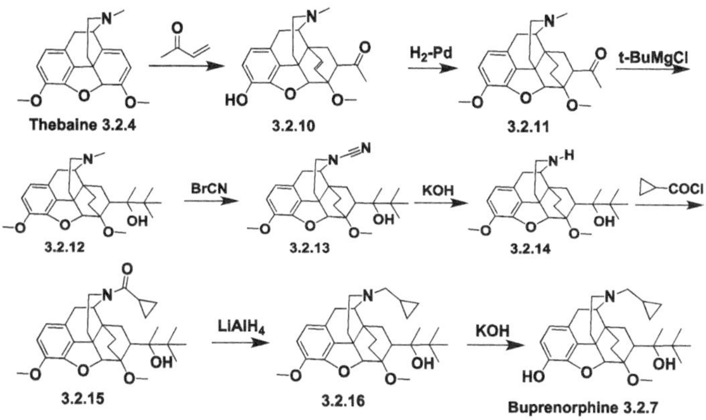

SCHEME 3.6 Synthesis of buprenorphine.

Nalorphine (**3.2.17**) reverses the effects of morphine and other narcotics. Nalorphine acts at two opioid receptors, at the κ receptor as an antagonist and at the μ receptors as an agonist. It is used to reverse opioid overdose. Naloxone (**3.2.18**) is an opioid antagonist used to counter the effects of opioid overdose. Naltrexone (**3.2.19**) helps patients overcome opioid addiction. Sometimes it is used for rapid detoxification regimens for opioid-dependent persons. The main use of naltrexone is for the treatment of alcohol dependence. Nalmefene (**3.2.20**) is an opioid receptor antagonist that is used in the management of alcohol dependence and for the treatment of other addictions. As with other drugs of this type, nalmefene can precipitate acute withdrawal symptoms in opioid-drug-dependent patients, and, more rarely, it can counteract the effects of strong opioids.

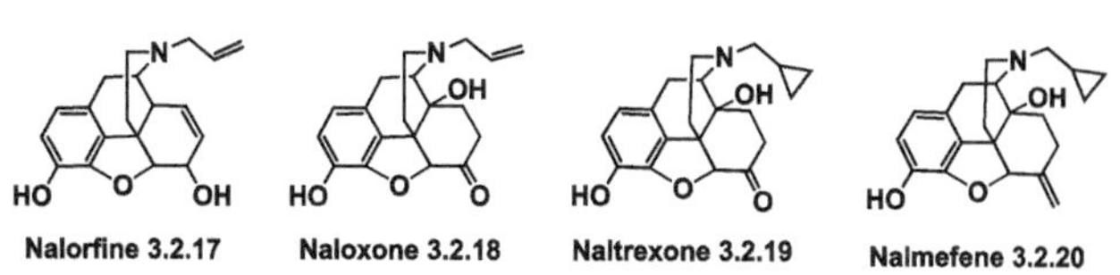

FIG. 3.13 Semisynthetic 4,5-epoxymorphinan derivatives (antagonists).

Morphinans

Morphinan skeleton has been chemically manipulated to obtain compounds with agonistic and antagonistic activity and selectivity toward different populations of opioid receptors. Affinities of these compounds as opioid analgesics at the μ receptor are lower than of their prototypes, but they possess high affinity at the κ receptor. Among morphinan derivatives are opioid analgesics, psychoactive drugs, dissociative hallucinogens, and cough suppressants. For example, levorphanol (**3.2.21**) is a medication used to treat severe pain and levallorphan (**3.2.22**) is used as an opioid antagonist.

Cyclorphan (**3.2.23**) is an opioid analgesic with δ- and κ-receptor agonistic and m-receptor antagonistic properties that provides strong pain relief. However, it is not marketed because of hallucinogenic side effects (Fig. 3.14.).

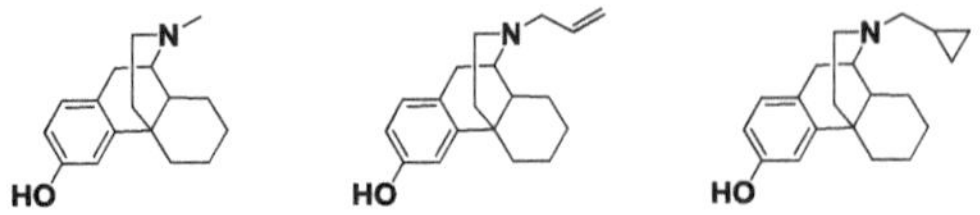

Levorphanol 3.2.21 Levallorphan 3.2.22 Cyclorphan 3.2.23

FIG. 3.14 **Structure of morphinans.**

Decahydroisoquinolines

These compounds (**3.2.24** to **3.2.27**) (Fig. 3.15.) possess high affinity at μ and δ receptors. Agonist or antagonist activity depends on the substituent at piperidine N-atom and the attached heterocyclic ring system pyrrole/pyridine [40]. No decahydroisoquinoline derivative has entered the pharmaceutical market as a drug.

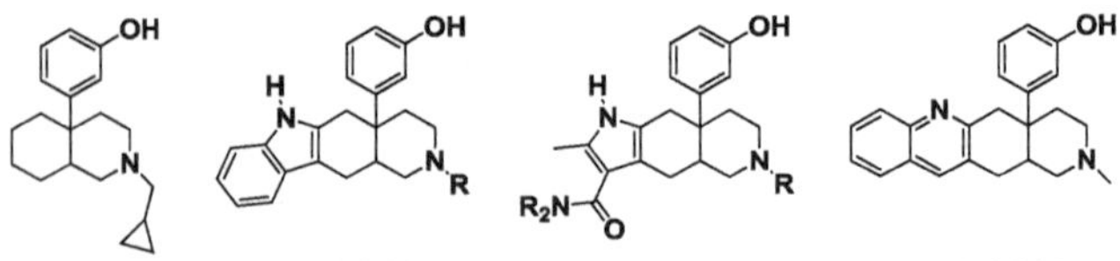

LY 113878 3.2.24 3.2.25 3.2.26 TAN-67 3.2.27

FIG. 3.15 **Structure of decahydroisoquinolines.**

Benzomorphans

Affinities of these compounds (**3.2.28** to **3.2.30**) (Fig. 3.16.) at the μ receptor are lower than of their prototypes, but they possess high affinity at the κ receptor. Pentazocine (**3.2.28**), metazocine (**3.2.29**), and cyclazocine (**3.2.30**) are representatives of this class of opioids, which class is used to treat moderate to moderately severe pain. Pentazocine (**3.2.28**), in particular, is used to treat moderate to severe pain. In general, the clinical use of benzomorphans is limited because of dysphoric and hallucinogenic effects, which are most likely caused by their activity at κ receptors.

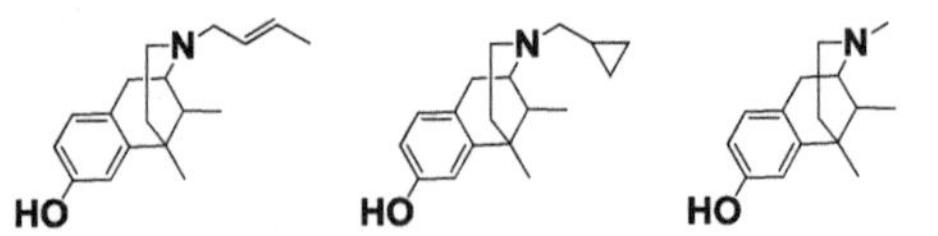

Pentazocine 3.2.28 Cyclazocine 3.2.29 Metazocine 3.2.30

FIG. 3.16 **Structure of benzomorphans.**

4-Phenylpiperidines

Pethidine (**3.2.31**) is the ancestor of a variety of opioid analgesics of the 4-phenylpiperidine series (Fig. 3.17.)—allylprodine, alphameprodine, alphaprodine, anileridine, betameprodine, benzethidine, diphenoxylate, etoxeridine, furethidine, morpheridine, phenoperidine, piminodine, trimeperidine—and was the first and simplest synthetic opioid analgesic synthesized for the treatment of moderate to severe pain. Prodine (**3.2.32**) is structurally closely related to phetidine (**3.2.31**) in which the propionyl fragment in the fourth position of the piperidine ring is replaced by a propionyloxy group. This compound is a strong analgesic with side effects that are common for morphine. Alvimopan (**3.2.33**) is a novel peripherally acting μ-opioid receptor antagonist in the 4-phenylpiperidine series and indicated for patients with large or small bowel resection to avoid postoperative ileus [41-43]. A closely related compound, JDTic (**3.2.34**) [44-47], is a new chemical entity, that exerts highly potent and selective opioid κ-receptor antagonist activity, has antidepressant-like effects, and has the potential to become a drug for the treatment of persons with physical dependence on morphine and cocaine. (Studies in humans are ongoing.)

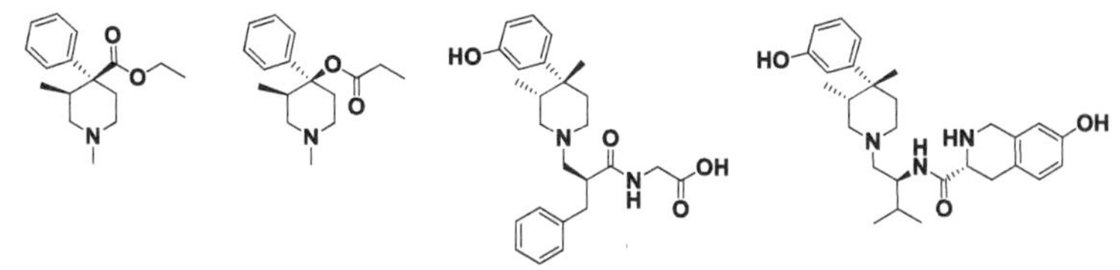

Pethidine 3.2.31 **Prodine 3.2.32** **Alvimopan 3.2.33** **JDTic 3.2.34**

FIG. 3.17 **Agonists and antagonists of 4-phenylpiperidine series.**

4-Anylinopiperidines

The 4-anilidopiperidines (Fig. 3.18.) class is the most potent class of opioid analgesics known. The prototype of this class, fentanyl (**3.2.35**), is approximately 300 times more potent than morphine in mice and rats, compared to 50 to 100 times in humans [48-58]. A very large number of fentanyl analogues (mefentanyl (**3.2.36**), phenaridine (**3.2.37**), lofentanil (**3.2.38**), ocfentanil (**3.2.39**), ohmefentanyl (**3.2.40**), carfentanil (**3.2.41**), sufentanil (**3.2.42**), thiofentanil (**3.2.43**), alfentanil (**3.2.44**), trefentanil (**3.2.45**), brifentanil (**3.2.46**), mirfentanil (**3.2.47**), and remifentanil (**3.2.48**)) have been synthesized since 1964, as have the acyclic compounds diampromide (**3.2.49**), and 2,3-seco-fentanyl (**3.2.50**).

Fentanyl–Fentora

Fentanyl (**3.2.35**), which is included on the list of Top 200 Drugs by sales for the 2010s, is a potent analgesic that was first synthesized more than 40 years ago. It is still the most popular opioid used in the perioperative period throughout the

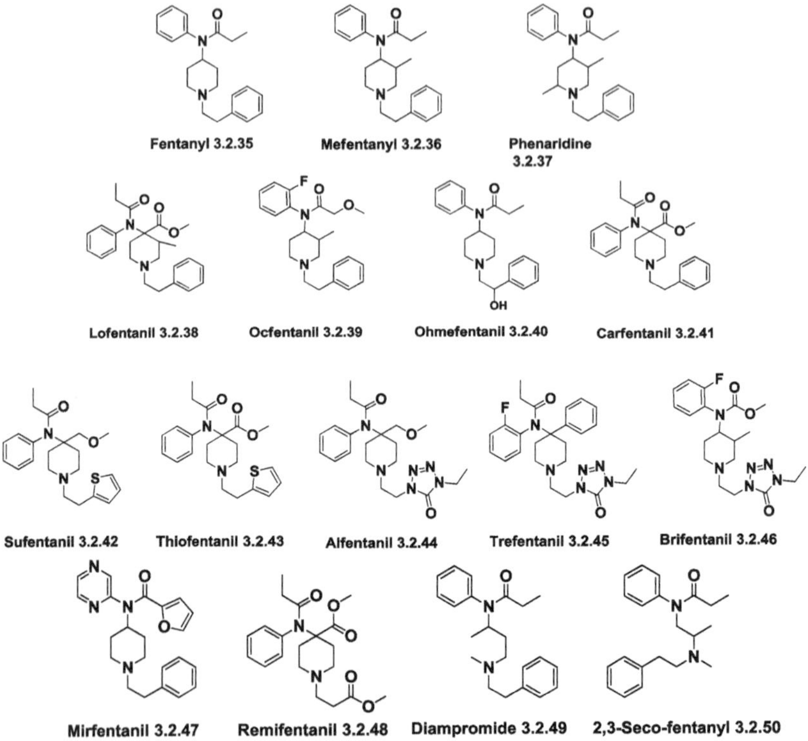

FIG. 3.18 Analgesics of 4-anilidopiperidine series.

world. In spite of the development of more potent, faster-onset opioids, fentanyl remains the mainstay of anesthesiologists. Fentanyl's popularity has resulted in its use in many acute and chronic pain conditions. Fentanyl, a pure opioid agonist, acts primarily through interaction with opioid μ receptors. The most clinically useful pharmacologic effects of the interaction of fentanyl are analgesia and sedation. Other opioid effects at clinically relevant doses may include somnolence, hypoventilation, bradycardia, postural hypotension, pruritus, dizziness, nausea, diaphoresis, flushing, euphoria, and confusion or difficulty in concentrating. Fentanyl is metabolized in the liver and in the intestinal mucosa to norfentanyl. It is primarily (more than 90%) eliminated by biotransformation to *N*-dealkylated and hydroxylated inactive metabolites. Fentanyl is a Schedule II controlled substance that can produce drug dependence pain. The story of fentanyl begins in late 1950s [59-62].

The first synthesis of fentanyl [59] (Scheme 3.7.) was accomplished beginning with 1-benzypiperidin-4-one (**3.2.51**), which was condensed with aniline to form the corresponding Schiff base (**3.2.52**). The double bond in this product was reduced by lithium aluminum hydride, and the resulting 1-benzyl-4-anilinopiperidine (**3.2.53**) was acylated using propionic anhydride. The resulting 1-benzyl-4-*N*-propinoylanilinopiperidine (**3.2.54**) underwent debenzylation using hydrogen and a palladium on carbon catalyst, to give 4-*N*-propanoylanilinopiperidine (**3.2.55**), which was *N*-alkylated by 2-phenylethylchloride, or tosylate to give fentanyl (**3.2.25**) [59-62].

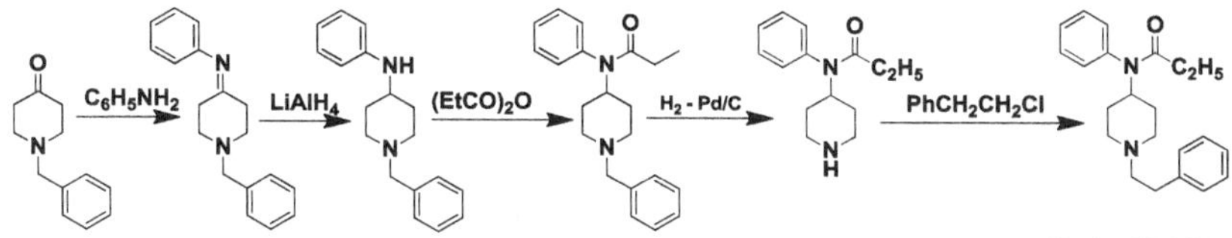

3.2.51 3.2.52 3.2.53 3.2.54 3.2.55 Fentanyl 3.2.35

SCHEME 3.7 **Synthesis of fentanyl.**

Later, a modified, shorter synthesis of fentanyl by the same scheme, but starting directly from 1-(2-phenethyl)piperidin-4-one, was proposed by group of chemists from Poland [63].

Another synthesis, starting from 4-anilinopyridine, was proposed [64] (Scheme 3.8.). 4-Anilinopyridine (**3.2.56**) was acylated with propionic anhydride to produce 4-*N*-propinoylanilinopyridine (**3.2.57**), which was alkylated with 2-phenylethylbromide to produce pyridinium salt (**3.2.58**), hydrogenation of which over PtO_2 resulted in a 4-*N*-anilinopiperidine derivative (**3.2.59**) that was acylated with propionic anhydride to produce fentanyl (**3.2.35**).

3.2.56 3.2.57 3.2.58 3.2.59 Fentanyl 3.2.35

SCHEME 3.8 **Alternative synthesis of fentanyl.**

The three most popular compounds in medicinal practice are fentanyl, sufentanil, and alfentanil. The analgesic potency of fentanyl, found to be 300 times higher than that of morphine in the tail withdrawal test in rats, had been enhanced up to 10,000 times than that of morphine (carfentanil) by the introduction of appropriate substituents the structure of fentanyl.

A huge amount of work had been published. The main changes of practical meaning in the structures of the compounds of these series are summarized in Fig. 3.19 and include:

a. replacement of the piperidine ring by a pyrrolidine, tropine, or azepine ring, as well as synthesis of open chain compounds;
b. replacement of the phenyl group in phenethyl moiety, for some aromatic heterocycles (changes of practical meaning are replacements of phenyl group for thiophene (sufentanil) and tetrazole rings (alfentanil, brifentanil, trefentanil), or carbomethoxy group as in remifentanil).

c. sensible change was done via attaching additional methoxymethyl (alfentanil, sufentanil) or metoxycarbonyl groups (lofentanil, carfentanil) to the fourth position of the piperidine ring.

d. changes of practical meaning had been done and in aniline part of the fentanyl molecule via introduction of fluorine atom into the ortho-position of aniline-(ocfentanil, brifentanil, trefentanil).

e. an additional one methyl group (lofentanil, brifentanil), as well as two methyl groups (phenaridine) had been introduced into the second, third or fifth positions of the piperidine ring.

Among other changes a lot of work had been done on replacement of propionyl group in the 4-anilido fragment of several other acyl groups. But the propionyl group remains preferable. The single successful case of replacement of the propionyl group is the replacement for the methoxyacetyl group (ocfentanil).

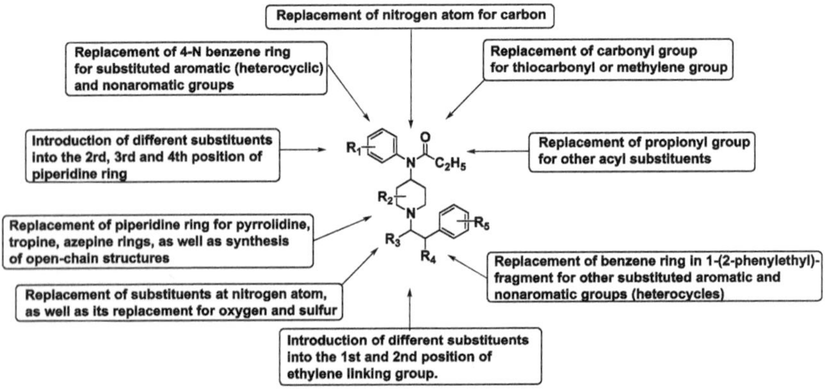

FIG. 3.19 Modifications of fentanyl structure.

More information concerning structure–activity relationships is found in reviews [65-72].

Table 3.1 compares the activities of opioid drugs that are available in the pharmaceutical market [73].

The target for the all of the above described opioid analgesics seems to be the μ receptor. But ligands specifically binding to each of three opioid receptors (μ, κ, and δ) have been created and which revealed unique pharmacological responses.

δ-Opioid Ligands

Agonists at the δ-opioid receptor were initially thought to be potential analgesic agents that may lack the undesired effects of μ opioids. Studies, however, uncovered that the δ-opioid system is involved in many other biological processes,

TABLE 3.1 Comparison of Opioid Drugs

Fentanyl Derivatives	Opioid Potency (Compared to Morphine)	Lipid Solubility*	Therapeutic Index†
Morphine	1	1,4	70
Fentanyl	300	800	300
Sufentanil	4500	1800	25,000
Alfentanil	75	150	1100
Remifentanil	220	18	33,000
Carfentanil	10,000		10,600

**Lipid solubility = octanol/water distribution coefficient.*
†Therapeutic index = median lethal dose (LD_{50})/lowest median effective dose (ED_{50}).

and generates undesirable side effects like convulsive properties (seizures in high doses). SNC-80 (**3.2.60**), and its analogues, like BW373U86 (**3.2.61**), DPI-287(**3.2.62**), and DPI-329 (**3.2.63**), are the accepted δ agonists. The δ-receptor selective antagonists are naltrindole (**3.2.64**) and naltriben (**3.2.65**), which are used in scientific research [74-77] (Fig. 3.20.).

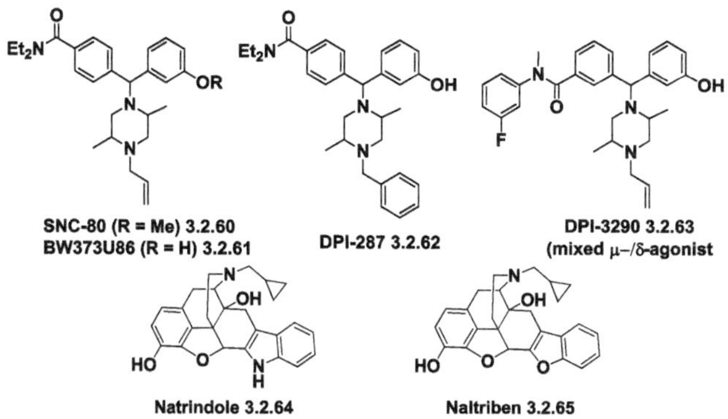

FIG. 3.20 δ-Opioid receptor ligands.

Functional association between μ- and δ-opioid receptors was first suggested by studies showing that μ activity could be modulated by δ ligands [78,79]. But the true role of δ ligands remains unrevealed. Some authors insist that δ agonists increase antinociceptive responses to μ- receptor agonists [80-82]; other authors insist that μ-agonist signaling can be enhanced by cotreatment with δ-selective antagonists [83,84].

κ-Opioid Ligands

κ-Opioid agonists are analgesics that lack the adverse side effects of constipation and respiratory depression, as well as the drug abuse and dependence liabilities associated with μ-opioid agonists. However, preclinical and clinical studies of κ agonists have revealed their own side effects profile, including sedation, diuresis, and dysphoria. Activation of the κ-opioid receptor appears to antagonize many of the effects of the μ receptor. A large body of evidence indicates that κ-opioid receptors may be involved in the modulation of some abuse-related effects of central nervous system (CNS) stimulants. κ Opioids attenuate stimulant self-administration in a variety of animal models. Selective agonists of this receptor may have therapeutic potential in the treatment of addiction. U-50488 (**3.2.66**) was one of the first selective κ agonists invented and research on its derivatives has led to the development of a large family of related compounds like U-69593 (**3.2.67**), and R-84760 (**3.2.68**). Norbinaltorphimine (nor-BNI) (**3.2.69**) is a prototypical κ antagonist, and is employed as a tool in opioid research [85-87] (Fig. 3.21.).

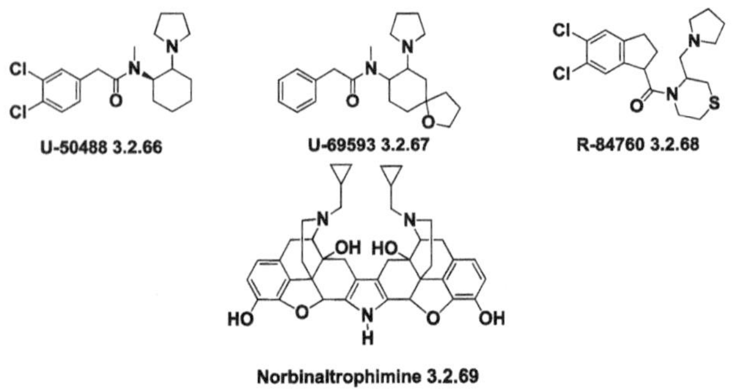

FIG. 3.21 **κ-Opioid receptor ligands.**

ORL-1–Opioid Ligands

ORL-1 receptor, or nociceptin/orphanin FQ peptide (NOP) receptor is the fourth member of the opioid peptide receptor family.

A large number of publications suggests that ORL-1 agonists may be useful for treatment of neuropathic pain/allodynia, substance abuse, stress, anxiety, cachexia, anorexia, asthma, and possibly hypertension. ORL-1 antagonists may be useful as analgesics and for learning ability and memory-enhancing and for treating locomotor disorders. They represented mainly as iperidine derivatives like aryl piperidines, spiperones benzimidazopiperidines (**3.2.70** to **3.2.79**), etc. Agonists and antagonists have been found in all of the classes. Currently, it is difficult to consider this compounds (3.2.70-3.2.79) and their analogues as potential drugs [88,89] (Fig. 3.22.).

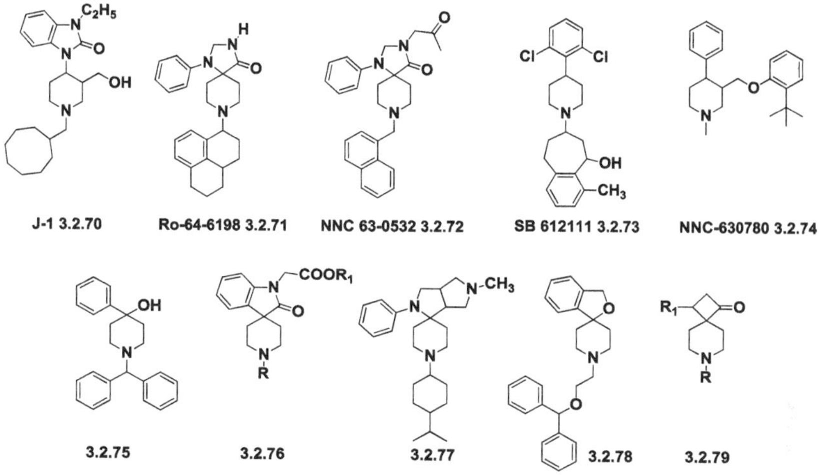

J-1 3.2.70 Ro-64-6198 3.2.71 NNC 63-0532 3.2.72 SB 612111 3.2.73 NNC-630780 3.2.74

3.2.75 3.2.76 3.2.77 3.2.78 3.2.79

FIG. 3.22 **ORL-1–opioid receptor ligands.**

3.3 ADJUNCTIVE AGENTS IN THE MANAGEMENT OF PAIN, OR "NONANALGESIC" PAIN KILLERS

Pain is a complex phenomenon that results from many of contributing factors. Development of chronic postsurgical pain syndromes, hyperalgesia and immunomodulation are some particular concerns as they may be related to opioid exposure. Of course the most important target for the opioid analgesia seems to be the μ receptor and COX. But the variety of types of pain and the plethora of possible pathways to produce pain, allow implementation of different strategies for creation of new analgesics. Acute pain is usually manageable with above described medications such as NSAIDs and opioids. Management of chronic pain, however, is much more difficult. Multiple classes of drugs, such as anticonvulsants, antidepressants, and antiarrhythmics, are known to be associated with pain. Anticonvulsant drugs (gabapentin, carbamazepine, oxcarbazepine), which typically are used to control seizures in epilepsy, may also be used to treat painful conditions, such as neuralgia and fibromyalgia. Tricyclic antidepressants (amitriptyline, imipramine, nortriptyline, desipramine) seem to have pain-relieving properties that are independent of their effects on reducing depression. Other antidepressants, other than the tricyclic group, such as venlafaxine and duloxetine, may also help with pain from nerve injury. Several local anesthetics and related antiarrhythmic drugs are potentially useful in treating diabetic neuropathy, including lidocaine, mexiletine, and tocainide.

Novel pain-regulating targets, such as neurokinin-1 (NK1) receptor modulators, cannabinoid (CB) receptor modulators, *N*-methyl-D-aspartate (NMDA) receptors, calcium channel modulators, transient receptor potential (TRPV1) channel modulators, and glial cell modulators, play a crucial role in the generation of pain and hyperalgesia in several pain states. The variety of types of possible pathways operating in the central and peripheral nervous system to produce pain allows implementation of different strategies to create new analgesics that target novel receptors. Currently, there is insufficient evidence to suggest that any one adjuvant analgesic has absolute advantages over another.

Antidepressants as Analgesics

Antidepressants are widely used for the treatment of chronic and neuropathic pain [90-96]. They exhibit a number of pharmacological actions: blocking reuptake of noradrenaline and serotonin, acting on opioid receptors, inhibiting histamine, blocking adenosine uptake, inhibiting cholinergic and NMDA receptors and ion channel activity. Antidepressants are effective in acute and neuropathic pain by diverse mechanisms independent of antidepressant effects.

Tricyclic antidepressants (TCAs)—amitriptyline (**3.3.1**), nortriptyline (**3.3.2**), desipramine (**3.3.3**), maprotiline (**3.3.4**) (Fig. 3.23.)—are effective in the treatment of neuropathic pain, fibromyalgia, low back pain, and headaches.

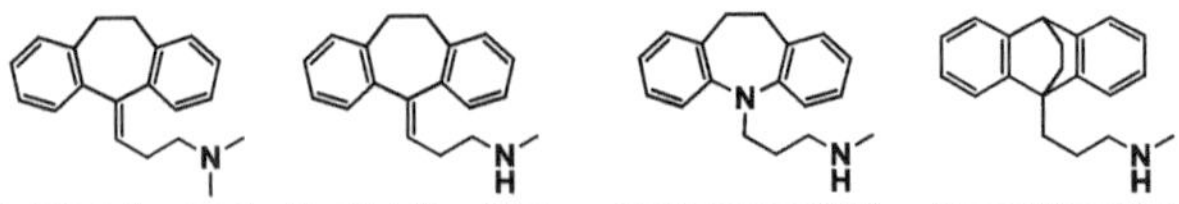

Amitriptyline 3.3.1 Nortriptyline 3.3.2 Desipramine 3.3.3 Maprotiline 3.3.4

FIG. 3.23 TCAs effective in the treatment of pain.

Selective serotonin reuptake inhibitors (SSRIs)—fluoxetine (Prozac) (**3.3.5**), citalopram (**3.3.6**), paroxetine (**3.3.7**) (Fig. 3.24.)—are better tolerated but are inferior to TCAs in controlling persistent pain. Fluoxetine has higher analgesic activity than either citalopram or paroxetine.

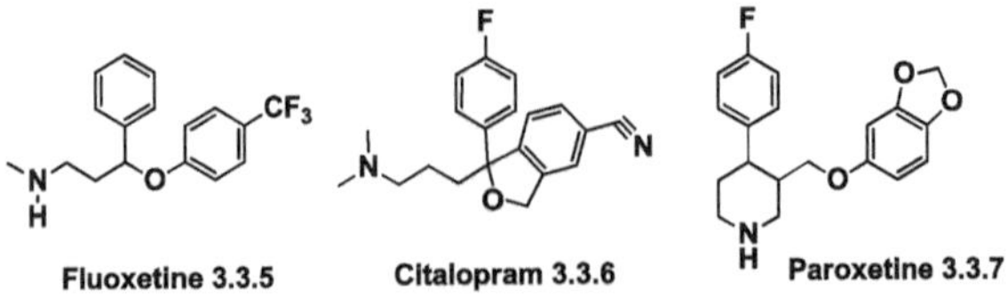

Fluoxetine 3.3.5 Citalopram 3.3.6 Paroxetine 3.3.7

FIG. 3.24 SSRIs effective in the treatment of pain.

Of the serotonin norepinephrine reuptake inhibitors (SNRIs)—venlafaxine (**3.3.8**), duloxetine (Cymbalta) (**3.3.9**), milnacipran (Savella) (**3.3.10**) (Fig. 3.25.)—venlafaxine is the most investigated drug for pain management. It is effective in the treatment of painful peripheral diabetic neuropathy. Duloxetine is proposed for use in diabetic peripheral neuropathic pain and for the safe treatment of many of the symptoms associated with fibromyalgia, particularly for women.

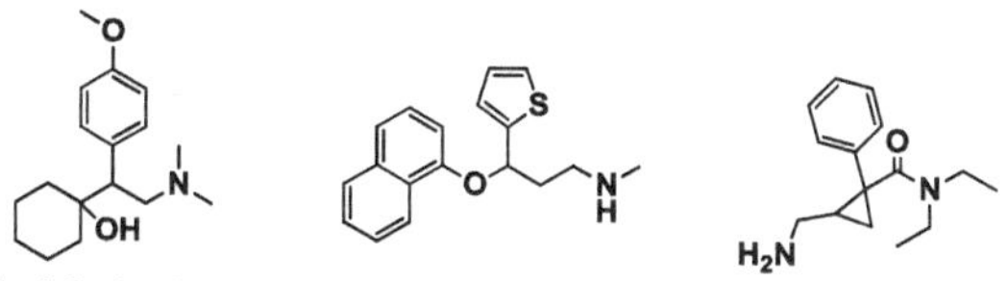

Venlafaxine 3.3.8 Duloxetine 3.3.9 Milnacipran 3.3.10

FIG. 3.25 SSNIs effective in the treatment of pain.

Miscellaneous Antidepressants

Mianserin (**3.3.11**) and its analog mirtazapine (**3.3.12**), which are noradrenergic and specific serotonergic antidepressants, increase the antinociceptive effect of various analgesic agents in an animal models. However, in humans, they show no analgesic effect. Trazodone (**3.3.13**), a SSRI and a selective 5-HT_2 antagonist, was ineffective in decreasing pain in patients with chronic low back pain. Nefazodone (**3.3.14**) (a serotonin–norepinephrine–dopamine reuptake inhibitor and 5-HT_2 antagonist) has only been tested in animals and was shown to produce analgesia and to potentiate the analgesic effect of morphine, but was not recommended because of potential hepatotoxicity (Fig. 3.26.).

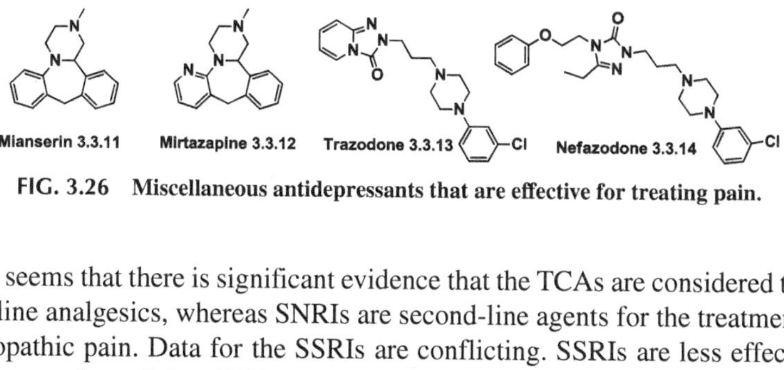

Mianserin 3.3.11 Mirtazapine 3.3.12 Trazodone 3.3.13 Nefazodone 3.3.14

FIG. 3.26 Miscellaneous antidepressants that are effective for treating pain.

It seems that there is significant evidence that the TCAs are considered to be first-line analgesics, whereas SNRIs are second-line agents for the treatment of neuropathic pain. Data for the SSRIs are conflicting. SSRIs are less effective, but better tolerated than TCAs.

Serotonin Norepinephrine Reuptake Inhibitors Combination with μ-Opioid Receptor Agonist (Tapentadol)

Recently, a new analgesic, tapentadol (**3.3.20**), with μ-opioid receptor agonist and noradrenaline reuptake inhibitor actions together with insignificant serotonin reuptake activity was approved by the FDA and presented on the pharmaceutical market for the treatment of moderate to severe acute pain.

Tapentadol represents a new pharmacological class ("MOR-NRI") that combines two mechanisms of action, μ-opioid receptor agonism (MOR) and noradrenaline reuptake inhibition (NRI) in a single molecule, and its overall analgesic effect is synergistic—that is, greater than the sum of the expected analgesia provided exclusively by noradrenaline reuptake inhibition and direct μ-receptor agonist effects. It is interesting, that on binding assays, tapentadol's μopioid binding was nearly 50 times lower than that of morphine, but tapentadol's analgesic potency was only two to three times lower than morphine's potency when tested in multiple preclinical pain models [97-101].

The compound was first disclosed in 1996 [102] and the synthetic pathway described below was proposed (Scheme 3.9.).

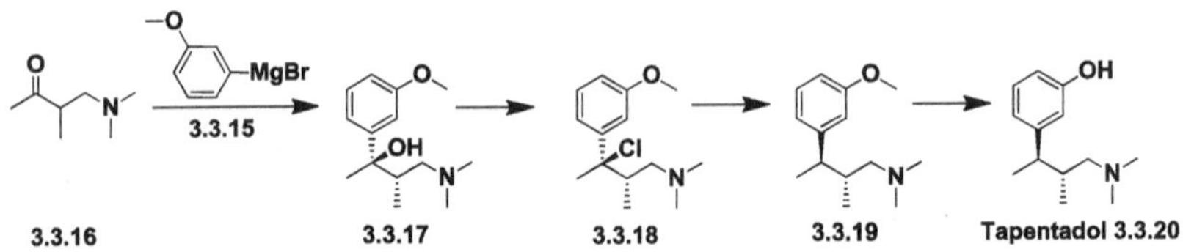

SCHEME 3.9 **Synthesis of tapentadol.**

(3-Methoxyphenyl) magnesium bromide (**3.3.15**) was reacted with 4-(dimethylamino)-3-methylbutan-2-one (**3.3.16**) to form racemic tertiary alcohol intermediate, separatedby chiral high-performance liquid chromatography (HPLC) to give (2S,3R)-1-(dimethylamino)-3-(3-methoxyphenyl)-2-methylbutan-3-ol (**3.3.17**), which was converted into the corresponding chloride (**3.3.18**). and was then Reduction of the last using either zinc borohydride, zinc cyanoborohydride, or tin cyanoborohydride to give (2R,3R)-3-(3-methoxyphenyl)-N,N,2-trimethylbutan-1-amine (**3.3.19**), which was demethylated with hydrobromic acid to produce the desired tapentadol (**3.3.20**). Other schemes of synthesis have been proposed [103-105].

Common side effects of tapentadol are nausea, vomiting, dizziness, somnolence, headache, fatigue, and dependence liability.

Gabapentinoids (Calcium Channel $\alpha_2\delta$ Ligands)

Some γ-aminobutyric (GABA) acid derivatives, particularly, gabapentin (**3.3.21**) and pregabalin (**3.3.22**) (Fig. 3.27.), are anticonvulsants used to treat disorders such as epilepsy, and are also effective for treatment of chronic pain, diabetic peripheral neuropathy, and fibromyalgia, although they are inactive in pain models designed for acute pain [106-111]. They interact with the $\alpha_2\delta$ subunit of the voltage-dependent calcium channels in the CNS, preventing the influx of calcium into the neuron, decreasing the release of neurotransmitters such as glutamate, noradrenaline, and substance P, and increasing neuronal levels of GABA by producing an increase of glutamic acid decarboxylase activity, which converts the excitatory neurotransmitter glutamate into the inhibitory GABA. But the mechanism of action that leads to their rapid analgesic effect is unknown. Additional $\alpha_2\delta$ ligands with α-amino acid structural matrices and their carboxylate bioisosters, as well as non–amino acid structures, have been disclosed (**3.3.23** to **3.3.28**) (Fig. 3.28.), but their analgesic activity was weaker than that of gabapentin [111].

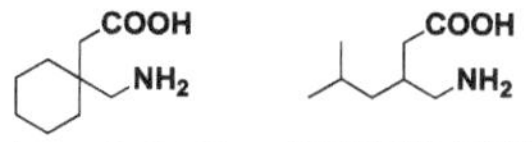

FIG. 3.27 Structures of gabapentin and pregabalin.

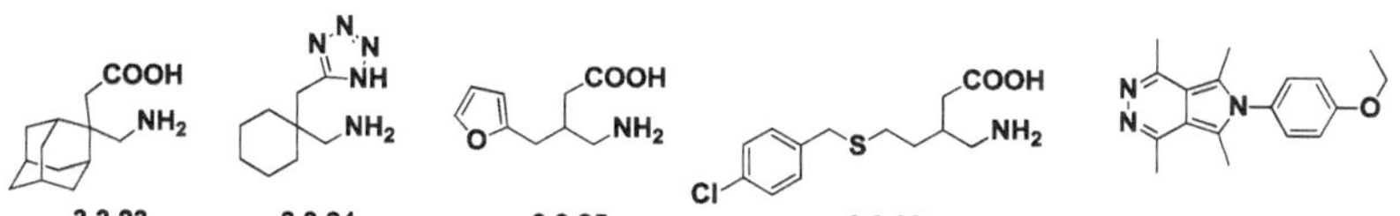

3.3.23 3.3.24 3.3.25 3.3.26 3.3.27

FIG. 3.28 **Structures of some newly disclosed gabapentinoids.**

Pregabalin–Lyrica

Pregabalin is used to treat epilepsy, anxiety disorder, and neuralgia- and fibromyalgia-generated pain. It is a medicine included on the list of Top 200 Drugs by sales for the 2010s. Side effects include severe allergic reactions; chest pain; confusion; arrhythmia and tachycardia; fever, chills; inability to control urination; loss of coordination; memory loss; mental or mood changes (anxiety, depression, restlessness, irritability, panic attacks, feeling "high," behavior changes, suicidal thoughts or attempts); seizures; shortness of breath; speaking problems; weight gain; swelling; tremor; trouble sleeping; trouble walking; unusual bruising or bleeding; tiredness or weakness; vision changes. This is not a complete list of all side effects that may occur.

The first none enantioselective synthesis of pregabalin (**3.3.22**) was developed starting with conjugate addition of nitromethane to ethyl 5-methylhex-2-enoate (**3.3.28**) followed by hydrogenation at atmospheric pressure, producing product (**3.3.30**) subjected to acid hydrolysis to produce the desired product [112] (Scheme 3.10.).

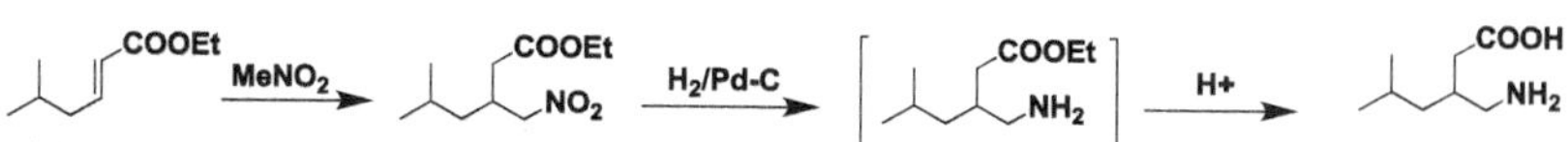

SCHEME 3.10 **Synthesis of pregabalin.**

The first synthesis enantioselective synthesis of pregabalin (**3.3.22**) was proposed [113] (Scheme 3.11.) starting from 4-methylpentanoic acid (**3.3.31**), which was converted into acid chloride (**3.3.32**), followed by acylation of the (4R,SS)-(+)-4-methyl-5-phenyl-2-oxazolidinone (**3.3.33**) to produce the acyloxazolidinone (**3.3.34**), which was alkylated with benzyl bromoacetate to produce the product (**3.3.35**) in >95% enantiomeric excess. The chiral auxiliary on the acyloxazolidinone (**3.3.35**) was removed by lithium hydroxide/hydrogen peroxide treatment, followed with reductive work-up with sodium bisulfite.

The resulting acid (**3.3.36**) was reduced to the alcohol (**3.3.37**) using borane dimethylsulfide complex. The last was converted to the azide (**3.3.38**), which was then converted to pregabalin (**3.3.22**) by hydrogenation on palladium catalyst.

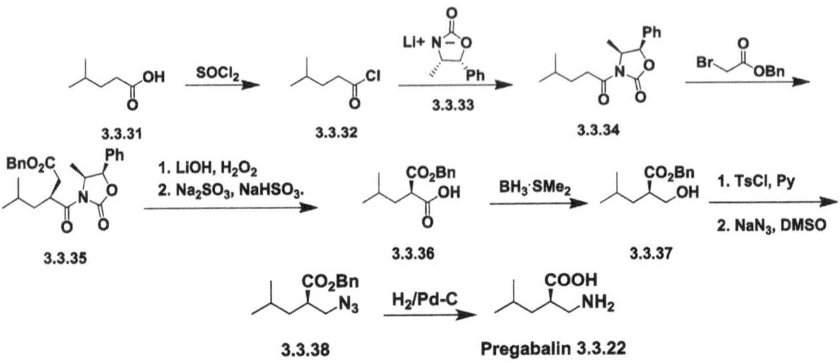

SCHEME 3.11 **Enantioselective synthesis of pregabalin.**

Several other schemes of synthesis are patented [113-119].

Pregabalin is used to relieve neuropathic pain, shingles, fibromyalgia, and seizures, and for pain management.

Calcium Channel Blockers

Voltage-gated Ca^{2+} channels (VGCC) play an important role in the development and maintenance of neuropathic pain. Currently available drugs that target the Ca^{2+} channel include gabapentin, pregabalin, and ziconotide [120], a synthetic version of the ω-conotoxin peptide derived from the toxin of the "marine cone snail", H-Cys-Lys-Gly-Lys-Gly-Ala-Lys-Cys-Ser-Arg-Leu-Met-Tyr-Asp-Cys-Cys-Thr-Gly-Ser-Cys-Arg-Ser-Gly-Lys-Cys-NH_2.

There are no VGCC blockers in clinical trials, but new compounds were reported recently [121,122]. Compounds currently in development include N-type Ca^{2+} channel blockers directly linked to pain transmission. Only one of them, NMED-160 (MK-6721), whose structure is not disclosed, achieved phase II clinical trials, but "does not demonstrate the ideal pharmaceutical characteristics considered necessary to advance the compound further in development" [123]. However, N-type blockers are represented by dihydropyridine derivatives like cilnidipine (**3.3.39**), an analog of nifedipine, a Ca^{2+} blocker for treatment of hypertension and ischemic heart disease, cyproheptadine (**3.3.40**), a potent L-type Ca^{2+} blocker, and histamine and serotonin receptors.

N-type blockers are also represented by anthranyl amide derivatives (GVIA) (**3.3.41**), some series of piperazine (**3.3.42**), piperidine (**3.3.43**), and dihydroquinazoline derivatives (**3.3.44**), and are excellently reviewed in [124,125] (Fig. 3.29.).

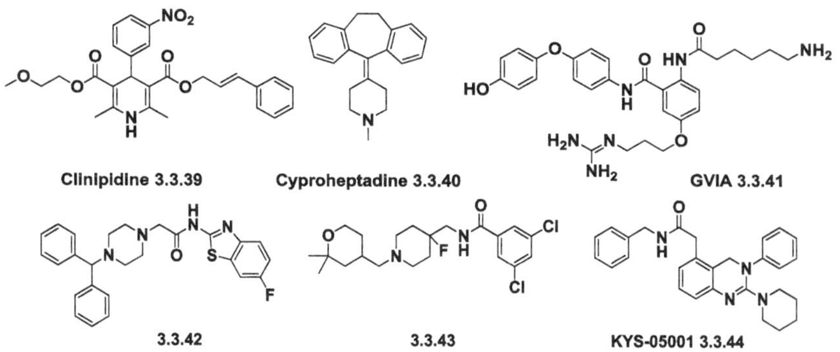

FIG. 3.29 Calcium channel blockers effective in the treatment of pain.

Sodium Channel Blockers

The therapeutic use of sodium channel blockers can be traced to 1884, when cocaine was introduced as the first local anesthetic drug after which the "caine" class of local anesthetics as sodium channel blocking drugs became the subject of research for more than a century. The aminoester class of "caines" includes as benzocaine (**3.3.1**) and procaine (**3.3.1**). The aminoamide class includes lidocaine (**3.3.1**) and bupivacaine (**3.3.1**). Local aesthetics, blocking voltage-gated sodium channels work as analgesics while retaining consciousness. However, their use is limited by their adverse cardiac and CNS effects (dizziness, light-headedness, somnolence) [126-129]. In addition to the local anesthetics designed for neural blockade, compounds of other "old" pharmacological classes such as antiarrhythmic mexiletine (**3.3.45**), tricyclic antidepressant amitriptyline (**3.3.1**), and the anticonvulsants diphenylhydantoin (**3.3.46**), carbamazepine (**3.3.47**), and lamotrigine (**3.3.48**) are used for the treatment of long-term "neuropathic" pain.

Some newly developed sodium channel blockers—oxcarbazepine (**3.3.49**), crobenetine (**3.3.50**), and ralfinamide (**3.3.51**)—also have proven effective. All of the clinical agents, however, have other diseases as their primary indication. Some novel sodium channel blockers, such as V102862 (**3.3.52**) and PPPA (**3.3.53**), possess binding affinities on the order of 10 nanomolar (nM), improved state-dependence of action, and good in vivo analgesic effects, accompanied by a low side-effect profile. Zonisamide (**3.3.54**), proposed for treatment of chronic cancer pain, has also been studied for the treatment of neuropathic pain. Ralfinamide (**3.3.52**) and BW-4030W92 (**3.3.55**) are in Phase II trials in patients with neuropathic pain [130-136] (Fig. 3.30.).

Potassium Channel Modulators

Opening of some K^+ channels plays an important role in the antinociception induced by opioids, NSAIDs, TCAs, agonists of serotonin 5-HT_{1A}, adrenoceptors agonists, CB receptors, GABA agonists, muscarinic M_2 agonists, and adenosine A_1. A limited number of potential K^+ channel modulators, like the Kv7 and K_{ATP} channels, have been identified as targets for the development of antinociceptive therapies.

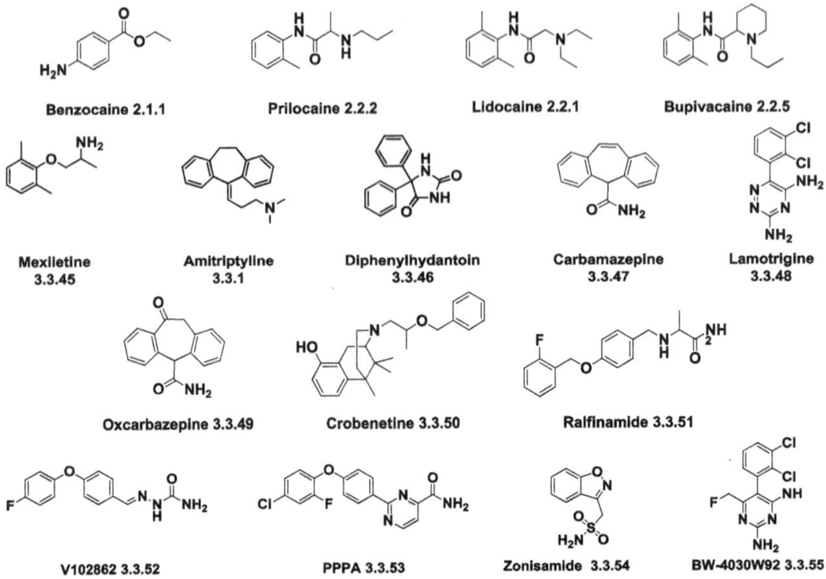

FIG. 3.30 Sodium channel blockers effective in pain treatment.

Data obtained from animal pain models and clinical observations of flupirtine (**3.3.56**) support the concept, that drugs capable of targeting the Kv7 channels could be effective analgesic agents for neuropathic pain and chronic pain conditions, like musculoskeletal pain, osteoporosis-related pain, cancer pain, and migraine pain. Examples of some significant first generation Kv7 potassium channel openers (**3.3.56** to **3.3.58**) are presented in Fig. 3.31.

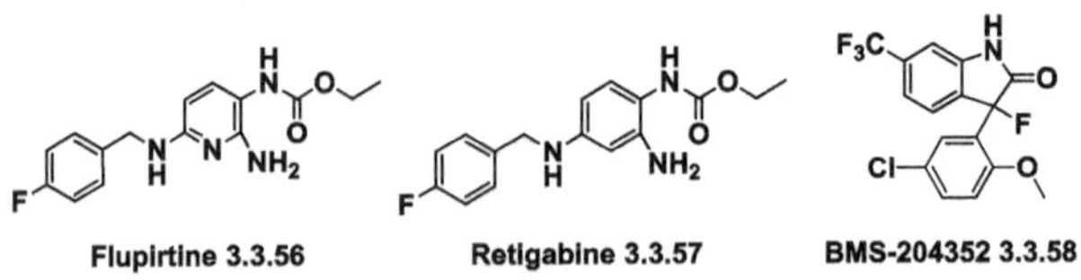

FIG. 3.31 First-generation Kv7 potassium channel openers.

Second-generation Kv7 potassium channel openers (**3.3.59** to **3.3.62**) are presented in Fig. 3.32.

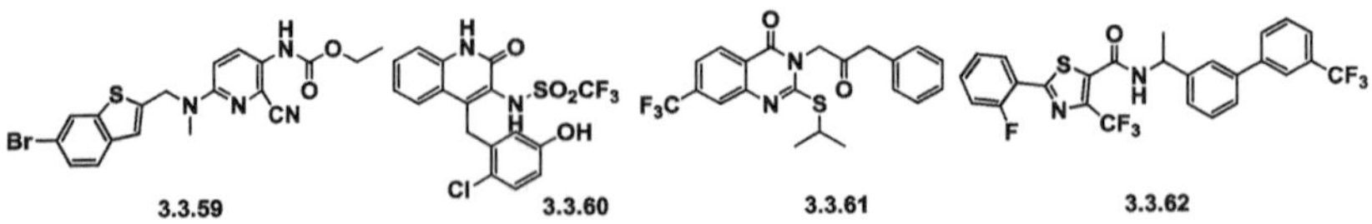

FIG. 3.32 Second-generation Kv7 potassium channel openers. Potassium channel modulators effective in the treatment of pain.

Selective Kv7 channels blockers (negative modulators), such as linopirdine (**3.3.63**) and XE-991 (**3.3.64**), are presented in Fig. 3.33. Big pharma does not show interest in further development of Kv7 blockers, but Kv7 openers still remain in focus as compounds with a novel mechanism of analgesic action [137-139].

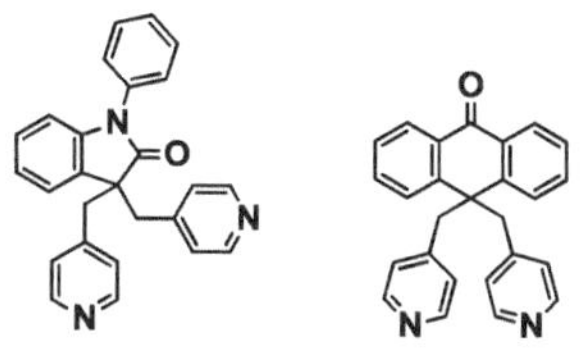

3.3.63 XE-991 3.3.64

FIG. 3.33 Selective Kv7 channels blockers (negative modulators).

A major limiting factor in assessing success in in vivo pain models for Kv7 blockers is their neurotoxic profile. Retigabine, for example, impairs motor function on testing.

Glutamate Receptor Modulators

Glutamate is the main excitatory neurotransmitter in the CNS. Glutamatergic transmission is primarily mediated by a family of metabotropic G-protein–coupled glutamate receptors, which include NMDA, α-amino-3-hydroxy-5-methyl-4-isoxazolepropionic acid (AMPA), and kainate receptors.

Excitatory amino acid glutamate plays a significant role in nociceptive processing, in pain sensation and transmission. Glutamate acts at several types of receptors, including ionotropic (directly coupled to ion channels) and metabotropic (directly coupled to intracellular second messengers). It also interacts with the opioid system, and coadministration of glutamate receptor antagonists with opioids may enhance analgesia while reducing the development of opioid tolerance and dependence. Application of glutamate, or agonists selective for one of the several types of glutamate receptor, induces nociceptive behaviors [140-142].

N-Methyl-D-Aspartate Antagonists

NMDA receptor antagonists such as ketamine (**3.3.65**), dizocilpine (**3.3.66**), bicifadine (**3.3.67**), dextromethorphan (**3.3.68**), and phencyclidine (**3.3.69**) (Fig. 3.34.) have shown positive activity in a number of pain models. Although some clinical data, search has yet failed to produce a truly effective drug for the relief of pain. These compounds will probably never see widespread clinical use because of their psychotomimetic side effects.

The most optimistic results in terms of overall pain relief have been obtained with compounds that provide potent blockade of the NMDA receptor and readily penetrate the CNS, such as ketamine. Moreover, preclinical data suggest that combinations of morphine and NMDA antagonists may exert synergistic effects in neuropathic pain models [143-145].

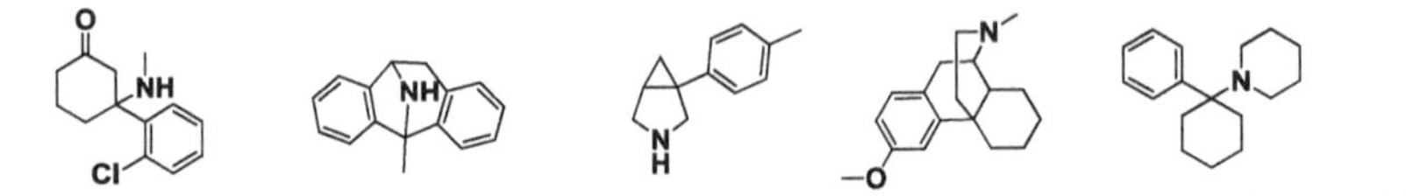

Ketamine 3.3.65 Dizocilpine 3.3.66 Bicifadine 3.3.67 Dextromethorphan 3.3.68 Phencyclidine 3.3.69

FIG. 3.34 **NMDA receptor antagonists effective in the treatment of pain.**

α-Amino-3-Hydroxy-5-Methyl-4-Isoxazolepropionic Acid Receptor Antagonists

Several pharmacological studies have confirmed role for AMPA receptors in pain states [146-148], but the most important clinical application for the AMPA receptor antagonists will probably be as neuroprotectants in neurodegenerative diseases, such as epilepsy, for the treatment of patients who are not responding to current therapies. Examples of AMPA/kainate receptor antagonists that are used in pain research are NBQX (**3.3.70**) and CNQX (**3.3.71**), which are shown in Fig. 3.35.

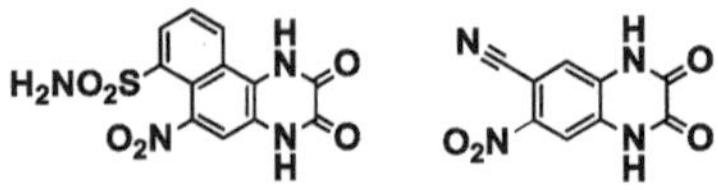

NBQX 3.3.70 CNQX 3.3.71

FIG. 3.35 **AMPA receptor antagonists effective in the treatment of pain.**

Kainate Receptor Antagonists

Preclinical and clinical and data indicate that kainate receptors play a role in pain signaling. Selective high-affinity decahydroisoquinoline derivative-kainate receptor antagonists LY-382884 (**3.3.72**), LY-294486 (**3.3.73**), LY-293558 (**3.3.74**), and LY-466195 (**3.3.75**) (Fig. 3.36.) display efficacy in the formalin model of persistent pain and provide evidence for the involvement of kainate receptors in migraine treatment. Tezampanel (LY-293558) (**3.3.74**), proposed as an agent to suppress the withdrawal symptoms from opioids, has anxiolytic effect [149-154].

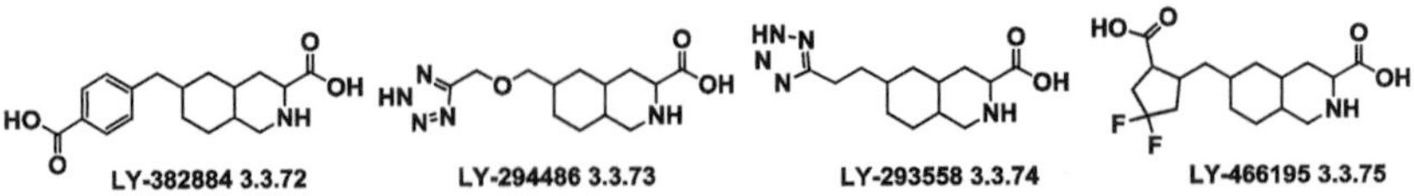

LY-382884 3.3.72 LY-294486 3.3.73 LY-293558 3.3.74 LY-466195 3.3.75

FIG. 3.36 **Kainate receptor antagonists effective in the treatment of pain.**

TRPV1 Antagonists (Transient Receptor Potential Cation Channel, Subfamily V, Member 1 Antagonists)

The largest group of noxious stimulus detectors is the transient receptor potential (TRP) channel family. Transient receptor potential channels (TRP channels) are a group of ion channels located in numerous human and animal cell types. TRP channels respond to mechanical, thermal, chemical, and other stimuli. There are six divisions of TRP channels: TRPC (canonical), TRPV (vanilloid), TRPM (melastatin), TRPA (ankyrin), TRPP (policystin), and TRPML (mucolipin). The potential of TRPV1 antagonists as analgesics has been widely reviewed [155-163].

TRPV1 receptors are involved in transmission and modulation of pain sensitization associated with tissue injury and inflammation. TRPV1 is involved in both afferent (sensation of pain) and efferent (neurotransmitter and neuropeptide release) functions and can mediate both inflammation and pain. Attempts to implement TRPV1 antagonists as potential analgesics is an alternative strategy for creation of novel drugs targeting the pain pathway at the very beginning of noxious stimuli at the peripheral terminals, and blockade of TRPV1 by antagonists and desensitization of TRPV1 by agonists represent alternative therapeutic strategies for the treatment of chronic pain. The most popular, studied, and characterized TRPV1 antagonist ligand is capsaicin (**3.3.76**) (Fig. 3.37.), the pungent ingredient found in the hot chilli pepper.

Capsaicin is currently used in topical ointments to relieve neuralgia (shingles), joint pain from osteoarthritis, and minor pain associated with rheumatoid arthritis or muscle sprains.

TRPV1 antagonists hold big promise in pain relief. However, other physiological functions that are mediated by TRPV1 channels also may be affected. Some studies have shown that TRPV1 antagonists generate a long-lasting hyperthermia, and the body temperature rises as much as 3°C. TRPV1 antagonists may also produce unwanted cardiovascular side effects. Some examples of the diverse chemical structures of TRPV1 antagonists (**3.3.77** to **3.3.81**) as promising analgesics are presented in Fig. 3.35 [155-163].

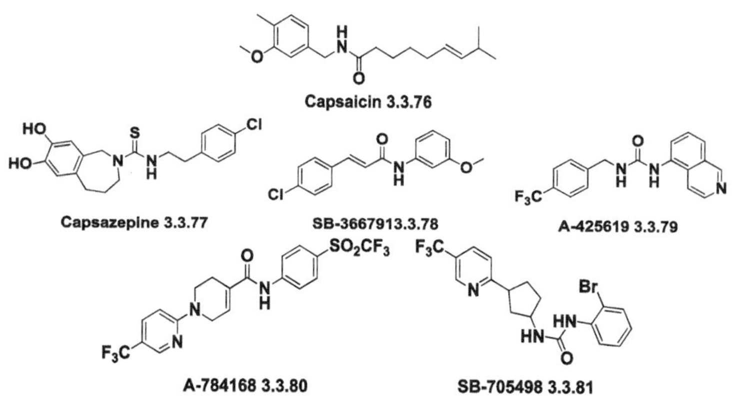

FIG. 3.37 TRPV1 antagonists that are effective in the treatment of pain.

α_2-Adrenergic Agonists

Coactivation of spinal α_2-adrenergic receptors and opioid receptors produces a well-documented antinociceptive synergy that has become part of the anesthesiologist's armamentarium, but the mechanism underlying this synergy remains unclear.

Clonidine, an α_2-adrenergic agonist, has a well-established analgesic profile in clinics and since the 1970s, has found wider application, particularly as an adjunct to anesthetics and analgesics in perioperative settings. There is not much new by way of chemical structures after the prototype clonidine (**3.3.82**). Clinically useful compounds are dexmedetomidine (**3.3.83**) and tizanidine (**3.3.84**) (Fig. 3.38.), but except for significant analgesic effect, all α_2 agonists produce diverse side effects, including anxiolysis, sedation, and sympatholysis [164-167].

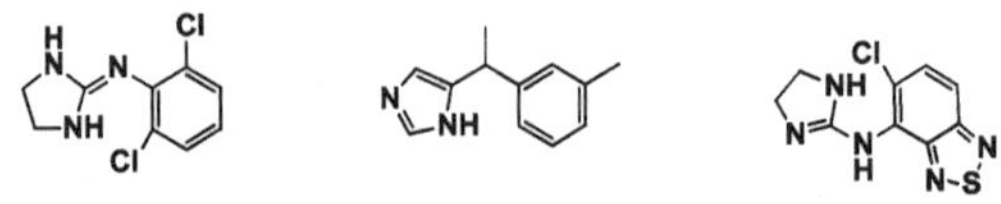

FIG. 3.38 **α_2-Adrenergic agonists effective in the treatment of pain.**

Some derivatives of imidazoles, imidazolines, and hydroxyethyl-thioureas that have α_2-adrenergic agonists with analgesic activity have been published (**3.3.85** to **3.3.88**) (Fig. 3.39.) [168,169].

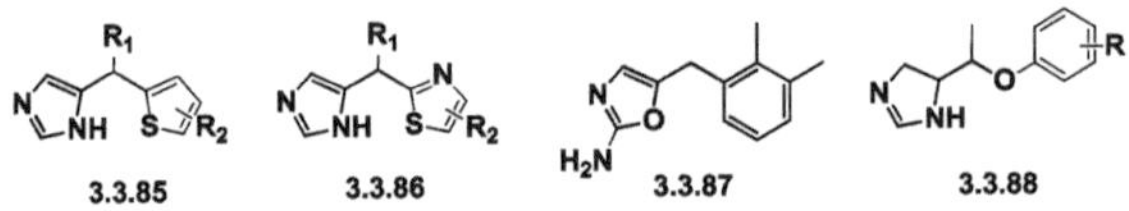

FIG. 3.39 **Some α_2-adrenergic agonist compounds with analgesic activity.**

Modulators of Nicotinic Acetylcholine Receptors

The analgesic properties of nicotine have generated attempts to develop compounds targeting nicotinic acetylcholine receptors (nAChRs). Acetylcholine mediates effects through both the nAChR (ligand-gated ion channels) and the G-protein–coupled muscarinic receptors. Only nAChR agonists have been reported as possible analgesics, although nAChR antagonists could also have an analgesic action. The effects of receptor agonists, termed *cholinomimetics* in analgesia, are well established. nAChR agonists exhibit antinociceptive, antihyperalgesic, and antiallodynic effects. These compounds successfully inhibit pain in different preclinical and clinical pain models without acting through an opioid mechanism, although suggesting a definite therapeutic potential. Various problems associated with the use of nAChR agonists as analgesics have been identified. Several nAChR agonist compounds, like tebanicline (ABT-594)

(**3.3.89**), sofinicline (ABT-894) (**3.3.90**) alkaloid epibatidine (3.3,91) and its analogues A-85380 (**3.3.92**), SIB-1663 (**3.3.93**) , ABT-202 (**3.3.94**), and ABT-366833 (**3.3.95**). (Fig. 3.40.), have been proposed as analgesics, but in general, efforts to create new analgesics targeting the cholinergic system have been largely unsuccessful. Tebanicline, in particular, has unacceptable gastrointestinal side effects. Other compounds have many undesirable effects, including addictive properties [170-175].

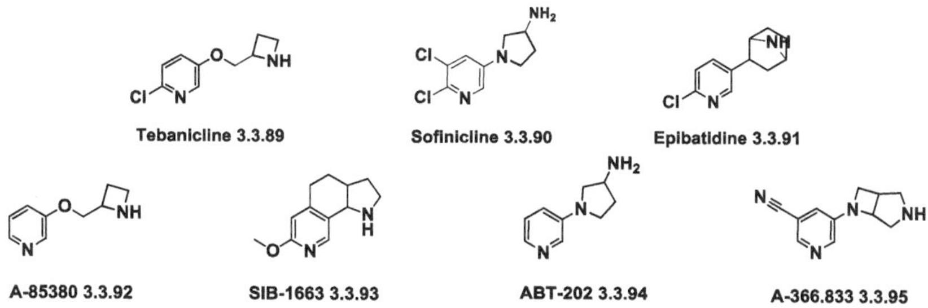

FIG. 3.40 Modulators of nAChRs with analgesic activity.

Neurokinin-1 Antagonists

Based on recent studies that substance P contributes to pain transmission in the CNS and in inflammatory processes [176,177], it was proved that NK1 antagonists could be developed as analgesic drugs. Much research has focused on the development of substance P antagonists as analgesics. Compounds like dapitant (RPR-100893) (**3.3.96**), lanepitant (LY-303870) (**3.3.97**), aprepitant (MK-869) (**3.3.98**), CP-99,994 (**3.3.99**), CI-1021 (**3.3.100**), TKA731 (**3.3.101**) (Fig. 3.41.), and others showed good activity in different models of pain and were developed up to Phase II clinical trials for the treatment of neuropathic pain, but showed no significant effects [178]. "Despite the identification of high affinity and selective substance P (NK1) receptor antagonists and a plethora of preclinical data supporting an analgesic profile of these agents, the outcome from clinical trials has been extremely disappointing with no clear analgesic efficacy being observed in a variety of pain states. This has led the pain community to seriously question the predictability and utility of preclinical pain assays, especially for novel targets" [179].

However, some literature data proved that definite NK1 antagonists (**3.3.102, 3.3.103**) (Fig. 3.42.) could be further developed as analgesics [180,181].

Cannabinoid Receptor Modulators

"Cannabis (marijuana) has been used anecdotally for more than 5000 years to treat a variety of conditions including hysteria, delirium, insomnia, nausea, anorexia, glaucoma, and pain." [182]. Cannabis contains more than

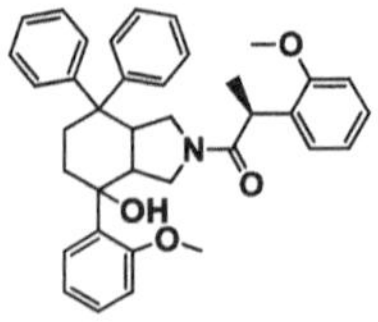

Dapitant 3.3.96

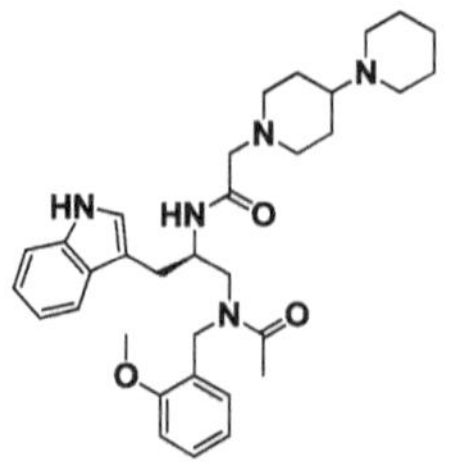

Lanepitant 3.3.97

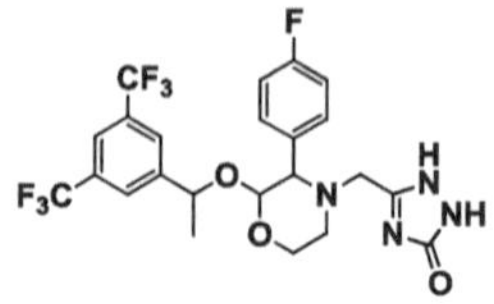

Aprepitant 3.3.98

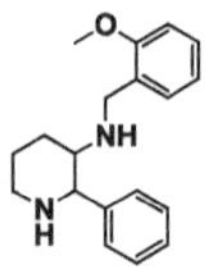

CP-99,994 3.3.99

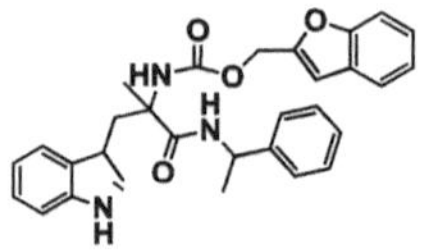

CI-1021 3.3.100

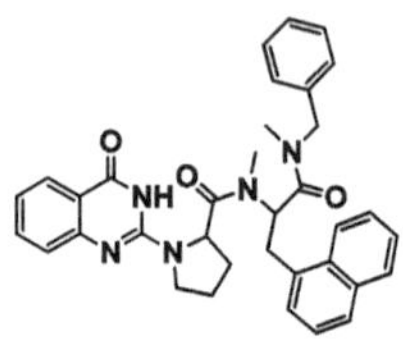

TKA731 3.3.101

FIG. 3.41 NK1 antagonists with analgesic activity.

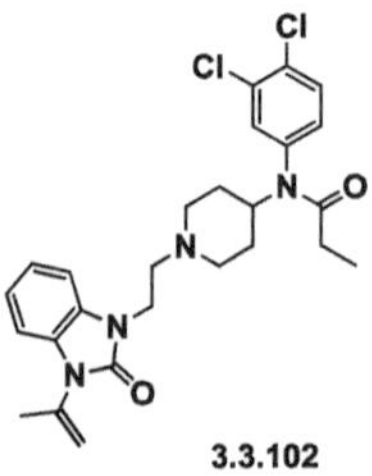

3.3.102

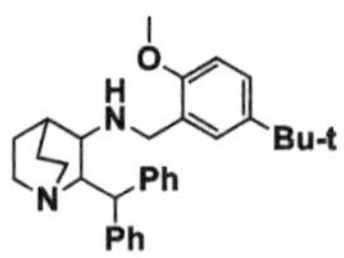

Maropitant 3.3.103

FIG. 3.42 NK1 antagonists with analgesic activity as potential new analgesics.

400 chemical components and produces many effects—both beneficial and harmful. While generating many psychoactive states in human behavior, Cannabis also has a potent analgesic effect on a variety of pain types. A number of Cannabis constituents provide therapeutic pain relief. Although cannabis products have been used for thousands of years to treat pain, it was not until the discovery of the "cannabis receptor" in the late 1980s that modern medicine started to take cannabis seriously. Cannabinoids produce their effect by interacting with CB1 and CB2 G-protein–coupled receptors. The CB1 receptor is recognized as an important therapeutic target for pain. An increasing amount of data demonstrate that CB receptor agonists produce relief of pain and neuroinflammation in a variety of animal models. However, the psychotropic side effects put a brake on CB receptor agonist use. The analgesic activity in chronic pain states may be mediated via spinal receptors, as well as peripheral CB1 and, potentially, CB2 receptors, and there is increasing evidence that selective CB1 agonists or mixed CB1/ CB2 agonists could be devoid of CNS side effects. There is a considerable

interest in developing cannabinoid agonists. Many new cannabinoid ligands have been synthesized and studied as analgesics, covering a wide variety of novel structural scaffolds, separating them into Δ^9-tetrahydrocannabinol (Δ^9-THC) (**3.3.104**) and its derivatives. New cannabinoids could be classified as derivatives of classical cannabinoids, for example, dexanabinol (HU-211) (**3.3.105**) and ajulemic acid (CT-3) (**3.3.106**), an analogue of Δ^9-THC metabolite that shows promise as an analgesia without psychoactive effect and nonclassical cannabinoids, in which the pyran ring is opened (CP-55,940) (**3.3.107**). Other cannabinoid agonists represented by indole – (JWH-151) (**3.3.108**), AM-1241 (**3.3.109**), WIN-55,212 (**3.3.110**)), pyrazole– (rimonabant (SR-141716 () (**3.3.111**), surinabant (SR-147778) (**3.3.112**), and benzimidazole derivatives represented by compound (**3.3.113**) (Fig. 3.43.). There are a number of excellent recent reviews that discuss these CB receptors and their ligands in considerable detail [183-190]. The design of novel compounds that either specifically target peripheral CB1 receptors or display high selectivity for CB2 receptors may offer new analgesics that avoid the adverse events seen with cannabinoid structures.

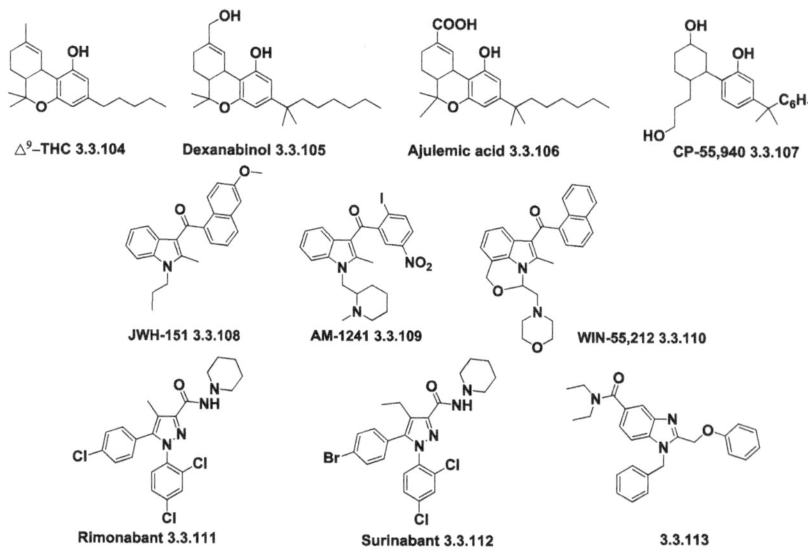

FIG. 3.43 Cannabinoid receptor modulators effective in the treatment of pain.

New structures for antagonists or inverse agonists of the cannabinoid CB1 receptor, among which are compounds (**3.3.114** to **3.3.117**), have been proposed and reviewed [191-193] (Fig. 3.44.).

The CB2 agonist annabinor (**3.3.118**), which is structurally related to CP-55,940, is currently in Phase II clinical trials for treatment of chronic pain (Fig. 3.45.).

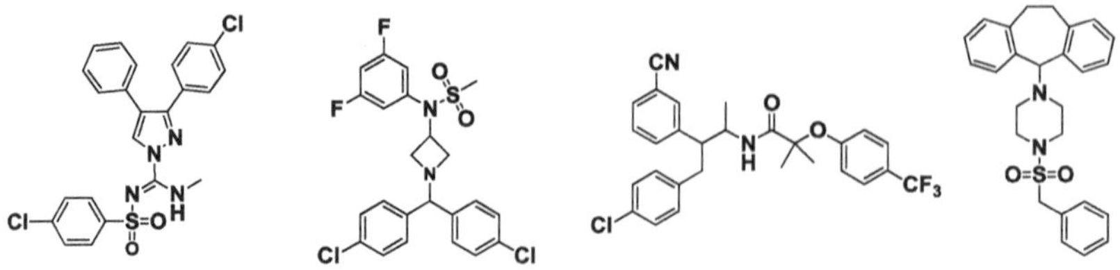

SLV-319 3.3.114 AVE-1625 3.3.115 Taranabant 3.3.116 3.3.117

FIG. 3.44 Structures for antagonists or inverse agonists of the cannabinoid CB1 receptor.

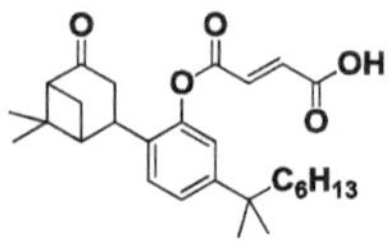

Cannabinor 3.3.118

FIG. 3.45 Structure of cannabinor.

The structures of the alkylindole family of compounds of CB2 agonists series AM-1241 (**3.3.119**), WIN-55,212 (**3.3.120**), JVH-015 (**3.3.121**) (and it's substituted analogues JVH-151, JVH-120), A-796260 (**3.3.122**), and L-768242 (**3.3.123**) are shown in Fig. 3.46.

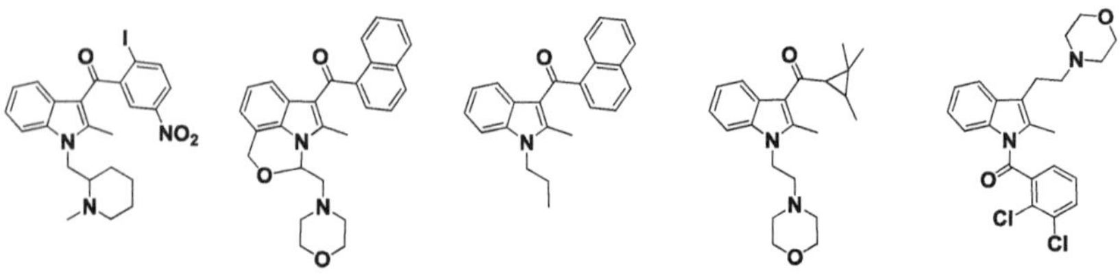

AM-1241 3.3.119 WIN 55,212 3.3.120 JVH-015 3.3.121 A-796260 3.3.122 L-768242 3.3.123

FIG. 3.46 Alkylindole family of CB2 agonists effective in the treatment of pain.

Glial Inhibitors

Modulation of resident immune cells, glia in the CNS plays an important role in creating and maintaining pain facilitation in chronic pain [194-206]. CNS glia cells become activated upon nerve injury and inflammation and release excitatory gliotransmitters, leading to pain sensitization. Glial inhibitors, such as fluorocitrate (**3.3.124**) [207-210] and minocycline (**3.3.125**) [207,211,212], were found to alleviate pain behavior in animal models of inflammatory, neuropathic, and postoperative pain. According to accumulated data, multiple other classes of compounds, like antimetabolite and antifolate drugs methotrexate (**3.3.126**), nonselective phosphodiesterase inhibitor ibudilast (**3.3.127**), inhibitors of proinflammatory cytokines propentofylline (**3.3.128**) and pentoxifylline (**3.3.129**), and thalidomide (**3.3.130**) derivatives with their potent antiinflammatory effects, are able to reduce glial activation [194-206,213-216] (Fig. 3.47.). Recently, it was found that CB2 agonists acting at the peripheral CB receptor agonist efficiently attenuate both inflammatory and neuropathic pain facilitation [217-226] and glial activity in the spinal cord [227,228].

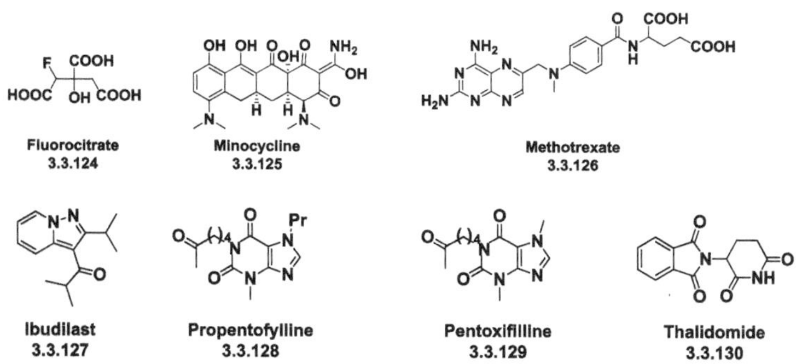

FIG. 3.47 Glial inhibitors effective in the treatment of pain.

Cholecystokinin Antagonists

Cholecystokinin (CCK) receptor ligands, especially the selective CCK antagonists, are promising compound drugs for use in the management of pain. CCK is a peptide originally discovered in the gastrointestinal tract. It is responsible for stimulating the digestion, but also for mediating a number of physiological processes in the CNS, producing behavioral changes such as anxiety, perturbation of memory, dysfunctioning neural pathways involved in neuropsychiatric disorders, and antagonizing opioid analgesia. CCK is able to attenuate the antinociceptive effect of morphine. Pain modulation appears to be one of the interesting effects of CCK antagonists.

CCK receptors have been divided into two subtypes, namely, the CCK1 and CCK2 receptors belonging to the class of G-protein–coupled receptors and represented by various chemical structures, including dipeptoids, benzodiazepine, quinazolinone, pyrazolidinone, thiazole, and amino acid derivatives that have both excellent selectivity and high affinity for either CCK1 or CCK2 receptors.

CCK antagonists are of certain interest in the management of pain. Two compounds, CCK2 antagonist L-365 260 (**3.3.131**) and CCK1 antagonist devazepide (**3.3.132**) (Fig. 3.48.), achieved Phase II of clinical trials. Combined administration of CCK1 antagonist PD-134 308 (**3.3.133**) with the enkephalinase inhibitor RB-101 (**3.3.134**) (Fig. 3.48.) did not induce development of tolerance to antinociception. CCK antagonists as adjuvants are usually administered alongside an opioid analgesic. CCK antagonists may reduce or mitigate the commonest side effects of opioid analgesics, such as nausea, vomiting, constipation, and drowsiness [229-232].

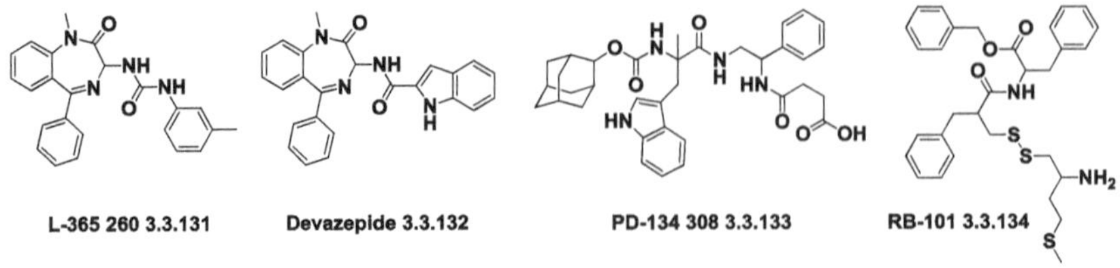

L-365 260 3.3.131 Devazepide 3.3.132 PD-134 308 3.3.133 RB-101 3.3.134

FIG. 3.48 CCK receptor ligands effective in the treatment of pain.

Triptans

Migraine is a common type of moderate to severe headache. It is a neurovascular disease, that affects millions of people and is more common in women than in men; its pathogenesis is still unclear. Episodes of pain usually follow a migraine.

Triptans are agonists of serotonin 5-HT_{1B} and 5-HT_{1D} receptors that are generally effective for treatment of migraine sufferers. It seems that triptans work in three different ways: peripheral inhibition of release of CGRP (calcitonin gene-related peptide) and substance P; modulation of second-order neurons centrally in the trigeminocervical pathway and via vasoconstriction; and narrowing of blood vessels in the brain to relieve swelling and thereby relieving the headache. Triptans have revolutionized the treatment of migraine and are currently the drugs of choice to treat it. They share many similarities, but also have important differences from one another [233-237]. Seven triptans are available worldwide and have proven efficacy in the relief of the pain and symptoms accompanying migraine crises. Triptans are classified to three generations, concerning their chemical structures and receptor subtype selectivity: first generation—sumatriptan (**3.3.136**); second generation—almotriptan (**3.3.136**), frovatriptan (**3.3.137**), donitriptan (**3.3.138**), avitriptan (**3.3.139**), and eletriptan(**3.3.140**); and third generation—rizatriptan (**3.3.141**) (Fig. 3.49.) and other analogues as selective 5-HT_{1D} receptor agonists. Eletriptan (**3.3.131**) and rizatriptan (**3.3.132**) have emerged as the most efficacious and are medications that are included on the list of Top 200 Drugs by sales for the 2010s.

Eletriptan–Relpax

Eletriptan (**3.3.140**) is an effective medication in the treatment of moderate to severe migraine attacks in adults [238-242]. The current manufacturing route to eletriptan [243] is based on well-developed chemistry (Scheme 3.11.). Reacting 5-bromoindole (**3.3.142**) with the Grignard reagent to get N-MgBr compound (**3.3.143**) and then with *N*-benzyloxycarbonyl-D-proline acid chloride (**3.3.144**) to produce the corresponding benzyl 2-(5-bromo-1H-indole-3-carbonyl)-pyrrolidine-1-carboxylate (**3.3.145**). Reducing the last with lithium aluminium hydride, 5-bromo-3-((1-methylpyrrolidin-2-yl)methyl)-1H-indole (**3.3.146**) was prepared, reaction of which with phenyl vinyl sulfone in the presence of

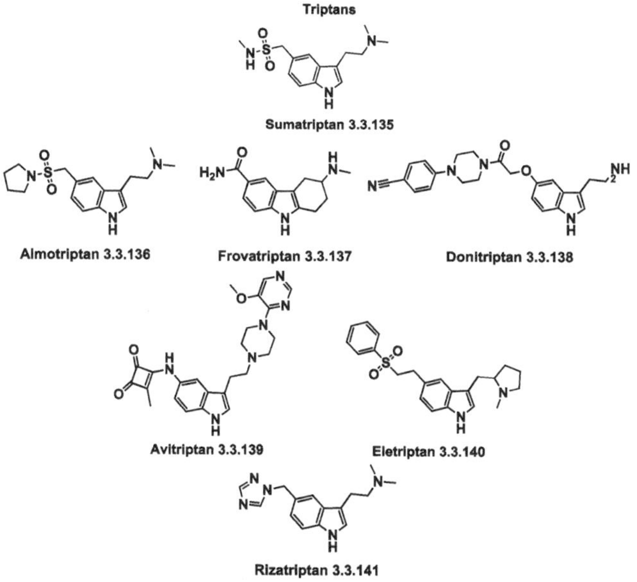

FIG. 3.49 Structures of triptans that provide relief for migraine sufferers.

$Pd(OAc)_2$ and tri-p-tolylphosphine produced 3-((1-methylpyrrolidin-2-yl) methyl)-5-(2-(phenylsulfonyl)vinyl)-1H-indole (**3.3.147**), the double bond of which was hydrogenated on Pd/C to produce the desired eletriptan (**3.3.140**). Other approaches are described in other papers [244,245] and in more than 60 patents.

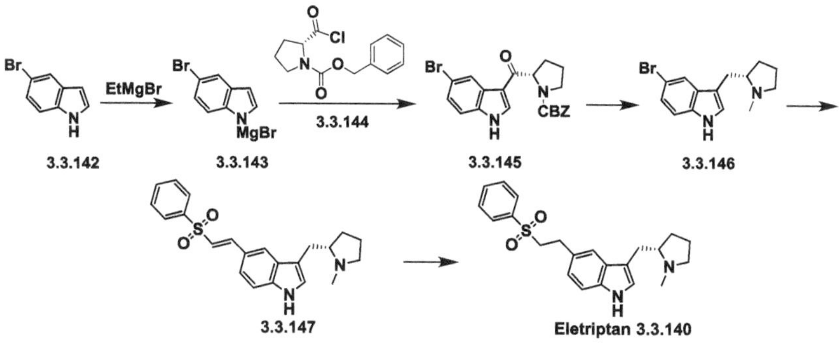

SCHEME 3.12 Synthesis of eletriptan.

Rizatriptan–Maxalt

Rizatriptan (**3.3.141**) is used to treat acute migraine attacks [246,247]. Its first proposed manufacturing consisted of indole Fischer cyclization of intermediate hydrazones obtained by the reaction of 1-(4-hydrazino phenyl) methyl-1,2,4-triazole (**3.3.148**) with 4-N,N,dimethylamino- (**3.3.149**) or 4-chlorobutanal dimethyl acetal (**3.3.1--50**), directly to rizatriptan (**3.3.141**), or via synthesis of intermediate 5-((1H-1,2,4-triazol-1-yl)methyl)-3-(2-chloroethyl)-1H-indole (**3.3.151**) followed by transformation to the desired rizatriptan (**3.3.141**) [248,249]. Alternative synthetic methods are also proposed [250-255] (Scheme 3.13.).

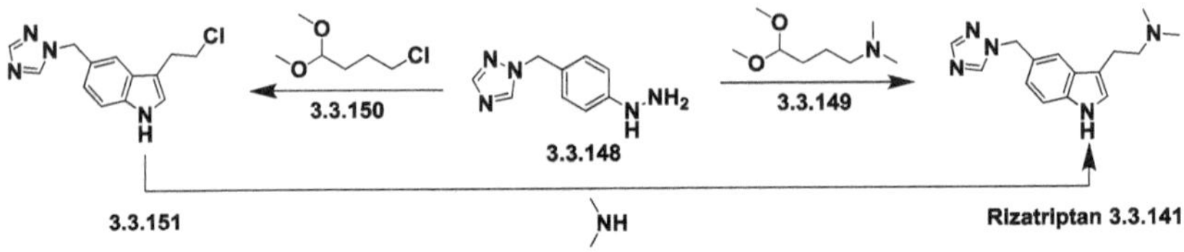

SCHEME 3.13 Synthesis of rizatriptan.

Thus, pain and pain syndromes are a very complex phenomena that are the result of many of contributing factors, and the variety of types of pain and the plethora of possible pathways to produce pain allow the implementation of different strategies to create of new analgesics.

REFERENCES

1. Woolf, C. J.; Borsook, D.; Koltzenburg, M. Mechanism-based classification of pain and analgesic drug discovery. In *Pain: Current Understanding, Emerging Therapies and Novel Approaches to Drug Discovery;* Bountra, C., Munglani, R., Schmidt, W. K., Eds.; Marcel Dekker, 2003.
2. Melnikova, I. Pain market. *Nat. Rev. Drug Discovery* **2010,** *9,* 589–590.
3. Vardanyan, R. S.; Hruby, V. J. *Synthesis of Essential Drugs;* Elsevier, 2006.
4. http://www.drugs.com/top200.html.
5. Henke, C. O.; Vaughen, J. V. Reduction of aryl nitro compounds, US 2198249 (1940).
6. Weil, H.; Traun, M.; Marcel, S. *Ber.,* **1922,** *55B,* 2664–2674.
7. Schaefer, W.; Winfried, H. S. Preparation of 5-aminosalicylic acid as a drug, DD 255941 (1988).
8. Sjoestrand, U. A method of preparing 5-aminosalicyclic acid, EP 253788 (1988).
9. Penning, T. D.; Talley, J. J.; Bertenshaw, S. R.; Carter, J. S.; Collins, P. W.; Docter, S.; Graneto, M. J.; Lee, L. F.; Malecha, J. W.; Miyashiro, J. M.; Rogers, R. S.; Rogier, D. J.; Yu, S. S.; Anderson, G. D.; Burton, E. G.; Cogburn, J. N.; Gregory, S. A.; Koboldt, C. M.; Perkins, W. E.; Seibert, K.; Veenhuizen, A. W.; Zhang, Y. Y.; Isakson, P. C. Synthesis and biological evaluation of the 1.5 diarylpyrazole class of cyclooxygenase-2 inhibitors: identification of 4-[5-(4-methylphenyl)-3-(trifluoromethyl)-1h-pyrazole-1-yl]benzenesulfonamide (SC-58634, Celecoxib). *J. Med. Chem.* **1997,** *40* (9), 1347–1265.
10. Fries, D. S. Opioid analgesics. In *Foye's Principles of Medicinal Chemistry,* 6th ed.; Lemke, T. L., Williams, D. A., Eds. Lippincott Williams & Wilkins, 2008; pp 652–678.
11. Mather, L. E. Trends in the pharmacology of opioids: Implications for the pharmacotherapy of pain. *Eur. J. Pain* **2001,** *5* (Suppl. A), 49–57.
12. Pasternak, G. W.; Letchworth, Sh. R. Future opioid analgesics: targeting the old and the new. *Curr. Opin. Centr. Periph. Nerv. Syst. Invest. Drugs* **1999,** *1* (1), 54–64.

13. Casy, A. F. Analgesics and their antagonists: recent developments. *Arzneim. Forsch.* **1978,** *22,* 149–227.
14. Dhawan, B. N.; Cesselin, F.; Raghubir, R.; Reisine, T.; Bradley, P. B.; Portoghese, P. S.; Hamon, M. International Union of Pharmacology. XII. Classification of opioid receptors. *Pharm. Rev.* **1966,** *48* (4), 567–592.
15. Gates, M.; Woodward, R. B.; Newhall, W. F.; Kunzli, R. Synthesis of ring systems related to morphine. IV. N-Methylisomorphinane. *J. Am. Chem. Soc.* **1950,** *72* (228), 1141–1146.
16. Gates, M.; Tschudi, G. The synthesis of morphine. *J. Am. Chem. Soc.* **1952,** *74,* 1109–1110.
17. Gates, M.; Tschudi, G. Synthesis of morphine. *J. Am. Chem. Soc.* **1956,** *78,* 1380–1393.
18. Iijima, I.; Minakuikava, J.; Jacobson, A. E.; Brossi, A.; Rice, K. C. Studies in the (+)-morphinan series. *J. Org. Chem.* **1978,** *43,* 1462–1463.
19. Beyerman, H. C.; Lie, T. S.; Maat, L.; Bosman, H. H.; Buurman, E.; Bijsterveld, H. J. M.; Sinnige, H. J. M. A convenient synthesis of codeine and morphine. *Rec. Trav. Chim.* **1976,** *95,* 24–25.
20. Beyerman, H. C.; van Berkel, J.; Lie, T. S.; Maat, L.; Wessels, I. C. M.; Buurman, H. H.; Bijsterveld, E. J. M.; Sinnigie, H. J. M. Chemistry of opium alkaloids. *Rec. Trav. Chim.* **1978,** *97,* 127–130.
21. Rice, K. C. Synthetic opium alkaloids and derivatives. *J. Org. Chem.* **1980,** *45* (15), 3135–3137.
22. Evans, D. A.; Mitch, C. H. Studies directed towards the total synthesis of morphine alkaloids. *Tetrahedron Lett.* **1982,** *23* (3), 285–288.
23. Toth, J. E.; Hamann, P. R.; Fuchs, P. L. Studies culminating in the total synthesis of (dl)-morphine. *J. Org. Chem.* **1988,** *53* (20), 4694–4708.
24. Parker, K. A.; Fokas, D. Convergent synthesis of (±)-dihydroisocodeine in 11 steps by the tandem radical cyclization strategy. A formal total synthesis of (±)-morphine. *J. Am. Chem. Soc.* **1992,** *114,* 9688–9689.
25. Hong, C. Y.; Kado, N.; Overman, L. E. Asymmetric synthesis of either enantiomer of opium alkaloids and morphinans. Total synthesis of (-)- and (+)-dihydrocodeinone and (-)- and (+)-morphine. *J. Am. Chem. Soc.* **1993,** *115* (23), 11028–11029.
26. Mulzer, J.; Dürner, G.; Trauner, D. Formal total synthesis of (-)-morphine by cuprate conjugate addition. *Angew. Chem., Int. Ed.* **1996,** *35* (23–24), 2830–2832.
27. White, J. D.; Hrnciar, P.; Stappenbeck, F. Asymmetric total synthesis of (+)-codeine via intramolecular carbenoid insertion. *J. Org. Chem.* **1999,** *64* (21), 7871–7884.
28. Taber, D. F.; Neubert, T. D.; Rheingold, A. L. Synthesis of (-)-morphine. *J. Am. Chem. Soc.* **2002,** *124* (42), 12416–12417.
29. Trost, B. M.; Tang, W. Enantioselective synthesis of (-)-codeine and (-)-morphine. *J. Am. Chem. Soc.* **2002,** *124* (49), 14542–14543.
30. Uchida, K.; Yokoshima, S.; Kan, T.; Fukuyama, T. Total synthesis of (+/-)-morphine. *Org. Lett.* **2006,** *8* (23), 5311–5313.
31. Varin, M.; Barré, E.; Iorga, B.; Guillou, C. Diastereoselective total synthesis of (±)-codeine. *Chem. - Eur. J.* **2008,** *14* (22), 6606–6608.
32. Stork, G.; Yamashita, A.; Adams, J.; Schulte, G. R.; Chesworth, R.; Miyazaki, Y.; Farmer, J. J. Regiospecific and stereoselective syntheses of (±)-morphine, codeine, and thebaine via a highly stereocontrolled intramolecular 4 + 2 cycloaddition leading to a phenanthrofuran system. *J. Am. Chem. Soc.* **2009,** *131* (32), 11402–11406.
33. Metzger, H. Dihydromorphinones, DE 623821 (1935).
34. Mannich, C.; Lovenheim, H. Two new reduction products of codeine. *Arch. Pharm. (Weinheim)* **1920,** *258,* 295–316.
35. Krauss, W. Verfahren zur Darstellung von Dihydrokodeinon, DE 415097 (1925).

36. Lewenstein M. J.; Weiss, U. 14-Hydroxydihydromorphinone, US 2806033 (1957).
37. Weiss, U. Derivatives of morphine. I. 14-Hydroxydihydromorphinone. *J. Am. Chem. Soc.* **1955,** *77,* 5891.
38. Bentley, K. W. Thebaine and oripavine derivatives, US 3433791 (1969).
39. Werner, L.; Machara, A.; Adams, D. R.; Cox, P. D.; Hudlicky, T. Synthesis of buprenorphine from oripavine via N-demethylation of oripavine quaternary salts. *J. Org. Chem.* **2011,** *76* (11), 4629–4634.
40. Colle, R.; Vecchietti, V.; Dondio, G.; Ronzoni, S. Preparation of hydroindoloisoquinoline derivatives and analogs as δ-receptor agonists, WO 9301186 (1993).
41. Curran, M. P.; Robyns, G. W.; Scott, L. J.; Perry, C. M. Alvimopan. *Drugs* **2008,** *68* (14), 2011–2019.
42. Delaney, C. P.; Yasothan, U.; Kirkpatrick, P. Alvimopan. *Nat. Rev. Drug Discovery* **2008,** *7* (9), 727–728.
43. Leslie, J. B. Alvimopan; a peripherally acting mu-opioid receptor antagonist. *Drugs Today* **2007,** *43* (9), 611–625.
44. Thomas, J. B.; Atkinson, R. N.; Rothman, R. B.; Fix, S. E.; Mascarella, S. W.; Vinson, N. A.; Xu, H.; Dersch, C. M.; Lu, Y.-F.; Cantrell, B. E.; Zimmerman, D. M.; Carroll, F. I. Identification of the first trans-(3R,4R)-dimethyl-4-(3-hydroxyphenyl)piperidine derivative to possess highly potent and selective opioid kappa receptor antagonist activity. *J. Med. Chem.* **2001,** *44* (17), 2687–2690.
45. Knoll, A. T.; Meloni, E. G.; Thomas, J. B.; Carroll, F. I.; Carlezon, W. A., Jr. Anxiolytic-like effects of kappa-opioid receptor antagonists in models of unlearned and learned fear in rats. *J. Pharmacol. Exp. Ther.* **2007,** *323* (3), 838–845.
46. Beardsley, P. M.; Howard, J. L.; Shelton, K. L.; Carroll, F. I. Differential effects of the novel kappa opioid receptor antagonist, JDTic, on reinstatement of cocaine-seeking induced by footshock stressors vs cocaine primes and its antidepressant-like effects in rats. *Psychopharmacology (Berl)* **2005,** *183* (1), 118–126.
47. Carroll, F. I.; Harris, L. S.; Aceto, M. D. Effects of JDTic, a selective kappa-opioid receptor antagonist, on the development and expression of physical dependence on morphine using a rat continuous-infusion model. *Eur. J. Pharmacol.* **2005,** *524* (1–3), 89–94.
48. Bagley, J. R.; Kudzma, L. V.; Lalinde, N. L.; Colapret, J. A.; Huang, B.; Lin, B.; Jerussi, T. P.; Benvenga, M. J.; Doorley, B. M.; Ossipov, M. H.; Spaulding, T. C.; Spencer, H. K.; Rudo, F. G.; Wynn, R. L. Evolution of the 4-anilidopiperidine class of opioid analgesics. *Med. Res. Rev.* **1991,** *11* (4), 403–436.
49. Stanley, T. H.; Fentanyl, J. *Pain Symptom Manage.* **2005,** *29* (5S), S67–S71.
50. Davis, M. P. Fentanyl for breakthrough pain: a systematic review. *Expert Rev. Neurother.* **2011,** *11* (8), 1197–1216.
51. Stanley, T. H. The history and development of the fentanyl series. *J. Pain Symptom Manage.* **1992,** *7* (3 Suppl), S3–S7.
52. Casy, A. F. Opioid receptors and their ligands: recent developments. *Adv. Drug Res.* **1989,** *18,* 177–289.
53. Casy, A. F.; Huckstep, M. R. Structure-activity studies of fentanyl. *J. Pharm. Pharmacol.* **1988,** *40* (9), 606–608.
54. Casy, A. F.; Hassan, M. M. A.; Simmonds, A. B.; Staniforth, D. Structure-activity relations in analgesics based on 4-anilinopiperidine. *J. Pharm. Pharmacol.* **1969,** *21* (7), 434–440.
55. Casy, A. F. Analgesics and their antagonists: recent developments. *Prog. Drug Res.* **1978,** *22,* 149–227.

56. Famini, G. R.; Ashman, W. P.; Mickiewicz, A. P.; Wilson, L. Y. Using theoretical descriptors in quantitative structure-activity relationships: Opiate receptor activity by fentanyl-like compounds. *Quant. Struct.-Act. Relat.* **1992,** *11* (2), 162–170.
57. Vuckovic, S.; Prostran, M.; Ivanovic, M.; Dosen-Micovic, L. j.; Todorovic, Z.; Nesic, Z.; Stojanovic, R.; Divac, N.; Mikovic, Z. Fentanyl analogs: structure-activity-relationship study. *Curr. Med. Chem.* **2009,** *16* (19), 2468–2474.
58. Dosen-Micovic, L. j. Molecular modeling of fentanyl analogs. *J. Serb. Chem. Soc.* **2004,** *69* (11), 834–854.
59. Janssen, P. A. J.; Gardocki, J. F. Method for producing analgesia, US 3141823 (1964).
60. Janssen, P. A. J. N-(1-Aralkyl-4-piperidyl)alkanoic acid anilides, FR M2430 (1964).
61. Janssen, P. A. J.; Niemegeers, C. J.; Dony, J. G. The inhibitory effect of fentanyl and other morphine-like analgesics on the warm water induced tail withdrawal reflex in rats. *Arzneim. Forsch.* **1963,** *13,* 502–507.
62. Janssen, P. A. J. 1-(g-Aroylpropyl)-4-(N-arylacylamino)piperidines. FR 1344366 (1963).
63. Jonczyk, A.; Jawdosiuk, M.; Makosza, M.; Czyzewski, J. Search for a new method for synthesis of the analgesic agent "Fentanyl". *Przem. Chem.* **1978,** *57* (3), 131–134.
64. Zee, S.-H.; Wang, W.-K. A new process for the synthesis of fentanyl. *J. Chin. Chem. Soc. (Weinheim, Ger.)* **1980,** *27* (4), 147–149.
65. Vardanyan, R. S.; Hruby, V. J. Fentanyl-related compounds and derivatives: current status and future prospects for pharmaceutical applications. *Future Med. Chem.* **2014,** *6* (4), 385–412.
66. Eguchi, M. Recent advances in selective opioid receptor agonists and antagonists. *Med. Res. Rev.* **2004,** *24* (2), 182–212.
67. Fuerst, S.; Hosztafi, S.; Friedmann, T. Structure-activity relationships of synthetic and semi-synthetic opioid agonists and antagonists. *Curr. Med. Chem.* **1995,** *1* (6), 423–440.
68. Schmidhammer, H. Opioid receptor antagonists. *Prog. Med. Chem.* **1998,** *35,* 83–132.
69. Goodman, A. J.; Le Bourdonnec, B.; Dolle, R. E. Mu opioid receptor antagonists: recent developments. *ChemMedChem* **2007,** *2* (11), 1552–1570.
70. Kaczor, A.; Matosiuk, D. Non-peptide opioid receptor ligands-recent advances. Part I. Agonists. *Curr. Med. Chem.* **2002,** *9* (17), 1567–1589.
71. Kaczor, A.; Matosiuk, D. Non-peptide opioid receptor ligands-recent advances. Part II. Antagonists. *Curr. Med. Chem.* **2002,** *9* (17), 1591–1603.
72. Meert, T. F. Pharmacotherapy of opioids: Present and future developments. *Pharm. World Sci.* **1996,** *18* (1), 1–15.
73. Wax, P. M.; Becker, Ch. E.; Curry, S. C. Unexpected "gas" casualties in Moscow: a medical toxicology perspective. *Ann. Emerg. Med.* **2003,** *41* (5), 700–705.
74. Putz, C. The δ opioid receptor. In *Analgesics: From Chemistry and Pharmacology to Clinical Application;* Buschmann, H., Christoph, T., Friderichs, E., Maul, C., Sundermann, B., Eds.; Wiley-VCH, 2002; pp 455–466.
75. Calderon, S. N.; Coop, A. SNC 80 and related δ opioid agonists. *Curr. Pharm. Des.* **2004,** *10* (7), 733–742.
76. Coop, A.; Rice, K. C. Role of δ-opioid receptors in biological processes. *Drug News Perspect.* **2000,** *13* (8), 481–487.
77. Micovic, V. I.; Ivanovic, M. D.; Dosen-Micovic, L. j. Structural requirements for ligands of the δ-opioid receptor. *J. Serb. Chem. Soc.* **2009,** *74* (11), 1207–1217.
78. Lee, N. M.; Leybin, L.; Chang, J. K.; Loh, H. H. Opiate and peptide interaction: effect of enkephalins on morphine analgesia. *Eur. J. Pharmacol.* **1980,** *68,* 181–185.

79. Schiller, P. W.; Fundytus, M. E.; Merovitz, L.; Weltrowska, G.; Nguyen, T. M.; Lemieux, C.; Chung, N. N.; Coderre, T. J. The opioid m agonist/d antagonist DIPP-NH2[J] produces a potent analgesic effect, no physical dependence, and less tolerance than morphine in rats. *J. Med. Chem.* **1999,** *42,* 3520–3526.
80. Porreca, F.; Takemori, A. E.; Sultana, M.; Portoghese, P. S.; Bowen, W. D.; Mosberg, H. I. Modulation of mu-mediated antinociception in the mouse involves opioid delta-2 receptors. *J. Pharmacol. Exp. Ther.* **1992,** *263* (1), 147–152.
81. Heyman, J. S.; Vaught, J. L.; Mosberg, H. I.; Haaseth, R. C.; Porreca, F. Modulation of μ-mediated antinociception by δ agonists in the mouse: selective potentiation of morphine and normorphine by [D-Pen2,D-Pen5]encephalin. *Eur. J. Pharmacol.* **1989,** *165* (1), 1–10.
82. Heyman, J. S.; Jang, Q.; Rothman, R. B.; Mosberg, H. I.; Porreca, F. Modulation of μ-mediated antinociception by δ agonists: characterization with antagonists. *Eur. J. Pharmacol.* **1989,** *169* (1), 43–52.
83. Rozenfeld, R.; Devi, L. A. Receptor heterodimerization leads to a switch in signaling: β-arrestin2-mediated ERK activation by μ-δ opioid receptor heterodimers. *FASEB J.* **2007,** *21* (10), 2455–2465.
84. Gomes, I.; Gupta, A.; Filipovska, J.; Szeto, H. H.; Pintar, J. E.; Devi, L. A. A role for heterodimerization of μ and δ opiate receptors in enhancing morphine analgesia. *Proc. Natl. Acad. Sci. U. S. A.* **2004,** *101* (14), 5135–5139.
85. Prisinzano, T. E.; Tidgewell, K.; Harding, W. W. Kappa opioids as potential treatments for stimulant dependence. *AAPS J.* **2005,** *7* (3), E592–E599.
86. Portoghese, Ph. P. Molecular recognition at kappa opioid receptors. *Pure Appl. Chem.* **2001,** *73* (9), 1387–1391.
87. Husbands, S. M. Kappa-opioid receptor ligands. *Expert Opin. Ther. Pat* **2004,** *14* (12), 1725–1741.
88. Bignan, G. C.; Connolly, P. J.; Middleton, S. A. Recent advances towards the discovery of ORL-1 receptor agonists and antagonists. *Expert Opin. Ther. Pat.* **2005,** *15* (4), 357–388.
89. Mustazza, C.; Bastanzio, G. Development of nociceptin receptor (NOP) agonists and antagonists. *Med. Res. Rev.* **2011,** *31* (4), 605–648.
90. Dharmshaktu, P.; Tayal, V.; Kalra, B. S. Efficacy of antidepressants as analgesics: a review. *J. Clin. Pharmacol.* **2012,** *52,* 6–17.
91. Mattia, C.; Coluzzi, F. Antidepressants in chronic neuropathic pain. *Mini-Rev. Med. Chem.* **2003,** *3* (7), 773–784.
92. Miller, A.; Rabe-Jablonska, J. The effectiveness of antidepressants in the treatment of chronic non-cancer pain. *Minerva Anestesiol.* **2002,** *68* (3), 105–114.
93. Mattia, C.; Paoletti, F.; Coluzzi, F.; Boanelli, A. New antidepressants in the treatment of neuropathic pain, a review. *Basic Clin. Pharmacol. Toxicol.* **2005,** *96* (6), 399–409.
94. Sindrup, S. H.; Otto, M.; Finnerup, N. B.; Jensen, T. S. Antidepressants in the treatment of neuropathic pain. *J. Psychiatry Neurosci.* **2001,** *26* (1), 30–36.
95. Lynch, M. E. Antidepressants as analgesics: a review of randomized controlled trials. *J. Psychiatry Neurosci.* **2001,** *26* (1), 30–36.
96. Arnold, L. M. Duloxetine and other antidepressants in the treatment of patients with fibromyalgia. *Pain Med.* **2007,** *8* (Suppl. 2), 63–74.
97. Ramanath, R. B. Pharmacological aspects of tapentadol. *Int. J. Pharma Bio Sci.* **2012,** *3* (1), 479–484.
98. Hartrick, C. T.; Rozek, R. J. Tapentadol in pain management: a μ-opioid receptor agonist and noradrenaline reuptake inhibitor. *CNS Drugs* **2011,** *25* (5), 359–370.
99. Kress, H. G. Tapentadol and its two mechanisms of action: Is there a new pharmacological class of centrally-acting analgesics on the horizon? *Eur. J. Pain (Oxford, U. K.)* **2010,** *14* (8), 781–783.

100. Tzschentke, T. M.; De Vry, J.; Terlinden, R.; Hennies, H.-H.; Lange, C.; Strassburger, W.; Haurand, M.; Kolb, J.; Schneider, J.; Buschmann, H.; Finkam, M.; Jahnel, U.; Friderichs, E. Tapentadol hydrochloride: analgesic mu-opioid receptor agonist noradrenaline reuptake inhibitor. *Drugs Future* **2006,** *31* (12), 1053–1061.
101. Frampton, J. E. Tapentadol immediate release: a review of its use in the treatment of moderate to severe acute pain. *Drugs* **2010,** *70* (13), 1719–1743.
102. Buschmann, H.; Strassburger, W.; Friderichs, E. Preparation of 1-phenyl-3-dimethylamino-propane derivatives as analgesics, EP 693475 (1996).
103. Vlasakova, R.; Hajicek, J.; Zezula, J. Process for preparing o-substituted (2r,3r)-3-(3-hydroxyphenyl)-2-methyl-4-pentenoic acids, CZ 303116 (2012).
104. Bhirud, S. B.; Johar, P. S.; Mishra, S.; Jamshad, D. Process for preparing 1-phenyl-3-dimethylaminopropane derivative, WO 2012038974 (2012).
105. Rajadhyaksha, M. N.; Nair; R.; Deshmukh, S. K.; Khabale, S. A.; Somnath, A.; Panandikar, A. M. Process for the preparation of tapentadol, WO 2012023147 (2012).
106. Froestl, W. An historical perspective on GABAergic drugs. *Future Med. Chem.* **2011,** *3* (2), 163–175.
107. Baidya, D. K.; Agarwal, A.; Khanna, P.; Arora, M. K. Pregabalin in acute and chronic pain. *J. Anaesthesiol., Clin. Pharmacol.* **2011,** *27* (3), 307–314.
108. Field, M. K.; Singh, L.; Gonzalez, I. M. Gabapentin and related compounds. In *Pain: Current Understanding, Emerging Therapies and Novel Approaches to Drug Discovery;* Bountra, C., Munglani, R., Schmidt, W. K., Eds.; Marcel Dekker, 2003; pp 775–779.
109. Maul, C.; Buschmann, H.; Sundermann, B. Gabapentin and gabapentinoids. In *Analgesics: From Chemistry and Pharmacology to Clinical Application;* Buschmann, H., Christoph, T., Friderichs, E., Maul, C., Sundermann, B., Eds.; Wiley-VCH, 2002; pp 287–295.
110. Thorpe, A. J.; Taylor, C. P. Calcium channel α2-δ ligands: gabapentin and pregabalin. In *Comprehensive Medicinal Chemistry II,* 8th ed.; Taylor, J. B., Triggle, D. J., Eds. Elsevier, 2006; pp 227–246.
111. Field, M. J.; Li, Z.; Schwarz, J. B. Ca2+ channel α2-δ ligands for the treatment of neuropathic pain. *J. Med. Chem.* **2007,** *50* (11), 2569–2575.
112. Andruszkiewicz, R.; Silverman, R. B. A convenient synthesis of 3-alkyl-4-aminobutanoic acids. *Synthesis* **1989,** *12*. 953–935.
113. Yuen, P.; Kanter, G. D.; Taylor, C. P.; Vartanian, M. G. Enantioselective synthesis of PD144723: a potent stereospecific anticonvulsant. *Bioorg. Med. Chem. Lett.* **1994,** *4* (6), 823–826.
114. Hoekstra, M. S.; Sobieray, D. M.; Schwindt, M. A.; Mulhern, T. A.; Grote, T. M.; Huckabee, B. K.; Hendrickson, V. S.; Franklin, L. C.; Granger, E. J.; Karrick, G. L. Chemical development of CI-1008, an enantiomerically pure anticonvulsant. *Org. Process Res. Dev.* **1997,** *1* (1), 26–38.
115. Huckabee, B. K.; Sobieray, D. M. Stereoselective synthesis of (S)-3-(aminomethyl)-5-methylhexanoic acid, WO 9638405 (1996).
116. Grote, T. M.; Huckabee, B. K.; Mulhern, T.; Sobieray, D. M.; Titus, R. D. Method of making (S)-3-(aminomethyl)-5-methylhexanoic acid, WO 9640617 (1996).
117. Hu, S.; Martinez, C. A.; Tao, J.; Tully, W. E.; Kelleher, P.; Dumond, Y. Preparation of pregabalin and related compounds, US 20050283023 (2005).
118. Kiran, C.; Mahendra, Y.; Chandrashekar, B.; Rakesh, R.; Sundeep, A. A novel process for synthesis of a substituted cyclopropane intermediate and a process for synthesis of pregabalin, IN 2007MU02055 (2009).
119. Kansal, V. K.; Chaurasia, B. P.; Shelke, S. H.; Tiwari, A. P. Synthesis of (s)-(+)-3-(aminomethyl)-5-methylhexanoic acid, WO 2008118427 (2008).
120. Skov, M. J.; Beck, J. C.; de Kater, A. W.; Shopp, G. M. Nonclinical safety of ziconotide: an intrathecal analgesic of a new pharmaceutical class. *Int. J. Toxicol.* **2007,** *26* (5), 411–421.

121. Pexton, T.; Moeller-Bertram, T.; Schilling, J. M.; Wallace, M. S. Targeting voltage-gated calcium channels for the treatment of neuropathic pain: a review of drug development. *Expert Opin. Invest. Drugs* **2011,** *20* (9), 1277–1284.
122. Hennies, H.-H.; Sundermann, B. Calcium channels. In *Analgesics: From Chemistry and Pharmacology to Clinical Application;* Buschmann, H., Christoph, T., Friderichs, E., Maul, C., Sundermann, B., Eds.; Wiley-VCH, 2002; pp 353–378.
123. http://www.bizjournals.com/philadelphia/stories/2007/08/06/daily17.html.
124. Bear, B.; Asgian, J.; Termin, A.; Zimmermann, N. Small molecules targeting sodium and calcium channels for neuropathic pain. *Curr. Opin. Drug Discovery Dev.* **2009,** *12* (4), 543–561.
125. Yamamoto, T.; Takahara, A. Recent updates of N-type calcium channel blockers with therapeutic potential for neuropathic pain and stroke. *Curr. Top. Med. Chem.* **2009,** *9* (4), 377–395.
126. Backonja, M. M. Local anesthetics as adjuvant analgesics. *J. Pain Symptom Manage.* **1994,** *9* (8), 491–499.
127. Kong, V. K.; Irwin, M. G. Systemic local anesthetics in pain control. *Eur. J. Anaesthesiol.* **2009,** *26* (2), 96–100.
128. Glazer, S.; Portenoy, R. K. Adjuvant analgesics. *J. Pain Symptom Manage.* **1991,** *6* (1), 30–39.
129. Knotkova, H.; Pappagallo, M. Adjuvant analgesics. *Anesthesiol. Clin.* **2007,** *25* (4), 775–786.
130. Mathie, A. Ion channels as novel therapeutic targets in the treatment of pain. *J. Pharm. Pharmacol.* **2010,** *62* (9), 1089–1095.
131. Kyle, D. J.; Ilyin, V. I. Sodium channel blockers. *J. Med. Chem.* **2007,** *50* (11), 2583–2588.
132. Roberson, D. P.; Binshtok, A. M.; Blasl, F.; Bean, B. P.; Woolf, C. J. Targeting of sodium channel blockers into nociceptors to produce long-duration analgesia: a systematic study and review. *Br. J. Pharmacol.* **2011,** *164* (1), 48–58.
133. Wang, G. K.; Strichartz, G. R. Therapeutic Na+ channel blockers beneficial for pain syndromes. *Drug Dev. Res.* **2001,** *54* (3), 154–158.
134. Zuliani, V.; Patel, M. K.; Fantini, M.; Rivara, M. Recent advances in the medicinal chemistry of sodium channel blockers and their therapeutic potential. *Curr. Top. Med. Chem.* **2009,** *9* (4), 396–415.
135. Hargus, N. J.; Patel, M. K. Voltage-gated Na+ channels in neuropathic pain. *Expert Opin. Invest. Drugs* **2007,** *16* (5), 635–646.
136. Tarnawa, I.; Bolcskei, H.; Kocsis, P. Blockers of voltage-gated sodium channels for the treatment of central nervous system diseases. *Recent Pat. CNS Drug Discovery* **2007,** *2,* 57–78.
137. Lawson, K. Potassium channels as targets for the management of pain. *Cent. Nerv. Syst. Agents Med. Chem.* **2006,** *6* (2), 119–128.
138. Ocanaa, M.; Cruz, C. M. C.; Cobos, R. J.; Entrena, J. M.; Baeyens, J. M. Potassium channels and pain: present realities and future opportunities. *Eur. J. Pharmacol.* **2004,** *500* (1–3), 203–219.
139. Munro, G.; Dalby-Brown, W. Kv7 modulators and neuropathic pain. *J. Med. Chem.* **2007,** *50* (11), 2576–2582.
140. Hovelsoe, N.; Sotty, F.; Montezinho, L. P.; Pinheiro, P. S.; Herrik, K. F.; Moerk, A. Therapeutic potential of metabotropic glutamate receptor modulators. *Curr. Neuropharmacol.* **2012,** *10* (1), 12–48.
141. Chiechio, S.; Nicoletti, F. Metabotropic glutamate receptors and the control of chronic pain. *Curr. Opin. Pharmacol.* **2012,** *12* (1), 28–34.
142. Bleakman, D.; Alt, A.; Nisenbaum, E. S. Glutamate receptors and pain. *Semin. Cell Dev. Biol.* **2007,** *17* (5), 592–604.
143. Childers, W. E., Jr.; Baudy, R. B. N-methyl-d-aspartate antagonists and neuropathic pain: the search for relief. *J. Med. Chem.* **2007,** *50* (11), 2557–2562.

144. Brown, D. G.; Krupp, J. J. N-methyl-d-aspartate receptor (NMDA) antagonists as potential pain therapeutics. *Curr. Top. Med. Chem.* **2006,** *6* (8), 749–770.
145. Ceber, M.; Salihoglu, T. Ketamine may be the first choice for anesthesia in burn patients. *J. Burn Care Res.* **2006,** *27* (5), 760–762.
146. Pogatzki, E. M.; Zahn, P. K.; Brennan, T. J. Effect of pretreatment with intrathecal excitatory amino acid receptor antagonists on the development of pain behavior caused by plantar incision. *Anesthesiology* **2000,** *93,* 489–496.
147. Nozaki-Taguchi, N.; Yaksh, T. L. Pharmacology of spinal glutamatergic receptors in post-thermal injury-evoked tactile allodynia and thermal hyperalgesia. *Anesthesiology* **2002,** *96,* 617–626.
148. Catarzi, D.; Colotta, V.; Varano, F. Competitive AMPA receptor antagonists. *Med. Res. Rev.* **2007,** *27* (2), 239–278.
149. Simmons, R. M.; Li, D. L.; Hoo, K. H.; Deverill, M.; Ornstein, P. L.; Iyengar, S. Kainate GluR5 receptor subtype mediates the nociceptive response to formalin in the rat. *Neuropharmacology* **1998,** *37,* 25–36.
150. Filla, S. A.; Winter, M. A.; Johnson, K. W.; Bleakman, D.; Bell, M. G.; Bleisch, T. J.; Castano, A. M.; Clemens-Smith, A.; del Prado, M.; Dieckman, D. K.; Dominguez, E.; Escribano, A.; Ho, K. H.; Hudziak, K. J.; Katofiasc, M. A.; Martinez-Perez, J. A.; Mateo, A.; Mathes, B. M.; Mattiuz, E. L.; Ogden, A. M. L.; Phebus, L. A.; Stack, D. R.; Stratford, R. E.; Ornstein, P. L. Ethyl (3S,4aR,6S 8aR)-6-(4-ethoxycar-bonylimidazol-1-ylmethyl)decahydroiso-quinoline-3-carboxylic ester: a prodrug of a GluR5 kainate receptor antagonist active in two animal models of acute migraine. *J. Med. Chem.* **2002,** *45,* 4383–4386.
151. Weiss, B.; Alt, A.; Ogden, A. M.; Gates, M.; Dieckman, D. K.; Clemens-Smith, A.; Ho, K. H.; Jarvie, K.; Rizkalla, G.; Wright, R. A.; Calligaro, D. O.; Schoepp, D.; Mattiuz, E. L.; Stratford, R. E.; Johnson, B.; Salhoff, C.; Katofiasc, M.; Phebus, L. A.; Schenck, K.; Cohen, M.; Filla, S. A.; Ornstein, P. L.; Johnson, K. W.; Bleakman, D. Pharmacological characterization of the competitive GLUK5 receptor antagonist decahydroisoquinoline LY466195 in vitro and in vivo. *J. Pharmacol. Exp. Ther.* **2006,** *318,* 772–781.
152. Ruscheweyh, R.; Sandkuhler, J. Role of kainate receptors in nociception. *Brain Res. Rev.* **2002,** *40,* 215–222.
153. Gilron, I. LY-293558.Eli Lilly & Co. *Curr. Opin. Invest. Drugs (BioMed Cent.)* **2001,** *2* (9), 1273–1278.
154. Weiss, B.; Alt, A.; Ogden, A. M.; Gates, M.; Dieckman, D. K.; Clemens-Smith, A.; Ho, K. H.; Jarvie, K.; Rizkalla, G.; Wright, R. A.; Calligaro, D. O.; Schoepp, D.; Mattiuz, E. L.; Stratford, R. E.; Johnson, B.; Salhoff, C.; Katofiasc, M.; Phebus, L. A.; Schenck, K.; Cohen, M.; Filla, S. A.; Ornstein, P. L.; Johnson, K. W.; Bleakman, D. Pharmacological characterization of the competitive GLUK5 receptor antagonist decahydroisoquinoline LY466195 in vitro and in vivo. *J. Pharmacol. Exp. Ther.* **2006,** *318* (2), 772–781.
155. Szallasi, A. New developments in the medicinal chemistry of vanilloid TRPV1 and related receptors. *Curr. Top. Med. Chem.* **2011,** *11* (17), 2116–2117.
156. Patapoutian, A.; Tate, S.; Woolf, C. J. Transient receptor potential channels: targeting pain at the source. *Nat. Rev. Drug Discovery* **2008,** *8* (1), 55–68.
157. Moran, M. M.; McAlexander, M. A.; Biro, T.; Szallasi, A. Transient receptor potential channels as therapeutic targets. *Nat. Rev. Drug Discovery* **2011,** *10* (8), 601–620.
158. Cortright, D. N.; Szallasi, A. TRP channels and pain. *Curr. Pharm. Des.* **2009,** *15* (15), 1736–1749.
159. Gharat, L. A.; Szallasi, A. Advances in the design and therapeutic use of capsaicin receptor TRPV1 agonists and antagonists. *Expert Opin. Ther. Pat.* **2008,** *18* (2), 159–209.

160. Premkumar, L. S. Targeting TRPV1 as an alternative approach to narcotic analgesics to treat chronic pain conditions. *AAPS J.* **2010,** *12* (3), 361–370.
161. Palazzo, E.; Luongo, L.; de Novellis, V.; Rossi, F.; Marabese, I.; Maione, S. Transient receptor potential vanilloid type 1 and pain development. *Curr. Opin. Pharmacol.* **2012,** *12* (1), 9–17.
162. Immke, D. C.; Gavva, N. R. The TRPV1 receptor and nociception. *Semin. Cell Dev. Biol.* **2006,** *17* (5), 582–591.
163. Kyle, D. J.; Tafesse, L. TRPV1 antagonists: a survey of the patent literature. *Expert Opin. Ther. Pat.* **2006,** *16* (7), 977–996.
164. Giovannoni, M. P.; Ghelardini, C.; Vergelli, C.; Dal Piaz, V. α2-Agonists as analgesic agents. *Med. Res. Rev.* **2009,** *29* (2), 339–368.
165. Kamibayashi, T.; Maze, M. Clinical uses of α2-adrenergic agonists. *Anesthesiology* **2000,** *93,* 1345–1349.
166. Kingery, W. S.; Davies, M. F.; Maze, M. Molecular mechanisms for the analgesic properties of alpha-2 adrenergic agonists. In *Molecular Neurobiology of Pain;* Borsook, D., Ed.; IASP Press, 1997; pp 275–304.
167. Overland, A. C.; Kitto, K. F.; Chabot-Dore, A.-J.; Rothwell, P. E.; Fairbanks, C. A.; Stone, L. S.; Wilcox, G. L. Protein kinase C mediates the synergistic interaction between agonists acting at α2-adrenergic and delta-opioid receptors in spinal cord. *J. Neurosci.* **2009,** *29* (42), 13264–13275.
168. Boyd, R. E. α2-Adrenergic receptor agonists as analgesics. *Curr. Top. Med. Chem.* **2001,** *1* (3), 193–197.
169. Jnoff, E.; Christophe, B.; Collart, P.; Coloretti, F.; Debeuckelaere, A.; De Ryck, M.; Fuks, B.; Genicot, C.; Gillard, M.; Guyaux, M.; Price, N.; Vandergeten, M.-C.; Vermeiren, C. Discovery of selective alpha2C adrenergic receptor agonists. *ChemMedChem* **2012,** *7* (3), 385–390.
170. Bartolini, A.; Mannelli, L. D. C.; Ghelardini, C. Analgesic and antineuropathic drugs acting through central cholinergic mechanisms. *Recent Pat. CNS Drug Discovery* **2011,** *6* (2), 119–140.
171. Lippiello, P. M.; Bencherif, M.; Hauser, T. A.; Jordan, K. G.; Letchworth, S. R.; Mazurov, A. A. Nicotinic receptors as targets for therapeutic discovery. *Expert Opin. Drug Discovery* **2007,** *2* (9), 1185–1203.
172. Meyer, M. D. Neuronal nicotinic acetylcholine receptors as a target for the treatment of neuropathic pain. *Drug Dev. Res.* **2006,** *67* (4), 355–359.
173. Jain, K. K. Modulators of nicotinic acetylcholine receptors as analgesics. *Curr. Opin. Invest. Drugs (BioMed Cent.)* **2004,** *5* (1), 76–81.
174. Decker, M. W.; Rueter, L. E.; Bitner, R. S. Nicotinic acetylcholine receptor agonists: A potential new class of analgesics. *Curr. Top. Med. Chem.* **2004,** *4* (3), 369–384.
175. Jones, P. G.; Dunlop, J. Targeting the cholinergic system as a therapeutic strategy for the treatment of pain. *Neuropharmacology* **2007,** *53* (2), 197–206.
176. Munoz, M.; Covenas, R. NK-1 receptor antagonists: a new paradigm in pharmacological therapy. *Curr. Med. Chem.* **2011,** *18* (12), 1820–1831.
177. Humphrey, J. M. Medicinal chemistry of selective neurokinin-1 antagonists. *Curr. Top. Med. Chem.* **2003,** *3* (12), 1423–1435.
178. Quartara, L.; Altamura, M. Tachykinin receptors antagonists: from research to clinic. *Curr. Drug Targets* **2006,** *7* (8), 975–992.
179. Boyce, S.; Hill, R. G. Substance P (NK1) receptor antagonists-analgesics or not?. *Handb. Exp. Pharmacol.* **2004,** *164,* 441–457 (Tachykinins).
180. Remond, G.; Portevin, B.; Bonnet, J.; Canet, E.; Regoli, D. De Nanteuil, G., Pharmacological profile of a novel series of NK1 antagonists. In vitro and in vivo potency of benzimidazolone derivatives. *Eur. J. Med. Chem.* **1997,** *32* (11), 843–868.

181. Boscan, P.; Twedt, D. Use of neurokinin type 1 receptor (NK-1) receptor antagonists in management of visceral pain, US 20120028980, (2012).
182. Burns, T. L.; Ineck, J. R. Cannabinoid analgesia as a potential new therapeutic option in the treatment of chronic pain. *Ann. Pharmacother.* **2006,** *40* (2), 251–260.
183. Fox, A.; Bevan, S. Therapeutic potential of cannabinoid receptor agonists as analgesic agents. *Expert Opin. Invest. Drugs* **2005,** *14* (6), 695–703.
184. Cheng, Y.; Hitchcock, S. A. Targeting cannabinoid agonists for inflammatory and neuropathic pain. *Expert Opin. Invest. Drugs* **2007,** *16* (7), 951–965.
185. Thakur, G. A.; Nikas, S. P.; Makriyannis, A. CB1 cannabinoid receptor ligands. *Mini-Rev. Med. Chem.* **2005,** *5* (7), 631–640.
186. Huffman, J. W. CB2 receptor ligands. *Mini-Rev. Med. Chem.* **2005,** *5* (7), 641–649.
187. Padgett, L. W. Recent developments in cannabinoid ligands. *Life Sci.* **2005,** *77* (14), 1767–1798.
188. Pertwee, R. G. Cannabinoid receptor ligands: clinical and neuropharmacological considerations, relevant to future drug discovery and development. *Expert Opin. Invest. Drugs* **2000,** *9,* 1553–1571.
189. Lange, J. H. M.; Kruse, Ch. G. Medicinal chemistry strategies to CB1 cannabinoid receptor antagonists. *Drug Discovery Today* **2005,** *10* (10), 693–702.
190. Hanus, L. O.; Mechoulam, R. Novel natural and synthetic ligands of the endocannabinoid system. *Curr. Med. Chem.* **2010,** *17* (14), 1341–1359.
191. Thakur, G. A.; Tichkule, R.; Bajaj, S.; Makriyannis, A. Latest advances in cannabinoid receptor agonists. *Expert Opin. Ther. Pat.* **2009,** *19* (12), 1647–1673.
192. Seltzman, H. H. Recent CB1 cannabinoid receptor antagonists and inverse agonists. *Drug Dev. Res.* **2009,** *70* (8), 601–615.
193. Xiong, W.; Cheng, K.; Cui, T.; Godlewski, G.; Rice, K. C.; Xu, Y.; Zhang, L. Cannabinoid potentiation of glycine receptors contributes to cannabis-induced analgesia. *Nat. Chem. Biol.* **2011,** *7,* 296–303.
194. Gosselin, R. D.; Marc, S. R.; Ji, R.-R.; Decosterd, I. Glial cells and chronic pain. *Neuroscientist* **2010,** *16* (5), 519–531.
195. McMahon, S. B.; Malcangio, M. Current challenges in glia-pain biology. *Neuron* **2009,** *64* (1), 46–54.
196. Cunha, T. M.; Quintino, Q. D. Glial modulation of pain: a step beyond. *J. Neurosci.* **2009,** *29* (11). 3340–2334.
197. Watkins, L. R.; Hutchinson, M. R.; Ledeboer, A.; Wieseler-Frank, J.; Milligan, E. D.; Maier, S. F. Glia as the "bad guys": implications for improving clinical pain control and the clinical utility of opioids. *Brain, Behav., Immun.* **2007,** *21* (2), 131–146.
198. Suter, M. R.; Wen, Y.-R.; Decosterd, I.; Ji, R. R. Do glial cells control pain? *Neuron Glia Biol.* **2007,** *3* (3), 255–268.
199. Scholz, J.; Woolf, C. J. The neuropathic pain triad: neurons, immune cells and glia. *Nat. Neurosci.* **2007,** *10* (11), 1361–1368.
200. Mika, J. Modulation of microglia can attenuate neuropathic pain symptoms and enhance morphine effectiveness. *Pharmacol. Rep.* **2008,** *60* (3), 297–307.
201. Watkins, L. R.; Maier, S. F. Targeting glia to control clinical pain: an idea whose time has come. *Drug Discovery Today: Ther. Strategies* **2004,** *1* (1), 83–88.
202. Hutchinson, M. R.; Bland, S. T.; Johnson, K. W.; Rice, K. C.; Maier, S. F.; Watkins, L. R. Opioid-induced glial activation: mechanisms of activation and implications for opioid analgesia, dependence, and reward. *Sci. World J.* **2007,** *7* (Suppl. 2), 98–111.
203. Watkins, L. R.; Hutchinson, M. R.; Milligan, E. D.; Maier, S. F. "Listening" and "talking" to neurons: implications of immune activation for pain control and increasing the efficacy of opioids. *Brain Res. Rev.* **2007,** *56* (1), 148–169.

204. Watkins, L. R.; Maier, S. F. Glia and pain: past, present, and future. In *Paths of Pain 1975-2005;* Merskey, H. J., Loeser, D., Dubner, R., Eds.; IASP Press, 2005; pp 165–175.
205. Watkins, L. R.; Hutchinson, M. R.; Johnston, I. N.; Maier, S. F. Glia: novel counter-regulators of opioid analgesia. *Trends Neurosci.* **2005,** *28* (12), 661–669.
206. Scholz, J.; Woolf, C. J. Can we conquer pain? *Nat. Neurosci* **2002,** *5* (Suppl.), 1062–1067.
207. Clarke, D. D. Fluoroacetate and fluorocitrate: mechanism of action. *Neurochem. Res.* **1991,** *16* (9), 1055–1058.
208. Qin, M.; Wang, J.-J.; Cao, R.; Zhang, H.; Duan, L.; Gao, B.; Xiong, Y.-F.; Chen, L.-W.; Rao, Z.-R. The lumbar spinal cord glial cells actively modulate subcutaneous formalin induced hyperalgesia in the rat. *Neurosci. Res.* **2006,** *55* (4), 442–450.
209. Chen, J.; Zhang, J.; Zhao, Y.; Yuan, L.; Nie, X.; Li, J.; Ma, Z.; Zhang, Y.; Wang, Q.; Chen, Y.; Jin, Y.; Rao, Z. Hyperalgesia in response to traumatic occlusion and GFAP expression in rat parabrachial nucleus: modulation with fluorocitrate. *Cell Tissue Res.* **2007,** *329* (2), 231–237.
210. Guo, W.; Wang, H.; Watanabe, M.; Shimizu; Zou, K. S.; LaGraize, S. C.; Wei, F.; Dubner, R.; Ren, K. Glial -cytokine-neuronal interactions underlying the mechanisms of persistent pain. *J. Neurosci.* **2007,** *27* (22), 6006–6018.
211. Mika, J.; Osikowicz, M.; Makuch, W.; Przewlocka, B. Minocycline and pentoxifylline attenuate allodynia and hyperalgesia and potentiate the effects of morphine in rat and mouse models of neuropathic pain. *Eur. J. Pharmacol.* **2007,** *560* (2–3), 142–149.
212. Ledeboer, A.; Sloane, E. M.; Milligan, E. D.; Frank, M. G.; Mahony, J. H.; Maier, S. F.; Watkins, L. R. Minocycline attenuates mechanical allodynia and proinflammatory cytokine expression in rat models of pain facilitation. *Pain* **2005,** *115* (1–2), 71–83.
213. Clark, A. K.; Gentry, C.; Bradbury, E. J.; McMahon, S. B.; Malcangio, M. Role of spinal microglia in rat models of peripheral nerve injury and inflammation. *Eur. J. Pain (Oxford, U. K.)* **2007,** *11* (2), 223–230.
214. Obata, H.; Eisenach, J. C.; Hussain, H.; Bynum, T.; Vincler, M. Spinal glial activation contributes to postoperative mechanical hypersensitivity in the rat. *J. Pain* **2006,** *7* (11), 816–822.
215. Ledeboer, A.; Hutchinson, M. R.; Watkins, L. R.; Johnson, K. W. Ibudilast (AV-411): A new class therapeutic candidate for neuropathic pain and opioid withdrawal syndromes. *Expert Opin. Invest. Drugs* **2007,** *16* (7), 935–950.
216. Johnson, K. W.; Watkins, L. R.; Hutchinson, M. Methods for potentiating opioid-induced analgesia in patients by administering phosphodiesterase inhibitor or glial attenuator, ibudilast, US 2008181876 (2008).
217. Guindon, J.; Hohmann, A. G. Cannabinoid CB2 receptors: a therapeutic target for the treatment of inflammatory and neuropathic pain. *Br. J. Pharmacol.* **2008,** *153* (2), 319–334.
218. Hoegenauer, E. K. Latest advances in the cannabinoids. *Expert Opin. Ther. Pat.* **2007,** *17* (12), 1457–1476.
219. Cheng, Y.; Hitchcock, S. A. Targeting cannabinoid agonists for inflammatory and neuropathic pain. *Expert Opin. Invest. Drugs* **2007,** *16* (7), 951–965.
220. Whiteside, G. T.; Lee, G. P.; Valenzano, K. J. The role of the cannabinoid CB2 receptor in pain transmission and therapeutic potential of small molecule CB2 receptor agonists. *Curr. Med. Chem.* **2007,** *14* (8), 917–936.
221. Poso, A.; Huffman, J. W. Targeting the cannabinoid CB2 receptor: modeling and structural determinants of CB2 selective ligands. *Br. J. Pharmacol.* **2008,** *153* (2), 335–346.
222. Xu, J. J.; Diaz, P.; Astruc-Diaz, F.; Craig, S.; Munoz, E.; Naguib, M. Pharmacological characterization of a novel cannabinoid ligand, MDA19, for treatment of neuropathic pain. *Anesth. Analg.* **2010,** *111* (1), 99–109.

223. Massi, P.; Valenti, M.; Bolognini, D.; Parolaro, D. Expression and function of the endocannabinoid system in glial cells. *Curr. Pharm. Des.* **2008,** *14* (23), 2289–2298.
224. Jhaveri, M. D.; Sagar, D. R.; Elmes, S. J. R.; Kendall, D. A.; Chapman, V. Cannabinoid CB2 receptor-mediated anti-nociception in models of acute and chronic pain. *Mol. Neurobiol.* **2007,** *36* (1), 26–35.
225. Malan, T. P.; Ibrahim, M. M.; Lai, J.; Vanderah, T. W.; Makriyannis, A.; Porreca, F. CB2 cannabinoid receptor agonists: pain relief without psychoactive effects? *Curr. Opin. Pharmacol.* **2003,** *3* (1), 62–67.
226. Pacher, P.; Batkai, S.; Kunos, G. The endocannabinoid system as an emerging target of pharmacotherapy. *Pharmacol. Rev.* **2006,** *58* (3), 389–462.
227. Galve-Roperh, I.; Aguado, T.; Palazuelos, J.; Guzman, M. Mechanisms of control of neuron survival by the endocannabinoid system. *Curr. Pharm. Des.* **2008,** *14* (23), 2279–2288.
228. Marchalant, Y.; Rosi, S.; Wenk, G. L. Anti-inflammatory property of the cannabinoid agonist WIN-55212-2 in a rodent model of chronic brain inflammation. *Neuroscience* **2007,** *144* (4), 1516–1522.
229. Herranz, R. Cholecystokinin antagonists: pharmacological and therapeutic potential. *Med. Res. Rev.* **2003,** *23* (5), 559–605.
230. Garcia-Lopez, T. M.; Gonzalez-Muniz, R.; Martin-Martinez, M.; Herranz, R. Strategies for design of non peptide CCK1R agonist/antagonist ligands. *Curr. Top. Med. Chem.* **2007,** *7* (12), 1180–1194.
231. Cawston, E. E.; Miller, L. J. Therapeutic potential for novel drugs targeting the type 1 cholecystokinin receptor. *Br. J. Pharmacol.* **2010,** *159,* 1009–1021.
232. Noble, F.; Roques, B. P. CCK-B receptor: chemistry, molecular biology, biochemistry and pharmacology. *Prog. Neurobiol.* **1999,** *58,* 349–379.
233. Johnston, M. M.; Rapoport, A. M. Triptans for the management of migraine. *Drugs* **2010,** *70* (12), 1505–1518.
234. Bigal, M. E.; Krymchantowski, A. V.; Hargreaves, R. The triptans. *Expert Rev. Neurother.* **2009,** *9* (5), 649–659.
235. Andreas, A. L.; Link, B. In the pipeline: triptans-newer developments. *Pharmazie* **2002,** *31* (5), 486–493.
236. Goadsby, P. J. Can we develop neurally acting drugs for the treatment of migraine? *Nat. Rev. Drug Discovery* **2005,** *4* (9), 741–750.
237. Tepper, S. J.; Rapoport, A. M. The triptans: a summary. *CNS Drugs* **1999,** *12* (5), 403–417.
238. McCormack, P. L.; Gillian, K. G. M. Eletriptan: a review of its use in the acute treatment of migraine. *Drugs* **2006,** *66* (8), 1129–1149.
239. Faerkkilae, M.; Kallela, M. Eletriptan review. *Expert Opin. Pharmacother.* **2005,** *6* (4), 625–630.
240. Sandrini, G.; Perrotta, A.; Nappi, G. Eletriptan: a review and new perspectives. *Expert Rev. Neurother.* **2006,** *6* (10), 1413–1421.
241. Sandrini, G.; Perrotta, A.; Tassorelli, C.; Nappi, G. Eletriptan. *Exp. Opin. Drug Metabolism Toxicol.* **2009,** *5* (12), 1587–1598.
242. Diener, H. Ch Eletriptan in migraine. *Expert Rev. Neurother.* **2005,** *5* (1), 43–53.
243. Ogilvie, R. J., New process for the preparation of the anti-migraine drug, eletriptan, WO 2002050063 (2002).
244. Ashcroft, Ch. P.; Hellier, P.; Pettman, A.; Watkinson, S. Second-generation process research towards eletriptan: a Fischer indole approach. *Org. Process Res. Dev.* **2011,** *15* (1), 98–103.

245. Pramanik, C.; Bhumkar, R.; Karhade, G.; Khairnar, P.; Tripathy, N. M.; Gurjar, M. K. Efficient synthesis of impurity-C of antimigraine agent rizatriptan benzoate. *Org. Process Res. Dev.* **2012,** *16* (3), 507–511.
246. Krymchantowski, A. V.; Marcelo, M. E. B. Rizatriptan in migraine. *Expert Rev. Neurother.* **2005,** *5* (5), 597–603.
247. Mannix, L. K. A review of the 5-HT1B/1D agonist rizatriptan: update on recent research and implications for the future. *Expert Opin. Pharmacother.* **2008,** *9* (6), 1001–1011.
248. Street, L. J.; Baker, R.; Davey, W.; Guiblin, A. R.; Jelly, R. A.; Reeve, A. J.; Routledge, H.; Sternfled, F.; Watt, A. P.; Beer, M. S.; Middlemiss, D. N.; Noble, A. J.; Stanton, J. A.; Scholey, K.; Hargreaves, R. J.; Sohal, B.; Graham, M. I.; Matassa, G. Synthesis and serotonergic activity of N,N-dimethyl-2-[5-(1,2,4-triazol-1-ylmethyl)-1H-indol-3-yl]ethyamine and analogues. Potent agonists for 5-HT1D receptors. *J. Med. Chem.* **1995,** *38,* 1799–1810.
249. Baker, R.; Hadham, M.; Matassa, V. G.; Pelham, F.; Street, L. J. Triazole containing indole derivatives, US 5298520 (1994).
250. Chen, C.-Y. Rizatriptan (Maxalt): a 5–HT1D receptor agonist. In *The Art of Process Chemistry;* Yasuda, N., Ed.; Wiley VCH, 2010; pp 117–142.
251. Chen, C.-Y.; Senanayake, C. H.; Billy, T. J.; Larsen, R. D.; Verhoeven, T. R.; Redider, P. J. Improved Fischer indole reaction for the preparation of N,N-dimethyltryptamines: synthesis of L-695,894, a potent 5-HT1D receptor agonist. *J. Org. Chem.* **1994,** *59* (13), 3738–3741.
252. Chem, C.-Y.; Lieberman, D. R.; Larsen, R. D.; Reamer, R. A.; Verhoeven, T. R.; Reider, P. J. Synthesis of the 5-HT1D receptor agonist MK-0462 via a Pd-catalyzed coupling reaction. *Tetrahedron Lett.* **1994,** *35* (38), 6981–6984.
253. Purna, C.; Bandari, M.; Qadeeruddin, M.; Ramanjaneyulu, G. S. Process for the large-scale production of high-purity rizatriptan benzoate, WO 2007054979 (2007).
254. Rabasseda, X.; Mealy, N.; Castaner, J. J. Synthesis of MK-0462. *Drugs Future* **1995,** *20* (7), 676–679.
255. Remuzon, P.; Dunny, C.; Jacquet, J. P.; Soumeillant, M.; Bouzard, D. 4-Hydroxy-2-methyl-3(2H)-isothiazolone-1,1-dioxide as protecting group of 4-(N-methylaminousulfonyl) methylphenyl-hydrazine during Fischer indole synthesis. *Tetrahedron Lett.* **1995,** *36* (35), 6227–6230.

Chapter 4

Soporific Agents (Hypnotics and Sedative Drugs)

Insomnia is the inability to obtain an adequate amount or quality of sleep. Insomnia is a symptom, and its proper treatment depends on finding the cause of sleeplessness and treating the underlying etiology. Numerous medical disorders can cause insomnia and many drugs have been implicated as causing insomnia.

Soporific agents are drugs that facilitate the development and normalization of sleep. However, the sleep that is induced by the majority of drugs differs from natural sleep.

For approximately 100 years, bromides, followed by chloral hydrate, and subsequently by barbiturates, were the only drugs capable of relieving patient conditions of insomnia and neurological disorders. However, today there are many known compounds of various chemical classes that can be classified as hypnotics and sedative drugs, which are capable of causing various degrees of central nervous system depression, relieving anxiety, and causing sleep. Sedation is an intermediate degree of central nervous system depression, whereas hypnosis is a degree of central nervous system depression similar to natural sleep. From the chemical point of view, soporific, sedative, and hypnotic drugs are classified as barbiturates, histamine H1 receptor antagonists, benzodiazepine hypnotics, which became available in the 1970s, nonbenzodiazepine hypnotics, so-called Z-drugs, melatonin receptor agonists, 5-HT_{2A} and 5-HT_{2C} antagonists/inverse agonists, NK1 receptor agonists, melatonergic (MT1/MT2) agonists, orexin receptor (OX1/OX2) antagonists, and so on. There are many excellent reviews covering different problems concerning treatment of insomnia and some of most recent are listed in the references [1-8].

4.1 BARBITURATES

Barbiturates, which first appeared in 1903, continue to be used today. Barbiturates act by depressing the central nervous system and slowing down many areas of the brain, assisting the induction of sleep. Perhaps the greatest danger with barbiturates is that, as with many drugs, they are addictive. Widespread concern about the safety of barbiturates eventually led to the development of alternative medicines and, since the 1970s, barbiturates have been widely replaced by reputedly "safer" benzodiazepine drugs. Phenobarbital (**4.1.1**), pentobarbital

Synthesis of Best-Seller Drugs. http://dx.doi.org/10.1016/B978-0-12-411492-0.00004-3

(**4.1.2**), amobarbital (**4.1.3**), secobarbital (**4.1.4**), and sodium thiopental (**4.1.5**) (Fig. 4.1.) are the more commonly used barbiturates for insomnia. The synthesis of these drugs is described in our previous book [9].

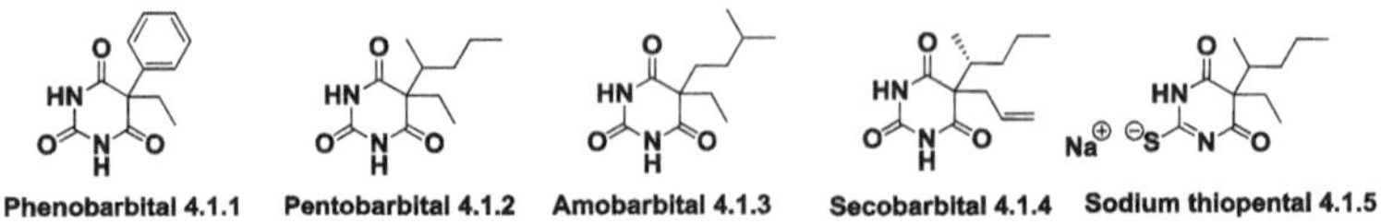

FIG. 4.1 **Barbiturates used as hypnotics and sedative drugs.**

4.2 HISTAMINE H1 RECEPTOR ANTAGONISTS

Histamine H1 receptor antagonists work by blocking the histamine H1 receptor, a mechanism of action different than that of any other medication for the treatment of insomnia. Administration of first-generation H1 receptor antagonists—chlorpheniramine (**4.2.1**), diphenhydramine (**4.2.2**), pyrilamine (**4.2.3**), and triprolidine (**4.2.4**)—produces somnolence, an increased likelihood of falling asleep. These effects led to the use of these drugs as nonprescription medications to promote sleep (Fig. 4.2.).

The tricyclic antidepressant doxepin (**4.2.5**) exhibits highly selective H1 antagonism in low doses and now is seeking approval for use in the treatment of insomnia.

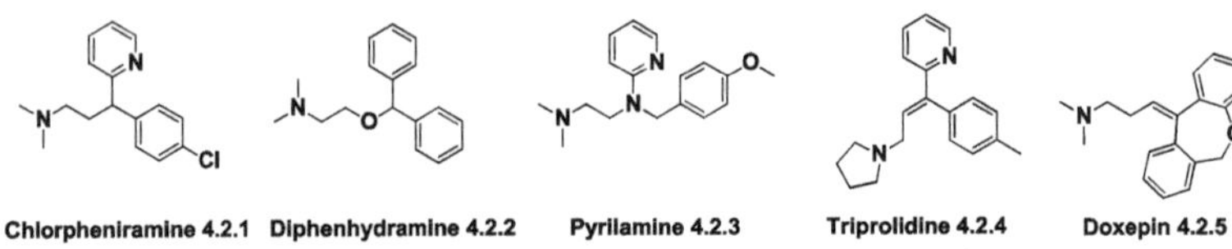

FIG. 4.2 **Histamine H1 receptor antagonists used as hypnotics and sedative drugs.**

4.3 BENZODIAZEPINES

The main actions of benzodiazepines (hypnotic, anxiolytic, anticonvulsant, myorelaxant, and amnesic), which appeared on the market in the late 1950s, confer a therapeutic value in a wide range of conditions. The pharmacological effects of benzodiazepines are believed to be realized by enhancement of the action of γ-aminobutyric acid (GABA) at the $GABA_A$ receptor and there are large pharmacological differences between particular benzodiazepines. Benzodiazepines commonly used for the treatment of insomnia include lorazepam (**4.3.1**), temazepam (**4.3.2**), estazolam (**4.3.3**), triazolam (**4.3.4**), flurazepam (**4.3.5**), and quazepam (**4.3.6**) (Fig. 4.3.). The syntheses of these drugs are described in our previous book [9].

The $GABA_A$ receptor is the site of action of benzodiazepines and Z-drugs. But different classes of hypnotics modulate GABAergic function through different $GABA_A$ subtypes.

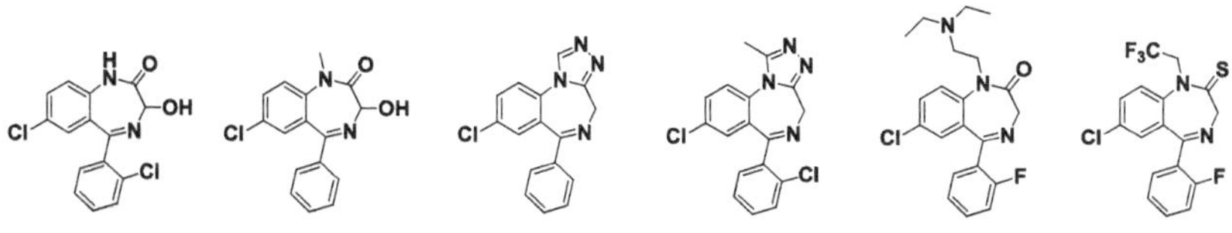

FIG. 4.3 Benzodiazepines used as hypnotics and sedative drugs.

4.4 Z-DRUGS

The Z-drugs, so named because of the initial letter of zolpidem, are zolpidem (Ambien) (**4.4.1**), eszopiclone (Lunesta) (**4.4.2**), which is the active isomer of zopiclone (Imovane) (**4.4.2a**), and zaleplon (Sonata) (**4.4.3**) (Fig. 4.4.). All are $GABA_A$ modulators that bind selectively α_1 subunits, thus, exhibiting mechanisms of action similar to those of the benzodiazepines. These compounds are generally well tolerated compared to the older benzodiazepines and barbiturates, and initially they were thought to show better safety profiles and be less addictive and/or habit-forming.

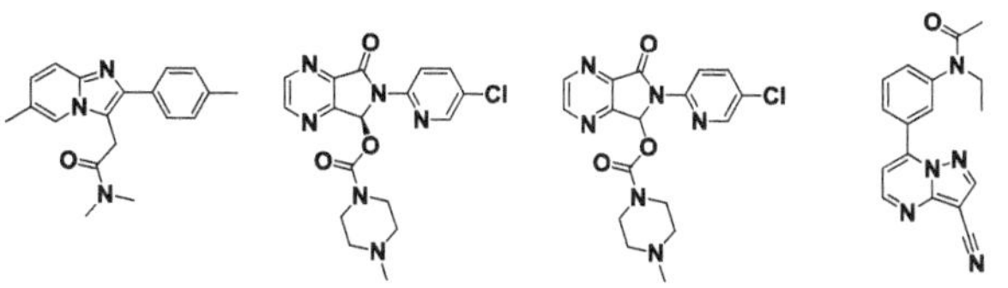

FIG. 4.4 Z-drugs used as hypnotics and sedative drugs.

Zolpidem (**4.4.1**) and eszopiclone (**4.4.2**) are included in the list of Top 200 Drugs by sales for the 2010s [10]. Several synthetic schemes have been developed for their synthesis.

Zolpidem–Ambien

The first synthesis of zolpidem (**4.4.1**) [11-14] involves reaction of 2-amino-5-methylpyridine (**4.4.4**) with 2-bromo-4-methylacetophenone (**4.4.5**) to produce imidazopyridine (**4.4.6**), which undergoes aminomethylation to form the amine (**4.4.7**). The obtained dimethylamino derivative (**4.4.7**) was methylated with methyl iodide and converted into its quaternary ammonium salt (**4.4.8**).

The reaction of quaternary ammonium salt (**4.4.8**) with the sodium cyanide produced nitrile (**4.4.9**), which, after acidic hydrolysis, was transformed into the corresponding acid chloride that on heating with dimethylamine produces the desired dimethylamide, zolpidem (**4.4.1**) (Scheme 4.1.).

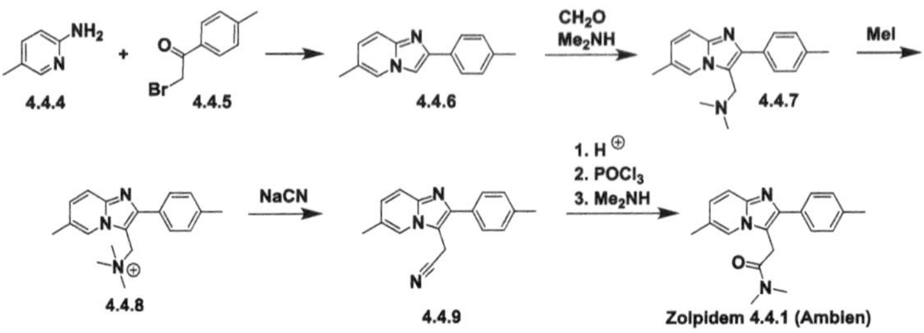

SCHEME 4.1 Synthesis of zolpidem.

An alternative preparation by the same authors involves a Vilsmeier-Haack formylation of (**4.4.6**) with dimethylformamide/phosphorus oxychloride that produces aldehyde (**4.4.10**), a reduction of the obtained aldehyde to the corresponding alcohol (**4.4.11**), the tosylation of the alcohol to (**4.4.12**), and the subsequent reaction with cyanide ion produce (**4.4.9**) [15] (Scheme 4.2.).

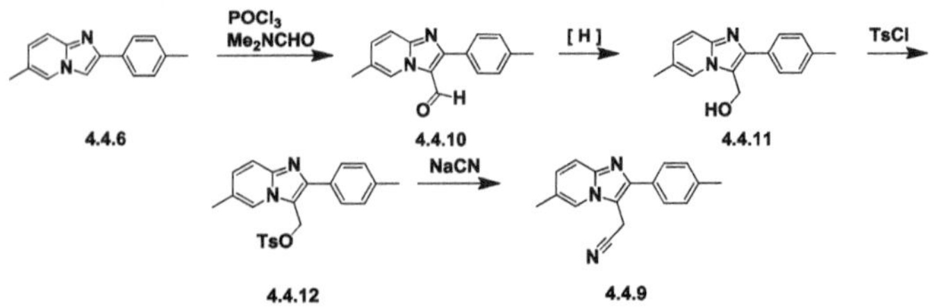

SCHEME 4.2 Alternative synthesis of zolpidem.

Another method involves bromination [16,17] or tosyloxylation [18] of (4-methylbenzoyl) propionamide (**4.4.13**) to produce 3-bromo-(or tosyloxy) (4-methylbenzoyl)propionamides (**4.4.14**) and cyclization of the resultant derivatives with 2-amino-5-methylpyridine (**4.4.4**) to produce imidazopyridine (zolpidem) (**4.4.1**) (Scheme 4.3.).

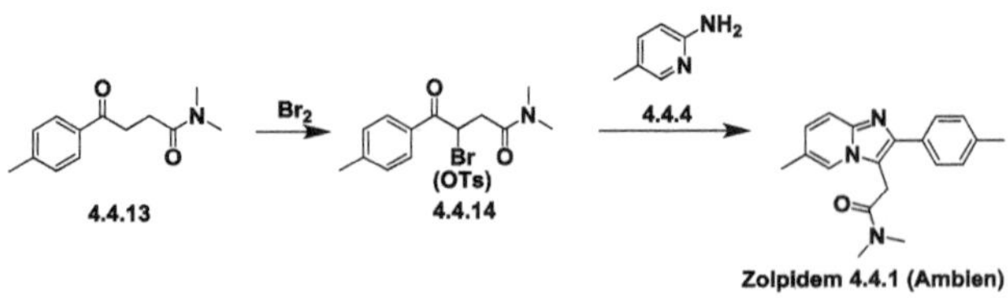

SCHEME 4.3 Alternative synthesis of zolpidem.

In another patent, imidazopyridine (**4.4.6**) was proposed to react with *N,N*-dimethylacetamide dimethyl acetal (**4.4.15**) to produce a hydroxy derivative (**4.4.16**). The hydroxy group of (**4.4.16**) was removed by chlorination with thionyl chloride followed by reduction with sodium hydrosulfite to produce zolpidem (**4.4.1**) [19] (Scheme 4.4.).

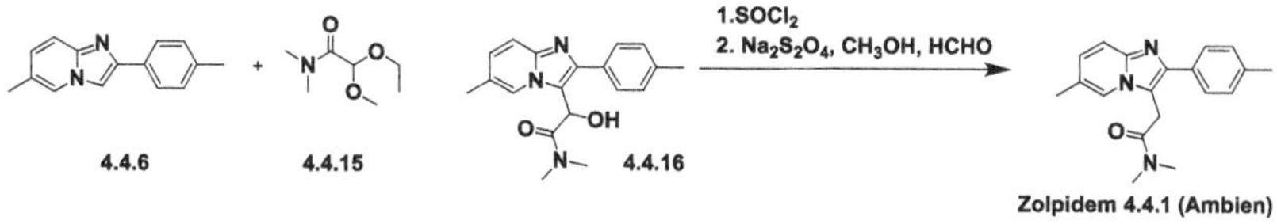

SCHEME 4.4 Alternative synthesis of zolpidem.

Zolpidem is an effective hypnotic. It improves sleep in patients with insomnia with a low propensity to cause clinical residual effects such as withdrawal, dependence, and tolerance, and minimal rebound insomnia [20-22].

Eszopiclone–Lunesta

Preparation of eszopiclone (**4.4.2**) includes the following steps, which involve synthesis of racemic zopiclone (**4.4.2a**) which was first proposed in 1972 [23], followed by resolution of racemic zopiclone with an optically active acid in suitable solvent. The formed (S)-zopiclone salt selectively peptidized from the solution, which was then treated with a base to form enriched (S)-zopiclone (eszopiclone) (**4.4.2**). Approximately 60 patents devoted to improvement of the synthesis process proposed new acids for formation of (S)-zopiclone salts. The general scheme of synthesis remains the same and is perfectly described in Guzzo [24] and includes the reaction of pyrazine-2,3-dicarboxylic anhydride (**4.4.17**) with 2-amino-5-chloropyridine (**4.4.18**), which produces pyrazine-2-carboxylic acid amide (**4.4.19**), followed by ring closure, and the use of thionyl chloride to produce the appropriate 5,7-dioxopyrrolopyrazine derivative imide (**4.4.20**). Selective KBH_4 reduction of only one of the carbonyl groups results in (**4.4.20**) producing the chiral 7-hydroxy-pyrazin-5-one derivative (**4.4.21**), reaction of which with 1-chloro-carbonyl-4-methylpiperazine (**4.4.22**) produces the racemic mixture zopiclone (**4.4.2.a**). Chiral resolution of racemic zopiclone with the appropriate acid (L-malic or L-tartaric acids) in a suitable solvent produces the (S)-zopiclone (eszopiclone) salt, which on workup with an appropriate base produces the desired eszopiclone (**4.4.2**) free base in crystalline form (Scheme 4.5.).

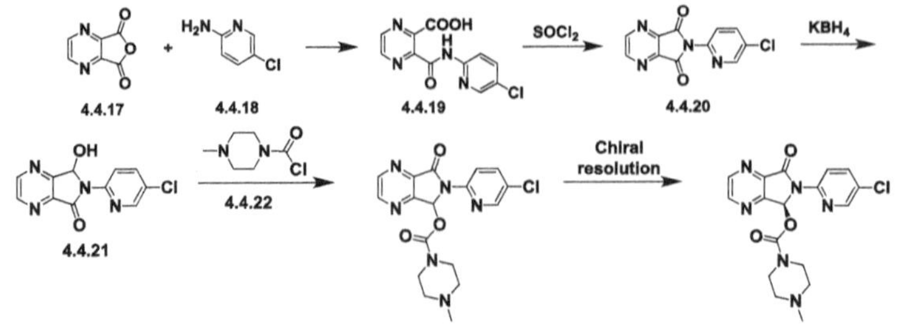

Zopiclone 4.4.2a (Imovane) Eszopiclone 4.4.2 (Lunesta)

SCHEME 4.5 **Synthesis of zopiclone and eszopiclone.**

Obtained eszopiclone (Lunesta), the S-enantiomer of racemic zopiclone, is an oral hypnotic agent for once-nightly therapy for insomnia, which significantly improves sleep onset and sleep maintenance [25-27].

4.5 NEW GABA RECEPTOR AGONISTS

New GABA receptor agonists of different structures are under development. They include indiplon (**4.5.1**) [28,29], adipiplon (**4.5.2**) [30] (in Phase II trials), which work by enhancing the action of GABA, and gaboxadol (**4.5.3**), a selective extrasynaptic GABA receptor agonist. Gaboxadol could be useful on sleep disorders, acting via a fundamentally different mechanism from that of the benzodiazepine receptor agonists and Z-drugs. The clinical probations of gaboxadol were cancelled as a result of revealed hallucinogenic effects on and disorientation of patients in efficacy trials [31]. The final GABA agonist under development is a fused γ-carboline compound, EVT-201 (**4.5.4**), which is a partial allosteric modulator of $GABA_A$ subtypes [32] (Fig. 4.5.).

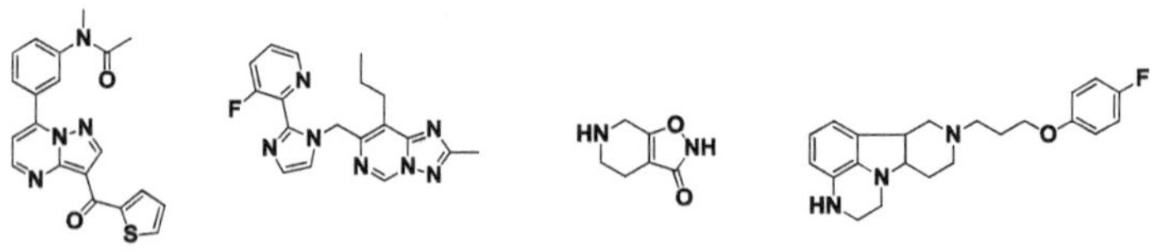

Indipilon 4.5.1 Adipiplon 4.5.2 Gaboxadol 4.5.3 EVT-201 4.5.4

FIG. 4.5 **GABA receptor agonists as sedatives under development.**

4.6 MELATONERGIC (MT1/MT2) AGONISTS

Melatonin (**4.6.1**) is the pineal hormone in the regulation of mammalian circadian rhythms. Melatonin receptors (MT1 and MT2) are thought to be involved in the maintenance circadian rhythms. Melatonin itself is successfully used for the treatment of sleep problems, but because of its short half-life (<30 min), development of prolonged-release medications (Circadin) [33] or melatonin analogues like ramelteon (**4.6.2**), agomelatine (**4.6.3**), tasimelteon (**4.6.4**), and TIK301 (**4.6.5**), are possible approaches for treatment of insomnia [34,35] (Fig. 4.6.).

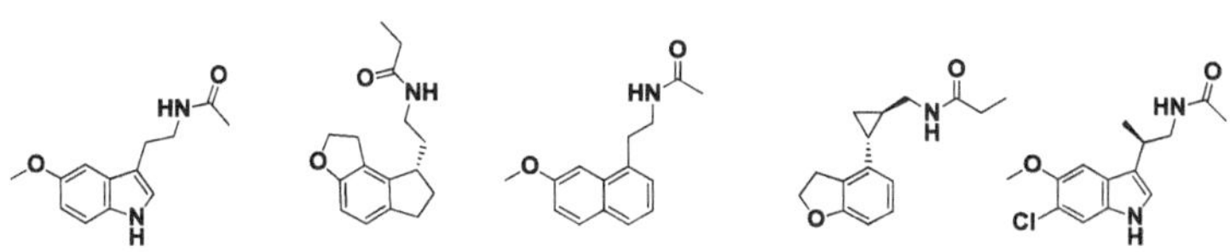

Melatonin 4.6.1 Ramelteon 4.6.2 Agomelatine 4.6.3 Tasimelteon 4.6.4 TIK301 4.6.5

FIG. 4.6 Melatonergic agonists as sedatives.

4.7 5-HT ANTAGONISTS AND ANTAGONISTS/ INVERSE AGONISTS

The serotonin system is implicated in the maintenance of wakefulness and, at least theoretically, serotonin antagonism remains a possible approach for insomnia treatment. The neuroleptic paliperidone (**4.7.1**)—which is an atypical antipsychotic, is the first agent approved for the treatment of schizoaffective disorder, is available for the treatment of schizophrenia, and is seeking approval for treatment of indications of depression and autism—is now being proposed as an insomnia remedy [36]. Clinical trials of some other piperidine derivatives, such as the selective 5-HT_{2A} antagonists ritanserin (**4.7.2**), volinanserin (**4.7.3**), have been discontinued. Pimavanserin (ACP-103) (**4.7.4**) is still in a phase III clinical trial for indications other than insomnia, as well as under development for insomnia (Fig. 4.7.).

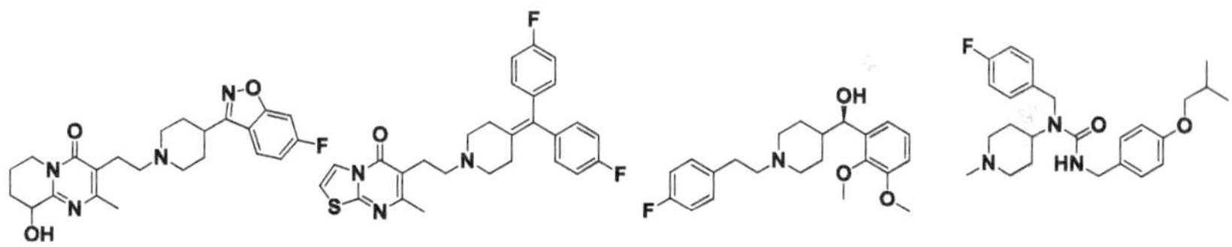

Paliperidone 4.7.1 Ritanserin 4.7.2 Volinanserin 4.7.3 Pimavanserin 4.7.4

FIG. 4.7 5-HT antagonists as sedatives.

Clinical trials of some piperazine derivatives (Fig. 4.8.), like pruvanserin (**4.7.5**) and fananserin (**4.7.6**), also have been discontinued [37,38].

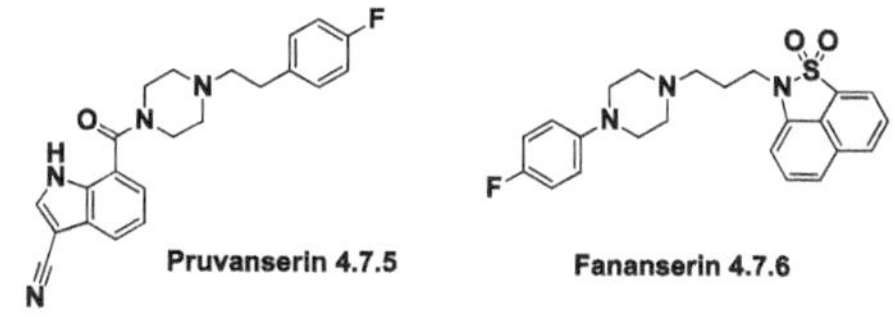

Pruvanserin 4.7.5 Fananserin 4.7.6

FIG. 4.8 5-HT antagonists proposed as sedatives whose clinical trials have been discontinued.

There are compounds of different chemical classes, like esmirtazapine (**4.7.7**) and eplivanserin (**4.7.8**) (Fig. 4.9.), whose clinical trials have been discontinued but for different reasons. Very interesting are compounds with mixed activities. Nelotanserin (**4.7.9**), which is a 5-HT_{2A} inverse agonist, was proposed for insomnia treatment but also was discontinued.

Another compound with mixed activity, H1 receptor antagonist/5-HT_{2A} inhibitor LY-2624803 (**4.7.10**). (Fig. 4.9.), is in phase II of clinical trials for the treatment of chronic insomnia [39].

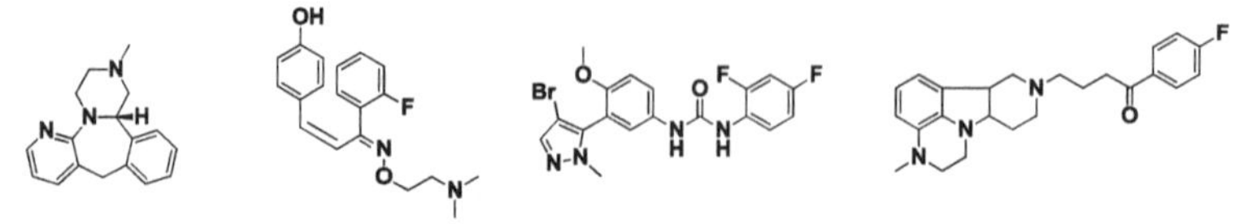

FIG. 4.9 **5-HT antagonists of different chemical classes proposed as sedatives.**

4.8 NK1-RECEPTOR ANTAGONISTS

A potent NK1-receptor antagonist, casopitant (**4.8.1**), was previously shown to be a potent medication for chemotherapy-induced and postoperative nausea and vomiting, and for the treatment of chronic anxiety and depression. Now it is in Phase II trials for insomnia [40]. Some other NK antagonists (**4.8.2** to **4.8.4**) [41-43] of the piperidine and piperazine series also display antiinsomnia properties (Fig. 4.10.).

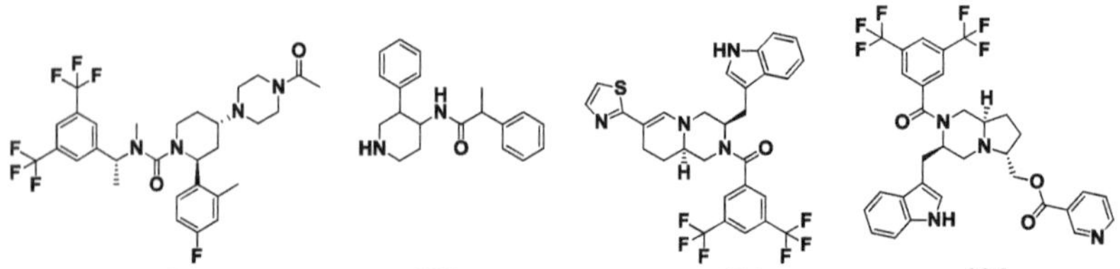

FIG. 4.10 **NK1-receptor antagonists proposed as sedatives.**

4.9 OREXIN RECEPTOR (OX1/OX2) ANTAGONISTS

The hypocretin–orexin system is a new area in the fight against insomnia and the use of an orexin receptor antagonist for the treatment of sleep disorders appears to be a new approach [44-46]. Proof-of-concept clinical studies in primary insomnia were reported with five structurally diverse dual orexin receptor antagonists—almorexant (**4.9.1**), MK-4305 (**4.9.2**), suvorexant (**4.9.3**), SB-649868 (**4.9.4**), and MK-6096 (**4.9.5**).

A potent orexin receptor (OX1/OX2) antagonist, almorexant (**4.9.1**) [47,48] is in phase III clinical trials. MK-4305 (**4.9.2**) [49,50] are under development for indications of insomnia. Suvorexant (**4.9.3**) [51], SB-649868 (**4.9.4**) [52], and MK-6096 (**4.9.5**) [53] also are on different phases of clinical trials (Fig. 4.11.).

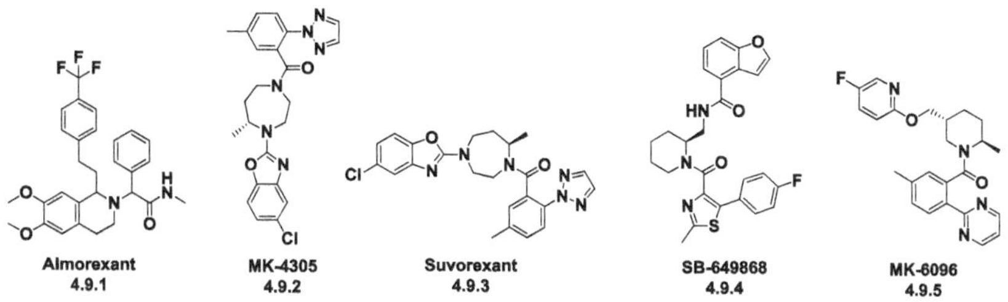

FIG. 4.11 Orexin receptor antagonists for insomnia under development.

4.10 SODIUM OXYBATE

Sodium oxybate (**4.10.1**) or γ-hydroxybutyrate is classified as a psychostimulant and was launched for the treatment of catalepsy associated with narcolepsy. Although it also improves overall insomnia, it is not officially approved for treatment of insomnia [54,55] (Fig. 4.12.).

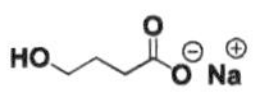

FIG. 4.12 Structure of sodium oxybate.

Since the early 2000s the number of soporific agents has changed significantly. Some new entities were approved for the treatment of insomnia (melatonin receptor agonists, histamine, serotonin, orexin antagonists). Newer targets, such as adenosine modulators, 5-HT_7 receptor antagonists, and *N*-acetyltransferase inhibitors, have been found useful for the treatment of insomnia [7].

REFERENCES

1. William, S. Sedative-hypnotics. In *Foye's Principles of Medicinal Chemistry,* 6th ed.; Lemke, T. L., Williams, D. A., Eds. Lippincott Williams & Wilkins, 2008; pp 504–520.
2. Richey, S. M.; Krystal, A. D. Pharmacological advances in the treatment of insomnia. *Curr. Pharm. Des.* **2011,** *17* (15), 1471–1475.
3. Sullivan, S. S.; Guilleminault, C. Emerging drugs for insomnia: new frontiers for old and novel targets. *Expert Opin. Emerg. Drugs* **2009,** *14* (3), 411–422.
4. Zammit, G. Comparative tolerability of newer agents for insomnia. *Drug Saf.* **2009,** *32* (9), 735–748.
5. Monti, J. M.; Monti, D. New directions in the treatment of insomnia. *Expert Opin. Ther. Pat.* **2005,** *15* (10), 1353–1359.
6. Bhat, A.; Shafi, F.; El Solh, A. A. Pharmacotherapy of insomnia. *Expert Opin. Pharmacother.* **2008,** *9* (3), 351–362.
7. Palomer, A.; Princep, M.; Guglietta, A. Recent advances in the treatment of insomnia. *Annu. Rep. Med. Chem.* **2007,** *42,* 63–80.
8. Zammit, G. Comparative tolerability of newer agents for insomnia. *Drug Saf.* **2009,** *32* (9), 735–748.

9. Vardanyan, R. S.; Hruby, V. J. *Synthesis of Essential Drugs;* Elsevier, 2006.
10. http://www.drugs.com/top200.html. Pharmaceutical Sales, 2010.
11. Kaplan, J. P.; George, P. Imidazo[1,2-a]pyridine derivatives and their therapeutic use, EP 50563 (1982).
12. George, P.; Rossey, G.; Depoortere, H.; Allen, J.; Wick, A. Zolpidem: a new hypnotic with the imidazo[1,2-a]pyridine structure. *Actual. Chim. Ther.* **1991,** *18,* 215–239.
13. Kaplam. J. P.; George, P. Imidazo[1,2-a] pyridine derivatives and their application as pharmaceuticals, US 4382938 (1983).
14. Jasty, A. M.; Tamma, R. R.; Mohanarangam, S.; Yasareni, S.; Rupakala G. S.; Debashish, G. Process for preparing zolpidem, US 2007/0027180 (2007).
15. Kaplan, J. P., George, P.; Bernandon J.M. Imidazo[1,2-a]pyridines and their therapeutic use, EP 92459 (1983).
16. Markus, S.; Wolfgang, W. Process for preparing zolpidem, US 2002/0183522 (2002).
17. Trapani, G.; Franco, M.; Ricciardi, L.; Latrofa, A.; Genchi, G.; Sanna, E.; Tuveri, F.; Cagetti, E.; Biggio, G.; Liso, G. J. Synthesis and binding affinity of 2-phenylimidazo[1,2-a]pyridine derivatives for both central and peripheral benzodiazepine receptors. A new series of high-affinity and selective ligands for the peripheral type. *J. Med. Chem.* **1997,** *40* (19), 3109–3118.
18. Patil, S. S.; Patil, S. V.; Bobade, V. D. An efficient synthesis of zolpidem. *Org. Prep. Proced. Int.* **2011,** *43* (2), 260–264.
19. Rossey, G.; Long, D. Process for the preparation of (2-phenylimidazopyridinyl) acetamides, FR 2600650 (1987).
20. Greenblatt, D. J.; Roth, T. Zolpidem for insomnia. *Expert Opin. Pharmacother.* **2012,** *13* (6), 879–893.
21. Moen, M. D.; Plosker, G. L. Zolpidem extended-release. *CNS Drugs* **2006,** *20* (5), 419–426.
22. Harrison, T. S.; Keating, G. M. Zolpidem: a review of its use in the management of insomnia. *CNS Drugs* **2005,** *19* (1), 65–89.
23. Cotrel, C.; Jeanmart, C.; Messer, M. N. Anticonvulsive and tranquilizing pyrrolopyrazines, DE 2300491 (1973).
24. Guzzo, P. R. Advances in the development of methods for the synthesis of GABAA receptor agonists for the treatment of insomnia [zolpidem (Ambien), zaleplon (Sonata), Eszopiclone (Estorra, Lunesta), indiplon]. In *Art of Drug Synthesis;* Johnson, D. S., Li, J. J., Eds. Wiley-Interscience, 2007; pp 215–223.
25. Melton, S. T.; Wood, J. M.; Kirkwood, C. K. Eszopiclone for insomnia. *Ann. Pharmacother.* **2005,** *39* (10), 1659–1666.
26. Owen, R. T. Eszopiclone: an update on its use in insomnia. *Drugs Today* **2011,** *47* (4), 263–275.
27. Hair, P. I.; McCormack, P. L.; Curran, M. P. Eszopiclone: a review of its use in the treatment of insomnia. *Drugs* **2008,** *68* (10), 1415–1434.
28. Marrs, J. C. Indiplon: A nonbenzodiazepine sedative–hypnotic for the, treatment of insomnia. *Ann. Pharmacother.* **2008,** *42* (7/8), 1070–1079.
29. Neubauer, D. N. Indiplon: the development of a new hypnotic. *Expert Opin. Invest. Drugs* **2005,** *14* (10), 1269–1276.
30. Xie, L.; Han, B.; Xu, Y.; Maynard, G.; Chenard, B. L.; Shaw, K.; Gao, Y. Imidazo-pyrimidines and triazolo-pyrimidines: benzodiazepine receptor ligands, WO 2005012306 (2005).
31. Wafford, K. A.; Ebert, B. Gaboxadol—a new awakening in sleep. *Curr. Opin. Pharmacol.* **2006,** *6,* 30–36.
32. Mates, S.; Fienberg, A. A.; Wennogle, L. P. Methods and compositions containing gamma-carboline compounds for treatment of sleep disorders and other disorders, WO 2009145900 A1 (2009).

33. Lemoine, P.; Zisapel, N. Prolonged-release formulation of melatonin (Circadin) for the treatment of insomnia. *Expert Opin. Pharmacother.* **2012,** *13* (6), 895–905.
34. Cardinali, D. P.; Srinivasan, V.; Brzezinski, A.; Brown, G. M. Melatonin and its analogs in insomnia and depression. *J. Pineal Res.* **2012,** *52* (4), 365–375.
35. Hardeland, R. New approaches in the management of insomnia: weighing the advantages of prolonged-release melatonin and synthetic melatoninergic agonists. *Neuropsychiatr. Dis. Treat.* **2009,** *5,* 341–354.
36. Sullivan, S. S.; Guilleminault, C. Emerging drugs for insomnia: new frontiers for old and novel targets. *Expert Opin. Emerg. Drugs* **2009,** *14* (3), 411–422.
37. Monti, J. M. Serotonin 5-HT2A receptor antagonists in the treatment of insomnia: present status and future prospects. *Drugs Today* **2010,** *46* (3), 183–193.
38. Vanover, K. E.; Davis, R. E. Role of 5-HT2A receptor antagonists in the treatment of insomnia. *Nat. Sci. Sleep* **2010,** *2,* 139–150.
39. Mates, S.; Fienberg, A. A.; Wennogle, L. P. Methods and compositions containing gamma-carboline compounds for treatment of sleep disorders and other disorders, WO 2009145900 (2009).
40. Navari, R. M. Casopitant a neurokinin-1 receptor antagonist with anti-emetic and antinausea activities. *Curr. Opin. Invest. Drugs (BioMed Cent.)* **2008,** *9* (7), 774–785.
41. O'Neill, B. T.; Parikh, V. D.; Welch, W. M. Phenylpiperidines as NK1 and NK3 antagonists, their preparation, pharmaceutical compositions and use in the treatment of neurokinin-mediated diseases, US 20050256164 (2005).
42. Iwema, B.; Wouter, I.; Van Scharrenburg, G. J. M.; Van Den Hoogenband, A.; McCreary, A. C. Preparation of hexa-and octahydro-pyrido[1,2-a]pyrazine derivatives with NK1 antagonistic activity, US 20050070548 (2005).
43. De Boer, D.; Coolen, H. K. A. C.; Hesselink, M. B.; Iwema B., Wouter I.; Kuil, G. D.; Van Maarseveen, J. H.; McCreary, A. C.; Van Scharrenburg, G. J. M. Preparation of aroylpyrrolo-pyrazines and related compounds as tachykinin NK1 antagonists, 2003084955 (2003).
44. Dugovic, C.; Shelton, J.E.; Aluisio, L. E.; Fraser, I. C.; Jiang, X.; Sutton, S. W.; Bonaventure, P.; Yun, S.; Li, X.; Lord, B.; Dvorak, C. A.; Carruthers, N. I.; Lovenberg, T. W. Blockade of orexin-1 receptors attenuates orexin-2 receptor antagonism-induced sleep promotion in the rat. *J. Pharmacol. Exp. Ther.* **2009,** *330* (1), 142–151.
45. Hoever, P.; Dorffner, G.; Benes, H.; Penzel, T.; Danker-Hopfe, H.; Barbanoj, M. J.; Pillar, G.; Saletu, B.; Polo, O.; Kunz, D.; Zeitlhofer, J.; Berg, S.; Partinen, M.; Bassetti, C. L.; Hoegl, B.; Ebrahim, I. O.; Holsboer-Trachsler, E.; Bengtsson, H.; Peker, Y.; Hemmeter, U.-M.; Chiossi, E.; Hajak, G.; Dingemanse, J. Orexin receptor antagonism, a new sleep-enabling paradigm: a proof-of-concept clinical trial. *Clin. Pharmacol. Ther. (Hoboken, NJ, U. S.)* **2012,** *91* (6), 975–985.
46. Coleman, P. J.; Renger, J. J. Orexin receptor antagonists: a review of promising compounds patented since 2006. *Expert Opin. Ther. Pat.* **2010,** *20* (3), 307–324.
47. Neubauer, D. N. Almorexant, a dual orexin receptor antagonist for the treatment of insomnia. *Curr. Opin. Invest. Drugs (BioMed Cent.)* **2010,** *11* (1), 101–110.
48. Hoever, P.; de Haas, S.; Winkler, J.; Schoemaker, R. S.; Chiossi, E.; van Gerven, J.; Dingemanse, J. Orexin receptor antagonism, a new sleep-promoting paradigm: an ascending single-dose study with almorexant. *Clin. Pharmacol. Ther. (Hoboken, NJ, U. S.)* **2010,** *87* (5), 593–600.
49. Baxter, C. A.; Cleator, E.; Brands, K. M. J.; Edwards, J. S.; Reamer, R. A.; Sheen, F. J.; Stewart, G. W.; Strotman, N. A.; Wallace, D. J. The first large-scale synthesis of MK-4305: a dual orexin receptor antagonist for the treatment of sleep disorder. *Org. Process Res. Dev.* **2011,** *15* (2), 367–375.

50. Cox, C. D.; Breslin, M. J.; Whitman, D. B.; Schreier, J. D.; McGaughey, G. B.; Bogusky, M. J.; Roecker, A. J.; Mercer, S. P.; Bednar, R. A.; Lemaire, W.; Bruno, J. G.; Reiss, D. R.; Harrell, C. M.; Murphy, K. L.; Garson, S. L.; Doran, S. M.; Prueksaritanont, T.; Anderson, W. B.; Tang, C.; Roller, S.; Cabalu, T. D.; Cui, D.; Hartman, G. D.; Young, S. D.; Koblan, K. S.; Winrow, C. J.; Renger, J. J.; Coleman, P. J. Discovery of the dual orexin receptor antagonist [(7R)-4-(5-Chloro-1,3-benzoxazol-2-yl)-7-methyl-1,4-diazepan-1-yl][5-methyl-2-(2H-1,2,3-triazol-2-yl)phenyl]methanone (MK-4305) for the treatment of insomnia. *J. Med. Chem.* **2010,** *53* (14), 5320–5332.
51. Hopkins, C. R. ACS chemical neuroscience—molecule spotlight on suvorexant. *ACS Chem. Neurosci.* **2012,** *3* (9), 647–648.
52. Bettica, P.; Nucci, G.; Pyke, C.; Squassante, L.; Zamuner, S. Phase I studies on the safety, tolerability, pharmacokinetics and pharmacodynamics of SB-64(9868), a novel dual orexin receptor antagonist. *J. Neuropsychopharmacol. (Oxford)* **2012,** *37* (5), 1224–1233.
53. Coleman, P. J.; Schreier, J. D.; Cox, C. D.; Breslin, M. J.; Whitman, D. B.; Bogusky, M. J.; McGaughey, G. B.; Bednar, R. A.; Lemaire, W.; Doran, S. M.; Fox, S. V.; Garson, S. L.; Gotter, A. L.; Harrell, C. M.; Reiss, D. R.; Cabalu, T. D.; Cui, D.; Prueksaritanont, T.; Stevens, J.; Tannenbaum, P. L.; Ball, R. G.; Stellabott, J.; Young, S. D.; Hartman, G. D.; Winrow, C. J.; Renger, J. J. Discovery of [(2R,5R)-5-{[(5-fluoropyridin-2-yl)oxy]methyl}-2-methylpiperidin-1-yl][5-methyl-2-(pyrimidin-2-yl)phenyl]methanone (MK-6096): A dual orexin receptor antagonist with potent sleep-promoting properties. *ChemMedChem* **2012,** *7* (3), 415–424.
54. Brown, M. A.; Guilleminault, C. A review of sodium oxybate and baclofen in the treatment of sleep disorders. *Curr. Pharm. Des.* **2011,** *17* (15), 1430–1435.
55. Mamelak, M.; Scharf, M. B.; Woods, M. Treatment of narcolepsy with gamma-hydroxybutyrate. A review of clinical and sleep laboratory findings. *Sleep* **1986,** *9* (1 Pt 2), 285–289.

Chapter 5

Anxiolytics (Tranquilizers)

Anxiety disorders, like generalized anxiety disorder, panic disorder, posttraumatic stress disorder, and obsessive-compulsive disorder, are the most common mental health conditions of modern life, but they remain underdiagnosed and undertreated. Historically, the primary drugs used for correcting psychoemotional disorders were barbiturates and hypnotic drugs, in particular, phenobarbital and glutetimide-piperidine-2,6-dione derivative.

Because barbiturates are now practically absent from the doctor's armamentarium, benzodiazepines dominate the treatment picture; even the term *anxiolytic* has become nearly synonymous with benzodiazepine. However, the tricyclic antidepressants, the monoamine oxidase inhibitors, and the selective serotonin reuptake inhibitors have shown efficacy in some of the anxiety disorders. Antihistamines are frequently used as hypnotics and anxiolytics (particularly in the primary care setting) as are newer categories of novel sedating nonbenzodiazepine hypnotics such as zaleplon and zolpidem and the nonsedating, nonaddictive anxiolytic azapirones. Treatments for anxiety disorders are varied and depend on each individual's case. In general, therapy includes benzodiazepines or tricyclic antidepressants targeting 5-hydroxytryptamine, dopamine, γ-aminobutyric acid (GABA), β-adrenoceptors, metabotropic glutamate, cholecystokinin, *N*-methyl-D-aspartate (NMDA), and opioid receptors [1,2].

Benzodiazepines were once called "minor tranquilizers." First-generation antipsychotics were considered "major tranquilizers." The calming and sedating properties of these classes of drugs led to their use as anxiolytics. (It is necessary to note that trifluoperazine was the single "major tranquilizer" allowed for use as an anxiolytic. Its use as an anxiolytic was limited because of its extrapyramidal side effects and risk of dyskinesia.) So-called atypical antipsychotics, also called second-generation antipsychotics, which are typically used to treat schizophrenia and bipolar disorder, are frequently used for agitation associated with anxiety disorder. Five of the atypical antipsychotics—aripiprazole (**5.1.1**), olanzapine (**5.1.2**), quetiapine (**5.1.3**), ziprasidone (**5.1.4**), and risperidone (**5.1.5**) (Fig. 5.1.)—were reported to be successfully used for augmentation treatment in patients with anxiety disorders, and the general conclusion was that "the use of atypical antipsychotics should be reserved for patients in whom other treatment failed" [3].

Synthesis of Best-Seller Drugs. http://dx.doi.org/10.1016/B978-0-12-411492-0.00005-5

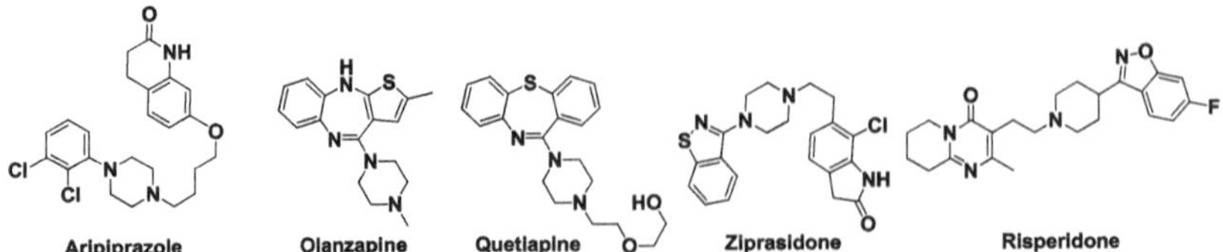

FIG. 5.1 Atypical antipsychotics.

5.1 BENZODIAZEPINES

Benzodiazepines have been in clinical use for decades and are still among the most widely prescribed drugs for the treatment anxiety disorders. However, their use is limited by side effects and the risk of drug dependence. Benzodiazepines nonselectively target GABA, the major inhibitory neurotransmitter in the brain and spinal cord and widely distributed in peripheral autonomic terminals allosterically, and nonselectively act via $GABA_A$, $GABA_B$, and $GABA_C$ receptors, which modulate emotions, cognition, and pain.

The most popular $GABA_A$ receptor modulators, classic 1,4-benzodiazepines, act via α_1, α_2, α_3, or α_5 subunits of $GABA_A$ receptors, and are among the most widely prescribed drugs for the treatment of insomnia and anxiety disorders.

The 1,4-benzodiazepines are a very effective class of medications and are commonly used for tranquilizing and improving anxiety conditions. They also display sedative, anticonvulsant, and muscle relaxant effects [4-8]. The syntheses of diazepam (**5.1.6**), clonazepam (**5.1.7**), lorazepam (**5.1.8**), and alprazolam (**5.1.9**) are described in our previous book [9] (Fig. 5.2.).

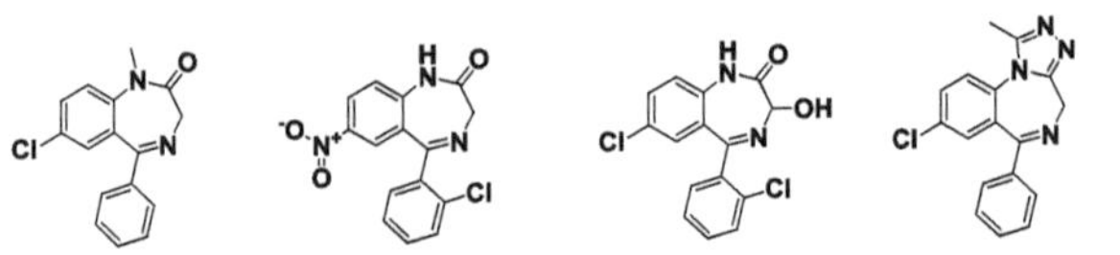

FIG. 5.2 Structures of diazepam, clonazepam, lorazepam and alprazolam.

For a long time it was not clear if anxiolytic and sedative effects are pharmacologically separable, but it seems that α_2-containing $GABA_A$ receptors are responsible for mediating of anxiolytic action of of 1,4-benzodiazepines, whereas α_1-containing $GABA_B$ receptors are responsible for mediating of sedative action [10,11]. Consequently, it is possible to hypothesize that selective α_2- or α_3-$GABA_A$ agonists would be nonsedating anxiolytics. This hypothesis is confirmed by preclinical testing of compounds like the imidazopyridine derivative TP-003 (**5.1.10**) (selective α_3 agonist), which, in preclinical studies, has shown anxiolytic properties without sedation. Other compounds from the series of triazolopyridazines are TPA-023 (**5.1.11**) (α_2, α_3 partial agonist, $\alpha 1$, $\alpha 5$ antagonist) and MK-0343 (**5.1.12**) (partial agonist for α_3, α_1, α_5, and α_2, respectively, but with greater agonist efficacy at the α_3 compared with α_1 subtypes), pyrazolotriazine derivative L-838417 (**5.1.13**) (partial agonist for α_2, α_3, α_5 subtypes), and the pyrido[3,4-b]indol-1-one derivative SL-651498 (**5.1.14**) (α_2, α_3 agonist, α_1, $\alpha5$ partial agonist) (Fig 5.3.).

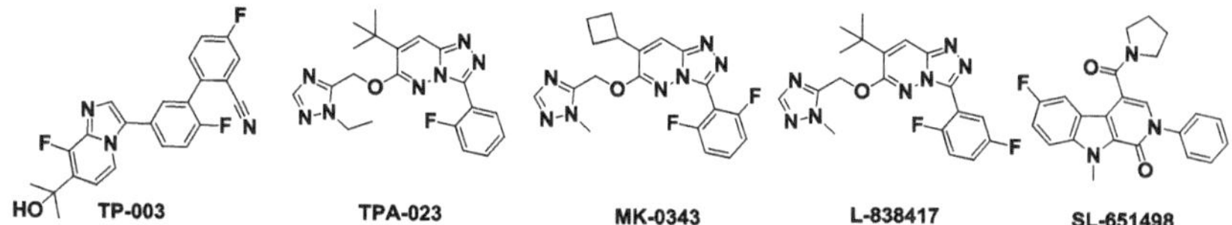

TP-003 5.1.10 TPA-023 5.1.11 MK-0343 5.1.12 L-838417 5.1.13 SL-651498 5.1.14

FIG. 5.3 Selective α_2- or α_3-GABA$_A$ agonists on preclinical testing.

There are other chemical classes of compounds that predominantly act on different "anxiolytically tuned" subtypes of GABA$_A$ receptors. Among them are benzimidazol derivative NS-11394 (**5.1.15**), pyrazolopyrimidin derivative - ocinaplon (**5.1.16**), benzoimidazopyrrolodiazepine derivative L-655708 (**5.1.17**), which are are ligands of GABA$_A$ receptors. On the other side, α_1-selective compounds of other classes, like the Z-drug for treatment of insomnia zolpidem (Ambien) (**5.1.18**), are sedative-hypnotics, which confirms the abovementioned hypothesis (Fig. 5.4.).

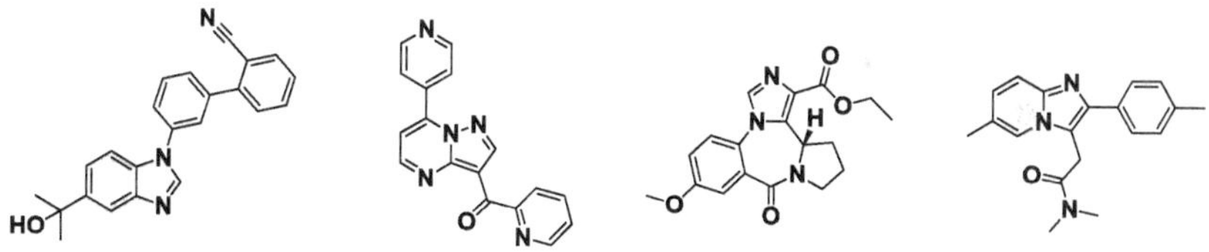

NS-11394 5.1.15 Ocinaplon 5.1.16 L-655708 5.1.17 Zolpidem (Ambien) 5.1.18

FIG. 5.4 Representatives of ligands of different "subtypes" of GABA$_A$ receptors belonging to other chemical classes.

New approaches to GABA pharmacology are evident [12], and search for novel targets for therapeutic intervention on anxiety disorders, actively continue [13-16].

5.2 ANXIOLYTIC DRUGS ACTING ON SEROTONIN RECEPTORS

Serotonin receptor subtypes are implicated in pathogenesis of many psychiatric and neurological disorders, including anxiety, and among 14 identified serotonin receptors, which are presently grouped into seven families (5-HT$_1$ to 5-HT$_7$), at least four subtypes—5-HT$_{1A}$, 5-HT$_{2A}$, 5-HT$_{2C}$, and 5-HT$_3$—are involved in the control of anxiety. Studies on anxiety disorders have shown dysregulation of serotonergic neurotransmission [17-19]. Conflicting hypotheses have been proposed regarding 5-HT excess and deficit and the mode of action of 5-HT. A reduced level of serotonin associates with anxiety of patients [20], although serotonin excess can also generate anxiety disorders [21].

The first generation of agonists for 5-HT_{1A} is usually represented as the azaperone series, but also include even more 1-(2-pyrimidinyl)piperazine derivatives like buspirone (**5.2.1**) (partial 5-HT_{1A} agonist), gepirone (**5.2.2**), ipsapirone (**5.2.3**), tandospirone (**5.2.4**), and others, from which only buspirone became available to the pharmaceutical market (Fig. 5.5.).

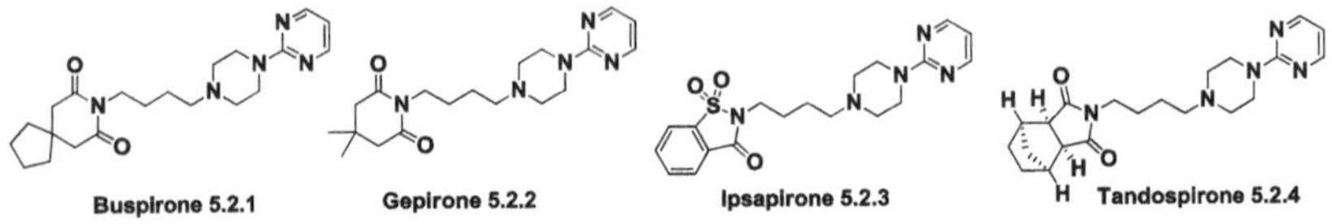

Buspirone 5.2.1 Gepirone 5.2.2 Ipsapirone 5.2.3 Tandospirone 5.2.4

FIG. 5.5 **Anxiolytic drugs acting on serotonin receptors.**

Representative of the second generation of 5-HT_{1A} agonists are selective compounds targeting only the 5-HT_{1A} receptor, such as flesinoxan (**5.2.5**), which did not show anxiolytic efficacy on animals. However, another new selective 5-HT_{1A} agonist, osemozotan (**5.2.6**), displays potentially useful properties as anxiolytic. The mixed 5-$HT_{1A/B/D}$-receptor antagonist GSK-163090 (**5.2.7**) shows a potent preclinical testing profile and was selected as a drug candidate as antidepressant/anxiolytic (Fig. 5.6.).

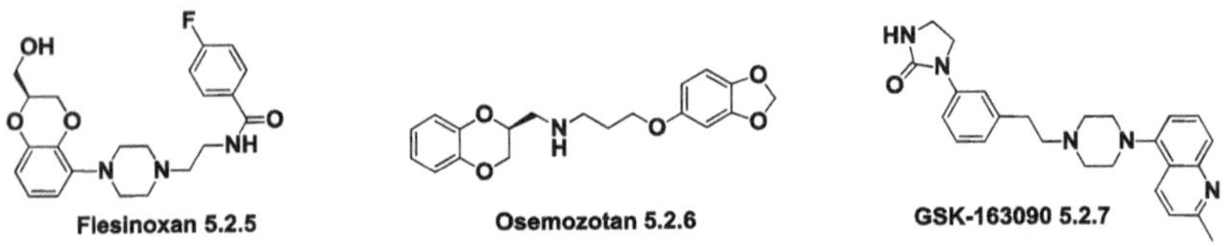

Flesinoxan 5.2.5 Osemozotan 5.2.6 GSK-163090 5.2.7

FIG. 5.6 **New 5-HT compounds targeting antidepressant/anxiolytic antidepressant/anxiolytic drug candidates.**

Compounds with antagonist activity like pirenperone (**5.2.8**), a nonselective serotonin 5-$HT_{2A/2C}$ antagonist, ketanserin (**5.2.9**), moderate selectivity for 5-HT_{2A} over 5-HT_{2B}, and ritanserin (**5.2.10**), a mixed 5-$HT_{2A/2C}$, have a weak anxiolytic effect. Agomelatine (**5.2.11**) (mixed 5-HT_{2C} antagonist/MT1 agonist), which represents a novel class of antidepressants and has been licensed for the treatment of depression, has potential indications for anxiety disorders (Fig. 5.7.).

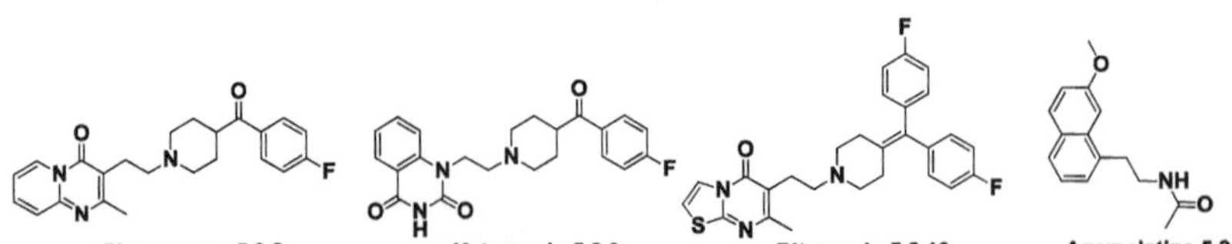

Pirenperone 5.2.8 Ketanserin 5.2.9 Ritanserin 5.2.10 Agomelatine 5.2.11

FIG. 5.7 **Compounds with 5-$HT_{2A/2C}$ antagonist activity indicated for anxiety disorders.**

The mixed 5-$HT_{2A/2C}$ antagonist deramciclane (**5.2.12**) was discontinued during Phase III trials due to clinically insignificant results.

5-HT_{1A} Agonist–5-$HT_{2A/3}$ antagonist CSP-2503 (**5.2.13**) show promising data in be useful in a number of conditions associated with anxiety and probably should be proved in clinical trials (Fig. 5.8.).

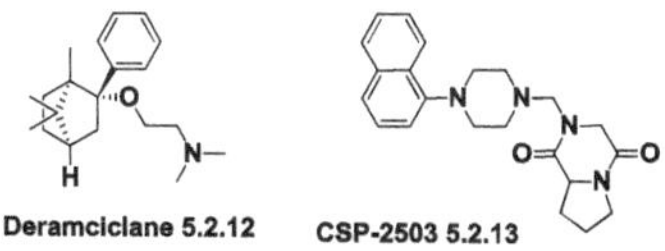

Deramciclane 5.2.12 CSP-2503 5.2.13

FIG. 5.8 **Mixed 5-HT receptors agonist–antagonists.**

Ondansetron (**5.2.14**) and zacopride (**5.2.15**) have been developed as 5-HT_3 antagonists, which works by blocking the action of serotonin also have been developed as antianxiety drugs (Fig. 5.9.).

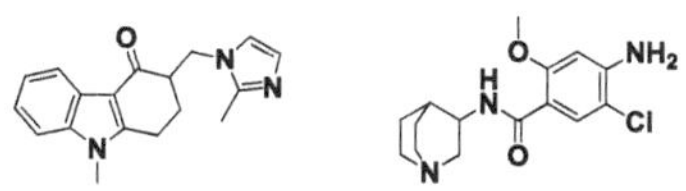

Ondansetron 5.2.14 Zacopride 5.2.15

FIG. 5.9 **5-HT_3 antagonists developed as antianxiety drugs.**

Selective serotonin reuptake inhibitors (SSRIs), which work via by blocking the serotonin transporter, an integral membrane protein that transports the neurotransmitter serotonin from synaptic spaces into presynaptic neurons, are primarily classified as antidepressants. SSRIs have anxiolytic properties, but they are also anxiogenics when first initiating treatment. SSRIs now available in the pharmaceutical market are fluoxetine, dapoxetine, sertraline, citalopram, fluvoxamine, and paroxetine. Four serotonin-norepinephrine reuptake inhibitors (SNRIs), which work by decreasing levels of serotonin and norepinephrine in the synaptic cleft, are alternate first-line agents to SSRIs, are also available. Effective anxyolytics of SSRI series are duloxetine, milnacipran, venlafaxine, and desvenlafaxine, but the side effects of SNRIs are similar to those of SSRI's, which limits their use. Tricyclic antidepressants—imipramine, doxepin, amitriptyline, and the unrelated trazodone—which act primarily as SSRIs and also have high affinity as antagonists at 5-HT_{2A}, 5-HT_{2c}, 5-HT_6, 5-HT_7, inhibit noradrenaline and other catecholamines reuptake, are anxiolytics as well, but they have serious side effects.

5.3 ANXIOLYTIC DRUGS ACTING ON ADRENORECEPTORS

Norepinephrine modulates activity in brain regions involved in anxiety and a nice hypothesis exists to support the evidence that norepinephrine is also involved in anxiety [22]. It is well known that the α_2-adrenoreceptor agonist clonidine (**5.3.1**) and the nonspecific β blockers propranolol (**5.3.2**) and oxprenolol (**5.3.3**) inhibit anxiety (Fig. 5.10.).

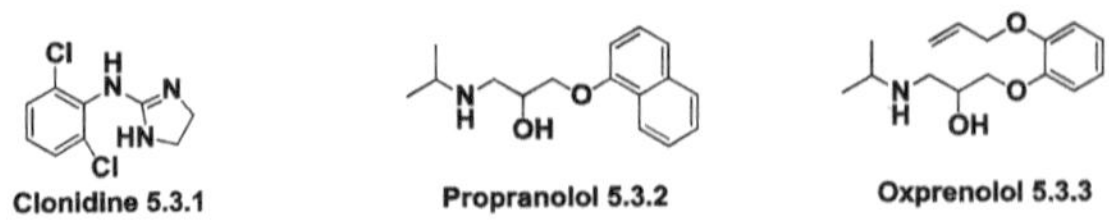

FIG. 5.10 **Anxiolytic drugs acting on adrenoreceptors.**

5.4 ANXIOLYTIC DRUGS ACTING ON GLUTAMATE RECEPTORS

Glutamate is one of the major excitatory neurotransmitters and there is consistent evidence that glutamate overactivity plays a major role in anxiety. Anticonvulsant pregabalin (Lyrica) (**5.4.1**), which acts as a presynaptic inhibitor of the release of excitatory transmitters, including glutamate, noradrenalin, substance P, and calcitonin gene-related peptide, is used to relieve neuropathic pain, fibromyalgia, and insomnia, and is used with other medications to treat certain types of epilepsy seizures. Pregabalin is an effective medication in anxiety disorders, which believed to be because of its inhibition of glutamate release in the brain [23] (Fig. 5.11.).

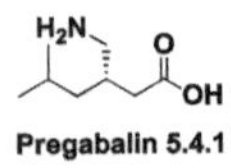

FIG. 5.11 **Structure of pregabalin.**

Under serious investigation are compounds AP-7 (**5.4.2**) (competitive) and MK-801 (Dizocilpine) (**5.4.3**) (noncompetitive) antagonists on the ionotropic NMDA receptor, which showed anxyolityic properties on preclinical models. Another example of anxiolytic drugs acting on glutamate receptors is a glycine site partial agonist d-cycloserine (**5.4.4**), which showed anxyolityic properties on preclinical models. But on human subjects all of the listed compoundshave serious side effects, such as sedation, memory impairment, and psychosis. Topiramate (**5.4.5**), an antiepilepsy drug, is another ionotropic α-amino-3-hydroxy-5-methyl-4-isoxazolepropionic acid (AMPA) receptor antagonist that showed anxiolytic properties. The metabotropic glutamate receptor group I $mGlu_5$ receptor antagonist MPEP (**5.4.6**) showed anxyolityic activity in preclinical trials. Group II mGlu nonselective $mGlu_2$ and $mGlu_3$ receptor agonist LY354740 (eglumegad) (**5.4.7**) and its analogue LY544344 (**5.4.8**) with improved bioavailability reduced anxiety and fear in experimental models [2,14,24,25] (Fig. 5.12.).

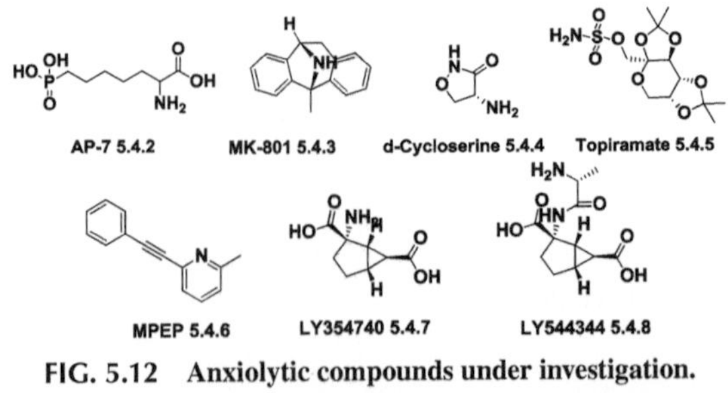

FIG. 5.12 **Anxiolytic compounds under investigation.**

5.5 ANXIOLYTIC COMPOUNDS ACTING ON OTHER RECEPTORS

Lines of evidence suggest that the cholecystokinin (CCK) system is also involved in anxiety disorders, and that CCK_2 receptor antagonists such as itriglumide (**5.5.1**) exert anxiolytic-like properties in models of anxiety [26,27] (Fig. 5.13.).

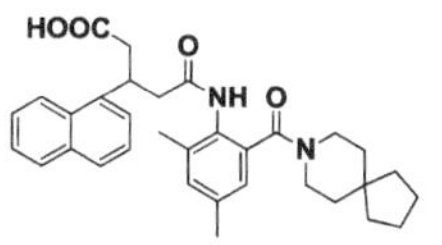

Itriglumide 5.5.1

FIG. 5.13 **Structure of itriglumide.**

Corticotropin-releasing factor (CRF) plays an important role in the development anxiogenic effects and CRF receptor antagonists such as pivagabine (**5.5.2**) may be effectve in the treatment of anxiety disorders [28,29] (Fig. 5.14.).

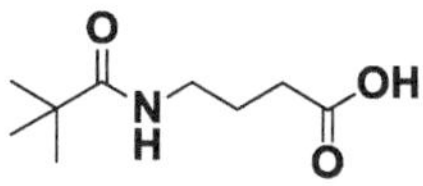

Pivagabine 5.5.2

FIG. 5.14 **Structure of pivagabine.**

In anxiety disorders, dopamine pathways, hypo- or hyperactivation of D_2 and D_3 receptors, and dopaminergic treatments need more research [30]. Dopamine receptors were employed comparatively less often in regard to anxiety. Nevertheless, two structural analogues—the D_2 dopamine receptor antagonist fluphenazine (**5.5.3**) and the D_1 and D_2 receptors antagonist trifluoperazine (**5.5.4**), as well as sulpiride (**5.5.5**), an antagonist at dopamine D_2 and D_3 receptors and pramipexole (**5.5.6**) with D_3-preferring receptor binding profile—have been shown to exert properties to treat anxiety disorders (Fig. 5.15.).

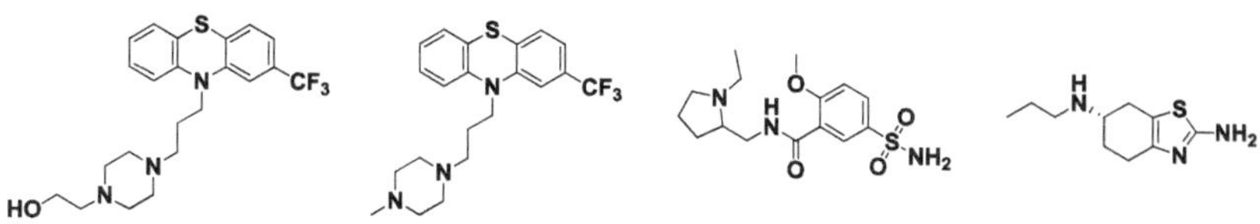

Fluphenazine 5.5.3 Trifluoperazine 5.5.4 Sulpiride 5.5.5 Pramipexole 5.5.6

FIG. 5.15 **Dopamine receptor antagonist drugs for anxiety disorders.**

Monoamine oxidases have been implicated in several neurological and psychiatric disorders, significantly increasing levels of dopamine, serotonin, and norepinephrine, attenuating depression and anxiety [15,31,32]. A new monoamine oxidase type A inhibitor, CX-157 (**5.5.7**), is now in the Phase I clinical investigations (Fig. 5.16.).

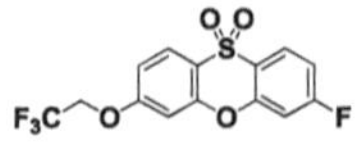

CX-157 5.5.7

FIG. 5.16 **Structure of CX-157.**

It has been shown in a wide range of models of anxiolytic activity that δ-opioid receptor agonists such as AZD-2327 (**5.5.8**) [33] may possess unique antidepressant and anxiolytic activities that could make a novel contribution to the pharmacotherapy of psychiatric disorders (Fig. 5.17.).

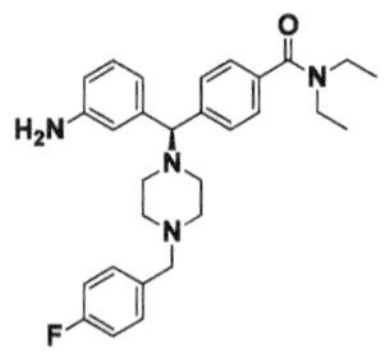

AZD-2327 5.5.8

FIG. 5.17 **Structure of AZD-2327.**

There is clear functional evidence that in patients who suffer from anxiety, tachykinin transmission is upregulated and NK1 receptor antagonists, such as L-822429 (**5.5.9**), vofopitant (**5.5.10**), and orvepitant (**5.5.11**) (Fig. 5.18.), with low side-effect profiles are currently under clinical investigation and may be effective in treatment of anxiety [26,34-36].

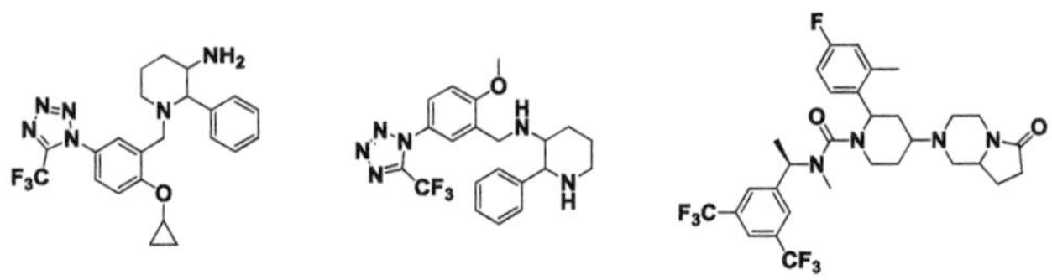

L-822429 5.5.9 Vofopitant 5.5.10 Orvepitant 5.5.11

FIG. 5.18 **NK1 receptor antagonists under clinical investigation for treatment of anxiety.**

The synthesis of antianxiety drugs is entering a new era and the abovementioned drug targets are currently undergoing major and intensive investigations.

None of the described drugs is included in the list of Top 200 Drugs by sales for the 2010s.

REFERENCES

1. Ravindran, L. N.; Murray, B. S. The pharmacologic treatment of anxiety disorders: a review of progress. *J. Clin. Psychiatry (Memphis, TN, U. S.)* **2010,** *71* (7), 839–854.
2. Sorbera, L. A.; Dulsat, C.; Rosa, E. Therapeutic targets for anxiety disorders. *Drugs Future* **2011,** *36* (6), 473–484.
3. Lorenz, R. A.; Cherry, W. J.; Saitz, M. Adjunctive use of atypical antipsychotics for treatment-resistant generalized anxiety disorder. *Pharmacotherapy* **2010,** *30* (9), 942–951.

4. Walser, A.; Fryer, R. I. 1,4-Benzodiazepines, in Chemistry of Heterocyclic Compounds. *(Chichester, UK)* **1991,** *50,* 431–543 (Bicyclic Diazepines).
5. Sternbach, L. H. The benzodiazepine story. *J. Med. Chem.* **1979,** *22* (1), 1–7.
6. Randall, L. O.; Schallek, W.; Sternbach, L. H.; Ning, R. Y. Chemistry and Pharmacology of the 1,4-Benzodiazepines. In *Medicinal Chemistry – A Series of Monographs;* Stevans, F., Ed.; Psychopharmacological Agents III, Part C, Vol. 4, 1974, Academic Press: New York/ N.Y.; pp 175–281.
7. Sternbach, L. H. Chemistry of 1,4-Benzodiazepines and Some Aspects of the Structure-Activity Relation. In *Benzodiazepines;* Garattini, S., Mussini, E.; Randall, L. O., Eds.; Raven Press: New York; 1973; pp 1–26.
8. Sternbach, L. H. 1,4-Benzodiazepines. Chemistry and some aspects of the structure-activity relation. *Angew. Chem., Int. Ed.* **1971,** *10* (1), 34–43.
9. Vardanyan, R. S.; Hruby, V. J. *Synthesis of Essential Drugs;* Elsevier, 2006.
10. Rudolph, U.; Crestani, F.; Benke, D.; Brunig, I.; Benson, J. A.; Fritschy, J.-M.; Martin, J. R.; Bluethmann, H.; Mohler, H. Benzodiazepine actions mediated by specific γ-aminobutyric acid receptor A subtypes. *Nature (London, U. K.)* **1999,** *401,* 796–800.
11. McKernan, R. M.; Rosahl, T. W.; Reynolds, D. S.; Sur, C.; Wafford, K. A.; Atack, J. R.; Farrar, S.; Myers, J.; Cook, G.; Ferris, P.; Garrett, L.; Bristow, L.; Marshall, G.; Macaulay, A.; Brown, N.; Howell, O.; Moore, K. W.; Carling, R. W.; Street, L. J.; Castro, J. L.; Ragan, C. I.; Dawson, G. R.; Whiting, P. J. Sedative but not anxiolytic properties of benzodiazepines are mediated by the $GABA_A$ receptor α1 subtype. *Nat. Neurosci.* **2000,** *3,* 587–592.
12. Mohler, H. The rise of a new GABA pharmacology. *Neuropharmacology* **2011,** *60,* 1042–1049.
13. Rudolph, U.; Knoflach, F. Beyond classical benzodiazepines: novel therapeutic potential of $GABA_A$ receptor subtypes. *Nat. Rev. Drug Discovery* **2011,** *10* (9), 685–697.
14. Christmas, D.; Hood, S.; Nutt, D. Potential novel anxiolytic drugs. *Curr. Pharm. Des.* **2008,** *14,* 3534–3546.
15. Solanki, G. Anti - anxiety drugs - an overview. *Int. J. Biomed. Res.* **2013,** *4* (1), 1–4.
16. Jainer, A. K.; Bharadwaj, A.; Marzanski, M. Developing new anti-anxiety drugs. *Int. Med. J.* **2007,** *14* (4), 291–294.
17. Hirschfeld, R. M. A. History and evolution of the monoamine hypothesis of depression. *J. Clin. Psychiatry (Memphis, TN, U. S.)* **2000,** *61,* 4–6.
18. Lucki, I. The spectrum of behaviors influenced by serotonin. *Biol. Psychiatry* **1998,** *44,* 151–162.
19. Naughton, M.; Mulrooney, J. B.; Leonard, B. E. A review of the role of serotonin receptors in psychiatric disorders. *Hum. Psychopharmacol.* **2000,** *15,* 397–415.
20. Baldwin, D. S.; Anderson, I. M.; Nutt, D. J.; Bandelow, B.; Bond, A.; Davidson, J. R. T.; den Boer, J. A.; Fineberg, N. A.; Knapp, M.; Scott, J.; Wittchen, H.-U. Evidence-based guidelines for the pharmacological treatment of anxiety disorders: recommendations from the British Association for Psychopharmacology. *J. Psychopharmacol. (London, U. K.)* **2005,** *19* (6), 567–596.
21. Charney, D. S.; Woods, S. W.; Goodman, W. K.; Heninger, G. R. Serotonin function in anxiety. II. Effects of the serotonin agonist MCPP in panic disorder patients and healthy subjects. *Psychopharmacology (Berl)* **1987,** *92* (1), 14–24.
22. Ressler, K. J.; Nemeroff, C. B. The role of serotonergic and noradrenergic systems in depression and anxiety disorders. *Depression Anxiety* **2000,** *12* (Suppl. 1), 2–19.
23. Strawn, J. R.; Geracioti Jr, T. D. The treatment of generalized anxiety disorder with pregabalin, an atypical anxiolytic. *Neuropsychiatr. Dis. Treat.* **2007,** *3* (2), 237–243.
24. Bergink, N.; van Megen, U. J.; Westenberg, H. G. Glutamate and anxiety. *Eur. Neuropsychopharmacol.* **2004,** *14* (3), 175–183.

25. Kaplan, G. B.; Moore, K. A. The use of cognitive enhancers in animal models of fear extinction,. *Pharmacol. Biochem. Behav.* **2011,** *99* (22), 217–228.
26. Steckler, T. Developing small molecule nonpeptidergic drugs for the treatment of anxiety disorders: is the challenge still ahead? *Curr. Top. Behav. Neurosci.* **2010,** *2,* 469–485.
27. Chung, L.; Moore, S. D. Cholecystokinin enhances GABAergic inhibitory transmission in basolateral amygdala. *Neuropeptides (Oxford, U. K.)* **2007,** *41,* 453–463.
28. Arborelius, L.; Owens, M. J.; Plotsky, P. M.; Nemeroff, C. B. The role of corticotropin-releasing factor in depression and anxiety disorders. *J. Endocrinol.* **1999,** *160,* 1–12.
29. Bale, T. L.; Lee, K.-F.; Vale, W. W. The role of corticotropin-releasing factor receptors in stress and anxiety. *Integr. Comp. Biol.* **2002,** *42* (3), 552–555.
30. Schneier, F. R.; Abi-Dargham, A.; Martinez, D.; Slifstein, M.; Hwang, D.-R.; Liebowitz, M. R.; Laruelle, M. Dopamine transporters, D2 receptors, and dopamine release in generalized social anxiety disorder. *Depression Anxiety* **2009,** *26* (5), 411–418.
31. Schneier, F. R. Pharmacotherapy of social anxiety disorder. *Expert Opin. Pharmacother.* **2011,** *12* (4), 615–625.
32. Farooq, M. U.; Moore, P. W.; Bhatt, A.; Aburashed, R.; Kassab, M. Y. Therapeutic role of zonisamide in neuropsychiatric disorders. *Mini-Rev. Med. Chem.* **2008,** *8* (10), 968–975.
33. Hudzik, T. J.; Maciag, C.; Smith, M. A.; Caccese, R.; Pietras, M. R.; Bui, K. H.; Coupal, M.; Adam, L.; Payza, K.; Griffin, A.; Smagin, G.; Song, D.; Swedberg, M. D. B.; Brown, W. Preclinical pharmacology of AZD2327: a highly selective agonist of the δ-opioid receptor. *J. Pharmacol. Exp. Ther.* **2011,** *338* (1), 195–204.
34. Sheehan, D. V.; Sheehan, K. H. Current approaches to the pharmacologic treatment of anxiety disorders. *Psychopharmacol. Bull.* **2007,** *40* (1), 98–109.
35. Ebner, K.; Sartori, S. B.; Singewald, N. Tachykinin receptors as therapeutic targets in stress-related disorders. *Curr. Pharm. Des.* **2009,** *15* (14), 1647–1674.
36. Singewald, N.; Chicchi, G. G.; Thurner, C. C.; Tsao, K.-L.; Spetea, M.; Schmidhammer, H.; Sreepathi, H. K.; Ferraguti, F.; Singewald, G. M.; Ebner, K. Modulation of basal and stress-induced amygdaloid substance P release by the potent and selective NK1 receptor antagonist L-82(2429). *J. Neurochem.* **2008,** *106* (6), 2476–2488.

Chapter 6

Antipsychotics

Antipsychotics are drugs that are used to control mental health conditions such as schizophrenia and other psychoses; hallucinations; delusional and bipolar disorders; agitation; severe anxiety; mania and violent or dangerously impulsive behavior; a state of apathy; lack of initiative; limited range of emotion; reduction in confusion and agitation; and normalization of psychomotor activity in psychotic patients.

Antipsychotics work by increasing or reducing the effects of neurotransmitters such as dopamine, serotonin, noradrenaline, and acetylcholine in the brain, which regulate numerous aspects of behavior, mood, emotions, control of sleeping and feeding, etc.

The antipsychotics are divided into three groups: classic or typical or first-, second-, or third-generation antipsychotics, according to their mechanism of action. They evolved from selective D_2 agents, to mixed D_2 and $5\text{-}HT_{2A}$ therapeutics, and then to those whose actions have been ascribed alternately to either D_2 partial agonism or D_2 functional selectivity. There is only one approved third-generation drug, aripiprazole, on the market. The original typical antipsychotic drugs, such as chlorpromazine and haloperidol, have been called typical or first-generation antipsychotics. These antipsychotics were developed in the 1950s and act primarily to reduce the effect of dopamine in the brain, causing both antipsychotic actions and side effects that are ascribed to their dopamine D_2 receptor antagonism. Discovery of chlorpromazine as an antipsychotic, which initially was proposed for use in medicine as an anesthetic is one of the most important milestones in the history of pharmacology.

Drugs such as clozapine, olanzapine, risperidone, and others (second-generation antipsychotics) were developed from the 1970s onward and avoided the neurological side effects. Although there is currently no general agreement on what makes them atypical, the new antipsychotics are thought to combine a number of clinical criteria (efficacy against positive and negative psychotic symptoms, efficacy in patients refractory to the antagonist action of dopamine, and the relatively low incidence of extrapyramidal symptoms, late onset dyskinesia, and secondary depressive or negative symptoms These compounds are divided into those that are high affinity D_2 and $5\text{-}HT_{2A}$ antagonists and those that also bind to other neuroreceptors. The single approved third-generation drug is aripiprazole.

Synthesis of Best-Seller Drugs. http://dx.doi.org/10.1016/B978-0-12-411492-0.00006-7

The effects of various chemical classes of typical and atypical antipsychotic drugs are divided into three groups: positive (hallucination and delusion), negative (social withdrawal and poverty of speech), and cognitive (reduction of working memory and attention) symptoms. More than 40 different antipsychotic medications have been introduced around the world. A number of reviews on antipsychotics have been published [1-16].

6.1 FIRST-GENERATION ANTIPSYCHOTICS

First-generation antipsychotics also can be classified by their chemical structure into phenothiazines, thioxanthenes, and diphenylbutylpiperidines.

Very few studies have found any significant differences among the various typical first-generation antipsychotic drugs in terms of their efficacy. In this sense, they are all thought to be equally effective and the drug of choice depends on the profile of adverse events and presenting situation. But the action mechanism of each is different. For example, typical antipsychotic drugs chlorpromazine and thioridazine act on D_1, D_2, 5-HT_2, α_1, H, and acetylcholine receptors (AChRs); levomepromazine on D_2, 5-HT_2, α_1, H, and ACH receptors; perphenazine and zuclopenthixol on D_2, 5-HT_2, α_1, H receptors; fluphenazine on 5-HT_2 and α_1; pimozide on D_2; sulpiride on D_1; clozapine on D_1, D_2, 5-HT_{2A}, norepinephrine (NE), α_1, α_2, H_1, and ACh; olanzapine on D_1, D_2, D_4, 5-HT_{2A}, NE, α_1, and H_1; quetiapine on D_1, D_2, 5-HT_{2A}, NE, α_1, α_2, and ACH; risperidone on D_2, 5-HT_{2A}, NE, α_1, α_2, and H_1; sertindole and ziprasidone on D_1, D_2, 5-HT_{2A}, NE, and α_1 [3].

Phenothiazines

The phenothiazine group of drugs is among the most widely prescribed psychotropic drugs in the world [17]. The phenothiazines are classified as typical antipsychotics because of their high potency to blockade the D_2 receptor. These drugs can be effective in treating the positive symptoms but not the negative signs of schizophrenia. Phenothiazines are highly sedating but with time, some tolerance to this effect develops. They have α-adrenergic blocking activity and can cause orthostatic hypotension. They have also moderate anticholinergic activity. The most popular phenothiazines (Fig. 6.1.) are the simple phenothiazin-10-yl-propan-1-amines-chlorpromazine (**6.1.1**), promazine (**6.1.2**), and levomepromazine (**6.1.3**); a little bit more complex are piperazin-1-yl-propyl-10H-phenothiazines-prochlorperazine (**6.1.4**), trifluoperazine (**6.1.5**), perphenazine (**6.1.6**), fluphenazine (**6.1.7**); piperidin-2-yl-ethyl-10H-phenothiazines-thioridazine (**6.1.8**), mesoridazine (**6.1.9**), pericyazine (**6.1.10**), and cyamemazine (**6.1.11**).

The cornerstone of this series of compounds, chlorpromazine (**6.1.1**), as well as all other compounds of the series, block the receptors that dopamine acts on and prevents its overactivity in the brain, thereby producing sedative

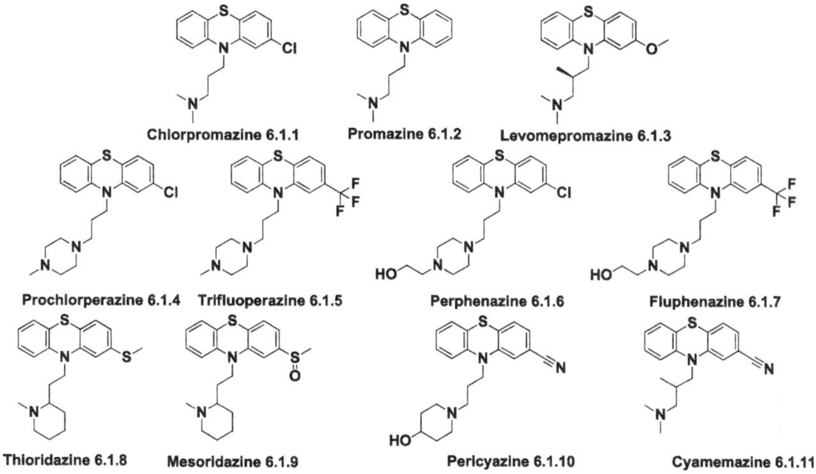

FIG. 6.1 Phenothiazines antipsychotics.

and calming effects. They are used to treat the symptoms of schizophrenia and other psychotic disorders, and to treat the symptoms of mania in people who have bipolar disorder and other abnormal moods. Cyamemazine (**6.1.11**) differs from other phenothiazine neuroleptics in that aside from the usual profile of D_2 antagonism it additionally produces potent blockade of 5-HT_{2A}, 5-HT_{2C}, and 5-HT_7 receptors. Despite being classified as a typical antipsychotic, it actually behaves like an atypical antipsychotic. Methods for the synthesis of phenothiazines, as well as of the below described thiothiexenes, butyrophenones, and diphenylbutylpiperidines, are described in our previous book [18].

Thioxanthenes

The thioxanthenes—chlorprothixene (**6.1.12**), clopenthixol (**6.1.13**), flupentixol (**6.1.14**), and thiothiexene (**6.1.15**) (Fig. 6.2.)—are typical first-generation antipsychotic drugs that are structurally and pharmacologically closely related to phenothiazines. They work like phenothiazines by blocking the D_2 receptor and suppressing the effect of dopamine in the brain. They also are interacting with α-adrenergic receptors, thereby blocking actions of endogenous or exogenous α-adrenergic agonists.

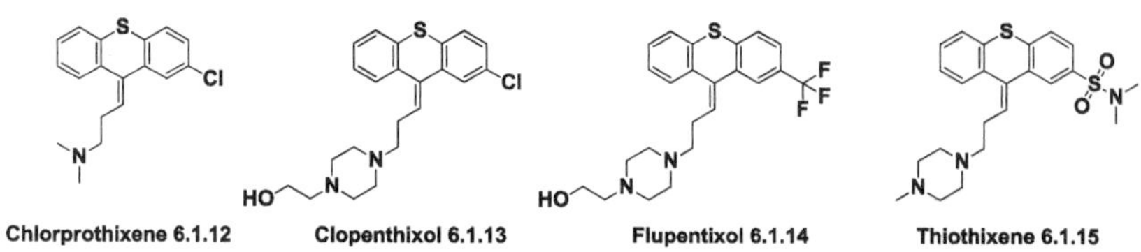

FIG. 6.2 Thioxanthene antipsychotics.

It seems that antipsychotic effects of flupentixol (**6.1.14**) provide an alleviating effect on the negative signs of schizophrenia, which is likely caused by D_2 and/or 5-HT_{2A} antagonism, whereas its antidepressant effects at lower doses may be mediated by preferential D_2/D_3 receptor blockade.

Butyrophenones

The concept of the central role of the dopamine receptor in schizophrenia was developed with the discovery of haloperidol and then was expanded into the family of butyrophenone antipsychotics and later to series of diphenylbutylpiperidines. Haloperidol (**6.1.16**) (Fig. 6.3.) is commonly prescribed in emergency cases for fast-acting treatment of the "positive symptoms" of psychosis and has been included in the World Health Organization's list of essential medicines. Haloperidol became the prototype drug for more efficient dopamine antagonists (particularly at the D_2 receptors). Haloperidol and some of its structural analogues are used to treat various psychiatric disorders such as schizophrenia; they also act as antiemetics. They are reported to provide effective migraine relief. The most usable butyrophenones, except for haloperidol itself, are trifluperidol (**6.1.17**), droperidol (**6.1.18**), benperidol (**6.1.19**), and the less popular melperone (**6.1.20**) and lenperone (**6.1.21**) (Fig. 6.3.).

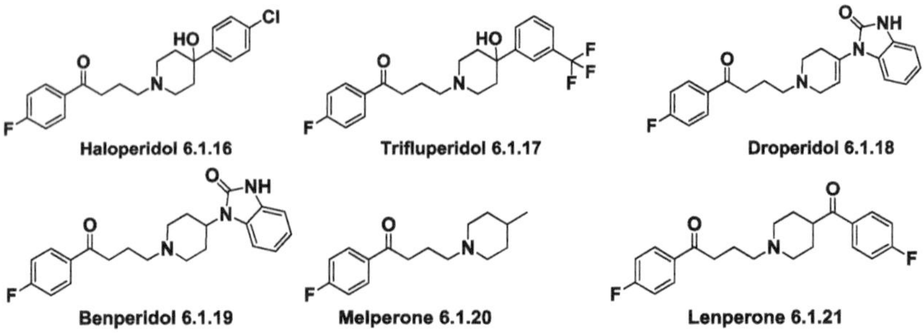

FIG. 6.3 Butyrophenone antipsychotics.

Butyrophenones are considered to be high-potency antipsychotics. In general, it is accepted to use the "chlorpromazine equivalence" concept to compare the relative effectiveness of antipsychotics. It is the amount in milligrams of a drug that must be administered to achieve the desired effects equivalent to 100 milligrams of chlorpromazine. Agents with "a chlorpromazine equivalence" of 5 to 10 milligrams are considered medium potency, and compounds in the range of approximately 2 milligrams are considered high potency.

Diphenylbutylpiperidines

Interestingly, diphenylbutylpiperidines, a class of typical antipsychotic drugs of another series, improve the negative manifestations of schizophrenia, which is explained by the considerable calcium channel–blocking properties of this class of compounds. This class includes pimozide (**6.1.22**), fluspirilene (**6.1.23**), and penfluridol (**6.1.24**) (Fig. 6.4.).

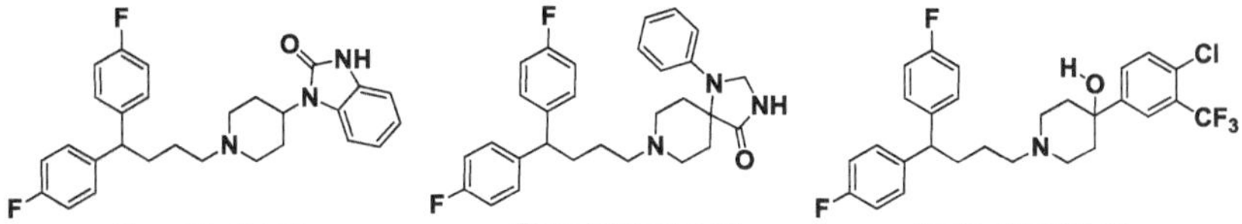

Pimozide 6.1.22 Fluspirilene 6.1.23 Penfluridol 6.1.24

FIG. 6.4 **Diphenylbutylpiperidine antipsychotics.**

Pimozide and other diphenylbutylpiperidines are highly selective D_2 receptor blockers. They are widely used in psychiatry for chronic psychoses, schizophrenia, treatment of Tourette syndrome (a condition characterized by motor or verbal tics). Diphenylbutylpiperidines do not differ greatly from other antipsychotics with respect to treatment efficacy, response or tolerability.

Butyrophenones and diphenylbutylpiperidines have been found to cause extrapyramidal side-effects. The most common are parkinsonian-like syndrome, dystonia, dyskinesia, sleepiness, headache, muscle tightness, muscle weakness, and difficulty moving. Other possible adverse effects include seizures and impotence.

6.2 SECOND-GENERATION ANTIPSYCHOTICS

The atypical antipsychotics are now considered to be first-line medications for schizophrenia and are replacing the typical antipsychotics. Based on the pharmacological principle of 5-HT_2/D antagonism, atypical antipsychotics have decreased propensity to cause extrapyramidal side effects; that is, they are not only capable of controlling the positive symptoms of schizophrenia, but they are also believed to improve the negative signs and cognitive abnormalities. They are the best medications available today in non–treatment-resistant and treatment-resistant schizophrenia, suicide risk in schizophrenia, aggressiveness or violence in psychiatric patients, psychosis in Parkinson disease, and prevention and treatment of tardive dyskinesia. Atypical antipsychotic agents differ in their receptor action and side-effect profiles. Since the development of clozapine, the prototype and the oldest atypical antipsychotic, the atypical antipsychotics are now considered to be first-line medications for schizophrenia and are replacing the atypical antipsychotics. Chemically, atypical antipsychotics belong to very-close-to-each-other classes: dibenzodiazepines, dibenzoxazepines, dibenzothiazepines, benzothienodiazepines, represented by clozapine, olanzapine, quetiapine and others; piperidinylindoles (sertindole); piperidinyl benzisoxazols (risperidone, paliperidone, iloperidone); piperazinyl benzoisothiazoles (ziprasidone, perospirone, and lurasidone); as well as of sulfamoylbenzamides: sulpiride, amisulpride, remoxipride or piperazinylbutoxyquinolin-one aripiprazole and cyclooctapyridine-blonanserin.

Dibenzodiazepines, Dibenzoxazepines, Dibenzothiazepines, and Benzothienodiazepines

Clozapine (**6.2.1**) (Fig. 6.5.), a dibenzodiazepine derivative, was the first atypical agent to become widely available and it has stimulated design and development of newer compounds with similar pharmacological profiles and less-adverse effects [19-24]. Clozapine has a complex pharmacology that depends on a high affinity for numerous receptors, including dopaminergic (D_1, D_4), serotonergic (5-HT_{2A}, 5-HT_{2C}, 5-HT_{3C}), α-adrenergic (α_1, α_2), histaminergic (H_1), and muscarinic (M_1) receptors, a fact that plays a big role in both its therapeutic and adverse effects. Clozapine was discovered as a result of structure-activity investigations that were made for a series of piperazinyl dibenzodiazepines [25,26]. The first example compounds were prepared by implementing the Vilsmeier-type reaction to oxodibenzodi(thia)azepines followed by a reaction with secondary amines in the presence of $POCl_3$ or PCl_5 (Scheme 6.1.).

Continuation of this fruitful work led to the discovery of several active compounds [26]. Substitutions at positions 2 and 8 of the benzene rings modified the pharmacological activity of this series of compounds. As a result, clozapine (**6.2.1**) was discovered. The replacement of the nitrogen atom in the 5-position of diazepine ring with sulfur resulted in the thiazepines clotiapine (**6.2.2**) and quetiapine (**6.2.3**), compounds with typical antipsychotic actions. Replacement of one benzene ring in dibenzodiazepine system for thiophene resulted in olanzapine (**6.2.4**) (Fig. 6.5.).

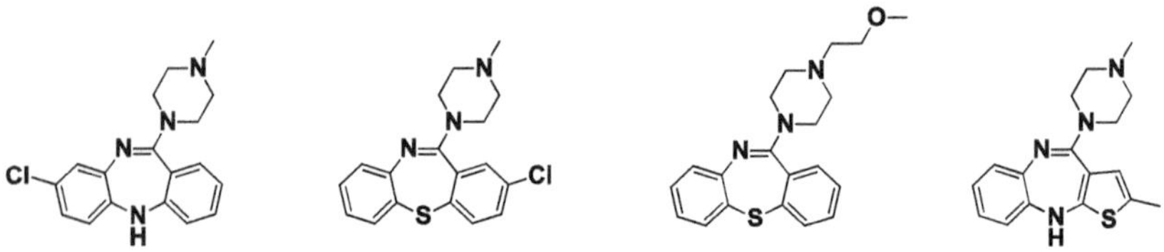

Clozapine 6.2.1 Clotiapine 6.2.2 Quetiapine 6.2.3 Olanzapine 6.2.4

FIG. 6.5 Structures of clozapine, clotiapine, quetiapine, and olanzapine.

Replacement of the same five-nitrogen atom with an oxygen produced the dibenzoxazepine antipsychotics loxapine (**6.2.5**) and amoxapine (**6.2.6**) (Fig. 6.6.). Omitting the diazepine nitrogen atom in the 5 position produced a hypnotic compound, perlapine (**6.2.7**). Replacement with sulfur produced zotepine (**6.2.8**). Beside clozapine, only the 2-Cl-dibenzoxazepine loxapine (**6.2.5**) and 2-Cl-dibenzothiepine zotepine (**6.2.8**) were introduced into clinical practice. In contrast with other compounds, all of which have a piperazinyl substituent at position 11, zotepine (**6.2.8**) has a 2-dimethylamino-ethoxy- substitution at the same position (Fig. 6.6.).

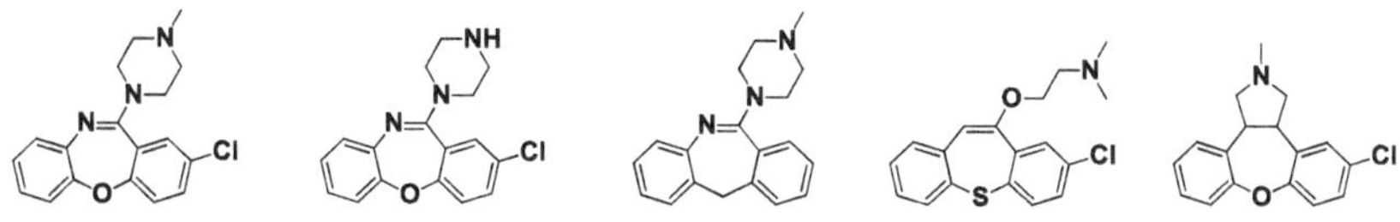

Loxapine 6.2.5 Amoxapine 6.2.6 Perlapine 6.2.7 Zotepine 6.2.8 Asenapine 6.2.9

FIG. 6.6 **Structures of loxapine, amoxapine, perlapine, zotepine and asenapine.**

Previously it was possible to consider the dibenzooxepine derivative asenapine (**6.2.9**) and the dibenzoazepine derivative mosapramine (**6.2.10**) (Fig. 6.7.), a potent dopamine antagonist with high affinity to the D_2, D_3, and D_4, and moderate activity to 5-HT_{2A}, receptors, in the same series. Clocapramine (**6.2.11**) is a D_2, 5-HT_{2A}, α_1, α_2 receptors antagonist, as is carpipramine (**6.2.12**) (Fig. 6.7.).

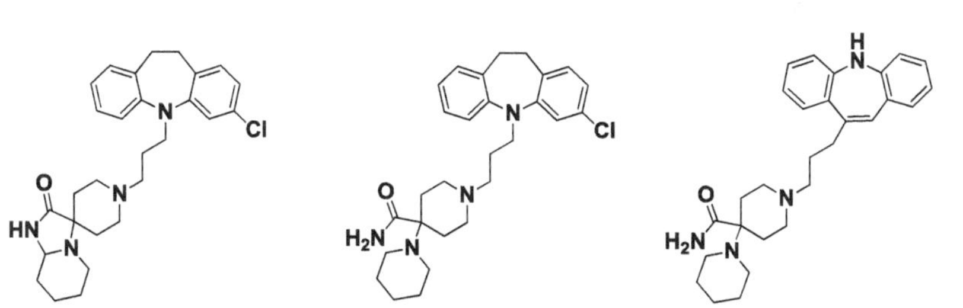

Mosapramine 6.2.10 Clocapramine 6.2.11 Carpipramine 6.2.12

FIG. 6.7 **Structures of mosapramine, clocapramine, and carpipramine.**

Clozapine (**6.2.1**) has been in use for more than 30 years and its first synthesis, as well as the synthesis of loxapine (**6.2.5**), are described in our book [18]. Quetiapine (Seroquel) (**6.2.3**) and olanzapine (Zyprexa) (**6.2.4**) are included in the list of Top 200 Drugs by sales for the 2010s and their syntheses are described below.

Quetiapine–Seroquel

The synthesis of quetiapine (**6.2.3**) is based on the intramolecular Leuckart amide synthesis of isocyanatodiphenyl sulfide (**6.2.17**), which was prepared starting from the o-nitrodiphenyl sulfide (**6.2.15**), which was prepared from o-chloronitrobenzene (**6.2.13**) and thiophenol (**6.2.14**) (Scheme 6.1.). The obtained o-nitrodiphenyl sulfide (**6.2.5**) was reduced with hydrogen using as a catalyst Raney Ni to the corresponding amine (**6.2.16**). Reaction of the last with phosgene gives the isocyanate (**6.2.17**), which on heating in the presence of $AlCl_3$ in o-dichlorobenzene forms the desired dibenzo[b,f][1,4]thiazepine-11(10H)-one (**6.2.18**) [19]. It was heated in the presence of $POCl_3$, and dimethylaniline to give the intermediate iminochloride (**6.2.19**), which was reacted with 2-(2-(piperazin-1-yl)ethoxy)ethanol (**6.2.20**) to produce the final product, quetiapine (**6.2.3**) [27,28].

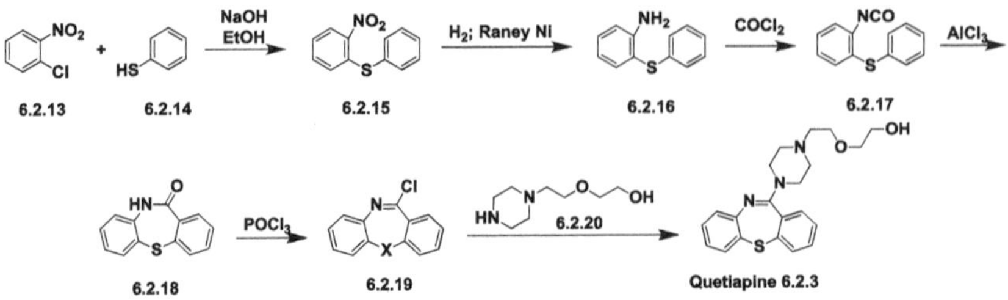

SCHEME 6.1 **Synthesis of quetiapine.**

A very close alternative route [29] suggests formation of dibenzo[b,f][1,4] thiazepine-11(10H)-one (**6.2.18**) via synthesis of the benzoylated intermediate (**6.2.21**), transferred into desired thiazepine-one (**6.2.18**) by cyclization with polyphosphoric acid, followed by the same method for the synthesis of iminochloride (**6.2.19**). The reaction of (**6.2.19**) with piperazine (**6.2.22**) produces a derivative (**6.2.23**), which on reaction with 2-(2-chloroethoxy)ethanol (**6.2.24**) produces the desired quetiapine (**6.2.3**) (Scheme 6.2.).

Quetiapine is the most recently approved atypical antipsychotic. The drug has a fairly broad spectrum of receptor affinity. It demonstrates moderate affinity for 5-HT_{2A}, α_1, M_1, and H_1 receptors, minor affinity for D_2 and 5-HT_{1A} receptors, low affinity for 5-HT_{2C}, α_2, and D_1 receptors, and no appreciable affinity for D_4 or benzodiazepine receptors. D_2 antagonism is similar to that of clozapine.

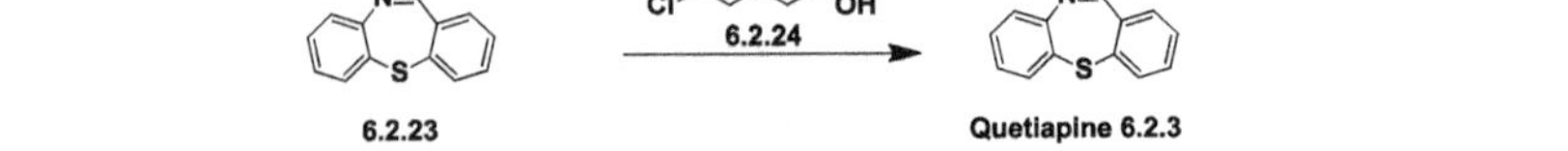

SCHEME 6.2 **Synthesis of quetiapine.**

Olanzapine–Zyprexa

The synthesis of olanzapine (**6.2.4**) is very similar to that of quetiapine (**6.2.3**) (Scheme 6.3.). The same o-chloronitrobenzene (**6.2.13**) was used for N-alkylation of 2-amino-5-methylthiophene-3-carbonitrile (**6.2.25**) and the nitro group in obtained 2-phenylamino-thiophene-3-carbonitrile derivative (**6.2.26**) was reduced to the amino group. Synthesized 4-amino-2-methyl-10H-thieno[2,3-b][1,5]benzodiazepine (**6.2.27**) on reaction with N-methylpiperazine (**6.2.28**) gave the desired amidine-olanzapine (**6.2.4**) [30-33].

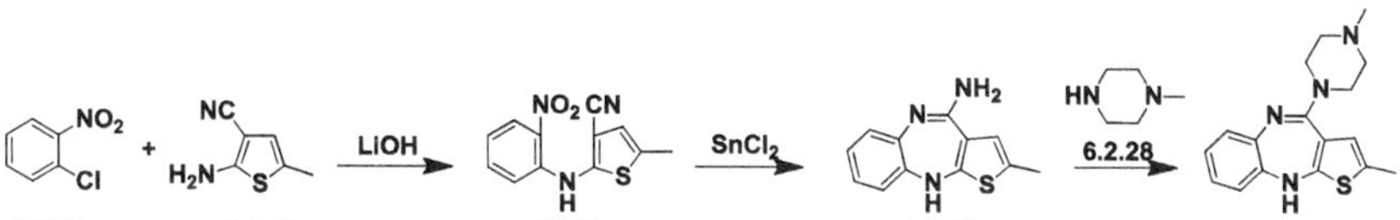

6.2.13 6.2.25 6.2.26 6.2.27 Olanzapine 6.2.4

SCHEME 6.3 **Synthesis of olanzapine.**

Olanzapine has a broad receptor binding profile for the D_1, D_2, D_4, 5-HT_{2C}, 5-HT_{2A}, 5-HT_3, α_1-adrenergic, H_1, and M_1 receptors.

Piperidinyl Isoxazoles, Piperazinyl Benzoisothiazoles, Piperidinylindoles

Piperidinyl isoxazoles, piperazinyl benzoisothiazoles, and piperidinylindoles are structurally related to each other and are members of a new group of "atypical" antipsychotics that cause no extrapyramidal symptoms and are effective against the negative symptoms of schizophrenia. Risperidone is one of the most promising drugs of this class. Risperidone (Scheme 6.4.) appears to be a unique alternative to clozapine in combining the known antipsychotic effects of conventional D_2 antagonist antipsychotics with the clozapine-like 5-HT antagonists.

Risperidone–Risperdal

Risperidone (**6.2.41**) has been synthesized (Scheme 6.4.) via condensation of separately prepared 3-(2-chloroethyl)-6,7,8,9-tetrahydro-2-methyl-4H-pyrido [1,2-a]-pyrimidin-4-one (**6.2.40**) with 6-fluoro-3-(piperidin-4-yl)benzo[d]isoxazole (**6.2.35**) in the presence of potassium iodide and sodium carbonate.

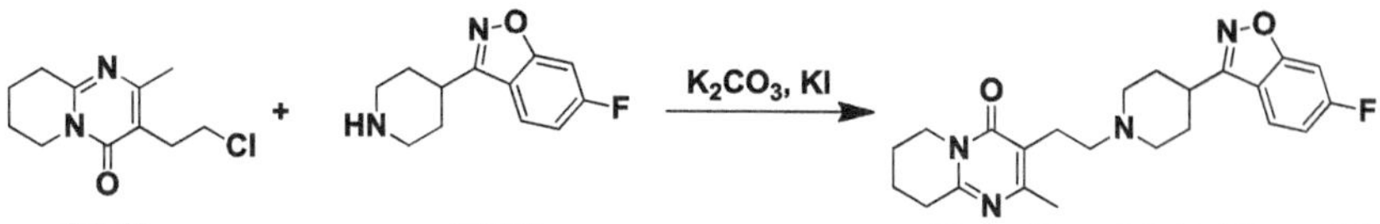

6.2.40 6.2.35 Risperidone 6.2.41 (Risperdal)

SCHEME 6.4 **Synthesis of risperidone.**

The piperidine component (**6.2.35**) was prepared by condensing 1-acetylpiperidine-4-carbonyl chloride (**6.2.29**) with 1,3-difluorobenzene (**6.2.30**) in the presence of aluminum chloride. Obtained compound (**6.2.31**) was hydrolyzed by hydrochloric acid to produce ketone (**6.2.31**), which was converted to the oxime by standard methods (**6.2.32**). Dehydrohalogenation of the product with strong inorganic bases gave 6-fluoro-3-(piperidin-4-yl)benzo[d]isoxazole (**6.2.35**) (Scheme 6.5.).

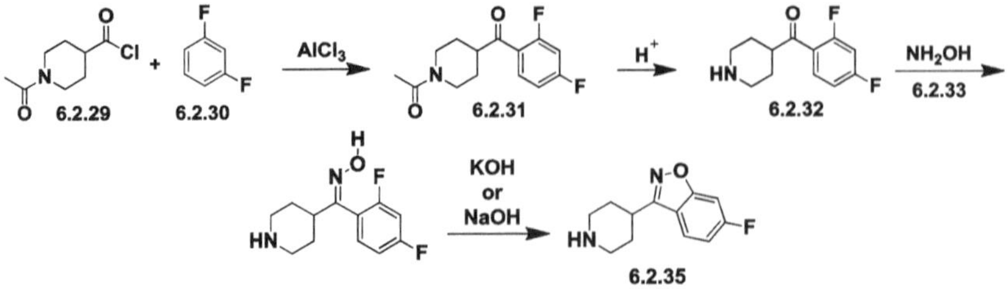

SCHEME 6.5 **Synthesis of 6-fluoro-3-(piperidin-4-yl)benzo[d]isoxazole (6.2.35).**

The second component, the derivative of 2-chloroethyl-tetrahydropyrimidin-4-one (**6.2.40**) was prepared by the cyclocondensation of 2-aminopyridine (**6.2.36**) with 2-acetylbutiro-lactone (**6.2.37**) (Scheme 6.6.), resulting in formation of the pyridopyrimidine (**6.2.38**), which was hydrogenated on Pd/C catalyst to the tetrahydropyridopyimidine derivative (**6.2.39**). The hydroxyl group in which was substituted to the chlorine with the use of thionyl chloride to give the desired (**6.2.40**) [34,35].

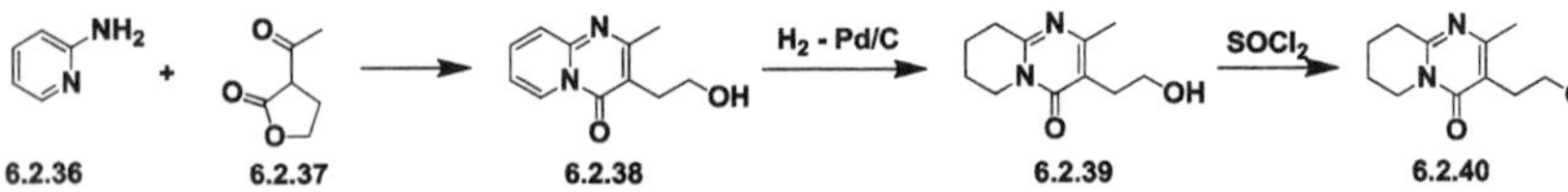

SCHEME 6.6 **Synthesis of 2-chloroethyl-tetrahydropyrimidin-4-one (6.2.40).**

Risperidone, is a widely prescribed antipsychotic agent used for the treatment of schizophrenia and bipolar disorder [36-39]. It has significant affinity for D_2, 5-HT_{2A}, 5-HT_{2C}, H_2, α_1, and α_2 receptors, and a somewhat lower affinity for D_1, 5-HT_{1D}, 5-HT_{1C}, and 5-HT_{1A} receptor subtypes [40].

It's interesting that the major route for metabolizing risperidone is its oxidation to enantiomeric 9-hydroxyrisperidone, which is also a powerful antipsychotic agent, and the racemate, paliperidone (**6.2.44**), was recently commercialized under the name Invega. Several approaches for the synthesis of paliperidone were described by its discoverers [41], but the industrial synthesis of paliperidone is very similar to that employed for risperidone. Just one of the key intermediates (**6.2.43**) was synthesized from pyridopyrimidinones carrying different protective groups, such as the benzyl derivative (**6.2.41**), the protective group of which was removed during simultaneous hydrogenation and hydrogenolysis procedure (Scheme 6.7.).

A very short synthesis of paliperidone via oxidation of risperidone with air oxygen under basic conditions [42] has been proposed (Scheme 6.8.).

Iloperidone (**6.2.52**) (Scheme 6.9.) is a relatively new drug from this series. It was recently approved for the acute treatment of schizophrenia. Processes for its preparation are described in a number of patents, which differ in details, but in general are very close to the methods described above. The first step in the synthesis of iloperidone is formylation of isonipecotic acid (**6.2.45**), to produce formyl protected piperidine-4-carboxylic acid (**6.2.46**) followed by conversion of the obtained derivative

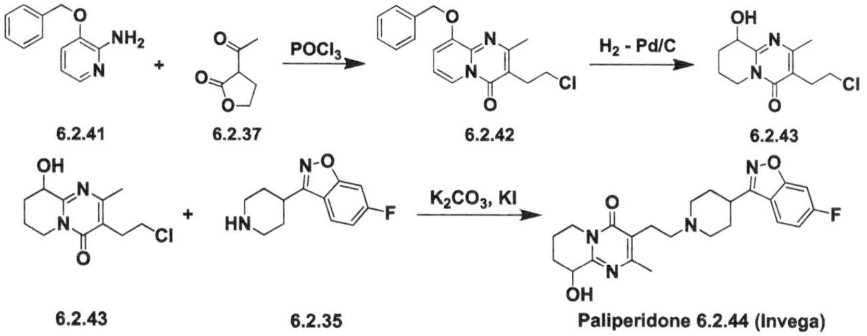

SCHEME 6.7 Synthesis of paliperidone.

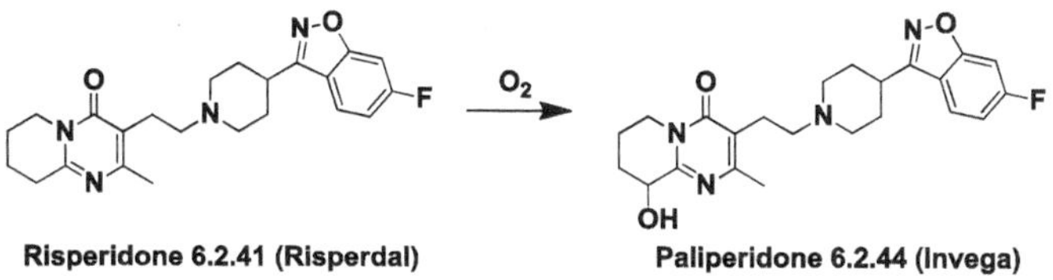

SCHEME 6.8 Synthesis of paliperidone.

to the acid chloride (**6.2.47**). The acid chloride (**6.2.47**) was used in Friedel-Crafts acylation of 1,3-difluorobenzene (**6.2.30**) to produce ketone (**6.2.48**), the key intermediate. The 4-(2,4-difluorobenzoyl)piperidine-1-carbaldehyde (**6.2.48**) was transformed to oxime (**6.2.49**), which underwent dehydrofluorination in presence of a strong base, such as KOH with concomitant loss of the *N*-formyl group in the first position of piperidine ring to give the known compound (**6.2.35**). Alkylation of the obtained oxazolo-piperidine derivative (**6.2.35**) with 1-chloro-3-bromopropane in the presence of K_2CO_3 produces an alkylated chloro derivative (**6.2.50**), which on reaction with 4-hydroxy-3-methoxyacetophenone (**6.2.51**) under basic conditions gives the desired iloperidone (**6.2.52**) [42-44] (Scheme 6.9.).

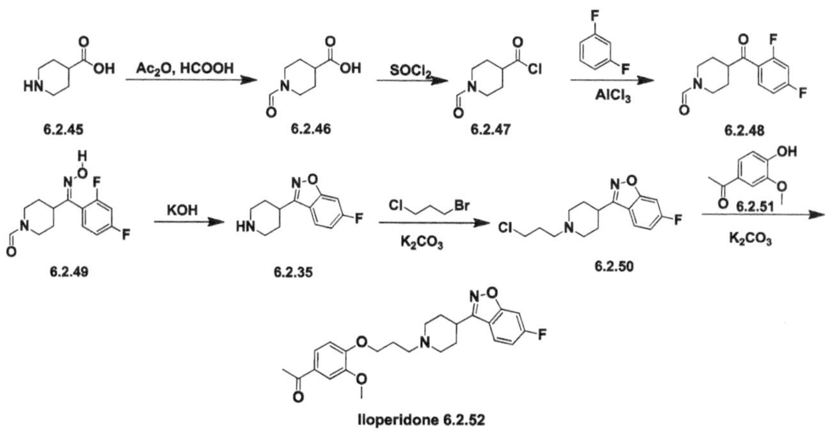

SCHEME 6.9 Synthesis of iloperidone.

Iloperidone targets a set of dopamine, norepinephrine, and serotonin receptor subtypes, such as 5-HT_{2A} and 5-HT_{2C}, as well as 5-HT_{1A}, 5-HT_{1B}, 5-HT_6, and 5-HT_7, dopamine D_1, D_{2A}, D_4, and D_5 receptors; norepinephrine α_{2A}, α_{2B}, β_1, and β_2; muscarine M_1 to M_5; histamine H_1; and CCK_A and CCK_B [45-47].

It appears that iloperidone is effective in treating schizophrenia. Common adverse effects include dizziness, dry mouth, orthostasis, and weight gain. Additionally, there are concerns about cardiac abnormalities. The pharmacology of iloperidone is well reviewed [48-55].

Sertindole–Serdolect

Sertindole (**6.2.60**), which can be considered as the closest structural analogue of the three above-described drugs risperidone (**6.2.41**), paliperidone (**6.2.44**) and iloperidone (**6.2.52**), is synthesized as outlined in Scheme 6.10. It involves the "Ullmann-type" Cu(I) catalyzed oxidative coupling *N*-arylation of 5-chloroindole (**6.2.53**) with 4-fluorobromobenzene (**6.2.54**). The resulting 1-(4-fluorophenyl)indole (**6.2.55**), on heating with 4-piperidinone (**6.2.56**) in a mixture of trifluoroacetic and acetic acids affords 5-chloro-1-(4-fluorophenyl)-3-(1,2,3,6-tetrahydropyridin-4-yl)-1H-indole (**6.2.57**). Hydrogenation of the double bond in the piperidine ring using hydrogen on a platinum oxide catalyst yields compound (**6.2.58**), which, on condensation with 1-(2-chloroethyl)imidazolidinone (**6.2.59**) in the presence of K_2CO_3 and KI in methyl isobutyl ketone, produces the desired sertindole (**6.2.60**) [56-61] (Scheme 6.10.).

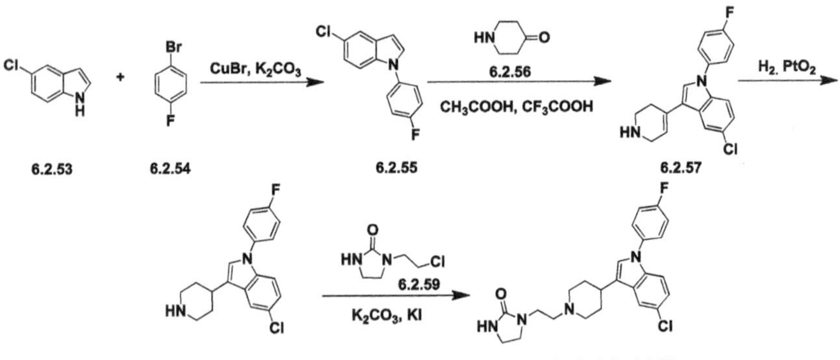

SCHEME 6.10 Synthesis of sertindole.

Sertindole has a high affinity for D_2, 5-HT_{2A}, 5-HT_{2C}, and α_1-adrenergic receptors, and moderate affinity for D_1 and D_4 receptors [62]. Sertindole's efficacy is comparable to risperidone but because of cardiovascular safety concerns, it is recommended as a second-line choice for patients intolerant to other antipsychotic agents [63-65].

The next series of compounds structurally related to the above-described benzisoxazolyl-piperidines risperidone (**6.2.41**), paliperidone (**6.2.44**), and iloperidone (**6.2.52**), are represented by ziprasidone (**6.2.61**) (Geodon), perospirone (**6.2.62**), and lurasidone (**6.2.63**), which represent piperazinyl-benzisothiazole derivatives. They also have combined serotonin (5-HT_2) and dopamine (D_2) receptor antagonist properties, and are used to treat the symptoms of schizophrenia and bipolar disorder, and episodes of mania and depression (Fig. 6.8.).

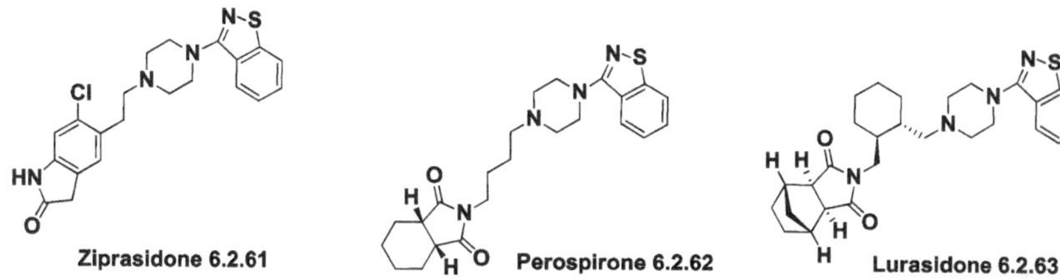

FIG. 6.8 Antipsychotics of piperazinyl-benzisothiazole series.

Ziprasidone–Geodon

Synthesis of ziprasidone (**6.2.61**) has been accomplished via alkylation of 1-(1,2-benzisothiazole-3-yl)piperazine (**6.2.69**) with 6-chloro-5-(2-chloroethyl) indolin-2-one (**6.2.67**) prepared separately [66-70] (Scheme 6.11.).

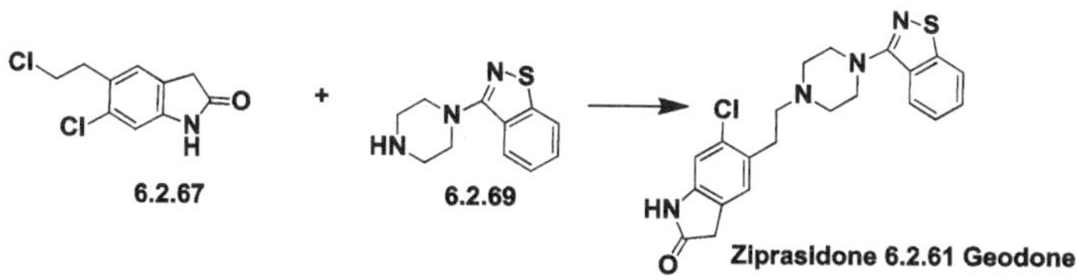

SCHEME 6.11 Synthesis of ziprasidone.

To prepare compound (**6.2.67**), the indolinone (**6.2.64**) was treated with chloroacetyl chloride (**6.2.65**) to produce the chloroacetylindolinone derivative (**6.2.66**), which was reduced to the desired (**6.2.67**) by using triethylsilane in trifluoroacetic acid as a hydride reductant (Scheme 6.12.).

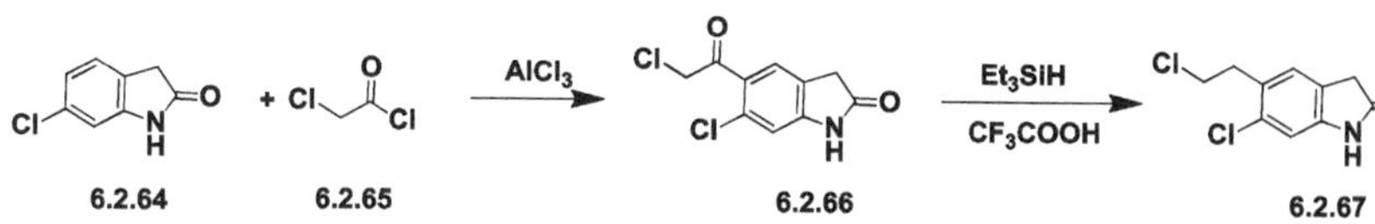

SCHEME 6.12 Synthesis of 6-chloro-5-(2-chloroethyl)indolin-2-one (6.2.67).

The synthesis of the second reaction component, 1-(1,2- benzisothiazole-3-yl)piperazine (**6.2.69**), was accomplished via direct interaction of 1,2-benzisothiazolin-3-one (**6.2.68**) with phosphorous oxychloride and excess of piperazine which afforded requested (**6.2.67**) (Scheme 6.13.).

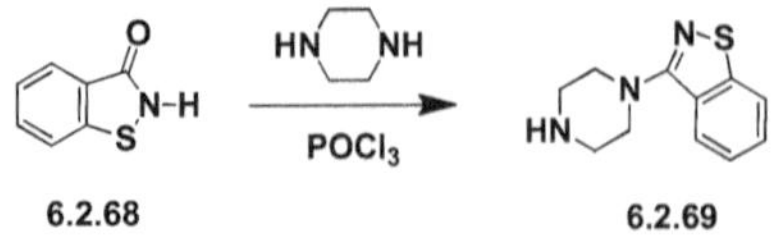

SCHEME 6.13 **Synthesis of 1-(1,2- benzisothiazole-3-yl)piperazine (6.2.69).**

Ziprasidone has a unique combination of pharmacological binding at human receptors. It has high affinity for human 5-HT and dopamine D_2 receptors. Ziprasidone is a 5-HT_{1A} receptor agonist, and an antagonist at 5-HT_{2A}, 5-HT_{2C}, and 5-$HT_{1B/1D}$ receptors. Additionally, ziprasidone inhibits neuronal uptake of 5-hydroxytryptamine and norepinephrine comparable to imipramine. This unique pharmacological profile may be related to its clinical effectiveness for treatment of the symptoms of schizophrenia with a low propensity for extrapyramidal side effects and weight gain [71]. Ziprasidone is approved for the treatment of acute agitation in schizophrenic patients; acute psychosis related to schizophrenia; schizoaffective disorder; and bipolar, delusional, or brief psychotic disorders in patients who need rapid control of agitation. Several reviews are devoted to the medicinal implementation of ziprasidone [72-75].

Structural analogues of ziprasidone, the compounds perospirone (**6.2.62**) [76-78] and lurasidone (**6.2.63**) [79-83], which show very close pharmacological profiles, have been synthesized in a similar manner to ziprasidone as described in tens of patents and as summarized in an excellent paper [84].

Benzamides Sulpiride, Amisulpride

A series of salicylamides, that is, o-hydroxy–substituted benzamides, which at could be named sulfamoylbenzamide derivatives and which in pharmacology belong to the class of antipsychotics titled substituted benzamides, includes compounds able to modulate dopaminergic neurons selectively and specifically. The first synthetic substituted benzamide was sulpiride (**6.2.70**) (Fig. 6.9.), followed by sultopride (**6.2.71**), tiapride (**6.2.72**), and metoclopramide (**6.2.73**), which were gradually replaced by the newer amisulpride (**6.2.74**) [85] (Fig. 6.9.). The last compound, remoxipride (**6.2.75**), with the absence of the aminosulfonyl group in the benzamide moiety, was examined in the 1990s, but removed from the market because of lethal side effects, specifically aplastic anemia, which occurred in as many as 1 in 10,000 cases.

Benzamide derivatives, are structurally distinct from other antipsychotics, and bind selectively to central and peripheral dopamine D_2 and D_3 receptors. At low doses they preferentially block presynaptic autoreceptors, producing an increase in dopamine release, and therefore acting as a dopaminergic compounds able to resolve the dopaminergic hypoactivity that characterizes depression. At higher doses they exert activity on postsynaptic D_2 and D_3 receptors located in the limbic region, producing selective dopaminergic inhibition, eliciting antipsychotic effects [86, 87].

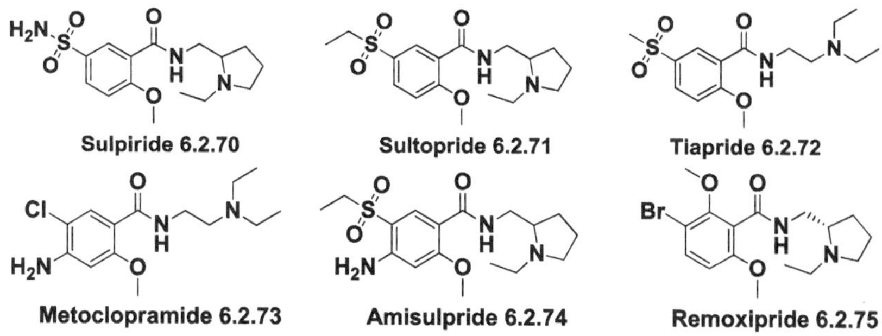

FIG. 6.9 Structure of antipsychotics titled "substituted benzamides."

Sulpiride was prepared with an excellent yields by methylation of 5-aminosulfosalicylic acid (**6.2.76**) with dimethylsulfate (**6.2.77**) followed with the reaction of obtained 5-aminosulfonyl-2-methoxybenzoic acid (**6.2.78**) with 2-aminomethyl-1-ethylpyrrolidine (**6.2.80**), using carbonyldiimidazole (**6.2.79**) as a condensing agent [88-92] (Scheme 6.14.).

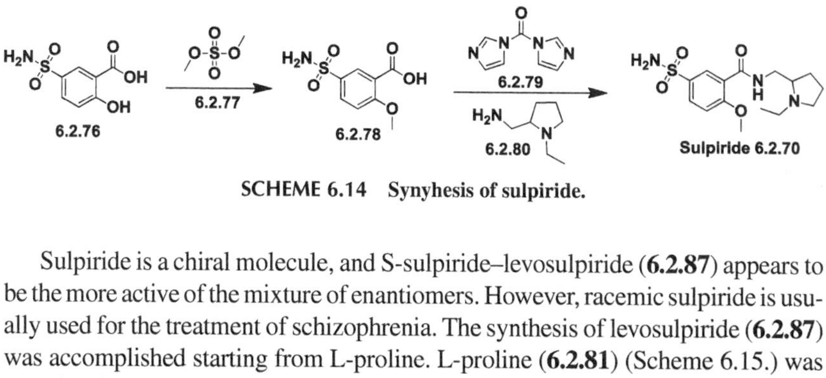

SCHEME 6.14 Synyhesis of sulpiride.

Sulpiride is a chiral molecule, and S-sulpiride–levosulpiride (**6.2.87**) appears to be the more active of the mixture of enantiomers. However, racemic sulpiride is usually used for the treatment of schizophrenia. The synthesis of levosulpiride (**6.2.87**) was accomplished starting from L-proline. L-proline (**6.2.81**) (Scheme 6.15.) was acetylated with acetic anhydride or acetyl chloride to give L-N-acetylproline (**6.2.82**). This product was reduced to L-N-ethyl-2-pyrrolidinomethanol (**6.2.83**). Then the hydroxyl group was replaced by chlorine with thionyl chloride to produce L-N-ethyl-2-pyrrolidinomethyl chloride (**6.2.84**). The chlorine atom, in turn, was replaced by an amino group by using excess ammonia in an alcohol solution. The obtained amine (**6.2.85**) was heated with 5-aminosulfonyl-2-methoxybenzoic acid ethyl ester (**6.2.86**) to produce the desired levosulpiride (**6.2.87**) [92].

Amisulpride (**6.2.74**) was prepared in a very similar manner, but starting from 4-amino-5-mercapto-2-methoxybenzoic acid (**6.2.88**) (Scheme 6.16.), the mercapto group was ethylated with diethylsulfate (**6.2.89**) to produce (**6.2.90**). The obtained product was oxidized with hydrogen peroxide in acetic acid to produce 4-amino-5-(ethylsulfonyl)-2-methoxybenzoic acid (**6.2.91**). This was transformed to the corresponding acid chloride (**6.2.92**), which on interaction with 2-aminomethyl-1-ethylpyrrolidine (**6.2.80**) produced the desired amisulpride (**6.2.74**) [93].

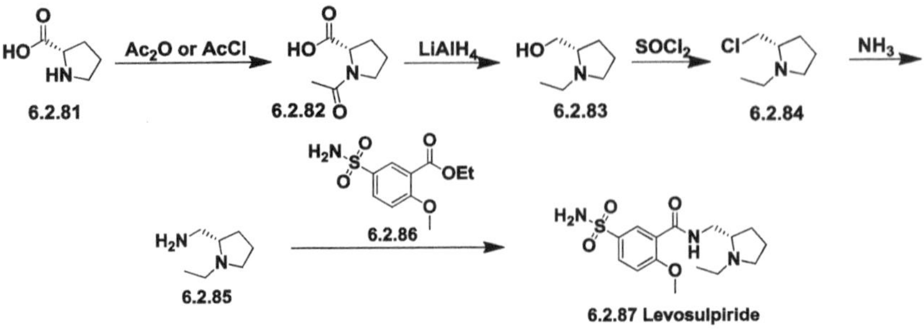

SCHEME 6.15 **Synyhesis of levosulpiride.**

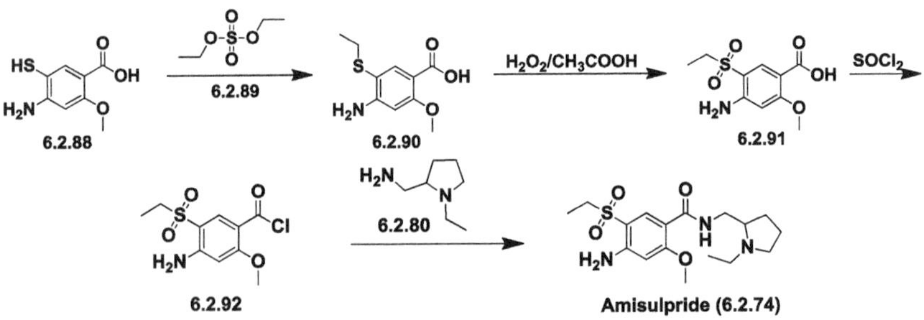

SCHEME 6.16 **Synyhesis of amisulpride.**

Amisulpride is an effective and well-tolerated option for the first-line treatment of patients with acute schizophrenia. The clinical properties of amisulpride appear more "atypical" than other members of this family despite having similar receptor profiles [94-96].

6.3 THIRD-GENERATION ANTIPSYCHOTICS

Third-generation antipsychotic agents are compounds that that have both D_2 antagonist and partial D_2 agonist qualities. The D_2 agonism/antagonism balance probably is critical for these series of compounds [97]. Aripiprazole (**6.3.1**) is the first representative on the market; several more, such as bifeprunox (**6.3.2**) and norclozapine (**6.3.3**), are in clinical development. Others are direct analogues of aripiprazole, such as OPC-4392 (**6.3.4**), SDZ 208-912 (**6.3.5**), and ACR16 (**6.3.6**) (Fig. 6.10.), are in various trial phases.

Aripiprazole–Abilify

A quinolinone derivative, aripiprazole (**6.3.1**) has been extoled as a new generation of antipsychotic drug. The extent to which its effects differ from other atypical antipsychotics is debated and the distinction between second- and third-generation antipsychotics seems not well proven. Aripiprazole is considered the first representative of a third-generation of atypical antipsychotics and

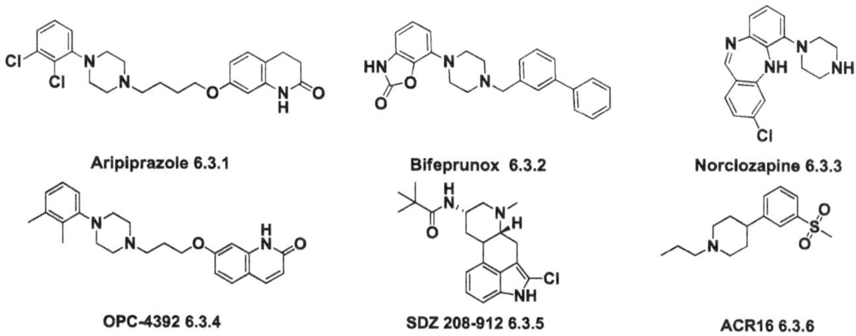

FIG. 6.10 **Third-generation antipsychotics.**

is a dopamine–serotonin system stabilizer. It is "functionally selective" and acts as an antagonist and partial agonist at D_2 receptors. In addition, it acts as a partial agonist at the 5-HT_{1A} receptor, and displays an antagonist profile at 5-HT_{2A} and 5-HT_7 receptors. Based on its D_2 partial agonist properties, pharmacologists have labeled it the "dopamine stabilizer" and a "mood stabilizer" [98-100]. In situations with high extracellular dopamine concentrations, the partial agonist properties of aripiprazole compete with dopamine and cause partial antagonism, offering a clinical benefit. Conversely, in situations where extracellular dopamine concentrations are low, the drug occupies additional receptors and causes partial activation [101]. But there is another opinion: the antagonistic action of these third-generation antipsychotics at 5-HT_7 receptors may be the main importance in their therapeutic efficacy [102].

Aripiprazole can be synthesized by various methods that were first described in its patent [103], and which later underwent improvements [104-107]. Scheme 6.17 represents what is probably the simplest approach to synthesizing aripiprazole. The starting material, 7-hydroxy-3,4-dihydroquinolin-2(1H)-one (**6.3.7**), was refluxed with 1,4-dibromobutane (**6.3.8**) in aqueous K_2CO_3 solution to produce (**6.3.9**). Adding acetonitrile to the obtained solution as a cosolvent—NaI, 1-(2,3-dichlorophenyl)piperazine (**6.3.10**)—and refluxing the whole mixture produced the desired aripiprazole (**6.2.93**).

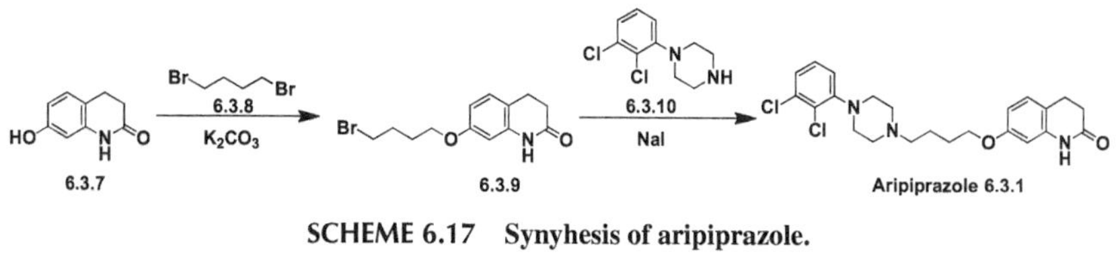

SCHEME 6.17 **Synyhesis of aripiprazole.**

Aripiprazole is approved for the treatment of schizophrenia and bipolar disorder in adults and adolescents. It has a low propensity for extrapyramidal side effects and causes minimal sedation and weight gain. Its action are described in a number of excellent reviews [108-112].

Over the last couple of decades, new drugs, known as atypical antipsychotics, have been synthesized. Some of them have covered certain aspects of the disorder that had previously been considered unmanageable, such as negative and depressive symptoms and cognitive function, with a lower risk of the patient suffering extrapyramidal symptoms. These have made them the first-line therapy for the treatment of the psychotic symptoms of schizophrenia, as well as other psychiatric and neurological disorders.

The development of antipsychotics has come full circle—from the development of selective agents with a single therapeutic action to multifunctional therapeutics. This progression reflects the development of our understanding of the mechanisms underlying the actions of drugs and their side effects. If we want to treat these diseases rather than try to prove suspicious concepts of the necessity of receptor selectivity or "magic bullets" for these diseases, it is likely that we will stop ignoring nonselective or "dirty" drugs with multiple mechanisms of therapeutic action.

Undoubtedly, new atypical agents with novel mechanisms of antipsychotic action that will have advantages over existing atypical antipsychotics will be commercialized. Among them could be some compounds with diverse chemical structures, such as those now in Phase II trials. Some of the disclosed chemical structures, such as GSK 742457 (**6.3.11**), a 5-HT_6 antagonist; BL-1020 (**6.3.12**), a dopamine antagonist/γ-aminobutyric acid (GABA) agonist; EVP-6124 (**6.3.13**), a partial agonist of α_7 neuronal nicotinic acetylcholine receptors; and LY-2140023 (**6.3.14**), glutamate 2/3 agonist, are presented in Fig. 6.11. Among them could be some of the compounds with undisclosed chemical structures such as cariprazine, a D_2/D_3 antagonist; L-83092, a GABA $\alpha_{2/3}$ agonist; MEM-3454, a partial nicotinic α_7 agonist; and others.

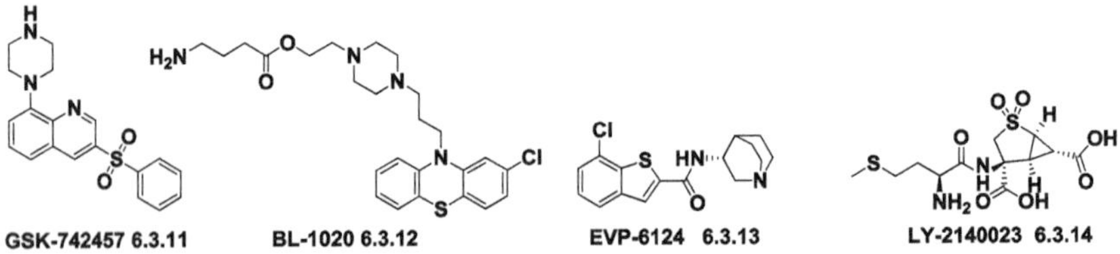
GSK-742457 6.3.11 BL-1020 6.3.12 EVP-6124 6.3.13 LY-2140023 6.3.14

FIG. 6.11 Atypical antipsychotics in Phase II trials.

Also among them could be some of the compounds that are now in Phase I trials. The structures of compounds with disclosed chemical structures are presented in Fig. 6.12. They are GSK1034702 (**6.3.15**), a M1 receptor agonist; ABT-107 (**6.3.16**), an α_7 nicotinic receptor agonist; neboglamine (**6.3.17**), a glycine, *N*-methyl-D-aspartate (NMDA) receptor agonist; PRX-07034 (**6.3.18**), a 5-HT_6 antagonist; and compounds with undisclosed chemical structures such as ACE-325, an AMPA receptor antagonist; AV-965, a 5-HT_{1A} antagonist; GSK1018921, a glycine transporter 1 inhibitor; ITI-007, a 5-HT_{2A} antagonist, dopamine phosphoprotein modulator; and MEM-63908, a partial nicotinic α_7 agonist.

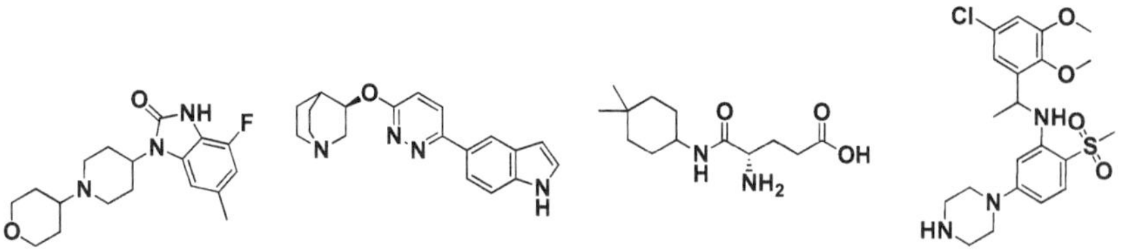

GSK1034702 6.3.15 **ABT-107 6.3.16** **Neboglamine 6.3.17** **PRX-07034 6.3.18**

FIG. 6.12 Atypical antipsychotics in Phase I trials.

REFERENCES

1. Kim, D. H.; Maneen, M. J.; Stahl, S. M. Building a better antipsychotic: receptor targets for the treatment of multiple symptom dimensions of schizophrenia. *Neurotherapeutics* **2009,** *6* (1), 78–85.
2. Awouters, F. H. L.; Lewi, P. J. Forty years of antipsychotic drug research—from haloperidol to paliperidone—with Dr. Paul Janssen. *Arzneim. Forsch.* **2007,** *57* (10), 625–632.
3. Hormaechea, J. A.; Garcia, L. E.; Prieto, L.; Gomez, J. C. Comparison of older and newer neuroleptics for the treatment of schizophrenia. *Expert Rev. Neurother.* **2001,** *1* (2), 161–170.
4. Jafari, S.; Fernandez-Enright, F.; Huang, X.-F. Structural contributions of antipsychotic drugs to their therapeutic profiles and metabolic side effects. *J. Neurochem.* **2012,** *120* (3 & 4), 371–384.
5. Emsley, R. Drugs in development for the treatment of schizophrenia. *Expert Opin. Invest. Drugs* **2009,** *18* (8), 1103–1118.
6. Huffman, J. C.; Alpert, J. E. An approach to the psychopharmacologic care of patients: antidepressants, antipsychotics, anxiolytics, mood stabilizers, and natural remedies. *Med. Clin. North Am.* **2010,** *94* (6), 1141–1160.
7. Margolis, R. L. The choice of antipsychotics in schizophrenia. *Nat. Rev. Neurol.* **2009,** *5* (6), 308–310.
8. Tandon, R. Antipsychotics in the treatment of schizophrenia: an overview. *J. Clin. Psychiatry (Memphis, TN, U. S.)* **2011,** *72* (Suppl. 1), 4–8.
9. Remington, G.; Foussias, G.; Agid, O. Progress in defining optimal treatment outcome in schizophrenia. *CNS Drugs* **2010,** *24* (1), 9–20.
10. Holmes, J. C.; Zacher, J. L. Second-generation antipsychotics: a review of recently approved agents and drugs in the pipeline. *Formulary* **2012,** *47* (3), 106–112, 119–121.
11. Newman-Tancredi, A.; Kleven, M. S. Comparative pharmacology of antipsychotics possessing combined dopamine D2 and serotonin 5-HT1A receptor properties. *Psychopharmacology (Berl).* **2011,** *216* (4), 451–473.
12. Charlton, B. G. If "atypical" neuroleptics did not exist, it wouldn't be necessary to invent them: Perverse incentives in drug development, research, marketing and clinical practice. *Med. Hypotheses* **2005,** *65* (6), 1005–1009.
13. Brooke, N. S.; Wiersgalla, M.; Salzman, C. Atypical uses of atypical antipsychotics. *Harv. Rev. Psychiatry* **2005,** *13,* 317–339.
14. Gao, K.; Muzina, D.; Gajwani, P.; Calabrese, J. R. Efficacy of typical and atypical antipsychotics for primary and comorbid anxiety symptoms or disorders: a review. *J. Clin. Psychiatry (Memphis, TN, U. S.)* **2006,** *67,* 1327–1340.
15. Jeste, D. V.; Dolder, C. R. Treatment of non-schizophrenic disorders: focus on atypical antipsychotics. *J. Psychiatr. Res.* **2004,** *38,* 73–103.

16. Nemeroff, C. B. Use of atypical antipsychotics in refractory depression and anxiety. *J. Clin. Psychiatry (Memphis, TN, U. S.)* **2005,** *66* (Suppl. 8), 13–21.
17. Fourrier, A.; Gasquet, I.; Allicar, M. P.; Bouhassira, M.; Lepine, J. P.; Begaud, B. Patterns of neuroleptic drug prescription: a national cross-sectional survey of a random sample of French psychiatrists. *Br. J. Clin. Pharmacol.* **2000,** *49,* 80–86.
18. Vardanyan, R. S.; Hruby, V. J. *Synthesis of Essential Drugs;* Elsevier, 2006.
19. Bruhwyler, J.; Chleide, E.; Mercier, M. Clozapine: an atypical neuroleptic. *Neurosci. Biobehav. Rev.* **1990,** *14* (4), 357–363.
20. Kiss, B.; Bitter, I. Structural analogues of clozapine. In *Analogue-based Drug Discovery;* Fischer, J., Ganellin, C. R., Eds.; Wiley-VCH, 2006; pp 297–313.
21. Schmutz, J. Neuroleptic piperazinyldibenzoazepines. Chemistry and structure-activity relations. *Arzneim. Forsch.* **1975,** *25* (5), 712–720.
22. Ananth, J.; Parameswaran, S.; Hara, B. Drug therapy in schizophrenia. *Curr. Pharm. Des.* **2004,** *10* (18), 2205–2217.
23. Markowitz, J. S.; Brown, C. S.; Moore, T. R. Atypical antipsychotics. Part I: pharmacology, pharmacokinetics, and efficacy. *Ann. Pharmacother.* **1999,** *33* (1), 73–85.
24. Brown, C. S.; Markowitz, J. S.; Moore, T. R.; Parker, N. G. Atypical antipsychotics: part II: adverse effects, drug interactions, and costs. *Ann. Pharmacother.* **1999,** *33* (2), 210–217.
25. Schmutz, J.; Hunziker, F. Preparation of 11-basic substituted dibenzodiazepines and dibenzothiazepines as pharmaceutically active compounds, US 3539573 (1970).
26. Hunziker, F.; Fischer, E.; Schmutz, J. Seven-membered heterocycles. X. 11-Amino-5H-dibenzo[b,e]-1,4-diazepines. *Helv. Chim. Acta* **1967,** *50* (6), 1588–1599.
27. Warawa, E. J.; Migler, B. M. Preparation of 11-[4-[2-(2-hydroxyethoxy)ethyl]-1-piperazinyl] dibenzo [b,f][1,4]thiazepine as a neuroleptic and antipsychotic, EP 240228 (1987).
28. Schmutz, J.; Kuenzle, F.; Hunziker, F.; Buerki, A. Seven-membered heterocycles. IV. New synthesis of dibenzo[b,f]-1,4-thiazepine, -oxazepine, and dibenzo[b,e]azepine lactams. *Helv. Chim. Acta* **1965,** *48* (2), 336–347.
29. Barker, A. C.; Copeland, R. J. Process for the preparation of a piperazinodibenzothiazepine with antidopaminergic activity, EP 282236 (1988),
30. Chakrabarti, J. K.; Horsman, L.; Hotten, T. M.; Pullar, I. A.; Tupper, D. E.; Wright, F. C. 4-Piperazinyl-lOH-thien0[2,3b-] [1,5]benzodiazepines as potential neuroleptics. *J. Med. Chem.* **1980,** *23,* 878–884.
31. Chakrabarti, J. K.; Hotten, T. M.; Tupper, D. E. Preparation of 2-methyl-4-(4-methyl-1-piperazinyl)-10H-thieno-[2,3-b][1,5]benzodiazepine, EP 454436 (1991).
32. Bunnell, C. A.; Hendriksen, B. A.; Hotten, T. M.; Larsen, S. D.; Tupper, D. E. Preparation of olanzapine solvates, EP 733634 (1996).
33. Beasley, C. M., Jr.; Chakrabarti, J. K.; Hotten, T. M.; Tupper, D. E. 2-Methyl-4-(4-methyl-1-piperazinyl)-10H-thieno-[2,3-b][1,5]benzodiazepine for treatment of psychoactive substance disorders, US 5817657 (1988).
34. Kennis, L. E. J.; Vandenberk, J. Preparation of 1,2-benzisoxazol-3-yl and 1,2-benzisothiazol-3-yl derivatives as antipsychotics, EP 196132 (1986).
35. Kennis, L. E. J.; Vandenberk, J. 3-piperidinyl-substituted 1,2-benzisoxazoles and 1,2-benzisothiazoles, US 4804663 (1989).
36. Janssen, P. A. J.; Niemegeers, C. J. E.; Awouters, F.; Schellekens, K. H. L.; Megens, A. A. H. P.; Meert, T. F. Pharmacology of risperidone (R 64 766), a new antipsychotic with serotonin-S2 and dopamine-D2 antagonistic properties. *J. Pharmacol. Exp. Ther.* **1988,** *244* (2), 685–693.
37. Livingston, M. G. Risperidone. *Lancet* **1994,** *343* (8895), 457–460.

38. Grant, S.; Fitton, A. Risperidone. A review of its pharmacology and therapeutic potential in the treatment of schizophrenia. *Drugs* **1994,** *48* (2), 253–273.
39. Germann, D.; Kurylo, N.; Han, F. Risperidone. *Profiles Drug Subst., Excipients, Relat. Methodol.* **2012,** *37,* 313–361.
40. Leysen, J. E.; Gommeren, W.; Eens, A.; De Chaffoy de Courcelles, D.; Stoof, J. C.; Jenssen, P. A. J. Biochemical profile of risperidone, a new antipsychotic. *J. Pharmacol. Exp. Ther.* **1988,** *247* (2), 661–670.
41. Janssen, C. G. M.; Knaeps, A. G.; Kennis, L. E. J.; Vandenberk, J. 3-[2-[4-(6-fluoro-1,2-benzisoxozol-3-yl)-1-piperidinyl]ethyl]-6,7,8,9 tetrahydro-9-hydroxy-2-methyl-4H-pyrido [1,2-a]pyrimidin-4-one, compositions and method of use, US 5158952 (1989).
42. Riva, R.; Banfi, L.; Castaldi, G.; Ghislieri, D.; Malpezzi, L.; Musumeci, F.; Tufaro, R.; Rasparini, M. Selective chemical oxidation of risperidone: a straightforward and cost-effective synthesis of paliperidone. *Eur. J. Org. Chem.* **2011,** 2319–2325.
43. Strupczewski, J. T.; Bordeau, K. J.; Chiang, Y.; Glamkowski, E. J.; Conway, P. G.; Corbett, R.; Hartman, H. B.; Szewczak, M. R.; Wilmot, C. A.; Helsley, G. C. 3-[[(Aryloxy)alkyl] piperidinyl]-1,2-benzisoxazoles as D2/5-HT2 antagonists with potential atypical antipsychotic activity: antipsychotic profile of iloperidone (HP 873). *J. Med. Chem.* **1995,** *38* (7), 1119–1131.
44. Strupczewski, J. T.; Helsley, G. C.; Glamkowski, E. J.; Chiang, Y.; Bordeau, K. J.; Nemoto, P. A.; Tegeler, J. J. 3-(Heteroaryl)-1-[(2,3-dihydro-1H-isoindol-2-yl)alkyl]pyrrolidines and 3-(heteroaryl)-1-[(2,3-dihydro-1H-indol-1-yl)alkyl]pyrrolidines and related compounds and their use as analgesics and antipsychotics, US 5776963 (1988).
45. Kongsamut, S.; Roehr, J. E.; Cai, J.; Hartman, H. B.; Weissensee, P.; Kerman, L. L.; Tang, L.; Sandrasagra, A. Iloperidone binding to human and rat dopamine and 5-HT receptors. *Eur. J. Pharmacol.* **1996,** *317* (2/3), 417–423.
46. Kalkman, H. O.; Subramanian, N.; Hoyer, D. Extended radioligand binding profile of iloperidone: a broad spectrum dopamine/serotonin/norepinephrine receptor antagonist for the management of psychotic disorders. *Neuropsychopharmacology* **2001,** *25* (6), 904–914.
47. Kalkman, H. O.; Feuerbach, D.; Lotscher, E.; Schoeffter, P. Functional characterization of the novel antipsychotic iloperidone at human D_2, D_3, a_{2C}, 5-HT_6, and 5-HT_{1A} receptors. *Life Sci.* **2003,** *73,* 1151–1159.
48. Szewczak, M. R.; Corbett, R.; Rush, D. K.; Wilmot, C. A.; Conway, P. G.; Strupczewski, J. T.; Cornfeldt, M. The pharmacological profile of iloperidone, a novel atypical antipsychotic agent. *J. Pharmacol. Exp. Ther.* **1995,** *274* (3), 1404–1413.
49. Citrome, L. Iloperidone: chemistry, pharmacodynamics, pharmacokinetics and metabolism, clinical efficacy, safety and tolerability, regulatory affairs, and an opinion. *Expert Opin. Drug Metab. Toxicol.* **2010,** *6* (12), 1551–1564.
50. Corbett, R.; Griffiths, L.; Shipley, J. E.; Shukla, U.; Strupczewski, J. T.; Szczepanik, A. M.; Szewczak, M. R.; Turk, D. J.; Vargas, H. M.; Kongsamut, S. the Iloperidone Project Team, Iloperidone: preclinical profile and early clinical evaluation. *CNS Drug Rev.* **1997,** *3* (2), 120–147.
51. Hesselink, J. M. K. Iloperidone Hoechst Marion Roussel Inc. *Curr. Opin. Cent. Peripher. Nerv. Syst. Invest. Drugs* **2000,** *2* (1), 71–78.
52. Arif, S. A.; Mitchell, M. M. Iloperidone: a new drug for the treatment of schizophrenia. *Am. J. Health-Syst. Pharm.* **2011,** *68* (4), 301–308.
53. Hale, K. S. Iloperidone-a second-generation antipsychotic for the treatment of acute schizophrenia. *J. Pharm. Technol.* **2010,** *26* (4), 193–202.
54. Mucke, H. A. M.; Castaner, J. Iloperidone: antipsychotic, dopamine D2 antagonist, 5-HT2A antagonist. *Drugs Future* **2000,** *25* (1), 29–40.

55. Marino, J.; Caballero, J. Iloperidone for the treatment of schizophrenia. *Ann. Pharmacother.* **2010,** *44* (5), 863–870.
56. Perregaard, J.; Arnt, J.; Bogeso, K. P.; Hyttel, J.; Sanchez, C. Noncataleptogenic, centrally acting dopamine D-2 and serotonin 5-HT2 antagonists within a series of 3-substituted l-(4-fluorophenyl)-lH-indoles. *J. Med. Chem.* **1992,** *35,* 1092–1101.
57. Perregaard, J. K. Indole derivatives and their antipsychotic activity, EP 200322 (1986).
58. Perregaard, J. K. 1-(4′-fluorophenyl)-3,5-substituted indoles useful in the treatment of psychic disorders and pharmaceutical compositions thereof, US 4710500 (1987).
59. Perregaard, J. K.; Skarsfeldt, T. Use of sertindole for the treatment of schizophrenia. EP 0392959 (1990).
60. Zanon, J.; Villa, M.; Ciardella, F. Method for manufacture of sertindole, WO03/080597 (2003).
61. Suni Kumar, I. V.; Anjaneyulu, G. S. R.; Hima Bind, V. Identification and synthesis of impurities formed during sertindole preparation. *Beilstein J. Org. Chem.* **2011,** *7,* 29–33.
62. Hietala, J.; Kuonnamaki, M.; Palvimaki, M.; Laakso, A.; Majasuo, H.; Syvalahti, E. Sertindole is a serotonin 5-HT_{2C} inverse agonist and decreases agonist but not antagonist binding to 5-HT_{2C} receptors after chronic treatment. *Psychopharmacology (Berl).* **2001,** *157* (2), 180–187.
63. Juruena, M. F.; Ponde de Sena, E.; Reis de Oliveira, I. Sertindole in the management of schizophrenia. *J. Cent. Nerv. Syst. Dis.* **2011,** *3,* 75–85.
64. Azorin, J.-M.; Kaladjian, A.; Fakra, E.; Adida, M. Sertindole for the treatment of schizophrenia. *Expert Opin. Pharmacother.* **2010,** *11* (18), 3053–3064.
65. Brown, L. A.; Levin, G. M. Sertindole, a new atypical antipsychotic for the treatment of schizophrenia. *Pharmacotherapy* **1998,** *18* (1), 69–83.
66. Yevich, J. P.; New, J. S.; Smith, D. W.; Lobeck, W. G.; Catt, J. D.; Minielli, J. L.; Eison, M. S.; Taylor, D. P.; Riblet, L. A.; Temple, D. L., Jr. Synthesis and biological evaluation of I-(1,2-benzisothiazol-3-yl)- and (1,2-benzisoxazol-3-yl)piperazine derivatives as potential antipsychotic agents. *J. Med. Chem.* **1986,** *29,* 359–369.
67. Bowles, P. Process for preparing aryl piperazinyl-heterocyclic compounds, US 5206366 (1994).
68. Lowe, J. A., III; Nagel, A. A. Neuroleptic arylpiperazinylalkyl-substituted heterocycles and their pharmaceutical compositions and use, US 4831031 (1989).
69. Bowles, P.; Busch, F. R.; Allen, D. J. M.; Diroma, S. A.; Godek, D. M. Process for preparing aryl piperazinyl-heterocyclic compounds useful as neuroleptics, CA 2095587 (1994).
70. Howard, H. R.; Shenk, K. D.; Smolarek, T. A.; Marx, M. H.; Windels, J. H.; Roth, R. W. Synthesis of 3H- and 14C-labeled CP-88,059: a potent atypical antipsychotic agent. *J. Labelled Compd. Radiopharm.* **1994,** *34* (2), 117–125.
71. Schmidt, A. W.; Lebel, L. A.; Howard, H. R., Jr.; Zorn, S. H. Ziprasidone: a novel antipsychotic agent with a unique human receptor binding profile. *Eur. J. Pharmacol.* **2001,** *425* (3), 197–201.
72. Rosa, A. R.; Franco, C.; Torrent, C.; Comes, M.; Cruz, N.; Horga, G.; Benabarre, A.; Vieta, E. Ziprasidone in the treatment of affective disorders: a review. *CNS Neurosci. Ther.* **2008,** *14* (4), 278–286.
73. Greenberg, W. M.; Citrome, L. Ziprasidone for schizophrenia and bipolar disorder: a review of the clinical trials. *CNS Drug Rev.* **2007,** *13* (2), 137–177.
74. Warrington, L.; Lombardo, I.; Loebel, A.; Ice, K. Ziprasidone for the treatment of acute manic or mixed episodes associated with bipolar disorder. *CNS Drugs* **2007,** *21* (10), 835–849.
75. Caley, C. F.; Cooper, C. K. Ziprasidone: the fifth atypical antipsychotic. *Ann. Pharmacother.* **2002,** *36* (5), 839–851.

76. de Paulis, T. Perospirone. *Curr. Opin. Invest. Drugs (BioMed Cent.),Drugs* **2002,** *3* (1), 121–129.
77. Onrust, S. V.; McClellan, K. Perospirone. *CNS Drugs* **2001,** *15* (4), 329–337.
78. Ishibashi, T.; Ohno, Y. Perospirone hydrochloride: the novel atypical antipsychotic agent with high affinities for 5-HT2, D2 and 5-HT1A receptors. In *Advances in Neuroregulation and Neuroprotection;* Collin, C., Ed.; CRC Press, 2005; pp 347–357.
79. Yasui-Furukori, N. Update on the development of lurasidone as a treatment for patients with acute schizophrenia. *Drug Des., Dev. Ther.* **2012,** *6,* 107–115.
80. Kane, J. M. Lurasidone: a clinical overview. *J. Clin. Psychiatry (Memphis, TN, U. S.)* **2011,** *72* (Suppl. 1), 24–28.
81. Meyer, J. M.; Loebel, A. D.; Schweizer, E. Lurasidone: a new drug in development for schizophrenia. *Expert Opin. Invest. Drugs* **2009,** *18* (11), 1715–1726.
82. Risbood, V.; Lee, J. R.; Roche-Desilets, J.; Fuller, M. A. Lurasidone: an atypical antipsychotic for schizophrenia. *Ann. Pharmacother.* **2012,** *46,* 1033–1046.
83. Ishibashi, T.; Horisawa, T.; Tokuda, K.; Ishiyama, T.; Ogasa, M.; Tagashira, R.; Matsumoto, K.; Nishikawa, H.; Ueda, Y.; Toma, S.; Oki, H.; Tanno, N.; Saji, I.; Ito, A.; Ohno, Y.; Nakamura, M. Pharmacological profile of lurasidone, a novel antipsychotic agent with potent 5-hydroxytryptamine 7 (5-HT_7) and 5-HT_{1A} receptor activity. *J. Pharmacol. Exp. Ther.* **2010,** *334* (1), 171–181.
84. Ishizumi, K.; Kojima, A.; Antoku, F.; Saji, I.; Yoshigi, M. Succinimide derivatives. II. Synthesis and antipsychotic activity of N-[4-[4-(1,2-benzisothiazol-3-yl)-1-piperazinyl]butyl]-1,2-cis-cyclohexanedicarboximide (SM-9018) and related compounds. *Chem. Pharm. Bull.* **1995,** *43* (12), 2139–2151.
85. Jenner, P.; Maraden, C. D. The substituted benzamides–a novel class of dopamine antagonists. *Life Sci.* **1979,** *25,* 479–486.
86. Racagni, G.; Canonico, P. L.; Ravizza, L.; Pani, L.; Amore, M. Consensus on the use of substituted benzamides in psychiatric patients. *Neuropsychobiology* **2004,** *50* (2), 134–143.
87. Green, B. Focus on amisulpride. *Curr. Med. Res. Opin.* **2002,** *18* (3), 113–117.
88. Engelhardt, E. L.; Miller, Ch. S. Verfahren zur Herstellung von heterocyclischen Benzamiden, DE 1595915 (1965).
89. Engelhardt, E. L.; Miller, Ch. S. 2-Alkoxybenzamide, DE 1795723 (1965).
90. Engelhardt, E. L.; Thominet, M. L. Heterocyclic aminoalkyl benzamides, US 3342826 (1969).
91. Bulteau, G.; Acher, J. Enamines, their derivatives and processes of production, US 4077976 (1978).
92. Mauri, F. Optically active benzamides, DE 2903891 (1979).
93. Thominet, M.; Acher, J.; Monier, J. C. 4-Amino-5-alkylsulfonyl o-anisamide derivatives useful as psychotropic agents, BE 872585 (1979).
94. Scatton, B.; Claustre, Y.; Cudennec, A.; Oblin, A.; Perrault, G.; Sanger, D. J.; Schoemaker, H. Amisulpride: from animal pharmacology to therapeutic action. *Int. Clin. Psychopharmacol.* **1997,** *12* (Suppl. 2), S29–S36.
95. Perrault, Gh.; Depoortere, R.; Morel, E.; Sanger, D. J.; Scatton, B. Psychopharmacological profile of amisulpride: an antipsychotic drug with presynaptic D2/D3 dopamine receptor antagonist activity and limbic selectivity. *J. Pharmacol. Exp. Ther.* **1997,** *280* (1), 73–82.
96. McKeage, K.; Plosker, G. L. Amisulpride: a review of its use in the management of schizophrenia. *CNS Drugs* **2004,** *18* (13), 933–956.
97. Tadori, Y.; Kitagawa, H.; Forbes, R. A.; McQuade, R. D.; Stark, A.; Kikuchi, T. Differences in agonist/antagonist properties at human dopamine D2 receptors between aripiprazole, bifeprunox and SDZ 208-912. *Eur. J. Pharmacol.* **2007,** *574* (2–3), 103–111.

98. Tamminga, C. A. Partial dopamine agonists in the treatment of psychosis. *J. Neural Transm.* **2002,** *109,* 411–420.
99. Stahl, S. M. Dopamine system stabilizers, aripiprazole, and the next generation of antipsychotics, part 1, "Goldilocks" actions at dopamine receptors. *J. Clin. Psychiatry (Memphis, TN, U. S.)* **2001,** *62,* 841–842.
100. Lieberman, J. A. Dopamine partial agonists: a new class of antipsychotic. *CNS Drugs* **2004,** *18,* 251–267.
101. Swainston, H. T.; Perry, C. M. Aripiprazole: a review of its use in schizophrenia and schizoaffective disorder. *Drugs* **2004,** *64* (15), 1715–1736.
102. Rauly-Lestienne, I.; Boutet-Robinet, E.; Ailhaud, M.-C.; Newman-Tancredi, A.; Cussac, D. Differential profile of typical, atypical and third generation antipsychotics at human 5-HT7a receptors coupled to adenylyl cyclase: detection of agonist and inverse agonist properties. *Naunyn-Schmiedeberg's Arch. Pharmacol.* **2007,** *376* (1–2), 93–105.
103. Oshiro, Y.; Sato, S.; Kurahashi, N. Preparation and formulation of 7-[(4-phenylpiperazino) butoxy]carbostyrils as dopaminergic neurotransmitter antagonists, EP 367141 (1990).
104. Kowalski, P.; Jaskowska, J. An efficient synthesis of aripiprazole, buspirone and NAN-190 by the reductive alkylation of amines procedure. *Arch. Pharm. (Weinheim, Ger.)* **2012,** *345* (1), 81–85.
105. Pettersson, F.; Ponten, H.; Waters, N.; Waters, S.; Sonesson, C. Synthesis and evaluation of a set of 4-phenylpiperidines and 4-phenylpiperazines as D2 receptor ligands and the discovery of the dopaminergic stabilizer 4-[3-(methylsulfonyl)phenyl]-1-propylpiperidine (Huntexil, Pridopidine, ACR16). *J. Med. Chem.* **2010,** *53* (6), 2510–2520.
106. Les, A.; Szelejewski, W. Optimization and scale-up of pharmaceutical synthesis. *Przem. Chem.* **2007,** *86* (12), 1174–1177.
107. Les, A.; Badowska-Roslonek, K.; Laszcz, M.; Kamienska-Duda, A.; Baran, P.; Kaczmarek, L. Optimization of aripiprazole synthesis. *Acta Pol. Pharm.* **2010,** *67* (2), 151–157.
108. McGavin, J. K.; Goa, K. L. Aripiprazole. *CNS Drugs* **2002,** *16* (11), 779–786.
109. Bowles, T. M.; Levin, G. M. Aripiprazole: a new atypical antipsychotic drug. *Ann. Pharmacother.* **2003,** *37* (5), 687–694.
110. Burris, K. D.; Molski, T. F.; Xu, C.; Ryan, E.; Tottori, K.; Kikuchi, T.; Yocca, F. D.; Molinoff, P. B. Aripiprazole, a novel antipsychotic, is a high-affinity partial agonist at human dopamine D2 receptors. *J. Pharmacol. Exp. Ther.* **2002,** *302* (1), 381–389.
111. Vergne, D. E.; Anton, R. F. Aripiprazole: a drug with a novel mechanism of action and possible efficacy for alcohol dependence. *CNS Neurol. Disord.: Drug Targets* **2010,** *9* (1), 50–54.
112. Mailman, R. B.; Murthy, V. Third generation antipsychotic drugs: partial agonism or receptor functional selectivity? *Curr. Pharm. Des.* **2010,** *16* (5), 488–501.

Chapter 7

Antidepressants

Depression is defined as a mental disorder characterized by general emotional dejection and withdrawal, sadness greater and more prolonged than that warranted by an objective reason, extreme gloom, feelings of inadequacy, and an inability to concentrate. Conditions characterized by the term *depression* include affective disorders, which are frequently accompanied by a number of other disturbances such as unmotivated sorrow, sleep disorders, changes in appetite, various psychomotor disturbances, loss of interest in things once pleasurable, loss of libido, feelings of worthlessness, and suicidal thoughts. There are sufficiently acceptable, although not universally accepted, classifications of depression.

There is extensive comorbidity between depression and anxiety disorders.

Stress is not a medical diagnosis, but may lead to a diagnosis of depression or anxiety and mood disorders.

The emotional response to stress is a key element of depression and many antidepressants are effective in the treatment of anxiety. In general, the noradrenergic, serotonergic, and dopaminergic systems regulate and modulate many of the same behavioral peculiarities, and comorbidity between depression and anxiety disorders is evident.

Depression is associated with an imbalance of serotonin, norepinephrine, and dopamine.

There exists an hypothesis or model, that "pure" anxiety states is a result of excessive serotonergic activity, and "pure" depressive state, is a result of deficient serotonergic activity [1-5].

Drugs that modulate activity of biogenic amines are used to treat depression. Tricyclic antidepressants, monoamine oxidase inhibitors, selective serotonin reuptake inhibitors (SSRIs), and selective noradrenaline reuptake inhibitors (SNRIs) are used as antidepressants. Recently, inhibitors of both serotonin and noradrenaline uptake (dual uptake inhibitors SSRI/SNRI) appeared in the market. The newest invention is the triple uptake inhibitors (SSRI/SNRI/selective dopamine reuptake inhibitors [SDARIs]), which were proposed for preclinical and clinical trials. It seems that the SDARIs possess greater efficacy than known antidepressants.

The first-generation of antidepressants appeared in the 1950s and were presented as tricyclic antidepressant (TCA) series and monoamine oxidase

Synthesis of Best-Seller Drugs. http://dx.doi.org/10.1016/B978-0-12-411492-0.00007-9

inhibitor (MAOI) series (monoamines are considered catecholamines, that is, serotonin, dopamine, epinephrine, and norepinephrine), which nonspecifically increased the level of monoamines by reducing their metabolism or inhibiting their reuptake.

The second-generation agents (approximately coinciding with the 1970s and 1980s) are considered the SSRIs such as mianserin and moclobemide and are characterized primarily by the era of their introduction rather than by their chemical structure or pharmacological effect.

Third-generation antidepressants used to refer to newer antidepressants. They have a variable mode of action and include venlafaxine, mirtazapine, reboxetine, nefazodone, and others and are not SSRIs. For example, venlafaxine is a combined inhibitor of the reuptake of serotonin, noradrenaline, and dopamine (to a lesser extent); reboxetine is more selective serotonin reuptake inhibitor than noradrenaline and dopamine; mirtazapine has no effect on noradrenaline reuptake but has high affinity for 5-HT_2 and 5-HT_3 receptors; and nefazodone's action is not fully understood.

Existing antidepressants are separable into six major classes:

- TCAs (imipramine, trimipramine, amitriptyline, doxepin, dothiepin, desipramine, protriptyline, nortriptyline, amoxapine, maprotiline, mirtazapine)
- MAOIs (phenelzine, isocarboxazid, tranylcypromine, moclobemide)
- SSRIs (fluoxetine, citalopram, escitalopram, fluvoxamine, paroxetine, sertraline)
- SNRIs (desipramine, reboxetine, maprotiline, viloxazine, atomoxetine)
- Dual uptake inhibitors SSRI/SNRI, (duloxetine, venlafaxine, desvenlafaxine milnacipran, sibutramine)
- Dual uptake inhibitors SNRI/SDARI (bupropion)

7.1 TRICYCLIC ANTIDEPRESSANTS

It is believed that TCAs binding to 5-HT and noradrenaline reuptake transporters block the reuptake of these monoamines from the synaptic cleft and their concentration returns to the normal range. For many years the TCAs were the first choice for and still are occasionally used for treatment of depression. The history of TCAs started with imipramine (**7.1.1**). The list of current TCAs includes the following medications which are modifications of the imipramine template. The most widely used TCAs are imipramine (**7.1.1**), clomipramine (**7.1.2**), desipramine (**7.1.3**), trimipramine (**7.1.4**), lofepramine (**7.1.5**), imipraminoxide (**7.1.6**), amitriptyline (**7.1.7**), nortriptyline (**7.1.8**), amitriptylinoxide (**7.1.9**), noxiptiline (**7.1.10**), protriptyline (**7.1.11**), quinupramine (**7.1.12**), doxepin (**7.1.13**), dothiepin (**7.1.14**), metapramine (**7.1.15**), dibenzepin (**7.1.16**), nitroxazepine (**7.1.17**), propizepine (**7.1.18**), amoxapine (**7.1.19**), mirtazapine (**7.1.20**), maprotiline (**7.1.21**), dimetacrine (**7.1.22**), melitracen (**7.1.23**), and pipofezine (**7.1.24**) [6,7] (Fig. 7.1.).

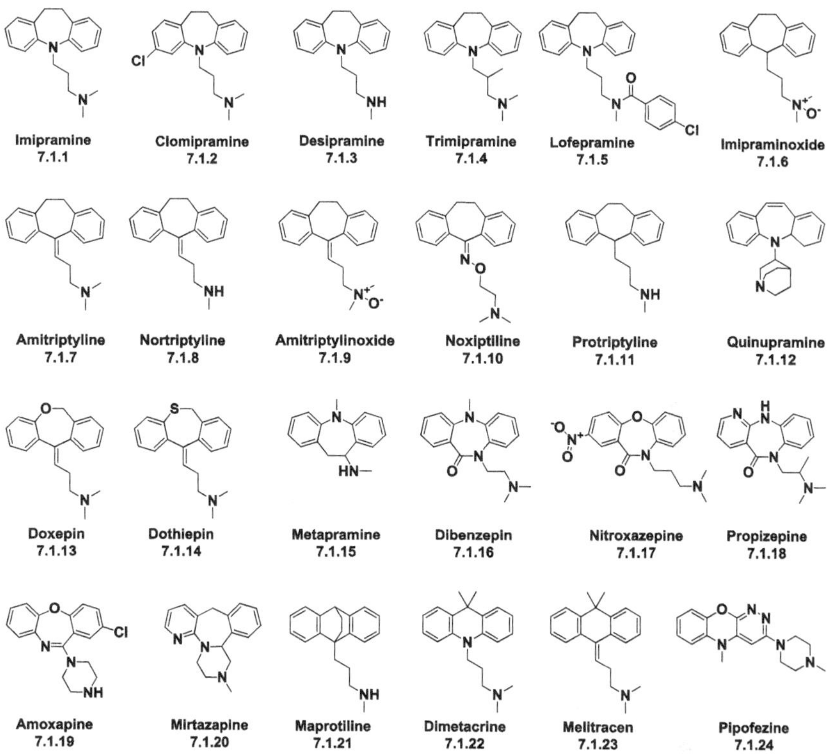

FIG. 7.1 Tricyclic antidepressants.

The synthesis of most of these compounds is described in our earlier book [8].

7.2 MONOAMINE OXIDASE INHIBITORS

MAOIs were the first class of drugs proposed as antidepressants mostly between the years 1957 and 1970. They were and are very effective antidepressants, but have many side effects, such as generation of liver dysfunctions, low blood pressure, sleepiness/insomnia, nervousness, difficulty urinating, and erectile dysfunction, which initiated their withdrawal from the medicinal practice. Monoamine oxidase inhibitors generally have been replaced by other antidepressants, but in certain cases, they are able to relieve depression when other treatments have failed. MAOIs act by inhibiting the activity of monoamine oxidase enzyme family preventing the breakdown of monoamines and thereby increasing their availability. The early MAOIs inhibited monoamine oxidase irreversibly, which meant that they permanently deactivated it, and the enzyme could not function until a mammalian organism replaced it. Some of newer MAOIs (e.g., moclobemide) are reversible and able to detach from the enzymes to facilitate its usual functioning. It's accepted to classify MAOIs to nonselective, MAO-A, and MAO-B inhibitors, considering that MAO-A inhibition reduces the breakdown of primarily serotonin, adrenaline, and dopamine, and MAO-B inhibitors reduce the breakdown of dopamine and phenethylamine. MAOIs approved for use today include phenelzine (**7.2.1**), isocarboxazid (**7.2.2**), selegiline (**7.2.3**), tranylcypromine (**7.2.4**), and moclobemide (**7.2.5**) [9-12] (Fig. 7.2.).

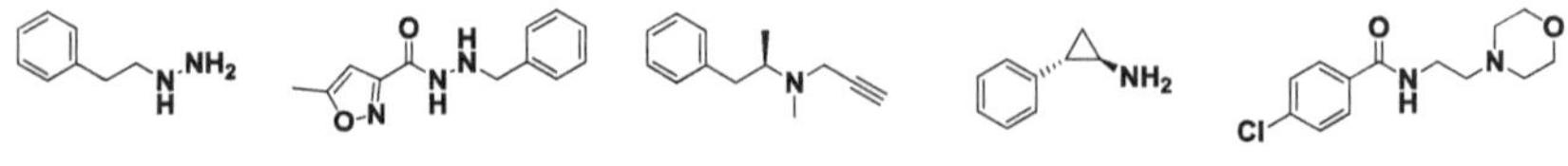

Phenelzine 7.2.1 Isocarboxazid 7.2.2 Selegiline 7.2.3 Tranylcypromine 7.2.4 Moclobemide 7.2.5

FIG. 7.2 Structure of MAOIs.

It is very instructive to glance over structures of compounds that have entered the pharmaceutical market and later have been withdrawn for different reasons, to see how minor changes could change a drug's properties (Fig. 7.3.), as well as to view some of the MAOIs accepted in other countries (Fig. 7.4.).

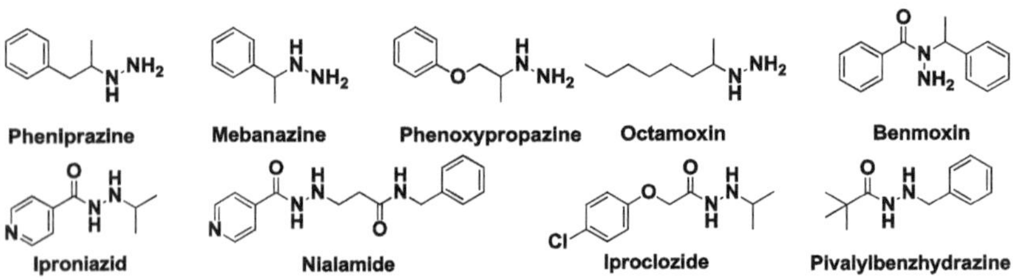

FIG. 7.3 Structure of withdrawn MAOIs.

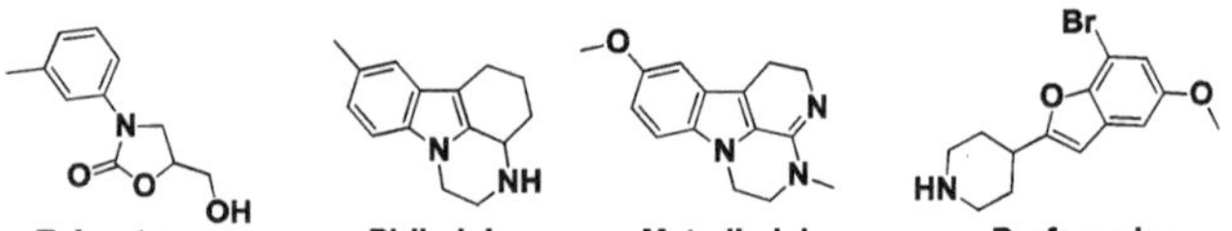

Toloxatone Pirlindole Metralindole Brofaromine

FIG. 7.4 Structure of MAOIs accepted in other countries.

7.3 SELECTIVE SEROTONIN REUPTAKE INHIBITORS

SSRIs are the most prescribed and widely used drugs for the treatment of depression.

It is believed that they act by increasing extracellular serotonin in the brain by blocking its uptake via the serotonin transporter, but the basis of action of these agents is poorly understood. They have better clinical efficacy, good tolerability, and relative safety in comparison to "first-generation antidepressants"— the classic TCAs and MAOIs. SSRIs belong to different chemical classes with wide structural variation. Approved agents in this pharmacological class include fluoxetine (**7.3.6**) (Prozac), paroxetine (**7.3.35**) (Paxil), sertraline (**7.3.53**) (Zoloft), fluvoxamine (**7.3.54**) (Luvox), citalopram (**7.3.60**) (Celexa), escitalopram (Lexapro)

(**7.3.60a**), the active isomer of racemic citalopram, and, the last representative of the series, vilazodone (**7.3.72**). The profile of action of each of these medications differ [13-18] (Fig. 7.5.).

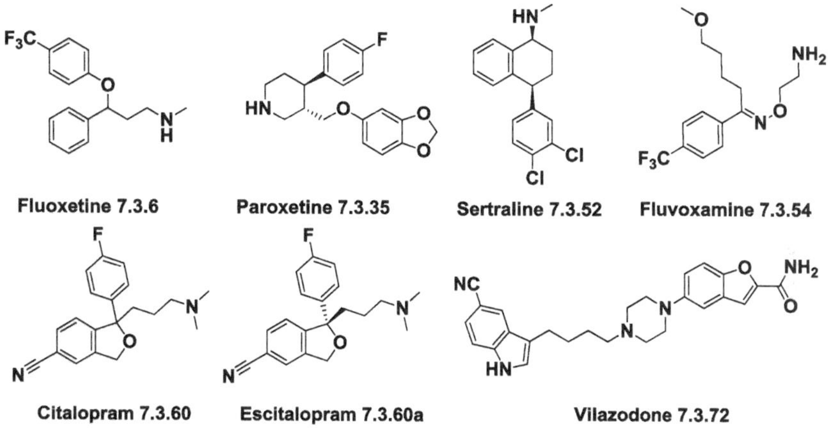

FIG. 7.5 Structure of SSRIs.

Fluoxetine–Prozac

Fluoxetine (**7.3.6**), a potent and specific SSRI, is an antidepressant approved for the treatment of major depressive disorder, obsessive compulsive disorder, panic disorder, bulimia nervosa, and premenstrual dysphoric disorder. Side effects include sexual dysfunction, dry mouth, nausea, headache, diarrhea, nervousness, restlessness, agitation, increased sweating, weight gain, insomnia, and drowsiness.

Fluoxetine (**7.3.6**) has been synthesized by different methods. The first patents [19-21] describe the synthesis starting from a Mannich reaction of acetophenone (**7.3.1**) with paraformaldehyde and methylamine, to produce dimethylaminopropiophenone (**7.3.2**), which is followed by reduction of the carbonyl group with diborane to produce alcohol (**7.3.3**), and nucleophilic substitution of the generated hydroxyl group for chlorine with thionyl chloride, which produced 3-chloro-3-phenylpropan-1-amine (**7.3.4**). The obtained product was treated with sodium 4-(trifluoromethyl)phenolate forming secondary amine (**7.3.5**), which was desmethylated by reaction with BrCN to produce fluoxetine (**7.3.6**). Another patent [22] proposes an industrial-scale process for the synthesis of amine (**7.3.5**) by direct phenylation of the alcohol (**7.3.2**) with 1-chloro-4-(trifluoromethyl)benzene in the presence of NaOH in a dipolar aprotic solvent (Scheme 7.1.).

Fluoxetine also was synthesized starting from benzyloxybenzene (**7.3.7**), which was easily prepared by the etherification of benzyl alcohol with 1-chloro-4-(trifluoromethyl)benzene. The obtained ether (**7.3.7**) was next alkylated with

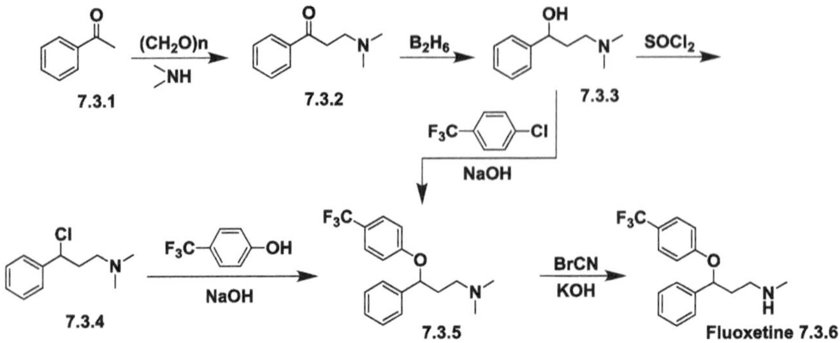

SCHEME 7.1 **Synthesis of fluoxetine.**

allyl bromide to produce 4-phenyl-4-(4-triflurormethylphenoxy)-1-butene (**7.3.8**), the double bond of which was then subjected to ozonolysis to produce aldehyde (**7.3.9**). After the forming of the corresponding methylimine (**7.3.10**) which was reduced with sodium cyanoborohydride to the desired fluoxetine (**7.3.6**) [23] (Scheme 7.2.).

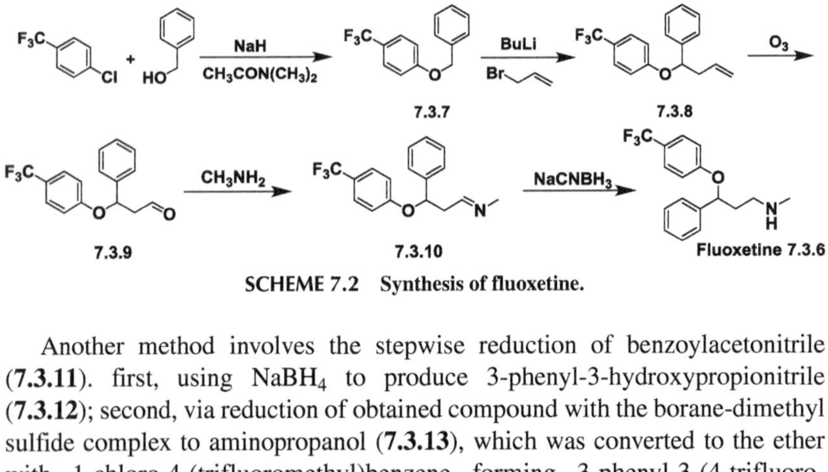

SCHEME 7.2 **Synthesis of fluoxetine.**

Another method involves the stepwise reduction of benzoylacetonitrile (**7.3.11**). first, using $NaBH_4$ to produce 3-phenyl-3-hydroxypropionitrile (**7.3.12**); second, via reduction of obtained compound with the borane-dimethyl sulfide complex to aminopropanol (**7.3.13**), which was converted to the ether with 1-chloro-4-(trifluoromethyl)benzene forming 3-phenyl-3-(4-trifluoromethylphenoxy)propanamine (**7.3.14**). The last was acylated with methyl chloroformate giving the carbamate (**7.3.15**), which was reduced with $LiAlH_4$ to obtain desired fluoxetine (**7.3.6**) [24] (Scheme 7.3.).

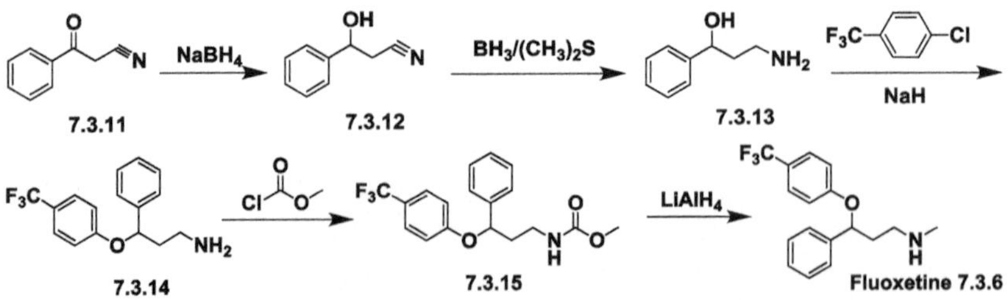

SCHEME 7.3 **Synthesis of fluoxetine.**

Fluoxetine was prepared by another interesting method. According to the patent [25], the 1,3-dipolar cycloaddition of 1-methyl nitrone (**7.3.16**), prepared via reaction of N-methylhydroxylamine with formaldehyde, to styrene produced 2-methyl-5-phenylisoxazolidine (**7.3.17**), which was hydrogenated on Pd/C catalyst to produce N-methyl-3-phenyl-3-hydroxypropylamine (**7.3.18**). The last, in turn, was reacted with 4-chlorobenzotrifluoride to form fluoxetine (**7.3.6**) (Scheme 7.4.).

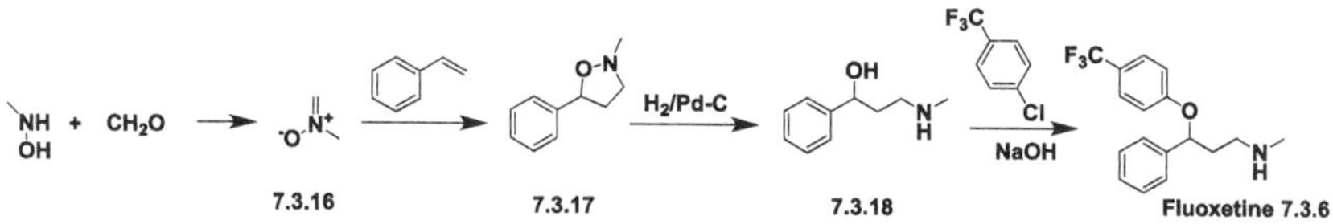

7.3.16 7.3.17 7.3.18 Fluoxetine 7.3.6

SCHEME 7.4 Synthesis of fluoxetine.

Fluoxetine is used therapeutically as a racemic mixture (50/50) of R-fluoxetine and S-fluoxetine enantiomers. In animal models, both enantiomers are potent serotonin uptake inhibitors with essentially equivalent pharmacologic activity.

The resolution of enantiomers of fluoxetine was performed by fractional crystallization of the D- and L-mandelic acid salts.

One of the first attempts to synthesize the two enantiomeric forms of fluoxetine was performed starting from (S)-(-)-3-chloro-1-phenylpropanol (**7.3.19**), which was transformed to the iodo derivative via a Finkelstein reaction. Obtained iodide (**7.3.20**) was reacted with aqueous methylamine to produce (S)-3-(methylamino)-l-phenylpropanol (**7.3.21**), which, when reacted with 1-chloro-4-(trifluoromethyl)benzene, produced (S)-fluoxetine (**7.3.6a**) pharmacological properties of which have been compared with the (R)-fluoxetine (**7.3.6b**) obtained via resolution of racemic fluoxetine and which showed that "there is little enantiospecificity regarding interactions of fluoxetine with the serotonin uptake carrier" [26] (Scheme 7.5.).

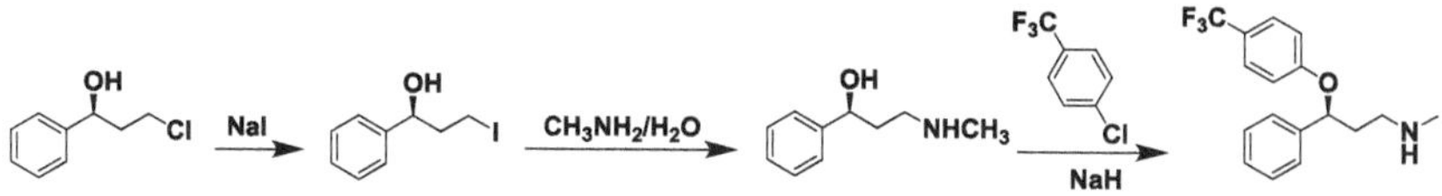

7.3.19 7.3.20 7.3.21 (S)-Fluoxetine 7.3.6a

SCHEME 7.5 Synthesis of (S)-fluoxetine.

Another paper proposes the same reaction sequence except for the source of (S)-(-)-3-chloro-1-phenylpropanol (**7.3.21**). An efficient synthetic route for its synthesis uses a stereospecific reduction of β-chloropropiophenon with BH_3 as reductant and (S)- or (R)-oxazoborolidine as catalyst [27].

The enantiomers of both compounds (S)- (**7.3.21a**) and (R)- (**7.3.21b**) have been synthesized starting with catalytic asymmetric epoxidation of cinnamyl alcohol (**7.3.22**) implementing Sharpless epoxidation reaction (winner of the 2001 Nobel Prize in chemistry). The oxidizing agent is t-butyl hydroperoxide. The stereochemistry of the resulting epoxides is determined by the catalyst formed from the chiral (L)-(+)- or (D)-(-)-dialkyl tartrate diesters and titanium tetra(isopropoxide). The obtained (+)-(2R,3R)-epoxy- and (-)-(2S,3S)-epoxy-alcohols (**7.3.23a,b**) correspondingly underwent regioselective reduction with sodium bis(2-methoxyethoxy)aluminum hydride (Red-Al) to give diols (**7.3.21a,b**), which were mesylated with methanesulfonyl chlorides. Treatment of obtained mesylates (**7.3.24a,b**) with methylamine in aqueous tetrahydrofuran afforded aminoalcohols (**7.3.25a,b**). Alkoxides of the last were prepared with the use of sodium hydride in dimethylacetamide on reaction with 1-chloro-4-(trifluoromethyl) benzene gave desired enantiomers of fluoxetine (**7.3.6a,b**) [28] (Scheme 7.6.).

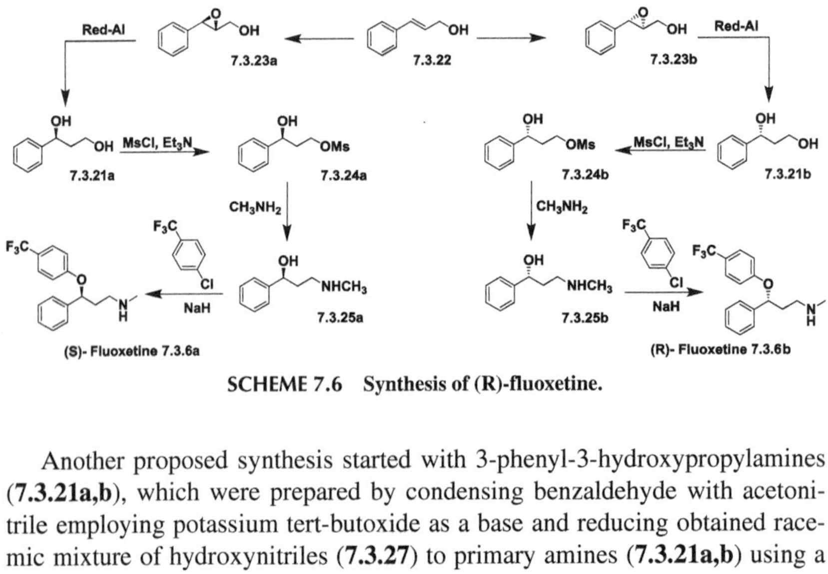

SCHEME 7.6 **Synthesis of (R)-fluoxetine.**

Another proposed synthesis started with 3-phenyl-3-hydroxypropylamines (**7.3.21a,b**), which were prepared by condensing benzaldehyde with acetonitrile employing potassium tert-butoxide as a base and reducing obtained racemic mixture of hydroxynitriles (**7.3.27**) to primary amines (**7.3.21a,b**) using a variety of reductants. Enantiomerically pure amino alcohols (**7.3.21a,b**) were obtained by classical resolution with (S)-(+)- and (R)-(-)-mandelic acid. For the further N-methylation enantiomeric amino alcohol, for example (**7.3.21a**), was acylated with methylchloroformate to form carbamate (**7.3.15a**) that was reduced to amino alcohol (**7.3.25a**), which was transformed to desired enantiomer of fluoxetine (**7.3.6a**), for example [29] (Scheme 7.7.).

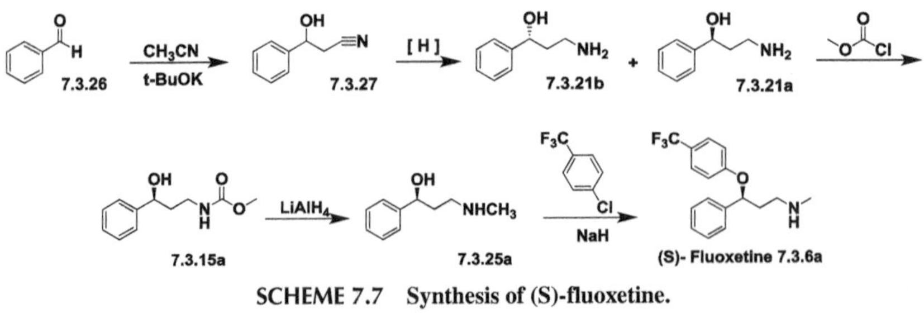

SCHEME 7.7 **Synthesis of (S)-fluoxetine.**

The history of the discovery of Prozac and its properties as a SSRI and as an antidepressant drug are discussed in Wong et al. [30], Childers and Rotella [31], Stanford [32], and Wong et al. [33].

Paroxetine–Paxil

Paroxetine is another specific serotonin reuptake inhibitor that is used to treat depression, obsessive-compulsive disorder, panic disorder, social anxiety disorder, premenstrual dysphoric disorder, generalized anxiety disorder, and post-traumatic stress disorder.

Several strategies for the synthesis of paroxetine have been proposed and are nicely described and analyzed in a recent review [34]. The relatively large-scale synthesis is based on a method for the preparation of the key starting material—the (+) enantiomer of the N-benzyl trans-4-(4-fluorophenyl)-3-piperidinemethanol (**7.3.32**). It was prepared starting from 1-benzyl-4-piperidone (**7.3.28**), which on Grignard reaction with 4-fluorophenyl magnesium bromide produced tertiary alcohol (**7.3.29**) dehydration of which, promoted by p-toluenesulfonic acid, produced tetrahydropyridine derivative (**7.3.30**). The Prins reaction of (**7.3.30**) produced the racemic mixture of tetrahydropyridine-3-methanol, which was resolved with (-)-L-dibenzoyltartaric acid to produce the (-) (**7.3.31**).The stereoselective reduction of (-) (**7.3.31**) on Pd/C catalyst under acidic conditions in water and the retention of the N-benzyl protective group and led to cis- (3R,4R isomer of piperidine-3-methanol (**7.3.32**). The obtained cis-alcohol (**7.3.32**) was converted into its cis-mesylate (**7.3.33**) with methanesulfonyl chloride. The reaction of the methanesulfonyl chloride with sodium salt of sesamol (3,4-methylenedioxyphenol) resulted in the formation of trans-N-benzylparoxetine (**7.3.34**), which was debenzylated on hydrogenation by Pd/C catalyst to produce the desired paroxetine (**7.3.35**) [35] (Scheme 7.8.).

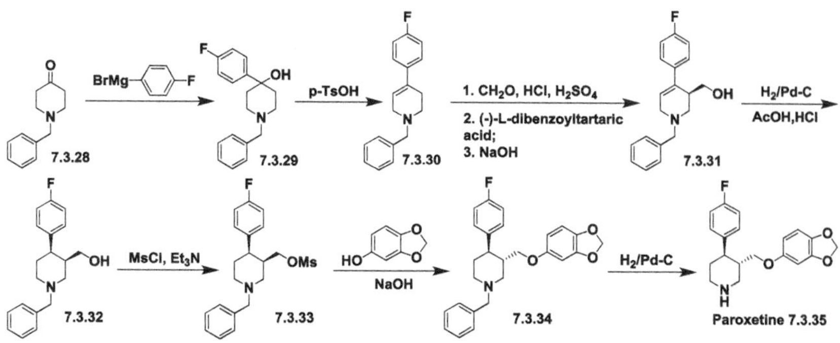

SCHEME 7.8 Synthesis of paroxetine.

Another method for the synthesis of paroxetine is based on a conjugate addition reaction of 4-fluorophenylmagnesium bromide to arecoline. This addition reaction produced the racemic esters (**7.3.37**) (+/-)-cis and (+/-)-trans. In addition, it was found that the cis form could be converted to a thermodynamically more stable trans form on heating with a strong base [36]. The obtained trans form of 1-methylpiperidine-3-carboxylate was hydrolyzed to acid (**7.3.38**) with concentrated hydrochloric acid and then converted to the corresponding acid chloride (**7.3.39**). The acid chloride was esterified with (-)-menthol and pure (-)-trans-(3S,4R) menthol ester (**7.3.40**) was separated as hydrobromide. The base liberated from hydrobromide was reduced with lithium aluminum hydride to give (-)-(3S,4R)-piperidin-3-yl-methanol (**7.3.41**), which was transferred to the (3S,4R)-3-(chloromethyl)-piperidine (**7.3.42**) which reaction with sodium sesamolate resulted in formation of (3S,4R)-3-(1,3-benzdioxolyl-3-oxymethyl)-piperidine. The last was desmethylated with phenylchloroformate to produce the desired paroxetine (**7.3.35**) [36] (Scheme 7.9.).

Very similar variations of proposed method based on the 4-arylpiperidine-3-carboxylates synthesis from arecoline are published [37-41].

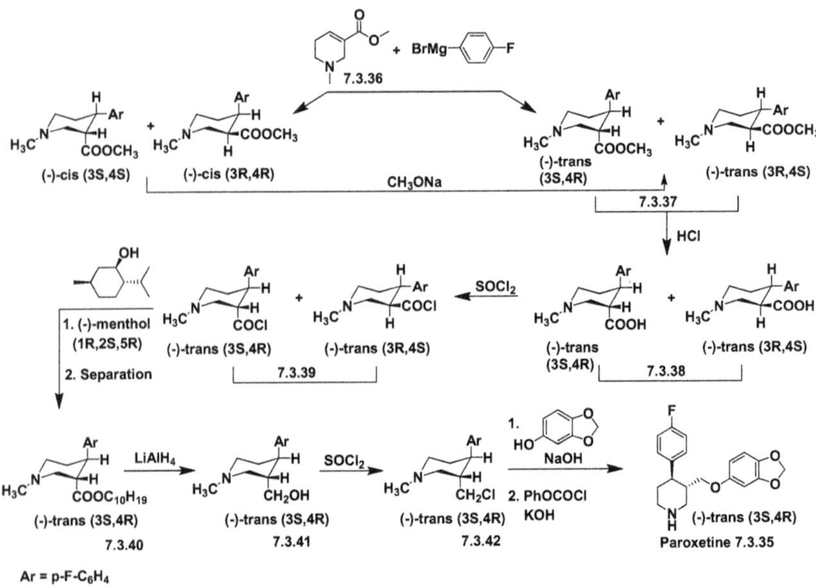

SCHEME 7.9 Synthesis of paroxetine.

Several other methods for the synthesis of paroxetine have been proposed. They generally concluded with preparation of ((3S,4R)-4-phenylpiperidin-3-yl) methanols from different starting materials. They are partly represented in Fig. 7.6 [34,42-45].

Paroxetine is effective for the psychic symptoms of anxiety, which include worry, tension, irritability, and concentration difficulties, and carry tolerable and safe side-effect profiles [46,47].

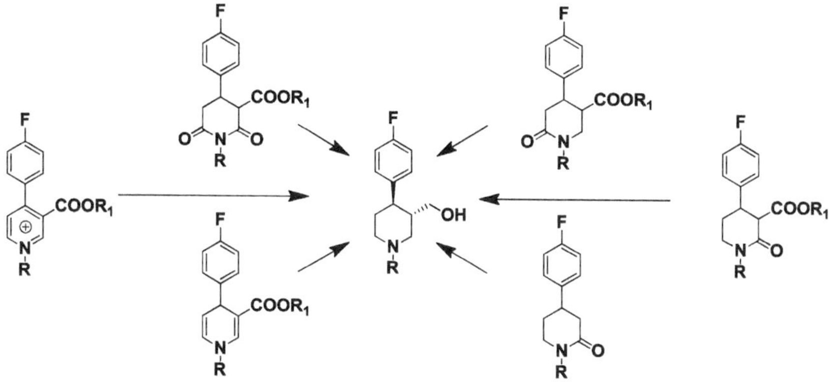

FIG. 7.6 **Preparation of ((3S,4R)-4-phenylpiperidin-3-yl)methanols from different starting materials.**

Sertraline–Zoloft

Sertraline is primarily used to treat depression. It can also be used to assist with the treatment of panic attacks, obsessive-compulsive disorder, posttraumatic stress disorder, and social anxiety disorder. It may also be prescribed to treat premenstrual dysphoric disorder, bloating, and breast tenderness. There is a high risk of side effects compared to other medications.

The original synthesis of sertraline (**7.3.52**) [48,49] started with a five-step (**7.3.43** → **7.3.50**) synthesis of tetralone (**7.3.49**), which later was replaced by a one-step procedure—direct condensation of α-naphthol (**7.3.48**) with o-dichlorobenzene in the presence of $AlCl_3$ [50]. The obtained tetralone (**7.3.49**) was condensed with methylamine in the presence of titanium chloride to form the imine (**7.3.50**). A 1:1 ratio of the rac-cis and rac-trans amines (**7.3.51**) resulted from the reduction of the imine double bond. In addition, the D-mandelic acid resolution at the end of the synthesis produced sertraline (**7.3.52**) (Scheme 7.10.) Many other schemes of synthesis of sertraline are proposed [51-61]; they are reviewed by Bhanja and Jena [62]. Reviews of pharmacology and clinical efficacy also are available [63-71].

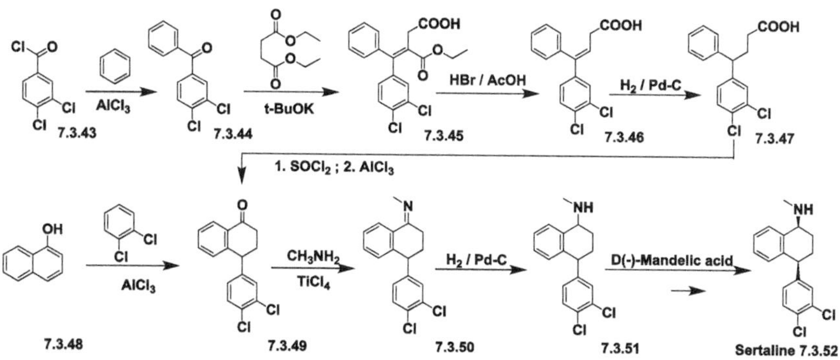

SCHEME 7.10 **Synthesis of sertraline.**

Fluvoxamine–Luvox

Fluvoxamine is used in the treatment of obsessive-compulsive disorder, social anxiety disorder, and major depression. Common side effects may include nausea, diarrhea, loss of appetite, increased sweating, mild skin rash, dizziness, drowsiness, weakness, anxiety, insomnia, dry mouth, heavy menstrual periods, muscle pain, or decreased sex drive.

Synthesis of fluvoxamine (**7.3.54**) is very simple and starts from 5-methoxy-4′-trifluoromethylvalerophenone (**7.3.53**), which is condensed with 2-aminooxyethylamine to produce the desired oxime fluvoxamine [72,73]. Another method [73,74] employs condensation of (**7.3.53**) with hydroxylamine, treating the obtained product with alkali and then with 2-chloroethamine (Scheme 7.11.)

Many reviews on the therapeutic potential of fluvoxamine are available [75-82].

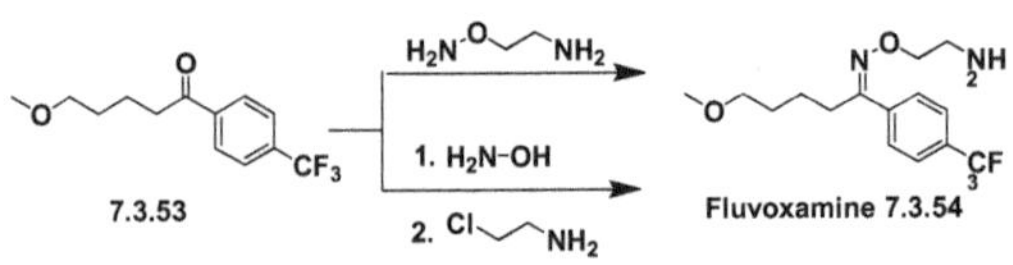

SCHEME 7.11 Synthesis of fluvoxamine.

Citalopram–Celexa and Escitalopram–Lexapro

Citalopram is the one of most prescribed antidepressants in the world. It is "first-line medication" for treating depression. It is also prescribed to treat anxiety panic disorders, panic attacks, and obsessive-compulsive disorder, and the behavioral disturbances of dementia.

Common side effects are numerous and affect the GI, integumentary, respiratory, cardiovascular, genitourinary, and central nervous systems. Life-threatening adverse effects include convulsions, hemorrhage, first-degree atrioventricular block, and myocardial infarction, angina pectoris, tachycardia, bradycardia, thrombophlebitis. It could cause hallucinations, psychosis, asthma, bronchitis, hypertension, cystitis, urine retention, or vomiting.

Several approaches have been reported for the preparation of citalopram [83-99] and escitalopram [99-102].

Synthesis of citalopram is proposed to occur by two general routes, usually called the anion strategy and the double Grignard strategy (Scheme 7.12.).

In both cases, synthesis is started from a reaction of phthalide (**7.3.56**) with p-fluorophenyl magnesium bromide giving a 2-hydroxymethyl benzophenone (**7.3.57**). In the anion strategy, the carbonyl group of (**7.3.56**) is reduced, producing diol (**7.3.57**), the ring closure of which occurred by using an acid to produce (**7.3.58**). Treatment of (**7.3.58**) with a strong base affords anion, which is alkylated with a N,N-dimethylaminopropyl chloride to produce the desired citalopram (**7.3.60**).

In the alternative double Grignard route, 2-hydroxymethyl benzophenone (**7.3.56**) in situ is directly treated with the Grignard reagent derived from (3-chloropropyl)-dimethylamine to produce diols (**7.3.61**), which is followed by acid to promote ring closure. In commercial production of citalopram the double Grignard strategy is applied. Chiral resolution of the diols (**7.3.61**) with (+)- and (-)-di-p-toluoyl-D-tartaric acid gave diols (**7.3.61a,b**), which separately underwent further ring closure under methanethiosulfonyl chloride/ triethylamine conditions to produce (S)-citalopram (**7.3.60a**) or escitalopram and (R)-citalopram (**7.3.60b**). Escitalopram inhibits serotonin uptake by two orders of magnitude better than the other enantiomer. There are contradictory data on (R)-citalopram. According to some data (R)-citalopram is inactive or has a significant affinity only for the allosteric site of the transporter [103].

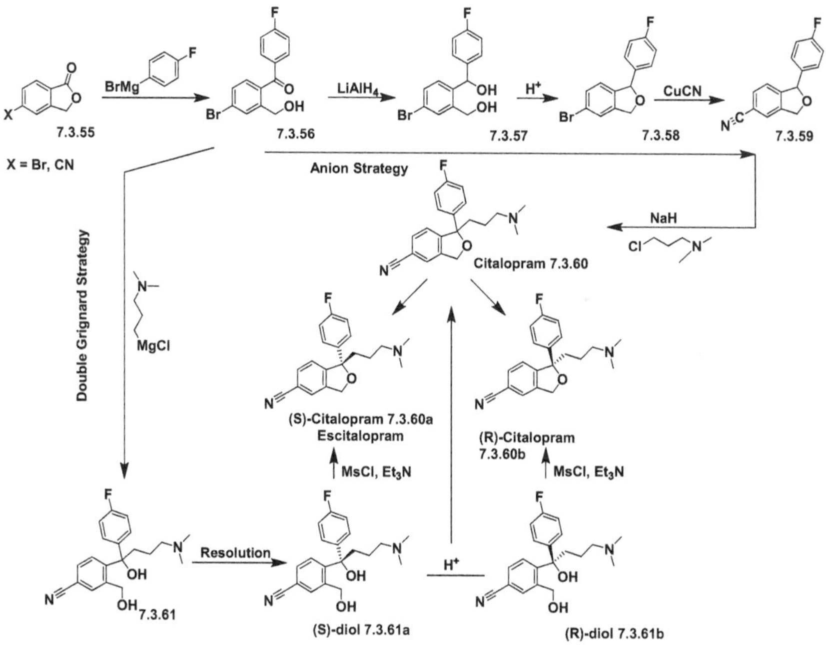

SCHEME 7.12 **Synthesis of citalopram and escitalopram.**

Many reviews on the use and management of citalopram [104-112] and escitalopram [113-119] have been published.

Vilazodone–Viibryd

Vilazodone was recently approved for the treatment of major depressive disorder in adults. It is a dual-acting serotonergic agent that combines the antidepressant effects of a SSRI with partial serotonin $(5\text{-HT})_{1A}$-receptor agonist activity.

Originally, vilazodone (**7.3.72**) was prepared starting from of 5-cyanoindole (**7.3.62**) acylation which, with 4-chlorobutyryl chloride (**7.3.63**) and in the presence of isobutyl-$AlCl_2$, led to the chlorobutanoyl indole (**7.3.64b**) alongside

(**7.3.64a**). After selective desoxygenation of the keto function with sodium bis(2-methoxyethoxy)aluminum hydride (Red-Al)), the building block (**7.3.65**), in moderate yield, was obtained (Scheme 7.13.).

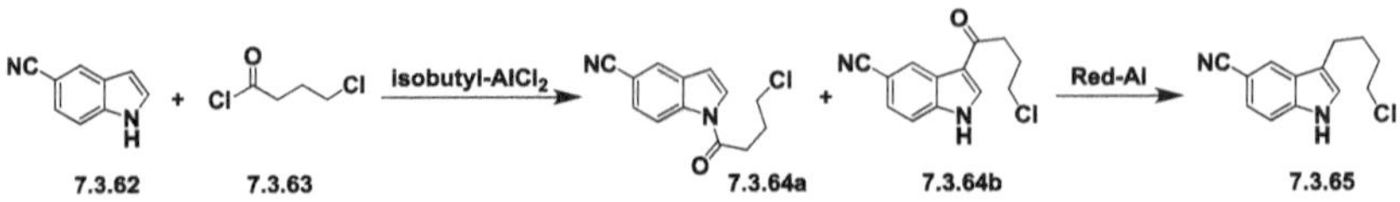

SCHEME 7.13 **Synthesis of 3-(4-chlorobutyl)-1H-indole-5-carbonitrile (7.3.65).**

The second block—ethyl 5-(piperazin-1-yl)benzofuran-2-carboxylate (**7.3.69**)—was synthesized starting from ethyl 5-nitrobenzofuran-2-carboxylate (**7.3.66**), the nitro group of which was hydrogenated with hydrogen on the Pd/C catalyst to the amine (**7.3.67**). The compound (**7.3.69**) was formed by reaction of (**7.3.67**) with bis(2-chloroethyl)ammonium chloride (**7.3.68**) (Scheme 7.14.).

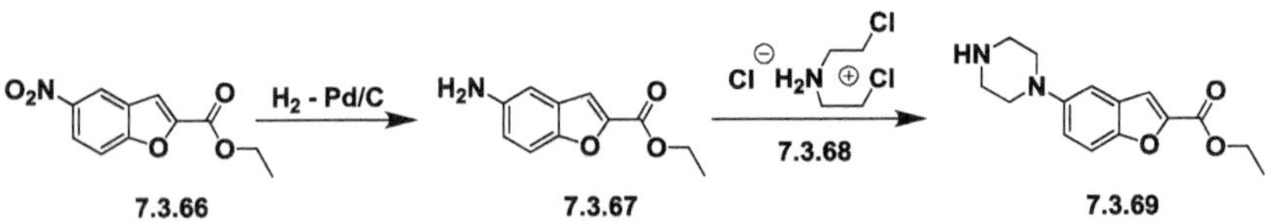

SCHEME 7.14 **Synthesis of ethyl 5-(piperazin-1-yl)benzofuran-2-carboxylate (7.3.69).**

The reaction of the chloride (**7.3.65**) with the amine (**7.3.69**) yielded product (**7.3.70**). The obtained ester was saponified with potassium hydroxide. The resulting acid (**7.3.71**) was activated with the Mukaiyama reagent (2-chloro-1-methylpyridinium iodide) and transferred into the corresponding amide—the desired vilazodone (**7.3.72**)—with gaseous ammonia [120] (Scheme 7.15.).

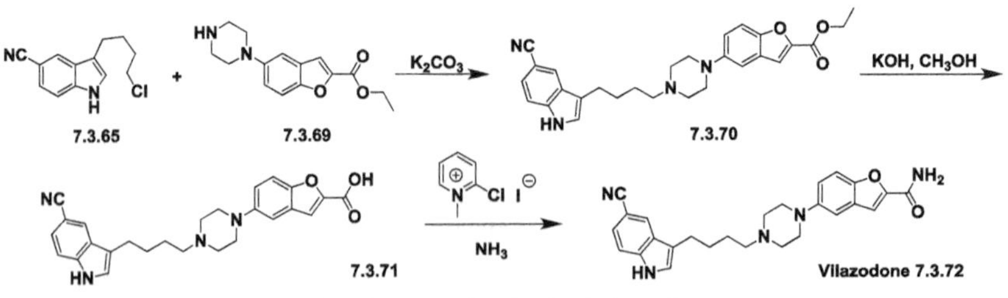

SCHEME 7.15 **Synthesis of vilazodone.**

A more efficient synthesis route, starting from N-protected 5-cyanoindole-1-tosyl-1H-indole-5-carbonitrile, implementation of selective deoxygenation in the $NaBH_4/CF_3COOH$ system instead of (Red-Al) reduction, and other improvements, enabled the proposing of a convenient method for scale-up production of vilazodone [121]. Many patents describe variations of vilazodone synthesis [122-126]. The therapeutic potential of vilazodone is widely discussed in different reviews [127-138].

7.4 SELECTIVE NOREPINEPHRINE (NORADRENALINE) REUPTAKE INHIBITORS

Selective blockers of norepinephrine reuptake inhibitors may be used as antidepressants. Among drugs currently approved for treatment of depression and attention deficit-hyperactivity disorder (ADHD) are some selective norepinephrine reuptake inhibitors such as desipramine (**7.1.3**), maprotiline (**7.1.20**), which is related to TCA series, and atomoxetine (**7.4.1**), reboxetine (**7.4.2**), viloxazine (**7.4.3**), which are structurally close to each other [139,140] (Fig. 7.7.) In Phase III clinical trials is a new norepinephrine reuptake inhibitor, edivoxetine (**7.4.4**).

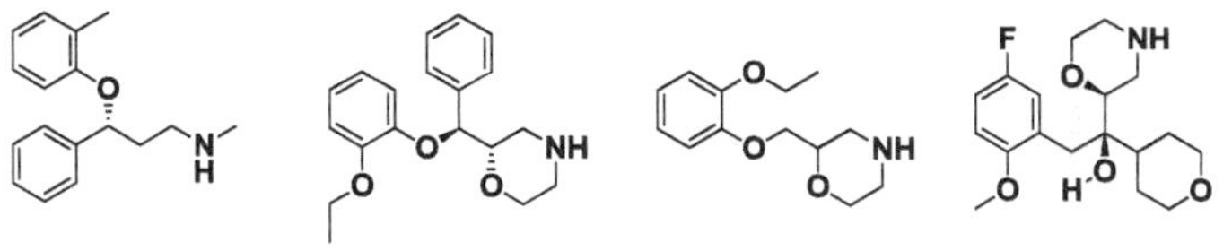

Atomoxetine 7.4.1 **Reboxetine 7.4.2** **Viloxazine 7.4.3** **Edivoxetine 7.4.4**

FIG. 7.7 **Selective norepinephrine reuptake inhibitors.**

Atomoxetine–Strattera

Atomoxetine is a selective norepinephrine reuptake inhibitor that is structurally closely related to fluoxetine. It primarily acts to increase noradrenalin levels; however, at higher doses it can increase dopamine levels. It is commonly used as a treatment for ADHD. Its synthesis is described in the same papers presenting fluoxetine synthesis [28,29] and differs only on O-substituent on 3-amino-1-phenylpropan-1-ol general carrier. According to Gao and Sharpless [28], methanesulfonate (**7.3.24b**) underwent a fairly complex Mitsunobu reaction with o-cresol using triphenyl phosphine and diethyl azodicarboxylate as catalysts, which inverts the stereochemistry at the chiral center and produces aryl ether (**7.4.1a**). Treatment of (**7.4.1a**) with methylamine in water/tetrahydrofuran solution produces the desired atomoxetine (**7.4.2**). In the paper by Koenig and Mitchell [29], alkoxide of (R)-3-amino-1-phenylpropan-1-ol (**7.3.21a**) was treated with o-fluorotoluene to produce (R)-(-)-atomoxetine (**7.4.1**) (Scheme 7.16.).

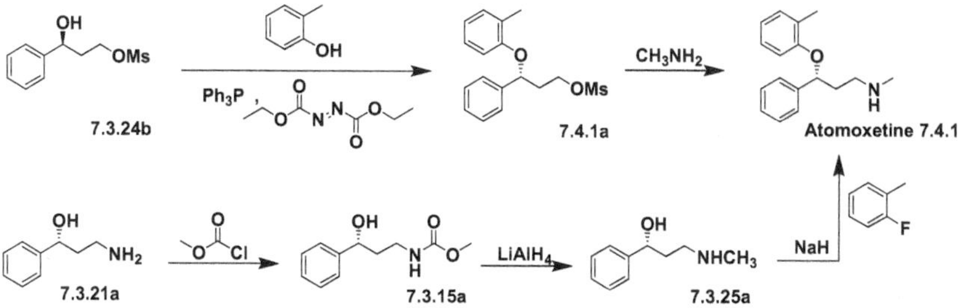

SCHEME 7.16 **Synthesis of atomoxetine.**

Because of the expense and difficulty of the Mitsunobu reaction at large scale, the aromatic displacement route is preferred. But in this method, epimerization of the chiral center was sometimes observed. An excellent review on atomoxetine synthesis is presented in Gray [141]. Modern industrial method of the synthesis of atomoxetine is based on (R)-3-chloro-1-phenylpropanol (**7.4.4**), which is usually obtained by selective reduction of 3-chloropropiophenone (**7.4.3**) with diisopinocampheylchloroborane ($^{d}Ipc_2BC1$) [142], whose chlorine atom was displaced with the dimethylamine moiety to produce (**7.4.5**). On reaction with protected o-fluorobenzaldehydes, such us 1-(dialkoxymethyl)-2-fluorobenzene (**7.4.6**) or 1-(dialkoxymethyl)-2-fluorobenzene (**7.4.7**), the obtained compound gave excellent yield of the desired ether (**7.4.8**) without significant racemization. The formyl protective group was removed by hydrolysis in acid media, and the obtained aldehyde was hydrogenated to benzylic alcohol and converted to chloride (**7.4.9**) with thionyl chloride. The synthesized (R)-3-(2-(chloromethyl)phenoxy)-N,N-dimethyl-3-phenylpropan-1-amine (**7.4.9**) underwent reductive dehalogenation using zinc in acetic acid to produce compound (**7.4.10**), demethylation of which with phenyl chloroformate gave the desired atomoxetine (**7.4.1**) [143] (Scheme 7.17.).

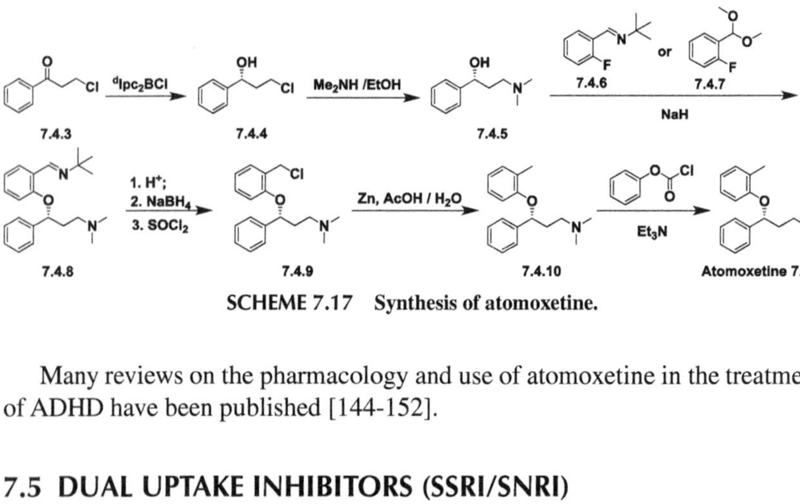

SCHEME 7.17 **Synthesis of atomoxetine.**

Many reviews on the pharmacology and use of atomoxetine in the treatment of ADHD have been published [144-152].

7.5 DUAL UPTAKE INHIBITORS (SSRI/SNRI)

Specific dual serotonin and noradrenaline reuptake inhibitors that increase the concentration in the synaptic cleft of both norepinephrine and serotonin are a relatively recent class of compounds with unique characteristics. The dual reuptake inhibition ensures efficacy that is comparable to the TCAs. The absence of interaction with monoamine oxidase, and with opioid, histaminic, muscarinic, α_1-adrenergic, and GABAergic receptors sharply limits the adverse effects. Currently, four drugs of this class are available: duloxetine (**7.5.5**), venlafaxine (**7.5.34**), desvenlafaxine (**7.5.38**), milnacipran (**7.5.39**). Combination of mechanisms of action in a single active agent is an important development in pharmacology. SNRIs are effective at easing depression symptoms and sometimes are used also to treat other mental health conditions such as anxiety (Fig. 7.8.).

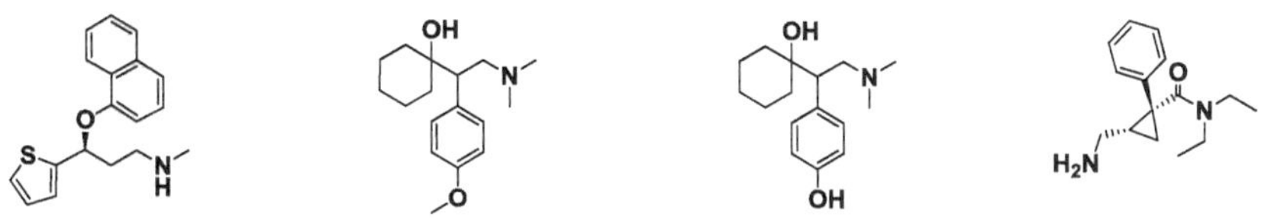

Duloxetine 7.5.5 (Cymbalta) Venlafaxine 7.5.34 (Effexor) Desvenlafaxine 7.5.38 (Pristiq) Milnacipran 7.5.39 (Savella)

FIG. 7.8 Dual serotonin and noradrenaline reuptake inhibitors.

Duloxetine–Cymbalta

Duloxetine originally was prepared by a four-step sequence starting from the readily available 2-acetylthiophene (**7.5.1**) that underwent Mannich aminomethylation, followed by reduction of the carbonyl group, etherification of the obtained alcohol with 1-fluoronaphthalene, and further demethylation of the tertiary amino group. The synthesis of duloxetine was disclosed in patents [153,154], and later preparation in kilogram amounts was discussed in detail [155]. In contrast to the preceding fluoxetine (Prozac), which is approved as a racemate, duloxetine is marketed in enantiomerically pure form. According to the patents [153,154], which differs in small details, enantioselective synthesis of duloxetine starts from 2-acetylthiophene (**7.5.1**), which underwent a Mannich aminomethylation reaction to produce β-aminoketone (**7.5.2**), reduction of which with sodium borohydride gives racemic alcohol (**7.5.3a,b**). Resolution with (S)-(+)-mandelic acid allows the separate (S)-(-)-alcohol (**7.5.3a**) alkylation with I-fluoronaphthalene to produce ether (**7.5.4**). Its N-methyl demethylation with phenyl chloroformate produces the desired product, duloxetine (**7.5.5**).

Reduction of some ketones with the Yamaguchi-Mosher-Poland complex [156], prepared by adding the (+)-(2S,3R)-4-dimethylamino-3-methyl-1,2-diphenyl-2-butanol to lithium aluminum hydride, was implemented using β-aminoketone (**7.5.2**) to produce (**7.5.3a**) with unexpectedly high stereoselectivity [155]. In this paper [156], demethylation of (**7.5.4**) was realized with 2,2,2-trichloroethyl chloroformate instead of phenyl chloroformate. Some improvements of the technical details of this general scheme have been proposed [157] (Scheme 7.18.).

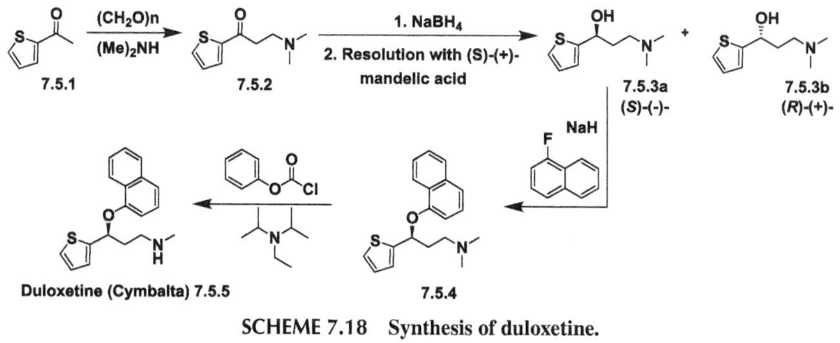

SCHEME 7.18 Synthesis of duloxetine.

Other routes to the synthesis of duloxetine have been developed. One started from 2-(3-chloropropionyl)thiophene (**7.5.6**), which was enantioselectively reduced using, separately, (R-) and (S-) 1-methyl-3,3-diphenyltetrahydropyrrolo[1,2-c][1,3,2]oxazaborole in the presence of borane resulted in the chloroalcohols (**7.5.7a,b**) (S)-(-)-(**7.5.7a**) was then transformed via the iodide to S-3-methylamino-1-thiophen-2-yl-propan-1-ol (**7.5.8**), which in a nucleophilic displacement reaction with 1-fluoronaphthalene to produce duloxetine (**7.5.5**). The isomeric alcohol-(R)-(+)-(**7.5.7b**) was subjected to a Mitsunobu reaction with 1-naphthol, diethylazodicarboxylate and triphenylphosphine and the resulting chloroether (**7.5.9**) was converted to duloxetine (**7.5.5**) by reaction with methylamine in THF [158,159] (Scheme 7.19.).

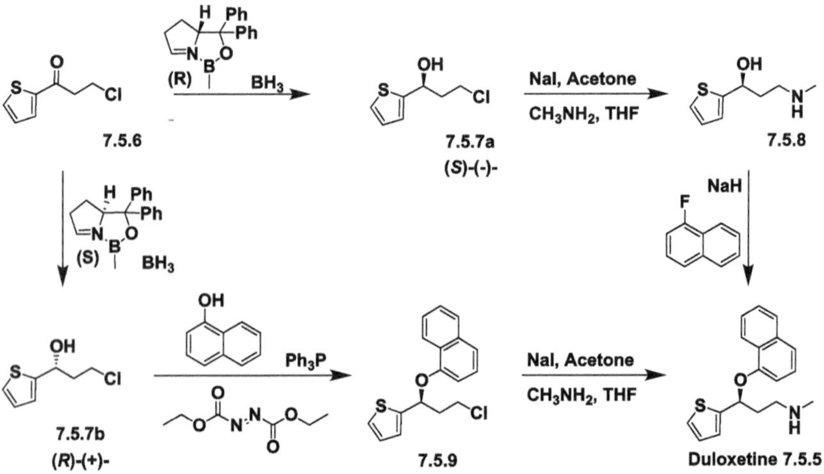

SCHEME 7.19 Synthesis of duloxetine.

Another approach to the synthesis of duloxetine was accomplished using asymmetric hydrogenation of 3-oxo-3-thienyl methylpropanoate (**7.5.14**). The required β-keto ester (**7.5.14**) was prepared in high yield from thiophene-2-carboxylic (**7.5.10**) using Masamune's simple procedure [160], which includes acylation of acids with 1,1′-carbonyldiimidazole (**7.5.11**) to obtain the corresponding imidazol-1-yl ketones (**7.5.12**), which by reacting with the magnesium salt of acetoacetic acid and further acidic workup, produced the desired β-keto ester (**7.5.14**). The β-keto ester was hydrogenated using a chiral (R)-MeO-BiphepRuBr$_2$ catalyst, affording the corresponding β-hydroxy ester (S)- (**7.5.15**) in quantitative yield and 90% enantiomeric excess. Lithium aluminium hydride reduction of the prepared methyl (S)-3-hydroxy-3-(thiophen-2-yl)propanoate (**7.5.15**) led to the diol (**7.5.16**), which was selectively mesylated at the primary position, yielding the monomesylate (**7.5.17**). Treatment of the (**7.5.17**) with methylamine in H_2O/THF mixture afforded the known amine (**7.4.16**), which was O-alkylated with fluoronaphthalene using sodium hydride to give duloxetine (**7.5.14**) [161] (Scheme 7.20.).

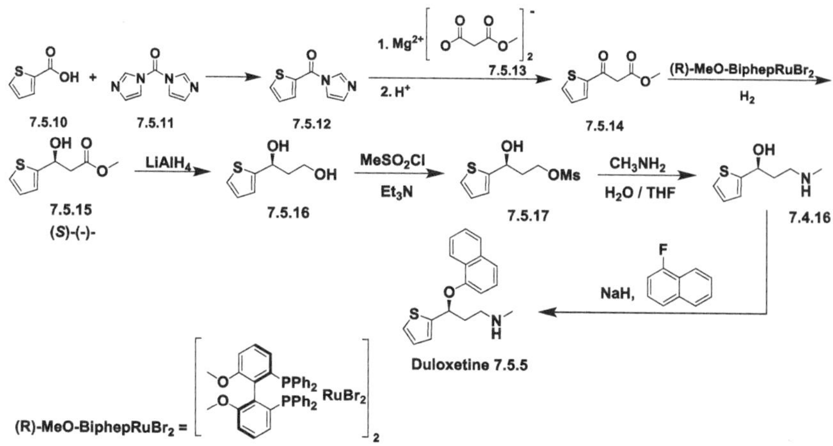

SCHEME 7.20 Synthesis of duloxetine using asymmetric hydrogenation.

A similar approach was applied after the discovery of direct catalytic aldol reaction of thioamides in presence of a soft Lewis acid/hard Bronsted base cooperative catalyst comprising $[Cu(CH_3CN)_4]PF_6$, (S,S)-Ph-BPE, and $Li(OC_6H_4$-p-OMe). This method offers a new entry to the concise enantioselective synthesis of duloxetine. The synthesis started from thenaldehyde (**7.5.18**), which on reaction with N,N-diallylethanethioamide (**7.5.19**) gave aldol product (**7.5.20**) in 92% enantiomeric excess, which after $LiAlH_4$ reduction gives allyl protected aminopropanol (**7.5.21**). The aminopropanol was deprotected using tetrakis(triphenylphosphine)palladium(0) and N,N-dimethylbarbituric acid as an allyl group scavenger to give the amine, which was acylated in situ with methyl chloroformate in presence of potassium carbonate to produce carbamate (**7.5.22**). The carbamate was hydrogenated to aminopropanol (**7.4.16**) and finally converted to duloxetine (**7.5.5**) by the reaction with fluoronaphthalene in the presence of sodium hydride [162] (Scheme 7.21.).

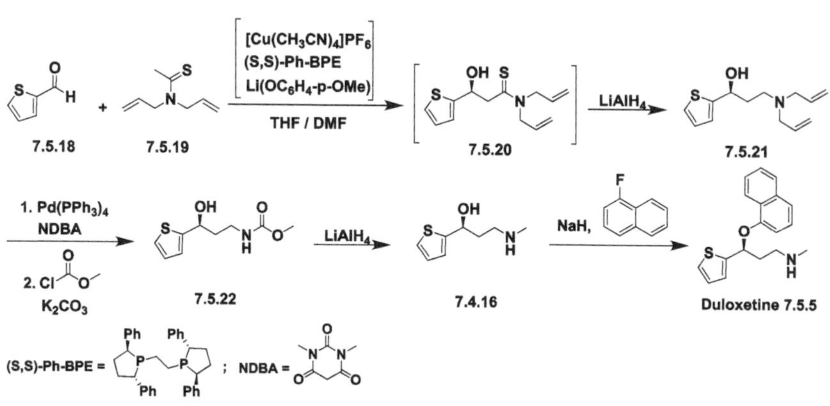

SCHEME 7.21 Synthesis of duloxetine.

Recently, a highly enantioselective Friedel-Crafts reaction of variously substituted thiophenes with glyoxylates using a catalytic amount of an easy accessible 6,6-dibromo-BINOL/Ti(IV) complex and yielding hydroxy(thiophene-2-yl) acetates was published [163]. Application of the proposed reaction to tert-butyl (S)-2-hydroxy-2-(thiophen-2-yl)acetate (**7.5.25**) was used for the formal synthesis of duloxetine according to the protocol proposed earlier [164].

Reaction of thiophene (**7.5.23**) with n-butyl glyoxylate (**7.5.24**) and 1 mol % of mentioned catalyst provides α-hydroxyacid (**7.5.25**) with a moderate yield and 92% enantioselectivity. The obtained product was reduced to the diol (**7.5.26**), the primary hydroxy group in which it was protected with the tosyl group to produce (**7.5.27**). The tosyl group in which was substituted with cyanide. The obtained nitrile (**7.5.28**) was then transformed to the respective 3-amino-1-(2-thienyl)-1-propanol derivative (**7.5.29**) employing $BH_3{\cdot}Me_2S$ as a reductant and converting the obtained amino alcohol in situ to the corresponding ethyl carbamate (**7.5.29**) upon treatment with ethyl chloroformate in aqueous K_2CO_3. The last was subjected to $LiAlH_4$ reduction to produce the N-monomethylated amino alcohol (**7.4.16**). O-Arylation of the last was carried out in usual way using 1-fluoronaphthalene employing NaH as the base to afford duloxetine (**7.5.5**) (Scheme 7.22.).

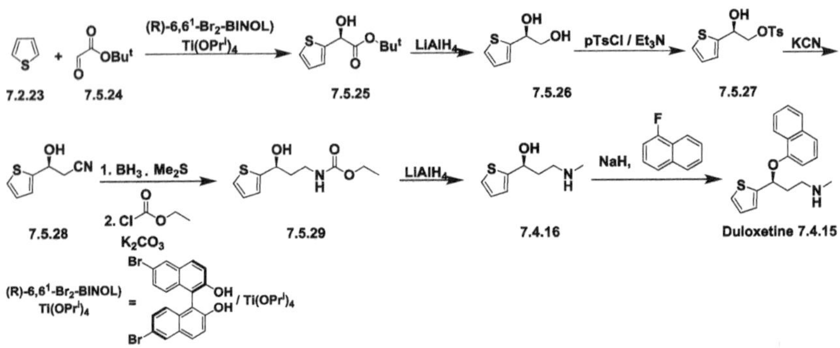

SCHEME 7.22 **Synthesis of duloxetine.**

Reviews on the synthesis of duloxetine have been published [165-168].

Duloxetine is used to treat depression and anxiety. It is also used for neuropathic pain caused by nerve damage associated with diabetes as well as to treat fibromyalgia and chronic pain that is related to muscles and bones. Duloxetine was approved for use of stress urinary incontinence. Multiple informative reviews on pharmacology of duloxetine have been published [169-188].

Venlafaxine–Effexor

Venlafaxine (**7.5.34**) was first synthesized by the nucleophilic addition of 4-methoxyphenyl acetonitrile (**7.5.30**) with cyclohexanone (**7.5.31**) using lithium diisopropylamide or butyl lithium to afford (RS)-1-[cyano-(4methoxyphenyl)methyl] cyclohexanol (**7.5.32**), catalytic hydrogenation of which over rhodium on alumina gave (RS)-1-[2-amino-1-(4-methoxypheny) ethyl]-cyclohexanol (**7.5.33**). Then (**7.5.33**) was dimethylated employing reductive amination Eschweiler-Clarke procedure to get desired venlafaxine (**7.5.34**), with overall yield of approximately 25% [189,190] (Scheme 7.23.).

This procedure was later improved by excluding industrially unattractive pyrophoric reagents such as n-BuLi and LDA and replacing them with sodium methoxide, and replacing the expensive Rh/Al_2O_3 catalyst for Raney Ni, simplifying the reductive amination process and scalable process for the production of venlafaxine to produce an overall yield of 55% and a highly pure state [191]. Another method proposed the use of sodium hydroxide instead of sodium methoxide [192].

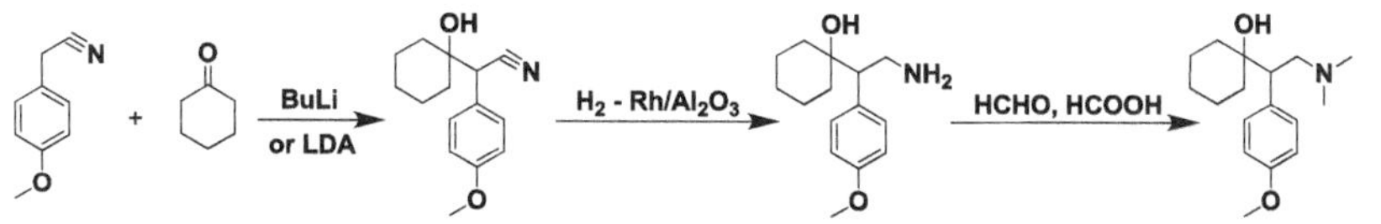

7.5.30 **7.5.31** **7.5.32** **7.5.33** **Venlafaxine 7.5.34**

LDA = $[(CH_3)_2CH]_2NLi$.

SCHEME 7.23 **Synthesis of venlafaxine.**

The reaction of 2-(4-methoxyphenyl)acetic acid ethyl ester (**7.5.37**) with paraformaldehyde, in the presence of K_2CO_3 and tetrabutylammonium iodide produces acrylate (**7.5.36**), which upon treatment with solution of Me_2NH in the presence of a catalytic amount of $FeCl_3$ rendered β-aminoester (**7.5.37**). The reaction of the obtained β-aminoester with the Grignard reagent prepared from 1,5-dibromopentane in THF, afforded venlafaxine (**7.5.34**) a 50% yield [193] (Scheme 7.24.).

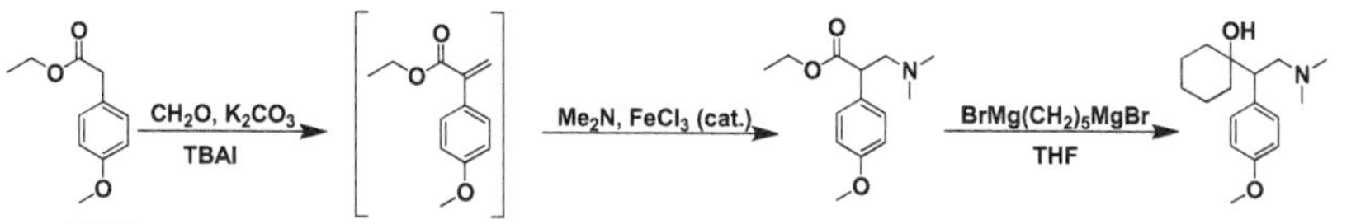

7.5.35 7.5.36 7.5.37 Venlafaxine 7.5.34

SCHEME 7.24 **Synthesis of venlafaxine.**

Other improved methods for the synthesis of venlafaxine have been proposed [194,195]. On the assumption that both enantiomers of venlafaxine have a role in its antidepressant activity (it had been hypothesized that (+)-enantiomer inhibits serotonin reuptake and the (-)-enantiomer inhibits norepinephrine reuptake), asymmetric synthesis of both the enantiomers has been proposed [196,197]. Multiple reviews on pharmacology of venlafaxine have been published [198-207].

7.6 DUAL UPTAKE INHIBITORS (SNRI/SDARI)

Bupropion (Wellbutrin) is the single approved antidepressant that acts as a dopamine and norepinephrine reuptake inhibitor. The synthesis of bupropion (**7.6.4**) starts with the reaction of 3-chlorobenzonitrile (**7.6.1**), with ethylmagnesium bromide to give 3-chloropropiophenone (**7.6.2**). Direct bromination produces 3α-bromopropiophenone (**7.6.3**), which on reaction with *tert*-butylamine produces the desired bupropion (**7.6.4**) [208-213] (Scheme 7.25.). The synthesis of the enantiomers of bupropion is described. The enantiomers were compared with the racemate. No significant differences were found in their potencies [214]. Many bupropion analogues were synthesized [215].

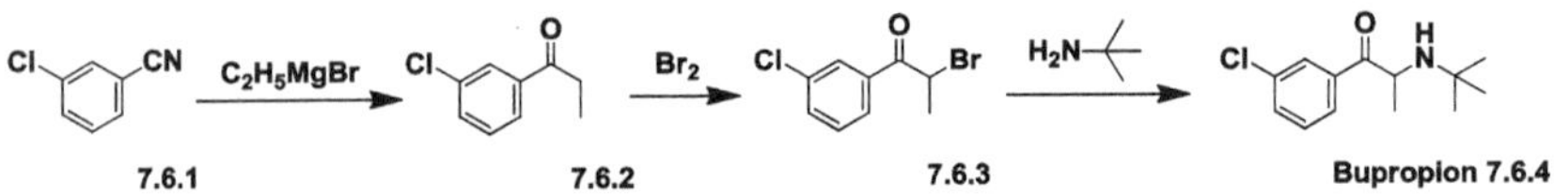

SCHEME 7.25 **Dual dopamine and norepinephrine reuptake inhibitor bupropion.**

Bupropion is used to treat attention deficit disorder, bipolar depression, chronic fatigue syndrome, nicotine addiction, cocaine addiction, and lower back pain. It causes seizures at high doses. The pharmacology of bupropion is widely reviewed [216-224].

7.7 TRIPLE UPTAKE INHIBITORS (SSRI/SNRI/SDARI)

There is a new series of psychoactive medications that work by inhibiting the neuronal reuptake of serotonin, norepinephrine, and dopamine [225-230]. However, there are no triple monoamine reuptake inhibitors available on the pharmaceutical market. Preclinical studies and clinical trials indicate that a drug inhibiting the uptake of all three of these neurotransmitters could produce a more rapid onset of action and possess greater efficacy than traditional antidepressants and could be the next generation of drugs for the treatment of depression. Pharma has developed different chemical classes of compounds as possible triple reuptake inhibitors; they are presented in Fig. 7.9. Among them are derivatives of substituted phenylpyrrolidines (**7.7.1**), 1-phenyl-3-azabicyclo[3.1.0]-hexanes (**7.7.2**), (1-phenylcyclobutyl)-alkanamines (**7.7.3**), 4-phenylpiperidines (**7.7.4**), 8-azabicyclo[3.2.1]octanes (**7.7.5**), substituted tetrahydroisoquinolines (**7.7.6**). Amitifadine (**7.1.1**), also known as DOV 21,947 or EB-1010, and is currently in clinical trials [229,230].

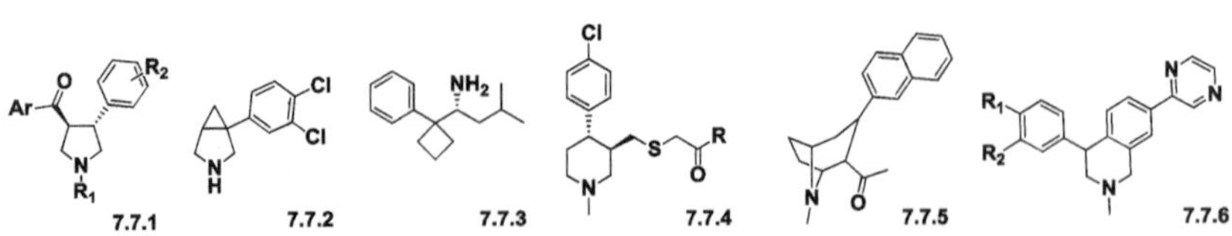

FIG. 7.9 **Structure of triple uptake inhibitors (SSRI/SNRI/SDARI).**

REFERENCES

1. Casacalenda, N.; Boulenger, J.-P. Pharmacologic treatments effective in both generalized anxiety disorder and major depressive disorder: clinical and theoretical implications. *Can. Psychiatr. Assoc. J.* **1998,** *43,* 722–730.
2. Holenz, J.; Diaz, J. L.; Buschmann, H. Chemistry. Marketed drugs and drugs in development. In Buschmann, H., Holenz, J., Parraga, A., Torrens, A., Vela, J. M., Diaz, J. L., Eds.; *Antidepressants, Antipsychotics, Anxiolytics: From Chemistry and Pharmacology to Clinical Application, 2*; Wiley-VCH, 2007; pp 1183–1196.
3. Pacher, P.; Kohegyi, E.; Kecskemeti, V.; Furst, S. Current trends in the development of new antidepressants. *Curr. Med. Chem.* **2001,** *8* (2), 89–100.
4. Ravindran, L. N.; Stein, M. B. The pharmacologic treatment of anxiety disorders: a review of progress. *J. Clin. Psychiatry (Memphis, TN, U. S.)* **2010,** *71* (7), 839–854.
5. Schubert-Zsilavecz, M.; Stark, H. Modern antidepressant medicinal chemistry-targets and drugs. *Pharm. Unserer Zeit* **2004,** *33* (4), 282–287.
6. Zettl, H.; Schubert-Zsilavecz, M.; Siebert, C. D. Medicinal chemistry of tricyclic Antidepressants. *Pharm. Unserer Zeit* **2008,** *37* (3), 206–213.
7. Hollister, L. E. Tricyclic antidepressants (first of two parts). *N. Engl. J. Med.* **1978,** *299* (20), 1106–1109.
8. Vardanyan, R. S.; Hruby, V. J. *Synthesis of Essential Drugs;* Elsevier, 2006.
9. Bolasco, A.; Carradori, S.; Fioravanti, R. Focusing on new monoamine oxidase inhibitors. *Expert Opin. Ther. Pat.* **2010,** *20* (7), 909–939.
10. Krishnan, K. R. Revisiting monoamine oxidase inhibitors. *J. Clin. Psychiatry (Memphis, TN, U. S.)* **2007,** *68* (Suppl. 8), 35–41.
11. Rajput, A.; Zesiewicz, T. A.; Hauser, R. A. Monoamine oxidase inhibitors, Neurologic. Disease Ther. In *Handbook of Parkinson's Disease,* 4th ed.; Pahwa, R., Lyons, K. E., Eds. CRC Press, 2007; pp 349–363.
12. Youdim, M. B. H.; Edmondson, D.; Tipton, K. F. The therapeutic potential of monoamine oxidase inhibitors. *Nat. Rev. Neurosci.* **2006,** *7* (4), 295–309.
13. Childers, W. E., Jr.; Rotella, D. P. Selective serotonin reuptake inhibitors for the treatment of depression. In *Analogue-based Drug Discovery II;* Fischer, J., Ganellin, C. R., Eds.; Wiley-VCH, 2010; pp 269–295.
14. Mandrioli, R.; Mercolini, L.; Saracino, M. A.; Raggi, M. A. Selective serotonin reuptake inhibitors (SSRIs): therapeutic drug monitoring and pharmacological interactions. *Curr. Med. Chem.* **2012,** *19* (12), 1846–1863.
15. Artigas, F. Future directions for serotonin and antidepressants. *ACS Chem. Neurosci.* **2013,** *4* (1), 5–8.
16. Daws, L. C.; Koek, W.; Mitchell, N. C. Revisiting serotonin reuptake inhibitors and the therapeutic potential of "uptake-2" in psychiatric disorders. *ACS Chem. Neurosci.* **2013,** *4* (1), 16–21.
17. Berton, O.; Nestler, E. J. New approaches to antidepressant drug discovery: beyond monoamines. *Nat. Rev. Neurosci.* **2006,** *7* (2), 137–151.
18. Quesseveur, G.; Nguyen, H. T.; Gardier, A. M.; Guiard, B. P. 5-HT2 ligands in the treatment of anxiety and depression. *Expert Opin. Invest. Drugs* **2012,** *21* (11), 1701–1725.
19. Molloy, B. B.; Schmiegel, K. K., 3-Aryloxy-3-phenylpropylamines, DE 2500110 (1975).
20. Molloy, B. B.; Schmiegel, K. K. Aryloxyphenylpropylamines in treating depression, US 4018895 (1977).
21. Molloy, B. B.; Schmiegel, K. K. Treatment of obesity with aryloxyphenylpropylamines, US 4626549 (1986).

22. Arosio, R.; Beratto, S. G. V.; Rossetti, V. Phenylation process for the industrial-scale preparation of high-purity fluoxetine and its acid-addition salts, US 5847214 (1998).
23. Dominguez, C. Process for the preparation of fluoxetine, ES 2103680 (1997).
24. Ezquerra, J. An industrially advantageous preparation of fluoxetine, ES 2101654 (1997).
25. Theriot, K. J. Preparation and use of 2-methyl-5-phenylisoxazolidine, US 5760243 (1998).
26. Robertson, D. W.; Krushinski, J. H.; Fuller, R. W.; Leander, J. D. The absolute configurations and pharmacological activities of the optical isomers of fluoxetine, a selective serotonin-uptake inhibitor. *J. Med. Chem.* **1988,** *31* (7), 1412–1417.
27. Corey, E. J.; Reichard, G. A. Enantioselective and practical syntheses of R- and S-fluoxetines. *Tetrahedron Lett.* **1989,** *30* (39), 5207–5210.
28. Gao, Y.; Sharpless, K. B. Asymmetric synthesis of both enantiomers of tomoxetine and fluoxetine. Selective reduction of 2,3-epoxycinnamyl alcohol with red-Al. *J. Org. Chem.* **1988,** *53* (17), 4081–4084.
29. Koenig, T. M.; Mitchell, D. A convenient method for preparing enantiomerically pure norfluoxetine, fluoxetine and tomoxetine. *Tetrahedron Lett.* **1994,** *35* (9), 1339–1342.
30. Wong, D. T.; Perry, K. W.; Bymaster, F. P. Case history: the discovery of fluoxetine hydrochloride (Prozac). *Nat. Rev. Drug Discovery* **2005,** *4* (9), 764–774.
31. Childers, W. E., Jr.; Rotella, D. P. Selective serotonin reuptake inhibitors for the treatment of depression. In *Analogue-based Drug Discovery II;* Fischer, J., Ganellin, C. R., Eds.; Wiley-VCH, 2010; pp 269–295.
32. Stanford, S. C. Prozac: panacea or puzzle? *Trends Pharmacol. Sci.* **1996,** *17* (4), 150–154.
33. Wong, D. T.; Bymaster, F. P.; Engleman, E. A. Prozac (fluoxetine, Lilly 110140), the first selective serotonin uptake inhibitor and an antidepressant drug: twenty years since its first publication. *Life Sci.* **1995,** *57* (5), 411–441.
34. De Risi, C.; Fanton, G.; Pollini, G. P.; Trapella, C.; Valente, F.; Zanirato, V. Recent advances in the stereoselective synthesis of trans-3,4-disubstituted-piperidines: applications to (-)-paroxetine. *Tetrahedron: Asymmetry* **2008,** *19* (2), 131–155.
35. Czibula, L.; Nemes, A.; Seboek, F.; Szantay, C., Jr.; Mak, M. A convenient synthesis of (-)-paroxetine. *Eur. J. Org. Chem.* **2004,** *15,* 3336–3339.
36. Christensen, J. A.; Squires, R. F. 4-Phenylpiperidine compounds, US 4007196 (1977).
37. Murthy, K. S. K.; Rey, A. W. Stereoselective preparation of 3-acyl-4-arylpiperidines, WO 9907680 (1999).
38. Engelstoft, M.; Hansen, J. B. Synthesis and 5HT modulating activity of stereoisomers of 3-phenoxymethyl-4-phenylpiperidines. *Acta Chem. Scand.* **1996,** *50* (2), 164–169.
39. Murthy, K. S. K.; Allan, W.; Rey, A. W.; Tjepkema, M. Enantioselective synthesis of 3-substituted-4-aryl piperidines useful for the preparation of paroxetine. *Tetrahedron Lett.* **2003,** *44* (28), 5355–5358.
40. Shinji, Y.; Jahan, I. A new route to 3,4-disubstituted piperidines: formal synthesis of (−)-paroxetine and (+)-femoxetine. *Tetrahedron Lett.* **2005,** *46* (50), 8673–8676.
41. Engelstoft, M.; Hansen, J. B. Synthesis and 5HT Modulating Activity of Stereoisomers of 3-Phenoxymethyl-4-phenylpiperidines. *Acta Chem. Scand.* **1996,** *50* (2), 164–169.
42. Cossy, J. Selective methodologies for the synthesis of biologically active piperidinic compounds. *Chem. Rec.* **2005,** *5* (2), 70–80.
43. Kim, M.; Park, Y.; Jeong, B.; Park, H.; Jew, S. Synthesis of (-)-paroxetine via enantioselective phase-transfer catalytic monoalkylation of malonamide ester. *Org. Lett.* **2010,** *12* (12), 2826–2829.
44. Bonifacio, F.; Mancinetti, D.; Crescenzi, C.; De Iasi, G.; Donnarumma, M.; Mastrangeli, C. Asymmetric hydrogenation applied to industrial processes: a convenient synthesis of paroxetine. *PharmaChem* **2003,** *2* (11–12), 13–15.

45. Lawrie, K. W. M.; Rustidge, D. C. The synthesis of [methylenedioxy-14C] paroxetine BRL 29060A. J. Labelled. *Compd. Radiopharm.* **1993,** *33* (8), 777–781.
46. Snyderman, S. H.; Rynn, M. A.; Bellew, K.; Rickels, K. Paroxetine in the treatment of generalised anxiety disorder. *Expert Opin. Pharmacother.* **2004,** *5* (8), 1799–1806.
47. Caley, C. F.; Weber, S. S. Paroxetine: a selective serotonin reuptake inhibiting antidepressant. *Ann. Pharmacother.* **1993,** *27* (10), 1212–1222.
48. Welch, W. M.; Herbert, C.A.; Koe, K.; Kraska, A. R. Antidepressant derivatives of cis-4-phenyl-1,2,3,4-tetrahydro-1-naphthalenamine, US 4536518 (1985).
49. Welch, W. M.; Kraska, A. R.; Sarges, R.; Koe, B. K. Nontricyclic antidepressant agents derived from cis- and trans-1-amino-4-aryltetralins. *J. Med. Chem.* **1984,** *27* (11), 1508–1515.
50. Adrian, G. Preparation of 4-(disubstituted aryl)-1-tetralones as intermediates for serotonin antagonists, EP 346226 (1989).
51. Lautens, M.; Rovis, T. Selective functionalization of 1,2-dihydronaphthalenols leads to a concise, stereoselective synthesis of sertraline. *Tetrahedron* **1999,** *55* (29), 8967–8976.
52. Quallich, G. J. Development of the commercial process for Zoloft/sertraline. *Chirality* **2005,** *17* (Suppl.), S120–S126.
53. Vukics, K.; Fodor, T.; Fischer, J.; Fellegvari, I.; Levai, S. Improved industrial synthesis of antidepressant sertraline. *Org. Process Res. Dev.* **2002,** *6* (1), 82–85.
54. Taber, G. P.; Pfisterer, D. M.; Colberg, J. C. A new and simplified process for preparing N-[4-(3,4-dichlorophenyl)-3,4-dihydro-1(2H)-naphthalenylidene]methanamine and a telescoped process for the synthesis of (1S-cis)-4-(3,4-dichlorophenol)-1,2,3,4-tetrahydro-N-methyl-1-naphthalenamine mandelate: key intermediates in the synthesis of sertraline hydrochloride. *Org. Process Res. Dev.* **2004,** *8* (3), 385–389.
55. Lee, S. H.; Kim, I. S.; Li, Q. R.; Dong, G. R.; Jeong, L. S.; Jung, Y. H. Stereoselective amination of chiral benzylic ethers using chlorosulfonyl isocyanate: total synthesis of (+)-sertraline. *J. Org. Chem.* **2011,** *76* (24), 10011–10019.
56. Wang, G.; Zheng, C.; Zhao, G. Asymmetric reduction of substituted indanones and tetralones catalyzed by chiral dendrimer and its application to the synthesis of (+)-sertraline. *Tetrahedron: Asymmetry* **2006,** *17* (14), 2074–2081.
57. Williams, Michael Quallich, George Sertraline: development of a chiral inhibitor of serotonin uptake. *Chem. Ind. (Chichester, U. K.)* **1990,** *10,* 315–319.
58. Zinnen, H. A.; Gattuso, M. J. Process for preparation of pharmaceutically desired chiral tetralone useful, e.g., in the preparation of sertraline, US (2002).
59. Roesner, S.; Casatejada, J. M.; Elford, T. G.; Sonawane, R. P.; Aggarwal, V. K. Enantioselective synthesis of (+)-sertaline and (+)-indatraline using lithiation/borylation-protodeboronation methodology. *Org. Lett.* **2011,** *13* (21), 5740–5743.
60. Garcia, A. E.; Ouizem, S.; Cheng, X.; Romanens, P.; Kuendig, E. P. Efficient enantioselective synthesis of sertraline, 2-epicatalponol and catalponol from tetralin-1,4-dione. *Adv. Synth. Catal.* **2010,** *352* (13), 2306–2314.
61. Chandrasekhar, S.; Reddy, M. V. An expedient total synthesis of cis-(+)- sertraline from D-phenylglycine. *Tetrahedron* **2000,** *56* (8), 1111–1114.
62. Bhanja, C.; Jena, S. Synthesis design of top-selling anti-depressant drug "sertraline": a retrosynthetic approach. *Asian J. Biochem. Pharm. Res.* **2012,** *2* (4), 163–179.
63. Murdoch, D.; McTavish, D. Sertraline. A review of its pharmacodynamic and pharmacokinetic properties, and therapeutic potential in depression and obsessive-compulsive disorder. *Drugs* **1992,** *44* (4), 604–624.
64. Welch, W. M. Discovery and preclinical development of the serotonin reuptake inhibitor sertraline. *Adv. Med. Chem.* **1995,** *3,* 113–148.

65. Tolbert, S. R.; Fuller, M. A. Sertraline: a new serotonin reuptake inhibitor. *J. Pharm. Technol.* **1992,** *8* (6), 238–241.
66. Perry, C. M.; Benfield, P. Sertraline. An overview of its pharmacological properties and a review of its therapeutic efficacy in obsessive-compulsive disorder. *CNS Drugs* **1997,** *7* (6), 480–500.
67. McRae, A. L.; Brady, K. T. Review of sertraline and its clinical applications in psychiatric disorders. *Expert Opin. Pharmacother.* **2001,** *2* (5), 883–892.
68. Mandrioli, R.; Mercolini, L.; Saracino, M. A.; Raggi, M. A. Selective serotonin reuptake inhibitors (SSRIs): therapeutic drug monitoring and pharmacological interactions. *Curr. Med. Chem.* **2012,** *19* (12), 1846–1863.
69. Hobgood, C. D.; Clayton, A. H. Sertraline in the treatment of panic disorder. *Drugs Today* **2009,** *45* (5), 351–361.
70. Warrington, S. J. Clinical implications of the pharmacology of sertraline. *Int. Clin. Psychopharmacol.* **1991,** *6* (Suppl. 2), 11–21.
71. Pastre, J. C.; Correia, C. R. D. Remarkable electronic effect on the diastereoselectivity of the Heck reaction of methyl cinnamate with arenediazonium salts: formal total synthesis of (±)-indatraline and (±)-sertraline. *Adv. Synth. Catal.* **2009,** *351* (9), 1217–1223.
72. Welle, H. B. A.; Claassen, V. Oxime ether compounds, CH 629761 (1982).
73. No Inventor data available, Substituted 4′-trifluoromethylvalerophenone O-(2-aminoethyl) oxime derivatives with antidepressive action, NL 7503310 (1976).
74. Chitturi, R.; Rajamannar, T.; Jadav, K. J.; Shah, H. A. Etherification and salification process for the industrial-scale manufacture of fluvoxamine maleate, CH 691124 (2001).
75. Irons, J. Fluvoxamine in the treatment of anxiety disorders. *Neuropsychiatr. Dis. Treat.* **2005,** *1* (4), 289–299.
76. Dell'Osso, B.; Allen, A.; Hollander, E. Fluvoxamine: a selective serotonin re-uptake inhibitor for the treatment of obsessive-compulsive disorder. *Expert Opin. Pharmacother.* **2005,** *6* (15), 2727–2740.
77. Silver, H. Fluvoxamine as an adjunctive agent in schizophrenia. *CNS Drug Rev.* **2001,** *7* (3), 283–304.
78. Figgitt, D. P.; McClellan, K. J. Fluvoxamine: An updated review of its use in the management of adults with anxiety disorders. *Drugs* **2000,** *60* (4), 925–954.
79. Goodman, W. K.; Ward, H.; Kablinger, A.; Murphy, T. Fluvoxamine in the treatment of obsessive-compulsive disorder and related conditions. *J. Clin. Psychiatry (Memphis, TN, U. S.)* **1997,** *58* (Suppl. 5), 32–49.
80. Palmer, K. J.; Benfield, P. Fluvoxamine: an overview of its pharmacological properties and review of its therapeutic potential in non-depressive disorders. *CNS Drugs* **1994,** *1* (1), 57–87.
81. Omori, I. M.; Watanabe, N.; Nakagawa, A.; Akechi, T.; Cipriani, A.; Barbui, C.; McGuire, H.; Churchill, R.; Furukawa, T. A. Efficacy, tolerability and side-effect profile of fluvoxamine for major depression: meta-analysis. *J. Psychopharmacol. (London, U. K.)* **2009,** *23* (5), 539–550.
82. Cheer, S. M.; Figgitt, D. P. Fluvoxamine: a review of its therapeutic potential in the management of anxiety disorders in children and adolescents. *Paediatr. Drugs* **2001,** *3* (10), 763–781.
83. Bogeso, K. P.; Toft, A. S. Anti-depressive substituted 1-dimethylaminopropyl-1-phenyl phtalans, US 4136193 (1979).
84. Bogeso, K. P. Intermediate in the preparation of 1-(3-dimethylaminopropyl)-1-(4-p-fluorophenyl)-1,3-dihydroisobenzofuran-5-carbonitrile, EP 171943 (1986).
85. Bogeso, K. P.; Perregaard, J. Preparation and isolation of antidepressant drug citalopram enantiomers and their pharmaceutical compositions, EP 347066 (1989).
86. Boegesoe, K. P., Novel intermediate and method for its preparation, US 4650884 (1987).

87. Boegesoe, K. P.; Jaegerspris, J.P. Pharmaceutically useful (+)-1-(3-dimethylaminopropyl)-1-(4′-fluorophenyl)-1,3-dihydrosobenzofuran, US 4943590 (1990).
88. Petersen, H.; Bogeso, K. P.; Bech S. M. Method for the preparation of citalopram, WO 9819511 (1998).
89. Petersen, H.; Bregnedal, P.; Bogeso, K. P. Method for the preparation of citalopram, WO 9819512 (1998).
90. Petersen, H.; Bregnedal, P.; Bogeso, K. P. Method for the preparation of citalopram, WO 9819513 (1998).
91. Petersen, H. Method for the preparation of citalopram, WO 9930548 (1999).
92. Rock, M. H.; Petersen, H.; Ellegaard, P. Method for the preparation of citalopram, WO 0012044 (2000).
93. Petersen, H.; Rock, M. H.; Svane, H., Method for the preparation of citalopram, WO 0013648 (2000).
94. Babu, N. A.; Vuddamari, S. G.; Laxman, G. S.; Manjunath, S. G.; Kulkarni, A. K. One pot synthesis of citalopram from 5-cyano phthalide, WO 2005/077927 (2005).
95. Greenhood, A. K.; McHattie, D.; Rechka, J. A.; Hedger, P. C. M.; Gamble, M. P. Process for the preparation of citalopram, US 7002025 (2006).
96. Humble, R. E.; Christensen, T. V.; Rock, M. H.; Nielsen, O.; Petersen, H.; Dancer, R. Process for the preparation of racemic citalopram and/or S-or R-citalopram by separation of a mixture of R-and S-citalopram, US 7112686 (2006).
97. Reddy, M. P.; Rao, A. K. S.B.; Usharani, V.; Dubey, P. K. Novel and improved process for the preparation of citalopram. *Asian J. Chem.* **2011,** *23* (4), 1829–1832.
98. Eildal, J. N. N.; Andersen, J.; Kristensen, A. S.; Jorgensen, A. M.; Bang-Andersen, B.; Jorgensen, M.; Stromgaard, K. From the selective serotonin transporter inhibitor citalopram to the selective norepinephrine transporter inhibitor talopram: synthesis and structure-activity relationship studies. *J. Med. Chem.* **2008,** *51* (10), 3045–3048.
99. Elati, C. R.; Kolla, N. R.; Mathad, V. T. Substrate modification approach to achieve efficient resolution: didesmethylcitalopram: a key intermediate for escitalopram, response to comments. *Org. Process Res. Dev.* **2009,** *13* (1), 34–37.
100. Vipin, K. K.; Umar, K. M.; Narsimha, R. B.; Ranjith, K. S.; Ramesh, D.; Sivakumaran, M. Process for the preparation of escitalopram, EP 2017271 (2009).
101. Sommer, M. B.; Nielsen, O.; Petersen, H.; Ahmadian, H.; Pedersen, H.; Brosen, P.; Geiser, F.; Lee, J.; Cox, G.; Dapremont, O.; Suten, C.; Assenza, S. P.; Hariharan, S.; Nair, U. Method for the preparation of escitalopram, US 2011/0065938 (2011).
102. Giridhar, T.; Srinivasulu, G.; Rao, K. S. Preparation of Escitalopram, Its Salts and Intermediates, US 2011/0092719 (2011).
103. Jacquot, C.; David, D. J.; Gardier, A. M.; Sanchez, C. Escitalopram and citalopram: the unexpected role of the R-enantiomer. *Encephale* **2007,** *33* (2), 179–187.
104. Pollock, B. G. Citalopram a comprehensive review. *Expert Opin. Pharmacother.* **2001,** *2* (4), 681–698.
105. Joubert, A. F.; Sanchez, C.; Larsen, F. Citalopram. *Hum. Psychopharmacol.* **2000,** *15* (6), 439–451.
106. Parker, N. G.; Brown, C. S. Citalopram in the treatment of depression. *Ann. Pharmacother.* **2000,** *34* (6), 761–771.
107. Tan, J. Y.; Levin, G. M. Citalopram in the treatment of depression and other potential uses in psychiatry. *Pharmacotherapy* **1999,** *19* (6), 675–689.
108. Joubert, A. F.; Stein, D. J. Citalopram and anxiety disorders. *Rev. Contemp. Pharmacother.* **1999,** *10* (2), 79–131.

109. Noble, S.; Benfield, P. Citalopram: a review of its pharmacology, clinical efficacy and tolerability in the treatment of depression. *CNS Drugs* **1997,** *8* (5), 410–431.
110. Montgomery, S. A.; Johnson, F. N. Citalopram in the treatment of depression. *Rev. Contemp. Pharmacother.* **1995,** *6* (6), 297–306.
111. Milne, R. J.; Goa, K. L. Citalopram. A review of its pharmacodynamic and pharmacokinetic properties, and therapeutic potential in depressive illness. *Drugs* **1991,** *41* (3), 450–477.
112. Hyttel, J. Citalopram-pharmacological profile of a specific serotonin uptake inhibitor with antidepressant activity. *Prog. Neuro-Psychopharmacol. Biol. Psychiatry* **1982,** *6* (3), 277–295.
113. Ahn, J.-H.; Patkar, A. A. Escitalopram for the treatment of major depressive disorder in youth. *Expert Opin. Pharmacother.* **2011,** *12* (14), 2235–2244.
114. Zhong, H.; Haddjeri, N.; Sanchez, C. Escitalopram, an antidepressant with an allosteric effect at the serotonin transporter-a review of current understanding of its mechanism of action. *Psychopharmacology (Berl)* **2012,** *219* (1), 1–13.
115. Garnock-Jones, K. P.; McCormack, P. L. Escitalopram: a review of its use in the management of major depressive disorder in adults. *CNS Drugs* **2010,** *24* (9), 769–796.
116. Waugh, J.; Goa, K. L. Escitalopram. A review of its use in the management of major depressive and anxiety disorders. *CNS Drugs* **2003,** *17* (5), 343–362.
117 Baldwin, D. S.; Nair, R. V. Escitalopram in the treatment of generalized anxiety disorder. *Expert Rev. Neurother.* **2005,** *5* (4), 443–449.
118. Aronson, S.; Delgado, P. Escitalopram. *Drugs Today* **2004,** *40* (2), 121–131.
119. Burke, W. J. Escitalopram. *Expert Opin. Invest. Drugs* **2002,** *11* (10), 1477–1486.
120. Heinrich, T.; Boettcher, H.; Gericke, R.; Bartoszyk, G. D.; Anzali, S.; Seyfried, C. A.; Greiner, H. E.; van Amsterdam, C. Synthesis and structure-activity relationship in a class of indolebutylpiperazines as dual 5-HT1A receptor agonists and serotonin reuptake inhibitors. *J. Med. Chem.* **2004,** *47* (19), 4684–4692.
121. Hu, B.; Song, Q.; Xu, Y. Scale-up synthesis of antidepressant drug vilazodone. *Org. Process Res. Dev.* **2012,** *16* (9), 1552–1557.
122. Andreas, B. Method for the production of 5-[4-[4-(5-cyano-3-indolyl)butyl]-1-piperazinyl] benzofuran-2-carboxamide, WO 2006114202 (2006).
123. Li, J. Q.; Wang, G.; Wang, C.; Wang, J. J. 4-(5-Cyano-1H-indole-3-yl)butyl substituted sulfonate-like compound and its application as key intermediate for preparing vilazodone or its pharmaceutically acceptable salt, CN 102267932 (2011).
124. Xu, W.; Zhang, R. J.; Zhu B. Method for preparing 3-(4-chloro-1-oxobutyl)-1H-indole-5-carbonitrile, CN 102249979 (2011).
125. Chen, M. Process for preparation of antidepressive vilazodone, CN 10218(0868) (2011).
126. Li, J. Q.; Wang, G.; Wang, C.; Huang, L. Method for preparing vilazodone or its hydrochloride, CN 10226(7985) (2011).
127. Choi, E.; Zmarlicka, M.; Ehret, M. J. Vilazodone: a novel antidepressant. *Am. J. Health-Syst. Pharm.* **2012,** *69* (18), 1551–1557.
128. Citrome, L. Vilazodone for major depressive disorder: a systematic review of the efficacy and safety profile for this newly approved antidepressant-what is the number needed to treat, number needed to harm and likelihood to be helped or harmed? *Int. J. Clin. Pract.* **2012,** *66* (4), 356–368.
129. Reed, C. R.; Kajdasz, D. K.; Whalen, H.; Athanasiou, M. C.; Gallipoli, S.; Thase, M. E. The efficacy profile of vilazodone, a novel antidepressant for the treatment of major depressive disorder. *Curr. Med. Res. Opin.* **2012,** *28* (1), 27–39.

130. Laughren, T. P.; Gobburu, J.; Temple, R. J.; Unger, E. F.; Bhattaram, A.; Dinh, P. V.; Fossom, L.; Hung, H. M. J.; Klimek, V.; Lee, J. E.; Levin, R. L.; Lindberg, C. Y.; Mathis, M.; Rosloff, B. N.; Wang, S.-J.; Wang, Y.; Yang, P.; Yu, B.; Zhang, H.; Zhang, L.; Zineh, I. Vilazodone: clinical basis for the US Food and Drug Administration's approval of a new antidepressant. *J. Clin. Psychiatry (Memphis, TN, U. S.)* **2011,** *72,* 1166–1173.
131. Hopkins, C. R. ACS chemical neuroscience molecule spotlight on Viibryd (vilazodone). *ACS Chem. Neurosci.* **2011,** *2* (10), 554.
132. Frampton, J. E. Vilazodone: in major depressive disorder. *CNS Drugs* **2011,** *25* (7), 615–627.
133. Dawson, L. A.; Watson, J. M. Vilazodone: a 5-HT1A receptor agonist/serotonin transporter inhibitor for the treatment of affective disorders. *CNS Neurosci. Ther.* **2009,** *15* (2), 107–117.
134. Dawson, L. A. The discovery and development of vilazodone for the treatment of depression: a novel antidepressant or simply another SSRI? *Exp. Opin. Drug Disc.* **2013,** *8* (12), 1529–1539.
135. Sorbera, L. A.; Rabasseda, X.; Silvestre, J.; Castaner, J. Vilazodone hydrochloride. Antidepressant 5-HT1A partial agonist 5-HT reuptake inhibitor. *Drugs Future* **2001,** *26* (3), 247–252.
136. dePaulis, T. Drug evaluation: vilazodone—a combined SSRI and 5-HT1A partial agonist for the treatment of depression. *IDrugs* **2007,** *10* (3), 193–201.
137. Robinson, D. S.; Kajdasz, D. K.; Gallipoli, S.; Whalen, H.; Wamil, A.; Reed, C. R. J. A 1-year, open-label study assessing the safety and tolerability of vilazodone in patients with major depressive disorder. *J. Clin. Psychopharmacol.* **2011,** *31*. 643–636.
138. Reed, C. R.; Kajdasz, D. K.; Whalen, H.; Athanasiou, M. C.; Gallipoli, S.; Thase, M. E. The efficacy profile of vilazodone, a novel antidepressant for the treatment of major depressive disorder. *Curr. Med. Res. Opin.* **2012,** *28* (1), 27–39.
139. Babu, R. P. K.; Maiti, S. N. Norepinephrine reuptake inhibitors for depression, ADHD and other neuropsychiatric disorders. *Heterocycles* **2006,** *69,* 539–567.
140. Frazer, A. Norepinephrine involvement in antidepressant action. *J. Clin. Psychiatry (Memphis, TN, U. S.)* **2000,** *61* (Suppl. 10), 25–30.
141. Gray, D. L. Approved treatments for attention deficit hyperactivity disorder: amphetamine (Adderall), methylphenidate (Ritalin), and atomoxetine (Strattera). In *The Art of Drug Synthesis;* Johnson, D. S., Li, J. J., Eds.; Wiley-Interscience, 2007; pp 241–259.
142. Srebnik, M.; Ramachandran, P. V.; Brown, H. C. Chiral synthesis via organoboranes. 18. Selective reductions. 43. Diisopinocampheylchloroborane as an excellent chiral reducing reagent for the synthesis of halo alcohols of high enantiomeric purity. A highly enantioselective synthesis of both optical isomers of tomoxetine, fluoxetine, and nisoxetine. *J. Org. Chem.* **1988,** *53* (13), 2916–2920.
143. Heath, P. C.; Ratz, A. M.; Weigel, L. O., Preparation of tomoxetine, WO 2000058262 (2000).
144. Wong, D. T.; Threlkeld, P. G.; Best, K. L.; Bymaster, F. P. A new inhibitor of norepinephrine uptake devoid of affinity for receptors in rat brain. *J. Pharmacol. Exp. Ther.* **1982,** *222,* 61–65.
145. Zerbe, R. L.; Rowe, H.; Enas, G. G.; Wong, D.; Farid, N.; Lemberger, L. Clinical pharmacology of tomoxetine, a potential antidepressant. *J. Pharmacol. Exp. Ther.* **1985,** *232,* 139–143.
146. Weiss, M. D.; Virani, A. L.; Wasdell, M.; Faulkner, L.; Rea, K.; Freeman, R. D.; Weiss, G.; Jokhani, V. Atomoxetine in clinical practice. *Future Neurol.* **2006,** *1* (3), 249–258.
147. Christman, A. K.; Fermo, J. D.; Markowitz, J. S. Atomoxetine, a novel treatment for attention-deficit-hyperactivity disorder. *Pharmacotherapy* **2004,** *24* (8), 1020–1036.
148. Kratochvil, C. J.; Vaughan, B. S.; Daughton, J. M.; Mayfield-Jorgensen, M. L.; Burke, W. J. Atomoxetine in the treatment of attention deficit hyperactivity disorder. *Expert Rev. Neurother.* **2004,** *4* (4), 601–611.

149. Thomason, C.; Michelson, D. Atomoxetine: treatment of attention deficit hyperactivity disorder: beyond stimulants. *Drugs Today* **2004,** *40* (5), 465–473.
150. Simpson, D.; Plosker, G. L. Atomoxetine: a review of its use in adults with attention deficit hyperactivity disorder. *Drugs* **2004,** *64* (2), 205–222.
151. Eiland, L. S.; Guest, A. L. Atomoxetine treatment of attention-deficit/hyperactivity disorder. *Ann. Pharmacother.* **2004,** *38* (1), 86–90.
152. Preti, A. T. Tomoxetine (Eli Lilly & Co). *Curr. Opin. Invest. Drugs (BioMed Cent.)* **2002,** *3* (2), 272–277.
153. Robertson, D. W.; Wong, D. T.; Krushinski, J. H., Preparation of 3-aryloxy-3-substituted-propanamines as antidepressants, EP 273658 (1988).
154. Berglund, R. A., Asymmetric synthesis of key intermediate in the synthesis of duloxetine via arylation of (S)-(-)-N,N-dimethyl-3-(2-thienyl)-3-hydroxypropanamine with 1-fluoronaphthalene in presence of potassium compound, US 5362886 (1994).
155. Deeter, J.; Frazier, J.; Staten, G.; Staszak, M.; Weigel, L. Asymmetric synthesis and absolute stereochemistry of LY24(8686). *Tetrahedron Lett.* **1990,** *31* (49), 7101–7104.
156. Yamaguchi, S.; Mosher, H. Asymmetric reductions with chiral reagents from lithium aluminum hydride and (3-)-(2S,3R)-4-dimethylamino-3-methyl-1,2-diphenyl-2-buta. *J. Org. Chem.* **1973,** *38* (10), 1870–1877.
157. Fujima, Y.; Ikunaka, M.; Inoue, T.; Matsumoto, J. Synthesis of (S)-3-(N-methylamino)-1-(2-thienyl)propan-1-ol: revisiting Eli Lilly's resolution-racemization-recycle synthesis of duloxetine for its robust processes. *Org. Process Res. Dev.* **2006,** *10* (5), 905–913.
158. Bymaster, F. P.; Beedle, E. E.; Findlay, J.; Gallagher, P. T.; Krushinski, J. H.; Mitchell, S.; Robertson, D. W.; Thompson, D. C.; Wallace, L.; Wong, D. T. Duloxetine (Cymbalta), a dual inhibitor of serotonin and norepinephrine reuptake. *Bioorg. Med. Chem. Lett.* **2003,** *13* (24), 4477–4480.
159. Wheeler, W. J.; Kuo, F. An asymmetric synthesis of duloxetine hydrochloride, a mixed uptake inhibitor of serotonin and norepinephrine, and its C-14 labeled isotopomers. *J. Labelled Compd. Radiopharm.* **1995,** *36* (3), 213–223.
160. Brooks, D. W.; Lu, L. D. L.; Masamune, S. Carbon acylation under practically neutral conditions. *Angew. Chem.* **1979,** *91* (1), 76–77.
161. Ratovelomanana-Vidal, V.; Girard, C.; Touati, R.; Tranchier, J. P.; Ben Hassine, B.; Genet, J. P. Enantioselective hydrogenation of β-keto esters using chiral diphosphine-ruthenium complexes: optimization for academic and industrial purposes and synthetic applications. *Adv. Synth. Catal.* **2003,** *345* (1+2), 261–274.
162. Suzuki, Y.; Iwata, M.; Yazaki, R.; Kumagai, N.; Shibasaki, M. Concise enantioselective synthesis of duloxetine via direct catalytic asymmetric Aldol reaction of thioamide. *J. Org. Chem.* **2012,** *77* (9), 4496–4500.
163. Majer, J.; Kwiatkowski, P.; Jurczak, J. Highly enantioselective Friedel-Crafts reaction of thiophenes with glyoxylates: formal synthesis of duloxetine. *Org. Lett.* **2009,** *11* (20), 4636–4639.
164. Ahmed Kamal, A.; Ramesh Khanna, G. B.; Ramu, R.; Krishnaji, T. Chemoenzymatic synthesis of duloxetine and its enantiomer: lipase-catalyzed resolution of 3-hydroxy-3-(2-thienyl) propane nitrile. *Tetrahedron Lett.* **2003,** *44* (25), 4783–4787.
165. Yang, A.; Chen, C. Review of the synthetic methods for antidepressants of duloxetine. *Guangdong Huagong* **2012,** *39* (4), 39–40. 54.
166. Pineiro-Nunez, M. Advances in development of methods for the synthesis of dual selective serotonin and norepinephrine reuptake inhibitors (SSNRIs) [venlafaxine hydrochloride (Effexor), milnacipran hydrochloride (Ixel, Dalcipran) and duloxetine hydrochloride (Cymbalta)]. In *The Art of Drug Synthesis;* Johnson, D. S., Li, J. J., Eds.; Wiley-Interscience, 2007; pp 199–213.

167. Jiang, Z. Development in synthesis research of duloxetine hydrochloride (to be continued). *Huagong Shengchan Yu Jishu* **2009,** *16* (4), 51–55.
168. Jiang, Z. Development in synthesis research of duloxetine hydrochloride (end). *Huagong Shengchan Yu Jishu* **2009,** *16* (5), 36–40.
169. Torres-Sanchez, S.; Perez-Caballero, L.; Mico, J. A.; Elorza, J.; Berrocoso, E. Preclinical discovery of duloxetine for the treatment of depression. *Expert Opin. Drug Discovery* **2012,** *7* (8), 745–755.
170. Monteleone, F.; Caputo, M.; Tecce, M. F.; Capasso, A. Duloxetine in the treatment of depression: an overview. *Cent. Nerv. Syst. Agents Med. Chem.* **2011,** *11* (3), 174–183.
171. Mancini, M.; Perna, G.; Rossi, A.; Petralia, A. Use of duloxetine in patients with an anxiety disorder, or with comorbid anxiety and major depressive disorder: A review of the literature. *Exp. Opin. Pharmacother.* **2010,** *11* (7), 1167–1181.
172. Norman, T. R.; Olver, J. S. Continuation treatment of major depressive disorder: is there a case for duloxetine? *Drug Des., Dev. Ther.* **2010,** *4,* 19–31.
173. Muneoka, K. Pharmacotherapy of major depressive disorder: focus on duloxetine. *Clin. Med.: Ther.* **2009,** *1,* 1541–1556.
174. Volz, H.-P. Duloxetine-current data and practical use. *Arzneimitteltherapie* **2009,** *27* (9), 273–278.
175. Carter, N. J.; McCormack, P. L. Duloxetine, a review of its use in the treatment of generalized anxiety disorder. *CNS Drugs* **2009,** *23* (6), 523–541.
176. Kornstein, S. G.; Russell, J. M.; Spann, M. E.; Crits-Christoph, P.; Ball, S. G. Duloxetine in the treatment of generalized anxiety disorder. *Expert Rev. Neurother.* **2009,** *9* (2), 155–165.
177. De Berardis, D.; Serroni, N.; Carano, A.; Scali, M.; Valchera, A.; Campanella, D.; D'Albenzio, A.; Di Giuseppe, B.; Moschetta, F. S.; Salerno, R. M.; Ferro, F. M. The role of duloxetine in the treatment of anxiety disorders. *Neuropsychiatr. Dis. Treat.* **2008,** *4* (5), 929–935.
178. Frampton, J. E.; Plosker, G. L. Duloxetine. A review of its use in the treatment of major depressive disorder. *CNS Drugs* **2007,** *21* (7), 581–609.
179. Crippa, J. A. S.; Zuardi, A. W. Duloxetine in the treatment of panic disorder. *Int. J. Neuropsychopharmacol.* **2006,** *9* (5), 633–634.
180. Bauer, M.; Moeller, H.-J.; Schneider, E. Duloxetine: a new selective and dual-acting antidepressant. *Expert Opin. Pharmacother.* **2006,** *7* (4), 421–427.
181. Westanmo, A. D.; Gayken, J.; Haight, R. Duloxetine: a balanced and selective norepinephrine- and serotonin-reuptake inhibitor. *Am. J. Health-Syst. Pharm.* **2005,** *62* (23), 2481–2490.
182. Bymaster, F. P.; Lee, T. C.; Knadler, M. P.; Detke, M. J.; Iyengar, S. The dual transporter inhibitor duloxetine: a review of its preclinical pharmacology, pharmacokinetic profile, and clinical results in depression. *Curr. Pharm. Des.* **2005,** *11* (12), 1475–1493.
183. Dugan, S. E.; Fuller, M. A. Duloxetine: a dual reuptake inhibitor. *Ann. Pharmacother.* **2004,** *38* (12), 2078–2085.
184. Dunner, D. L.; Goldstein, D. J.; Mallinckrodt, C.; Lu, Y.; Detke, M. J. Duloxetine in treatment of anxiety symptoms associated with depression. *Depression Anxiety* **2003,** *18* (2), 53–61.
185. Anttila, S.; Leinonen, E. Duloxetine Eli Lilly. *Curr. Opin. Invest. Drugs (BioMed Cent.)* **2002,** *3* (8), 1217–1221.
186. Pitsikas, N. Duloxetine (Eli Lilly & Co.). *Curr. Opin. Invest. Drugs (BioMed Cent.)* **2000,** *1* (1), 116–121.
187. Karpa, K. D.; Cavanaugh, J. E.; Lakoski, J. M. Duloxetine pharmacology: profile of a dual monoamine modulator. *CNS Drug Rev.* **2002,** *8* (4), 361–376.
188. Wong, D. T.; Bymaster, F. P. Dual serotonin and noradrenaline uptake inhibitor class of antidepressants-potential for greater efficacy or just hype? *Prog. Drug Res.* **2002,** *58,* 169–222.

189. Husbands, G. E. M.; Yardley, J. P.; Muth, E. A. Phenethylamine derivatives and intermediates, EP 112669 (1984).
190. Yardley, J. P.; Husbands, G. E. M.; Stack, G.; Butch, J.; Bicksler, J.; Moyer, J. A.; Muth, E. A.; Andree, T.; Fletcher, H., III; James, M. N. G.; Sieleckit, A. R. 2-Phenyl-2-(1-hydroxycycloalkyl)ethylamine derivatives: synthesis and antidepressant activity. *J. Med. Chem.* **1990,** *33* (10), 2899–2905.
191. Saravanan, M.; Satyanarayana, B.; Reddy, P. P. An improved and impurity-free large-scale synthesis of venlafaxine hydrochloride. *Org. Process Res. Dev.* **2011,** *15* (6), 1392–1395.
192. Basappa; Kavitha, C. V.; Rangappa, K. S. Simple and an efficient method for the synthesis of 1-[2-dimethylamino-1-(4-methoxyphenyl)-ethyl]-cyclohexanol hydrochloride: (±)venlafaxine racemic mixtures. *Bioorg. Med. Chem. Lett.* **2004,** *14* (12), 3279–3281.
193. Chavan, S. P.; Khobragade, D. A.; Thakkar, M.; Kalkote, U. R. Practical synthesis of (±)-venlafaxine. *Synthetic Communications* **2007,** *37* (22), 3901–3906.
194. Zhou, J.; Zhang, H.; Huang, X.; Huang, W. An improved novel method of venlafaxine synthesis. *Zhongguo Yaoke Daxue Xuebao* **1999,** *30* (4), 249–250.
195. Lu, X.; Wu, W.-L.; Zhang, Q.; Dong, D. Synthesis of venlafaxine hydrochloride. *Zhejiang Huagong* **2011,** *42* (5), 9–11.
196. Chavan, S. P.; Garai, S.; Pawar, K. P. Asymmetric total synthesis of (-)-venlafaxine using an organocatalyst. *Tetrahedron Lett.* **2013,** *54* (17), 2137–2139.
197. Bhuniya, R.; Nanda, S. Asymmetric synthesis of both the enantiomers of antidepressant venlafaxine and its analogues. *Tetrahedron Lett.* **2012,** *53* (15), 1990–1992.
198. Katzman, M. A.; Jacobs, L. Venlafaxine in the treatment of panic disorder. *Neuropsychiatr. Dis. Treat.* **2007,** *3* (1), 59–67.
199. Thase, M. E. Treatment of anxiety disorders with venlafaxine XR. *Expert Rev. Neurother.* **2006,** *6* (3), 269–282.
200. Katzman, M. Venlafaxine in the treatment of anxiety disorder. *Expert Rev. Neurother.* **2004,** *4* (3), 371–381.
201. Gutierrez, M. A.; Stimmel, G. L.; Aiso, J. Y. Venlafaxine: a 2003 update. *Clin. Ther.* **2003,** *25* (8), 2138–2154.
202. Balfour, J. A. B.; Jarvis, B. Venlafaxine extended-release: A review of its clinical potential in the management of generalized anxiety disorder. *CNS Drugs* **2000,** *14* (6), 483–503.
203. Burnett, F. E.; Dinan, T. G. The clinical efficacy of venlafaxine in the treatment of depression. *Rev. Contemp. Pharmacother.* **1998,** *9* (5), 303–320.
204. Burnett, F. E.; Dinan, T. G. Venlafaxine. Pharmacology and therapeutic potential in the treatment of depression. *Hum. Psychopharmacol.* **1998,** *13* (3), 153–162.
205. Benkert, O.; Gruender, G.; Wetzel, H. Is there an advantage to venlafaxine in comparison with other antidepressants? *Hum. Psychopharmacol.* **1997,** *12* (1), 53–64.
206. Schweizer, E.; Thielen, R. J.; Frazer, A. Venlafaxine: a novel antidepressant compound. *Expert Opin. Invest. Drugs* **1997,** *6* (1), 65–78.
207. Holliday, S. M.; Benfield, P. Venlafaxine: a review of its pharmacology and therapeutic potential in depression. *Drugs* **1995,** *49* (2). 280–294.
208. Mehta, N. B.; Yeowell, D. A. m-Chloro(or fluoro)-α-(tert-butylamino)propiophenone hydrochloride, DE 2059618 (1971).
209. Yeowell, D. A. α-Halopropiophenones, DE 2064934 (1971).
210. Mehta, N. B.; Yeowell, D. A. α-(tert-Butylamino)-m-chloropropiophenone, CA 97778 (1975).
211. Hill, J. A.; Scharver, J. D. Synthesis of carbon-14- and tritium-labeled forms of bupropion hydrochloride-a novel antidepressant. *J. Labelled Compd. Radiopharm.* **1988,** *25* (10), 1095–1104.

212. Perrine, D. M.; Ross, J. T.; Nervi, S. J.; Zimmerman, R. H. A short, one-pot synthesis of bupropion (Zyban, Wellbutrin). *J. Chem. Educ.* **2000,** *77* (11), 1479–1480.
213. Xu, Z.; Zhou, W.; Liu, H.; Li, X. Studies on the synthesis of bupropion hydrochloride, Anhui Daxue Xuebao. *Ziran Kexueban* **2006,** *30* (5), 87–90.
214. Musso, D. L.; Mehta, N. B.; Soroko, F. E.; Ferris, R. M.; Hollingsworth, E. B.; Kenney, B. T. Synthesis and evaluation of the antidepressant activity of the enantiomers of bupropion. *Chirality* **1993,** *5* (7), 495–500.
215. Carroll, F. I.; Blough, B. E.; Mascarella, S. W.; Navarro, H. A.; Eaton, J. B.; Lukas, R. J.; Damaj, M. I. Synthesis and biological evaluation of bupropion analogues as potential pharmacotherapies for smoking cessation. *J. Med. Chem.* **2010,** *53* (5), 2204–2214.
216. Dhillon, S.; Yang, L. P. H.; Curran, M. P. Bupropion a review of its use in the management of major depressive disorder. *Drugs* **2008,** *68* (5), 653–689.
217. Moreira, R. The efficacy and tolerability of bupropion in the treatment of major depressive disorder. *Clin. Drug Invest.* **2011,** *31* (Suppl. 1), 5–17.
218. Clayton, A. H. Extended-release bupropion: an antidepressant with a broad spectrum of therapeutic activity? *Expert Opin. Pharmacother.* **2007,** *8* (4), 457–466.
219. Wilkes, S. Bupropion. *Drugs Today* **2006,** *42* (10), 671–681.
220. Foley, K. F.; DeSanty, K. P.; Kast, R. E. Bupropion: pharmacology and therapeutic applications. *Expert Rev. Neurother.* **2006,** *6* (9), 1249–1265.
221. West, R. Bupropion SR for smoking cessation. *Expert Opin. Pharmacother.* **2003,** *4* (4), 533–540.
222. Bryant, S. G.; Guernsey, B. G.; Ingrim, N. B. Review of bupropion. *Clin. Pharm.* **1983,** *2* (6), 525–537.
223. Maxwell, R.; Mehta, N. B.; Tucker, W. E.; Schroeder, D. H.; Stern, W. C. Bupropion. *Pharmacol. Biochem. Prop. Drug Subst.* **1981,** *3,* 1–55.
224. Dwoskin, L. P.; Rauhut, A. S.; King-Pospisil, K. A.; Bardo, M. T. Review of the pharmacology and clinical profile of bupropion, an antidepressant and tobacco use cessation agent. *CNS Drug Rev.* **2006,** *12* (3–4), 178–207.
225. Chen, Z.; Skolnick, P. L. Triple uptake inhibitors: therapeutic potential in depression and beyond. *Expert Opin. Invest. Drugs* **2007,** *16* (9), 1365–1377.
226. Capelli, A. M.; Micheli, F. Triple monoamine uptake inhibitors. *Pharm. Pat. Anal.* **2012,** *1* (4), 469–481.
227. Skolnick, P. Triple-uptake inhibitors (broad-spectrum antidepressants). In *Polypharmacology in Drug Discovery;* Peters, J.-U., Ed.; Wiley, 2012; pp 363–382.
228. Weikert, R. J. The rational design of triple reuptake inhibitors for the treatment of depression. In *Designing Multi-Target Drugs;* Morphy, J. R., Harris, C. J., Eds.; Royal Society of Chemistry, 2012; pp 270–289.
229. Marks, D. M. Amitifadine hydrochloride: Triple reuptake inhibitor treatment of depression. *Drugs Future* **2012,** *37* (4), 241–246.
230. Gallou, I. Amitifadine hydrochloride: triple reuptake inhibitors treatment of depression. *Org. Prep. Proced. Int.* **2007,** *39* (4), 355–383.

Chapter 8

Central Nervous System Stimulants

CNS stimulants are agents that increase physical and motor activity, elevate mood and euphoria, improve concentration, decrease appetite, improve task performance, and increase alertness and attention span. CNS stimulants speed up mental and physical processes in the body, which can be useful in the treatment of certain medical conditions, but, they have many side effects, a high potential for addiction, and the risk of destroying mental and physical health [1-3].

Historically, CNS stimulants were used to treat respiratory problems, obesity, and neurological disorders. Now, stimulants are prescribed for treating only a few health conditions, including ADHD, depression that has not responded to other treatments, and narcolepsy. They may also be used for short-term treatment of obesity and some cases of asthma.

Most CNS stimulants are classified as controlled substances.

There are different categories of stimulants possessing different chemical structures and employing different pharmacological mechanisms.

Alkaloid nicotine (**8.1.1**) (Fig. 8.1.) in tobacco products is the legal stimulant with an important public health impact.

Other alkaloids-methylxanthines, such as caffeine (**8.1.2**) and theophylline (**8.1.2**) (Fig. 8.1.), are present in tea, coffee, and "energy" beverages. They are potent stimulants whose use is widely accepted by society.

Cocaine (**8.1.4**) (Fig. 8.1.), another alkaloid, is the most potent naturally occurring stimulant. It is the most commonly abused drug after cannabis and is highly addictive.

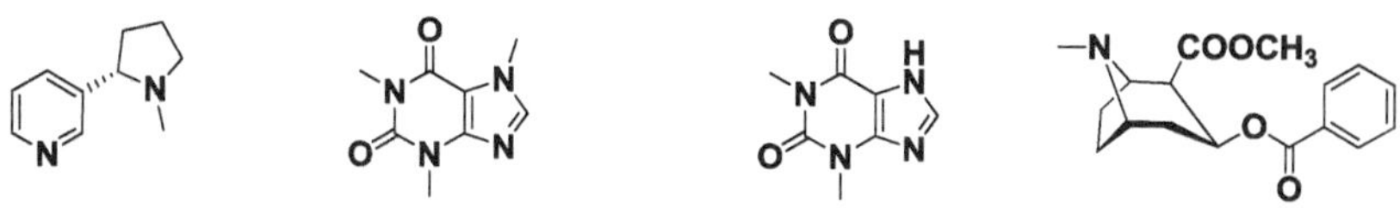

Nicotine 8.1.1 **Caffeine 8.1.2** **Theophylline 8.1.3** **Cocaine 8.1.4**

FIG. 8.1 **Alkaloid CNS stimulants.**

Amphetamines (**8.1.5** to **8.1.8**) and their precursors and derivatives are structurally related to key brain neurotransmitters, such as dopamine and norepinephrine (Fig. 8.2.).

Synthesis of Best-Seller Drugs. http://dx.doi.org/10.1016/B978-0-12-411492-0.00008-0

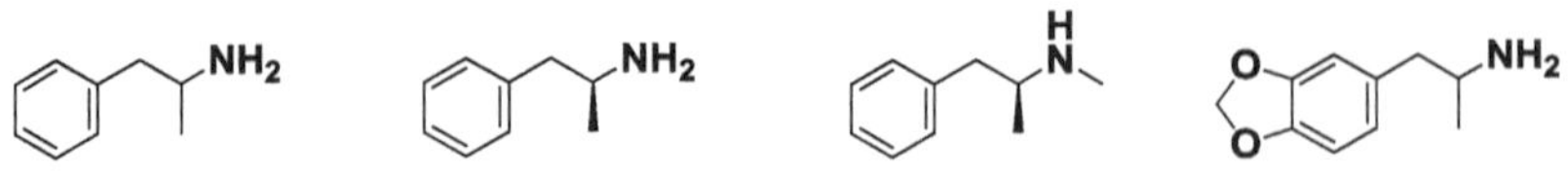

Amphetamine 8.1.5 (Benzedrine) | Dextroamphetamine 8.1.6 | Methamphetamine 8.1.7 | MDMA 8.1.8 (Ecstasy)

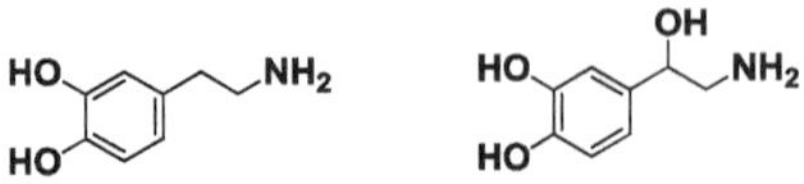

Dopamine | Norepinephrine

FIG. 8.2 **Drugs structurally related to key brain neurotransmitters.**

Compounds such as phentermine (**8.1.9**), fenfluramine (Pondimin) (**8.1.10**), methylphenidate (Ritalin) (**8.1.11**), phenmetrazine (Preludin) (**8.1.12**), in which the terminal amine is incorporated into a morpholine ring, and pemoline (Cylert) (**8.1.13**) represent a broad range of stimulant agents including variety of drugs formerly used to treat obesity (Fig. 8.3.).

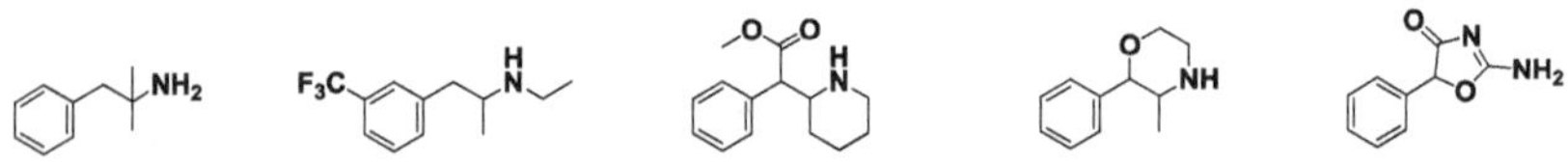

Phentermine 8.1.9 | Fenfluramine 8.1.10 (Pondimin) | Methylphenidate 8.1.11 (Ritalin) | Phenmetrazine 8.1.12 (Preludin) | Pemoline 8.1.13 (Cylert)

FIG. 8.3 **Variety of stimulant drugs structurally related to the amphetamine core.**

All amphetamines derive from the β-phenethylamine core structure and are compounds that easily cross the blood–brain barrier. Although amphetamines are widely acknowledged as synthetic drugs, humans have used natural amphetamines for several millenniums, through the consumption of amphetamines produced in various plants, namely from the plants *Catha edulis* (khat) and *Ephedra sinica*. Recently, a wave of new amphetamines has emerged in the market, including mephedrone (**8.1.14**), methylone (**8.1.15**), methedrone (**8.1.16**), and butylone (**8.1.17**), among others (Fig. 8.4.).

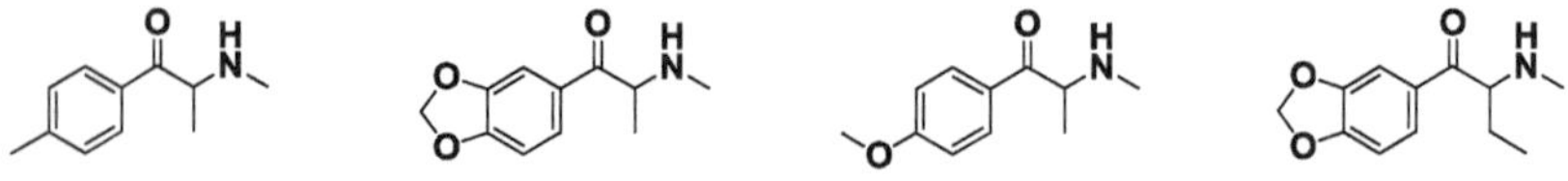

Mephedrone 8.1.14 | Methylone 8.1.15 | Methedrone 8.1.16 | Butylone 8.1.17

FIG. 8.4 **New amphetamines that have recently emerged in the market.**

The amphetamines block the reuptake of norepinephrine and dopamine from the synapse, thus increasing their amount circulating in the cerebral cortex and reticular activating system. They inhibit the action of monoamine oxidase and cause catecholamines to be released. The amphetamines produce CNS and respiratory stimulation, a pressor response, mydriasis, bronchodilation, and contraction of the urinary sphincter. The anorexigenic effect of the amphetamines is probably secondary to the CNS stimulating effect.

Eugeroics, which simply means "good arousal," is a unique class of drugs belonging to the family of diphenylmethylsulfinylacetamides, which contain only two representatives—modafinil (Provigil) (**8.1.18**) and adrafinil (Olmifon) (**8.1.19**)—and are relatively nonaddictive and relatively non–dependence-forming CNS stimulants. They are prescribed for the treatment of narcolepsy (Fig. 8.5.).

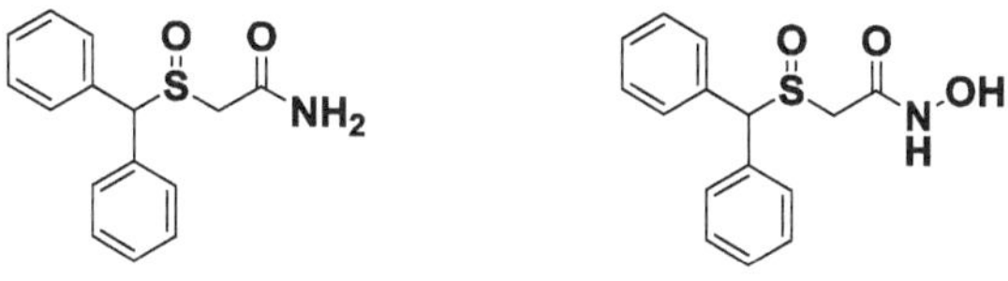

Modafinil 8.1.18
(Provigil)

Adrafinil 8.1.19
(Olmifon)

FIG. 8.5 **Structure of diphenylmethylsulfinylacetamides.**

A broad group of chemicals that possess stimulant and hallucinogenic activity and have a high potential for addiction and abuse are included in the Controlled Substances List.

The bestselling CNS stimulants on the pharmaceutical market today are a mixture of amphetamine salts (Adderall) (**8.1.5** to **8.1.8**), lisdexamfetamine (Vyvanse) (**8.1.24**), methylphenidate (Ritalin) (**8.1.30**), and dexmethylphenidate (Focalin) (**8.1.30a**). They all belong to the amphetamine class of compounds. Another bestselling CNS stimulant, armodafinil (Nuvigil) (**8.1.18**) which is the (-)-(R)-enantiomer of the racemic drug modafinil (Provigil) (**8.1.18**).

Adderall is a brand of pharmaceutical psychostimulant created and prescribed to correct mental disorders, most commonly ADHD. Adderall a combination of four amphetamine salts and contains sulfate salts of amphetamine (**8.1.5**) and dextroamphetamine (**8.1.6**), with the dextroamphetamine saccharate and D,L-amphetamine aspartate monohydrate [4,5].

Lisdexamfetamine–Vyvanse

Lisdexamfetamine (**8.1.24**) dimesylate is a long-acting amphetamine prodrug indicated for the treatment of ADHD [6-8]. Lisdexamfetamine is an amphetamine prodrug, comprising an L-lysine covalently bonded to dextroamphetamine. Lisdexamfetamine itself is a therapeutically inactive molecule, butafter the oral intake, during ingestion, due to enzymic hydrolysis, D-amphetamine released, which is responsible for the pharmacological effect of the drug in a time-dependent manner (sustained release). Because of the need for enzymic hydrolysis, there is a decreased likelihood of misuse and diversion with lisdexamfetamine.

L-lysine-D-amphetamine synthesis was disclosed in patents [9-12]. The synthesis process includes preparation of protected Lys-hydroxysuccinimide ester such as (**8.1.22**). First, di-BOC-protected L-lysine (**8.1.21**) was prepared using a standard procedure, starting from L-lysine dihydrochloride (**8.1.20**) and di(tert-butyl) dicarbonate, which were reacted in sodium hydroxide aqueous solution to produce (**8.1.21**). The last was then reacted with hydroxysuccinimide in the presence of N,N′-dicyclohexylcarbodiimide to produce ester (**8.1.22**). The obtained compound was coupled with D-amphetamine (**8.1.6**) and the product (**8.1.23**) was deprotected with an acid—HCl [9] or methane sulfonic acid [10]—to produce the desired product (**8.1.24**) as dihydrochloride or dimesylate (Scheme 8.1.).

Another patent describes synthesis of the analogues, but employs benzyloxycarbonyl (Cbz) protective groups [11] instead of amino groups in lysine.

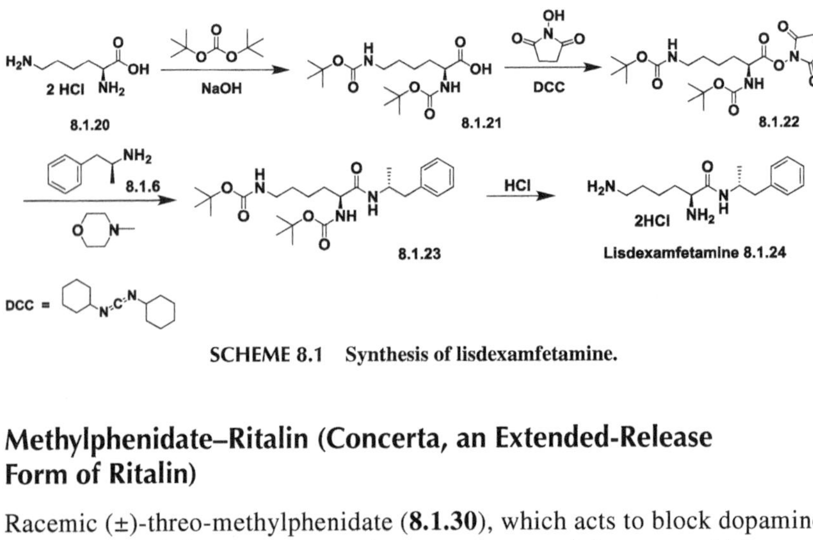

SCHEME 8.1 Synthesis of lisdexamfetamine.

Methylphenidate–Ritalin (Concerta, an Extended-Release Form of Ritalin)

Racemic (±)-threo-methylphenidate (**8.1.30**), which acts to block dopamine and norepinephrine transporters, is currently one of the most widely used drugs for the treatment of ADHD [13-18]. It has been demonstrated that pharmacological specificity of methylphenidate resides entirely in the (2R,2′R)-(+)-threo-methylphenidate [19-24]. Methylphenidate is synthesized and marketed as a mixture of two racemates: (±)-erythro and (±)-threo. Original synthesis of this medication starts from the condensation of phenylacetonitrile (**8.1.25**) with 2-chloro-pyridine (**8.1.26**) in the presence of sodium amide. Obtained phenyl-2-pyridylacetonitrile (**8.1.27**) was hydrolysed with sulfuric acid to give α-phenyl-2-pyridylacetamide (**8.1.28**), which reacted with dimethyl sulfate in MeOH to give methyl 2-phenyl-2-(pyridin-2-yl) acetate (**8.1.29**). The last was hydrogenated in acetic acid in the presence of Pt catalyst at room temperature to produce the desired methylphenidate (**8.1.30**) [25] (Scheme 8.2.).

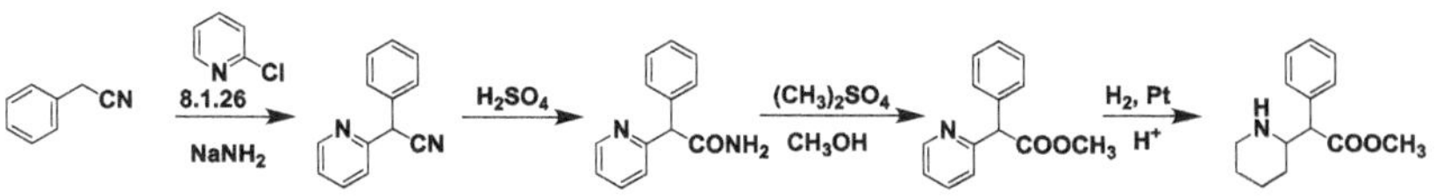

8.1.25 8.1.27 8.1.28 8.1.29 Methylphenidate 8.1.30

SCHEME 8.2 Synthesis of methylphenidate.

Dexmethylphenidate–Focalin (D-Threo-Methylphenidate)

Methylphenidate that is synthesized as described above exists as four possible stereoisomers (**8.1.30a-d**) (Fig. 8.6.).

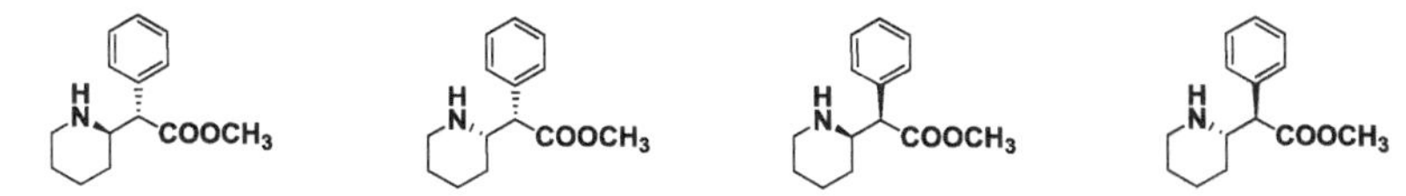

L-erythro 2S,2'R-Methylphenidate 8.1.30d L-threo (2S, 2'S)-Methylphenidate 8.1.30b D-threo (2R, 2'R)-Methylphenidate 8.1.30a D-erythro 2R,2'S-Methylphenidate 8.1.30c

FIG. 8.6 Four possible stereoisomers of methylphenidate.

Experiments with the individual D- and L-threo-enantiomers have shown that the biological activity resides predominantly in the D-threo enantiomer (**8.1.30a**). The erythro isomers (**8.1.30c,d**) have been shown to exhibit little therapeutic effect.

Various methods and approaches, such as classical resolution [19,21-24] use of enantiomerically pure precursors and enantioselective synthesis approaches, and as well as others, to the preparation of enantiomerically pure (2R,2′R)-(+)-threo-methylphenidate are proposed [26-30]. Resolution via enzymatic hydrolysis, which afford (2S,2′S)-(-)-threo and (2R,2′R)-(+)-threo methylphenidate, is also described [31] and all have been reviewed [32]. Process for the racemization of L-threo-methylphenidate product is reported [33]. One method for stereoselective synthesis of dexmethylphenidate with the use of D-pipecolic acid as the chiral precursor is described below [30].

The synthesis starts from D-pipecolic acid (**8.1.31**), which was BOC-protected with di-t-butyl dicarbonate to produce BOC-acid (**8.1.32**), and on treatment with N-methylhydroxylamine was transformed to N-methoxy-N-methylamide (**8.1.33**). The reaction of N-methoxy-N-methylamide with phenyllithium produced optically pure aromatic amino ketone (**8.1.34**). Employing the Wittig reaction with methylene triphenylphosphonium ylide prepared from methyl triphenylphosphonium bromide and potassium tert-butoxide obtained ketone (**8.1.34**), which was converted to the chiral aromatic alkene (**8.1.35**). Only threo alcohol (**8.1.36**) was isolated when olefin (**8.1.35**) underwent an hydroboration–oxidation reaction followed by hydroperoxide anion treatment, which gave the high yield of threo isomeric alcohol. The obtained alcohol (**8.1.36**) was oxidized with pyridinium chlorochromate in N,N-dimethylformamide (DMF) to produce acid (**8.1.37**), which was esterified using diazomethane to produce the desired (2R,2′R)-(+)-threo-methylphenidate–dexmethylphenidate (**8.1.30a**) (Scheme 8.3.).

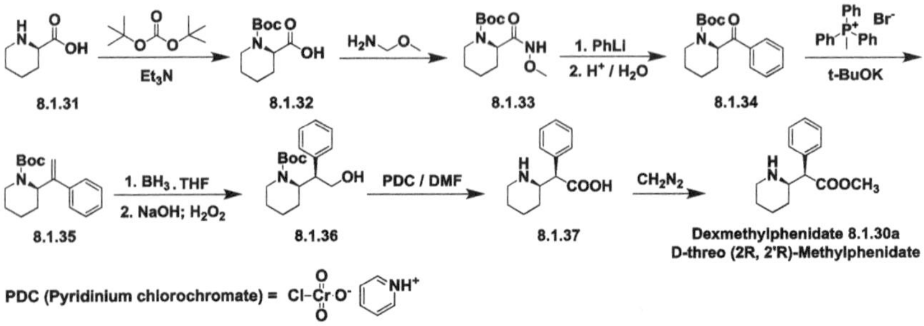

SCHEME 8.3 **Synthesis of dexmethylphenidate.**

The pharmacological profile of dexmethylphenidate-the single isomer version of racemic methylphenidate (Ritalin) is well presented in reviews [34,35].

Armodafinil–Nuvigil

Armodafinil (Nuvigil) R- is the longer-lasting isomer of the racemic compound modafinil, a wakefulness-promoting medication that is indicated to improve wakefulness in patients with excessive sleepiness associated with obstructive sleep apnea, shift work disorder, and narcolepsy. The most commonly reported adverse events are headache, nausea, dizziness, and insomnia. Serious rash, requiring hospitalization and discontinuation of treatment, has been reported with the use of armodafinil and modafinil.

Armodafinil (**8.1.18a**) was synthesized starting from acetylation of benzhydrol (**8.1.38**), with acetic anhydride and catalytic sulfuric acid to prepare acetate (**8.1.39**). Obtained crude acetate was reacted with methyl thioglycolate (**8.1.40**) to produce 2-(benzhydrylthio)acetamide (**8.1.41**). The last was aminated with ammonia to give 2-(benzhydrylthio)acetamide (**8.1.42**) and then oxidized to the desired sulfoxide (**8.1.18a**). In another method, benzhydrol (**8.1.38**), via reaction with thiourea and further acidic workup, was transformed to diphenylmethanethiol (**8.1.43**). The obtained thiol then reacted with chloroacetic acid ester (**8.1.44**) or chloroacetamide (**8.1.45**) to produce the requested 2-(benzhydrylthio)acetamide (**8.1.42**). Prepared in different ways, this amide then underwent the asymmetric oxidation step, with hydrogen peroxide, tert-butylhydroperoxide or cumene hydroperoxide in the presence of chiral complex obtained from (S,S)-(-)-diethyl tartrate, titanium (IV) isopropoxide, and water to produce the desired (+)-modafinil-armodafinil (**8.1.18a**) [36,37] (Scheme 8.4.)

Numerous other catalytic and stoichiometric methods of oxidation with enantioselective reagents [38] and microbial oxidation [39] are described. The racemic form was approved earlier as modafinil (**8.1.18**) was used for preferential crystallization of racemate to provide Phase I trials of armodafinil. Enantiospecific resolution of armodafinil on cellulose-based high-performance liquid chromatography (HPLC) columns, as well as large-scale chromatography was implemented for enantiomers of armodafinil [40,41].

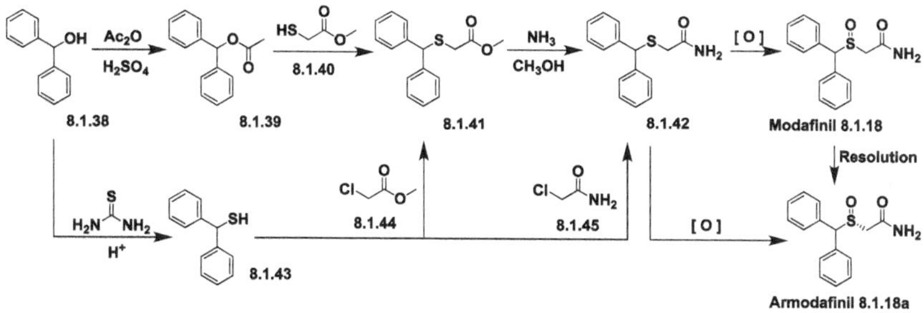

SCHEME 8.4 Synthesis of armodafinil.

REFERENCES

1. Favrod-Coune, T.; Broers, B. The health effect of psychostimulants: a literature review. *Pharmaceuticals* **2010,** *3,* 2333–2361.
2. Pliszka, S. R. Psychostimulants. In *Pharmacotherapy of Child and Adolescent Psychiatric Disorders,* 3rd ed.; Rosenberg, D., Gershon, S., Eds. Wiley-Blackwell, 2012; pp 65–104.
3. Fernandez-Espejo, E.; Rodriguez-Espinosa, N. Psychostimulant drugs and neuroplasticity. *Pharmaceuticals* **2011,** *4,* 976–991.
4. Sallee, F. R.; Smirnoff, A. V. Adderall XR: long acting stimulant for single daily dosing. *Expert Rev. Neurother.* **2004,** *4* (6), 927–934.
5. McKeage, K.; Scott, L. J. SLI-381 (Adderall XR). *CNS Drugs* **2003,** *17* (9), 669–675.
6. Faraone, S. V. Lisdexamphetamine dimesylate: the first long-acting prodrug stimulant treatment for attention deficit/hyperactivity disorder. *Expert Opin. Pharmacother.* **2008,** *9* (9), 1565–1574.
7. Childress, A. C.; Sallee, F. R. The use of lisdexamphetamine dimesylate for the treatment of ADHD. *Expert Rev. Neurother.* **2012,** *12* (1), 13–26.
8. Blick, S. K. A.; Keating, G. M. Lisdexamfetamine. *Paediatr. Drugs* **2007,** *9* (2), 129–135.
9. Mickle, T.; Krishnan, S.; Bishop, B.; Lauderback, C.; Moncrief, J. S.; Oberlender, R.; Piccariello, T. Abuse-resistant amphetamine prodrugs, US 20070042955 (2007).
10. Mickle, T.; Krishnan, S.; Bishop, B.; Lauderback, C.; Moncrief, J. S.; Oberlender, R.; Piccariello, T.; Paul, B. J.; Verbicky, C. A. Abuse-resistant amphetamine prodrugs, US 20090234002 (2009).
11. Bhirud, S. B.; Sarin, G. S.; Kumar, R., Process for preparation of lisdexamphetamine and salts thereof, WO 2013011526 (2013).
12. Bauer, M. J.; Callen, G. R.; Humphrey, J. C.; Johnson, T. J.; Schiesher, M. W. Methods for preparing lisdexamphetamine and its salts, US 20120157706 (2012).
13. Capp, P. K.; Pearl, P. L.; Conlon, C. Methylphenidate HCl: therapy for attention deficit hyperactivity disorder. *Expert Rev. Neurother.* **2005,** *5* (3), 325–331.

14. Leonard, B. E.; McCartan, D.; White, J.; King, D. J. Methylphenidate: a review of its neuropharmacological, neuropsychological and adverse clinical effects. *Hum. Psychopharmacol.* **2004,** *19* (3), 151–180.
15. Lyseng-Williamson, K. A.; Keating, G. M. Extended-release methylphenidate (Ritalin LA). *Drugs* **2002,** *62* (15), 2251–2259.
16. Keating, G. M.; Figgitt, D. P. Dexmethylphenidate. *Drugs* **2002,** *62* (13), 1899–1908.
17. Challman, T. D.; Lipsky, J. J. Methylphenidate: its pharmacology and uses. *Mayo Clin. Proc.* **2000,** *75* (7), 711–721.
18. Perel, J. M.; Dayton, P. G. Methylphenidate. *Psychopharmacology (Berl).* **1977,** *2* (Pt. 2), 1287–1316.
19. Patrick, K. S.; Caldwell, R. W.; Ferris, R. M.; Breese, G. R. Pharmacology of the enantiomers of threo-methylphenidate. *J. Pharmacol. Exp. Ther.* **1987,** *241* (1), 152–158.
20. Ding, Y. S.; Fowler, J. S.; Volkow, N. D.; Dewey, S. L.; Wang, G. J.; Logan, J.; Gatley, S. J.; Pappas, N. Chiral drugs. Comparison of the pharmacokinetics of [11C]d-threo and l-threo-methylphenidate in the human and baboon brain. *Psychopharmacology (Berl).* **1997,** *131* (1), 71–78.
21. Hartmann, M.; Panizzon, L. Pyridine and piperidine derivatives, US 2507631 (1950).
22. Rometsch, R. Conversion of stereoisomers, US 2957880 (1960).
23. Maxwell, R. E.; Chaplin, E.; Eckhardt, S. B.; Soares, J. R.; Hite, G. Conformational similarities between molecular models of phenethylamine and of potent inhibitors of the uptake of tritiated norepinephrine by adrenergic nerves in rabbit aorta. *J. Pharmacol. Exp. Ther.* **1970,** *173* (1), 158–165.
24. Froimowitz, M.; Patrick, K. S.; Cody, V. Conformational analysis of methylphenidate and its structural relationship to other dopamine reuptake blockers such as CFT. *Pharm. Res.* **1995,** *12,* 1430–1434.
25. Panizzon, L. Preparation of pyridyl- and piperidylarylacetonitriles and some derivatives. *Helv. Chim. Acta* **1944,** *27,* 1748–1756.
26. Deutsch, H. M.; Shi, Q.; Gruszecka-Kowalik, E.; Schweri, M. M. Synthesis and pharmacology of potential cocaine antagonists. 2. Structure-activity relationship studies of aromatic ring-substituted methylphenidate analogs. *J. Med. Chem.* **1996,** *39* (6), 1201–1209.
27. Axten, J. M.; Krim, L.; Kung, H. F.; Winkler, J. D. A stereoselective synthesis of dl-threo-methylphenidate: preparation and biological evaluation of novel analogs. *J. Org. Chem.* **1998,** *63* (2), 9628–9629.
28. Russowsky, D.; Amaro da Silveira Neto, B. A concise and stereoselective synthesis of (+/-)-erythro-methylphenidate. *Tetrahedron Lett.* **2003,** *44* (14), 2923–2926.
29. Lapinsky, D. J.; Yarravarapu, N.; Nolan, T. L.; Surratt, C. K.; Lever, J. R.; Tomlinson, M.; Vaughan, R. A.; Deutsch, H. M. Evolution of a compact photoprobe for the dopamine transporter based on (±)-threo-methylphenidate. *ACS Med. Chem. Lett.* **2012,** *3* (5), 378–382.
30. Thai, D. L.; Sapko, M. T.; Reiter, C. T.; Bierer, D. E.; Perel, J. M. Asymmetric synthesis and pharmacology of methylphenidate and its para-substituted derivatives. *J. Med. Chem.* **1998,** *41* (4), 591–601.
31. Prashad, M.; Har, D.; Repic, O.; Blacklock, T. J.; Giannousis, P. Enzymic resolution of (±)-threo-methylphenidate. *Tetrahedron: Asymmetry* **1998,** *9* (12), 2133–2136.
32. Prashad, M. Approaches to the preparation of enantiomerically pure (2R,2′R)-(+)-threo-methylphenidate hydrochloride. *Adv. Synth. Catal.* **2001,** *343* (5), 379–392.
33. Chavan, A. B.; Gundecha, S. S.; Kadam, P. N.; Maikap, G. C.; Gurjar, M. K. An efficient process for the racemization of unwanted (2s,2′s or l-threo)-α-phenyl-α-(2-piperidyl)acetamide. *Org. Process Res. Dev.* **2010,** *14* (6), 1473–1475.

34. Anonymous, Dexmethylphenidate-Novartis/Celgene: Focalin, D-MPH, D-methylphenidate hydrochloride, D-methylphenidate, dexmethylphenidate, dexmethylphenidate hydrochloride. *Drugs R&D* **2002,** *3* (4), 279–282.
35. Liu, F.; Minami, H.; Silva, R. R. Dexmethylphenidate hydrochloride in the treatment of attention deficit hyperactivity disorder. *Neuropsychiatric Disease Treatment* **2006,** *2* (4), 467–473.
36. Rebiere, F.; Duret, G.; Prat, L. Process for enantioselective synthesis of single enantiomers of modafinil by asymmetric oxidation, EP 1516869 (2005).
37. Rebiere, F.; Duret, G.; Prat, L. Process for enantioselective synthesis of single enantiomers of modafinil and related compounds by asymmetric oxidation of the corresponding sulfides in the presence of chiral metal complexes, WO 2005028428 (2005).
38. Ternois, J.; Guillen, F.; Plaquevent, J.-C.; Coquerel, G. Asymmetric synthesis of modafinil and its derivatives by enantioselective oxidation of thioethers: comparison of various methods including synthesis in ionic liquids. *Tetrahedron: Asymmetry* **(2008) 2007,** *18* (24), 2959–2964.
39. Olivo, H. F.; Osorio-Lozada, A.; Peeples, T. L. Microbial oxidation/amidation of benzhydrylsulfanyl acetic acid. Synthesis of (+)-modafinil. *Tetrahedron: Asymmetry* **2005,** *16* (21), 3507–3511.
40. Nageswara, R. R.; Shinde, D. D.; Kumar Talluri, M. V. N. Enantioselective HPLC resolution of synthetic intermediates of armodafinil and related substances. *J. Sep. Sci.* **2008,** *31* (6–7), 981–989.
41. Hauck, W.; Adam, P.; Bobier, C.; Landmesser, N. Use of large-scale chromatography in the preparation of armodafinil. *Chirality* **2008,** *20* (8), 896–899.

Chapter 9

Antiepileptic Drugs

Epilepsy is defined as a neurological condition that is characterized by recurrent, unprovoked seizures.

The International League Against Epilepsy and the International Bureau for Epilepsy have come to consensus definitions for the term *epilepsy*: "Epilepsy is a disorder of the brain characterized by an enduring predisposition to generate epileptic seizures and by the neurobiologic, cognitive, psychological, and social consequences of this condition" [1]. Epilepsy is one of the most common neurological conditions, with an estimated prevalence of approximately 1% of the total population of North America.

Pharmacological treatment of epileptic seizures practically started in the middle of the 19th century with the introduction of potassium bromide, the earliest effective treatment for epilepsy. There would not be a better drug until 1910 to 1912 when phenobarbital (**9.1.1**), along with other barbiturates, entered the pharmaceutical market followed by mephobarbital (**9.1.2**) and primidone (**9.1.3**). In the 1940s, another effective drug from the imidazolidine-2,4-dione class of compounds, phenytoin (**9.1.4**), was proposed for the treatment of epilepsy. Then followed next structural simplifications, and the pyrrolidine-2,5-dione derivative ethosuximide (**9.1.5**) appeared toward the end of the 1950s. These events were followed by the discovery of valproic acid's (**9.1.6**) anticonvulsant properties in the 1960s, the sodium salt of which became the drug of choice for treatment of primary generalized epilepsies. Carbamazepine (**9.1.7**) was introduced in 1974. Also in the mid-1970s some benzodiazepines, such as diazepam (**9.1.8**), clonazepam (**9.1.9**), and lorazepam (**9.1.10**), which act as central nervous system depressants, were proposed for the treatment of epileptic seizures (Fig. 9.1.). The synthesis of all of them is described in our previous book [2].

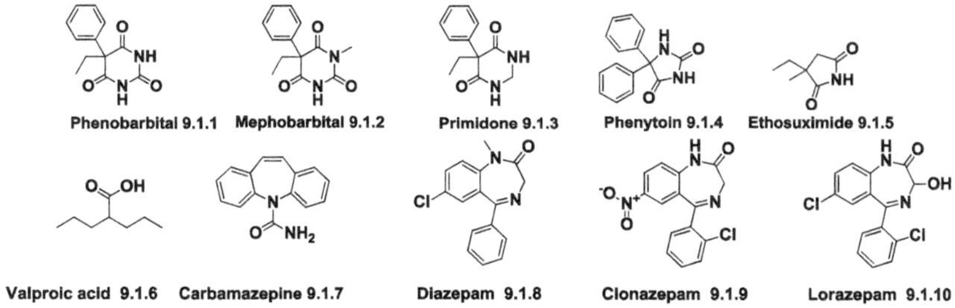

FIG. 9.1 First generation of antiepileptic drugs.

Synthesis of Best-Seller Drugs. http://dx.doi.org/10.1016/B978-0-12-411492-0.00009-2

The listed medications were the mainstays of seizure treatment until the 1990s when a newer, second generation of antiepileptic drugs (AEDs) appeared, which are listed here in chronological order: felbamate (**9.1.11**) (1993) [3-6], gabapentin (**9.1.12**) (1993) [7-9], lamotrigine (**9.1.13**) (1994) [10-13], topiramate (**9.1.14**) (1996) [14-18], tiagabine (**9.1.15**) (1997) [19-23], levetiracetam (**9.1.16**) (1999) [24-27], oxcarbazepine (**9.1.17**) (2000) [28-31], zonisamide (**9.1.18**) (2000) [32-38], and pregabalin (**9.1.19**) (2005) [39-44] appeared in doctor's armory (Fig. 9.2.).

This series of compounds is accepted as the second generation of AEDs and are well reviewed [45-55].

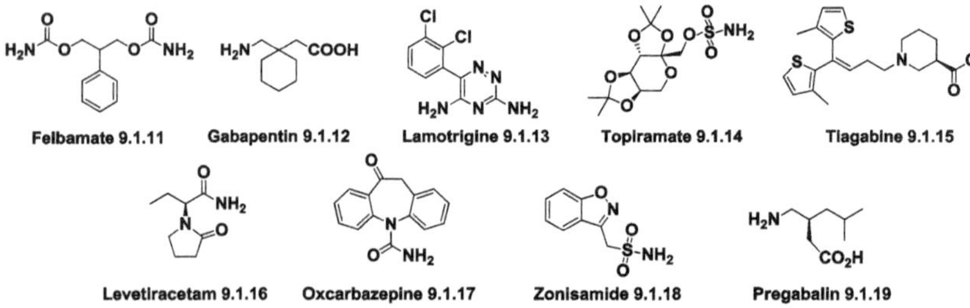

FIG. 9.2 Second generation of AEDs.

These newer antiepileptics offer the potential advantages of fewer drug interactions, unique mechanisms of action, and a broader spectrum of activity.

There is a group of compounds formally named the third generation of AEDs. They should be divided into two categories: second generation of the first and second generation compounds such as new derivatives of valproic acid (**9.1.6**) like valrocemide (**9.1.20**) [56-60] and DP-valproic acid (**9.1.21**) [59-62]; eslicarbazepine (**9.1.22**) [63-70], a new derivative of oxcarbazepine (**9.1.17**); brivaracetam (**9.1.23**) [71-78] and seletracetam (**9.1.24**) [79-83], which are new derivatives of levetiracetam (**9.1.16**); fluorofelbamate (**9.1.25**) [84,85], a new derivative of felbamate (**9.1.11**); and fosphenytoin (**9.1.26**) [86-89], a new derivative of phenytoin (**9.1.4**) (Fig. 9.3.).

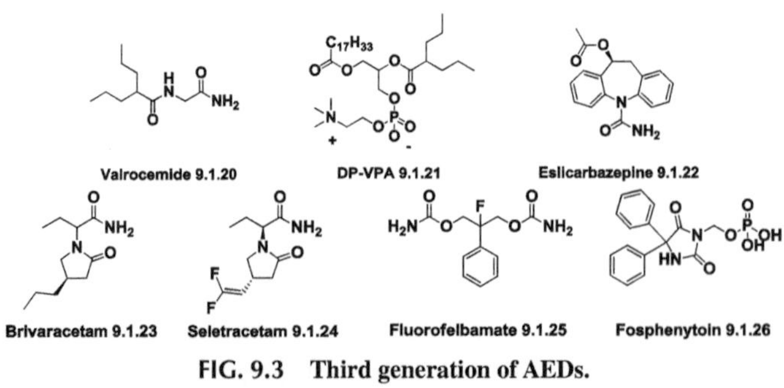

FIG. 9.3 Third generation of AEDs.

There are also compounds with completely new chemical structures. One group belongs to the 2-amino-N-benzylacetamide series. Among this group are lacosamide (**9.1.27**) [90-104] and remacemide (**9.1.28**) [105-111], and the 2-(benzylamino)acetamide series drug safinamide (**9.1.29**) [112-117], which are believed to act mainly through voltage-gated sodium channel inactivation. Other new compounds proposed as AEDs are represented as individual chemical entities belonging to different chemical classes. They are retigabine (**9.1.30**) [118-128], a potassium channel opener; rufinamide (**9.1.31**) [129-140], which probably is working via stabilizing a sodium channel inactive state; soretolide (**9.1.32**) [141] whose action is close to that of carbamazepine; losigamone (**9.1.33**) [142-144] whose mechanism of action is not known; carisbamate (**9.1.34**) [145,146] whose exact mechanism of action is also unknown; stiripentol (**9.1.35**) [147-151], which increases γ-aminobutyric acid (GABA) transmission; perampanel (**9.1.36**) [152-156] and talampanel (**9.1.37**) [157-163], which are noncompetitive antagonists of ionotropic transmembrane α-amino-3-hydroxy-5-methyl-4-isoxazolepropionic acid (AMPA) receptors; carabersat (**9.1.38**) [164], which has some novel mechanism of action; vigabatrin (**9.1.39**) [165-176], which inhibits the catabolism of GABA by irreversibly inhibiting GABA transaminase; and first-in-class synthetic novel class neurosteroid ganaxolone (**9.1.40**) [177-181], which is devoid of any hormonal activity and allosterically modulates the $GABA_A$ receptor complex (Fig. 9.4.).

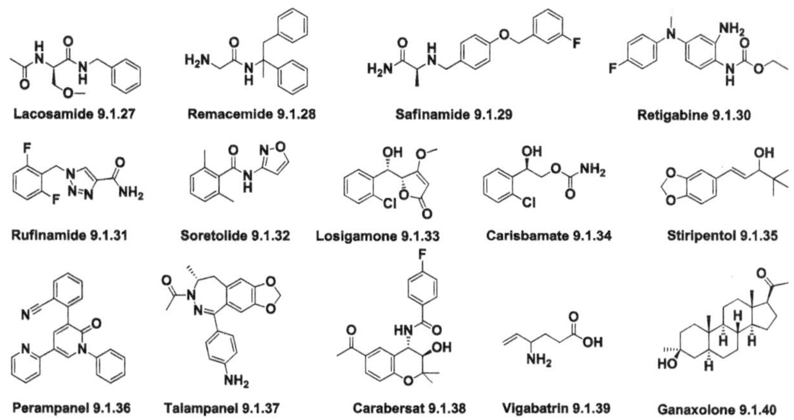

FIG. 9.4 Antiepileptic drugs with completely new chemical structures.

Unlike other therapeutic classes, detailed actions of AED at the molecular level are not completely understood. Most of them have more than one mechanism of action, which is why they are not classified according to modes of action. Nevertheless, a number of mechanisms have been identified. Many AEDs act at various subtypes of voltage-gated channels. One pharmacological mechanisms is blockade of the voltage-dependent sodium channels, which has been established as the primary action of phenytoin (**9.1.4**), carbamazepine

(**9.1.7**), oxcarbazepine (**9.1.17**), lamotrigine (**9.1.13**), topiramate (**9.1.14**), zonisamide (**9.1.18**), and felbamate (**9.1.11**). Blockade of T-type calcium channels is a primary mechanism for ethosuximide (**9.1.5**) and zonisamide (**9.1.18**). Modulation of excitatory transmission release through blockade of N-type and P/Q-type calcium channels seems to be responsible for the action of gabapentin (**9.1.12**) and pregabalin (**9.1.19**) in epilepsy. Lamotrigine (**9.1.13**) and levetiracetam (**9.1.16**) differentially affect potassium channels. Some AEDs suppress epileptic seizures by GABAergic inhibition. Tiagabine (**9.1.15**) exerts such an effect. Topiramate (**9.1.14**) acts on AMPA/kainate receptors; felbamate (**9.1.11**) acts on N-methyl-D-aspartate (NMDA) receptors; and lamotrigine (**9.1.13**) may modulate serotoninergic transmission. More likely, the clinical efficacy of many AEDs is related to multifactorial mechanisms. Progress reports on new AEDS are summarized for the regular conferences on progress regarding new AEDS in different stages of development, as well as new findings [182-187]. Recent advances and trends also are carefully summarized in the reviews [188-196].

Pregabalin–Lyrica

Pregabalin (**9.1.19**) is among the bestsellers drugs for the treatment of epilepsy. It is structurally related to both the inhibitory neurotransmitter ΓABA and gabapentin (**9.1.12**). The S-(+) enantiomer is more potent than the R-(-)-enantiomer. The two enantiomers were prepared by chiral oxazolidinone alkylation chemistry. The synthetic route of discovery synthesis (Scheme 9.1.) [197] was based on asymmetric alkylation of a chiral oxazolidinone auxiliary (**9.1.43**). Thus the chloride of 4-methylpentanoic acid (**9.1.42**) prepared from the corresponding acid (**9.1.41**) was used for acylation of the anion of chiral (4R,SS)-(+)-4-methyl-5-phenyl-2-oxazolidinone (**9.1.43**) to produce the acyloxazolidinone (**9.1.44**), which was then alkylated with benzyl bromoacetate to produce the product (**9.1.45**) with greater than 95% enantiomeric excess (ee). The chiral auxiliary was removed by lithium hydroxide/hydrogen peroxide treatment followed by a reductive workup with a solution of sodium bisulfite, preadjusted to neutral pH with sodium sulfite. The resulting acid (**9.1.46**) was then selectively reduced to the corresponding alcohol (**9.1.47**) using borane dimethylsulfide complex. The obtained alcohol (**9.1.47**) was converted to the tosylate (**9.1.48**) with tosyl chloride in pyridine and then to azide (**9.1.49**) with sodium azide in dimethyl sulfoxide. The obtained azide (**9.1.49**) was finally converted to pregabalin (**9.1.19**) by simultaneous hydrogenation of azide and benzyl ester functions using palladium on charcoal catalyst. The (R)-enantiomer, was prepared by the same method using (4S,SR)-(-)-4-methyl-5-phenyl-2-oxazolidinone as the chiral auxiliary. The optical purity of the product from this synthesis was excellent. However the route is limited use for scale-up, owing principally to low temperature (−78 °C) reactions, column chromatography for isolation of intermediates, some side reactions, and low overall yield (10%).

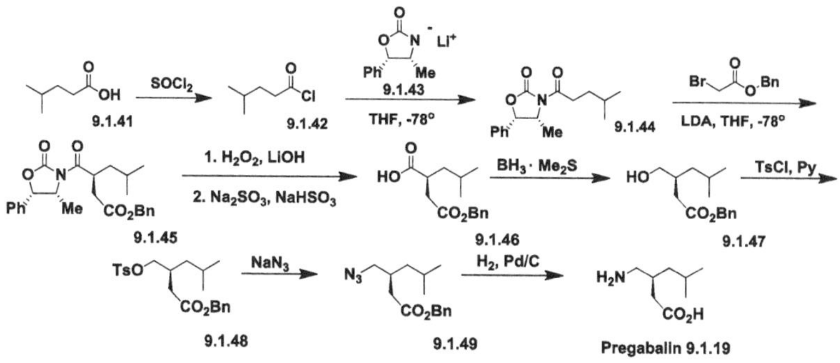

SCHEME 9.1 **Synthesis of pregabalin.**

Use of the tert-butyl ester in place of the benzyl ester showed improved results and allowed to use hydrolysis of t-butyl analogue of (**9.1.49**) under basic conditions instead of hydrogenation [41]. Other alternate processes, such as L-leucine approach, Stobbe condensation, Hofmann rearrangement, malonate route, and others, were found by the same group [41]. For example, the route for the synthesis of pregabalin (**9.1.19**) starting from L-leucine (**9.1.50**) is illustrated in Scheme 9.2.

The bromo acid (**9.1.51**) was prepared from L-leucine by treatment with sodium nitrite and sodium bromide in aqueous sulfuric acid. This reaction has been shown to proceed with retention of configuration. The obtained bromo acid (**9.1.19**) was esterified using tert-butyl acetate to produce the ester (**9.1.52**). Displacement of bromide with diethylsodiomalonate (**9.1.53**) produced the compound (**9.1.54**), which was hydrolyzed to acid (**9.1.55**). Borane dimethylsulfide reduction of produced the chiral lactone (**9.1.56**), which was opened with trimethylsilyl iodide, producing compound (**9.1.57**) and elaborated to pregabalin via the preparation of azide (**9.1.58**) followed by its hydrolysis to acid (**9.1.59**) and hydrogenation to pregabalin (**9.1.19**).

The L-leucine route was found not to be low enough in cost, and was set aside in favor of other, more advantageous sequences.

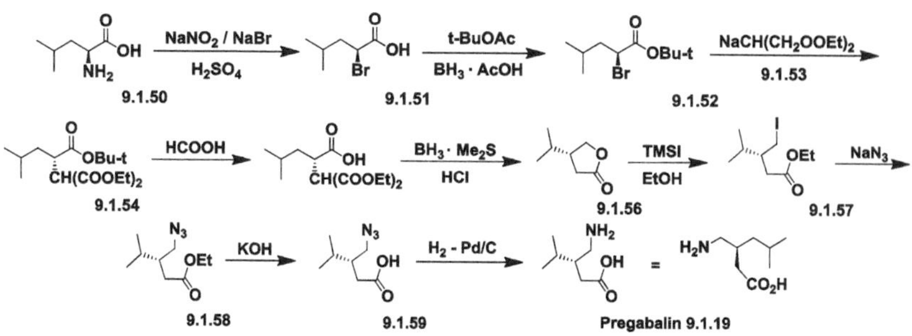

SCHEME 9.2 **Synthesis of pregabalin.**

In another route, Stobbe condensation of isovalderaldehyde (**9.1.60**) with diethylmalonate (**9.1.61**) was implemented using di-n-propylamine/acetic acid as catalyst to produce the diester (**9.1.62**) (Scheme 9.3.) [41]. Conjugate hydrocyanation reaction has been used to obtain cyano diester (**9.1.63**), which was converted to racemic compound (**9.1.65**) by a series of reactions, including hydrolysis and decarboxylation in high yield. The preferred isomer, pregabalin (**9.1.19**), emerged from the obtained racemate with (S)-(+)-mandelic acid followed by two recrystallizations.

Benzyl ester, t-butyl and malonate routs are acceptable for industrial implementation and, it may come as a surprise, that the lowest cost process is to use classical resolution rather than enantioselective synthesis [41]. Many other ingenious ways for the synthesis of pregabalin and its analogues have been proposed [98-203].

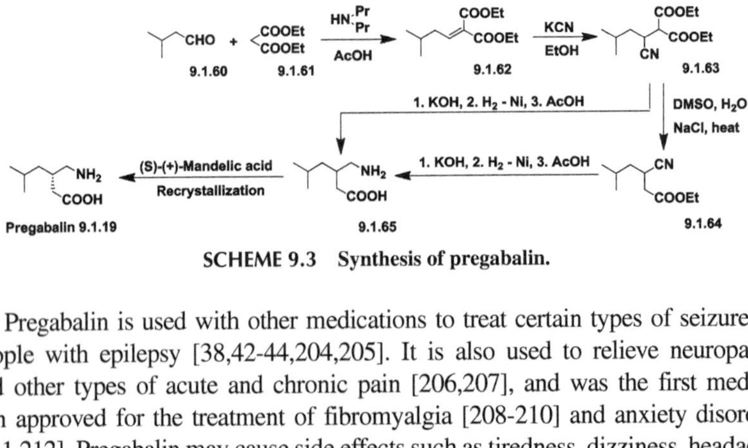

SCHEME 9.3 **Synthesis of pregabalin.**

Pregabalin is used with other medications to treat certain types of seizures in people with epilepsy [38,42-44,204,205]. It is also used to relieve neuropathic and other types of acute and chronic pain [206,207], and was the first medication approved for the treatment of fibromyalgia [208-210] and anxiety disorders [211,212]. Pregabalin may cause side effects such as tiredness, dizziness, headache, speech problems, difficulty remembering or forgetfulness, anxiety, lack of coordination, uncontrollable shaking or jerking of a part of the body, and changes in libido.

Lamotrigine–Lamictal

Two key methods for the synthesis of lamotrigine (**9.1.13**) have been reported. The first method [10,11] is based on condensation of 2,3-dichlorobenzoyl cyanide (**9.1.68**), which is obtained by conversion of 2,3-dichlorobenzoic acid (**9.1.66**) to its acid chloride (**9.1.67**), which is treated with copper cyanide to produce 2,3-dichlorobenzoyl cyanide (**9.1.68**). On condensation with aminoguanidine (**9.1.69**), 2,3-dichlorobenzoyl cyanide (**9.1.68**) proceeds to yield approximately 16% of the desired lamotrigine (**9.1.13**) (Scheme 9.4.).

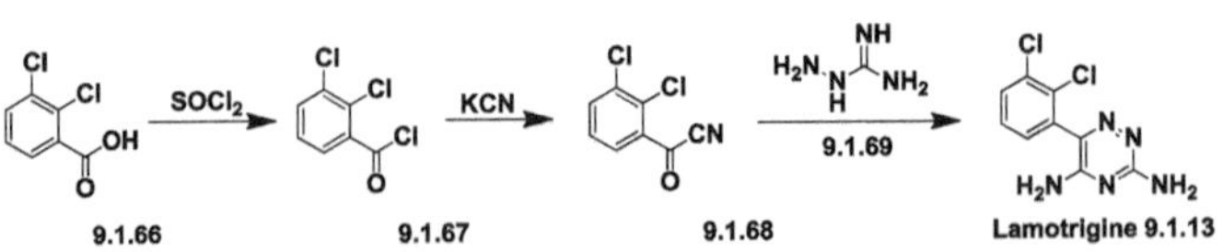

SCHEME 9.4 **Synthesis of lamotrigine.**

The other method used to prepare lamotrigine (**9.1.13**) [213] includes condensation of 2-(2,3-dichlorophenyl)-2-oxoacetic acid (**9.1.70**) with thiosemicarbazide (**9.1.71**) to produce 3-thioxo-3,4-dihydro-1,2,4-triazin-5(6H)-one derivative (**9.1.72**), which on reaction with methyl iodide in aqueous sodium hydroxide solution produces an S-methylated product (**9.1.73**). On interaction with any chlorinating agent, particularly with phosphorous oxychloride, the S-methylated product (**9.1.73**) undergoes replacement of its thiomethyl and hydroxyl groups for chlorine, thereby producing compound (**9.1.74**), which, on reaction with ammonia, produces the desired lamotrigine (**9.1.13**) (Scheme 9.5.).

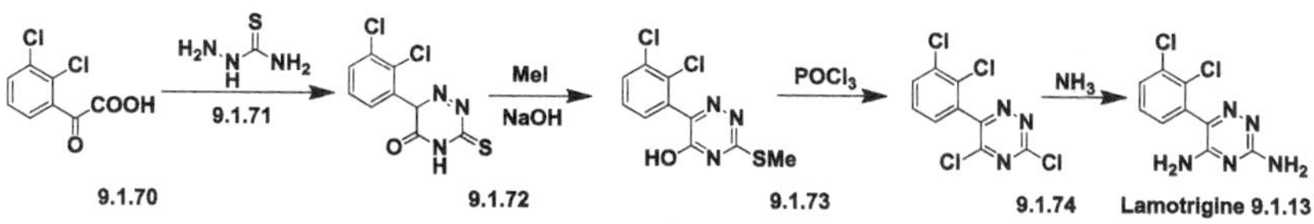

SCHEME 9.5 Synthesis of lamotrigine.

The third, not industrial, but conceptually new method for the synthesis of lamotrigine uses as a starting materials tetrazole diazonium salt (**9.1.75**) and 3-oxo-2-phenylpropanenitrile derivative (**9.1.76**) was proposed recently [214]. Tetrazole diazonium salt (**9.1.75**) does not enter into azo coupling reactions with regular arylacetonitriles because of their low CH acidity. But reaction with 3-oxo-2-phenylpropanenitrile derivative, where the formyl group plays double roles, and of electron withdrawing group, and of easily leaving group gave possibility to synthesize hydrazone (**9.1.77**), which by a series of transformations (**9.1.77** → **9.1.78a,b**) is isomerized to (**9.1.78b**), and when refluxed in acetic acid or DMF is converted into tetrazolo[1,5-b][1,2,4]triazine (**9.1.79**). Finally, the obtained tetrazolotriazine compound (**9.1.79**) was decomposed on heating in acetic acid medium in the presence of a freshly prepared powdered copper to produce lamotrigine (**9.1.13**) (Scheme 9.6.).

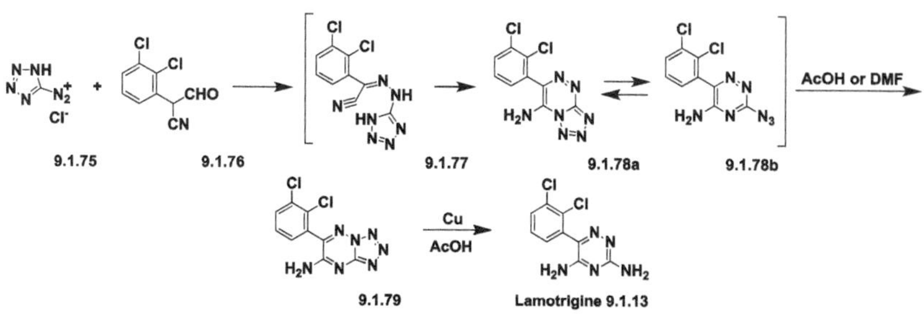

SCHEME 9.6 Synthesis of lamotrigine.

Some related approaches for the synthesis of lamotrigine derivatives can be found in the reviews [215,216].

Lamotrigine is an anticonvulsant drug used in the treatment of epilepsy and bipolar disorder. In addition, it is used to reduce neuropathic pain, trigeminal neuralgia, and cluster headaches [217-224].

Lamotrigine can cause serious side effects such as rashes, fever, swollen glands, body aches, flu symptoms, headache, increased sensitivity to light, pain, muscle weakness, chest pain, irregular heart rhythm, and feeling short of breath.

Levetiracetam–Keppra

Levetiracetam is another bestselling drug included in the list of Top 200 Drugs by sales for the 2010s.

Levetiracetam belongs to class of racetams, which are nootropics capable of reversing amnesia induced by scopolamine, electroconvulsive shock, hypoxia, memory enhancers, neuro enhancers, cognitive enhancers, and intelligence enhancers. There is no universally accepted mechanism of action for racetams. Piracetam (**9.1.80**), aniracetam (**9.1.81**), nefiracetam (**9.1.82**), pramiracetam (**9.1.83**), oxiracetam (**9.1.84**), rolziracetam (**9.1.85**), etiracetam (**9.1.86**), and their structural analogues are representatives of racetam's family [225] (Fig. 9.5.).

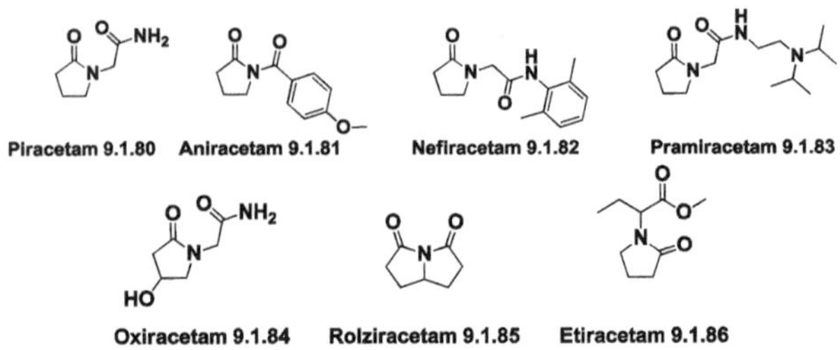

FIG. 9.5 Nootropic racetams.

Several alternative processes for the preparation of levetiracetam have been disclosed. But the first synthesis of racemic levetiracetam–etiracetam (**9.1.86**) probably was proposed in patent [226]. The protocol describes reaction of the sodium derivative of the pyrrolidin-2-one (**9.1.87**) with the ethyl 2-bromobutanoate (**9.1.88**), hydrolysis of obtained ester (**9.1.89**) to the corresponding acid (**9.1.90**), its conversion into the ammonium salt (**9.1.91**), which on heating undergoes thermal dehydration to the desired amide—etiracetam (**9.1.86**) (Scheme 9.7.).

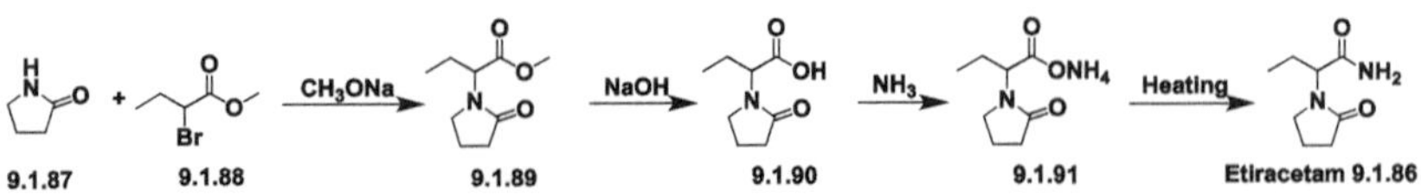

SCHEME 9.7 Synthesis of etiracetam.

The protocol of the synthesis of levetiracetam (**9.1.16**) was first described in the patent [227] (Scheme 9.8.). The method proposed for the separation of the racemate (**9.1.90**) is the use of (S)-1-phenylethanamine which give separable crystalline salt with (S)-2-(2-oxopyrrolidin-1-yl)butanoic acid (**9.1.90a**). (S)-2-(2-Oxopyrrolidin-1-yl)butanoic acid (**9.1.90a**) the product obtained after workup of the separated salt with a strong base and then with acid, on treatment with methyl chloroformate was converted to mixed anhydride (**9.1.92**), which on reaction with ammonia produced the desired levetiracetam (**9.1.16**).

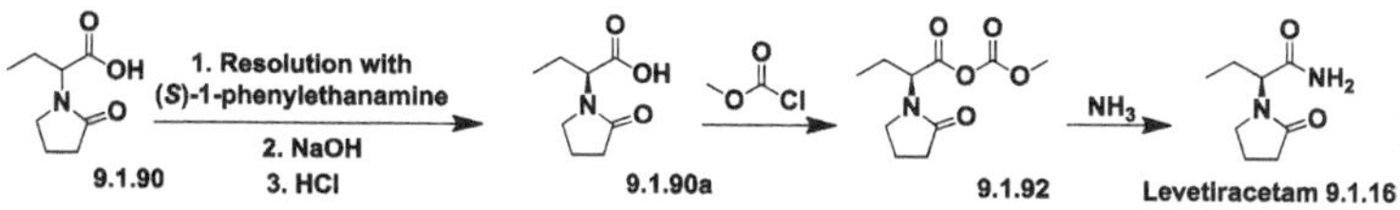

SCHEME 9.8 **Preparation of levetiracetam using (S)-1-phenylethanamine.**

The same patent [227] demonstrated another approach for the synthesis of levetiracetam (**9.1.16**). For that purpose (S)-2-aminobutanamide (**9.1.92**) was acylated with 4-chlorobutanoyl chloride (**9.1.93**) to produce amide (**9.1.95**), or it was alkylated with methyl 4-chlorobutanoate (**9.1.94**) to produce amine (**9.1.96**). Cyclization of the obtained amide (**9.1.95**) in the presence of a base and Bu4NBr, as well as cyclization of amine (**9.1.96**) in the presence of 2-hydroxyperidine produced the desired levetiracetam (**9.1.16**) (Scheme 9.9.).

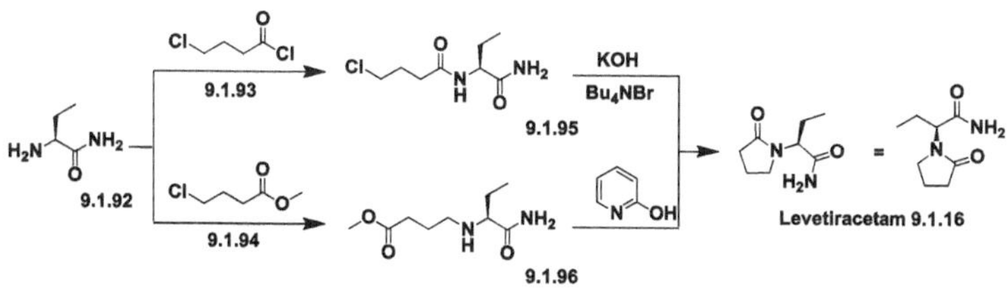

SCHEME 9.9 **Synthesis of levetiracetam using (S)-2-aminobutanamide.**

In another patent [228], levetiracetam (**9.1.16**) was synthesized via hydrogenolysis of (S)-α-[2-(methylthio)ethyl]-2-oxo-1-pyrrolidineacetamide (**9.1.101**) (Scheme 9.10.), which was prepared in a fashion similar to the previous method, but starting with methioninamide (**9.1.97**) followed by desulfurization via hydrogenation with the Raney Ni catalyst. The compound (**9.1.101**) itself was prepared either by cyclization of the products obtained on reaction of (S)-methioninamide (**9.1.97**) with 4-chloro-butyryl chloride (**9.1.93**), or by its alkylation with ethyl 4-bromobutanoate (**9.1.98**) and further cyclization of the obtained products **9.1.99** and **9.1.101**.

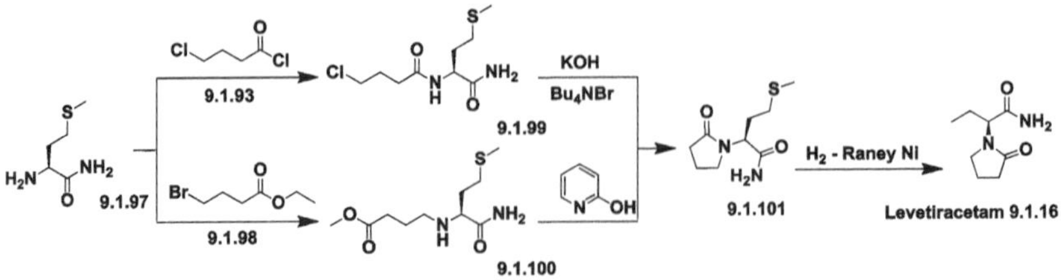

SCHEME 9.10 **Synthesis of levetiracetam.**

An interesting approach for the synthesis of levetiracetam was proposed in another patent [229] (Scheme 9.11.). α-Ketobutyric acid (**9.1.103**) was condensed with pyrrolidinone (**9.1.102**) to produce (**9.1.104**), which has the geometry of a double bond assigned to be Z. The obtained acid was converted to acid chloride with the use of phosphorous pentachloride and then to amide with gaseous ammonia to produce the Z-isomer of amide (**9.1.105**). The Z-isomer of amide (**9.1.105**) underwent asymmetric hydrogenation by a Rh(I) or Ru(II) catalyst such as Rh(I)$(MeOH)_2$[(R)-Binap] or [RuCl(R-Binap)(C6H6)]$^+$Cl$^-$ at a pressure 4 atm to produce the desired levetiracetam (**9.1.16**).

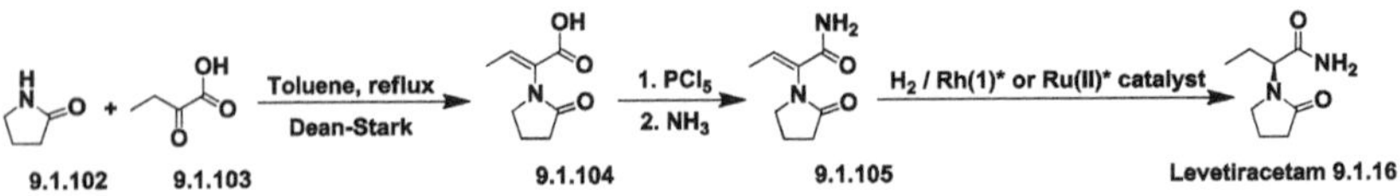

SCHEME 9.11 **Synthesis of levetiracetam.**

A patent [230] describes a method of preparation of levetiracetam from the aminomethyl derivatives of racemic α-ethyl-2-oxo-l-pyrrolidineacetamide, its resolution followed by deaminomethylation. Papers [231-238] describe more sophisticated approaches. Patents [239-241] describe preparation end separation of isomeric mixtures of levetiracetam.

Levetiracetam is a broad-spectrum AED that is effective against a variety of seizure types in adults and children [242-253]. Levetiracetam may cause side effects such as drowsiness, weakness, coordination problems, headache, pain, anxiety, loss of appetite, vomiting, diarrhea, and constipation.

Topiramate–Topamax

Topiramate (**9.1.14**) is among the bestselling drugs for the treatment of epilepsy and is indicated for monotherapy and for adjunctive therapy in patients with partial onset or primary generalized tonic–clonic seizures, and for patients with seizures associated with Lennox-Gastaut syndrome.

Topiramate is synthesized by two methods: D-fructose (**9.1.106**), is reacted with acetone to produce bisacetonide (**9.1.107**). This compound is then condensed with sulfamoyl chloride (**9.1.108**) in the presence of sodium hydride to produce the desired topiramate (**9.1.14**) [254] (Scheme 9.12.).

Alternatively, intermediate (**9.1.107**) is reacted with sulfuryl chloride (**9.1.109**) to produce sulfochloridate (**9.1.110**), which, with sodium azide, produces the azido derivative (**9.1.111**). The azido derivative (**9.1.111**) was reduced either by copper powder in methanol or by catalytic hydrogenation with palladium on carbon to produce topiramate [254-258] (Scheme 9.12.).

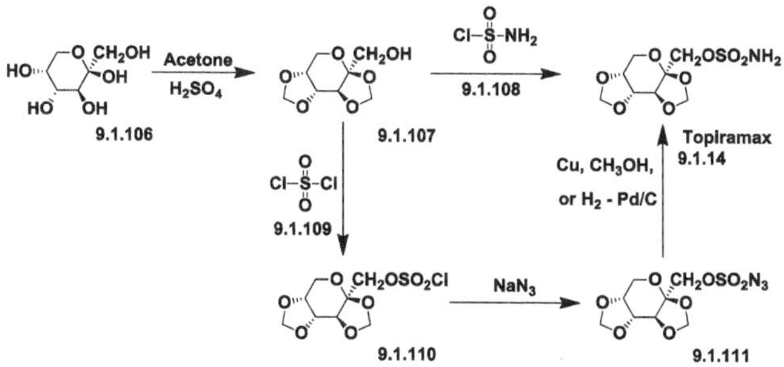

SCHEME 9.12 Synthesis of topiramate.

Topiramate has drawn much attention in the treatment of many diseases other than epilepsy, such as bipolar disorder, bulimia nervosa, alcohol dependence, cocaine dependence, schizophrenia, and personality disorder. Topiramate appears to be a promising drug for the prevention and treatment of migraine and obesity. Moreover, it could be also used to treat Prader-Willi syndrome, essential tremor, and nerve injury [259-268].

Common side effects of topiramate include drowsiness, dizziness, fatigue, lack of coordination, loss of appetite, inability to concentrate or speak clearly, and nervousness.

REFERENCES

1. Fisher, R. S.; van Emde Boas, W.; Blume, W.; Elger, C.; Genton, P.; Lee, P.; Engel, J., Jr. Epileptic seizures and epilepsy: definitions proposed by the International League Against Epilepsy (ILAE) and the International Bureau for Epilepsy (IBE). *Epilepsia* **2005,** *46* (4), 470–472.
2. Vardanyan, R. S.; Hruby, V. J. *Synthesis of Essential Drugs;* Elsevier, 2006.
3. Berger, F. M.; Ludwig, B. J. 2-Phenyl-1,3-propanediol dicarbamate, US 2884444 (1959).
4. Ludwig, B. J.; Powell, L. S.; Berger, F. M. Carbamate derivatives related to meprobamate. *J. Med. Chem.* **1969,** *12* (3), 462–472.
5. Sofia, R. D.; Kramer, L.; Perhach, J. L.; Rosenberg, A. Felbamate. *Epilepsy Res., Suppl.* **1991,** *3,* 103–108.

6. Leppik, I. E.; White, J. R. Felbamate. In *The Treatment of Epilepsy,* 3rd ed.; Shorvon, S., Perucca, E., Engel, J., Jr., Eds. Wiley-Blackwell, 2009; pp 511–518.
7. Satzinger, G.; Hartenstein, J.; Herrmann, M.; Heldt, W. Cyclic amino acids, DE 2460891 (1976).
8. Goa, K. L.; Sorkin, E. M. Gabapentin: a review of its pharmacological properties and clinical potential in epilepsy. *Drugs* **1993,** *46* (3), 409–427.
9. Honarmand, A.; Safavi, M.; Zare, M. Gabapentin: an update of its pharmacological properties and therapeutic use in epilepsy. *J. Res. Med. Sci.* **2011,** *16* (8), 1062–1069.
10. Baxter, M. G.; Elphick, A. R.; Miller, A. A.; Sawyer, D. A. 1,2,4-Triazine derivatives, pharmaceutical compositions and intermediates utilized for their preparation, EP 21121 (1981).
11. Baxter, M. G.; Elphick, A. R.; Miller, A. A.; Sawyer, D. A. Substituted aromatic compounds, CA 1133938 (1982).
12. Choi, H.; Morrell, M. J. Review of lamotrigine and its clinical applications in epilepsy. *Expert Opin. Pharmacother.* **2003,** *4* (2), 243–251.
13. Matsuo, F.; Riaz, A. Lamotrigine. In *The Treatment of Epilepsy,* 3rd ed.; Shorvon, S., Perucca, E., Engel, J., Jr., Eds. Wiley-Blackwell, 2009; pp 535–558.
14. Maryanoff, B. E.; Gardocki, J. F. Anticonvulsant sulfamate derivatives, US 4513006 (1985).
15. Maryanoff, B. E.; Nortey, S. O.; Gardocki, J. F.; Shank, R. P.; Dodgson, S. P. Anticonvulsant O-alkyl sulfamates. 2,3:4,5-bis-O-(1-methylethylidene)-β-D-fructopyranose sulfamate and related compounds. *J. Med. Chem.* **1987,** *30* (5), 880–887.
16. Silberstein, S. D.; Ben-Menachem, E.; Shank, R. P.; Wiegand, F. Topiramate monotherapy in epilepsy and migraine prevention. *Clin. Ther.* **2005,** *27* (2), 154–165.
17. Sommer, B. R.; Fenn, H. H. Review of topiramate for the treatment of epilepsy in elderly patients. *Clin. Interventions Aging* **2010,** *5,* 89–99.
18. Cross, J. H.; Riney, C. J. Topiramate. In *The Treatment of Epilepsy,* 3rd ed.; Shorvon, S., Perucca, E., Engel, J., Jr., Eds. Wiley-Blackwell, 2009; pp 673–683.
19. Groenvald, F. C.; Braestrup, C. Diheterocyclylbutenylamino acids as GABA uptake inhibitors, WO 8700171 (1987).
20. Schwartz, T. L.; Nihalani, N. Tiagabine in anxiety disorders, *Exp. Opin. Pharrmacother.,* (**2006**), *7* (14), 1977–1987.
21. Leach, J. P.; Brodie, M. J. Tiagabine. *Lancet* **1998,** *351* (9097), 203–207.
22. Kalviainen, R. Tiagabine. In *The Treatment of Epilepsy,* 3rd ed.; Shorvon, S., Perucca, E., Engel, J., Jr., Eds. Wiley-Blackwell, 2009; pp 663–672.
23. Bauer, J.; Cooper-Mahkorn, D. Tiagabine: efficacy and safety in partial seizures-current status. *Neuropsychiatr. Dis. Treat* **2008,** *4* (4), 731–736.
24. Gobert, J.; Geerts, J. P.; Bodson, G. (S)-α-Ethyl-2-oxo-1-pyrrolidineacetamide, EP 162036 (1985).
25. Lyseng-Williamson, K. A. Levetiracetam: a review of its use in epilepsy. *Drugs* **2011,** *71* (4), 489–514.
26. French, J. A.; Tonner, F. Levetiracetam. In *The Treatment of Epilepsy,* 3rd ed.; Shorvon, S., Perucca, E., Engel, J., Jr., Eds. Wiley-Blackwell, 2009; pp 559–573.
27. Crepeau, A. Z.; Treiman, D. M. Levetiracetam: a comprehensive review. *Expert Rev. Neurother.* **2010,** *10* (2), 159–171.
28. Schindler, W. Central suppressive 10-oxo-10,11-dihydeo-5H-dibenz[b,f]azepine-5-carboxamide, DE 2011087 (1970).
29. Kaufmann, D.; Fuenfschilling, P. C.; Beutler, U.; Hoehn, P.; Lohse, O.; Zaugg, W. A new synthesis of oxcarbazepine using a Friedel-Crafts cyclization strategy. *Tetrahedron Lett.* **2004,** *45* (27), 5275–5278.

30. Faught, E.; Limdi, N. Oxcarbazepine. In *The Treatment of Epilepsy,* 3rd ed.; Shorvon, S., Perucca, E., Engel, J., Jr., Eds. Wiley-Blackwell, 2009; pp 575–584.
31. Beydoun, A.; Kutluay, E. Oxcarbazepine, Expert Opin. *Pharmacother.* **2002,** *3* (1), 59–71.
32. Uno, H.; Kurokawa, M.; Masuda, Y. Methane-sulfonamide derivatives, US 4172896 (1979).
33. Uno, H.; Kurokawa, M.; Masuda, Y.; Nishimura, H. Studies on 3-substituted 1,2-benzisoxazole derivatives. 6. Syntheses of 3-(sulfamoylmethyl)-1,2-benzisoxazole derivatives and their anticonvulsant activities. *J. Med. Chem.* **1979,** *22* (2), 180–183.
34. Masuda, Y.; Karasawa, T.; Shiraishi, Y.; Hori, M.; Yoshida, K.; Shimizu, M. 3-Sulfamoylmethyl-1,2-benzisoxazole, a new type of anticonvulsant drug. Pharmacological profile. *Arzneim. Forsch* **1980,** *30* (3), 477–483.
35. Dupont, S.; Stefan, H. Zonisamide in clinical practice. *Acta Neurol. Scand.* **2012,** *126* (Suppl. 194), 29–35.
36. Brodie, M. J.; Ben-Menachem, E.; Chouette, I.; Giorgi, L. Zonisamide: its pharmacology, efficacy and safety in clinical trials. *Acta Neurol. Scand.* **2012,** *126* (Suppl. 194), 19–28.
37. Holder, J. L., Jr.; Wilfong, A. A. Zonisamide in the treatment of epilepsy. *Expert Opin. Pharmacother.* **2011,** *12* (16), 2573–2581.
38. Wroe, S. J. Zonisamide. In *The Treatment of Epilepsy,* 3rd ed.; Shorvon, S., Perucca, E., Engel, J., Jr., Eds. Wiley-Blackwell, 2009; pp 713–720.
39. Andruszkiewicz, R.; Silverman, R. B. A convenient synthesis of 3-alkyl-4-aminobutanoic acids. *Synthesis* **1989,** *12*. 953–935.
40. Yuen, P. W.; Kanter, G. D.; Taylor, C. P.; Vartanian, M. G. Enantioselective synthesis of PD144723: a potent stereospecific anticonvulsant. *Bioorg. Med. Chem. Lett.* **1994,** *4* (6), 823–826.
41. Hoekstra, M. S.; Sobieray, D. M.; Schwindt, M. A.; Mulhern, T. A.; Grote, T. M.; Huckabee, B. K.; Hendrickson, V. S.; Franklin, L. C.; Granger, E. J.; Karrick, G. L. Chemical development of CI-(1008), an enantiomerically pure anticonvulsant. *Org. Process Res. Dev.* **1997,** *1* (1), 26–38.
42. Huckle, R. Pregabalin (Pfizer). *Curr. Opin. Invest. Drugs (BioMed Cent.)* **2004,** *5* (1), 82–89.
43. Rheims, S.; Ryvlin, P. Pregabalin. In *The Treatment of Epilepsy,* 3rd ed.; Shorvon, S., Perucca, E., Engel, J., Jr., Eds. Wiley-Blackwell, 2009; pp 627–635.
44. Arain, A. M. Pregabalin in the management of partial epilepsy. *Neuropsychiatr. Dis. Treat.* **2009,** *5,* 407–413.
45. Perucca, E. An Introduction to antiepileptic drugs. *Epilepsia* **2005,** *46* (Suppl. 4), 31–37.
46. Fattore, C.; Perucca, E. Novel medications for epilepsy. *Drugs* **2011,** *71* (16), 2151–2178.
47. Perucca, E.; Berlowitz, D.; Brinbaum, A.; Cloyd, J. C.; Garrard, J.; Hanlon, J. T.; Levy, R. H.; Pugh, M. J. Pharmacological and clinical aspects of antiepileptic drug use in the elderly. *Epilepsy Res.* **2006,** *68* (Suppl. 1), S49–S63.
48. Perucca, E.; French, J.; Bialer, M. Development of new antiepileptic drugs: challenges, incentives, and recent advances. *Lancet Neurol.* **2007,** *6* (9), 793–804.
49. Ahsan, W.; Safhi, M. M.; Siddiqui, N.; Javed, S.; Alam, M. S.; Azad, B.; Akhtar, J. An insight into the new anticonvulsant agents. *Curr. Top. Med. Chem.* **2012,** *12* (9), 1072–1092.
50. Unverferth, K.; Rundfeldt, C. Antiepileptic drugs. *Pharmaceuticals* **2000,** *2,* 469–488.
51. Le Duc, B. Antiseizure drugs. In *Foye's Principles of Medicinal Chemistry,* 6th ed.; Lemke, T. L., Williams, D. A., Eds. Lippincott Williams & Wilkins, 2008; pp 521–546.
52. LaRoche, S. M.; Hetners, S. L. The new antiepileptic drugs. Scientific review. *JAMA, J. Am. Med. Assoc.* **2004,** *291* (5), 605–614.
53. Duncan, J. S. The promise of new antiepileptic drugs. *Br. J. Clin. Pharmacol.* **2002,** *53* (2), 123–131.

54. Gerlach, A. C.; Krajewski, J. L. Antiepileptic drug discovery and development: what have we learned and where are we going? *Pharmaceuticals* **2010,** *3,* 2884–2899.
55. Das, N.; Dhanawat, M.; Shrivastava, S. K. An overview on antiepileptic drugs. *Drug Discoveries Ther.* **2012,** *6* (4), 178–193.
56. Patsalos, P. N.; Sander, J. W. Antiepileptic drugs in early clinical development. In *The Treatment of Epilepsy,* 3rd ed.; Shorvon, S., Perucca, E., Engel, J., Jr., Eds. Wiley-Blackwell, 2009; pp 733–740.
57. Gavernet, L.; Elvira, J. E.; Samaja, G. A.; Pastore, V.; Sella, C. M.; Enrique, A.; Estiu, G.; Bruno-Blanch, L. E. Synthesis and anticonvulsant activity of amino acid-derived sulfamides. *J. Med. Chem.* **2009,** *52* (6), 1592–1601.
58. Hovinga, C. A. Valrocemide (Teva/Acorda). *Curr. Opin. Invest. Drugs (BioMed Cent.)* **2004,** *5* (1), 101–106.
59. Bialer, M.; Yagen, B. Valproic acid: second generation. *Neurotherapeutics* **2007,** *4* (1), 130–137.
60. Trojnar, M. K.; Wierzchowska-Cioch, E.; Krzyzanowski, M.; Jargiello, M.; Czuczwar, S. J. New generation of valproic acid. *Pol. J. Pharmacol.* **2004,** *56* (3). 283–238.
61. Rosenberg, G.; Friedman, J. E.; Shapiro, I.; Kozak, A. DP-VPA: antiepileptic drug. *Drugs Future* **2005,** *30* (12), 1212–1218.
62. Labiner; David, M. DP-VPA (D-Pharm). *Curr. Opin. Invest. Drugs (BioMed Cent.)* **2002,** *3* (6), 921–923.
63. Ravinder, B.; Rajeshwar, R. S.; Sridhar, M.; Murali, M. M.; Srinivas, K.; Panasa, R. A.; Bandichhor, R. An efficient synthesis for eslicarbazepine acetate, oxcarbazepine, and carbamazepine. *Tetrahedron Lett.* **2013,** *54* (22), 2841–2844.
64. Jain, S. K.; Vasantharaju, S. G.; Muddukrishna, B. S. Eslicarbazepine acetate: a new promising antiepileptic agent. *Elixir Int. J.* **Dec. 2011,** 5736–5740.
65. Singh, R. P.; Asconape, J. J. A review of eslicarbazepine acetate for the adjunctive treatment of partial-onset epilepsy. *J. Cent. Nerv. Syst. Dis.* **2011,** *3,* 179–187.
66. Almeida, L.; Bialer, M.; Soares-da-Silva, P. Eslicarbazepine acetate. In *The Treatment of Epilepsy,* 3rd ed.; Shorvon, S., Perucca, E., Engel, J., Jr., Eds. Wiley-Blackwell, 2009; pp 485–498.
67. Dulsat, C.; Mealy, N.; Castaner, R.; Bolos, J. Eslicarbazepine acetate: sodium channel blocker antiepileptic drug. *Drugs Future* **2009,** *34* (3), 189–195.
68. McCormack, P. L.; Robinson, D. M. Eslicarbazepine acetate. *CNS Drugs* **2009,** *23* (1), 71–79.
69. Almeida, L.; Soares-da-Silva, P. Eslicarbazepine acetate (BIA 2-093). *Neurotherapeutics* **2007,** *4* (1), 88–96.
70. Owen, R. T. Eslicarbazepine acetate: a novel agent for the adjunctive treatment of epilepsy. *Drugs Today* **2010,** *46* (1), 23–31.
71. Differding, E.; Kenda, B.; Lallemand, B.; Matagne, A.; Michel, P.; Pasau, P.; Talaga, P. Preparation of 2-oxo-1-pyrrolidine derivatives and their anticonvulsant activity, WO 2001062726 (2001).
72. Kenda, B. M.; Matagne, A. C.; Talaga, P. E.; Pasau, P. M.; Differding, E.; Lallemand, B. I.; Frycia, A. M.; Moureau, F. G.; Klitgaard, H. V.; Gillard, M. R.; Fuks, B.; Michel, P. Discovery of 4-substituted pyrrolidone butanamides as new agents with significant antiepileptic activity. *J. Med. Chem.* **2004,** *47* (3), 530–549.
73. Schulze-Bonhage, A. Brivaracetam for the treatment of epilepsy. *Expert Opin. Pharmacother.* **2011,** *12* (12), 1959–1966.
74. von Rosenstiel, P.; Perucca, E. Brivaracetam. In *The Treatment of Epilepsy,* 3rd ed.; Shorvon, S., Perucca, E., Engel, J., Jr., Eds. Wiley-Blackwell, 2009; pp 447–457.

75. Bestha, D. P.; Yang, B.; Madaan, V. Brivaracetam: SV2A ligand antiepileptic drug. *Drugs Future* **2010,** *35* (3), 165–172.
76. Rogawski, M. A. Brivaracetam: a rational drug discovery success story. *Br. J. Pharmacol.* **2008,** *154* (8), 1555–1557.
77. von Rosenstiel, P. Brivaracetam (UCB 34714). *Neurotherapeutics* **2007,** *4* (1), 84–87.
78. Malawska, B.; Kulig, K. Brivaracetam (UCB). *Curr. Opin. Invest. Drugs (BioMed Cent.)* **2005,** *6* (7), 740–746.
79. Matagne, A.; Margineanu, D.-G.; Potschka, H.; Loescher, W.; Michel, P.; Kenda, B.; Klitgaard, H. Profile of the new pyrrolidone derivative seletracetam (UCB 44212) in animal models of epilepsy. *Eur. J. Pharmacol.* **2009,** *614* (1–3), 30–37.
80. Pollard, J. R. Seletracetam, a small-molecule SV2A modulator for the treatment of epilepsy. *Curr. Opin. Invest. Drugs (BioMed Cent.)* **2008,** *9* (1), 101–107.
81. Wang, Y.; Serradell, N.; Bolos, J. Seletracetam: antiepileptic drug. *Drugs Fut.* **2006,** *31* (12), 1048–1052.
82. Bennett, B.; Matagne, A.; Michel, P.; Leonard, M.; Cornet, M.; Meeus, M.-A.; Toublanc, N. Seletracetam (UCB 44212). *Neurotherapeutics* **2007,** *4* (1), 117–122.
83. Pollard, J. R. Seletracetam, a small molecule SV2A modulator for the treatment of epilepsy. *Curr. Opin. Invest. Drugs (BioMed Cent.)* **2008,** *9* (1), 101–107.
84. Roecklein, B. A.; Sacks, H. J.; Mortko, H.; Stables, J. Fluorofelbamate. *Neurotherapeutics* **2007,** *4* (1), 97–101.
85. Mortko, H.; He, W.; Andersen, M. W.; Dotse, A. K.; Li, J. Methods for the synthesis of dicarbamate compounds such as fluorofelbamate, US 20060241298 (2006).
86. Ramsay, R. E.; Pryor, F. M. Phenytoin and fosphenytoin. In *Status Epilepticus: Mechanisms and Management;* Wasterlain, C. G., Treiman, D. M., Eds.; MIT Press, 2006; pp 545–552.
87. Luer, M. S. Fosphenytoin. *Neurol. Res.* **1998,** *20* (2), 178–182.
88. Browne, T. R. Fosphenytoin (Cerebyx). *Clin. Neuropharmacol.* **1997,** *20* (1), 1–12.
89. Boucher, B. A. Fosphenytoin a novel phenytoin prodrug. *Pharmacotherapy* **1996,** *16* (5), 777–791.
90. Kohn, H. Preparation of anticonvulsant enantiomeric amino acid derivatives, WO 9733861 (1997).
91. King, A. M.; Salome, C.; Salome-Grosjean, E.; De Ryck, M.; Kaminski, R.; Valade, A.; Stables, J. P.; Kohn, H. Primary amino acid derivatives: substitution of the 4′-N′-benzylamide site in (R)-N′-benzyl 2-amino-3-methylbutanamide, (R)-N′-benzyl 2-amino-3,3-dimethylbutanamide, and (R)-N′-benzyl 2-amino-3-methoxypropionamide provides potent anticonvulsants with pain-attenuating properties. *J. Med. Chem.* **2011,** *54* (19), 6417–6431.
92. Muthukrishnan, M.; Mujahid, M.; Sasikumar, M.; Mujumdar, P. First asymmetric synthesis of the antiepileptic drug Lacosamide (Vimpat) based on a hydrolytic kinetic resolution strategy. *Tetrahedron: Asymmetry* **2011,** *22* (12), 1353–1357.
93. Salome, C.; Salome-Grosjean, E.; Stables, J. P.; Kohn, H. Merging the structural motifs of functionalized amino acids and α-amino amides: compounds with significant anticonvulsant activities. *J. Med. Chem.* **2010,** *53* (9), 3756–3771.
94. Morieux, P.; Stables, J. P.; Kohn, H. Synthesis and anticonvulsant activities of N-benzyl (2R)-2-acetamido-3-oxysubstituted propionamide derivatives. *Bioorg. Med. Chem.* **2010,** *16* (19), 8968–8975.
95. Beyreuther, B. K.; Freitag, J.; Heers, C.; Krebsfaenger, N.; Scharfenecker, U.; Stoehr, T. Lacosamide: a review of preclinical properties. *CNS Drug Rev.* **2007,** *13* (1), 21–42.
96. Riedner, J. Improved synthesis scheme for Lacosamide, EP 1642889 (2006).
97. Andurkar, S. V.; Stables, J. P.; Kohn, H. Synthesis and anticonvulsant activities of (R)-(O)-methylserine derivatives. *Tetrahedron: Asymmetry* **1998,** *9* (21), 3841–3854.

98. Uysal, S.; Calis, U.; Soyer, Z. Synthesis and anticonvulsant activity of some 2/3-benzoylaminopropion-anilide derivatives. *Arzneim. Forsch.* **2012,** *62* (6), 295–300.
99. Fernandez, E. M.; Franck, A. J. Lacosamide for the treatment of refractory status epilepticus. *Ann. Pharmacother.* **2011,** *45* (11), 1445–1449.
100. Salome, C.; Salome-Grosjean, E.; Park, K. D.; Morieux, P.; Swendiman, R.; DeMarco, E.; Stables, J. P.; Kohn, H. Synthesis and anticonvulsant activities of (R)-N-(4′-substituted)benzyl 2-acetamido-3-methoxypropionamides. *J. Med. Chem.* **2010,** *53* (3), 1288–1305.
101. Park, K. D.; Morieux, P.; Salome, C.; Cotton, S. W.; Reamtong, O.; Eyers, C.; Gaskell, S. J.; Stables, J. P.; Liu, R.; Kohn, H. Lacosamide isothiocyanate-based agents: novel agents to target and identify lacosamide receptors. *J. Med. Chem.* **2009,** *52* (21), 6897–6911.
102. Rogawski, M. A.; Tofighy, A.; White, H. S.; Matagne, A.; Wolff, C. Current understanding of the mechanism of action of the antiepileptic drug lacosamide. *Epilepsy Res.* **2015,** *110,* 189–205.
103. Ben-Menachem, E. Lacosamide: an investigational drug for adjunctive treatment of partial-onset seizures. *Drugs Today* **2008,** *44* (1), 35–40.
104. Andurkar, S. V.; Stables, J. P.; Kohn, H. The anticonvulsant activities of N-benzyl 3-methoxypropionamides. *Bioorg. Med. Chem.* **1999,** *7* (11), 2381–2389.
105. Griffith, R. C.; Napier, J. J. N-(Phenylalkyl)acetamide derivatives as antiepileptics and sedatives and their preparation, EP 279937 (1988).
106. Griffith, R. C.; Napier, J. J. Use of arylalkylamides in the treatment of neurodegenerative diseases, EP 427427 (1991).
107. Malek, R.; Borowicz, K. K.; Kimber-Trojnar, Z.; Sobieszek, G.; Piskorska, B.; Czuczwar, S. J. Remacemide—a novel potential antiepileptic drug. *Pol. J. Pharmacol.* **2003,** *55* (5), 691–698.
108. Schachter, S. C.; Tarsy, D. Remacemide: current status and clinical applications, Expert Opin. Invest. *Drugs* **2000,** *9* (4), 871–883.
109. Davies, J. A. Remacemide hydrochloride: a novel antiepileptic agent. *Gen. Pharmacol.* **1997,** *28* (4), 499–502.
110. Muir, K. T.; Palmer, G. C. Remacemide. *Epilepsy Res., Suppl.* **1991,** *3,* 147–152.
111. Leach, J. P.; Marson, A. G.; Hutton, J. L. Remacemide for drug-resistant localization related epilepsy. *The Cochrane database of systematic reviews* **2002,** *4,* 1–23 (CD001900).
112. Dostert, P.; Pevarello, P.; Heidempergher, F.; Varasi, M.; Bonsignori, A.; Roncucci, R. Preparation of α-(phenylalkylamino)carboxamides as drugs, EP 400495 (1990).
113. Pevarello, P.; Bonsignori, A.; Dostert, P.; Heidempergher, F.; Pinciroli, V.; Colombo, M.; McArthur, R. A.; Salvati, P.; Post, C.; Fariello, R. G.; Varasi, M. Synthesis and anticonvulsant activity of a new class of 2-[(Arylalkyl)amino]alkanamide derivatives. *J. Med. Chem.* **1998,** *41* (4), 579–590.
114. Chazot, P. L. Safinamide (Newron Pharmaceuticals). *Curr. Opin. Invest. Drugs (BioMed Cent.)* **2001,** *2* (6), 809–813.
115. Fariello, R. G. Safinamide. *Neurotherapeutics* **2007,** *4* (1), 110–116.
116. Malek, N. M.; Grosset, D. G. Investigational agents in the treatment of Parkinson's disease: focus on safinamide. *J. Exp. Pharmacol.* **2012,** *4,* 85–90.
117. Sorbera, L. A.; Leeson, P. A.; Castaner, J. Safinamide mesilate Prop INNM NW-1015 PNU-151774E FCE-2(6743). *Drugs Future* **2001,** *26* (8), 745–749.
118. Blackburn-Munro, G.; Dalby-Brown, W.; Mirza, N. R.; Mikkelsen, J. D.; Blackburn-Munro, R. E., Retigabine: chemical synthesis to clinical application, CNS Drug Rev., 11(1), 1–20.
119. Czuczwar, S. J.; Patsalos, P. N. The new generation of GABA enhancers. Potential in the treatment of epilepsy. *CNS Drugs* **2001,** *15* (5), 339–350.
120. Tikoo, D.; Gupta, M. Retigabine (Ezogabine) for management of partial onset seizures: a mini review. *Int. J. Pharmacol. Clin. Sci.* **2012,** *1* (3), 85–90.

121. Rheims, S.; Ryvlin, P. Retigabine for partial onset seizures. *Expert Rev. Neurother.,Neurotherapeutics* **2012,** *12* (5), 509–517.
122. Gunthorpe, M. J.; Large, C. H.; Sankar, R. The mechanism of action of retigabine (ezogabine), a first-in-class K+ channel opener for the treatment of epilepsy. *Epilepsia* **2012,** *53* (3), 412–424.
123. Deeks, E. D. Retigabine (ezogabine): in partial-onset seizures in adults with epilepsy. *CNS Drugs* **2011,** *24* (10), 887–900.
124. Stafstrom, C. E.; Grippon, S.; Kirkpatrick, P. Ezogabine (retigabine). *Nat. Rev. Drug Discovery* **2011,** *10* (10), 729–730.
125. Mansbach, H.; Baulac, M. Retigabine. In *The Treatment of Epilepsy,* 3rd ed.; Shorvon, S., Perucca, E., Engel, J., Jr., Eds. Wiley-Blackwell, 2009; pp 637–646.
126. Czuczwar, P.; Wojtak, A.; Cioczek-Czuczwar, A.; Parada-Turska, J.; Maciejewski, R.; Czuczwar, S. J. Retigabine: the newer potential antiepileptic drug. *Pharmacol. Rep.* **2010,** *62* (2), 211–219.
127. Chung, S. S. Retigabine: could it be the next broad-spectrum antiepileptic drug? *Curr. Drug Ther.* **2010,** *5* (1), 10–16.
128. Porter, R. J.; Nohria, V.; Rundfeldt, C. Retigabine. *Neurotherapeutics* **2007,** *4* (1), 149–154.
129. Meier, R. Preparation of fluorinated phenylalkyltriazoles as anticonvulsants and pharmaceutical compositions containing them, EP 199262 (1986).
130. Heaney, D.; Walker, M. C. Rufinamide. *Drugs Today* **2007,** *43* (7), 455–460.
131. Hakimian, S.; Cheng-Hakimian, A.; Anderson, G. D.; Miller, J. W. Rufinamide: a new antiepileptic medication. *Expert Opin. Pharmacother.* **2007,** *8* (12), 1931–1940.
132. Palhagen, S.; Canger, R.; Henriksen, O.; van Parys, J. A.; Riviere, M.-E.; Karolchyk, M. A. Rufinamide: a double-blind, placebo-controlled proof of principle trial in patients with epilepsy. *Epilepsy Res.* **2001,** *43* (2), 115–124.
133. Deeks, E. D.; Scott, L. J. Rufinamide. *CNS Drugs* **2006,** *20* (9), 751–760.
134. Arroyo, S. Rufinamide. *Neurotherapeutics* **2007,** *4* (1), 155–162.
135. Wisniewski, C. S. Rufinamide: a new antiepileptic medication for the treatment of seizures associated with Lennox-Gastaut syndrome. *Ann. Pharmacother.* **2010,** *44* (4), 658–667.
136. Sorbera, L. A.; Leeson, P. A.; Rabasseda, X.; Castaner, J. Rufinamide. *Drugs Future* **2000,** *25* (11), 1145–1149.
137. Patel, D. S.; Mehta, H. R.; Goswami, H. J.; Sheth, A. A.; Patel, D. S.; Mehta, M. M.; Shanker, N.; Patel, K. J.; Mehta, A. A.; Deshpande, S. Rufinamide: a novel antiepileptic drug. *Res. J. Pharm. Biol. Chem. Sci.* **2011,** *2* (1), 855–865.
138. Biton, V. Rufinamide. In *The Treatment of Epilepsy,* 3rd ed.; Shorvon, S., Perucca, E., Engel, J., Jr., Eds. Wiley-Blackwell, 2009; pp 647–655.
139. Ferrie, C. D. Rufinamide: a new antiepileptic drug treatment for Lennox-Gastaut syndrome. *Expert Rev. Neurother.* **2010,** *10* (6), 851–860.
140. Patel, D. S.; Mehta, H. R.; Goswami, H. J.; Sheth, A. A.; Patel, D. S.; Mehta, M. M.; Shanker, N.; Patel, K. J.; Mehta, A. A.; Deshpande, S. Rufinamide: a novel antiepileptic drug. *Res. J. Pharm., Biol. Chem. Sci.* **2011,** *2* (1), 855–865.
141. Fatope, M. O. Soretolide (Laboratoires Biocodex). *Curr. Opin. Invest. Drugs (BioMed Cent.)* **2001,** *2* (6), 824–827.
142. Kimber-Trojnar, Z.; Borowicz, K. K.; Malek, R.; Sobieszek, G.; Piskorska, B.; Czuczwar, S. J. Perspectives of losigamone in epilepsy treatment. *Pol. J. Pharmacol.* **2003,** *55* (5), 675–682.
143. Willmore, L. J. Losigamone. Dr Willmar Schwabe. *Curr. Opin. Invest. Drugs (BioMed Cent.)* **2001,** *2* (12), 1763–1766.
144. Stein, U.; Klessing, K.; Chatterjee, S. S. Losigamone. *Epilepsy Res., Suppl.* **1991,** *3,* 129–133.
145. Novak, G. P.; Brodie, M. J. Carisbamate. In *The Treatment of Epilepsy,* 3rd ed.; Shorvon, S., Perucca, E., Engel, J., Jr., Eds. Wiley-Blackwell, 2009; pp 475–484.

146. Novak, G. P.; Kelley, M.; Zannikos, P.; Klein, B. Carisbamate (RWJ-333369). *Neurotherapeutics* **2007,** *4* (1), 106–109.
147. Plosker, G. L. Stiripentol: in severe myoclonic epilepsy of infancy (Dravet syndrome). *CNS Drugs* **2012,** *26* (11), 993–1001.
148. Eriksson, K.; Keranen, T. Stiripentol. In *The Treatment of Epilepsy,* 3rd ed.; Shorvon, S., Perucca, E., Engel, J., Jr., Eds. Wiley-Blackwell, 2009; pp 657–661.
149. Czuczwar, S. J.; Trojnar, M. K.; Gergont, A.; Kroczka, S.; Kacinski, M. Stiripentol—characteristic of a new antiepileptic drug. *Expert Opin. Drug Discovery* **2008,** *3* (4), 453–460.
150. Trojnar, M. K.; Wojtal, K.; Trojnar, M. P.; Czuczwar, S. J. Stiripentol. A novel antiepileptic drug. *Pharmacol. Rep.* **2005,** *57* (2), 154–160.
151. Chiron, C. Stiripentol, Expert Opin. Invest. *Drugs* **2005,** *14* (7), 905–911.
152. Dhir, A. Perampanel: AMPA-receptor antagonist antiepileptic agent. *Drugs Future* **2012,** *37* (1), 13–17.
153. Steinhoff, B. J. The new antiepileptic drug perampanel. *Arzneimitteltherapie* **2012,** *30* (12), 375–380.
154. Shvarts, V.; Chung, S. Perampanel: newly approved, novel antiepileptic medication for partial-onset seizures. *Expert Rev. Neurother.* **2013,** *13* (2), 131–134.
155. Plosker, G. L. Perampanel: as adjunctive therapy in patients with partial-onset seizures. *CNS Drugs* **2012,** *26* (12), 1085–1096.
156. Owen, R. T. Perampanel: a novel antiepileptic for the adjunctive treatment of refractory partial onset seizures. *Drugs Today* **2013,** *49* (1), 23–31.
157. Andrasi, F.; Berzsenyi, P.; Botka, P.; Farkas, S.; Goldschmidt, K.; Hamori, T.; Korosi, J.; Moravcsik, I.; Tarnawa, I. Preparation of 1-(4-aminophenyl)-7,8-methylenedioxy-2,3-benzodiazepines as muscle relaxants, anticonvulsants, and cerebral antiischemics, EP492485 (1992).
158. Andrasi, F.; Berzsenyi, P.; Botka, P.; Farkas, S.; Goldschmidt, K.; Hamori, T.; Korosi, J.; Moravcsik, I.; Tarnawa, I. N-acylated 1-(4-aminophenyl)-4-methyl-7,8-methylenedioxy-3,4-dihydro-5H-2,3-benzodiazepines as anticonvulsants, muscle relaxants, and neuroprotective agents, US 5459137 (1995).
159. Ling, I.; Podanyi, B.; Hamori, T.; Solyom, S. Asymmetric reduction of a carbon-nitrogen double bond: enantioselective synthesis of 4,5-dihydro-3H-2,3-benzodiazepines. *J. Chem. Soc., Perkin Trans. 1 (1972–1999)* **1995,** (11), 1423–1427.
160. Anderson, B. A.; Hansen, M. M.; Vicenzi, J. T.; Zmijewski, M. J. In *Chemistry, biocatalysis and engineering: an interdisciplinary approach to the manufacture of the benzodiazepine drug candidate LY30(0164);* Gadamasetti, K. G., Ed.; 1999; pp 263–282. (Process Chem. Pharm. Ind.
161. Andrasi, F. Talampanel Prop INN GYKI-53773 IDR-53773 LY-30(0164). *Drugs Future* **2001,** *26* (8), 754–756.
162. Meng, C. Q. Talampanel Eli Lilly & Co. *Curr. Opin. Cent. Peripher. Nerv. Syst. Invest. Drugs* **1999,** *1* (5), 637–643.
163. Howes, J. F.; Bell, C. Talampanel. *Neurotherapeutics* **2007,** *4* (1), 126–129.
164. Crespi, F. Carabreast. *Curr. Opin. Cent. Peripher. Nerv. Syst. Invest. Drugs* **1999,** *1* (5), 644–648.
165. Metcalf, B. W.; Jung, M. Olefinic derivatives of amino acids, US 3960927 (1976).
166. Seiler, N.; Sarhan, S., Treatment of seizure disorders with γ-vinyl GABA and glycines, EP 124091 (1984).
167. Goralski, C. T.; Hoops, J. F.; Ramanarayanan, K. A. Process for the production of vinyl GABA, EP 427197 (1991).
168. Al-Majed, A. Vigabatrin. *Profiles Drug Subst., Excipients, Relat. Methodol.* **2010,** *35,* 309–345.

169. French, J. A. Vigabatrin. *Epilepsia* **1999,** *40* (Suppl. 5), S11–S16.
170. Connelly, J. F. Vigabatrin. *Ann. Pharmacother.* **1993,** *27* (2), 197–204.
171. Richens, A. Pharmacology and clinical pharmacology of vigabatrin. *J. Child Neurol.* **1991,** (Suppl 2), S7–S10.
172. Gaily, E. Vigabatrin monotherapy for infantile spasms. *Expert Rev. Neurother.* **2012,** *12* (3), 275–286.
173. Pesaturo, K. A.; Spooner, Li. M.; Belliveau, P. Vigabatrin for infantile spasms. *Pharmacotherapy* **2011,** *31* (3), 298–311.
174. Kramer, G.; Wohlrab, G. Vigabatrin. In *The Treatment of Epilepsy,* 3rd ed.; Shorvon, S., Perucca, E., Engel, J., Jr., Eds. Wiley-Blackwell, 2009; pp 699–712.
175. Tolman, J. A.; Faulkner, M. A. Vigabatrin: a comprehensive review of drug properties including clinical updates following recent FDA approval. *Expert Opin. Pharmacother.* **2009,** *10* (18), 3077–3089.
176. Camposano, S. E.; Major, P.; Halpern, E.; Thiele, E. A. Vigabatrin in the treatment of childhood epilepsy: a retrospective chart review of efficacy and safety profile. *Epilepsia* **2008,** *49* (7), 1186–1191.
177. Cook, M. C.; Lawrence, R.; Phillipps, G. H.; Hunter, A. C.; Newall, C. E.; Stephenson, L.; Weir, N. G., 3α-Hydroxy-5α-pregnan-20-one derivatives, DE 2162555 (1972).
178. Hogenkamp, D. J.; Tahir, S. H.; Hawkinson, J. E.; Upasani, R. B.; Alauddin, M.; Kimbrough, C. L.; Acosta-Burruel, M.; Whittemore, E. R.; Woodward, R. M.; Lan, N. C.; Gee, K. W.; Bolger, M. B. Synthesis and in vitro activity of 3β-substituted-3α-hydroxypregnan-20-ones: allosteric modulators of the GABAA receptor. *J. Med. Chem.* **1997,** *40* (1), 61–72.
179. Nohria, V.; Giller, E. G. Ganaxolone. *Neurotherapeutics* **2007,** *4* (1), 102–105.
180. Reddy, D. S.; Woodward, R. G. Ganaxolone: A prospective overview. *Drugs Future* **2004,** *29* (3), 227–242.
181. Monaghan, E. P.; McAuley, J. W.; Data, J. L. Ganaxolone: a novel positive allosteric modulator of the GABAA receptor complex for the treatment of epilepsy. *Expert Opin. Invest. Drugs* **1999,** *8* (10), 1663–1671.
182. Bialer, M.; Johannessen, S. I.; Kupferberg, H. J.; Levym, R. H.; Loiseau, P.; Perucca, E. Progress report on new antiepileptic drugs: a summary of the Fifth Eilat Conference (EILAT V). *Epilepsy Res.* **2001,** *43* (1), 11–58.
183. Bialer, M.; Johannessen, S. I.; Kupferberg, H. J.; Levym, R. H.; Loiseau, P.; Perucca, E. Progress report on new antiepileptic drugs: a summary of the Sixth Eilat Conference (EILAT VI). *Epilepsy Res.* **2002,** *51* (1), 31–71.
184. Bialer, M.; Johannessen, S. I.; Kupferberg, H. J.; Loiseau, P.; Perucca, E.; Tomson, T. Progress report on new antiepileptic drugs: a summary of the Seventh Eilat Conference (EILAT VII). *Epilepsy Res.* **2004,** *61* (1), 1–48.
185. Bialer, M.; Johannessen, S. I.; Kupferberg, H. J.; Loiseau, P.; Perucca, E.; Tomson, T. Progress report on new antiepileptic drugs: a summary of the Eighth Eilat Conference (EILAT VIII). *Epilepsy Res.* **2007,** *73* (1), 1–52.
186. Bialer, M.; Johannessen, S. I.; Kupferberg, H. J.; Loiseau, P.; Perucca, E.; Tomson, T. Progress report on new antiepileptic drugs: a summary of the Ninth Eilat Conference (EILAT IX). *Epilepsy Res.* **2009,** *83* (1), 1–43.
187. Bialer, M.; Johannessen, S. I.; Kupferberg, H. J.; Perucca, E.; Tomson, T.; White, H. S. Progress report on new antiepileptic drugs: a summary of the Tenth Eilat Conference (EILAT X). *Epilepsy Res.* **2010,** *92* (2–3), 89–124.
188. Luszczki, J. J. Third-generation antiepileptic drugs: mechanisms of action, pharmacokinetics and interactions. *Pharmacol. Rep.* **2009,** *61* (2), 197–216.

189. LaRoche, S. M. A new look at the second-generation antiepileptic drugs: a decade of experience. *Neurologist* **2007,** *13* (3), 133–139.
190. Pollard, J. R.; French, J. Antiepileptic drugs in development. *Lancet Neurol.* **2006,** *5* (12), 1064–1067.
191. Sirven, J. I.; Noe, K.; Hoerth, M.; Drazkowski, J. Antiepileptic drugs 2012: recent advances and trends. *Mayo Clin. Proc.* **2012,** *87* (9), 879–889.
192. Patsalos, P. N. The new generation of anti-epileptic drugs. *Emerging Drugs (London, U. K.)* **1999,** *4,* 87–106.
193. Bialer, M. How did phenobarbital's chemical structure affect the development of subsequent antiepileptic drugs (AEDs)? *Epilepsia* **2012,** *53* (Suppl. 8), 3–11.
194. Bialer, M.; White, H. S. Key factors in the discovery and development of new antiepileptic drugs. *Nat. Rev. Drug Discovery* **2010,** *9* (1), 68–82.
195. Prunetti, P.; Perucca, E. New and forthcoming anti-epileptic drugs. *Curr. Opin. Neurol.* **2011,** *24* (2), 159–164.
196. Klitgaard, H. Antiepileptic drug discovery: lessons from the past and future challenges. *Acta Neurol. Scand., Suppl.* **2005,** *181,* 68–72.
197. Yuen, P.; Kanter, G. D.; Taylor, C. P.; Vartanian, M. G. Enantioselective synthesis of PD144723: a potent stereospecific anticonvulsant. *Bioorg. Med. Chem. Lett.* **1994,** *4* (6), 823–826.
198. Belliotti, T. R.; Capiris, T.; Ekhato, I. V.; Kinsora, J. J.; Field, M. J.; Heffner, T. G.; Meltzer, L. T.; Schwarz, J. B.; Taylor, C. P.; Thorpe, A. J.; Vartanian, M. G.; Wise, L. D.; Ti, Z.-S.; Weber, M. L.; Wustrow, D. J. Structure-activity relationships of pregabalin and analogues that target the α2-δ protein. *J. Med. Chem.* **2005,** *48* (7), 2294–2307.
199. Shelke, S. H.; Mhaske, P. C.; Bobade, V. D. An efficient total synthesis of (±)-pregabalin Indian. *J. Chem., Sect. B: Org. Chem. Incl. Med. Chem.* **2012,** *51B* (4), 631–634.
200. Shi, W.; Liu, H.; Zhang, Y.; Zhong, B.; Yang, H. Design, synthesis, and preliminary evaluation of gabapentin- pregabalin mutual prodrugs in relieving neuropathic pain. *Arch. Pharm. (Weinheim, Ger.)* **2005,** *338* (8), 358–364.
201. Hoge, G. Synthesis of both enantiomers of a P-chirogenic 1,2-bisphospholanoethane ligand via convergent routes and application to rhodium-catalyzed asymmetric hydrogenation of CI-1008 (pregabalin). *J. Am. Chem. Soc.* **2003,** *125* (34), 10219–10227.
202. Martin, L.; Rabasseda, X.; Leeson, P.; Castaner, J. Pregabalin: antiepileptic. *Drugs Future* **1999,** *24* (8), 862–870.
203. Horvat, S.; Hamersak, Z.; Stipetic, I.; Jolas, T. Synthesis, characterization and in vitro pharmacology of novel pregabalin derivatives. *Eur. J. Med. Chem.* **2010,** *45* (4), 1447–1452.
204. Arain, A. M. Pregabalin in the management of partial epilepsy. *Neuropsychiatr. Dis. Treat.* **2009,** *5,* 407–413.
205. Ryvlin, P.; Perucca, E.; Rheims, S. Pregabalin for the management of partial epilepsy. *Neuropsychiatr. Dis. Treat.* **2008,** *4* (6), 1211–1224.
206. Baidya, D. K.; Agarwal, A.; Khanna, P.; Arora, M. K. Pregabalin in acute and chronic pain. *J. Anaesthesiol., Clin. Pharmacol* **2011,** *27* (3), 307–314.
207. Chiechio, S.; Zammataro, M.; Caraci, F.; Rampello, L.; Copani, A.; Sabato, A. F.; Nicoletti, F. Pregabalin in the treatment of chronic pain: an overview. *Clin. Drug Invest.* **2009,** *29* (3), 203–213.
208. Smith, M. T.; Moore, B. J. Pregabalin for the treatment of fibromyalgia. *Expert Opin. Pharmacother.* **2012,** *13* (10), 1527–1533.
209. Boomershine, C. S. Pregabalin for the management of fibromyalgia syndrome. *J. Pain Res.* **2010,** *3,* 81–88.
210. Lawson, K. Pregabalin and fibromyalgia syndrome: a treatment option. *Clin. Med.: Ther.* **2009,** *1,* 809–824.

211. Wensel, T. M.; Powe, K. W.; Cates, M. E. Pregabalin for the treatment of generalized anxiety disorder. *Ann. Pharmacother.* **2012,** *46* (3), 424–426.
212. Montgomery, S. A.; Kasper, S. Pharmacotherapy update: pregabalin in the treatment of generalized anxiety disorder. *Clin. Med. Insights: Ther.* **2010,** *2,* 189–202.
213. Lee, G. R. Preparation of lamotrigine, WO 9620935 (1996).
214. Ulomskii, E. N.; Shestakova, T. S.; Deev, S. L.; Rusinov, V. L.; Chupakhin, O. N. A new approach to the synthesis of lamotrigine and other 3,5-diamino-1,2,4-triazine derivatives. *Russ. Chem. Bull.* **2005,** *54* (3), 726–732.
215. Qian, Y.; Lv, P.-C.; Shi, L.; Fang, R.-Q.; Song, Z.-C.; Zhu, H.-L. Synthesis and antimicrobial activity of lamotrigine and its ammonium derivatives. *J. Chem. Sci. (Amritsar, India)* **2009,** *121* (4), 463–470.
216. Hlavac, J.; Buchtik, R.; Slouka, J.; Hradil, P.; Wiedermannova, I. Synthesis of oxo analogs of Lamotrigine and related compounds. *ARKIVOC (Gainesville, FL, U. S.)* **2003,** *1,* 22–28.
217. Beattie, K.; Phadke, G.; Novakovic, J. Lamotrigine. *Profiles Drug Subst., Excipients, Relat. Methodol.* **2012,** *37,* 245–285.
218. Vajda, F. J. E.; Dodd, S.; Horgan, D. Lamotrigine in epilepsy, pregnancy and psychiatry—a drug for all seasons? *J. Clin. Neurosci.* **2013,** *20* (1), 13–16.
219. Matsuo, F.; Riaz, A. Lamotrigine. In *The Treatment of Epilepsy,* 3rd ed.; Shorvon, S., Perucca, E., Engel, J., Jr., Eds. Wiley-Blackwell, 2009; pp 535–558.
220. Blaszczyk, B.; Czuczwar, S. J. Efficacy, safety, and potential of extended-release lamotrigine in the treatment of epileptic patients. *Neuropsychiatr. Dis. Treat.* **2010,** *6,* 145–150.
221. Malik, S.; Arif, H.; Hirsch, L. J. Lamotrigine and its applications in the treatment of epilepsy and other neurological and psychiatric disorders. *Expert Rev. Neurother.* **2006,** *6* (11), 1609–1627.
222. Eisenberg, E.; Shifrin, A.; Krivoy, N. Lamotrigine for neuropathic pain. *Expert Rev. Neurother.* **2005,** *5* (6), 729–735.
223. Bhagwagar, Z.; Goodwin, G. M. Lamotrigine in the treatment of bipolar disorder. *Expert Opin. Pharmacother.* **2005,** *6* (8), 1401–1408.
224. Herman, E. Lamotrigine: a depression mood stabilizer. *Eur. Neuropsychopharmacol.* **2004,** *14* (Suppl. 2), S89–S93.
225. Gouliaev, A. H.; Senning, A. Piracetam and other structurally related nootropics. *Brain Res. Rev.* **1994,** *19* (2), 180–222.
226. Strubbe, J. H. L.; Linz, R. A. Pharmaceutical (2-oxopyrrolidine)acetamide derivatives, DE 2106418 (1971).
227. Gobert, J.; Geerts, J. P.; Bodson, G. (S)-α-Ethyl-2-oxo-1 pyrrolidineacetamide, EP 162036 (1985).
228. Cossement, E.; Motte, G.; Geerts, J. P.; Gobert, J. Preparation of S-α-ethyl-2-oxo-1-pyrrolidineacetamide via desulfurization/hydrogenolysis, GB 2225322 (1990).
229. Surtees, J.; Marmon, V; Differding, E.; Zimmermann, V. Process for preparation of 2-oxo-1-pyrrolidine derivatives, WO 2001064637 (2001).
230. Artus Surroca, J. J.; Rafecas, J. L.; Garriga, S. L.; Pericas, B. M. A.; Sola, C., L. Improved process for the preparation of levetiracetam, an antiepileptic agent, from etiracetam, via N-aminomethylation, resolution, and deaminomethylation, WO 2004076416 (2004).
231. Imahori, T.; Omoto, K.; Hirose, Y.; Takahata, H. Asymmetric synthesis of the antiepileptic drug levetiracetam. *Heterocycles* **2008,** *76* (2), 1627–1632.
232. Das, S. K.; Zhang, J.; Huang, Y.; Davidson, J. G. Amino acid esters and amides for reductive amination of mucochloric acid: synthesis of novel γ-lactams, short peptides and antiseizure agent Levetiracetam (Keppra). *Eur. J. Org. Chem.* **2006,** *16,* 3730–3737.
233. Kotkar, S. P.; Sudalai, A. A short enantioselective synthesis of the antiepileptic agent levetiracetam based on proline-catalyzed asymmetric α-aminooxylation. *Tetrahedron Lett.* **2006,** *47* (38), 6813–6815.

234. Boschi, F.; Camps, P.; Comes-Franchini, M.; Munoz-Torrero, D.; Ricci, A.; Sanchez, L. A synthesis of levetiracetam based on (S)-N-phenylpantolactam as a chiral auxiliary. *Tetrahedron: Asymmetry* **2005,** *16* (22), 3739–3745.
235. Mujahid, M.; Mujumdar, P.; Sasikumar, M.; Kunte, S. S.; Muthukrishnan, M. An alternate synthesis of enantiomerically pure levetiracetam (Keppra). *Tetrahedron: Asymmetry* **2012,** *23* (20–21), 1512–1515.
236. Chandra, B. K.; Buchi, R. R.; Mukkanti, K.; Suresh, K.; Madhusudhan, G.; Nigam, S. Enantioselective synthesis of antiepileptic agent, (-)-levetiracetam, through Evans asymmetric strategy. *J. Chem.* **2013,** *17* (6512), 1–5.
237. Mandal, A. K.; Mahajan, S. W.; Ganguly, P.; Chetia, A.; Chauhan, N. D.; Bhatt, D. N.; Baria, R. R. Process for the preparation of levetiracetam and racemization of (R)- and (S)-2-aminobutyric acids and amides, WO 2006103696 (2006).
238. Mylavarapu, R.; Anand, R. V.; Kondaiah, G. C. M.; Reddy, L. A.; Reddy, G. S.; Roy, A.; Bhattacharya, A.; Mukkanti, K.; Bandichhor, R. An alternate synthesis of levetiracetam. *Green Chem. Lett. Rev.* **2010,** *3* (3), 225–230.
239. Futagawa, T.; Canvat, J.-P.; Cavoy, E.; Deleers, M.; Hamende, M.; Zimmermann, V. Process for the preparation of levetiracetam, US 6107492 (2000).
240. Cavoy, E.; Hamende, M.; Deleers, M.; Canvat, J.-P.; Zimmermann, V. Process for preparing (s)- and (r)-α-ethyl-2-oxo-1-pyrrolidineacetamide using a simulated mobile bed chromatography with chiral phase of silica gel supported polymer, US 6124473 (2000).
241. Li, Y. Q.; Wang, Z.-X.; Guntoori, B. Reddy improved process for the preparation of (S)- and (R)-α-ethyl-2-oxo-1-pyrrolidineacetamide, WO 2006053441 (2006).
242. Lyseng-Williamson, K. A. Levetiracetam: a review of its use in epilepsy. *Drugs* **2011,** *71* (4), 489–514.
243. French, J. A.; Tonner, F. Levetiracetam. In *The Treatment of Epilepsy,* 3rd ed.; Shorvon, S., Perucca, E., Engel, J., Jr., Eds. Wiley-Blackwell, 2009; pp 559–573.
244. Crepeau, A. Z.; Treiman, D. M. Levetiracetam: a comprehensive review. *Expert Rev. Neurother.* **2010,** *10* (2), 159–171.
245. Carreno, M. Levetiracetam. *Drugs Today* **2007,** *43* (11), 769–794.
246. Klitgaard, H.; Verdru, P. Levetiracetam: the first SV2A ligand for the treatment of epilepsy. *Expert Opin. Drug Discovery* **2007,** *2* (11), 1537–1545.
247. De Smedt, T.; Raedt, R.; Vonck, K.; Boon, P. Levetiracetam: the profile of a novel anticonvulsant drug—part I: preclinical data. *CNS Drug Rev.* **2007,** *13* (1), 43–56.
248. De Smedt, T.; Raedt, R.; Vonck, K.; Boon, P. Levetiracetam: part II, the clinical profile of a novel anticonvulsant drug. *CNS Drug Rev.* **2007,** *13* (1), 57–78.
249. Safdieh, J. E.; Harden, C. L. Levetiracetam: past, present and future. *Future Neurol.* **2006,** *1* (4), 365–371.
250. Patsalos, P. N. Levetiracetam: pharmacology and therapeutics in the treatment of epilepsy and other neurological conditions. *Rev. Contemp. Pharmacother.* **2004,** *13* (1,2), 1–168.
251. Zelano, J.; Kumlien, E. Levetiracetam as alternative stage two antiepileptic drug in status epilepticus: a systematic review. *Seizure* **2012,** *21* (4), 233–236.
252. Nash, E. M.; Sangha, K. S. Levetiracetam. *Am. J. Health-Syst. Pharm.* **2001,** *58* (13), 1195–1199.
253. Hovinga, C. A. Levetiracetam: a novel antiepileptic drug. *Pharmacotherapy* **2001,** *21* (11), 1375–1388.
254. Maryanoff, B. E.; Gardocki, J. F. Anticonvulsant sulfamate derivatives, US 4513006 (1985).
255. Maryanoff, B. E.; Nortey, S. O.; Gardocki, J. F.; Shank, R. P.; Dodgson, S. P. Anticonvulsant O-alkyl sulfamates. 2,3:4,5-Bis-O-(1-methylethylidene)-beta-D-fructopyranose sulfamate and related compounds. *J. Med. Chem.* **1987,** *30* (5), 880–887.

256. Maryanoff, B. E. Sugar sulfamates for seizure control: discovery and development of topiramate, a structurally unique antiepileptic drug. *Curr. Top. Med. Chem.* **2009,** *9* (11), 1049–1052.
257. Maryanoff, B. E. Pharmaceutical "gold" from neurostabilizing agents: topiramate and successor molecules. *J. Med. Chem.* **2009,** *52* (11), 3431–3440.
258. Parker, M. H.; Maryanoff, B. E.; Reitz, A. B. Synthesis of carba-β-L-fructopyranose and carbacyclic analogs of topiramate, an anticonvulsant agent, agent. *Synlett* **2004,** (12), 2095–2098.
259. Cross, J. H.; Riney, C. J. Topiramate. In *The Treatment of Epilepsy,* 3rd ed.; Shorvon, S., Perucca, E., Engel, J., Jr., Eds. Wiley-Blackwell, 2009; pp 673–683.
260. Lyseng-Williamson, K. A.; Yang, L. P. H. Topiramate: a review of its use in the treatment of epilepsy. *Drugs* **2007,** *67* (15), 2231–2256.
261. Guerrini, R.; Parmeggiani, L. Topiramate and its clinical applications in epilepsy. *Expert Opin. Pharmacother* **2006,** *7* (6), 811–823.
262. Sommer, B. R.; Fenn, H. H. Review of topiramate for the treatment of epilepsy in elderly patients. *Clin. Interventions Aging* **2010,** *5,* 89–99.
263. Naegel, S.; Obermann, M. Topiramate in the prevention and treatment of migraine: efficacy, safety and patient preference. *Neuropsychiatr. Dis. Treat.* **2010,** *6,* 17–28.
264. Kenna, G. A.; Lomastro, T. L.; Schiesl, A.; Leggio, L.; Swift, R. M. Review of topiramate: an antiepileptic for the treatment of alcohol dependence. *Curr. Drug Abuse Rev.* **2009,** *2* (2), 135–142.
265. Lyseng-Williamson, K. A.; Yang, L. P. H. Spotlight on topiramate in epilepsy. *CNS Drugs* **2008,** *22* (2), 171–174.
266. Faught, E. Topiramate in the treatment of partial and generalized epilepsy. *Neuropsychiatr. Dis. Treat.* **2007,** *3* (6), 811–821.
267. van Passel, L.; Arif, H.; Hirsch, L. J. Topiramate for the treatment of epilepsy and other nervous system disorders. *Expert Rev. Neurother.* **2006,** *6* (1), 19–31.
268. Shank, R. P.; Gardocki, J. F.; Vaught, J. L.; Davis, C. B.; Schupsky, J. J.; Raffa, R. B.; Dodgson, S. J.; Nortey, S. O.; Maryanoff, B. E. Topiramate: preclinical evaluation of a structurally novel anticonvulsant. *Epilepsia* **1994,** *35* (2), 450–460.

Chapter 10

Antiparkinsonian Drugs

Parkinson disease is a neurodegenerative disorder of the brain that affects movement. It develops gradually, starting with a tremor followed by bradykinesia or slowing of movement, rigidity of the limbs and trunk, postural instability or impaired balance and coordination. Symptoms worsen and progress over time. Parkinson disease is characterized by an imbalance between acetylcholine and dopamine, which probably results from the degeneration of a dopaminergic nigrostriatal pathway.

Breakdown of acetylcholine–dopamine balance hampers proper functioning of the corticobasal ganglia–thalamocortical loop circuits. In this disease, dopamine depletion blocks autoinhibition of acetylcholine release through muscarinic autoreceptors, leading to excessive acetylcholine release that eventually prunes spines of the indirect pathway's projection of neurons of the striatum, thereby interrupting information transfer from motor command centers in the cerebral cortex.

Many medications are available to treat the symptoms of Parkinson disease; however, the disease can't be cured and medications just help control its symptoms.

The first effective drugs for Parkinson disease were anticholinergics. Since the introduction of levodopa many new drugs have emerged for the treatment of Parkinson disease such as dopamine agonists, monoamine oxidase (MAO) B inhibitors, amantadine, and catechol O-methyltransferase inhibitors. In all stages of the disease, levodopa remains the most effective drug for improving motor symptoms. However, long-term treatment with levodopa is accompanied by the development of adverse effects such as motor fluctuations and dyskinesia.

Levodopa (**10.1.1**) was the first drug approved specifically for Parkinson disease in the 1970s. Levodopa is the immediate precursor of dopamine. In the brain, levodopa is metabolized to dopamine by enzymes, replacing the depleted endogenous neurotransmitter.

Levodopa and the levodopa–carbidopa (**10.1.2**) combination is an effective Parkinson disease medication. In this combination, carbidopa, a decarboxylase inhibitor, protects levodopa from premature conversion to dopamine outside of the brain. In Europe, levodopa is often combined with benserazide (**10.1.3**), another decarboxylase inhibitor (Fig. 10.1.).

Synthesis of Best-Seller Drugs. http://dx.doi.org/10.1016/B978-0-12-411492-0.00010-9

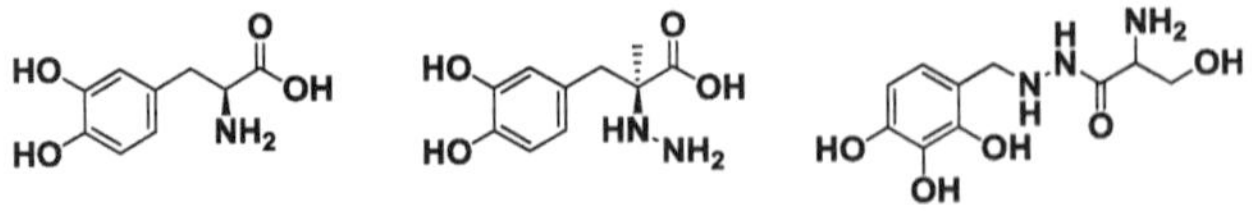

FIG. 10.1 **Structure of levodopa, carbidopa, and benserazide.**

10.1 LEVODOPA (LEVODOPA–CARBIDOPA)

Levodopa has been the mainstay of treatment for Parkinson disease for more than 40 years and attempts to optimize levodopa formulations to minimize side effects and enhance central nervous system has been carefully reviewed [1].

The synthesis of levodopa has been described in many publications and patents. It is proposed to synthesize starting from 3-nitro-L-tyrosine [2], or from 3-(3,4-methylenedioxyphenyl)-L-alanine [3], or from L-tyrosine [4]. Synthesis based on catalytic asymmetric hydrogenation of a racemic mixture of alfa-acetamido-4-hydroxy-3-alkoxy-cinnamic acids also was proposed [5]. A series of processes by enzymatic and fermentation [6], by fermentation of L-tyrosine [7], by stereoselective hydrolysis of N-acyl derivatives with *Escherichia coli* acylase [8], and by reaction between protected 3,4-dihydroxyphenyl pyruvic acids and glutamic or aspartic acid in presence of aminotransferase produced by genetically modified *E. coli* [8] are described. A method for the optical resolution of precursors of the levodopa has been described in Krubiner [9], for the resolution of the d,l-N-acetyl-3-(3,4-dimethoxyphenyl)alanine with d-α-methylbenzylamine is described in Schubel et al. [10], resolution of the d,l-N-benzoyl-3,4-dihydroxyphenylalanine with cinchonidine in Stoeck et al. [11], and in many others.

Levodopa and levodopa–carbidopa are not included in the list of Top 200 Drugs by sales for the 2010s, but because of their extreme importance in the treatment of Parkinson disease, in addition to methods described in our previous book [12], one of the more recent synthetic Schemes of industrial meaning for the synthesis [13] is presented in Scheme 10.1.

The process is based on Strecker amino-acid synthesis starting with condensation of 3,4-dimethoxyphenyl-acetaldehyde (**10.1.4**) with sodium cyanide in ammonium hydroxide media to form the racemic α-aminonitrile (**10.1.5**). The obtained aminonitrile, is subjected to an optical resolution by selective precipitation and separation of the d-diastereoisomeric salt (**10.1.7**) made by the d-camphorsulfonic acid (10.1.6) and D-2-amino-3-(3,4-dimethoxyphenyl)-propionitrile (**10.1.8**). By using ammonium hydroxide to transform the camphorsulfonic salt (**10.1.7**) into the base D-2-amino-3-(3,4-dimethoxyphenyl)propionitrile (**10.1.8**) and with the subsequent insufflation of gaseous hydrogen chloride into the solution, the D-2-amino-3-(3,4-dimethoxyphenyl)-propionitrile hydrochloride (**10.1.9**) was prepared. The hydrolysis of nitrile group and the demethylation of methoxy groups in the obtained compound was accomplished on reflux of the obtained hydrochloride (**10.1.9**) with 48% HBr solution. Bringing the pH to 4.5 with ammonium hydroxide gave the desired L-(-)-2-amino-3-(3,4-dihydroxyphenyl)propionic acid-levodopa (**10.1.1**) (Scheme 10.1.).

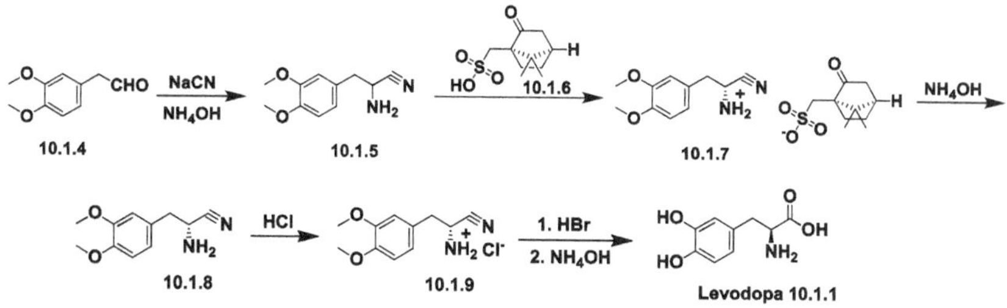

SCHEME 10.1 **Synthesis of levodopa.**

Several methods have been provided in the literature for the preparation of carbidopa. According to one of them [14-16] (Scheme 10.2.), the racemic carbidopa (**10.1.2**) has been prepared by the use of a Strecker reaction in which the source of ammonia was replaced with aqueous hydrazine. This condensation of 1-(4′-hydroxy-3′-methoxyphenyl)-2-propanone, vanillyl methyl ketone (**10.1.10**) with potassium cyanide in aqueous hydrazine resulted in the hydrazinonitrile (**10.1.1**) which was hydrolyzed in two stages. First, to an amide (**10.1.12**) by fortified hydrochloric acid at −10 °C to 0 °C; second, to the desired racemic hydrazine acid (**10.1.3**) by refluxing with boiling 48% hydrobromic acid. Two serious limitations make this simple method not very acceptable: (a) The starting ketone (**10.1.10**) readily reacts with obtained hydrazinonitrile; and (b) the hydrazinonitrile itself readily loses hydrogen cyanide [17].

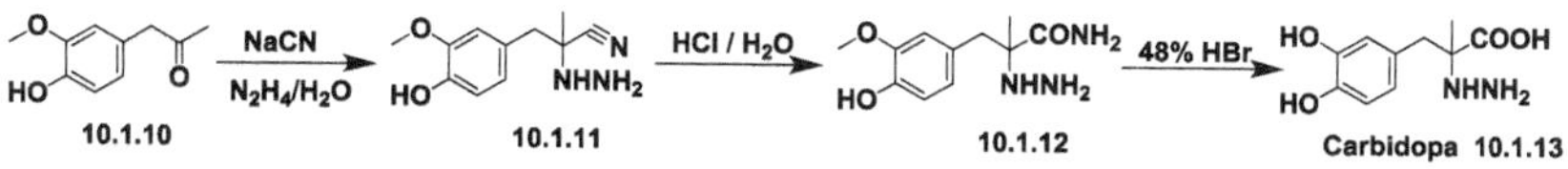

SCHEME 10.2 **Synthesis of racemic carbidopa.**

Another disadvantage of this method is the practical impossibility of resolving the racemate, except when 1-menthoxyacetyl chloride was implemented to synthesize the mixture of corresponding racemic hydrazides (**10.1.14**). Surprisingly, crystals of mainly levorotatory hydrazide (**10.1.14a**) precipitated and after hydrolysis allow to obtain desired levorotatory hydrazine acid, carbidopa (**10.1.12**) [17] (Scheme 10.3.).

The ready availability of L(-)-α-methyldopa and its derivatives provided attempts for its direct N-amination, that is, direct synthesis of carbidopa, when formal proof of the absolute configuration is constituted. Variations of reactions hydroxylamine-O-sulfonic acid with chloramine, intermediate formation of hydantoic acid by the use of cyanates with further oxidation to hydrazino acid and other were proposed [18-21].

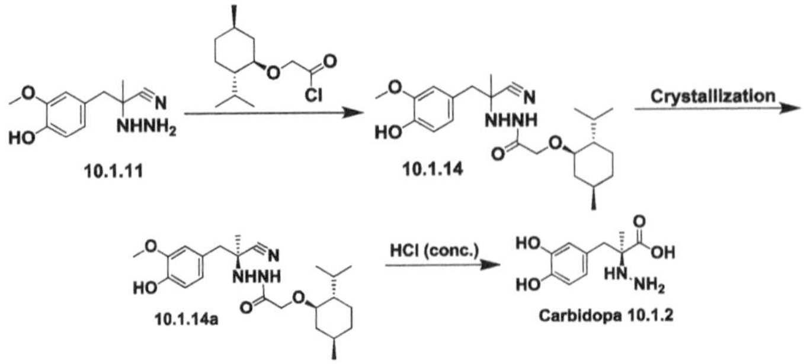

SCHEME 10.3 **Synthesis of carbidopa.**

The "state of the art" elegant synthesis of L-carbidopa, but unlikely applicable at industrial scale, was demonstrated recently [22] (Scheme 10.4.).

Reaction of 3,4-dimethoxyacetophenone (**10.1.15**) with dimethylcarbonate in the presence of sodium hydride produced the corresponding β-keto ester, which underwent transesterification with isopropanol using catalytic amounts of ZnO. After that classical alkylation of obtained product with methyl iodide using potassium carbonate in acetone, compound (**10.1.16**) was formed. Obtained 3-oxo-3-phenylpropanoate derivative (**10.1.16**) then underwent enantioselective R-hydra-zination with di-tert-butyl azodicarboxylate using a mixture of $Eu(OTf)_3$/bis(diphenyl-oxazoline) derivative (**10.1.17**) as a catalyst. The prepareded in 95% yield adduct (**10.1.18**) was 98% enantiomeric excess (ee). The Boc groups were easily removed with trifluoroacetic acid, and the free hydrazine intermediate was protected with the use of benzyl chloroformate to produce (**10.1.19**). The keto group in (**10.1.19**) underwent a smooth reduction using the $Et(Me)_2SiH_3$ (**10.1.20**)–TFA combination to produce (**10.1.21**) in a 72% yield. Finally, the global deprotection of (**10.1.21**) with PBr_3 led directly to L-carbidopa (**10.1.2**) as a hydrobromide.

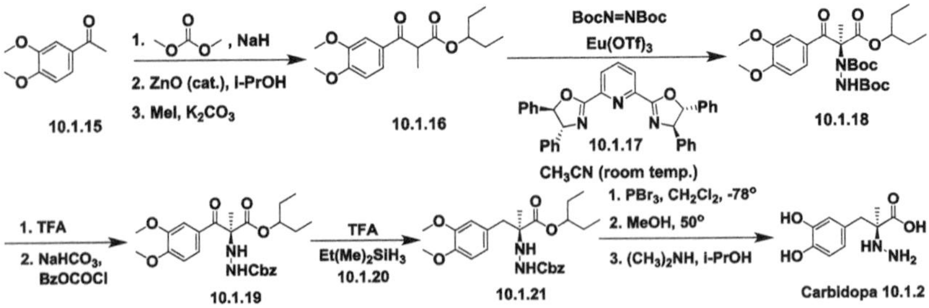

SCHEME 10.4 **Synthesis of carbidopa.**

10.2 DOPAMINE AGONISTS

Dopamine agonists exert their antiparkinsonian effects by acting directly on dopamine receptors and mimicking the endogenous neurotransmitter. Unlike levodopa, dopamine agonists don't transfer into dopamine. They just mimic dopamine effects in brain. Although they aren't as effective as levodopa, they last longer and may be used with levodopa to smooth the sometimes off-and-on effect of levodopa. A variety of dopamine agonists are available that are classified as ergoline and nonergoline dopamine agonists. These differences relate to the chemical structure of the medications rather than their action.

Ergoline itself (**10.2.1**) (Fig. 10.2.) is a chemical compound whose structural skeleton is contained in a diverse range of alkaloids and synthetic drugs.

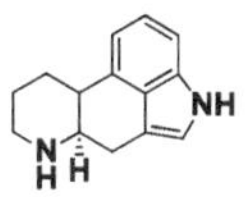

Ergoline 10.2.1

FIG. 10.2 Structure of ergoline.

Some ergoline alkaloids cause convulsive and gangrenous symptoms. Others have psychedelic properties. As well as acting on dopamine receptors, ergoline dopamine agonists—bromocriptine (**10.2.2**), cabergoline (**10.2.3**), lisuride (**10.2.4**), and pergolide (**10.2.5**) (Fig. 10.3.)—may also act on other chemical receptors in the brain. Hallucinations are a frequent complication of pharmacotherapy of Parkinson disease and occur more often with dopamine agonists than with levodopa. Pergolide was withdrawn from the U.S. market in 2007 because of reported cases of valvular heart disease.

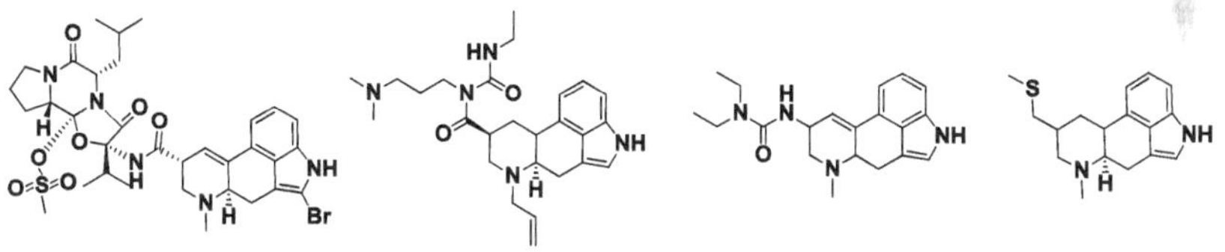

Bromocriptine 10.2.2 Cabergoline 10.2.3 Lisuride 10.2.4 Pergolide 10.2.5

FIG. 10.3 Structure of bromocriptine, cabergoline, lisuride, and pergolide.

Nonergoline dopamine agonists—apomorphine (**10.2.6**), pramipexole (**10.2.7**), ropinirole (**10.2.8**), and rotigotine (**10.2.9**) (Fig. 10.4.)—are selective and work on specific (D_1, D_2, D_3) dopamine receptors.

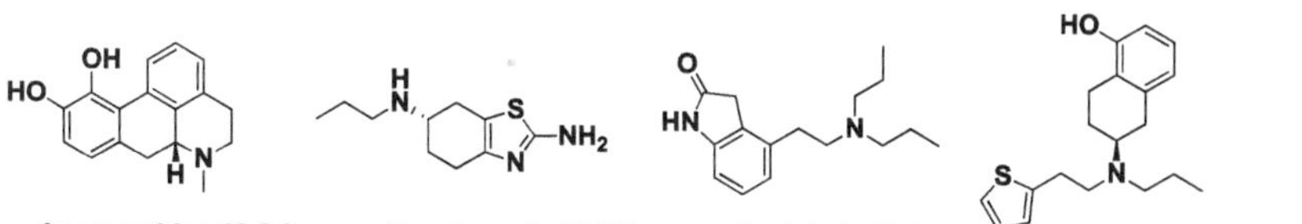

Apomorphine 10.2.6 Pramipexole 10.2.7 Ropinirole 10.2.8 Rotigotine 10.2.9

FIG. 10.4 Nonergoline dopamine agonists.

None of the mentioned compounds is included in the list of Top 200 Drugs by sales for the 2010s.

10.3 ANTICHOLINERGICS

Anticholinergic medications were used for many years to help control the tremor associated with Parkinson disease. Activation of muscarinic acetylcholine receptors (mAChRs) has an excitatory effect, opposite to that of dopaminergic activation, causing increased parasympathetic activity, that is, vasodilation; constriction of pupils in the eyes; increased secretion of sweat, saliva, and tears; slow heart rate; mucus secretion in the respiratory tract; and constriction of bronchioles; and so on. Anticholinergic antiparkinson agents or acetylcholine antagonists block the mAChRs and cholinergic nerve activity. Some of mAChRs blockers are currently approved for clinical use include nonselective antagonists for the treatment of Parkinson disease via blocking the mAChRs and cholinergic nerve activity.

Several anticholinergic medications are available, including diphenhydramine (**10.3.1**), benztropine (**10.3.2**), biperiden (**10.3.3**), procyclidine (**10.3.4**), and trihexyphenidyl (**10.3.5**) (Fig. 10.5.) and their synthesis is described in our previous book [12].

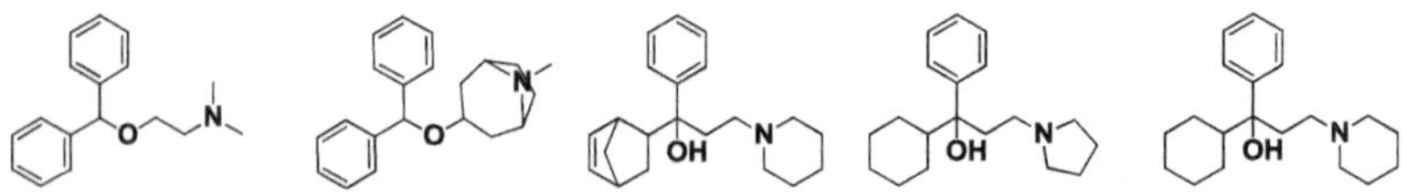

Diphenhydramine 10.3.1 Benztropine 10.3.2 Biperiden 10.3.3 Procyclidine 10.3.4 Trihexyphenidyl 10.3.5

FIG. 10.5 **Anticholinergic medications to help control the tremor associated with Parkinson disease.**

10.4 MAO INHIBITORS

MAO inhibitors inhibit MAO B, which metabolizes brain dopamine. When added to carbidopa-levodopa, these medications can increase the risk of hallucinations. These medications include selegiline (**10.4.1**) and rasagiline (**10.4.2**) (Fig. 10.6.).

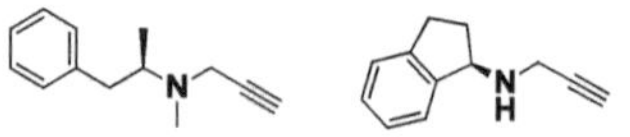

Selegiline 10.4.1 Rasagiline 10.4.2

FIG. 10.6 **Structure of selegiline and rasagiline.**

10.5 CATECHOL-O-METHYLTRANSFERASE INHIBITORS

Catechol-O-methyltransferase (COMT) inhibitors are a new therapeutic option in the treatment of patients with Parkinson disease. They mildly prolong the effect of levodopa therapy via blocking the enzyme COMT, allowing a larger amount of levodopa to reach the brain, thus raising dopamine levels there. The representatives of COMT are members of the class of drugs known as nitrocatechols and two of them, entacapone (**10.5.1**) and tolcapone (**10.5.2**) (Fig. 10.7.), are only used in conjunction with levodopa. Their action is somewhat similar to that of carbidopa and benserazide in that they inhibit the action of the enzyme that converts levodopa into a compound that cannot cross the blood–brain barrier. Side effects are primarily those attributable to an enhanced levodopa effect, including an increased risk of involuntary movements (dyskinesias). Another limitation of their implementation is the risk of serious liver damage.

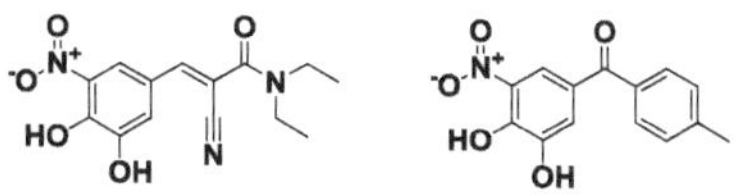

Entacapone 10.5.1 Tolcapone 10.5.2

FIG. 10.7 **Structure of entacapone and tolcapone.**

10.6 AMANTADINE

Amantadine (**10.6.1**) (Fig. 10.8.) is another pharmacological adjuvant used in conjunction with levodopa–carbidopa and dopamine agonists. It reduces symptoms of fatigue, tremor, and bradykinesia and dyskinesia in early and in more advanced Parkinson disease. Side effects may include a purple mottling of the skin, ankle swelling, or hallucinations.

Amantadine 10.6.1

FIG. 10.8 **Structure of amantadine.**

10.7 NEW DRUGS IN DEVELOPMENT FOR PARKINSON DISEASE

Numerous drugs that target the primary motor disorder in Parkinson disease are in development. Among them are modulators of mitochondrial function coenzyme Q10, also known as ubiquinone (**10.7.1**), and creatine (**10.7.2**) (Fig. 10.9.), which are thought to act as an antioxidant modulating mitochondrial function of patients with Parkinson disease. Creatine causes modest increases in strength in people with Parkinson disease. It has the ability to increase muscle stores of phosphocreatine, potentially increasing the muscle's ability to resynthesize adenosine triphosphate (ATP) from adenosine diphosphate (ADP) via transferring a phosphoryl group to ADP to meet increased energy demands.

Other neuroprotective strategies employ neuroimmunophilin ligands such as immunosuppressant FK506 (**10.7.3**) (tacrolimus), a macrolide antibiotic with immunosuppressive properties isolated from the fungus *Streptomyces tsukubaensis*, and which is used as an immunosuppressant to prevent the rejection of organ transplants. CEP-1347 (**10.7.4**) is an inhibitor of mixed lineage kinases that elicits neuroprotective and antiinflammatory responses in models of neurodegenerative diseases and others, and has variable efficacy in reversing neuronal degeneration and in preventing cell death. In neuroprotective animal models of brain trauma and neurodegenerative diseases, a second-generation tetracycline derivative, minocycline (**10.7.5**), showed expressed neuroprotective properties.

Several strategies have been proposed and developed to improve existing dopaminergic agents and it has been shown that the methyl ester of levodopa, melevodopa (**10.7.6**), is as effective as levodopa, and might have the advantage of faster absorption. Innovations can be expected from the development of new selective dopamine agonists such as the highly selective D_2 receptor full agonist sumanirole (**10.7.7**), and D_2 and D_3 partial agonist pardoprunox (SLV308) (**10.7.8**).

Safinamide (**10.7.9**), with multiple sites of action, is a novel experimental drug that combines several pharmacological properties and potentially is useful in the treatment of Parkinson disease.

Istradefylline (**10.7.10**) is a novel adenosine A2A receptor antagonist designed to treat patients with motor fluctuations and dyskinesias, particularly in Parkinson disease.

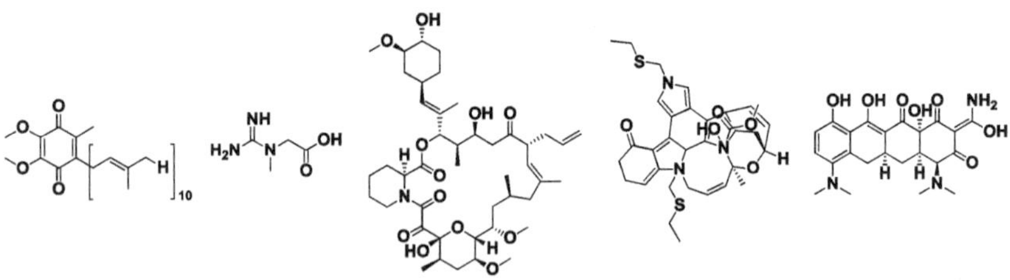

Coenzyme Q10 10.7.1 Creatine 10.7.2 FK506 (tacrolimus) 10.7.3 CEP-1347 (10.7.4) Minocycline 10.7.5

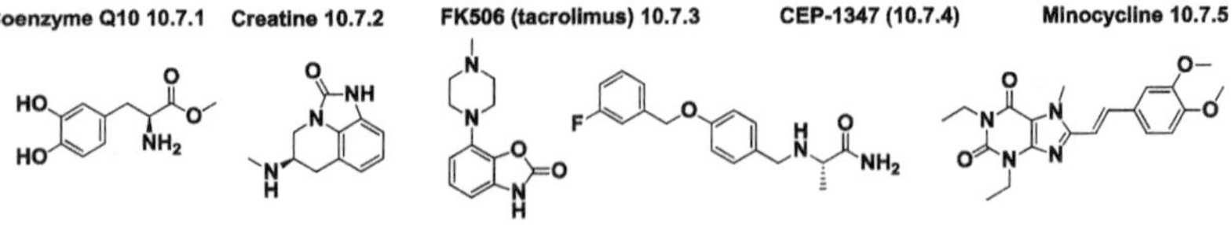

Melevodopa 10.7.6 Sumanirole 10.7.7 Pardoprunox 10.7.8 Safinamide 10.7.9 Istradefylline 10.7.10

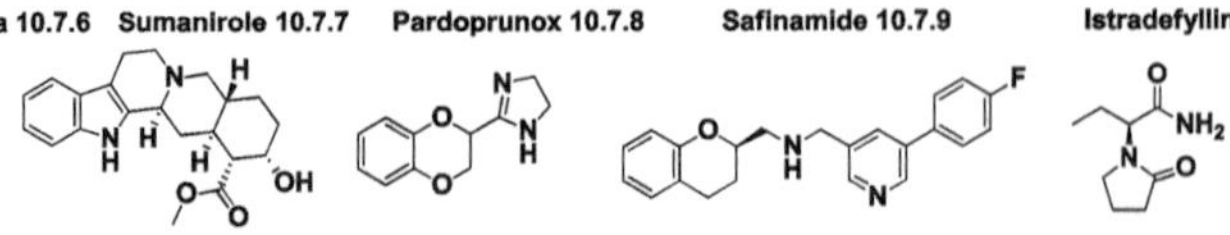

Yohimbine 10.7.11 Idazoxan 10.7.12 Sarizotan 10.7.13 Levetiracetam 10.7.14

FIG. 10.9 New drugs in development for Parkinson disease.

Several studies have shown that motor fluctuations and levodopa-induced dyskinesia can be modulated using drugs acting on neurotransmitters other than dopamine, including glutamate, γ-aminobutyric acid (GABA), epinephrine, acetylcholine, 5-HT, adenosine, and cholecystokinin receptors.

It was only recently found that amantadine (**10.6.1**), an antiviral compound, acts as a noncompetitive *N*-methyl-D-aspartate (NMDA) glutamatergic antagonist.

The coadministration of yohimbine (**10.7.11**), which has a variety of pharmacological properties, although it primarily acts as α-adrenergic receptor antagonist, treats levodopa-induced dyskinesia.

Idazoxan (**10.7.12**), a selective α_2-adrenoreceptor antagonist, was developed to reduce the dyskinetic side effects of levodopa.

Sarizotan (**10.7.13**), a selective 5-HT_{1A} receptor agonist and D_2 receptor antagonist, induced a significant reduction in dyskinesia.

The antiepileptic drug levetiracetam (**10.7.14**) significantly reduces levodopa-induced dyskinesia in animal models [23-27].

REFERENCES

1. Hauser, R. A. Levodopa: past, present, and future. *Eur. Neurol.* **2009,** *62* (1), 1–8.
2. Waser, E.; Lewandowski, M. Phenylalanine series. I. Synthesis of I-3,4-dihydroxyphenylalanine. *Helv. Chim. Acta* **1921,** *4,* 657–666.
3. Yamada, S.; Fujii, T.; Shioiri, T. Optically active amino acids. III. Preparation of 3-(3,4-dihydroxyphenyl)-DL-, -D-, and -L-alanine alanine. *Chem. Pharm. Bull.* **1962,** *10,* 693–697.
4. Vorbrueggen, H.; Krolikiewicz, K. New synthesis of 3-(3,4-dihydroxyphenyl)-L-alanine (L-DOPA). *Chem. Ber.* **1972,** *105* (4), 1168–1173.
5. Knowles, W, S.; Sabacky, M. J.; Vineyard, B. D, L-Dopa and intermediates, US 4005127 (1977).
6. Florent, J.; Lunel, J.; Renaut, J. L-β-(3,4-Dihydroxyphenyl)-α-alanine, DE 2102793 (1971).
7. Bamberg, P.; Sjoberg, B. O. H. L-Dopa and L-m-tyrosine for treating parkinsonism, and their intermediates, DE 2100445 (1971).
8. Dyer, R. L.; Lewis, D. J. L-Dopa, EP 189938 (1986).
9. Krubiner, A. M. L-Dopa by optical resolution with (+)-α-methylbenzylamine, DE 1964422 (1970).
10. Schubel, H.; Janssen, P.; Rutz, H. Pure L-N-benzoyl-3,4-dihydroxyphenylalanine, DE 1963991 (1971).
11. Stoeck, G.; Budka, H. G.; Topfmeier, F.; Gradel, W. L-Dopa from D, L-N-benzoyl-3,4-dimethoxyphenylalanine, DE 2039253 (1972).
12. Vardanyan, R. S.; Hruby, V. J. *Synthesis of Essential Drugs;* Elsevier, 2006.
13. Cannata, V.; Tamerlani, G.; Morotti, M. New process for the synthesis of levodopa, EP 357565 (1990).
14. Sletzinger, M.; Chemerda, J. M.; Bollinger, F. W. Potent decarboxylase inhibitors. Analogs of methyldopa. *J. Med. Chem.* **1963,** *6,* 101–103.
15. Pfister, K. III. α-Hydrazino-β-phenylpropionic acid-α-alkylphenylalanine mixtures, FR M1 1553 (1962).
16. Inventor data not available. α-Hydrazino acids, GB 940596 (1963).

17. Karady, S.; Ly, M. G.; Pines, S. H.; Sletzinger, M. Synthesis of D- and L-alpha-(3,4-dihydroxybenzyl)-alpha-hydrazinopropionic acid via resolution. *J. Org. Chem.* **1971,** *36* (14), 1946–1948.
18. Karady, S.; Ly, M. G.; Pines, S. H.; Sletzinger, M. Synthesis of L-alpha-(3,4-dihydroxybenzyl)-alpha-hydrazinopropionic acid from optically active precursors by N-homologization. *J. Org. Chem.* **1971,** *36* (14), 1949–1951.
19. Ribalta, B. M.; Mirallas G. S. Industrial process for preparation of hydrazine derivative, L-α-hydrazino-3,4-dihydroxy-α-methylbenzenepropanoic acid monohydrate, ES 2026759 (1992).
20. Poslinska-Bucewka, H.; Rudnicki, A.; Paszkowski, S. Method of preparing L-α-methyl-α-hydrazino-β-(3,4-dihydroxyphenyl)propionic acid [carbidopa], PL 162497 (1993).
21. Katai, F. K.; Farkas, J.; Lukacs, G.; Lang, F.; Jurak, F.; Csala, I.; Koroknai, T.; Porcs-Makkay, M. Process for the preparation of carbidopa WO 2007042848 (2007).
22. Pericas, A.; Shafir, A.; Vallribera, A. Asymmetric synthesis of L-carbidopa based on a highly enantioselective α-amination. *Org. Lett.* **2013,** *7* (15), 1448–1451.
23. Colosimo, C.; Fabbrini, G.; Berardelli, A. Drug insight: new drugs in development for Parkinson's disease. *Nat. Clin. Pract. Neurol.* **2006,** *2* (11), 600–610.
24. Bockler, Frank Competition for the "gold standard" L-dopa? Receptor mediated Parkinson's therapy. *Pharmazie* **2006,** *35* (3), 204–216.
25. Djaldetti, R.; Melamed, E. New drugs in the future treatment of Parkinson's disease. *J. Neurol.* **2002,** *249* (Suppl. 2), 30–35.
26. Gottwald, M. D.; Bainbridge, J. L.; Dowling, G. A.; Aminoff, M. J.; Alldredge, B. K. New pharmacotherapy for Parkinson's disease. *Ann. Pharmacother.* **1997,** *31* (10), 1205–1217.
27. Almeida, Q. J.; Hyson, H. C. Pharmacological treatment for Parkinson's disease: a continuing evolution. *Front. CNS Drug Discovery* **2010,** *1,* 147–155.

Chapter 11

Adrenergic (Sympathomimetic) Drugs

Adrenaline is a hormone produced within the adrenal gland in response to stress that increases heart rate, strengthens the force of the heart's contraction and cardiac output, increases blood pressure and opens up the bronchioles in the lungs, and raises the blood levels of glucose and lipids among other effects. The secretion of adrenaline is a part of the human "fight or flight" response, the acute stress response to fear, perceived threat, or panic.

An excessive adrenaline level usually is observed in highly emotional and overaggressive persons.

On the other hand, there are some disorders of the adrenal glands that reduce the level of epinephrine to below normal, including Addison disease and other forms of hypoadrenalism.

Any drug that mimics the functioning of the sympathetic nervous system by affecting the release or action of epinephrine (adrenaline) (**11.1**), norepinephrine (noradrenaline) (**11.2**), and dopamine (**11.3**)—hormones that are secreted by the adrenal gland—is considered an adrenergic drug. The term *adrenergic* literally means "having to do with adrenaline (epinephrine) and/or noradrenaline (norepinephrine)" [1,2] (Fig. 11.1.).

The main function of the mentioned hormones is to adapt the body to stressful situations and they have a half-life of a few minutes when circulating in the blood. They undergo degradation either by catechol-O-methyltransferases or by monoamine oxidases.

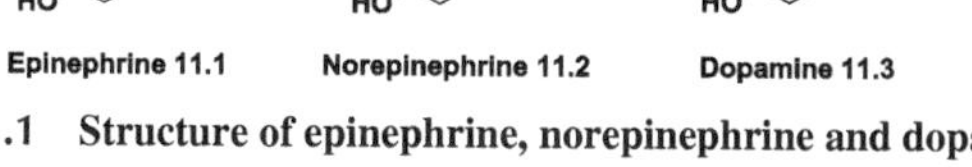

FIG. 11.1 Structure of epinephrine, norepinephrine and dopamine.

By definition, adrenergic agonists produce their effects by activating adrenergic receptors.

There are at least two adrenergic receptor types—α and β (α_1, α_2, β_1, and β_2 receptors)—by which adrenergic drugs exert their effects. Norepinephrine activates primarily α and epinephrine activates primarily β receptors, although epinephrine may also activate α receptors. Stimulation of α receptors is associated

Synthesis of Best-Seller Drugs. http://dx.doi.org/10.1016/B978-0-12-411492-0.00011-0

with constriction of small blood vessels in the bronchial mucosa and relaxation of smooth muscles of the intestinal tract. β-Receptor activation relaxes bronchial smooth muscles, which causes the bronchi to dilate, causing an increase in the rate and force of heart contractions.

Adrenergic agonists (sympathomimetic agents) raise blood pressure and increase heart rate. Therapeutically, these drugs are used to provide patients relief from disorders such as bronchial asthma, chronic obstructive pulmonary diseases, cardiac arrest, allergic reactions, and nasal decongestants; they are also used as appetite suppressants.

Adrenergic drugs increase the output of the heart, and used to be given as heart stimulants to raise blood pressure, reversing the drop in blood pressure, and to increase urine flow as part of the treatment of shock. They also may be used to stop bleeding by causing the blood vessels to constrict.

Adrenergic agonists are subdivided into three classes: direct acting, indirect acting, and dual acting.

1. *Direct-acting agonists* bind to and activate α_1, α_2, β_1, and β_2 receptors. Naturally occurring molecules that bind to these receptors include epinephrine (**11.1**), which binds to α_1,α_2, and β_1 receptors, norepinephrine (**11.2**), which binds to α_1, α_2, and β_1 receptors, and dopamine (**11.3**), which binds to dopamine receptors as well as to α_1 and β_1 receptors. Direct-acting agonists such as isoproterenol (**11.4**), dobutamine (**11.5**), phenylephrine (**11.6**), and clonidine (**11.7**) (Fig. 11.2.), are also adrenergic drugs.

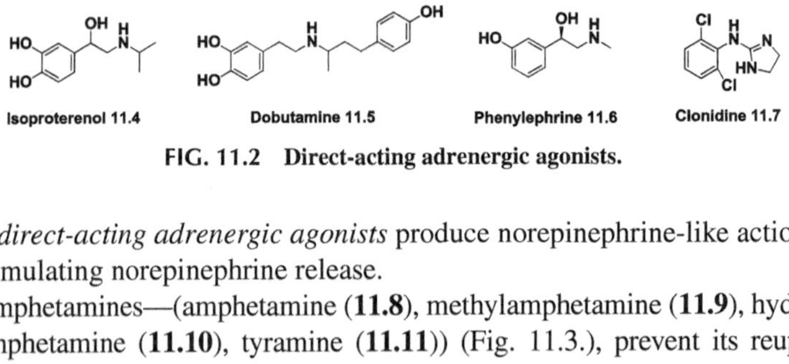

FIG. 11.2 **Direct-acting adrenergic agonists.**

2. *Indirect-acting adrenergic agonists* produce norepinephrine-like actions by stimulating norepinephrine release.
   Amphetamines—(amphetamine (**11.8**), methylamphetamine (**11.9**), hydroxyamphetamine (**11.10**), tyramine (**11.11**)) (Fig. 11.3.), prevent its reuptake. Tricyclic antidepressant (amitriptyline (**11.12**), cocaine (**11.13**)), which acts by multiple mechanisms on brain catecholaminergic neurons and compounds inhibiting of norepinephrine inactivation (monoamine oxidase [MAO] inhibitors) increase the amount of norepinephrine available for release.

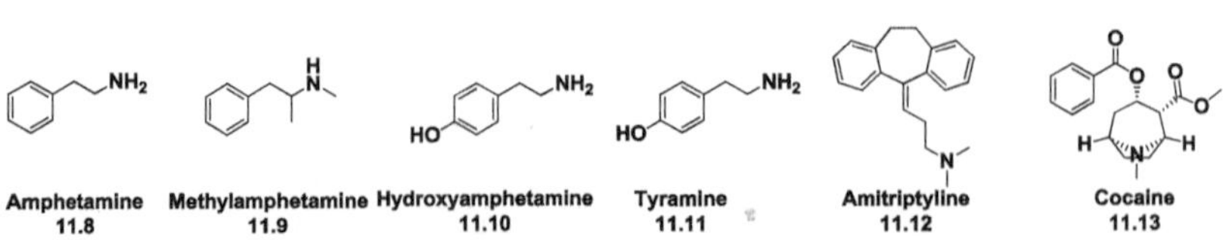

FIG. 11.3 **Indirect acting adrenergic agonists.**

3. *Dual-acting adrenergic agonists*, which bind to adrenergic receptors and stimulate norepinephrine release. Among them are drugs such as ephedrine (**11.14**) and metaraminol (**11.15**) (Fig. 11.4.).

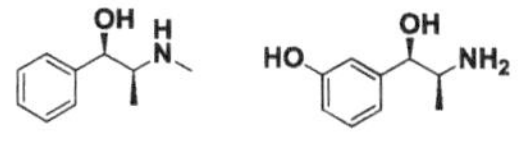

Ephedrine 11.14 Metaraminol 11.15

FIG. 11.4 Dual-acting adrenergic agonists.

According to their chemical structure, sympathomimetic drugs may be divided into catecholamines and noncatecholamines. Catecholamines are hormones produced by the adrenal glands. It is a group of sympathomimetic amines (including dopamine, epinephrine, and norepinephrine), the aromatic constituent of whose molecules is catechol (o-dihydroxybenzene).

Noncatecholamines used as adrenomimetics are phenylephrine (**11.6**), amphetamine (**11.8**), methylamphetamine (**11.9**), hydroxyamphetamine (**11.10**), tyramine (**11.11**), and ephedrine (**11.14**). Because they lack a catechol group, noncatecholamines are not substrates for metabolizing enzymes. As a result, their half-lives are much longer than those of catecholamines, they can be given to a patient orally, and they are more able to cross the blood–brain barrier.

The receptor mediating the vasoconstrictor actions of catecholamines is referred to as an α receptor. α Receptors are associated mainly with increased contractibility of vascular smooth muscle and intestinal relaxation.

α Receptors have been further subdivided into α_1 and α_2 receptors. Epinephrine and norepinephrine activate both receptors.

11.1 α_1-RECEPTOR ACTIVATING DRUGS

α_1 Receptors are characteristic of vascular smooth muscle, although their density varies throughout the body—eye, skin, viscera, mucous membranes, veins, sex organs, and bladder—but are quite abundant on blood vessels of the oral and nasal mucosae and urinary sphincters. α_1 Receptors are also found on muscles attached to hair follicles. Their contraction accounts for "goose bumps." They are also found on apocrine sweat glands in the armpits, groin, and palms, causing "nervous sweat" solely as a result of the hormones in the blood. Contraction of the dilator muscle of the iris is caused by α_1 binding, which causes the pupil to dilate and can facilitate eye examinations and ocular surgery. Dilator muscle contraction is the only clinical use of α_1 activation that is not based on vasoconstriction.

Activation of α_1 receptors elicits two responses that can be of therapeutic use: vasoconstriction and mydriasis. Vasoconstriction is the one for which α_1 agonists are used most often.

Because of their properties as vasoconstrictive agents, α_1 agonists are used to reduce edema and inflammation. They are used to stop bleeding primarily in the skin and mucous membranes, and as nasal decongestants. α_1-Agonists are frequently combined with local anesthetics to delay anesthetic absorption.

Common α_1-receptor activating drugs include phenylephrine (**11.6**), pseudoephedrine (**11.1.1**), methoxamine (**11.1.2**), and propylhexedrine (**11.1.3**), which are based on the phenylethylamine skeleton, and dihydroimidazole derivatives such as naphazoline (**11.1.4**), tetrahydrozoline (**11.1.5**), xylometazoline (**11.1.6**), oxymetazoline (**11.1.7**), and cirazoline (**11.1.8**) (Fig. 11.5.)

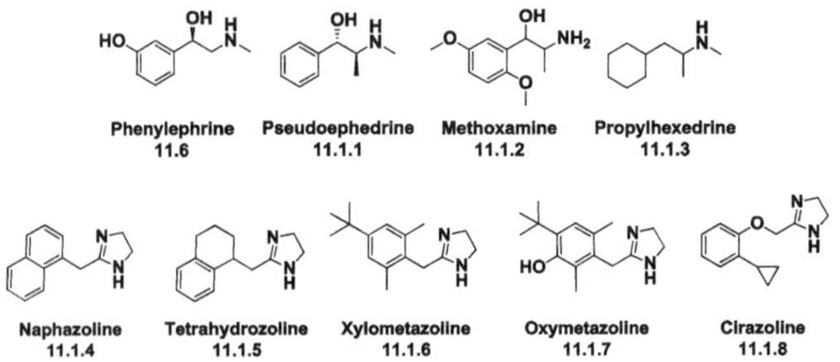

FIG. 11.5 α_1**-Receptor activating drugs.**

α_1-Receptor activating drugs are used for temporary relief of congestion in the nose caused by various conditions, including the common cold, sinusitis, hay fever, and allergies, and to relieve redness, puffiness, and itchy/watering eyes resulting from colds, allergies, or an eye irritation. α_1-Agonists can cause headache, reflex bradycardia, excitability, and restlessness.

11.2 α_2-RECEPTOR ACTIVATING DRUGS

α_2 Receptors are located in presynaptic nerve terminal. In the periphery, they are mainly located on vascular smooth muscle of veins, more so than on arteries. They are abundant in the brain and are associated with the pain perception. Activation of these receptors inhibits the release of norepinephrine. There are no therapeutic applications related to activation of peripheral α_2 receptors. In contrast, activation of α_2 receptors in the central nervous system (CNS) are of great clinical significance, producing two useful effects: reduction of sympathetic outflow to the heart and blood vessels and relief of severe pain. Several α_2 selective adrenergic agonists are known: methylepinephrine (**11.2.1**), α-methylnorepinephrine (**11.2.2**), clonidine (**11.2.3**), tizanidine (**11.2.4**), guanabenz (**11.2.5**), guanfacine (**11.2.6**), and guanethidine (**11.2.7**) (Fig. 11.6.).

Methylepinephrine 11.2.1   α-Methylnorepinephrine 11.2.2

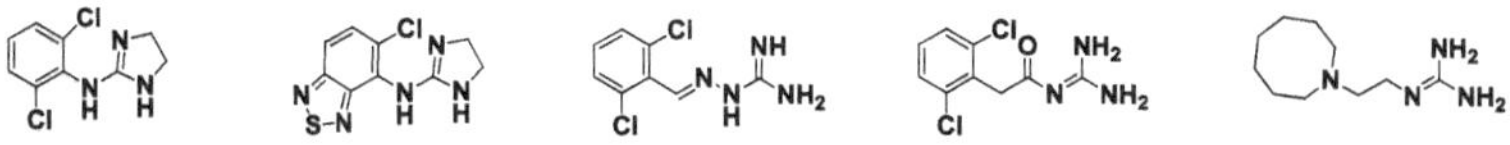

Clonidine 11.2.3   Tizanidine 11.2.4   Guanabenz 11.2.5   Guanfacine 11.2.6   Guanethidine 11.2.7

FIG. 11.6 **α_2 Receptor agonists.**

The α_2-adrenoceptor agonists are used very occasionally as centrally acting hypotensive agents.

Side effects of centrally acting α_2-adrenoceptor agonists include sedation, dry mouth and nasal mucosa, bradycardia, orthostatic hypotension, and impotence.

11.3 β_1-RECEPTOR ACTIVATING DRUGS

β_1-Receptor locations are heart muscle and kidney. These receptors are associated with the conducting system (e.g., pacemaker) and the ventricular musculature. They are also found in eccrine sweat and salivary glands. Epinephrine (**11.1.1**) or norepinephrine (**11.1.2**) causes excitatory responses in these tissues. Activation of β_1 receptors in the heart has a positive inotropic effect by increasing the force of contraction, thereby improving cardiac performance. The primary goal of treatment with β_1 agonists is to maintain blood flow to vital organs. By activating cardiac β_1 receptors, drugs can initiate contraction in a heart that has stopped beating. Selective β_1-activating drugs are dobutamine (**11.5**), denopamine (**11.3.1**), prenalterol (**11.3.2**), and xamoterol (**11.3.3**), which is considered third-generation β-adrenergic receptor partial agonist that provides cardiac stimulation at rest, but acts as a β blocker during exercise (Fig. 11.7.).

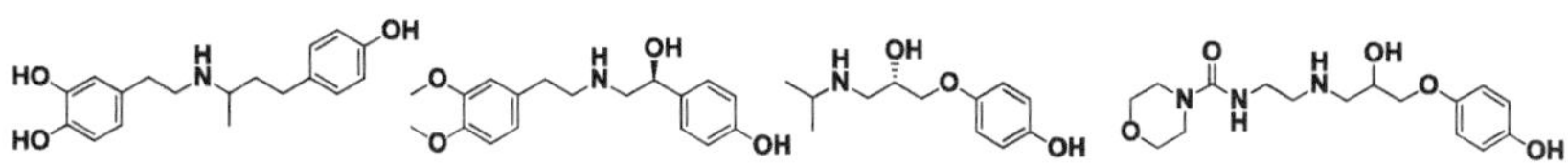

Dobutamine 11.5   Denopamine 11.3.1   Prenalterol 11.3.2   Xamoterol 11.3.3

FIG. 11.7 **β_1 Receptor agonists.**

Drugs that activate the β_1 receptor are used in heart failure to improve the contractile state of the failing heart. They also increase heart rate, but excess stimulation can induce significant increases in heart rate and arrhythmias

11.4 β_2-RECEPTOR ACTIVATING DRUGS

β_2 Receptors are found in bronchial smooth muscle and blood vessels of skeletal muscle, arterioles, heart, lung, uterus, and liver. Therapeutic applications of β_2 activation are limited to the lungs and the uterus. Because drugs that activate β_2

receptors cause smooth muscle relaxation in the lung promoting bronchodilation and help relieve or prevent asthma attacks, β_2 agonists are used for treatment of airways dysfunction such as bronchial asthma, chronic bronchitis, and emphysema, and in cases of premature labor for relaxing uterine smooth muscle.

Drugs used for their β_2-activating ability also can be classified as short-acting—albuterol (**11.4.1**), pirbuterol (**11.4.2**), procaterol (**11.4.3**), orciprenaline (**11.4.5**), terbutaline (**11.4.6**), fenoterol (**11.4.7**), and isoprenaline (**11.4.8**)—and long-acting— formoterol (**11.4.9**), bambuterol (**11.4.10**), clenbuterol (**11.4.11**), and salmeterol (**11.4.12**)—drugs (Fig. 11.8.).

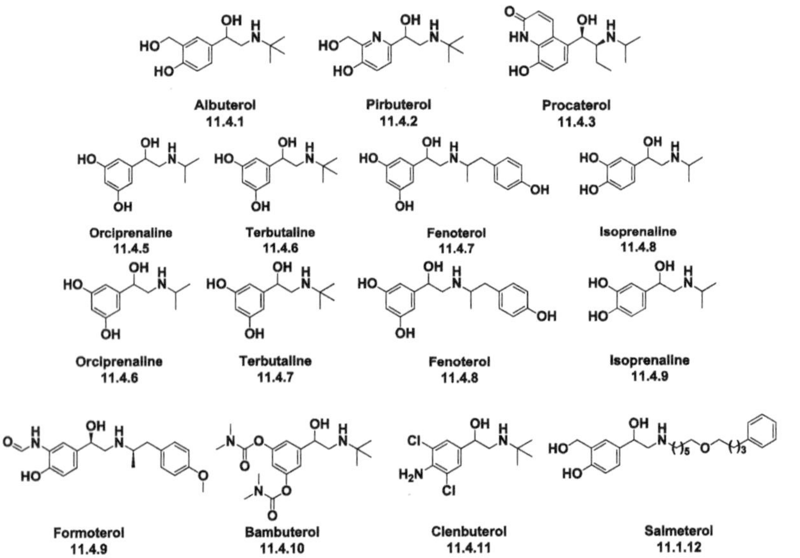

FIG. 11.8 **β_2 Receptor agonists.**

Most side effects of β_2 agonists result from their concurrent β_1 activity and include an increase in heart rate, a rise in systolic pressure, a decrease in diastolic pressure, chest pain, and arrhythmia.

11.5 DOPAMINE-RECEPTOR ACTIVATING DRUGS

Dopamine receptors are involved in many neurological functions, such as motivation, memory, motor control, endocrine signaling. Dysfunction of dopaminergic neurotransmission in the CNS could be a reason of some neuropsychiatric disorders. There are at least five subtypes of dopamine receptors, D_1 to D_5. Some dopamine agonists can treat the hypodopaminergic state of organism, which is viewed as one of the main causes that triggers drug-seeking and taking; they are typically used for relieving Parkinson disease symptoms and restless legs syndrome.

Activation of peripheral dopamine receptors causes dilation of the renal vasculature. This effect is exploited in the treatment of shock: by dilating renal blood vessels, renal perfusion is improved, thereby reducing the risk of renal failure. Dopamine itself is a drug that can activate dopamine receptors. It should be noted that, when dopamine is given to treat shock, the drug also enhances cardiac performance (because it activates β_1 receptors in the heart).

Common dopamine receptor agonists are dopamine (**11.3**) itself, apomorphine (**11.5.1**), bromocriptine (**11.5.2**), cabergoline (**11.5.3**), lisuride (**11.5.4**), pergolide (**11.5.5**), pramipexole (**11.5.6**), ropinirole (**11.5.7**), rotigotine (**11.5.8**), ciladopa (**11.5.9**), and derivatives of hexidine, namely, dihydrexidine (**11.5.10**), dinapsoline (**11.5.11**), and dinoxyline (**11.5.12**) (Fig. 11.9.).

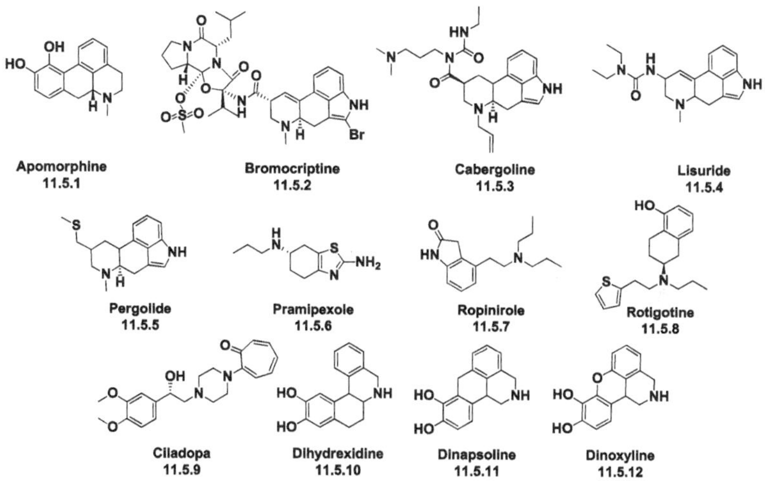

FIG. 11.9 **Dopamine receptor agonists.**

11.6 MULTIPLE ADRENORECEPTOR ACTIVATION

Anaphylactic shock is a manifestation of severe allergy. The reaction is characterized by hypotension (from widespread vasodilation), bronchoconstriction, and edema of the glottis. Epinephrine (**11.1**), is the treatment of choice for anaphylactic shock. Benefits derive from activating three types of adrenergic receptors: α_1, β_1, and β_2. By activating these receptors, epinephrine can reverse the most-severe manifestations of the anaphylactic reaction. Activation of β_1 receptors increases cardiac output, thereby helping elevate blood pressure. Blood pressure is also increased because epinephrine promotes α_1-mediated vasoconstriction. In addition to increasing blood pressure, vasoconstriction helps suppress glottal edema. By activating β_2 receptors, epinephrine can counteract bronchoconstriction. Mixed-acting drugs are epinephrine (**11.1**), norepinephrine (**11.2**), dobutamine (**11.5**), ephedrine (**11.14**), pseudoephedrine (**11.1.1**), and amphetamine (**11.8**).

Synthesis of the most of the discussed drugs is described in our previous book [3].

11.7 DOPAMINE-RECEPTOR AGONIST BESTSELLER DRUGS

On the list of Top 200 Drugs by sales for the 2010s are Guanfacine–Intuniv (**11.2.5**) and Albuterol (Salbutamol)–Ventolin (**11.4.1**)

Guanfacine–Intuniv

Guanfacine is a classic, centrally acting antihypertensive known to stimulate central α_2-adrenoceptors, which induces peripheral sympathoinhibition and hence reduces elevated blood pressure, predominantly as a result of vasodilation and a consequent decrease in peripheral vascular resistance. Its antihypertensive efficacy is beyond doubt, but the profile of adverse reactions is considered unfavorable [4,5]. Possible adverse reactions include blurred vision, confusion, dizziness, sweating, unusual tiredness, and weakness. Nevertheless guanfacine is one of the bestselling drugs for reducing high blood pressure. Moreover, after the discovery that the prefrontal cortex, which regulates behavior, thought, and emotion, is among the most evolved brain regions and that many cognitive disorders involve impairment of the prefrontal cortex [6], guanfacine was found to be capable of managing attention deficit hyperactivity disorder [7,8].

The synthesis of guanfacine is pretty simple and consists of an interaction of 2,6-dichlorophenylacetyl chloride or ethyl 2,6-dichlorophenyl acetate (**11.7.1**) with guanidine [9]. An alternate synthesis proposed by the same authors includes formylation of 2,6-dichlorophenylbenzyl cyanide (**11.7.2**) with ethyl formate to prepare a formylphenylacetonitrile derivative (**11.7.3**), which, after condensation with guanidine, produced (**11.7.4**), which was hydrated using hydrochloric acid to produce the desired guanfacine (**11.2.5**) [10,11] (Scheme 11.1.).

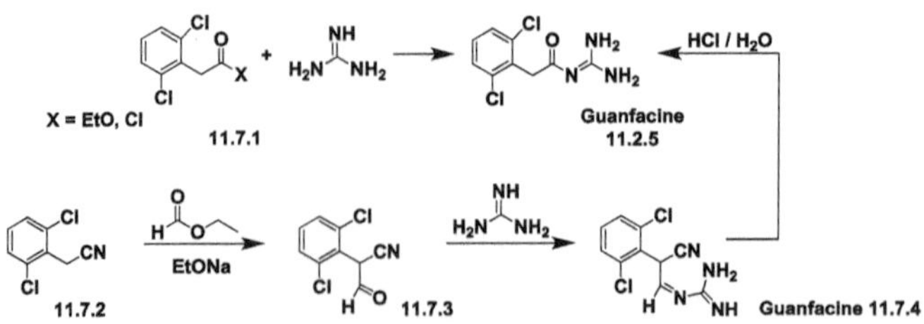

SCHEME 11.1 **Synthesis of guanfacine.**

Albuterol–Salbutamol

Albuterol is a short-acting β_2-adrenergic receptor agonist and for more than 25 years was used to relieve bronchospasms in conditions such as asthma, bronchospasm, and other obstructive pulmonary diseases. Its side effects include restlessness, irritability, nervousness, sometimes tremor, and increased or irregular heart rate [12-14].

The two general methods for the synthesis of albuterol are based on two starting materials: substituted benzophenones and salicylic acid derivatives.

The general approach started from 4-hydroxyacetophenone is described as follows [15] (Scheme 11.2.): 4-Hydroxyacetophenone (**11.7.5**) Blanc chloromethylation using formaldehyde and hydrogen chloride produces 4-hydroxy-3-chloromethylyacetophenone (**11.7.6**), which was acylated with a mixture of acetic anhydride, acetic acid, and sodium acetate to yield ketone (**11.7.7**). Bromination of the obtained ketone produces bromoketone (**11.7.8**), which on reaction with N-benzyl-N-t-butyl amine produces the aminoketone (**11.7.9**). After hydrolysis of acetyl groups with hydrochloric acid, the keto group in the diol (**11.7.10**) was reduced with sodium borohydride to give the triol (**11.7.11**). The triol was debenzylated with hydrogen using a Pd-C catalyst to produce the desired albuterol (**11.4.1**). Different modifications of this general approach have been proposed [16-22].

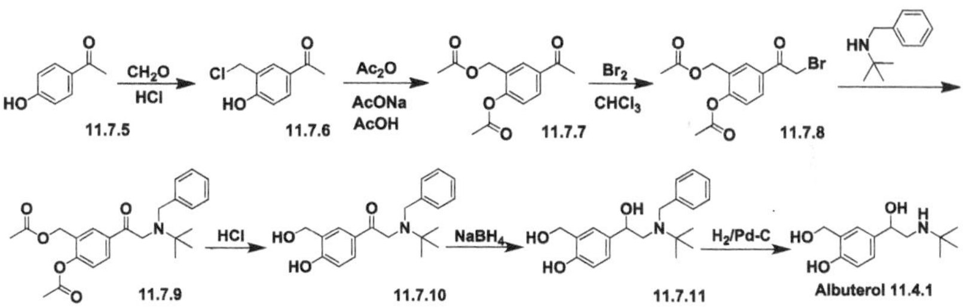

SCHEME 11.2 Synthesis of albuterol.

An approach started from methyl salicylate is outlined in Scheme 11.3 [16,17].

Methyl salicylate (**11.7.12**) was acylated with bromoacetyl chloride and the obtained bromoketone (**11.7.13**), on reaction with N-benzyl-N-t-butyl amine, produced aminoketone (**11.7.14**). Hydrogenation of the aminoketone with lithium aluminium hydride produced the known triol (**11.7.11**), which on debenzylation with hydrogen by Pd-C catalyst produced the desired albuterol (**11.4.1**).

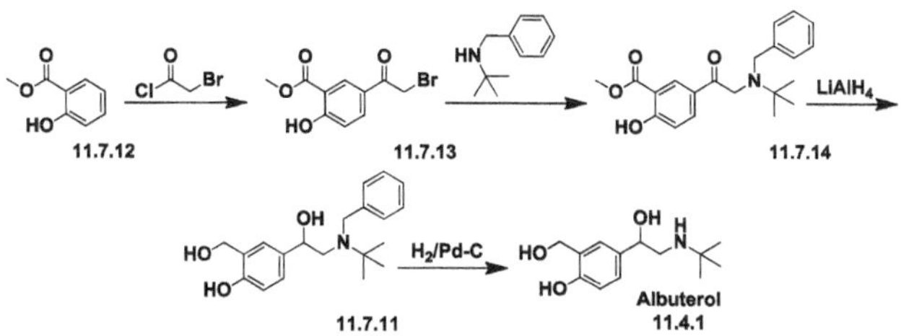

SCHEME 11.3 Synthesis of albuterol.

The details for both methods are discussed in the review [18].

The two enantiomers of the drugs are usually found to have significantly different potencies, but in the case of albuterol, effects of (R)-albuterol compared with (R,S)-albuterol, which have been documented in various cells and in vivo animal models, remain controversial [19-22].

REFERENCES

1. Brunton, L.; Lazo, J. S.; Parker, K.; Goodman, L. S.; Gilman, A. G. *Goodman & Gilman's The Pharmacological Basis of Therapeutics*, 11th ed.; McGraw-Hill, 2006.
2. Lemke, T. L.; David, A.; Williams, D. A. *Foye's Principles of Medicinal Chemistry*, 6th ed.; Lippincott Williams & Wilkins, 2008.
3. Vardanyan, R. S.; Hruby, V. J. *Synthesis of Essential Drugs;* Elsevier, 2006.
4. Scholtysik, G.; Jerie, P.; Picard, C. W. Guanfacine. In *Pharmacology of Antihypertension Drugs;* Scriabine, A., Ed.; Raven Press, 1980; pp 79–98.
5. Scholtysik, G.; Fetkovska, N. Pharmacology of guanfacine. *Cor Vasa* **1987,** *29* (4 Suppl. 1), S11–S16.
6. Arnsten, A. F. T.; Jin, L. E. Guanfacine for the treatment of cognitive disorders: a century of discoveries at Yale. *Yale J. Biol. Med.* **2012,** *85* (1), 45–58.
7. Posey, D. J.; McDougle, C. J. Guanfacine and guanfacine extended release: treatment for ADHD and related disorders. *CNS Drug Rev.* **2007,** *13* (4), 465–474.
8. Bukstein, O. G.; Head, J. Guanfacine ER for the treatment of adolescent attention-deficit/hyperactivity disorder. *Expert Opin. Pharmacother.* **2012,** *13* (15), 2207–2213.
9. Bream, J. B.; Picard, C. W. Acetylguanidine derivatives, FR 1584670 (1969).
10. Bream, J. B.; Picard, C. W. Phenylacetylguanidines, CH511816 (1971).
11. Bream, J. B.; Lauener, H.; Picard, C. W.; Scholtysik, G.; White, T. G. Substituted phenylacetylguanidines, a new class of antihypertensive agents. *Arzneim. Forsch.* **1975,** *25* (10), 1477–1482.
12. Ahrens, R. C.; Smith, G. D. Albuterol: an adrenergic agent for use in the treatment of asthma. *Pharmacology, pharmacokinetics and clinical use, Pharmacotherapy* **1984,** *4* (3), 105–121.
13. Price, A. H.; Clissold, S. P. Salbutamol in the 1980s. *A reappraisal of its clinical efficacy, Drugs* **1989,** *38* (1), 77–122.
14. Colice, G. L. Albuterol HFA for the management of obstructive airway disease. *Expert Rev. Respir. Med.* **2008,** *2* (2), 149–159.
15. Lunts, L. H. C.; Toon, P. Hydroxy-αγ-(aminomethyl)-m-xylene-α',α3-diols as stimulators, US 3644353 (1972).
16. Lunts, L. H. C.; Toon, P. 1-Phenyl-2-aminoethanol derivatives as bronchodilators, US 3705233 (1972).
17. Collin, D. T.; Hartley, D.; Jack, D.; Lunts, L. H. C.; Press, J. C.; Ritchie, A. C.; Toon, P. Saligenin analogs of sympathomimetic catechol amines. *J. Med. Chem.* **1970,** *13* (4), 674–680.
18. Skachilova, S. Y.; Zueva, E. F.; Muravskaya, I. D.; Goncharenko, L. V.; Smirnov, L. D. Procedures for preparing salbutamol (a review). *Khim.-Farm. Zh.* **1991,** *25* (10), 59–65.
19. Hartley, D.; Middlemiss, D. Absolute configuration of the optical isomers of salbutamol. *J. Med. Chem.* **1971,** *14* (9), 895–896.

20. Hawkins, C. J.; Klease, G. T. Relative potency of (-)- and (plus-)-salbutamol on guinea pig tracheal tissue. *J. Med. Chem.* **1973,** *16* (7), 856–857.
21. Bakale, R. P. The development of routes to (R)-albuterol hydrochloride. *Spec. Chem.* **1995,** *15* (6), 249–250. 253.
22. Barnes, P. J. Treatment with (R)- albuterol has no advantage over racemic albuterol. *Rebuttal by Dr. Barnes, Am. J. Respir. Crit. Care Med.* **2006,** *174* (9), 974.

Chapter 12

Adrenoblockers

Adrenergic antagonists (adrenoblockers) are compounds that inhibit the action of adrenaline (epinephrine), noradrenaline (norepinephrine), and other catecholamines that control autonomic outflow and some functions of the central nervous system at the adrenergic receptors or inhibit their release. These agents produce their effects through various mechanisms, including direct binding to the α or β adrenoceptors. Adrenoblockers are highly effective pharmaceuticals that are used broadly in the treatment of cardiac diseases. Their physiological effects include the dilation of blood vessels, which lowers blood pressure, and slows heart rate [1-5].

Two types of adrenoreceptors—α_1 and α_2—are known. The α_1 adrenoreceptors are predominantly located on vascular smooth muscle and regulate their contraction. α_2 Adrenoreceptors are also found in the smooth muscles including vascular. They inhibit the release of norepinephrine and modulate the feedback mechanism for its release.

β Adrenoblockers are divided into β_1, β_2, and β_3 blockers. β_1 Receptors are predominantly present in heart tissues, β_2 receptors exist in vascular and bronchial smooth muscle, and β_3 receptors are located in the adipocytes.

The α and β adrenoblockers are subdivided into selective and nonselective groups. Nonselective α and β adrenoblockers exhibit affinity correspondingly for α_1 and α_2 and β_1 and β_2 adrenoreceptors [5,6].

12.1 α ADRENOBLOCKERS

Nonselective α Adrenoblockers

The nonselective α antagonists are usually reserved for use in hypotensive emergencies caused by a pheochromocytoma. The nonselective α antagonists are phenoxybenzamine (**12.1.1**), phentolamine (**12.1.2**), tolazoline (**12.1.3**), and trazodone (**12.1.4**) (Fig. 12.1.).

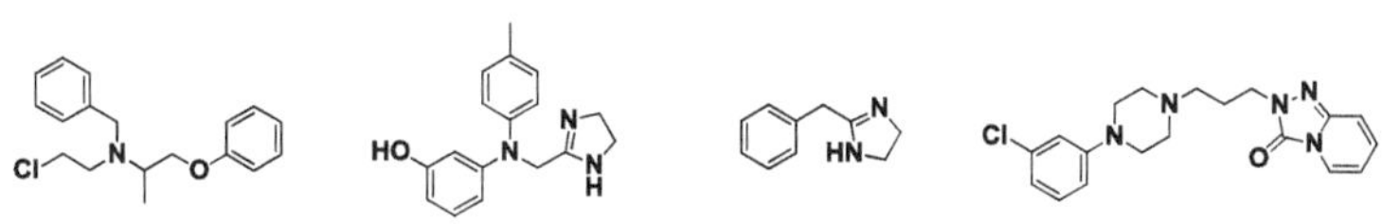

Phenoxybenzamine 12.1.1 Phentolamine 12.1.2 Tolazoline 12.1.3 Trazodone 12.1.4

FIG. 12.1 Nonselective α adrenoblockers.

Synthesis of Best-Seller Drugs. http://dx.doi.org/10.1016/B978-0-12-411492-0.00012-2

Selective α_1 Adrenoblockers

Selective α_1 adrenoblockers inhibit the ability of catecholamines to constrict blood vessels. Consequently, blocking these adrenoceptors causes a widespread relaxation of the blood vessels. These drugs are sometimes used to treat primary hypertension, although their use is not as widespread as other antihypertensive dugs. They can also be used in cardiac failure and in the treatment of some urinary bladder conditions because they block the contraction of the sphincter at the bladder outlet, which is mediated by α_1 adrenoceptors [3,7]. Examples of α_1 antagonists include prazosin (**12.1.5**), terazosin (**12.1.6**), doxazosin (**12.1.7**), alfuzosin (**12.1.8**), silodosin (**12.1.9**), tamsulosin (**12.1.10**), and urapidil (**12.2.11**) (Fig. 12.2.).

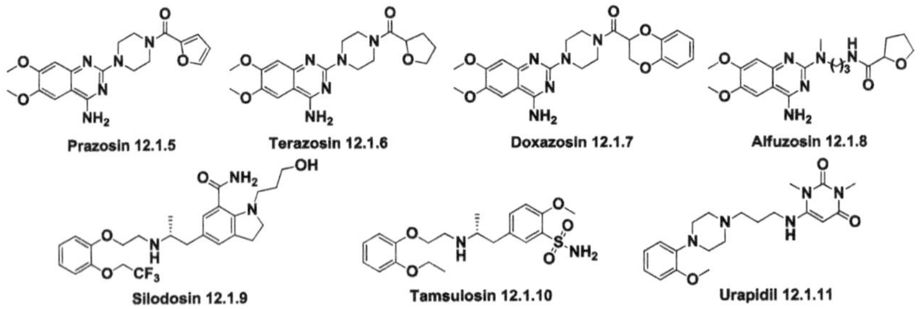

FIG. 12.2 **Selective α_1 adrenoblockers.**

Uroselective α_1 adrenoblockers (doxazosin, terazosin, tamsulosin, alfuzosin) are effective in the treatment of benign prostatic hyperplasia (BPH), a common benign tumor in men [8,9].

Selective α_2 Adrenoblockers

α_2 Adrenoblockers are mainly used in research, and have limited clinical application. They sometimes are used for reversing the effects of α_2 agonists. Idazoxan (**12.1.12**), atipamezole (**12.1.13**), and yohimbine (**12.1.14**) (Fig. 12.3.) provide an alternative for discontinuing sedation or anaesthesia in veterinary medicine [10]. There are few human studies indicating the usefulness of this approach in clinical medicine. Global downregulation of neurotransmitters induced by α_2 adrenoblockers significantly increases the level of adrenergic, dopaminergic, and serotonergic neurotransmitters and insulin secretion, hence decreasing blood sugar levels, which may cause symptoms of depression and other neurological problems. The alkaloid yohimbine (**12.1.14**), which is a nonspecific inhibitor of the presynaptic α_2 adrenoreceptors, has little therapeutic use. It has been proposed in the treatment of erectile dysfunction and postural hypotension.

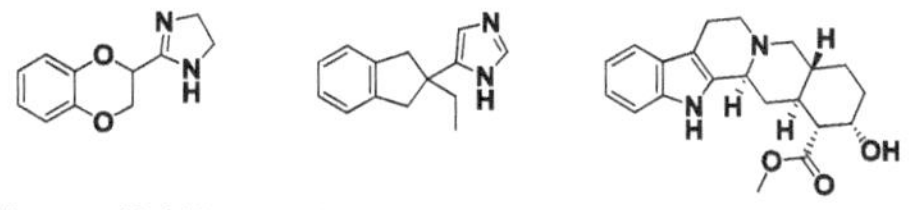

FIG. 12.3 Selective α_2 adrenoblockers.

Alfuzosin–Uroxatral

Alfuzosin is an uroselective α_1 adrenoblocker and is used in the treatment of BPH. It is included in the list of Top 200 Drugs by sales for the 2010s. It used in men to treat symptoms of BPH, which include difficulty and painful urinating, urinary frequency and urgency, dribbling, weak stream, and incomplete bladder emptying. It works by blocking α_1 adrenoreceptors in the lower urinary tract, causing smooth muscles in the bladder neck and prostate to relax, resulting urine in the bladder flowing more easily [11-14].

Alfuzosin (**12.1.8**) was synthesized via reaction of 4-amino-2-chloroquinazoline (**12.1.21**) with N-(3-(methylamino)propyl)tetrahydrofuran-2-carboxamide (**12.1.25**), which were prepared separately [15,16] (Scheme 12.1.).

The synthesis of the quinazoline compound (**12.1.21**) started from 3,4-dimethoxybenzaldehyde or methylvanillin (**12.1.15**), which was nitrated to produce nitrobenzaldehyde (**12.1.16**). The obtained aldehyde was oxidized to nitrobenzoic acid using potassium permanganate, which was stepwise converted to acid chloride and then to nitrobenzamide (**12.1.17**). After reduction with iron in acetic acid the nitrobenzamide (**12.1.17**) produced aminoamide (**12.1.18**), which, on reaction with urea, was cyclized to quinazoline-2,4-dione (**12.1.19**). Phosphorus oxychloride treatment of the obtained compound (**12.1.19**), followed by reaction of the resulting 2,4-dichloroquinazoline (**12.1.20**) with ammonia, yielded the 4-amino-2-chloroquinazoline (**12.1.21**) [17].

The second component N-(3-(methylamino)propyl)tetrahydrofuran-2-carboxamide (**12.1.25**) was synthesized starting from

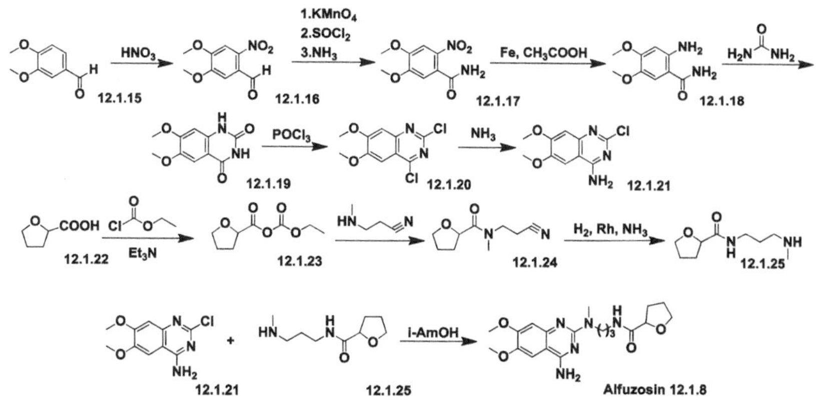

SCHEME 12.1 Synthesis of alfuzosin.

tetrahydrofuran-2-carboxylic acid (**12.1.22**), which was transformed to a mixed anhydride (**12.1.23**) via interaction with ethyl chloroformate in the presence of triethylamine. The obtained compound (**12.1.23**) on reaction with 3-(methylamino)propanenitrile produced tetrahydrofuroyl amide (**12.1.24**), which, on hydrogenation by Rh catalyst in methanol saturated with ammonia, provided the secondary amine (**12.1.25**). The secondary amine (**12.1.25**) via reaction with 4-amino-2-chloroquinazoline (**12.1.21**) in i-AmOH produced the desired alfuzosin (**12.1.8**). Some improvements for described Scheme 12.1. have been proposed [18-20].

Tamsulosin–Flomax

Tamsulosin (**12.1.10**), another bestselling uroselective α_1 adrenoblocker, works in the same way as alfuzosin (**12.1.8**)—it causes the bladder neck muscles and muscle fibers in the prostate itself to relax, making it easier to urinate in the presence of of BPH [21-27]. Tamsulosin also is used to assist in the passage of kidney stones by the same mechanism of smooth muscle relaxation via α antagonism [28,29].

The key intermediate in the preparation of tamsulosin (**12.1.10**) is the 1-methyl-2-(4-methoxy-3-sulfonamido phenyl) ethyl amine (**12.1.29**). Synthesis of this amine began with 4-methoxyphenylacetone (**12.1.26**), which underwent chlorosulfonation followed by treatment with obtained sulfonyl chloride with ammonia, a standard method in the preparation of sulfonamides for the pharmaceutical industry to produce 5-acetonyl-2-methoxybenzene sulfonamide (**12.1.27**). The mixture of the (**12.1.27**) and (R)-(+)-α-methylbenzylaminewas hydrogenated at 50 to 52 °C under normal pressure in the presence of PtO_2 or Raney nickel in methanol to produce 2-methoxy-5-((R)-2-(((R)-1-phenylethyl) amino)propyl)benzenesulfonamide (**12.1.28**), which on debenzylation under hydrogenation over Pd/C catalyst produced the requested (R)-5-(2-aminopropyl)-2-methoxybenzene-sulfonamide (**12.1.29**). The obtained compound was refluxed with 1-(2-bromoethoxy)-2-ethoxybenzene in ethanol to produce the desired tamsulosin (**12.1.10**) [30] (Scheme 12.2.).

Another patent [31] describes the synthesis of tamsulosin starting with the reaction of (R)-(-)-2-(4-methoxyphenyl)-1-methylethylamine (**12.1.30**) with

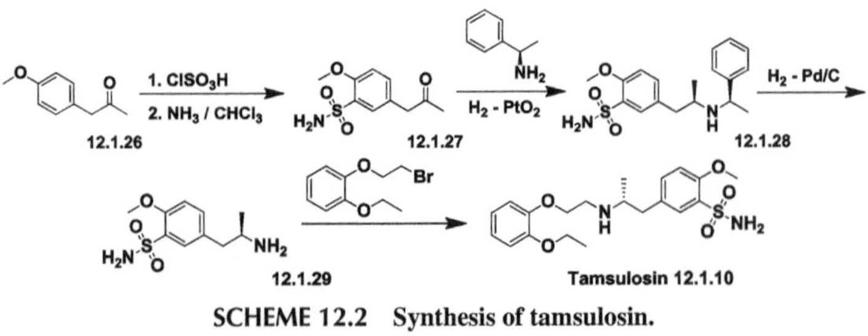

SCHEME 12.2 Synthesis of tamsulosin.

bromoacetic acid in the presence of pivaloyl chloride and triethylamine, which produced (R)-(+)-2-bromo-N-[2-(4-methoxyphenyl)-1-methylethyl]acetamide (**12.1.31**), which was chlorosulfonated with chlorosulphonic acid at room temperature and the obtained product was treated with ammonia to produce (R)-(+)-N-[2-(3-aminosulfonyl-4-methoxyphenyl)-1-methylethyl]-2-bromoacetamide (**12.1.32**). The last compound was aryloxylated with 2-ethoxyphenol in the presence of potassium carbonate and the carbonyl group in the obtained product was reduced with lithium aluminum hydride to produce the desired tamsulosin (**12.1.10**) [31] (Scheme 12.3.).

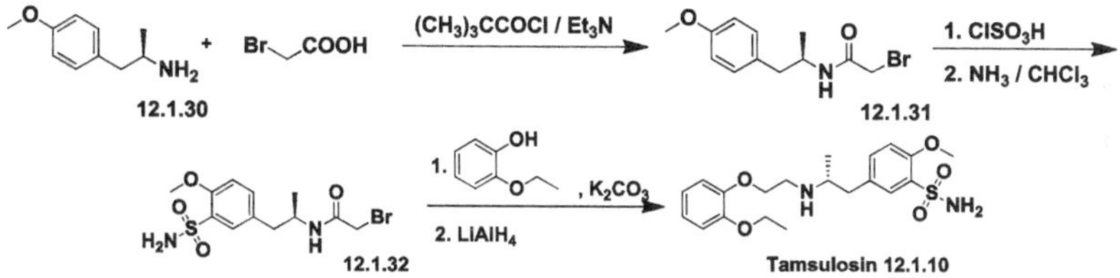

SCHEME 12.3 **Synthesis of tamsulosin.**

Many other patents and paper describe similar approaches for the synthesis of tamsulosin [32-37].

12.2 β ADRENOBLOCKERS

Drugs that exhibit reversible competitive blocking action on β-adrenoreceptive receptors and that counteract effects of catecholamines are called β adrenoblockers. They are extremely useful in treating hypertension, cardiac failure, angina, arrhythmias, chest pain, and migraines, in controlling muscle tremors and anxiety, for the treatment of glaucoma, all of which result from overactivity of the sympathetic system. The main effect of β adrenoblockers is a slowing of the release of sympathetic, tonic impulses [38].

β Adrenoblockers are divided into β_1, β_2, and β_3 blockers, according to three different types of β receptors identified by molecular pharmacology. In turn, β_1, β_2, and β_3 blockers are subdivided into selective and nonselective groups. β_1 Receptors are predominantly present in heart tissues and act on heart rate by interacting with cardiac pacemaker cells. β_2 Receptors exist in vascular and bronchial smooth muscle, and cause vasodilation and bronchodilation on activation. β_3 Receptors are located in the adipocytes where they are supposed to be involved in fatty acid metabolism.

Nonselective β adrenoblockers (propranolol, nadolol, timolol, and labetalol, which is a combined α and β adrenoblocker) exhibit affinity for both β_1 and β_2 adrenoreceptors. Selective β_1 blockers (acebutolol, atenolol, esmolol, metoprolol) in therapeutic doses predominantly bind to β_1 adrenoreceptors. Nonselective β blockers usually are classified as first-generation drugs, the cardioselective β_1

blockers as second-generation drugs, and recently discovered β blockers with vasodilator action (selective or nonselective) as third-generation drugs (nebivolol, carvedilol, labetalol) [39].

All mentioned β blockers belong to the 1-aryloxy-3-aminopropanol-2 class, and their action results in reduction of heart rate and strength of cardiac beats, slowing of atrioventricular conductivity, reduction of the level of renin in the plasma, and reduction of blood pressure. Structure activity relationships in β adrenoblockers are well reviewed [40].

There are no therapeutically useful selective β_2 adrenoblockers, although a number of laboratory tools with expressed β_2-adrenoblocking activity already exist (butoxamine, ICI-118,551).

The β-adrenergic blockade as a therapeutic approach first emerged in the 1960s, and places it far ahead of many competing treatments, not only in the cardiovascular area, but in all of therapeutics. β blockers were first discovered in 1962 by Sir James Black [41]. The development of propranolol was considered one of the most important contributions to medicine and pharmacology. Black was awarded the Nobel prize in 1988 and the Nobel committee called his discovery "the greatest breakthrough when it comes to pharmaceuticals against heart illness since the discovery of digitalis 200 years ago."

All β blockers cause a competitive inhibition of the β receptors and block effects of catecholamines. β Blockers are used effectively for the treatment of heart failure, angina pectoris, postmyocardial infarction, hypertension, hyperdynamic circulatory syndrome, sinus, ventricular and supraventricular tachycardias, chronic atrial fibrillation, and mitral stenosis [42-50].

Nonselective β Adrenoblockers (First-Generation β Blockers)

The syntheses of most of the first-generation β blockers—propranolol (**12.2.1**), alprenolol (**12.2.2**), oxprenolol (**12.2.3**), penbutolol (**12.2.4**), nadolol (**12.2.5**), carteolol (**12.2.6**), sotalol (**12.2.7**), timolol (**12.2.8**), pindolol (**12.2.9**), bopindolol (**12.2.10**), bucindolol (**12.2.11**), nipradilol (**12.2.12**), and tilisolol (**12.2.13**) (Fig. 12.4.)—are described in our previous book [51], and some structure–activity relationships of β blockers in general have been comprehensively reviewed [40]. Noncardioselective β blockers cause an equal blockade of β_1- and β_2-adrenergic receptors.

Cardioselective β_1 Blockers (Second-Generation β Blockers)

Syntheses of metoprolol (**12.2.14**), atenolol (**12.2.15**), esmolol (**12.2.16**), acebutolol (**12.2.17**), bisoprolol (**12.2.18**), betaxolol (**12.2.19**), and celiprolol (**12.2.20**) (Fig. 12.5.) are also described in our previous book [51]. Cardioselective β blockers have a higher affinity for the β_1 receptor. Nevertheless, with increasing doses some β_2 effects may appear.

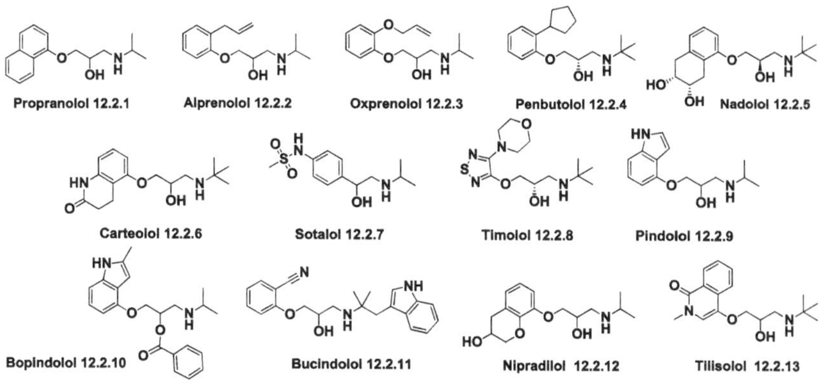

FIG. 12.4 Nonselective β adrenoblockers (first-generation β blockers).

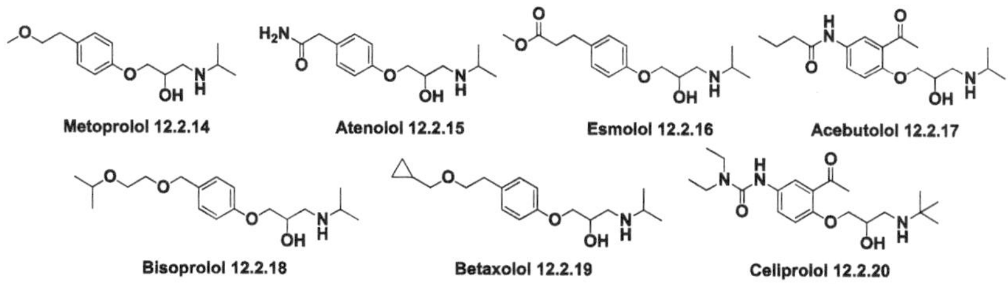

FIG. 12.5 Cardioselective β_1 blockers (second-generation β blockers).

β Blockers with Vasodilator Action (Third-Generation β Blockers)

Noncardioselective (first-generation) and cardioselective (second-generation) β blockers have no vasodilatory properties. The noncardioselective β-blocker carvedilol (**12.2.21**) simultaneously blocks the β_1- and β_2-adrenergic receptors, and also exhibits vasodilating properties that result from α_1-receptor blockade. Nebivolol (**12.2.22**), another β blocker, is cardioselective and has demonstrated the highest cardioselectivity of any currently available β blocker. In addition, nebivolol (**12.2.22**) exhibits vasodilatory properties unrelated to α_1-receptor blockade but mediated by nitric oxide (NO). Labetalol (**12.2.23**) combines nonselective, competitive, β-blocking with selective, competitive, α_1-blocking activities. This causes a decrease in blood pressure and vascular resistance without a substantial reduction in heart rate or cardiac output (Fig. 12.6.).

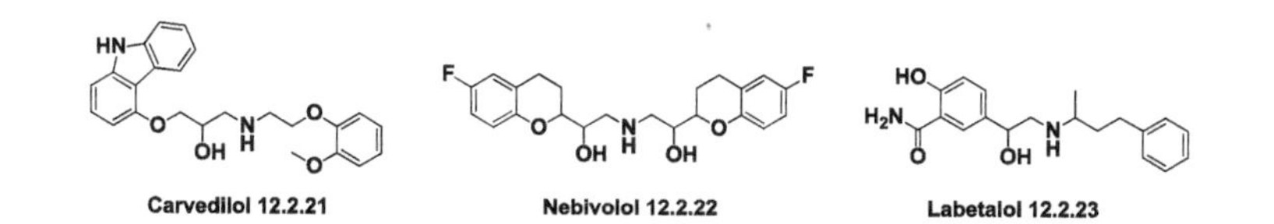

FIG. 12.6 β Blockers with vasodilator action (third-generation β blockers).

β_2 Adrenoblockers

There are no therapeutically useful selective β_2 adrenoblockers, although a number of laboratory tools with expressed β_2-adrenoblocking activity, such as butoxamine (**12.2.24**) and ICI-118,551 (**12.2.25**), already exist (Fig. 12.7.).

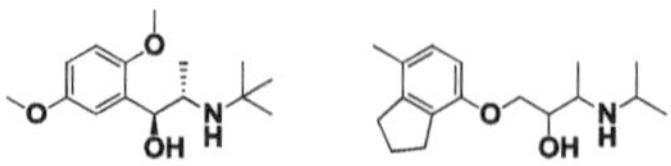

Butoxamine 12.2.24 ICI-118,551 12.2.25

FIG. 12.7 β_2 **Adrenoblockers.**

β_3 Adrenoblockers

β_3 Adrenoceptors mediate some of the effects of catecholamines on tissues such as blood vessels or the urinary bladder. β_3 Receptors are located in various species and tissues, including the heart, but mainly in brown adipose tissue, gallbladder, and urinary bladder, and are involved in the regulation of thermogenesis, lipolysis, and stress. β_3 Agonists could be used for treatment of overactive bladder and weight loss. Studies suggest that β_3 receptors might play an important role in the pathophysiology of heart failure by counterbalancing the effects of a β_1 and β_2 stimulation. The known β_3 adrenoblockers are L-748,337 (**12.2.26**) and SR 59230A (**12.2.27**) and they are used as laboratory tools (Fig. 12.8.).

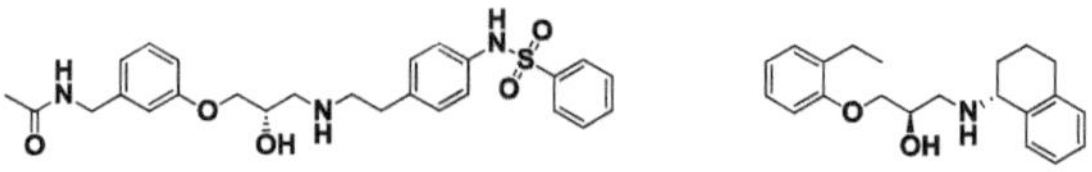

L-748,337 12.2.26 SR 59230A 12.2.27

FIG. 12.8 β_3 **Adrenoblockers.**

The adverse effects of most of β-blockers are predictable. Fatigue, headache, dizziness, upset stomach, constipation, and diarrhea are the common side effects.

Bronchospasms resulting from blockade of β_2 receptors, which mediate dilation in the bronchi, are possible. Asthma is an absolute contraindication for all β blockers.

The syntheses of most of the mentioned β blockers are described in our previous book [51]. Carvedilol (Coreg) and Nebivolol (Bystolic) are the two β blockers included in the list of Top 200 Drugs by sales for the 2010s.

Carvedilol–Coreg

Carvedilol is a third-generation β_1 and β_2 blocker that also possesses α_1-adrenergic–blocking properties. It is indicated for the treatment of hypertension, stable angina pectoris, and congestive heart failure. This compound has shown superior efficacy to certain other β blockers in heart failure [52-60]. Some side effects such as chest pain, tightness, or heaviness dizziness, lightheadedness, or fainting, swelling, shortness of breath can occur on taking carvedilol.

Carvedilol originally was synthesized in two steps via condensing 4-hydroxycarbazole (**12.2.28**) with epichlorohydrin to produce 4-(2,3-epoxypropoxy)carbazole (**12.2.29**), which then was reacted with 2-(2-methoxyphenoxy)ethanamine (**12.2.30**) to produce the desired carvedilol (**12.2.21**) [61] (Scheme 12.4.).

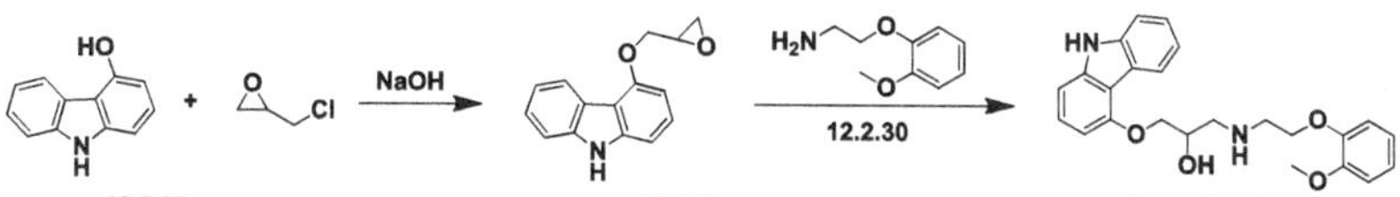

SCHEME 12.4 **Synthesis of carvedilol.**

Alternate methods for the synthesis of carvedilol have been proposed [62-67].

Nebivolol–Bystolic

Nebivolol is another third-generation, highly β_1-specific β blocker, included in the list of Top 200 Drugs by sales for the 2010s and recommended for the treatment of hypertension. In addition to its β-blocking effects, nebivolol activates NO synthase, increasing endothelin-dependent NO, giving it a unique peripheral vasodilatory action [68-78].

The synthesis process of nebivolol is described in detail in the first patent [79] (Scheme 12.5.) and several modifications have been proposed, but keeping the general approach the same [80-83]. The synthesis started from 6-fluoro-4-oxo-4H-1-benzopyran-2-carboxylic acid (**12.2.31**), which was hydrogenated by Pd-C catalyst to 6-fluoro-3,4-dihydro-2H-1-benzopyran-2-carboxylic acid (**12.2.32**). Several attempts to transform the carboxylic group to aldehyde have been made and one was the implementation of the coupling reactions of carbonyldiimidazole with acids to produce the corresponding ketone, which was followed by its reduction with bis-(2-methylpropyl)-aluminum hydride to aldehyde (**12.2.33**). The Corey–Chaykovsky reaction of obtained aldehyde with trimethylsulfoxonium iodide enabled the synthesis of oxiranyl chromane (**12.2.34**), which, on reflux with benzylamine in ethanol, underwent two consequent ring-opening reactions to produce bis-product (**12.2.35**), which was debezylated on hydrogenation by Pd-C catalyst to produce the desired nebivolol **(12.2.22)** [79] (Scheme 12.5.).

Nebivolol is a racemic mixture of (S,R,R,R) and (R,S,S,S) enantiomers and is the most β_1-selective adrenoceptor antagonist currently available for clinical use. However, its hemodynamic effects differ from those of classical β-adrenoceptor antagonists as a result of vasodilating action. Common and rare side effects are mild headache, rash, and sleeplessness.

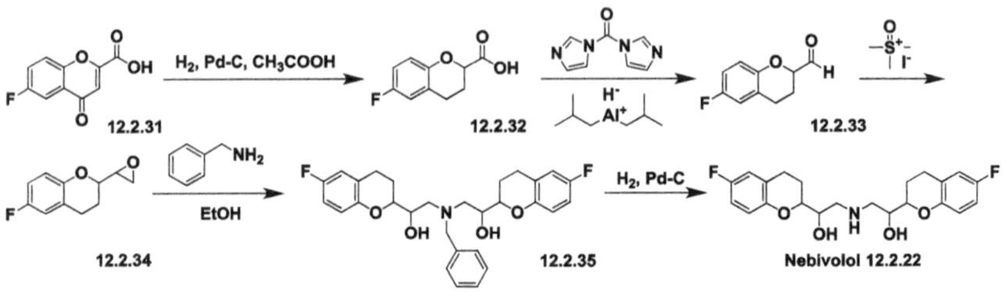

SCHEME 12.5 Synthesis of nebivolol.

REFERENCES

1. Pupo, A. S.; Minneman, K. P. Adrenergic pharmacology: focus on the central nervous system. *CNS Spectr.* **2001,** *6* (8), 656–662.
2. Kobilka, B. K. Structural insights into adrenergic receptor function and pharmacology. *Trends Pharmacol. Sci.* **2011,** *32* (4), 213–218.
3. Timmermans, P. B. M. W. M.; Chiu, A. T.; Thoolen, M. J. M. C. α-Adrenergic receptors. In Hansch, C., Sammes, P. G., Taylor, J. B., Eds.; *Comprehensive Medicinal Chemistry*, Vol. 3; Pergamon Press, 1990; pp 133–185.
4. Main, B. G. β-Adrenergic receptors. In Hansch, C., Sammes, P. G., Taylor, J. B., Eds.; *Comprehensive Medicinal Chemistry*, Vol. 3; Pergamon Press, 1990; pp 187–228.
5. Muscholl, E.; Rahn, K. H. Adrenergic α- and β-receptors and their specific inhibitors. *Klin. Wochenschr.* **1968,** *46* (3), 113–119.
6. Philipp, M.; Hein, L. Adrenergic receptor knockout mice: distinct functions of 9 receptor subtypes. *Pharmacol. Ther.* **2004,** *101* (1), 65–74.
7. Sica, D. A. Alpha1-adrenergic blockers: current usage considerations. *J. Clin. Hypertens. (Hoboken, NJ, U. S.)* **2005,** *7* (12), 757–762.
8. Kumar, V. L.; Dewan, S. Alpha adrenergic blockers in the treatment of benign hyperplasia of the prostate. *Int. Urol. Nephrol.* **2000,** *32* (1), 67–71.
9. Yuan, J. Q.; Liu, Y.; Yang, Z. Y.; Qin, X.; Yang, K. H.; Mao, C. The efficacy and safety of alpha-1 blockers for benign prostatic hyperplasia: an overview of 15 systematic reviews. *Curr. Med. Res. Opin.* **2013,** *29* (3), 279–287.
10. Berlan, M.; Montastruc, J.; Lafontan, M. Pharmacological prospects for α_2-adrenoreceptor antagonist therapy. *Trends Pharmacol. Sci.* **1992,** *13* (7), 277–282.
11. Weiner, D. M.; Lowe, F. Alfuzosin for the management of benign prostate hyperplasia. *Expert Opin. Pharmacother.* **2003,** *4* (11), 2057–2063.
12. Hofner, K.; Jonas, U. Alfuzosin: A clinically uroselective α_1-blocker. *World J. Urol.* **2002,** *19* (6), 405–412.
13. Cho, K. J.; Kim, J. C. Alfuzosin for the treatment of storage symptoms suggestive of overactive bladder. *Expert Opin. Pharmacother.* **2012,** *13* (8), 1143–1151.

14. Elhilali, M. M. Alfuzosin: an α_1-receptor blocker for the treatment of lower urinary tract symptoms associated with benign prostatic hyperplasia. *Expert Opin. Pharmacother.* **2006,** *7* (5), 583–596.
15. Manoury, P. M. 4-Amino-6,7-dimethoxyquinazolin-2-ylalkylenediamines, US 4315007 (1982)
16. Manoury, P. M.; Binet, J. L.; Dumas, A. P.; Lefevre-Borg, F.; Cavero, I. Synthesis and antihypertensive activity of a series of 4-amino-6,7-dimethoxyquinazoline derivatives. *J. Med. Chem.* **1986,** *29* (1), 19–25.
17. Althuis, T. H.; Hess, H. J. Synthesis and identification of the major metabolites of prazosin formed in dog and rat. *J. Med. Chem.* **1977,** *20* (1), 146–149.
18. Sadanand, N. S.; Bhivsan, A. P.; Chandrakantrao, J. U. Preparation of alfuzosin, US 20070105880 (2007).
19. Sadanand, N.S. Process for the preparation of alfuzosin hydrochloride, IN 2004MU01229 (2006).
20. Zhao, Z.; Peng, L. Process for preparation of alfuzosin hydrochloride, CN 1616438 (2005).
21. Wilde, M. I.; McTavish, D. Tamsulosin: a review of its pharmacological properties and therapeutic potential in the management of symptomatic benign prostatic hyperplasia. *Drugs* **1996,** *52* (6), 883–898.
22. Noble, A. J.; Chess-Williams, R.; Couldwell, C.; Furukawa, K.; Uchyiuma, T.; Korstanje, C.; Chapple, C. R. The effects of tamsulosin, a high affinity antagonist at functional alpha 1A- and alpha 1D-adrenoceptor subtypes. *Br. J. Pharmacol.* **1997,** *120* (2), 231–238.
23. Dong, Z. L.; Wang, Z. P.; Yang, K. H.; Liu, Y. L.; Gao, W. H.; Chen, W. Y. Tamsulosin versus terazosin for benign prostatic hyperplasia: a systematic review. *Syst. Biol. Reprod. Med.* **2009,** *55* (4), 129–136.
24. Michel, M. C.; de la Rosette, J. J. Efficacy and safety of tamsulosin in the treatment of urological diseases. *Expert Opin. Pharmacother.* **2004,** *5* (1), 151–160.
25. Dunn, C. J.; Matheson, A.; Faulds, D. M. Tamsulosin: a review of its pharmacology and therapeutic efficacy in the management of lower urinary tract symptoms. *Drugs Aging* **2002,** *19* (6), 135–161.
26. Chapple, C.; Andersson, K.-E. Tamsulosin: An overview. *World J. Urol.* **2002,** *19* (6), 397–404.
27. Lyseng-Williamson, K. A.; Jarvis, B.; Wagstaff, A. J. Tamsulosin: An update of its role in the management of lower urinary tract symptoms. *Drugs* **2002,** *62* (1), 135–167.
28. Lu, Z.; Dong, Z.; Ding, H.; Wang, H.; Ma, B.; Wang, Z. Tamsulosin for ureteral stones: a systematic review and meta-analysis of a randomized controlled trial. *Urol. Int.* **2012,** *89* (1), 107–115.
29. Zheng, S.; Liu, L. R.; Yuan, H. C.; Wei, Q. Tamsulosin as adjunctive treatment after shockwave lithotripsy in patients with upper urinary tract stones: a systematic review and meta-analysis. *Scand. J. Urol. Nephrol.* **2010,** *44* (6), 425–432.
30. Okada, M.; Yoshida, K.; Takanobu, K. Process for producing optically active m-(aminoalkyl) benzenesulfonamide derivatives useful as antihypotensives and as drug intermediates, EP 257787 (1988).
31. Ito, Y.; Kato, H.; Etsuchu, E.; Mitani, K.; Yagi, N.; Yagi, N., Phenoxyacetamides as α-blockers, JP 02306958 (1990).
32. Imai, K.; Niigata, K.; Fujikura, T.; Hashimoto, S.; Takenaka, T. Sulfamoyl-substituted phenethylamine derivatives, and pharmaceutical compositions, containing them, EP 34432 (1981).
33. Niigata, K.; Fujikura, T. Sulfamoyl-substituted phenethylamine derivatives with α-adrenergic blocking activity for use as antihypertensives, and their pharmaceutical compositions, US 4731478 (1988).

34. Gizur, T.; Fogassy, E.; Balint, J.; Egri, G.; Torley, J.; Demeter, A.; Greiner, I. New practical synthesis of tamsulosin. *Chirality* **2008,** *20* (6), 790–795.
35. Sagratini, G.; Angeli, P.; Buccioni, M.; Gulini, U.; Marucci, G.; Melchiorre, C.; Poggesi, E.; Giardina, D. Synthesis and α_1-adrenoceptor antagonist activity of tamsulosin analogues. *Eur. J. Med. Chem.* **2010,** *45* (12), 5800–5807.
36. Reddy, A. V.; Rao, S. U. B.; Narasimha, G. L.; Dubey, P. K. Improved process for the preparation of tamsulosin hydrochloride. *Synth. Commun.* **2009,** *39* (8), 1451–1456.
37. Reddy, A. V.; Amasa, Reddy, A. S.; Bhatthula, G. B. K.; Kompella, L. S.; Matta, V. P.; Subha, M. C. S. Asymmetric synthesis of unnatural amino acids and tamsulosin chiral intermediate. *Synth. Commun.* **2013,** *43* (21), 2892–2897.
38. Crowther, A. F. The discovery of the first β-adrenergic blocking agents. *Drug Des. Delivery* **1990,** *6* (2), 149–156.
39. Lopez-Sendon, J.; Swedberg, K.; McMurray, J.; Tamargo, J.; Maggioni, A. P.; Dargie, H.; Tendera, M.; Waagstein, F.; Kjekshus, J.; Lechat, P.; Torp-Pedersen, C. Expert consensus document on beta-adrenergic receptor blockers. *Eur. Heart J.* **2004,** *25* (15), 1341–1362.
40. Yuzhakov, S. D.; Glushkov, R. G.; Mashkovskii, M. D. Structure activity relationships in beta-adrenoblockers. *Pharm. Chem. J.* **1991,** *25* (5), 283–295.
41. Black, J. W.; Stephenson, J. S. Pharmacology of a new adrenergic beta-receptor-blocking compound (Nethalide). *Lancet* **1962,** *2* (7251), 311–314.
42. Benovic, J. L.; Bouvier, M.; Caron, M. G.; Lefkowitz, R. J. Regulation of adenylyl cyclase-coupled beta-adrenergic receptors. *Annu. Rev. Cell Biol.* **1988,** *4,* 405–428.
43. Chrysant, S. G.; Chrysant, G. S. Current status of β blockers for the treatment of hypertension: an update. *Drugs Today* **2012,** *48* (5), 353–366.
44. Hollenberg, N. K. The role of β blockers as a cornerstone of cardiovascular therapy. *Am. J. Hypertens.* **2005,** *18* (12 Pt. 2), 165S–168S.
45. Toda, N. Vasodilating β-adrenoceptor blockers as cardiovascular therapeutics. *Pharmacol. Ther.* **2003,** *100* (3), 215–234.
46. Ladagd, D.; Schwinger, R. H. G.; Brixius, K. Cardio-selective beta-blocker: pharmacological evidence and their influence on exercise capacity. *Cardiovasc. Ther.* **2013,** *31* (2), 76–83.
47. Bielecka-Dabrowa, A.; Aronow, W. S.; Rysz, J.; Banach, M. Current place of beta-blockers in the treatment of hypertension. *Curr. Vascular Pharmacol.* **2010,** *8* (6), 733–741.
48. Ram, C.; Venkata, S. Beta-blockers in hypertension. *Am. J. Cardiol.* **2010,** *106* (12), 1819–1825.
49. Pedersen, M. E.; Cockcroft, J. R. The vasodilatory beta-blockers. *Curr. Hypertens. Rep.* **2007,** *9* (4), 269–277.
50. Di Nicolantonio, J. J.; Hackam, D. G. Carvedilol: a third-generation β-blocker should be a first-choice β-blocker. *Expert Rev. Cardiovasc. Ther.* **2012,** *10* (1), 13–25.
51. Vardanyan, R. S.; Hruby, V. J. *Synthesis of Essential Drugs;* Elsevier, 2006.
52. Hirohashi, M.; Takasuna, K.; Tamura, K.; Yamaguchi, K.; Maekawa, K.; Yamada, S.; Iwasaki, S.; Yoshida, M.; Nomura, M.; Taguchi, K. General pharmacological profiles of the new β-adrenoceptor antagonist carvedilol. *Arzneim. Forsch.* **1990,** *40* (7), 735–746.
53. Ruffolo, R. R.; Feuerstein, G. Z. Carvedilol. In Taylor, J. B., Triggle, D. J., Eds.; *Comprehensive Medicinal Chemistry II,* Vol. 8; Elsevier, 2006; pp 137–147 .
54. Ruffolo, R. R.; Feuerstein, G. Z. Carvedilol case history: the discovery and development of the first β-blocker for the treatment of congestive heart failure. *Expert Opin. Drug Discovery* **2006,** *1* (1), 85–89.
55. Weir, R. A. P.; Dargie, H. J. Carvedilol in chronic heart failure: past, present and future. *Future Cardiol.* **2005,** *1* (6), 723–734.

56. Di Nicolantonio, J. J.; Hackam, D. G. Carvedilol: a third-generation β-blocker should be a first-choice β-blocker. *Expert Rev. Cardiovasc. Ther.* **2012,** *10* (1), 13–25.
57. Tsochatzis, E. A.; Triantos, C. K.; Burroughs, A. K. Carvedilol-the best β-blocker for primary prophylaxis? *Nat. Rev. Gastroenterol. Hepatol.* **2009,** *6* (12), 692–694.
58. Stafylas, P.; Sarafidis, P. A. Carvedilol in hypertension treatment. *Vasc. Health Risk Manage.* **2008,** *4* (1), 23–30.
59. Doughty, R. N.; White, H. D. Carvedilol: use in chronic heart failure. *Expert Rev. Cardiovasc. Ther.* **2007,** *5* (1), 21–31.
60. Carreira, R. S.; Monteiro, P.; Goncalves, L. M.; Providencia, L. A. Carvedilol: just another beta-blocker or a powerful cardioprotector? *Cardiovasc. Hematol. Disord.: Drug Targets* **2006,** *6* (4), 257–266.
61. Wiedemann, F.; Kampe, W.; Thiel, M.; Sponer, G.; Roesch, E.; Dietmann, K. Carbazolyl-4-oxypropanolamine derivatives, DE 2815926 (1979).
62. Anandkumar, B.; Reddy, R. Buchi; Gangaiah, L.; Madhusudhan, G.; Mukkanti, K. A new and alternate synthesis of carvedilol: an adrenergic receptor. *Pharma Chem.* **2011,** *3* (6), 620–626.
63. Kumar, B. A.; Babu, K. V.; Rao, R. K.; Srinivas, K.; Madhusudhan, G.; Mukkanti, K. Synthesis of racemic and chiral carvedilol starting from corresponding 5-(chloromethyl)oxazolidin-2-one. *Indian J. Chem., Sect. B: Org. Chem. Incl. Med. Chem.* **2012,** *51B* (9), 1430–1435.
64. Kumar, B. A.; Ashrafuddin, Md.; Rajesh, V.; Parveen, S.; Madhusudhan, G. Convenient synthesis of carvedilol utilizing 3-(9H-carbazol-4-yloxy)-1-chloropropan-2-yl phenyl carbonate as a key intermediate. *Indian J. Chem., Sect. B: Org. Chem. Incl. Med. Chem* **2012,** *51B* (9), 780–784.
65. Rao, S. N.; Sitaramaiah, D.; Srimannarayana, K.; Rao, C. N.; Rao, P. S.; Babu, K. S. Synthesis and characterization of potential impurities of carvedilol, an antihypertensive drug. *Synth. Commun.* **2011,** *41* (1), 85–93.
66. Naidu, K. C.; Kumar, V. K.; Naresh, E.; Kameswararao, Ch.; Madhusudhan, G.; Mukkanti, K. A new and facile synthesis of 5-[(9H-carbazol-4-yloxy)methyl]oxazolidin-2-one intermediate towards the synthesis of carvedilol, a β-adrenergic blocking agent. *Org. Chem.: Indian J.* **2010,** *6* (2), 171–177.
67. Kumar, B. A.; Vysabhattar, R.; Ramadasu, G.; Mukkanti, K.; Madhusudhan, G. Synthesis of carvedilol via 1-(9H-carbazol-4-yloxy)-3-(N-(2-(2-methoxy phenoxy)ethyl)-N-(4-methoxybenzyl)amino)propan-2-ol. *Org. Chem.: Indian J.* **2010,** *6* (1), 70–73.
68. McNeely, W.; Goa, K. L. Nebivolol in the management of essential hypertension: a review. *Drugs* **1999,** *57* (4), 633–651.
69. Mangrella, M.; Rossi, F.; Fici, F.; Rossi, F. Pharmacology of nebivolol. *Pharmacol. Res.* **1998,** *38* (6), 419–431.
70. Dery, A. S.; Hamilton, L. A.; Starr, J. A. Nebivolol for the treatment of heart failure. *Am. J. Health-Syst. Pharm.* **2011,** *68* (10), 879–886.
71. Gao, Y.; Vanhoutte, P. M. Nebivolol: An endothelium-friendly selective β_1-adrenoceptor blocker. *J. Cardiovasc. Pharmacol.* **2012,** *59* (1), 16–21.
72. Karter, Y. Nebivolol: more than a highly selective Beta blocker. *Recent Pat. Cardiovasc. Drug Discovery* **2007,** *2* (2), 152–155.
73. Riva, N.; Lip, G. Y. H. Nebivolol for the treatment of heart failure. *Expert Opin. Invest. Drugs* **2011,** *20* (12), 1733–1746.
74. Lipsic, E.; van Veldhuisen, D. J. Nebivolol in chronic heart failure: current evidence and future perspectives. *Expert Opin. Pharmacother.* **2010,** *11* (6), 983–992.
75. Gupta, S.; Wright, H. M. Nebivolol: a highly selective β_1-adrenergic receptor blocker that causes vasodilation by increasing nitric oxide. *Cardiovasc. Ther.* **2008,** *26* (3), 189–202.

76. Prisant, L. M. Nebivolol: pharmacologic profile of an ultraselective, vasodilatory β_1-blocker. *J. Clin. Pharmacol.* **2008,** *48* (2), 225–239.
77. Veverka, A.; Nuzum, D. S.; Jolly, J. L. Nebivolol: a third-generation β-adrenergic blocker. *Ann. Pharmacother.* **2006,** *40* (7/8), 1353–1360.
78. Cockcroft, J. Nebivolol: a review. *Expert Opin. Pharmacother.* **2004,** *5* (4), 893–899.
79. Van Lommen, G. R. E.; De Bruyn, M. F. L.; Schroven, M. F. J. 2,2′-Iminobisethanol derivatives, EP 145067 (1985).
80. Mauro, S.; Fattori, D.; D'Andrea, P.; Cipollone, A. Process for the preparation of nebivolol, WO 2012095707 (2012).
81. Haldar, P.; Golla, C. M.; Swapna, A.; Kumar, K. V.; Jawlekar, S.; Tummala, A. K.; Patro, D.; Yarraguntla, S. R.; Ramakrishnan, S.; Chimala, V. R. R. Process for the preparation of nebivolol and its salts, US 20110250454 (2011).
82. Motaleb, M. A.; Moustapha, M. E.; Ibrahim, I. T. Synthesis and biological evaluation of 125I-nebivolol as a potential cardioselective agent for imaging β_1-adrenoceptors. *J. Radioanal. Nucl. Chem.* **2011,** *289* (1), 239–245.
83. Wang, N.-X.; Xing, Y.; Wang, Y.-J. Asymmetric synthesis of chiral flavanone and chromanone derivatives. *Curr. Org. Chem.* **2013,** *17* (14), 1555–1562.

Chapter 13

Cholinomimetics

Acetylcholine (ACh), the first neurotransmitter to be characterized, modulates many physiological processes in the central and peripheral nervous systems. In the central nervous system (CNS), ACh regulates motor function, nociception, sensory perception, cognitive processing, sleep/wake cycles, while in the peripheral nervous system (PNS) it controls heart rate, smooth muscle activity, and gastrointestinal tract motility [1-8].

Any drug that can reproduce the effects and mimic the actions of the neurotransmitter ACh, the primary transmitter of nerve impulses within the parasympathetic nervous system, that dilates blood vessels, contracts smooth muscles (eye pupils), slows the heart rate, and increases bodily secretions (salvia, sweat, tears, mucus in the respiratory tract, intestinal cramps) is considered a cholinomimetic.

The basis for the classification of compounds named cholinomimetics [9] could be their mechanism of action.

Cholinomimetics are classified as direct-acting drugs, that is, drugs that activate the receptor site by mimicking the effects of Ach, and indirectly acting drugs that prevent the enzyme cholinesterase (AChE) from hydrolyzing ACh at the receptor site. In turn, indirectly acting drugs can be reversible or irreversible.

Cholinomimetic drugs could also be classified based on the spectrum of their actions.

ACh acts on two vastly functionally and structurally distinct classes of receptors—nicotinic receptors (nAChRs) (with two subtypes, one at the neuromuscular junction of skeletal muscle, the other within ganglia and the CNS), and muscarinic receptors (mAChRs) (widely distributed within both peripheral and central nervous systems). These receptors originally were distinguished from each other by the selectivity of the alkaloid agonists, muscarine and nicotine, respectively.

The Nobel Prize in Physiology or Medicine 1936 was awarded to Sir Henry Hallett Dale and Otto Loewi "for their discoveries relating to chemical transmission of nerve impulses." Dale was the first, in 1914, to show that cholinomimetics produce two distinct types of effects: muscarinic effects produced by the alkaloid muscarine, which are blocked selectively by atropine, and nicotinic effects, which are produced by the alkaloid nicotine and blocked by *d*-tubocurarine but not by atropine.

Synthesis of Best-Seller Drugs. http://dx.doi.org/10.1016/B978-0-12-411492-0.00013-4

The nAChRs are members of the ligand-gated ion channel superfamily that includes γ-aminobutyric acid A ($GABA_A$) and $GABA_C$ receptors, and mediate fast synaptic neurotransmission throughout the nervous system. The mAChRs are members of the G-protein–coupled receptors and provide responses through second-messenger systems. Both muscarinic and nicotinic receptors exist in various subtypes.

Neuronal nAChRs are pentameric ligand-gated ion channels. Molecular cloning has identified nine α subunits (α_2 to α_{10}) and three β subunits (β_2 to β_4). Homomeric nAChRs are composed only of α subunits, whereas heteromeric nAChRs have different combinations of α and β subunits. The most abundant neuronal subunits are α_4, α_7, and β_2. More than 90% of all neuronal nAChRs are made up of the heteromeric $\alpha_4\beta_2$ receptor subtype.

Five molecularly distinct mammalian subtypes of mAChRs, M1 to M5, have been cloned. Each of the five mAChR subtypes is a seven-transmembrane protein that can be further divided into two major functional classes based on G-protein–coupling.

Most frequently used central AChR agonists are used as key structures and research tools in experimental studies.

13.1 DIRECT-ACTING CHOLINOMIMETICS

Mixed AChR Agonists

ACh (**13.1.1**) and carbachol (**13.1.2**) are mixed direct-acting cholinomimetics (Fig. 13.1.).

Acetylcholine 13.1.1 Carbachol 13.1.2

FIG. 13.1 **Structure of ACh and carbachol.**

Nicotinic AChR Agonists

Nicotine (**13.1.3**), cytisine (**13.1.4**), anatoxin (**13.1.5**), epibatidine (**13.1.6**), ferruginine (**13.1.7**) epiquinamide (**13.1.8**), and 5-iodo-3-[2(S)-2-azetidinyl-methoxy]pyridine (**13.1.9**) are more or less selective nicotinic AChR agonists. Varenicline (**13.1.10**), a medication used to treat smoking addiction, is a partial agonist of nicotinic AChR (Fig. 13.2.).

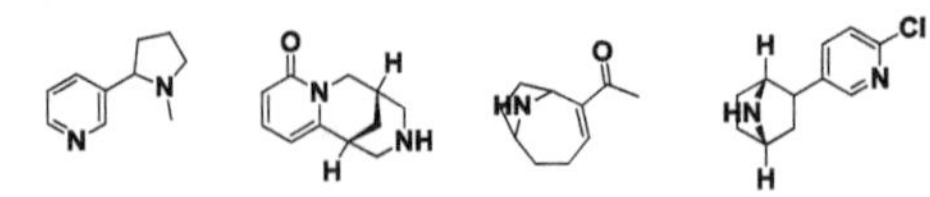

Nicotine 13.1.3 Cytisine 13.1.4 Anatoxin 13.1.5 Epibatidine 13.1.6

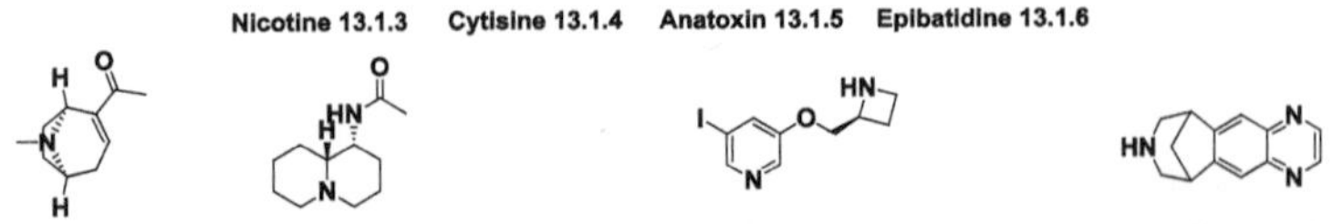

Ferruginine 13.1.7 Epiquinamide 13.1.8 5-Iodo-3-[2(S)-azetidinylmethoxy]pyridine 13.1.9 Varenicline 13.1.10

FIG. 13.2 **Nicotinic AChR agonists.**

Muscarinic AChR Agonists

Methacholine (**13.1.11**), bethanechol (**13.1.12**), muscarine (**13.1.13**), pilocarpine (**13.1.14**), arecoline (**13.1.15**), milameline (**13.1.16**), xanomeline (**13.1.17**), RS-86 (**13.1.18**), aceclidine (**13.1.19**), cevimeline (**13.1.20**), talsaclidine (**13.1.21**), sabcomeline (**13.1.22**), WAY-132983 (**13.1.23**), oxotremorine (**13.1.24**), β-acetoxynortropane (**13.1.25**), and bao gong teng A (BGT-A) (**13.1.26**) are representatives of muscarinic AChR agonists (Fig. 13.3.).

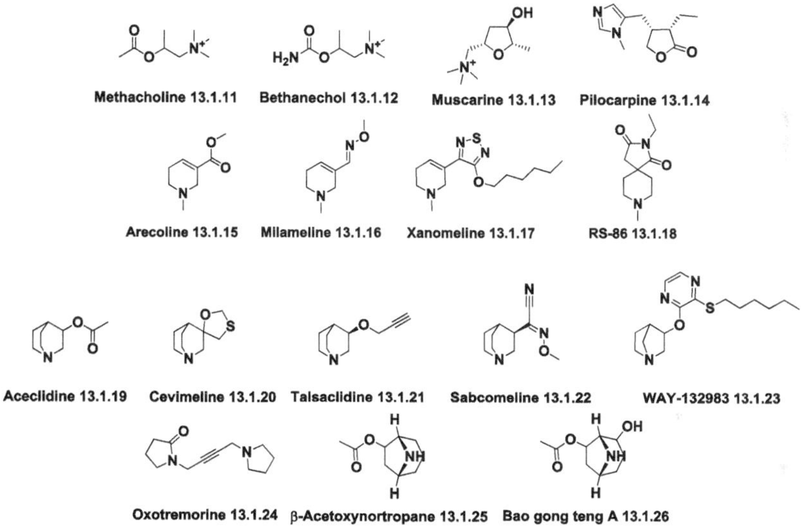

FIG. 13.3 Muscarinic ACh receptor agonists.

The selectivities, binding assays and pharmacological effects of variety of different compounds belonged to ACh (**13.1.1**), nicotine (**13.1.3**), anatoxin (**13.1.5**), epibatidine (**13.1.6**), ferruginine (**13.1.7**) derivatives are documented in reviews [10-18].

Muscarinic receptors are located in many areas of the body. The major targets of muscarinic agonists include blood vessels, heart, eyes, lungs, and the smooth muscle of the gastrointestinal and urinary tracts. As a result, muscarinic agonists have diverse effects. Nevertheless, because of their inability to activate nicotinic receptors, as compared to ACh, the muscarinic agonists are pharmacologically preferred. It is believed that muscarinic receptors are important in learning and memory. Currently, direct muscarinic receptor agonists are not used for treatment of any CNS disease, but they produce CNS effects, including learning and memory-improving agents, and medications for the treatment of Alzheimer and Parkinson diseases and for symptoms associated with schizophrenia [19-27]. Some muscarinic receptor ligands, such as xanomeline and cevimeline, have been investigated in preclinical or in clinical studies for the treatment of nervous system diseases (Alzheimer and Sjögren diseases). Muscarinic agonists such as vedaclidine

(**13.1.27**), CMI-936 (**13.1.28**), and CMI-1145 (**13.1.29**) have been demonstrated to have analgesic effects in animal models comparable to those produced by known opioids [28] (Fig. 13.4.).

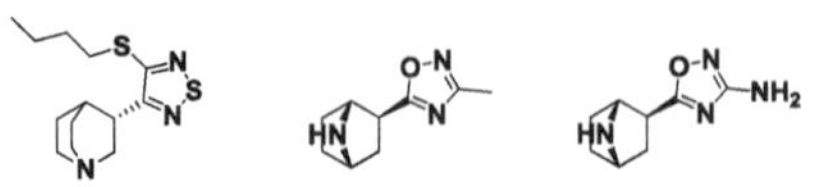
Vedaclidine 13.1.27 CMI-936 13.1.28 CMI-1145 13.1.29

FIG. 13.4 **Muscarinic agonists with analgesic effects in animal models.**

ACh itself has no practical therapeutic use because it does not differentiate between nicotinic and muscarinic receptors and is rapidly degraded. Nicotine also has no clinical use because it stimulates both sympathetic and parasympathetic systems. Muscarine is not used clinically because it causes symptoms of mushroom poisoning.

The major therapeutic uses of the cholinomimetics are to treat diseases of the eye (glaucoma), the diseases of the gastrointestinal and urinary tracts (postoperative atony, neurogenic bladder), and the neuromuscular junction (myasthenia gravis, curare-induced neuromuscular paralysis). Their last implementation involves attempts to treat patients with Alzheimer and Parkinson diseases and schizophrenia.

ACh (**13.1.1**) is used very rarely and only as an intraocular solution for obtaining miosis during ocular surgery. It has no other therapeutic use because it does not differentiate between nicotinic and muscarinic receptors and is rapidly degraded. Nicotine also has no clinical use because it stimulates both sympathetic and parasympathetic systems. Muscarine is not used clinically because it causes symptoms of mushroom poisoning.

Carbachol (**13.1.2**), which is synthesized by replacement of acetic acid fragment of ACh by carbamic acid, is resistant to cholinesterases and more active, but also more toxic, than ACh. It has both muscarinic and nicotinic actions and rarely is used as an ophthalmic solution for treatment of glaucoma.

Bethanechol (**13.1.12**) or carbamylmethylcholine has primarily muscarinic effects and stimulates gastrointestinal and vesical motricity. It is used in treatment of gastroesophageal reflux in infants and for vesical atonicity to facilitate micturition.

Pilocarpine (**13.1.13**) is used to constrict pupils and reduce pressure caused by glaucoma. It contracts the ciliary muscle with causes the iris to be withdrawn. This action permits drainage of the aqueous humor and thus relieves the pressure caused by a glaucoma condition. It is also used, by general route, to treat hyposialism.

Aceclidine (**13.1.16**) is also used in the form of ophthalmic solution to treat glaucoma.

Cevimeline (**13.1.17**) is used to treat symptoms of dry mouth that accompanies Sjögren syndrome.

Cholinomimetics generally are not found on lists of the "Top 200" prescription medications, but they do serve as a useful model class.

13.2 CHOLINESTERASE INHIBITORS (INDIRECT-ACTING CHOLINOMIMETICS)

Stimulation of cholinoreceptors can be achieved by an alternate mechanism—inhibiting the acetyl and butyryl cholinesterases. The enzyme inactivation, induced by various inhibitors, leads to ACh accumulation, and in clinical conditions this approach can be therapeutically advantageous.

Cholinesterases are involved in the termination of impulse transmission metabolizing choline esters in numerous cholinergic pathways in the central and peripheral nervous systems. The enzyme inactivation, leads to choline esters accumulation, hyperstimulation of nicotinic and muscarinic receptors, and disrupted neurotransmission. Pharmacological effects of cholinesterase inhibitors include CNS (increased alertness, convulsions, seizures, coma), cardiovascular system (decreased heart rate and cardiac output), respiratory system (increased bronchial tone and mucosal secretions), neuromuscular junction system (increased muscle strength, ataxia, tremors), gastrointestinal system (increased motility, digestive secretions, cramps, defecation), and the eyes (miosis, decreased far vision, changes of intraocular pressure).

Cholinesterase inhibitors are a wide group of chemical compounds that are used as drugs, insecticides, and chemical warfare agents.

Cholinesterase inhibitors are classified into two groups: reversible, such as physostigmine (**13.2.1**), neostigmine (**13.2.2**) pyridostigmine (**13.2.3**), rivastigmine (**13.2.4**), edrophonium (**13.2.5**), demecarium (**13.2.6**), ambenonium (**13.2.7**), donepezil (**13.2.8**), tacrine (**13.2.9**), the prototypical cholinesterase inhibitors for treatment of Alzheimer's disease, naturally occurring sesquiterpene alkaloid huperzine A (**13.2.10**), and galantamine (**13.2.11**), another alkaloid that is obtained synthetically or from the bulbs and flowers of *Galanthus caucasicus*, as well as ladostigil (**13.2.12**), a dual acetylcholine-butyrylcholinesterase and brain selective monoamine oxidase (MAO) A and B inhibitor, all of which lead to carbamoylation of cholinesterases and prevent binding and metabolism of choline esters (Fig. 13.5.).

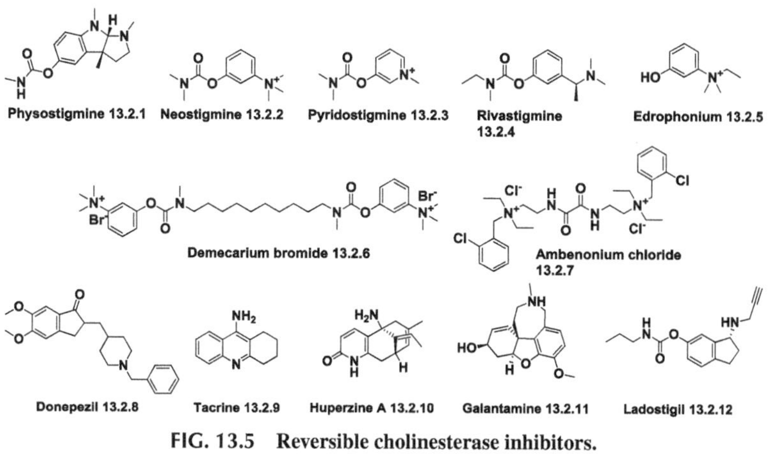

FIG. 13.5 Reversible cholinesterase inhibitors.

Irreversible cholinesterase inhibitors are mainly organophosphorus compounds used as insecticides, such as diisopropylfluorophosphate (**13.2.13**), echothiophate (**13.2.14**), which is rarely used in the treatment of chronic glaucoma, and many others, such as malathion (**13.2.15**), dichlorvos (**13.2.16**), chlorpyrifos (**13.2.17**), and parathion (**13.2.18**), which are widely used as an organophosphorus insecticides (Fig. 13.6.) and which lead to phosphorylation of serine residue in the cholinesterase active site, making it incapable of metabolizing choline esters.

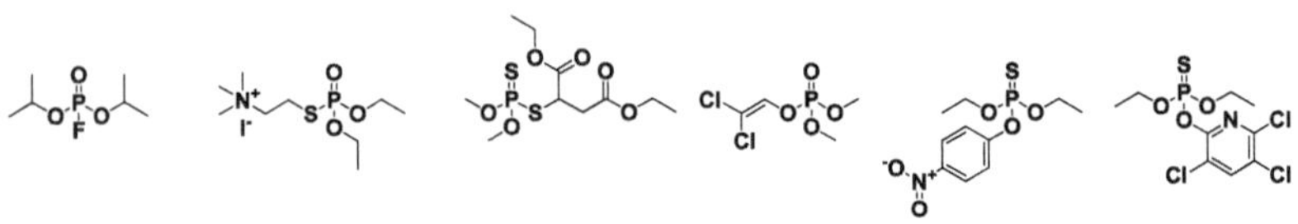

Diisopropylfluoro-phosphate 13.2.13 Echothiophate 13.2.14 Malathion 13.2.15 Dichlorvos 13.2.16 Chlorpyrifos 13.2.17 Parathion 13.2.18

FIG. 13.6 Irreversible cholinesterase inhibitors.

Nerve gases such as sarin (**13.2.19**), soman (**13.2.20**), tabun (**13.2.21**), a group called the V-agents, namely, VX-gas (**13.2.22**), VE-gas(**13.2.23**), VG-gas (**13.2.24**), and others, which are among the most potent synthetic toxic agents known and are used as warfare agents (Fig. 13.7.).

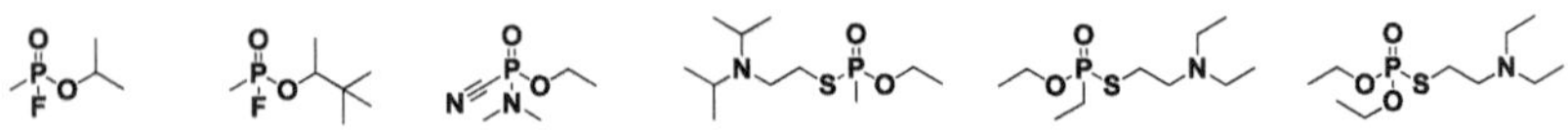

Sarin 13.2.19 Soman 13.2.20 Tabun 13.2.21 VX-Gas 13.2.22 VE-Gas 13.2.23 VG-Gas 13.2.24

FIG. 13.7 Nerve gases used as warfare agents.

So, cholinesterase inhibitors are used not only in the treatment of human diseases, but for the control of insect pests and as chemical warfare agents and weapons. Antidotal treatment in cases of this type of poisoning includes general supportive measures, and management with atropine for the treatment of muscarinic symptoms and pralidoxime for the regeneration of the enzyme.

Therapeutic uses of cholinesterase inhibitors [29-40] in general include improvement of gastrointestinal and urinary motility, neuromuscular transmission in myasthenia gravis, and treatment of glaucoma that results from increased intraocular pressure. Inhibition of cholinesterase is associated also with treatment of some Alzheimer disease symptoms [41-43] although the etiology of Alzheimer disease remains unclear and seems to involve multiple factors. Moreover, cholinergic transmission in the brain is an essential component of cognitive function. In addition to Alzheimer disease, cholinesterase inhibitors have been examined in the treatment of a variety of other neurological disorders associated with cognitive decline, such as mild cognitive impairment, Down syndrome, Parkinson disease, Huntington disease, multiple sclerosis, epilepsy, psychotic disorders, and pain [44-48]. Cholinesterase inhibitors are occasionally used in the treatment of atropine overdosage and certain types of arrhythmias.

Physostigmine (**13.2.1**) enhances cholinergic responses to the iris increases aqueous flow and decreases intraocular pressure, making it useful in the treatment of glaucoma.

Neostigmine (**13.2.2**), used for the control of myasthenia gravis, prevention and treatment of postoperative distention, and urinary http://www.rxlist.com/script/main/art.asp?articlekey=5912 retention after mechanical injuries, and reversal of effects of nondepolarizing neuromuscular blocking agents (tubocurarine, metocurine, gallamine, and pancuronium) after surgery.

Pyridostigmine (**13.2.3**) is used to decrease muscle weakness resulting from myasthenia gravis.

Edrophonium (**13.2.4**) is used as diagnostic test for the myasthenic condition and for dose assessment.

Rivastigmine (**13.2.5**) is in multiple clinical major Phase III trials as a pseudoirreversible acetylcholinesterase inhibitor that improves cognitive functions.

Demecarium (**13.2.6**) is used in the management of glaucoma.

Ambenonium (**13.2.7**) is used to improve muscle strength in patients with myasthenia gravis.

Donepezil (**13.2.8**) significantly improves cognition.

Tacrine (**13.2.9**) appears to improve cognitive function and behavioral deficits in patients with Alzheimer disease.

Huperzine A (**13.2.10**) is used for Alzheimer disease, memory and learning enhancement, and age-related memory impairment. It is also used for treating myasthenia gravis, for increasing alertness and energy, and for protecting against agents that damage the nerves, such as nerve gases.

Galantamine (**13.2.11**) is one of the last drugs approved for the treatment of Alzheimer disease.

Ladostigil (**13.2.12**) is a multitarget neuroprotective and neurorestorative anti-Parkinson and anti-Alzheimer drug.

Donepezil (**13.2.8**) is the single cholinesterase inhibitor found in the list of Top 200 Drugs by sales for the 2010s.

Donepezil–Aricept

Donepezil (**13.2.8**) is the most widely prescribed of the cholinesterase inhibitors. It improves cognitive performance and stabilizes functional abilities in people with mild-to-moderate Alzheimer disease by increasing the amount of the ACh in the brain, the deficit of which is thought to play a major role in the clinical presentation of Alzheimer disease. It shows good safety and long-term tolerability [49-59]. The synthesis of donepezil involves the condensation of 5,6-dimethoxy-1-indanone (**13.2.25**) and 1-benzyl-4-piperidinecarbox-aldehyde (**13.2.26**) in the presence of n-BuLi as base, at −78°C in tetrahydrofuran/

hexamethylphosphoric triamide solvent mixture. The resulting compound (**13.2.27**) is hydrogenated over palladium carbon to produce the desired donepezil (**13.2.8**) [60,61] (Scheme 13.1.). Improved methods [62,63] and other approaches [64-68] for the synthesis of donepezil (**13.2.8**) have been proposed.

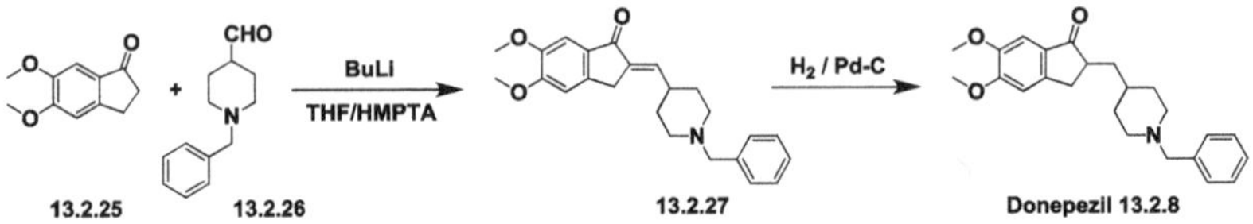

SCHEME 13.1 **Synthesis of donepezil.**

REFERENCES

1. Jensen, A. A.; Krogsgaard-Larsen, P. Acetylcholine. In *Textbook of Drug Design and Discovery,* 4th ed.; Krogsgaard-Larsen, P., Stroemgaard, K., Madsen, U., Eds. CRC Press, 2010; pp 263–281.
2. Dani, J. A.; Bertrand, D. Nicotinic acetylcholine receptors and nicotinic cholinergic mechanisms of the central nervous system. *Annu. Rev. Pharmacol. Toxicol.* **2007,** *47,* 699–729.
3. Wess, J.; Buhl, T.; Lambrecht, G.; Mutschler, E. Cholinergic receptors. In Hansch, C., Sammes, P. G., Taylor, J. B., Eds.; *Comprehensive Medicinal Chemistry*, Vol. 3; Pergamon Press, 1990; pp 423–491.
4. Van der Zee, E. A.; Platt, B.; Riedel, G. Acetylcholine: future research and perspectives. *Behav. Brain Res.* **2011,** *221* (2), 583–586.
5. Abreu-Villaca, Y.; Filgueiras, C. C.; Manhaes, A. C. Developmental aspects of the cholinergic system. *Behav. Brain Res.* **2011,** *221* (2), 367–378.
6. Tiwari, P.; Dwivedi, S.; Singh, M. P.; Mishra, R.; Chandy, A. Basic and modern concepts on cholinergic receptor: a review. *Asian Pac. J. Trop. Dis.* **2013,** *3* (5), 413–420.
7. Hollenhorst, M.; Clauss, W.; Fronius, M. Acetylcholine as nonneuronal cholinergic system. *Biol. Unserer Zeit* **2012,** *42* (6), 390–395.
8. Contestabile, A. The history of the cholinergic hypothesis. *Behav. Brain Res.* **2011,** *221* (2), 334–340.
9. Caulfield, M. P.; Birdsall, N. J. M. International Union of Pharmacology. XVII. Classification of muscarinic acetylcholine receptors. *Pharmacol. Rev.* **1998,** *50* (2), 279–290.
10. Daly, J. W. Nicotinic agonists, antagonists, and modulators from natural sources. *Cell. Mol. Neurobiol.* **2005,** *25* (3/4), 513–552.
11. Romanelli, M. N.; Gualtieri, F. Cholinergic nicotinic receptors: competitive ligands, allosteric modulators, and their potential applications. *Med. Res. Rev.* **2003,** *23* (4), 393–426.
12. Hurst, R.; Rollema, H.; Bertrand, D. Nicotinic acetylcholine receptors: from basic science to therapeutics. *Pharmacol. Ther.* **2013,** *137* (1), 22–54.
13. Toma, L.; Barlocco, D.; Gelain, A. Neuronal nicotinic acetylcholine receptor agonists. *Expert Opin. Ther. Pat.* **2004,** *14* (7), 1029–1040.
14. Tonder, J. E.; Olesen, P. H. Agonists at the $\alpha_4\beta_2$ nicotinic acetylcholine receptors: structure-activity relationships and molecular modeling. *Curr. Med. Chem.* **2001,** *8* (6), 651–674.
15. Nirogi, R.; Goura, V.; Abraham, R.; Jayarajan, P. $\alpha_4\beta_2$* Neuronal nicotinic receptor ligands (agonist, partial agonist and positive allosteric modulators) as therapeutic prospects for pain. *Eur. J. Pharmacol.* **2013,** *712* (1–3), 22–29.

16. Bunnelle, W. H.; Dart, M. J.; Schrimpf, M. R. Design of ligands for the nicotinic acetylcholine receptors: the quest for selectivity. *Curr. Top. Med. Chem.* **2004,** *4* (3), 299–334.
17. Glennon, R. A.; Dukat, M. Central nicotinic receptor ligands and pharmacophores. *Pharm. Acta Helv.* **2000,** *74* (2–3), 103–114.
18. Holladay, M. W.; Dart, M. J.; Lynch, J. K. Neuronal nicotinic acetylcholine receptors as targets for drug discovery. *J. Med. Chem.* **1997,** *40* (26), 4169–4194.
19. Eglen, R. M. Muscarinic receptor subtype pharmacology and physiology. *Prog. Med. Chem.* **2005,** *43,* 105–136.
20. Fisher, A.; Karton, Y.; Heldman, E.; Gurwitz, D.; Haring, R.; Meshulam, H.; Brandeis, R.; Pittel, Z.; Segall, Y.; Marciano, D. Progress in medicinal chemistry of novel selective muscarinic agonists. *Drug Des. Discovery* **1993,** *9* (3–4), 221–235.
21. Broadley, K. J.; Kelly, D. R. Muscarinic receptor agonists and antagonists. *Recent Res. Dev. Org. Chem.* **2002,** *6* (Pt. 2), 747–792.
22. Bubser, M.; Byun, N.; Wood, M. R.; Jones, C. K. Muscarinic receptor pharmacology and circuitry for the modulation of cognition. *Handb. Exp. Pharmacol.* **2012,** *208,* 121–166 (Muscarinic Receptors).
23. Tata, A. M. Muscarinic acetylcholine receptors as novel therapeutic targets. *Front. CNS Drug Discovery* **2010,** *1,* 362–377.
24. Kuduk, S. D.; Beshore, D. C. Novel M1 allosteric ligands: a patent review. *Expert Opin. Ther. Pat.* **2012,** *22* (12), 1385–1398.
25. Messer, W. S., Jr. Drugs that target muscarinic cholinergic receptors. In Buccafusco, J. J., Ed.; *Cognitive Enhancing Drugs*; Springer, 2004; pp 37–48.
26. Ellis, J. Muscarinic receptors. In Pangalos, M. N., Davies, C. H., Eds.; *Understanding G Protein-Coupled Receptors and Their Role in the CNS*; Oxford University Press, 2002; pp 349–371.
27. Jones, C., K.; Byun, N.; Bubser, M. Muscarinic and nicotinic acetylcholine receptor agonists and allosteric modulators for the treatment of schizophrenia. *Neuropsychopharmacology* **2012,** *37* (1), 16–42.
28. Greig, N. H.; Reale, M.; Tata, A. M. New pharmacological approaches to the cholinergic system: an overview on muscarinic receptor ligands and cholinesterase inhibitors. *Recent Pat. CNS Drug Discovery* **2013,** *8* (2), 123–141.
29. Colovic, M. B.; Krstic, D. Z.; Lazarevic-Pasti, T. D.; Bondzic, A. M.; Vasic, V. M. Acetylcholinesterase inhibitors: pharmacology and toxicology. *Curr. Neuropharmacol.* **2013,** *11* (3), 315–335.
30. Racchi, M.; Mazzucchelli, M.; Porrello, E.; Lanni, C.; Govoni, S. Acetylcholinesterase inhibitors: novel activities of old molecules. *Pharmacol. Res.* **2004,** *50* (4), 441–451.
31. Amenta, F.; Di Tullio, M. A.; Parnetti, L.; Tayebati, S. K. Cholinesterase inhibitors: from weapons, to pesticides, to cognition enhancing drugs. *Curr. Enzyme Inhib.* **2006,** *2* (3), 249–259.
32. Pope, C.; Karanth, S.; Liu, J. Pharmacology and toxicology of cholinesterase inhibitors: uses and misuses of a common mechanism of action. *Environ. Toxicol. Pharmacol.* **2005,** *19* (3), 433–446.
33. Karczmar, A. Anticholinesterases: dramatic aspects of their use and misuse. *Neurochem. Int.* **1998,** *32* (5–6), 401–411.
34. Karczmar, A. G. Cholinesterases (ChEs) and the cholinergic system in ontogenesis and phylogenesis, and non-classical roles of cholinesterases—a review. *Chem.-Biol. Interact.* **2010,** *187* (1–3), 34–43.
35. Silman, I.; Sussman, J. L. Acetylcholinesterase: "classical" and "non-classical" functions and pharmacology. *Curr. Opin. Pharmacol.* **2005,** *5* (3), 293–302.
36. Pohanka, M. Cholinesterases, a target of pharmacology and toxicology. *Biomed. Pap.* **2011,** *155* (3), 219–230.

37. Hostettmann, K.; Borloz, A.; Urbain, A.; Marston, A. Natural product inhibitors of acetylcholinesterase. *Curr. Org. Chem.* **2006,** *10* (8), 825–847.
38. Houghton, P. J.; Ren, Y.; Howes, M.-J. Acetylcholinesterase inhibitors from plants and fungi. *Nat. Prod. Rep.* **2006,** *23* (2), 181–199.
39. Pohanka, M. Acetylcholinesterase inhibitors: a patent review (2008-present). *Expert Opin. Ther. Pat.* **2012,** *22* (8), 871–886.
40. de los Rios, C. Cholinesterase inhibitors: a patent review (2007-2011). *Expert Opin. Ther. Pat.* **2012,** *22* (8), 853–869.
41. Van Beek, A. H. E. A.; Claassen, J. A. H. R. The cerebrovascular role of the cholinergic neural system in Alzheimer's disease. *Behav. Brain Res.* **2011,** *221* (2), 537–542.
42. Small, D. H. Acetylcholinesterase inhibitors for the treatment of dementia in Alzheimer's disease: do we need new inhibitors? *Expert Opin. Emerg. Drugs* **2005,** *10* (4), 817–825.
43. Mehta, M.; Adem, A.; Sabbagh, M. New acetylcholinesterase inhibitors for Alzheimer's disease. *Int. J. Alzheimer's Dis.* **2012,** *60* (6), 1–8.
44. Bohnen, N. I.; Albin, R. L. The cholinergic system and Parkinson disease. *Behav. Brain Res.* **2011,** *221* (2), 564–573.
45. Woolf, N. J.; Butcher, L. L. Cholinergic systems mediate action from movement to higher consciousness. *Behav. Brain Res.* **2011,** *221* (2), 488–498.
46. Schliebs, R.; Arendt, T. The cholinergic system in aging and neuronal degeneration. *Behav. Brain Res.* **2011,** *221* (2), 555–563.
47. Jones, C. K.; Byun, N.; Bubser, M. Muscarinic and nicotinic acetylcholine receptor agonists and allosteric modulators for the treatment of schizophrenia. *Neuropsychopharmacology* **2012,** *37* (1), 16–42.
48. Dagyte, G.; Den, B. J. A.; Trentani, A. The cholinergic system and depression. *Behav. Brain Res.* **2011,** *221* (2), 574–582.
49. Cheewakriengkrai, L.; Gauthier, S. A 10-year perspective on donepezil. *Expert Opin. Pharmacother.* **2013,** *14* (3), 331–338.
50. Doody, R. S.; Cummings, J. L.; Farlow, M. R. Reviewing the role of donepezil in the treatment of Alzheimer's disease. *Curr. Alzheimer Res.* **2012,** *9* (7), 773–781.
51. Jelic, V.; Darreh-Shori, T. Donepezil: A review of pharmacological characteristics and role in the management of Alzheimer disease. *Clin. Med. Insights: Ther.* **2010,** *2,* 771–788.
52. Asiri, Y. A.; Mostafa, G. A. E. Donepezil, Profiles Drug Subst., Excipients. *Relat. Methodol* **2010,** *35,* 117–150.
53. Tsuno, N. Donepezil in the treatment of patients with Alzheimer's disease. *Expert Rev. Neurother.* **2009,** *9* (5), 591–598.
54. Benjamin, B.; Burns, A. Donepezil for Alzheimer's disease. *Expert Rev. Neurother.* **2007,** *7* (10), 1243–1249.
55. Seltzer, B. Donepezil: an update. *Expert Opin. Pharmacother.* **2007,** *8* (7), 1011–1023.
56. Sugimoto, H. Donepezil hydrochloride: a treatment drug for Alzheimer's disease. *Chem. Rec.* **2001,** *1* (1), 63–73.
57. Dooley, M.; Lamb, H. M. Donepezil: a review of its use in Alzheimer's disease. *Drugs Aging* **2000,** *16* (3), 199–226.
58. Whitehouse, P. J. Donepezil. *Drugs Today* **1998,** *34* (4), 321–326.
59. Bryson, H. M.; Benfield, P. Donepezil. *Drugs Aging* **1997,** *10* (3), 234–239.
60. Sugimoto, H.; Tsuchiya, Y.; Higurashi, K.; Karibe, N.; Iimura, Y.; Sasaki, A.; Yamanashi, Y.; Ogura, H.; Araki, S.; Kashiwa, M.; Kosasa, T.; Kusota, A.; Kozasa, M.; Yamatsu, K. Preparation of 1-benzyl-4-(substituted alkyl)piperidines and analogs as acetylcholinesterase inhibitors. EP 296560 (1988).

61. Iimura, Y.; Mishima, M.; Sugimoto, H. Synthesis of 1-benzyl-4-[(5,6-dimethoxy[2-14C]-1-indanon)-2-yl]methylpiperidine hydrochloride (E2020-14C). *J. Labelled Compd. Radiopharm.* **1989,** *27* (7), 835–839.
62. Niphade, N.; Mali, A.; Jagtap, K.; Ojha, R. C.; Vankawala, P. J.; Mathad, V. T. An improved and efficient process for the production of donepezil hydrochloride: substitution of sodium hydroxide for n-butyl lithium via phase transfer catalysis. *Org. Process Res. Dev* **2008,** *12* (4), 731–735.
63. Rao, R. J. R.; Rao, A. K. S.B.; Murthy, Y. L. N. Efficient and industrially viable synthesis of donepezil. *Synth. Commun.* **2007,** *37* (17), 2847–2853.
64. Bolea, I.; Juarez-Jimenez, J.; de los Rios, C.; Chioua, M.; Pouplana, R.; Luque, F. J.; Unzeta, M.; Marco-Contelles, J.; Samadi, A. Synthesis, biological evaluation, and molecular modeling of donepezil and N-[(5-(benzyloxy)-1-methyl-1H-indol-2-yl)methyl]-N-methylprop-2-yn-1-amine hybrids as new multipotent cholinesterase/monoamine oxidase inhibitors for the treatment of Alzheimer's disease. *J. Med. Chem.* **2011,** *54* (24), 8251–8270.
65. Andreani, A.; Cavalli, A.; Granaiola, M.; Guardigli, M.; Leoni, A.; Locatelli, A.; Morigi, R.; Rambaldi, M.; Recanatini, M.; Roda, A. Synthesis and screening for antiacetylcholinesterase activity of (1-benzyl-4-oxopiperidin-3-ylidene)methylindoles and -pyrroles related to donepezil. *J. Med. Chem.* **2001,** *44* (23), 4011–4014.
66. Mulla, S.; Mukhopadhyay, R. N. Synthesis, isolation, and characterization of Donepezil open ring impurity. *Int. J. Pharm. Pharm. Sci.* **2013,** *5* (3), 826–829.
67. Chavakula, R.; Mutyala, N. R.; Chennupati, S. R. Industrially viable preparation of n-benzyl-4-formylpiperidine, a key intermediate of donepezil. *Org. Prep. Proced. Int.* **2013,** *45* (2), 168–170.
68. Dubey, S. K.; Kharbanda, M.; Dubey, S. K.; Mathela, C. S. A new commercially viable synthetic route for donepezil hydrochloride: anti-Alzheimer's drug. *Chem. Pharm. Bull.* **2010,** *58* (9), 1157–1160.

Chapter 14

Anticholinergic Drugs

Anticholinergic drugs are medications that reduce the effect of acetylcholine, a neurotransmitter that is involved in many major bodily functions in the central and peripheral nervous systems. These drugs are commonly used in critical care medicine.

Anticholinergic agents are classified into two categories according to the receptors that they act on: antinicotinic and antimuscarinic.

14.1 ANTINICOTINIC AGENTS (NEUROMUSCULAR BLOCKING DRUGS)

Antinicotinic agents act on the nicotinic acetylcholine receptors. They prevent the transmission of signals from motor nerves to neuromuscular structures of the skeletal muscle. The main therapeutic use of neuromuscular blockers is in surgical procedures as muscle relaxants, either as adjuvants or as premedication drugs for anesthesia.

Muscle relaxants are typically classified by their mechanism of action (depolarizing versus nondepolarizing), chemical structure, and duration of action (short, intermediate, and long acting).

Curare, the prototype of antinicotinic agents, is a common name for various South American poisons. It causes paralysis of mammalians by blocking transmission between nerve and muscle, without affecting nerve conduction or muscle contraction directly. The main toxin of curare is the isoquinoline derivative D-tubocurarine (**14.1.1**), a classical example of a nondepolarizing agent that competitively blocks the action of acetylcholine on nicotinic receptors. D-tubocurarine is the first anticholinergic drug to have been used in anesthesia to produce the necessary level of muscle relaxation. Currently, D-tubocurarine is rarely used, because much safer alternatives are available.

Nondepolarizing muscle relaxants prevent access of acetylcholine to the receptor, interrupting transmission at the skeletal neuromuscular junction without causing depolarization of the motor end plate. They prevent acetylcholine from triggering muscle contraction and are used as muscle relaxants in convulsive states, as well as anesthesia adjuvants. The action of nondepolarizing drugs can be reversed by cholinesterase inhibitors like neostigmine. Nondepolarizing drugs can be divided into chemical groups of tetrahydroisoquinoline derivatives—tubocurarine (**14.1.1**) itself, atracurium (**14.1.2**), doxacurium

Synthesis of Best-Seller Drugs. http://dx.doi.org/10.1016/B978-0-12-411492-0.00014-6

(**14.1.3**), mivacurium (**14.1.4**), and cisatracurium (**14.1.5**) (Fig. 14.1.)—and aminosteroids—vecuronium (**14.1.6**), pancuronium (**14.1.7**), rocuronium (**14.1.8**), and pipecuronium (**14.1.9**) (Fig. 14.2.)—and other chemical classes, which include compounds such as gallamine (**14.1.10**) and β-erythroidine (**14.1.11**) (Fig. 14.3.).

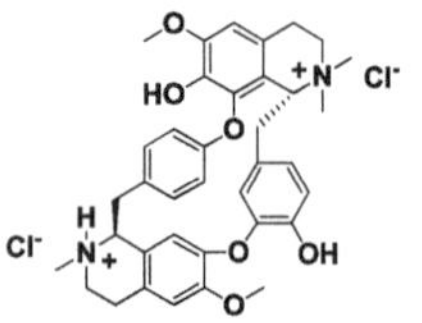

(D)-Tubocurarine 14.1.1

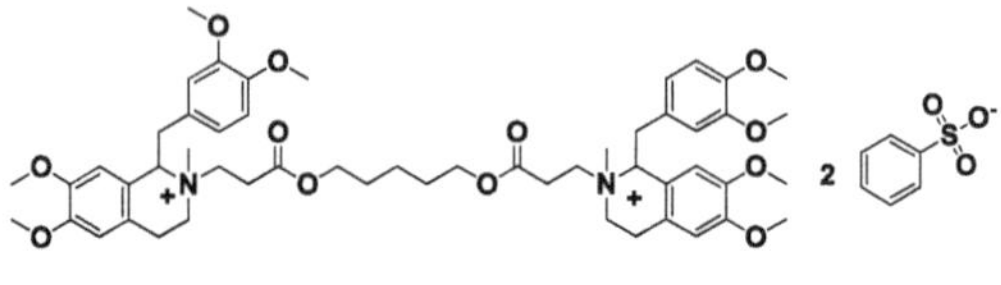

Atracurium besilate 14.1.2

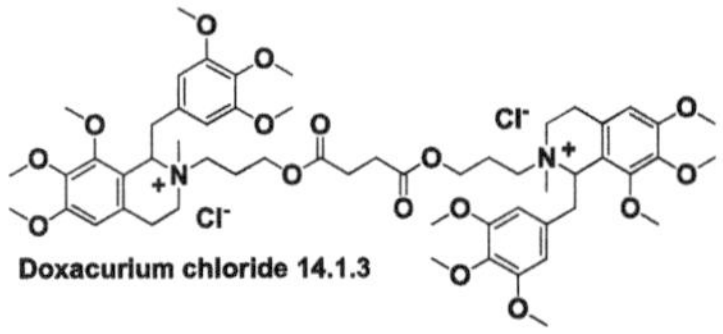

Doxacurium chloride 14.1.3

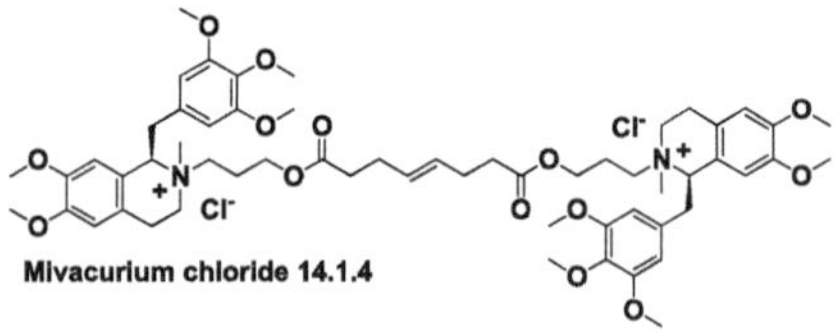

Mivacurium chloride 14.1.4

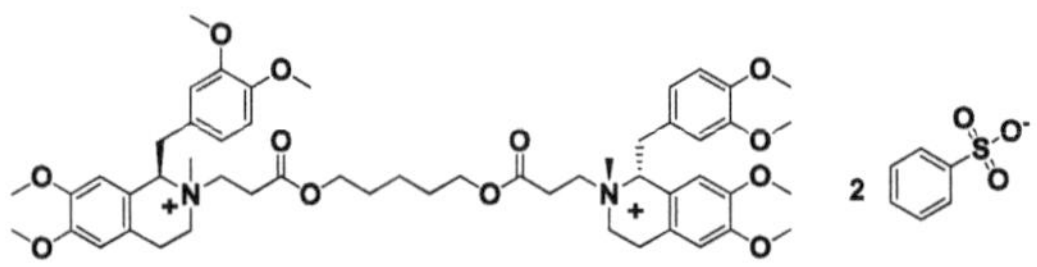

Cisatracurium besilate 14.1.5

FIG. 14.1 Nondepolarizing drugs that are classified as tetrahydroisoquinoline derivatives.

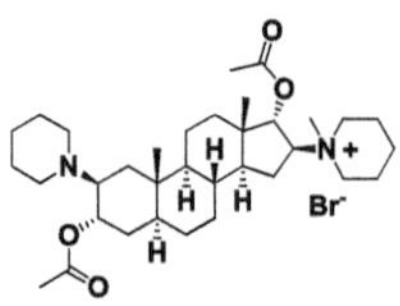

Vecuronium bromide 14.1.6

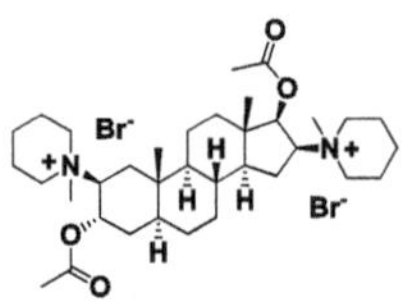

Pancuronium bromide 14.1.7

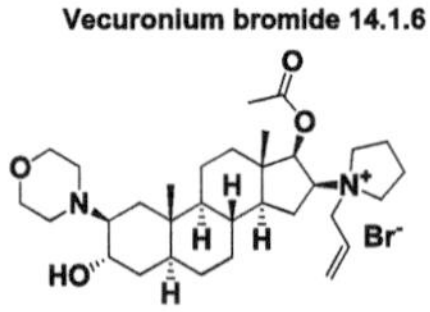

Rocuronium bromide 14.1.8

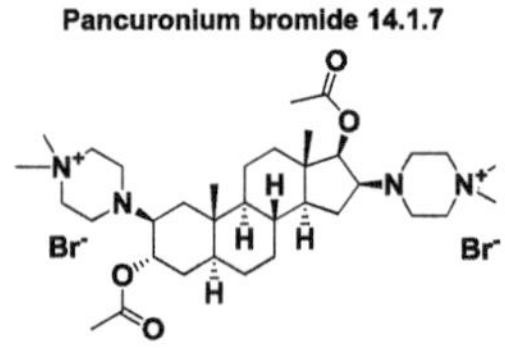

Pipecuronium bromide 14.1.9

FIG. 14.2 Nondepolarizing drugs that are classified as aminosteroids.

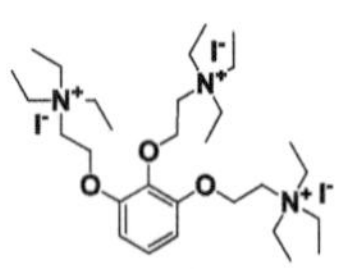

Gallamine triethiodide 14.1.10

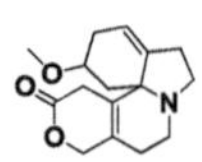

Dihydro-beta-erythroidine 14.1.11

FIG. 14.3 Structure of representatives of other chemical classes of nondepolarizing drugs.

Nondepolarizing neuromuscular block is monitored throughout surgery and antagonized at the end of anaesthesia to restore muscle tone rapidly and completely and is reversed by acetylcholinesterase inhibitors.

Mivacurium is a short-action medication; atracurium, cisatracurium, vecuronium, and rocuronium are intermediate-action medications; and doxacurium and pancuronium are long-action medications.

Depolarizing neuromuscular blocking agents, that is, drugs that interrupt transmission at the skeletal neuromuscular junction by causing sustained depolarization of the motor end plate, are more resistant to degradation by acetylcholinesterase, and thus can more persistently depolarize the muscle fibers, having a prolonged effect. They do not require reversal with cholinesterase inhibitors. These agents are primarily used as adjuvants in surgical anesthesia to cause skeletal muscle relaxation. The most popular depolarizing agent is suxamethonium (succinylcholine) (Fig. 14.4.). Structurally, succinylcholine (**14.1.12**) is made up of two acetylcholine molecules. It is basically the only depolarizing muscle blocker used clinically and in surgical procedures, despite a number of disadvantages.

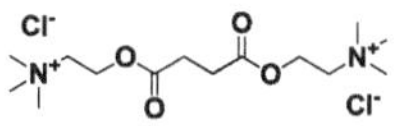

Succinylcholine 14.1.12

FIG. 14.4 Structure of succinylcholine.

Succinylcholine was synthesized in 1906. Daniel Bovet was the first to describe succinylcholine-induced paralysis. The discovery and development of succinylcholine pharmacological properties lead to the awarding of a Nobel Prize in Physiology or Medicine in 1957 "for discoveries relating to synthetic compounds that inhibit the action of certain body substances, and especially their action on the vascular system and the skeletal muscles."

14.2 GANGLIONIC BLOCKERS

Agents having as their major action the interruption of neural transmission at nicotinic receptors on postganglionic autonomic neurons without sensitive changes in membrane potentials are called *ganglionic blockers*. They occupy nicotinic receptors and stabilize postsynaptic membrane against the action of acetylcholine. Ganglioblockers may be used, and historically have been used, for management of hypertensive cardiovascular disease and hypertensive crises; that is, to control a patient's blood pressure and for the induction of hypotension in surgery.

Examples of ganglioblocking drugs are hexamethonium (**14.2.1**), pentolonium tartrate (**14.2.2**), trimethaphan (**14.2.3**), and mecamylamine (**14.2.4**) (Fig. 14.5.).

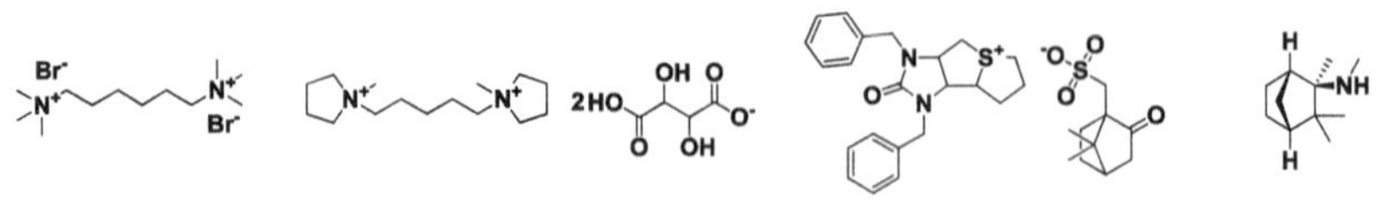

Hexamethonium 14.2.1 — Pentolonium tartrate 14.2.2 — Trimethaphan 14.2.3 — Mecamylamine 14.2.4

FIG. 14.5 Structure of ganglioblocking drugs.

14.3 ANTIMUSCARINIC AGENTS

Drugs that bind to muscarinic receptors but block the actions of endogenous acetylcholine or other cholinomimetics are called *antimuscarinic agents, antimuscarinics,* or *muscarine antagonists*. They have widespread effects, including actions on the iris and ciliary muscle of the eye, the heart and blood vessels, act on secretions of the respiratory tract, gastrointestinal system, and salivary glands, central nervous system. Centrally active antimuscarinics are generally used in psychiatry to treat the extrapyramidal side effects of antipsychotic medications. They may have antidepressant and mood-elevating properties, and are liable to abuse, have hallucinogenic features.

Anticholinergic medications have been used in various preparations from plants since antiquity.

The naturally occurring antimuscarinics atropine and scopolamine were extracted from plants of the family Solanaceae the best known of which probably is the plant belladonna, which in Italian means "beautiful woman," because the herb was used in eyedrops by women to dilate the pupils of the eyes to make them appear seductive. Belladonna has a long history of use as a medicine, cosmetic, and poison. It was used as an anesthetic for surgery. Historically, anticholinergic agents were known more for their toxicity than for their therapeutic effects.

Atropine (**14.3.1**) and scopolamine (**14.3.2**) are the earlier tropane alkaloids isolated from plant sources. The name "atropa" comes from Atropos, one of the three Fates in Greek mythology. No other drug class can claim as long a history with so many therapeutic applications and significant developments in pharmacology.

Antimuscarinic agents constitute the vast majority of anticholinergic drugs.

A basic example is atropine itself, which blocks acetylcholine receptor sites, opposes the actions of the vagus nerve, increases firing of the sinoatrial node and conduction through the atrioventricular node of the heart, and decreases bronchiole secretions. The overall effect of atropine is to lower the parasympathetic activity of all muscles and glands regulated by the parasympathetic nervous system.

The effects of scopolamine on the peripheral nervous system are similar to the effects of atropine. However, scopolamine is a central nervous system depressant and constitutes a highly effective treatment to prevent motion sickness.

Antimuscarinic agents can be classified based on their chemical structure or by their therapeutic implementation.

From a chemical point of view, one possible classification could be done by dividing them into natural belladonna alkaloids—atropine (dl-hyoscyamine) (**14.3.1**) and scopolamine (**14.3.2**) (Fig. 14.6.).

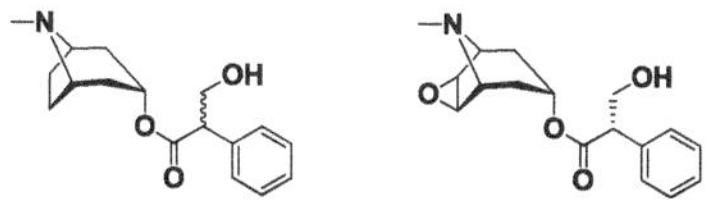

Atropine 14.3.1 Scopolamine 14.3.2

FIG. 14.6 Structure of atropine and scopolamine.

The semisynthetic class of antimuscarinics is represented by ipratropium bromide (**14.3.3**), oxitropium bromide (**14.3.4**), tiotropium bromide (**14.3.5**), trospium bromide (**14.3.6**), benzatropine (**14.3.7**), homatropine (**14.3.8**), methylscopolamine (**14.3.9**) (Fig. 14.7.).

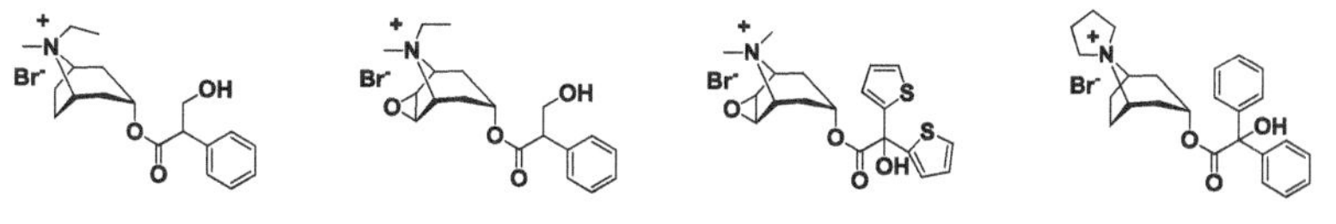

Ipratropium bromide 14.3.3 Oxitropium bromide 14.3.4 Tiotropium bromide 14.3.5 Trospium bromide 14.3.6

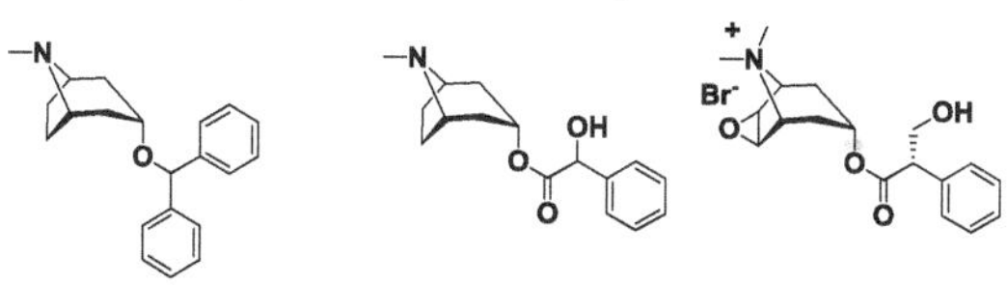

Benzatropine 14.3.7 Homatropine 14.3.8 Methylscopolamine 14.3.9

FIG. 14.7 Semisynthetic antimuscarinics.

Synthetic antimuscarinics belong to different chemical classes. Procyclidine (**14.3.10**), tridihexethyl (**14.3.11**), trihexyphenidyl (**14.3.12**), and biperiden (**14.3.13**) belong to the 1-phenyl-3-amino-propan-1-ol series; dicycloverine (**14.3.14**) to the bi(cyclohexane)]-1-carboxylates; cyclopentolate (**14.3.15**) and glycopyrronium bromide (**14.3.16**) are possible to rank as derivatives 1-hydroxycyclopentyl-2-phenylacetates; oxyphencyclimine (**14.3.17**) and oxybutynin (**14.3.18**) to the 2-cyclohexyl-2-hydroxy-2-phenylacetates; and mepenzolate (**14.3.19**) and propiverine (**14.3.20**) to the 2-cyclohexyl-2-hydroxy-2-phenylacetic acid series (Fig. 14.8.).

A large number of antimuscarinic drugs are impossible to classify according to chemical structure, including clidinium bromide (**14.3.21**), aclidinium bromide (**14.3.22**), darifenacin (**14.3.23**), propantheline (**14.3.24**), flavoxate (**14.3.25**), solifenacin (**14.3.26**), tolterodine (**14.3.27**), tropicamide (**14.3.28**), pirenzepine (**14.3.29**), telenzepine (**14.3.30**), hydroxyzine (**14.3.31**), and mebeverine (**14.3.32**) (Fig. 14.9.).

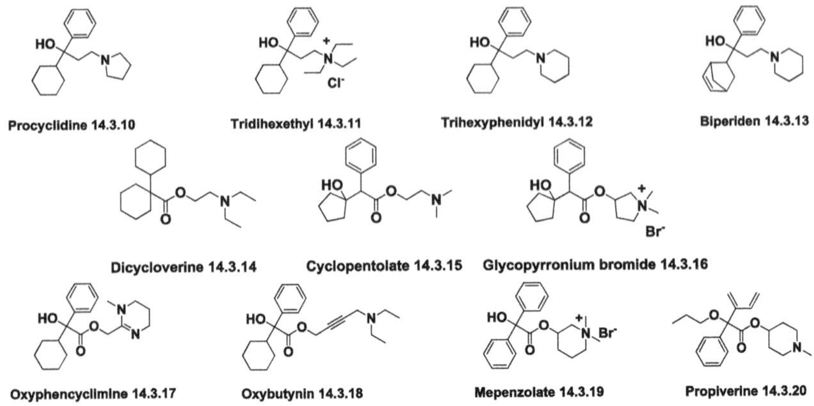

FIG. 14.8 Synthetic antimuscarinics.

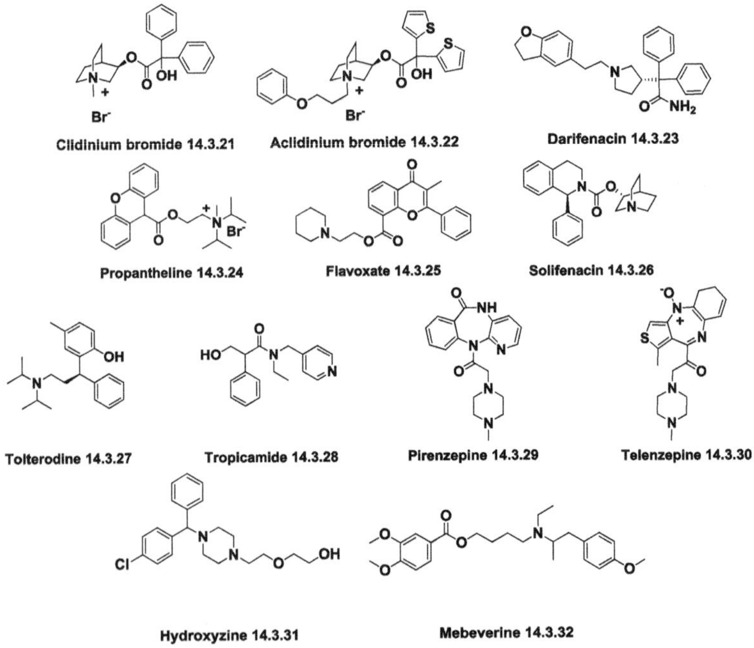

FIG. 14.9 Unclassified antimuscarinics.

Therapeutic classification of antimuscarinic drugs have been used for the treatment of different diseases [1-12] and can be presented as:

1. *Mydriatics and cycloplegics:* atropine (**14.3.1**), scopolamine (**14.3.2**), cyclopentolate (**14.3.15**), tropicamide (**14.3.28**);
2. *Antispasmodics/antidiarrheals;* atropine (**14.3.1**), mepenzolate (**14.3.19**), clidinium (**14.3.21**);

3. *Preanesthetic medications:* atropine (**14.3.1**), scopolamine (**14.3.2**);
4. *Medication for cardiovascular disorders*: atropine (**14.3.1**);
5. *Drugs used for motion sickness*: atropine (**14.3.1**), scopolamine (**14.3.2**);
6. *Drugs used for peptic ulcer*: pirenzepine (**14.3.29**), telenzepine (**14.3.30**);
7. *Drugs for bronchial asthma*: ipratropium (**14.3.3**), tiotropium (**14.3.5**), benzatropine (**14.3.7**);
8. *Medications used to treat urinary incontinence*: trospium (**14.3.6**), oxybutynin (**14.3.18**); propiverine (**14.3.20**), darifenacin (**14.3.23**), solifenacin (**14.3.26**), tolterodine (**14.3.27**); inhaled anticholinergics constitute the cornerstone of therapy in chronic obstructive pulmonary disease (COPD);
9. *Drugs used for Parkinson disease*: benztropine (**14.3.7**), procyclidine (**14.3.10**); trihexyphenidyl (**14.3.12**), biperiden (**14.3.13**);
10. *Antidote to cholinergic poisoning*: atropine (**14.3.1**).

Another classification of antimuscarinic drugs can be done based on their selectivity to distinct subtypes of muscarine acetylcholine receptors:

1. *M_1 antagonists*: pirenzepine (**14.3.26**), telenzepine (**14.3.30**), dicycloverine (**14.3.14**);
2. *M_2 antagonists*: analogue of pirenzepine (**14.3.29**), telenzepine (**14.3.30**), experimental compound AF-DX 116 (**14.3.33**), alkaloid himbacine (**14.3.34**), bis(hexane-1,6-diamine) derivatives methoctramine (**14.3.35**), and tripitramine (**14.3.36**) (Fig. 14.10.);
3. *M_3 antagonists*: oxybutynin (**14.3.18**), darifenacin (**14.3.23**), solifenacin (**14.3.26**), and tolterodine (**14.3.27**).

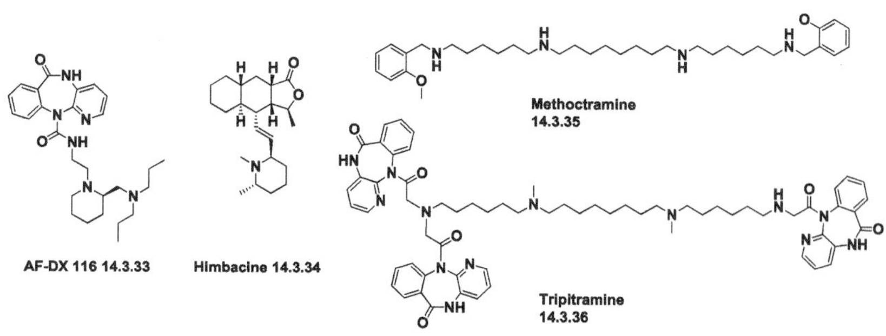

FIG. 14.10 Structure of some M_2 antimuscarinic agents.

In addition to the listed antimuscarinic agents, drugs from other categories may have antimuscarinic effects. Examples of such drug classes are antidepressants, antiemetics, antipsychotics, drugs used in cardiovascular and gastrointestinal diseases, muscle relaxants, antiparkinsonian drugs, and some others.

Four antimuscarinic drugs are included in the list of Top 200 Drugs by sales for the 2010s: tiotropium bromide (**14.3.5**) (Spiriva); darifenacin (**14.3.23**) (Enablex); Solifenacin (**14.3.26**) (VESIcare), and tolterodine (**14.3.27**) (Detrol).

Tiotropium Bromide–Spiriva

Tiotropium bromide (**14.3.5**) is an anticholinergic agent that has gained worldwide acceptance as a first-line, once-daily maintenance therapy for patients with moderate-to-severe COPD [13-20]. It is an antimuscarinic compound that binds to all receptor subtypes with similar affinity, but it dissociates from the M_1 and M_3 muscarinic receptors approximately 100-fold more slowly than from the M_2 receptor [21]. Adverse events associated with the use of tiotropium bromide may include coughing, influenza-like symptoms, sinusitis, dry mouth, and arthritis. More serious possible side effects include high pressure in the eye (glaucoma), blurred vision, difficulty and pain in passing urine, and heartburn.

Synthesis of tiotropium bromide (**14.3.5**) was first described in a patent [22] by transesterification of methyl di-(2-thienyl)glycolate thienylcarboxylate (**14.3.38**) with scopine (**14.3.37**) to produce the appropriate scopine ester (**14.3.39**), which on treatment with MeBr was quaternized to the desired product, tiotropium bromide (**14.3.5**). Alternate patents describe synthesis of tiotropium bromide starting from an already quaternized scopine salt [23], and starting from tropenol (**13.3.40**) and then oxidizing the obtained tropenol ester (**14.3.41**) to scopine ester (**14.3.39**) [24].

An interesting approach is demonstrated in a patent [25], in which by reaction of scopine methobromide (**14.3.43**) with 5,5-di(thiophen-2-yl)-1,3-dioxolane-2,4-dione (**14.3.42**) prepared from lithium salt of oxalic acid monomethyl ester and thiophene-2-magnesium bromide the desired tiotropium bromide (**14.3.5**) was synthesized (Scheme 14.1.). Many other methods with minor changes in described methods are also patented.

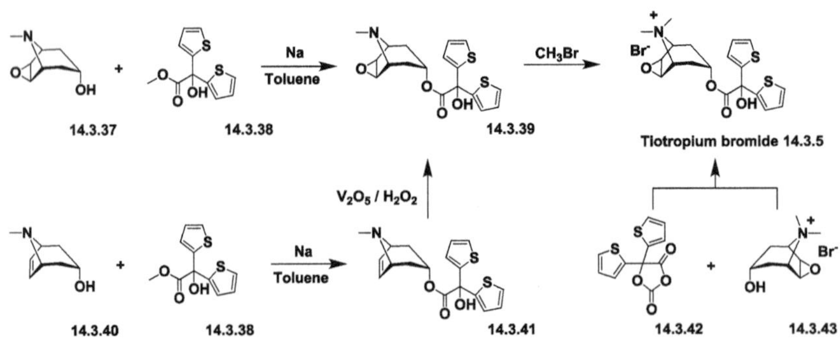

SCHEME 14.1 Synthesis of tiotropium bromide.

Darifenacin–Enablex

Darifenacin (**14.3.23**) works by blocking the M_3 muscarinic acetylcholine receptor, which is primarily responsible for bladder muscle contractions. It reduces muscle spasms of the bladder and urinary tract relaxing the bladder muscles to prevent urgent, frequent, or uncontrolled urination. More common side effects of darifenacin are acid or sour stomach, heartburn, indigestion discomfort, upset, and pain [26-28].

Synthesis of darifenacin was first proposed in a patent [29]. The proposed method consists of coupling of (S)-2,2-diphenyl-2-(pyrrolidin-3-yl)acetamide (**14.3.50**) with 5-(2-bromoethyl)-2,3-dihydrobenzofuran.

The synthesis started with commercially available (2S,4R)-4-hydroxypyrrolidine-2-carboxylic acid (**14.3.44**), which, on heating in cyclohexanol in the presence of a catalytic amount of cyclohex-2-en-1-one, has been decarboxylated to produce (R)-pyrrolidin-3-ol (**14.3.45**). Obtained cyclic amino alcohol was N-tosylated in the presence of pyridine to produce (R)-1-tosylpyrrolidin-3-ol (**14.3.46**), which was O-tosylated in Mitsunobu reaction conditions using triphenylphosphine, p-toluenesulfonic acid methyl ester, and diethyl azodicarboxylate to obtain (S)-enantiomer of 1-tosylpyrrolidin-3-yl 4-methylbenzenesulfonate (**14.3.47**), which, when coupled with diphenylacetonitrile in the presence of sodium hydride, produced the desired (S)-enantiomer of 2,2-diphenyl-2-(1-tosylpyrrolidin-3-yl)acetonitrile (**14.3.48**). The obtained product was detosylated under reflux in 48% hydrobromic acid to produce the compound (**14.3.49**) and undergo hydrolysis of the cyano group using 95% sulfuric acid to produce (S)-2,2-diphenyl-2-(pyrrolidin-3-yl)acetamide (**14.3.50**). The last was alkylated with 5-(2-bromoethyl)-2,3-dihydrobenzofuran to produce the desired darifenacin (**14.3.23**) (Scheme 14.2.).

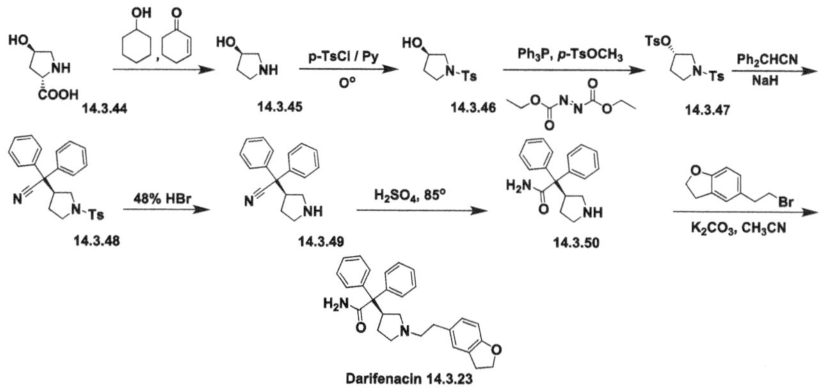

SCHEME 14.2 Synthesis of darifenacin.

Alternative methods to synthesize darifenacin are proposed in the same patent [29], as well as in some others [30-32] that are focused on alkylation, acylation, or reductive amination reactions using as starting material compounds of the series (**14.3.49**) or (**14.3.50**), which are summarized in Fig. 14.11.

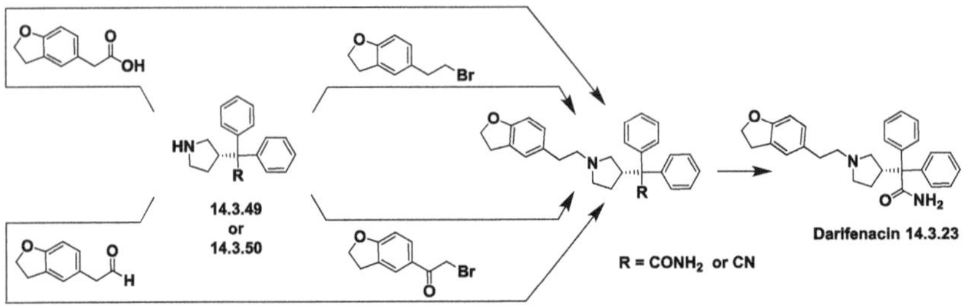

FIG. 14.11 **Alternative ways to synthesize darifenacin.**

Solifenacin–VESIcare

Overactive bladder (OAB) syndrome is the term used to describe the symptom complex of urinary urgency with or without urge incontinence, usually with frequency and nocturia. Solifenacin (**14.3.26**) [33-37] is a novel muscarinic receptor antagonist for the treatment of men and women with OAB. The common side effects include dry mouth, blurred vision, indigestion, nausea, and stomach pain.

The first publications [38,39] and the strategy used for industrial scale synthesis of solifenacin (**14.3.26**) is based on Bischler-Napiralski cyclization of N-phenethylbenzamide (**14.3.51**) synthesized from benzoic acid and phenethylamine using a mixture of $POCl_3$ and P_2O_5 as a condensation reagent to produce 1-phenyl-3,4-dihydroisoquinoline (**14.3.52**). Resulting imine was hydrogenated with $NaBH_4$ to 1-phenyl-1,2,3,4-tetrahydroisoquinoline (**14.3.53**). After this point, two different strategies have been employed. According to the first strategy, the racemic (**14.3.53**) was reacted with ethyl chloroformate to yield dihydroisoquinoline-carboxylate (**14.3.54**). The resulting compound was reacted with (R)-quinuclidin-3-ol in the presence of sodium hydride in toluene to produce (3R)-quinuclidyl carboxylates (**14.3.26 rac**), which were separated using preparative high-performance liquid chromatography (HPLC) to produce the desired solifenacin (**14.3.26**). According to the second strategy, racemic 1-phenyl-1,2,3,4-tetrahydroisoquinoline (**14.3.53 rac**) was separated via resolution of the tartrate salt, which provided (S)-1-phenyl-1,2,3,4-tetrahydroisoquinoline intermediate (**14.3.53**), which, after consequent reactions with ethyl chloroformate and (R)-quinuclidin-3-ol, produced the desired solifenacin (**14.3.26**) (Scheme 14.3.).

Other published patents are based on these two approaches [40-49] with minor changes, like asymmetric hydrogenation of 1-phenyl-3,4-dihydroisoquinoline [47] or enantioselective synthesis of1-phenyl-tetrahydroisoquinoline [48,49].

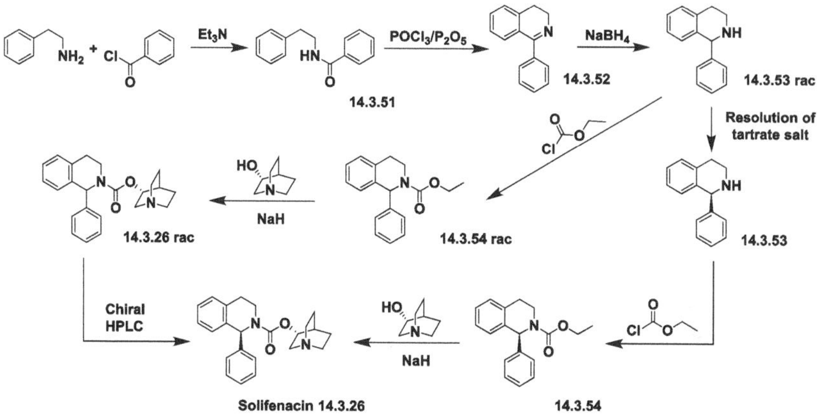

SCHEME 14.3 **Synthesis of solifenacin.**

Tolterodine–Detrol

Tolterodine is a potent, competitive muscarinic receptor antagonist that exhibits in vivo selectivity for the bladder over other tissues that contain muscarinic receptors (such as the salivary glands and the eye), and represents the first drug specifically developed for the treatment of OAB [50-54].

The first method for the synthesis of tolterodine (**14.3.27**) was reported in a patent [55] and a paper in *Drugs* [56], and was based on the condensation of *p*-cresol with cinnamic acid, which was conducted in tetralone media in the presence of sulfuric acid and at high temperature to produce 6-methyl-4-phenylchroman-2-one (**14.3.55**). The next step consisted in simultaneous lactone ring opening with methanol and phenolic hydroxyl etherification which occurred in refluxing methanol/acetone solution containing K_2CO_3 and MeI giving 2-hydroxy-3-phenylpropanoate (**14.3.56**). The last was reduced to corresponding 2-(3-hydroxy-1-phenylpropyl)-4-methylphenol (**14.3.57**) which was tosylated with tosyl chloride in pyridine to give product (**14.3.58**). This product was reacted with diisopropylamine in acetonitrile to give amine (**14.3.59**), the methyl ether fragment of which was cleaved with phosphorous tribromide in dichloromethane giving racemic tolterodine (**14.3.27rac**) which was resolved with L-(+)-tartaric acid giving the desired tolterodine (**14.3.27**) (Scheme 14.4.).

Other publications patents are based on very close two approaches [57-65] which propose asymmetric hydrogenation of [57-64] or enantioselective synthesis of [65] (Scheme 14.4.).

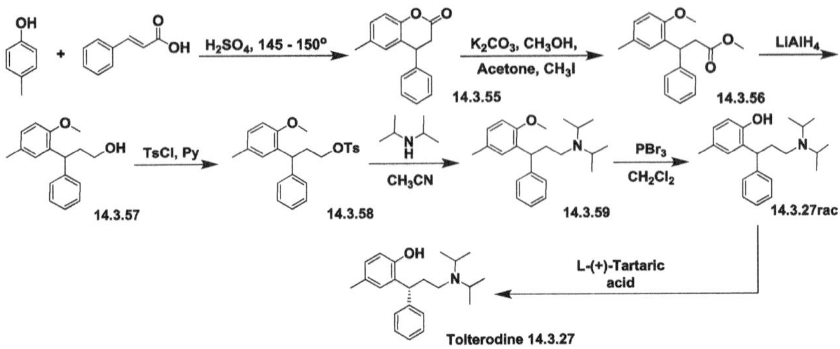

SCHEME 14.4 **Synthesis of tolterodine.**

REFERENCES

1. Weiner, M. F.; Davis, K. L. Anticholinergic drugs. In *Drugs in Psychiatry;* Burrows, G. D., Norman, T. R., Davies, B., Eds.; *Antimanics, Anticonvulsants Other Drugs Psychiatry*, Vol. 4; Elsevier, 1986; pp 191–205.
2. Gross, N. J. Anticholinergic drugs. In Barnes, P. J., Grunstein, M. M., Leff, A. R., Woolcock, A. J., Eds.; *Asthma*, Vol. 2; Lippincott Williams and Wilkins, 1997; pp 1555–1568.
3. Stockbruegger, R. W. Antimuscarinic drugs. *Methods Find. Exp. Clin. Pharmacol.* **1989,** *11* (Suppl. 1), 79–86.
4. Celli, B. R. The clinical use of anticholinergics. In *Therapeutic Strategies in COPD;* Cazzola, M., Ed.; CRC Press, 2005; pp 93–105.
5. Gross, N. J. Anticholinergic agents in asthma and COPD. *Eur. J. Pharmacol.* **2006,** *533* (1–3), 36–39.
6. Guyer, A. C.; Long, A. A. Long-acting anticholinergics in the treatment of asthma. *Curr. Opin. Allergy Clin. Immunol.* **2013,** *13* (4), 392–398.
7. Plusa, T. Rationale basis for new anticholinergic drugs for chronic obstructive pulmonary disease. *Int. Rev. Allergol. Clin. Immunol.* **2011,** *17* (3–4), 49–52.
8. Novelli, F.; Malagrino, L.; Dente, F. L.; Paggiaro, P. Efficacy of anticholinergic drugs in asthma. *Expert Rev. Respir. Med.* **2012,** *6* (3), 309–319.
9. Virchow, J. C.; Lommatzsch, M. Anticholinergic agents in asthma. In *Therapeutic Strategies in Asthma: Current Treatments;* Polosa, R., Holgate, S. T., Eds.; Clinical Publishing, 2007; pp 79–90.
10. Shvarts, G. Y.; Shvarts, P. G.; Plotnikov, A. N.; Savvin, D. Y. Drugs for the treatment of overactive bladder syndrome: present and future (a review). *Pharm. Chem. J.* **2013,** *46* (12), 699–706.
11. Chancellor, M.; Boone, T. Anticholinergics for overactive bladder therapy: central nervous system effects. *CNS Neurosci. Ther.* **2012,** *18* (2), 167–174.
12. Buccafusco, J. J. The cholinergic hypothesis—past and present. In *Cognitive Enhancing Drugs;* Buccafusco, J. J., Ed.; Birkhäuser, 2004; pp 1–10.
13. Yohannes, A. M.; Connolly, M. J.; Hanania, N. A. Ten years of tiotropium: clinical impact and patient perspectives. *Int. J. Chronic Obstruct. Pulm. Dis.* **2013,** *8,* 117–125.
14. Keating, G. M. Tiotropium bromide inhalation powder: a review of its use in the management of chronic obstructive pulmonary disease. *Drugs* **2012,** *72* (2), 273–300.
15. Mamary, A. J.; Criner, G. J. Tiotropium bromide for chronic obstructive pulmonary disease. *Expert Rev. Respir. Med.* **2009,** *3* (3), 211–220.
16. Rodrigo, G. J.; Nannini, L. J. Tiotropium for the treatment of stable chronic obstructive pulmonary disease: A systematic review with meta-analysis. *Pulm. Pharmacol. Ther.* **2007,** *20* (5), 495–502.

17. Lipson, D. A. Tiotropium bromide. *Int. J. Chronic Obstruct. Pulm. Dis.* **2006,** *1* (2), 107–114.
18. Hvizdos, K. M.; Goa, K. L. Tiotropium bromide. *Drugs* **2002,** *62* (8), 1195–1203.
19. Barnes, P. J. Tiotropium bromide. *Expert Opin. Invest. Drugs* **2001,** *10* (4), 733–740.
20. Barnes, P. J. The pharmacological properties of tiotropium. *Chest* **2000,** *117* (2 Suppl.), 63S–66S.
21. Mundy, C.; Kirkpatrick, P. Fresh from the pipeline: tiotropium bromide. *Nat. Rev. Drug Discovery* **2004,** *3* (8), 643–644.
22. Banholzer, R.; Bauer, R.; Reichl, R. Preparation of anticholinergic scopine, (nor)tropine, and granatoline esters of thienylcarboxylic acids and their quaternary salts, EP 418716 (1991).
23. Belzer, W.; Hamm, R.; Hofmann, M.; Lock, R. New procedure for the production of tiotropium salts from scopine methobromide, DE 102004041253 (2006).
24. Banholzer, R.; Graulich, M.; Luettke, S.; Mathes, A.; Meissner, H.; Specht, P.; Broeder, W. Procedure for the production of the anticholinergic tiotropium bromide from tropenol, DE 10064816 (2002).
25. Soukup, M. Manufacturing process for tiotropium bromide, US 20120123125 (2012).
26. Haab, F. Darifenacin in the treatment of overactive bladder. *Drugs Today* **2005,** *41* (7), 441–452.
27. Chapple, C. R. Darifenacin a novel M3 muscarinic selective receptor antagonist for the treatment of overactive bladder. *Expert Opin. Invest. Drugs* **2004,** *13* (11), 1493–1500.
28. Zinner, N. Darifenacin: a muscarinic M3-selective receptor antagonist for the treatment of overactive bladder. *Expert Opin. Pharmacother.* **2007,** *8* (4), 511–523.
29. Cross, P. E.; Mackenzie, A. R. Preparation of pyrrolidine derivatives as muscarinic receptor antagonists, EP 388054 (1990).
30. Lorente B.-L. A.; Rodriguez L. M.; Fernandez S. Y.; Gutierrez F. L. G. Process for preparation of 1,3-difunctionalized pyrrolidines (e.g. Darifenacin) from 3-hydroxypyrrolidines, WO 2010112488 (2010).
31. Soldevilla M. N. Process for preparation (3S)-substituted pyrrolidines from (S)-malic acid, protected amines, and diphenylacetonitrile or diphenylacetamide, WO 2009153649 (2009).
32. Merli, V.; Canavesi, A.; Daverio, P. Processes for preparing darifenacin hydrobromide, WO 2007076157 (2007).
33. Robinson, D.; Cardozo, L. Solifenacin for the treatment of overactive bladder. *Therapy* **2011,** *8* (6), 691–701.
34. Robinson, D.; Cardozo, L. Solifenacin: pharmacology and clinical efficacy. *Expert Rev. Clin. Pharmacol.* **2009,** *2* (3), 239–253.
35. Payne, C. K. Solifenacin in overactive bladder syndrome. *Drugs* **2006,** *66* (2), 175–190.
36. Chapple, C. R.; Cardozo, L.; Steers, W. D.; Govier, F. E. Solifenacin significantly improves all symptoms of overactive bladder syndrome. *Int. J. Clin. Pract.* **2006,** *60* (8), 959–966.
37. Luo, D.; Liu, L.; Han, P.; Wei, Q.; Shen, H. Solifenacin for overactive bladder: a systematic review and meta-analysis. *Int. Urogynecol. J.* **2012,** *23* (8), 983–991.
38. Mealy, N.; Castaner, J. YM-905: treatment of urinary incontinence, muscarinic M3 antagonist. *Drugs Future* **1999,** *24* (8), 871–874.
39. Naito, R.; Yonetoku, Y.; Okamoto, Y.; Toyoshima, A.; Ikeda, K.; Takeuchi, M. Synthesis and antimuscarinic properties of quinuclidin-3-yl 1,2,3,4-tetrahydroisoquinoline-2-carboxylate derivatives as novel muscarinic receptor antagonists. *J. Med. Chem.* **2005,** *48* (21), 6597–6606.
40. Takeuchi, M.; Naito, R.; Hayakawa, M.; Okamoto, Y.; Yonetoku, Y.; Ikeda, K.; Isomura, Y., Preparation of new quinuclidine derivatives as muscarinic M3 receptor antagonists, WO 9620194 (1996).
41. Perlman, N.; Nidam, T. Processes for preparation of solifenacin, WO 2007076116 (2007).

42. Puig, J.; Sanchez, L.; Masllorens, E.; Auger, I.; Bosch, J. Process for the preparation of solifenacin from (R)-3-quinuclidinol and (S)-1-phenyl-1,2,3,4-tetrahydroisoquinoline containing an activated acyl group in the presence of Lewis acid, WO 2008062282 (2008).
43. Dave, M. G.; Pandey, B.; Kothari, H. M.; Patel, P. R. Process for preparing chemically and chirally pure solifenacin base and its salts, WO 2009087664 (2009).
44. Mathad, V. T.; Lilakar, J. D.; Gilla, G.; Kikkuru, S.; Chinta, R. R.; Dudipala, S. Drug delivery systems containing solifenacin with high purity, WO 2008011462 (2008).
45. Ruzic, M.; Prudic, D.; Pecavar, A.; Zanotti-Gerosa, A.; Stropnik, T. Preparation of solifenacin salts and inclusion into pharmaceutical dosage forms, WO 2010012459 (2010).
46. Bolchi, C.; Pallavicini, M.; Fumagalli, L.; Straniero, V.; Valoti, E. One-pot racemization process of 1-phenyl-1,2,3,4-tetrahydroisoquinoline: a key intermediate for the antimuscarinic agent solifenacin. *Org. Process Res. Dev.* **2013,** *17* (3), 432–437.
47. Ruzic, M.; Pecavar, A.; Prudic, D.; Kralj, D.; Scriban, C.; Zanotti-Gerosa, A. The development of an asymmetric hydrogenation process for the preparation of solifenacin. *Org. Process Res. Dev.* **2012,** *16* (7), 1293–1300.
48. Niphade, N. C.; Jagtap, K. M.; Mali, A. C.; Solanki, P. V.; Jachak, M. N.; Mathad, V. T. Efficient and single pot process for the preparation of enantiomerically pure solifenacin succinate, an antimuscarinic agent. *Monatsh. Chem.* **2011,** *142* (11), 1181–1186.
49. Wang, S.; Onaran, M. B.; Seto, C. T. Enantioselective synthesis of 1-aryltetrahydroiso-quinolines. *Org. Lett.* **2010,** *12* (12), 2690–2693.
50. Nilvebrant, L.; Hallen, B.; Larsson, G. Tolterodine - a new bladder-selective muscarinic receptor antagonist: preclinical pharmacological and clinical data. *Life Sci.* **1997,** *60* (13/14), 1129–1136.
51. Hills, C. J.; Winter, S. A.; Balfour, J. A. Tolterodine. *Drugs* **1998,** *55* (6), 813–820.
52. Wefer, J.; Truss, M. C.; Jonas, U. Tolterodine: An overview. *World J. Urol.* **2001,** *19* (5), 312–318.
53. Rovner, E. S. Tolterodine for the treatment of overactive bladder: a review. *Exp. Opin. Pharmacother.* **2005,** *6* (4), 653–666.
54. Nilvebrant, L. The mechanism of action of tolterodine. *Rev. Contempor. Pharmacother.* **2000,** *11* (1), 13–27.
55. Joensson, N. A.; Sparf, B. A.; Mikiver, L.; Moses, P.; Nilvebrant, L.; Glas, G. 3,3-Diphenylpropylamines as drugs, especially anticholinergic agents, and their preparation and formulations containing them, EP 325571 (1998).
56. Graul, A.; Martel, A. M.; Castaner, J. Tolterodine. Agent for urinary incontinence muscarinic receptor antagonist. *Drugs Future* **1997,** *22* (7), 733–737.
57. Barancelli, D. A.; Salles, A. G.; Taylor, J. G.; Correia, C. R. D. Coumarins from free ortho-hydroxy cinnamates by Heck-Matsuda arylations: a scalable total synthesis of (R)-tolterodine. *Org. Lett.* **2012,** *14* (23), 6036–6039.
58. De Castro, K. A.; Ko, J.; Park, D.; Park, S.; Rhee, H. Reduction of ethyl benzoylacetate and selective protection of 2-(3-hydroxy-1-phenylpropyl)-4-methylphenol: a new and facile synthesis of tolterodine. *Org. Process Res. Dev.* **2007,** *11* (5), 918–921.
59. Turner, H. M.; Patel, J.; Niljianskul, N.; Chong, J. M. Binaphthol-catalyzed asymmetric conjugate arylboration of enones. *Org. Lett.* **2011,** *13* (21), 5796–5799.
60. Gallagher, B. D.; Taft, B. R.; Lipshutz, B. H. Asymmetric conjugate reductions of coumarins. a new route to tolterodine and related coumarin derivatives. *Org. Lett.* **2009,** *11* (23), 5374–5377.
61. Yoo, K.; Kim, H.; Yun, J. Enantioselective synthesis of (R)- tolterodine via CuH-catalyzed asymmetric conjugate reduction. *J. Org. Chem.* **2009,** *74* (11), 4232–4235.
62. Ulgheri, F.; Marchetti, M.; Piccolo, O. Enantioselective synthesis of (s)- and (r)-tolterodine by asymmetric hydrogenation of a coumarin derivative obtained by a Heck reaction. *J. Org. Chem.* **2007,** *72* (16), 6056–6059.

63. Hedberg, C.; Andersson, P. G. Catalytic asymmetric total synthesis of the muscarinic receptor antagonist (R)-tolterodine. *Adv. Synth. Catal.* **2005,** *347* (5), 662–666.
64. Andersson, P. G.; Schink, H. E.; Oesterlund, K. Asymmetric total synthesis of (+)-tolterodine, a new muscarinic receptor antagonist, via copper-assisted asymmetric conjugate addition of aryl grignard reagents to 3-phenyl-prop-2-enoyloxazolidinones. *J. Org. Chem.* **1998,** *63* (22), 8067–8070.
65. Selenski, C.; Pettus, T. R. R. Enantioselective [4 + 2] cycloadditions of o-quinone methides: total synthesis of (+)-mimosifoliol and formal synthesis of (+)-tolterodine. *J. Org. Chem.* **2004,** *69* (26), 9196–9203.

Chapter 15

Centrally Acting Skeletal Muscle Relaxants

Skeletal muscle relaxants (myorelaxants) are a group of chemical compounds that act both centrally and peripherally and have the ability to relax skeletal muscle, which reduces muscle contractility by blocking the transmission of nerve impulses, or by decreasing the excitability of the motor end plate, or by other actions. There are two major groups of muscle relaxants: Neuromuscular blockers act peripherally as skeletal muscle relaxants (tubocurarine, atracurium, doxacurium, mivacurium, pancuronium, rocuronium, succinylcholine, vecuronium), which block neuromuscular junction function. They are generally used in anesthesia to prevent spontaneous movement of muscle during surgical intervention and act by blocking the nicotinic acetylcholine receptor, thus inhibiting neuron transmission to muscle, which is described in Chapter 14.

Centrally acting skeletal muscle relaxants are commonly indicated to alleviate two different conditions: spasticity and muscular pain or spasms. Spasticity can be associated with a number of diseases and represents a condition in which muscles are continuously contracted, causing stiffness or tightness interfering with movements and speech. Muscular pain associated with muscle spasms include low back pain, neck pain, tension headaches, fibromyalgia, and myofascial pain syndrome. Although widely used, there appear to be gaps in understanding of the action, efficacy and safety of this heterogeneous group of medications of this series [1-7].

Skeletal muscle relaxants consist of two types of drugs: antispasticity and antispasmodic agents. The antispasticity agents—baclofen (**15.1.1**), tizanidine (**15.1.2**), dantrolene (**15.1.3**), and benzodiazepines of the diazepam series (**15.1.4**) (Fig. 15.1.)—are used in medicinal practice for improving muscle hypertonicity and involuntary jerks.

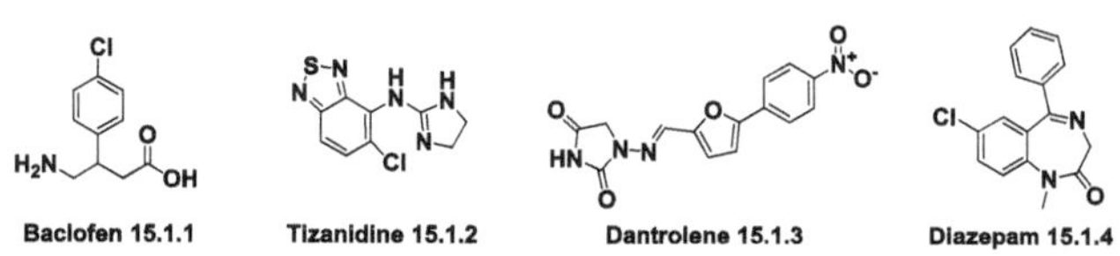

FIG. 15.1 Antispasticity skeletal muscle relaxants.

Synthesis of Best-Seller Drugs. http://dx.doi.org/10.1016/B978-0-12-411492-0.00015-8

Antispasmodic agents, such as mephenesin (**15.1.5**), methocarbamol (**15.1.6**), meprobamate (**15.1.7**), carisoprodol, (**15.1.8**), orphenadrine (**15.1.9**), cyclobenzaprine (**15.1.10**), chlorzoxazone (**15.1.11**), and metaxalone (**15.1.12**) (Fig. 15.2.), are primarily used to treat musculoskeletal conditions.

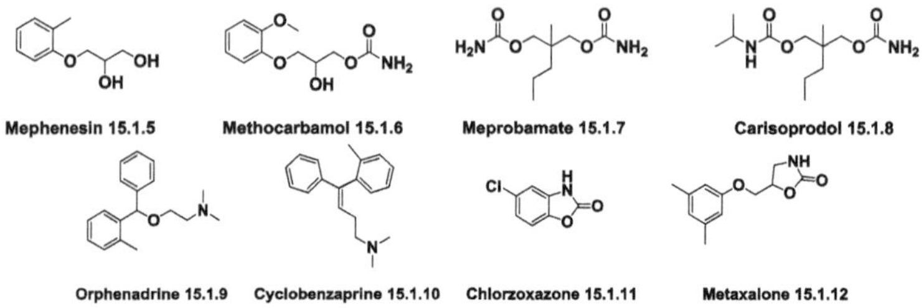

FIG. 15.2 **Antispasmodic skeletal muscle relaxants.**

Many classes of chemical compounds exhibit centrally acting skeletal muscle relaxant properties, including glycerol ether analogues of mephenesin (**15.1.5**) and methocarbamol (**15.1.6**); derivatives of 1,3-propandiols such as meprobamate (**15.1.7**) and carisoprodol (**15.1.8**); derivatives of diols; drugs that are structurally close to styramate (**15.1.13**); aminoalcohol derivatives such as orphenadrine (**15.1.9**) or compounds of the series (**15.1.14**) that are also considered as thiazole derivatives; heterocyclic compounds such as pyrrole derivatives (**15.1.15**); a variety of pyrazoles (**15.1.16**); oxazoles, of which the most prominent representative is metaxalone (**15.1.12**); thiadiazoles (**15.1.17**); oxadiazoles (**15.1.18**); benzofuran derivatives (**15.1.19**), which can be considered cyclized mephenesin derivatives; a variety of benzazoles (**15.1.20**), one representative of which is chlorzoxazone (**15.1.8**); benzoxazoles (**15.1.21**); quinazolines (**15.1.22**); and benzodiazepines such as diazepam (**15.1.8**), well reviewed decades ago [8] (Fig. 15.3.).

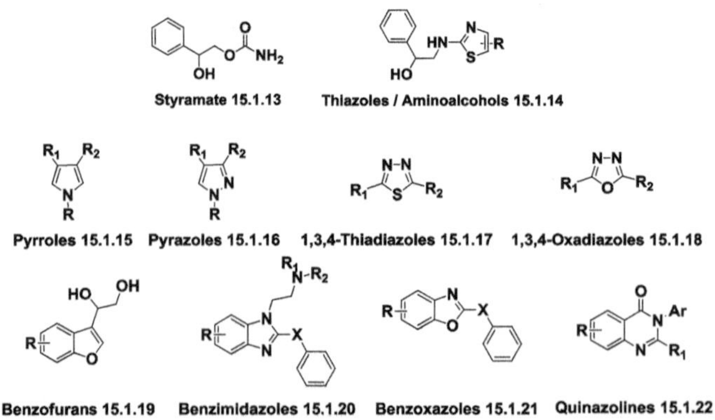

FIG. 15.3 **Different classes of chemical compounds which exhibit centrally acting skeletal muscle relaxants properties.**

Metaxalone (**15.1.12**) is the only muscle relaxant drug included in the list of Top 200 Drugs by sales for the 2010s.

Metaxalone–Skelaxin

Metaxalone (**15.1.12**), is a skeletal muscle relaxant, indicated as an adjunct to rest physical therapy to relax muscles and relieve pain and discomfort caused by muscle injuries (strains, sprains, and other) [9-12].

The most frequent reactions to Skelaxin (metaxalone) are drowsiness, dizziness, headache, nervousness, nausea, vomiting, and rash.

Metaxalone (**15.1.12**) was prepared firstly by the condensation of 3,5-dimethylphenol (**15.1.23**) with glycidol (**15.1.24**) to produce 3-(3,5-dimethylphenoxy)-1,2-propanediol (**15.1.25**). The obtained diol was rapidly heated with urea to the temperature of 180 to 200 °C and maintained there for several hours to produce the desired product, metaxalone (**15.1.12**) [13-15]. The close alternate procedure was proposed via the etherification of starting 3,5-dimethylphenol (**15.1.23**) with epichlorohydrin (**15.1.26**) which gives a mixture of 1-(3,5-dimethylphenoxy)-3-chloro-2-propanol (**15.1.27**) and 1-(3,5-dimethylphenoxy)-2,3-epoxypropane (**15.1.28**). The obtained mixture was reacted with benzylamine to produce a benzylamino alcohol derivative (**15.1.29**). The last was debenzylated by hydrogenation using Pd/C to produce the product (**15.1.30**). Cyclocondensation of (**15.1.30**) with dimethyl carbonate in the presence of a strong base produced the desired metaxalone (**15.1.12**) [16] (Scheme 15.1.).

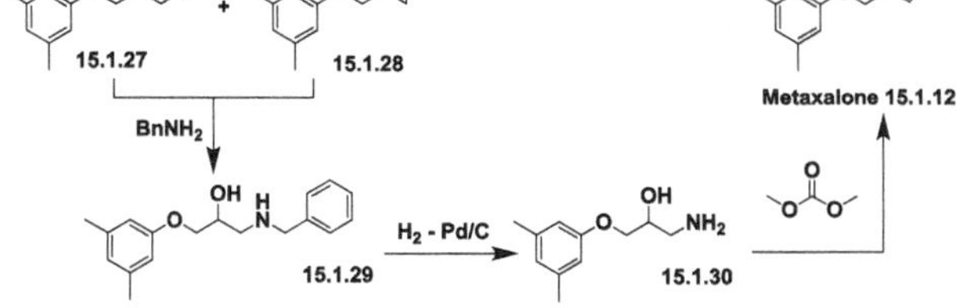

SCHEME 15.1 **Synthesis of metaxalone.**

Metaxalone (**15.1.12**) was also prepared by epoxide ring opening of triglycidyl isocyanurate (**15.1.31**) with 3,5-dimethylphenol (**15.1.23**) followed by rearrangement of intermediate (**15.1.32**) [17,18] (Scheme 15.2.).

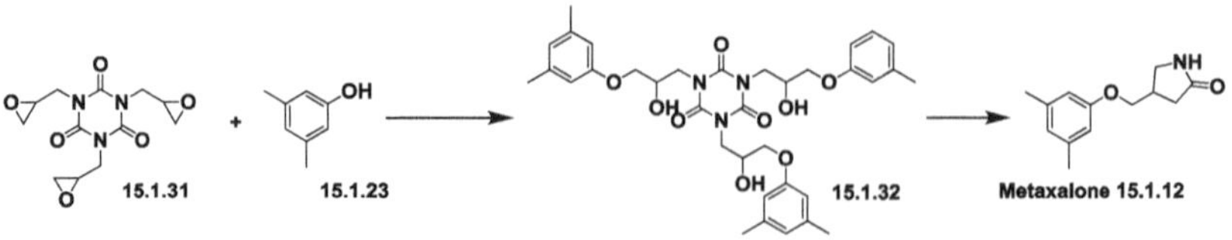

SCHEME 15.2 Synthesis of metaxalone by epoxide ring opening.

REFERENCES

1. Toth, P. P.; Urtis, J. Commonly used muscle relaxant therapies for acute low back pain: a review of carisoprodol, cyclobenzaprine hydrochloride, and metaxalone. *Clin. Ther.* **2004,** *26* (9), 1355–1367.
2. Booth, R. G. Drugs used to treat neuromuscular disorders: antiparkinsonian and spasmolytic agents. In *Foye's Principles of Medicinal Chemistry,* 6th ed.; Lemke, T. L., Williams, D. A., Eds. Lippincott Williams & Wilkins, 2008; pp 679–697.
3. Chou, R.; Peterson, K.; Helfand, M. Comparative efficacy and safety of skeletal muscle relaxants for spasticity and musculoskeletal conditions: a systematic review. *J. Pain Symptom Manage.* **2004,** *28* (2), 140–175.
4. Elenbaas, J. K. Centrally acting oral skeletal muscle relaxants. *Am. J. Hosp. Pharm.* **1980,** *37* (10), 1313–1323.
5. Waldman, H. J. Centrally acting skeletal muscle relaxants and associated drugs. *J. Pain Symptom Manage.* **1994,** *9* (7), 434–441.
6. See, S.; Ginzburg, R. Skeletal muscle relaxants. *Pharmacotherapy* **2008,** *28* (2), 207–213.
7. Meleger, A. L. Muscle relaxants and antispasticity agents. *Phys. Med. Rehabil. Clin. N. Am.* **2006,** *17* (2), 401–413.
8. Donahoe, H. B.; Kimura, K. K. Synthetic centrally acting skeletal muscle relaxants. In *Drugs Affecting the Central Nervous System;* Bueger, A., Ed.; Vol. 2, Marcel Dekker, 1968; pp 265–326.
9. Nicholson, B. Metaxalone in the therapy of muscle spasm. *Int. Cong. Symp. Series-Royal Soc. Med.* **2000,** *245,* 45–53 (Medical Management of Selected Neurological Disorders: Epilepsy, Spasticity and Pain).
10. Xiao, G. Metaxalone. In *Handbook of Metabolic Pathways of Xenobiotics;* Lee, P. W., Ed.; John Wiley and Sons, Inc, 2014; pp 1767–1768.
11. Bosak, A. R.; Skolnik, A. B. Serotonin syndrome associated with metaxalone overdose. *J. Med. Toxico.* **2014,** *10* (4), 402–405.
12. Bruce, R. B.; Turnbull, L.; Newman, J.; Pitts, J. Metabolism of metaxalone. *J. Med. Chem.* **1966,** *9* (3), 2868.
13. Lunsford, C. D.; Mays, R. P.; Richman, J. A., Jr.; Murphey, R. S. 5-Aryloxymethyl-2-oxazolidinones. *J. Am. Chem. Soc.* **1960,** *82,* 1166–1171.
14. Lunsford, C. D. 5-(o-Methoxyphenoxy)-2-oxazolidone, US 2895960 (1959).
15. A. H. Robins Inc. (Polysubstituted phenoxymethyl)-2-oxazolidones, GB 888594 (1962).
16. Breviglieri, G.; Contrini, S.; Bruno, G.; Assanelli, C. Four-step process for the preparation of metaxalon, US 6538142 (2003).
17. Lee, F.-Y.; Huang, T.-M.; Chung, C.-H. Preparation of 5-aryloxymethyl-2-oxazolidinones by reaction of triglycidyl isocyanurate with phenols, US 6562980 (2003).
18. Iacoangeli, T.; Chiavarini, M.; Fazio, A.; Marchetti, M.; Ciottoli, G. B. Method for the preparation of metaxalone, WO 2012104139 (2012).

Chapter 16

Antihistamine Drugs

From the labyrinth of intricate views on histamine it is possible to define histamine as an important neurotransmitter and mediator, formed by enzymatic decarboxylation of histidine in multiple processes in central and peripheral tissues. Histamine plays a pivotal role in allergic inflammation. It is released by the human immune system during allergic reactions, causing itching, swelling, and congestion, stimulating gastric secretion and constriction of bronchial smooth muscle, and dilating blood vessels, which causes a fall in blood pressure. It is involved also in the regulation of basic body functions, including energy and endocrine homeostasis, cognition and memory, and the sleep–wake cycle. The term *antihistamine* refers to drugs that antagonize the actions of histamine (**16.1.1**) [1-11] (Fig. 16.1.).

The name "histamine" means "an amine present in all tissues." It was identified as a major pathogenic mediator of allergic disorders in 1910s and 1920s, and the very first antihistamine—piperoxan (**16.1.2**) (Fig. 16.1.)—was introduced in beginning of the 1930s by Daniel Bovet and Ernest Fourneau at the Pasteur Institute. Bovet won the 1957 Nobel Prize in Physiology or Medicine "for his discoveries relating to synthetic compounds that inhibit the action of certain body substances, and especially their action on the vascular system and the skeletal muscles."

Allergies are caused by a hypersensitivity reaction of the antibody class immunoglobulin E (IgE). When an allergen is encountered, it binds to IgE, which excessively activates the mast cells or basophils, leading them to release massive amounts of histamine, which lead to inflammatory responses ranging from runny nose to anaphylactic shock.

FIG. 16.1 Structure of histamine and the first antihistamine.

Histamine exerts its effects through its interaction with four major types of receptors (H_1, H_2, H_3, H_4) belonging to the superfamily of G-protein–coupled receptors (GPCRs). H_1 and H_2 receptors are widely expressed, in contrast to H_3 and H_4 receptors, which are not. All four are structurally similar and are classified as GPCRs [12-15].

Synthesis of Best-Seller Drugs. http://dx.doi.org/10.1016/B978-0-12-411492-0.00016-X

H_1 receptors have been identified in central nervous system (CNS) neurons, in endothelial and epithelial cells, and in the bronchial, vascular, and gastrointestinal smooth muscle. They are responsible for the constriction of bronchial, vascular, and gastrointestinal smooth muscle; for the activation of the afferent vagal nerves of the airways and the cough receptors; for the increase in vascular permeability; and for local irritative manifestations. H_1 histamine receptor blockers are widely used in allergic rhinitis, conjunctivitis, urticarial, and other allergic diseases.

H_2 receptors are present in the gastric mucosa, the uterus, and the brain. They also increase vascular permeability and stimulate secretion of the gastric acid. Compounds of this class are used in ulcer diseases and gastroesophageal reflux disease.

H_3 receptors are located in the brain and in the bronchial smooth muscle. They are responsible for cerebral vasodilation and might be involved in the feedback system, inhibiting their own synthesis and release of histamine from the nerve endings. The H_3 antagonists, currently in the phase of laboratory experimentation, and could find their use in the therapy of processes affecting the CNS such as Alzheimer disease, Parkinson disease, attention-deficit hyperactivity disorder (ADHD), schizophrenia, epilepsy, and obesity.

No H_3 antagonists have been approved for use as of this writing, but several compounds for allergic rhinitis are undergoing clinical trials.

H_4 receptors are present in the CNS: neutrophils, monocytes, Langerhans cells, mast cells, and endocrine cells.

Several compounds that may be potentially useful for treatment of atopic dermatitis/eczema, allergic rhinitis, and other inflammatory disorders are in the laboratory experimentation phase.

Since the discovery of piperoxan (**16.1.2**), more than 40 antihistamine compounds have entered the market.

16.1 H_1 ANTIHISTAMINES

H_1 antihistamines were and remain the first-line medications for the treatment of allergic diseases, rhinoconjunctivitis, and urticarial [16-25].

H_1 antihistamines, in turn, are classified as being of the first-, second-, or third-generation class [26-29].

The first-generation antihistamines do cause marked sedation, CNS dysfunction, and anticholinergic adverse effects, resulting in performance or cognitive function impairment and therapy nonadherence.

According to the World Health Organization (WHO) classification, first-generation antihistamines are divided into six main chemical groups:

1. *Aminoalkyl ethers*—diphenhydramine (**16.1.3**), orphenadrine (**16.1.4**), chlorodiphenhydramine (**16.1.5**), bromazine (**16.1.6**), carbinoxamine (**16.1.7**), doxylamine (**16.1.8**), embramine (**16.1.9**), clemastine (**16.1.10**), and phenyltoloxamine (**16.1.11**) (Fig. 16.2.).

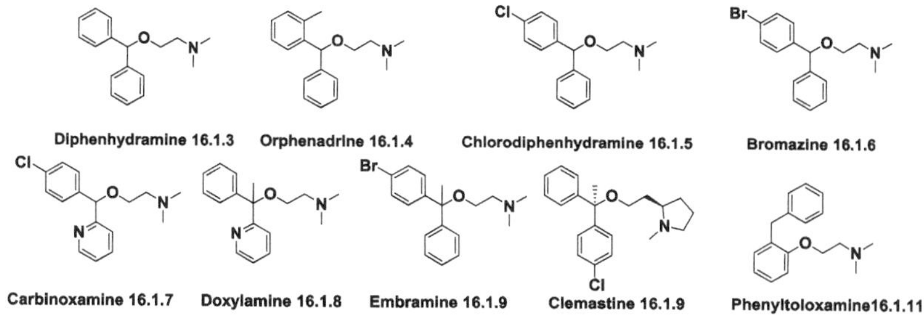

FIG. 16.2 **Structure of first-generation antihistamines–aminoalkyl ethers.**

2. *Substituted alkylamines*—pheniramine (**16.1.12**), chlorpheniramine (**16.1.13**), brompheniramine (**16.1.14**), dimetindene (**16.1.15**), triprolidine (**16.1.16**), and acrivastine (**16.1.17**) (Fig. 16.3.).

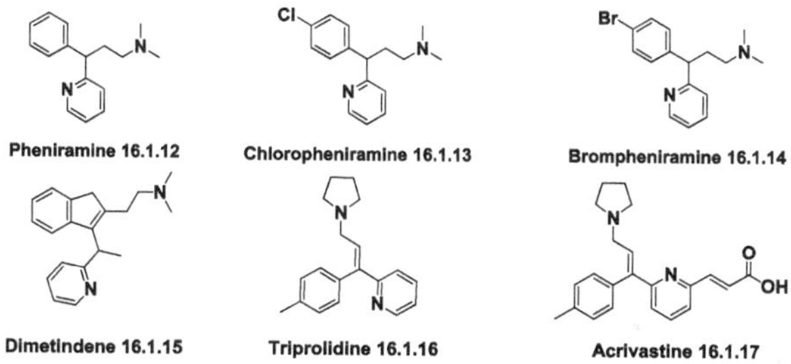

FIG. 16.3 **Structure of first-generation antihistamines–substituted alkylamines.**

3. *Substituted ethylenediamines*—tripelennamine (**16.1.18**), chloropyramine (**16.1.19**), mepyramine (**16.1.20**), methapyrilene (**16.1.21**), and antazoline (**16.1.22**) (Fig. 16.4.).

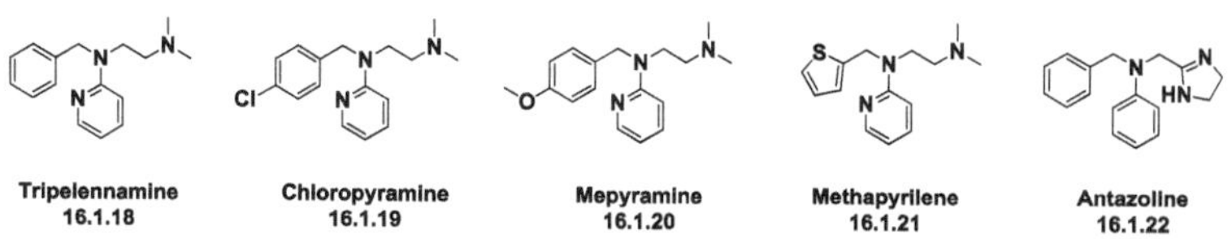

FIG. 16.4 **Structure of first-generation antihistamines–ethylenediamines.**

4. *Piperazine derivatives*—cyclizine (**16.1.23**), meclozine (**16.1.24**), buclizine (**16.1.25**), cetirizine (**16.1.26**), hydroxyzine (**16.1.27**), levocetirizine (**16.1.28**), and quetiapine (**16.1.29**) (Fig. 16.5.).
5. *Phenothiazine derivatives*—promethazine (**16.1.30**), alimemazine (**16.1.31**), and methdilazine (**16.1.32**) (Fig. 16.6.).
6. *Others,* including cyproheptadine (**16.1.33**), desloratadine (**16.1.34**), loratadine (**16.1.35**), rupatadine (**16.1.36**) alcaftadine (**16.1.37**), ketotifen

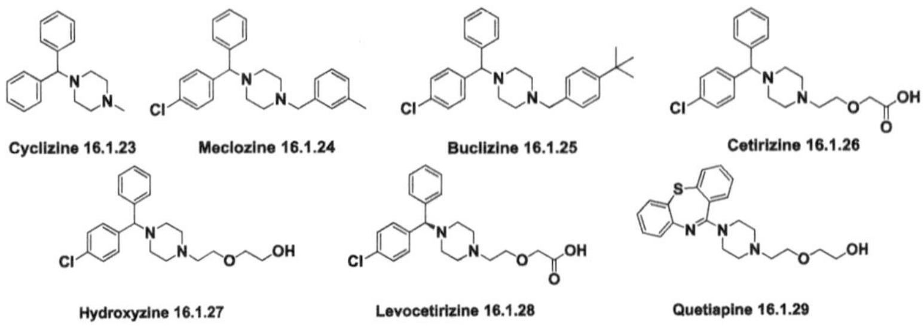

FIG. 16.5 **Structure of first-generation antihistamines–piperazine derivatives.**

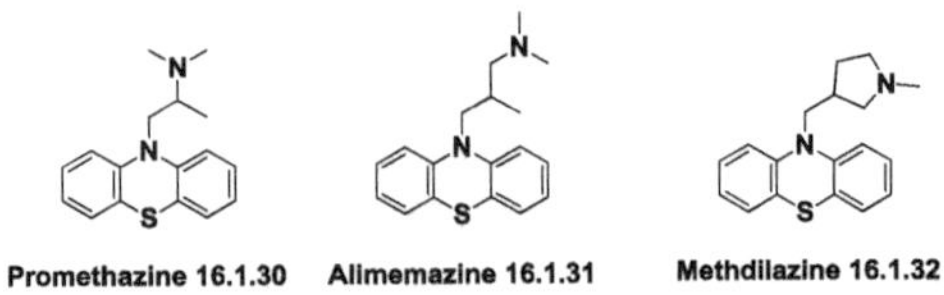

FIG. 16.6 **Structure of first-generation antihistamines–phenothiazine derivatives.**

(**16.1.38**), ebastine (**16.1.39**), bepotastine (**16.1.40**), fexofenadine (**16.1.41**), terfenadine (**16.1.42**), levocabastine (**16.1.43**), bilastine (**16.1.44**), astemizole (**16.1.45**), mizolastine (**16.1.46**), phenindamine (**16.1.47**), azelastine (**16.1.48**), mirtazapine (**16.1.49**), olopatadine (**16.1.50**), and quifenadine (**16.1.51**). The WHO classification of first-generation antihistamines probably could be divided into seven main chemical groups, adding piperidine derivatives group for compounds (**16.1.31** to **16.1.47**), leaving only compounds (**16.1.48** to **16.1.51**) to be included in the "others" group (Fig. 16.7.).

H_1-receptor antagonists are effective in treatment of acute allergic urticaria; anaphylactic or anaphylactoid shock; allergic angioedema; prevention and treatment of allergic and pseudoallergic reactions caused by drugs; seasonal allergic rhinitis; severe allergic reactions to alimentary products; and serum sickness. In addition, some antihistamines block cholinergic, muscarinic, and serotonin receptors, and can be used in sleep disorders, motion sickness, nausea, and dizziness. First-generation antihistamines easily permeate the blood–brain barrier, which explains the sedative effect.

The common adverse reactions related to these preparations are sedation, headache, dizziness, weakness, somnolence/insomnia, anxiety, irritability, tremor, seizures, paresthesia, retardation, dystaxia, dry mouth, nausea, diarrhea, upset stomach, nasal congestion, breathing disorders, hypotension, tachycardia, agranulocytosis, thrombocytopenia, and photosensitivity.

The second-generation antihistamines, including dimetindene (**16.1.15**), acrivastine (**16.1.17**), cetirizine (**16.1.26**), loratadine (**16.1.35**), ketotifen (**16.1.38**), bepotastine (**16.1.40**), fexofenadine (**16.1.41**), terfenadine (**16.1.42**), levocabastine (**16.1.43**), astemizole (**16.1.44**), mizolastine (**16.1.45**),

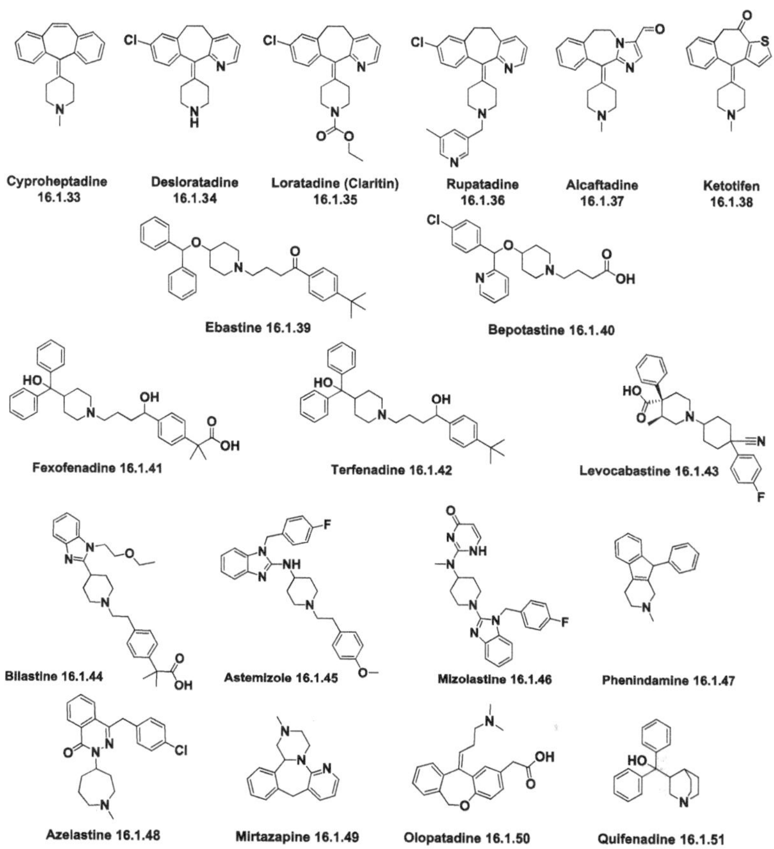

FIG. 16.7 Structure of unclassified first-generation antihistamines.

azelastine (**16.1.48**), mirtazapine (**16.1.49**), olopatadine (**16.1.50**), and quifenadine (**16.1.51**) have been known as the "nonsedating antihistamines." They were developed in the 1980s to minimize these side effects. They have almost no sedative and cholinolytic effects, however, they may express cardiotoxic effects (Fig. 16.8.).

The second-generation H_1-receptor antagonists and are used for treatment of bronchial asthma, atopic dermatitis, pollen fever, and allergic rhinitis. Terfenadine and astemizole are now withdrawn from the market because of serious cardiovascular side effects.

The third-generation antihistamines—fexofenadine (**16.1.41**), desloratadine (**16.1.52**), and levocetirizine (**16.1.53**)—are the newest antihistamines. Desloratadine is an active metabolite of loratadine. These agents have been used clinically to treat various allergic disorders such as seasonal or perennial allergic rhinitis and chronic urticaria. Their main characteristic is the inability to express cardiotoxic effects [29] (Fig. 16.9.).

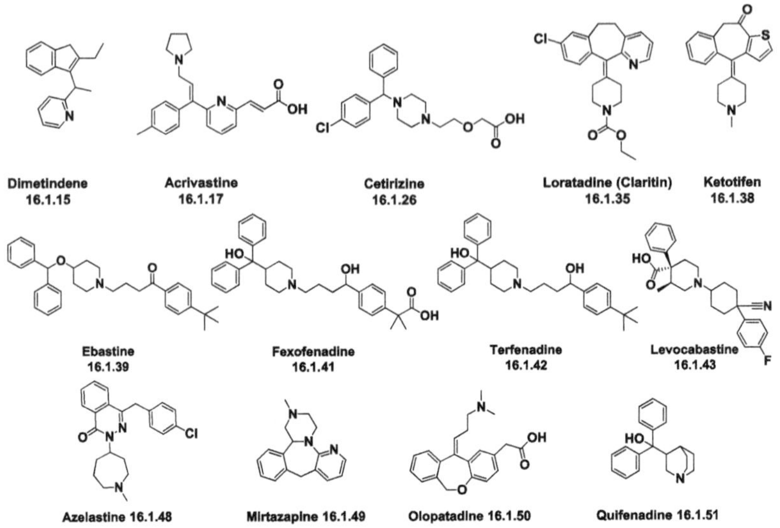

FIG. 16.8 The second-generation antihistamines.

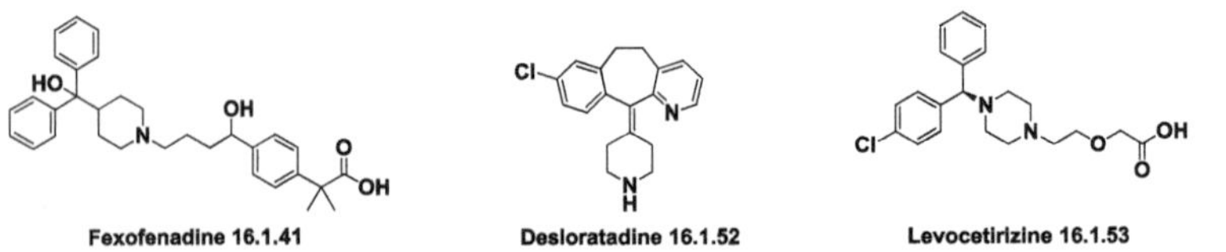

FIG. 16.9 The third-generation antihistamines.

The third-generation H_1-receptor antagonists are used in patients with perennial allergic rhinitis, seasonal allergic rhinitis (conjunctivitis) with a long duration chronic urticaria, atopic dermatitis, and allergic contact dermatitis. An important benefit of this group of antihistamine drugs is the absence of sedative effects.

The common adverse effects of second- and third-generation H_1-receptor antagonists include xerostomia, nausea, vomiting, and constipation. Sometimes they may cause mild headache, dizziness, insomnia, agitation, lethargy, fatigue, mood disturbance, paresthesia, etc.

Two H_1-antihistamine drugs, the second-generation olopatadine (**16.1.50**) and the third-generation levocetirizine (**16.1.53**), are included in the list of Top 200 Drugs by sales for the 2010s.

Olopatadine–Patanol

Olopatadine (**16.1.50**) was proposed to synthesize by different alternative methods, all of which were based on the previously described ketones 2-(11-oxo-6,11-dihydrodibenzo[b,e]oxepin-2-yl)acetic acid (**16.1.54**) and appropriate methyl ester (**16.1.55**). Formation of a double bond in position 11 of the mentioned oxepines was accomplished by means of Grignard or Wittig reactions. For the implementation of the Grignard reaction the carboxyl group of acid (**16.1.54**) was protected via transformation to an oxazoline derivative (**16.1.57**) using an excess of 2-amino-2-methylpropan-1-ol and thionyl chloride. The obtained product was allowed to react with 3-(dimethylamino) propyl magnesium chloride to produce alcohol (**16.1.58**), which was dehydrated using a thionyl chloride–pyridine combination to prepare the compound (**16.1.59**). Olopatadine (**16.1.50**) was prepared from this compound or via transformation to olopatadine methyl ester (**16.1.60**) in methanol TsOH mixture followed by basic hydrolysis with sodium hydroxide to desired olopatadine (**16.1.50**), or by direct hydrolysis of the oxazoline derivative (**16.1.57**) in acidic water media.

Two closely related schemes were employed for the synthesis of olopatadine (**16.1.50**). Both of them implement Wittig reaction to produce olopatadine starting from methyl ester (**16.1.55**). Wittig olefination of this compound with [N,N-(dimethylamino)propyl]triphenylphosphonim bromide produced the olopatadine methyl ester (**16.1.60**), which was hydrolyzed to olopatadine (**16.1.50**). Olefination of methyl ester (**16.1.53**) in Wittig conditions with [3-[(tetrahydro-2H-pyran-2-yl)oxy]propyl]triphenylphosphonium bromide produced the compound (**16.1.61**), which, on treatment with a mixture of TsOH in water and dioxane, was hydrolyzed to alcohol (**16.1.62**). The obtained alcohol was mesylated to produce compound (**16.1.63**), which on amination with dimethylamine with subsequent basic hydrolysis produced the desired olopatadine (**16.1.50**) [30,31] (Scheme 16.1.).

In all methods described, a mixture of (E/Z) in different ratios of isomers were obtained. Separation was done on different stages of synthesis by chromatography or by fractional crystallization. Difference between each geometrical isomers of olopatadine was small. The activities of the E isomer both in the bronchoconstriction and in the H_1 receptor binding higher than those of the Z isomer. Structurally, olopatadine-dibenz[*b,e*]oxepin-2-acetic acid, 11-[3-(dimethylamino)propylidene]-6,11-dihydro-(11Z) (**16.1.60**) is a compound bearing a Z-configured (dimethylamino)propylidene substituent and an acetic acid chain at the C-11 and C-2 positions, respectively, of the tricyclic system. It is usually obtained as a precipitate on desalination of products after saponification of the crude mixture (E/Z = 1/2) of olopatadine methyl ester (**16.1.60**). Stereoselective syntheses of olopatadine have been proposed [32].

Olopatadine is a novel antiallergic drug [33-41] that is a selective histamine H_1-receptor antagonist possessing inhibitory effects on the release of inflammatory lipid mediators, such as leukotriene and thromboxane, from human polymorphonuclear leukocytes and eosinophils. Olopatadine also inhibits the tachykininergic contractions in guinea pig bronchi by prejunctional inhibition of peripheral sensory nerves.

Possible side effects are blurred vision, burning/stinging/redness/dryness of the eye, and headache.

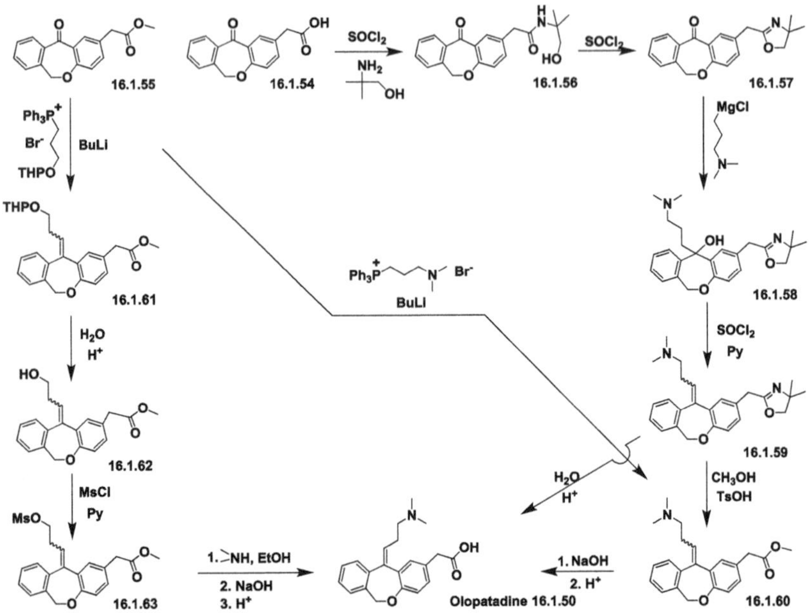

SCHEME 16.1 Synthesis of olopatadine.

Levocetirizine–Xyzal

The levorotatory enantiomer of a second-generation H_1-receptor antagonist cetirizine (**16.1.26**) [42,43] displays an improved pharmacological profile compared to the racemate. Levocetirizine (**16.1.53**) was obtained by a classical resolution of racemic cetirizine using D-(-)-tartaric acid. The early syntheses of racemic cetirizine [44,45] and its isomer's separation [45] are based on a general scheme of synthesis started from 4-chlorobenzhydrylchloride (**16.1.64**), which, on reaction with ethyl piperazine-1-carboxylate (**16.1.65**), produced compound (**16.1.66**), which underwent acidic hydrolysis to produce the benzhydrylpiperazine derivative (**16.1.67**). The last was reacted with methyl 2-(2-chloroethoxy) acetate to produce the product (**16.1.68**). The obtained ester was easily converted into a free acid cetirizine (**16.1.26**) and the desired product, levocetirizine (**16.1.53**), was isolated via crystallization of D-(-)-tartaric salt from the racemic mixture (Scheme 16.2.).

Alternate enantioselective syntheses of each isomer of cetirizine have been proposed [46-49]. One alternate [46] started from each isomer of 4-chlorobenzhydrylamine (**16.1.69**), separated with the use of (-)-, or (+)-tartaric acids (R)-(4-chlorophenyl)(phenyl)methanamine (**16.1.70**) was reacted with N,N-bis(2-chloroethyl)-4-methylbenzenesulfonamide (**16.1.71**) in boiling diisopropylethylamine to produce a tosyl derivative (**16.1.72**), which easily crystallized from ethanol. Reductive removal of the N-tosyl group occurred with the use of 4-hydroxybenzoic acid (phenol component) in HBr/CH_3COO. After the deprotection reaction was complete, a very clean product (**16.1.73**) was obtained. The last was alkylated with 2-(2-chloroethoxy)acetamide to produce (**16.1.74**), which was hydrolyzed in hydrochloric acid to produce the desired levocetirizine (**16.1.53**).

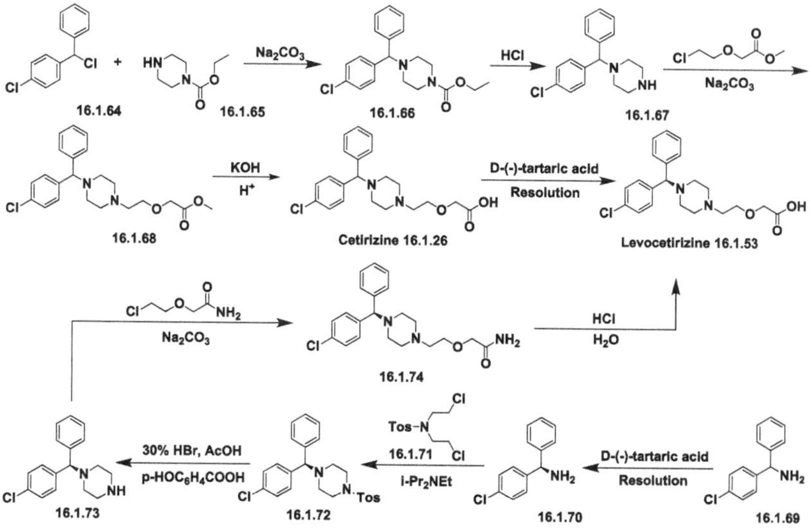

SCHEME 16.2 Synthesis of levocetirizine.

Levocetirizine [50-53] represents a third generation of antihistamines, which was developed from the second-generation antihistamine cetirizine with a higher affinity and selectivity to H_1 receptors than its racemate cetirizine. It is rapidly absorbed and minimally metabolized and has been found effective for relieving symptoms of allergic rhinitis, including nasal congestion, sneezing, itching, and redness and tearing of the eyes caused by hay fever, with minor side effects. It is specifically indicated for the relief of symptoms associated with allergic rhinitis. The most common side effects are drowsiness, low energy, and feeling weak.

16.2 H_2 ANTIHISTAMINES

Histamine plays a big role in regulating the secretion of hydrochloric acid in the stomach and since the 1970s a new class of synthetic drugs that blocks the action of histamine at H_2 receptors entered the pharmaceutical market. H_2 receptors have specific locations in the body, mainly in gastric parietal cells; a low level can be found in vascular smooth muscle, neutrophils, CNS, heart, and uterus.

H_2-histamine drugs antagonize the action of histamine in stimulating acid secretion and in blocking other stimulants of acid secretion. They are used to treat peptic ulcer disease and gastroesophageal reflux disease.

H_2-receptor blockers are amongst the most commonly prescribed medications in the world and available as over-the-counter preparations in some dosage forms. Their most important action is a reduction in gastric acid secretion as a result of H_2-receptor blockade [54-57].

The prototypical H_2 antagonist is cimetidine (**16.2.1**). Ranitidine (**16.2.2**), nizatidine (**16.2.3**), and famotidine (**16.2.4**) are the most used structural analogues of cimetidine. Famotidine is approximately eight times more potent than ranitidine and 40 times more potent than cimetidine. Other representatives of H_2 antagonists—burimamide (**16.2.5**), metiamide (**16.2.6**), and tiotidine (**16.2.7**)—are used much less frequently (Fig. 16.10.).

Unlike the earlier agents in this class, certain of the newer H_2-receptor antagonists (so-called second-generation H_2-receptor antagonists) have a multimodal mechanism of action. It seems that they not only suppress gastric acid secretion, but by activating mucosal defense mechanisms, induce collagen synthesis in the gastric mucosa. Roxatidine (**16.2.8**) and lafutidine (**16.2.9**) represent second-generation H_2 histamine antagonists and belong to absolutely no other chemical group of compounds.

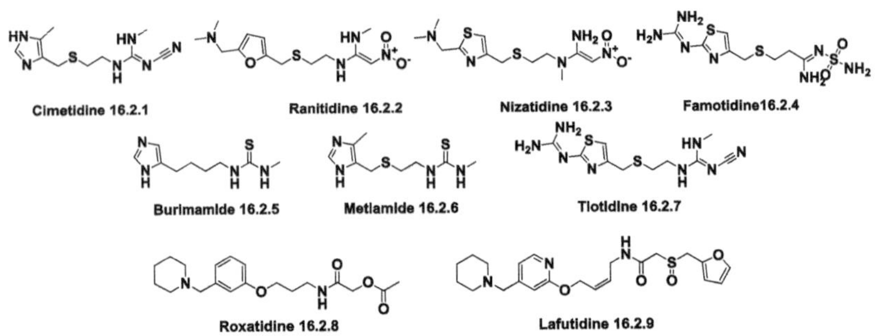

FIG. 16.10 **H_2 antihistamines.**

H_2-receptor blockers are used to treat conditions associated with excess amounts of stomach acid, and are used to treat the duodenal and gastric ulcers, and for therapy and maintenance of gastroesophageal reflux disease, pathological hypersecretory conditions such as Zollinger-Ellison syndrome, upper gastrointestinal bleeding, heartburn, and sour stomach.

H_2-receptor block side effects include abdominal stomach pain; blistering; burning; redness; scaling or tenderness of skin; blurred vision; fast, pounding, or irregular heartbeat; fever and/or flu-like symptoms; and a general feeling of discomfort or illness.

16.3 H_3 ANTIHISTAMINES

After the immense success of the H_1 and H_2 antihistamine blockbuster drugs developed in the 1960s and 1970s, the histamine receptor family was enriched with the discovery of the third histamine receptor, H_3, in 1983 [58], which belonged to the family of GPCRs. It turned out to be an interesting target for the modulation of a variety of important processes.

The early H_3 antagonists were based on the 4-substituted imidazole motif. In 1987, the first H_3R antagonist thioperamide (**16.3.1**), which did not exhibit significant interactions with H_1R and H_2R receptor subtypes, was synthesized. Thereafter, compounds of diverse structures came: clobenpropit (**16.3.2**) and ciproxifan (**16.3.3**). Cipralisant (**16.3.4**) was one of the first H_3 compounds to advance to the clinic, reaching Phase II enrollment for ADHD before progress was halted in 2002 (Fig. 16.11.).

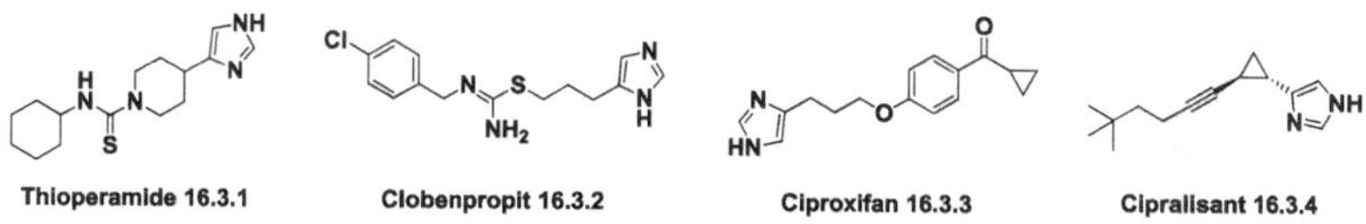

FIG. 16.11 **H_3 antihistamines.**

Many other ligands with nanomolar affinity binding parameters to human H_3 receptors were designed, synthesized, and proposed as H_3 antagonists (**16.3.5-16.3.16**) (Fig. 16.12.).

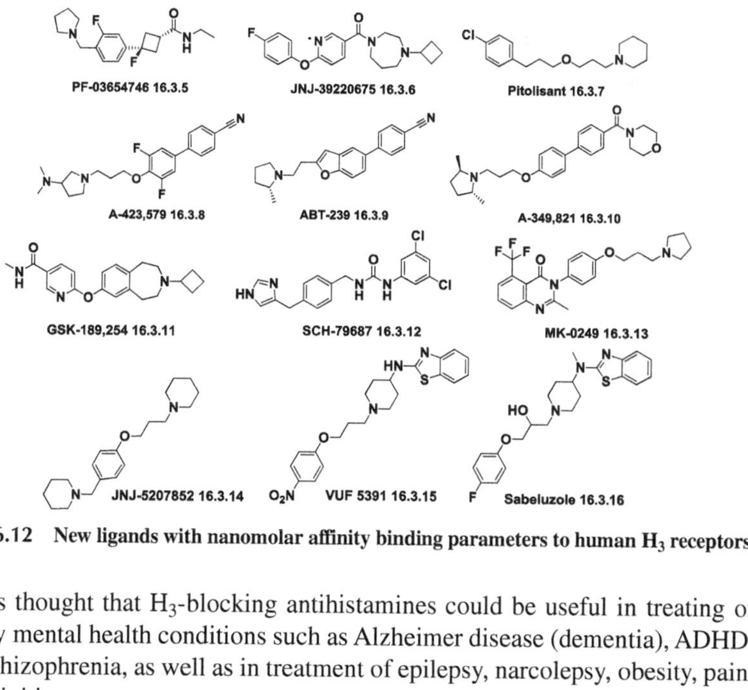

FIG. 16.12 **New ligands with nanomolar affinity binding parameters to human H_3 receptors.**

It's thought that H_3-blocking antihistamines could be useful in treating of variety mental health conditions such as Alzheimer disease (dementia), ADHD, and schizophrenia, as well as in treatment of epilepsy, narcolepsy, obesity, pain, and rhinitis.

Although some compounds have reached and passed different stages of clinical trials, there are no H_3-antagonist drugs on the pharmaceutical market. For example compounds AZD-5213 and SAR-110894 (structures undisclosed) are both in Phase I trials, ABT-288, MK-0249 (structures undisclosed), and PF-03654746 (**16.3.5**) are in Phase II trials as remedies for Alzheimer disease.

ABT-288 and PF-03654746, as well as GSK-239512 and MK-0249 (structures undisclosed), are in Phase II trials as medications for use in schizophrenia.

PF-03654746 is in Phase II trials as a medication for treatment of ADHD. Pitolisant (**16.3.7**) has advanced to several Phase III trials for treatment of sleep-related problems.

Drug discovery efforts targeting the histamine H_3 receptor as an attractive GPCR drug target that regulates neurotransmission in the CNS and plays a role in cognitive and homeostatic functions is still the focus of academic researchers and numerous pharmaceutical companies. The results of their research are summarized in multiple reviews [59-75].

16.4 H_4 ANTIHISTAMINES

Further investigations in histamine receptor research resulted in the discovery and disclosure of the fourth human histamine receptor, H_4, which is preferentially expressed in mast cells and various cells of the immune system [76]. It also has been identified on dendritic cells, basophils, and T lymphocytes, suggesting that it may play an important role in inflammatory responses. The discovery of the H_4 receptor by several groups lead to the reevaluation of the physiological role for histamine.

Steady output of patents and publications directed to search for H_4R ligands started to demonstrate the therapeutic potential of such compounds. Thioperamide (**16.3.1**), originally developed as an H_3 antagonist, has an H_4R affinity, acting as an H_4R inverse agonist. The H_4R high-throughput screening led to identification of JNJ-7777120 (**16.4.1**). Benzimidazole (**16.4.2, 16.4.3**) and thienopyrrole (**16.4.4**) derivatives were prepared as bioisosteres of the indole carboxamides, which showed affinities close to that of indole (**16.4.1**). In the search for novel H_4 antagonists or inverse agonists, a variety of ligand classes have been identified. Among them are series of derivatives of aminopyrimidines (**16.4.5**) and benzofused aminopyrimidines–quinazolines and related compounds (**16.4.6** to **16.4.10**), quinoxalines and related compounds (**16.4.11**), oxazepine (**16.4.12**), analogues of the H_2 agonist S-(2-guanidylethyl)isothiourea derivative dimaprit (**16.4.13**), and analogues compounds (**16.4.14-16.4.16**), series of imidazolyl cyclopropanes (**16.4.17**) (Fig. 16.13.).

The discovery of selective ligands for the H_4 receptor was crucial to the uncovering of the receptor's function. These compounds show that the H_4 receptor plays a role in mast cell and eosinophil chemotaxis, as well as cytokine production in dendritic and T cells.

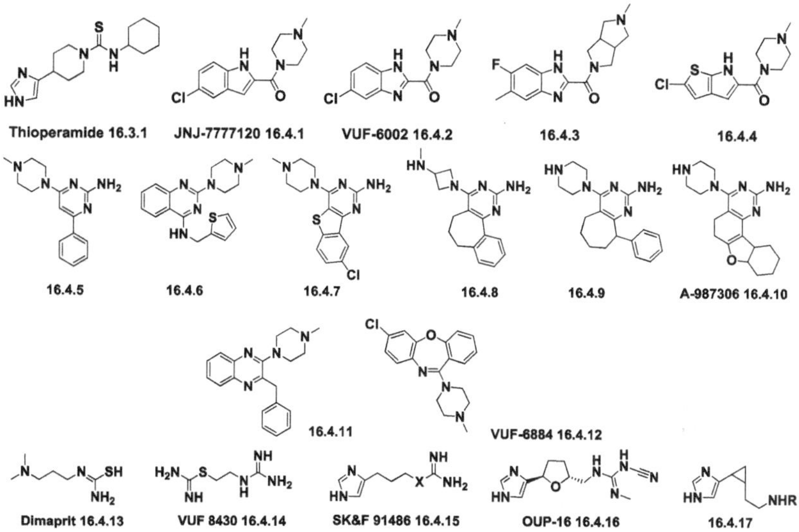

FIG. 16.13 **H_4 antihistamines.**

H_4-receptor antagonists have shown efficacy in a variety of inflammatory animal models, including in peritonitis, colitis, and airway inflammation models. These data suggest that the H_4 receptor is an attractive target for possible treatment of inflammation, allergy, and asthma.

Preliminary biological results in animal models suggest that H_4-receptor antagonists have definite clinical potential, especially for allergic and inflammatory conditions in diseases for which H_1-receptor antagonists are ineffective. H_4-receptor antagonists could become remedies for the treatment of chronic inflammatory disorders, such as allergic diseases of the respiratory and gastrointestinal tracts and of the skin.

The majority of the pharmacological research that has been published in the literature on the H_4 receptors has been carefully reviewed [77-89].

REFERENCES

1. Parsons, M. E.; Ganellin, C. R. Histamine and its receptors. *Br. J. Pharmacol.* **2006,** *147* (Suppl. 1), S127–S135.
2. Hough, L. B.; Leurs, R. Histamine receptors. In *Understanding G Protein-Coupled Receptors and Their Role in the CNS;* Pangalos, M. N., Davies, C. H., Eds.; Oxford University Press, 2002; pp 307–348.
3. de Esch, I.; Timmerman, H.; Leurs, R. Histamine receptors. In *Textbook of Drug Design and Discovery,* 4th ed.; Krogsgaard-Larsen, P., Stroemgaard, K., Madsen, U., Eds. CRC Press, 2010; pp 283–297.
4. Simons, F.; Estelle, R.; Simons; Keith, J. Antihistamines. In 2nd ed.; Kay, A. B., Bousquet, J., Holt, P. G., Kaplan, A. P., Eds.; *Allergy and Allergic Diseases*, Vol. 1; Wiley-Blackwell, 2008; pp 551–565.

5. Bovill, J. G. Histamine and histamine antagonists. In *Pharmacology for Anaesthesiologists;* Fee, J. P. H., Bovill, J. G., Eds.; Taylor & Francis, 2005; pp 223–230.
6. Leurs, R.; Vischer, H. F.; Wijtmans, M.; de Esch, I. J. P. En route to new blockbuster anti-histamines: surveying the offspring of the expanding histamine receptor family. *Trends Pharmacol. Sci.* **2011,** *32* (4), 250–257.
7. Keller, G. A.; Di Girolamo, G. Antihistamines: past answers and present questions. *Curr. Drug Saf.* **2010,** *5* (1), 58–64.
8. Centanni, S.; Santus, P. The role of antihistamine drugs in allergic disorders. *Curr. Med. Chem.: Anti-Inflammatory Anti-Allergy Agents* **2003,** *2* (3), 284–295.
9. Bielory, L.; Lien, K. W.; Bigelsen, S. Efficacy and tolerability of newer antihistamines in the treatment of allergic conjunctivitis. *Drugs* **2005,** *65* (2), 215–228.
10. Toemoeskoezi, Z. Histamine agonists, antagonists, and inverse agonists. In *Histamine: Biology and Medical Aspects;* Falus, A., Ed.; Karger, 2004; pp 78–88.
11. Akdis, C. A.; Simons, F.; Estelle, R. Histamine receptors are hot in immunopharmacology. *Eur. J. Pharmacol.* **2006,** *533* (1–3), 69–76.
12. Walter, M.; Stark, H. Histamine receptor subtypes: a century of rational drug design. *Front. Biosci., Scholar Ed.* **2012,** *S4* (2), 461–488.
13. Bongers, G.; de Esch, I.; Leurs, R. Molecular pharmacology of the four histamine receptors Adv. *Exp. Med. Biol.* **2010,** *709* (Histamine in Inflammation), 11–19.
14. Strasser, A.; Wittmann, H.-J.; Buschauer, A.; Schneider, E. H.; Seifert, R. Species-dependent activities of G-protein-coupled receptor ligands: lessons from histamine receptor orthologs. *Trends Pharmacol. Sci.* **2013,** *34* (1), 13–32.
15. Oppenheimer, J. J.; Casale, T. B. Next generation antihistamines: therapeutic rationale, accomplishments and advances. *Expert Opin. Invest. Drugs* **2002,** *11* (6), 807–817.
16. Passalacqua, G.; Canonica, G. W.; Bousquet, J. Structure and classification of H_1-antihistamines and overview of their activities. *Clin. Allergy Immunol.* **2002,** *17,* 65–100.
17. Vena, G. A.; Cassano, N.; Buquicchio, R.; Ventura, M. T. Antiinflammatory effects of H_1-antihistamines: clinical and immunological relevance. *Curr. Pharm. Des.* **2008,** *14* (27), 2902–2911.
18. Howarth, P. H. The choice of an H_1-antihistamine for the 21st century. *Clin. Exp. Allergy Rev.* **2002,** *2* (1), 18–25.
19. Simons, F. E. R. Advances in H_1-Antihistamines. *N. Engl. J. Med.* **2004,** *351* (21), 2203–2217.
20. Sharma, A.; Hamelin, B. A. Classic histamine H_1 receptor antagonists: a critical review of their metabolic and pharmacokinetic fate from a bird's eye view. *Curr. Drug Metab.* **2003,** *4* (2), 105–129.
21. Kalpaklioglu, F.; Baccioglu, A. Efficacy and safety of H_1-antihistamines: an update. *Anti-Inflammatory Anti-Allergy Agents Med. Chem.* **2012,** *11* (3), 230–237.
22. Simons, F.; Estelle, R.; Simons, K. J. Histamine and H_1-antihistamines: celebrating a century of progress. *J. Allergy Clin. Immunol.* **2011,** *128* (6), 1139–1150.
23. Thurmond, R. L.; Gelfand, E. W.; Dunford, P. J. The role of histamine H_1 and H_4 receptors in allergic inflammation: the search for new antihistamines. *Nat. Rev. Drug Discovery* **2008,** *7* (1), 41–53.
24. Beaton, G.; Moree, W. J. The expanding role of H_1 antihistamines: a patent survey of selective and dual activity compounds 2005–2010. *Expert Opin. Ther. Pat.* **2010,** *20* (9), 1197–1218.
25. H_1 Antihistamines: a review. *I., J. Investig. Allergol. Clin. Immunol.* **1999,** *14* (5), 300–312.
26. Slater, J. W.; Zechnich, A. D.; Haxby, D. G. Second-generation antihistamines: a comparative review. *Drugs* **1999,** *57* (1), 31–47.
27. Walsh, G. M.; Annunziato, L.; Frossard, N.; Knol, K.; Levander, S.; Nicolas, J.-M.; Taglialatela, M.; Tharp, M. D.; Tillement, J. P.; Timmerman, H. New insights into the second-generation antihistamines. *Drugs* **2001,** *61* (2), 207–236.

28. Tillement, J.-P. Pharmacological profile of the new antihistamines. *Clin. Exp. Allergy Rev.* **2005,** *5* (1), 7–11.
29. Handley, D. A.; Magnetti, A.; Huggins, A. J. Therapeutic advantages of third generation antihistamines. *Expert Opin. Invest. Drugs* **1998,** *7* (7), 1045–1054.
30. Oshima, E.; Kumazawa, T.; Otaki, S.; Obase, H.; Ohmori, K.; Ishii, H.; Manabe, H.; Tamura, T.; Shuto, K. Dibenz[b,e]oxepin derivatives, procedure for their preparation, and their use as antiallergic and antiinflammatory agents, EP 235796 (1987).
31. Ohshima, E.; Otaki, S.; Sato, H.; Kumazawa, T.; Obase, H.; Ishii, A.; Ishii, H.; Ohmori, K.; Hirayama, N. Synthesis and antiallergic activity of 11-(aminoalkylidene)-6,11-dihydrodibenz[b,e]oxepin derivatives. *J. Med. Chem.* **1992,** *35* (11), 2074–2084.
32. Bosch, J.; Bachs, J.; Gomez, A. M.; Griera, R.; Ecija, M.; Amat, M. Stereoselective syntheses of the antihistaminic drug olopatadine and its e-isomer. *J. Org. Chem.* **2012,** *77* (14), 6340–6344.
33. Ohmori, K.; Hayashi, K.-I.; Kaise, T.; Ohshima, E.; Kobayashi, S.; Yamazaki, T.; Mukouyama, A. Pharmacological, pharmacokinetic and clinical properties of olopatadine hydrochloride, a new antiallergic drug. *Jpn. J. Pharmacol.* **2002,** *88* (4), 379–397.
34. Ohmori, K.; Hasegawa, K.; Tamura, T.; Miyake, K.; Matsubara, M.; Masaki, S.; Karasawa, A.; Urayama, N.; Horikoshi, K.; Kajita, J.; Hasegawa, M.; Taniguchi, K.; Komada, T.; Kawamoto, Y. Properties of olopatadine hydrochloride, a new antiallergic/antihistaminic drug. *Arzneim. Forsch.* **2004,** *54* (12), 809–829.
35. Kaliner, M. A.; Oppenheimer, J.; Farrar, J. R. Comprehensive review of olopatadine: the molecule and its clinical entities. *Allergy Asthma Proc.* **2010,** *31* (2), 112–119.
36. Kurt, R. A.; Ucakhan-Guenduez, O.; Guenduez, K. Olopatadine 0.1% and 0.2% ophthalmic solution for the management of ocular allergy. *Expert Rev. Ophthalmol.* **2010,** *5* (3), 287–296.
37. Leonardi, A.; Quintieri, L. Olopatadine: a drug for allergic conjunctivitis targeting the mast cell. *Expert Opin. Pharmacother.* **2010,** *11* (6), 969–981.
38. Uchio, E. Treatment of allergic conjunctivitis with Mast cell stabilization and anti-histamine effects of olopatadine ophthalmic solution: a review pre-clinical and clinical research. *Clin. Ophthalmol.* **2008,** *2* (3), 525–531.
39. Rosenwasser, L. J.; O'Brien, T.; Weyne, J. Mast cell stabilization and anti-histamine effects of olopatadine ophthalmic solution: a review pre-clinical and clinical research. *Curr. Med. Res. Opin.* **2005,** *21* (9), 1377–1387.
40. Abelson, M. B. A review of olopatadine for the treatment of ocular allergy. *Expert Opin. Pharmacother.* **2004,** *5* (9), 1979–1994.
41. Wade, L.; Bielory, L.; Rudner, S. Ophthalmic antihistamines and H_1-H_4 receptors. *Curr. Opin. Allergy Clin. Immunol.* **2012,** *12* (5), 510–516.
42. Curran, M. P.; Scott, L. J.; Perry, C. M. Cetirizine: a review of its use in allergic disorders. *Drugs* **2004,** *64* (5), 523–561.
43. Portnoy, J. M.; Dinakar, C. Review of cetirizine hydrochloride for the treatment of allergic disorders. *Expert Opin. Pharmacother.* **2004,** *5* (1), 125–135.
44. Baltes, E.; De Lannoy, J.; Rodriguez, L. 2-[4-(Diphenylmethyl)-1-piperazinyl]acetic acids and their amides and pharmaceutical compositions, EP 58146 (1982).
45. Cossement, E.; Motte, G.; Bodson, G.; Gobert, J. Process for preparation of cetirizine, its dihydrochloride, and optical isomers via hydrolysis and corresponding nitriles, GB 2225321 (1990).
46. Opalka, C. J.; D'Ambra, T. E.; Faccone, J. J.; Bodson, G.; Cossement, E. A novel synthesis of the enantiomers of an antihistamine drug by piperazine formation from a primary amine. *Synthesis* **1995,** *7,* 766–768.

47. Cossement, E.; Bodson, G.; Gobert, J., Enantiomers of 1-[(4-chlorophenyl)phenylmethyl]-4-[(4-methylphenyl)sulfonyl]piperazine and their preparation and use as intermediates for antihistaminics, EP 617028 (1994).
48. Pflum, D. A.; Wilkinson, H. S.; Tanoury, G. J.; Kessler, D. W.; Kraus, H. B.; Senanayake, C. H.; Wald, S. A. A large-scale synthesis of enantiomerically pure cetirizine dihydrochloride using preparative chiral HPLC. *Org. Process Res. Dev.* **2001,** *5* (2), 110–115.
49. Pflum, D. A.; Krishnamurthy, D.; Han, Z.; Wald, S. A.; Senanayake, C. H. Asymmetric synthesis of cetirizine dihydrochloride. *Tetrahedron Lett.* **2002,** *43* (6), 923–926.
50. Klimek, L. Levocetirizine: from scientific evidence to a potent modern-day treatment of today's allergic patients. *Drugs Today* **2009,** *45* (3), 213–225.
51. Walsh, G. M. Levocetirizine: an update. *Curr. Med. Chem.* **2006,** *13* (22), 2711–2715.
52. Hair, P. I.; Scott, L. J. Levocetirizine: a review of its use in the management of allergic rhinitis and skin allergies. *Drugs* **2006,** *66* (7), 973–996.
53. Day, J. H.; Ellis, A. K.; Rafeiro, E. Levocetirizine: a new selective H_1 receptor antagonist for use in allergic disorders. *Drugs Today* **2004,** *40* (5), 415–421.
54. Ganellin, C. R. Development of anti-ulcer H_2-receptor histamine antagonists. In *Analogue-based Drug Discovery;* Fischer, J., Ganellin, C. R., Eds.; Wiley-VCH, 2006; pp 71–80.
55. Kubas, H.; Stark, H. Medicinal chemistry of histamine-H_2-receptor antagonists. *Pharm. Unserer Zeit* **2007,** *36* (1), 24–32.
56. Ichikawa, T.; Hotta, K.; Ishihara, K. Second-generation histamine H_2 receptor antagonists with gastric mucosal defensive properties. *Mini-Rev. Med. Chem.* **2009,** *9* (5), 581–589.
57. Eriksson, S.; Langstrom, G.; Rikner, L.; Carlsson, R.; Naesdal, J. Omeprazole and H_2-receptor antagonists in the acute treatment of duodenal ulcer, gastric ulcer and reflux oesophagitis: a meta-analysis. *Eur. J. Gastroenterol. Hepatol.* **1995,** *7* (5), 467–475.
58. Arrang, J. M.; Garbarg, M.; Schwartz, J. C. Autoinhibition of brain histamine release mediated by a novel class (H_3) of histamine receptor. *Nature (London, U. K.)* **1983,** *302,* 832–837.
59. Leurs, R.; Bakker, R. A.; Timmerman, H.; de Esch, I. J. P. The histamine H_3 receptor: from gene cloning to H_3 receptor drugs. *Nat. Rev. Drug Discovery* **2005,** *4* (2), 107–120.
60. Berlin, M.; Boyce, C. W.; de Lera Ruiz, M. Histamine H_3 receptor as a drug discovery target. *J. Med. Chem.* **2011,** *54* (1), 26–53.
61. Berlin, M.; Boyce, C. W. Recent advances in the development of histamine H_3 antagonists. *Expert Opin. Ther. Pat.* **2007,** *17* (6), 675–687.
62. Celanire, S.; Wijtmans, M.; Talaga, P.; Leurs, R.; de Esch, I. J. P. Keynote review: histamine H_3 receptor antagonists reach out for the clinic. *Drug Discovery Today* **2005,** *10* (23/24), 1613–1627.
63. Cowart, M.; Altenbach, R.; Black, L.; Faghih, R.; Zhao, C.; Hancock, A. A. Medicinal chemistry and biological properties of non-imidazole histamine H_3 antagonists. *Mini-Rev. Med. Chem.* **2004,** *4* (9), 979–992.
64. Hudkins, R. L.; Raddatz, R. Recent advances in drug discovery of histamine H_3 antagonists. *Annu. Rep. Med. Chem.* **2007,** *42,* 49–62.
65. Singh, M.; Jadhav, H. R. Histamine H_3 receptor function and ligands: recent developments. *Mini-Rev. Med. Chem.* **2013,** *13* (1), 47–57.
66. Kuhne, S.; Wijtmans, M.; Lim, H. D.; Leurs, R.; de Esch, I. J. P. Several down, a few to go: histamine H_3 receptor ligands making the final push towards the market. *Expert Opin. Invest. Drugs* **2011,** *20* (12), 1629–1648.
67. Plancher, J.-M. The histamine H_3 receptor as a therapeutic drug target for metabolic disorders: status, challenges and opportunities. *Curr. Top. Med. Chem.* **2011,** *11* (12), 1430–1446.
68. Lebois, E. P.; Jones, C. K.; Lindsley, C. W. The evolution of histamine H_3 antagonists/inverse agonists. *Curr. Top. Med. Chem.* **2011,** *11* (6), 648–660.

69. Vohora, D.; Bhowmik, M. Histamine H_3 receptor antagonists/inverse agonists on cognitive and motor processes: relevance to Alzheimer's disease, ADHD, schizophrenia and drug abuse. *Front. Syst. Neurosci.* **2012,** *6* (Oct.), 72.
70. Lazewska, D.; Kiec-Kononowicz, K. Recent advances in histamine H_3 receptor antagonists/inverse agonists. *Expert Opin. Ther. Pat.* **2010,** *20* (9), 1147–1169.
71. Letavic, M. A.; Barbier, A. J.; Dvorak, C. A.; Carruthers, N. I. Recent medicinal chemistry of the histamine H_3 receptor. *Prog. Med. Chem.* **2006,** *44,* 181–206.
72. Aslanian, R.; Shih, N.-Y. Recent progress in histamine H_3 receptor chemistry. *Annu. Rep. Med. Chem.* **2004,** *39,* 57–66.
73. Wijtmans, M.; Leurs, R.; de Esch, I. Histamine H_3 receptor ligands break ground in a remarkable plethora of therapeutic areas. *Expert Opin. Invest. Drugs* **2007,** *16* (7), 967–985.
74. Stark, H.; Schunack, W. Histamine H_3-receptor agonists and antagonists: chemical, pharmacological, and clinical aspects. In *Chemistry and Molecular Aspects of Drug Design and Action;* Rekka, E. A., Kourounakis, P. N., Eds.; CRC Press, 2008; pp 199–214.
75. Brioni, J. D.; Esbenshade, T. A.; Garrison, T. R.; Bitner, S. R.; Cowart, M. D. Discovery of histamine H_3 antagonists for the treatment of cognitive disorders and Alzheimer's disease. *J. Pharmacol. Exp. Ther.* **2011,** *336* (1), 38–46.
76. Jablonowski, J. A.; Carruthers, N. I.; Thurmond, R. L. The histamine H_4 receptor and potential therapeutic uses for H_4 ligands. *Mini-Rev. Med. Chem.* **2004,** *4* (9), 993–1000.
77. Marson, C. M. Targeting the histamine H_4 receptor. *Chem. Rev. (Washington, DC, U. S.)* **2011,** *111* (11), 7121–7156.
78. Kiss, R.; Keseru, G. M. Histamine H_4 receptor ligands and their potential therapeutic applications. *Expert Opin. Ther. Pat.* **2009,** *19* (2), 119–135.
79. Dunford, P. J.; Thurmond, R. L. Histamine H_4 antagonists. *Prog. Respir. Res.* **2010,** *39,* 187–191.
80. Smits, R. A.; Leurs, R.; de Esch, I. J. P. Major advances in the development of histamine 4 receptor ligands. *Drug Discovery Today* **2009,** *14* (15–16), 745–753.
81. Venable, J. D.; Thurmond, R. L. Development and chemistry of histamine H_4 receptor ligands as potential modulators of inflammatory and allergic responses. *Anti-Inflammatory Anti-Allergy Agents Med. Chem.* **2006,** *5* (4), 307–322.
82. Engelhardt, H.; Smits, R. A.; Leurs, R.; Haaksma, E.; de Esch, I. J. P. A new generation of anti-histamines: histamine H_4 receptor antagonists on their way to the clinic. *Curr. Opin. Drug Discovery Dev.* **2009,** *12* (5), 628–643.
83. Saravanan, C.; Bharti, S. K.; Jaggi, S.; Singh, S. K. Histamine H_4 receptor: a novel target for inflammation therapy. *Mini-Rev. Med. Chem.* **2011,** *11* (2), 143–158.
84. Walter, M.; Kottke, T.; Stark, H. The histamine H_4 receptor: targeting inflammatory disorders. *Eur. J. Pharmacol.* **2011,** *668* (1–2), 1–5.
85. Salcedo, C.; Pontes, C.; Merlos, M. Is the H_4 receptor a new drug target for allergies and asthma? *Front. Biosci., Elite Ed.* **2013,** *E5* (1), 178–187.
86. Bhatt, H. G.; Agrawal, Y. K.; Raval, H. G.; Manna, K.; Desai, P. R. Histamine H_4 receptor: a novel therapeutic target for immune and allergic responses. *Mini-Rev. Med. Chem.* **2010,** *10* (14), 1293–1308.
87. Kiss, R.; Keseru, G. M. Histamine H_4 receptor ligands and their potential therapeutic applications: an update. *Expert Opin. Ther. Pat.* **2012,** *22* (3), 205–221.
88. Fung-Leung, W.-P.; Thurmond, R. L.; Ling, P.; Karlsson, L. Histamine H_4 receptor antagonists: the new antihistamine? *Curr. Opin. Invest. Drugs (BioMed Cent.)* **2004,** *5* (11), 1174–1183.
89. Yu, F.; Bonaventure, P.; Thurmond, R. L. The future antihistamines: histamine H_3 and H_4 receptor ligands. *Adv. Exp. Med. Biol.* **2010,** *709* (Histamine in Inflammation), 125–140.

Chapter 17

Cardiotonic Inotropic Drugs

The term "heart failure" is used to describe the failure of physiological cardiac function responsible for pumping blood and characterized by exercise intolerance, fatigue, and dyspnea (shortness of breath).

Inotropic agents, are medicines that alter the force or energy of muscular contraction. Cardiotonic inotropic drugs, which are pharmacological agents that have a strengthening effect on the heart or that can increase cardiac output, represent the most frequent treatment for heart failure. Sometimes they are called positive inotropes; negative inotropes weaken the force of the heartbeat.

Positive inotropic drugs in general increase the Ca^{2+} level in the heart muscle, increase its contractility, both increase renin production and prevent activation of the renin–angiotensin system, and increase renal blood flow and urine output.

The drug therapy for "heart failure" includes positive inotropic drugs such as digoxin, angiotensin-converting enzyme inhibitors, angiotensin receptor antagonists, β-receptor blockers, calcium antagonists, and aldosterone receptor antagonist, as well as other medications. The nondrug therapy includes cardiac pacemaker, cardiac defibrillator therapy, heart transplantation, and others.

A classification system for inotropic agents was proposed more than 20 years ago. The system categorizes inotropic agents according to their mechanisms of action. Agents are classified as those that increase the cyclic adenosine monophosphate level (Class I); regulate ion channels or pumps in the myocardial cell membrane (Class II); modulate intracellular calcium regulation (Class III); and augment contractility through various mechanisms (Class IV). This classification system does not suggest that some classes of inotropic agents might be more effective than others [1].

Although there are many agents that improve the symptoms of heart failure, the practical guidelines for the diagnosis, assessment, and treatment of heart failure given in the United States [2] and in Europe [3] recommend digoxin (**17.2.1**), the only inotropic drug, which was introduced in medicinal practice in the 18th century.

The group of cardiotonic glycosides to which digoxin belongs is extracted mostly from leaves of digitalis plants. Their scientific name comes from Latin word *digitus*, which means finger and refers to the ease with which a flower of *Digitalis purpurea* can be fitted over a human fingertip.

Synthesis of Best-Seller Drugs. http://dx.doi.org/10.1016/B978-0-12-411492-0.00017-1

The mechanism of action of digoxin is not completely understood and is a little bit controversial [4], however, after 200 years of use, digoxin remains both well tolerated and the only oral inotrope. It is one of the most commonly prescribed cardiac medications. The main indications for digoxin include heart failure and atrial fibrillation [5,6].

17.1 CLASS I CARDIOTONIC INOTROPIC AGENTS (DRUGS THAT INCREASE THE CYCLIC ADENOSINE MONOPHOSPHATE LEVEL)

Drugs That Increase Cyclic Adenosine Monophosphate Level By β-Receptor Stimulation

Increased cyclic adenosine monophosphate (cAMP) content in a patient's body has inotropic and vasodilatory effects. β-Adrenoreceptor agonists—isoproterenol (**17.1.1**), denopamine (**17.1.2**), prenalterol (**17.1.3**), and docarpamine (**17.1.4**)—are effective β_1 stimulants with positive inotropic effect for heart failure short-term therapy. The same effect displays dobutamine (**17.1.5**), a potent β_1 agonist and α_1 antagonist, and dopamine and dopamine-receptor stimulant. Noradrenaline (norepinephrine) (**17.1.6**) increases blood pressure through stimulation of α receptors of peripheral vessels (Fig. 17.1.).

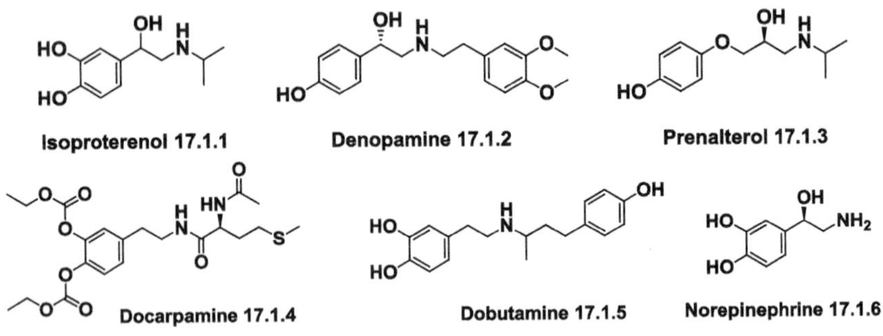

FIG. 17.1 **Drugs that increase blood pressure through stimulation of α receptors of peripheral vessels.**

Drugs That Increase Cyclic Adenosine Monophosphate Level By Stimulating Adenylyl Cyclase

Forskolin (**17.1.7**) is an example of a direct stimulator of adenylyl cyclase. Forskolin is a labdane diterpene produced by the plant *Coleus forskohlii*. It activates adenylate cyclase, the enzyme that increases the levels of intracellular cAMP, which exert a positive inotropic effect (Fig. 17.2.).

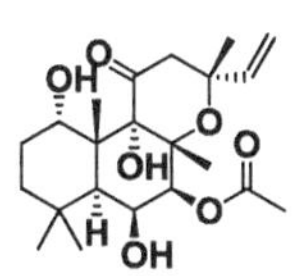

Forskolin 17.1.7

FIG. 17.2 Structure of forskolin.

Phosphodiesterase Inhibitors

Among the five isoenzymes of phosphodiesterase (PDE) inhibitors, selective PDE3 inhibitors amrinone (**17.1.8**) and milrinone (**17.1.9**) specifically degrade cAMP and are recognized to be good cardiotonic drugs.

There is great interest in developing new drugs in this category. A second generation of PDEs, such as CI-930 (**17.1.10**), meribendan (**17.1.11**), and pimobendan (**17.1.12**), levosimendan (**17.1.13**), and structurally unrelated to this raw enoximone (**17.1.14**), which have a higher selectivity as PDE3 inhibitors in comparison with the parent compounds, have been synthesized and are in different stages of trials (Fig. 17.3.).

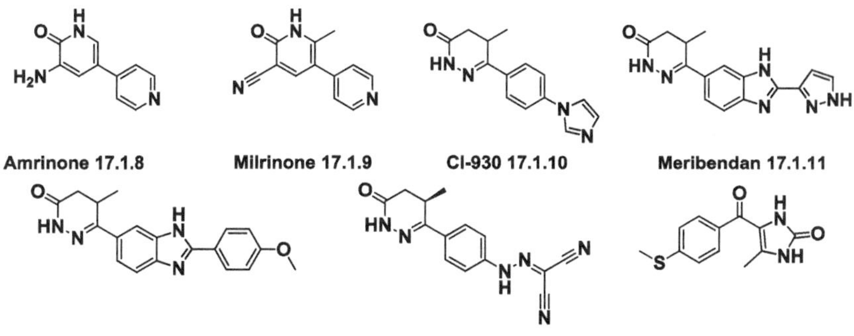

Pimobendan 17.1.12 Levosimendan 17.1.13 Enoximone 17.1.14

FIG. 17.3 Structure of selective PDE3 inhibitors.

17.2 CLASS II CARDIOTONIC INOTROPIC AGENTS (DRUGS THAT REGULATE ION CHANNELS OR PUMPS IN THE MYOCARDIAL CELL MEMBRANE)

Digitalis, which exerts a variety of effects on heart failure, including myocardial contractile force, is used for preparation of drugs that contain cardiac glycosides, extracted from various plants of the *Digitalis* genus. Digoxin (**17.2.1**) and digitoxin (**17.2.2**) are the two most used digitalis cardiac inotropes. Several hundreds of partial-synthetic derivatives have been prepared and evaluated and all of which were poorer in their therapeutic indices than digoxin and digitoxin (Fig. 17.4.).

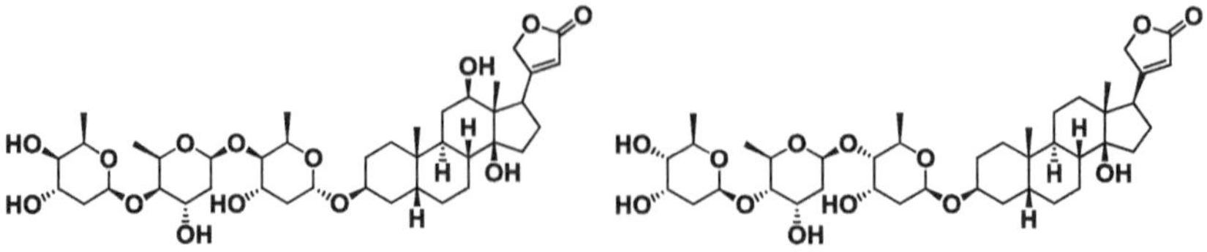

Digoxin 17.2.1

Digitoxin 17.2.2

FIG. 17.4 **Structure of the cardiac inotropes digoxin and digitoxin.**

The mechanism of their action is not completely understood, however, it is believed that they inhibit the sodium-potassium adenosine triphosphatase (ATPase), which, in turn, causes an increase in the level of Na^+ ions in the myocytes, leading to a rise in the level of the Ca^{2+} ion concentration in the myocardiocytes and pacemaker cells, which leads to increased contractility of the heart. However, the clinical use of digoxin and digitoxin is limited by a narrow therapeutic index [6].

17.3 CLASS III CARDIOTONIC INOTROPIC AGENTS (DRUGS THAT MODULATE INTRACELLULAR CALCIUM REGULATION)

This group of compounds increases the contractile response to any given level of intracellular Ca^{2+}. The molecular mechanism of action of intracellular calcium regulators is supposed to be at the calcium-binding protein troponin C level of the troponin complex, or by activation of phosphoinositide cascade. The representatives of this group of compounds are the already-mentioned pimobendan (**17.1.12**) and the compound MCI-154 (**17.3.1**) (Fig. 17.5.).

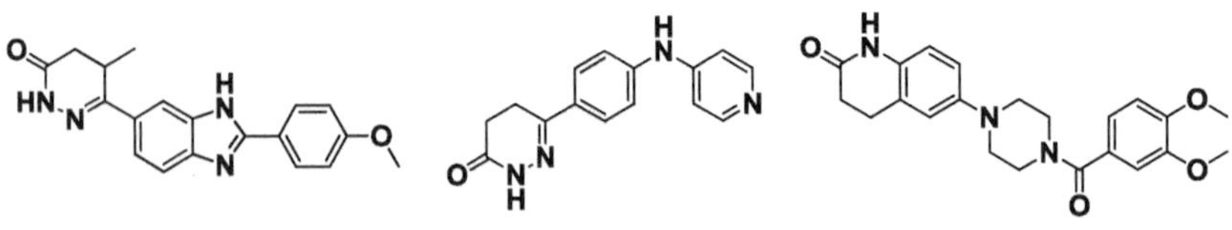

Pimobendan 17.1.12

MCI-154 17.3.1

Vesnarinone 17.4.1

FIG. 17.5 **Class III and Class IV drugs that modulate intracellular calcium regulation.**

17.4 CLASS IV DRUGS THAT MODULATE INTRACELLULAR CALCIUM REGULATION AGENTS (DRUGS THAT AUGMENT CONTRACTILITY THROUGH VARIOUS MECHANISMS)

This group represents compounds that augment heart muscle contractility through two or more mechanisms and includes pimobendan (**17.1.12**), and vesnarinone (**17.4.1**) (see Fig. 17.5.).

None of the described cardiotonic inotropic drugs has been shown to be capable of preventing absolutely all manifestations of congestive heart failure. None is included in the list of Top 200 Drugs by sales for the 2010s. Recent literature on cardiotonic inotropic drugs is well reviewed [7-15].

REFERENCES

1. Feldman, A. M. Classification of positive inotropic agents. *J. Am. Coll. Cardiol.* **1993,** *22* (4), 1223–1227.
2. Williams, J. F., Jr.; Bristow, M. R.; Fowler, M. ,B.; Francis, G. S.; Garson, A., Jr.; Gersh, B. ,J.; Hammer, D. F.; Hlatky, M. A.; Leier, C.; Packer, M.; Pitt, B.; Ullyot, D.; Wexler, L. F.; Winters, W., Jr.; Ritchie, J.; Cheitlin, M. D.; Eagle, K.; Gardner, T. J.; Garson, A., Jr.; Gibbons, R. J.; Lewis, R. P.; O'Rourke, R.; Ryan, T. J. Guidelines for the evaluation and management of heart failure. *Report of the American College of Cardiology/American Heart Association Task Force on Practice Guidelines (Committee on Evaluation and Management of Heart Failure), Circulation* **1995,** *92* (9), 2764–2784.
3. Dickstein, K.; Cohen-Solal, A.; Filippatos, G.; McMurray, J. J. V.; Ponikowski, P.; Poole-Wilson, P. A.; Stroemberg, A.; van Veldhuisen, D. J.; Atar, D.; Hoes, A. W.; Keren, A.; Mebazaa, A.; Nieminen, M.; Priori, S. G.; Swedberg, K. ESC guidelines for the diagnosis and treatment of acute and chronic heart failure (2008). *Eur. Heart J.* **2008,** *29* (19), 2388–2442.
4. Kalman, J. J. Digoxin therapy: does it still have a role in the management of heart failure. In *Therapeutic Strategies in Heart Failure;* Yancy, C. W., Young, J. B., Eds.; Clinical Publishing, 2007; pp 39–45.
5. Eichhorn, E. J.; Gheorghiade, M. Digoxin. *Prog. Cardiovasc. Dis.* **2002,** *44* (4), 251–266.
6. Albrecht, H. P.; Geiss, K.-H. Cardiac glycosides and synthetic inotropic drugs. In McGuire, J. L., Ed.; *Pharmaceuticals: Classes, Therapeutic Agents, Areas of Application, vol. 1*; Wiley-VCH, 2000; pp 173–200.
7. Fraga, C. A. M.; Barreiro, E. J. Cardiotonic drugs: history and perspectives on an old and important class of therapeutic agents. *Quim. Nova* **1996,** *19* (2), 182–189.
8. Repke, K. R. H.; Megges, R. Status and prospect of current inotropic agents. *Expert Opin. Ther. Pat.* **1997,** *7* (11), 1297–1306.
9. Chen, S. Y.; Tang, W. H. W. Emerging drugs for acute and chronic heart failure: current and future developments. *Expert Opin. Emerg. Drugs* **2007,** *12* (1), 75–95.
10. Hasenfuss, G.; Teerlink, J. R. Cardiac inotropes: current agents and future directions. *Eur. Heart J.* **2011,** *32* (15), 1838–1845.
11. Triposkiadis, F.; Parissis, J. T.; Starling, R. C.; Skoularigis, J.; Louridas, G. Current drugs and medical treatment algorithms in the management of acute decompensated heart failure. *Expert Opin. Invest. Drugs* **2009,** *18* (6), 695–707.
12. Hamad, E.; Mather, P. J.; Srinivasan, S.; Rubin, S.; Whellan, D. J.; Feldman, A. M. Pharmacologic therapy of chronic heart failure. *Am. J. Cardiovasc. Drugs* **2007,** *7* (4), 235–248.
13. Balakumar, P.; Singh, M. Recent advances in pharmacotherapy for heart failure: future directions. *Trends Med. Res.* **2007,** *2* (2), 61–71.
14. Swynghedauw, B.; Charlemagne, D. What is wrong with positive inotropic drugs? Lessons from basic science and clinical trials. *Eur. Heart J. Suppl.* **2002,** *4* (Suppl. D), D43–D49.
15. Repke, K. R. H.; Megges, R. Status and prospect of current inotropic agents. *Expert Opin. Ther. Pat.* **1997,** *7* (11), 1297–1306.

Chapter 18

Antiarrhythmic Drugs

The regular beating and contraction of the heart, which is remarkably constant and robust, supplies the whole body with necessary blood.

Cardiac arrhythmia is an irregular rate, force, or rhythm of muscle contractions in the heart.

Most often arrhythmias are the result of an underlying cardiac disease. They can arise from the morphological and structural alterations of cardiac tissues. Heart is very vulnerable to different stressors such as psychological, physiological, and pharmacological, and can be many causes of an arrhythmia, including cardiac diseases, structural alterations of cardiac tissues, such as scarring of heart tissue, blocking arteries in the heart, morphological changes in the heart's structure, high blood pressure, stress, and different medications.

Generally, it is believed that arrhythmias occur when the electrical impulses in the heart, primarily driven by cellular Na^+, K^+, and Ca^{2+} ions through cell membranes currents, are disrupted. According to a model proposed in the 1950s [1-3], voltage-gated ion channels, which directionally propagate electrical signals, play an exceptional role for pharmacological therapy in general and for arrhythmias in particular, coordinating heartbeats [4].

Depending on their origin, arrhythmias can occur in the upper (atrial) or lower (ventricular) chambers of the heart and in the junctional atrioventricular tissue.

The term *cardiac arrhythmia* covers a very large number of very different conditions that are classified by rate, site of origin, or mechanism.

Arrhythmias as an irregular heart beating are classified as tachycardia, (if the heart beating is too fast), bradycardia (if it is too slow), fibrillation (if it is too irregular), or premature contraction (if it is too early).

There are hundreds of different types of cardiac arrhythmias. The normal rhythm of the heart (sinus rhythm), can be disturbed through failure of automaticity, or overactivity, or premature atrial, or ventricular contractions, or fibrillation.

The most common example of a relatively benign arrhythmia is atrial fibrillation. Similarly common are premature atrial and premature ventricular arrhythmias.

Drugs used for preventing and treating irregular heart rate and heart beat are called *antiarrhythmic drugs*.

Synthesis of Best-Seller Drugs. http://dx.doi.org/10.1016/B978-0-12-411492-0.00018-3

Antiarrhythmic medicines help return the heart to its normal sinus rhythm and rate and maintain it after it has been achieved, stabilizing the heart muscle tissue.

Arrhythmia mechanisms includes numerous pathways such as ion channels, membrane adaptor proteins, transcription factors, and cytoplasmic signaling cascades. There are different classifications of antiarrhythmic drugs based on different principles: location of the drug action, which means that substances can act directly on the myocardium, or substances that have an effect on the efferent innervation of the heart; or classifications based on effectiveness of drugs for definite types of arrhythmia, such as effective for supraventricular arrhythmia, or those effective for ventricular arrhythmia.

Based on knowledge regarding the origin of arrhythmias, a variety of classifications for antiarrhythmic drugs have been suggested. The first classification was proposed by Vaughan Williams in 1970 [5], and then was successively accepted [6-8].

Other classfication schemes exist [9-16], such as Touboul's classification [9], which takes into account data from in vivo electrophysiological explorations, the Goldberger and Curtis classification [10], and "The Sicilian Gambit classification" [11], but the Vaughan Williams classification [5-8] remains the most acceptable and continues to be used.

According to the Vaughan Williams classification, antiarrhythmic drugs can be subdivided into five main classes.

18.1 CLASS I

The first class represents sodium channel blockers (quinidine, procainamide, disopyramide, lidocaine, tocainide, mexiletine, flecainide, encainide, and phenytoin (antiepileptic). It is necessary to mention that procainamide, disopyramide, lidocaine, tocainide, flecainide, and encainide "at 10 to 100 times their antiarrhythmic concentrations behaved as local anesthetics in nerves" [7].

The first class is, in turn, subdivided into three, based on specific kinetic characteristics and their effect on the length of the action potential duration (association/dissociation) within the group.

Class Ia (Sodium Channel Blockers with Intermediate Association/Dissociation Kinetics and Action Potential Duration)

This class of compounds is represented by quinidine (**18.1.1**), procainamide (**18.1.2**), and disopyramide (**18.1.3**) (Fig. 18.1.).

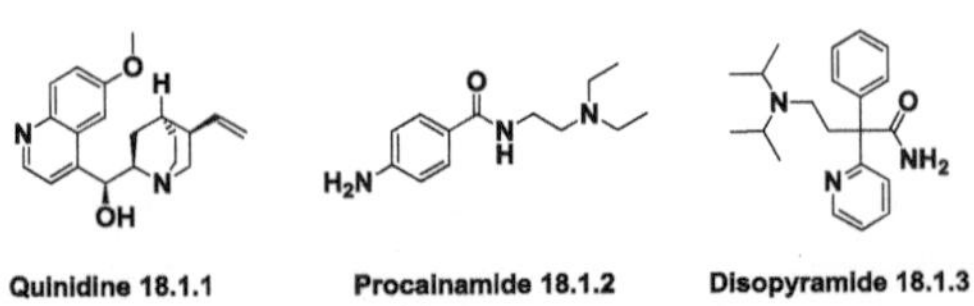

FIG. 18.1 Class Ia antiarrhythmic drugs.

Class Ib (Sodium Channel Blockers with Rapid Association/Dissociation Kinetics)

This class of compounds is represented by mexiletine (**18.1.4**), tocainide (**18.1.5**), which is no longer sold in the United States, lidocaine (**18.1.6**), a local anesthetic and antiarrhythmic drug, and phenytoin (**18.1.7**), an antiepileptic drug (Fig. 18.2.).

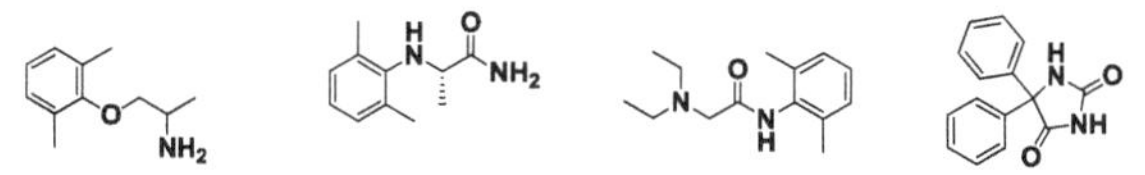

Mexiletine 18.1.4 Tocainide 18.1.5 Lidocaine 18.1.6 Phenytoin 18.1.7

FIG. 18.2 **Class Ib antiarrhythmic drugs.**

Class Ic (Sodium Channel Blockers with Slow Association/Dissociation Kinetics)

This class of compounds is represented by propafenone (**18.1.8**), flecainide (**18.1.9**), encainide (**18.1.10**), which is no longer in use because of its frequent proarrhythmic side effects, and moricizine (**18.1.1**), which was withdrawn in 2007 for commercial reasons (Fig. 18.3.).

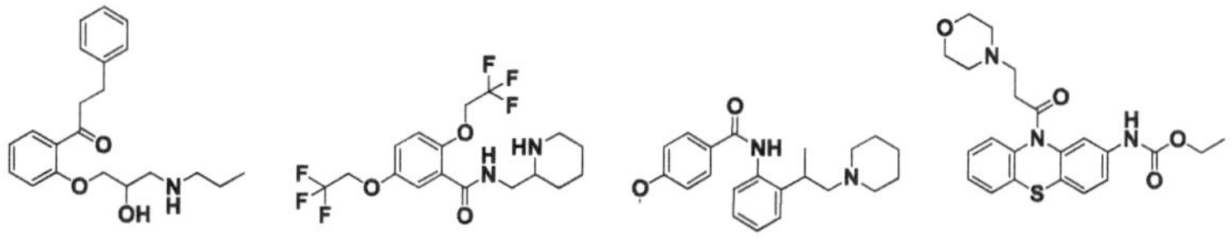

Propafenone 18.1.8 Flecainide 18.1.9 Encainide 18.1.10 Moricizine 18.1.11

FIG. 18.3 **Class Ic antiarrhythmic drugs.**

Ranolazine (**18.1.12**) is a new inhibitor of Na(+) current that was initially developed as an antianginal agent and found to exert antiarrhythmic actions without affecting heart rate or blood pressure. It improves myocardial ischemia and cardiac diastolic function, and tends to prevent related cardiac arrhythmias. It would probably best be placed in the Class Ib antiarrhythmics.

The antiarrhythmic efficacy of ranolazine in humans is unknown, but it represents intriguing possibilities in arrhythmia treatment [17] (Fig. 18.4.).

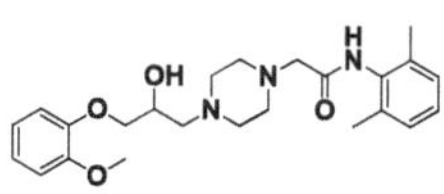

Ranolazine 18.1.12

FIG. 18.4 **Structure of ranolazine.**

18.2 CLASS II

The second class represents drugs that block action of endogenous catecholamines in the heart that have certain significance in the pathogenesis of arrhythmia (β blockers: propranolol (**18.2.1**), esmolol (**18.2.2**), metoprolol (**18.2.3**), bisoprolol (**18.2.4**), atenolol (**18.2.5**), and timolol (**18.2.6**) (Fig. 18.5.).

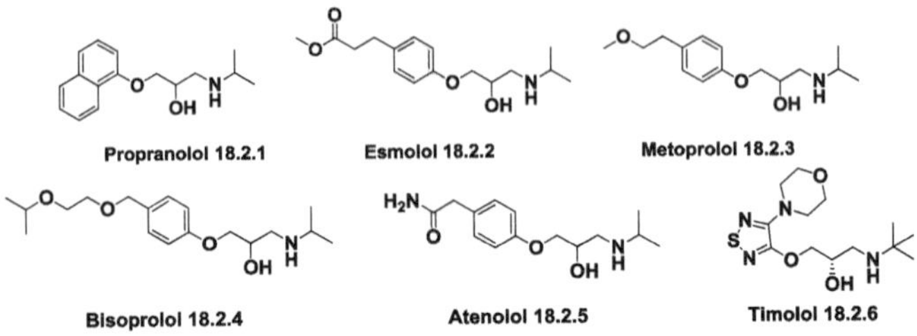

FIG. 18.5 **Class II antiarrhythmic drugs.**

18.3 CLASS III

The third class consists of drugs that predominantly block the potassium channels. Dofetilide (**18.3.1**) and ibutilide (**18.3.2**) are pure potassium channel blockers. Sotalol (**18.3.3**), which is the leading molecule in this group, is a combined β- and potassium channel blocker. Amiodarone (**18.3.4**) is a multichannel blocker, having all properties of Class I to IV drugs, and used to treat and prevent certain types of serious, life-threatening ventricular arrhythmias. Amiodarone is a structural analogue of thyroid hormone, and some of its toxicity, such as induced hypothyroidism, which occurs in 15 to 20% of patients, is attributable to its interaction with nuclear receptors of thyroid hormones [18].

Dronedarone (**18.3.5**) is the latest multichannel antiarrhythmic drug with a different mix of effects than amiodarone. Dronedarone was developed to treat atrial fibrillation and then started to be used as an antiarrhythmic drug. It is a close analogue of amiodarone, but because it lacks the iodine moieties of amiodarone, it does not induce thyroid dysfunctions in patients [19-21].

Vernakalant (**18.3.6**) is a new antiarrhythmic agent that was recently approved in Europe. It works by blocking K+ and Na+ channels, prolonging atrial refractory periods [22] (Fig. 18.6.).

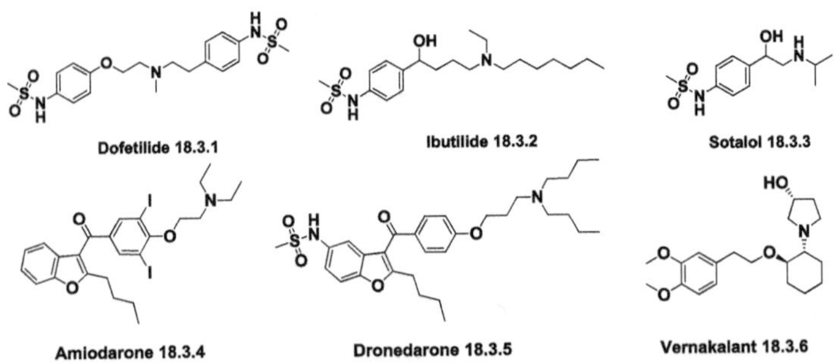

FIG. 18.6 **Class III antiarrhythmic drugs.**

18.4 CLASS IV

The fourth class of antiarrhythmic drugs is represented by antianginal drugs, the Ca^{2+} channel blockers verapamil (**18.4.1**) and diltiazem (**18.4.2**) (Fig. 18.7.).

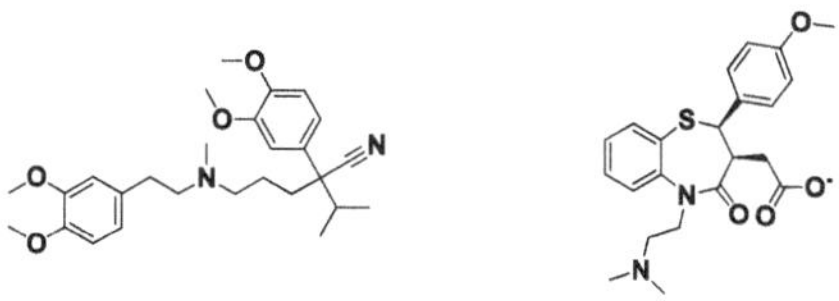

Verapamil 18.4.1 **Diltiazem 18.4.2**

FIG. 18.7 **Class IV antiarrhythmic drugs.**

18.5 CLASS V

Some researchers adhere to a system of dividing antiarrhythmic drugs into five classes. The fifth class contains those antiarrhythmics that do not fit in any of the four standard classes and are not part of the original Vaughan Williams [7] classification scheme and include digoxin (**18.5.1**) and adenosine (**18.5.2**) (Fig. 18.8.).

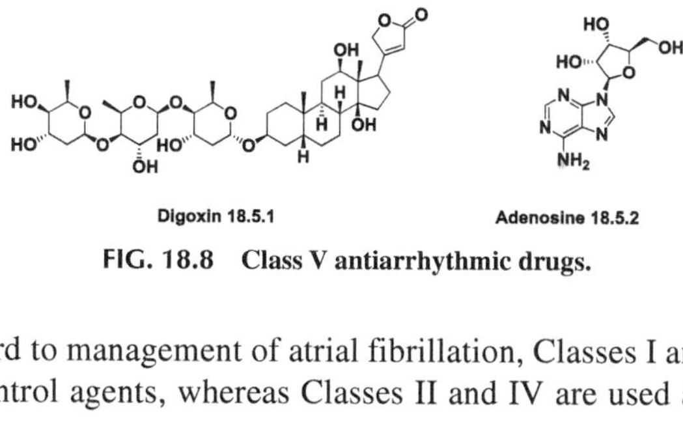

Digoxin 18.5.1 **Adenosine 18.5.2**

FIG. 18.8 **Class V antiarrhythmic drugs.**

With regard to management of atrial fibrillation, Classes I and III are used as rhythm control agents, whereas Classes II and IV are used as rate control agents.

Syntheses of all presented drugs are described in our previous book [23]. In the last decade no novel antiarrhythmics entered the pharmaceutical market, and none is presented in the list of Top 200 Drugs by sales for the 2010s.

Recent advances and progress in antiarrhythmic drugs have been critically reviewed [24-38].

REFERENCES

1. Hodgkin, A. L.; Huxley, A. F.; Katz, B. Measurement of current-voltage relations in the membrane of the giant axon of Loligo. *J. Physiol. (Oxford, U. K.)* **1952,** *116* (4), 424–448.
2. Hodgkin, A. L.; Huxley, A. F. Currents carried by sodium and potassium ions through the membrane of the giant axon of Loligo. *J. Physiol. (Oxford, U. K.)* **1952,** *116* (4), 449–472.
3. Hodgkin, A. L.; Huxley, A. F. A quantitative description of membrane current and its application to conduction and excitation in nerve. *J. Physiol. (Oxford, U. K.)* **1952,** *117* (4), 500–544.

4. Clarkson, C. W.; Hondeghem, L. M. Electrophysiological evidence that local anesthetics and antiarrhythmic drugs bind to a specific receptor site in cardiac sodium channels: displacement of bupivacaine by lidocaine. *Proc. West. Pharmacol. Soc.* **1984,** *27,* 23–25.
5. Vaughan Williams, E. M. The experimental basis for the choice of an antiarrhythmic drug. *Adv. Cardiol.* **1970,** *4,* 275–289.
6. Singh, B. N.; Hauswirth, O. Comparative mechanisms of action of antiarrhythmic drugs. *Am. Heart J.* **1974,** *87,* 367–382.
7. Vaughan Williams, E. M. A classification of antiarrhythmic actions reassessed after a decade of new drugs. *J. Clin. Pharmacol.* **1984,** *24* (4), 129–147.
8. Vaughan Williams, E. M. Classification of Antiarrhythmic Agents. In *Antiarrhythmic Drugs;* Vaughan Williams, E. M., Ed.; Springer-Verlag, 1989; p 45.
9. Touboul, P.; Atallah, G.; Gressard, A.; Michelon, G.; Chatelain, M. T.; Delahaye, J. P. Electrophysiological effects of anti-arrhythmia agents in man. Attempt at classification. *Arch. Mal. Coeur Vaiss.* **1979,** *72,* 72–81.
10. Goldberger, A. L.; Curtis, G. P. An "autonomic" classification of antiarrhythmic drugs. *J. Electrocardiol.* **1982,** *15,* 397–400.
11. Gambit, T. S. The Sicilian gambit. A new approach to the classification of antiarrhythmic drugs based on their actions on arrhythmogenic mechanisms. Task Force of the Working Group on Arrhythmias of the European Society of Cardiology. *Circulation* **1991,** *84,* 1831–1851.
12. Harrison, D. C.; Winkle, R. A.; Sami, M.; Mason, J. Encainide: a new and potent antiarrhythmic agent. *Am. Heart J.* **1980,** *100,* 1046–1054.
13. Hondeghem, L. M. Development of class III antiarrhythmic agents. *J. Cardiovasc. Pharmacol.* **1992,** *20* (Suppl. 2), S17–S22.
14. Cosio, F. G.; Delpon, E. New antiarrhythmic drugs for atrial flutter and atrial fibrillation. A conceptual breakthrough at last? *Circulation* **2002,** *105* (3), 276–278.
15. Hondeghem, L. M. Classification of antiarrhythmic agents and the two laws of pharmacology. *Cardiovasc. Res.* **2000,** *45* (1), 57–60.
16. Carnes, C. A. Antiarrhythmic drug classification. In *Novel Therapeutic Targets for Antiarrhythmic Drugs;* Billman, G. E., Ed.; Wiley, 2010; pp 155–170.
17. Keating, G. M. Ranolazine. *Drugs* **2013,** *73* (1), 55–73.
18. van Erven, L.; Schalij, M. J. Amiodarone: an effective antiarrhythmic drug with unusual side effects. *Heart (London, U. K.)* **2010,** *96* (19), 1593–1600.
19. Oyetayo, O. O.; Rogers, C. E.; Hofmann, P. O. Dronedarone: a new antiarrhythmic agent. *Pharmacotherapy* **2010,** *30* (9), 904–915.
20. Naccarelli, G. V.; Wolbrette, D. L.; Levin, V.; Samii, S.; Banchs, J. E.; Penny-Peterson, E.; Gonzalez, M. D. Safety and efficacy of dronedarone in the treatment of atrial fibrillation/flutter. *Clin. Med. Insights: Cardiol.* **2011,** *5,* 103–119.
21. Lin, L.; Bai, R. Dronedarone: is it time to turn it down? *Expert Opin. Drug Saf.* **2013,** *12* (1), 5–8.
22. Vizzardi, E.; Salghetti, F.; Bonadei, I.; Gelsomino, S.; Lorusso, R.; D'Aloia, A.; Curnis, A. A new antiarrhythmic drug in the treatment of recent-onset atrial fibrillation: vernakalant. *Cardiovasc. Ther.* **2013,** *31* (5), e55–e62.
23. Vardanyan, R. S.; Hruby, V. J. *Synthesis of Essential Drugs;* Elsevier, 2006.
24. Polovina, M. M.; Potpara, T. S. Recent advances in antiarrhythmic drug treatment of atrial fibrillation. *Recent Pat. Cardiovasc. Drug Discovery* **2013,** *8* (2), 112–126.
25. Hegyi, B.; Komaromi, I.; Nanasi, P. P.; Szentandrassy, N. Selectivity problems with drugs acting on cardiac Na+ and Ca2+ channels. *Curr. Med. Chem.* **2013,** *20* (20), 2552–2571.
26. MacRae, C. A. Recent advances in in vivo screening for antiarrhythmic drugs. *Expert Opin. Drug Discovery* **2013,** *8* (2), 131–141.

27. Saklani, P.; Skanes, A. Novel anti-arrhythmic medications in the treatment of atrial fibrillation. *Curr. Cardiol. Rev.* **2012,** *8* (4), 302–309.
28. Yang, X. S.; Sun, J. P.; Yu, C. M. Contemporary therapy of atrial fibrillation. *World J. Cardiovasc. Dis.* **2012,** *2* (3), 111–117.
29. Camm, A. J.; Camm, C. F.; Savelieva, I. Medical treatment of atrial fibrillation. *J. Cardiovasc. Med. (London, U. K.)* **2012,** *13* (2), 97–107.
30. Kozlowski, D.; Budrejko, S.; Lip, G. Y. H.; Mikhailidis, D. P.; Rysz, J.; Raczak, G.; Banach, M. Dronedarone: an overview. *Ann. Med.* **2012,** *44* (1), 60–72.
31. Burashnikov, A.; Antzelevitch, C. Novel pharmacological targets for the rhythm control management of atrial fibrillation. *Pharmacol. Ther.* **2011,** *132* (3), 300–313.
32. Thireau, J.; Pasquie, J.-L.; Martel, E.; Le Guennec, J.-Y.; Richard, S. New drugs vs. old concepts: a fresh look at antiarrhythmics. *Pharmacol. Ther.* **2011,** *132* (2), 125–145.
33. Dagres, N.; Sommer, P.; Anastasiou-Nana, M.; Hindricks, G. Treating arrhythmias: an expert opinion. *Expert Opin. Pharmacother.* **2011,** *12* (9), 1359–1367.
34. Marinelli, A.; Capucci, A. Antiarrhythmic drugs for atrial fibrillation. *Expert Opin. Pharmacother.* **2011,** *12* (8), 1201–1215.
35. Hohnloser, S. H. Novel antiarrhythmic approaches to treatment of atrial fibrillation. *Hot Top. Cardiol.* **2009,** *18,* 15–23.
36. Antoons, G.; Willems, R.; Sipido, K. R. Targeting Na+Ca2+ exchange as an antiarrhythmic strategy. In *Novel Therapeutic Targets for Antiarrhythmic Drugs;* Billman, G. E., Ed.; Wiley, 2010; pp 313–338.
37. Burashnikov, A.; Antzelevitch, C. New developments in atrial antiarrhythmic drug therapy. *Nat. Rev. Cardiol.* **2010,** *7* (3), 139–148.
38. Kakadiya, J. Novel anti-arrhythmic agents. *Pharmacologyonline* **2009,** (3, News Letters), 221–243.

Chapter 19

Antianginal Drugs

Angina pectoris was first clearly described in the 18th century, although it was recognized centuries ago. Later clear links were found proving its origins to abnormalities of the coronary arteries such as platelet accumulation on the vessel walls, narrowing of the epicardial arteries leading to thrombosis in the coronary arteries, which, in turn, generates acute chest pain caused by lack of sufficient oxygen.

Angina pectoris is a common disease (ischemic heart disease) that affects middle-aged persons, usually males. Angina pectoris is chest pain arising from the heart, sometimes radiating into the left shoulder and down the inner side of the left arm. It is believed that the dull, tight chest pain of angina occurs when the heart's muscular wall is not getting enough oxygen.

The primary aim of stable angina management is symptomatic relief; the secondary aim is prevention.

Antianginal drugs are a group of medicines used to prevent and relieve angina attacks. The therapeutic effect of this group is associated with the member drugs' antiischemic and analgesic effects.

The medical treatment of angina has changed little in the past few years.

Antianginal drugs are divided into medications that decrease the demand of cardiac tissue in oxygen and simultaneously increase blood supply to the heart (nitrates, calcium channel blockers), medications that reduce the effects of catecholamines (β blockers), and drugs that increase blood supply to the heart (β adrenomimetics), all of which basically reduce the oxygen demand of the heart.

Nitrates for treatment of the angina pectoris condition were first described in 1867. Beta blockers were introduced in the 1960s, and calcium antagonists were introduced in the 1970s [1-6].

19.1 NITRATES

Nitroglycerine (**19.1.1**) was first used for angina pectoris in 1876. A hundred years later, in 1977, release of nitric oxide from nitroglycerine and its action on vascular smooth muscle was discovered. Three years later, in 1980, nitric oxide was identified as an endothelial-derived relaxing factor [7].

Other organic and inorganic nitrates are prodrugs that undergo enzymatic denitration releasing nitric oxide, and it is thought that it acts by the same mechanism, by relaxing vascular and, in particular, venous smooth muscle, thus decreasing the return of blood to the heart, which lowers oxygen demand.

Synthesis of Best-Seller Drugs. http://dx.doi.org/10.1016/B978-0-12-411492-0.00019-5

Nitroglycerine (**19.1.1**) remains the treatment of choice for relieving angina. Other nitrates, such as erythrityl tetranitrate (**19.1.2**), pentaerythritol tetranitrate (**19.1.3**), isosorbide mononitrate (**19.1.4**), isosorbide dinitrate (**19.1.5**) also are available as drugs. These agents all exert their effect by being converted to the potent natural vasodilator nitric oxide [8-10] (Fig. 19.1.).

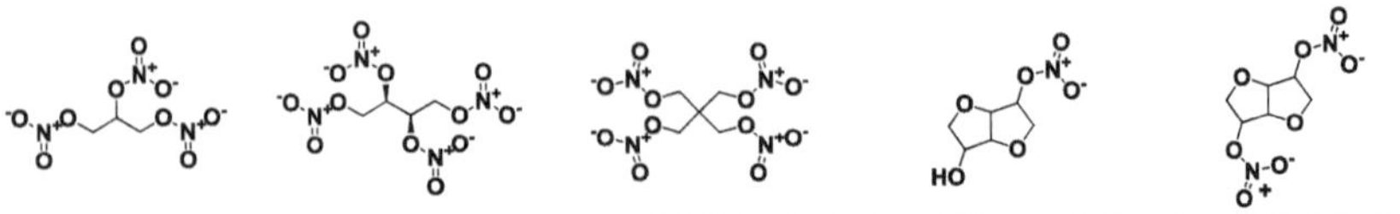

Nitroglycerin 19.1.1 | Erythrityl tetranitrate 19.1.2 | Pentaerythritol tetranitrate 19.1.3 | Isosorbide mononitrate 19.1.4 | Isosorbide dinitrate 19.1.5

FIG. 19.1 **Antianginal nitrates.**

19.2 CALCIUM-CHANNEL BLOCKERS

Calcium-channel blockers are presumed to have their major impact by increasing coronary flow as a consequence of dilatation of the coronary vessels inhibiting myocyte calcium influx. They enhance relaxation of vascular smooth muscle, decrease systemic vascular resistance, reduce heart rate and nodal conduction.

They are classified according to chemical structure as 1,4-dihydropyridines which has served as an excellent nucleus for the creation of ligands for voltage-gated Ca^{2+} channel such as nifedipine (**19.2.1**), felodipine (**19.2.2**), amlodipine (**19.2.3**), and nicardipine (**19.2.4**) [11,12]. Nondihydropyridine calcium-channel blockers as antianginal drugs are represented as phenylalkylamines, such as verapamil [13,14] (**19.2.5**), and benzothiazepines, such as diltiazem (**19.2.6**) [15], further separated according to generations (first, second) (Fig. 19.2.).

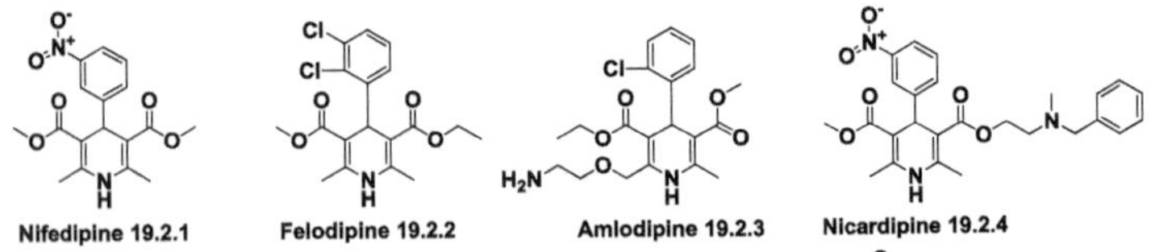

Nifedipine 19.2.1 | Felodipine 19.2.2 | Amlodipine 19.2.3 | Nicardipine 19.2.4

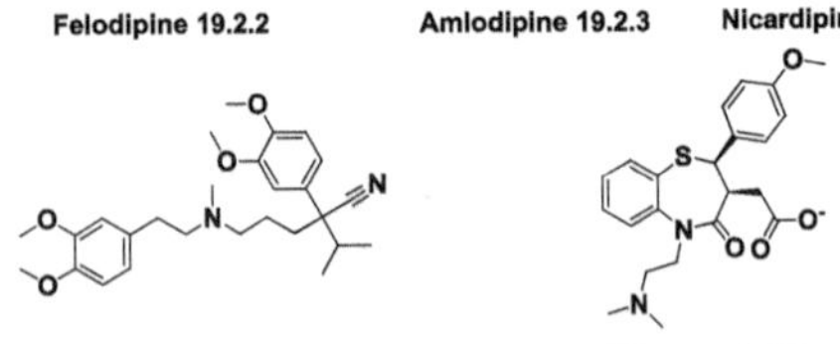

Verapamil 19.2.5 | Diltiazem 19.2.6

FIG. 19.2 **Antianginal calcium-channel blockers.**

19.3 β-ADRENERGIC BLOCKERS

β Blockers, one of the most useful groups of drugs in use today. They are being used not only for their original purpose to treat angina and cardiac arrhythmias, but also as effective therapeutics for hypertension, cardiac failure, anxiety, migraine, and glaucoma [16,17]. β Adrenergic blockers with different pharmacological properties are usually indicated as first-line therapy as antianginal drugs. These drugs lower heart rate and systolic blood pressure by inhibition of β-adrenergic receptors, which are activated by circulating endogenous norepinephrine and epinephrine in response to exercise or stress, cause vasodilation. β Blockers as antianginal drugs are represented by propranolol (**19.3.1**), atenolol (**19.3.2**), metoprolol (**19.3.3**), bisoprolol (**19.3.4**), nadolol (**19.3.5**), celiprolol (**19.3.6**), labetalol (**19.3.7**), timolol (**19.3.8**), carvedilol (**19.3.9**), and nebivolol (**19.3.10**) (Fig. 19.3.).

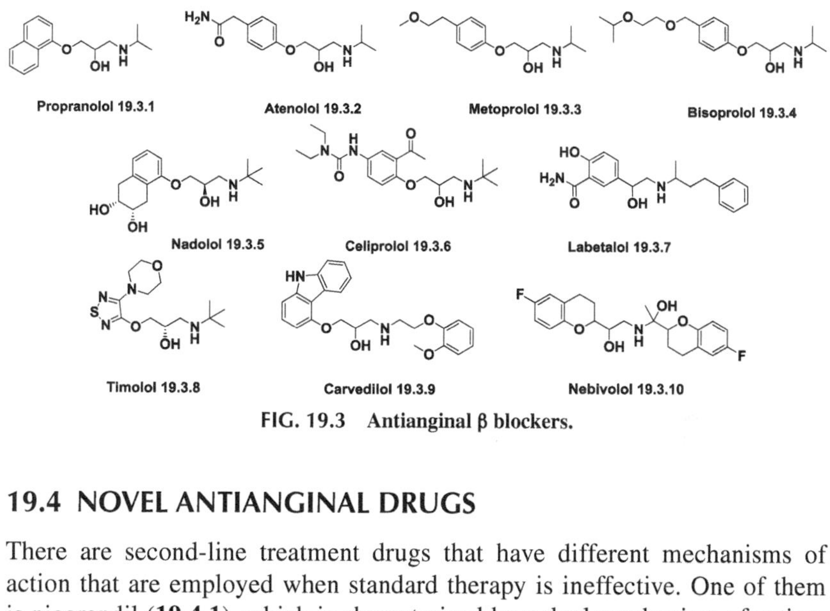

FIG. 19.3 **Antianginal β blockers.**

19.4 NOVEL ANTIANGINAL DRUGS

There are second-line treatment drugs that have different mechanisms of action that are employed when standard therapy is ineffective. One of them is nicorandil (**19.4.1**), which is characterized by a dual mechanism of action. The nicotinamide moiety acts as an opener of adenosine triphosphate (ATP)-sensitive potassium channels, whereas the presence of NO_2 group explains its nitrate-like properties, which significantly improve symptoms of angina pectoris [18,19].

Another drug is ivabradine (**19.4.2**), which specifically and selectively inhibits mixed sodium–potassium inward I_f current, with the sole action of heart rate reduction and no impact on any other cardiac parameter [20,21].

Trimetazidine (**19.4.3**) is one of the drugs used for optimization of myocardial energy metabolism. The mechanism of trimetazidine effects is believed to improve "fuel efficiency" via inhibition of myocardial free fatty acid β-oxidation, thereby allowing better inotropy and lusitropy [22,23].

Ranolazine (**19.4.4**) is specifically indicated for the treatment of chronic angina in patients who have failed to respond to prior angina therapy. It improves exercise tolerance and reduces the frequency of angina attacks in patients with ischemic heart disease [24-34] (Fig. 19.4.).

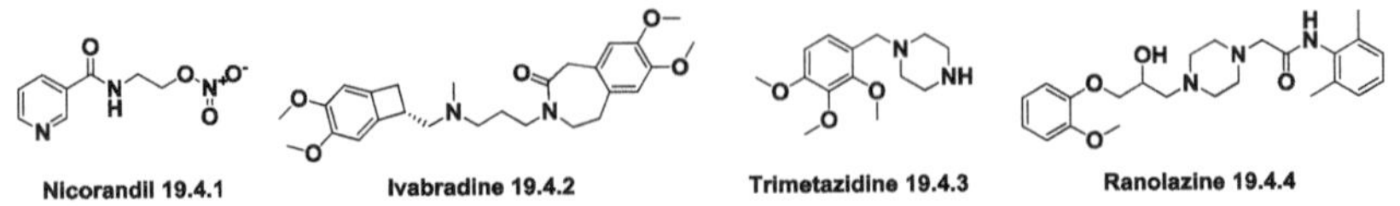

FIG. 19.4 **Novel antianginal drugs.**

Ranolazine–Ranexa

Ranolazine (**19.4.4**) was approved in 2006 for the treatment of chronic angina pectoris, and it was the first approved agent from a new class of antianginal drugs in almost 25 years.

The mechanism of action of ranolazine is not totally clear. Initially, ranolazine was thought to exert its therapeutic efficacy through partial inhibition of fatty acid oxidation. More recent evidence suggests that ranolazine reduces calcium overload in the ischemic myocyte through inhibition of the late sodium current. The most frequently reported adverse reactions are dizziness, headache, constipation, and nausea.

Ranolazine is included in the list of Top 200 Drugs by sales for the 2010s.

The synthesis of ranolazine was first disclosed in the first patent [35].

The process for preparation of ranolazine is simple enough and consists of condensation of (piperazin-1-yl)acetamide (**19.4.5**) with phenoxyoxirane (**19.4.6**) in boiling toluene/methanol solvent mixture and further purification by column chromatography. In another patent for the same synthesis, purification was carried out via transformation of the crude product to fumaric acid salt [36] or hydrochloride, its crystallization, and then further basification [37]. Some optimizations of the synthetic scheme were proposed recently [38]. Another approach includes a cyclization reaction between bis(2-chloroethyl)amino)-N-phenylacetamide (**19.4.7**) and 1-amino-3-phenoxypropan-2-ol derivative (**19.4.8**) forming in acetone/water media a piperazine ring and giving desired ranolazine (**19.4.4**) [39] (Scheme 19.1.).

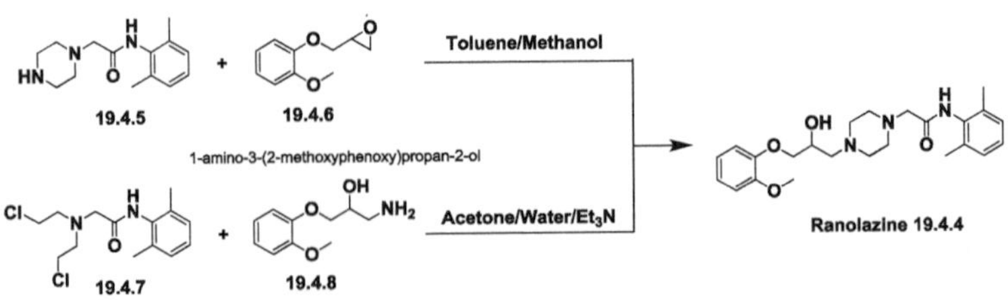

SCHEME 19.1 **Synthesis of ranolazine.**

REFERENCES

1. Poole-Wilson, P. A.; Jacques, A.; Lyon, A. Treatment of angina: a commentary on new therapeutic approaches. *Eur. Heart J. Suppl.* **2006,** *8* (Suppl. A), A20–A25.
2. Siama, K.; Tousoulis, D.; Papageorgiou, N.; Siasos, G.; Tsiamis, E.; Bakogiannis, C.; Briasoulis, A.; Androulakis, E.; Tentolouris, K.; Stefanadis, C. Stable angina pectoris: current medical treatment. *Curr. Pharm. Des.* **2013,** *19* (9), 1569–1580.
3. Chaitman, B. R.; Laddu, A. A. Stable angina pectoris: antianginal therapies and future directions. *Nat. Rev. Cardiol.* **2012,** *9* (1), 40–52.
4. Ben-Dor, I.; Battler, A. Treatment of stable angina. *Heart (London, U. K.)* **2007,** *93* (7), 868–874.
5. Gupta, S. P. Quantitative structure-activity relationships of antianginal drugs. *Prog. Drug Res.* **2001,** *56,* 121–154.
6. Denktas, A.; Bayes-Genis, A.; Schwartz, R. S. New approaches to the pharmacological treatment of angina. *Curr. Opin. Pharmacol.* **2001,** *1* (2), 151–158.
7. Marsh, N.; Marsh, A. A short history of nitroglycerine and nitric oxide in pharmacology and physiology. *Clin. Exp. Pharmacol. Physiol.* **2000,** *27* (4), 313–319.
8. Kojima, J. Evidence of acute myocardial infarction: efficacy of nitrate drugs. *Ther. Res.* **2007,** *28* (7), 1271–1283.
9. Dezsi, L. Development of organic nitrates for coronary heart disease. In *Analogue-based Drug Discovery;* Fischer, J., Ganellin, C. R., Eds.; Wiley-VCH, 2006; pp 247–258.
10. Sneader, W. Organic nitrates. *Drug News Perspect.* **1999,** *12* (1), 58–63.
11. Swarnalatha, G.; Prasanthi, G.; Sirisha, N. Madhusudhana Chetty, C. 1,4-Dihydropyridines: a multifunctional molecule—a review. *Int. J. ChemTech Res.* **2011,** *3* (1), 75–89.
12. Triggle, D. J. 1,4-Dihydropyridines as calcium channel ligands and privileged structures. *Cell. Mol. Neurobiol.* **2003,** *23* (3), 293–303.
13. Duchene-Marullaz, P. Verapamil anti-anginal effects. *Therapie* **1984,** *39* (2), 153–165.
14. Singh, B. N.; Chew, C. Y. C.; Josephson, M. A.; Packer, M. Pharmacologic and hemodynamic mechanisms underlying the antianginal actions of verapamil. *Am. J. Cardiol.* **1982,** *50* (4), 886–893.
15. Chaffman, M.; Brogden, R. N.; Diltiazem A review of its pharmacological properties and therapeutic efficacy. *Drugs* **1985,** *29* (5), 387–454.
16. Baker, J. G.; Hill, S. J.; Summers, R. J. Evolution of β-blockers: from anti-anginal drugs to ligand-directed signaling. *Trends Pharmacol. Sci.* **2011,** *32* (4), 227–234.
17. Erdmann, E. Clinical pharmacology of beta-blockers in cardiology: trial results and clinical applications. *Hot Top. Cardiol.* **2008,** *10,* 1–48.
18. Schmid, J.-P.; Schroeder, V. Nicorandil—review of pharmacological properties and clinical applications. *HeartDrug* **2005,** *5* (4), 220–229.
19. Hiremath, J. G.; Valluru, R.; Jaiprakash, N.; Katta, S. A.; Matad, P. P. Pharmaceutical aspects of nicorandil. *Int. J. Pharm. Pharm. Sci.* **2010,** *2* (4), 24–29.
20. Guglin, M. Heart rate reduction in heart failure: ivabradine or beta blockers? *Heart Failure Rev.* **2013,** *18* (4), 517–528.
21. Riccioni, G. Ivabradine: an intelligent drug for the treatment of ischemic heart disease. *Molecules* **2012,** *17,* 13592–13604.
22. Onay-Besikci, A.; Ozkan, S. A. Trimetazidine revisited: a comprehensive review of the pharmacological effects and analytical techniques for the determination of trimetazidine. *Cardiovasc. Ther.* **2008,** *26* (2), 147–165.

23. Marzilli, M. Cardioprotective effects of trimetazidine: a review. *Curr. Med. Res. Opin.* **2003,** *19* (7), 661–672.
24. Mathew, M.; Sajeeth, C. I.; Santhi, K.; Madhu, E. N. Ranolazine, a new addition to angina treatment. *Int. J. Pharma Bio Sci.* **2012,** *2* (1), 157–165.
25. Sossalla, S.; Maier, L. S. Role of ranolazine in angina, heart failure, arrhythmias, and diabetes. *Pharmacol. Ther.* **2012,** *133* (3), 311–323.
26. Tamargo, J.; Caballero, R.; Delpoon, E. Ranolazine: an antianginal drug with antiarrhythmic properties. *Expert Rev. Cardiovasc. Ther.* **2011,** *9* (7), 815–827.
27. Vadnais, D. S.; Wenger, N. K. Emerging clinical role of ranolazine in the management of angina. *Ther. Clin. Risk Manage.* **2010,** *6,* 517–530.
28. Aslam, S.; Gray, D. Ranolazine (Ranexa) in the treatment of chronic stable angina. *Adv. Ther.* **2010,** *27* (4), 193–201.
29. Keating, G. M. Ranolazine: a review of its use in chronic stable angina pectoris. *Drugs* **2008,** *68* (17), 2483–2503.
30. Nash, D. T.; Nash, S. D. Ranolazine for chronic stable angina. *Lancet* **2008,** *372* (9646), 1335–1341.
31. Siddiqui, M. A. A.; Keam, S. J. Ranolazine: a review of its use in chronic stable angina pectoris. *Drugs* **2006,** *66* (5), 693–710.
32. Tafreshi, M. J.; Fisher, E. Ranolazine: a new approach to management of patients with angina. *Ann. Pharmacother.* **2006,** *40* (4), 689–693.
33. Stanley, W. C. Ranolazine: new approach for the treatment of stable angina pectoris. *Expert Rev. Cardiovasc. Ther.* **2005,** *3* (5), 821–829.
34. Tavazzi, L. Ranolazine, a new antianginal drug. *Future Cardiol.* **2005,** *1* (4), 447–455.
35. Kluge, A. F.; Clark, R. D.; Strosberg, A. M.; Pascal, J. C.; Whiting, R. L. Cardioselective aryloxy- and arylthiohydroxypropylpiperazinyl acetanilides which affect calcium entry, EP 126449 (1984).
36. Kompella, A. K.; Adibhatla K. S.; Venkaiah C. A process for the preparation of highly pure ranolazine base, WO 2008139492 (2008).
37. Siyan, R. S.; Gohel, S. V.; Walunj, R.; Anurag, S.; Singh, G. P. Process for the preparation of ranolazine from 2,6-xylidine, chloroacetyl chloride, piperazine, and 2-[(2-methoxyphenoxy) methyl]oxirane, WO 2010097805 (2010).
38. Madivada, L. R.; Anumala, R. R.; Gilla, G.; Kagga, M.; Bandichhor, R. An efficient synthesis of 1-(2-methoxyphenoxy)-2,3-epoxypropane: key intermediate of β-adrenoblockers. *Org. Process Res. Dev.* **2012,** *16* (10), 1660–1664.
39. gai-Csongor, E.; Gizur, T.; Harsanyi, K.; Trischler, F.; Demeter-Szabo, A.; Csehi, A.; Vajda, E.; Szabo-Komlosi, G. Process for preparation of piperazine derivatives, EP 483932 (1992).

Chapter 20

Hypolipidemic and Antihyperlipidemic Drugs

Hyperlipidemia is a metabolic (mainly inherited) disorder characterized by elevated lipid and/or lipoprotein levels in the blood and is an important risk factor in development of atherosclerosis and heart disease.

Hyperlipidemias are divided in primary and secondary subtypes. Primary hyperlipidemia is usually caused by genetic factors, whereas secondary hyperlipidemia arises from other metabolic disorders (e.g., diabetes).

It is hypothesized and believed that cholesterol is the main constituent of the fatty lumps in the arteries walls, which the walls to narrow. In addition, there is a direct link between the concentration of lipoproteins in blood plasma and expressed atherosclerotic changes in arteries.

Lipoproteins are particles that transport triglycerides and cholesterol in the blood between all the tissues of the body; lipoproteins are circulated in the blood. Cholesterol is transported via particles of various size that are made up of triglycerides, cholesterol esters, and phospholipids, each of which plays a specific role.

Lipoproteins are divided into five basic types. The largest particles are chylomicrons, low-density proteins (LDLs), followed by intermediate-density lipoproteins (IDLs), very-low-density lipoproteins (VLDLs), and, finally, the smallest–high-density lipoproteins (HDLs).

Cholesterol, triglycerides, and LDLs, sometimes referred to as the "bad cholesterol," carry cholesterol from the liver to cells of the body are different forms of lipids, which are responsible for possible complications in the human body. HDLs, which collect cholesterol from the body's tissues and take it back to the liver, are referred to as the "good cholesterol" lipoprotein. Cholesterol is essential for the functioning of mammalians, but high levels of cholesterol increase the risk of cardiovascular disease and represent a major cardiovascular risk factor.

Compounds that lower the concentration of lipoproteins in the plasma by inhibiting their production in the organism or by removing them from the plasma are called hypolipidemics, or antihyperlipidemic agents.

A diverse group of pharmaceuticals, called lipid-lowering drugs, promote the reduction of lipid concentrations in the serum and are used in the treatment of hyperlipidemias.

Synthesis of Best-Seller Drugs. http://dx.doi.org/10.1016/B978-0-12-411492-0.00020-1

Despite significant improvements in lipid-lowering therapy, it still remains a substantial contributor to the incidence of cardiovascular disease.

Drugs therapies available for the treatment of hyperlipidemia include use of drugs like niacin, fibrates (clofibrate, gemfibrozil), bile acid–binding resins (cholestyramine and colestipol), probucol, and statins, such as 3-hydroxy-3-methylglutaryl-CoA (HMG-CoA) reductase inhibitors, which are used to reduce cholesterol levels.

20.1 STATINS

The HMG-CoA reductase catalyzes the formation of mevalonic acid, an early intermediate in the biosynthesis of cholesterol, and efficiently lower blood serum cholesterol. The Nobel Prize in Physiology or Medicine 1985 was awarded to Michael S. Brown and Joseph L. Goldstein "for their discoveries concerning the regulation of cholesterol metabolism." Their findings [1], as well as much earlier obtained data that some derivatives of mevalonic acid and their ring closed lactone form - mevalonolactones inhibit the biosynthesis of cholesterol [2,3] led to the development of a variety of statin drugs.

Mevalonic acid is very soluble in water and exists in equilibrium with the mevalonolactone. In biological media, the anion form of mevalonic acid is the predominant form (Fig. 20.1.).

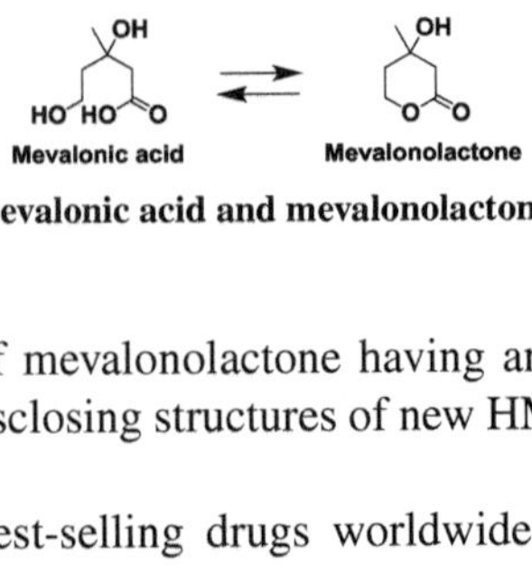

FIG. 20.1 Mevalonic acid and mevalonolactone equilibrium.

Several derivatives of mevalonolactone having antilipidemic activity have been initially patented disclosing structures of new HMG-CoA reductase inhibitors [4-6].

Statins, which are best-selling drugs worldwide with a market value of approximately $30 billion, have demonstrated multiple beneficial effects in lowering levels of LDL and in the prevention of cardiovascular disease.

Since the introduction of statins in the 1980s, no other class of lipid modulators has entered the market.

Mevastatin (**20.1.1**) was the first agent identified in 1970 and is considered to be the first statin drug. Lovastatin (**20.1.2**) is another naturally occurring compound of this series and was the first statin approved by the FDA (in 1987) as a hypolipidemic agent for lowering LDL cholesterol.

Commonly prescribed drugs available worldwide today are lovastatin (**20.1.2**), simvastatin (**20.1.3**) (Zocor), pravastatin (**20.1.4**) (Pravachol), atorvastatin (**20.1.5**) (Lipitor), fluvastatin (**20.1.6**) (Lescol), pitavastatin (**20.1.7**) (Livalo), cerivastatin (**20.1.8**) (Baycol), and rosuvastatin (**20.1.9**) (Crestor) (Fig. 20.2.) It is believed that first three compounds—mevastatin (**20.1.1**), lovastatin (**20.1.2**) and simvastatin (**20.1.3**)—are administered as prodrug lactones, which, over a period of time, convert in vivo to their respective active hydroxyl acid forms, such as in structures (**20.1.4** to **20.1.9**).

Atorvastatin and rosuvastatin are the most prescribed drugs. Pitavastatin is licensed in a few countries. Cerivastatin was voluntarily withdrawn from the market worldwide in 2001, as a result of reports of fatal rhabdomyolysis. These drugs display different pharmacokinetic properties and are metabolized by different pathways.

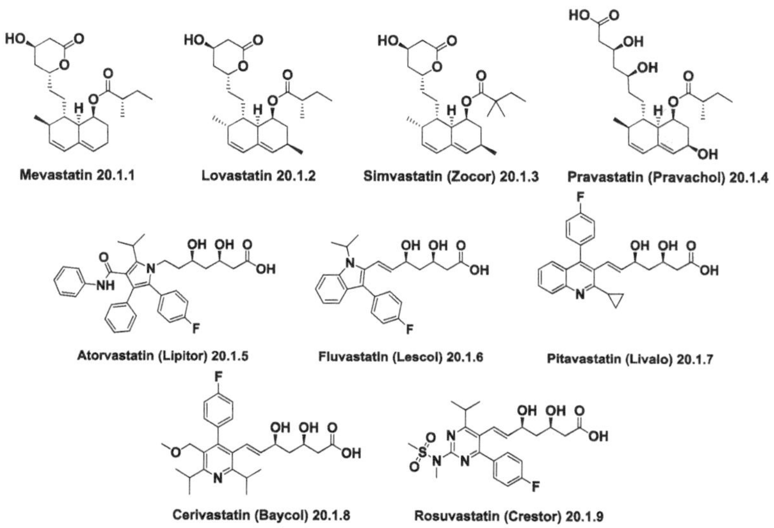

FIG. 20.2 Statins proposed for the pharmaceutical market.

Mevastatin, also known as ML-236B or compactin, (**20.1.1**) was first discovered during the search for a new antimicrobial agent in the fermentation broth of *Penicillium citrinum.* It was shown to have antimicrobial activity and, at the same time, cholesterol biosynthesis pathway inhibitor properties by Akira Endo [7,8] and presented in review papers [9-12]. Mevastatin was never marketed as an HMG-CoA reductase inhibitor because of its multiple adverse effects, including tumors, muscle deterioration, and death of laboratory animals.

Lovastatin (**20.1.2**) is another naturally occurring statin isolated from *Aspergillus terreus*, and was the first statin approved by the FDA (in 1987) as a hypolipidemic agent for lowering LDL cholesterol. Available today are two semisynthetic statins (simvastatin and pravastatin) and four synthetic statins (atorvastatin, fluvastatin, pitavastatin, and rosuvastatin). Two of them, atorvastatin (**20.1.5**) and rosuvastatin (**20.1.9**), are included in the list of Top 200 Drugs by sales for the 2010s.

Atorvastatin–Lipitor

Atorvastatin was introduced under the trade name Lipitor in 1997. Since 2001, Lipitor has been the leading pharmaceutical product in global sales, reaching $13.6 billion in 2006. Several reviews provide the story of discovery and synthesis of atorvastatin [13-15].

Racemic atorvastatin (**20.1.17**), with further separation of diastereomers, was first synthesized in mid-1980s via condensation of ethyl 2-bromo-2-(4-fluorophenyl)acetate (**20.1.10**) with 2-(1,3-dixolan-2-yl)ethylamine in acetonitrile in the presence of triethylamine, which produced ethyl 2-[2-(1,3-dioxolan-2-yl)ethylamino]-2-(4-fluorophenyl) acetate (**20.1.11**). The last was acylated with isobutyryl chloride, which, after hydrolysis of the ester group with NaOH in methanol/water, yielded free acid (**20.1.12**), which was cyclized with propynamide (**20.1.13**) by a 3 + 2 cycloaddition reaction, forming only one of the two possible regioisomeric products. This reaction, followed by extrusion of CO_2, occurred by heating the reagents at 90°C in acetic anhydride to produce a pyrrole derivative (**20.1.14**), which underwent a two-step procedure for the hydrolysis of the dioxolane group. First, it was converted to diethylacetal on refluxing in ethanol/HCl. The final hydrolysis to aldehyde group occurs on refluxing the obtained diethylacetal in p-TsOH, acetone/water system, yielding the corresponding aldehyde (**20.1.15**). Obtained aldehyde was condensed with methyl acetoacetate under Weiler dianion condensation reaction conditions by means of NaH and n-BuLi in THF, yielding the racemic 5-hydroxy-3-oxoheptanoic acid methyl ester (**20.1.16**). Diastereoselective reduction of the carbonyl group of (**20.1.16**) with tributylborane and $NaBH_4$ in THF produces the corresponding dihydroxy ester as a 9:1 mixture of syn- ant anti- diols, respectively. The implemented method previously was shown to be effective for stereoselective synthesis of 1,3-diols from β-hydroxyketones [16]. Most probably in the first stage of the reaction, triethylborane converts the β-hydroxyketone to the corresponding sufficiently rigid diethylborinic acid ester, which, on reduction with $NaBH_4$, gives predominantly the syn-diol isomer, ester group in which underwent base hydrolysis ($NaOH/H_2O$), yielding the acid (**20.1.17**). The lactonization of the obtained acid in refluxing toluene produces lactone (**20.1.18**), which was submitted to reaction with (R)-1-phenylethylamine to produce the mixture of diastereomeric amides (**20.1.19a,b**), which was followed by their chromatographic separation, and further base hydrolysis to appropriate acids. This was followed by heating in refluxing toluene with azeotropic removal of water to obtain the optically pure (+)-lactone (**20.1.20b**), which was finally hydrolyzed with using $NaOH/CH_3OH/H_2O$ mixture, and then treated with calcium chloride or calcium acetate to give desired biologically active (+)-stereoisomer of atorvastatin (**20.1.5**) as a calcium salt [17-20] (Scheme 20.1.).

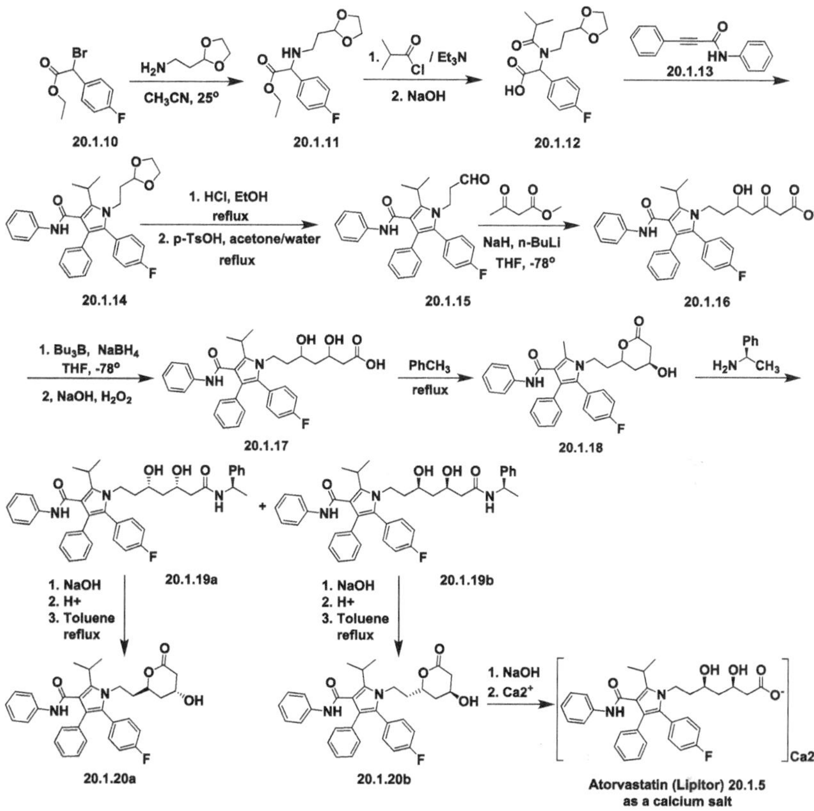

SCHEME 20.1 Synthesis of atorvastatin.

Several schemes for enantioselective synthesis of (+)-stereoisomer of atorvastatin have been proposed, and one of the earliest [20] was based on the aldehyde (**20.1.15**) available in gram quantities.

In the first enantioselective synthesis of atorvastatin magnesium enolate of (S)-(+)-2-acetoxy-l,1,2-triphenylethanol (**20.1.21**) obtained on interaction of (**20.1.21**) with lithium diisopropyl amide (LDA) followed by cation exchange with $MgBr_2$ was reacted with aldehyde (**20.1.15**), producing alcohol (**20.1.22**) in good yield and high optical purity (the enantiomeric excess 97%). After transesterification of (**20.1.22**) using sodium methoxide in CH_3OH/THF media, ester (**20.1.23**) was obtained, which underwent Claisen condensation with an excess of lithio tert-butylacetate (**20.1.24**) to produce β-keto ester (**20.1.25**) in high yield. Reduction of the last with triethylborane and $NaBH_4$, as it is described in Scheme 20.1., and further base hydrolysis (NaOH) of the ester group, after lactonization in boiling toluene with azeotropic, removal of water produced a 98:2 mixture of stereoisomers of (**20.1.20a,b**), which on crystallization from ethyl acetate-hexanes, produced the desired pure (+)-(**20.1.20b**) as a single isomer in gram quantities (Scheme 20.2.).

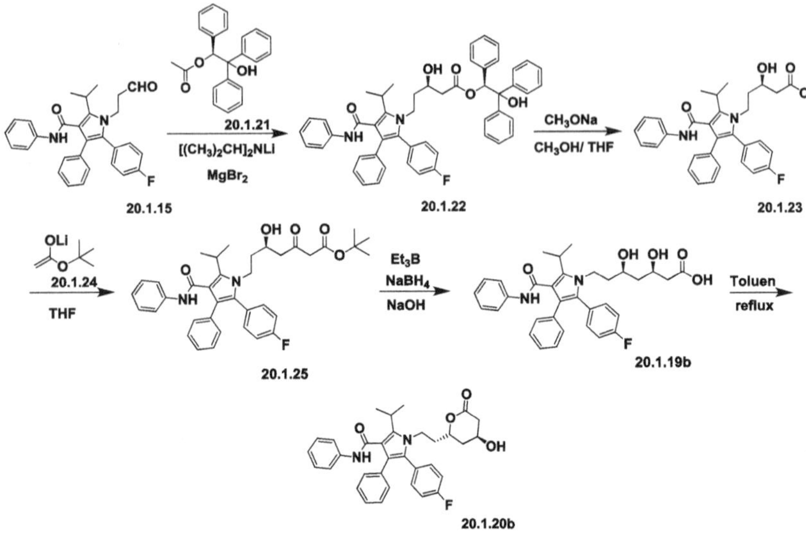

SCHEME 20.2 **Enantioselective synthesis of (+)-stereoisomer of atorvastatin.**

Another approach has been demonstrated for the kilogram scale production of atorvastatin; it implements the Paal-Knorr pyrrole synthesis reaction via condensing of appropriate 1,4-diketones with amines for the synthesis of one of two key intermediates—aldehyde (**20.1.15**).

To begin, 1,4-diketone (**20.1.31**) was prepared in two steps [21,22]. Commercially available isobutyrylacetanilide (**20.1.26**) and benzaldehyde (**20.1.27**) were coupled in Knoevenagle reaction conditions using acetic acid and β-alanine as catalysts to produce product (**20.1.28**) with high yield. The enone (**20.1.28**) and p-fluorobenzaldehyd were involved in a Stetter reaction, a conjugate addition of an aldehyde, in the given case benzaldehyde (**20.1.28**), to produce an α,β-unsaturated compound in the presence of thiazolium salt (**20.1.29**) as a catalyst, which produced 1,4-dicarbonyl compound (**20.1.31**). The last was cyclized to desired the pyrrole derivative (**20.1.15**) on reflux in boiling toluene-heptane-THF 1:4:1 solvent mixture using pivalic acid as a catalyst.

Several other methods based on Paal-Knorr pyrrole synthesis via direct condensation of 1,4-diketone (**20.1.31**) with a variety of second key intermediates—different derivatives of 7-amino-3,5-dihydroxyheptanoic acid like compound (**20.1.32**)—have been employed [23-27] (Scheme 20.3.). Methods for the synthesis of atorvastatin have been reviewed [14,28].

In dosages of 10 to 80 mg/day, atorvastatin reduces levels of total cholesterol, LDL—cholesterol, triglyceride, and VLDL—cholesterol and increases HDL cholesterol in patients with a wide variety of dyslipidemias. Pharmacology and implementations of atorvastatin are reviewed [29-33].

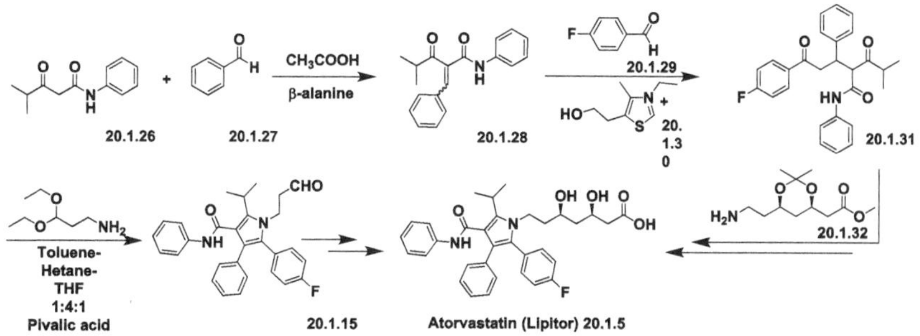

SCHEME 20.3 **Synthesis of atorvastatin.**

Rosuvastatin–Crestor

Rosuvastatin is the latest statin to receive approval by the FDA. It is indicated to reduce elevated levels of total cholesterol, LDL cholesterol, and triglycerides, and to increase levels of HDL cholesterol in patients with hypercholesterolemia and dyslipidemia. It possesses significant (HMG-CoA) inhibiting activity, and at daily doses of 5 to 40 mg, produces a reduction in plasma LDL cholesterol of 45 to 63%, which is statistically greater than those achieved with equivalent doses of atorvastatin, simvastatin, and pravastatin.

Synthesis of rosuvastatin starts from Knoevenagel condensation of ethyl isobutyrylacetate (**20.1.33**) with p-fluorobenzaldehyde (**20.1.9**), which provided the benzylidene keto ester (**20.1.34**), cyclocondensation of which with S-methylisothiourea in hexamethylphosphoramide (HMPA) at 100°C produced the intermediate (**20.1.35**), which, without further purification, underwent dehydrogenation-aromatization with 2,3-dichloro-5,6-dicyano-p-benzoquinone (DDQ) to produce the pyrimidine derivative (**20.1.36**). The sulfur moiety in the synthesized S-methyl pyrimidine (**20.1.36**) was oxidized by m-chloroperbenzoic acid (m-CPBA) in chloroform to produce methylsulfonylpyrimidine (**20.1.37**). This compound was transformed to a methylamino derivative (**20.1.38**) via interaction with methylamine ethanol solution and then to sulfonamide (**20.1.39**) by treatment with sodium hydride followed by methanesulfonyl chloride. Reduction of the ester group of compound (**20.1.39**) with diisobutylaluminium hydride (DIBAL) in toluene, followed by tetrapropylammonium perruthenate (TPAP) oxidation of synthesized primary alcohol (**20.1.40**) produced the corresponding aldehyde (**20.1.41**). The last underwent Wittig reaction with the optically active ylide-(3R)-3-(tert-butyldimethylsilyloxy)-5-oxo-6-triphenylphosphoranylidenehexanoate (**20.1.42**) in boiling MeCN to produce heptanoate (**20.1.43**), which was deprotected with HF in MeCN, and stereoselectively chelation controlled with diethylmethoxyborane intermediate was reduced to syn-diol with the use of $NaBH_4$ in THF to produce the ester (**20.1.44**).The dihydroxyheptanoate (**20.1.44**) reacted with aqueous NaOH to produce the sodium salt, which was transformed to desired calcium salt, rosuvastatin (**20.1.9**) [34,35] (Scheme 20.4.).

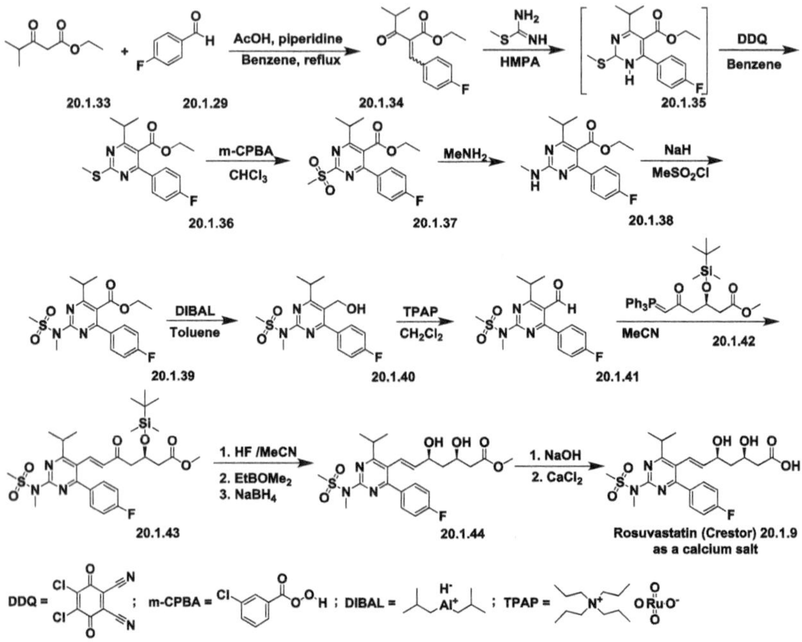

SCHEME 20.4 Synthesis of rosuvastatin.

Another approach for the synthesis of rosuvastatin has been proposed. Using the Wittig reaction between in situ prepared phosporhane (**20.1.45**) and (2S,4R)-4-(tert-butyldimethylsilyloxy)-6-oxotetrahydro-2H-pyran-2-carbaldehyde (**20.1.46**) followed with subsequent one-pot deprotection/hydrolysis step without any detectable epimerization gave an excellent overall yield of desired rosuvastatin (**20.1.9**) [36] (Scheme 20.5.).

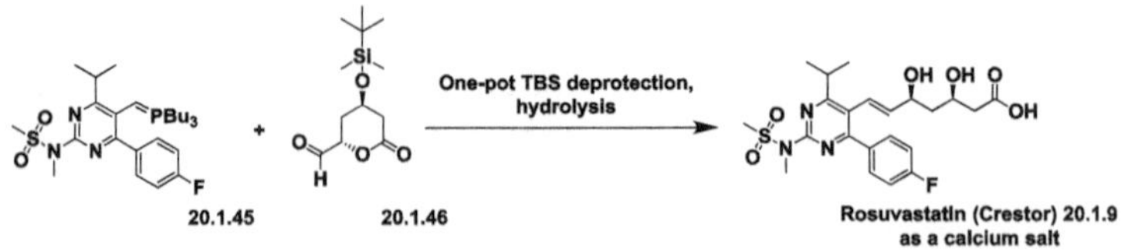

SCHEME 20.5 Synthesis of rosuvastatin.

Rosuvastatin (Crestor) is prescribed to lower cholesterol and triglyceride levels and to prevent heart attacks and strokes. The usual dose ranges from 5 to 20 mg per day. In some patients, side effects, such as muscle and abdominal pain, diarrhea, or allergic symptoms (hives, shortness of breath, tissue swelling) can develop. Pharmacology and medicinal use of atorvastatin have been reviewed [37-51] and it can be considered the drug of choice for the management of dyslipidemia because it has a good safety profile, but the risk of development of gastrointestinal disturbances, myopathy [52-54], and diabetogenic effects [55-58] raises some questions.

20.2 FIBRATES

Fibrates are a class of lipid-lowering drugs used primarily for hypertriglyceridemia [59-75]. In contrast to statins, this group of drugs does not inhibit cholesterol biosynthesis. The effects of fibrates are mediated through the activation of the transcription factor peroxisome proliferator-activated receptors (PPARs), that regulates the expression of many enzymes and genes.

The PPARs appear to play a significant role as sensors and regulators of lipid metabolism. There are three receptor isoforms including PPAR-α, -γ, and -δ, which are encoded by different genes, and normalization of lipid metabolism is achieved via pharmacological modulation of the PPAR-α with corresponding agonists.

It is suggested that fibrates induce lipoprotein lipolysis, reduce hepatic triglyceride formation, and induce hepatic fatty acid uptake. Fibrates intensify removal of LDLs and increase HDL production. Moreover, fibrates also target aldose reductase, a major player in the development of diabetes.

Fibrates promote a shift from small, dense LDL to larger particles, which are less susceptible to oxidation and possess higher binding affinity for removal by the nonatherogenic LDL receptor pathway, correcting lipid abnormalities commonly observed in patients at high risk of cardiovascular disease.

These combined actions are demonstrated by a reduction of elevated triglyceride levels by up to 50% and a rise in HDL cholesterol concentrations by 5 to 15%. Fibrates are reasonable second-line therapies for dyslipidemia and diabetes. They are safe in combination therapy with statins.

Fibrates have been used clinically since the late 1960s. All of them except for gemfibrozil are derivatives of 2-hydroxy-2-methylpropanoic acid. This drug group includes clofibrate (**20.2.1**), ciprofibrate (**20.2.2**), bezafibrate (**20.2.3**), fenofibric acid (**20.2.4**), fenofibrate (**20.2.5**), etofibrate (**20.2.6**), etofylline (**20.2.7**), and gemfibrozil (**20.2.8**). Prolonged use of some of these drugs, for example, clofibrate and ciprofibrate, can cause peroxisome proliferation leading to hepatomegaly and tumor formation in the liver of some laboratory animals (Fig. 20.3.) Only gemfibrozil and fenofibrate, because of their milder effect on peroxisome proliferation, are being used as lipid-lowering drugs in humans.

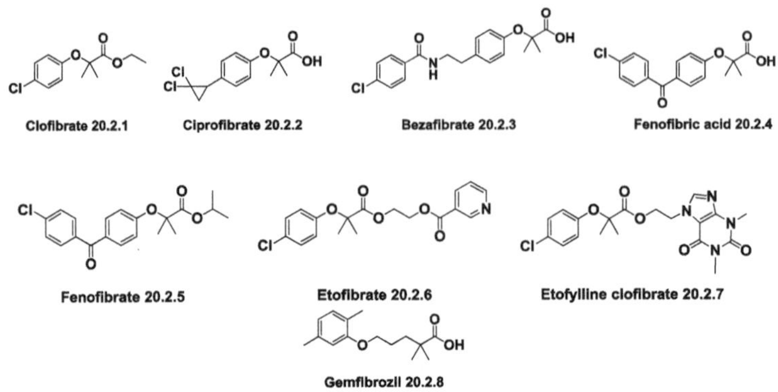

FIG. 20.3 Structure of fibrates.

The syntheses of clofibrate and gemfibrozil have been described in our previous book [76]. Fenofibrate is included in the list of Top 200 Drugs by sales for the 2010s.

Fenofibrate–Tricor

Several protocols for the preparation of fenofibrate are known. The first patents [77-79] and later published paper [80] disclosed the synthesis of fenofibrate starting from 4-chloro-4′-hydroxy-benzophenone (**20.2.12**), which was obtained from condensation of 4-chlorobenzoyl chloride (**20.2.9**) and anisole (**20.2.10**), which was demethylated with the use of hydrobromic acid to produce the desired phenol derivative (**20.2.12**).

The obtained compound was dissolved in acetone, to which was added NaOH followed by $CHCl_3$, which gave the appropriate fenofibric acid (**20.2.4**), which was esterified on reflux with isopropanol in presence of sulfuric acid. Another method suggests direct etherification of phenol (**20.2.12**) with isopropyl 2-bromo-2-methylpropanoate (**20.2.12**) in refluxing isopropanol in presence of potassium bicarbonate [81,82], or on reflux with 2-halogen-2-methylpropanoic acid (**20.2.14**) in aqueous NaOH followed with esterification with isopropanol [83,84] (Scheme 20.6.).

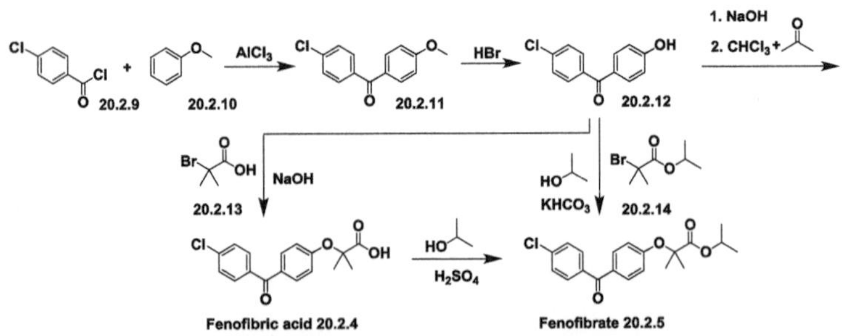

SCHEME 20.6 Synthesis of fenofibrate.

The lipid-modifying effects of fenofibrate are mediated by activation of PPAR-α. It is indicated for use in the treatment of primary hypercholesterolemia, mixed dyslipidemia, and hypertriglyceridemia. Fenofibrate improves triglyceride and HDL-cholesterol levels. Greater improvements in lipid levels are seen when fenofibrate is administered in combination with statins or in combination with ezetimibe. Fenofibrate also has nonlipid (i.e., pleiotropic) effects. It reduces fibrinogen, C-reactive protein, and uric acid levels, and improves flow-mediated dilation. Serious side effects include severe stomach/abdominal pain, persistent nausea/vomiting, yellowing eyes/skin, and dark urine. Rare, but serious, side effects could appear as unusual muscle pain, tenderness, or weakness [85-91].

20.3 BILE ACID SEQUESTRANTS

Bile acid sequestrants, whose mechanism of action is based on the contradictory hypothesis that disruption of bile acid reabsorption from the gut will remove bile acid molecules from the gastrointestinal tract, thus decreasing cholesterol levels are the oldest lipid-lowering agents, having been used for more than 50 years in the treatment of hypercholesterolemia [92-97]. This hypothesis is based on the evidence that bile acids—amphipathic molecules that are synthesized in the liver via a multistep metabolic pathway—from cholesterol (**20.3.1**), which is first converted to the primary bile acids, chenodeoxycholic acid (**20.3.2**) and cholic acid (**20.3.3**), which, in turn, are transformed to secondary bile acids, lithocholic acid (**20.3.4**) and deoxycholic acid (**20.3.5**) (Fig. 20.4.), will lower cholesterol levels. But cholesterol homeostasis is critical because it plays a role not only in hypercholesterolemia but in the synthesis of biologic membranes, steroid hormones, and vitamin D, and modulate glycemic control, and its maintenance must take into account all these factors.

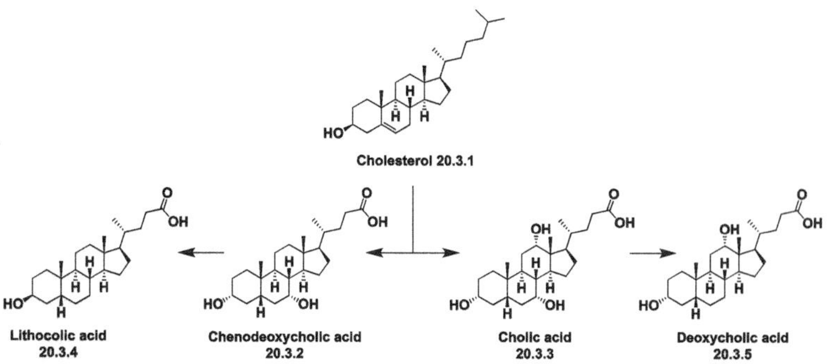

FIG. 20.4 Schematic metabolic pathway of the synthesis of bile acids from cholesterol.

Nonetheless, bile acid sequestrants are a group of polymers that bear positively charged nitrogen cations that bind to bile acids anions in the small intestine and then are excreted with feces. This approach has been developed as a strategy to treat hypercholesterolemia.

Because of their high molecular weight, the field of action of bile sequestrant polymers is limited to the gastrointestinal tract, which allow avoiding systemic exposure. It is considered an advantage over conventional small-molecule drugs.

Different vinyl, acrylic, and allyl polymers were used to synthesize potential bile acid sequestrants based on conventional polymerization techniques.

Three compounds are available on the market: cholestyramine (**20.3.6**), colestipol (**20.3.7**) (first-generation bile acid sequestrants), and colesevelam (**20.3.8**) (Fig. 20.5.) Cholestyramine and colestipol have greater affinity for dihydroxy than for trihydroxy bile salts, which creates an imbalance in the bile salt pool, increasing the amount of trihydroxy bile salt fraction. In contrast, colesevelam binds to entire bile salts equally.

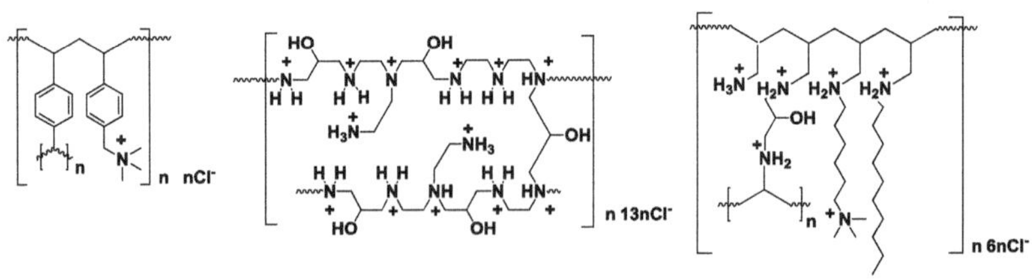

Cholestyramine **20.3.6** | Colestipol **20.3.7** | Colesevelam **20.3.8**

FIG. 20.5 **Bile acid sequestrants.**

Colesevelam–Welchol

Colesevelam (**20.3.8**) is included in the list of Top 200 Drugs by sales for the 2010s.

Colesevelam (**20.3.8**) is a crosslinked polyallylamine polymer composed of polyallylamine crosslinked with epichlorohydrin in water solution at room temperature. The obtained crosspolymer was then alkylated with the mixture of 1-bromodecane and 6-bromohexyltrimethylammonium bromide. The separated hydrobromide salt was treated with aqueous sodium hydroxide and then with concentric hydrochloric acid to produce the desired colesevelam as a hydrochloride [98-104].

Colesevelam is a new bile acid sequestrant. It binds to bile acids with higher affinity than traditional sequestrants with fewer side effects and drug interactions. Colesevelam is efficacious alone or in combination with statins in reducing LDL cholesterol levels. Colesevelam improves glycemic control when added to existing antidiabetes therapy in patients with type 2 diabetes [105-113].

20.4 EZETIMIBE–VYTORIN

Another approach for the treatment of hyperlipidemias is inhibition of absorption of cholesterol from the intestines. The majority of the cholesterol that is absorbed from the intestines comes from the liver and was excreted into the intestines in bile; only some of intestinal cholesterol comes from the diet. It is believed that drugs that could selectively inhibit the intestinal absorption of cholesterol and related phytosterols via binding to a critical mediator of cholesterol absorption, the polytopic transmembrane protein Niemann-Pick C1-like 1 (NPC1L1) [114], which is mediated extracellular sterol transport across the brush border in the gastrointestinal tract epithelial cells, could reduce the growth of fatty plaques in arteries.

Ezetimibe[115-134]representsthefirstexampleofanewclassoflipid-lowering drugs such as (**20.4.1** to **20.4.3**) which inhibit cholesterol absorption from the intestines (Fig. 20.6.) [116, 135-137] and belong to derivatives of β-lactams, which in general are a part of the core structure of widest range of antibiotics such as penams, cephems, monobactams, carbapenems.

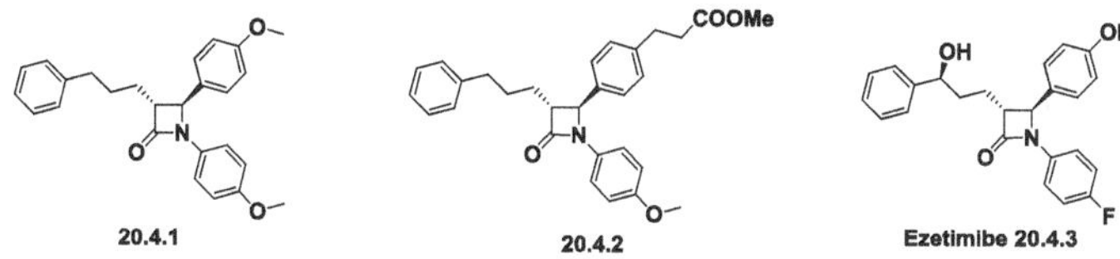

FIG. 20.6 Inhibitors of absorption of cholesterol from the intestines.

Ezetimibe reduces LDL cholesterol when administered either alone or in combination with statins. However, this drug has been the subject of a number of reports where its effect remains under question because it failed to demonstrate significant changes on the cardiovascular mortality according to the published studies [138-141].

Ezetimibe (**20.4.8**) is included in the list of Top 200 Drugs by sales for the 2010s.

A process for preparation of ezetimibe has been disclosed [142-145]. The key step in the synthesis of ezetimibe is formation of four-membered azetidinone (**20.4.8**), which was synthesized via a 2 + 2 ketene-imine cycloaddition reaction with high trans diastereoselectivity. Imine (**20.4.6**) was prepared simply by condensation of p-benzyloxybenzaldehyde (**20.4.4**) with p-fluoroaniline (**20.4.6**). By adding methyl glutaryl chloride in toluene to a mixture of imine and double excess of tributhylamine in boiling heptane, the intermediate ketene was generated, which on reaction with imine (**20.4.6**) gave desired azetidinone (**20.4.8**). The reaction proceeds to produce almost exclusively the trans isomer (15:1 trans:cis) of (**20.4.8**) which was additionally purified via chiral chromatography (Chiracel OD). The ester group in the obtained compound was hydrolyzed with lithium hydroxide to produce acid (**20.4.9**), which was transformed to acid chloride (**20.4.10**) with oxalyl chloride. The acid chloride was condensed with the zinc reagent formed in situ from p-fluoromagnesium bromide and zinc chloride in the presence of catalytic amount of palladium–tetrakis(triphenylphosphine) to produce the ketone (**20.4.11**). The carbonyl group of the obtained ketone was then reduced with diborane or dimethyl sulfide borane to produce the equal mixture of diastereomeric alcohols (**20.4.12a,b**) separation of which on Chiracel OD column produced the desired (3S) alcohol (**20.4.12a**). Removal of the benzyl protecting group from compound (**20.4.12a**) by hydrogenolysis over palladium catalyst finally produced ezetimibe (**20.4.3**) [142-145]. (Scheme 20.7.).

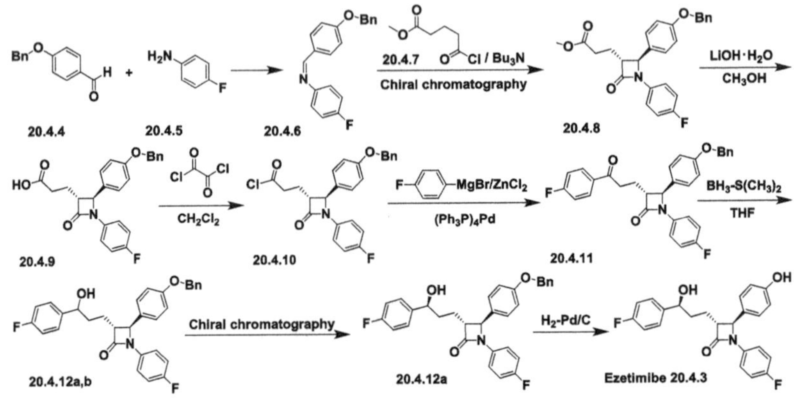

SCHEME 20.7 **Synthesis of ezetimibe.**

Another alternate way for the synthesis of the azetidinones of the series (**20.4.8**) was proposed as a one-step reaction between commercially available (S)-3-hydroxy-γ-lactone (**20.4.13**) and imine (**20.4.6**). This reaction represents a one-step enantio- and diastereoselective practical synthesis for azetidinones from γ-lactones.

For preparation of ezetimibe (**20.4.3**) a dianion of lactone (**20.4.13**) was generated with LDA in THF/N,N-dimethylpropyleneurea/(DMPU) mixture and added to imine (**20.4.6**) dissolved in DMF. The reaction was maintained at −25 to −30°C for approximately 20 to 22 hours and then the LiCl dissolved in DMF was added to the reaction mixture to accelerate the cyclization reaction (**20.4.14** to **20.4.15**), which proceeded with high diastereoselectivity (95:5 trans:cis ratio), which was further enhanced after crystallization. The mixture of diols (**20.4.15**) underwent an oxidative cleavage with sodium periodate in acetonitrile to form aldehyde (**20.4.16**). Interestingly, the cis isomer (**20.4.15a**) under the reaction conditions was epimerized at C-3 to give the enantiomer of the desired product, resulting in a 100% trans β-lactam aldehyde. This compound was converted to β-hydroxyketone by condensation with 4-fluoroacetophenone implementing Mukaiyama aldol condensation conditions. For that purpose 4-fluoroacetophenone first was enolized by the use of LDA and then the obtained lithium enolate was transferred to trimethylsilyl ether, reacting with trimethylsilyl chloride (TMSCl). Further reaction of enol trimethylsilyl ether with aldehyde (**20.4.16**) catalyzed by $TiCl_4$ produced diastereomers of β-hydroxy ketone (**20.4.17**), which were dehydrated with p-TsOHin toluene to give enone

(**20.4.18**). The crude (**20.4.18**) was subjected to hydrogenation under hydrogen pressure of 60 psi using the Wilkinson catalyst, $(Ph_3P)_3RhCl$, to produce ketone (**20.4.19**). The ketone (**20.4.19**) was hydrogenated with borane dimethylsulfide complex using the Corey-Bakshi-Shibata chiral catalyst prepared in situ from trimethylboroxine and diphenylprolinol, obtaining benzyl-protected ezetimibe (**20.4.12a**), which was deprotected by standard palladium-charcoal catalyzed hydrogenation procedure to produce the desired ezetimibe (**20.4.3**) [145,146] (Scheme 20.8.).

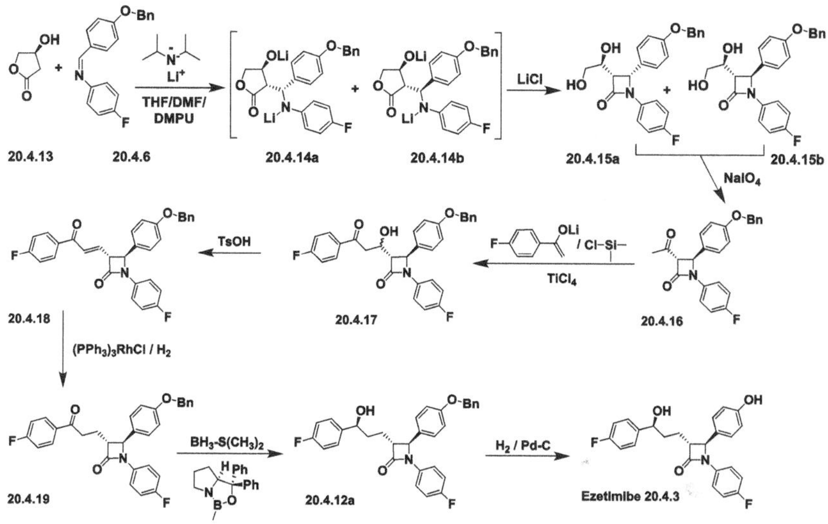

SCHEME 20.8 Synthesis of ezetimibe.

Further attempts for the scalable methodology for ezetimibe synthesis devoid of chromatography were based on Evans chemistry of oxazolidinones.

Commercially available 5(S)-phenyloxazolidinone (**20.4.20**) was acylated with methyl glutaryl chloride (**20.4.7**) to provide oxazolidinone (**20.4.21**). The oxazolidinone (**20.4.21**) was converted to titanium enolate with either $TiCl_4$ or $TiCl_3O$-i-Pr in presence of two equivalents of a base, such as tertiary amines and, particularly, N,N-diisopropylethylamine (DIPEA), and the obtained titanium enolate was reacted with imine (**20.4.6**) to produce, after recrystallization, the desired aminoamide (**20.4.22**) with very high stereochemical purity. Silylation of (**20.4.22**) with bis(trimethylsilyl)acetamide (BSA) followed by fluoride ion catalyzed (tetrabutylammonium fluoride [TBAF], LiF, KF, CsF) cyclization produced the previously described 2-azetidinone (**20.4.8**) in a one-pot operation. After the ester group hydrolysis with lithium hydroxide, transformation to acid chloride with oxalyl chloride and condensation with p-fluorophenylzinc bromide in presence of palladium-tetrakis(triphenylphosphine), the known ketone (**20.4.19**) was prepared. The carbonyl group of the (**20.4.19**) was reduced with diborane using the Corey-Bakshi-Shibata chiral catalyst to produce the desired (3S) alcohol (**20.4.12a**). Removal of the benzyl protecting group from which by standard hydrogenolysis over palladium catalyst produced ezetimibe (**20.4.3**) [147-149] (Scheme 20.9.). Some modified approaches of described methodology have been disclosed [150].

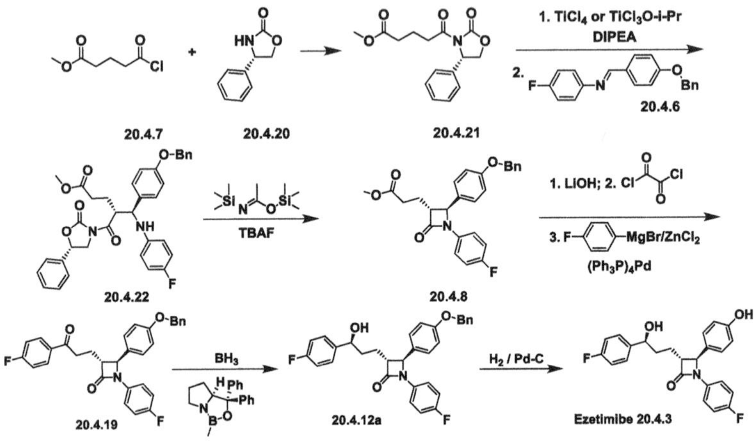

SCHEME 20.9 Synthesis of ezetimibe.

Ezetimibe may be used alone or in combination with any HMG-CoA reductase inhibitors (statins) together with lifestyle changes (diet, weight loss, exercise). The most common side effects are fever, headache, muscle pain, runny nose, sore throat.

20.5 NIACIN

Niacin, vitamin B_3, or pyridine-3-carboxylic acid (**20.5.1**) is the oldest lipid-lowering medication with unique antiatherosclerotic property and is referred as vitamin B_3 because it was the third compound on discovery of eight compounds in the vitamin B series. Sometimes it is referred as vitamin PP, which comes from the definition "pellagra-preventive factor."

Niacin has favorable effects on all aspects of an abnormal lipid profile. It reduces LDL-C, triglycerides and lipoprotein and increases HDL-C. The mechanisms by which niacin exerts its effects are complex and as yet incompletely characterized, but it is clear that it decreases the flux of fatty acids from adipose tissue to the liver by inhibiting hormone-sensitive lipase activity; inhibits triglyceride formation within hepatocytes by inhibiting diacylglycerol acyl transferase, an enzyme that esterifies fatty acid to glycerol; reduces serum triglyceride levels; increases LDL particle size; and decreases LDL particle number.

Niacin is not included in the list of Top 200 Drugs by sales for the 2010s but its world demand is more than 40,000 tons in recent years. Figure 20.7 presents a brief description of its methods of synthesis.

The main industrial method consists in oxidation of 2-methyl-5-ethylpyridine with HNO_3 [151,152]. 2-Methyl-5-ethylpyridine for this purpose is industrially produced in large scale from paraldehyde and ammonia [153]. Another method of synthesis starts from pyridine that has been brominated in the third position to produce 3-bromopyidine, which, in turn, was converted to 3-cyanopyidine and then hydrolyzed to the desired pyridine-3-carboxylic acid [154]. Oxidation of nicotine, β-picoline, and quinolone is also proposed as a method for synthesizing pyridine-3-carboxylic acid–niacin (**20.5.1**) [155,156].

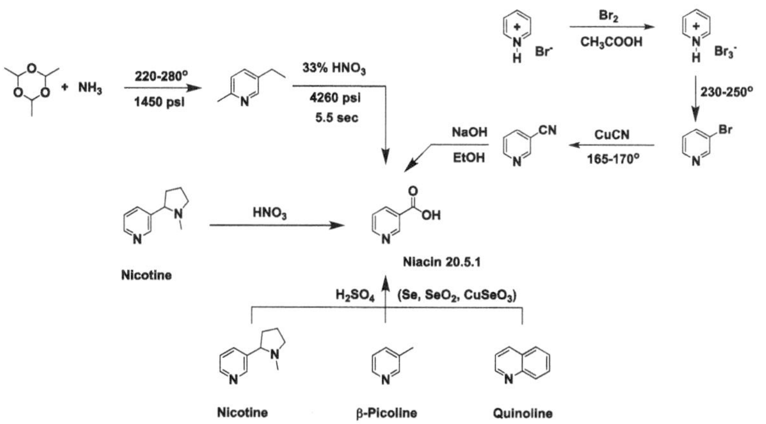

FIG. 20.7 **Synthesis of niacin.**

Since the 1950s, niacin has been used to lower elevated LDL cholesterol and triglyceride levels in the blood.

It seems that niacin can be considered one the most effective and the most contradictory medications in clinical use for increasing HDL cholesterol and lowering triglycerides and LDL cholesterol. Opposing opinions can be found in recent publications [157-166].

The most frequent side effects of niacin are rapid heartbeat, flushed skin, itching, nausea and vomiting, abdominal pain and other gastrointestinal problems, and liver damage.

20.6 OMEGA-3 FATTY ACIDS (FISH OILS)

The omega-3 fatty acids, essential fatty acids in the human diet, are polyunsaturated fatty acids with a double bond at the third carbon atom from the end of the carbon chain. Their list includes many important compounds, such as eicosapentaenoic acid, docosahexaenoic acid, α-linolenic acid, stearidonic acid, and eicosapentaenoic acid. Fish is the main supplier of omega-3 fatty acids that control a wide array of bodily functions and are vital for normal metabolism. Fish oil contains a complex mixture of omega-3 fatty acids, which are predominantly eicosapentaenoic acid (**20.6.1**), docosapentaenoic acid (**20.6.2**), and docosahexaenoic acid (**20.6.3**) (Fig. 20.8.). Each of them has distinct biological effects.

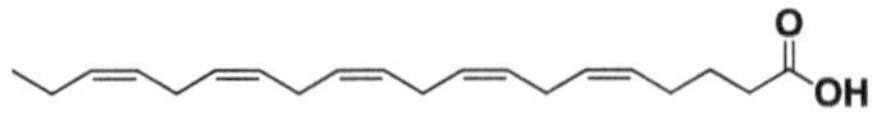

Eicosapentaenoic acid 20.6.1

Docosahexaenoic acid 20.6.2

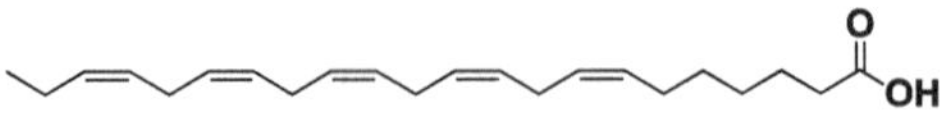

Docosapentaenoic acid 20.6.3

FIG. 20.8 **Structure of omega-3 fatty acids.**

Eicosapentaenoic (**20.6.1**) and docosahexaenoic (**20.6.1**) acids are thought to be beneficial in treating hypertriglyceridemia. Their mechanism of action is complex and not fully understood.

The suggested mechanism for their action includes suppression of lipogenesis, decreased triglycerides synthesis through inhibition of diacylglycerol acyltransferase-2 (DGAT2), and an increase of mitochondrial β-oxidation. Moreover, they may be involved in the regulation of genes in lipid metabolism. The most recent reviews on omega-3 fatty acids.

Fish oil is derived from tissues of oily fish [167]. But the fish themselves do not produce omega-3 fatty acids; they accumulate them from feeding sources. Marine and freshwater fish oil vary in content of different fatty acids.

Omega-3 fatty acids are effective in lowering serum triglyceride, usually in combination with other lipid-lowering medications such as statins, fibrates, and niacin. The most recent reviews on action and application of omega-3 fatty acids on LDL cholesterol and triglyceride levels give conflicting results [168-177].

20.7 NEW AGENTS FOR THE PHARMACOTHERAPY OF DYSLIPIDEMIA

Current lipid-altering agents—statins, fibrates, bile acid sequestrants, niacin, and omega-3 fatty acids—are used to rationally treat hyperlipidemias. A considerable number of new approaches for dyslipidemia therapies are in development [178-180].

Current agents in development include microsomal triglyceride transfer protein inhibitors, acyl-cholesterol acyl transferase inhibitors, squalene synthase inhibitors, preprotein convertase subtilisin kexin-9 inhibitors peroxisomal, antisense oligonucleotide therapies, cholesteryl ester transfer protein inhibitors, proliferator activating receptor α and γ agonists, liver X receptor and farnesoid X receptor (bile acid receptor) targeting agents, and apolipoprotein B synthesis inhibitors.

Microsomal Triglyceride Transfer Protein Inhibitors

Inhibition of triglyceride transfer protein results in an increase in the size of hepatic cholesterol pools. Two representatives of this class of compounds lomitapide (**20.7.1**) [181-184] and SLx-4090 (**20.7.2**) [185] (Fig. 20.9.), first-generation microsomal triglyceride transfer protein inhibitors, effectively block the formation of chylomicrons and still continue to remain on trials.

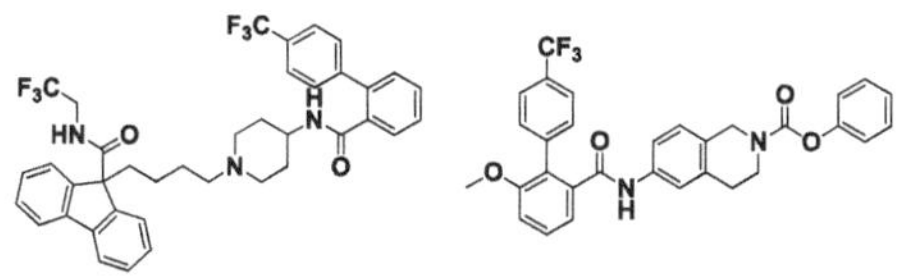

Lomitapide 20.7.1 SLx-4090 20.7.2

FIG. 20.9 Microsomal triglyceride transfer protein inhibitors.

Squalene Synthase Inhibitors

Inhibition of squalene synthase, which is further downstream in the synthesis of cholesterol, leads to a reduction in cholesterol synthesis [186]. Lapaquistat (**20.7.3**) (Fig. 20.10.), which is representative of squalene synthase inhibitors, unfortunately caused hepatic dysfunction and its trials were discontinued.

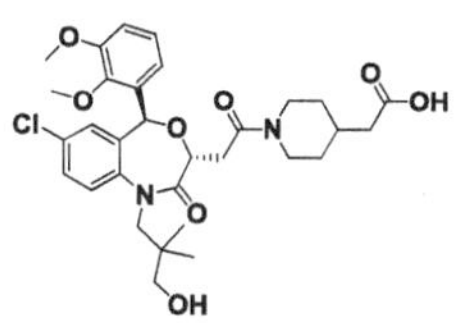

Lapaquistat 20.7.3

FIG. 20.10 Structure of lapaquistat.

Preprotein Convertase Subtilisin Kexin-9 Inhibitors

Preprotein convertase subtilisin kexin-9 (PCSK-9) is a newly discovered protein involved in regulation of LDL receptor expression. Antibodies to PCSK-9 like AMG-145, REGN-727/SAR-236553, NVP-LGT-209, REGN-727, ALN-PCS, and SPC 5001, deliver up to 70% reduction in LDL-C when added to other therapies [178-180,187-190].

Antisense Oligonucleotide Therapies

It has been shown that specifically reducing or silencing gene expression by infusing a short complementary antisense oligonucleotide sequence to messenger RNA has potential for treatment of lipid disorders. Mipomersen, an apolipoprotein B (apoB) synthesis inhibitor intended to lower LDL-C, is a synthetic oligonucleotide that was approved by the FDA in 2013 for the treatment of homozygous familial hypercholesterolemia [191-197].

Cholesterol Ester Transfer Protein Inhibitors

Cholesterol esters are transferred in exchange for triglycerides in plasma by cholesteryl ester transfer protein (CETP). Inhibition of CETP is one approach to increasing HDL-C concentrations. CETP is a plasma glycoprotein that mediates the transfer of cholesteryl esters from HDL to the apoB-containing lipoproteins, with a balanced transfer of triglycerides. Inhibition of CETP results in an accumulation of cholesteryl esters in HDL, thus resulting in increased HDL-C [198-201].

Torcetrapib (**20.7.4**) [202-204] (Fig. 20.11.) was the first CETP inhibitor administered in humans for inhibition of the process of transfer cholesterol from HDL form ("good" cholesterol) to LDL form ("bad" cholesterol). Torcetrapib's Phase III trials were cut off because of "an imbalance of mortality and cardiovascular events" associated with its use. Other CETP inhibitors, such as anacetrapib (**20.7.5**) [205,206] and evacetrapib (**20.7.6**) [207-209] (Fig. 20.11.), are still in different clinical trials phases.

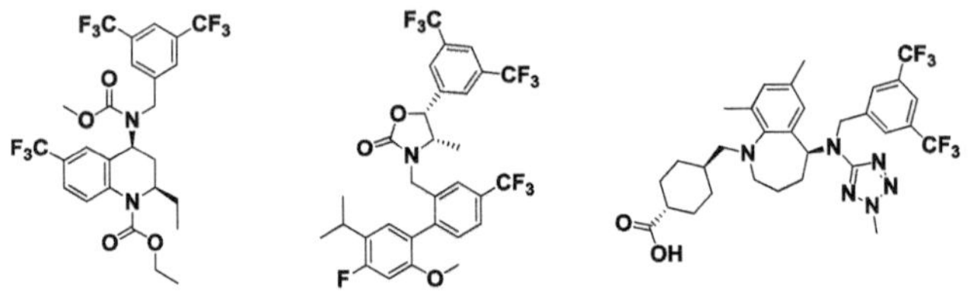

FIG. 20.11 Cholesterol ester transfer protein inhibitors.

Peroxisomal Proliferator Activating Receptor-α and -γ Agonists

PPAR-α agonists (fibrates) can reduce triglycerides by 70%, raise HDL by 20%, and reduce LDL by 10 to 25%. PPAR-γ agonists rosiglitazone (**20.7.7**) and pioglitazone (**20.7.8**) display controversial data.

PPAR-α–γ coagonists aleglitazar (**20.7.9**), muraglitazar (**20.7.10**), tesaglitazar (**20.7.11**), ragaglitazar (**20.7.12**), naveglitazar (**20.7.13**), and MK-0767 (**20.7.14**) (Fig. 20.12.) reached clinical trials, but were abandoned after bladder cancers were detected in animal models with ragaglitazar. The same cancer-generation effect that seems to be PPAR-γ driven has been suggested for pioglitazone in man. Only aleglitazar (**20.7.9**) remains in trials.

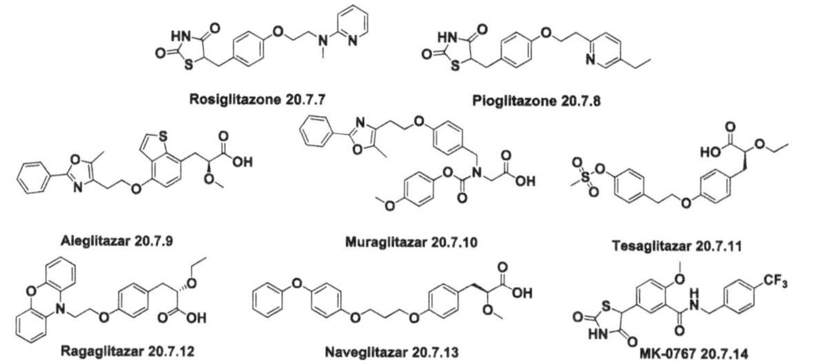

FIG. 20.12 Peroxisomal proliferator activating receptor-α and -γ agonists.

Probably, one of these newer agents will have to show a significant advantage in efficacy, tolerability, and safety over existing agents to be used in treatment of hyperlipidemias.

REFERENCES

1. Brown, M. S.; Goldstein, J. L. Lowering plasma cholesterol by raising LDL receptors. *N. Engl. J. Med.* **1981,** *305* (9), 515–517.
2. Singer, F. M.; Januszka, J. P.; Borman, A. New inhibitors of in vitro conversion of acetate and mevalonate to cholesterol. *Proc. Soc. Exp. Biol. Med.* **1959,** *102,* 370–373.
3. Hulcher, F. H. Inhibition of hepatic cholesterol biosynthesis by 3,5-dihydroxy-3,4,4-trimethylvaleric acid and its site of action. *Arch. Biochem. Biophys.* **1971,** *146* (2), 422–427.
4. Oka, H.; Terahara, A.; Endo, A. 4-Hydroxy-2-pyrone derivatives, and their pharmaceutical preparations, EP 10951 (1980).
5. Mitsui, S.; Ogiso, A.; Endo, A. Mevalonolactone derivatives, DE 2822848 (1978).
6. Willard, A. K.; Novello, F. C.; Hoffman, W. F.; Cragoe, E. J., Jr. Substituted pyranone inhibitors of cholesterol synthesis, US 4567289 (1986).
7. Endo, A.; Kuroda, M.; Tsujita, Y. ML-236A, ML-236B, and ML-236C, new inhibitors of cholesterogenesis produced by *Penicillium citrinium. J. Antibiot.* **1976,** *29,* 1346–1348.
8. Endo, A.; Tsujita, Y.; Kuroda, M.; Tanzawa, K. Inhibition of cholesterol synthesis in vitro and in vivo by ML-236A and ML-236B, competitive inhibitors of 3-hydroxy-3-methylglutaryl-coenzyme A reductase. *Eur. J. Biochem.* **1977,** *77,* 31–36.
9. Endo, A. A historical perspective on the discovery of statins. *Proc. Jpn. Acad., Ser. B* **2010,** *86* (5), 484–493.
10. Stossel, T. P. The discovery of statins. *Cell* **2008,** *134* (6), 903–905.
11. Lyons, K. S.; Harbinson, M. Statins: in the beginning. *J. R. Coll. Physicians Edinb.* **2009,** *39* (4), 362–364.
12. Patchett, A. A. 2002 Alfred Burger award address in medicinal chemistry. Natural products and design: interrelated approaches in drug discovery. *J. Med. Chem.* **2002,** *45* (26), 5609–5616.
13. Roth, B. D. The discovery and development of atorvastatin, a potent novel hypolipidemic agent. *Prog. Med. Chem.* **2002,** *40,* 1–22.
14. Jack, J.; Johnson, D. S.; Sliskovic, D. R.; Roth, B. D. *Contemporary Drug Synthesis;* John Wiley, 2004; pp 113–124.

15. Hajkova, M.; Kratochvil, B.; Radl, S. Atorvastatin-the world's bestselling drug. *Chem. Listy* **2008,** *102* (1), 3–14.
16. Narasaka, K.; Pai, H. C. Stereoselective synthesis of meso (or erythro) 1,3-diols from β-hydroxyketones. *Chem. Lett.* **1980,** *9* (11), 1415–1418.
17. Roth, B. D. Preparation of trans-6-[(carbamoylpyrrolyl)alkyl)]-4-hydroxypyranones as hypocholesterolemics, US 4681893 (1987).
18. Roth, B. D. Preparation of anticholesteremic (R-(R*R*))-2-(4-fluorophenyl)-β, δ-dihydroxy-5-(1-methylethyl-3-phenyl-4((phenylamino)carbonyl)-1H-pyrrolyl-1-heptanoic acid, its lactone form and salts thereof, EP 409281 (1991).
19. Roth, B. D.; Ortwine, D. F.; Hoefle, M. L.; Stratton, C. D.; Sliskovic, D. R.; Wilson, M. W.; Newton, R. S. Inhibitors of cholesterol biosynthesis. 1. trans-6-(2-Pyrrol-1-ylethyl)-4-hydroxypyran-2-ones, a novel series of HMG-CoA reductase inhibitors. 1. Effects of structural modifications at the 2- and 5-positions of the pyrrole nucleus. *J. Med. Chem.* **1990,** *33* (1), 21–31.
20. Roth, B. D.; Blankley, C. J.; Chucholowski, A. W.; Ferguson, E.; Hoefle, M. L.; Ortwine, D. F.; Newton, R. S.; Sekerke, C. S.; Sliskovic, D. R.; Wilson, M. W. Inhibitors of cholesterol biosynthesis. 3. Tetrahydro-4-hydroxy-6-[2-(1H-pyrrol-1-yl)ethyl]-2H-pyran 2-one inhibitors of HMG-CoA reductase. 2. Effects of introducing substituents at positions three and four of the pyrrole nucleus. *J. Med. Chem.* **1991,** *34* (1), 357–366.
21. Brower, P. L.; Butler, D. E.; Deering, C. F.; Le, T. V.; Millar, A.; Nanninga, T. N.; Roth, B. D. The synthesis of (4R-cis)-1,1-dimethylethyl 6-cyanomethyl-2,2-dimethyl-1,3-dioxane-4-acetate, a key intermediate for the preparation of CI-981, a high potent, tissue selective inhibitor of HMG-CoA reductase. *Tetrahedron Lett.* **1992,** *33* (17), 2279–2282.
22. Baumann, K. L.; Butler, D. E.; Deering, C. F.; Mennen, K. E.; Millar, A.; Nanninga, T. N.; Palmer, C. W.; Roth, B. D. The convergent synthesis of CI-981, an optically active, highly potent, tissue-selective inhibitor of HMG-CoA reductase. *Tetrahedron Lett.* **1992,** *33* (17), 2283–2284.
23. Beck, G.; Jendralla, H.; Kesseler, K. Practical large scale synthesis of tert-butyl (3R,5S)-6-hydroxy-3,5-O-isopropylidene-3,5-dihydroxyhexanoate: Essential building block for HMG-CoA reductase inhibitors. *Synthesis* **1995,** (8), 1014–1018.
24. Sletzinger, M.; Verhoeven, T. R.; Volante, R. P.; McNamara, J. M.; Corley, E. G.; Liu, T. M. H. A diastereospecific, non-racemic synthesis of a novel β-hydroxy-δ-lactone HMG-CoA reductase inhibitor. *Tetrahedron Lett.* **1985,** *26* (25), 2951–2954.
25. Chen, K.-M.; Hardtmann, G. R.; Prasad, K.; Repic, O.; Shapiro, M. J. 1,3-Syn diastereoselective reduction of β-hydroxyketones utilizing alkoxydialkylboranes. *Tetrahedron Lett.* **1987,** *28* (2), 155–158.
26. Radl, S. A new way to tert-butyl [(4R,6R)-6-(2-aminoethyl)-2,2-dimethyl-1,3-dioxan-4-yl]acetate, a key intermediate of atorvastatin synthesis. *Synth. Commun.* **2003,** *33* (13), 2275–2283.
27. Choi, H.-W.; Shin, H. Efficient synthesis of (3R,5S)-3,5,6-trihydroxyhexanoic acid derivative as a chiral side chain of statins. *Synlett* **2008,** (10), 1523–1525.
28. Casar, Z. Historic overview and recent advances in the synthesis of super-statins. *Curr. Org. Chem.* **2010,** *14* (8), 816–845.
29. Lea, A. P.; Mctavish, D. Atorvastatin a review of its pharmacology and therapeutic potential in the management of hyperlipidemias. *Drugs* **1997,** *53* (5), 828–847.
30. Graul, A.; Castaner, J. Atorvastatin calcium. *Drugs Future* **1997,** *22* (9), 956–968.
31. Haque, T.; Khan, B. V. Atorvastatin: a review of its pharmacological properties and use in familial hypercholesterolemia. *Clin. Lipidol.* **2010,** *5* (5), 615–625.
32. Poli, A. Atorvastatin: pharmacological characteristics and lipid-lowering effects. *Drugs* **2007,** *67* (Suppl. 1), 3–15.

33. Malhotra, H. S.; Goa, K. L. Atorvastatin: An updated review of its pharmacological properties and use in dyslipidemia. *Drugs* **2001,** *61* (12), 1835–1881.
34. Watanabe, M.; Koike, H.; Ishiba, T.; Okada, T.; Seo, S.; Hirai, K. Synthesis and biological activity of methanesulfonamide pyrimidine- and n-methanesulfonyl pyrrole-substituted 3,5-dihydroxy-6-heptenoates, a novel series of hmg-coa reductase inhibitors. *Bioorg. Med. Chem.* **1997,** *5* (2), 437–444.
35. Pfefferkorn, J. A. Advances in the development of methods for the synthesis of second-generation HMG-CoA reductase inhibitors [fluvastatin sodium, (Lescol), rosuvastatin calcium (Crestor), pitavasstatin calcium (Livalo)]. In *Art of Drug Synthesis;* Johnson, D. S., Li, J. J., Eds.; Wiley, 2007; pp 169–182.
36. Casar, Z.; Steinbucher, M.; Kosmrlj, J. Lactone pathway to statins utilizing the Wittig reaction. The synthesis of rosuvastatin. *J. Org. Chem.* **2010,** *75* (19), 6681–6684.
37. White, C. M. A review of the pharmacologic and pharmacokinetic aspects of rosuvastatin. *J. Clin. Pharmacol.* **2002,** *42* (9), 963–970.
38. Carswell, C. I.; Plosker, G. L.; Jarvis, B. Rosuvastatin. *Drugs* **2002,** *62* (14), 2075–2085.
39. Rosenson, R. S. Rosuvastatin: a new inhibitor of HMG-CoA reductase for the treatment of dyslipidemia. *Expert Rev. Cardiovasc. Ther.* **2003,** *1* (4), 495–505.
40. Olsson, A. G.; McTaggart, F.; Raza, A. Rosuvastatin: A highly effective new HMG-CoA reductase inhibitor. *Cardiovasc. Drug Rev.* **2002,** *20* (4), 303–328.
41. McTaggart, F. Comparative pharmacology of rosuvastatin. *Atheroscler. Suppl.* **2003,** *4* (1), 9–14.
42. Cheng, J. W. M. Rosuvastatin in the management of hyperlipidemia. *Clin. Ther.* **2004,** *26* (9), 1368–1387.
43. McKenney, J. M. Efficacy and safety of rosuvastatin in treatment of dyslipidemia. *Am. J. Health-Syst. Pharm.* **2005,** *62* (10), 1033–1047.
44. Kostapanos, M. S.; Milionis, H. J.; Elisaf, M. S. An overview of the extra-lipid effects of rosuvastatin. *J. Cardiovasc. Pharmacol. Ther.* **2008,** *13* (3), 157–174.
45. Soran, H.; Durrington, P. Rosuvastatin: efficacy, safety and clinical effectiveness. *Expert Opin. Pharmacother.* **2008,** *9* (12), 2145–2160.
46. Scott, L. J.; Curran, M. P.; Figgitt, D. P. Rosuvastatin: a review of its use in the management of dyslipidemia. *Am. J. Cardiovasc. Drugs* **2004,** *4* (2), 117–138.
47. Luvai, A.; Mbagaya, W.; Hall, A. S.; Barth, J. H. Rosuvastatin: a review of the pharmacology and clinical effectiveness in cardiovascular disease. *Clin. Med. Insights: Cardiol.* **2012,** *6,* 17–33.
48. Carter, N. J. Rosuvastatin: A review of its use in the prevention of cardiovascular disease in apparently healthy women or men with normal LDL-C levels and elevated hsCRP levels. *Am. J. Cardiovasc. Drugs* **2010,** *10* (6), 383–400.
49. Kapur, N. K. Rosuvastatin: A highly potent statin for the prevention and management of coronary artery disease. *Expert Rev. Cardiovasc. Ther.* **2007,** *5* (2), 161–175.
50. Culhane, N. S.; Lettieri, S. L.; Skae, J. R. Rosuvastatin for the treatment of hypercholesterolemia. *Pharmacotherapy* **2005,** *25* (7), 990–1000.
51. Schuster, H. Rosuvastatin—a highly effective new 3-hydroxy-3-methylglutaryl coenzyme A reductase inhibitor: Review of clinical trial data at 10-40 mg doses in dyslipidemic patients. *Cardiology* **2003,** *99* (3), 126–139.
52. Tobert, J. A. Lovastatin and beyond: the history of the HMG-CoA reductase inhibitors. *Nat. Rev. Drug Discovery* **2003,** *2* (7), 517–526.
53. Armitage, J. The safety of statins in clinical practice. *Lancet* **2007,** *370* (9601), 1781–1790.
54. Joy, T. R.; Hegele, R. A. Narrative review: statin-related myopathy. *Ann. Intern. Med.* **2009,** *150* (12), 858–868.

55. Sattar, N.; Preiss, D.; Murray, H. M.; Welsh, P.; Buckley, B. M.; de Craen, A. J. M.; Seshasai, S. R. K.; McMurray, J. J.; Freeman, D. J.; Jukema, J. W.; MacFarlane, P. W.; Packard, C. J.; Stott, D. J.; Westendorp, R. G.; Shepherd, J.; Davis, B. R.; Pressel, S. L.; Marchioli, R.; Marfisi, R. M.; Maggioni, A. P.; Tavazzi, L.; Tognoni, G.; Kjekshus, J.; Pedersen, T. R.; Cook, T. J.; Gotto, A. M.; Clearfield, M. B.; Downs, J. R.; Nakamura, H.; Ohashi, Y.; Mizuno, K.; Ray, K. K.; Ford, I. Statins and risk of incident diabetes: a collaborative meta-analysis of randomized statin trials. *Lancet* **2010,** *375* (9716), 735–742.
56. Kolich, M. N.; Howard, P. A. Statins: is there an increased risk of diabetes? *Hosp. Pharm.* **2012,** *47* (7), 513–517.
57. Sattar, N.; Taskinen, M.-R. Statins are diabetogenic—myth or reality? *Atheroscler. Suppl.* **2012,** *13* (1), 1–10.
58. Law, M.; Rudnicka, A. R. Statin safety: a systematic review. *Am. J. Cardiol.* **2006,** *97* (8A), 52C–60C.
59. Schoonjans, K.; Staels, B.; Auwerx, J. Role of the peroxisome proliferator-activated receptor (PPAR) in mediating the effects of fibrates and fatty acids on gene expression. *J. Lipid Res.* **1996,** *37* (5), 907–925.
60. Watts, G. F.; Dimmitt, S. B. Fibrates, dyslipoproteinemia and cardiovascular disease. *Curr. Opin. Lipidol.* **1999,** *10* (6), 561–574.
61. Gervois, P.; Fruchart, J.-C.; Staels, B. Drug insight: mechanisms of action and therapeutic applications for agonists of peroxisome proliferator-activated receptors. *Nat. Clin. Pract. Endocrinol. Metab.* **2007,** *3* (2), 145–156.
62. Fruchart, J.-C.; Duriez, P.; Staels, B. Molecular mechanism of action of fibrates. *J. Soc. Biol.* **1999,** *193* (1), 67–75.
63. Rader, D. J.; Haffner, S. M. Role of fibrates in the management of hypertriglyceridemia. *Am. J. Cardiol.* **1999,** *83* (9B), 30F–35F.
64. Chapman, M. J. Fibrates: therapeutic review. *Br. J. Diabetes Vasc. Dis.* **2006,** *6* (1), 11–18.
65. Katsiki, N.; Nikolic, D.; Montalto, G.; Banach, M.; Mikhailidis, D. P.; Rizzo, M. The role of fibrate treatment in dyslipidemia: an overview. *Curr. Pharm. Des.* **2013,** *19* (17), 3124–3131.
66. Drexel, H. Statins, fibrates, nicotinic acid, cholesterol absorption inhibitors, anion-exchange resins, omega-3 fatty acids: which drugs for which patients? *Fundam. Clin. Pharmacol.* **2009,** *23* (6), 687–692.
67. Toth, P. P.; Dayspring, T. D.; Pokrywka, G. S. Drug therapy for hypertriglyceridemia: fibrates: and omega-3 fatty acids. *Curr. Atheroscler. Rep.* **2009,** *11* (1), 71–79.
68. Filippatos, T.; Milionis, H. J. Treatment of hyperlipidemia with fenofibrate and related fibrates. *Expert Opin. Invest. Drugs* **2008,** *17* (10), 1599–1614.
69. Florentin, M.; Liberopoulos, E. N.; Mikhailidis, D. P.; Elisaf, M. S. Fibrate-associated adverse effects beyond muscle and liver toxicity. *Curr. Pharm. Des.* **2008,** *14* (6), 574–587.
70. Backes, J. M.; Gibson, C. A.; Ruisinger, J. F.; Moriarty, P. M. Fibrates: what have we learned in the past 40 years? *Pharmacotherapy* **2007,** *27* (3), 412–424.
71. Steinhilber, D.; Schubert-Zsilavecz, M. Molecular pharmacology and medicinal chemistry of fibrates. *Pharm. Unserer Zeit* **2007,** *36* (2), 108–113.
72. Balendiran, G. K.; Verma, M.; Perry, E. Chemistry of fibrates. *Curr. Chem. Biol.* **2007,** *1* (3), 311–316.
73. Keating, G. M.; Croom, K. F. Fenofibrate: a review of its use in primary dyslipidemia, the metabolic syndrome and type 2 diabetes mellitus. *Drugs* **2007,** *67* (1), 121–153.
74. McKeage, K.; Keating, G. M. Fenofibrate: a review of its use in dyslipidaemia. *Drugs* **2011,** *71* (14), 1917–1946.

75. Ewang-Emukowhate, M.; Wierzbicki, A. S. Lipid-lowering agents. *J. Cardiovasc. Pharmacol. Ther.* **2013,** *18* (5), 401–411.
76. Vardanyan, R. S.; Hruby, V. J. *Synthesis of Essential Drugs;* Elsevier, 2006.
77. Mieville, A. Phenoxyalkylcarboxylic acid derivatives, DE 2003430 (1970).
78. Mieville, A. Phenoxycarboxylic acid derivatives and pharmaceutical preparations containing them, DE 2250327 (1973).
79. Mieville, A. Pharmaceutical p-acyloximephenoxyacetic acids and their derivatives, DE 2065956 (1977).
80. Sornay, R.; Gurrieri, J.; Tourne, C.; Renson, F. J.; Majoie, B.; Wulfert, E. Antilipidemic drugs. Part 1: synthesis and structure-activity relation of new alkyl- and benzoylphenoxycarboxylic acids. *Arzneim. Forsch.* **1976,** *26* (5), 885–889.
81. Tubertini, P.; Vecchio, E. Process for preparation of pure fenofibrate by reaction of 4-chloro-4′-hydroxybenzophenone with isopropyl α-bromoisobutyrate followed by recrystallization of crude product from alcohols and ketones to remove polymeric impurity, EP 1837327 (2007).
82. Gignier, J. P.; Bourrelly, J. p-Chlorobenzoylphenoxyisobutyric acid esters, EP 2151 (1979).
83. Guazzi, G. An etherification process for the preparation of hypocholesteremic fibrates, WO 2002062743 (2002).
84. Bourgogne, J. P.; Sornay, R. Preparation of 2-phenoxy-2-methylpropionate derivatives as hypolipemics and hypocholesteremics, EP 245156 (1987).
85. Keating, G. M.; Croom, K. F. Fenofibrate: a review of its use in primary dyslipidemia, the metabolic syndrome and type 2 diabetes mellitus. *Drugs* **2007,** *67* (1), 121–153.
86. Filippatos, T.; Milionis, H. J. Treatment of hyperlipidemia with fenofibrateand related fibrates. *Expert Opin. Invest. Drugs* **2008,** *17* (10), 1599–1614.
87. Goto, M. Fenofibrate: panacea for aging-related conditions?. In *Dyslipidemia: From Prevention to Treatment;* Kelishadi, R., Ed.; InTech, 2012; pp 447–458.
88. McKeage, K.; Keating, G. M. Fenofibrate: a review of its use in dyslipidaemia. *Drugs* **2011,** *71* (14), 1917–1946.
89. Keating, G. M. Fenofibrate: a review of its lipid-modifying effects in dyslipidemia and its vascular effects in type 2 diabetes mellitus. *Am. J. Cardiovasc. Drugs* **2011,** *11* (4), 227–247.
90. Tsimihodimos, V.; Liberopoulos, E.; Elisaf, M. Pleiotropic effects of fenofibrate. *Curr. Pharm. Des.* **2009,** *15* (5), 517–528.
91. Farnier, M. Update on the clinical utility of fenofibrate in mixed dyslipidemias: mechanisms of action and rational prescribing. *Vasc. Health Risk Manage.* **2008,** *4* (5), 991–1000.
92. Dhal, P. K.; Huval, C. C.; Holmes-Farley, S. R. Biologically active polymeric sequestrants: design, synthesis, and therapeutic applications. *Pure Appl. Chem.* **2007,** *79* (9), 1521–1530.
93. Out, C.; Groen, A. K.; Brufau, G. Bile acid sequestrants: more than simple resins. *Curr. Opin. Lipidol.* **2012,** *23* (1), 43–55.
94. Insull, W., Jr. Clinical utility of bile acid sequestrants in the treatment of dyslipidemia: a scientific review. *South. Med. J.* **2006,** *99* (3), 257–273.
95. Benson, G.; Hickey, D. Bile acid sequestrants. *Expert Opin. Invest. Drugs* **1994,** *3* (5), 493–500.
96. Princen, H. M. G.; Post, S. M.; Twisk, J. Regulation of bile acid biosynthesis. *Curr. Pharm. Des.* **1997,** *3* (1), 59–84.
97. Mandeville, W. Harry; Goldberg, Dennis I. The sequestration of bile acid, a non-absorbed method for cholesterol reduction. *A review, Curr. Pharm. Des.* **1997,** *3* (1), 15–28.
98. Mandeville, W. H., III; Holmes-Farley, S. R. Process for removing bile salts from a patient using alkylated crosslinked polyamines, and compositions therefor, WO 9534585 (1995).
99. Mandeville, W. H., III; Holmes-Farley, S. R. Alkylated amine polymers, US 5693675 (1997).

100. Holmes-Farley, S. R.; Petersen, J. S. Hydrophilic nonamine-containing and amine-containing copolymers and their use as bile acid sequestrants, US 5929184 (1999).
101. Holmes-Farley, S. R.; Mandeville, W. H., III. Water-insoluble noncrosslinked bile acid-sequestrant and hypocholesterolemic amine polymers, and therapeutic use, US 6129910 (2000).
102. Gaboardi, M.; Baruto, A.; Castaldi, M. Methods for production of Colesevelam, WO 2015092669 (2015).
103. Mendonca, P. V.; Serra, A. C.; Silva, C. L.; Simoes, S.; Coelho, J. F. J. Polymeric bile acid sequestrant-Synthesis using conventional methods and new approaches based on "controlled"/living radical polymerization. *Prog. Polym. Sci.* **2013,** *38* (3–4), 445–461.
104. Dhal, P. K.; Huval, C. C.; Holmes-Farley, S. R. Biologically active polymeric sequestrants: design, synthesis, and therapeutic applications. *Pure Appl. Chem.* **2007,** *79* (9), 1521–1530.
105. Wong, N. N. Colesevelam a new bile acid sequestrant. *Heart Dis.* **2001,** *3* (1), 63–70.
106. Melian, E.; Balmori, P.; Greg, L. Colesevelam. *Am. J. Cardiovasc. Drugs* **2001,** *1* (2), 141–146.
107. Aldridge, M. A.; Ito, M. K. Colesevelam hydrochloride: A novel bile acid-binding resin. *Ann. Pharmacother.* **2001,** *35* (7/8), 898–907.
108. Steinmetz, K. L.; Schonder, K. S. Colesevelam: potential uses for the newest bile resin. *Cardiovasc. Drug Rev.* **2005,** *23* (1), 15–30.
109. Tziomalos, K.; Karagiannis, A.; Mikhailidis, D. P.; Athyros, V. G. Colesevelam: a new and improved bile acid sequestrant? *Curr. Pharm. Des.* **2013,** *19* (17), 3115–3123.
110. Levy, P. Review of studies on the effect of bile acid sequestrants in patients with type 2 diabetes mellitus. *Metab. Syndr. Relat. Disord.* **2010,** *8* (Suppl. 1), S9–S13.
111. Manghat, P.; Wierzbicki, A. S. Colesevelam hydrochloride: a specifically engineered bile acid sequestrant. *Future Lipidol.* **2008,** *3* (3), 237–255.
112. Base, H.; Jones, P. J. Colesevelam hydrochloride: reducing atherosclerotic coronary heart disease risk factors. *Vasc. Health Risk Manage.* **2007,** *3* (5), 733–742.
113. Steinmetz, K. L. Colesevelam hydrochloride. *Am. J. Health-Syst. Pharm.* **2002,** *59* (10), 932–939.
114. Betters, J. L.; Yu, L. NPC1L1 and cholesterol transport. *FEBS Lett.* **2010,** *584* (13), 2740–2747.
115. Burnett, D. A. β-Lactam cholesterol absorption inhibitors. *Curr. Med. Chem.* **2004,** *11* (14), 1873–1887.
116. Clader, J. W. Ezetimibe and other azetidinone cholesterol absorption inhibitors. *Curr. Top. Med. Chem.* **2005,** *5* (3), 243–256.
117. Gupta, E. K.; Ito, M. K. Ezetimibe: the first in a novel class of selective cholesterol-absorption inhibitors. *Heart Dis.* **2002,** *4* (6), 399–409.
118. Clader, J. W. Ezetimibe. In Taylor, J. B., Triggle, D. J., Eds.; *Comprehensive, Medicinal Chemistry II*, Vol. 8; Elsevier, 2006; pp 65–82.
119. Jeu, L. A.; Cheng, J. W. M. Pharmacology and therapeutics of ezetimibe (SCH 58235), a cholesterol-absorption inhibitor. *Clin. Ther.* **2003,** *25* (9), 2352–2387.
120. Mikhailidis, D. P.; Wierzbicki, A. S.; Daskalopoulou, S. S.; Al-Saady, N.; Griffiths, H.; Hamilton, G.; Monkman, D.; Patel, V.; Pittard, J.; Schachter, M. The use of ezetimibe in achieving low density lipoprotein lowering goals in clinical practice: Position statement of a United Kingdom consensus panel. *Curr. Med. Res. Opin.* **2005,** *21* (6), 959–969.
121. Toth, P. P.; Davidson, M. H. Cholesterol absorption blockade with ezetimibe. *Curr. Drug Targets: Cardiovasc. & Haematol. Disord.* **2005,** *5* (6), 455–462.
122. Harris, M.; Davis, W.; Brown, W. V. Ezetimibe. *Drugs Today* **2003,** *39* (4), 229–247.
123. Davidson, M. H. Ezetimibe: a novel option for lowering cholesterol. *Expert Rev. Cardiovasc. Ther.* **2003,** *1* (1), 11–21.

124. Darkes, M. J. M.; Poole, R. M.; Goa, K. L. Ezetimibe. *Am. J. Cardiovasc. Drugs* **2003,** *3* (1), 67–76.
125. Katsiki, N.; Theocharidou, E.; Karagiannis, A.; Athyros, V. G.; Mikhailidis, D. P. Ezetimibe therapy for dyslipidemia: an update. *Curr. Pharm. Des.* **2013,** *19* (17), 3107–3114.
126. Husain, A.; Alam, M. M.; Azim, M. S.; Mitra, M.; Bhasin, P. S. A review on pharmacological and pharmaceutical properties of ezetimibe. *J. Pharm. Res. (Bangalore, India)* **2012,** *5* (8), 4056–4059.
127. Kostapanos, M. S.; Elisaf, M. S.; Mikhailidis, D. P. Ezetimibe—a new approach in hypercholesterolemia management. *Pharmacol. Rep.* **2012,** *64* (4), 997–998.
128. Oswald, S.; Siegmund, W. Ezetimibe and cholesterol absorption. In *Development of Therapeutic Agents Handbook;* Gad, S. C., Ed.; Wiley, 2012; pp 723–741.
129. Phan, B. A. P.; Dayspring, T. D.; Toth, P. P. Ezetimibe therapy: mechanism of action and clinical update. *Vasc. Health Risk Manage.* **2012,** *8,* 415–427.
130. Bays, H. Ezetimibe. *Expert Opin. Invest. Drugs* **2002,** *11* (11), 1587–1604.
131. Suchy, D.; Labuzek, K.; Stadnicki, A.; Okopien, B. Ezetimibe; more than a low density lipoprotein cholesterol lowering drug? An update after 4 years—a new approach in hypercholesterolemia management. *Pharmacol. Rep.* **2011,** *63* (6), 1335–1348.
132. Lestari, M. L. A.D.; Ardiana, F.; Indrayanto, G. Ezetimibe. *Profiles Drug Subst., Excipients, Relat. Methodol.* **2011,** *36,* 103–149.
133. Lioudaki, E.; Ganotakis, E. S.; Mikhailidis, D. P. Ezetimibe; more than a low density lipoprotein cholesterol lowering drug? An update after 4 years. *Curr. Vasc. Pharmacol.* **2011,** *9* (1), 62–86.
134. Sarigianni, M.; Katsiki, N.; Mikhailidis, D. P. Ezetimibe in diabetes: more than cholesterol lowering? *Curr. Med. Res. Opin.* **2010,** *26* (10), 2517–2520.
135. Burnett, D. A.; Caplen, M. A.; Davis, H. R., Jr.; Burrier, R. E.; Clader, J. W. 2-Azetidinones as inhibitors of cholesterol absorption. *J. Med. Chem.* **1994,** *37* (12), 1733–1736.
136. Ross, J. S.; Frazee, S. G.; Garavaglia, S. B.; Levin, R.; Novshadian, H.; Jackevicius, C. A.; Stettin, G.; Krumholz, H. M. Trends in use of Ezetimibe after the ENHANCE trial, 2007 through 2010. *JAMA Internal Med.* **2014,** *174* (9), 1486–1493.
137. Clader, J. W.; Burnett, D. A.; Caplen, M. A.; Domalski, M. S.; Dugar, S.; Vaccaro, W. S. R.; Browne, M. E.; Zhao, H.; Burrier, R. E.; Salisbury, B.; Davis, H. R., Jr. 2-Azetidinone cholesterol absorption inhibitors: structure–activity relationships on the heterocyclic nucleus. *J. Med. Chem.* **1996,** *39* (19), 3684–3893.
138. Doggrell, S. A. The ezetimibe controversy—can this be resolved by comparing the clinical trials with simvastatin and ezetimibe alone and together? *Expert Opin. Pharmacother.* **2012,** *13* (10), 1469–1480.
139. Kalogirou, M.; Tsimihodimos, V.; Elisaf, M. Pleiotropic effects of ezetimibe: do they really exist? *Eur. J. Pharmacol.* **2010,** *633* (1–3), 62–70.
140. Gouni-Berthold, I.; Mikhailidis, D. P.; Rizzo, M. Clinical benefits of ezetimibe use: is absence of proof, proof of absence? *Expert Opin. Pharmacother.* **2012,** *13* (14), 1985–1988.
141. Khanderia, U.; Regal, R. E.; Rubenfire, M.; Boyden, T. The ezetimibe controversy: implications for clinical practice. *Ther. Adv. Cardiovasc. Dis.* **2011,** *5* (4), 199–208.
142. Rosenblum, S. B.; Dugar, S.; Burnett, D. A.; Clader, J. W.; Mckittrick, B. A. Preparation of hydroxy-substituted azetidinone compounds useful as hypocholesterolemic agents, US 5846966 (1997).
143. Rosenblum, S. B.; Dugar, S.; Burnett, D. A.; Clader, J. W.; McKittrick, B. A. Preparation of hydroxy-substituted azetidinone compounds as HMG-CoA reductase inhibitors, US 5846966 (1998).

144. Rosenblum, S. B.; Huynh, T.; Afonso, A.; Davis, H. R., Jr.; Yumibe, N.; Clader, J. W.; Burnett, D. A. Discovery of 1-(4-fluorophenyl)-(3R)-[3-(4-fluorophenyl)-(3S)-hydroxypropyl]-(4S)-(4-hydroxyphenyl)-2-azetidinone (SCH 58235): a designed, potent, orally active inhibitor of cholesterol absorption. *J. Med. Chem.* **1998,** *41* (6), 973–980.
145. Rosenblum, S. B. Advances in the development of methods for the synthesis of cholesterol absorption inhibitors [ezetimibe (Zetia, Ezetrol)]. In *Art of Drug Synthesis;* Johnson, D. S., Li, J. J., Eds.; Wiley, 2007; pp 183–196.
146. Wu, G.; Wong, Y. S.; Chen, X.; Ding, Z. A novel one-step diastereo- and enantioselective formation of trans-azetidinones and its application to the total synthesis of cholesterol absorption inhibitors. *J. Org. Chem.* **1999,** *64* (10), 3714–3718.
147. Vaccaro, W. D.; Sher, R.; Davis, H. R. Carboxy-substituted 2-azetidinones as cholesterol absorption inhibitors. *Bioorg. Med. Chem. Lett.* **1998,** *8* (3), 319–322.
148. Thiruvengadam, T. K.; Tann, C. H.; Mcallister, T. L. Process for the stereospecific synthesis of azetidinones, US 5561227 (1996).
149. Fu, X.; McAllister, T. L.; Thiruvengadam, T.; Tann, C. H. Process for enantioselective synthesis of oxazolidinone deriv. as an intermediate for hydroxyalkyl substituted azetidinone, WO 2002079174 (2002).
150. Collier, S. J.; Liang, J.; Fu F. J.; Wilson, R. J. Synthesis of ezetimibe, WO 2010141494 (2010).
151. Illich, G. M., Jr. Continuous production of nicotinic acid, US 2905688 (1959).
152. Stocker, A.; Marti, O.; Pfammatter, T.; Schreiner, G.; Brander, S. Nicotinic acid, DE 2046556 (1971).
153. Nenz, A.; Pieroni, M. Commercial synthetic pyridine bases. 1. MEP [methyl ethyl pyridine] manufacture, chemistry, and uses. *Hydrocarb. Process.* (1966–2001) **1968,** *47* (11), 139–144.
154. McElvain, S. M.; Goese, M. A. Preparation of nicotinic acid from pyridine. *J. Am. Chem. Soc.* **1941,** *63,* 2283–2284.
155. McElvain, S. M. Nicotinic acid. *Org. Synth. Coll.Vol.* **1941,** *1,* 385.
156. Woodward, C. F.; Badgett, C. O.; Kaufman, J. G. Chemical-catalytic liquid-phase oxidation of nicotine, β-picoline and quinoline to nicotinic acid. *Ind. Eng. Chem.* **1944,** *36,* 544–546.
157. Brooks, E. L.; Kuvin, J. T.; Karas, R. H. Niacin's role in the statin era. *Expert Opin. Pharmacother.* **2010,** *11* (14), 2291–2300.
158. Al-Mohaissen, M. A.; Pun, S. C.; Frohlich, J. J. Niacin: from mechanisms of action to therapeutic uses. *Mini-Rev. Med. Chem.* **2010,** *10* (3), 204–217.
159. Kamanna, V. S.; Kashyap, M. L. Mechanism of action of niacin. *Am. J. Cardiol.* **2008,** *101* (8a), 20B–26B.
160. Ginsberg, H. N.; Reyes-Soffer, G. Niacin: a long history, but a questionable future. *Curr. Opin. Lipidol.* **2013,** *24* (6), 475–479.
161. Song, W.-L.; FitzGerald, G. A. Niacin, an old drug with a new twist. *J. Lipid Res.* **2013,** *54* (10), 2586–2594.
162. Gouni-Berthold, I.; Berthold, H. K. The role of niacin in lipid-lowering treatment: are we aiming too high? *Curr. Pharm. Des.* **2013,** *19* (17), 3094–3106.
163. Kamanna, V. S.; Ganji, S. H.; Kashyap, M. L. Recent advances in niacin and lipid metabolism. *Curr. Opin. Lipidol.* **2013,** *24* (3), 239–245.
164. Wierzbicki, A. S. Failure to THRIVE: the end for niacin? *Nat. Rev. Cardiol.* **2013,** *10* (5), 246–247.
165. Hochholzer, W.; Berg, D. D.; Giugliano, R. P. The facts behind niacin. *Ther. Adv. Cardiovasc. Dis.* **2011,** *5* (5), 227–240.
166. Chen, J.; Chopp, M. Niacin, an old drug, has new effects on central nervous system disease. *Open Drug Disc. J.* **2010,** *2,* 181–186.

167. Hernandez, E. M. Processing of omega-3 oils. In *Omega-3 Oils;* Hernandez, E. M., Hosokawa, M., Eds.; AOCS Press, 2011; pp 107–128.
168. Mozaffarian, D.; Wu, J. H. Y. Omega-3 fatty acids and cardiovascular disease: effects on risk factors, molecular pathways, and clinical events. *J. Am. Coll. Cardiol.* **2011,** *58* (20), 2047–2067.
169. Poudyal, H.; Panchal, S. K.; Diwan, V.; Brown, L. Omega-3 fatty acids and metabolic syndrome: effects and emerging mechanisms of action. *Prog. Lipid Res.* **2011,** *50* (4), 372–387.
170. Saravanan, P.; Davidson, N. C.; Schmidt, E. B.; Calder, P. C. Cardiovascular effects of marine omega-3 fatty acids. *Lancet* **2010,** *376* (9740), 540–550.
171. Weylandt, K. H.; Chiu, C.-Y.; Gomolka, B.; Waechter, S. F.; Wiedenmann, B. Omega-3 fatty acids and their lipid mediators: Towards an understanding of resolvin and protectin formation. *Prostaglandins Other Lipid Mediators* **2012,** *97* (3-4), 73–82.
172. Swanson, D.; Block, R.; Mousa, S. A. Omega-3 fatty acids EPA and DHA: health benefits throughout life. *Adv. Nutr.* **2012,** *3* (1), 1–7.
173. Kromhout, D.; de Goede, J. Update on cardiometabolic health effects of w-3 Fatty Acids. *Curr. Opin. Lipidol.* **2014,** *25* (1), 85–90.
174. Kromhout, D.; Yasuda, S.; Geleijnse, J. M.; Shimokawa, H. Fish oil and omega-3 fatty acids in cardiovascular disease: do they really work? *Eur. Heart J.* **2012,** *33* (4), 436–443.
175. Davidson, M. H. Omega-3 fatty acids: new insights into the pharmacology and biology of docosahexaenoic acid, docosapentaenoic acid and eicosapentaenoic acid. *Curr. Opin. Lipidol.* **2013,** *24* (6), 467–474.
176. Gerber, P. A.; Gouni-Berthold, I.; Berneis, K. Omega-3 fatty acids: role in metabolism and cardiovascular disease. *Curr. Pharm. Des.* **2013,** *19* (17), 3074–3093.
177. Jump, D. B.; Depner, C. M.; Tripathy, S. Omega-3 fatty acid supplementation and cardiovascular disease. *J. Lipid Res.* **2012,** *53* (12), 2525–2545.
178. Bays, H.; Stein, E. A. Pharmacotherapy for dyslipidaemia-current therapies and future agents. *Expert Opin. Pharmacother.* **2003,** *4* (11), 1901–1938.
179. Ewang-Emukowhate, M.; Wierzbicki, A. S. Lipid-lowering agents. *J. Cardiovasc. Pharmacol. Ther.* **2013,** *18* (5), 401–411.
180. Wierzbicki, A. S.; Hardman, T. C.; Viljoen, A. New lipid-lowering drugs: an update. *Int. J. Clin. Pract.* **2012,** *66* (3), 270–280.
181. Panno, M. D.; Cefalu, A. B.; Averna, M. R. Lomitapide: a novel drug for homozygous familial hypercholesterolemia. *Clin. Lipidol.* **2014,** *9* (1), 19–32.
182. Rizzo, M. Lomitapide, a microsomal triglyceride transfer protein inhibitor for the treatment of hypercholesterolemia. *IDrugs* **2010,** *13* (2), 103–111.
183. Cuchel, M.; Bloedon, L.-A. T.; Szapary, P. O.; Kolansky, D. M.; Wolfe, M. L.; Sarkis, A.; Millar, J. S.; Ikewaki, K.; Sigelman, E. S.; Gregg, R. E.; Rader, D. J. Inhibition of microsomal triglyceride transfer protein in familial hypercholesterolemia. *N. Engl. J. Med.* **2007,** *356* (2), 148–156.
184. Robl, J. A.; Sulsky, R.; Sun, C.-Q.; Simpkins, L. M.; Wang, T.; Dickson, J. K., Jr.; Chen, Y.; Magnin, D. R.; Taunk, P.; Slusarchyk, W. A.; Biller, S. A.; Lan, S.-J.; Connolly, F.; Kunselman, L. K.; Sabrah, T.; Jamil, H.; Gordon, D.; Harrity, T. W.; Wetterau, J. R. A novel series of highly potent benzimidazole-based microsomal triglyceride transfer protein inhibitors. *J. Med. Chem.* **2001,** *44* (6), 851–856.
185. Kim, E.; Campbell, S.; Schueller, O.; Wong, E.; Cole, B.; Kuo, J.; Ellis, J.; Ferkany, J.; Sweetnam, P. A small-molecule inhibitor of enterocytic microsomal triglyceride transfer protein, SLx-4090: biochemical, pharmacodynamic, pharmacokinetic, and safety profile. *J. Pharmacol. Exp. Ther.* **2011,** *337* (3), 775–785.
186. Davidson, M. H. Squalene synthase inhibition: a novel target for the management of dyslipidemia. *Curr. Atheroscler. Rep.* **2007,** *9* (1), 78–80.

187. Wierzbicki, A. S.; Hardman, T. C.; Viljoen, A. Inhibition of pro-protein convertase subtilisin kexin-9 (PCSK-9) as a treatment for hyperlipidaemia. *Expert Opin. Invest. Drugs* **2012,** *21* (5), 667–676.
188. Lee, P.; Hegele, R. A. Current phase II proprotein convertase subtilisin/kexin 9 inhibitor therapies for dyslipidemia. *Expert Opin. Invest. Drugs* **2013,** *22* (11), 1411–1423.
189. Seidah, N. G. Proprotein convertase subtilisin kexin 9 (PCSK) inhibitors in the treatment of hypercholesterolemia and other pathologies. *Curr. Pharm. Des.* **2013,** *19* (17), 3161–3172.
190. Shen, L.; Peng, H.; Xu, D.; Zhao, S. The next generation of novel low-density lipoprotein cholesterol-lowering agents: proprotein convertase subtilisin/kexin 9 inhibitors. *Pharmacol. Res.* **2013,** *73,* 27–34.
191. Raal, F. J.; Santos, R. D.; Blom, D. J.; Marais, A. D.; Charng, M.-J.; Cromwell, W. C.; Lachmann, R. H.; Gaudet, D.; Tan, J. L.; Chasan-Taber, S.; Tribble, D. L.; Flaim, J.-A. D.; Crooke, S. T. Mipomersen an apolipoprotein B synthesis inhibitor, for lowering of LDL cholesterol concentrations in patients with homozygous familial hypercholesterolaemia: a randomized, double-blind, placebo-controlled trial. *Lancet* **2010,** *375* (9719), 998–1006.
192. Merki, E.; Graham, M. J.; Mullick, A. E.; Miller, E. R.; Crooke, R. M.; Pitas, R. E.; Witztum, J. L.; Tsimikas, S. Antisense oligonucleotide directed to human apolipoprotein B-100 reduces lipoprotein(a) levels and oxidized phospholipids on human apolipoprotein B-100 particles in lipoprotein(a) transgenic mice. *Circulation* **2008,** *118* (7), 743–753.
193. Thomas, G. S.; Cromwell, W. C.; Ali, S.; Chin, W.; Flaim, J.-A. D.; Davidson, M. Mipomersen an apolipoprotein B synthesis inhibitor, reduces atherogenic lipoproteins in patients with severe hypercholesterolemia at high cardiovascular risk. *J. Am. Coll. Cardiol.* **2013,** *62* (23), 2178–2184.
194. Hair, P.; Cameron, F.; McKeage, K. Mipomersen sodium: first global approval. *Drugs* **2013,** *73* (5), 487–493.
195. McGowan, M. P.; Tardif, J.-C.; Ceska, R.; Burgess, L. J.; Soran, H.; Gouni-Berthold, I.; Wagener, G.; Chasan-Taber, S. Randomized, placebo-controlled trial of mipomersen in patients with severe hypercholesterolemia receiving maximally tolerated lipid-lowering therapy. *PLoS One* **2012,** *7* (11), e49006.
196. Bell, D. A.; Hooper, A. J.; Burnett, J. R. Mipomersen, an antisense apolipoprotein B synthesis inhibitor. *Expert Opin. Invest. Drugs* **2011,** *20* (2), 265–272.
197. Visser, M. E.; Akdim, F.; Tribble, D. L.; Nederveen, A. J.; Kwoh, T. J.; Kastelein, J. J. P.; Trip, M. D.; Stroes, E. S. G. Effect of mipomersen, an apolipoprotein B synthesis inhibitor, on low-density lipoprotein cholesterol in patients with familial hypercholesterolemia. *J. Lipid Res.* **2010,** *51* (5), 1057–1062.
198. Barter, P. J.; Kastelein, J. J. P. Targeting cholesteryl ester transfer protein for the prevention and management of cardiovascular disease. *J. Am. Coll. Cardiol.* **2006,** *47* (3), 492–499.
199. Joy, T.; Hegele, R. A. Is raising HDL a futile strategy for atheroprotection? *Nat. Rev. Drug Discovery* **2008,** *7* (2), 143–155.
200. Chapman, M. J.; Le Goff, W.; Guerin, M.; Kontush, A. Cholesteryl ester transfer protein: at the heart of the action of lipid-modulating therapy with statins, fibrates, niacin, and cholesteryl ester transfer protein inhibitors. *Eur. Heart J.* **2010,** *31* (2), 149–164.
201. Hunt, J. A.; Lu, Z. Cholesteryl ester transfer protein (CETP) inhibitors. *Curr. Top. Med. Chem.* **2009,** *9* (5), 419–427.
202. Eckardstein, A. Implications of torcetrapib failure for the future of HDL therapy: is HDL-cholesterol the right target? *Expert Rev. Cardiovasc. Ther.* **2010,** *8* (3), 345–358.
203. Kontush, A.; Guerin, M.; Chapman, M. J. Spotlight on HDL-raising therapies: insights from the torcetrapib trials. *Nat. Clin. Pract. Cardiovasc. Med.* **2008,** *5* (6), 329–336.

204. Johns, D. G.; Duffy, J.; Fisher, T.; Hubbard, B. K.; Forrest, M. J. On- and off-target pharmacology of torcetrapib: current understanding and implications for the structure activity relationships (SAR) discovery and development of cholesteryl ester-transfer protein (CETP) inhibitors. *Drugs* **2012,** *72* (4), 491–507.
205. Aperis, G.; Paliouras, C.; Tsampikaki, E.; Papakonstantinou, N.; Alivanis, P. Anacetrapib: a new weapon against dyslipidemia. *Curr. Clin. Pharmacol.* **2011,** *6* (4), 227–235.
206. Masson, D. Anacetrapib, a cholesterol ester transfer protein (CETP) inhibitor for the treatment of atherosclerosis. *Curr. Opin. Invest. Drugs (BioMed Cent.)* **2009,** *10* (9), 980–987.
207. Guoqing, B.; Thomas, P.; Zhang, Y.; Schmidt, R. J.; Chen, Y. Q.; Cockerham, S. L.; Zimmerman, K. M.; Karathanasis, S. K.; Cannady, E. A.; Fields, T.; Mantlo, N. B. Evacetrapib is a novel, potent, and selective inhibitor of cholesteryl ester transfer protein that elevates HDL cholesterol without inducing aldosterone or increasing blood pressure. *J. Lipid Res.* **2011,** *52* (12), 2169–2176.
208. Nicholls, S. J. Evacetrapib. *Curr. Cardiol. Rep.* **2012,** *14* (3), 245–250.
209. Nicholls, S. J.; Brewer, H. B.; Kastelein, J. J. P.; Krueger, K. A.; Wang, M.-D.; Shao, M.; Hu, B.; McErlean, E.; Nissen, S. E. Effects of the CETP inhibitor evacetrapib administered as monotherapy or in combination with statins on HDL and LDL cholesterol: a randomized controlled trial, JAMA. *J. Am. Med. Assoc.* **2011,** *306* (19), 2099–2109.

Chapter 21

Diuretics

Diuretic therapy is generally used to treat edematous states in the cases of renal insufficiency, nephrotic syndrome, liver cirrhosis, and heart failure. Edema is an abnormal accumulation of fluid in the intercellular spaces of connective tissue. Interestingly, there are minor differences in composition of fluid generated with various diseases. It is believed that edema is mostly a symptom of disease rather than a disease, which is why treatment of edema generally consists of treatment of the underlying disease, such as insufficient functioning of the kidney or heart.

Nonedematous states in which diuretic therapy is useful include hypertension, acute renal failure, diabetes insipidus, hypercalcemia, and hypercalciuria.

In general, diuretics work by causing the kidneys to excrete increased amounts of salt and water from the body, resulting in a decrease in plasma volume and the volume of blood within the arteries, which, in turn, reduces the "pushing" on the artery walls, reducing stroke volume, and, consequently, blood pressure.

Diuretics are medications that are designed to increase the flow of urine, promoting the removal of excess of water, salts, metabolic products and toxins from the body.

Diuresis can be generated in different ways by choosing different types of drugs.

There are a few disorders that are treated with diuretics itself, but diuretics changed the approach to many diseases and have turned once fatal conditions into tolerable ones. Some of the recent reviews on the action and classification of diuretics are noted in the references [1-9].

Diuretic properties of hundreds of plants have been known since antiquity. Among the plants known to have diuretic properties are xanthine-containing (caffeine, theophylline, theobromine) herbs and watermelon, grapes, olives, fennel, and garlic [10].

Mercury salts had been implemented as a diuretic in the 16th century. Organomercurials such as meralluride (**21.1**), novasurol (**21.2**), and salyrgan (**21.3**) (Fig. 21.1.) were introduced to pharmaceutical market before World War II.

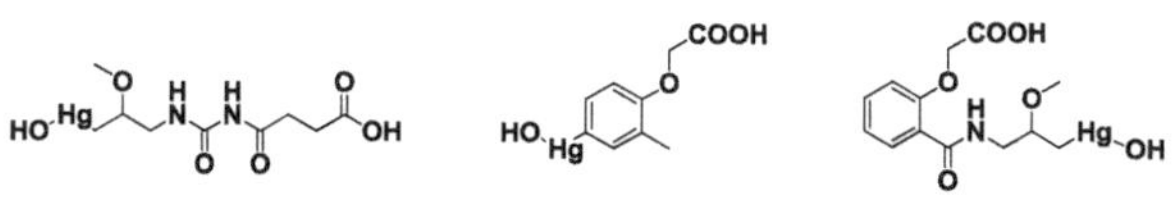

FIG. 21.1 Organomercurial diuretics.

Synthesis of Best-Seller Drugs. http://dx.doi.org/10.1016/B978-0-12-411492-0.00021-3

Carbonic anhydrase inhibitors were the first diuretic medications to replace mercurials and to have significantly less-toxic effects. The carbonic anhydrase inhibitors, which were developed in 1930s, were based on the observation that patients treated with newly discovered antibacterial sulfanilamides had increased volumes of urine along with increased amounts of sodium ions in the urine. In the mid-1950s, acetazolamide (**21.1.4**), the first carbonic anhydrase inhibitor diuretic, was developed. The late 1950s and early 1960s became a time of vigorous creation of new generations of diuretics, which started from 1,2,4-benzothiadiazine derivatives, later called thiazides, loop diuretics, or "high-ceiling" diuretics, such as furosemide (Lasix) and ethacrynic acid. They were called loop diuretics because they act in the thick ascending limb of the loop of Henle and the discovery that on their implementation, diuresis increases almost linearly with dosage increase.

Later, aldosterone antagonists and other diuretics appeared on the pharmaceutical market. Among them are polyvalent carbonic anhydrase inhibitors, diuretics that influence blood glucose and blood pressure, uric acid, lipids, and, generally speaking, compounds that provide additional improvement of renal function.

These compounds inhibit carbonic anhydrase, an enzyme that converts carbon dioxide and water to carbonic acid, which causes an alkalization of the urine and diuresis. In the eye, this property of anhydrase inhibitors manifested as decrease in intraocular pressure making these agents valuable drugs for the treatment of glaucoma.

Diuretics are among the most important drugs in the therapeutic armamentarium.

Although all diuretics, except spironolactone, exert their effects from the lumen of the nephron, acting primarily by impairing sodium reabsorption in the renal tubules, they differ in their fine mechanism and site of action and, therefore, in their specific pharmacological properties and clinical indications.

There is no single system for diuretics classification that could be considered as the best. Diuretics can be classified according to their diuretic potency, their mechanism and site of action, or their chemical structure.

On the basis of potency, diuretics could be formally classified as:

1. High-efficacy diuretics—inhibitors of $Na^+/K^+/2Cl^-$ cotransport—such as furosemide, bumetanide, ethacrynic acid, torsemide.
2. Moderately potent diuretics, which include drugs of the thiazide series such as chlorothiazide, hydrochlorothiazide, benzthiazide, hydroflumethiazide, and of the thiazide-like series such as chlorthalidone, metolazone, xipamide, indapamide.
3. Weak diuretics such as:
   a. Carbonic anhydrase inhibitors: acetazolamide, methazolamide.
   b. Potassium-sparing diuretics: triamterene, spironolactone, amiloride
   c. Osmotic diuretics: glycerol, mannitol, isosorbide.
   d. Xanthines: theophylline, caffeine.

According to mechanism and site of action, diuretics can be and are most often classified as:

1. Carbonic anhydrase inhibitors, acting at proximal convoluted tubules and suppressing the action of carbonic anhydrase enzyme. These compounds belong to the series of acetazolamide, dichlorphenamide, methazolamide.
2. Loop diuretics, acting on the loop of Henle, such as furosemide, bumetanide, torsemide, ethacrynic acid.
3. Diuretics acting at acting at the distal convoluted tubule, which include:
   a. Thiazide diuretics.
   b. Thiazide-like drugs (sulfonamides).
4. Diuretics acting at the collecting duct system (potassium-sparing diuretics):
   a. Epithelial sodium channel blockers such as amiloride and triamterene.
   b. Competitive aldosterone antagonists such as spironolactone, canrenone, and eplerenone.
   c. Vasopressin antagonists: "vaptans"—tolvaptan and conivaptan—lithium salts, demeclocycline.
5. Osmotic diuretics acting at the proximal convoluted tubules and limiting the reabsorption of water in the tubule: glycerol, mannitol, isosorbide.

21.1 CARBONIC ANHYDRASE INHIBITORS

Carbonic anhydrases are zinc-containing metalloenzymes that catalyze CO_2 hydration to bicarbonate and hydrogen ions. Carbonic anhydrases are present in many nephron sites, with the predominant location in the luminal membrane of the proximal tubule cells. Carbonic anhydrase inhibitors suppress the activity of these enzymes, preventing reabsorption of bicarbonate ion in the proximal tubule, which leads to several effects, including bicarbonate and potassium retention in the urine and decreased sodium absorption, acting, in general, as mild diuretics.

Carbonic anhydrase inhibitors are considered to be natriuretic (increase the excretion of sodium ions), kaluretic (increase the excretion of potassium), and bicarbonaturetic (increase the excretion of bicarbonate).

The carbonic anhydrase inhibitors were the forerunners of modern diuretics and were discovered and proposed as diuretics in the mid-1930s, the time of the boom around the discovery of bacteriostatic sulfonamides, when it was found that they cause an alkaline diuresis and hyperchloremic metabolic acidosis.

Carbonic anhydrase inhibitors are now rarely used as diuretics, but they still have several specific applications particularly for the treatment of glaucoma. The prototypical diuretic carbonic anhydrase inhibitor was acetazolamide (**21.1.1**); other diuretic carbonic anhydrase inhibitors, such as methazolamide (**21.1.2**) and dichlorphenamide (**21.1.3**) (Fig. 21.2.), were created later. An important limiting factor in the use of acetazolamide is its excessive bicarbonate excretion, which is accompanied by a comparably less-severe chloruresis [11-13].

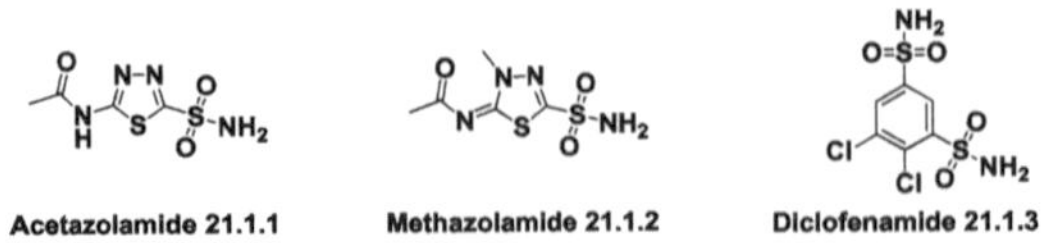

FIG. 21.2 Carbonic anhydrase inhibitors.

21.2 LOOP DIURETICS (HIGH-CEILING DIURETICS)

Loop diuretics are the most commonly used diuretics. Because of their high diuretic potential, they often are described as "high-ceiling" diuretics.

Loop diuretics inhibit the $Na^+/K^+/2Cl^-$ cotransporter, which was recently demonstrated to be an aldosterone-induced protein in the thick ascending loop of Henle, in the kidneys and stop the transport of sodium chloride out of the tubule into the interstitial tissue, causing a decrease in Na^+, K^+, and Cl^- reabsorption. Inhibition of this transporter leads to a significant increase in concentration of ions in the tubule and reduced hypertonicity in the surrounding interstitium, causing less water to be reabsorbed into the blood. This causes more urine to be produced and a decrease in blood volume. Examples of the most used loop diuretics include ethacrynic acid (**21.2.1**), furosemide (**21.2.2**), bumetanide (**21.2.3**), and torsemide (**21.2.4**). Examples of a-little-bit-less-used loop diuretics, include azosemide (**21.2.5**), piretanide (**21.2.6**), tienilic acid (**21.1.7**), and etozolin (**21.2.8**). Among them, ethacrynic acid (**21.2.1**) and tienilic acid (**21.2.7**) are derivatives of phenoxyacetic acid, etozolin (**21.2.8**) is a thiazolidinone derivative, and all others can formally be classified as sulfonamides (Fig. 21.3.).

Loop diuretics are indicated in the treatment of fluid overload/edema in congestive heart failure, acute pulmonary edema, hepatic ascites, nephrotic syndrome, renal failure, and other edematous disorders. Loop diuretics are especially useful in emergencies. They do not lower blood pressure significantly [14,15].

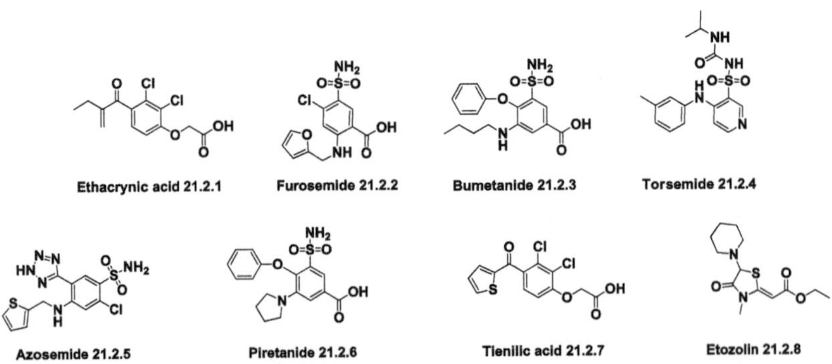

FIG. 21.3 Loop diuretics.

Furosemide was the first approved loop diuretic in 1966, followed by ethacrynic acid in 1967, bumetanide in 1983, and torsemide in 1993.

Like any drug, diuretics come with potential side effects. They can include frequent urination, arrhythmia, dizziness, electrolyte abnormalities, extreme tiredness or weakness, and muscle cramps.

21.3 DIURETICS ACTING AT THE DISTAL CONVOLUTED TUBULE

The distal convoluted tubule is a part of kidney nephron located between the loop of Henle and the collecting duct system.

Thiazides and Thiazide-Like Diuretics

Thiazides

Thiazides and thiazide-like diuretics inhibit Na^+/Cl^- cotransporter located on the apical membrane of the early segment of the distal convoluted tubule. They have been widely prescribed for more than 60 years for the treatment of hypertension and various edematous states.

Some members of this series retain significant carbonic anhydrase inhibitory activity.

They are mild diuretics and effective antihypertensives.

Since the introduction of the first representatives of thiazides into medicinal practice in the mid-1950s, decades of criticism and controversy about them has not ceased. But 60 years later, they still remain one of the most important classes of drugs.

The thiazide diuretics emerged from efforts to synthesize more potent carbonic anhydrase inhibitors by structural variations of sulfanilamides. A newly synthesized compound, 6-chloro-7-sulfamoyl-2H-1,2,4-benzothiadiazine 1,1-dioxide, displayed much weaker activity as a carbonic anhydrase suppressant than acetazolamide (**21.1.1**), but it produced a remarkable sodium and chloride excretion with relatively little bicarbonate output. The compound was named chlorothiazide (**21.3.1**) and became the prototype for a series of effective thiazide series diuretics that include hydrochlorothiazide (**21.3.2**), hydroflumethiazide (**21.3.3**), trichlormethiazide (**21.3.4**), mebutizide (**21.3.5**), cyclopenthiazide (**21.3.6**), cyclothiazide (**21.3.7**), bendroflumethiazide (**21.3.8**), polythiazide (**21.3.9**), and methycyclothiazide (**21.3.10**) (Fig. 21.4.).

Thiazide diuretics were the first well-established and tolerated, efficient antihypertensive drugs. As diuretics they are usually used in combination with a loop diuretic to augment the diuresis in patients with refractory edema.

The most common adverse effects associated with the thiazide diuretics include skin rashes, interstitial nephritis, pancreatitis, gout, alkalosis, volume depletion, hypokalemia, hypomagnesemia, hypercholesterolemia, hypertriglyceridemia, hyperglycemia in diabetics, and azotemia [16-19].

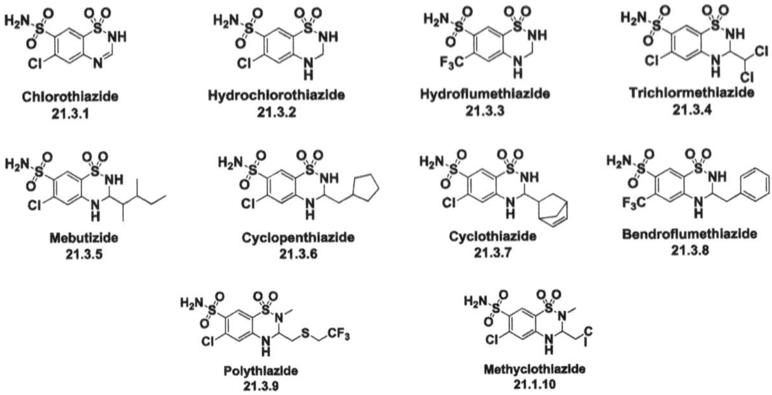

FIG. 21.4 **Thiazide diuretics.**

Thiazide-Like Diuretics

Multiple attempts of modifications in the thiadiazine nucleus resulted in creation of series of thiazide-like diuretics such as clofenamide (**21.3.11**), mefruside (**21.3.12**), meticrane (**21.3.13**), clorexolone (**21.3.14**), xipamide (**21.3.15**), clopamide (**21.3.16**), indapamide (**21.3.17**), chlortalidone (**21.3.18**), quinethazone (**21.3.19**), fenquizone (**21.3.20**), and metolazone (**21.3.21**) (Fig. 21.5.).

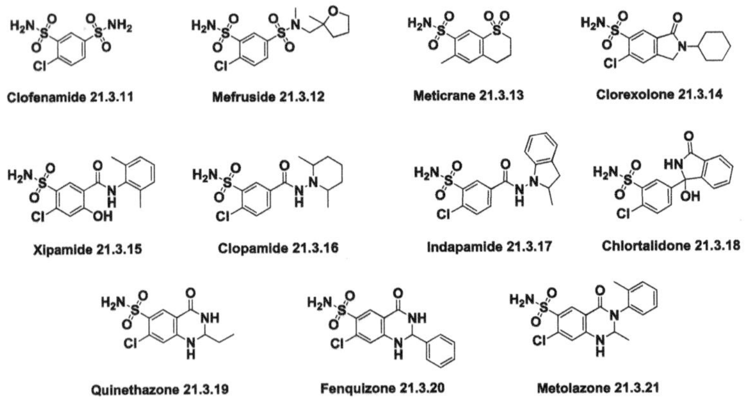

FIG. 21.5 **Thiazide-like diuretics.**

These drugs represent different chemical categories, but they all have pharmacological effects similar to those of the classical thiazides. Their common, general feature formally could be considered the sulfonamide moiety and sometimes they bring united under the name sulfonamides [20].

21.4 DIURETICS ACTING AT THE COLLECTING DUCT SYSTEM

The collecting duct system is another part of the kidney nephron, consisting of a series of tubules and ducts that are involved in the physiologic regulation of renal electrolytes and fluid balance. This part of the kidney is regulated by the aldosterone and vasopressin hormones.

There exist a variety of diuretics that block exchange of sodium for potassium and hydrogen ions in the distal tubule, increasing sodium and chloride excretion without increasing potassium excretion.

Diuretics that act by inhibiting the Na^+/K^+ exchange in the collecting duct system and cause a mild diuresis and potassium retention are called potassium-sparing diuretics.

In general, potassium-sparing diuretics increase diuresis by interfering on the Na^+/K^+ exchange in in three ways: via blocking the epithelial sodium channel in the kidneys; acting as an antagonist at the aldosterone receptors; or at the vasopressin receptors.

Potassium-sparing diuretics are divided into the (a) cycloamidine-type epithelial sodium channel blockers, (b) competitive aldosterone antagonists, and (c) vasopressin antagonists.

Some Li salts and demeclocycline, an antibiotic of tetracycline series, are also considered potassium-sparing diuretics.

Epithelial Sodium Channel Blockers

Epithelial sodium channel blockers include the cycloamidine-type compounds amiloride (**21.4.1**), triamterene (**21.4.2**), and benzamil (**21.4.3**), which act by blocking the Na^+ channels in the luminal membrane of the principal cells of the cortical collecting ducts. This reduces the Na^+ entry through the luminal membrane and hence the net reabsorption of NaCl. Potassium-sparing diuretics are used to prevent hypokalemia induced by loop or thiazide diuretics [21] (Fig. 21.6.).

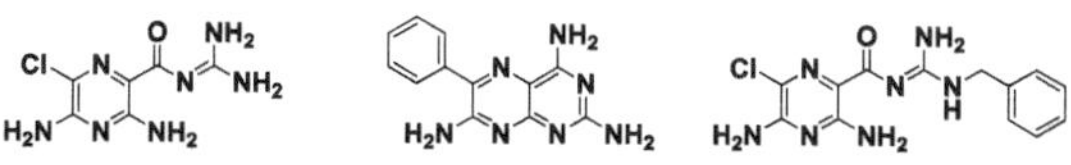

Amiloride 21.4.1 Triamterene 21.4.2 Benzamil 21.4.3

FIG. 21.6 **Epithelial sodium channel blockers.**

Aldosterone Antagonists

Aldosterone is a hormone secreted by the cortical region of the adrenal glands, which are located on top of the kidneys. Aldosterone plays a pivotal role in stimulating reabsorption of Na^+ in the distal tubule and collecting duct in exchange for K^+ and H^+. Derivatives of spirolactone—spironolactone (**21.4.4**), canrenone (**21.4.5**), and eplerenone (**21.4.6**) (Fig. 21.7.)—competitively block the aldosterone receptor at the cytosolic receptor level [22-24].

Vasopressin Antagonists

Vasopressin-2 receptor antagonists ("vaptans") provide a new approach to the treatment of hyponatremia. Effects of vasopressin via both V1- and V2-receptors

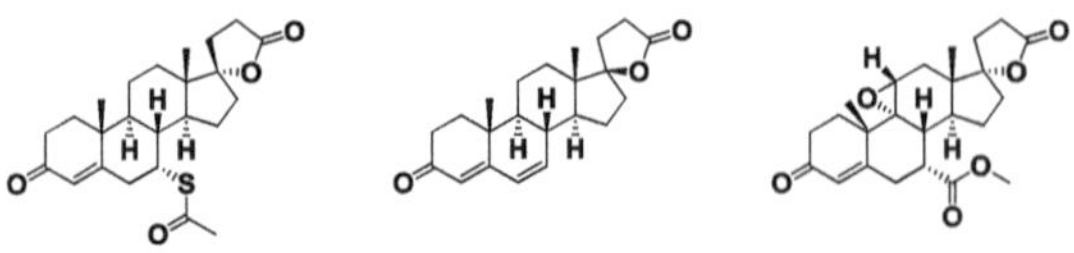

Spironolactone 21.4.4 Canrenone 21.4.5 Eplerenone 21.4.6

FIG. 21.7 **Aldosterone antagonists.**

are closely implicated in a variety of water-retaining and cardiovascular diseases, including heart failure, hyponatremia, hypertension, and renal diseases [25-30].

Among this series, tolvaptan (**21.4.7**) is an approved diuretic, a new drug indicated for the treatment of severe hyponatremia in most countries. Conivaptan (**21.4.8**) and lixivaptan (**21.4.8**) (Fig. 21.8.) are investigational agents.

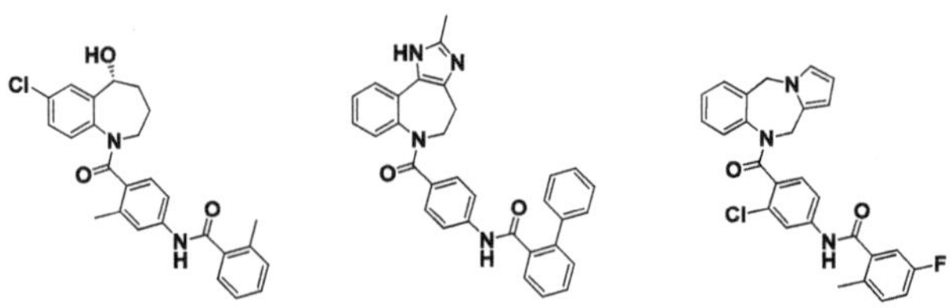

Tolvaptan 21.4.7 Conivaptan 21.4.8 Lixivaptan 21.4.9

FIG. 21.8 **Vasopressin antagonists ("vaptans").**

Potassium-sparing diuretics are used to prevent hypokalemia induced by loop or thiazide diuretics, to secondary hyperaldosteronism attributable to hepatic cirrhosis and ascites, and to treat primary hyperaldosteronism (Conn syndrome), Bartter and Liddle syndromes, and hirsutism.

21.5 OSMOTIC DIURETICS

Osmotic diuretics, which include glycerol (**21.5.1**), mannitol (**21.5.2**), and isosorbide (**21.5.3**) (Fig. 21.9.), are substances that have a low molecular weight and are filtered through the glomerulus. Their presence limits the reabsorption of water in the tubule and leads to an increase in the osmolarity of the filtrate. To maintain osmotic balance, water is retained in the urine [31].

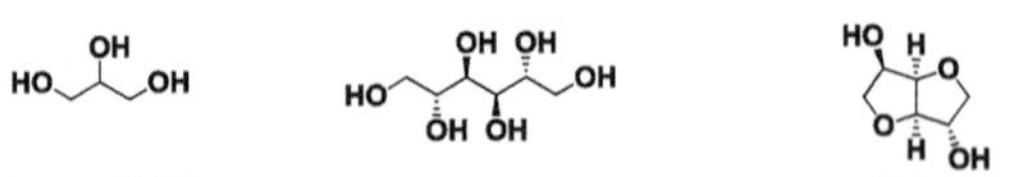

Glycerol 21.5.1 Mannitol 21.5.2 Isosorbide 21.5.3

FIG. 21.9 **Osmotic diuretics.**

21.6 NOVEL DIURETIC TARGETS

Few academic and pharma researchers are involved in the search for novel diuretics, but in the late 20th century, novel, clinically unexploited diuretic targets, such as renal ion-transport proteins, which participate in the regulation of fluid volume homeostasis, have been discovered.

Renal outer medullary potassium channel nanomolar inhibitors such as VU-590 (**21.6.1**), VU-591 (**21.6.2**), and BNBI (**21.6.3**), chloride channels antagonists such as MT-189 (**21.6.4**) [32], urea transporters inhibitors such as PU-14 (**21.6.5**), UTBinh-14 (**21.6.6**), and compound (**21.6.7**) [33-35], and others from very different chemical classes, have been discovered (Fig. 21.10.). Therefore unpredicted opportunities for translating these discoveries into novel diuretic therapies still exist.

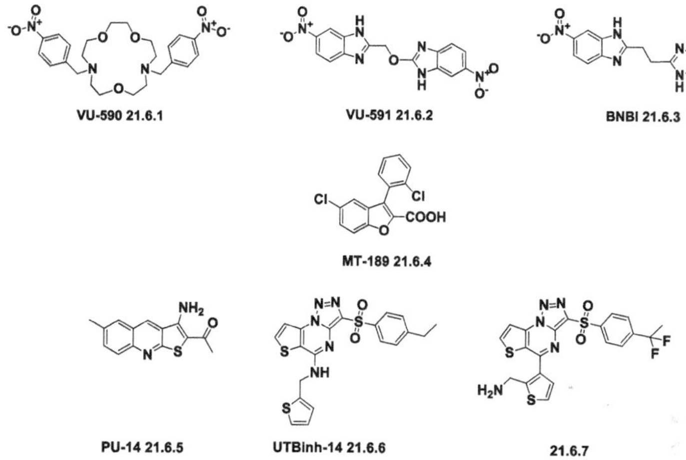

FIG. 21.10 Novel experimental chemical classes of diuretics.

No diuretic is included in the list of Top 200 Drugs by sales for the 2010s. Synthesis of generic, essential diuretics is described in our previous book [36].

REFERENCES

1. Lang, H.-J.; Hropot, M. Discovery and development of diuretic agents. In *Handbook of Experimental Pharmacology*; Greger, R. F., Knauf, H., Mutschler, E., Eds.; Vol. 117; Springer, 1995; pp 141–172.
2. Rankin, G. O. Diuretics. In *Foye's Principles of Medicinal Chemistry,* 6th ed.; Lemke, T. L., Williams, D. A., Eds. Lippincott Williams & Wilkins, 2008; pp 722–737.
3. Wile, D. Diuretics: a review. *Ann. Clin. Biochem.* **2012,** *49* (5), 419–431.
4. Puschett, J. B. Pharmacological classification and renal actions of diuretics. *Cardiology* **1994,** *84* (2), 4–13.
5. Sica, D. A. Diuretic use in renal disease. *Nat. Rev. Nephrol.* **2012,** *8* (2), 100–109.
6. Sarafidis, P. A.; Georgianos, P. I.; Lasaridis, A. N. Diuretics in clinical practice, part I: mechanisms of action, pharmacological effects and clinical indications of diuretic compounds. *Expert Opin. Drug Saf.* **2010,** *9* (2), 243–257.

7. Sarafidis, P. A.; Georgianos, P. I.; Lasaridis, A. N. Diuretics in clinical practice. Part II: electrolyte and acid-base disorders complicating diuretic therapy. *Expert Opin. Drug Saf.* **2010,** *9* (2), 259–273.
8. Presne, C.; Monge, M.; Mansour, J.; Oprisiu, R.; Choukroun, G.; Achard, J. M.; Fournier, A. Diuretic-based therapy. *Nephrol. Ther.* **2007,** *3* (6), 392–426.
9. Wang, D. J.; Gottlieb, S. S. Diuretics: still the mainstay of treatment. *Crit. Care Med.* **2008,** *36* (1 Suppl.), S89–S94.
10. Wright, C. I.; Van-Buren, L.; Kroner, C. I.; Koning, M. M. G. Herbal medicines as diuretics: a review of the scientific evidence. *J. Ethnopharmacol.* **2007,** *114* (1), 1–31.
11. Vullo, D.; Innocenti, A.; Supuran, C. T. Diuretics with carbonic anhydrase inhibitory activity: toward novel applications for sulfonamide drugs. In *Design of Zinc-Enzyme Inhibitors: Functional, Structural, and Disease Applications;* Supuran, C. T., Winum, J.-Y., Drug, Eds.; Wiley, 2009; pp 155–170.
12. Supuran, C. T. Diuretics: from classical carbonic anhydrase inhibitors to novel applications of the sulfonamides. *Curr. Pharm. Des.* **2008,** *14* (7), 641–648.
13. Carta, F.; Supuran, C. T. Diuretics with carbonic anhydrase inhibitory action: a patent and literature review (2005–2013). *Expert Opin. Ther. Pat.* **2013,** *23* (6), 681–691.
14. Wargo, K. A.; Banta, W. M. A comprehensive review of the loop diuretics: should furosemide be first line? *Ann. Pharmacother.* **2009,** *43* (11), 1836–1847.
15. Rossignol, P.; Zannad, F. Loop diuretics and ultrafiltration in heart failure. *Expert Opin. Pharmacother.* **2013,** *14* (12), 1641–1648.
16. Tamargo, J.; Segura, J.; Ruilope, L. M. Diuretics in the treatment of hypertension. Part 1: thiazide and thiazide-like diuretics. *Expert Opin. Pharmacother.* **2014,** *15* (4), 527–547.
17. Messerli, F. H.; Bangalore, S. Half a century of hydrochlorothiazide: facts, fads, fiction, and follies. *Am. J. Med.* **2011,** *124* (10), 896–899.
18. Dinges, J. Thiazide-based diuretics for the treatment of hypertension and genitourinary disorders. In *Bioactive Heterocyclic Compound Classes: Pharmaceuticals;* Dinges, J., Lamberth, C., Eds.; Wiley-VCH, 2012; pp 169–182.
19. Ernst, M. E.; Grimm, R. H., Jr. Thiazide diuretics: 50 years and beyond. *Curr. Hypertens. Rev.* **2008,** *4* (4), 256–265.
20. Hughes, A. D. How do thiazide and thiazide-like diuretics lower blood pressure? *JRAAS* **2004,** *5* (4), 155–160.
21. Teiwes, J.; Toto, R. D. Epithelial sodium channel blockers inhibition in cardiovascular disease. A potential role for amiloride. *Am. J. Hypertens.* **2007,** *20* (1), 109–117.
22. Ponda, M. P.; Hostetter, T. H. Aldosterone antagonism in chronic kidney disease. *Clin. J. Am. Soc. Nephrol.* **2006,** *1* (4), 668–677.
23. Wenzel, U. Aldosterone antagonists: silver bullet or just sodium excretion and potassium retention? *Kidney Int.* **2007,** *71* (5), 374–376.
24. Roongsritong, C.; Kumar, A.; Jenkins, L. A. Hypertensive heart disease and the role of aldosterone antagonists. *Curr. Hypertens. Rev.* **2007,** *3* (2), 137–142.
25. Kumar, S.; Berl, T. Vasopressin antagonists in the treatment of water-retaining disorders. *Semin. Nephrol.* **2008,** *28* (3), 279–288.
26. Decaux, G. V2-antagonists for the treatment of hyponatraemia. *Nephrol. Dial. Transplant.* **2007,** *22* (7), 1853–1855.
27. Kim, C. S. Pharmacologic management of the cardio-renal syndrome. *Electrolyte Blood Pressure* **2013,** *11* (1), 17–23.
28. Jovanovich, A. J.; Berl, T. Where vaptans do and do not fit in the treatment of hyponatremia. *Kidney Int.* **2013,** *83* (4), 563–567.

29. Rozen-Zvi, B.; Yahav, D.; Gheorghiade, M.; Korzets, A.; Leibovici, L.; Gafter, U. Vasopressin receptor antagonists for the treatment of hyponatremia: systematic review and meta-analysis. *Am. J. Kidney Dis.* **2010,** *56* (2), 325–337.
30. Costello-Boerrigter, L. C.; Boerrigter, G.; Burnett, J. C., Jr. Pharmacology of vasopressin antagonists. *Heart Failure Rev.* **2009,** *14* (2), 75–82.
31. Better, O. S.; Rubinstein, I.; Winaver, J. Osmotic diuretics: mannitol. *Handb. Exp. Pharmacol.* **1995,** *117,* 423–441.
32. Denton, J. S.; Pao, A. C.; Maduke, M. Novel diuretic targets. *Am. J. Physiol.* **2013,** *305* (4, Pt. 2), F931–F942.
33. Anderson, M. O.; Zhang, J.; Liu, Y.; Yao, C.; Phuan, P.-W.; Verkman, A. S. Nanomolar potency and metabolically stable inhibitors of kidney urea transporter UT-B. *J. Med. Chem.* **2012,** *55* (12), 5942–5950.
34. Li, F.; Lei, T.; Zhu, J.; Wang, W.; Sun, Y.; Chen, J.; Dong, Z.; Zhou, H.; Yang, B. A novel small-molecule thienoquinolin urea transporter inhibitor acts as a potential diuretic. *Kidney Int.* **2013,** *83* (6), 1076–1086.
35. Yao, C.; Anderson, M. O.; Zhang, J.; Yang, B.; Phuan, P.-W.; Verkman, A. S. Triazolothienopyrimidine inhibitors of urea transporter UT-B reduce urine concentration. *J. Am. Soc. Nephrol.* **2012,** *23* (7), 1210–1220.
36. Vardanyan, R. S.; Hruby, V. J. *Synthesis of Essential Drugs;* Elsevier, 2006.

Chapter 22

Antihypertensive Drugs

The blood pressure is the pressure of the blood within the arteries that originates in the pumping action of the heart. It is produced primarily by the contraction of the heart muscle. It is measured in two indexes: the systolic pressure, which shows the force that blood exerts on the artery walls at the heart's contraction to pump the blood to the peripheral organs and tissues, and the diastolic pressure, which remains on the artery walls as the heart relaxes between beats. It is postulated that the normal value of systolic pressure for healthy individuals must vary between 100 and 140 millimeters of mercury (mm Hg) and the diastolic pressure value varies between 60 and 100 mm Hg [1-3].

For 90 to 95% of patients, the causes of high blood pressure are unknown; this hypertension is classified as essential or primary hypertension. The remaining 5 to 10% are cases of secondary hypertension caused by underlying heart, kidney, or endocrinal diseases, certain cancers, or use of cocaine, amphetamines, thyroid supplements, or corticosteroids.

Hypertension (high blood pressure) can lead to an increased risk of heart attack, stroke, and renal failure. Statistically, on average approximately 20% of the adult population is hypertensive.

Antihypertensive drugs are medications that help lower blood pressure [4-10].

Venesection or bloodletting and the use of leeches historically have been the first attempted methods to decrease a patient's blood pressure. Prior to World War II there were no effective antihypertensive drugs. Sodium thiocyanate (**22.1**) was in limited use by 1900 and sporadically thereafter. But its use needed permanent control and blood level measurements to keep the dosage within a safe range. Mainly for these reason the drug never became popular [11].

Another attempt to lower blood pressure by means of intravenously administered drug was the infusion of sodium nitroprusside (**22.2**), a compound that dilates both arterial and venous resistance and which still remains an effective drug for the rapid reduction of significant arterial hypertension regardless of the etiology [12,13].

The ganglion blockers were the first drugs applied in long-term treatment of hypertension.

Hexamethonium (**22.3**) was one of the first drugs used for treating hypertensive patients but because of its adverse effects, it is no longer used.

Synthesis of Best-Seller Drugs. http://dx.doi.org/10.1016/B978-0-12-411492-0.00022-5

Other drugs belonging to different pharmacological classes, such as antipsychotic and antihypertensive reserpine (**22.4**), a direct-acting smooth muscle relaxant, vasodilator hydralazine (**22.5**), the antimalarial pentaquine (**22.6**), the antihypertensive guanethidine (**22.7**), and the α_2-adrenergic agonist methyldopa (**22.8**) (Fig. 22.1.), were implemented to control hypertension, but their antihypertensive action was associated with adverse effects such as dizziness, sedation effect, impotency, blurred vision, dry mouth, and constipation that were so significant that relatively few patients complied with these medications.

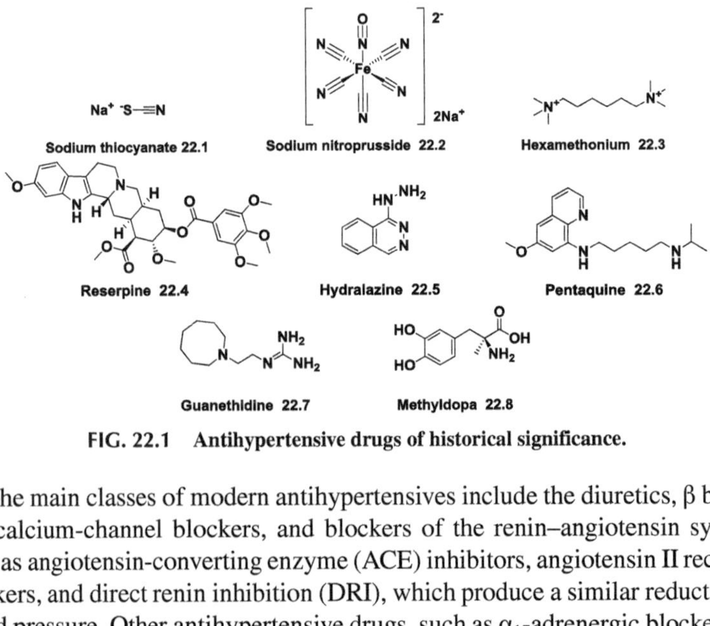

FIG. 22.1 Antihypertensive drugs of historical significance.

The main classes of modern antihypertensives include the diuretics, β blockers, calcium-channel blockers, and blockers of the renin–angiotensin system, such as angiotensin-converting enzyme (ACE) inhibitors, angiotensin II receptor blockers, and direct renin inhibition (DRI), which produce a similar reduction in blood pressure. Other antihypertensive drugs, such as α_1-adrenergic blockers, α_2 agonists, direct-acting vasodilators, and aldosterone receptor antagonists (Aldactone or eplerenone) are reserved for some special indications. State-of-the-art treatment of hypertension was presented recently in an excellent review [14].

The currently used antihypertensive agents are listed in the two major guidelines [1,2].

Few new drug therapies for hypertension have become available in recent years.

Synthesis of generic, essential antihypertensives is described in our previous book [15]. Synthesis of antihypertensives such as valsartan, olmesartan, irbesartan, losartan, and elmisartan, which are included in the list of Top 200 Drugs by sales for the 2010s are described below.

22.1 DIURETICS

The diuretics are listed as the current first-line antihypertensives in the major guidelines [1,2] for the treatment of hypertension. They are used as monotherapy or in combination with medications of other antihypertensive classes. Their therapeutic effect consists in decreasing the plasma and extracellular fluid volumes, which leads to a decrease in cardiac output, and consequently blood pressure [16,17].

The thiazide diuretics [chlorothiazide (**22.1.1**), hydrochlorothiazide (**22.1.2**), and polythiazide (**22.1.9**)], thiazide-like diuretics [indapamide (**22.1.4**), chlortalidone (**22.1.5**), and metolazone (**22.1.6**)], loop diuretics [furosemide (**22.1.7**), bumetanide (**22.1.8**), and torsemide (**22.1.9**)], potassium-sparing diuretics–epithelial sodium channel blockers [amiloride (**22.1.10**), triamterene (**22.1.11**)], and aldosterone antagonists [spironolactone (**22.1.12**) and eplerenone (**22.1.13**)] (Fig. 22.2.) are the most commonly used diuretics for treatment of hypertension. The prescriptions for chlorothiazide (**22.1.1**) outnumbers those written for any other single diuretic [14].

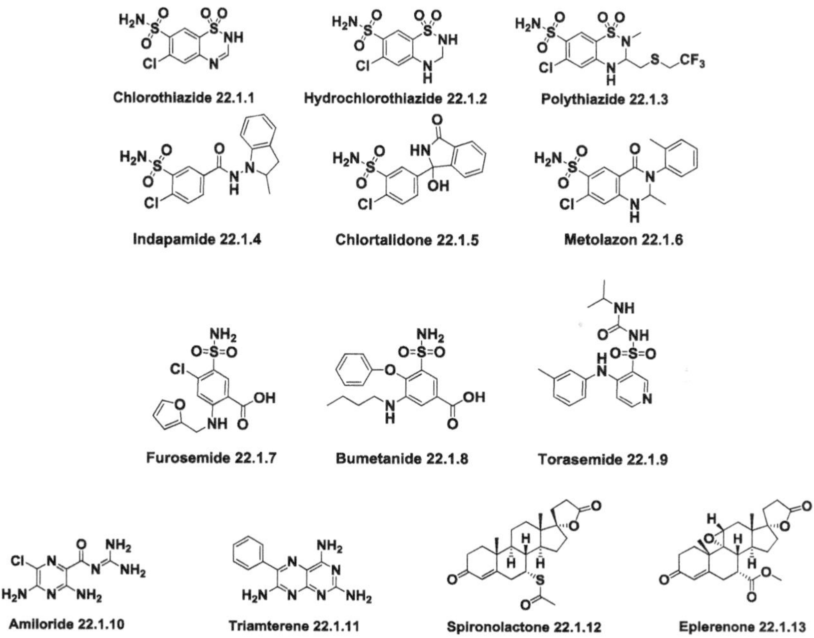

FIG. 22.2 **Diuretics are the current first-line antihypertensives.**

22.2 β BLOCKERS

β Blockers announced a new era in pharmacology and have been used in the treatment of cardiovascular conditions for decades. β Blockers have long been used as a first-line therapy for hypertension. They act via blocking the effects of the sympathetic nervous system and specifically on β receptors in the heart; they reduce the heart rate, the heart's workload, and the heart's output of blood, which lowers blood pressure [18-21]. Traditional β blockers—propranolol (**22.2.1**), atenolol (**22.2.2**), metoprolol (**22.2.3**), acebutolol (**22.2.4**), bisoprolol

(**22.2.5**), betaxolol (**22.2.6**), penbutolol (**22.2.7**), nadolol (**22.2.8**), pindolol (**22.2.9**), and timolol (**22.2.10**) (Fig. 22.3.)—are the most recommended [1,2] and commonly used β blockers for treatment of hypertension.

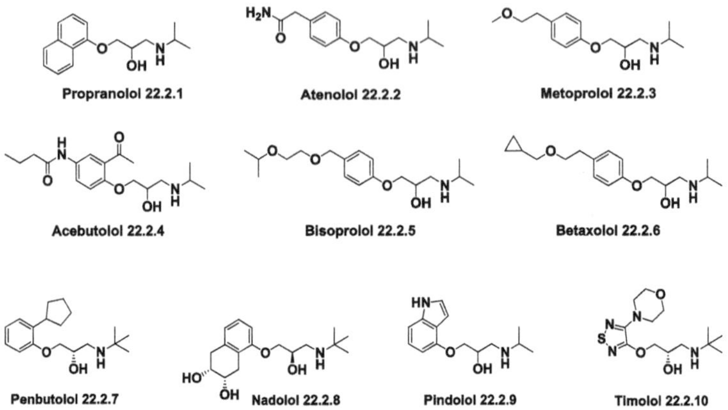

FIG. 22.3 Commonly used β blockers for treatment of hypertension.

Moreover, it is shown that combined α/β blockers such as carvedilol (**22.2.11**), nebivolol (**22.2.12**) and labetalol (**22.2.13-**) (Fig. 22.4.) can reduce blood pressure by reducing systemic vascular resistance rather than by decreasing cardiac output, as is observed with regular β blockers, yet not have the same vascular and metabolic effects as atenolol or metoprolol.

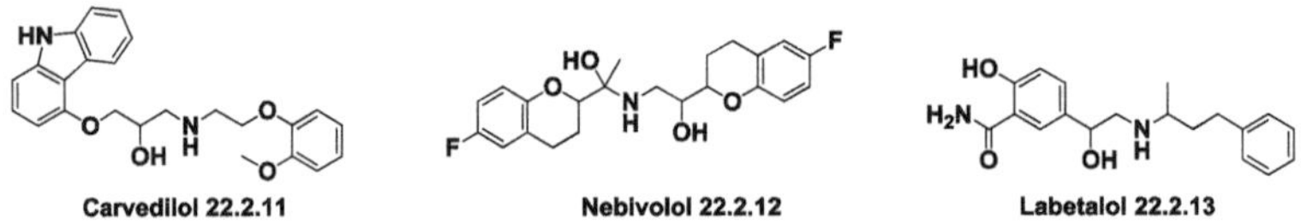

FIG. 22.4 Combined α/β blockers for treatment of hypertension.

22.3 CALCIUM-CHANNEL BLOCKERS

Calcium-channel blockers, also called calcium antagonists, inhibit transmembrane calcium influx in cardiac and vascular smooth muscle, leading to vasodilation and lowering of blood pressure. By decreasing the calcium influx, the heart's contraction becomes less forceful, reducing heart rate, relaxing and opening up narrowed blood vessels, and lowering blood pressure [22,23]. The Joint National Committee on Prevention, Detection, Evaluation, and Treatment of High Blood Pressure [1], as well as the European Society of Hypertension [2], recommend calcium-channel blockers of the dihydropyridine series—nifedipine (**22.3.1**), felodipine (**22.3.2**), lacidipine (**22.3.3**), amlodipine (**22.3.4**), nicardipine (**22.3.5**), and isradipine (**22.3.6**)—and nondihydropyridine series—verapamil (**22.3.7**) and diltiazem (**22.3.8**) (Fig. 22.5.)—for hypertension prevention and management.

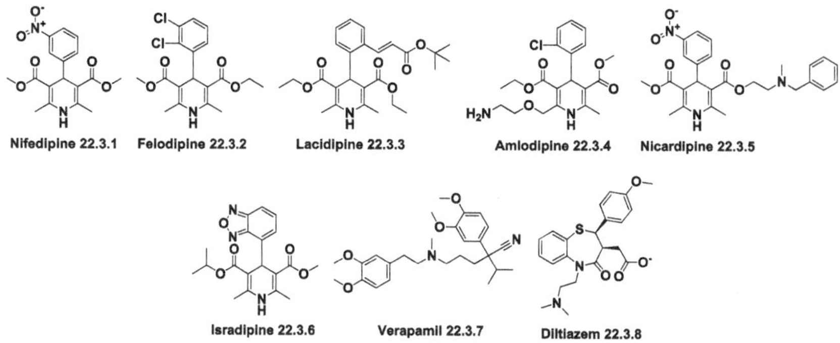

FIG. 22.5 Calcium-channel blockers recommend for hypertension prevention and management.

It seems, that the long-acting second-generation drug amlodipine (**22.3.4**) has taken the lead in the number of prescriptions issued [14].

New third-generation calcium-channel blockers, such as lercanidipine (**22.3.9**), lacidipine (**22.3.10**), and manidipine (**22.3.11**) (Fig. 22.6.), have improved drug persistence in comparison with other calcium-channel blockers, with values close to those of renin–angiotensin system inhibitors.

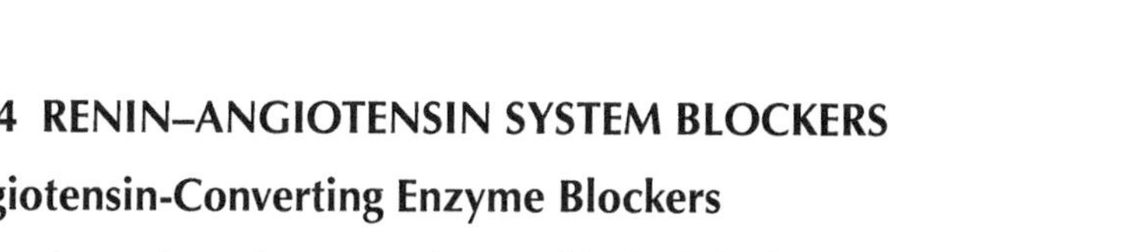

FIG. 22.6 Third-generation calcium-channel blockers.

22.4 RENIN–ANGIOTENSIN SYSTEM BLOCKERS

Angiotensin-Converting Enzyme Blockers

The renin–angiotensin system plays a critical role in the regulation of blood pressure. Renin is an enzyme which is secreted by kidneys in response on decrease in arterial blood pressure, decrease of sodium chloride level and decrease of blood pressure controlled by β_1 receptors. It cleaves angiotensinogen, a γ-globulin, to yield angiotensin I, which is further converted into angiotensin II by another enzyme—ACE. Angiotensin II, in turn, constricts blood vessels and turns on a series of other mechanisms that lead to an increase in blood pressure.

Angiotensin-I is a decapeptide hormone that can be cleaved to octapeptide angiotensin-II, which raises blood pressure in response to signals from the kidneys. Angiotensin-I converting enzyme is a key regulator of blood pressure.

ACE catalyzes the conversion of inactive angiotensin-I into a potent vasoconstrictor, angiotensin-II, through cleavage of terminal dipeptide (His-Leu) of angiotensin-I. ACE also inactivates the vasodilator bradykinin. Inhibition of ACE mainly results in an overall antihypertensive effect.

Inhibitors of ACE are often used to treat hypertension, and other cardio-related diseases. The influences of ACE on blood pressure make it a good target in the treatment of hypertension. Therefore, ACE inhibitors are one of the major classes of antihypertensive drugs [24-32]. Captopril (**22.4.1**) was the first orally active ACE inhibitory antihypertensive drug, discovered in 1977. Several other synthetic ACE inhibitors, such as enalapril (**22.4.2**), lisinopril (**22.4.3**), fosinopril (**22.4.4**), ramipril (**22.4.5**), trandolapril (**22.4.6**), perindopril (**22.4.7**), quinapril (**22.4.8**), moexipril (**22.4.9**), and benazepril (**22.4.10**), are on the list of recommended drugs in the Guidelines of Joint National Committee on Prevention, Detection, Evaluation, and Treatment of High Blood Pressure [1,2] and are in clinical use for the treatment of hypertension (Fig. 22.7.).

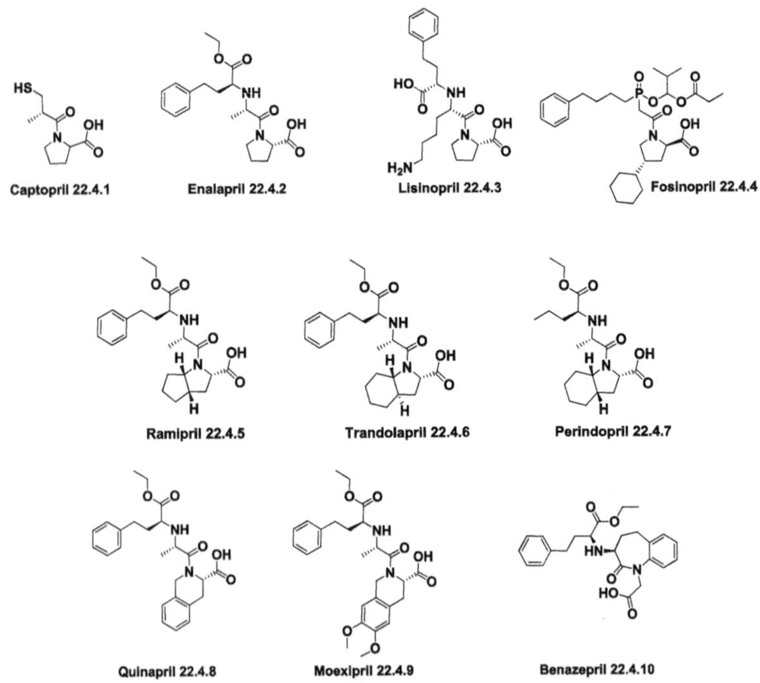

FIG. 22.7 Angiotensin-converting enzyme blockers.

Angiotensin II Receptor Blockers

Angiotensin-II is a major player of the renin-angiotensin system which acts via two receptor subtypes: AT1 and AT2. The AT1 receptor mediates modulations of angiotensin-II on the vasoconstriction, aldosterone and vasopressin release, sodium and water retention, insulin resistance, inflammation, and other causes of modulation.

Consequently, the AT1 receptor antagonists block angiotensin-II–dependent effects, resulting in a reduction in blood pressure and some other physiological effects.

The sartans are a class of new antihypertension drugs of angiotensin-II receptor blockers. Losartan (**22.4.11**) is the prototype of this class of drugs. Since it first came to market in 1994, the sartans have been developed rapidly for their unique clinical efficacy and good safety.

Renin–angiotensin–aldosterone system blockade is currently viewed as the best treatment strategy for hypertension; it will delay progression of chronic proteinuric nephropathies and has the best tolerability profile of all antihypertensive drug classes [33-45].

Today, the sartans class consists of seven compounds: irbesartan (**22.4.12**), candesartan (**22.4.13**), olmesartan (**22.4.14**), valsartan (**22.4.15**), telmisartan (**22.4.16**), eprosartan (**22.4.17**) (Fig. 22.8.).

The newest angiotensin receptor blocker marketed for the treatment of arterial hypertension is azilsartan (**22.4.18**) (Fig. 22.8.).

Azilsartan is the eighth sartan to join the pharmaceutical market. It is probably the first representative of next-generation angiotensin-II receptor blockers. This new agent induces a potent and long-lasting antihypertensive effect, insulin resistance improvement and albuminuria inhibitory effects.

Sartans are also used in the treatment of diabetic nephropathy an hypertensive patients with type 2 diabetes mellitus, treatment of congestive heart failure. Most angiotensin receptor blockers are available also in combination with adjunctive medications.

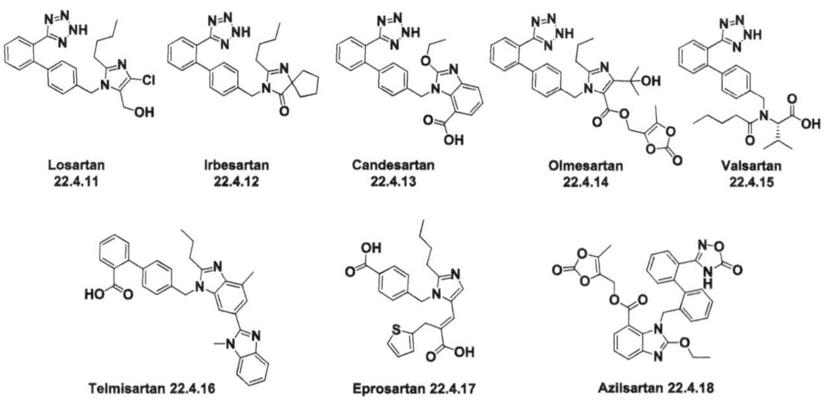

FIG. 22.8 Angiotensin II receptor blockers.

Losartan (**22.4.11**), irbesartan (**22.4.12**), olmesartan (**22.4.14**), valsartan (**22.4.15**), and telmisartan (**22.4.16**) are included in the list of Top 200 Drugs by sales for the 2010s.

Losartan–Cozaar

The first approach employed in the synthesis of losartan (**22.4.11**) included alkylation of (2-butyl-4-chloro-1H-imidazol-5-yl)methanol (**22.4.20**) with 4′-(bromomethyl)-[1,1′-biphenyl]-2-carbonitrile (**22.4.19**) to give intermediate (**22.4.21**) followed by cyclization of its nitrile moiety to tetrazole (**22.4.11**) using the known reaction of sodium azide with a nitrile, in the presence of an ammonium salt, or trimethyltin azide as sodium azide surrogate instead of sodium azide usually implemented on large scale synthesis for safety considerations. The last method needs further workup to remove trimethyltin moiety attached to two isomeric tetrazoles formed. The trimethyltin group was removed employing sodium hydroxide, and alkylating with bulky triphenylmethyl chloride. The requested product was isolated as triphenylmethyl derivative, which deprotection was easily achieved by hydrolysis using 10% hydrochloric acid in tetrahydrofuran producing the desired losartan (**22.4.11**) [46,47] (Scheme 22.1.)

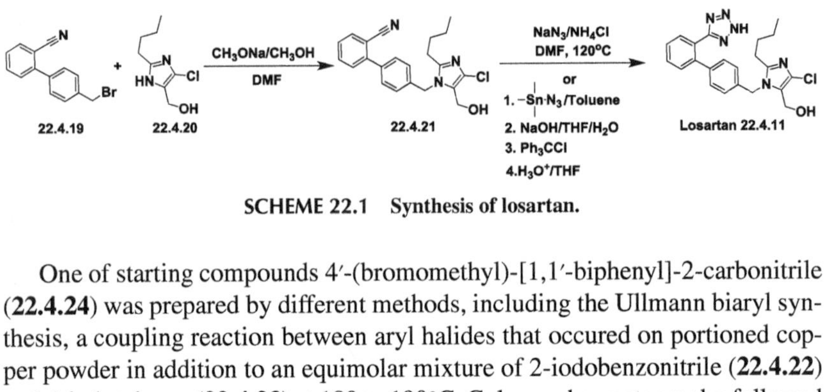

SCHEME 22.1 Synthesis of losartan.

One of starting compounds 4′-(bromomethyl)-[1,1′-biphenyl]-2-carbonitrile (**22.4.24**) was prepared by different methods, including the Ullmann biaryl synthesis, a coupling reaction between aryl halides that occured on portioned copper powder in addition to an equimolar mixture of 2-iodobenzonitrile (**22.4.22**) and 4-iodotoluene (**22.4.23**) at 180 to 190°C. Column chromatography followed by Kugelrohr distillation produced 4′-methyl-[1,1′-biphenyl]-2-carbonitrile (**22.4.24**) with low yields. The last was converted to the desired (**22.4.19**) on reaction with N-bromosuccinimide in presence of dibenzoyl peroxide.

Another method for the preparation of the biphenyl intermediate (**22.4.24**) uses the Meyers chemistry displacement of an o-methoxy group with organometallics on the reaction of oxazolidine (**22.4.26**), which in turn, was obtained by a two-step reaction of 2-methoxybenzoyl chloride (**22.4.25**) with 2-amino-2-methyl-1-propanol to give an intermediate amide and further cyclizing the obtained amide with thionyl chloride. The oxazolidine (**22.4.26**) on reaction with (4-methylpheny1)magnesium bromide produced the biphenylyloxazolidine (**22.4.27**). The oxazolidine moiety was removed under acidic conditions, providing corresponding acid, which was converted to biphenylnitrile (**22.4.24**) by stepwise transformation to an acid chloride, then primary amide, and, finally, converting the carboxamido group into the cyano group using thionyl chloride. Alternatively, the oxazoline ring can be fragmented by employing phosphorus oxychloride in pyridine to furnish the desired nitrile (**22.4.24**) and, finally, converting it to the desired (**22.4.19**) [46,47] (Scheme 22.2.).

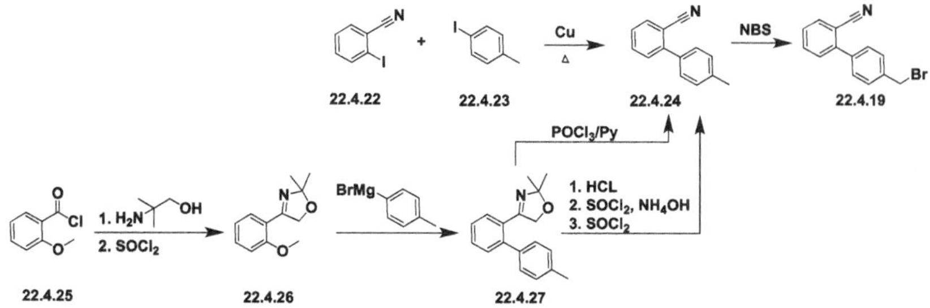

SCHEME 22.2 **Synthesis of 4′-(bromomethyl)-[1,1′-biphenyl]-2-carbonitrile (22.4.24).**

The second starting compound (2-butyl-4-chloro-1H-imidazol-5-yl)methanol (**22.4.20**) have been prepared from ethyl valerimidate (**22.4.28**), dihydroxyacetone (**22.4.29**), and ammonia, which on reaction in autoclave under elevated temperature and pressure produce (2-butyl-1H-imidazol-5-yl)methanol (**22.4.30**), which on further chlorination with N-chlorosuccinimide [48] produces the desired (**22.4.20**). Some further improvements in synthetic details were been done later [49,50] (Scheme 22.3.)

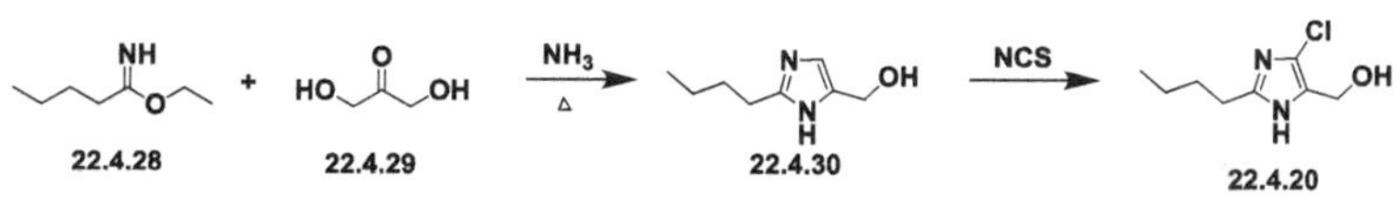

SCHEME 22.3 **Synthesis of (2-butyl-4-chloro-1H-imidazol-5-yl)methanol (22.4.20).**

Another, and highly efficient, convergent approach avoiding the use of azides in the synthesis of losartan (**22.4.11**) has been described. It implements the Suzuki palladium-catalyzed cross-coupling reaction of phenylboronic acid (**22.4.31**) and 4-bromobenzyl-imidazolyl (**22.4.32**) derivatives with further deprotection from trityl group with sulfuric acid in water/acetonitrile mixture [51] (Scheme 22.4.).

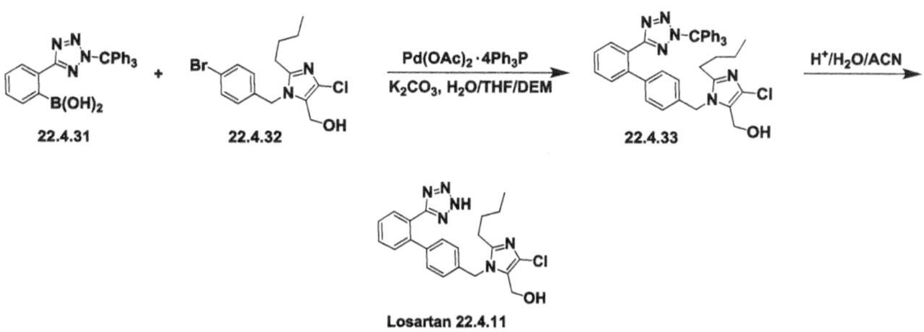

SCHEME 22.4 **Synthesis of losartan.**

The starting phenylboronic acid (**22.4.32**) was prepared from commercially available 5-phenyltetrazole (**22.4.34**), which was easily alkylated with trityl chloride in THF to produce (**22.4.35**). BuLi was then added to protected tritylphenyltetrazole (**22.4.36**) followed by triisopropyl borate, which gave phenylboronic acid isopropyl ester (**22.4.36**). The obtained ester was hydrolyzed with saturated aqueous NH_4Cl to produce the desired boronic acid (**22.4.31**) (Scheme 22.5.). (Preliminary protection of the tetrazole moiety was necessary for a number of reasons among which are: forming specifically one isomer on the N-2 position in the synthesis of (**22.4.31**) in a series of further transformations; trityl group easy addition and removal under mild conditions; prevention of tetrazole group action as a poison on palladium reagent.)

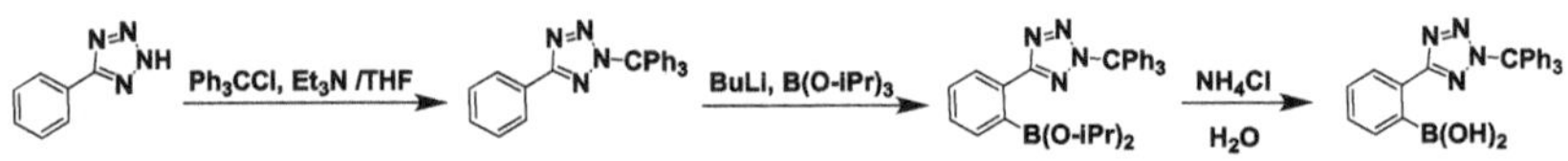

SCHEME 22.5 **Synthesis of the starting phenylboronic acid (22.4.32).**

The second starting material—the 4-bromobenzyl-imidazolyl (**22.4.32**) derivative—was synthesized in two ways. First was direct oxidation of (2-butyl-4-chloro-1H-imidazol-5-yl)methanol (**22.4.20**) with manganese dioxide was proposed by the same authors in the same publications [48-50]. Other elegant approaches were demonstrated starting from the easily accessible methyl valerimidate (**22.4.37**).

Methyl valerimidate (**22.4.37**), and glycine methyl ester in a methanolic solution formed 2-butyl-3,5-dihydro-4H-imidazol-4-one (**22.4.38**) [51]. The last under Vilsmeyer reaction conditions—treatment with $POCl_3$/DMF at 40°C—produced (**22.4.39**) as the major product, which was heated with $POCl_3$ followed with aqueous workup to give 2-butyl-4-chloro-1H-imidazole-5-carbaldehyde (**22.4.41**). Efficient "one-pot" procedures for the synthesis of the same aldehyde have been developed and optimized. The product of the reaction of methyl valerimidate (**22.4.38**) and glycine-(1-iminopentyl)glycine (**22.4.40**) under Vilsmeyer conditions, but at 100°C, produced the desired (**22.4.41**) of high purity [52] (Scheme 22.6.).

Aldehyde (**22.4.41**) was benzylated with 1-bromo-4-(bromomethyl)benzene in dimethylacetamide (DMAC) with K_2CO_3 as base to provide a 97:3 mixture of regioisomers - products of two imidazole nitrogen atoms alkylation. The use of aldehyde was necessary for providing the high ratio of the desired regioisomer and a suitable rate of reaction [51]. (Alcohol (**22.4.20**) alkylated 100 times slower and provided a higher ratio of the undesired isomer.)

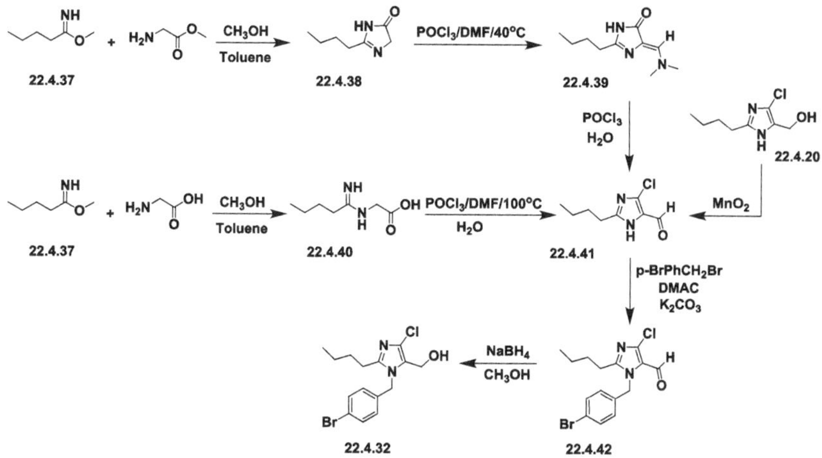

SCHEME 22.6 **Synthesis of (1-(4-bromobenzyl)-2-butyl-4-chloro-1H-imidazol-5-yl)methanol (22.4.32).**

Slightly different approaches also have been proposed [53-55].

Losartan is an angiotensin-II receptor blocker. It is indicated for the treatment of hypertension and may be used alone or in combination with other antihypertensive agents, including diuretics [56-65]. Side effects of losartan include diarrhea, muscle cramps, dizziness, insomnia, and nasal congestion.

Olmesartan–Benicar

Olmesartan (**22.4.14**), another angiotensin receptor blocker [66-78], is a close structural analogue of losartan (**22.4.11**), irbesartan (**22.4.12**), and candesartan (**22.4.13**), having the general 4-(imidazol-1-yl)methyl)-[1,1-biphenyl]-2-yl)-2H-tetrazole skeleton.

Synthesis of olmesartan started from preparation of imidazole-dicarbonitrile (**22.4.45**) via cyclocondensation of diaminomaleonitrile (**22.4.44**) and trimethyl orthobutyrate (**22.4.43**) by coherent heating in boiling CH_3CN and then in toluene, which results in a 96% yield of product. The obtained dinitrile (**22.4.45**) was hydrolyzed in acid media to the corresponding diacid (**22.4.46**), which, in turn, was esterified to diethyl ester (**22.4.47**) in a standard manner. The diethyl ester (**22.4.47**) was treated with four equivalents of MeMgBr to produce 4-hydroxyalkyl imidazole (**22.4.48**).

The obtained imidazole was alkylated with biphenyl-tetrazolyl bromide derivative (**22.4.49**). Hydrolysis of the ester group in the obtained compound (**22.4.50**) was furnished with LiOH, and the obtained lithium salt of acid (**22.4.51**) was alkylated with medoxomil chloride (4-chloromethyl-5-methyl-2-oxo-1,3-dioxole) (**22.4.52**) to produce trityl-protected olmesartan (**22.4.53**). The protective group was removed using 25% aqueous acetic acid to produce the desired olmesartan (**22.4.14**) [79-83] (Scheme 22.7.).

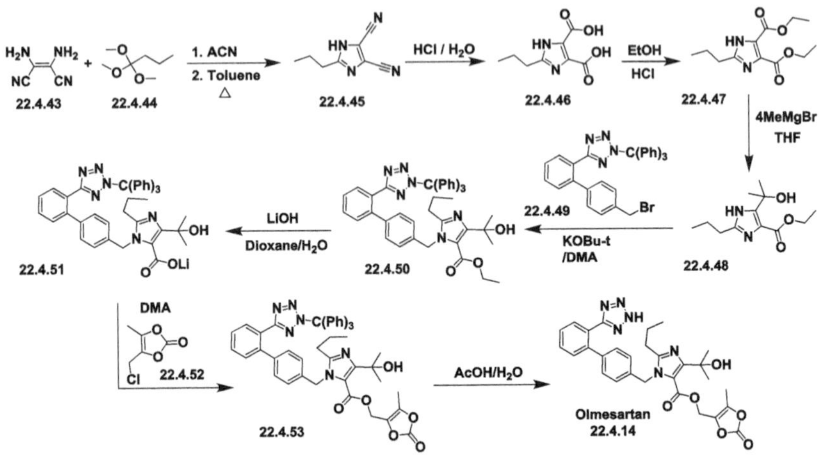

SCHEME 22.7 **Synthesis of olmesartan.**

The synthesis of biphenyl-tetrazolyl bromide derivative (**22.4.9**) was proposed via heterocyclization of 4′-methyl-[1,1′-biphenyl]-2-carbonitrile (**22.4.24**) with trimethyltin producing isomeric tetrazoles with trimethyltin moiety attached to two isomeric postions (**22.4.54**). The predominant isomer was separated and trimethyltin group was removed employing sodium hydroxide to produce (**22.4.55**). After alkylation with triphenylmethyl chloride, the desired product was isolated and brominated with N-bromosuccinimide producing (**22.4.9**) [47] (Scheme 22.8.).

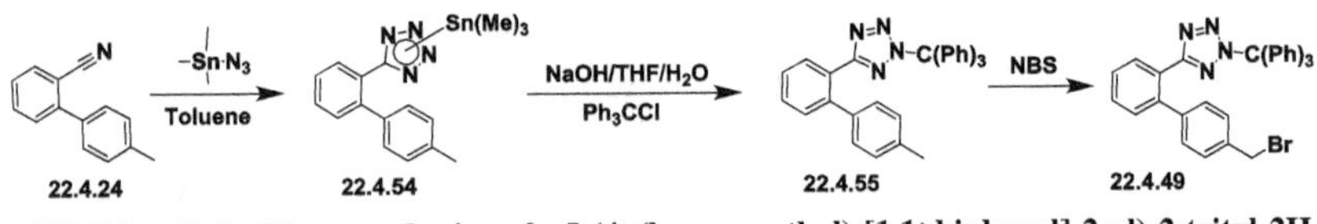

SCHEME 22.8 **The synthesis of 5-(4′-(bromomethyl)-[1,1′-biphenyl]-2-yl)-2-trityl-2H-tetrazole (22.4.49).**

Olmesartan is reported to be an effective agent for the treatment of hypertension with blood pressure-lowering effects comparable to other antihypertensive agents and other [66-78]. Effects were seen as early as 2 weeks and persisted when olmesartan medoxomil was administered long term. Common side effects include dizziness, bronchitis, back pain, headache, and flu-like symptoms.

Valsartan–Diovan

In the first patent [84] disclosing valsartan (**22.4.15**) and in a later paper by the same authors [85], different approaches for its synthesis are demonstrated.

According to one of them, L-valine methyl or benzyl esters were alkylated with biphenylbromomethylnitrile (**22.4.19**) to produce amines (**22.4.56**), which was followed by acylation with valeroyl chloride to produce N-valeryl derivatives (**22.4.57**). The obtained product was refluxed in xylene with tributyltin azide, producing tributyltin terazole derivatives, which then were hydrolyzed to terazole derivatives (**22.4.58**) by coherent use of base and acid. Finally, in the case of methyl ester, it underwent hydrolysis in base conditions or, in the case of benzyl ester, it was hydrogenated over the palladium catalyst, to produce the desired valsartan (**22.4.15**).

Another method also started from the same biphenylbromomethylnitrile (**22.4.59**), which was stepwise converted to acetate (**22.4.59**), hydrolyzed to the corresponding benzyl alcohol (**22.4.60**), which underwent Swern oxidation using oxalyl chloride in DMSO and in the presence of triethylamine to produce aldehyde (**22.4.61**). Reductive amination of the last with amino component L-valine methyl ester and hydrogenation agent sodium cyanoborohydride produced biphenyl nitrile (**22.4.56**), which by the same sequence of reactions (**22.4.56** to **22.4.57** to **22.4.58**) produced the desired valsartan (**22.4.15**) (Scheme 22.9.).

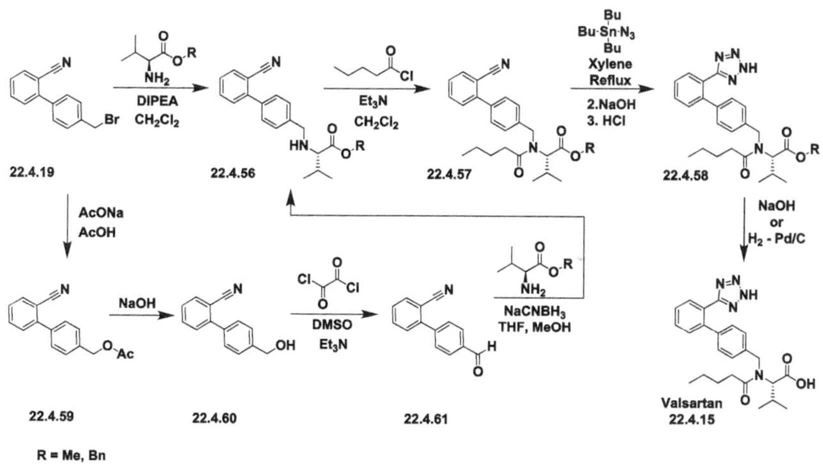

SCHEME 22.9 Synthesis of valsartan.

Many other attempts to improve details of the described synthetic scheme have been launched [86-95].

Valsartan is a highly effective, orally available angiotensin receptor blocker. It is indicated for treatment of mild to moderate essential hypertension. Valsartan reduces unwanted effects of angiotensin-II, such as aldosterone, vasopressin and endothelin secretion, vasoconstriction, diuresis, endothelial cell hyperplasia, mitogenesis, induction of growth factors, and production of collagen [96-106]. Side effects include chest pain, vomiting, shortness of breath, palpitations, anorexia, and angioedema.

Irbesartan–Avapro

According to the first publication [107] and first disclosed patents [108,109], synthesis of irbesartan starts (**22.4.12**) from the cyclocondensation reaction between ethyl ester of 1-amino-cyclopentanecarboxylic acid (**22.4.62**) and ethyl valerimidate (**22.4.28**), which, on reflux in toluene in the presence of the catalytic amount of acetic acid, produced imidazolone derivative (**22.4.58**). The imidazolone derivative (**22.4.63**) was alkylated with a known biphenyl-bromomethylnitrile (**22.4.9**) in DMF and in the presence of sodium methylate, easily producing compound (**22.4.64**). After reaction with tributyltin azide, the obtained product was hydrolyzed to produce the desired irbesartan (**22.4.12**) (see Scheme 22.10.).

Another ingenious approach for the synthesis of irbesartan was demonstrated recently [110].

According to that publication, the starting compound 4′-methyl-biphenyl-2-carboxylic acid (**22.4.65**) was transferred to benzylamide (**22.4.66**) in usual way (SOC12, benzylamine). Reaction of the obtained amide with triphenylphosphine/diethyl azodicaroxylate/trimethylsilylazide mixture allowed easily to converted it mild conditions and in high yield to tetrazole derivative (**22.4.67**). Benzylic bromination with 1,3-dibromo-5,5-dimethylhydantoin in ethylacetate and in presence of free radical initiator (2,2′-azobis(2-methylpropionitrile) furnished the intermediate (**22.4.68**) in high yield. The last was reacted with imidazolone derivative (**22.4.63**) in the presence of potassium hydroxide in acetone to afford the benzyl-protected irbesartan (**22.4.69**), which on debenzylation using a hydrogenation Pd/C catalyst produced the desired irbesartan (**22.4.12**) (Scheme 22.10.).

A series of other publications [111-115] and patents [116-118] have proposed several improvements in described scheme of synthesis.

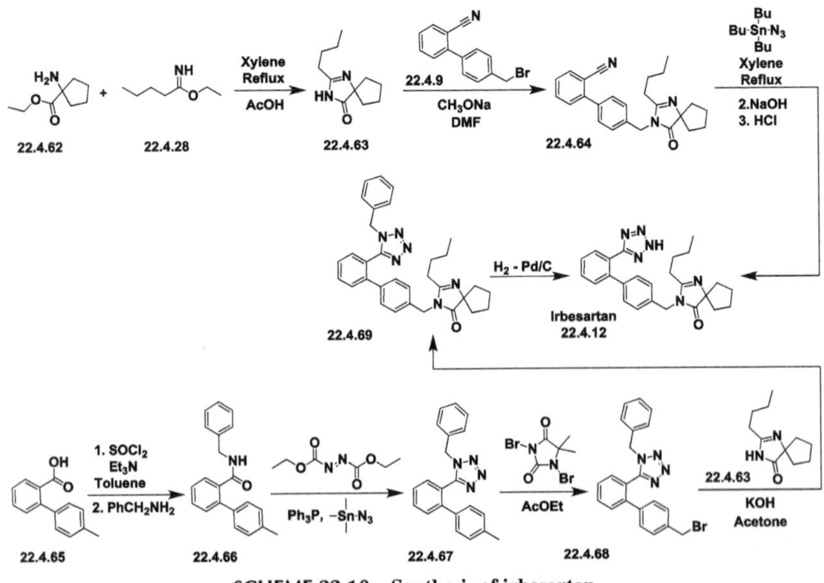

SCHEME 22.10 Synthesis of irbesartan.

Irbesartan is used to treat mild to moderate hypertension alone or with other medicines. It is effective in the elderly and nonelderly, men and women, and has attractive pharmacokinetic and clinical features [119-132].

Irbesartan's most frequent side effects are diarrhea, dizziness, and tiredness.

Telmisartan–Micardis

The first total synthesis of telmisartan (**22.4.16**) started with the acylation of 4-amino-3-methyl benzoic acid methyl ester (**22.4.70**) with butyryl chloride producing amide (**22.4.71**), followed by nitration with the mixture of HNO_3 and H_2SO_4 at 0°C, which gave compound (**22.4.72**). Reduction of the nitro group with hydrogen on Pd catalyst to produce a phenylenediamine derivative (**22.4.68**), which on subsequent cyclization with acetic acid gave the benzimidazole derivative (**22.4.69**).

After base hydrolysis of the carbomethoxyl group, the acid (**22.4.75**) was obtained. It was condensed with N-methyl-1,2-phenylenediamine to produce the bis-benzimidazole (**22.4.76**), which was then alkylated with the 4′-(bromomethyl)-2-biphenylcarboxylic acid tert-butyl ester (**22.4.77**) to produce, after deprotection of acid group, the desired telmisartan (**22.4.16**) [133] (Scheme 22.11.).

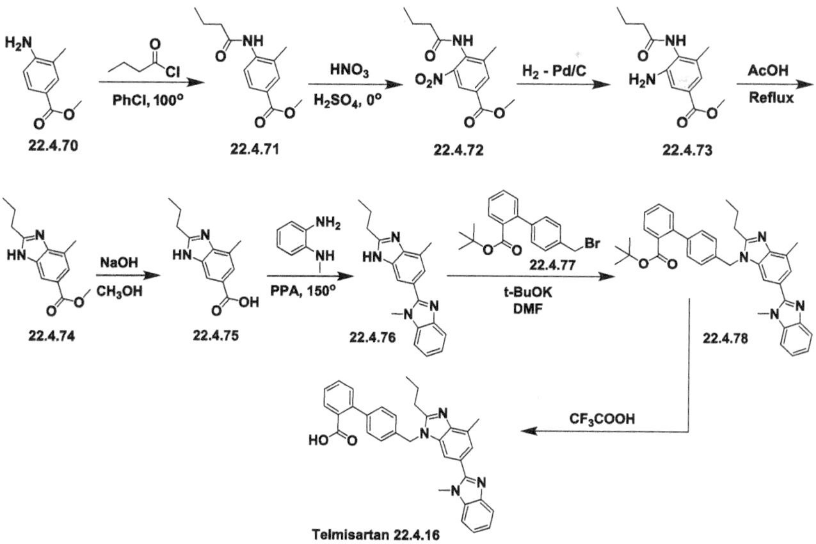

SCHEME 22.11 Synthesis of telmisartan.

A principial change in the synthesis of telmisartan was done using as a starting material 2′-(4,4-dimethyl-4,5 dihydro-1,3-oxazol-2-yl) biphenyl-4-carbaldehyde (**22.4.79**) prepared on Suzuki coupling of

2-(2-bromophenyl)-4,4-dimethyl-2-oxazoline with 4-formylphenylboronic acid. The reductive amination of this aldehyde with phenylenediamine derivative (**22.4.73**) was carried out in refluxing toluene in presence of p-TsOH, followed by hydrogenation of the formed Schiff base with hydrogen on Pd catalyst to produce the intermediate compound (**22.4.80**), which was cyclized in refluxing acetic acid to oxazoline-protected telmisartan (**22.4.76**). Protective oxazoline moiety was removed using boiling concentric hydrochloric acid to produce telmisartan (**22.4.81**) [134,135] (Scheme 22.12.).

Other minor changes in these two principal schemes were proposed [136-142].

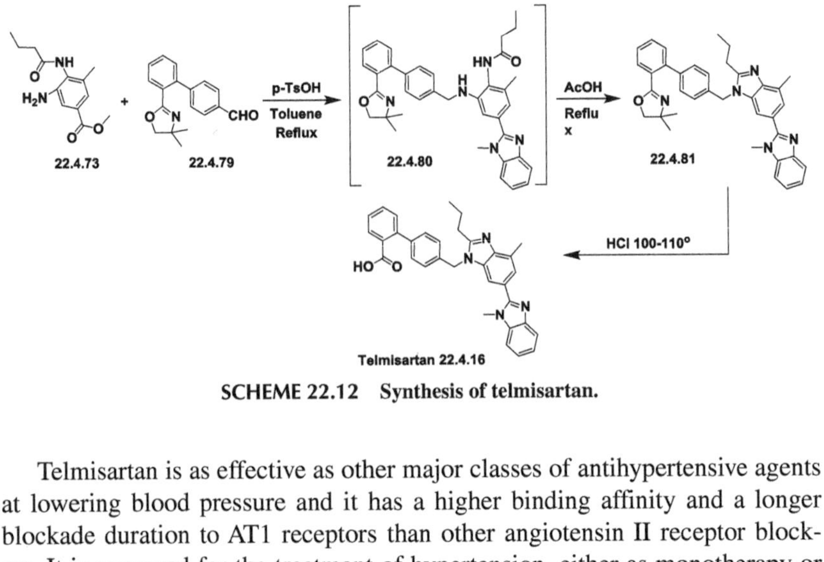

SCHEME 22.12 Synthesis of telmisartan.

Telmisartan is as effective as other major classes of antihypertensive agents at lowering blood pressure and it has a higher binding affinity and a longer blockade duration to AT1 receptors than other angiotensin II receptor blockers. It is approved for the treatment of hypertension, either as monotherapy or in combination with other antihypertensive agents. It seems that telmisartan lowered serum levels of glucose, insulin and triglycerides, and body weight [143-159].

Rare side effects can occur while taking telmisartan, including dizziness, lightheadedness or fainting, fast heartbeat, changes in vision, painful urination or changes in urinary frequency, and swelling in the hands, lower legs, and feet.

Direct Renin Inhibitors

More recently, a third approach to the inhibition of the renin–angiotensin system–DRI has become available. This new way of blocking the renin–angiotensin–aldosterone system provides a unique opportunity to block the renin–angiotensin system at its initial step was very promising as it aliskiren (**22.4.82**) was proposed [160-167] (Fig. 22.9.).

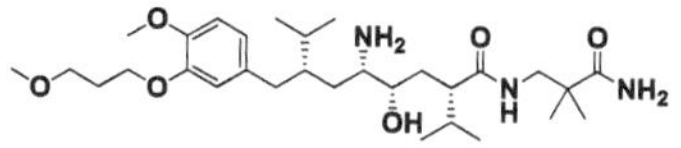

Aliskiren 22.4.82

FIG. 22.9 **Structure of aliskiren.**

Aliskiren (**22.4.82**) is the first drug representing a new generation of DRIs with a unique structure, becoming the first new therapy since 2007 for the treatment of hypertension. Aliskiren is the first member of the new class of DRIs approved by the U.S. Food and Drug Administration for the treatment of hypertension. Treatment with aliskiren lowers blood pressure effectively in patients with mild-to-moderate hypertension; it has a good safety and tolerability profile. Possible side effects could be signs of an allergic reaction, hives, severe stomach pain, difficulty breathing, and swelling of face, lips, tongue, or throat.

Aliskiren (**22.4.82**) was designed by molecular modeling [168,169].

DRIs block the conversion of angiotensinogen to angiotensin-I, which subsequently results in a reduction in angiotensin-II concentrations. Unlike the ACE inhibitors and the angiotensin-II receptor blockers, which reactively stimulate an increase in plasma renin activity, DRIs suppress the effects of renin and lead to a reduction in plasma renin activity.

The first reported DRI was a decapeptide, labeled based on its structure. Several peptide-based peptidomimetics, such as enalkiren (**22.4.83**), zankiren (**22.4.84**) and remikiren (**22.4.85**) (Fig. 22.10.), were created as investigational drugs, which, unfortunately, had limited druglike properties.

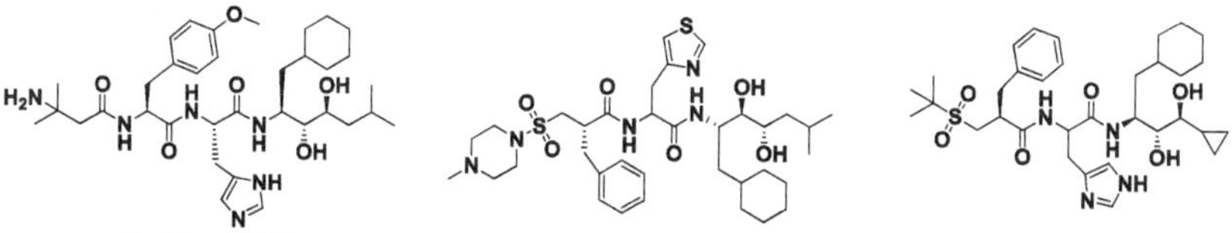

Enalkiren 22.4.83 Zankiren 22.4.84 Remikiren 22.4.85

FIG. 22.10 **Direct renin inhibitors.**

After identification of ligand-binding sites in the renin, the first small molecule DRIs, such as (**22.4.86** to **22.4.89**) (Fig. 22.11.), were designed. The literature on existing and experimental DRIs is reviewed by Webb et al [170] and in other reviews [170-176].

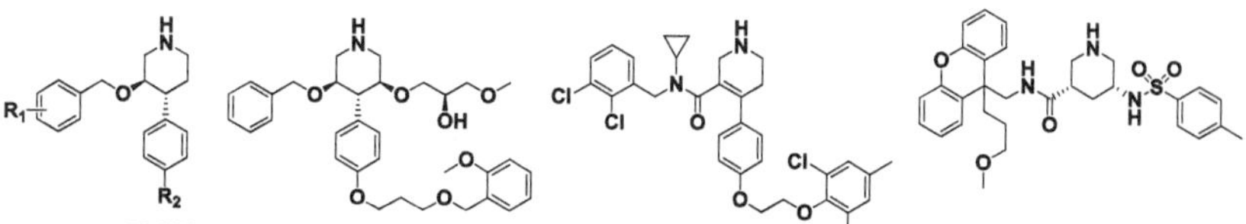

22.4.86 22.4.87 22.4.88 22.4.89

FIG. 22.11 **Small molecule DRIs.**

22.5 MISCELLANEOUS

α_1-Adrenergic Receptor Antagonists

α_1 Antagonists are drugs that block adrenaline and noradrenaline binding on α adrenoreceptors, thereby reducing arteriolar resistance and increasing venous capacitance. Some α-adrenergic agonists selective for α_2-adrenoreceptors, such as prazosin (**22.5.1**), terazosin (**22.5.2**), and doxazosin (**22.5.3**) (Fig. 22.12.), are sometimes used to treat high blood pressure.

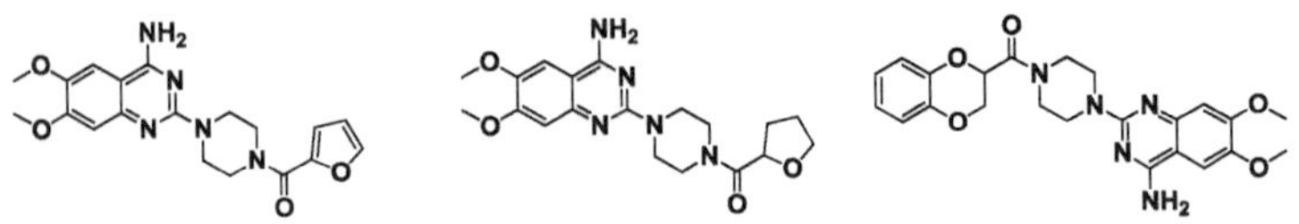

Prazosin 22.5.1 Terazosin 22.5.2 Doxazosin 22.5.3

FIG. 22.12 **α_1-Adrenergic receptor antagonists.**

α_2-Adrenergic Receptor Agonists

The α_2-adrenergic agonists are centrally acting vasodilators used in the treatment of hypertension.

α_2 Agonists lower blood pressure by stimulating α receptors to open peripheral arteries easing blood flow. Clonidine (**22.5.4**), guanabenz (**22.5.5**), guanfacine (**22.5.6**), and moxonidine (**22.5.7**) are structurally related compounds and have similar antihypertensive profiles. α-Methyldopa (**22.5.8**) is a structural analogue of dopa and functions as a prodrug. After administration, α-methyldopa is converted to α-methylnorepinephrine, which then serves as the α_2-adrenoceptor agonist (Fig. 22.13).

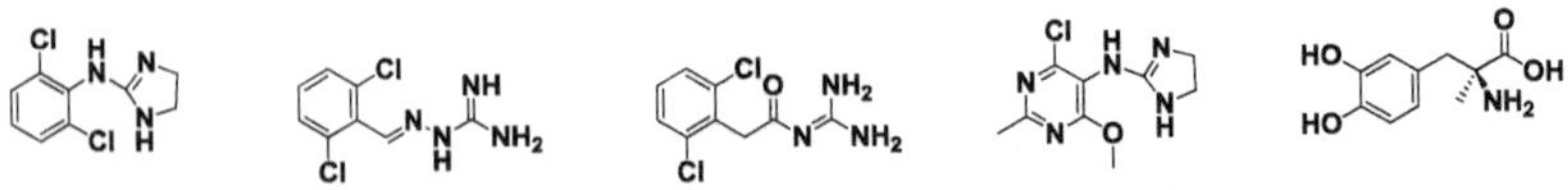

Clonidine 22.5.4 Guanabenz 22.5.5 Guanfacine 22.5.6 Moxonidine 22.5.7 Methyldopa 22.5.8

FIG. 22.13 **α_2-Adrenergic receptor antagonists**

Direct- and Indirect-Acting Vasodilators

Direct-acting vasodilators treat hypertension by acting primarily in arteries to relax vascular smooth muscle, thereby lowering blood pressure.

Drugs in this group, hydralazine (**22.5.9**) and minoxidil (**22.5.10**) (Fig. 22.14.), do not fit neatly into the other mechanistic classes, in part, because their mechanism of action is not entirely clear and it appears they have multiple direct effects on the vascular smooth muscle.

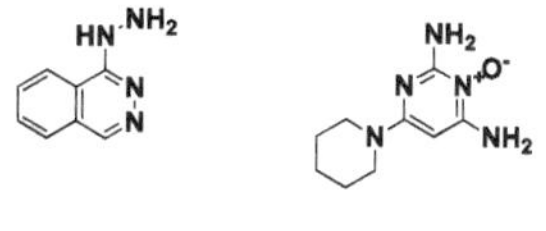

Hydralazine 22.5.9 Minoxidil 22.5.10

FIG. 22.14 **Direct-acting vasodilators.**

The category of indirect-acting drugs includes guanethidine, mecamylamine, and reserpine.

Guanethidine (**22.5.11**) is an antihypertensive drug that reduces the release of adrenaline and noradrenaline. It relaxes the blood vessels so that blood passes through them more easily. This helps to lower blood pressure.

Mecamylamine (**22.5.12**) is a nonselective and noncompetitive antagonist of nicotinic acetylcholine receptors. The hypotensive effect of mecamylamine is attributed to reduction in sympathetic tone, vasodilation, and reduced cardiac output, and is primarily postural.

The antihypertensive action of reserpine (**22.5.13**) is explained by its ability to deplete the adrenaline, noradrenaline, and dopamine involved in controlling peripheral vascular resistance (Fig. 22.15.).

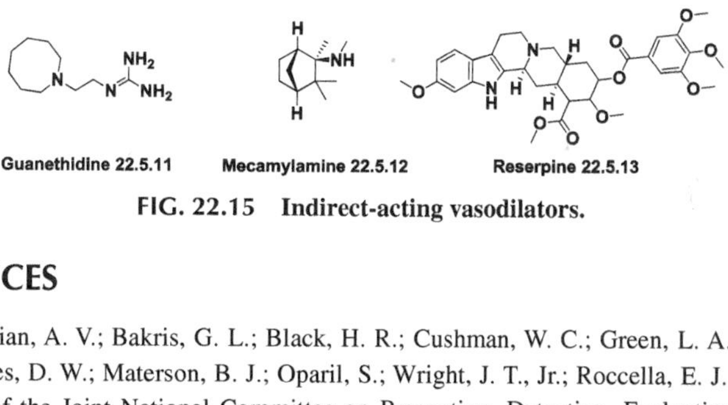

Guanethidine 22.5.11 Mecamylamine 22.5.12 Reserpine 22.5.13

FIG. 22.15 **Indirect-acting vasodilators.**

REFERENCES

1. Chobanian, A. V.; Bakris, G. L.; Black, H. R.; Cushman, W. C.; Green, L. A.; Izzo, J. L., Jr.; Jones, D. W.; Materson, B. J.; Oparil, S.; Wright, J. T., Jr.; Roccella, E. J. The seventh report of the Joint National Committee on Prevention, Detection, Evaluation, and Treatment of High Blood Pressure. *The JNC 7 report, JAMA. J. Am. Med. Assoc.* **2003,** *289* (19), 2560–2572.
2. Guidelines Committee 2003 European Society of Hypertension-European Society of Cardiology guidelines for the management of arterial hypertension. *J. Hypertens.* **2003,** *21* (6), 1011–1053.
3. Carretero, O. A.; Oparil, S. Essential hypertension. Part I: definition and etiology. *Circulation* **2000,** *101* (3), 329–335.
4. Smith, R. E. T.; Ashiya, M. Antihypertensive therapies. *Nat. Rev. Drug Discovery* **2007,** *6* (8), 597–598.
5. Laurent, S.; Schlaich, M.; Esler, M. New drugs, procedures, and devices for hypertension. *Lancet* **2012,** *380* (9841), 591–600.
6. Grassi, G.; Seravalle, G.; Trevano, F. Q.; Dell'Oro, R.; Mancia, G. Blood pressure control and antihypertensive treatment. *Curr. Vasc. Pharmacol.* **2012,** *10* (4), 506–511.
7. Borghi, C.; Cicero, A. F. G. Hypertension: management perspectives. *Expert Opin. Pharmacother.* **2012,** *13* (14), 1999–2003.

8. Carpino, P. A.; Flynn, D. Review of companies and drug classes in the 2007-2011 antihypertensive patent literature. *Pharm. Pat. Anal.* **2012,** *1* (1), 45–64.
9. Liamis, G.; Milionis, H.; Elisaf, M. Blood pressure drug therapy and electrolyte disturbances. *Int. J. Clin. Pract.* **2008,** *62* (10), 1572–1580.
10. Van Zwieten, P. A. Do we need new antihypertensive treatments? *Blood Pressure* **2007,** *16* (5), 291–300.
11. Hines, E. A., Jr. The thiocyanates in the treatment of hypertensive disease. *Med. Clin. North Am.* **1946,** *30* (4), 869–877.
12. Friederich, J. A.; Butterworth, J. F. 4th. Sodium nitroprusside: twenty years and counting. *Anesth. Analg. (Hagerstown, MD, U. S.)* **1995,** *81* (1), 152–162.
13. Johnson, C. C. Actions and toxicity of sodium nitroprusside. *Arch. Int. Pharmacodyn. Ther.* **1929,** *35,* 480–496.
14. Burnier, M.; Vuignier, Y.; Wuerzner, G. State-of-the-art treatment of hypertension: established and new drugs. *Eur. Heart J.* **2014,** *35* (9), 557–562.
15. Vardanyan, R. S.; Hruby, V. J. *Synthesis of Essential Drugs;* Elsevier, 2006.
16. De Stevens, G. Diuretics in the clinical treatment of hypertension. In *Medicinal Chemistry,* Vol. 7; Academic Press, 1967; pp 263–277.
17. Vogel, L. Diuretics for the treatment of hypertension and edema-cost-effective therapeutic alternatives according to ∮115c SGBV. *Arzneimitteltherapie* **2006,** *24* (3), 90–99.
18. Ram, C. V. S. Beta-blockers in hypertension. *Am. J. Cardiol.* **2010,** *106* (12), 1819–1825.
19. Frishman, W. H.; Saunders, E. β-Adrenergic blockers. *J. Clin. Hypertens. (Hoboken, NJ, U. S.)* **2011,** *13* (9), 649–653.
20. Baker, J. G.; Hill, S. J.; Summers, R. J. Evolution of β-blockers: from anti-anginal drugs to ligand-directed signaling. *Trends Pharmacol. Sci.* **2011,** *32* (4), 227–234.
21. Chrysant, S. G.; Chrysant, G. S. Current status of β-blockers for the treatment of hypertension: an update. *Drugs Today* **2012,** *48* (5), 353–366.
22. Punzi, H. Calcium antagonists and systemic hypertension. In *Calcium Channel Blockers and Renal Disease;* Robles, N. R., Ed.; Nova Science, 2009; pp 27–45.
23. Weber, M. A. Calcium channel antagonists in the treatment of hypertension. *Am. J. Cardiovasc. Drugs* **2002,** *2* (6), 415–431.
24. Ramirez-Sanchez, M.; Prieto, I.; Wangensteen, R.; Banegas, I.; Segarra, A. B.; Villarejo, A. B.; Vives, F.; Cobo, J.; de Gasparo, M. The renin-angiotensin system: new insight into old therapies. *Curr. Med. Chem.* **2013,** *20* (10), 1313–1322.
25. Redon, J.; Trenkwalder, P. R. A.; Barrios, V. Efficacy of combination therapy with angiotensin-converting enzyme inhibitor and calcium channel blocker in hypertension. *Expert Opin. Pharmacother.* **2013,** *14* (2), 155–164.
26. Bruno, Seva P.; van der Lubbe, N.; Verdonk, K.; Roks, A. J. M.; Hoorn, E. J.; Danser, A. H. J. Key developments in renin-angiotensin-aldosterone system inhibition. *Nat. Rev. Nephrol.* **2013,** *9* (1), 26–36.
27. Ruschitzka, F.; Taddei, S. Angiotensin-converting enzyme inhibitors: first-line agents in cardiovascular protection? *Eur. Heart J.* **2012,** *33* (16), 1996–1998.
28. Verdecchia, P.; Gentile, G.; Angeli, F.; Reboldi, G. Beyond blood pressure: evidence for cardiovascular, cerebrovascular, and renal protective effects of renin-angiotensin system blockers. *Ther. Adv. Cardiovasc. Dis.* **2012,** *6* (2), 81–91.
29. Anthony, C. S.; Masuyer, G.; Sturrock, E. D.; Acharya, K. R. Structure based drug design of angiotensin-I converting enzyme inhibitor. *Curr. Med. Chem.* **2012,** *19* (6), 845–855.
30. Bhuyan, B. J.; Mugesh, G. Angiotensin-converting enzyme inhibitors in the treatment of hypertension. *Curr. Sci.* **2011,** *101* (7), 881–887.

31. Izzo, J. L., Jr.; Weir, M. R. Angiotensin-converting enzyme inhibitors. *J. Clin. Hypertens. (Hoboken, NJ, U. S.)* **2011,** *13* (9), 667–675.
32. Pattan, S. R.; Chavan, P. A.; Dighe, N. S.; Kothiwale, V. A.; Hiremath, S. V.; Sanikop, M. B.; Bhosle, S. P.; Godge, R. K.; Bagi, J. G. The recent developments on ACE inhibitors—a review. *Pharmacologyonline* **2010,** 213–220 (1, Newsletter).
33. Tylicki, L.; Lizakowski, S.; Rutkowski, B. Renin-angiotensin-aldosterone system blockade for nephroprotection: current evidence and future directions. *J. Nephrol.* **2012,** *25* (6), 900–910.
34. Volpe, M. Angiotensin II receptor blockers: clinical relevance and new opportunities. *Hot Top. Hypertens.* **2012,** (14), 7–13.
35. Shrikrishna, D.; Astin, R.; Kemp, P. R.; Hopkinson, N. S. Renin-angiotensin system blockade: a novel therapeutic approach in chronic obstructive pulmonary disease. *Clin. Sci.* **2012,** *123* (8), 487–498.
36. Volpe, M. Preventing cardiovascular events with angiotensin II receptor blockers: a closer look at telmisartan and valsartan. *Expert Rev. Cardiovasc. Ther.* **2012,** *10* (8), 1061–1072.
37. Verdecchia, P.; Gentile, G.; Angeli, F.; Reboldi, G. Beyond blood pressure: evidence for cardiovascular, cerebrovascular, and renal protective effects of renin-angiotensin system blockers. *Ther. Adv. Cardiovasc. Dis.* **2012,** *6* (2), 81–91.
38. Khairnar, A. K.; Baviskar, D. T.; Jain, D. K. Angiotensin II receptor blockers: an overview. *Int. J. Pharm. Pharm. Sci.* **2012,** *4* (Suppl. 3), 50–56.
39. Cheung, B. M. Y. Therapeutic potential of angiotensin II receptor blockers in hypertension. *Expert Opin. Invest. Drugs* **2006,** *15* (6), 625–635.
40. Kurtz, T. W. The antidiabetic effect of angiotensin II receptor antagonists: Evidence, mechanisms, and clinical significance. In *Angiotensin II Receptor Antagonists,* 2nd ed.; Mancia, G., Ed. CRC Press, 2006; pp 85–98.
41. Mavromoustakos, T.; Agelis, G.; Durdagi, S. AT1 antagonists: a patent review (2008-2012). *Expert Opin. Ther. Pat.* **2013,** *23* (11), 1483–1494.
42. Muszalska, I.; Sobczak, A.; Dolhan, A.; Jelinska, A. Analysis of sartans: a review. *J. Pharm. Sci.* **2014,** *103* (1), 2–28.
43. Dina, R.; Jafari, M. Angiotensin II-receptor antagonists: an overview. *Am. J. Health-Syst. Pharm.* **2000,** *57* (13), 1231–1241.
44. Smith, R. D.; Cunningham, G.; Kivlighn, S. D. Angiotensin II antagonists. *Emerging Drugs (London, U. K.)* **1998,** *3,* 81–93.
45. Ashton, W. T. Nonpeptide angiotensin II receptor antagonists. *Expert Opin. Invest. Drugs* **1994,** *3* (11), 1105–1142.
46. Carini, D. J.; Wong, P. C. B.; Duncia, J. J. V. Preparation of (biphenylylmethyl)imidazoles and analogs as antihypertensive agents, EP 324377 (1989).
47. Carini, D. J.; Duncia, J. V.; Aldrich, P. E.; Chiu, A. T.; Johnson, A. L.; Pierce, M. E.; Price, W. A.; Santella, J. B., III; Wells, G. J.; Wexler, R. R.; Wong, P. C.; Yoo, S.-E.; Timmermans, P. B. M. W. M. Nonpeptide angiotensin II receptor antagonists: the discovery of a series of N-(biphenylylmethyl)imidazoles as potent, orally active antihypertensives. *J. Med. Chem.* **1991,** *34* (8), 2525–2547.
48. Yoshiyasu, R.; Shoji, K.; Kohei, N. Imidazole derivatives, EP 28833 (1981).
49. Watson, S. P. A convenient synthesis of 2-butyl-4(5)-chloro-1H-imidazole-5(4)-carboxaldehyde. *Synth. Commun.* **1992,** *22* (20), 2971–2977.
50. Shi, Y. J.; Frey, L. F.; Tschaen, D. M.; Verhoeven, T. R. A practical synthesis of 2-butyl-4(5)-chloro-5(4)-hydroxymethyl-1H-imidazole. *Synth. Commun.* **1993,** *23* (18), 2623–2630.

51. Larsen, R. D.; King, A. O.; Chen, C. Y.; Corley, E. G.; Foster, B. S.; Roberts, F. E.; Yang, C.; Lieberman, D. R.; Reamer, R. A.; Tschaen, D. M.; Verhoeven, T. R.; Reider, P. J. Efficient synthesis of losartan, a nonpeptide angiotensin ii receptor antagonist. *J. Org. Chem.* **1994,** *59* (21), 6391–6394.
52. Griffiths, G. J.; Hauck, M. B.; Imwinkelried, R.; Kohr, J.; Roten, C. A.; Stucky, G. C. Novel syntheses of 2-butyl-5-chloro-3*H*-imidazole-4-carbaldehyde: a key intermediate for the synthesis of the angiotensin II antagonist losartan. *J. Org. Chem.* **1999,** *64,* 8084–8089.
53. Carini, D. J.; Duncia, J. J. V. Preparation of angiotensin II receptor-blocking (phenylalkyl) imidazoles, EP 253310 (1988).
54. Lo, Y. S.; Rossano, L. T.; Larsen, R. D.; King, A. O. Preparation of [(heterocyclylmethyl) biphenylyl]tetrazoles as angiotensin II receptor antagonist intermediates, US 5310928 (1994).
55. Lo, Y. S.; Rossano, L. T. Preparation of tetrazolylphenylboronic acid intermediates for the synthesis of angiotensin II receptor antagonists, US 5130439 (1992).
56. Siegl, P. K. S. Discovery of losartan, the first specific non-peptide angiotensin II receptor antagonist. *J. Hypertens.* **1993,** *11* (3), S19–S22.
57. Wong, P. C.; Barnes, T. B.; Chiu, A. T.; Christ, D. D.; Duncia, J. V.; Herblin, W. F.; Timmermans, P. B. M.W.M. Losartan (DuP 753), an orally active nonpeptide angiotensin II receptor antagonist. *Cardiovasc. Drug Rev.* **1991,** *9* (4), 317–339.
58. Goa, K. L.; Wagstaff, A. J. Losartan potassium: a review of its pharmacology, clinical efficacy and tolerability in the management of hypertension. *Drugs* **1996,** *51* (5), 820–845.
59. Carr, A. A.; Prisant, L. M. Losartan: first of a new class of angiotensin antagonists for the management of hypertension. *J. Clin. Pharmacol.* **1996,** *36* (1), 3–12.
60. Johnston, C. I. Angiotensin receptor antagonists: focus on losartan. *Lancet* **1995,** *346* (8987), 1404–1407.
61. Mcintyre, M.; Caffe, S. E.; Michalak, R. A.; Reid, J. L. Losartan, an orally active angiotensin (AT1) receptor antagonist: a review of its efficacy and safety in essential hypertension. *Pharmacol. Ther.* **1997,** *74* (2), 181–194.
62. Smith, R. D.; Sweet, C. S.; Goldberg, A.; Timmermans, P. B. M.W.M. Losartan potassium (Cozaar): a nonpeptide antagonist of angiotensin II. *Drugs Today* **1996,** *32* (Suppl. F), 1–42.
63. Ramirez, F.; Rifkin, D. B. Is losartan the drug for all seasons? *Curr. Opin. Pharmacol.* **2012,** *12* (2), 223–224.
64. Tallant, E. A.; Ferrario, C. M. Biology of angiotensin II receptor inhibition with a focus on losartan: a new drug for the treatment of hypertension. *Expert Opin. Invest. Drugs* **1996,** *5* (9), 1201–1214.
65. Xu, F.; Mao, C.; Liu, Y.; Wu, L.; Xu, Z.; Zhang, L. Losartan chemistry and its effects via AT1 mechanisms in the kidney. *Curr. Med. Chem.* **2009,** *16* (28), 3701–3715.
66. Warner, G. T.; Jarvis, B. Olmesartan medoxomil. *Drugs* **2002,** *62* (9), 1345–1353.
67. Mire, D. E.; Silfani, T. N.; Pugsley, M. K. A review of the structural and functional features of olmesartan medoxomil, an angiotensin receptor blocker. *J. Cardiovasc. Pharmacol.* **2005,** *46* (5), 585–593.
68. Unger, T.; McInnes, G. T.; Neutel, J. M.; Boehn, M. The role of olmesartan medoxomil in the management of hypertension. *Drugs* **2004,** *64* (24), 2731–2739.
69. Koike, H.; Konse, T.; Sada, T.; Ikeda, T.; Hyogo, A.; Hinman, D.; Saito, H.; Yanagisawa, H. Olmesartan medoxomil, a novel potent angiotensin II blocker. *Ann. Rep. Sankyo Res. Lab.* **2004,** *55,* 1–91.
70. Chrysant, S. G.; Chrysant, G. S. Antihypertensive efficacy of olmesartan medoxomil alone and in combination with hydrochlorothiazide. *Expert Opin. Pharmacother.* **2004,** *5* (3), 657–667.

71. Brousil, J. A.; Burke, J. M. Olmesartan medoxomil: an angiotensin II-receptor blocker. *Clin. Ther.* **2003,** *25* (4), 1041–1055.
72. de la Sierra, A.; Volpe, M. Olmesartan-based therapies: an effective way to improve blood pressure control and cardiovascular protection. *J. Hypertens.* **2013,** *31* (Suppl. 1), S13–S17.
73. Yanagisawa, H. i; Koike, H.; Miura, S. Olmesartan medoxomil: an angiotensin II receptor blocker. In *Case Studies in Modern Drug Discovery and Development;* Huang, X., Aslanian, R. G., Eds.; Wiley, 2012; pp 45–66.
74. Grassi, G.; Mancia, G. Olmesartan medoxomil: as monotherapy and in combination treatment in hypertension. *High Blood Pressure Cardiovasc. Prev.* **2010,** *17* (1), 1–14.
75. Redon, J.; Fabia, M. J. Efficacy in angiotensin receptor blockade: a comparative review of data with olmesartan. *JRAAS* **2009,** *10* (3), 147–156.
76. Schindler, C.; Ferrario, C. M. Olmesartan for the treatment of arterial hypertension. *Future Cardiol.* **2008,** *4* (4), 357–372.
77. Brunner, H. R. Olmesartan medoxomil: current status of its use in monotherapy. *Vasc. Health Risk Manage.* **2006,** *2* (4), 327–340.
78. Gardner, S. F.; Franks, A. M. Olmesartan medoxomil: the seventh angiotensin receptor antagonist. *Ann. Pharmacother.* **2003,** *37* (1), 99–105.
79. Yanagisawa, H.; Amemiya, Y.; Kanazaki, T.; Shimoji, Y.; Fujimoto, K.; Kitahara, Y.; Sada, T.; Mizuno, M.; Ikeda, M.; Miyamoto, S.; Furukawa, Y.; Koike, H. Nonpeptide angiotensin II receptor antagonists: synthesis, biological activities, and structure-activity relationships of imidazole-5-carboxylic acids bearing alkyl, alkenyl, and hydroxyalkyl substituents at the 4-position and their related compounds. *J. Med. Chem.* **1996,** *39* (1), 323–338.
80. Yanagisawa, H.; Shimoji, Y.; Fujimoto, K.; Kanazaki, T.; Anemiya, Y.; Koike, H.; Sada, T. Preparation of 1-[(carboxy-biphenylyl)methyl]imidazole-5-carboxylates and analogs as angiotensin II antagonists, EP 503785 (1992).
81. Yanagisawa, H.; Fujimoto, K.; Amemiya, Y.; Shimoji, Y.; Kanazaki, T.; Koike, H.; Sada, T. Preparation of 4′-(imidazolomethyl)biphenyl-2-carboxylates as angiotensin II receptor antagonists, US 5616599 (1997).
82. Firke, R. V.; Sisodia, U.; Bhangale, C.; Shivdavkar, R. B.; Godbole, H. M.; Singh, G. P. Preparation of olmesartan medoxomil, WO 2013021312 (2013).
83. Hedvati, L.; Pilarsky, G. Process for the preparation of olmesartan medoxomil, WO 2006029056 (2006).
84. Buehlmayer, P.; Ostermayer, F.; Schmidlin, T. Preparation of [(tetrazolylbiphenyl)methyl] amines and analogs as angiotensin II antagonists, EP 443983 (1991).
85. Buhlmayer, P.; Furet, P.; Criscione, L.; de Gasparo, M.; Whitebread, S.; Schmidlin, T.; Lattmann, R.; Wood, J. Valsartan, a potent, orally active angiotensin II antagonist developed from the structurally new amino acid series. *Bioorg. Med. Chem. Lett.* **1994,** *4* (1), 29–34.
86. Patel, R. N.; Patel, D. S.; Patel, R. B.; Patel, K. S. A novel and industrial approach for the synthesis of valsartan. *Het. Lett.* **2013,** *3* (4), 513–518.
87. Senthil, K. N.; Reddy, S. B.; Sinha, B. K.; Mukkanti, K.; Dandala, R. New and improved manufacturing process for valsartan. *Org. Process Res. Dev.* **2009,** *13* (6), 1185–1189.
88. Pandarus, V.; Desplantier-Giscard, D.; Gingras, G.; Beland, F.; Ciriminna, R.; Pagliaro, M. Greening the valsartan synthesis: scale-up of key Suzuki-Miyaura coupling over SiliaCat DPP-Pd. *Org. Process Res. Dev.* **2013,** *17* (12), 1492–1497.
89. Aalla, S.; Gilla, G.; Bojja, Y.; Anumula, R. R.; Vummenthala, P. R.; Padi, P. R. An efficient and telescopic process for valsartan, an angiotensin II receptor blocker. *Org. Process Res. Dev.* **2012,** *16* (4), 682–686.

90. Penikelapati, H. R.; Ambati, S.; Maruthikumar, T. V.; Ambati, N. B. New and improved synthesis of valsartan: an antihypertensive drug. *Res. J. Pharm. Biol. Chem. Sci.* **2011,** *2* (4), 632–639.
91. Wang, G.; Sun, B.; Peng, C. An improved synthesis of valsartan. *Org. Process Res. Dev.* **2011,** *15* (5), 986–988.
92. Reddy, C. R.; Shankar, N. G. B.; Reddy, N. M.; Somannavar, Y. S.; Reddy, B. S.; Islam, A.; Sivakumaran, M. An improved process for the preparation of valsartan, WO 2012001484 (2012).
93. Rao, M. N.; Ravilla, L.; Kanakamajalu, S.; Rao, S. K.; Nagarajan, K. Improved process for the preparation of angiotensin II antagonists, WO 2013072924 (2013).
94. Radl, S.; Stach, J.; Dedinova, E. A method for the preparation of N-(1-oxopentyl)-N-[[2′-(1H-tetrazol-5-y1)[1,1′-biphenyl]-4-y1]methyl]-L-valine (valsartan), WO 2004101534 (2004).
95. Cosme, G. A.; Palomo, N.; Francisco, E. Process for the preparation of valsartan from oxazolidinone derivative, WO 2008138871 (2008).
96. Kumar, Y.; Prasad, M.; Lahiri, S.; Maheshwari, N.; Saxena, I. Process for preparation of pure valsartan and its barium salt, WO 2005049587 (2005).
97. Markham, A.; Goa, K. L. Valsartan: a review of its pharmacology and therapeutic use in essential hypertension. *Drugs* **1997,** *54* (2), 299–311.
98. Mann, J.; Julius, S. The valsartan antihypertensive long-term use evaluation (VALUE) trial of cardiovascular events in hypertension. Rationale and design. *Blood Pressure* **1998,** *7* (3), 176–183.
99. Criscione, L.; Bradley, W. A.; Buhlmayer, P.; Whitebread, S.; Glazer, R.; Lloyd, P.; Mueller, P.; De Gasparo, M. Valsartan: preclinical and clinical profile of an antihypertensive angiotensin-II antagonist. *Cardiovasc. Drug Rev.* **1995,** *13* (3), 230–250.
100. Black, H. R.; Bailey, J.; Zappe, D.; Samuel, R. Valsartan: more than a decade of experience. *Drugs* **2009,** *69* (17), 2393–2414.
101. Mistry, N. B.; Westheim, A. S.; Kjeldsen, S. E. The angiotensin receptor antagonist valsartan: a review of the literature with a focus on clinical trials. *Expert Opin. Pharmacother.* **2006,** *7* (5), 575–581.
102. Kober, L.; Torp-Pedersen, C. Valsartan: the past, present and future. *Future Cardiol.* **2005,** *1* (5), 591–598.
103. Kjeldsen, S. E.; Brunner, H. R.; McInnes, G. T.; Stolt, P. Valsartan in the treatment of hypertension. *Aging Health* **2005,** *1* (1), 27–36.
104. Thomas, M. C.; Johnston, C. I. Valsartan. *J. Drug Eval.* **2004,** *2* (3), 67–101.
105. Remuzzi, A.; Perico, N.; Remuzzi, G. Pharmacological and clinical profile of valsartan. *Drugs Today* **1998,** *34* (11), 973–986.
106. Chiolero, A.; Burnier, M. Pharmacology of valsartan, an angiotensin II receptor antagonist. *Expert Opin. Invest. Drugs* **1998,** *7* (11), 1915–1925.
107. Bernhart, C. A.; Perreaut, P. M.; Ferrari, B. P.; Muneaux, Y. A.; Assens, J. L. A.; Clement, J.; Haudricourt, F.; Muneaux, C. F.; Taillades, J. E.; Vignal, M.-A.; Gougat, J.; Guiraudou, P. R.; Lacour, C. A.; Roccon, A.; Cazaubon, C. F.; Jean-Claude Brelihre, J.-C.; Le Fur, G.; Nisato, D. A new series of imidazolones: highly specific and potent nonpeptide AT1 angiotensin II receptor antagonists. *J. Med. Chem.* **1993,** *36* (22), 3371–3380.
108. Bernhart, C.; Breliere, J. C.; Clement, J.; Nisato, D.; Perreaut, P. Preparation of N-(carboxybiphenylylmethyl)spiro[cycloalkane-imidazolinone] derivatives and analogs as angiotensin II inhibitors, WO 9114679 (1991).
109. Perreaut, P.; Muneaux, C.; Muneaux, Y. N-substituted heterocyclic derivatives as angiotensin II inhibitors, EP 532410 (1993).

110. Rao, S. N.; Babu, K. S. An improved and efficient synthesis of Irbesartan, an antihypertensive drug. *Org. Commun.* **2011,** *4* (4), 105–111.
111. Kumar, M. R.; Park, K.; Lee, S. Synthesis of amido-N-imidazolium salts and their applications as ligands in Suzuki-Miyaura reactions: coupling of hetero-aromatic halides and the synthesis of milrinone and irbesartan. *Adv. Synth. Catal.* **2010,** *352* (18), 3255–3266.
112. Patel, F. A.; Abed, S. Green chemical approach towards Irbesartan synthesis. *Org. Chem.: Indian J.* **2010,** *6* (1), 116–118.
113. Satyanarayana, B.; Anjaneyulu, Y.; Veerasomaiah, P.; Reddy, P. P. Synthesis and characterization of Irbesartan impurities. *Heterocycl. Commun.* **2007,** *13* (4), 223–228.
114. Satyanarayana, B.; Sumalatha, Y.; Sridhar, C.; Venkatraman, S.; Reddy, P. P. New synthesis of analogs of the antihypertensive active pharmaceutical ingredient irbesartan. *Synth. Commun.* **2006,** *36* (14), 2079–2086.
115. Ekhato, I. V.; Bonacorsi, S., Jr. The synthesis of radiolabeled irbesartan using N,N-dimethyl[14C]formamide as a source of carbon-14 isotope. *J. Labelled Compd. Radiopharm.* **2011,** *54* (4), 202–205.
116. Nisnevich, G.; Rukhman, I.; Pertsikov, B.; Kaftanov, J.; Dolitzky, B. Synthesis of irbesartan, WO 2004065383 (2004).
117. Selic, L. Process for the preparation of pure irbesartan, WO 2007115990 (2007).
118. Deshpande, P. B.; Luthra, P.; Rathod, D. M.; Patel, H. K.; Parikh, P. T. Process for preparation of irbesartan, WO 2007052301 (2007).
119. Gillis, J. C.; Markham, A. Irbesartan: a review of its pharmacodynamic and pharmacokinetic properties and therapeutic use in the management of hypertension. *Drugs* **1997,** *54* (6), 885–902.
120. Casas, A.; Merlos, M.; Castaner, J. Irbesartan. Antihypertensive, treatment of congestive heart failure, angiotensin II AT1 antagonist. *Drugs Future* **1997,** *22* (5), 481–491.
121. Markham, A.; Spencer, C. M.; Jarvis, B. Irbesartan: an updated review of its use in cardiovascular disorders. *Drugs* **2000,** *59* (5), 1187–1206.
122. Powell, J. R.; Reeves, R. A.; Marino, M. R.; Cazaubon, C.; Nisato, D. A review of the new angiotensin II-receptor antagonist irbesartan. *Cardiovasc. Drug Rev.* **1998,** *16* (3), 169–194.
123. Borghi, C.; Cicero, A. F. G. The role of irbesartan in the treatment of patients with hypertension: a comprehensive and practical review. *High Blood Pressure Cardiovasc. Prev.* **2012,** *19* (1), 19–31.
124. Husain, A.; Azim, M. S.; Mitra, M.; Bhasin, P. S. A review of pharmacological and pharmaceutical profile of irbesartan. *Pharmacophore* **2011,** *2* (6), 276–286.
125. Morishita, R.; Kanda, Y.; Nakajima, M. Irbesartan: second generation of ARB as metabosartan. *Curr. Hypertens. Rev.* **2010,** *6* (3), 173–179.
126. Bramlage, P.; Durand-Zaleski, I.; Desai, N.; Pirk, O.; Hacker, C. The value of irbesartan in the management of hypertension. *Expert Opin. Pharmacother.* **2009,** *10* (11), 1817–1831.
127. Croom, K. F.; Plosker, G. L. Irbesartan: a review of its use in hypertension and diabetic nephropathy. *Drugs* **2008,** *68* (11), 1543–1569.
128. Borghi, C.; Ertek, S.; Cicero, A. F. G. Irbesartan: a review of its use alone and in combination with hydrochlorothiazide. *Therapy* **2006,** *3* (6), 733–749.
129. Waeber, B. A review of irbesartan in antihypertensive therapy: comparison with other antihypertensive agents. *Curr. Ther. Res.* **2001,** *62* (7), 505–523.
130. Waeber, B. Blood pressure control: a review on irbesartan. *Eur. Heart J. Suppl.* **2000,** *2* (Suppl. B), B2–B7.
131. Johnston, C. I. Pharmacology of irbesartan. *Expert Opin. Invest. Drugs* **1999,** *8* (5), 655–670.

132. Pouleur, H. G. Clinical overview of irbesartan: a new angiotensin II receptor antagonist. *Am. J. Hypertens.* **1997,** *10* (12, Pt. 2), 318S–324S.
133. Ries, U. J.; Mihm, G.; Narr, B.; Hasselbach, K. M.; Wittneben, H.; Entzeroth, M.; van Meel, J. C. A.; Wienen, W.; Hauel, N. H. 6-Substituted benzimidazoles as new nonpeptide angiotensin II receptor antagonists: synthesis, biological activity, and structure-activity relationships. *J. Med. Chem.* **1993,** *36* (25), 4040–4051.
134. Kumar, A. S.; Ghosh, S.; Mehta, G. N. A modification to the synthesis of telmisartan: an antihypertensive drug. *J. Chem. Res.* **2010,** *34* (2), 95–97.
135. Sanjeev, K. A.; Ghosh, S.; Mehta, G. N. Efficient and improved synthesis of telmisartan. *Beilstein J. Org. Chem.* **2010,** *6* (25), 1–5.
136. Reddy, K. S.; Srinivasan, N.; Reddy, C. R.; Kolla, N.; Anjaneyulu, Y.; Venkatraman, S.; Bhattacharya, A.; Mathad, V. T. An efficient and impurity-free process for telmisartan: an antihypertensive drug. *Org. Process Res. Dev.* **2007,** *11* (1), 81–85.
137. Venugopal, S.; Ramanatham, J.; Devanna, N.; Kumar, A. S.; Ghosh, S.; Soundararajan, R.; Kale, B.; Mehta, G. N. New strategy for the synthesis of telmisartan: an antihypertensive drug. *Asian J. Chem.* **2010,** *22* (4), 2767–2773.
138. Sharma, M. C.; Kohli, D. V.; Sharma, S.; Sharma, A. D. Synthesis and antihypertensive activity of 4′-{2-[4-[2-(Substitutedphenyl)-4-oxo-thiazolidin-3-yl]-phenyl} benzoimidazol-1-yl-methyl}-biphenyl-2-carboxylic acids. *Pharm. Sin.* **2010,** *1* (1), 58–73.
139. Kumar, A. S.; Ghosh, S.; Mehta, G. N.; Soundararajan, R.; Sarma, P. S. R.; Bhima, K. New and improved synthesis of telmisartan: an antihypertensive drug. *Synth. Commun.* **2009,** *39* (23), 4149–4157.
140. Singh, S.; Joshi, A.; Newadkar, R. Process for preparing telmisartan, EP 2277866 (2011).
141. Amarnath, U.; Suryakiran, U. One pot process for the preparation of telmisartan, WO 2014027280 (2014).
142. Khamar, B. M.; Siddiqui, I. H.; Ponnaiah, R.; Modi, I. A. An improved process for the preparation of substantially pure telmisartan, WO 2010018441 (2010).
143. Wienen, W.; Entzeroth, M.; Van Meel, J. C. A.; Stangier, J.; Busch, U.; Ebner, T.; Schmid, J.; Lehmann, H.; Matzek, K.; Kempthorne-Rawson, J.; Gladigau, V.; Hauel, N. H. A review on telmisartan: a novel, long-acting angiotensin II-receptor antagonist. *Cardiovasc. Drug Rev.* **2000,** *18* (2), 127–156.
144. Sharpe, M.; Jarvis, B.; Goa, K. L. Telmisartan: a review of its use in hypertension. *Drugs* **2001,** *61* (10), 1501–1529.
145. Battershill, A. J.; Scott, L. J. Telmisartan: a review of its use in the management of hypertension. *Drugs* **2006,** *66* (1), 51–83.
146. Rosario, B. H.; Hendra, T. J. Telmisartan in the treatment of hypertension. *Expert Opin. Drug Metab. Toxicol.* **2008,** *4* (4), 485–492.
147. Yamagishi, S.; Takeuchi, M. Telmisartan is a promising cardiometabolic sartan due to its unique PPAR-γ-inducing property. *Med. Hypotheses* **2005,** *64* (3), 476–478.
148. Doggrell, S. A. Telmisartan-killing two birds with one stone. *Expert Opin. Pharmacother.* **2004,** *5* (11), 2397–2400.
149. Goebel, M.; Clemenz, M.; Unger, T. Effective treatment of hypertension by AT1 receptor antagonism: the past and future of telmisartan. *Expert Rev. Cardiovasc. Ther.* **2006,** *4* (5), 615–629.
150. McClellan, K. J.; Markham, A. Telmisartan. *Drugs* **1998,** *56* (6), 1039–1044.
151. Merlos, M.; Casas, A.; Castaner, J. Telmisartan. BIBR-277. 4′-[4-methyl-6-(1-methyl-1H-benzimidazol-2-yl)-2-propyl-1H-benzimidazol-1-ylmethyl]biphenyl-2-carboxylic acid. *Drugs Future* **1997,** *22* (10), 1112–1116.

152. Destro, M.; Cagnoni, F.; Dognini, G. P.; Galimberti, V.; Taietti, C.; Cavalleri, C.; Galli, E. Telmisartan: just an antihypertensive agent? *A literature review, Expert Opin. Pharmacother.* **2011,** *12* (17), 2719–2735.
153. Yamagishi, S.; Nakamura, K.; Matsui, T. Potential utility of telmisartan, an angiotensin II type 1 receptor blocker with peroxisome proliferator-activated receptor-γ (PPAR-γ)-modulating activity for the treatment of cardiometabolic disorders. *Curr. Mol. Med.* **2007,** *7* (5), 463–469.
154. Costa, F. V. Telmisartan: standing out in a crowded contest? *High Blood Pressure Cardiovasc. Prev.* **2006,** *13* (3), 85–94.
155. Yamagishi, S.; Matsui, T.; Nakamura, K. Telmisartan, a unique angiotensin II type 1 receptor blocker with selective peroxisome proliferator-activated receptor-γ-modulating activity. *Adipocytes* **2006,** *2* (2), 47–50.
156. Ruilope, L. M. Telmisartan for the management of patients at high cardiovascular risk. *Curr. Med. Res. Opin.* **2011,** *27* (8), 1673–1682.
157. Frampton, J. E. Telmisartan: a review of its use in cardiovascular disease prevention. *Drugs* **2011,** *71* (6), 651–677.
158. Deppe, S.; Boeger, R. H.; Weiss, J.; Benndorf, R. A. Telmisartan: a review of its pharmacodynamic and pharmacokinetic properties. *Expert Opin. Drug Metab. Toxicol.* **2010,** *6* (7), 863–871.
159. Rizos, C. V.; Elisaf, M. S.; Liberopoulos, E. N. Are the pleiotropic effects of telmisartan clinically relevant? *Curr. Pharm. Des.* **2009,** *15* (24), 2815–2832.
160. Rueger, H.; Stutz, S.; Goschke, R.; Spindler, F.; Maibaum, J. A convergent synthesis approach towards CGP60536B, a non-peptide orally potent renin inhibitor, via an enantiomerically pure keto lactone intermediate. *Tetrahedron Lett.* **2000,** *41* (51), 10085–10089.
161. Dondoni, A.; De Lathauwer, G.; Perrone, D. A convergent synthesis of the renin inhibitor SPP-100 using a nitrone intermediate. *Tetrahedron Lett.* **2001,** *42* (29), 4819–4823.
162. Goeschke, R.; Maibaum, J. K.; Schilling, W.; Stutz, S.; Rigollier, P.; Yamaguchi, Y.; Cohen, N. C.; Herold, P. Preparation of δ-amino-γ-hydroxy-ω-arylalkanoic acid amides as renin inhibitors, EP 678503 (1995).
163. Gradman, A. H.; Schmieder, R. E.; Lins, R. L.; Nussberger, J.; Chiang, Y.; Bedigian, M. P. Aliskiren, a novel orally effective renin inhibitor, provides dose-dependent antihypertensive efficacy and placebo-like tolerability in hypertensive patients. *Circulation* **2005,** *111* (8), 1012–1018.
164. Daugherty, K. K. Aliskiren. *Am. J. Health-Syst. Pharm.* **2008,** *65* (14), 1323–1332.
165. Saleem, T. S. M.; Jain, A.; Tarani, P.; Ravi, V.; Gauthaman, K. Aliskiren: a novel, orally active renin inhibitor. *Syst. Rev. Pharm.* **2010,** *1* (1), 93–98.
166. Verdecchia, P.; Angeli, F.; Reboldi, G. Good news from aliskiren? *Clin. Med.: Ther.* **2009,** *1,* 999–1002.
167. Sanoski, C. A. Aliskiren: an oral direct renin inhibitor for the treatment of hypertension. *Pharmacotherapy* **2009,** *29* (2), 193–212.
168. Maibaum, J.; Feldman, D. L. Case history on Tekturna/Rasilez (aliskiren), a highly efficacious direct oral renin inhibitor as a new therapy for hypertension. *Annu. Rev. Med. Chem.* **2009,** *44,* 105–127.
169. Wood, J. M.; Maibaum, J.; Rahuel, J.; Grutter, M. G.; Cohen, N. C.; Rasetti, V.; Rueger, H.; Goschke, R.; Stutz, S.; Fuhrer, W.; Schilling, W.; Rigollier, P.; Yamaguchi, Y.; Cumin, F.; Baum, H. P.; Schnell, C. R.; Herold, P.; Mah, R.; Jensen, C.; O'Brien, E.; Stanton, A.; Bedigian, M. P. Structure-based design of aliskiren, a novel orally effective renin inhibitor. *Biochem. Biophys. Res. Commun.* **2003,** *308,* 698–705.

170. Webb, R. L.; Schiering, N.; Sedrani, R.; Maibaum, J. Direct renin inhibitors as a new therapy for hypertension. *J. Med. Chem.* **2010,** *53* (21), 7490–7520.
171. Scott, B. B.; McGeehan, G. M.; Harrison, R. R. Development of inhibitors of the aspartyl protease renin for the treatment of hypertension. *Curr. Protein Pept. Sci.* **2006,** *7,* 241–254.
172. Ostermann, N.; Ruedisser, S.; Ehrhardt, C.; Breitenstein, W.; Marzinzik, A.; Jacoby, E.; Vangrevelinghe, E.; Ottl, J.; Klumpp, M.; Hartwieg, J. C. D.; Cumin, F.; Hassiepen, U.; Trappe, J.; Sedrani, R.; Geisse, S.; Gerhartz, B.; Richert, P.; Francotte, E.; Wagner, T.; Kromer, M.; Kosaka, T.; Webb, R. L.; Rigel, D. F.; Maibaum, J.; Baeschlin, D. K. A novel class of oral direct renin knhibitors: highly potent 3,5-disubstituted piperidines bearing a tricyclic p3-p1 pharmacophore. *J. Med. Chem.* **2013,** *56* (6), 2196–2206.
173. Robles, N. R.; Cerezo, I.; Hernandez-Gallego, R. Renin-angiotensin system blocking drugs. *J. Cardiovasc. Pharmacol. Ther.* **2014,** *19* (1), 14–33.
174. Yokokawa, F. Recent progress on the discovery of non-peptidic direct renin inhibitors for the clinical management of hypertension. *Expert Opin. Drug Discovery* **2013,** *8* (6), 673–690.
175. Volpe, M.; Pontremoli, R.; Borghi, C. Direct renin inhibition: from pharmacological innovation to novel therapeutic opportunities. *High Blood Pressure Cardiovasc. Prev.* **2011,** *18* (3), 93–105.
176. Chaudhary, K.; Nistala, R.; Whaley-Connell, A. Is there a future for direct renin inhibitors? *Expert Opin. Invest. Drugs* **2010,** *19* (5), 653–661.

Chapter 23

Drugs for Treating Respiratory System Diseases

Respiratory disease is considered any disorder that affects nasal cavities, the throat, the trachea, the bronchi and bronchioles, or lungs, generating shortness of breath (dyspnea), cough, or production of sputum.

Any pathological problem in the lungs that prevents them from working properly is considered lung disease. Lung diseases in general could be classified as:

- Airway diseases, such as asthma, bronchitis, and emphysema, which are the result of narrowing or blocking of the airways.
- Lung circulation diseases, which are caused by clotting, scarring, or inflammation of the blood vessels, pulmonary embolism, pulmonary venoocclusive disease, pulmonary edema, or chronic thromboembolic disease.
- Lung tissue diseases such as fibrosis and lung cancer, which are diseases characterized by uncontrolled cell growth in tissues of the lung and sarcoidosis.

Major symptoms of lung disease are coughing, dyspnea, and chest pain.

Chronic obstructive pulmonary disease (COPD) by definition is a type of obstructive lung disease that is characterized by chronically poor airflow, mucus hypersecretion, loss of body mass, and cardiovascular effects. It typically worsens over time; the main symptoms include shortness of breath, cough, and sputum production. Emphysema and chronic bronchitis are the most common forms of COPD. COPD represents a variety of diseases that are characterized by differing phenotypes and overlapping features. It is the fifth leading cause of death in the world.

Although COPD is a major medical problem worldwide, unfortunately, it has no cure; however, there are remedies that relieve some of a patient's symptoms. Treatment options include β_2 agonists, anticholinergics, particularly muscarinic receptor blockers, xanthines, and glucocorticoids, which are by far the most effective therapy for controlling asthma and which could be used as monotherapy or in combination [1-16].

Synthesis of Best-Seller Drugs. http://dx.doi.org/10.1016/B978-0-12-411492-0.00023-7

23.1 BRONCHODILATORS

Bronchodilators are the mainstay of the current symptomatic management of asthma and COPD. Bronchodilators work through their direct relaxation effect on airway smooth muscles. Their use via an inhaled route is currently preferred [17-19].

β_2-Adrenoceptor Agonists

β_2 Adrenoceptors are located throughout the airways. Acting on airway smooth muscle, β_2 agonists reduce airflow limitations. Bronchodilators do not significantly modify disease progression, but they provide symptomatic relief via dilation of distal airways.

Historically, epinephrine (adrenaline) (**23.1.1**) was the first nonselective α- and β-adrenoceptor agonist to be introduced into clinical practice for the treatment of asthma and COPD in the 20th century. The next compound that appeared for the same purpose in the 1940s was the nonselective β-adrenoceptor agonist isoproterenol (**23.1.2**). Then in 1960s to 1970s came metaproterenol (**23.1.3**), a mixed β_1 and β_2 agonist, and later appeared the selective β_2-agonists albuterol (salbutamol) (**23.1.4**), terbutaline (**23.1.5**), and fenoterol (**23.1.6**), all of them short-acting compounds (Fig. 23.1.). Fast- and short-acting agents are best used for rescue of symptoms, whereas long-acting agents are best used for maintenance therapy.

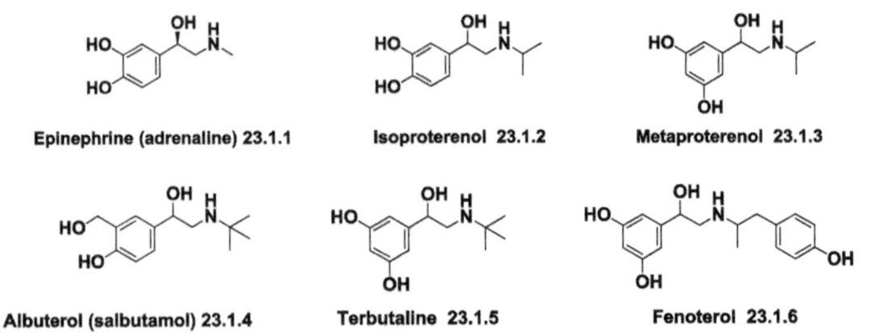

FIG. 23.1 Nonselective adrenoceptor agonists for the treatment of asthma and COPD.

Later in the 1980s salmeterol (**23.1.7**), and formoterol (**23.1.8**) were invented as long-acting, selective, β_2 agonists. Clenbuterol (**23.1.9**), bambuterol (**23.1.10**), also belong to this series of long-acting compounds.

These compounds exhibit sufficient duration of action to support twice daily dosing regimens following inhaled delivery. Abediterol (**23.1.11**), is an experimental compound of the same series of long-acting compounds (Fig. 23.2.).

Several ultralong-acting β_2-adrenoreceptor agonists for once-a-day treatment, which produce sustained bronchodilation lasting for at least 12 to 24 hours, such as indacaterol (**23.1.12**), carmoterol (**23.1.13**), olodaterol (**23.1.14**),

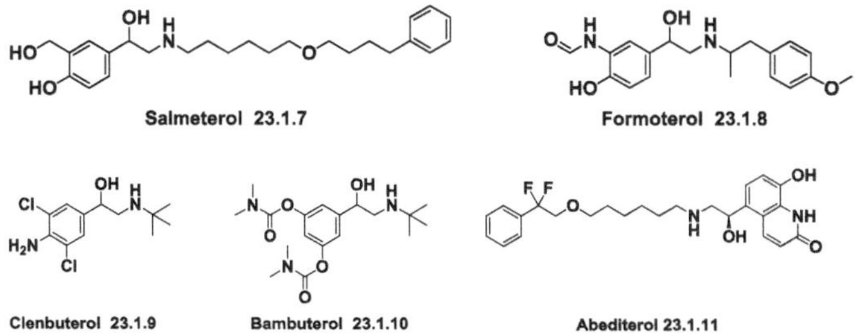

FIG. 23.2 Selective long-acting adrenoceptor β_2-agonist compounds for the treatment of asthma and COPD.

and milveterol (**23.1.15**), vilanterol (**23.1.16**) have been proposed as COPD relievers. Several reviews summarize progress in the field of β_2 agonists with extended action [20-27]. All of the drugs except indacaterol (**23.1.12**) [28], which is approved for medicinal use, are now in different stages of clinical trials (Fig. 23.3.).

It seems that discovery of long-acting bronchodilators made the shift from symptomatic drugs to "disease modifiers" and long-acting bronchodilators can not be simply ranked as symptomatic drugs. Moreover, COPD probably no longer should be considered an irreversible disease as it can be treated effectively [29,30].

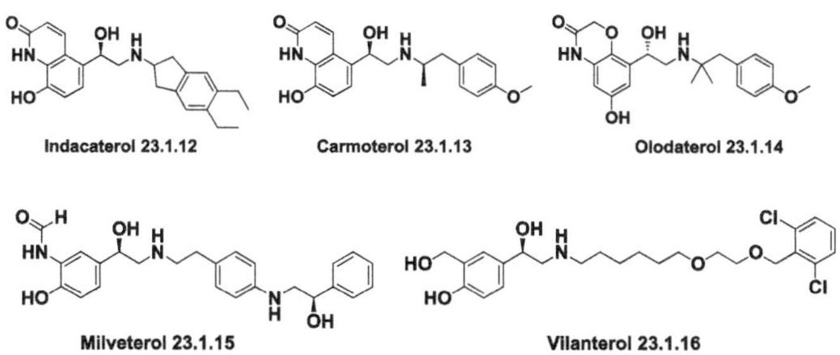

FIG. 23.3 Ultralong-acting β_2-adrenoreceptor agonists.

Of the above-described β-adrenoceptor agonists, albuterol (salbutamol) (**23.1.4**), salmeterol (**23.1.7**), and formoterol (**23.1.8**) are included in the list of Top 200 Drugs by sales for the 2010s, being the constituent part of combination preparations for management of asthma and COPD such as:

- Combivent—a combination preparation of albuterol (salbutamol) (**23.1.4**) and ipratropium bromide.
- Advair—a combination preparation of salmeterol (**23.1.7**) and a corticosteroid fluticasone.
- Symbicort—a combination formulation containing formoterol (**23.1.8**) and budesonide.

Albuterol–Salbutamol

The first proposed synthesis of albuterol (**23.1.4**) started from chloromethylation of 4-hydroxyacetophenone (**23.1.17**) in formaldehyde/hydrochloric acid solution on passing hydrogen chloride gas to produce 3-(chloromethyl)-4-hydroxyacetophenone (**23.1.18**) [31]. Two hydroxyl groups of the obtained compound were acetylated by heating at 100°C with sodium acetate and acetic anhydride in acetic acid to produce the requested 2-acetoxy-5-acetylbenzyl acetate (**23.1.19**). The last was directly brominated with bromine in chloroform at room temperature, providing 2-acetoxy-5-(2-bromo-acetyl)benzyl acetate (**23.1.20**), which was readily condensed with tret-butyl benzylamine on heating in benzene or methyl ethylketone to produce the amine (**23.1.21**). The product was hydrolyzed to (hydroxymethyl)phenol (**23.1.22**) using hydrochloric acid at room temperature. The carbonyl group of the (hydroxymethyl)phenol (**23.1.22**) was reduced with sodium borohydride to produce the product (**23.1.23**). Final debenzylation via hydrogenation on Pd/C catalyst produced the desired albuterol (**23.1.4**) [32-37] (Scheme 23.1.).

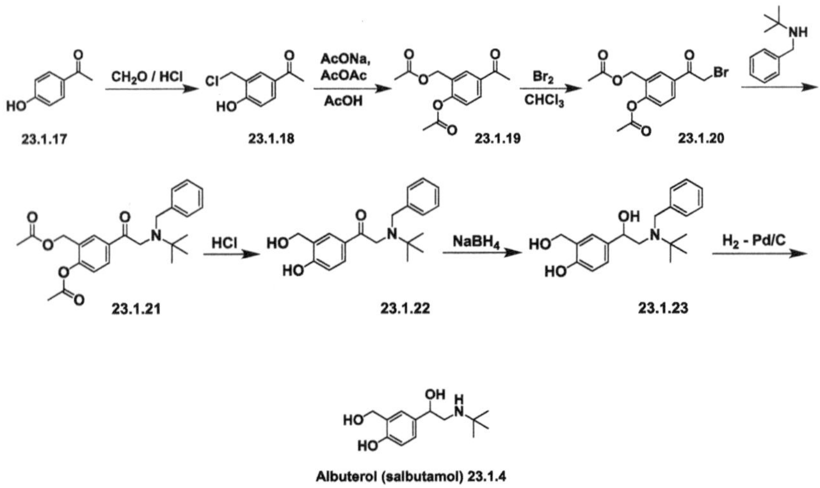

SCHEME 23.1 Synthesis of albuterol.

Albuterol (salbutamol) is extensively used in the treatment of reversible obstructive airways disease. It represents a first-line therapy because it offers rapid bronchodilation, usually relieving bronchospasm within minutes. It is particularly useful in those patients who are unable to coordinate the use of inhalers. Albuterol is generally well tolerated and any side effects observed, such as restlessness, irritability, or nervousness, are a predictable extension of its pharmacology [38-40].

Salmeterol–Serevent

Various approaches for the synthesis of salmeterol were described first in a patent [41] and many related patents [42-44] and papers [45-50]. Disclosed and published data are summarized by Scheme 23.2.

The essence of the proposed methods consists of two general approaches: (a) in interaction of 4-(2-amino-1-hydroxyethyl)-2-(hydroxymethyl)phenol (**23.1.24**) with (4-((6-bromohexyl)oxy)butyl)benzene (**23.1.25**), and (b) via interaction of bromoacetophenone (**23.1.26**) or its derivatives or precursors (**23.1.27** to **23.1.29**) with easily available N-benzyl-6-(4-phenylbutoxy)hexyl-1-amine (**23.1.32**) (Scheme 23.2. and Scheme 23.3.).

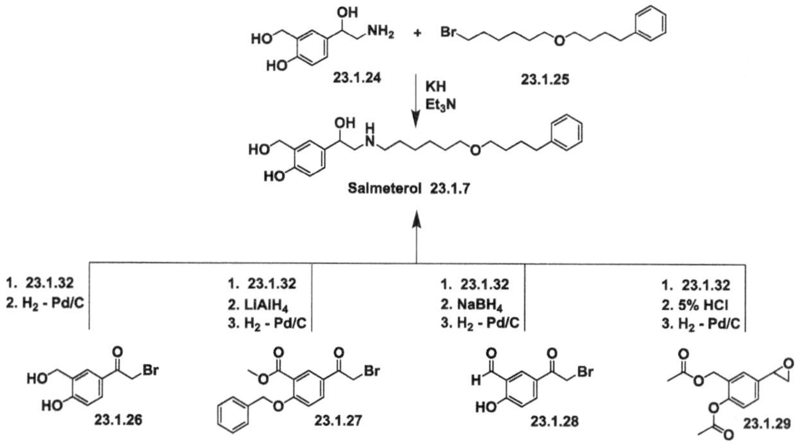

SCHEME 23.2 Synthesis of salmeterol.

Amine (**23.1.32**), in turn, was prepared starting from 4-phenylbutan-1-ol (**23.1.30**), according to Scheme 23.3.

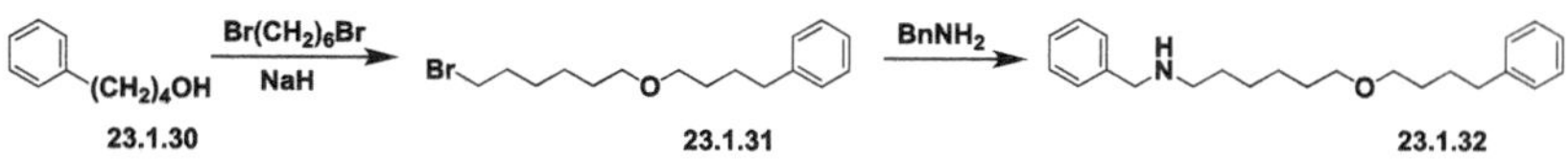

SCHEME 23.3 Synthesis of N-benzyl-6-(4-phenylbutoxy)hexyl-1-amine (23.1.32).

Salmeterol is a potent and highly β_2-selective adrenoceptor agonist with a duration of action longer than 12 hours. It is administrated via dry power inhalation and clinical studies show it has a good efficacy and a good safety profile. More common side effects of salmeterol are difficulty with breathing, irritation of the throat, cough producing mucus, headache, and runny nose [51-56].

Formoterol–Perforomist

Formoterol (**23.1.8**) is commercialized in its racemic form, but the different stereoisomers, corresponding to the two stereogenic centers of the molecule, have different pharmacological potencies. The order of potency is (R,R) » (R,S) = (S,R) > (S,S) [57]. It was found that the (R,R)-isomer is 1000-fold more potent than the (S,S)-isomer.

The first described synthesis of formoterol started from bromination of 4-benzyloxy-3-nitroacetophenone (**23.1.33**), which produced 4-benzyloxy-3-nitro-α-bromoacetophenone (**23.1.34**).

The 4-benzyloxy-3-nitro-α-bromoacetophenone (**23.1.34**) was reacted with N-benzyl-1-(4-methoxyphenyl)propan-2-amine (**23.1.35**) to produce aminoketone (**23.1.36**). The keto group of the obtained compound was hydrogenated with sodium borohydride, producing the amino alcohol (**23.1.37**), the nitro group of which was reduced to an amino group using iron powder in 50% aqueous acetic acid solution to produce (**23.1.38**). Formylation of an aromatic amino group in (**23.1.38**) was carried out with a mixture of acetic anhydride and formic acid, and just leaving the mixture of reagents to stand overnight produced the compound (**23.1.39**). Debenzylation of the compound (**23.1.39**) by a standard protocol using hydrogen on Pd/C produced the desired formoterol (**23.1.8**) [58,59] (Scheme 23.4.).

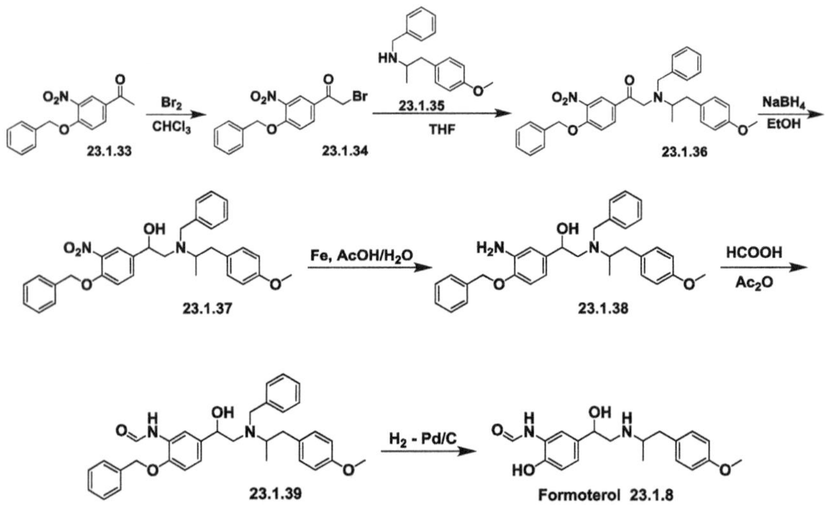

SCHEME 23.4 Synthesis of formoterol.

The (R,R)-enantiomer is more active than the other stereoisomers [57,58].

Resolution procedure of all four stereoisomers of formoterol has been described in the literature via a resolution, but it gave low overall yield [57,60].

Diastereo- and enantioselective syntheses of formoterol have been reported [61-67]. The starting point for them is enantioselective reduction of bromo ketone (**23.1.34**) to enantiomerically pure (R)-1-(4-(benzyloxy)-3-nitrophenyl)-2-bromoethanol (**23.1.40**), which was achieved in different ways, such as cis-l-amino-2-indanol catalyzed borane reduction [61,62], or using Rh-PEGBsDPEN catalyst prepared in situ by mixing η5-pentamethy lcyclopentadienylrhodium dimer ($([RhC12(Cp^*)]_2)$) and polyethylene glycol derivative of N-((1S,2S)-2-amino-1,2-diphenylethyl)-4-hydroxybenzene-sulfonamide (PEGBsDPEN) [63]. The nitro group of the secondary alcohol (**23.1.40**) was reduced by hydrogenation in the presence of 5% Pt/C, and then reacted with $HCOOH/Ac_2O$ mixture to produce the formamide derivative (**23.1.41**). The obtained compound was transformed into epoxide (**23.1.42**) quantitatively in the mixture of THF and MeOH as the solvent in presence of potassium carbonate at room temperature. Heating the mixture of epoxide (**23.1.42**) and amine (**23.1.43**), which was prepared by one-pot Pt/C, catalyzed hydrogenation of chiral 1-phenylethanamine and 1-(4-methoxyphenyl) propane-2-one, at 120°C for 24 hours to produce the ring-opening product (**23.1.44**). Removal of the protecting group by Pd/C catalyzed hydrogenolysis produced (R,R)-formoterol (**23.1.8**) [63] (Scheme 23.5.).

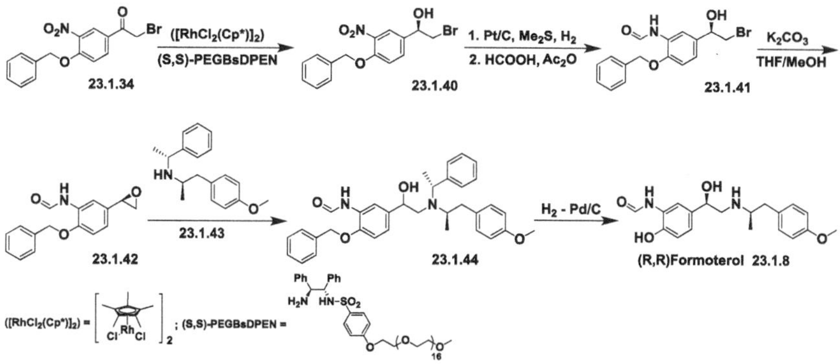

SCHEME 23.5 Diastereo- and enantioselective synthesis of formoterol.

Formoterol is the first highly selective β_2-agonist combining the clinical advantage of rapid bronchodilation with a long duration of action. It is effective for the treatment of COPD patients with moderate or greater severity of airflow obstruction. Formoterol is a safe and well-tolerated drug. Its side effects include difficulty with breathing, cough, nasal congestion, body aches or pain, headache, and unusual tiredness or weakness [68-76].

Other extremely β_2-selective and effective β-adrenoceptor agonists in which the phenyl ring was substituted for a heterocycle such as broxaterol (**23.1.45**), pirbuterol (**23.1.46**), ZK-90055 (**23.1.47**), and procaterol (**23.1.48**) were proposed for treatment of asthma and COPD, but they had no advantage compared to salbutamol. An extremely potent selective β_2-adrenoceptor agonist that was orders of magnitude more potent than adrenaline was found in marine sponge and coded as S-1319 (**23.1.49**) (Fig. 23.4.).

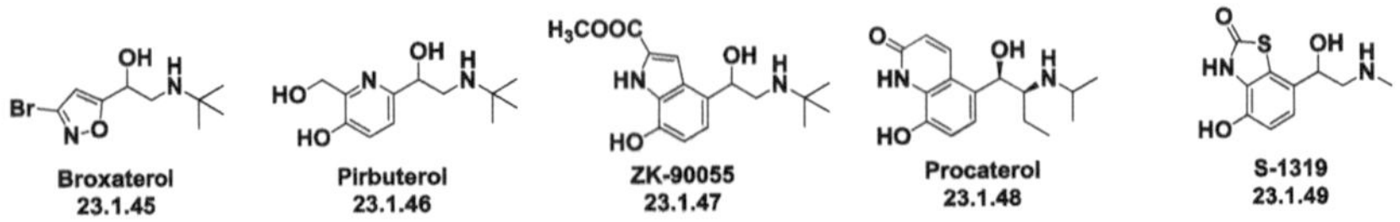

FIG. 23.4 New β_2-selective and effective β-adrenoceptor agonists.

An excellent review describing development of β_2-adrenoceptor agonists as bronchodilators during the last 100 years concludes that β_2-adrenoceptor agonists are the most effective and safe bronchodilators currently available, and have not been surpassed by any other bronchodilating agents [77,78].

Investigational β_2-adrenoceptor agonists—compounds PF-610355 (**23.1.50**) and bedoradrine (**23.1.51**), as well as LAS100977 and GSK-642444 whose structures are not displayed—are in clinical trials (Fig. 23.5.).

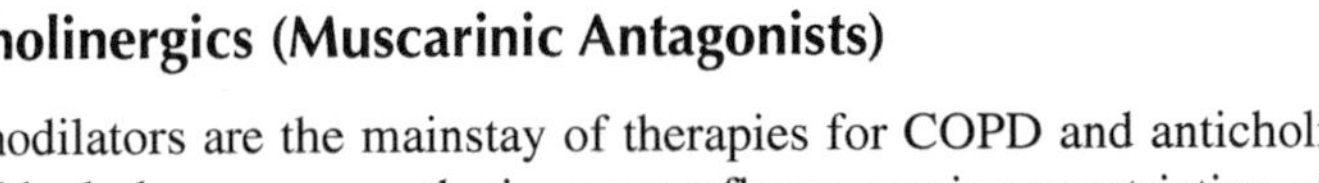

FIG. 23.5 Investigational β_2-adrenoceptor agonists.

Anticholinergics (Muscarinic Antagonists)

Bronchodilators are the mainstay of therapies for COPD and anticholinergics, which block the parasympathetic nerve reflexes causing constriction of the airways, allow the air passages to remain open.

Airway tone and hyperreactivity are mediated by the parasympathetic nerves that release acetylcholine onto muscarinic receptors. Only three of five muscarinic receptor subtypes (M1, M2, and M3) exert physiological effects on the lungs. Stimulation of M1 and M3 muscarinic receptors causes bronchoconstriction. The M2 receptor is an inhibitory prejunctional autoreceptor, thus the ideal anticholinergic drug for treatment of COPD would be a selective inhibitor of M1 and M3 muscarinic receptors that does not inhibit M2 muscarinic receptors. Subsequent findings show that M3 receptor selective blockers that do not antagonize M2 receptor function could have a greater therapeutic benefit than nonselective drugs [7,8,17,18,79].

Smoking the dried plant known as "belladonna" is a remedy for therapy of asthma known since antiquity.

The anticholinergic action of deadly nightshade, ("belladonna") from dried plant smoking was caused by the presence of alkaloids of the atropine (**23.1.52**) series and because its use was not limited to that of therapy for asthma; it was also used for preparation of other drugs, cosmetics, eyedrops, and poisons. Unfortunately, atropine easily penetrates the blood–brain barrier, generating multiple side effects.

Chemical modification of atropine yielded a number of congeners as effective as anticholinergics, which are used in treatment of respiratory diseases. Among them are nonselective muscarinic receptor antagonists such as methylatropinium nitrate (**23.1.53**) ipratropium (**23.1.54**), oxitropium (**23.1.55**), and bromides which are short acting (6 to 8 hours), none selective receptor antagonists (Fig. 23.6.).

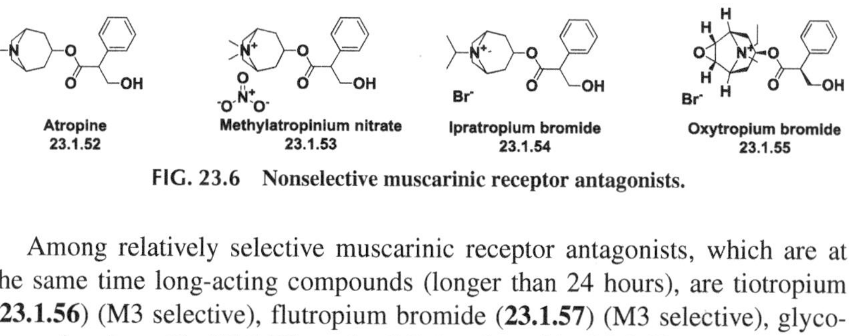

Atropine
23.1.52
Methylatropinium nitrate
23.1.53
Ipratropium bromide
23.1.54
Oxytropium bromide
23.1.55

FIG. 23.6 **Nonselective muscarinic receptor antagonists.**

Among relatively selective muscarinic receptor antagonists, which are at the same time long-acting compounds (longer than 24 hours), are tiotropium (**23.1.56**) (M3 selective), flutropium bromide (**23.1.57**) (M3 selective), glycopyrronium bromide (**23.1.58**), and aclidinium bromide (**23.1.59**) (M3/M2 selective) (Fig. 23.7.).

These compounds are as effective as atropine, but longer acting and with fewer side effects. Tiotropium bromide is included in the list of Top 200 Drugs by sales for the 2010s.

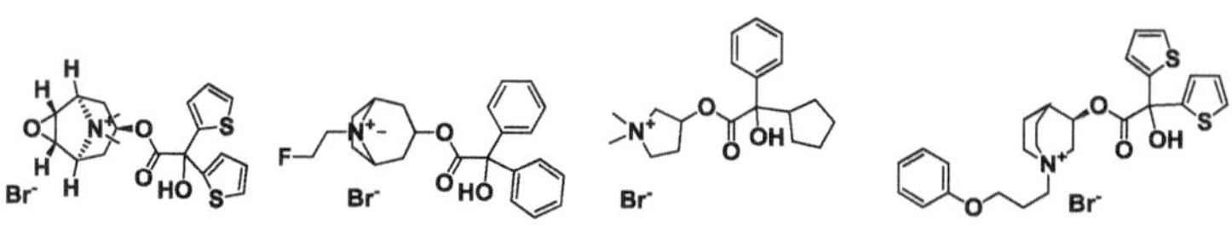

Tiotropium bromide
23.1.56
Flutropium bromide
23.1.57
Glycopyrronium bromide
23.1.58
Aclidinium bromide
23.1.59

FIG. 23.7 **Relatively selective muscarinic receptor antagonists.**

Several other muscarinic antagonists bronchodilators, such as umeclidinium bromide (**23.1.60**), previously known as GSK573719, compound (**23.1.61**) (Fig. 23.8.), trospium, dexpirronium, PF-4522971, TD-4208, CHF 5407, and QAT370 (structures of which are not disclosed) are under development.

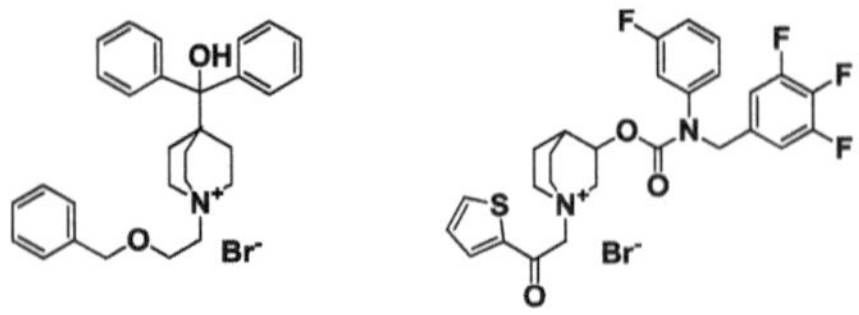

Umeclidinium bromide 23.1.60 23.1.61

FIG. 23.8 Muscarinic antagonists under development.

Tiotropium Bromide–Spiriva

Preparation of tiotropium bromide (**23.1.56**) was first published in a patent [80]. The first step consists of transesterification of methyl 2-hydroxy-2,2-di(thiophen-2-yl)acetate (**23.1.63**) with scopine (**23.1.65**) in presence of sodium or sodium methoxide in refluxing toluene to produce (**23.1.62**). The second step of the synthesis of the desired product (**23.1.57**) is quaternization with methyl bromide carried out in acetonitrile or an acetonitrile/dichloromethane mixture [80] (see Scheme 23.6.). Alternative ways of synthesis imply preparation of ester (**23.1.64**) directly from scopine (**23.1.63**) and 2-hydroxy-2,2-di(thiophen-2-yl) acetic acid or 2-hydroxy-2,2-di(thiophen-2-yl)acetic anhydride [81] (Scheme 23.6.). Further minor modifications of this general scheme have been reported [82-84].

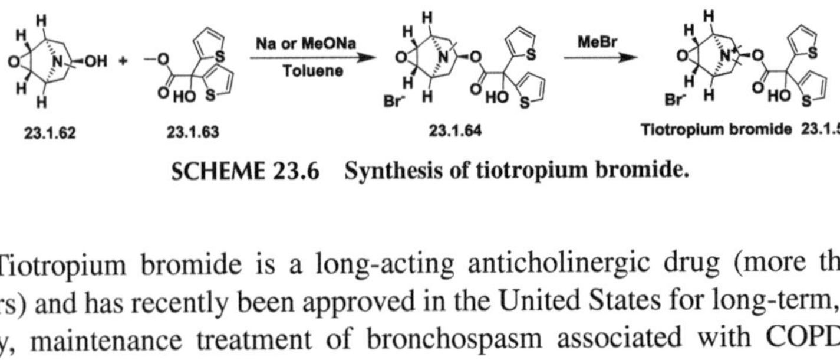

23.1.62 23.1.63 23.1.64 Tiotropium bromide 23.1.56

SCHEME 23.6 Synthesis of tiotropium bromide.

Tiotropium bromide is a long-acting anticholinergic drug (more than 24 hours) and has recently been approved in the United States for long-term, once-daily, maintenance treatment of bronchospasm associated with COPD. The potency and long duration of effect of this anticholinergic bronchodilator result primarily from a prolonged blockade of the M1 and M3 muscarinic receptors in the airways and a relatively more rapid dissociation from the M2 receptor, which provides inhibitory feedback [85-101].

Xanthines

Xanthines, compounds structurally related to caffeine (**23.1.66**), have been used in the treatment of asthma for approximately 150 years, and theophylline (1,3-dimethylxanthine), which was first prescribed for the treatment of asthma in 1937 [102], is still widely used for asthma and COPD and still remains a very useful add-on therapy. "One of the commonest and best reputed remedies of asthma…is strong coffee." This sentence was written in 1859 [103].

Xanthines don't open airways as well as the previously described inhaled bronchodilators. But xanthines, working via mechanisms of action that are still not well understood, produce several effects that are beneficial to patients with COPD and asthma. They decrease diaphragmatic muscle fatigue, increase mucociliary clearance, block centrally mediated hypoventilation, and decrease capillary leakage. At the same time, at high doses they stimulate heart rate and force of contraction, can generate cardiac arrhythmias, and can lead to convulsions.

Xanthines and methylxanthines, used in the treatment of lung diseases, include caffeine (**23.1.65**), theophylline (**23.1.66**), aminophylline (**23.1.67**), bamifylline (**23.1.68**), acebrophylline (**23.1.69**), and doxofylline (**23.1.70**) (Fig. 23.9.).

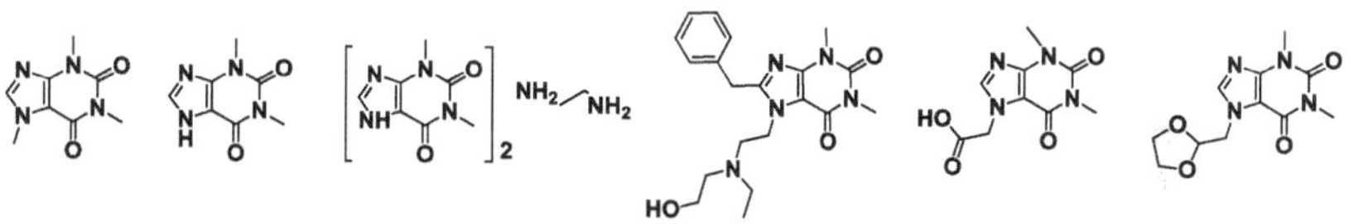

Caffeine 23.1.65 · Theophylline 23.1.66 · Aminophylline 23.1.67 · Bamifylline 23.1.68 · Acebrophylline 23.1.69 · Doxofylline 23.1.70

FIG. 23.9 Xanthines and methylxanthines used in the treatment of lung diseases.

Doxofylline [104] is a novel xanthine drug with similar efficacy to theophylline [105] and seems to provide a serious alternative to theophylline. It appears to be a bronchodilator and an antiinflammatory drug with a wider therapeutic window than other xanthines. Doxofylline has better efficacy and fewer side effects than theophylline.

None of the xanthines is included in the list of Top 200 Drugs by sales for the 2010s.

23.2 CORTICOSTEROIDS

Bronchodilators, including β_2-agonists and antimuscarinic receptor antagonists, are the mainstay of pharmacotherapy in COPD, but the central role in the pathophysiology of COPD is systemic and local inflammation. More than 60% of patients with COPD are treated with inhaled corticosteroids even though their use is subject to debate.

Corticosteroids (inhaled) are extensively used in the treatment of asthma and COPD. Because of their broad antiinflammatory and immunosuppressive effects, they are recommended and widely implemented in the treatment of asthma and COPD cases that are characterized by airway inflammation. Corticosteroids are the preferred agents for managing persistent asthma, improving lung function, and decreasing bronchial hyperresponsiveness, but they do not alter the natural progression of the disease.

When used appropriately, corticosteroids have few adverse events at low and medium doses. Potential systemic side effects include suppression of the hypothalamus–pituitary–adrenal axis, Cushing syndrome, osteoporosis, cataracts, dermal thinning and bruising, adrenal insufficiency, and growth suppression in children [106,107].

Corticosteroids currently available are triamcinolone (**23.2.1**), beclomethasone (**23.2.2**), budesonide (**23.2.3**), flunisolide (**23.2.4**), ciclesonide (**23.2.5**), fluticasone (**23.2.6**), and mometasone (**23.2.7**) (Fig. 23.10.).

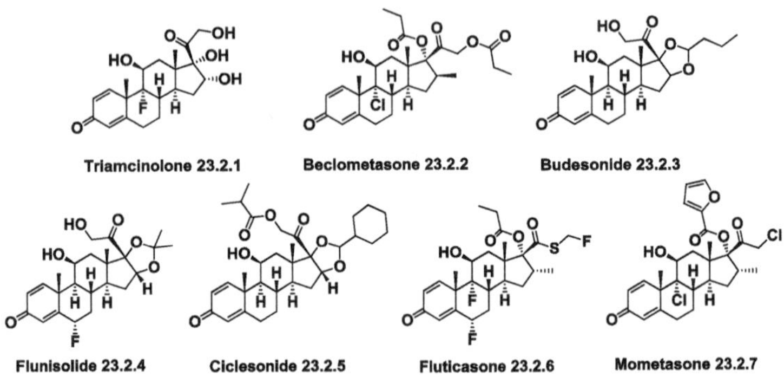

FIG. 23.10 **Corticosteroids used in the treatment of lung diseases.**

Triamcinolone (**23.2.1**) and budesonide (**23.2.3**) are included in the list of Top 200 Drugs by sales for the 2010s. The first – triamcinolon, for prevention of asthma attacks, the second – budesonide as a component of drug combinations. Current guidelines generally recommend use of corticosteroids in combination with long-acting bronchodilators for maintenance treatment of moderate-to-severe chronic COPD.

Triamcinolone–Nasacort

The synthesis of triamcinolone (**23.2.1**) starts from ketalization of cortisol 21-acetate (**23.2.8**) using ethylene glycol. Dehydration of the obtained compound (**23.2.9**) for creation of a double bond in position 16,17 of the steroid skeleton through the series of sequential reactions of chlorination, dehydrochlorination, hydrolysis, and acetylation produces 21-acetoxy-4,9(11),16-pregnatriene-3, 20-dione (**23.2.10**), treatment of which with osmium tetroxide in benzene and pyridine produced diol (**23.2.11**), the secondary hydroxyl group of which, in position 16, was acetylated with acetic anhydride in pyridine to produce the diacetate (**23.2.12**). The obtained compound in dioxane and water was treated with N-bromoacetamide and 10% perchloric acid to yield bromohydrine (**23.2.13**). Dehydrobromination of the bromohydrine (**23.2.13**) with anhydrous potassium acetate in refluxing ethanol produced the epoxy-derivative (**23.2.14**). Opening of the epoxide ring in (**23.2.14**) with anhydrous hydrogen fluoride in chloroform produced (**23.2.15**). Microbiological dehydrogenation of the obtained product with *Corynebacterium simplex* produced crude diacetate (**23.2.16**), saponification of which produced triamcinolone (**23.2.1**) [108-110] (Scheme 23.7.).

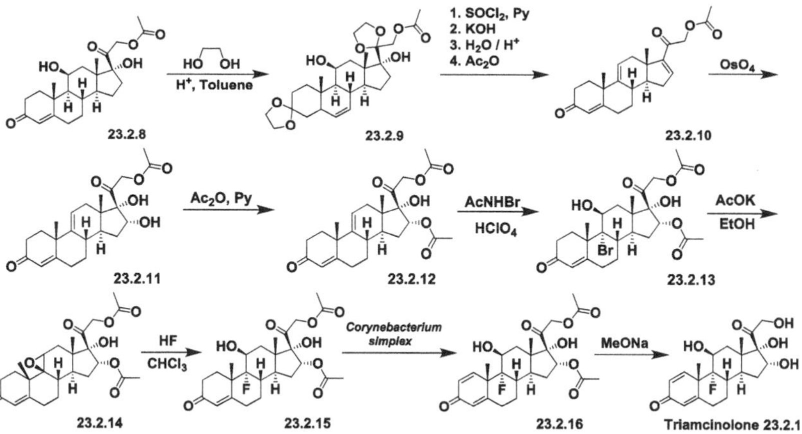

SCHEME 23.7 Synthesis of triamcinolone.

Triamcinolone is commonly used in the treatment of respiratory inflammation and improves airway reactivity, decreasing respiratory problems. Strangely, there are only few reviews of the pharmacotherapy of triamcinolone [111-113].

Budesonide–Pulmicort

Budesonide (**23.2.3**) was prepared simply by acetalization of 16a-hydroxyprednisolone (**23.2.17**) with butyraldehyde (**23.2.18**) in presence of anhydrous perchloric acid [114,115] (Scheme 23.8.).

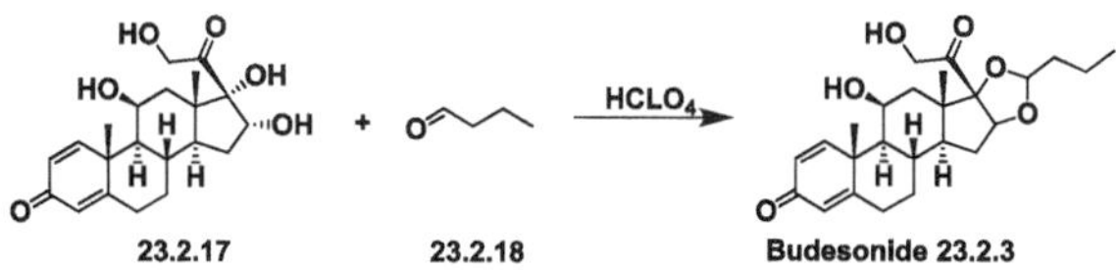

SCHEME 23.8 Synthesis of budesonide.

Budesonide is used to prevent wheezing, shortness of breath, and troubled breathing caused by severe asthma and other lung diseases [116].

23.3 COMBINATION AGENTS

Many drugs for treating COPD and respiratory system diseases in general are included in the list of Top 200 Drugs by sales for the 2010s. Among them are tiotropium bromide (**23.1.57**) under the trade name Spiriva; triamcinolone (**23.2.1**) under the trade name Nasacort; and budesonide (**23.2.3**) under the trade names Entocort and Pulmicort. Among the many combination agents used to treat COPD are Combivent, which is a combination of albuterol (**23.1.4**) and ipratropium bromide (**23.1.3**); Advair, which is a combination of fluticasone (**23.2.6**) and salmeterol (**23.1.7**); and Symbicort, which is a combination of formoterol (**23.1.8**) and budesonide (**23.2.3**). Dulera, which is a combination of formoterol (**23.1.8**) and mometasone (**23.2.7**), and DuoNeb, which is a combination of ipratropium bromide (**23.1.3**) and albuterol (**23.1.4**), are not included in the "list," but they are also very popular prescriptions written by pulmonologists.

23.4 MUCOACTIVE DRUGS

Mucus hypersecretion is a clinical feature of respiratory diseases such as COPD, asthma, and cystic fibrosis.

The main purpose of mucoactive drugs is to increase the ability to expectorate sputum and/or decrease mucus hypersecretion.

Mucolytic and related agents have been in use since antiquity. These agents belong to several distinct chemical classes. For many of them their mechanisms of action remain not well understood.

Compounds that affect mucus have been precisely categorized [114,115] and now can be classified as expectorants, mucoregulators, mucolytics, or mucokinetics.

Expectorants

Expectorants increase secretion of mucins and are defined as drugs that induce discharge or expulsion of mucus from the respiratory tract.

The most frequently used expectorants are hypertonic saline used as an aerosol, guaifenesin (**23.4.1**), and some glycerol iodination products, such as domiodol (**23.4.2**) and organidin (**23.4.3**). In general, iodine-containing agents are considered to be expectorants that promote the secretion of airway fluid and have long been used in medicine. Some purinergic receptor agonists, such as uridine triphosphate (**23.4.4**) and adenosine triphosphate (**23.4.5**), have been recently proposed and developed as expectorants and are currently undergoing clinical studies (Fig. 23.11.).

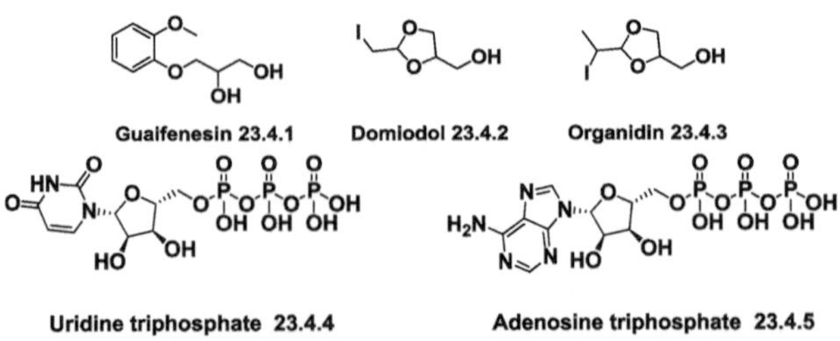

FIG. 23.11 Expectorants.

Mucoregulators

Mucoregulators are drugs that regulate mucus secretion by reducing the process of mucus hypersecretion. It is a diverse group of different compounds that includes cysteine derivatives, anticholinergics, glucocorticoids, and macrolide antibiotics.

Cysteine derivatives, mecysteine (**23.4.6**), acetylcysteine (**23.4.7**), carbocysteine (**23.4.8**), fudosteine (**23.4.9**), their analogue erdosteine (**23.4.10**), and Nacystelyn, a recently developed lysine salt of N-acetylcysteine. Cysteine derivatives are considered as compounds which restores viscoelastic properties of mucus, provides antiinflammatory, and antioxidant properties. All of which has the ability to increase the synthesis of sialomucins, important structural components of mucus (Fig. 23.12.).

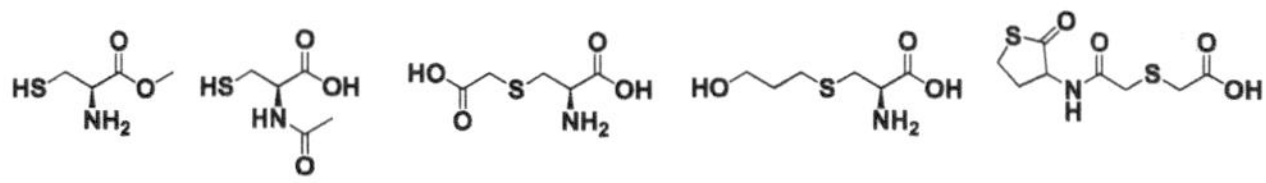

Mecysteine 23.4.6 Acetylcysteine 23.4.7 Carbocysteine 23.4.8 Fudosteine 23.4.9 Erdosteine 23.4.10

FIG. 23.12 Mucoregulators.

Anticholinergics that are used as mucoregulators include scopolamine (**23.4.11**), atropine (**23.1.52**), ipratropium (**23.1.54**), tiotropium bromide (**23.1.56**), and glycopyrronium bromide (**23.1.58**) (Fig. 23.13.). It is considered that they block secretory reflexes, and reduce glandular output and sputum volume.

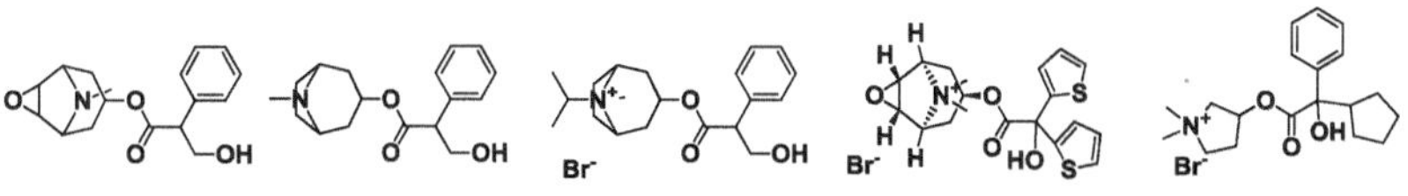

Scopolamine 23.4.11 Atropine 23.1.52 Ipratropium bromide 23.1.54 Tiotropium bromide 23.1.56 Glycopyrronium bromide 23.1.58

FIG. 23.13 Anticholinergic used as mucoregulators.

The glucocorticoids are potent antiinflammatory agents that are used in the management of acute exacerbations in COPD and asthma patients. Prednisolone (**23.4.12**) (Fig. 23.14.) is the most frequently used drug among glucocorticoids and provides improvement in lung clearance in stable asthmatics.

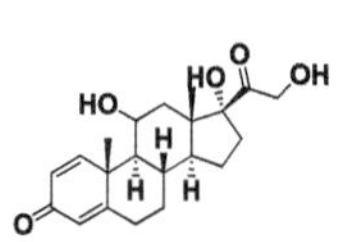

Prednisolone 23.4.12

FIG. 23.14 Structure of prednisolone.

Macrolide antibiotics, which include erythromycin (**23.4.13**), azithromycin (**23.4.14**), clarithromycin (**23.4.15**), and roxithromycin (**23.4.16**) (Fig. 23.15.), have been successfully used to treat a range of chronic, inflammatory lung disorders, significantly reducing sputum production.

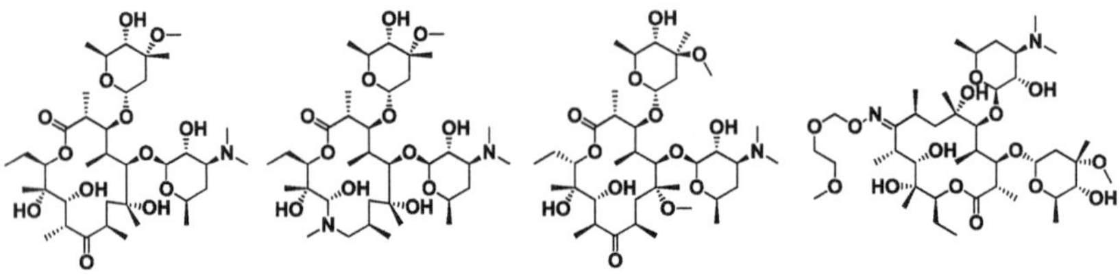

Erythromycin 23.4.13 Azithromycin 23.4.14 Clarithromycin 23.4.15 Roxithromycin 23.4.16

FIG. 23.15 **Macrolide antibiotics used to treat a range of lung disorders.**

Mucolytics

Mucolytics are drugs that decrease mucus viscosity. N-acetylcysteine (**23.4.7**) is a classic mucolytic drug. A little bit more modern drugs are fudosteine (**23.4.9**), erdosteine (**23.4.10**), and Nacystelyn (see Fig. 23.12.).

Mucokinetics

Mucokinetics are drugs that facilitate removal of mucus from the respiratory tract, effectively increasing the transportability of mucus by cough. These agents include adrenoceptor agonist bronchodilators such as albuterol and a surfactant. A special drug for that purpose is ambroxol (**23.4.17**) (Fig. 23.16.).

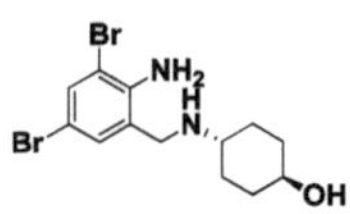

Ambroxol 23.4.17

FIG. 23.16 **Structure of ambroxol.**

Several reviews summarize the most clinically relevant mucoactive drugs used worldwide in the management of several acute and chronic respiratory diseases [116-119].

23.5 DEVELOPMENT OF NEW DRUGS FOR CHRONIC OBSTRUCTIVE PULMONARY DISEASE

COPD is characterized by a chronic inflammation in the lungs that leads to progressive airflow obstruction. Consequently, one of the strategies for treating COPD could be blocking of the synthesis of inflammatory mediators or corresponding receptors involved in pathophysiology of COPD.

Different research groups have been focused on the key inflammatory regulators, such as inhibitors of phosphodiesterase-4 (PDE4), p38 mitogen-activated protein kinase (p38), and Janus kinases [10-13].

Phosphodiesterase-4 Inhibitors

Phosphodiesterase-4 is expressed in neutrophils, T cells, and macrophages; consequently, its inhibition may have inhibitory effects on inflammation. The selective PDE4 inhibitor, roflumilast (**23.5.1**) (Fig. 23.17.) is an example of a PDE4 inhibitor and shows a promising inhibiting effect on lung inflammation and emphysema on model testing.

Cytokine Inhibitors

Tumor necrosis factor-α (TNF-α) is a key mediator of neutrophilic inflammation in COPD. The TNF-α inhibitor infliximab, which is an artificial chimeric monoclonal antibody, achieved clinical trials, but showed little effect in COPD patients.

Chemokine Antagonists

Chemokine receptor antagonists have also been proposed as drug targets in COPD. CXCR1/2 antagonists ADZ-4818 (**23.5.2**) (Fig. 23.17.), ADZ8309 (structure is not displayed), and CXCR2 antagonists SB-656933 (structure is not displayed) and SCH-527123 (**23.5.3**) (Fig. 23.17.) are currently being evaluated in clinical trials.

Toll-Like Receptor Inhibitors

Toll-like receptor (TLR) inhibitors, especially TLR2, TLR4, and TLR9 inhibitors are attractive targets for inhibition of inflammation on COPD. Different antibodies, such as TLR2 inhibitor OPN-305 and TLR4 inhibitor NI-0101, are undergoing trials.

Nuclear Factor-κB Inhibitors

The nuclear factor-κB (NF-κB) family of transcription factors regulates the expression of multiple proinflammatory mediators (cytokines, chemokines, adhesion molecules, matrix metalloproteinases, and cyclooxygenases) in response to inflammation. Numerous NF-κB inhibitor compounds are under development for potential antiinflammatory therapy: IMD-0354 (**23.5.4**), BMS-345541 (**23.5.5**) (Fig. 23.17.).

p38 Mitogen-Activated Protein Kinase Inhibitors

The p38 mitogen-activated protein kinase (MAPK) signaling pathway plays a central role in regulating inflammation. Activation of p38 MAPK in COPD patients correlates with the degree of inflammation. p38 inhibitors have been evaluated in clinical trials, including losmapimod (**23.5.6**), SB-681323 (**23.5.7**), and PF-03715455 (**23.5.8**) (Fig. 23.17.). Compounds PH-797804, ARRY-371797, GSK-681323, and GSK-856553, whose structures are not displayed, are currently under development.

Protease Inhibitors

More than 30 years ago human neutrophil elastase was identified as a target for COPD treatment. Sivelestat (**23.5.9**) has been approved for clinical use

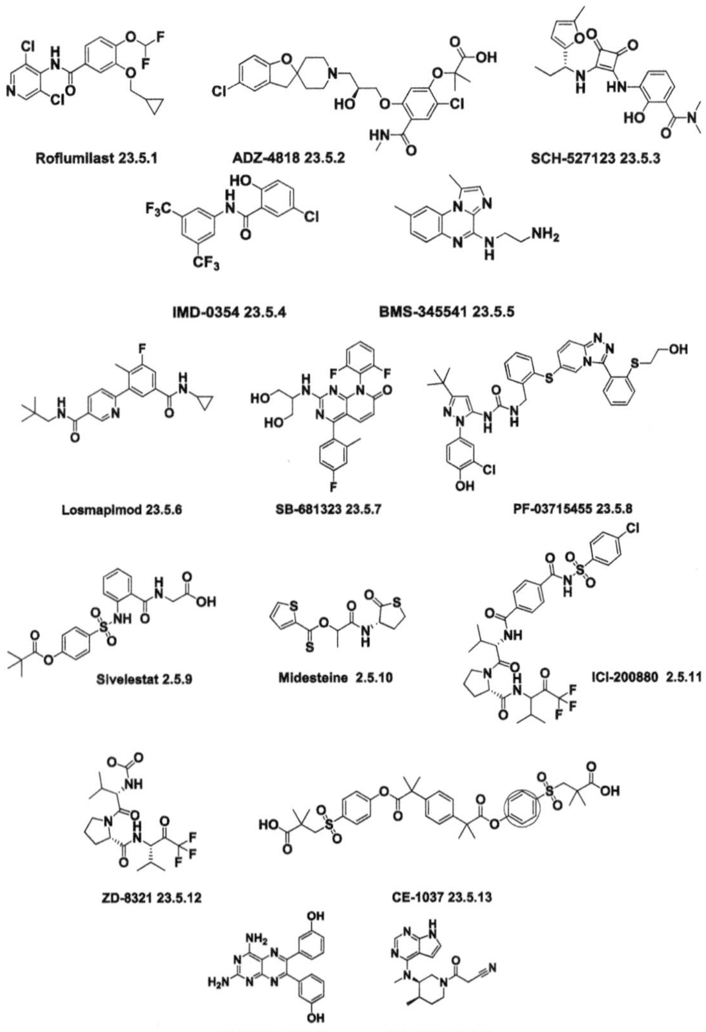

FIG. 23.17 **Structure of new drugs for COPD under investigation.**

as human neutrophil elastase inhibitor in some countries. Other inhibitors, including midesteine (**23.5.10**), ICI-200880 (**23.5.11**), ZD-8321 (**23.5.12**), and CE-1037 (**23.5.13**) have been discontinued at different phases of clinical trials (Fig. 23.17.).

Phosphoinositide-3 Kinase Inhibitors

Because phosphoinositide-3 kinase inhibitors are also involved in regulating inflammatory gene expression of experimental compounds TG-100115 (**23.5.14**) is now in early stages of drug development (Fig. 23.17.).

Janus Kinase/Signal Transducers and Activators of Transcription Inhibitors

Several cytokines and inflammatory mediators in COPD signal via the Janus kinase/signal transducers and activators of transcription pathway, which inhibition may represent definite interest as pulmonary inflammation remedy. One of these compounds, tofacitinib (**23.5.15**), is currently in clinical trials (Fig. 23.17.).

REFERENCES

1. Barnes, P. J. Chronic obstructive pulmonary disease (COPD). In *Chemical Biology: Approaches to Drug Discovery and Development to Targeting Disease;* Civjan, N., Ed.; Wiley, 2012; pp 245–265.
2. Cazzola, M.; Segreti, A.; Rogliani, P. Comparative effectiveness of drugs for chronic obstructive pulmonary disease. *Drugs Today* **2012,** *48* (12), 785–794.
3. Herdegen, T. Control and relief: drugs against asthma and COPD. *Dtsch. Apoth. Ztg.* **2013,** *153* (27), 52–62.
4. Reid, D. J.; Pham, N. T. Emerging therapeutic options for the management of COPD. *Clin. Med. Insights: Circ., Respir. Pulm. Med.* **2013,** *7,* 7–15.
5. Matera, M. G.; Calzetta, L.; Segreti, A.; Cazzola, M. Emerging drugs for chronic obstructive pulmonary disease. *Expert Opin. Emerg. Drugs* **2012,** *17* (1), 61–82.
6. Ejiofor, S.; Turner, A. M. Pharmacotherapies for COPD. *Clin. Med. Insights: Circ., Respir. Pulm. Med.* **2013,** *7,* 17–34.
7. Plusa, T. Rationale basis for new anticholinergic drugs for chronic obstructive pulmonary disease. *Int. Rev. Allergol. Clin. Immunol.* **2011,** *17* (3–4), 49–52.
8. Compton, C.; McBryan, D.; Bucchioni, E.; Patalano, F. The Novartis view on emerging drugs and novel targets for the treatment of chronic obstructive pulmonary diseases. *Pulm. Pharmacol. Ther.* **2013,** *26* (5), 562–573.
9. Page, C.; Pitchford, S.; Spina, D. Development of new drugs for the treatment of respiratory diseases: from concept to the clinic. *J. Drug Delivery Sci. Technol.* **2011,** *21* (4), 347–352.
10. Barnes, P. J. Development of new drugs for COPD. *Curr. Med. Chem.* **2013,** *20* (12), 1531–1540.
11. Ngkelo, A.; Adcock, I. M. New treatments for COPD. *Curr. Opin. Pharmacol.* **2013,** *13* (3), 362–369.

12. Restrepo, R. D.; Tate, A.; Coquat, J. Evaluation of for the treatment of chronic obstructive pulmonary diseases. *Expert Opin. Pharmacother.* **2013,** *14* (14), 1993–2002.
13. D'Urzo, A.; Vogelmeier, C. Future of chronic obstructive pulmonary diseases management. *Expert Rev. Respir. Med.* **2012,** *6* (3), 285–299.
14. Montuschi, P.; Ciabattoni, G. Bronchodilating Drugs for Chronic Obstructive Pulmonary Disease: Current Status and Future Trends. *J. Med. Chem.* **2015,** *58* (10), 4131–4164.
15. Ambrosino, N.; Paggiaro, P. The management of asthma and chronic obstructive pulmonary disease: current status and future perspectives. *Expert Rev. Respir. Med.* **2012,** *6* (1), 117–127.
16. Hudd, T. R.; Zaiken, K. Management of chronic obstructive pulmonary disease: an emphasis on recently approved medications and products in the pipeline. *Formulary* **2011,** *46* (9), 374–380, 389-393.
17. Spina, D. Current and novel bronchodilators in respiratory disease. *Curr. Opin. Pulm. Med.* **2014,** *20* (1), 73–86.
18. Cazzola, M.; Page, C. P.; Calzetta, L.; Matera, M. G. Pharmacology and therapeutics of bronchodilators. *Pharmacol. Rev.* **2012,** *64* (3), 450–504.
19. Mak, G.; Hanania, N. A. New bronchodilators. *Curr. Opin. Pharmacol.* **2012,** *12* (3), 238–245.
20. Bouyssou, T.; Rudolf, K.; Hoenke, C.; Lustenberger, P.; Schnapp, A.; Konetzki, I. Studies towards topical selective β_2-adrenoceptor agonists with a long duration of action. *Bioorg. Med. Chem. Lett.* **2009,** *19* (17), 5237–5240.
21. Hoenke, C.; Bouyssou, T.; Tautermann, C. S.; Rudolf, K.; Schnapp, A.; Konetzki, I. Use of 5-hydroxy-4H-benzo[1,4]oxazin-3-ones as β_2-adrenoceptor agonists. *Bioorg. Med. Chem. Lett.* **2009,** *19* (23), 6640–6644.
22. Bouyssou, T.; Hoenke, C.; Rudolf, K.; Lustenberger, P.; Pestel, S.; Sieger, P.; Lotz, R.; Heine, C.; Büttner, F. H.; Schnapp, A.; Konetzki, I. Discovery of olodaterol, a novel inhaled β_2-adrenoceptor agonist with a 24 h bronchodilatory efficacy. *Bioorg. Med. Chem. Lett.* **2010,** *20* (4), 1410–1414.
23. Glossop, P. A.; Price, D. A. Progress in the development of inhaled, long-acting β_2-adrenoceptor agonists. *Annu. Rep. Med. Chem.* **2006,** *41,* 237–248.
24. Brown, A. D.; Bunnage, M. E.; Glossop, P. A.; James, K.; Jones, R.; Lane, C. A. L.; Lewthwaite, R. A.; Mantell, S.; Perros-Huguet, C.; Price, D. A.; Trevethick, M.; Webster, R. The discovery of long acting β_2-adrenoreceptor agonists. *Bioorg. Med. Chem. Lett.* **2007,** *17* (14), 4012–4015.
25. Brown, A. D.; Bunnage, M. E.; Glossop, P. A.; Holbrook, M.; Jones, R. D.; Lane, C. A. L.; Lewthwaite, R. A.; Mantell, S.; Perros-Huguet, C.; Price, D. A.; Webster, R. The discovery of indole-derived long acting β_2-adrenoceptor agonists for the treatment of asthma and COPD. *Bioorg. Med. Chem. Lett.* **2007,** *17* (22), 6188–6191.
26. Brown, A. D.; Bunnage, M. E.; Glossop, P. A.; James, K.; Jones, R.; Lane, C. A. L.; Lewthwaite, R. A.; Mantell, S.; Perros-Huguet, C.; Price, D. A.; Trevethick, M.; Webster, R. The discovery of adamantyl-derived, inhaled, long acting β_2-adrenoreceptor agonists. *Bioorg. Med. Chem. Lett.* **2008,** *18* (4), 1280–1283.
27. Procopiou, P. A.; Barrett, V. J.; Bevan, N. J.; Biggadike, K.; Butchers, P. R.; Coe, D. M.; Conroy, R.; Edney, D. D.; Field, R. N.; Ford, A. J.; Guntrip, S. B.; Looker, B. E.; McLay, I. M.; Monteith, M. J.; Morrison, V. S.; Mutch, P. J.; Stephen, A.; Richards, S. A.; Sasse, R.; Smith, C. E. Synthesis and structure-activity relationships of long-acting β_2 adrenergic receptor agonists incorporating arylsulfonamide groups. *J. Med. Chem.* **2009,** *52* (8), 2280–2288.
28. Rossi, A.; Polese, G. Indacaterol: a comprehensive review. *Int. J. Chronic Obstruct. Pulm. Dis.* **2013,** *8,* 353–363.

29. Russell, R.; Anzueto, A.; Weisman, I. Optimizing management of chronic obstructive pulmonary disease in the upcoming decade. *Int. J. Chronic Obstruct. Pulm. Dis.* **2011,** *6,* 47–61.
30. Moen, M. D. Indacaterol in chronic obstructive pulmonary disease. *Drugs* **2010,** *70* (17), 2269–2280.
31. Trave, R. The chloromethylation of hydroxyacetophenones. II. *Gazz. Chim. Ital.* **1951,** *81,* 773–781.
32. Lunts, L. H. C.; Toon, P. 4-Hydroxy-$\alpha\gamma$-(aminomethyl)-m-xylene-α',α^3-diols as stimulators, US 3644353 (1972).
33. Collin, D. T. α^1-[tert-Butylamino)methyl]-4-hydroxy-m-xylene-α^1,α^3-diol, US 3642896 (1972).
34. Lunts, L. H. C.; Toon, P. 1-Phenyl-2-aminoethanol derivatives as bronchodilators, US 3705233 (1972).
35. Collin, D. T.; Hartley, D.; Jack, D.; Lunts, L. H. C.; Press, J. C.; Ritchie, A. C.; Toon, P. Saligenin analogs of sympathomimetic catechol amines. *J. Med. Chem.* **1970,** *13* (4), 674–680.
36. Skachilova, S. Y.; Zueva, E. F.; Muravskaya, I. D.; Goncharenko, L. V.; Smirnov, L. D. Procedures for preparing salbutamol (a review). *Khim.-Farm. Zh.* **1991,** *25* (10), 59–65.
37. Battram, C.; Charlton, S. J.; Cuenoud, B.; Dowling, M. R.; Fairhurst, R. A.; Farr, D.; Fozard, J. R.; Leighton-Davies, J. R.; Lewis, C. A.; McEvoy, L.; Turner, R. J.; Trifilieff, A. In vitro and in vivo pharmacological characterization of 5-[(R)-2-(5,6-diethyl-indan-2-ylamino)-1-hydroxy-ethyl]-8-hydroxy-1H-quinolin-2-one (indacaterol), a novel inhaled β_2 adrenoceptor agonist with a 24 h duration of action. *J. Pharmacol. Exp. Ther.* **2006,** *317,* 762–770.
38. Ahrens, R. C.; Smith, G. D. Albuterol: an adrenergic agent for use in the treatment of asthma. *Pharmacology, pharmacokinetics and clinical use, Pharmacotherapy* **1984,** *4* (3), 105–121.
39. Brittain, R. T.; Harris, D. M. Albuterol. *Pharmacol. Biochem. Prop. Drug Subst.* **1977,** *1,* 257–276.
40. Colice, G. L. Albuterol HFA for the management of obstructive airway disease. *Expert Rev. Respir. Med.* **2008,** *2* (2), 149–159.
41. Skidmore, I. F.; Lunts, L. H. C.; Finch, H.; Naylor, A. Phenethanolamine derivatives useful in the treatment of respiratory problems, DE 3414752 (1984).
42. Ariza Aranda, J.; Serra, M. J.; Monserrat, V. C. New process and intermediates for the preparation of α1-[6-(4-phenylbutoxy)hexylaminomethyl]-4-hydroxy-1,3-benzenedimethanol [salmeterol], ES 2065269 (1995).
43. Hong, H.; Ma, J.; Li, J.; Huang, J.; Zhang, L. Process for the preparation of salmeterol xinafoate from 2-(bromoethyl)benzene, CN 103420856 (2013).
44. Dwivedi, S. D.; Shah, N. S. Process for preparation of salmeterol, WO 2012032546 (2012).
45. Rong, Y.; Ruoho, A. E. A new synthetic approach to salmeterol. *Synth. Commun.* **1999,** *29* (12), 2155–2162.
46. Hett, R.; Stare, R.; Helquist, P. Enantioselective synthesis of salmeterol via asymmetric borane reduction. *Tetrahedron Lett.* **1994,** *35* (50), 9375–9378.
47. Guo, Z.-L.; Deng, Y.-Q.; Zhong, S.; Lu, G. Enantioselective synthesis of (R)- salmeterol employing an asymmetric Henry reaction as the key step. *Tetrahedron: Asymmetry* **2011,** *22* (13), 1395–1399.
48. Liu, J.; Zhou, D.; Jia, X.; Huang, L.; Li, X.; Chan, A. S. C. A convenient synthesis of (R)-salmeterol via Rh-catalyzed asymmetric transfer hydrogenation. *Tetrahedron: Asymmetry* **2008,** *19* (15), 1824–1828.

49. Procopiou, P. A.; Barrett, V. J.; Bevan, N. J.; Biggadike, K.; Box, P. C.; Butchers, P. R.; Coe, D. M.; Conroy, R.; Emmons, A.; Ford, A. J.; Holmes, D. S.; Horsley, H.; Kerr, F.; Li-Kwai-Cheung, A.-M.; Looker, B. E.; Mann, I. S.; McLay, I. M.; Morrison, V. S.; Mutch, P. J.; Smith, C. E.; Tomlin, P. Synthesis and structure-activity relationships of long-acting β_2 adrenergic receptor agonists incorporating metabolic inactivation: an antedrug approach. *J. Med. Chem.* **2010,** *53* (11), 4522–4530.
50. Molinski, T. F.; Stanley, S. D. Improved synthesis of 13C, 2H3- and 2H3-salmeterol by Cs2CO_3-mediated monoalkylation of a primary amine. *J. Labelled Compd. Radiopharm.* **2002,** *45* (9), 755–762.
51. Johnson, M. Salmeterol. *Med. Res. Rev.* **1995,** *15* (3), 225–257.
52. Jarvis, B.; Markham, A. Inhaled salmeterol: a review of its efficacy in chronic obstructive pulmonary disease. *Drugs Aging* **2001,** *18* (6), 441–472.
53. Johnson, M. The pharmacology of salmeterol. *Lung* **1990,** *168* (Suppl.), 115–119.
54. Hochhaus, G.; Buchwald, A.; Schulz, M. Salmeterol, a long acting β_2-agonist. *Pharmz. Zeit.* **1988,** *143* (27), 2318–2324.
55. Adkins, J.; McTavish, D. Salmeterol: a review of its pharmacological properties and clinical efficacy in the management of children with asthma. *Drugs* **1997,** *54* (2), 331–354.
56. Meyer, J. M.; Wenzel, C. L.; Kradjan, W. A. Salmeterol: a novel, long-acting beta2-agonist. *Ann. Pharmacother.* **1993,** *27* (12), 1478–1487.
57. Murase, K.; Mase, T.; Ida, H.; Takahashi, K.; Murakami, M. Absolute configurations of four isomers of 3-formamido-4-hydroxy-α-[[N-(p-methoxy-α-methylphenethyl)amino]methyl] benzyl alcohol, a potent β-adrenoreceptor stimulant. *Chem. Pharm. Bull.* **1978,** *26* (4), 1123–1129.
58. Murase, K.; Mase, T.; Ida, H.; Takahashi, K.; Murakami, M. New β-adrenoreceptor stimulants. Studies on 3-acylamino-4-hydroxy-α-(N-substituted aminomethyl)benzyl alcohols. *Chem. Pharm. Bull.* **1977,** *25* (6), 1368–1377.
59. Murakami, M.; Takahashi, K.; Mase, T.; Murase, K.; Ida, H. α-(Aminomethyl)benzyl alcohol derivatives, DE 2305092 (1973).
60. Trofast, J.; Oesterberg, K.; Kaellstroem, B. L.; Waldeck, B. Steric aspects of agonism and antagonism at β-adrenoceptors: synthesis and pharmacological experiments with the enantiomers of formoterol and their diastereomers. *Chirality* **1991,** *3* (6), 443–450.
61. Hett, R.; Fang, Q. K.; Gao, Y.; Hong, Y.; Butler, H. T.; Nie, X.; Wald, S. A. Enantioselective and diastereoselective synthesis of all four stereoisomers of formoterol. *Tetrahedron Lett.* **1997,** *38* (7), 1125–1128.
62. Hett, R.; Fang, Q. K.; Gao, Y.; Wald, S. A.; Senanayake, C. H. Large-scale synthesis of enantio- and diastereomerically pure (R,R)-formoterol. *Org. Process Res. Dev.* **1998,** *2* (2), 96–99.
63. Huang, L.; Liu, J.; Shan, W.; Liu, B.; Shi, A.; Li, X. The asymmetric synthesis of (R,R)-formoterol via transfer hydrogenation with polyethylene glycol bound Rh catalyst in PEG2000 and water. *Chirality* **2010,** *22* (2), 206–211.
64. Fan, W.; Chen, L.; Hai, L.; Wu, Y. New method in synthesizing an optical active intermediate for (R,R)-formoterol. *Chin. Chem. Lett.* **2008,** *19* (3), 279–280.
65. Mereyala, H. B.; Sambaru, K. Synthesis of N-[2-benzyloxy-5-(2-oxiranyl)phenyl]formamide. *Formal synthesis of formoterol, Indian J. Chem., Sect. B: Org. Chem. Incl. Med. Chem.* **2005,** *44B* (1), 167–169.
66. Wilkinson, H. S.; Tanoury, G. J.; Wald, S. A.; Senanayake, C. H. Diethylanilineborane: a practical, safe, and consistent-quality borane source for the large-scale enantioselective reduction of a ketone intermediate in the synthesis of (R,R)-formoterol. *Org. Process Res. Dev.* **2002,** *6* (2), 146–148.

67. Campos, F.; Bosch, M. P.; Guerrero, A. An efficient enantioselective synthesis of (R,R)-formoterol, a potent bronchodilator, using lipases. *Tetrahedron: Asymmetry* **2000,** *11* (13), 2705–2717.
68. Anderson, G. P. Formoterol: pharmacology, molecular basis of agonism, and mechanism of long duration of a highly potent and selective β_2-adrenoceptor agonist bronchodilator. *Life Sci.* **1993,** *52* (26), 2145–2160.
69. Cheer, S. M.; Scott, L. J. Formoterol: a review of its use in chronic obstructive pulmonary disease. *Am. J. Respir. Med.* **2002,** *1* (4), 285–300.
70. Friedman, M.; Della, C. G.; Kottakis, J. Formoterol therapy for chronic obstructive pulmonary disease: a review of the literature. *Pharmacotherapy* **2002,** *22* (9), 1129–1139.
71. Berger, W. E. The use of inhaled formoterol in the treatment of asthma. *Ann. Allergy, Asthma, Immunol.* **2006,** *97* (1), 24–33.
72. Steiropoulos, P.; Tzouvelekis, A.; Bouros, D. Formoterol in the management of chronic obstructive pulmonary disease. *Int. J. Chronic Obstruct. Pulm. Dis.* **2008,** *3* (2), 205–215.
73. Rodrigo, G. J.; Neffen, H.; Colodenco, F. D.; Castro-Rodriguez, J. A. Formoterol for acute asthma in the emergency department: a systematic review with met-analysis. *Ann. Allergy, Asthma, Immunol.* **2010,** *104* (3), 247–252.
74. Sears, M. R.; Ottosson, A.; Radner, F.; Suissa, S. Long-acting β-agonists: a review of formoterol safety data from asthma clinical trials. *Eur. Respir. J.* **2009,** *33* (1), 21–32.
75. Rubins, J. B. Formoterol fumarate inhalation solution (Perforomist) for COPD. *Expert Rev. Clin. Immunol.* **2008,** *4* (4), 415–423.
76. Antoniu, S. A. Formoterol as a rescue medication for asthma. *Expert Opin. Pharmacother.* **2006,** *7* (17), 2439–2441.
77. Waldeck, B. β-Adrenoceptor agonists and asthma—100 years of development. *Eur. J. Pharmacol.* **2003,** *445(1* (2)), 1–12.
78. Cazzola, M.; Page, C. P.; Rogliani, P.; Matera, M. G. β_2-agonist therapy in lung diseases. *Am. J. Respir. Crit. Care Med.* **2013,** *187* (7), 690–696.
79. Lee, A. M.; Jacoby, D. B.; Fryer, A. D. Selective muscarinic receptor antagonists for airway diseases. *Curr. Opin. Pharmacol.* **2001,** *1* (3), 223–229.
80. Banholzer, R.; Bauer, R.; Reichl, R. Preparation of anticholinergic scopine, (nor)tropine, and granatoline esters of thienylcarboxylic acids and their quaternary salts, EP 418716 (1991).
81. Bilgic, M. Methods for the synthesis of tiotropium bromide from scopine, US 20130030182 (2013).
82. Soukup, M. Manufacturing process for tiotropium bromide from oxalic acid derivative, WO 2013050929 (2013).
83. Soukup. Milan manufacturing process for tiotropium bromide, US 20120123125 (2012).
84. Cipollone, A.; Fattori, D.; Fincham, C. I. Process for the preparation of scopine esters, WO 2013046138 (2013).
85. Koumis, T.; Samuel, S. Tiotropium bromide: a new long-acting bronchodilator for the treatment of chronic obstructive pulmonary disease. *Clin. Ther.* **2005,** *27* (4), 377–392.
86. Gross, N. J. Tiotropium bromide. *Chest* **2004,** *126* (6), 1946–1953.
87. Keam, S. J.; Keating, G. M. Tiotropium bromide: a review of its use as maintenance therapy in patients with COPD. *Treat. Respir. Med.* **2004,** *3* (4), 247–268.
88. Mundy, C.; Kirkpatrick, P. Fresh from the pipeline: tiotropium bromide. *Nat. Rev. Drug Discovery* **2004,** *3* (8), 643–644.
89. ZuWallack, A. R.; ZuWallack, R. L. Tiotropium bromide, a new, once-daily inhaled anticholinergic bronchodilator for chronic-obstructive pulmonary disease. *Expert Opin. Pharmacother.* **2004,** *5* (8), 1827–1835.

90. Hvizdos, K. M.; Goa, K. L. Tiotropium bromide. *Drugs* **2002,** *62* (8), 1195–1203.
91. Barnes, P. J. Tiotropium bromide. *Expert Opin. Invest. Drugs* **2001,** *10* (4), 733–740.
92. Barnes, P. J. The pharmacological properties of tiotropium. *Chest* **2000,** *117* (2 Suppl.), 63S-66S.
93. Keating, G. M. Tiotropium bromide inhalation powder: a review of its use in the management of chronic obstructive pulmonary disease. *Drugs* **2012,** *72* (2), 273–300.
94. Bollmeier, S. G.; Lee, S.-Y. The emerging role of tiotropium for patients with asthma. *Ann. Pharmacother.* **2013,** *47* (5), 704–713.
95. Yohannes, A. M.; Connolly, M. J.; Hanania, N. A. Ten years of tiotropium: clinical impact and patient perspectives. *Int. J. Chronic Obstruct. Pulm. Dis.* **2013,** *8,* 117–125.
96. Bateman, E. D.; Rennard, S.; Barnes, P. J.; Dicpinigaitis, P. V.; Gosens, R.; Gross, N. J.; Nadel, J. A.; Pfeifer, M.; Racke, K.; Rabe, K. F.; Rubin, B. K.; Welte, T.; Wessler, I. Alternative mechanisms for tiotropium. *Pulm. Pharmacol. Ther.* **2009,** *22* (6), 533–542.
97. Mamary, A. J.; Criner, G. J. Tiotropium bromide for chronic obstructive pulmonary disease. *Expert Rev. Respir. Med.* **2009,** *3* (3), 211–220.
98. Oba, Y.; Zaza, T.; Thameem, D. M. Safety, tolerability and risk benefit analysis of tiotropium in COPD. *Int. J. Chronic Obstruct. Pulm. Dis.* **2008,** *3* (4), 575–584.
99. Rodrigo, G. J.; Nannini, L. J. Tiotropium for the treatment of stable chronic obstructive pulmonary disease: a systematic review with meta-analysis. *Pulm. Pharmacol. Ther.* **2007,** *20* (5), 495–502.
100. Lipson, David A. Tiotropium bromide. *Int. J. Chronic Obstruct. Pulm. Dis.* **2006,** *1* (2), 107–114.
101. Somand, H.; Remington, T. L. Tiotropium: a bronchodilator for chronic obstructive pulmonary disease. *Ann. Pharmacother.* **2005,** *39* (9), 1467–1475.
102. Barnes, P. J. Theophylline: new perspectives for an old drug. *Am. J. Respir. Crit. Care Med.* **2003,** *167,* 813–818.
103. Salter, H. On some points in the treatment and clinical history of asthma. *Edinburgh Med. J.* **1859,** *4,* 1109–1115.
104. Page, C. P. Doxofylline: a novofylline. *Pulm. Pharmacol. Ther.* **2010,** *23* (4), 231–234.
105. Barnes, P. J. Theophylline. *Am. J. Respir. Crit. Care Med.* **2013,** *188* (8), 901–906.
106. Cazzola, M.; Rogliani, P.; Novelli, L.; Matera, M. G. Inhaled corticosteroids for chronic obstructive pulmonary disease. *Expert Opin. Pharmacother.* **2013,** *14* (18), 2489–2499.
107. Raissy, H. H.; Kelly, H. W.; Harkins, M.; Szefler, S. J. Inhaled corticosteroids in lung disease. *Am. J. Respir. Crit. Care Med.* **2013,** *187* (8), 798–803.
108. Bernstein, S.; Lenhard, R. H.; Allen, W. S.; Heller, M.; Littell, R.; Stolar, S. M.; Feldman, L. I.; Blank, R. H. 16-Hydroxylated steroids. IV. The synthesis of the 16α-hydroxy derivatives of 9α-halo steroids. *J. Am. Chem. Soc.* **1956,** *78,* 5693–5694.
109. Bernstein, S.; Lenhard, R. H.; Allen, W. S.; Heller, M.; Littell, R.; Stolar, S. M.; Feldman, L. I.; Blank, R. H. 16-Hydroxylated steroids. VI. The synthesis of the 16α-hydroxy derivatives of 9α-substituted steroids. *J. Am. Chem. Soc.* **1959,** *81,* 1689–1696.
110. Allen, G. R., Jr.; Marx, M.; Weiss, M. J. 16,17-Dihydroxy steroids, US 3021347 (1962).
111. Florey, K. Triamcinolone. *Anal. Profiles Drug Subst.* **1972,** *1,* 367–396.
112. Sieh, D. H. Triamcinolone diacetate. *Anal. Profiles Drug Subst.* **1982,** *11,* 651–661.
113. Doggrell, S. A. Triamcinolone: new and old indications. *Expert Opin. Pharmacother.* **2001,** *2* (7), 1177–1186.
114. Braga, P. C. Mucus pharmacology. *Respiration* **1991,** *58* (Suppl. 1), 47–51.
115. Yuta, A.; Baraniuk, J. N. Therapeutic approaches to mucus hypersecretion. *Curr. Allergy Asthma Rep.* **2005,** *5* (3), 243–251.

116. Rogers, D. F. Mucoactive agents for airway mucus hypersecretory diseases. *Respir. Care* **2007,** *52* (9), 1176–1193.
117. Balsamo, R.; Lanata, L.; Egan, C. G. Mucoactive drugs. *Eur. Respir. Rev.* **2010,** *19* (116), 127–133.
118. Allegra, L. Mucoactive drugs. In *Therapeutic Strategies in COPD;* Cazzola, M., Ed.; CRC Press, 2005; pp 247–277.
119. Miyata, T.; Kai, H.; Isohama, Y.; Takahama, K. Current opinion of muco-active drug, research: strategies and problems. *Eur. Respir. J.* **1998,** *11* (2), 480–491.

Chapter 24

Antithrombotic Drugs (Anticoagulants, Antiplatelets, and Thrombolytics)

Thromboembolic disorders are one of the major disorders that causes morbidities and mortalities worldwide. (More than 1 million deaths occur each year in the United States alone.) Blood clots that form in the circulatory system often break off and travel to other areas of the system, causing major organ damage or death. Heart attacks and strokes usually are caused by these "emboli." Pathologic clotting or bleeding usually is the result of disbalance of hemostasis, which is a complex series of physiologic processes based on interactions among the blood vessels and supporting tissues, endothelial cells, platelets, fibrinolytic system, plasma coagulation proteins, and protease inhibitors that maintain the fluidity of the blood, stop bleeding when injury to a vessel occurs, and confine blood to the vascular spaces.

Thrombosis is the process of formation of thrombus in the circulation from constituents of flowing blood. At the same time hemostatic plugs are the blood clots formed in healthy individuals at the site of bleeding that stop the escape of blood and plasma, whereas thrombi developing in the unruptured blood vessels may be harmful.

Three primary events predispose to thrombus formation: alterations in or injury to the vascular endothelium, activation of the coagulation system, and/or reduction in flow (Virchow's triad) [1].

Under normal circumstances, platelets circulate in an inactive form and do not significantly interact with the vessel wall. But atherosclerotic plaque rupture during vascular injury, during percutaneous coronary interventions, and during acute coronary syndromes facilitates platelet adhesion and activation.

Following adhesion, platelets form a monolayer at the site of vessel wall injury and undergo activation, resulting in morphologic changes.

Blood clots can develop in the arterial or venous circulation. The clots are composed of polymerized fibrin, platelets, trapped neutrophils, and red blood cells. Fibrin is the major component of clots that form in veins, and platelets are the major component of clots that form in arteries where they can cause heart attacks and strokes by blocking the flow of blood in the heart and brain, respectively, although fibrin also plays an important role in arterial thrombosis.

Synthesis of Best-Seller Drugs. http://dx.doi.org/10.1016/B978-0-12-411492-0.00024-9

Arterial thrombi are commonly formed after endothelial injury, whereas venous thrombus is formed by blood stasis in veins.

By knowing the nature of thrombus, effective therapy can be chosen; that is, arterial thrombus should be treated by antiplatelet agents and venous thrombosis should be treated by anticoagulants mainly. These drugs prevent thrombus extension, recurrence and embolic complications, but they do not dissolve the already formed clot. For the lysis of already formed thrombus fibrinolytic drugs are used.

Antithrombotic drugs in general are divided into three types, which are based on a somewhat outmoded model of coagulation cascade but promulgated by "Big Pharma":

- Anticoagulant drugs whose primary target of action are enzymes such as thrombin (an enzyme of the blood plasma that catalyzes the conversion of fibrinogen to fibrin, the last step of the blood clotting process) or factor Xa (an enzyme in coagulation cascade). These include vitamin K antagonists and heparin (unfractionated, low-molecular-weight heparins, heparin mimetics, warfarin, and its analogues).
- Antiplatelet drugs, which act on platelet receptors. The main function of platelets, sometimes called "thrombocytes," is to contribute the process of stopping bleeding at the site of interruption receptors. These include aspirin, compounds of ticlopidine and clopidogrel series, and dipyridamole.
- Fibrinolytic drugs, which act on fibrinogen, a plasma glycoprotein that is converted into fibrin during clot formation. These include streptokinase, urokinase, alteplase, reteplase, tenecteplase, and tissue plasminogen activator [t-PA], among others.

The history of the founding antithrombotic agents—aspirin, heparin, and warfarin—is an interesting odyssey of vision, observation, and serendipity of smart personalities [2,3].

The introduction of heparin (**24.1.1**) for the treatment of thromboembolic disorders in the 1930s was a major breakthrough, and provided the first widely available anticoagulant agent [4]. Another drug of historical importance, introduced in 1940, is the first oral anticoagulant agent, warfarin (**24.1.2**), synthesized and named after the Wisconsin Alumni Research Foundation [5]. Warfarin and heparin have been in wide use for 60 and 80 years, respectively.

The efficacy of aspirin (**24.1.3**) for prevention of thrombotic events in cardiovascular patients is well established, demonstrating as early as the 1950s a reduction in vascular death of approximately 15% and a further reduction in nonfatal vascular events of 30%. Aspirin is widely used for the prevention of recurrent stroke in patients with transient ischemic attack (TIA) and ischemic stroke of arterial origin. Although the benefit of aspirin is undisputed, it is also known that aspirin is associated with the risk of bleeding. That is why most guidelines recommend using the lowest aspirin dose [6,7].

24.1 ANTICOAGULANT DRUGS

Anticoagulant drugs inhibit clot formation by blocking the action of clotting factors of blood, are used for a variety of conditions, including the avoidance of ischemic stroke in patients with atrial fibrillation, attenuation of thrombin generation in acute coronary syndromes, and prevention of thromboembolism and thrombus extension in cases of venous thrombosis or pulmonary embolism.

Anticoagulants generally are divided into two types: heparin, which is given by intravenous or subcutaneously injection because it is poorly absorbed by the intestine and is used primarily in hospitalized patients; and orally administered coumarin derivatives, such as warfarin, which is the most widely prescribed oral anticoagulant drug.

Heparin (**24.1.1**) is a key anticoagulant and has been used in medicine for more than 80 years. Heparin is used to prevent blood clotting during and after surgery and to treat various disorders in which there is an increased risk of blood clot formation. Discovered in 1920s, heparin represents a naturally occurring mixture of mucopolysaccharides that are present in liver and lung tissues. Because of its anticlotting effect, the drug creates a significant risk of excessive bleeding, which may be reversed with protamine. Another adverse effect of heparin is a reduction in the number of circulating platelets (thrombocytopenia).

Legend says that in 1916, Jay McLean, a second-year medical student at Johns Hopkins University in Baltimore, who was working under the guidance of William H. Howell, isolated from mammalian tissues fractions that inhibited blood coagulation. McLean left a beaker of cat's blood with extracted "antithrombin" on Howell's desk and asked the professor to call him when it clotted. Howell never called McLean, but extracted "antithrombin" himself and named it heparin. After some initial skepticism, the discovery was recognized as a possible remedy to treat coagulation disorders, and by the 1920s several groups started manufacturing heparin by extracting it from porcine intestines. Today, pharmaceutical heparins are produced in tons quantities and details of industrial protocol are tightly guarded.

Heparin is an animal tissue extract that is a highly acidic mucopolysaccharide, is polydisperse in nature, and which belongs to the glycosaminoglycan family. The molecular weight ranges from 6 to 20kDa.

Heparin consists of a disaccharide repeating unit of either iduronic acid or glucuronic acid and glucosamine residues, each capable of carrying sulfate groups. The locations of sulfate groups in iduronic acid and glucuronic acid dictate the anticoagulant activity of heparin.

Generalizing, heparin is possible to describe as a linear polymer, which belongs to the glycosaminoglycan family, and represents linear polymers consisting of repeating units of 1→4 linked pyranosyluronic acid and 2-amino-2-deoxyglucopyranose (glucosamine) residues and exists primarily as a helical structure. The uronic acid residues typically consist of 90% L- iduronic acid and 10% D-glucuronic acid. Heparin contains a high proportion of sulfo-groups, making it one of the strongest acids in nature. Simplifying schematically, heparin is possible to represent as a structure composed of a major (75 to 95%) trisulfated disaccharide repeating unit. The uronic acid can be modified by sulfation at the 2-O position, while the glucosamine residue can be unmodified, N-sulfated or N-acetylated, and can contain variable patterns of O-sulfation at the 3–O and 6–O positions [8-11] (Fig. 24.1.).

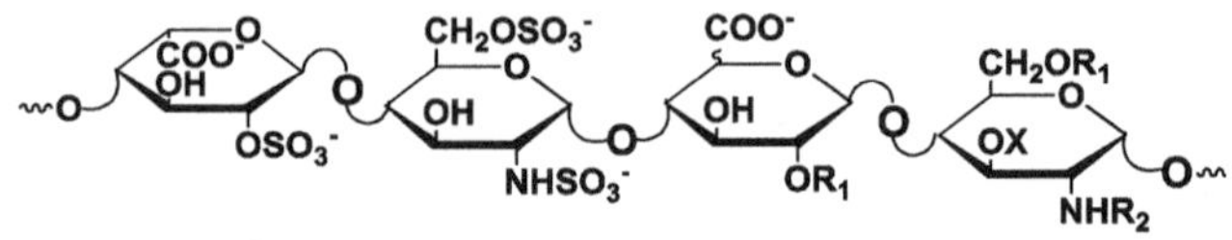

Heparin 24.1.1

R_1 = H, SO_3^-; R_2 = H, SO_3^-, Ac

FIG. 24.1 Generalized structure of heparin.

The typically used unfractionated heparin has an average molecular weight of 14 to 18kDa.

Heparin and the structurally related heparan sulfate participate in numerous important biological processes, such as blood anticoagulation, pathogen infection, cell differentiation, growth, migration, and inflammation. Heparin acts promoting the activity of antithrombin in blood plasma, which is, in turn, a protein that inactivates clotting promotor enzyme thrombin.

Heparin has been used as an anticoagulant drug since 1935 for the treatment and prevention of deep vein thrombosis during surgery, blood transfusion, and renal dialysis [12-19]. It produces a powerful anticoagulant effect via a high-affinity binding to antithrombin, thus inactivating thrombin and activated factor X (factor Xa). But heparin can cause bleeding episodes, even several weeks after its implementation. It can also cause allergic reactions, such as nausea, vomiting, sweating, hives, itching, and trouble breathing [20].

Another danger of heparin use is the heparin-induced thrombocytopenia, an immune response to heparin, the most common symptom of which is enlargement or extension of a previously diagnosed blood clot, and the development of a new blood clot elsewhere in the body that can progress to severe thrombosis, amputation, and, in some cases, death [21-23].

Another problem with heparin concerns contamination issues on its industrial production. In 2007, for example, more than 100 deaths and hundreds of adverse clinical effect cases concerning contamination were reported [24].

A variable dose–response relationship, which is a result of its structural heterogeneity, dictates a critical need for developing a safer substitute for heparin that has more predictable bioactivity and reduced side effects.

Different forms of heparin-based anticoagulants have been developed as alternatives to animal source heparin. One of the goals of these investigators was to create new anticoagulants that target a single coagulation factor and have a predictable dose–response relationship [25-34].

Low-molecular-weight heparins (LMWHs) were prepared by the cleaving of heparin using different depolymerization techniques and methods, yielding fragments with an average molecular weight of 4.5 to 6.0 kDa. Obtained preparations are widely used now as anticoagulants in a range of different clinical indications.

Among the new heparin-based anticoagulants that have entered into widespread clinical use are enoxaparin (prepared by alkaline depolymerization of heparin), ardeparin and Parnaparin (by oxidative depolymerization with Cu^{2+} and H_2O_2), Certoparin and sandoparin (by deaminative cleavage with isoamyl nitrite), dalteparin, reviparin, and nadroparin (by deaminative cleavage with nitrous acid), miniparin and bioparin (cleavage methods not disclosed), and fondaparinux and idraparinux (synthetic LMWHs). These are compounds with increased bioavailability and a reduced risk of heparin-induced thrombocytopenia.

Enoxaparin (Lovenox) and fondaparinux (Arixtra) are included in the list of Top 200 Drugs by sales for the 2010s.

Enoxaparin–Lovenox

Enoxaparin sodium (**24.1.2**) is obtained by alkaline depolymerization of the heparin benzyl ester derived from porcine intestinal mucosa. Its structure is characterized by a 2-O-sulfo-4-enepyranosuronic acid group at the nonreducing end and a 2-N,6-O-disulfo-D-glucosamine at the reducing end of the chain.

The process for preparation of LMWH in general is as follows: (a) transformation of heparin raw material to a quaternary ammonium salt by reacting with appropriate quaternary ammonium salt, particularly with benzethonium chloride, to generate heparin quaternary ammonium salt; (b) reaction of the carboxy groups of synthesized quaternary ammonium salt with benzyl halide to generate heparin benzyl ester; (c) hydrolysis of the obtained heparin benzyl ester under alkali conditions, usually NaOH, to obtain LMWH segment; (d) separation and purification of obtained product enoxaparin [35].

Approximately 20% of the enoxaparin structure contains a 1,6-anhydro derivative on the reducing end of the polysaccharide chain. The drug substance is the sodium salt (Fig. 24.2.) The average molecular weight is approximately

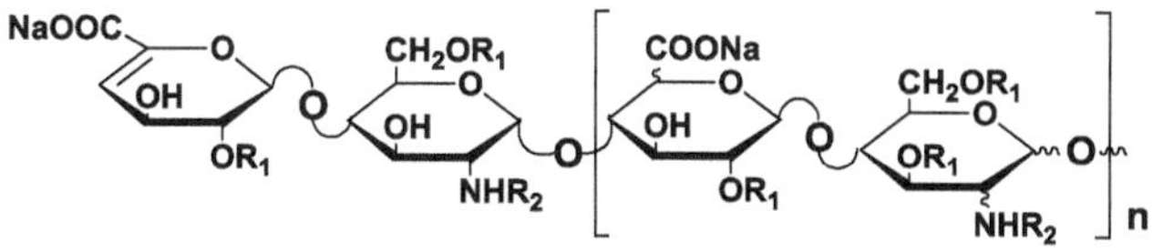

Enoxaparin 24.1.2

R_1 = H or SO_3Na, R_2 = SO_3Na, $COOCH_3$; n = 1-20

FIG. 24.2 Generalized structure of enoxaparin.

4.5 kDa. Other methods of preparation of enoxaparin, such as enzymatic [36] or chemoenzymatic synthesis [37], are proposed.

Enoxaparin has become the treatment of choice for various thromboembolic diseases, as well as to prevent clotting in the extracorporeal circulation during hemodialysis. Enoxaparin does not appear to be associated with an increased bleeding risk and can be used without the need for monitoring and adjustment of regimens. Empirical dose adjustment and biological monitoring seem to be necessary along with therapeutic doses. Enoxaparin also possesses antiproliferative properties, and reduces neointimal proliferation following vascular injury [38-43].

Fondaparinux–Arixtra

Understanding of the anticoagulant action of heparin led to the conclusion that short heparin epitopes may offer an attractive alternative to natural heparins and resulted in creation of synthetic mimetics. In 2001, fondaparinux, a synthetic and selective ultra-LMWH inhibitor of factor Xa, was registered in the United States and Europe as a new antithrombotic drug under the name Arixtra. The developed route for its synthesis consists of more than 50 steps with 0.1% overall yield [44-46]. Schematically, it could be described as follows (Scheme 24.1.). Three blocks for coupling—two disaccharide (**24.1.3**) and (**24.1.4**) and one monosaccharide (**24.1.5**)—were pr_epared separately by known carbohydrate chemistry methods and protocols. The later steps are described below.

The glycosyl bromide (**24.1.3**) and O-methyl α-L-idopyranosyluronate-α-D-glucopyranoside (**24.1.4**) were coupled using silver triflate as a catalyst and 2,4,6-collidine as an acid acceptor. The chloroacetyl group in the isolated the tetrasaccharide derivative (**24.1.6**) was then removed, implementing a known protocol with hydrazinecarbodithioic acid in 2,4,6-collidine-HOAc mixture to produce (**24.1.7**). Glycosylation of the (**24.1.7**) with (**24.1.5**) produced the pentasaccharide derivative (**24.1.8**). After hydrolysis of the acetyl and methoxy carbonyl groups with NaOH in H_2O/methanol/chloroform media, compound was obtained (**24.1.9**). O-sulfation of (**24.1.9**) in N,N-dimethylformamide, using SO_3 in the presence of Et_3N, produced the sulfated product (**24.1.10**), which was hydrogenated on Pd/C catalyst to debenzylate and to hydrogenate the azido group to an amino group simultaneously, which allowed synthesizing of the compound (**24.1.11**). Amino groups in the obtained compound were sulfated using pyridine/SO_3 complex to produce the desired fondaparinux (**24.1.12**).

The commercial drug fondaparinux is a methylated derivative of pentasaccharide, bearing short, well-defined heparin epitopes. Fondaparinux differs from heparin in that it is a specific anti–factor Xa agent, and does not have many of the pharmacological properties of the polycomponent drug heparin. Fondaparinux is now synthesized in kilogram scale and it is possible to suggest that further development of the synthetic methods [47-50] or a chemoenzymatic approach will result in cost-effective products, accelerating the modernization of LMWH therapeutics.

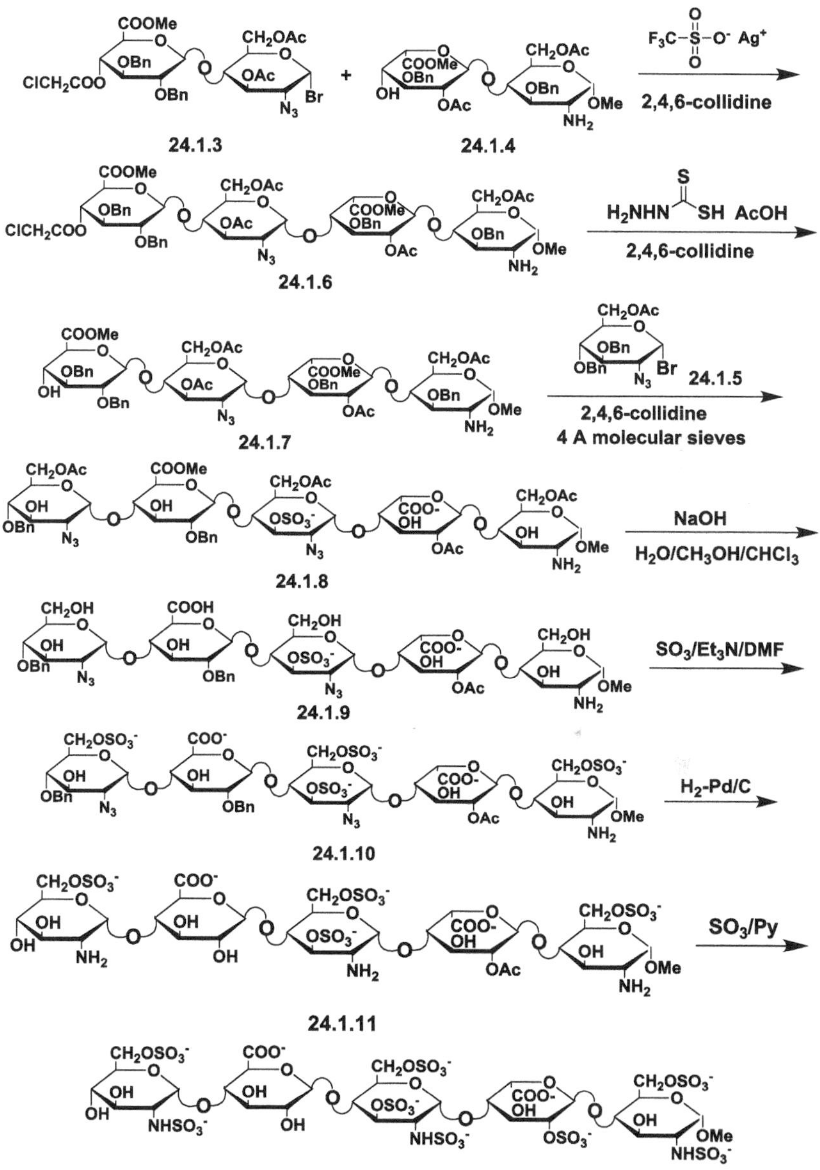

Fondaparinux 24.1.12

SCHEME 24.1 Synthesis of fondaparinux.

Fondaparinux is indicated for prophylaxis of deep vein thrombosis (DVT) in patients undergoing some types of surgery (including extended prophylaxis), treatment of DVT or acute pulmonary embolism. It reduces the risk of ischemic events, substantially reduces major bleeding and improves long-term mortality and morbidity [51-62].

Idraparinux (**24.1.13**) is a novel pentasaccharide that is under development. It is an analogue of fondaparinux in which the N-sulfates are replaced by O-sulfates and the hydroxyl groups are methylated was synthesized through a 39-step synthesis starting from D-glucose and methyl α-D-glucopyranoside. Up to now, there is no consensus on whether they are better than other anticoagulation methods for long-term thromboembolic events [63] (Fig. 24.3.).

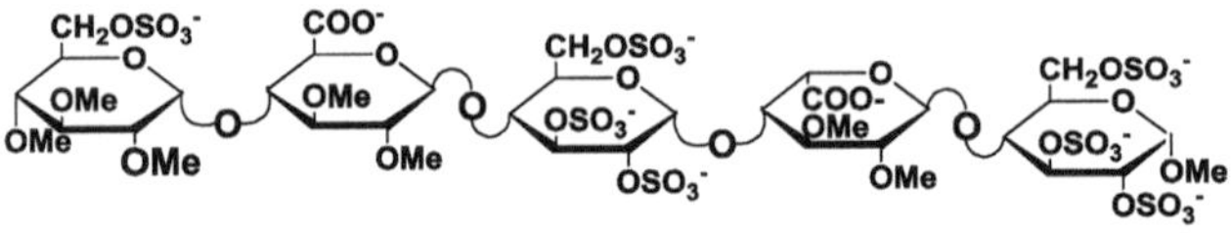

Idraparinux 24.1.13

FIG. 24.3 **Generalized structure of idraparinux.**

The third generation of synthetic pentasaccharides is represented by SSR-126517E whose structure is not yet disclosed.

Thus there are two type of drugs—heparin (**24.1.1**) and its congeners and warfarin (**24.1.15**) and its congeners—that form the mainstay in the prophylaxis of DVT, stroke prevention in atrial fibrillation, and treatment of thromboembolic disease.

Heparin and its modified analogues are parenteral drugs and act immediately.

The second type of anticoagulants, warfarin and its modifications, are orally taken drugs that require several hours for the onset of the anticoagulant effect and have longer duration of action. Hemorrhage is the principal toxic effect during oral anticoagulant therapy.

Warfarin

Warfarin (**24.1.15**) is another drug of historical importance that was discovered in the Wisconsin laboratory of Dr. Karl Paul Link when, in 1933, a farmer brought him a milk container full of "blood completely destitute of clotting capacity" from cattle with hemorrhagic "sweet clover disease," which happened after farmers imported sweet clover plants from Europe to America and Canada. This case led to the discovery of the first oral anticoagulant, dicoumarol (**24.1.14**), in 1941. In 1948, the coumarin derivative, warfarin (**24.1.15**), received its name from Wisconsin Alumni Research Foundation (WARF) and was thereafter promoted as rodent poison.

The structure of dicoumarol (**24.1.14**) became the prototype of the derivative anticoagulant class of drugs of the 4-hydroxycoumarin series. Dicoumarol itself, for a short time, was employed as an anticoagulant drug, but has been replaced by warfarin.

Coumaric oral anticoagulants today include warfarin (**24.1.15**), acenocoumarol (**24.1.16**), and phenprocoumon (**24.1.17**) (Fig. 24.4.). They have constituted the standard worldwide oral anticoagulant treatment for thromboembolic disorders. Despite their undisputable effectiveness, coumaric oral anticoagulants have a narrow therapeutic window and are associated with a high risk of major bleeding.

Recently, another coumarin compound, ethyl biscoumacetate (**24.1.18**), was investigated experimentally and clinically, but it showed no great advantages over the known triad.

The second-generation drugs of the 4-hydroxycoumarin series have larger lipid-soluble substituents at the 3-position, which increases their half-life in the body. This generates the idea of creation of rodenticides (rat poisons) (**24.1.19** to **24.1.24**) (Fig. 24.4.).

The oral anticoagulants warfarin, acenocoumarol, and phenprocoumon are universally accepted in medicine and work via inhibition of the synthesis of vitamin K-dependent γ-carboxylation of clotting factors, which include factors II, VII, IX, and X, and the anticoagulant proteins C and S.

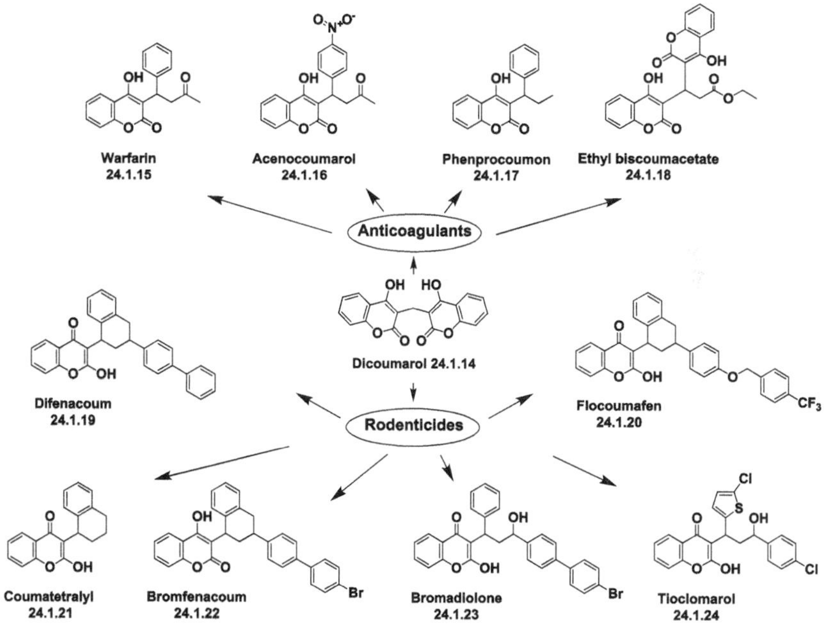

FIG. 24.4 Coumaric oral anticoagulants and rodenticides.

Despite the fact that warfarin has been used as an anticoagulant for many years, the safety profile for this drug has been poor. Because of its narrow therapeutic index there is greater than 10-fold interindividual variability in the dose required to attain a therapeutic response. Inappropriate dosing continues to contribute to significant problems. Warfarin may cause severe bleeding, especially in the stomach, esophagus, intestines, urinary tract or bladder, and lungs, that can be life-threatening and even cause death. Moreover, it can cause high blood pressure;

heart attack; angina; pericarditis and endocarditis; stroke; and aneurysm. In addition, it can cause cancer, chronic diarrhea, kidney, or liver diseases [64-74].

That is why there is a greatest unmet medical need in new anticoagulant drugs to replace warfarin in long-term treatment cases. Most of the attention has focused on the development of oral agents that target thrombin or factor Xa and there is no evidence that one target is any better than the other.

Thrombin Inhibitors

Thrombin is a logical target for new anticoagulants. Thrombin is the most potent platelet agonist; it converts fibrinogen to fibrin and also amplifies its own production by feedback activation of factors V and VIII, key cofactors for the prothrombinase and intrinsic tenase complex, respectively. In addition, as the most potent platelet agonist, thrombin coordinates the process of platelet activation and aggregation with coagulation. Because of its multiple roles in coagulation, thrombin inhibition can not only block fibrin formation but it also attenuates further thrombin formation and platelet activation and may become a novel class of anticoagulants [75-81].

The naturally occurring thrombin inhibitor is the 65-amino-acid polypeptide hirudin, which is obtained from medicinal leeches.

Historically, the first synthetic direct thrombin inhibitor, for which clinical trials were completed and which was found to be comparable to warfarin, was ximelagatran (**24.1.25**); however, it was not approved for medicinal use because of its hepatotoxicity.

Dabigatran (**24.1.26**) is the only commercially available direct thrombin inhibitor. Inhibition of thrombin attenuates formation of fibrin, reduces thrombin generation, and may limit platelet aggregation [82-89].

Atecegatran (AZD-0837) (**24.1.27**), tecarfarin (ATI-5923) (**24.1.28**), and odiparcil (**24.1.29**) are novel thrombin inhibitors that are under development.

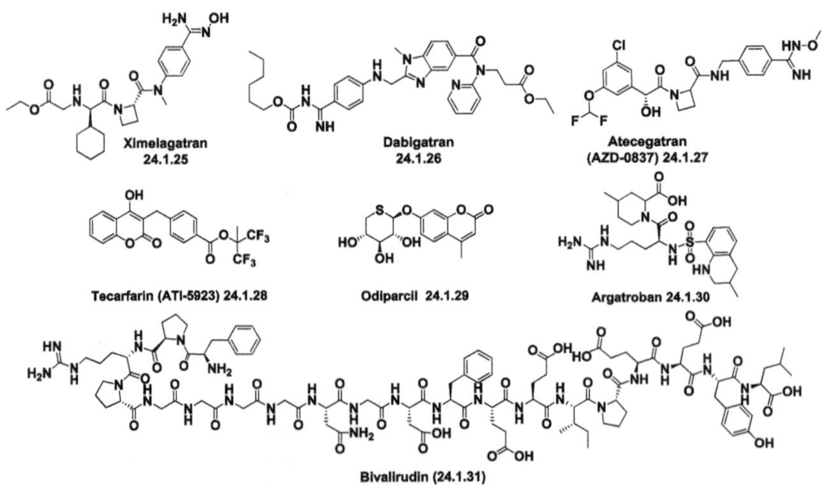

FIG. 24.5 **Thrombin inhibitors.**

The parenteral direct thrombin inhibitors argatroban (**24.1.30**), a synthetic 20-amino-acid peptide, and bivalirudin (**24.1.31**) are approved for use in cases of unstable angina and heparin-induced thrombocytopenia, respectively (Fig. 24.5.)

Factor Xa Inhibitors

Factor Xa plays a critical role in coagulation by regulating thrombin generation. Regulation of thrombin generation via inhibition of factor Xa is a highly significant advance [90-95].

Natural direct factor Xa inhibitors usually are represented by the 119-amino-acid peptide antistasin from leeches and the anticoagulant 60-amino-acid peptide TAP from ticks.

Synthetic antithrombotic agents that directly inhibit factor Xa have been synthesized. By directly binding to its active site, these compounds lead to interruption of the intrinsic and extrinsic coagulation cascade pathways and thus to inhibition of thrombin formation and thrombus development. These new compounds present a low variability in pharmacokinetic and pharmacodynamic profiles, which means that dose adjustments and biological monitoring of hemostasis are not required.

New antithrombotic direct factor Xa inhibitors ("xabans"), such as rivaroxaban (**24.1.32**), apixaban (**24.1.33**), TAK-442 (**24.1.34**), YM-466 (**24.1.35**), darexaban (**24.1.36**), eribaxaban (**24.1.37**), betrixaban (**24.1.38**), otamixaban (**24.1.39**), LY517717 (**24.1.40**), and edoxaban (**24.1.41**), are being developed in an attempt to overcome current therapeutic limitations (Fig. 24.6.). Among

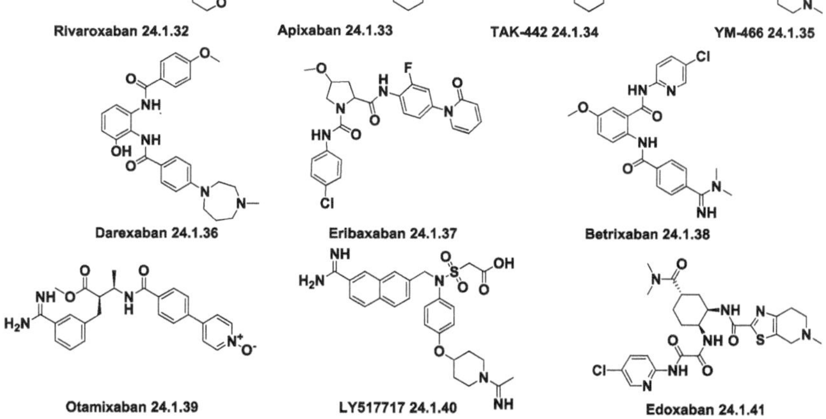

FIG. 24.6 Factor Xa inhibitors ("xabans").

these compounds, rivaroxaban (**24.1.32**) and apixaban (**24.1.33**), are undergoing Phase III trials.

Several of these drugs have reached various stages of clinical development and have shown promising results. They have the potential to either replace or act as alternatives to traditional anticoagulants. Rivaroxaban has been approved in many countries to reduce the risk of stroke in patients with atrial fibrillation and the treatment and prevention of venous thromboembolism.

Dual Thrombin/Factor Xa Inhibitors

Thrombosis models predict that inhibition of both thrombin and factor Xa simultaneously could produce an additive, even a synergistic, antithrombotic effect. The first compounds to demonstrate dual inhibitor properties were BIBM1015 (**24.1.42**) and tanogitran (BIBT986) (**24.1.43**) which belong to the methylbenzimidazole series (Fig. 24.7.).

Tanogitran achieved clinical trials but no further reports appeared in literature.

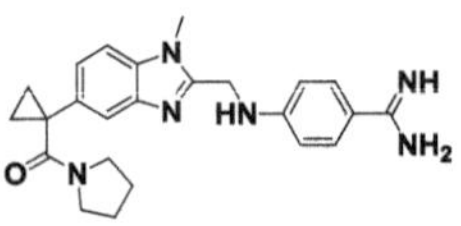

BIBM1015 24.1.42

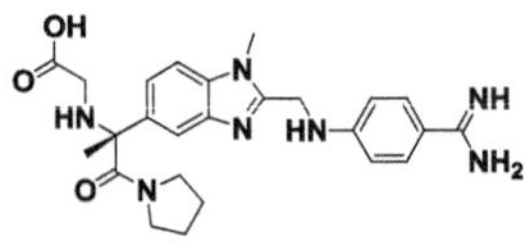

Tanogitran (BIBT986) 24.1.43

FIG. 24.7 **Dual thrombin/factor Xa inhibitors.**

An unbelievable amount of literature has been published about anticoagulant drugs. Some of latest reviews are found in the references [96-115].

24.2 ANTIPLATELET DRUGS

Platelets are major players in arterial thrombosis, and antiplatelet therapy is most important for the treatment and prevention of cardiovascular events. The clinical manifestations of these disorders include coronary heart disease, stroke, and peripheral vascular diseases, and antiplatelet drugs are the cornerstone of treatment of cardiovascular diseases.

There are strategies for inhibiting platelet functions: inhibition of cyclooxygenase, which also inhibits platelet aggregation with aspirin;, blockade of the P2Y12 adenosine diphosphate (ADP) receptor antagonists with clopidogrel or prasugrel; and blockade of the glycoprotein IIb/IIIa receptor (GPIIb/IIIa) with abciximab, eptifibatide, and tirofiban. These compounds are well-known prototypes of antiplatelet drugs.

Antiplatelet drugs also can be classified on the basis of the site of their action: drugs inhibiting platelet activation, drugs inhibiting platelet aggregation, and drugs inhibiting platelet adhesion.

Pharmacology, mechanisms of action, strategies of implementation, and uncertainties about antiplatelet drugs that are intended to prevent and/or reverse platelet aggregation in arterial thrombosis are summarized in several excellent books and reviews [116-128].

Drugs Inhibiting Platelet Activation

In general, there are three platelet activation pathways that can be triggered: thromboxane A_2 (TXA_2) synthesis and/or activation of the TXA_2 receptor, adenosine diphosphate P2Y12 receptor activation and thrombin activation of the protease activated receptor-1 (PAR-1). Accordingly, there will exist corresponding blockers or antagonists to inhibit these platelet activation pathways.

Inhibitors of the TXA_2 Pathway

Platelets possess two receptors for ADP, P2Y1 and P2Y12. The P2Y1 receptor is one of many platelet receptors coupled to Gq and initiates ADP-induced activation. The P2Y12 receptor plays a special role in the amplification of platelet activation initiated by numerous pathways. Platelet activation leads to a range of responses that play a critical role in arterial thrombosis and the inflammatory responses. P2Y12 receptor antagonists have dramatic inhibitory effects on platelet function regardless of the activating stimuli. This phenomenon, makes the receptor an ideal target for pharmaceutical therapy [129].

The single known drug of this series is aspirin (**24.2.1**), the most commonly used drug in antiplatelet therapy, which selectively inhibits synthesis of TXA_2-potent platelet aggregator selectively acetylating cyclooxygenase-1.

Among investigational TXA_2 pathway inhibitors are picotamide (**24.2.2**), which is a dual inhibitor of both TXA_2 receptors and TXA_2 synthase, terutroban (**24.2.3**), and selective inhibitors of the TXA_2 receptor on platelets, Z-335 (**24.2.4**) and BM-573 (**24.2.5**) (Fig. 24.8.).

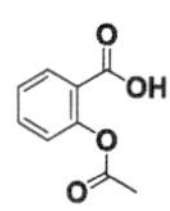

Aspirin 24.2.1

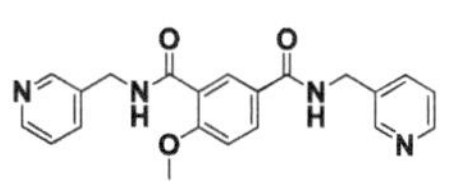

Picotamide 24.2.2

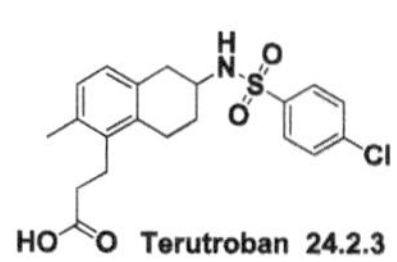

Terutroban 24.2.3

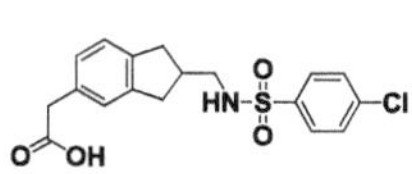

Z-335 24.2.4

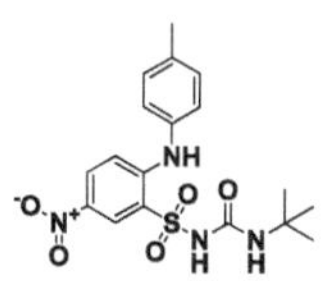

BM-573 24.2.5

FIG. 24.8 Inhibitors of the TXA_2 pathway.

Inhibitors of the Adenosine Diphosphate P2Y12 Receptor Pathway

The group of drugs that inhibit ADP receptors on platelet cell membranes involved in platelet aggregation are called P2Y12 receptor blockers.

They could be classified into three types:

- Irreversibly binding, orally active prodrugs (derivatives of thienopyridines) such as ticlopidine (**24.2.6**), formally a first-generation thienopyridine; clopidogrel (**24.2.7**), a second-generation compound; and prasugrel (**24.2.8**), which is representative of third-generation thienopyridines (Fig. 24.9.).

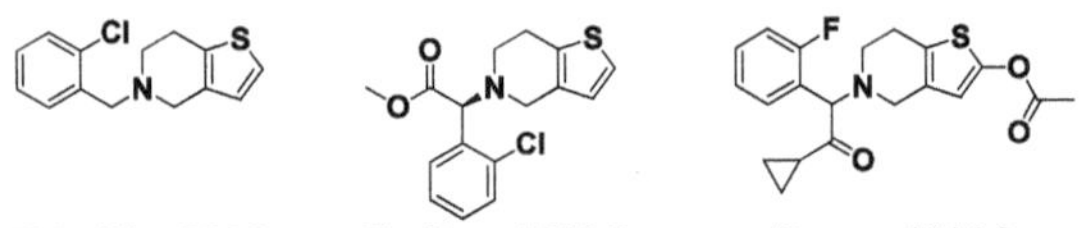

Ticlopidine 24.2.6 Clopidogrel 24.2.7 Prasugrel 24.2.8

FIG. 24.9 Irreversibly binding inhibitors of the adenosine diphosphate P2Y12 receptor pathway.

- Irreversibly binding, orally active drug ticagrelor (**24.2.9**), which is chemically distinct from the thienopyridine structure. It is an allosteric antagonist and has a binding site different from ADP, making the blockage reversible (Fig. 24.10.).

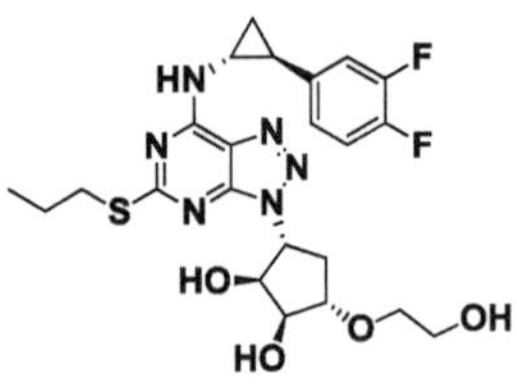

Ticagrelor 24.2.9

FIG. 24.10 Structure of ticagrelor.

- Reversibly binding, intravenously administered drugs that chemically represent a modified adenosine triphosphate derivative stable enough for enzymatic degradation. Cangrelor (**24.2.10**), which is in Phase III trials, is an example. It causes almost complete inhibition of ADP-induced platelet aggregation. Another novel experimental direct-acting and reversible platelet P2Y12 receptor blocker compound is elinogrel (**24.2.11**), which is chemically distinct from the above mentioned drugs (Fig. 24.11.).

Ticlopidine was the first approved thienopyridine antiplatelet drug, but it is rarely used now because of the risk of neutropenia and the availability of clopidogrel, which is included in the list of Top 200 Drugs by sales for the 2010s.

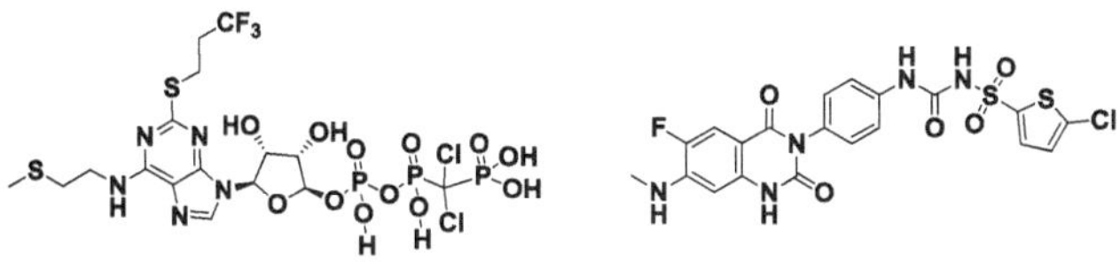

FIG. 24.11 Reversibly binding inhibitors of the adenosine diphosphate P2Y12 receptor pathway.

Clopidogrel, which has a similar mechanism of action but a safer pharmacological profile, is the widely used medication in cases of cerebrovascular disease, and in combination with aspirin is implemented after angioplasty and stenting.

Clopidogrel–Plavix

Various methods and strategies have been reported for the synthesis of clopidogrel. Product is usually obtained as a racemate and needs to be resolved through fractional crystallization with an appropriate resolving agent, such as tartaric or camphorsulfonic acid. Only the S-enantiomer is suitable for pharmaceutical use, as the R-enantiomer shows no antithrombotic activity and causes convulsions in animal experiments.

The first synthesis of clopidogrel [130] was carried out via rarely used intramolecular Mannich or Pictet–Spengler type reaction–annulation of arylethylamine after imine formation with formaldehyde. The role of β-arylethylamine moiety in the reaction played methyl (S)-2-(2-chlorophenyl)-2-((2-(thiophen-2-yl)ethyl)amino)acetate (**24.2.19**) obtained on reaction of methyl (S)-2-(2-chlorophenyl)glycinate (**24.2.15a**) and 2-(thiophen-2-yl)ethyl benzenesulfonate (**24.2.18**).

Methyl (S)-2-(2-chlorophenyl)glycinate (**24.2.15a**) was synthesized by Strecker condensation of 2-chlorobenzaldehyde (**24.2.12**) with KCN and $(NH_4)_2CO_3$ in water-ethanol to produce hydantoin (**24.2.13**), which was hydrolyzed in alkaline conditions to produce 2-(2-chlorophenyl)glycine (**24.2.15**). Its conversion to methyl (S)-2-(2-chlorophenyl)glycinate (**24.2.15a**) was carried out in two ways. First, transformation of the obtained racemic amino acid to methyl ester by passing HCl gas through its methanolic solution and further separation of (S)-enantiomer (**24.2.15a**) by crystallization of its salt formed with (+)-tartaric acid and further workup with base. Second, separation of the (S)-enantiomer of amino acid (**24.2.14a**) using the (+)-camphorsulfonic acid resolution method and then esterification to (**24.2.15a**) under the same conditions (methanol/HCl gas). The second reagent, 2-(thiophen-2-yl)ethyl benzenesulfonate (**24.2.18**), was prepared starting from thiophene (**24.2.16**), which was α-methylated using butyl lithium and the obtained Li derivative, upon reaction with ethyleneoxide and subsequent hydrolysis produced

2-thiophene ethanol (**24.2.17**), which was converted to benzenesulfonyl ester with benzenesulfonyl chloride in pyridine to produce the desired compound 2-(thiophen-2-yl)ethyl benzenesulfonate (**24.2.18**). Nucleophilic displacement reactions of benzenesulfonate (**24.2.18**) in ethylacetate in presence of sodium bicarbonate with methyl (S)-2-(2-chlorophenyl)glycinate (**24.2.15a**) produced the compound (**24.2.19**), which was cyclized to the desired clopidogrel (**24.2.7**) via the above-mentioned intramolecular Mannich or Pictet–Spengler type reaction (Scheme 24.2.).

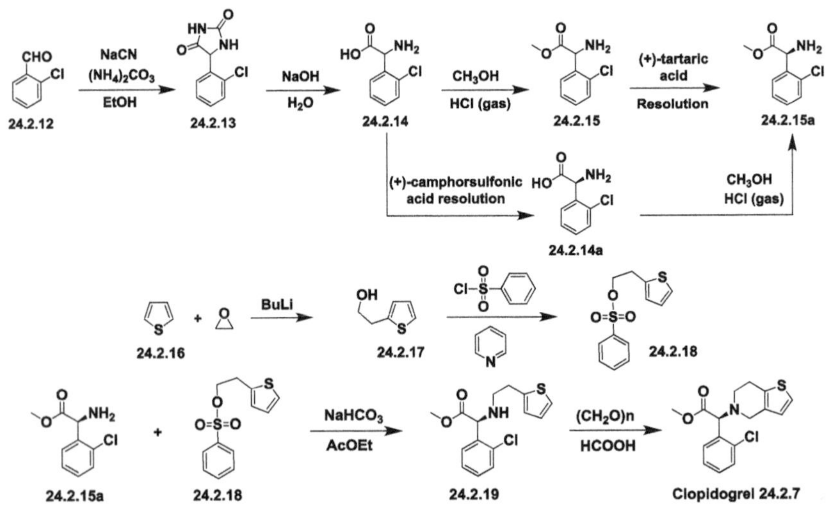

SCHEME 24.2 Synthesis of clopidogrel.

Another approach [131], one can say general approach, is that the methyl 2-bromo-2-(2-chlorophenyl)acetate (**24.2.24**) or other derivatives of methyl 2-halogen-2-(2-chlorophenyl)acetates were used for alkylation of 4,5,6,7-tetrahydrothieno[3,2-c] pyridine (**24.2.23**), which was prepared using the same above described Pictet-Spengler reaction applied to β-thienylethylamine (**24.2.23**), which was condensed with HCHO followed by cyclization of the imines in 20% HCl to tetrahydrothienopyridine in high yield.

β-Thienylethylamine (**24.2.22**) in turn was prepared via sequence of Vilsmeier's, Darzens' and Leuckart's reactions (**24.2.16** → **24.2.22**) by using thiophene (**24.2.16**) as raw material. Racemic clopidogrel (**24.2.25**) was resolved through fractional crystallization with camphorsulfonic acid. This general approach with peculiar nuances [132-138], including asymmetric synthesis [139-141], were implemented in different patents and papers (Scheme 24.3.).

Clopidogrel selectively and irreversibly inhibits ADP-induced platelet aggregation and reduces ischemic complications in a wide range of patients with coronary artery disease [142-157]. It is used alone or with aspirin to prevent serious or life-threatening problems with the heart and blood vessels in

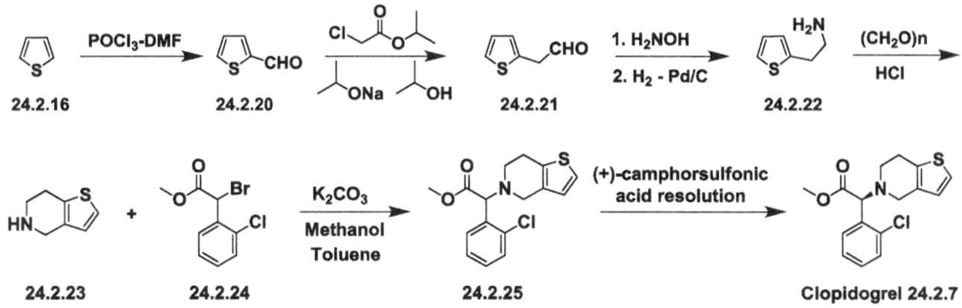

SCHEME 24.3 Synthesis of clopidogrel.

people who have had a stroke, heart attack, or severe chest pain. In combination with aspirin, clopidogrel is the current "gold standard" for reducing cardiovascular events.

Clopidogrel may cause side effects such as excessive tiredness, headache, dizziness, nausea and vomiting, stomach pain, and diarrhea.

Recent data suggest higher rates of adverse cardiovascular events in patients on proton pump inhibitors and clopidogrel. Coadministration of these drugs significantly decreases the effect of clopidogrel on platelet aggregation, probably because goth drugs are metabolized by similar pathways of the cytochrome P450 system and proton pump inhibitors may inhibit the metabolism of clopidogrel to its active metabolite.

Inhibitors of Protease Activated Receptor-1

PAR-1, which belongs to the family of G-protein–coupled receptors, is expressed on the surface of a wide variety of cells. PAR-1 is the major thrombin receptor on human platelets, and antagonists of PAR-1 have been developed as novel antiplatelet agents for the prevention of arterial thrombosis. Recently, a novel class of antiplatelet agents able to inhibit thrombin-mediated platelet activation was developed. The most advanced drugs are vorapaxar (**24.2.26**) and atopaxar (**24.2.27**), which are experimental compounds proposed for treatment of acute coronary syndrome chest pain caused by coronary artery disease [158,159] (Fig. 24.12.).

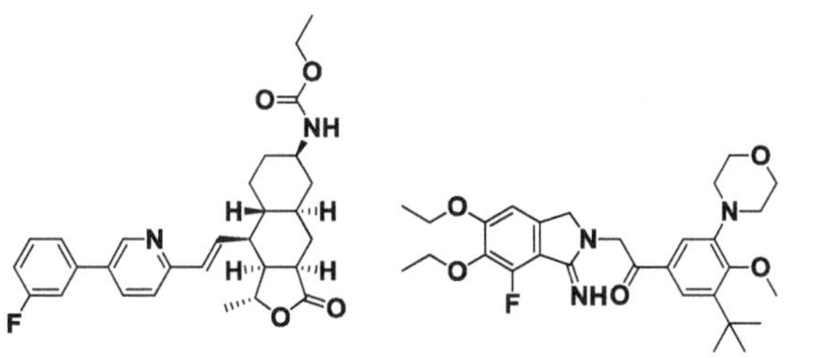

FIG. 24.12 Structure of experimental compounds vorapaxar and atopaxar.

Phosphodiesterase Inhibitors

Platelets express three phosphodiesterase (PDE) isoenzymes, 2, 3, and 5, which regulate the levels of 3,5′-cyclic adenosine monophosphate and 3′,5′-cyclic guanosine monophosphate, which, in turn, serve as intracellular signals to suppress platelet activation and subsequent aggregation. Modulation of these secondary messengers could be fundamental in regulating platelet activation and thrombosis.

Drugs, such as cilostazol (**24.2.28**) and dipyridamole (**24.2.29**) increase the levels of cyclic adenosine monophosphate by inhibiting PDE (Fig. 24.13.).

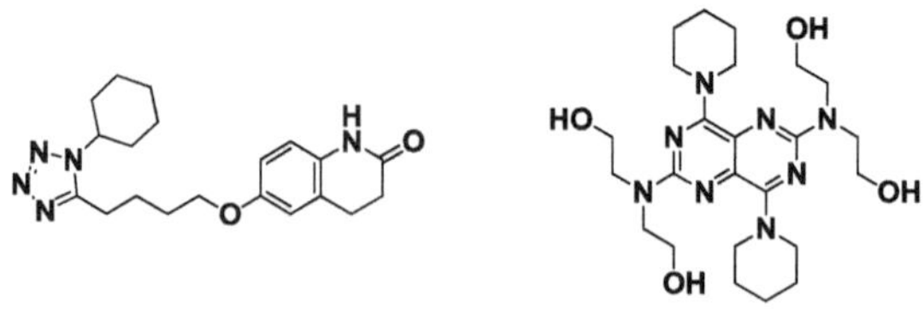

Cilostazol 24.2.28 **Dipyridamole 24.2.29**

FIG. 24.13 **Structure of cilostazol and dipyridamole.**

Although cilostazol is a safe and effective drug, it does not show superiority when compared with standard aspirin and clopidogrel treatment. Action of cilostazol sometimes accompanied with such side effects as headaches, gastrointestinal symptoms, and skin rash, which often lead to its discontinuation.

Dipyridamole, like cilostazol, inhibits cyclic nucleotide PDE and blocks adenosine uptake, which results in both antiplatelet and vasodilator properties. The antithrombotic activity of dipyridamole via inhibition of platelets was initially discovered in in vivo experiments about 50 years ago. Now it is clear that in addition to antiplatelet effects, dipyridamole also possesses beneficial vasculature properties, including direct and indirect effects on the endothelium, such as inhibition of proliferation of venous and arterial smooth muscle cell, antioxidant, and antiinflammatory properties, as well as their subsequent effect on cell signaling [160-165]. Dipyridamole is approved for stroke prevention and is included in the list of Top 200 Drugs by sales for the 2010s.

Dipyridamole–Aggrenox

Dipyridamole (**24.2.29**) was synthesized starting from 6-methylpyrimidine-2,4-diol (**24.2.31**), which, in turn, was prepared by treating ethyl acetoacetate with excess urea and Ac_2O to prepare intermediate ethyl-3-ureidobut-2-enoate (**24.2.30**) and further cyclization of formed intermediate with NaOH to (**24.2.31**). Obtained product was simultaneously oxidized and nitrated on dissolving in concentric H_2SO_4 and the careful adding of HNO_3 at room temp prepared 2,6-dihydroxy-5-nitropyrimidine-4-carboxylic acid (**24.2.32**). The nitro group of the last was reduced to an amino group by sodium dithionite

($Na_2S_2O_4$) in methanol or with hydrogen on Pd/C catalyst, producing the corresponding 5-amino-2,6-dihydroxypyrimidine-4-carboxylic acid (**24.2.33**). The obtained product was cyclized into pyrimido[5,4-d]pyrimidine-2,4,6,8-tetraol (**24.2.34**) by heating with urea or sodium or potassium cyanate at high temperatures. Refluxing the last with the PCl_5 and $POCl_3$ mixture, a tetrachloroderivative (**24.2.35**) was prepared. Reacting perchloropyrimido[5,4-d]pyrimidine (**24.2.35**) and piperidine at room temperature, 2,6-dichloro-4,8-diaminopyrimidopyrimidine (**24.2.36**) was synthesized. The prepared dichlorodiamino compound was heated for 1 hour with diethanolamine and a catalytic amount of $CuSO_4$ bomb-tube to produce the desired dipyridamole (**24.2.29**) [166,167] (Scheme 24.4.).

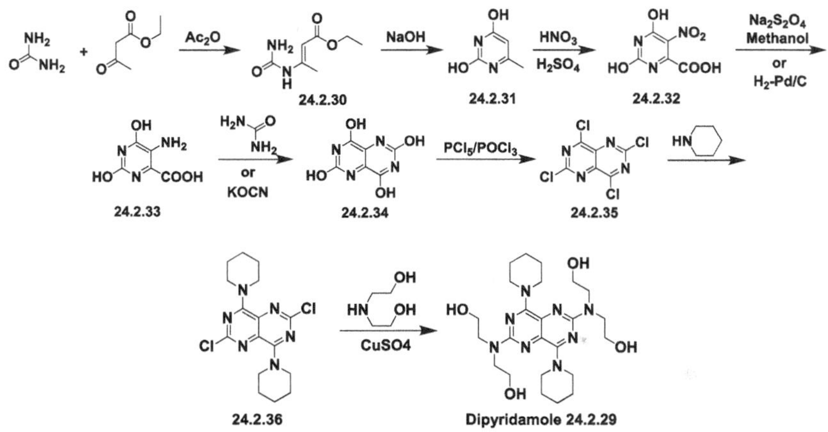

SCHEME 24.4 Synthesis of dipyridamole.

Some improvements of the process for the preparation of dipyridamole have been proposed later [168-171].

Inhibitors of Platelet Aggregation

Activation of receptor function of platelet membrane glycoprotein (GPIIb/IIIa) leads to the binding of fibrinogen and is the final common pathway to platelet aggregation.

Glycoprotein IIb/IIIa (GPIIb/IIIa) receptor antagonists inhibit the binding of fibrinogen to the platelet GPIIb/IIIa receptor, which prevents platelet-to-platelet aggregation [127,172,173].

Abciximab, eptifibatide, and tirofiban are three commercially available GPIIb/IIIa inhibitors of platelet aggregation intended for intravenous administration.

Abciximab (ReoPro) is a recombinant monoclonal antibody with unique properties that binds to the platelet GPIIb/IIIa receptor with high affinity for weeks and indicated for use in cases of percutaneous coronary intervention [174].

Eptifibatide (Integrilin) (**24.2.37**) is a truncated derivative of the naturally occurring rattlesnake venom protein known as barbourin. It is a cyclic heptapeptide containing 6 amino acids and mercaptopropionyl residue derived from snake venom protein. It is used to reduce the risk of acute cardiac ischemic events only in hospitalized patients in surgery or nonsurgery medical treatment because of the possible side effects [175].

Tirofiban (**24.2.38**) is a sulfonamide derivative of phenylpropanoic acid that selectively and reversibly blocks the GPIIb/IIIa receptor. Tirofiban indicated to reduce the cases of thrombotic cardiovascular events. It prevents the blood from clotting during heart attack, or while the patient is undergoing a procedure to treat a blocked coronary artery [176] (Fig. 24.14.).

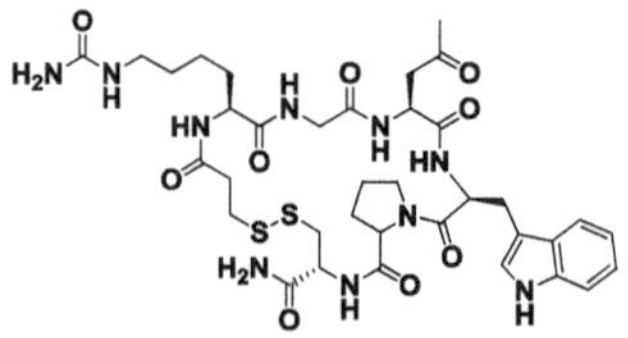

Eptifibatide 24.2.37

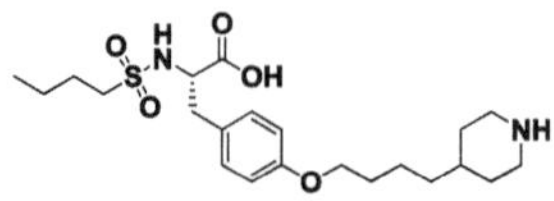

Tirofiban 24.2.38

FIG. 24.14 **Structures of eptifibatide and tirofiban used in medicine.**

Structures of (GPIIb/IIIa) receptor antagonists in development: sibrafiban (**24.2.39**), lamifiban (**24.2.40**), lotrafiban (**24.2.41**), xemilofiban (**24.2.42**), roxifiban (**24.2.43**), orbofiban (**24.2.44**), and lefradafiban (**24.2.45**) are presented on Fig. 24.15.

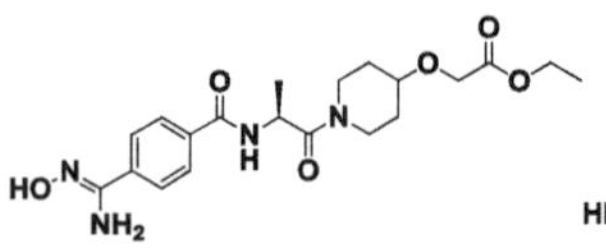

Sibrafiban 24.2.39

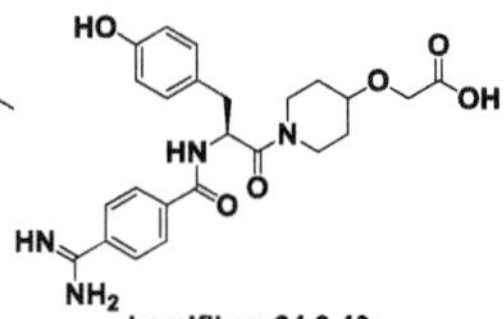

Lamifiban 24.2.40

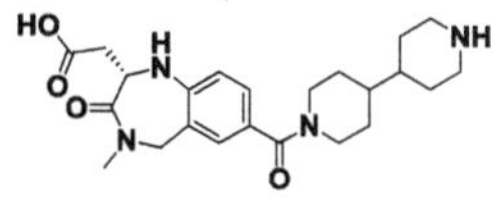

Lotrafiban 24.2.41

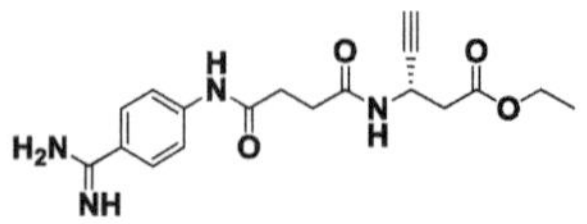

Xemilofiban 24.2.42

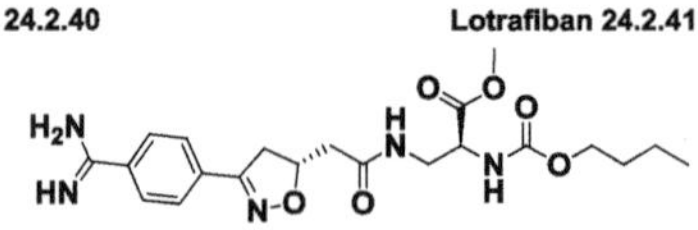

Roxifiban 24.2.43

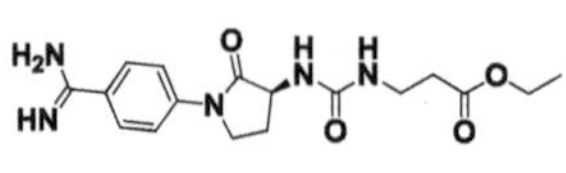

Orbofiban 24.2.44

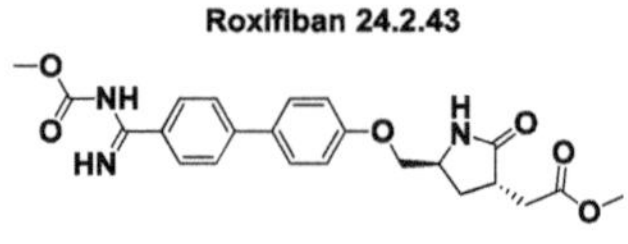

Lefradafiban 24.2.45

FIG. 24.15 **Structure of new, experimental platelet aggregation inhibitors.**

Inhibitors of Platelet Adhesion

Platelet adhesion is a highly coordinated process achieved by surface receptors, protein ligands, and matrix proteins operating at the platelet–subendothelium interface. This process can be inhibited by collagen–platelet interaction and include humanized monoclonal antibodies such as AJvW-2 and AJvW-200, 6B4, JAQ1, OM2, oligonucleotide aptamers such as R9.32, R9.14, ARC17792 (aptamers-single-stranded nucleic acids with unique three-dimensional structures), and pegnivacogin, proteins derived from the medicinal leech, and small molecules. None has reached Phase III [177].

24.3 FIBRINOLYTIC DRUGS

Thrombosis is a major cause of death worldwide. Thrombosis occurs when there is an imbalance between prothrombotic and antithrombotic mechanisms.

Fibrinolysis is a process that prevents growing of blood clots and is classified for two types: primary and secondary fibrinolysis. The primary type is a regular body process, whereas secondary fibrinolysis is the breakdown of clots caused by some medical disorder or some other reason.

Blood clots can occur in any vascular bed, and their appearance in coronary, cerebral, or pulmonary vessels can be life-threatening, causing myocardial infarctions, cerebrovascular strokes, and respiratory and cardiac failure.

Thrombolytic drugs are those used to dissolve (lyse) blood clots (thrombi).

Thrombolytic drugs dissolve blood clots by stimulating activation of plasminogen, which forms a cleaved product called plasmin. Plasmin is a proteolytic enzyme that is capable of breaking crosslinks between fibrin molecules, which provide the structural integrity of blood clots. Because of these actions, thrombolytic drugs are also called "fibrinolytic drugs."

Medical intervention with fibrinolytic drugs such as tissue plasminogen activator and streptokinase is the principal treatment for life-treating thromboembolic disorders [178,179].

There are three major classes of fibrinolytic drugs: tissue plasminogen activators (tPAs), streptokinase, and urokinase. While drugs in these three classes all have the ability to effectively dissolve blood clots, they differ in their detailed mechanisms in ways that alter their selectivity for fibrin clots.

tPA is a family of thrombolytic drugs used in cases of acute myocardial infarction, cerebrovascular thrombotic stroke, and pulmonary embolism, and include:

- Alteplase (Activase) (**24.3.1**) which is a recombinant form of human tPA.
- Alteplase is the most commonly administered thrombolytic agent. It is a 527-amino-acid tPA produced by recombinant DNA technology from a human melanoma cell line. However, because of the high cost of production and other considerations, the details of its production are not available [180,181,184,185].
- Reteplase (Retavase) (**24.3.2**) is a genetically engineered, smaller derivative of recombinant tPA that has increased potency and is faster acting than recombinant tPA [186].
- Tenecteplase (**24.3.3**) in a genetically engineered variant of alteplase [187].

The following drugs mimic endogenous tPA.

- Streptokinase and anistreplase are isolated and purified from streptococci bacteria production. Streptokinase (**24.3.4**) is one of the leading and one of the most-often employed, particularly of its lower cost when coded to others fibrinolytic agent widely used in thromboembolic conditions [182,183,188,189].
- Anistreplase (Eminase) (**24.3.5**) is a complex of streptokinase and plasminogen. Both are used in acute myocardial infarction, arterial and venous thrombosis, and pulmonary embolism.

Common adverse effects of all the thrombolytic drugs are bleeding complications related to systemic fibrinogenolysis and lysis of normal hemostatic plugs.

REFERENCES

1. del Zoppo, G. J. Virchow's triad: the vascular basis of cerebral injury. *Rev. Neurol. Dis.* **2008,** *5* (Suppl 1), S12–S21.
2. Mueller, R. L.; Scheidt, S. History of drugs for thrombotic disease: discovery, development, and directions for the future. *Circulation* **1994,** *89* (1), 432–449.
3. Galanaud, J.-P.; Laroche, J.-P.; Righini, M. The history and historical treatments of deep vein thrombosis. *J. Thromb. Haemostasis* **2013,** *11* (3), 402–411.
4. Hirsh, J.; Anand, S. S.; Halperin, J.; Fuster, V. Guide to anticoagulant therapy: heparin. A statement for healthcare professionals from the American Heart Association. *Circulation* **2001,** *103* (24), 2994–3018.
5. Link, K. P. The discovery of dicumarol and its sequels. *Circulation* **1959,** *19* (1), 97–107.
6. Lordkipanidze, M. Advances in monitoring of aspirin therapy. *Platelets* **2012,** *23* (7), 526–536.
7. Botting, R. M.; Gryglewski, R. J.; Vane, J. R. The anti-thrombotic and fibrinolytic actions of aspirin. In *Aspirin and Other Salicylates;* Vane, J. R., Bottling, R. M., Eds.; Chapman & Hall, 1992; pp 245–291.
8. Mulloy, B. Structure and physicochemical characterisation of heparin. *Handb. Exp. Pharmacol.* **2012,** *207* (Heparin), 77–98.
9. Casu, B. Structure and biological activity of heparin. *Adv. Carbohydr. Chem. Biochem.* **1985,** *43,* 51–134.
10. Lane, D. A., Lindahl, U., Eds. *Heparin: Chemical and Biological Properties, Clinical Applications;* Edward Arnold, 1989.
11. McDuffie, N. M., Ed. *Heparin: Structure, Cellular Functions, and Clinical Applications;* Elsevier, 1979.
12. Comper, W. D. *Heparin (and Related Polysaccharides): Structural and Functional Properties;* (Volume 7 of Polymer monographs). Gordon and Breach Science Publishers, 1981.
13. Carlsson, P.; Kjellen, L. Heparin biosynthesis. *Handb. Exp. Pharmacol.* **2012,** *207* (Heparin), 23–41.
14. Coyne, E. Heparin-past, present and future. *Dev. Biochem.* **1981,** *12* (Chem. Biol. Heparin), 9–17.
15. Lever, R., Mulloy, B., Page, C. P., Eds. *Heparin—a century of progress.* Handb. Exp. Pharmacol., 2012; p 207.
16. Pavao, M. S. G.; Mourao, P. A. S. Challenges for heparin production: artificial synthesis or alternative natural sources? *Glycobiol. Insights* **2012,** *3,* 1–6.

17. Szczubialka, K.; Kaminski, K.; Zasada, K.; Karewicz, A.; Nowakowska, M. Heparin—a key drug in the treatment of the circulatory degenerative diseases: controlling its action with polymers. *Curr. Pharm. Des.* **2012,** *18* (18), 2591–2606.
18. Linhardt, R. J.; Liu, J. Synthetic heparin. *Curr. Opin. Pharmacol.* **2012,** *12* (2), 217–219.
19. Gandhi, N. S.; Mancera, R. L. Heparin/heparan sulphate-based drugs. *Drug Discovery Today* **2010,** *15* (23/24), 1058–1069.
20. Alban, S. Adverse effects of heparin. *Handb. Exp. Pharmacol.* **2012,** *207* (Heparin), 211–263.
21. Warkentin, T. E. History of heparin-induced thrombocytopenia. *Fundam. Clin. Cardiol.* **2013,** *66,* 1–23.
22. Kyriakou, E. S.; Kokori, S. I.; Stylos, D. A.; Kardoulaki, A. P.; Tsantes, A. E. Heparin-induced thrombocytopenia: pathophysiology, diagnosis, and treatment monitoring. *Drug Dev. Res.* **2013,** *74* (8), 558–567.
23. Dasararaju, R.; Singh, N.; Mehta, A. Heparin-induced thrombocytopenia: review. *Expert Rev. Hematol.* **2013,** *6* (4), 419–428.
24. Turnbull, J. E. Getting the farm out of pharma for heparin production. *Science (Washington, DC, U. S.)* **2011,** *334* (6055), 462–463.
25. Barrowcliffe, T. W.; Johnson, E. A.; Thomas, D. P. *Low Molecular Weight Heparin;* Wiley, 1992.
26. Doutremepuich, C., Ed. *Low Molecular Weight Heparins in Clinical Practice;* CRC Press, 1992.
27. Coombe, D.; Kett, W. C. Heparin mimetics. *Handb. Exp. Pharmacol.* **2012,** *207* (Heparin), 361–383.
28. Oh, Y. I.; Sheng, G. J.; Chang, S.-K.; Hsieh-Wilson, L. C. Tailored glycopolymers as anticoagulant heparin mimetics. *Angew. Chem., Int. Ed.* **2013,** *52* (45), 11796–11799.
29. Xu, Y.; Cai, C.; Chandarajoti, K.; Hsieh, P.-H.; Li, L.; Pham, T. Q.; Sparkenbaugh, E. M.; Sheng, J.; Key, N. S.; Pawlinski, R.; Harris, E. N.; Linhardt, R. J.; Liu, J. Homogeneous low-molecular-weight heparins with reversible anticoagulant activity. *Nat. Chem. Biol.* **2014,** *10* (4), 248–250.
30. Driguez, P.-A. Synthesis of natural and nonnatural heparin fragments: optimizations and applications toward modulation of FGF2-mediated FGFR signaling. In *Modern Synthetic Methods in Carbohydrate Chemistry: From Monosaccharides to Complex Glycoconjugates;* Werz, D. B., Vidal, S., Eds.; Wiley-VCH, 2014; pp 191–220.
31. Becker, R. C. Optimizing heparin compounds: a working construct for future antithrombotic drug development. *J. Thromb. Thrombolysis* **2004,** *18* (1), 55–58.
32. Driguez, P.-A.; Lederman, I.; Strassel, J.-M.; Herbert, J.-M.; Petitou, M. Synthetic carbohydrate derivatives as low sulfated heparin mimetics. *J. Org. Chem.* **1999,** *64* (26), 9512–9520.
33. Petitou, M.; van Boeckel, C. A. A. Heparin: from the original "soup" to well-designed heparin mimetics. *Pure Appl. Chem.* **1997,** *69* (9), 1839–1846.
34. Linhardt, R. J.; Gunay, N. S. Production and chemical processing of low molecular weight heparins. *Semin. Thromb. Hemostasis* **1999,** *25* (Suppl. 3), 5–16.
35. Mardiguian, J. Sulfated polysaccharides and their use as medicines, EP 40144 (1981).
36. Munoz, E.; Xu, D.; Avci, F.; Kemp, M.; Liu, J.; Linhardt, R. J. Enzymatic synthesis of heparin related polysaccharides on sensor chips: Rapid screening of heparin-protein interactions. *Biochem. Biophys. Res. Commun.* **2006,** *339* (2), 597–602.
37. Zhang, Z.; McCallum, S. A.; Xie, J.; Nieto, L.; Corzana, F.; Jimenez-Barbero, J.; Chen, M.; Liu, J.; Linhardt, R. J. Solution structures of chemo-enzymatically synthesized heparin and its precursors. *J. Am. Chem. Soc.* **2008,** *130* (39), 12998–13007.
38. Noble, S.; Spencer, C. M. Enoxaparin: a review of its clinical potential in the management of coronary artery disease. *Drugs* **1998,** *56* (2), 259–272.

39. Hofmann, T. Clinical application of enoxaparin. *Expert Rev. Cardiovasc. Ther.* **2004,** *2* (3), 321–337.
40. Turpie, A. G. G.; Mason, J. A. Review of enoxaparin and its clinical applications in venous and arterial thromboembolism. *Expert Opin. Pharmacother.* **2002,** *3* (5), 575–598.
41. Ingle, R. G.; Agarwal, A. S. A world of low molecular weight heparins (LMWHs) enoxaparin as a promising moiety–a review. *Carbohydr. Polym.* **2014,** *106,* 148–153.
42. Lee, S.; Raw, A.; Yu, L.; Lionberger, R.; Ya, N.; Verthelyi, D.; Rosenberg, A.; Kozlowski, S.; Webber, K.; Woodcock, J. Scientific considerations in the review and approval of generic enoxaparin in the United States. *Nat. Biotechnol.* **2013,** *31* (3), 220–226.
43. Iqbal, Z.; Cohen, M. Enoxaparin: a pharmacologic and clinical review. *Expert Opin. Pharmacother.* **2011,** *12* (7), 1157–1170.
44. Petitou, M.; Jacquinet, J. C.; Sinay, P.; Choay, J.; Lormeau, J. C.; Nassr, M. Organic oligosaccharides, corresponding to fragments of natural mucopolysaccharides, and their biological applications. EP 84999 (1983).
45. Petitou, M.; Jacquinet, J-C.; Sinay, P.; Choay, J.; Lormeau, J-C.; Nassr, M. Preparation of oligosaccharide uronates, corresponding to fragments of natural mucopolysaccharides, and their biological applications. US 4818816 (1989).
46. Petitou, M.; Duchaussoy, P.; Lederman, I.; Choay, J.; Jacquinet, J. C.; Sinay, P.; Torri, G. Synthesis of heparin fragments: a methyl α-pentaoside with high affinity for antithrombin III. *Carbohydr. Res.* **1987,** *167,* 67–75.
47. Li, T.; Ye, H.; Cao, X.; Wang, J.; Liu, Y.; Zhou, L.; Liu, Q.; Wang, W.; Shen, J.; Zhao, W.; Wang, P. Total synthesis of anticoagulant pentasaccharide fondaparinux. *ChemMedChem* **2014,** *9,* 1071–1080.
48. Lin, F.; Lian, G.; Zhou, Y. Synthesis of fondaparinux: modular synthesis investigation for heparin synthesis. *Carbohydr. Res.* **2013,** *371,* 32–39.
49. Manikowski, A.; Koziol, A.; Czajkowska-Wojciechowska, E. An alternative route for fondaparinux sodium synthesis via selective hydrogenations and sulfation of appropriate pentasaccharides. *Carbohydr. Res.* **2012,** *361,* 155–161.
50. Reverter, J. C. Fondaparinux sodium. *Drugs Today* **2002,** *38* (3), 185–194.
51. Samama, M.-M.; Gerotziafas, G. T. Evaluation of the pharmacological properties and clinical results of the synthetic pentasaccharide (fondaparinux). *Thromb. Res.* **2003,** *109* (1), 1–11.
52. Petitou, M.; Duchaussoy, P.; Herbert, J.-M.; Duc, G.; El Hajji, M.; Branellec, J.-F.; Donat, F.; Necciari, J.; Cariou, R.; Bouthier, J.; Garrigou, E. The synthetic pentasaccharide fondaparinux: first in the class of antithrombotic agents that selectivity inhibit coagulation factor Xa. *Semin. Thromb. Hemostasis* **2002,** *28* (4), 393–402.
53. Bauer, K. A.; Hawkins, D. W.; Peters, P. C.; Petitou, M.; Herbert, J.-M.; van Boeckel, C. A. A.; Meuleman, D. G. The synthetic pentasaccharide fondaparinux: first in the class of antithrombotic agents that selectivity inhibit coagulation factor. *Cardiovasc. Drug Rev.* **2002,** *20* (1), 37–52.
54. Cheng, J. W. M. Fondaparinux: a new antithrombotic agent. *Clin. Ther.* **2002,** *24* (11), 1757–1769.
55. Giangrande, P. L. F. Fondaparinux (Arixtra): a new anticoagulant. *Int. J. Clin. Pract.* **2002,** *56* (8), 615–617.
56. Nadar, S. K.; Goyal, D.; Shantsila, E.; Banerjee, P.; Lip, G. Y. H. Fondaparinux: an overview. *Expert Rev. Cardiovasc. Ther.* **2009,** *7* (6), 577–585.
57. Robinson, D. M.; Wellington, K. Fondaparinux sodium: a review of its use in the treatment of acute venous thromboembolism. *Am. J. Cardiovasc. Drugs* **2005,** *5* (5), 335–346.

58. Blick, S. K. A.; Orman, J. S.; Wagstaff, A. J.; Scott, L. J. Fondaparinux sodium: a review of its use in the management of acute coronary syndromes. *Am. J. Cardiovasc. Drugs* **2008,** *8* (2), 113–125.
59. Toschi, V.; Lettino, M. Fondaparinux: pharmacology and clinical experience in cardiovascular medicine. *Mini-Rev. Med. Chem.* **2007,** *7* (4), 383–387.
60. Bauer, K. A. Fondaparinux: a new synthetic and selective inhibitor of factor Xa. *Best Pract. Res., Clin. Haematol.* **2004,** *17* (1), 89–104.
61. Turpie, A. G. G. Fondaparinux: a Factor Xa inhibitor for antithrombotic therapy. *Expert Opin. Pharmacother.* **2004,** *5* (6), 1373–1384.
62. Reverter, J. C. Fondaparinux sodium. *Drugs Today* **2002,** *38* (3), 185–194.
63. Diaz-Ricart, M.; del Fresno, M. Idraparinux sodium: anticoagulant factor Xa inhibitor. *Drugs Future* **2002,** *27* (7), 639–644.
64. Babhair, S. A.; Tariq, M.; Al-Badr, A. A. Warfarin. *Anal. Profiles Drug Subst.* **1985,** *14,* 423–452.
65. Park, B. K. Warfarin: metabolism and mode of action. *Biochem. Pharmacol. (Amsterdam, Neth.)* **1988,** *37* (1), 19–27.
66. Scully, M. Warfarin therapy: rat poison and the prevention of thrombosis. *Biochemist* **2002,** *24* (1), 15–17.
67. Keller, C.; Matzdorff, A. C.; Kemkes-Matthes, B. Pharmacology of warfarin and clinical implications. *Semin. Thromb. Hemostasis* **1999,** *25* (1), 13–16.
68. Porter, W. R. Warfarin: history, tautomerism and activity. *J. Comput.-Aided Mol. Des.* **2010,** *24* (6–7), 553–575.
69. Kimmel, S. E. Warfarin therapy: in need of improvement after all these years. *Expert Opin. Pharmacother.* **2008,** *9* (5), 677–686.
70. Miller, G. P. Warfarin therapy: how the less interesting half just got interesting. *J. Thromb. Haemostasis* **2010,** *8* (12), 2705–2707.
71. Ruff, C. T.; Braunwald, E. Will warfarin ever be replaced? *J. Cardiovasc. Pharmacol. Ther.* **2010,** *15* (3), 210–219.
72. Lin, P. J. Reviewing the reality: why we need to change. *Eur. Heart J. Suppl.* **2005,** *7* (Suppl. E), E15–E20.
73. Garcia, D. Rethinking warfarin reversal. *Blood* **2010,** *116* (5), 675–676.
74. Mohr, J. P. Update on antithrombotic therapies: warfarin. *Int. J. Clin. Pract., Suppl.* **2003,** *136,* 3–6.
75. Mehta, A. Y.; Jin, Y.; Desai, U. R. An update on recent patents on thrombin inhibitors (2010-2013). *Expert Opin. Ther. Pat.* **2014,** *24* (1), 47–67.
76. Coppens, M.; Eikelboom, J. W.; Gustafsson, D.; Weitz, J. I.; Hirsh, J. Translational success stories: development of direct thrombin inhibitors. *Circ. Res.* **2012,** *111* (7), 920–929.
77. Arsenault, K. A.; Hirsh, J.; Whitlock, R. P.; Eikelboom, J. W. Direct thrombin inhibitors in cardiovascular disease. *Nat. Rev. Cardiol.* **2012,** *9* (7), 402–414.
78. O'Brien, P. J.; Mureebe, L. Direct thrombin inhibitors. *J. Cardiovasc. Pharmacol. Ther.* **2012,** *17* (1), 5–11.
79. Lee, C. J.; Ansell, J. E. Direct thrombin inhibitors. *Br. J. Clin. Pharmacol.* **2011,** *72* (4), 581–592.
80. Mehta, R. S. Novel oral anticoagulants. Part II: direct thrombin inhibitors. *Expert Rev. Hematol.* **2010,** *3* (3), 351–361.
81. Schwienhorst, A. Direct thrombin inhibitors-a survey of recent developments. *Cell. Mol. Life Sci.* **2006,** *63* (23), 2773–2791.
82. Ganetsky, M.; Babu, K. M.; Salhanick, S. D.; Brown, R. S.; Boyer, E. W. Dabigatran: review of pharmacology and management of bleeding complications of this novel oral anticoagulant. *J. Med. Toxicol.* **2011,** *7* (4), 281–287.

83. Kaur, K.; Gupta, V. Dabigatran etexilate: a drug update. *Pharma Innov.* **2013,** *2* (3), 141–148.
84. Hauel, N.; Clemens, A.; Nar, H.; Priepke, H.; van Ryn, J.; Wienen, W. The discovery of dabigatran etexilate. In *Analogue-Based Drug Discovery III;* Fischer, J., Ganellin, C. R., Rotella, D. P., Eds.; Wiley-VCH, 2013; pp 243–267.
85. Nagarakanti, R.; Ellis, C. R. Dabigatran in clinical practice. *Clin. Ther.* **2012,** *34* (10), 2051–2060.
86. Mungall, D. BIBR-1048 (Boehringer Ingelheim). *Curr. Opin. Invest. Drugs (BioMed Cent.)* **2002,** *3* (6), 905–907.
87. Hauel, N.; Ries, U.; Priepke, H.; Wienen, W.; Stassen, J. M. Preparation of amidinophenylethylbenzimidazolylcarboxamides and related compounds as thrombin inhibitors, WO 9837075 (1998).
88. Hauel, N.; Ries, U.; Priepke, H.; Wienen, W.; Stassen, J. M. Preparation of 2-(amidinoanilinomethyl)benzimidazole-5-carboxamides and analogs as antithrombotics, DE 19706229 (1998).
89. Hauel, N. H.; Nar, H.; Priepke, H.; Ries, U.; Stassen, J.-M.; Wienen, W. Structure-based design of novel potent nonpeptide thrombin inhibitors. *J. Med. Chem.* **2002,** *45* (9), 1757–1766.
90. Pinto, D. J. P.; Smallheer, J. M.; Cheney, D. L.; Knabb, R. M.; Wexler, R. R. Factor Xa inhibitors: next-generation antithrombotic agents. *J. Med. Chem.* **2010,** *53* (17), 6243–6274.
91. Bondarenko, M.; Curti, C.; Montana, M.; Rathelot, P.; Vanelle, P. Efficacy and toxicity of Factor Xa inhibitors. *J. Pharm. Pharm. Sci.* **2013,** *16* (1), 74–88.
92. Yeh, C. H.; Fredenburgh, J. C.; Weitz, J. I. Oral direct factor Xa inhibitors. *Circ. Res.* **2012,** *111* (8), 1069–1078.
93. Pinto, D. J. P.; Qiao, J. X.; Knabb, R. M. The emergence of factor Xa inhibitors for the treatment of cardiovascular diseases: a patent review. *Expert Opin. Ther. Pat.* **2012,** *22* (6), 645–661.
94. Young, R. J. The successful quest for oral factor Xa inhibitors: learnings for all of medicinal chemistry? *Bioorg. Med. Chem. Lett.* **2011,** *21* (21), 6228–6235.
95. Lee, Y.-K.; Player, M. R. Developments in factor Xa inhibitors for the treatment of thromboembolic disorders. *Med. Res. Rev.* **2011,** *31* (2), 202–283.
96. Mueller, R. L.; Scheidt, S. History of drugs for thrombotic disease: discovery, development, and directions for the future. *Circulation* **1994,** *89* (1), 432–449.
97. Yates, S.; Sarode, R. Novel thrombin and factor Xa inhibitors: challenges to reversal of their anticoagulation effects. *Curr. Opin. Hematol.* **2013,** *20* (6), 552–557.
98. Duffull, S. B. Is the ideal anticoagulant a myth? *Expert Rev. Clin. Pharmacol.* **2012,** *5* (3), 231–236.
99. Ahrens, I.; Bode, C. New parenteral anticoagulants: focus on factor Xa and thrombin inhibitors. *Curr. Drug Discovery Technol.* **2012,** *9* (2), 129–136.
100. Gosavi, S.; Mukherjee, D. Review of newer anticoagulants and anti-platelet agents in acute coronary syndrome and cardiovascular diseases. *Cardiovasc. Hematol. Agents Med. Chem.* **2013,** *11* (3), 194–202.
101. DeWald, T. A.; Becker, R. C. The pharmacology of novel oral anticoagulants. *J. Thromb. Thrombolysis* **2014,** *37* (2), 217–233.
102. Page, C. Heparin and related drugs: beyond anticoagulant activity. *ISRN Pharmacol.* **2013,** *91* (0743), 14.
103. Makaryus, J. N.; Halperin, J. L.; Lau, J. F. Oral anticoagulants in the management of venous thromboembolism. *Nat. Rev. Cardiol.* **2013,** *10* (7), 397–409.
104. Eriksson, B. I.; Quinlan, D. J.; Eikelboom, J. W. Novel oral factor Xa and thrombin inhibitors in the management of thromboembolism. *Annu. Rev. Med.* **2011,** *62,* 41–57.

105. Weitz, J. I.; Eikelboom, J. W.; Samama, M. M. New antithrombotic drugs: Antithrombotic Therapy and Prevention of Thrombosis, 9th ed: American College of Chest Physicians Evidence-Based Clinical Practice Guidelines. *Chest* **2012,** *141* (2 Suppl.), e120S–e151S.
106. Mousa, S. A. The future of anticoagulant therapy. *Cardiovasc. Ther.* **2012,** *30* (5), 255–256.
107. Mousa, S. A. Highlights of latest advances in antithrombotics. *Methods Mol. Med.* **2004,** *93* (Anticoagulants, Antiplatelets, and Thrombolytics), 1–7.
108. Weitz, J. I.; Linkins, L.-A. Beyond heparin and warfarin: the new generation of anticoagulants. *Expert Opin. Invest. Drugs* **2007,** *16* (3), 271–282.
109. Hoffman, R.; Brenner, B. The promise of novel direct oral anticoagulants. *Best Pract. Res., Clin. Haematol.* **2012,** *25* (3), 351–360.
110. Goel, R.; Srivathsan, K. Newer oral anticoagulant agents: a new era in medicine. *Curr. Cardiol. Rev.* **2012,** *8* (2), 158–165.
111. Capodanno, D.; Giacchi, G.; Tamburino, C. Current status and ongoing development of reversing agents for novel oral anticoagulants (NOACs). *Recent Pat. Cardiovasc. Drug Discovery* **2013,** *8* (1), 2–9.
112. Khoo, C. W.; Tay, K.-H.; Shantsila, E.; Lip, G. Y. H. Novel oral anticoagulants. *Int. J. Clin. Pract.* **2009,** *63* (4), 630–641.
113. Miyares, M. A.; Davis, K. Newer oral anticoagulants: a review of laboratory monitoring options and reversal agents in the hemorrhagic patient. *Am. J. Health-Syst. Pharm.* **2012,** *69* (17), 1473–1484.
114. Harder, S.; Graff, J. Novel oral anticoagulants: clinical pharmacology, indications and practical considerations. *Eur. J. Clin. Pharmacol.* **2013,** *69* (9), 1617–1633.
115. Straub, A.; Roehrig, S.; Hillisch, A. Oral, direct thrombin and factor Xa inhibitors: the replacement for warfarin, leeches, and pig intestines?. *Angew. Chem., Int. Ed.* **2011,** *50* (20). 4574–45-90.
116. Peter, K. Antiplatelet drugs. In *Principles of Molecular Cardiology;* Runge, M. S., Patterson, C., Willerson, J. T., Eds.; Humana Press, 2005; pp 203–218.
117. Ruggeri, Z. M.; FitzGerald, G. A.; Shattil, S. J. Platelet thrombus formation and antiplatelet therapy. In *Molecular Basis of Cardiovascular Disease: A Companion to Braunwald's Heart Disease,* 1st ed.; Chien, K. R., Ed.; W. B. Saunders, 1999; pp 566–589.
118. Kalra, K.; Franzese, C. J.; Gesheff, M. G.; Lev, E. I.; Pandya, S.; Bliden, K. P.; Tantry, U. S.; Gurbel, P. A. Pharmacology of antiplatelet agents. *Curr. Atheroscler. Rep.* **2013,** *15* (12), 1–13.
119. Ungerer, M.; Muench, G. Novel antiplatelet drugs in clinical development. *Thromb. Haemostasis* **2013,** *110* (5), 868–875.
120. Garg, A.; Pandey, A.; Oza, P.; Chaturvedi, M. Newer antiplatelet and antithrombotic drugs on the horizon. *Natl. J. Physiol., Pharm. Pharmacol.* **2013,** *3* (2), 105–110.
121. Yeung, J.; Holinstat, M. Newer agents in antiplatelet therapy: a review. *J. Blood Med.* **2012,** *3,* 33–42.
122. Eikelboom, J. W.; Hirsh, J.; Spencer, F. A.; Baglin, T. P.; Weitz, J. I. Antiplatelet drugs: Antithrombotic Therapy and Prevention of Thrombosis, 9th ed: American College of Chest Physicians Evidence-Based Clinical Practice Guidelines. *Chest* **2012,** *141* (2 Suppl.), e89S–e119S.
123. Tello-Montoliu, A.; Jover, E.; Rivera, J.; Valdes, M.; Angiolillo, D. J.; Marin, F. New perspectives in antiplatelet therapy. *Curr. Med. Chem.* **2012,** *19* (3), 406–427.
124. Coccheri, S. Antiplatelet drugs—do we need new options?: With a reappraisal of direct thromboxane inhibitors. *Drugs* **2010,** *70* (7), 887–908.
125. Siddique, A.; Butt, M.; Shantsila, E.; Lip, G. Y. H. New antiplatelet drugs: beyond aspirin and clopidogrel. *Int. J. Clin. Pract.* **2009,** *63* (5), 776–789.

126. Grove, E. L.; Kristensen, S. D. Update on oral antiplatelet therapy: principles, problems and promises. *Future Cardiol.* **2009,** *5* (3), 247–258.
127. Mousa, S. A. Antiplatelet therapies: from aspirin to GPIIb/IIIa-receptor antagonists and beyond. *Drug Discovery Today* **1999,** *4* (12), 552–561.
128. Geiger, J. Anti-aggregatory drugs: I. Platelet receptor antagonists. *Expert Opin. Ther. Pat.* **1999,** *9* (10), 1389–1414.
129. Storey, R. F. Biology and pharmacology of the platelet P2Y12 receptor. *Curr. Pharm. Des.* **2006,** *12* (10), 1255–1259.
130. Aubert, D.; Ferrand, C.; Maffrand, J. P. Thieno[3,2-c]pyridine derivatives and their therapeutical use, EP 99802 (1984).
131. Bousquet, A.; Calet, S.; Heymes, A. Isopropyl 2-thienylglycidate, process for its preparation, and its use as synthetic intermediate for ticlopidine and clopidogrel, EP 465358 (1992).
132. Preparation of d-α-5-(4,5,6,7-tetrahydro[3,2-c]thienopyridyl)-2-(chlorophenyl) acetic acid methyl ester as an antithrombotic, JP 63203684 (1988).
133. Wang, L.; Shen, J.; Tang, Y.; Chen, Y.; Wang, W.; Cai, Z.; Du, Z. Synthetic improvements in the preparation of clopidogrel. *Org. Process Res. Dev.* **2007,** *11* (3), 487–489.
134. Castro, B.; Dormoy, J-R.; Previero, A. Improved method for preparing 2-thienylethylamine derivatives, including an intermediate for clopidogrel, WO 9839322 (1998).
135. Bakonyi, M.; Csatari N. M.; Molnar, L., Mrs.; Makovi, Z.; Jobb, P.; Bai, T. New 2-[(2-thienyl) ethylamino](2-halophenyl)acetamide intermediates for clopidogrel and analogs, and process for their preparation, WO 9851681 (1998).
136. Heymes, A.; Castro, B.; Bakonyi, M.; Csatari N. M.; Molnar, L. New 2-[(2-thienyl)ethylamino](2-halophenyl)acetonitrile intermediates for clopidogrel and analogs, and process for their preparation, WO 9851682 (1998).
137. Bousquet, A.; Musolino, A. Hydroxyacetic ester derivatives, namely (R)-methyl 2-(sulfonyloxy)-2-(chlorophenyl)acetates, preparation method, and use as synthesis intermediates for clopidogrel, WO 9918110 (1999).
138. Bouisset, M.; Radisson, J., Process for preparing phenylacetic derivatives of thienopyridines and intermediate α-bromophenylacetic acids, EP 420706 (1991).
139. Sashikanth, S.; Raju, V.; Somaiah, S.; Rao, P. S.; Reddy, K. V. An asymmetric synthesis of clopidogrel hydrogen sulfate. *Synthesis* **2013,** *45* (5), 621–624.
140. Sadhukhan, A.; Saravanan, S.; Khan, N. H.; Kureshy, R. I.; Abdi, S. H. R.; Bajaj, H. C. Modified asymmetric strecker reaction of aldehyde with secondary amine: a protocol for the synthesis of S-clopidogrel (an antiplatelet agent). *J. Org. Chem.* **2012,** *77* (16), 7076–7080.
141. Ferraboschi, P.; De Mieri, M.; Galimberti, F. Chemo-enzymatic approach to the synthesis of the antithrombotic clopidogrel. *Tetrahedron: Asymmetry* **2010,** *21* (17), 2136–2141.
142. Savi, P.; Herbert, J.-M. Clopidogrel and ticlopidine: P2Y12 adenosine diphosphate-receptor antagonists for the prevention of atherothrombosis. *Semin. Thromb. Hemostasis* **2005,** *31* (2), 174–183.
143. Quinn, M. J.; Fitzgerald, D. J. Ticlopidine and clopidogrel. *Circulation* **1999,** *100* (15), 1667–1672.
144. Herbert, J. M.; Frehel, D.; Vallee, E.; Kieffer, G.; Gouy, D.; Berger, Y.; Necciari, J.; Defreyn, G.; Maffrand, J. P. Clopidogrel, a novel antiplatelet and antithrombotic agent. *Cardiovasc. Drug Rev.* **1993,** *11* (2), 180–198.
145. Savi, P.; Nurden, P.; Nurden, A. T.; Levy-Toledano, S.; Herbert, J.-M. Clopidogrel: a review of its mechanism of action. *Platelets* **1998,** *9* (3/4), 251–255.
146. Coukell, A. J.; Markham, A. Clopidogrel. *Drugs* **1997,** *54* (5), 745–750.

147. Feuerstein, G.; Nichols, A. J.; Ruffolo, R. R., Jr. Clopidogrel: a novel antiplatelet drug for prevention and treatment of thrombotic disorders. *Expert Opin. Invest. Drugs* **1995,** *4* (5), 425–430.
148. Lerner, R. G.; Frishman, W. H.; Mohan, K. T. Clopidogrel: a new antiplatelet drug. *Heart Dis.* **2000,** *2* (2), 168–173.
149. Sadanandan, S.; Singh, I. M. Clopidogrel: the data, the experience, and the controversies. *Am. J. Cardiovasc. Drugs* **2012,** *12* (6), 361–374.
150. Huber, K. Clopidogrel in coronary artery disease: update (2012). *Adv. Cardiol* **2012,** *47,* 31–38.
151. Sarafoff, N.; Byrne, R. A.; Sibbing, D. Clinical use of clopidogrel. *Curr. Pharm. Des.* **2012,** *18* (33), 5224–5239.
152. Diener, H.-C.; Ringleb, P. A.; Savi, P. Clopidogrel for the secondary prevention of stroke. *Expert Opin. Pharmacother.* **2005,** *6* (5), 755–764.
153. Rossini, R.; Musumeci, G.; Nijaradze, T.; Gavazzi, A. Clopidogrel bisulfate: a review of its use in the management of acute coronary syndrome. *Clin. Med.: Ther.* **2009,** *1,* 899–910.
154. Gurbel, P. A.; Antonino, M. J.; Tantry, U. S. Recent developments in clopidogrel pharmacology and their relation to clinical outcomes. *Expert Opin. Drug Metab. Toxicol.* **2009,** *5* (8), 989–1004.
155. Chow, G.; Ziegelstein, R. C. Clopidogrel. The good, the bad, and the ugly. *Am. J. Cardiovasc. Drugs* **2007,** *7* (3), 169–171.
156. Fox, K. A. A.; Chelliah, R. Clopidogrel: an updated and comprehensive review. *Expert Opin. Drug Metab. Toxicol.* **2007,** *3* (4), 621–631.
157. Plosker, G. L.; Lyseng-Williamson, K. A. Clopidogrel: a review of its use in the prevention of thrombosis. *Drugs* **2007,** *67* (4), 613–646.
158. Cui, H.; Tan, W.; Shi, J.; Xia, Y. Recent development in thrombin receptor antagonist as novel antithrombotic agent. *Open J. Med. Chem.* **2012,** *2* (4), 112–118.
159. Leonardi, S.; Becker, R. C. PAR-1 inhibitors: a novel class of antiplatelet agents for the treatment of patients with atherothrombosis. *Handb. Exp. Pharmacol.* **2012,** *210* (Antiplatelet Agents), 239–260.
160. Eisert, W. G. Dipyridamole in antithrombotic treatment. *Adv. Cardiol.* **2012,** *47,* 78–86.
161. Schaper, W. Dipyridamole, an underestimated vascular protective drug. *Cardiovasc. Pharmacother., Proc. Int. Congr., 9th* **2005,** *19* (5), 357–363.
162. Eisert, W. G. *Dipyridamole. Platelets,* 3rd ed.; Michelson, A. D., Ed.; Academic Press, 2013; pp 1155–1170. Ed. Michelson, A. D.
163. Diener, H. C.; Forbes, C. Dipyridamole. In *Drug Therapy for Stroke Prevention;* Bogousslavsky, J., Ed.; CRC Press, 2001; pp 87–93.
164. Rivey, M. P.; Alexander, M. R.; Taylor, J. W. Dipyridamole: a critical evaluation. *Drug Intell. Clin. Pharm.* **1984,** *18* (11), 869–880.
165. Harker, L. A.; Kadatz, R. A. Mechanism of action of dipyridamole. *Thromb. Res.* **1983,** (Suppl. 4), 39–46.
166. Fischer, F. G.; Roch, J.; Kottler, A. Derivatives of pyrimido[5,4-d]pyrimidine, GB 807826 (1959).
167. Fischer, F. G.; Roch, J.; Kottler, A. Pyrimido[5,4-d]pyrimidines, US 3031450 (1962).
168. Rao, S. M.; Padmanabhan, R.; Senthilkumar, A. M. An improved process for the preparation of dipyridamole WO 2007080463 (2007).
169. Niegel, H.; Goldner, H.; Meyer, H. P.; Lorenz, D. Preparation of 2,6-dichloro-4,8-dipiperidinopyrimido[5,4-d]pyrimidine, DD 280008 (1990).

170. Krahnefeld, H.; Seifert, M.; Stutzriemer, S.; Wolf, J.; Rumler, E.; Weise, D. Preparation of pure dipyridamole monohydrochloride, DD 238975 (1986).
171. Murakami, M.; Takahashi, K.; Imai, K.; Kojima, T.; Tamazawa, K. Dipyridamole, DE 2437178 (1975).
172. Agah, R.; Plow, E. F.; Topol, E. J. GPIIb-IIIa antagonists. In *Platelets;* Michelson, A. D., Ed.; (Academic Press), 2002; pp 769–785.
173. Topol, E. J.; Byzova, T. V.; Plow, E. F. Platelet GPIIb-IIIa blockers. *Lancet* **1999,** *353* (9148), 227–231.
174. Gabriel, H. M.; Oliveira, E. I. Role of abciximab in the treatment of coronary artery disease. *Expert Opin. Biol. Ther.* **2006,** *6* (9), 935–942.
175. Phillips, D. R.; Scarborough, R. M. Clinical pharmacology of eptifibatide. *Am. J. Cardiol.* **1997,** *80* (4A), 11B–20B.
176. Winter, J. P.; Juergens, C. P. The role of tirofiban in the management of coronary artery disease. *Cardiovasc. Hematol. Disord.: Drug Targets* **2008,** *8* (2), 138–146.
177. Kiefer, T. L.; Becker, R. C. Inhibitors of platelet adhesion. *Circulation* **2009,** *120* (24), 2488–2495.
178. Keramati, M.; Mianroodi, R. A.; Memarnejadian, A.; Mirzaie, A.; Sazvari, S.; Aslani, M. M.; Roohvand, F. Towards a superior streptokinase for fibrinolytic therapy of vascular thrombosis. *Cardiovasc. Hematol. Agents Med. Chem.* **2013,** *11* (3), 218–229.
179. Wardlaw, J. M.; Murray, V.; Berge, E.; del Zoppo, G.; Sandercock, P.; Lindley, R. L.; Cohen, G. Recombinant tissue plasminogen activator for acute ischemic stroke: an updated systematic review and meta-analysis. *Lancet* **2012,** *379* (9834), 2364–2372.
180. Dhillon, S. Alteplase: a review of its use in the management of acute ischaemic stroke. *CNS Drugs* **2012,** *26* (10), 899–926.
181. Anderson, C. Thrombolysis with alteplase after stroke: extending outcomes. *Lancet Neurol.* **2013,** *12* (8), 731–732.
182. Simpson, D.; Siddiqui, M.; Asif, A.; Scott, L. J.; Hilleman, D. E. Reteplase: a review of its use in the management of thrombotic occlusive disorders. *Am. J. Cardiovasc. Drugs* **2006,** *6* (4), 265–285.
183. Behrouz, R. Intravenous tenecteplase in acute ischemic stroke: an updated review. *J. Neurol.* **2014,** *261* (16), 1069–1072.
184. Kumar, P. S.; Pulicherla, K. K.; Rao, K. R. S.S. Current status of production, clinical usage and market scenario of streptokinase. *J. Pharm. Res. (Bangalore, India)* **2012,** *5* (8), 4223–4229.
185. Vaishnavi, B.; Mohanasrinivasan, V.; Devi, Subathra C. Streptokinase: a novel clot buster. *J. Pharm. Res. (Bangalore, India)* **2011,** *4* (10), 3784–3788.
186. Simpson, D.; Siddiqui, M. A. A.; Scott, L. J.; Hilleman, D. E. Spotlight on reteplace in thrombotic occlusive disorders. *BioDrugs* **2007,** *21* (1), 65–68.
187. Rabasseda, X. Tenecteplase (TNK tissue plasminogen activator): A new fibrinolytic for the acute treatment of myocardial infarction. *Drugs of Today* **2001,** *37* (11), 749–760.
188. Kumar, P. Suresh; Pulicherla, K. K.; Rao, K. R. S. Sambasiva Current status of production, clinical usage and market scenario of streptokinase. *J. Pharm. Res.* **2012,** *5* (8), 4223–4229.
189. Munger, M. A.; Forrence, E. A. Anistreplase: a new thrombolytic for the treatment of acute myocardial infarction. *Clin. Pharm.* **1990,** *9* (7), 530–540.

Chapter 25

Thyroid and Antithyroid Drugs

Thyroid hormones have profound effects and control many "big time" physiologic processes, such as protein, lipid, and carbohydrate metabolism, development, and growth.

The effects of thyroid hormones are very complex. They act on basal metabolism and thermogenesis; skeletal musculature; cardiovascular, central nervous, and reproductive systems; kidneys; and liver, but not directly on the brain. Very important, however, is the action of thyroid hormones on maturation of the central nervous system and mental development. Their deficiency before a child's birth can cause cretinism and ossification defects; after the birth, deficiency causes growth retardation.

Thyroid hormones contain iodine atoms and for their normal balance and functioning in mammals, and certain amount of iodine, supplied in the diet in the form of iodide I^- or iodate IO^{3-} ions, is necessary for the synthesis and proper functioning of thyroid hormones.

Thyroxine (T4) (**25.2.1**) and triiodothyronine (T3) (**25.2.2**) are produced by the thyroid gland and are the only two endogenous hormones that contain iodine atoms.

Thyroid disorders are one of the most common endocrine disorders.

The thyroid gland produces two types of hormones: the tyrosine-based compounds T4 (T4) and T3 (Fig. 25.1.), and a 32-amino-acid linear polypeptide, calcitonin.

Secretion of thyroid hormones T4 and T3 is controlled by the hypothalamus, which secretes thyrotropin-releasing hormone, and the anterior pituitary gland, which secretes the thyrostimulin hormone thyrotropin.

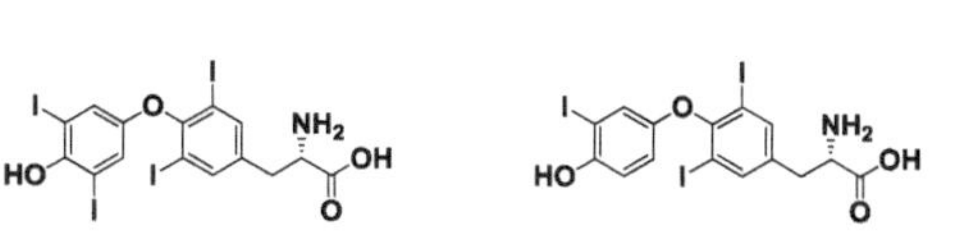

Thyroxine 25.2.1 (3,5,3',5'-tetraiodothyronine)   Triiodothyronine 25.2.2 (3,3',5-triiodotriiodothyronine)

FIG. 25.1 Endogenous hormones thyroxine (T4) and triiodothyronine (T3).

Thyroid hormones affect every cell and all the organs of the body. Too much thyroid hormone speeds things up and too little slows things down. Diseases are associated with both inadequate production and overproduction of thyroid

Synthesis of Best-Seller Drugs. http://dx.doi.org/10.1016/B978-0-12-411492-0.00025-0

hormones. Both types of disease are relatively common afflictions of humans and animals [1-3].

25.1 DRUGS FOR TREATMENT OF HYPERTHYROIDISM

Hyperthyroidism is a condition in which the thyroid produces more thyroid hormones than is needed by the body. It is also referred to as thyrotoxicosis, or as an overactive thyroid.

Hyperthyroidism is a chronic, even lifelong, disorder. It can be treated with an antithyroid drug, radioactive iodine (which is taken up by thyroid cells having as a result that the radiation destroys them, thereby reducing thyroid hormone production) or thyroidectomy (when thyroid gland is removed). Known drugs just block the production of thyroid hormone but have no effect or the underlying cause of the hyperthyroidism.

The clinical symptoms caused by an excess of circulating free T4, free T3, or both, include goiter, exophthalmos (bulging eyeballs), cardiac arrhythmias, weight loss in the presence of increased appetite, intolerance to heat, and profuse sweating. The clinical triad of hyperthyroidism was independently described by Basedow and Graves and included goiter, exophthalmos, and tachycardia. In continental European countries, the term *Basedow disease* is the more common name, whereas it is known as *Graves disease* in the English-speaking world.

It is a common disorder that affects approximately 2% of women and 0.2% of men [4-11].

The medication treatment of hyperthyroidism started at the end of 1920s with the first observation that Brassica plants—cabbage, cauliflowers, broccoli, and turnips—containing phenylthiourea acted positively on persons with thyrotoxicosis, which led to the introduction of thiourea and thiouracil as possible medical treatments. Further studies prompted the synthesis of methimazole and propylthiouracil. Since that time the simple molecules, called *thionamide* drugs, have been in routine use for the treatment of hyperthyroidism. Carbimazole (**25.1.1**), its active metabolite methimazole (**25.1.2**), and propylthiouracil (**25.1.3**) (Fig. 25.2.), synthesis of which was described in our previous book [12], are the mainstays of antithyroid drug therapy. These compounds decrease the synthesis of thyroid hormones via inhibiting thyroperoxidase, which results in impairment of iodide oxidation and coupling of iodotyrosyl residues to form the thyroid hormones T3 and T4.

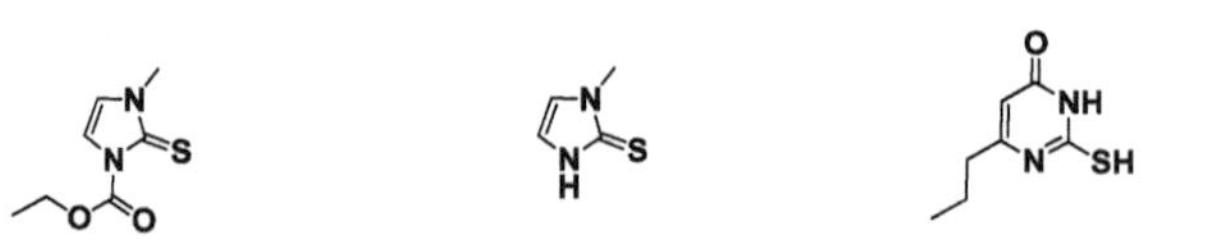

Carbimazole 25.1.1 Methimazole 25.1.2 Propylthiouracil 25.1.3

FIG. 25.2 Structure of carbimazole, methimazole, and propylthiouracil.

Most of the side effects of antithyroid drugs are minor (itching, rash, hives, joint pain and swelling). Fortunately, cases of agranulocytosis, which is the most severe adverse effect of antithyroid drug therapy, are very rare.

25.2 DRUGS FOR TREATMENT OF HYPOTHYROIDISM

Hypothyroidism is a condition in which the thyroid gland does not produce the necessary amount of thyroid hormone. This condition is often called *underactive thyroid.*

Hypothyroidism is the most common pathological hormone deficiency. It is a serious condition that is prevalent in older women. Severe hypothyroidism can lead to heart failure, psychosis, and coma.

Hypothyroidism can be classified as congenital or acquired, or, depending on the level of endocrine dysfunction, primary or secondary, depending on its severity and whether it is clinical or subclinical [13-16]. For cases of overt severe hypothyroidism, the extreme manifestation of hypothyroidism, the term *myxoedema* is usually used [17].

Treatment usually consists of T4 replacement with manufactured levothyroxine (**25.2.1**) therapy, which is highly effective and safe. Levothyroxine is the drug of choice for hypothyroidism [18-23]. It has a high degree of effectiveness and small risk of adverse reactions; in addition, it is a lifelong thyroid hormone replacement therapy. Levothyroxine is included in the list of Top 200 Drugs by sales for the 2010s.

Levothyroxine–Synthroid

The first synthesis of racemic T4 (**25.2.1a**), was carried out in 1927 [24]. According to that protocol, 4-methoxyphenol (**25.2.3**) was coupled with 3,4,5-triiodonitrobenzene (**25.2.4**) and then the nitro group in the obtained product (**25.2.5**) was reduced with stannous (II) chloride in acetic acid and a flow of dry hydrogen chloride to produce the aniline derivative (**25.2.6**). The aromatic amine group in the aniline derivative (**25.2.6**) easily underwent diazotization to produce an intermediate diazonium salt, which on reaction with KCN in hydrochloric acid solution was converted to a nitrile (**25.2.7**). The obtained nitrile was transformed to aldehyde, implementing the Stephen reaction protocol with the use of anhydrous stannous (II) chloride upon reduction in acetic acid and stream of gaseous hydrogen chloride, which produced the intermediate imine that was further hydrolyzed to 3,5-diiodo-4-(4′-methoxy-phenoxy) benzaldehyde (**25.2.8**). The obtained arylaldehyde was reacted with hippuric acid in the presence of sodium acetate in acetic anhydride and after work-up with sulfuric acid in ethanol, produced the derivative (**25.2.9**). The double bond in the derivative (**25.2.9**) was reduced using red phosphorous in hydrochloric acid, which proceeded with the simultaneous hydrolysis of the benzamide and ester groups. The resulting product was treated with iodine and potassium iodide in an aqueous ammonium hydroxide solution to produce racemic T4 (**25.2.1a**) (Scheme 25.1.).L-T4 (**25.2.1**) was later prepared via optical resolution of (±)-3,5-diiodothyronine [25].

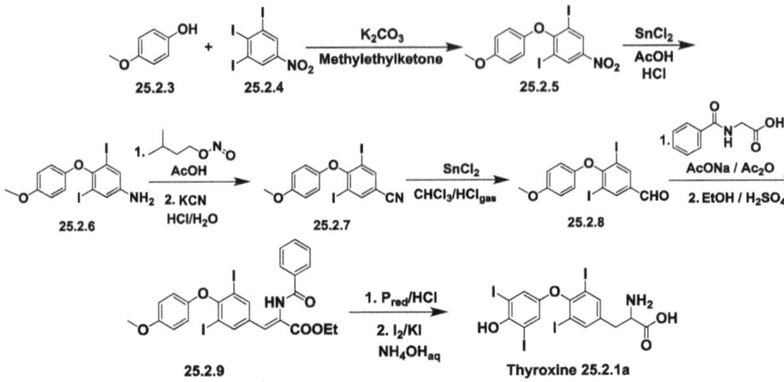

SCHEME 25.1 **Synthesis of racemic T4.**

Several other protocols have been performed and developed over the years for the preparation of L-thyroxine. There are several patents [26-32] and papers [33-36] describing them, which have been reviewed [37,38]. Two of the proposed protocols are used for large-scale adaptations -biomimetic approach [39] and chemical synthesis [40], both utilizing L-tyrosine (**25.2.10**). Chemical synthesis started from tyrosine (**25.2.10**) which was nitrated with a mixture of concentrated sulfuric acid and nitric acids at temperatures lower than 10°C to produce the corresponding 3,5-dinitrotyrosine (**25.2.11**). The amine group in 3,5-dinitrotyrosine (**25.2.11**) was protected by acetylation with acetic anhydride in a NaOH solution preparing (**25.2.12**). Carboxylic acid group, in (**25.2.12**) was protected by esterification with ethanol in chloroform, using toluenesulfonic acid as a catalyst to produce the ester (**25.2.13**). The ester was tosylated and coupled with 4-methoxyphenol (**25.2.3**) in a NaOH solution to produce a dinitrothyronine derivative (**25.2.14**). Reduction of both nitro groups took place with hydrogen on Pd/C catalyst and produced amine (**25.2.15**), which was by diazotized with sodium nitrite in sulfuric acid and further treated with sodium iodide/iodine mixture solution to produce the L-diiodothyronine derivative (**25.2.16**) in high yield. Subsequent hydrolysis of all protective groups simultaneously with 57% hydriodic acid in acetic acid produced the compound (**25.2.17**), which was followed by iodination using iodine and potassium iodide to produce L-thyroxine (**25.2.1**) in an 89% yield [40] (Scheme 25.2.).

Levothyroxine is one of the most prescribed drugs worldwide.

Levothyroxine is the overwhelming choice of clinicians for the treatment of hypothyroid states and for the suppression of goiter and thyroid nodules in selected cases. Levothyroxine is the preferred drug for treatment of myxedema

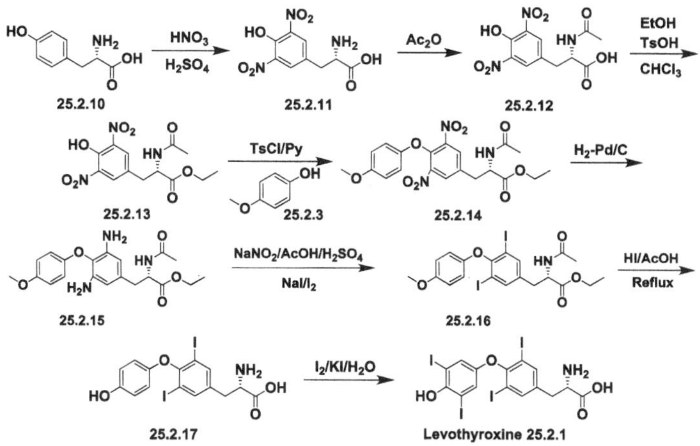

SCHEME 25.2 Synthesis of levothyroxine.

coma, neonatal hypothyroidism, and primary, secondary, and tertiary hypothyroidism. Intoxication with levothyroxine has a low prevalence.

REFERENCES

1. Murthy, M. B.; Jain, S. S.; Ramteke, K. B.; Raparti, G. T. Thyroid: disorders, disruptors and drugs. *Int. J. Nutr., Pharmacol., Neurol. Dis.* **2013,** *3* (2), 87–95.
2. Gessl, A.; Lemmens-Gruber, R.; Kautzky-Willer, A. Thyroid disorders. *Handb. Exp. Pharmacol.* **2012,** *214,* 361–386.
3. Fatourechi, M. M.; Fatourechi, V. An update on subclinical hypothyroidism and subclinical hyperthyroidism. *Expert Rev. Endocrinol. Metab.* **2014,** *9* (2), 137–151.
4. Mohanty, B. B.; Agrawal, D.; Rath, K.; Kumar, S.; Roy, D. K. Goitre: a complete review. *Int. J. Pharm. Biol. Sci.* **2012,** *3* (3), 33–48.
5. Fumarola, A.; Di Fiore, A.; Dainelli, M.; Grani, G.; Calvanese, A. Medical treatment of hyperthyroidism: state of the art. *Exp. Clin. Endocrinol. Diabetes* **2010,** *118* (10), 678–684.
6. Garcia-Mayor, R. V.; Larranaga, A. Treatment of Graves' hyperthyroidism with thionamides-derived drugs: review. *Med. Chem. (Sharjah, United Arab Emirates)* **2010,** *6* (4), 239–246.
7. Toft, A. D. Antithyroid drugs. In Martini, L., Ed.; *Encyclopedia of Endocrine Diseases*, Vol. 1; Elsevier, 2004; pp 261–264.
8. Cooper, D. S. Antithyroid drugs. *N. Engl. J. Med.* **2005,** *352* (9), 905–917.
9. Cooper, D. S. Antithyroid drugs for the treatment of hyperthyroidism caused by Graves' disease. *Endocrinol. Metab. Clin. North Am.* **1998,** *27* (1), 225–247.
10. Marchant, B.; Lees, J. F. H.; Alexander, W. D. Antithyroid drugs. In *International Encyclopedia of Pharmacology and Therapeutics;* Hershman, J. M., Bray, G. A., Eds.; Vol. 101, Elsevier, 1979; pp 209–252.
11. Gilman, A. G.; Murad, F. Thyroid and antithyroid drugs. In *Pharmacological Basis of Therapeutics,* 5th ed.; Goodman, L. S., Gilman, A., Eds. Macmillan, 1975; pp 1398–1422.
12. Vardanyan, R. S.; Hruby, V. J. *Synthesis of Essential Drugs;* Elsevier, 2006.
13. Beck-Peccoz, P.; Persani, L. Central hypothyroidism. In *Thyroid Diseases;* Monaco, F., Ed.; CRC Press, 2012; pp 215–223.

14. Deladoey, J.; Van Vliet, G. Treating congenital hypothyroidism-which levothyroxine? *Nat. Rev. Endocrinol.* **2013,** *9* (5), 257–258.
15. Gaitonde, D. Y.; Rowley, K. D.; Sweeney, L. B. Hypothyroidism: an update. *Am. Fam. Physician* **2012,** *86* (3), 244–251.
16. Jayakumar, R. V. Hypothyroidism. *J. Indian Med. Assoc.* **2006,** *104* (10), 557–560. 562.
17. Wall, C. R. Myxedema coma: diagnosis and treatment. *Am. Fam. Physician* **2000,** *62* (11), 2485–2490.
18. Sypniewski, E. Comparative pharmacology of the thyroid hormones. *Ann. Thorac. Surg.* **1993,** *56* (1 Suppl.), S6–S8.
19. Post, A.; Warren, R. J. Sodium levothyroxine. *Anal. Profiles Drug Subst.* **1976,** *5,* 225–281.
20. Bianco, A. C.; Casula, S. Thyroid hormone replacement therapy: three simple questions, complex answers. *Eur. Thyroid J.* **2012,** *1* (2), 88–98.
21. Hennessey, J. V. Levothyroxine a new drug? Since when? How could that be? *Thyroid* **2003,** *13* (3), 279–282.
22. Mandel, S. J.; Brent, G. A.; Larsen, P. R. Levothyroxine therapy in patients with thyroid disease. *Ann. Intern. Med.* **1993,** *119* (6), 492–502.
23. Wartofsky, L. Levothyroxine: therapeutic use and regulatory issues related to bioequivalence. *Expert Opin. Pharmacother.* **2002,** *3* (6), 727–732.
24. Harrington, C. R.; Barger, G. Thyroxine. III. Constitution and synthesis of thyroxine. *Biochem. J.* **1927,** *21,* 169–183.
25. Harington, C. R. Resolution of dl-thyroxine. *Biochem. J.* **1928,** *22,* 1429–1435.
26. Ginger, L. G.; Anthony, P. Z. Thyroxine, US 2889363 (1959).
27. Anthony, P. Z.; Ginger, L. G. Thyroxine, US 2889364 (1959).
28. Wegner, R.; Rudnick, K. L-3,3′,5-Triiodothyronine and L-thyroxine, DD 65933 (1969).
29. Every, C. E. Thyroxine, GB 598691 (1948).
30. Turner, C. W.; Reineke, E. P. Thyroxine from diiodotyrosine, US 2435947 (1948).
31. Khamar, B. M.; Gurusamy, R.; Ravi, M. N.; Reddy, V. M.; Edde, B.; Ponnaiah, R.; Modi, I. A. An improved process for the preparation of levothyroxine sodium with reduced levels of impurities, WO 2009136249 (2009).
32. Zhang, W.; Luo, F.; Ji, M.; Chen, H., Process for preparation of L-thyroxine sodium with L-tyrosine as raw material, CN 102199103 (2010).
33. Chalmers, J. R.; Dickson, G. T.; Elks, J.; Hems, B. A. Synthesis of thyroxine and related substances. V. A synthesis of L-thyroxine from L-tyrosine. *J. Chem. Soc.* **1949,** 3424–3433.
34. Bal'on, Y. G.; Simurov, O. V.; Stel'makh, A. M. Synthesis of thyroid hormones. *Farm. Zh. (Kiev, Ukr.)* **1999,** *4,* 51–55.
35. Salamonczyk, G. M.; Oza, V. B.; Sih, C. J. A concise synthesis of thyroxine (T4) and 3,5,3′-triiodo-L-thyronine (T3). *Tetrahedron Lett.* **1997,** *38* (40), 6965–6968.
36. Bell, N. V.; Bowman, W. R.; Coe, P. F.; Turner, A. T.; Whybrow, Del. Synthesis of thyroxine: biomimetic studies. *Can. J. Chem.* **1997,** *75* (6), 873–883.
37. Kochergin, P. M.; Palei, R. M.; Kravchenko, A. N.; Popova, E. V. Methods for synthesizing thyroxine and triiodothyronine. *Khim.-Farm. Zh.* **1990,** *24* (6), 43–49.
38. Chemburkar, S. R.; Deming, K. C.; Reddy, R. E. Chemistry of thyroxine: an historical perspective and recent progress on its synthesis. *Tetrahedron* **2010,** *66* (11), 1955–1962.
39. von Mutzenbecher, P. Formation of thyroxine from diiodotyrosine. *Z. Physiol. Chem.* **1939,** *261,* 253–256.
40. Chalmers, J. R.; Dickson, G. T.; Elks, J.; Hems, B. A. Synthesis of thyroxine and related substances. V. A synthesis of L-thyroxine from L-tyrosine. *J. Chem. Soc.* **1949,** 3424–3433.

Chapter 26

Hyperglycemic and Hypoglycemic Drugs

The most common disorder of carbohydrate metabolism can be classified as hyperglycemia (high blood glucose) which happens when the body has too little insulin or can't use insulin. Insulin, is a hormone that plays multiple roles in the body; it processes glucose, activates glucose transporters, facilitates the conversion of glucose to glycogen-polysaccharide $(C_6H_{10}O_5)x$, a form in which glucose is stored in tissues as energy source, and hypoglycemia (low level of blood glucose), which usually happens on increase of insulin release caused by external insulin injection, or use of hypoglycemic agents

Symptoms of hyperglycemia include fatigue, irritability, weight loss, increased thirst and hunger, frequent urination, recurrent infections, and blurred vision. Chronic hyperglycemia can contribute to many complications, such as poor wound healing, kidney failure, cardiac events, blindness, and need of amputation.

Symptoms of hypoglycemia are headache, dizziness, sweating, hunger, shakiness, even seizures and coma, and the immediate need for glucose consumption or glucagon injection.

Diabetes mellitus is a disorder of carbohydrate metabolism associated with hyperglycemia. According to the World Health Organization, at least 171 million people worldwide suffer from diabetes mellitus. It can be defined as a group of heterogeneous disorders with the common elements of hyperglycemia and glucose intolerance caused by insulin deficiency, impaired effectiveness of insulin action, or both. There are several types of diabetes: type 1 diabetes mellitus is a severe deficiency in insulin secretion associated with destruction of pancreatic β cells, resulting from atrophy of the islets of Langerhans and causing hyperglycemia; type 2 diabetes mellitus, which is also known as non–insulin-dependent diabetes mellitus. Type 2 diabetes mellitus consists of an array of dysfunctions resulting from the combination of resistance to insulin action, inadequate insulin secretion, and excessive or inappropriate glucagon secretion which brings to body's inability to compensate increased insulin production. A more complex form of type 2 diabetes—insulin-resistant diabetes—occurs when insulin is produced and circulated in an organism whose receptors have become insensitive. Special types of diabetes could be the result of complications of other diseases, or of the action

Synthesis of Best-Seller Drugs. http://dx.doi.org/10.1016/B978-0-12-411492-0.00026-2

of immunosuppressive drugs or chemicals. A final form of diabetes, called gestational diabetes, can occur with pregnancy.

Antidiabetic medications treat diabetes mellitus by lowering glucose levels in the blood. With the exceptions of insulin, all are administered orally and are called oral hypoglycemic agents or oral antihyperglycemic agents.

26.1 INSULIN AND INSULIN PREPARATIONS

Insulin

Diabetes mellitus type 1 is a disease caused by the lack of insulin and its treatment requires insulin supplementation.

Insulin produced by the pancreatic β cells and is one of the most important hormones in mammalians regulating carbohydrate and fat metabolism in the body for energy production and many other vital activities. Insulin performs also unique functions within the central nervous system (CNS). It crosses the blood–brain barrier, affecting feeding and cognition through CNS mechanisms largely independent of glucose utilization, and regulates lipid metabolism and cholesterol synthesis. Insulin also reduces myocardial O_2 consumption and increases cardiac efficiency. It seems to augment cardiomyocyte contraction, while it affects favorably myocardial relaxation, stimulates vascular endothelial growth factor and thereby angiogenesis, suppresses apoptosis, promotes cell survival, and ameliorates both myocardial microcirculation and coronary artery resistance, leading to increased blood perfusion of myocardium [1-3]. More than 250 books, 80,000 reviews, and half a million articles have been published about insulin.

Insulin was the first protein to have its amino acid sequence sequenced, in 1955 by Fred Sanger, bringing him a Nobel prize in 1958. Insulin was also the first peptide hormone to be measured by radioimmunoassay, bringing a Nobel Prize in 1977 to Rosalyn Yalow "for the development of radioimmunoassays of peptide hormones." It was also the first protein to be synthesized in microorganisms by recombinant DNA technology in the late 1970s.

For 30 years, diabetic patients were successfully treated with insulin without researchers knowing the chemical structure of the hormone, which is a 51-amino-acid polypeptide.

Insulin (pig) has two chains: A containing 21, and B containing 30 amino acid residues constrained by one intrachain and two interchain disulphide bonds [4-6]. Insulin produced in other organisms may have a slightly different amino acid sequence, or extra amino acids, but the next levels of structure are not greatly altered by these variations. Notably, the positions of these three disulfide bonds are invariant in mammalian forms of insulin (Fig. 26.1.).

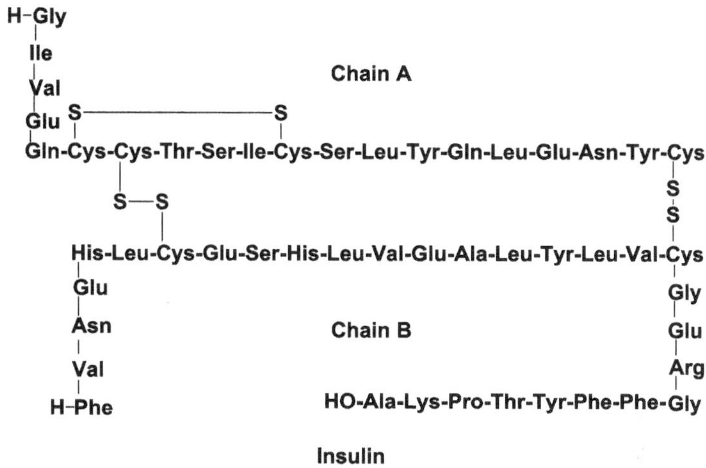

FIG. 26.1 Structure of insulin.

Two disulfide bridges (residues A7 to B7, and A20 to B19) covalently tether the chains, and chain A contains an internal disulfide bridge (residues A6 to A11).

The A chain, which is fairly compact, contains two sections of α helix (A2 Ile-A8 Thr and A13 Leu-A19 Tyr). The B chain consists of a larger section of α helix (B9 Ser-B19Cys) and the smaller glycine residues at B20 Gly and B23Gly. It appears to wrap around the A chain. Insulin can form into granules consisting of hexamers (six insulin molecules grouped around two Zn^{2+} ions) as a result of interactions between hydrophobic surfaces and the three-dimensional structure of insulin, which was ultimately solved by Dorothy Hodgkin and colleagues in 1976 [7]. This toroidal form is the one in which insulin is stored in the β cells and secreted into the bloodstream. Insulin may also form into dimers (double units); however, the active form is apparently a single unit.

At present, insulin is commercially available in different forms for subcutaneous, intramuscular, and intravenous injections.

Insulin Preparations

Until the 1980s, insulin that came in vials, was produced from pig and cow pancreases using a protocol according to which from 2 tons of animal material 200 to 250 g of purified insulin was prepared, and the product often caused allergic reactions, sometimes even infections. Today, insulin is produced mainly biotechnologically using a safe strain of *Escherichia coli* bacteria into which a human gene for insulin was introduced. Insulin is the first recombinant DNA drug product in the world.

Newly created pharmaceutical insulin preparations are classified as rapid, intermediate, or long acting according to their onset and duration of action.

Rapid-acting insulin preparations include Humulin R, which is the brand name of biosynthetic insulin and which is identical in chemical structure to human insulin. Others are lispro, aspart, and glulisine insulins.

Insulin lispro is the first human insulin analogue produced by recombinant DNA technology via site-directed mutation. Amino acids 28 and 29 of the B chain in lispro are switched (hence the name lispro) instead of the pro-lys sequence in regular human insulin.

Insulin aspart is another analogue prepared by replacing the B28 Pro with aspartic acid (hence the name aspart).

Glulisine is the newest analogue and differs from human insulin in the replacement of the B3 asparagine with lysine and the B29 lysine with glutamic acid (hence the name glulisine).

The importance of B28 Pro and B29 Lys on the insulin properties was established by systematically truncating the C terminus of the B chain. Insulin properties can be drastically altered by substitution of one or two key amino acids [8,9].

Intermediate-acting insulin preparations include Humulin N, Novolin N, and lente insulin.

Long-acting human insulin preparation is represented by glargine, a human insulin analogue developed by recombinant DNA technology that has multi-mutation sites in both the A and B chains of human insulin. In the A chain, the amino acid at position A21 (asparagine) is replaced with glycine, while the B chain has two additional arginine residues added to the C-terminus.

Biosynthetic insulin analogues: lispro (Humalog) [10-12], aspart (NovoLog) [13,14], and glulisine (Lantus) [15,16] are included in the list of Top 200 Drugs by sales for the 2010s.

26.2 ORAL ANTIDIABETIC DRUGS

Type 2 diabetes is the most common form of diabetes; it causes blood glucose levels to rise higher than normal. It occurs either from insufficient insulin production or the inability of cells to adequately respond to normal levels of insulin, which is called insulin resistance.

The primary etiology of diabetes is unknown. Diabetes could be a consequence of genetics, lifestyle, some other diseases, or the action of a drug or chemical agent that causes blood glucose levels to rise higher than normal. Type 2 diabetes, which is when the body does not use insulin properly and is the most common form, can be successfully controlled with suitable drugs. Current treatment includes insulin secretagogues, sulfonylureas and meglitinides (stimulate insulin secretion); biguanides (suppress hepatic glucose production); insulin sensitizers; thiazolidinediones–glitazones (improve insulin sensitivity and peripheral glucose uptake); α-glucosidase inhibitors (delay digestion and absorption of intestinal carbohydrate); glucagon-like peptide 1 agonists; dipeptidyl peptidase-IV inhibitors; and some newer drugs that are in development [17-28].

Insulin Secretagogues

Sulfonylureas

The story of hypoglycemic sulfonamides started in France in the spring of 1942, during the early years of World War II. As usually happens in the history of drug discovery, Dr. Marcel Janbon, at the Clinic of the Montpellier Medical School, noted that patients who were receiving the antibacterial sulfonamide drug 2254 RP (**26.2.1**) (Fig. 26.2.) developed symptomatic hypoglycemia and concluded that this compound probably stimulates insulin release. This observation and hypothesis became the moving force for the creation of hypoglycemic sulfonamides, later transformed to sulfonylureas class of insulin secretagogues.

Sulfonylureas stimulate insulin secretion from the pancreas; because of this, they are sometimes referred to as "insulin secretagogues." Over time, such therapy fails, and additional treatment is required.

In 1955, carbutamide (**26.2.2**) (Fig. 26.2.), the first effective hypoglycemic sulfonylurea, was introduced into clinical study, but because of toxic effects was withdrawn from further study.

2254 RP 26.2.1 Carbutamide 26.2.2

FIG. 26.2 Structure of 2254 RP and carbutamide.

Tolbutamide (**26.2.3**), the first safe and effective hypoglycemic sulfonylurea, was introduced into clinical practice in the United States in 1956 and was followed over the next 10 years by three additional drugs: chlorpropamide (**26.2.4**), acetohexamide (**26.2.5**), and tolazamide (**26.2.6**). These four drugs are known as the first-generation sulfonylureas (Fig. 26.3.).

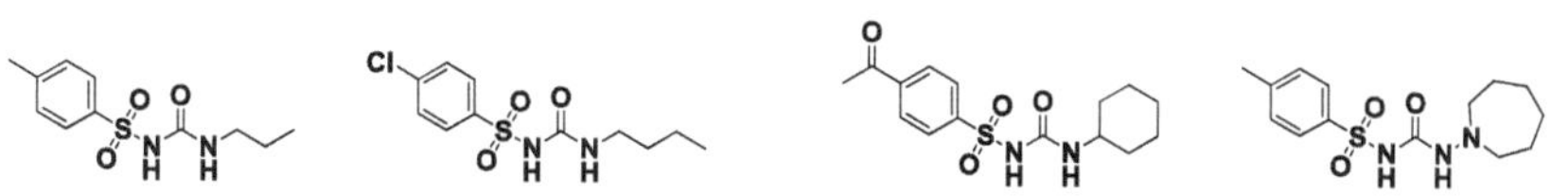

Tolbutamide 26.2.3 Chlorpropamide 26.2.4 Acetohexamide 26.2.5 Tolazamide 26.2.6

FIG. 26.3 The first-generation hypoglycemic sulfonylureas.

Glyburide (**26.2.7**), glipizide (**26.2.8**), and gliclazide (**26.2.9**), which have been called the second-generation sulfonylureas, were introduced into clinical practice in 1969 to 1979 decade. Pharmaceutical companies in European countries and Japan also produce other second-generation sulfonylureas such as glibornuride (**26.2.10**), gliquidone (**26.2.11**), glisoxepide, (**26.2.12**), and glyclopyramide (**26.2.13**) (Fig. 26.4.). These drugs, have 50 to 100 times greater hypoglycemic potency than the first-generation agents and a duration of action of longer than 24 hours.

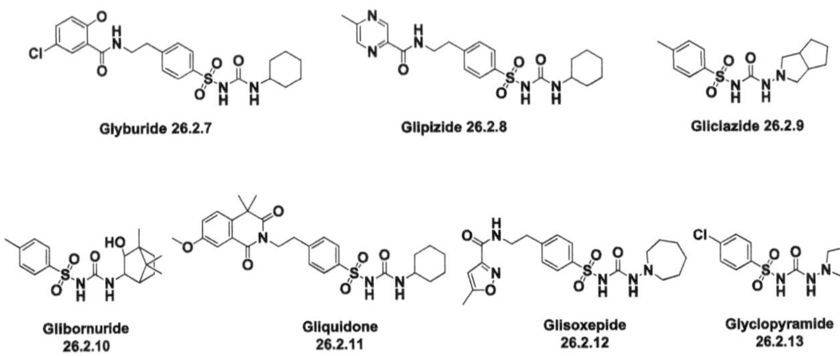

FIG. 26.4 The second-generation hypoglycemic sulfonylureas.

Glimepiride (**26.2.14**) (Fig. 26.5.), proposed in 1995, is considered a third-generation sulfonylurea.

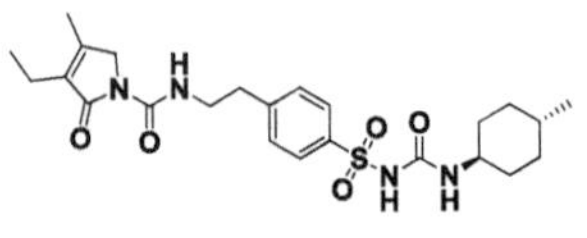

Glimepiride 26.2.14

FIG. 26.5 Structure of glimepiride, the third-generation sulfonylurea.

Glimepiride has some advantages over the members of the second-generation group, such as a lower risk of hypoglycemia, no interaction with cardiovascular adenosine triphosphate–sensitive potassium channels, responsible for various forms of cardiac stress, and a possibility that it may increase insulin sensitivity.

Sulfonylureas, and glyburide particularly, are the most commonly prescribed medications for type 2 diabetes mellitus worldwide. Although guidelines do not recommend the use of sulfonylureas for first-line treatment, the drugs are still commonly prescribed this way in many countries [29-35].

Interestingly, a huge amount of newly synthesized sulfonylureas demonstrated herbicidal properties and have been introduced to the pesticide market since 1995, or are currently in their later stages of development. These include flupyrsulfuron-methyl-sodium, sulfosulfuron, iodosulfuron-methyl-sodium, mesosulfuron-Me, tritosulfuron, monosulfuron, and monosulfuron-ester for use in cereals; ethoxysulfuron, azimsulfuron, cyclosulfamuron, flucetosulfuron, orthosulfamuron, propyrisulfuron, and metazosulfuron for use in rice; foramsulfuron for use in maize; oxasulfuron for use in soybeans; and trifloxysulfuron-sodium for use in sugarcane and cotton [36].

Meglitinides (Glinides)

Meglitinide (**26.2.15**) is the prototype of a chemically heterogeneous new class of insulin-secreting agents commonly called "glinides," characterized by a rapid onset and short duration of action. This class of "insulin secretagogues" is represented by three agents: repaglinide (**26.2.16**), nateglinide (**26.2.17**), and mitiglinide (**26.2.18**) (Fig. 26.6.).

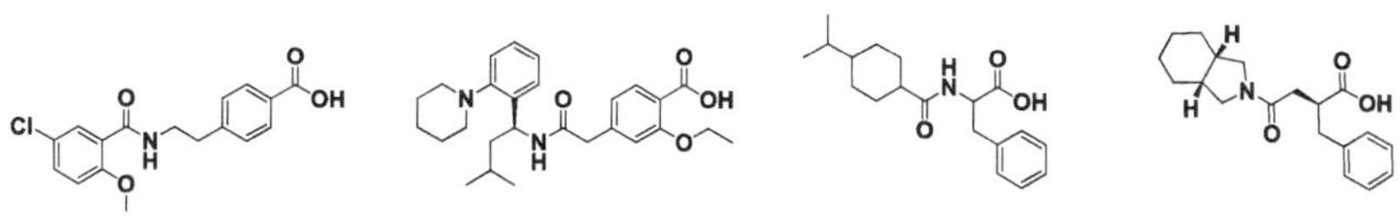

Meglitinide 26.2.15 Repaglinide 26.2.16 Nateglinide 26.2.17 Mitiglinide 26.2.18

FIG. 26.6 Structure of "glinides."

Meglitinides are rapid-acting insulin secretion–stimulating agents that are approved for the treatment of type 2 diabetes as adjuncts to metformin therapy for those patients with continued postprandial hyperglycemia. They act synergistically not only with metformin and thiazolidinediones (pioglitazone and rosiglitazone), but can be also combined with long-acting insulin [37,38].

These agents may be used as first-line, second-line, or adjunct therapy behind metformin for treatment of type 2 diabetes mellitus. Sulfonylureas and meglitinides are effective treatments, but cumulative data over decades of research raise concerns regarding universal prescribing because of reported, blunting of ischemic preconditioning, the incidence of hypoglycemia, modest weight gain and the unproven link to cancer.

Repaglinide is included in the list of Top 200 Drugs by sales for the 2010s.

Repaglinide–Prandin

Repaglinide is the first of a new class of oral antidiabetic agents, the glinides, which are designed to normalize glucose levels in patients with type 2 diabetes mellitus. Like the sulfonylureas, repaglinide reduces blood glucose by stimulating insulin release from pancreatic β cells, but differs from these and other antidiabetic agents in its structure, binding profile,duration of action, and mode of excretion [39-46].

Synthesis of repaglinide consists of a reaction of 2-(3-ethoxy-4-(ethoxycarbonyl)phenyl)acetic acid (**26.2.19**) with (S)-3-methyl-1-(2-(piperidin-1-yl)phenyl)butan-1-amine (**26.2.20**) using as a coupling agent dicyclohexylcarbodiimide with further hydrolysis of the ethoxycarbonyl group in the obtained product (**26.2.21**) with base (Scheme 26.1.).

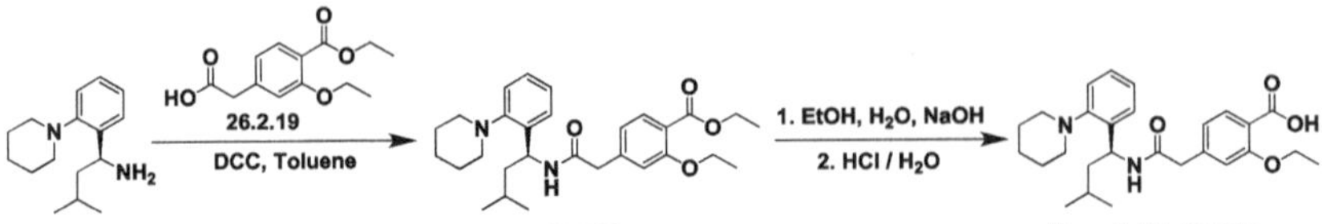

SCHEME 26.1 **Synthesis of repaglinide.**

To start, the benzoic acid derivative was prepared from commercially available 2-hydroxy-4-methylbenzoic acid (**26.2.22**), which was simultaneously etherified and esterified on heating in autoclave with ethyl bromide in presence of potassium carbonate in acetone to produce the product (**26.2.23**). The last was reacted with N-bromosuccinimide in CCl_4 to produce the benzylic bromination product (**26.2.24**), which, with sodium cyanide in phase-transfer conditions, produced nitrile (**26.2.25**). The obtained nitrile was converted to an ester derivative (**26.2.26**) and then selectively alcoholized to the requested acid (**26.2.19**) (Scheme 26.2.).

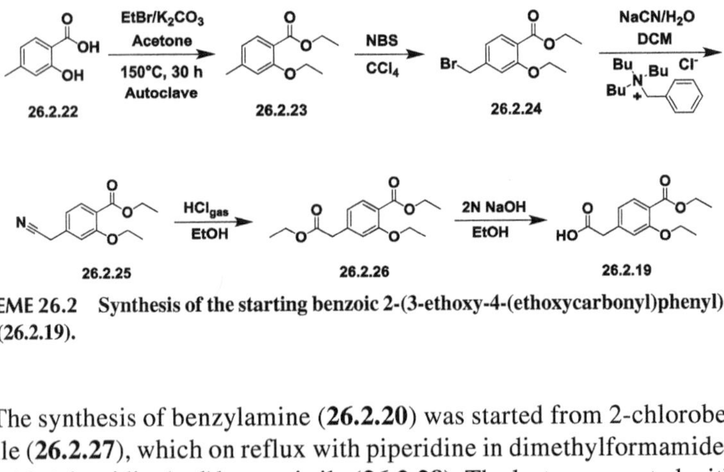

SCHEME 26.2 **Synthesis of the starting benzoic 2-(3-ethoxy-4-(ethoxycarbonyl)phenyl)acetic acid (26.2.19).**

The synthesis of benzylamine (**26.2.20**) was started from 2-chlorobenzonitrile (**26.2.27**), which on reflux with piperidine in dimethylformamide produced 2-(piperidin-1-yl)benzonitrile (**26.2.28**). The last was reacted with the Grignard reagent i-butylmagnesium bromide to produce ketone (**26.2.29**). The obtained ketone was converted to imine (**26.2.30**) with 100% yield by dropping into a cold mixture of concentrated ammonia and saturated aqueous ammonium chloride solution. Ketimine (**26.2.30**) was acylated with acetic anhydride, and the resulting enamide (E/Z)-17 was hydrogenated in methanol using (S)-Ru$(OAc)_2$(BINAP) as a catalyst to prepare the corresponding amide, which was hydrolyzed further with hydrochloric acid to (S)-3-methyl-1-(2-(piperidin-1-yl)phenyl)butan-1-amine (**26.2.20**) [47,48] (Scheme 26.3.).

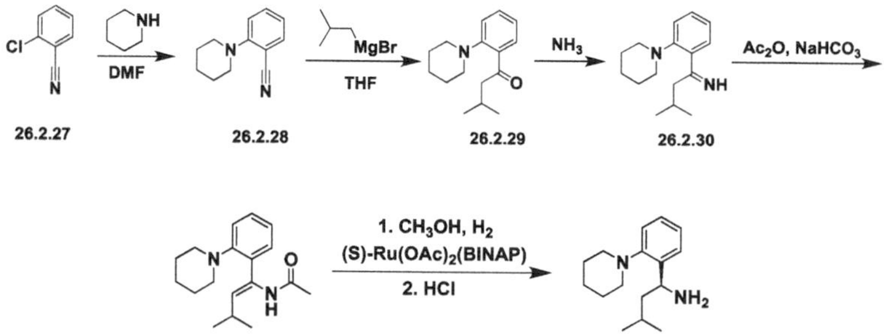

26.2.27 26.2.28 26.2.29 26.2.30 26.2.31 26.2.20

SCHEME 26.3 Synthesis of the starting (S)-3-methyl-1-(2-(piperidin-1-yl)phenyl)butan-1-amine (26.2.20).

Biguanides

The biguanides represent another class of hypoglycemic drugs. Their history started in the 1920s with the investigations of the sugar-lowering effects of a weed goat's rue (*Galega officinalis*), which has been known since medieval times to relieve the symptoms of diabetes. It turned out to contain an alkaloid galegine (**26.2.32**) that was evaluated in unsuccessful clinical trials in patients with diabetes in the 1920s and 1930s. In the 1950s, three biguanides—metformin (**26.2.33**), phenformin (**26.2.34**), and buformin (**26.2.35**)—were proposed as hypoglycemic drugs (Fig. 26.7.).

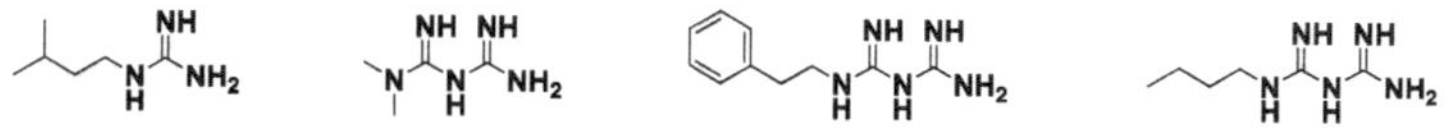

Galegine 26.2.32 Metformin 26.2.33 Phenformin 26.2.34 Buformin 26.2.35

FIG. 26.7 Biguanide hypoglycemic drugs.

Metformin and phenformin were withdrawn in 1978 because use of phenformin led to increased incidences of lactic acidosis. In 1995, metformin, was reapproved in the United States after being in use in Europe for 20 years. Currently, metformin represents the cornerstone of treatment of type 2 diabetes and has been used as the first-line treatment, either alone or in combination with other diabetes agents. Metformin exerts its prevailing, glucose-lowering effect by inhibiting hepatic gluconeogenesis and opposing the action of glucagon [49-54]. Furthermore, in recent years, an increasing number of studies have found that metformin has additional cardiovascular protective effects, including the inhibition of atherosclerosis, heart failure, and myocardial infarction. It has been described as a geroprotector, and several studies show that metformin can slow down the rate of aging [55,56]. Other research suggests metformin may help to prevent cancer [57,58]. Metformin is included in the list of Top 200 Drugs by sales for the 2010s.

Metformin–ACTOplus

The synthesis of metformin with high yields [59-64], involves the reaction of dicyanamide (**26.2.36**) dimethylamine on cooling followed with a slow flow of an equimolar amount of gaseous hydrogen chloride. The obtained required product, metformin (**26.2.33**) hydrochloride, recrystallized from methanol (Scheme 26.4.).

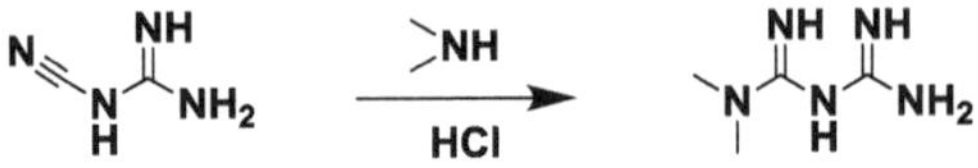

26.2.36 **Metformin 26.2.33**

SCHEME 26.4 **Synthesis of metformin.**

Thiazolidinediones

Peroxisome proliferator-activated receptor (PPAR) family of nuclear receptors plays a major regulatory role in energy homeostasis and metabolic function. PPAR-γ is a key regulator of glucose and lipid metabolism and therefore an important pharmacological target for metabolic diseases. Activation of PPAR-γ, particularly, causes insulin sensitization and enhances glucose metabolism [65-67]. Thiazolidinedione derivatives–glitazone-type drugs, are perfect antidiabetic agents that increase the insulin sensitivity, via activation of PPAR-γ. Moreover, recent findings supporting a role of PPAR-γ in inflammation and cell growth have promoted the investigation of PPAR-γ agonists as experimental drugs for some chronic diseases, such as atherosclerosis and cancer. There is strong evidence that insulin resistance is involved in the development of not only hyperglycemia, but also dyslipidemia, hypertension, hypercoagulation, vasculopathy, and, ultimately, atherosclerotic cardiovascular disease.

Thiazolidinediones (TZDs), are drugs universally used as antidiabetic agents in patients with type 2 diabetes [68-71].

The discovery of thiazolidinediones was the result of the search for new, more potent fibrates, which led to the discovery of the first thiazolidinedione, the prototypical compound ciglitazone (**26.2.37**), which was never marketed. In 1997, troglitazone (**26.2.38**), which improved insulin sensitivity and thus reduced blood glucose, became the first thiazolidinedione to be approved for clinical use, but it was withdrawn in 2000 because it caused liver damage. Two other thiazolidinediones, pioglitazone (**26.2.39**) and rosiglitazone (**26.2.40**), were approved in 1999. Both of are included in the list of Top 200 Drugs by sales for the 2010s.

The use of rosiglitazone is slightly restricted because of its potential to cause cardiovascular ischemia. It should be used with caution for patients with congestive heart failure. The most common side effect is edema. Frequent side effects are weight gain. Anemia and osteoporosis may also occur.

Numerous other TZDs manufactured to the highest quality standards, such as englitazone (**26.2.41**), darglitazone (**26.2.42**), netoglitazone (**26.2.43**), rivoglitazone (**26.2.44**), and others, were and are in trials (Fig. 26.8.).

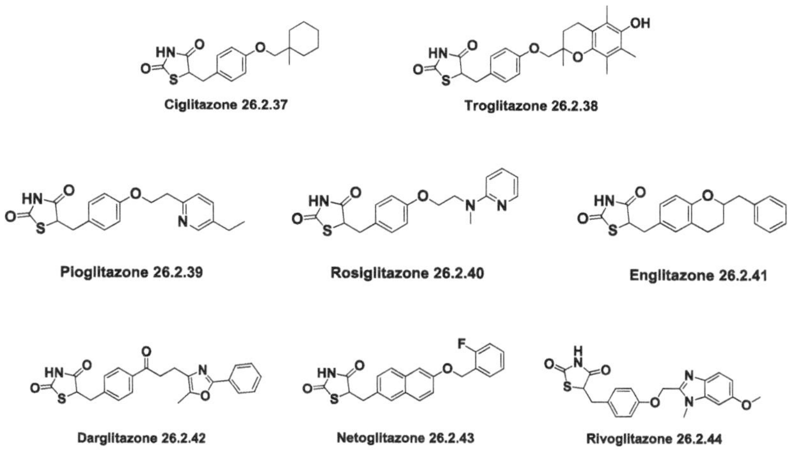

FIG. 26.8 **Antidiabetic thiazolidinediones.**

Pioglitazone–Actos

Pioglitazone is an oral antidiabetic agent that decreases insulin resistance in adipose tissue, liver and muscles which action is mediated by its link to PPAR-γ, which improves insulin sensitivity, decreases hepatic glucose production, and increases glucose uptake in the peripheral tissues. Beyond these effects on glucose metabolism, pioglitazone has positive effects on lipid metabolism, blood pressure, and endothelial function [72-83].

The first described synthesis of pioglitazone started with condensation of 1-fluoro-4-nitrobenzene (**26.2.45**) using 2-(5-ethyl-2-pyridyl)ethanol (**26.2.46**) to produce pyridylethoxybenzene (**26.2.47**), which was hydrogenated with palladium on charcoal catalyst to produce the aromatic amine (**26.2.48**). Diazotization of the aromatic amine (**26.2.48**) in acetone methanol mixture and continuing the workup with hydrobromic acid, and coupling with methylacrylate in the presence of copper (I) oxide (the Meerwein arylation) produced the methyl 2-bromo-propanoate derivative (**26.2.49**). This underwent cyclocondensation with thiourea, followed by hydrolysis of the intermediate imino compound to produce the desired pioglitazone (**26.2.39**) [84,85] (Scheme 26.5.).

Another set of the synthesis starts from 4-(2-(5-ethylpyridin-2-yl)ethoxy) benzaldehyde (**26.2.51**) prepared by different ways. The 2-(5-ethyl-2-pyridyl) ethyl toluene sulfonate (**26.2.50**) obtained from the reaction of 2-(5-ethyl-2-pyridyl)ethanol (**26.2.46**) with *p*-toluene sulfonic acid in phase transfer catalysis conditions (benzyltributyl ammonium chloride, sodium hydroxide)

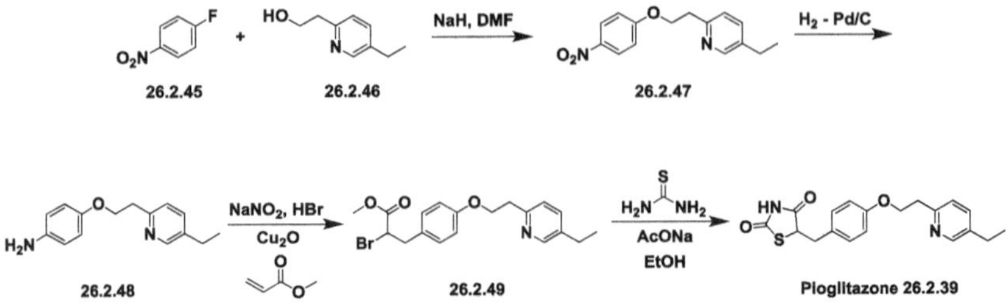

SCHEME 26.5 Synthesis of pioglitazone.

was condensed with 4-hydroxybenzaldehyde in aqueous sodium hydroxide and the obtained aldehyde (**26.2.51**) was condensed with thiazolidine-2,4-dione in Knoevenagel-type reaction conditions to produce the benzylidene derivative (**26.2.52**), which was hydrogenated using palladium on charcoal catalyst to produce the desired pioglitazone (**26.2.39**) [86,87] (Scheme 26.6.). Several attempts have been made to modify this synthesis [88].

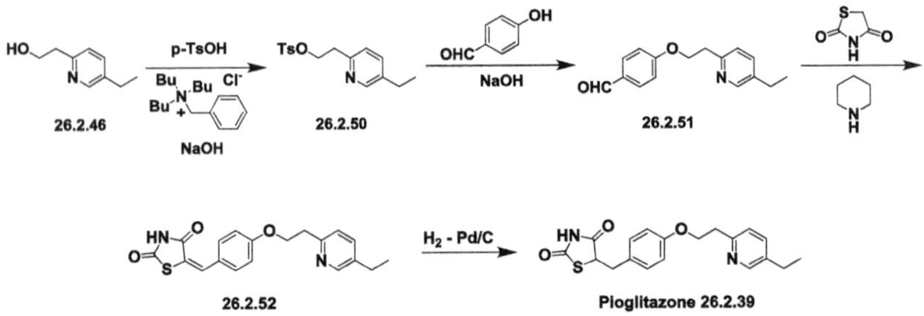

SCHEME 26.6 Synthesis of pioglitazone.

Rosiglitazone–Avandia

The synthesis of rosiglitazone (**26.2.40**) was proposed by a method closely related to that described above for pioglitazone (**26.2.39**).

Heating under reflux (120°C for 15 hours) a mixture of 2-chloropyridine (**26.2.54**) in excess amount of 2-(methylamino)ethanol (**26.2.53**) produced a good yield of amino alcohol (**26.2.55**). Transformation of the amino alcohol (**26.2.55**) to the sodium alcoholate with sodium hydride in DMF and further reaction with 4-fluorobenzaldehyde gave the corresponding aryl aldehyde (**26.2.56**), which on Knoevenagel condensation with 2,4-thiazolidinedione in refluxing toluene containing a catalytic amount of piperidinium acetate or benzoate, produced the benzylidene derivative (**26.2.57**). Because of poisoning of the Pd/C catalyst in standard reduction conditions, it was hydrogenated by using the magnesium–methanol electron transfer technique [89], which is preactivated with iodine magnesium turnings in methanol, to produce the rosiglitazone (**26.2.40**) [90-97] (Scheme 26.7.).

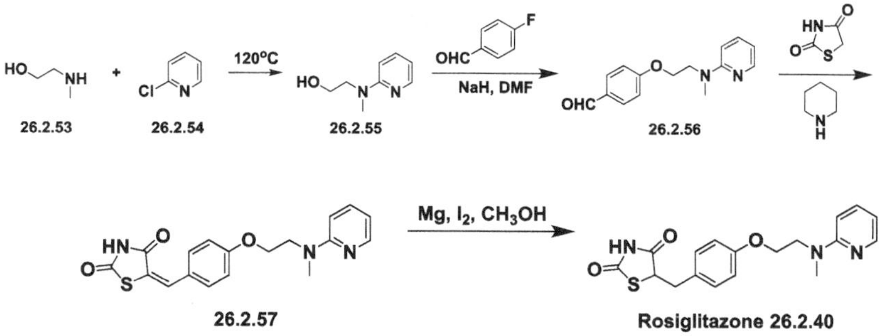

SCHEME 26.7 Synthesis of rosiglitazone.

Rosiglitazone is an effective antihyperglycemic agent of the thiazolidinedione class that significantly improves glycemic control primarily by increasing hepatic and peripheral insulin sensitivity. In general, rosiglitazone as monotherapy or in combination with other antihyperglycemic agents, improves glycemic control in adults with type 2 diabetes mellitus [98-103].

The drug has been extensively used and many warnings on its potential risks were ignored, mostly because of an aggressive commercial strategy. Available data show that rosiglitazone increases the risk for cardiac ischemia in patients. Some papers raised doubts on the safety of the drug and started a serious dispute which led to the unveiling of many other side effects and emphasizes the need for a more accurate evaluation of any drug [104-111].

Recently, a number of naturally occurring compounds, ligands of PPAR-γ, chemically unrelated to thiazolidinediones have been discovered [112].

New compounds of this series—dual PPAR-α/γ agonists, so-called "glitazars"—are reported for the management of hyperglycemia and hyperlipidemia. Compounds such as ragaglitazar (**26.2.58**), tesaglitazar (**26.2.59**), KRP-297 (**26.2.60**), muraglitazar (**26.2.61**), **and** DRF-2519 (**26.2.62**) were and are under development. Ragaglitazar, and KRP-297 were discontinued following cases of carcinogenicity. But muraglitazar and some others are still in trials, and research and development of new dual PPAR-α/γ agonists is continuing [113] (Fig. 26.9.).

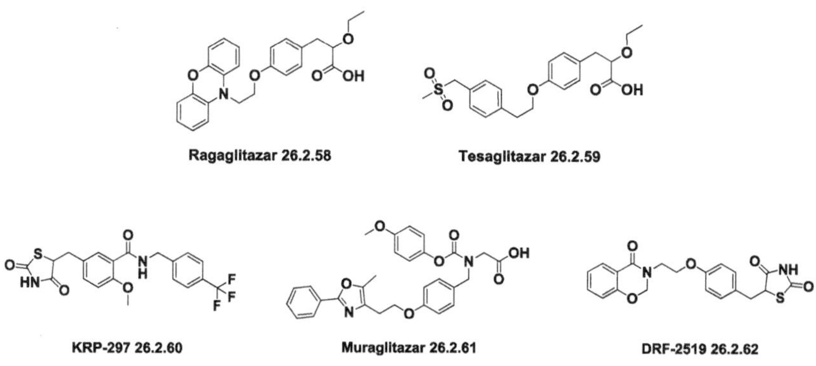

FIG. 26.9 Structure of "glitazars."

α-Glucosidase Inhibitors

Glycoside trimming enzymes are very important in a broad range of metabolic pathways, including carbohydrate digestion in the intestinal tract.

Glucosidases in general are a group of key intestinal enzymes involved in the digestion of carbohydrates, and α-glucosidase is an enzyme that cleaves the glycosidic bond of the oligosaccharides to liberate glucose and its inhibition retards the carbohydrate digestion.

α-Glucosidase inhibitors [114-124], which delay the digestion and absorption of carbohydrates, thereby inhibiting postprandial hyperglycemia and hyperinsulinemia, are of interest as promising therapeutic potential in the treatment of diabetes.

α-Glucosidase inhibitors are novel hypoglycemic drugs that have been used in the treatment of type 2 diabetes mellitus since the 1980s. These drugs significantly reduce the postprandial rise in glycemic and plasma insulin levels both in nondiabetics and in type 2 diabetic patients. They play a relatively minor role in the treatment of type 2 diabetes mellitus and currently only three drugs belonging to this category are in the market: acarbose (**26.2.63**), which was introduced in 1996 [125,126], miglitol (**26.2.64**) [127,128], and voglibose (**26.2.65**) [129] (Fig. 26.10.).

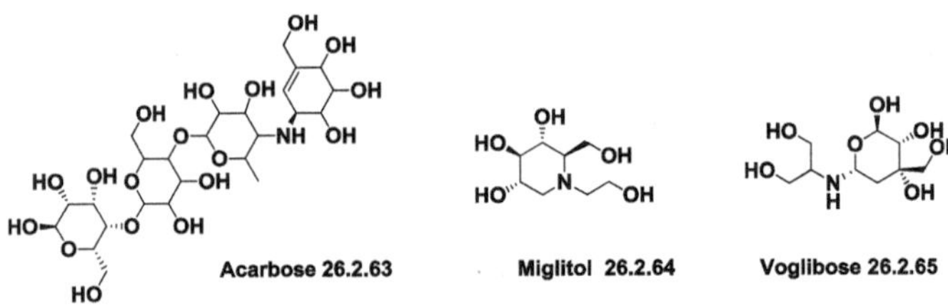

FIG. 26.10 α-Glucosidase inhibitors.

Glucagon-Like Peptide 1 Agonists and Dipeptidyl Peptidase-IV Inhibitors

In the first decade of the 21st century it is becoming evident that type 2 diabetes originates not only from a decline in insulin secretion capacity or insulin resistance, but also from disturbing the equilibria in the secretion or action of some other hormones. Recently, it was revealed that normal glucose homeostasis is controlled by insulin as well as by other glucoregulatory hormones, including glucagon, glucagon-like peptide-1 (GLP-1), insulinotropic peptide (GIP), amylin, and, probably, some others. In general, GLP-1, secreted by L cells, glucose-dependent GIP, secreted by K cells, and amylin, secreted from the pancreatic β cells, are a group of hormones that increase insulin release and thus stimulate a decrease of blood glucose levels and are defined as members of the secretin family of hormones called incretins [130-141].

Glucagon-Like Peptide 1 Agonists

Therapeutic agents mimicking the functions of GLP-1 (“incretin mimetics,” a group of hormones that stimulate a decrease of glucose levels in blood) are a relatively new group of drugs for treatment of type 2 diabetes; they are now commercially available.

GLP-1 itself is a peptide consisting of 37 amino acids. Its activation is mediated by the cleavage of six amino acids from the N-terminus. Because of the short plasma half-life, a therapy with GLP-1$_{(7-36)}$ itself is not possible [142-144].

Incretin mimetics offer a new and interesting treatment modality in diabetes and are currently implemented in medicinal practice with exenatide (**26.2.66**) and liraglutide (**26.2.67**), which act as agonists of the GLP-1 receptor, mimicking the effects of the physiological ligand. They stimulate the release of insulin by the pancreas, inhibiting the release of glucagon and slowing glucose absorption into the bloodstream. The most common adverse event associated with GLP-1 receptor agonists is mild nausea, which lessens over time.

Both exenatide and liraglutide are included in the list of Top 200 Drugs by sales for the 2010s.

Exenatide–Byetta

Exenatide (**26.2.66**) (Fig. 26.11.)is a synthetic version of the natural compound exendin-4, which is a GLP-1 receptor agonist (incretin mimetic) [145-156]. It is indicated as an adjunct to diet and exercise to improve glycemic control in adults with type 2 diabetes mellitus and works by stimulating the pancreas to secrete insulin.

Exenatide blunts the postprandial rise of plasma glucose by increasing glucose-dependent insulin secretion, suppressing inappropriately high glucagon secretion, and delaying gastric emptying.

Historically, exenatide was discovered as exendin-4, a protein originally isolated from the venom of *Heloderma suspectum* (Gila monster) that shares 53% sequence identity with GLP-1. Synthetically prepared exendin-4 (exenatide) has higher biological stability than GLP-1. As a side effect, exenatide use can cause nausea, vomiting, diarrhea, dizziness, headache, weakness, acid stomach, and weight loss.

Commercially, exenatide is produced by direct chemical synthesis using solid-phase peptide chemistry protocols.

His-Gly-Glu-Gly-Thr-Phe-Thr-Ser-Asp-Leu-Ser-Lys-Gln-Met-Glu-Glu-Glu-Ala-Val-Arg-Leu-Phe-Ile-Glu-Trp-Leu-Asn-Gly-Gly-Pro-Ser-Ser-Gly-Ala-Pro-Pro-SerNH_2

Exenatide 26.2.66

FIG. 26.11 Structure of exenatide.

Liraglutide–Victoza

Liraglutide (**26.2.67**) (Fig. 26.12.)is a full agonist of the GLP-1 receptor and shares 97% of its amino acid sequence identity with human GLP-1. Because of the high sequence identity with native GLP-1, the antibody response to liraglutide treatment is significantly lower than with exenatide. The Lys 34 in GLP-1 was substituted for by an arginine and a N-e(γ Glu[N-a-hexadecanoyl]) residue and was attached to Lys 26. This lipid anchor causes strong albumin binding, which protects the peptide against degradation.

It is indicated as an adjunct to diet and exercise to improve glycemic control in adults with type 2 diabetes and works by helping the pancreas to release the right amount of insulin when blood sugar levels are high. [157-162].

Liraglutide is produced by recombinant DNA technology and used in the treatment of type 2 diabetes mellitus.

His-Ala-Glu-Gly-Thr-Phe-Thr-Ser-Asp-Val-Ser-Ser-Tyr-Leu-Glu-
Gly-Gln-Ala-Ala-Lys-Glu-Phe-Ile-Ala-Trp-Leu-Val-Arg-Gly-Arg-Gly

Liraglutide 26.2.67

FIG. 26.12 Structure of liraglutide.

Some new GLP-1 analogues, such as taspoglutide and albiglutide, are in various stages of clinical development [163,164]. CJC-1134-PC, the modified exendin-4 analogue conjugated to recombinant human albumin is now in Phase II trials [165].

Attempts to discover nonpeptidic small molecule GLP-1 receptor (GLP-1R) agonists have been published [166-168].

Dipeptidyl Peptidase-IV Inhibitors

Dipeptidyl-peptidase-4 (DPP-4), is the one of nine known members of serine endopeptidases. DPP-4 selectively cleaves two amino acids ("dipeptidyl") from the N terminus of GIP and GLP-1—two peptide hormone incretins responsible for a significant part of postprandial insulin secretion. DPP-4 consists of 766 amino acids and is expressed in a large number of tissues.

Inhibition of DPP-4 leads to prolongation of the half-life of endogenous GLP-1, and becomes a therapeutic concept and a target for developing new drugs for treatment of diabetes mellitus.

The first developed DPP-4 inhibitor, a "gliptin" [169], sitagliptin (**26.2.68**), resulted from a high-throughput drug screening approach. It was followed by vildagliptin (**26.2.69**) and saxagliptin (**26.2.70**). Alogliptin (**26.2.71**), linagliptin (**26.2.72**), teneligliptin (**26.2.73**), and other DPP-4 compounds are awaiting approval by the FDA (Fig. 26.13.).

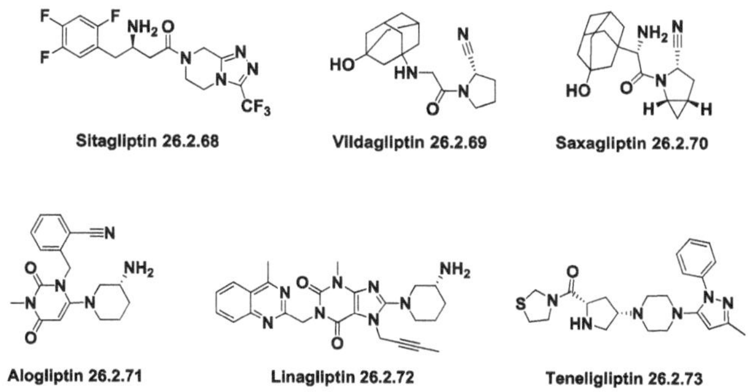

FIG. 26.13 The dipeptidyl peptidase-4 inhibitors "gliptins."

At present, DPP-4s have not been associated with any significant improvements in cardiovascular risk factors, such as blood pressure or lipid levels [170].

Sitagliptin–Januvia

Sitagliptin (**26.2.68**) is included in the list of Top 200 Drugs by sales for the 2010s.

It is an effective and generally well tolerated, orally administered, potent, and highly selective inhibitor of DPP-4 and was the first agent of its class to be approved for use in the management of type 2 diabetes. The drug increases the GLP-1 concentration, thereby improving insulin secretion from pancreatic p-cells, restoring glycemic control [171-179]. Sitagliptin may cause side effects such as stuffed or runny nose and sore throat, headache, diarrhea, nausea, and vomiting.

Several processes have been disclosed in the prior art for the preparation of sitagliptin and its analogues [180-192], but the general approach is that standard peptide coupling of β-amino acid-(R)-3-amino-4-(2,4,5-trifluorophenyl) butanoic acid (**26.2.74**) with 3-(trifluoromethyl)-5,6,7,8-tetrahydro-[1,2,4] triazolo[4,3-a]pyrazine (**26.2.75**) with further deprotection of obtained Boc-sitagliptin (**26.2.76**) (Scheme 26.8.).

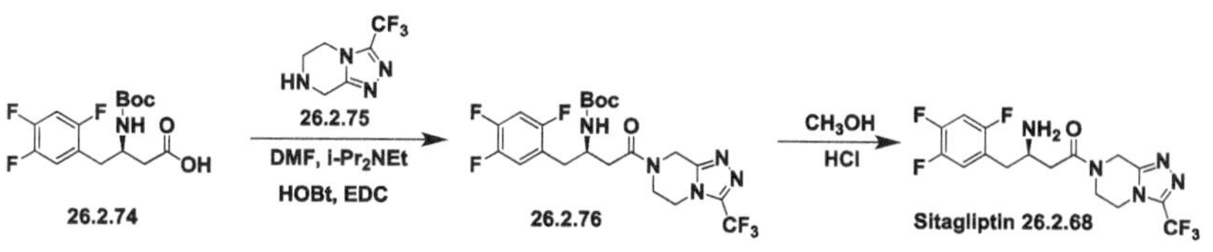

SCHEME 26.8 Synthesis of sitagliptin.

The starting β-amino acid (**26.2.74**) according to the first publication [180] was prepared via the Arndt–Eistert homologation of the corresponding α-amino acid, which was prepared using Schöllkopf's bis-lactam methodology.

The Schöllkopf reagent (2S)-(+)-2,5-dihydro-3,6-dimethoxyl-2-isopropylpyrazine (**26.2.78**) was alkylated with 1-(bromomethyl)-2,4,5-trifluorobenzene (**26.2.77**) to prepare the appropriate 2,5-dihydropyrazine (**26.2.79**), which was hydrolyzed with hydrochloric acid followed by di-tert-butyl dicarbonate protection of amino group in obtained aminoacid to produce the ester (**26.2.80**). Basic hydrolysis of the ester (**26.2.80**) produced the α-amino acid (**26.2.81**), which was treated with isobutyl chloroformate followed by diazomethane to produce the diazo ketone (**26.2.82**), which in the presence of silver salt and methanol as nucleophile, underwent Wolff rearrangement to ester (**26.2.83**) followed by basic hydrolysis to produce the desired β-amino acid (**26.2.74**) (enantiomeric excess >99%). The same acid (**26.2.74**) could also be prepared in one step by sonication of diazo ketone (**26.2.82**) in the presence of silver benzoate (Scheme 26.9.).

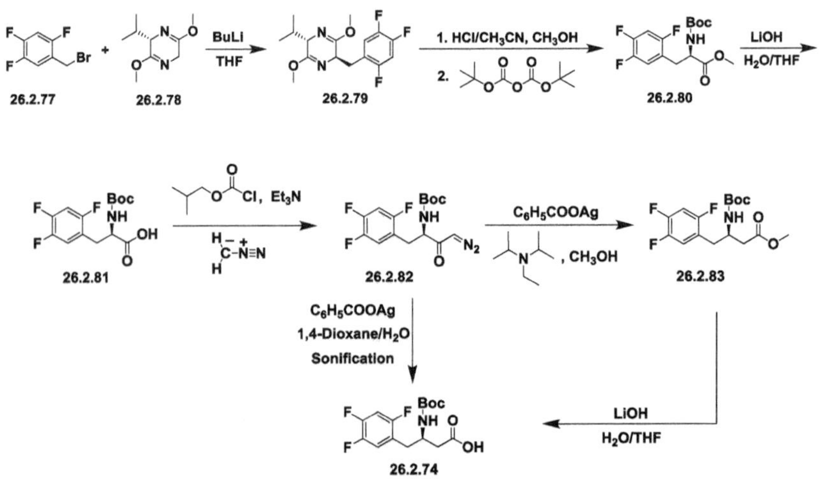

SCHEME 26.9 Preparation of the starting β-amino acid (26.2.74) for sitagliptin synthesis.

3-(Trifluoromethyl)-5,6,7,8-tetrahydro-[1,2,4]triazolo[4,3-a]pyrazine (**26.2.75**) for this synthesis was prepared according to the (Scheme 26.10.)

The starting compound was the commercially available chloropyrazine (**26.2.84**), which was converted to hydrazine (**26.2.85**) with hydrazine in refluxing ethanol, and then acylated with trifluoroacetic anhydride to produce amide (**26.2.86**), which was then condensed in polyphosphoric acid (PPA) to produce triazolopyrazine (**26.2.87**). Subsequent catalytic hydrogenation proceeded smoothly to produce the target piperazine derivative (**26.2.75**).

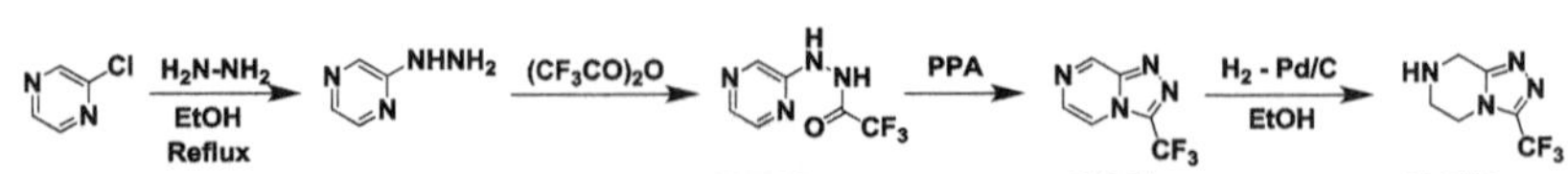

SCHEME 26.10 Preparation of the starting 3-(Trifluoromethyl)-5,6,7,8-tetrahydro-[1,2,4] triazolo[4,3-a]pyrazine (26.2.75) for sitagliptin synthesis.

Another approach [181] offers a possible industrial method for the preparation of sitagliptin, which differs in the method of the synthesis of a key β-amino acid: (R)-3-amino-4-(2,4,5-trifluorophenyl)butanoic acid (**26.2.74**). The method starts from the commercially available 2-(2,4,5-trifluorophenyl) acetic acid (**26.2.88**), which was reacted with Meldrum acid in THF in the presence of 1,1′-carbonyldiimidazole, producing adduct (**26.2.89**). Methanolysis of the adduct (**26.2.89**) produced β-ketoester (**26.2.90**), which, on reaction with anhydrous ammonium acetate in methanol and preferably at temperatures (60 to 65°C), produced a 3-aminobut-2-enoate derivative (**26.2.91**). Reducing the last with sodium cyanoborohydride in a methanol racemic methyl-(3RS)-3-amino-4-(2,4,5-trifluorophenyl)butanoate (**26.2.92**) was prepared. Treating the obtained butanoate (**26.2.92**) with (R)-(−)-mandelic acid, an enantiomerically enriched crude methyl(3R)-3-amino-4-(2,4,5-trifluorophenyl)butanoate salt of (**26.2.93**) was separated and further purified. Then the hydrolysis of (**26.2.93**) and protection of the amino group were carried out as a one-pot reaction using lithium hydroxide in THF–water mixture followed by di-tert-butyl dicarbonate work up. The isolated desired acid (**26.2.74**) was coupled with 3-trifluoromethyl-5,6,7,8-tetrahydro[1,2,4]triazolo[4,3-a]pyrazine hydrochloride (**26.2,75**) to obtain BOC-protected sitagliptin (**26.2.76**). The BOC-protected sitagliptin (**26.2.76**) was placed in methanol, and 25% methanolic hydrochloride solution was added at 25 to 30°C. After 10 hours the methanol was removed and the residue was worked up with 10% sodium carbonate in ethyl acetate to produce the desired sitagliptin (**26.2.68**) [181] (Scheme 26.11.).

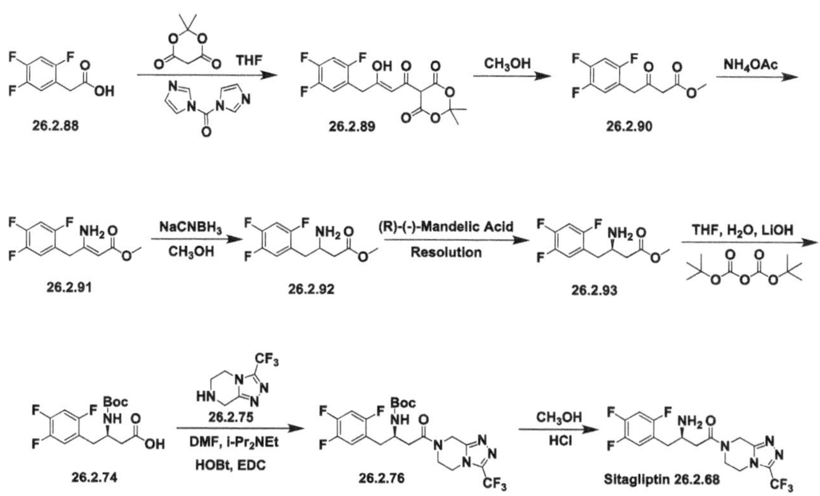

SCHEME 26.11 Synthesis of sitagliptin.

Another method [182], which differs from those described above, again focused on the method of preparation of (R)-3-((tert-butoxycarbonyl)amino)-4-(2,4,5-trifluorophenyl)butanoic acid (**26.2.74**).

The synthesis started with the L-homoserine (**26.2.94**), in which, alternately, the hydroxyl group was protected with the use of tert-butyldimethylsilyl chloride (TBSCl) in the presence of 1,8-diazabicyclo[5.4.0]undec-7-ene (DBU) in CH_3CN to form intermediate silyl ether and an amino group with di-tert-butyl dicarbonate to produce the compound (**26.2.95**). The obtained protected amino acid (**26.2.95**) was then condensed with N-hydroxysuccinimide (NHS) in presence of N, N′-dicyclohexylcarbodiimide (DCC) to produce the corresponding ester, which was then reduced with 1 equivalent of $NaBH_4$ to form the diol derivative (**26.2.96**). Mesylation of the diol derivative (**26.2.96**) with methanesulfonyl chloride, which was followed by ring closure under basic condition (NaH), formed the aziridine (**26.2.97**) with high enantio purity (>99% enantiomeric excess). The Grignard reaction using 2,4,5-trifluoro-phenyl magnesium bromide (**26.2.98**) with the obtained aziridine in the presence of the copper catalyst dimethyl sulfide complex produced compound (**26.2.99**). After deprotection of the hydroxyl group with tetrabutylammonium fluoride (TBAF) in MeOH, the obtained amino alcohol (**26.2.101**) was oxidized with NaClO and (2,2,6,6-tetramethyl-piperidin-1-yl)oxyl (TEMPO). The resulting 3-R-Boc-amino-4-(2,4,5-trifluorophenyl)butyric acid (**26.2.74**) was obtained in good yield (90%). The pyrazine derivative (**26.2.75**) was then coupled to the obtained amino acid using HOBT-EDCI, to provide intermediate (**26.2.76**) in 95% yield, and after the BOC group removal, the desired sitagliptin (**26.2.68**) was isolated in 90% yield [182] (Scheme 26.12.).

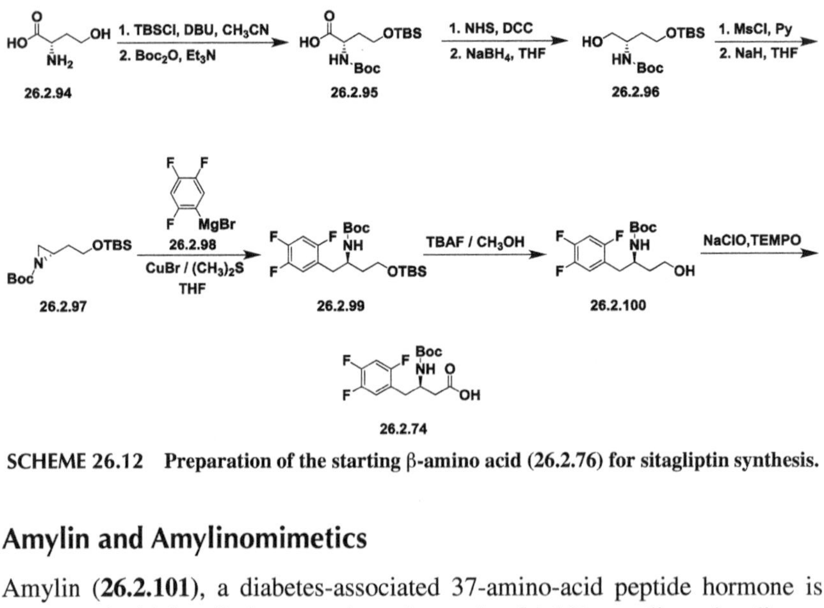

SCHEME 26.12 Preparation of the starting β-amino acid (26.2.76) for sitagliptin synthesis.

Amylin and Amylinomimetics

Amylin (**26.2.101**), a diabetes-associated 37-amino-acid peptide hormone is cosecreted with insulin in approximately a ratio of 1:100, amylin-to-insulin.

The amylin amino acid sequence is KCNTATCATQRLANFLVHSSN-NFGAILSSTNVGSNTY, with a disulfide bridge between cysteine residues 2 and 7 (Fig. 26.14.).

Lys-Cys-Asn-Thr-Ala-Thr-Cys-Ala-Thr-Gln-Arg-Leu-Ala-Asn-Phe-Leu-Val-His-Ser-Ser-Asn-Asn-Phe-Gly-Ala-Ile-Leu-Ser-Ser-Thr-Asn-Val-Gly-Ser-Asn-Thr-Tyr

Amylin 26.2.101

FIG. 26.14 **Structure of amylin.**

The actions of amylin agonists appear to be synergistic to insulin. The design of soluble derivatives of amylin, so called amylinomimetics, brought to creation of new drugs such as pramlintide (**26.2.102**), the first member of a new class of drugs, which is used as a supplement to insulin in the treatment of type 1 diabetes [193-198] (Fig. 26.15.).

Lys-Cys-Asn-Thr-Ala-Thr-Cys-Ala-Thr-Gln-Arg-Leu-Ala-Asn-Phe-Leu-Val-His-Ser-Ser-Asn-Asn-Phe-Gly-Pro-Ile-Leu-Pro-Pro-Thr-Asn-Val-Gly-Ser-Asn-Thr-Tyr-NH_2

Pramlintide 26.2.102

FIG. 26.15 **Structure of pramlintide.**

Sodium Glucose Cotransporter Inhibitors

The sodium-glucose cotransporter system (SGLT) plays an important role in glucose reabsorption process and has received considerable attention in recent years as a target for the treatment of type 2 diabetes mellitus. SGLTs are a group of proteins that exist in the small intestine and renal proximal tubules. Accordingly, two types of transporters (SGLT-1 [small intestine]) and (SGLT-2 [renal tubules]) are identified. SGLT-2 regulates the reuptake of the majority glucose filtered in urine, and inhibition of SGLT-2 leads to an increase in glucose exertion, providing a new mechanism to lower elevated blood glucose level. Consequently, inhibition of SGLT-2 represents an innovative therapeutic strategy for the treatment of patients with type 1 or type 2 diabetes by enhancing glucose loss through the urine [199-205].

Dapagliflozin (**26.2.103**) [206-209] and canagliflozin (**26.2.104**) [209-213] belong to a novel class of antidiabetic drugs known as SGLT-2 inhibitors. The analogues ipragliflozin (**26.2.105**) and empagliflozin (**26.2.106**) are under investigation (Fig. 26.16.).

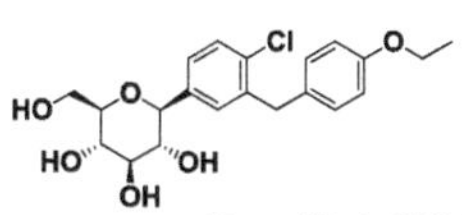

Dapagliflozin 26.2.103

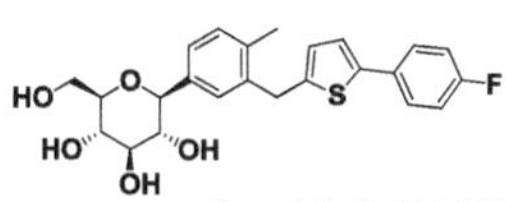

Canagliflozin 26.2.104

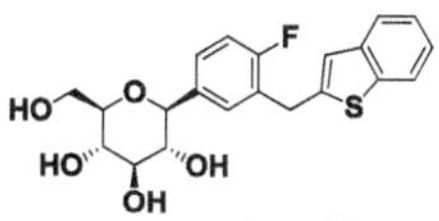

Ipragliflozin 26.2.105

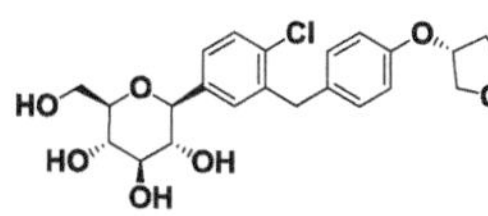

Empagliflozin 26.2.106

FIG. 26.16 **Sodium-glucose cotransporter system (SGLT) inhibitors.**

Drugs in Development

Glucagon Antagonists

Glucagon (**26.2.107**) is a crucial hormone in glucose homeostasis and hence has great potential for application as an essential regulator of hepatic glucose production.

Native glucagon is a 29-amino-acid peptide having the sequence shown in Fig. 26.17.

His-Ser-Glu-Gly-Thr-Phe-Thr-Ser-Asp-Tyr- Ser-Lys-Tyr-Leu-Asp-
Ser-Arg-Arg-Ala-Gln-Asp- Phe-Val-Gln-Trp-Leu-Met-Asn-Thr-OH

Glucagon 26.2.107

FIG. 26.17 **Structure of glucagon.**

There is strong evidence that excessive glucagon levels contribute to the hyperglycemia of non–insulin-dependent diabetes mellitus by inappropriately stimulating glucose output. Studies suggest that glucagon suppression or an action that antagonizes glucagon could be a useful adjunct to conventional treatment of hyperglycemia in diabetic patients. The antagonist can be peptidic or nonpeptidic in nature [214-221].

A number of publications have disclosed glucagon analogues that can act as antagonists [222-225]. Probably, the most thoroughly characterized antagonists are DesHis1[Glu9]-glucagon amide [226], des-His^1-[Nle^9-Ala^{11}-Ala^{16}] glucagon amide [227], [des His^1, des Phe^6, Glu^9] glucagon amide [228]. Other antagonists have also been characterized [229-233].

Peptide antagonists are quite potent. However, they are generally known not to be orally available because of degradation by enzymes, and poor distribution in vivo. As a result, orally available nonpeptide antagonists of peptide hormones are generally preferred. Multiple classes of small molecule glucagon receptor antagonists have been reported [234-237] and reviewed [238-240].

Among the nonpeptide glucagon antagonists, biaryl amides (**26.2.108**) [241], 1-phenyl pyrazole derivatives (**26.2.109**) [242], a quinoxaline derivative such as CP-99,711 (**26.2.110**) [243], substituted pyridylpyrroles (**26.2.111**) [236], imidazoles (**26.2.112**) [214,244-246], thiophene-derived compounds (**26.2.113**) [247], ureas (**26.2.114, 26.2.115**) [248-250], substituted pyrimidine and pyridone compounds (**26.2.116**) [251], benzimidazol-2-ylthio compounds such as NNC 92-1687 (**26.2.117**) [252], 4-aryl-pyridines (**26.2.118**) [253,254], alkylidene hydrazides (**26.2.119**) [255-259], biphenyl derivatives (**26.2.120**) [260], and even disilacyclohexanes (**26.2.121**) [261] (Fig. 26.18.).

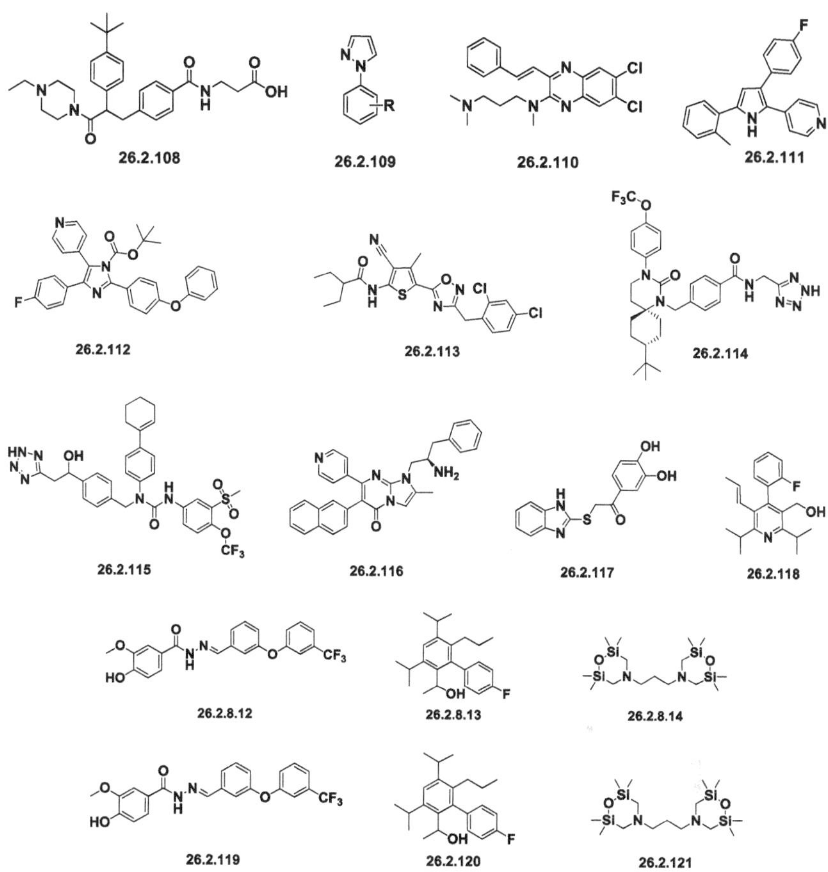

FIG. 26.18 Structural variety of nonpeptide glucagon antagonists.

Protein Tyrosine Phosphatase Inhibitors

Protein tyrosine phosphatase (PTP) is a large family of enzymes that are key players in many disease states. Dysfunction of PTP activity is associated with diabetes, cancer, autoimmune disorders, and neural diseases. Inhibition of protein tyrosine phosphatase 1B (PTP1B), considered to be one of the best validated biological targets for the treatment of type 2 diabetes, has been proposed as a novel therapy to treat this disease [262-268].

Many structurally diverse small-molecule inhibitors have been synthesized and evaluated as PTP inhibitors. Among them are naphthalene derivatives (**26.2.122**) [265], a series of benzotriazole phenyldifluoromethylphosphonic acids (**26.2.123**) [269], benzofuransulfonamides (**26.2.124**) [270], 2-amino-3-carboxylic acid derivatives of thieno[3,2-c]-tetrahydropyridine (**26.2.125**) [271],

2 thieno[2,3-c]-tetrahydropyridine (**26.2.126**) [272], derivatives of 1,2,3,4-tetrahydroisoquinoline sulfamic acid (**26.2.127**) [273], novel pyridazine analogues (**26.2.128**) [274], and many others (Fig. 26.19.).

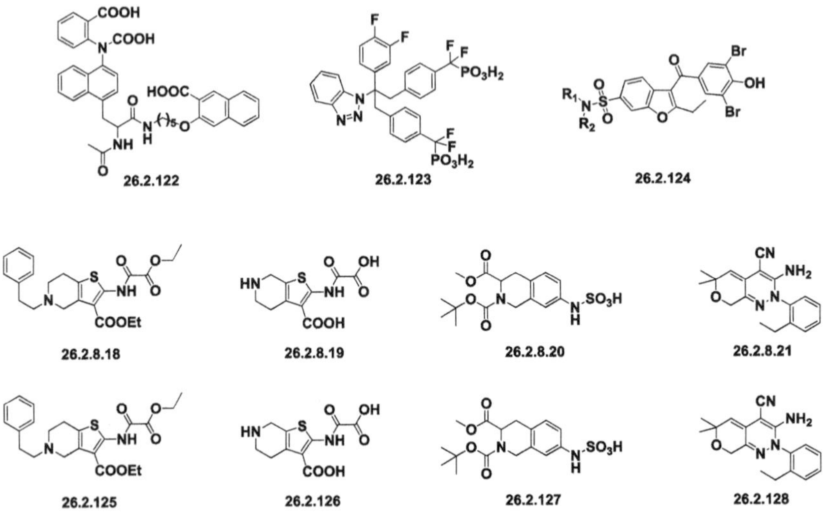

FIG. 26.19 Protein tyrosine phosphatase inhibitors.

Glycogen Phosphorylase Inhibitors

The alternative approach—inhibition of glycogenolysis via inhibition of hepatic glycogen phosphorylase—became a promising treatment strategy for attenuating hyperglycemia in type 2 diabetes.

Structurally diverse small-molecule inhibitors such as indole-2-carboxamide derivative (**26.2.129**) CP-91149 [275] and urea derivative (**26.2.130**) [276-282] have been synthesized and evaluated as PTP inhibitors (Fig. 26.20.).

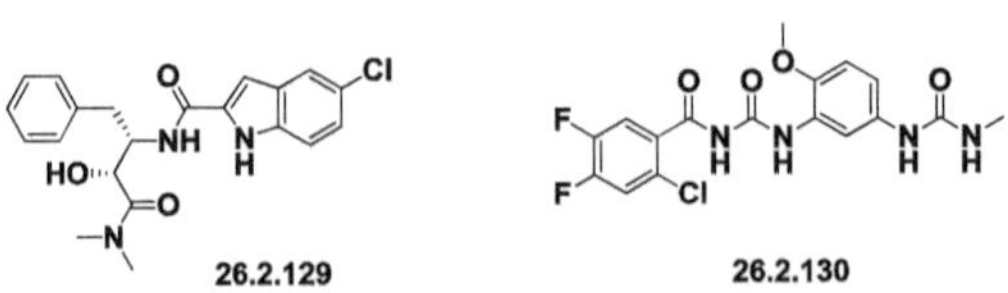

FIG. 26.20 Glycogen phosphorylase inhibitors.

Leptin

Leptin is an adipocyte-derived 146-amino-acid protein hormone that is also called "satiety hormone." It acts as a major regulator for food intake and energy homeostasis. It plays a crucial role in glucose homeostasis, including regulation of insulin secretion by pancreatic β cells. Studies suggest that leptin could be used as an adjunct of insulin therapy in insulin-deficient diabetes [283,284].

Glucokinase Inhibitors

Glucokinase is a member of the hexokinase family of enzymes that are responsible for the phosphorylation of glucose to glucose-6-phosphate. It was identified as a new drug target for developing antidiabetic medicines. The enzyme plays a key role in glucose homeostasis. Phosphorylation of glucose by glucokinase in the liver promotes glycogen synthesis, while in the β-cell it results in insulin release. Activators of glucokinase increase the sensitivity of the enzyme to glucose, leading to increased insulin secretion and liver glycogen synthesis and a decrease in liver glucose output [285,286]. Many small molecules have been identified as glucokinase activators. Among them are structurally diverse compounds like MK-0941 (**26.2.131**) [287], Ro 028-1675 (**26.2.132**) [288], (**26.2.133**) [289], (**26.2.134**) [290], piragliatin (**26.2.135**) [291], (**26.2.136**) [292], (**26.2.137**) [293], (**26.2.138**) and (**26.2.139**) [294], and (**26.2.140**) [295] (Fig. 26.21.).

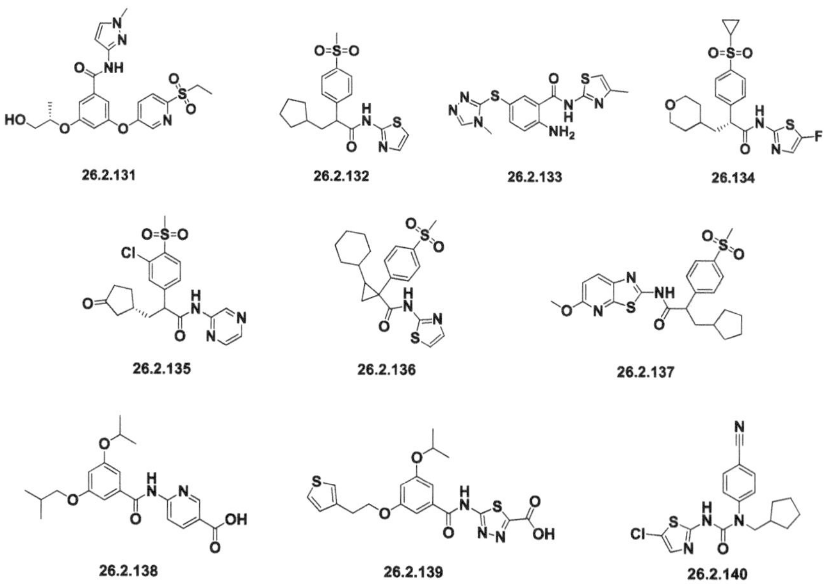

FIG. 26.21 Glucokinase inhibitors.

REFERENCES

1. Roth, J.; Qureshi, S.; Whitford, I.; Vranic, M.; Kahn, C. R.; Fantus, I. G.; Dirks, J. H. Insulin's discovery: new insights on its ninetieth birthday. *Diabetes/Metab. Res. Rev.* **2012,** *28* (4), 293–304.
2. Smith, G. D. Insulin. In Messerschmidt, A., Ed.; *Handbook of Metalloproteins*, Vol. 3; Wiley, 2004; pp 367–377.
3. Brandenburg, D. Insulin. Structure, function, design. *Exp. Clin. Endocrinol. Diabetes* **1999,** *107* (Suppl. 2), S6–S12.
4. Adams, M. J.; Blundell, T. L.; Dodson, E. J.; Dodson, G. G.; Vijayan, M.; Baker, E. N.; Harding, M. M.; Hodgkin, D. C.; Rimmer, B.; Sheats, S. Structure of rhombohedral 2 zinc insulin crystals. *Nature (London, U. K.)* **1969,** *224* (5218), 491–499.

5. Yip, C. C.; Ottensmeyer, P. Three-dimensional structural interactions of insulin and its receptor. *J. Biol. Chem.* **2003,** *278* (30), 27329–27332.
6. Mayer, J. P.; Zhang, F.; DiMarchi, R. D. Insulin structure and function. *Biopolymers* **2007,** *88* (5), 687–713.
7. Bentley, G.; Dodson, E.; Dodson, G.; Hodgkin, D.; Mercola, D. Structure of insulin in 4-zinc insulin. *Nature (London, U. K.)* **1976,** *261* (5556), 166–168.
8. Brems, D. N.; Alter, L. A.; Beckage, M. J.; Chance, R. E.; DiMarchi, R. D.; Green, L. K.; Long, H. B.; Pekar, A. H.; Shields, J. E.; Frank, B. H. Altering the association properties of insulin by amino acid replacement. *Protein Eng.* **1992,** *5* (6), 527–533.
9. DiMarchi, R. D.; Mayer, J. P.; Fan, L.; Brems, D. N.; Frank, B. H.; Green, L. K.; Hoffmann, J. A.; Howey, D. C.; Long, H. B.; Shaw, W. N.; Shields, J. E.; Slieker, L. J.; Su, K. S. E.; Sundel, K. L.; Chance, R. E. Synthesis of a fast-acting insulin based on structural homology with insulin-like growth factor I. In *Pept.: Chem. Biol., Proc. Am. Pept. Symp., 12th;* Smith, J. A., Rivier, J. E., Eds.; ESCOM Science, 1992; pp 26–28.
10. Esposito, K.; Capuano, A.; Giugliano, D. Humalog (lispro) for type 2 diabetes. *Expert Opin. Biol. Ther.* **2012,** *12* (11), 1541–1550.
11. Chance, R. E.; Glazer, N. B.; Wishner, K. L. Insulin lispro (Humalog). In *Biopharmaceuticals, an Industrial Perspective;* Walsh, G., Murphy, B., Eds.; Springer, 1999; pp 149–171.
12. Gale, E. A. M. Insulin lispro: a new quick-acting insulin analog. *Expert Opin. Invest. Drugs* **1997,** *6* (9), 1247–1256.
13. Reynolds, N. A.; Wagstaff, A. J. Insulin aspart: a review of its use in the management of type 1 or 2 diabetes mellitus. *Drugs* **2004,** *64* (17), 1957–1974.
14. Owens, D.; Vora, J. Insulin aspart: a review. *Expert Opin. Drug Metab. Toxicol.* **2006,** *2* (5), 793–804.
15. Garg, S.; Moser, E.; Dain, M.-P.; Rodionova, A. Clinical experience with insulin glargine in type 1 diabetes. *Diabetes Technol. Ther.* **2010,** *12* (11), 835–846.
16. Goykhman, S.; Drincic, A.; Desmangles, J. C.; Rendell, M. Insulin glargine: a review 8 years after its introduction. *Expert Opin. Pharmacother.* **2009,** *10* (4), 705–718.
17. Bueno, A. B.; Castano, A. M.; Rodriguez, A. Diabetes drugs. In *Drug Discovery: Practices, Processes, and Perspectives;* Li, J. J., Corey, E. J., Eds.; Wiley, 2013; pp 205–243.
18. Mehanna, A. Antidiabetic agents: past, present and future. *Future Med. Chem.* **2013,** *5* (4), 411–430.
19. Quianzon, C. C. L.; Cheikh, I. E. History of current non-insulin medications for diabetes mellitus. *J. Community Hosp. Intern. Med. Perspect.* **2012,** *2* (3).
20. Laliberte, B. K.; Neumiller, J. J. Review of medications used in the treatment of diabetes mellitus. *J. Pharm. Technol.* **2010,** *26* (3), 136–146.
21. Rathod, S.; Kakadiya, J. Oral hypoglycemic agent—overview. *Pharmacologyonline* **2009,** (1), 498–524.
22. Zito, S. W.; Shinde, J.; Chen, I.-C. S.; Taldone, T.; Barletta, M. Oral hypoglycemics: a review of chemicals used to treat type 2 diabetes. *Curr. Bioact. Compd.* **2008,** *4* (2), 68–85.
23. Nain, S.; Bansal, N. Anti-diabetic drugs for the treatment of diabetes: a review. *Asian J. Biochem. Pharm. Res.* **2012,** *2* (2), 148–153.
24. Laliberte, B. K.; Neumiller, J. J. Review of medications used in the treatment of diabetes mellitus. *J. Pharm. Technol.* **2010,** *26* (3), 136–146.
25. Krentz, A. J.; Patel, M. B.; Bailey, C. J. New drugs for type 2 diabetes mellitus: what is their place in therapy? *Drugs* **2008,** *68* (15), 2131–2162.
26. Nathan, D. M.; Buse, J. B.; Davidson, M. B.; Ferrannini, E.; Holman, R. R.; Sherwin, R.; Zinman, B. Medical management of hyperglycemia in type 2 diabetes: a consensus algorithm for the initiation and adjustment of therapy: a consensus statement of the American Diabetes Association and the European Association for the study of diabetes. *Diabetes Care* **2009,** *32* (1), 193–203.

27. Mehanna, A. Antidiabetic agents: past, present and future. *Future Med. Chem.* **2013,** *5* (4), 411–430.
28. Mizuno, C. S.; Chittiboyina, A. G.; Kurtz, T. W.; Pershadsingh, H. A.; Avery, M. A. Type 2 diabetes and oral antihyperglycemic drugs. *Curr. Med. Chem.* **2008,** *15* (1), 61–74.
29. Thule, P. M.; Umpierrez, G. Sulfonylureas: a new look at old therapy. *Curr. Diabetes Rep.* **2014,** *14* (4), 1–8.
30. Rendell, M. The role of sulfonylureas in the management of type 2 diabetes mellitus. *Drugs* **2004,** *64* (12), 1339–1358.
31. Del Prato, S.; Aragona, M.; Coppelli, A. Sulfonylureas and hypoglycaemia. *Diabetes, Nutr. Metab.* **2002,** *15* (6), 444–451.
32. Goyal, S.; Rai, J. K.; Rajesh, K. S.; Narang, R. K. Sulfonylureas for antidiabetic therapy, an overview for glipizide. *Int. J. Pharm. Pharm. Sci.* **2010,** *2* (Suppl. 2), 1–6.
33. Green, J. B.; Feinglos, M. N. Are sulfonylureas passe? *Curr. Diabetes Rep.* **2006,** *6* (5), 373–377.
34. Del Prato, S.; Pulizzi, N. The place of sulfonylureas in the therapy for type 2 diabetes mellitus. *Metab., Clin. Exp.* **2006,** *55* (5 Suppl. 1), S20–S27.
35. Chen, X.; Hussain, S.; Parveen, S.; Zhang, S.; Yang, Y.; Zhu, C. Sulfonyl group-containing compounds in the design of potential drugs for the treatment of diabetes and its complications. *Curr. Med. Chem.* **2012,** *19* (21), 3578–3604.
36. Ort, O. Newer sulfonylureas. In 2nd ed.; Kraemer, W., Schirmer, U., Eds.; *Modern Crop Protection Compounds*, Vol. 1; Wiley-VCH, 2012; pp 50–88.
37. Pfeiffer, A. F. H. Oral hypoglycemic agents: sulfonylureas and meglitinides. In *Textbook of Type 2 Diabetes;* Goldstein, B. J., Mueller-Wieland, D., Eds.; CRC Press, 2003; pp 77–85.
38. Lamos, E. L.; Stein, S. A.; Davis, S. N. Sulfonylureas and meglitinides: historical and contemporary issues. *Panminerva Med.* **2013,** *55* (3), 239–251.
39. Scott, L. J. Repaglinide: a review of its use in type 2 diabetes mellitus. *Drugs* **2012,** *72* (2), 249–272.
40. Culy, C. R.; Jarvis, B. Repaglinide: a review of its therapeutic use in type 2 diabetes mellitus. *Drugs* **2001,** *61* (11), 1625–1660.
41. Massi-Benedetti, M.; Damsbo, P. Pharmacology and clinical experience with repaglinide. *Expert Opin. Invest. Drugs* **2000,** *9* (4), 885–898.
42. Wolffenbuttel, B. H. R. Repaglinide-a new compound for the treatment of patients with type 2 diabetes. *Neth. J. Med.* **1999,** *55* (5), 229–234.
43. Owens, D. R. Repaglinide: a new short-acting insulinotropic agent for the treatment of type 2 diabetes. *Eur. J. Clin. Invest.* **1999,** *29* (Suppl. 2), 30–37.
44. Malaisse, W. J. Repaglinide, a new oral antidiabetic agent: a review of recent preclinical studies. *Eur. J. Clin. Invest.* **1999,** *29* (Suppl. 2), 21–29.
45. Guay, D. R. P. Repaglinide, a novel, short-acting hypoglycemic agent for type 2 diabetes mellitus. *Pharmacotherapy* **1998,** *18* (6), 1195–1204.
46. Balfour, J. A.; Faulds, D. Repaglinide. *Drugs Aging* **1998,** *13* (2), 173–180.
47. Grell, W.; Greischel, A.; Zahn, G.; Mark, M.; Knorr, H.; Rupprecht, E.; Mueller, U., Preparation and formulation of (S)-(+)-2-ethoxy-4-[N-[1-(2-piperidinophenyl)-3-methyl-1-butyl] aminocarbonylmethyl]benzoic acid, WO 9300337 (1993).
48. Grell, W.; Hurnaus, R.; Griss, G.; Sauter, R.; Rupprecht, E.; Mark, M.; Luger, P.; Nar, H.; Wittneben, H.; Mueller, P. Repaglinide and related hypoglycemic benzoic acid derivatives. *J. Med. Chem.* **1998,** *41* (26), 5219–5246.
49. Hermann, L. S. Clinical pharmacology of biguanides. *Handb. Exp. Pharmacol.* **1996,** *119,* 373–407.
50. Pernicova, I.; Korbonits, M. Metformin—mode of action and clinical implications for diabetes and cancer. *Nat. Rev. Endocrinol.* **2014,** *10* (3), 143–156.

51. Hajjar, J.; Habra, M. A.; Naing, A. Metformin: an old drug with new potential. *Expert Opin. Invest. Drugs* **2013,** *22* (12), 1511–1517.
52. Rena, G.; Pearson, E. R.; Sakamoto, K. Molecular mechanism of action of metformin: old or new insights? *Diabetologia* **2013,** *56* (9), 1898–1906.
53. Rojas, L. B. A.; Gomes, M. B. Metformin: an old but still the best treatment for type 2 diabetes. *Diabetol. Metab. Syndr* **2013,** 5–6.
54. Mahmood, K.; Naeem, M.; Rahimnajjad, N. A. Metformin: the hidden chronicles of a magic drug. *Eur. J. Intern. Med.* **2013,** *24* (1), 20–26.
55. Bulterijs, S. Metformin As a geroprotector. *Rejuvenation Res.* **2011,** *14* (5), 469–482.
56. Anisimov, V. N. Metformin: do we finally have an anti-aging drug? *Cell Cycle* **2013,** *12* (22), 3483–3489.
57. Belda-Iniesta, C.; Pernia, O.; Simo, R. Metformin: a new option in cancer treatment. *Clin. Transl. Oncol.* **2011,** *13* (6), 363–367.
58. Martin, M.; Marais, R. Metformin: a diabetes drug for cancer, or a cancer drug for diabetics? *J. Clin. Oncol.* **2012,** *30* (21), 2698–2700.
59. Bauerreis, R.; Eubel, J., Monosubstituted aliphatic biguanide salts, DE 1023757 (1958).
60. 1,1-Dimethylbiguanide hydrochloride, FR 2322860 (1977), Fr. Demande (1977), FR 2322860 A1 19770401.
61. Shapiro, S. L.; Parrino, V. A.; Freedman, L. Hypoglycemic agents. I., Chemical properties of β-phenethylbiguanide. A new hypoglycemic agent. *J. Am. Chem. Soc.* **1959,** *81* (9), 2220–2225.
62. Werner, E.; Bell, J. The preparation of methylguanidine, and of ββ-dimethylguanidine by the interaction of dicyanodiamide, and methylammonium and dimethylammonium chlorides respectively. *J. Chem. Soc., Trans.* **1922,** *121,* 1790–1795.
63. Chava, S.; Gorantla, S. R.; Indukuri, V. S. K.; Ketavarapu, N. R.; Gorantla, V. C., An improved process for the preparation of metformin hydrochloride, WO 2014041566 (2014).
64. Koduru, R. M.; Bhattacharyya, P. K.; Kale, U. R.; Bhuwan, R., Improved process for preparation of metformin hydrochloride, IN 2010MU01409 (2012).
65. Cho, N.; Momose, Y. Peroxisome proliferator-activated receptor γ agonists as insulin sensitizers: from the discovery to recent progress. *Curr. Top. Med. Chem.* **2008,** *8* (17), 1483–1507.
66. Chiarelli, F.; di Marzio, D. Peroxisome proliferator-activated receptor-γ agonists and diabetes: current evidence and future perspectives. *Vasc. Health Risk Manage.* **2008,** *4* (2), 297–304.
67. Chang, F.; Jaber, L. A.; Berlie, H. D.; O'Connell, M. B. Evolution of peroxisome proliferator-activated receptor agonists. *Ann. Pharmacother.* **2007,** *41* (6), 973–983.
68. Sood, V.; Colleran, K.; Burge, M. R. Thiazolidinediones: a comparative review of approved uses. *Diabetes Technol. Ther.* **2000,** *2* (3), 429–440.
69. Barnett, A. H. Thiazolidinediones and cardiovascular outcomes. *Br. J. Diabetes Vasc. Dis.* **2008,** *8* (1), 45–49.
70. Mudaliar, S.; Henry, R. R. New oral therapies for type 2 diabetes mellitus: the glitazones or insulin sensitizers. *Annu. Rev. Med.* **2001,** *52,* 239–257.
71. Hulin, B.; McCarthy, P. A.; Gibbs, E. M. The glitazone family of antidiabetic agents. *Curr. Pharm. Des.* **1996,** *2* (1), 85–102.
72. Gillies, P. S.; Dunn, C. J. Pioglitazone. *Drugs* **2000,** *60* (2), 333–343.
73. Waugh, J.; Keating, G. M.; Plosker, G. L.; Easthope, S.; Robinson, D. M. Pioglitazone: a review of its use in type 2 diabetes mellitus. *Drugs* **2006,** *66* (1), 85–109.
74. Betteridge, D. J. Effects of pioglitazone on lipid and lipoprotein metabolism. *Diabetes, Obes. Metab.* **2007,** *9* (5), 640–647.
75. Smith, U. Pioglitazone: mechanism of action. *Int. J. Clin. Pract., Suppl.* **2001,** *121,* 13–18.

76. Schernthaner, G.; Currie, C. J.; Schernthaner, G.-H. Do we still need pioglitazone for the treatment of type 2 diabetes? A risk-benefit critique in 2013. *Diabetes Care* **2013,** *36* (Suppl. 2), S155–S161.
77. Kostapanos, M. S.; Elisaf, M. S.; Mikhailidis, D. P. Pioglitazone and cancer: angel or demon? *Curr. Pharm. Des.* **2013,** *19* (27), 4913–4929.
78. Govindan, J.; Evans, M. Pioglitazone in clinical practice: where are we now? *Diabetes Ther.* **2012,** *3* (1), 1/1–1/8.
79. de Pablos-Velasco, P. Pioglitazone: beyond glucose control. *Expert Rev. Cardiovasc. Ther.* **2010,** *8* (8), 1057–1067.
80. Shah, P.; Mudaliar, S. Pioglitazone: side effect and safety profile. *Expert Opin. Drug Saf.* **2010,** *9* (2), 347–354.
81. Dormandy, J.; Bhattacharya, M.; van Troostenburg de Bruyn, A.-R. The PROactive investigators, Safety and tolerability of pioglitazone in high-risk patients with type 2 diabetes an overview of data from PROactive. *Drug Saf.* **2009,** *32* (3), 187–202.
82. Ryder, R. E. J. Pioglitazone: an agent which reduces stroke, myocardial infarction and death and is also a key component of the modern paradigm for the optimum management of type 2 diabetes. *Br. J. Diabetes Vasc. Dis.* **2011,** *11* (3), 113–120.
83. Papanas, N.; Katsiki, N.; Hatzitolios, A. I.; Maltezos, E. Pioglitazone: a valuable component of combination therapy for type 2 diabetes mellitus. *Expert Opin. Pharmacother.* **2011,** *12* (10), 1457–1461.
84. Meguro, K.; Fujita, T., Thiazolidinedione derivatives and their use, EP 193256 (1986).
85. Sohda, T.; Momose, Y.; Meguro, K.; Kawamatsu, Y.; Sugiyama, Y.; Ikeda, H. Studies on antidiabetic agents. Synthesis and hypoglycemic activity of 5-[4-(pyridylalkoxy)benzyl]-2,4-thiazolidinediones. *Arzneim. Forsch.* **1990,** *40* (1), 37–42.
86. Momose, Y.; Meguro, K.; Ikeda, H.; Hatanaka, C.; Oi, S.; Sohda, T. Studies on antidiabetic agents. X. Synthesis and biological activities of pioglitazone and related compounds. *Chem. Pharm. Bull.* **1991,** *39* (6), 1440–1445.
87. Arita, M.; Mizuno, Y., Preparation of ether-containing 2,4-thiazolinedione derivatives, EP 506273 (1992).
88. Ortiz, A.; Sansinenea, E. Synthetic thiazolidinediones, potential antidiabetic compounds. *Curr. Org. Chem.* **2011,** *15* (1), 108–127.
89. Watt, D. S.; Profitt, J. A.; Corey, E. J. A reagent for the reduction of conjugated nitriles. *J. Org. Chem.* **1975,** *40,* 127–128.
90. Cantello, B. C. C.; Cawthorne, M. A.; Cottam, G. P.; Duff, P. T.; Haigh, D.; Hindley, R. M.; Lister, C. A.; Smith, S. A.; Thurlby, P. L. [[ω-(Heterocyclylamino)alkoxy]benzyl]-2,4-thiazolidinediones as potent antihyperglycemic agents. *J. Med. Chem.* **1994,** *37* (23), 3977–3985.
91. Cantello, B. C. C.; Eggleston, D. S.; Haigh, D.; Haltiwanger, R. C.; Heath, C. M.; Hindley, R. M.; Jennings, K. R.; Sime, J. T.; Woroniecki, S. R. Facile biocatalytic reduction of the carbon-carbon double bond of 5-benzylidenethiazolidine-2,4-diones. Synthesis of (±)-5-(4-{2-[methyl(2-pyridyl)amino]ethoxy}benzyl)thiazolidine-2,4-dione (BRL 49653), its (R)-(+)-enantiomer and analogs. *J. Chem. Soc., Perkin Trans. 1 (1972–1999)* **1994,** (22), 3319–3324.
92. Cantello, B. C. C.; Cawthorne, M. A.; Haigh, D.; Hindley, R. M.; Smith, S. A.; Thurlby, P. The synthesis of BRL 49653-a novel and potent antihyperglycemic agent. *Bioorg. Med. Chem. Lett.* **1994,** *4* (10), 1181–1184.
93. Hindley, R. M., Substituted thiazolidinedione derivatives, their preparation and pharmaceutical compositions, and their use in medicaments for therapy of hyperglycemia and hyperlipidemia, EP 306228 (1989).

94. Li, J. Advances in the development of methods for the synthesis of peroxisome proliferator-activated receptor (PPAR) agonists [rosiglitazone maleate (Avandia), pioglitazone hydrochloride (Actos), muraglitazar (Pargluva)], Art of Drug Synthesis. In Johnson, D. S., Li, J. J., Eds.; Wiley, 2007; pp 117–127.
95. Giles, R. G.; Lewis, N. J.; Quick, J. K., Process for the preparation of thiazolidinedione derivatives, WO 9923095 (1999).
96. Giles, R. G.; Lewis, N. J.; Moore, S.; Pool, C. R.; Quick, J. K.; Urquhart, M., Preparation of 5-benzylthiazolidine-2,4-diones, WO 9837073 (1998).
97. Hindley, R. M.; Woroniecki, S. R., Process for the preparation of pharmaceutically active thiazolidine or oxazolidine compounds by a yeast reductase, WO 9310254 1993).
98. Wagstaff, A. J.; Goa, K. L. Rosiglitazone: a review of its use in the management of type 2 diabetes mellitus. *Drugs* **2002,** *62* (12), 1805–1837.
99. Deeks, E. D.; Keam, S. J. Rosiglitazone: a review of its use in type 2 diabetes mellitus. *Drugs* **2007,** *67* (18), 2747–2779.
100. Werner, A. L.; Travaglini, M. T. A review of rosiglitazone in type 2 diabetes mellitus. *Pharmacotherapy* **2001,** *21* (9), 1082–1099.
101. Wolffenbuttel, B. H. R.; Sels, J.-P.; Huijberts, M. S. P. Rosiglitazone. *Expert Opin. Pharmacother.* **2001,** *2* (3), 467–478.
102. Goldstein, B. J. Rosiglitazone. *Int. J. Clin. Pract.* **2000,** *54* (5), 333–337.
103. Balfour, J. A. B.; Plosker, G. L. Rosiglitazone. *Drugs* **1999,** *57* (6), 921–930.
104. Gilbert, R. E. Rosiglitazone: opening Pandora's black box? *Clin. J. Am. Soc. Nephrol.* **2007,** *2* (6), 1329–1331.
105. Krentz, A. J. Rosiglitazone: trials, tribulations and termination. *Drugs* **2011,** *71* (2), 123–130.
106. Nissen, S. E.; Wolski, K. Rosiglitazone revisited: an updated meta-analysis of risk for myocardial infarction and cardiovascular mortality. *Arch. Intern. Med.* **2010,** *170* (14), 1191–1201.
107. Gale, E. A. M. The second time as farce: rosiglitazone and the regulators. *Nat. Rev. Endocrinol.* **2011,** *7* (1), 5–6.
108. Cheung, B. M. Y. Behind the rosiglitazone controversy. *Expert Rev. Clin. Pharmacol.* **2010,** *3* (6), 723–725.
109. Gopala, A.; Sharma, V. K.; Ganti, S. S. Rosiglitazone—a journey that never completed. *Int. J. Pharm. Pharm. Sci.* **2010,** *2* (Suppl. 2), 7–9.
110. Tatti, P. The rise and fall of rosiglitazone. *Clin. Med.: Ther.* **2009,** *1,* 313–319.
111. Goldfine, A. B. The rough road for rosiglitazone. *Curr. Opin. Endocrinol., Diabetes Obes.* **2008,** *15* (2), 113–117.
112. Salam, N. K.; Huang, T. H. -W.; Kota, B. P.; Kim, M. S.; Li, Y.; Hibbs, D. E. Novel PPAR-gamma agonists identified from a natural product library: a virtual screening, induced-fit docking and biological assay study. *Chem. Biol. Drug Des.* **2008,** *71* (1), 57–70.
113. Madhavan, G. R.; Chakrabarti, R.; Reddy, K. A.; Rajesh, B. M.; Balraju, V.; Rao, P. B.; Rajagopalan, R.; Iqbal, J. Dual PPAR-α and -γ activators derived from novel benzoxazinone containing thiazolidinediones having antidiabetic and hypolipidemic potential. *Bioorg. Med. Chem.* **2006,** *14* (2), 584–591.
114. Bischoff, H. Pharmacology of a-glucosidase inhibition. *Eur. J. Clin. Invest.* **1994,** *24* (Suppl. 3), 3–10.
115. Lebovitz, H. E. Alpha-glucosidase inhibitors. *Endocrinol. Metab. Clin. North Am.* **1997,** *26* (3), 539–551.
116. Truscheit, E.; Frommer, W.; Junge, B.; Mueller, L.; Schmidt, D. D.; Wingender, W. Chemistry and biochemistry of microbial α-glucosidase inhibitors. *Angew. Chem.* **1981,** *93* (9), 738–755.

117. Scheen, A. J. Is there a role for α-glucosidase inhibitors in the prevention of type 2 diabetes mellitus? *Drugs* **2003,** *63* (10), 933–951.
118. Derosa, G.; Maffioli, P. α-Glucosidase inhibitors and their use in clinical practice. *Arch. Med. Sci.* **2012,** *8* (5), 899–906.
119. van de Laar, F. A. Alpha-Glucosidase inhibitors in the early treatment of type 2 diabetes. *Vasc. Health Risk Manage.* **2008,** *4* (6), 1189–1195.
120. Godbout, A.; Chiasson, J.-L. Who should benefit from the use of alpha-glucosidase inhibitors. *Curr. Diabetes Rep.* **2007,** *7* (5), 333–339.
121. Borges de Melo, E.; Gomes, A. S.; Carvalho, I. α- and β-Glucosidase inhibitors: chemical structure and biological activity. *Tetrahedron* **2006,** *62* (44), 10277–10302.
122. Tanaka, K. S. E.; Winters, G. C.; Batchelor, R. J.; Einstein, F. W. B.; Bennet, A. J. A new structural motif for the design of potent glucosidase inhibitors. *J. Am. Chem. Soc.* **2001,** *123* (5), 998–999.
123. Goke, B.; Herrmann-Rinke, C. Structure-activity relationship of a-glucosidase inhibitors Diabetes/Metab. *Rev.* **1998,** *14* (Suppl. 1), S31–S38.
124. Baron, A. D. Postprandial hyperglycemia and α-glucosidase inhibitors. *Diabetes Res. Clin. Pract.* **1998,** *40* (Suppl., Postprandial Hyperglycaemic State: Emerging Concepts), S51–S55.
125. Martin, A. E.; Montogomery, P. A. Acarbose: an α-glucosidase inhibitor. *Am. J. Health-Syst. Pharm.* **1996,** *53* (19), 2277–2290.
126. Clissold, S. P.; Edwards, C. Acarbose. A preliminary review of its pharmacodynamic and pharmacokinetic properties, and therapeutic potential. *Drugs* **1988,** *35* (3), 214–243.
127. Sels, J.-P.; Huijberts, M. S. P.; Wolffenbuttel, B. H. R. Miglitol, a new α-glucosidase inhibitor. *Expert Opin. Pharmacother.* **1999,** *1* (1), 149–156.
128. Campbell, L. K.; Baker, D. E.; Campbell, R. K. Miglitol: assessment of its role in the treatment of patients with diabetes mellitus. *Ann. Pharmacother.* **2000,** *34* (11), 1291–1301.
129. Chen, X.; Zheng, Y.; Shen, Y. Voglibose (Basen, AO-128), one of the most important α-glucosidase inhibitors. *Curr. Med. Chem.* **2006,** *13* (1), 109–116.
130. Drucker, D. J.; Nauck, M. A. The incretin system: glucagon-like peptide-1 receptor agonists and dipeptidyl peptidase-4 inhibitors in type 2 diabetes. *Lancet* **2006,** *368* (9548), 1696–1705.
131. Lovshin, J. A.; Drucker, D. J. Incretin-based therapies for type 2 diabetes mellitus. *Nat. Rev. Endocrinol.* **2009,** *5* (5), 262–269.
132. Holst, J. J.; Vilsboell, T.; Deacon, C. F. The incretin system and its role in type 2 diabetes mellitus. *Mol. Cell. Endocrinol.* **2009,** *297* (1-2), 127–136.
133. Green, B. D.; Flatt, P. R. Incretin hormone mimetics and analogues in diabetes therapeutics. *Best Pract. Res., Clin. Endocrinol. Metab.* **2007,** *21* (4), 497–516.
134. DeFronzo, R. A. From the triumvirate to the ominous octet: a new paradigm for the treatment of type 2 diabetes mellitus. *Diabetes* **2009,** *58* (4), 773–795.
135. Hansen, K. B.; Vilsboll, T.; Knop, F. K. Incretin mimetics: a novel therapeutic option for patients with type 2 diabetes—a review. *Diabetes, Metab. Syndr. Obes.: Targets Ther.* **2010,** *3,* 155–163.
136. Zettl, H.; Schubert-Zsilavecz, M.; Steinhilber, D. Medicinal chemistry of incretin mimetics and DPP-4 inhibitors. *ChemMedChem* **2010,** *5* (2), 179–185.
137. McIntosh, C. H. S. Incretin -based therapies for type 2 diabetes. *Can. J. Diabetes* **2008,** *32* (2), 131–139.
138. Frias, J. P.; Edelman, S. V. Incretins and their role in the management of diabetes. *Curr. Opin. Endocrinol., Diabetes Obes.* **2007,** *14* (4), 269–276.
139. Tasyurek, H. M.; Altunbas, H. A.; Balci, M. K.; Sanlioglu, S. Incretins: Their physiology and application in the treatment of diabetes mellitus. *Diabetes/Metabolism Res. Rev.* **2014,** *30* (5), 354–371.

140. Joy, S. V.; Rodgers, P. T.; Scates, A. C. Incretin mimetics as emerging treatments for type 2 diabetes. *Ann. Pharmacother.* **2005,** *39* (1), 110–118.
141. Aaboe, K.; Krarup, T.; Madsbad, S.; Holst, J. J. GLP-1: physiological effects and potential therapeutic applications. *Diabetes, Obes. Metab.* **2008,** *10* (11), 994–1003.
142. Garg, S. K. The role of basal insulin and glucagon-like peptide 1 agonists in the therapeutic management of type 2 diabetes—a comprehensive review. *Diabetes Technol. Ther.* **2010,** *12* (1), 11–24.
143. Minze, M. G.; Klein, M. S.; Jernigan, M. J.; Wise, S. L.; Fruge, K. Once-weekly exenatide: an extended-duration glucagon-like peptide agonist for the treatment of type 2 diabetes mellitus. *Pharmacotherapy* **2013,** *33* (6), 627–638.
144. Tzefos, M.; Harris, K.; Brackett, A. Clinical efficacy and safety of once-weekly glucagon-like peptide 1 agonists in development for treatment of type 2 diabetes mellitus in adults. *Ann. Pharmacother.* **2012,** *46* (1), 68–78.
145. Barnett, A. Exenatide. *Expert Opin. Pharmacother* **2007,** *8* (15), 2593–2608.
146. McCormack, P. L. Exenatide twice daily: a review of its use in the management of patients with type 2 diabetes mellitus. *Drugs* **2014,** *74* (3), 325–351.
147. Sennik, D.; Ahmed, F.; Russell-Jones, D. Exenatide, a GLP-1 agonist in the treatment of type 2 diabetes. *Expert Rev. Endocrinol. Metab.* **2012,** *7* (1), 15–26.
148. Norris, S. L.; Lee, N.; Thakurta, S.; Chan, B. K. S. Exenatide efficacy and safety: a systematic review. *Diabetic Med.* **2009,** *26* (9), 837–846.
149. Robles, G. I.; Singh-Franco, D. A review of exenatide as adjunctive therapy in patients with type 2 diabetes. *Drug Des., Dev. Ther.* **2009,** *3,* 219–240.
150. Gentilella, R.; Bianchi, C.; Rossi, A.; Rotella, C. M. Exenatide: a review from pharmacology to clinical practice. *Diabetes, Obes. Metab.* **2009,** *11* (6), 544–556.
151. Briones, Mariele; Bajaj, M. Exenatide: a GLP-1 receptor agonist as novel therapy for type 2 diabetes mellitus. *Expert Opin. Pharmacother.* **2006,** *7* (8), 1055–1064.
152. Nielsen, L. L.; Young, A. A.; Parkes, D. G. Pharmacology of exenatide (synthetic exendin-4): a potential therapeutic for improved glycemic control of type 2 diabetes. *Regul. Pept.* **2004,** *117* (2), 77–88.
153. Barnett, A. H. Exenatide. *Drugs Today* **2005,** *41* (9), 563–578.
154. Keating, G. M. Exenatide. *Drugs* **2005,** *65* (12), 1681–1692.
155. Amori, R. E.; Lau, J.; Pittas, A. G. Efficacy and safety of incretin therapy in type 2 diabetes: systematic review and meta-analysis. *JAMA, J. Am. Med. Assoc.* **2007,** *298,* 194–206.
156. van Genugten, R. E.; van Raalte, D. H.; Diamant, M. Does glucagon-like peptide-1 receptor agonist therapy add value in the treatment of type 2 diabetes? Focus on exenatide. *Diabetes Res. Clin. Pract.* **2009,** *86S,* S26–S34.
157. Perry, C. M. Liraglutide: a review of its use in the management of type 2 diabetes mellitus. *Drugs* **2011,** *71* (17), 2347–2373.
158. Ryan, G. J.; Foster, K. T.; Jobe, L. J. Review of the therapeutic uses of liraglutide. *Clin. Ther.* **2011,** *33* (7), 793–811.
159. Davies, M. J.; Kela, R.; Khunti, K. Liraglutide-overview of the preclinical and clinical data and its role in the treatment of type 2 diabetes. *Diabetes, Obes. Metab.* **2011,** *13* (3), 207–220.
160. Croom, K. F.; McCormack, P. L. Liraglutide: a review of its use in type 2 diabetes mellitus. *Drugs* **2009,** *69* (14), 1985–2004.
161. Russell-Jones, D. Molecular, pharmacological and clinical aspects of liraglutide, a once-daily human GLP-1 analogue. *Mol. Cell. Endocrinol.* **2009,** *297* (1–2), 137–140.
162. Vilsboll, T. Liraglutide: a once-daily GLP-1 analogue for the treatment of type 2 diabetes mellitus. *Expert Opin. Invest. Drugs* **2007,** *16* (2), 231–237.

163. Gupta, V. Glucagon-like peptide-1 analogues: an overview. *Indian J. Endocrinol. Metab.* **2013,** *17* (3), 413–421.
164. St. Onge, E. L.; Miller, S. A. Albiglutide: a new GLP-1 analog for the treatment of type 2 diabetes. *Expert Opin. Biol. Ther.* **2010,** *10* (5), 801–806.
165. Madsbad, S.; Kielgast, U.; Asmar, M.; Deacon, C. F.; Torekov, S. S.; Holst, J. J. An overview of once-weekly glucagon-like peptide-1 receptor agonists—-available efficacy and safety data and perspectives for the future. *Diabetes, Obes. Metab.* **2011,** *13* (5), 394–407.
166. Murphy, K. G.; Bloom, S. R. Nonpeptidic glucagon-like peptide 1 receptor agonists: a magic bullet for diabetes? *Proc. Natl. Acad. Sci. U. S. A.* **2007,** *104* (3), 689–690.
167. Liu, Q.; Li, N.; Yuan, Y.; Lu, H.; Wu, X.; Zhou, C.; He, M.; Su, H.; Zhang, M.; Wang, J.; Wang, B.; Wang, Y.; Na, D.; Ye, Y.; Weiss, H.-C.; Gesing, E. R. F.; Liao, J.; Wang, M.-W. Cyclobutane derivatives as novel nonpeptidic small molecule agonists of glucagon-like peptide-1 receptor. *J. Med. Chem.* **2012,** *55* (1), 250–267.
168. Moon, H.-S.; Kim, M.-K.; Son, M.-H. The development of non-peptide glucagon-like peptide 1 receptor agonists for the treatment of type 2 diabetes. *Arch. Pharmacal Res.* **2011,** *34* (7), 1041–1043.
169. Scheen, A. J. A review of gliptins in 2011. *Expert Opin. Pharmacother.* **2012,** *13* (1), 81–99.
170. Russell-Jones, D.; Gough, S. Recent advances in incretin-based therapies. *Clin. Endocrinol. (Oxford, U. K.)* **2012,** *77* (4), 489–499.
171. Thornberry, N. A.; Weber, A. E. Discovery of JANUVIA (sitagliptin), a selective dipeptidyl peptidase IV inhibitor for the treatment of type 2 diabetes. *Curr. Top. Med. Chem.* **2007,** *7* (6), 557–568.
172. Lyseng-Williamson, K. A. Sitagliptin. *Drugs* **2007,** *67* (4), 587–597.
173. Miller, S. A.; St. Onge, E. L. Sitagliptin: a dipeptidyl peptidase IV inhibitor for the treatment of type 2 diabetes. *Ann. Pharmacother.* **2006,** *40* (7/8), 1336–1343.
174. Subbarayan, S.; Kipnes, M. Sitagliptin: a review. *Expert Opin. Pharmacother.* **2011,** *12* (10), 1613–1622.
175. Engel, S. S.; Williams-Herman, D. E.; Golm, G. T.; Clay, R. J.; Machotka, S. V.; Kaufman, K. D.; Goldstein, B. J. Sitagliptin: review of preclinical and clinical data regarding incidence of pancreatitis. *Int. J. Clin. Pract.* **2010,** *64* (7), 984–990.
176. Parmee, E. R.; SinhaRoy, R.; Xu, F.; Givand, J. C.; Rosen, L. A. Discovery and development of the DPP-4 inhibitor Januvia (Sitagliptin). In *Case Studies in Modern Drug Discovery and Development;* Huang, X., Aslanian, R. G., Eds.; Wiley, 2012; pp 10–44.
177. Florentin, M.; Liberopoulos, E. N.; Mikhailidis, D. P.; Elisaf, M. S. Sitagliptin in clinical practice: a new approach in the treatment of type 2 diabetes. *Expert Opin. Pharmacother.* **2008,** *9* (10), 1705–1720.
178. Plosker, G. L. Sitagliptin: a review of its use in patients with type 2 diabetes mellitus. *Drugs* **2014,** *74* (2), 223–242.
179. Dhillon, S. Sitagliptin: a review of its use in the management of type 2 diabetes mellitus. *Drugs* **2010,** *70* (4), 489–512.
180. Kim, D.; Wang, L.; Beconi, M.; Eiermann, G. J.; Fisher, M. H.; He, H.; Hickey, G. J.; Kowalchick, J. E.; Leiting, B.; Lyons, K.; Marsilio, F.; McCann, M. E. P.; Reshma, A.; Petrov, A.; Scapin, G.; Patel, S. B.; Roy, R. S.; Wu, J. K.; Wyvratt, M. J.; Zhang, B. B.; Zhu, L.; Thornberry, N. A.; Weber, A. E. (2R)-4-Oxo-4-[3-(trifluoromethyl)-5,6-dihydro[1,2,4]triazolo[4,3-a]pyrazin-7(8H)-yl]-1-(2,4,5-trifluorophenyl)butan-2-amine: a potent, orally active dipeptidyl peptidase IV inhibitor for the treatment of type 2 diabetes. *J. Med. Chem.* **2005,** *48* (1), 141–151.
181. Gore, V.; Gadkar, M.; Priyanka, B.; Suresh, S., Novel processes for the preparation of sitagliptin, IN 2009KO00722 (2010).

182. Pan, X.; Li, X.; Lu, Q.; Yu, W.; Li, W.; Zhang, Q.; Deng, F.; Liu, F. Na-acyl derivatives of aminoacyl-2-cyanopyrrolidine-inhibitors of prolyl endopeptidase and dipeptidyl peptidase-IV, having hypoglycemic, antihypoxic, neuroprotective action and action of cognitive function improvement. *Tetrahedron Lett.* **2013,** *54* (50), 6807–6809.
183. Edmondson, S. D.; Xu, F.; Armstrong, J. D., III Sitagliptin (Januvia): a treatment for type 2 diabetes. In *Modern Drug Synthesis;* Li, J. J., Johnson, D. S., Eds.; Wiley, 2010; pp 125–140.
184. Desai, A. A. Sitagliptin manufacture: a compelling tale of green chemistry, process intensification, and industrial asymmetric catalysis. *Angew. Chem., Int. Ed.* **2011,** *50* (9), 1974–1976.
185. Balsells, J.; Hsiao, Y.; Hansen, K. B.; Xu, F.; Ikemoto, N.; Clausen, A.; Armstrong, J. D., III Synthesis of sitagliptin, the active ingredient in Januvia and Janumet. In *Green Chemistry in the Pharmaceutical Industry;* Dunn, P. J., Wells, A. S., Williams, M. T., Eds.; Wiley-VCH, 2010; pp 101–126.
186. Liu, Y.; Si, M.; Tang, L.; Shangguan, S.; Wu, H.; Li, J.; Wu, P.; Ma, X.; Liu, T.; Hu, Y. Synthesis and biological evaluation of novel benzyl-substituted (S)-phenylalanine derivatives as potent dipeptidyl peptidase 4. *Bioorg. Med. Chem.* **2013,** *21* (18), 5679–5687.
187. Hansen, K. B.; Balsells, J.; Dreher, S.; Hsiao, Y.; Kubryk, M.; Palucki, M.; Rivera, N.; Steinhuebel, D.; Armstrong, J. D., III; Askin, D.; Grabowski, E. J. J. First generation process for the preparation of the DPP-IV inhibitor sitagliptin. *Org. Process Res. Dev.* **2005,** *9* (5), 634–639.
188. Hansen, K. B.; Hsiao, Y.; Xu, F.; Rivera, N.; Clausen, A.; Kubryk, M.; Krska, S.; Rosner, T.; Simmons, B.; Balsells, J.; Ikemoto, N.; Sun, Y.; Spindler, F.; Malan, C.; Grabowski, E. J. J.; Armstrong, J. D. Highly efficient asymmetric synthesis of sitagliptin. *J. Am. Chem. Soc.* **2009,** *131* (25), 8798–8804.
189. Savile, C. K.; Janey, J. M.; Mundorff, E. C.; Moore, J. C.; Tam, S.; Jarvis, W. R.; Colbeck, J. C.; Krebber, A.; Fleitz, F. J.; Brands, J.; Devine, P. N.; Huisman, G. W.; Hughes, G. J. Biocatalytic asymmetric synthesis of chiral amines from ketones applied to sitagliptin manufacture. *Science (Washington, DC, U. S.)* **2010,** *329* (5989), 305–309.
190. Edmondson, S. D.; Fisher, M. H.; Kim, D.; MacCoss, M.; Parmee, E. R.; Weber, A. E.; Xu, J., Preparation of β-amino tetrahydroimidazo[1,2-a]pyrazines and tetrahydrotrioazolo[4,3-a] pyrazines as dipeptidyl peptidase inhibitors for the treatment or prevention of diabetes, WO 2003004498 (2003).
191. Dreher, S. D.; Ikemoto, N.; Njolito, E.; Rivera, N. R.; Tellers, D. M.; Xiao, Y., Process for preparation of chiral β-amino acid derivatives, WO 2004085661 (2004).
192. Angelaud, R.; Armstrong, J. D., III; Askin, D.; Balsells, J.; Hansen, K.; Lee, J.; Maligres, P. E.; Rivera, N. R.; Xiao, Y.; Zhong, Y-Li., Process for the preparation of β-amino acid amide dipeptidyl peptidase-IV inhibitors, WO 2004087650 (2004).
193. Adeghate, E.; Kalasz, H. Amylin analogues in the treatment of diabetes mellitus: medicinal chemistry and structural basis of its function. *Open Med. Chem. J.* **2011,** *5,* 78–81.
194. Day, C. Amylin analogue as an antidiabetic agent. *Br. J. Diabetes Vasc. Dis.* **2005,** *5* (3), 151–154.
195. McQueen, J. Pramlintide acetate. *Am. J. Health-Syst. Pharm.* **2005,** *62* (22), 2363–2372.
196. Ryan, G. J.; Jobe, L. J.; Martin, R. Pramlintide in the treatment of type 1 and type 2 diabetes mellitus. *Clin. Ther.* **2005,** *27* (10), 1500–1512.
197. Edelman, S.; Maier, H.; Wilhelm, K. Pramlintide in the treatment of diabetes mellitus. *BioDrugs* **2008,** *22* (6), 375–386.
198. Younk, L. M.; Mikeladze, M.; Davis, S. N. Pramlintide and the treatment of diabetes: a review of the data since its introduction. *Expert Opin. Pharmacother.* **2011,** *12* (9), 1439–1451.
199. Isaji, M. Sodium-glucose cotransporter inhibitors for diabetes. *Curr. Opin. Invest. Drugs (BioMed Cent.)* **2007,** *8* (4), 285–292.

200. Idris, I.; Donnely, R. Sodium-glucose co-transporter-2 inhibitors: an emerging new class of oral antidiabetic drug. *Diabetes, Obes. Metab.* **2009,** *11* (2), 79–88.
201. Bailey, C. J. Renal glucose reabsorption inhibitors to treat diabetes. *Trends Pharmacol. Sci.* **2011,** *32* (2), 63–71.
202. Nair, S.; Wilding, J. P. H. Sodium glucose cotransporter 2 inhibitors as a new treatment for diabetes mellitus. *J. Clin. Endocrinol. Metab.* **2010,** *95* (1), 34–42.
203. Bouchie, A. SGLT2 inhibitors enter crowded diabetes space. *Nat. Biotechnol.* **2013,** *31* (6), 469–470.
204. Washburn, W. N. SGLT2 inhibitors in development. *RSC Drug Discovery Ser.* **2012,** *27,* 29–87.
205. Ghosh, R. K.; Ghosh, S. M.; Chawla, S.; Jasdanwala, S. A. SGLT2 inhibitors: a new emerging therapeutic class in the treatment of type 2 diabetes mellitus. *J. Clin. Pharmacol.* **2012,** *52* (4), 457–461.
206. Plosker, G. L. Dapagliflozin: a review of its use in type 2 diabetes mellitus. *Drugs* **2012,** *72* (17), 2289–2312.
207. Shah, N. K.; Deeb, W. E.; Choksi, R.; Epstein, B. J. Dapagliflozin: a novel sodium-glucose cotransporter type 2 inhibitor for the treatment of type 2 diabetes mellitus. *Pharmacotherapy* **2012,** *32* (1), 80–94.
208. Katsiki, N.; Papanas, N.; Mikhailidis, D. P. Dapagliflozin: more than just another oral glucose-lowering agent? *Expert Opin. Invest. Drugs* **2010,** *19* (12), 1581–1589.
209. Kipnes, M. Dapagliflozin: an emerging treatment option in type 2 diabetes. *Expert Opin. Invest. Drugs* **2009,** *18* (3), 327–334.
210. Elkinson, S.; Scott, L. J. Canagliflozin. *Drugs* **2013,** *73* (9), 979–988.
211. Lamos, E. M.; Younk, L. M.; Davis, S. N. Canagliflozin, an inhibitor of sodium-glucose cotransporter 2, for the treatment of type 2 diabetes mellitus. *Expert Opin. Drug Metab. Toxicol.* **2013,** *9* (6), 763–775.
212. Dietrich, E.; Powell, J.; Taylor, J. R. Canagliflozin: a novel treatment option for type 2 diabetes. *Drug Des., Dev. Ther.* **2013,** *7,* 1399–1408.
213. Babu, A. Canagliflozin for the treatment of type 2 diabetes. *Drugs Today* **2013,** *49* (6), 363–376.
214. Livingston, J. N.; Schoen, W. R. Glucagon and glucagon-like peptide-1. *Annu. Rep. Med. Chem.* **1999,** *34,* 189–198.
215. Connell, R. D. Glucagon antagonists for the treatment of type 2 diabetes. *Expert Opin. Ther. Pat.* **1999,** *9* (6), 701–709.
216. Bagger, J. I.; Knop, F. K.; Holst, J. J.; Vilsboell, T. Glucagon antagonism as a potential therapeutic target in type 2 diabetes. *Diabetes, Obes. Metab.* **2011,** *13* (11), 965–971.
217. Hruby, V. J. Strategies in the development of peptide antagonists. *Prog. Brain Res.* **1992,** *92,* 215–224.
218. Zechel, C.; Trivedi, D.; Hruby, V. J. Synthetic glucagon antagonists and partial agonists. *Int. J. Pept. Protein Res.* **1991,** *38* (2), 131–138.
219. Unson, C. G.; Cypess, A. M.; Wu, C.-R.; Goldsmith, P. K.; Merrifield, R. B.; Sakmar, T. P. Antibodies against specific extracellular epitopes of the glucagon receptor block glucagon binding. *Proc. Natl. Acad. Sci. U. S. A.* **1996,** *93* (1), 310–315.
220. Kurukulasuriya, R.; Link, J. T. Progress towards glucagon receptor antagonist therapy for type 2 diabetes. *Expert Opin. Ther. Pat.* **2005,** *15* (12), 1739–1749.
221. Ling, A. L.; Wasserman, J. I. Approaches to glucagon receptor antagonists. *Expert Opin. Ther. Pat.* **2003,** *13* (1), 15–22.
222. Ahn, J.-M.; Medeiros, M.; Trivedi, D.; Hruby, V. J. Development of potent truncated glucagon antagonists. *J. Med. Chem.* **2001,** *44* (9), 1372–1379.

223. Bregman, M. D.; Hruby, V. J. Synthesis and isolation of a glucagon antagonist. *FEBS Lett.* **1979,** *101* (1), 191–194.
224. Johnson, D. G.; Goebel, C. U.; Hruby, V. J.; Bregman, M. D.; Trivedi, D. Hyperglycemia of diabetic rats decreased by a glucagon receptor antagonist. *Science (Washington, DC, U. S.)* **1982,** *215* (4536), 1115–1116.
225. Hruby, V. J.; Ahn, J.-M.; Trivedi, D. The design and biological activities of glucagon agonists and antagonists, and their use in examining the mechanisms of glucose action. *Curr. Med. Chem.: Immunol., Endocr. Metab. Agents* **2001,** *1* (3), 199–215.
226. Unson, C. G.; Gurzenda, E. M.; Merrifield, R. B. Biological activities of des-His1[Glu9] glucagon amide, a glucagon antagonist. *Peptides* **1989,** *10* (6), 1171–1177.
227. Post, S. R.; Rubinstein, P. G.; Tager, H. S. Mechanism of action of des-His1-[Glu9]glucagon amide, a peptide antagonist of the glucagon receptor system. *Proc. Natl. Acad. Sci. U. S. A.* **1993,** *90* (5), 1662–1666.
228. Unson, C. G.; Wu, C. R.; Fitzpatrick, K. J.; Merrifield, R. B. Multiple-site replacement analogs of glucagon. *A molecular basis for antagonist design J. Biol. Chem.* **1994,** *269* (17), 12548–12551.
229. Azizeh, B. Y.; Van Tine, B. A.; Sturm, N. S.; Hutzler, A. M.; David, C.; Trivedi, D.; Hruby, V. J. [des-His1, des-Phe6, Glu9]glucagon amide: a newly designed "pure" glucagon antagonist. *Bioorg. Med. Chem. Lett.* **1995,** *5* (16), 1849–1852.
230. Azizeh, B. Y.; Ahn, J.-M.; Caspari, R.; Shenderovich, M. D.; Trivedi, D.; Hruby, V. J. The role of phenylalanine at position 6 in glucagon's mechanism of biological action: multiple replacement analogues of glucagon. *J. Med. Chem.* **1997,** *40,* 2555–2562.
231. Cho, Y. M.; Merchant, C. E.; Kieffer, T. J. Targeting the glucagon receptor family for diabetes and obesity therapy. *Pharmacol. Therapeut.* **2012,** *135* (3), 247–278.
232. Post, S. R.; Rubinstein, P. G.; Tager, H. S. Mechanism of action of des-His1-[Glu9] glucagon amide, a peptide antagonist of the glucagon receptor system. *Proc. Natl. Acad. Sci. U. S. A.* **1993,** *90,* 1662–1666.
233. Unson, C. G.; Wu, C.-R.; Fizpatrick, K. J.; Merrifield, R. B. Multiple-site replacement analogues of glucagon. *J. Biol. Chem.* **1994,** *269,* 12548–12551.
234. Madsen, P.; Brand, C. L.; Holst, J. J.; Knudsen, B. Advances in non-peptide glucagon receptor antagonists. *Curr. Pharmaceut. Design* **1999,** *5* (9), 683–691.
235. Qureshi, S. A.; Candelore, M. R.; Xie, D.; Yang, X.; Tota, L. M.; Ding, V. D. H.; Li, Z.; Bansal, A.; Miller, C.; Cohen, S. M.; Jiang, G.; Brady, E.; Saperstein, R.; Duffy, J. L.; Tata, J. R.; Capman, K. T.; Moller, D. E.; Zhang, B. B. A novel glucagon receptor antagonist inhibits glucagon-mediated biological effects. *Diabetes* **2004,** *53,* 3267–3273.
236. Hasegawa, F.; Niidome, K.; Migihashi, C.; Murata, M.; Negoro, T.; Matsumoto, T.; Kato, K.; Fujii, A. Discovery of furan-2-carbohydrazides as orally active glucagon receptor antagonists. *Bioorg. Med. Chem. Lett.* **2014,** *24* (17), 4266–4270.
237. Kodra, J. T.; Jorgensen, A. S.; Andersen, B.; Behrens, C.; Brand, C. L.; Christensen, I. T.; Gulbrandt, M.; Jeppesen, C. B.; Knudsen, L. B.; Madsen, P.; Nishimura, E.; Sams, C. K.; Sidelmann, U. G.; Pedersen, R. A.; Lynn, F. C.; Lau, J. Novel glucagon receptor antagonists with improved selectivity over the glucose-dependent insulinotropic polypeptide receptor. *J. Med. Chem.* **2008,** *51,* 5387–5396.
238. Ling, A. L.; Wasserman, J. I. Approaches to glucagon receptor antagonists. *Expert Opin. Ther. Pat.* **2003,** *13* (1), 15–22.
239. Djuric, S. W.; Grihalde, N.; Lin, C. W. Glucagon receptor antagonists for the treatment of type II diabetes: current prospects. *Curr. Opin. Invest. Drugs (BioMed Cent.)* **2002,** *3* (11), 1617–1623.

240. Ling, A. Small-molecule glucagon receptor antagonists. *Drugs Future* **2002,** *27* (10), 987–993.
241. Kurukulasuriya, R.; Sorensen, B. K.; Link, J. T.; Patel, J. R.; Jae, H. S.; Winn, M. X.; Rohde, J. R.; Grihalde, N. D.; Lin, C. W.; Ogiela, C. A.; Adler, A. L.; Collins, C. A. Biaryl amide glucagon receptor antagonists. *Bioorg. Med. Chem. Lett.* **2004,** *14,* 2047–2050.
242. Anderson, P. L.; Paolella, N. A., 1-Phenylpyrazole derivatives as glucagon inhibitors, US 4359474 (1982).
243. Collins, J. L.; Dambek, P. J.; Goldstein, S. W.; Faraci, W. S. CP-99,711: a nonpeptide glucagon receptor antagonist. *Bioorg. Med. Chem. Lett.* **1992,** *2* (9), 915–918.
244. Chang, L. L.; Sidler, K. L.; Cascieri, M. A.; de Laszlo, S.; Koch, G.; Li, B.; MacCoss, M.; Mantlo, N.; O'Keefe, S.; Pang, M.; Rolando, A.; Hagmann, W. K. Substituted imidazoles as glucagon receptor antagonists. *Bioorg. Med. Chem. Lett.* **2001,** *11,* 2549–2553.
245. Chang, L. L., Triaryl substituted imidazoles, compositions containing such compounds and methods of use, WO 9821957 (1998).
246. Chang, L. L., Preparation of triaryl substituted imidazoles as glucagon antagonists, WO 9822108 (1998).
247. Duffy, J. L.; Kirk, B. A.; Konteatis, Z.; Campbell, E. L.; Liang, R.; Brady, E. J.; Candelore, M. R.; Ding, V. D. H.; Jiang, G.; Liu, F.; Qureshi, S. A.; Saperstein, R.; Szalkowski, D.; Tong, S.; Tota, L. M.; Xie, D.; Yang, X.; Zafian, P.; Zheng, S.; Chapman, K. T.; Zhang, B. B.; Tata, J. R. Discovery and investigation of a novel class of thiophene-derived antagonists of the human glucagon receptor. *Bioorg. Med. Chem. Lett.* **2005,** *15,* 1401–1405.
248. Shen, D. M.; Zhang, F.; Brady, E. J.; Candelore, M. R.; Yang, Q. D.; Ding, V. D. H.; Dragovic, J.; Feeny, W. P.; Jiang, G.; McCann, P.; Mock, S.; Quereshi, S. A.; Saperstein, R.; Shen, X.; Tamvakopoulos, C.; Tong, X.; Tota, L. M.; Wright, M. J.; Yang, X.; Zheng, S.; Chapman, K. T.; Zhang, B. B.; Tata, J. R.; Parmee, E. R. Discovery of novel, potent, and orally active spiro-urea human glucagon receptor antagonist. *Bioorg. Med. Chem. Lett.* **2005,** *15,* 4564–4569.
249. Liang, R.; Abrado, L.; Brady, E. J.; Candelore, M. R.; Ding, V.; Saperstein, R.; Tota, L. M.; Wright, M.; Mock, S.; Tamvakopoulos, C.; Tong, S.; Zheng, S.; Zhang, B. B.; Tata, J. R.; Parmee, E. R. Design and synthesis of conformationally constrained trisubstituted ureas as potent antagonists of the human glucagon receptor. *Bioorg. Med. Chem. Lett.* **2007,** *17,* 587–592.
250. Behrens, C.; Lau, J.; Madsen, P., Preparation of tetrazolylethylbenzylureas, oxadiazolylmethyl-aminocarbonylbenzylureas, and related compounds as glucagon antagonists/inverse agonists, US 20030203946 (2003).
251. Spohr, U. D.; Malone, M. J.; Mantlo, N. B.; Zablocki, J. A., Preparation of arylpyrimidinones and analogs as drugs, WO 9824780 (1998).
252. Jiang, G.; Zhang, B. B. Glucagon and regulation of glucose metabolism. *Am. J. Phisiol.* **2003,** *284* (4, Pt. 1), E671–E678.
253. Ladouceur, G. H.; Cook, J. H.; Doherty, E. M.; Schoen, W. R.; MacDougall, M. L.; Livingston, J. N. Discovery of 5-hydroxyalkyl 4-phenylpyridines as a new class of glucagon receptor antagonists. *Bioorg. Med. Chem. Lett.* **2002,** *12,* 461–464.
254. Smith, R. A.; Hertzog, D. L.; Osterhout, M. H.; Ladouceur, G. H.; Korpusik, M.; Bobko, M. A.; Howard Jones, J.; Phelan, K.; Romero, R. H.; Hundertmark, T.; MacDougall, M. L.; Livingston, J. N.; Schoen, W. R. Optimization of the 4-aryl group of 4-aryl-pyridine glucagon antagonists: development of an efficient, alternative synthesis. *Bioorg. Med. Chem. Lett.* **2002,** *12,* 1303–1306.
255. Ling, A.; Hong, Y.; Gonzalez, J.; Gregor, V.; Polinsky, A.; Kuki, A.; Shi, S.; Teston, K.; Porter, J.; Kiel, D.; Lakis, J.; Anderes, K.; May, J.; Knudsen, L. B.; Lau, J. Identification of alkylidene hydrazides as glucagon receptor antagonists. *J. Med. Chem.* **2001,** *44,* 3141–3149.

256. Ling, A.; Plewe, M.; Gonzalez, J.; Madsen, P.; Sams, C. K.; Lau, J.; Gregor, V.; Murphy, D.; Teston, K.; Kuki, A.; Shi, S.; Truesdale, L.; Kiel, D.; May, J.; Lakis, J.; Anderes, K.; Iatsimirskaia, E.; Sidelmann, U. G.; Knudsen, L. B.; Brand, C. L.; Polinsky, A. Human glucagon receptor antagonists based on alkylidene hydrazides. *Bioorg. Med. Chem. Lett.* **2002,** *12,* 663–666.
257. Madsen, P.; Ling, A.; Sams, C. K.; Knudsen, L. B.; Sidelmann, U. G.; Ynddal, L.; Brand, C. L.; Plewe, M.; Murphy, D.; Teng, M.; Truesdale, L.; Kiel, D.; May, J.; Kuki, A.; Shi, S.; Feng, J.; Johnson, M. D.; Teston, K. A.; Lakis, J.; Anderes, K.; Gregor, V.; Lau, J. Optimization of alkylidene hydrazide based human glucagon receptor antagonists. Discovery of the highly potent and orally available 3-cyano-4 hydroxybenzoic acid (1-(2,3,5,6-tetramethylbenzyl)-1H-indol-4-ylm ethylene-hydrazide. *J. Med. Chem.* **2002,** *45,* 5755–5775.
258. Madsen, P.; Ling, A.; Plewe, M.; Sams, C. K.; Knudsen, L. B.; Sidelmann, U. G.; Ynddal, L.; Brand, C. L.; Andersen, B.; Murphy, D.; Teng, M.; Truesdale, L.; Kiel, D.; May, J.; Kuki, A.; Shi, S.; Johnson, M. D.; Teston, K. A.; Feng, J.; Lakis, J.; Anderes, K.; Gregor, V.; Lau, J. Optimization of Alkylidene Hydrazide Based Human Glucagon Receptor Antagonists. Discovery of the Highly Potent and Orally Available 3-Cyano-4-hydroxybenzoic Acid [1-(2,3,5,6-Tetramethylbenzyl)-1H-indol-4-ylmethylene]hydrazide. *J. Med. Chem.* **2002,** *45* (26), 5755–5775.
259. Ling, A.; Gregor, V.; Gonzalez, J.; Hong, Y.; Kiel, D.; Kuki, A.; Shi, S.; Naerum, L.; Madsen, P.; Sams, C.; Lau, J.; Plewe, M. B.; Feng, J.; Teng, M.; Johnson, M. D.; Teston, K. A.; Sidelmann, U. G.; Knudsen, L. B., Preparation of aroyl hydrazides and related compounds as glucagon antagonists/inverse agonists, US 6613942 (2003).
260. Lee, E. C. Y.; Tu, M.; Stevens, B. D.; Bian, J.; Aspnes, G.; Perreault, C.; Sammons, M. F.; Wright, S. W.; Litchfield, J.; Kalgutkar, A. S.; Sharma, R.; Didiuk, M. T.; Ebner, D. C.; Filipski, K. J.; Brown, J.; Atkinson, K.; Pfefferkorn, J. A. Guzman-Perez, A.,l Identification of a novel conformationally constrained glucagon receptor antagonist. *Bioorg. Med. Chem. Lett.* **2014,** *24* (3), 839–844.
261. Barcza, S., 4,4′-(Alkanediyl)bis(2,2,6,6-tetraalkyl-1-oxa-4-aza-2,6-disilacyclohexanes), US 4374130 (1983).
262. Zhang, S.; Zhang, Z.-Y. PTP1B as a drug target: recent developments in PTP1B inhibitor discovery. *Drug Discovery Today* **2007,** *12* (9&10), 373–381.
263. Montalibet, J.; Kennedy, B. P. Therapeutic strategies for targeting PTP1B in diabetes. *Drug Discovery Today: Ther. Strategies* **2005,** *2* (2), 129–135.
264 Wei, Y.; Chen, Y.-T.; Shi, L.; Gao, L.-X.; Liu, S.; Cui, Y.-M.; Zhang, W.; Shen, Q.; Li, J.; Nan, F.-J. Discovery and structural modification of novel inhibitors of PTP1B inspired by the ACT fragment of scleritodermin A. *MedChemComm* **2011,** *2* (11), 1104–1109.
265. Szczepankiewicz, B. G.; Liu, G.; Hajduk, P. J.; Abad-Zapatero, C.; Pei, Z.; Xin, Z.; Lubben, T. H.; Trevillyan, J. M.; Stashko, M. A.; Ballaron, S. J.; Liang, H.; Huang, F.; Hutchins, C. W.; Fesik, S. W.; Jirousek, M. R. Discovery of a potent, selective protein tyrosine phosphatase 1B inhibitor using a linked-fragment strategy. *J. Am. Chem. Soc.* **2003,** *125* (14), 4087–4096.
266. Combs, A. P.; Zhu, W.; Crawley, M. L.; Glass, B.; Polam, P.; Sparks, R. B.; Modi, D.; Takvorian, A.; McLaughlin, E.; Yue, E. W.; Wasserman, Z.; Bower, M.; Wei, M.; Rupar, M.; Ala, P. J.; Reid, B. M.; Ellis, D.; Gonneville, L.; Emm, T.; Taylor, N.; Yeleswaram, S.; Li, Y.; Wynn, R.; Burn, T. C.; Hollis, G.; Liu, P. C. C.; Metcalf, B. Potent benzimidazole sulfonamide protein tyrosine phosphatase 1B inhibitors containing the heterocyclic (S)-isothiazolidinone phosphotyrosine mimetic. *J. Med. Chem.* **2006,** *49* (13), 3774–3789.
267. Yin, J.-P.; Tang, C.-L.; Gao, L.-X.; Ma, W.-P.; Li, J.-Y.; Li, Y.; Li, J.; Nan, F.-J. Design and synthesis of paracaseolide A analogues as selective protein tyrosine phosphatase 1B inhibitors. *Org. Biomol. Chem.* **2014,** *12* (21), 3441–3445.

268. Rakse, M.; Karthikeyan, C.; Deora, G. S.; Moorthy, N. S. H.N.; Rathore, V.; Rawat, A. K.; Srivastava, A. K.; Trivedi, P. Design, synthesis and molecular modelling studies of novel 3-acetamido-4-methyl benzoic acid derivatives as inhibitors of protein tyrosine phosphatase 1B. *Eur. J. Med. Chem.* **2013,** *70,* 469–476.
269. Lau, C. K.; Bayly, C. I.; Gauthier, J. Y.; Li, C. S.; Therien, M.; Asante-Appiah, E.; Cromlish, W.; Boie, Y.; Forghani, F.; Desmarais, S.; Wang, Q.; Skorey, K.; Waddleton, D.; Payette, P.; Ramachandran, C.; Kennedy, B. P.; Scapin, G. Structure based design of a series of potent and selective non peptidic PTP-1B inhibitors. *Bioorg. Med. Chem. Lett.* **2004,** *14* (4), 1043–1048.
270. Wiesmann, C.; Barr, K. J.; Kung, J.; Zhu, J.; Erlanson, D. A.; Shen, W.; Fahr, B. J.; Zhong, M.; Taylor, L.; Randal, M.; McDowell, R. S.; Hansen, S. K. Allosteric inhibition of protein tyrosine phosphatase 1B. *Nat. Struct. Mol. Biol.* **2004,** *11* (8), 730–737.
271. Andersen, H. S.; Olsen, O. H.; Iversen, L. F.; Sorensen, A. L. P.; Mortensen, S. B.; Christensen, M. S.; Branner, S.; Hansen, T. K.; Lau, J. F.; Jeppesen, L.; Moran, E. J.; Su, J.; Bakir, F.; Judge, L.; Shahbaz, M.; Collins, T.; Vo, T.; Newman, M. J.; Ripka, W. C.; Moller, N. P. H. Discovery and SAR of a novel selective and orally bioavailable nonpeptide classical competitive inhibitor class of protein-tyrosine phosphatase 1B. *J. Med. Chem.* **2002,** *45* (20), 4443–4459.
272. Iversen, L. F.; Andersen, H. S.; Branner, S.; Mortensen, S. B.; Peters, G. H.; Norris, K.; Olsen, O. H.; Jeppesen, C. B.; Lundt, B. F.; Ripka, W.; Moller, K. B.; Moller, N. P. H. Structure-based design of a low molecular weight, nonphosphorus, nonpeptide, and highly selective inhibitor of protein-tyrosine phosphatase 1B. *J. Biol. Chem.* **2000,** *275* (14), 10300–10307.
273. Klopfenstein, S. R.; Evdokimov, A. G.; Colson, A.-O.; Fairweather, N. T.; Neuman, J. J.; Maier, M. B.; Gray, J. L.; Gerwe, G. S.; Stake, G. E.; Howard, B. W.; Farmer, J. A.; Pokross, M. E.; Downs, T. R.; Kasibhatla, B.; Peters, K. G. 1,2,3,4-Tetrahydroisoquinolinyl sulfamic acids as phosphatase PTP1B inhibitors. *Bioorg. Med. Chem. Lett.* **2006,** *16* (6), 1574–1578.
274. Liljebris, C.; Martinsson, J.; Tedenborg, L.; Williams, M.; Barker, E.; Duffy, J. E. S.; Nygren, A.; James, S. Synthesis and biological activity of a novel class of pyridazine analogues as non-competitive reversible inhibitors of protein tyrosine phosphatase 1B (PTP1B). *Bioorg. Med. Chem.* **2002,** *10* (10), 3197–3212.
275. Martin, W. H.; Hoover, D. J.; Armento, S. J.; Stock, I. A.; McPherson, R. K.; Danley, D. E.; Stevenson, R. W.; Barrett, E. J.; Treadway, J. L. Discovery of a human liver glycogen phosphorylase inhibitor that lowers blood glucose in vivo. *Proc. Natl. Acad. Sci. U. S. A.* **1998,** *95* (4), 1776–1781.
276. Baker, D. J.; Timmons, J. A.; Greenhaff, P. L. Glycogen phosphorylase inhibition in type 2 diabetes therapy: a systematic evaluation of metabolic and functional effects in rat skeletal muscle. *Diabetes* **2005,** *54* (8), 2453–2459.
277. Li, W.-L.; Jian, C.; Luo, M.-H.; Han, L.; Li, M.; Li, W.-L. Recent advances in design of glycogen phosphorylase inhibitors. *Curr. Enzyme Inhib.* **2011,** *7* (4), 259–267.
278. Gaboriaud-Kolar, N.; Skaltsounis, A.-L. Glycogen phosphorylase inhibitors: a patent review. *Expert Opin. Ther. Pat.* **2013,** *23* (8), 1017–1032.
279. Somsak, L.; Czifrak, K.; Toth, M.; Bokor, E.; Chrysina, E. D.; Alexacou, K.-M.; Hayes, J. M.; Tiraidis, C.; Lazoura, E.; Leonidas, D. D. New inhibitors of glycogen phosphorylase as potential antidiabetic agents. *Curr. Med. Chem.* **2008,** *15* (28), 2933–2983.
280. Treadway, J. L.; Mendys, P.; Hoover, D. J. Glycogen phosphorylase inhibitors for treatment of type 2 diabetes mellitus. *Expert Opin. Invest. Drugs* **2001,** *10* (3), 439–454.
281. Klabunde, T.; Wendt, K. U.; Kadereit, D.; Brachvogel, V.; Burger, H.-J.; Herling, A. W.; Oikonomakos, N. G.; Kosmopoulou, M. N.; Schmoll, D.; Sarubbi, E.; Von Roedern, E.; Schoenafinger, K.; Defossa, E. Acyl ureas as human liver glycogen phosphorylase inhibitors for the treatment of type 2 diabetes. *J. Med. Chem.* **2005,** *48* (20), 6178–6193.

282. Defossa, E.; Kadereit, D.; Klabunde, T.; Burger, H-J.; Herling, A.; Wendt, K-U.; Von Roedern, E; Schoenafinger, K., Preparation of N-[(phenylamino)carbonyl]benzamides as glycogenphosphorylase-A inhibitors for the treatment of diabetes, WO 2004007437 (2004).
283. Grasso, P. Novel approaches to the treatment of obesity and type 2 diabetes mellitus: bioactive leptin-related synthetic peptide analogs. *Recent Pat. Endocr., Metab. Immune Drug Discovery* **2011,** *5* (3), 163–175.
284. Coppari, R.; Bjorbaek, C. Leptin revisited: its mechanism of action and potential for treating diabetes. *Nat. Rev. Drug Discovery* **2012,** *11* (9), 692–708.
285. Pal, M. Recent advances in glucokinase activators for the treatment of type 2 diabetes. *Drug Discovery Today* **2009,** *14* (15/16), 784–792.
286. Sarabu, R.; Berthel, S. J.; Kester, R. F.; Tilley, J. W. Glucokinase activators as new type 2 diabetes therapeutic agents. *Expert Opin. Ther. Pat.* **2008,** *18* (7), 759–768.
287. Meininger, G. E.; Scott, R.; Alba, M.; Shentu, Y.; Luo, E.; Amin, H.; Davies, M. J.; Kaufman, K. D.; Goldstein, B. J. Effects of MK-(0941), a novel glucokinase activator, on glycemic control in insulin-treated patients with type 2 diabetes. *Diabetes Care* **2011,** *34* (12), 25602566.
288. Grimsby, J.; Sarabu, R.; Corbett, W. L.; Haynes, N.-E.; Bizzarro, F. T.; Coffey, J. W.; Guertin, K. R.; Hilliard, D. W.; Kester, R. F.; Mahaney, P. E.; Marcus, L.; Qi, L.; Spence, C. L.; Tengi, J.; Magnuson, M. A.; Chu, C. A.; Dvorozniak, M. T.; Matschinsky, F. M.; Grippo, J. F. Allosteric activators of glucokinase: potential role in diabetes therapy. *Science (Washington, DC, U. S.)* **2003,** *301* (5631), 370–373.
289. Futamura, M.; Hosaka, H.; Kadotani, A.; Shimazaki, H.; Sasaki, K.; Ohyama, S.; Nishimura, T.; Eiki, J.; Nagata, Y. An allosteric activator of glucokinase impairs the interaction of glucokinase and glucokinase regulatory protein and regulates glucose metabolism. *J. Biol. Chem.* **2006,** *281* (49), 37668–37674.
290. Fyfe, M. C. T.; White, J. R.; Taylor, A.; Chatfield, R.; Wargent, E.; Printz, R. L.; Sulpice, T.; McCormack, J. G.; Procter, M. J.; Reynet, C.; Widdowson, P. S.; Wong-Kai-In, P. Glucokinase activator PSN-GK1 displays enhanced antihyperglycaemic and insulinotropic actions. *Diabetologia* **2007,** *50* (6), 1277–1287.
291. Daniewski, A. R.; Liu, W.; Radinov, R. N., Process for preparation of piragliatin, WO 2007115968 (2007).
292. Efanov, A. M.; Barrett, D. G.; Brenner, M. B.; Briggs, S. L.; Delaunois, A.; Durbin, J. D.; Giese, U.; Guo, H.; Radloff, M.; Gil, G. S.; Sewing, S.; Wang, Y.; Weichert, A.; Zaliani, A.; Gromada, J. A novel glucokinase activator modulates pancreatic islet and hepatocyte function. *Endocrinology* **2005,** *146* (9), 3696–3701.
293. Bebernitz, G. R., Preparation of N-2-thiazolylsulfonamides and related compounds as glycokinase activators for the treatment of type 2 diabetes, WO 2004050645 (2004).
294. Brocklehurst, K. J.; Payne, V. A.; Davies, R. A.; Carroll, D.; Vertigan, H. L.; Wightman, H. J.; Aiston, S.; Waddell, I. D.; Leighton, B.; Coghlan, M. P.; Agius, L. Stimulation of hepatocyte glucose metabolism by novel small molecule glucokinase activators. *Diabetes* **2004,** *53* (3), 535–541.
295. Castelhano, A. L.; Dong, H.; Fyfe, M. C. T.; Gardner, L. S.; Kamikozawa, Y.; Kurabayashi, S.; Nawano, M.; Ohashi, R.; Procter, M. J.; Qiu, L.; Rasamison, C. M.; Schofield, K. L.; Shah, V. K.; Ueta, K.; Williams, G. M.; Witter, D.; Yasuda, K. Glucokinase-activating ureas. *Bioorg. Med. Chem. Lett.* **2005,** *15* (5), 1501–1504.

Chapter 27

Steroid Hormones

Steroids are large group of chemical substances that share a 17-carbon-atom skeleton composed of four fused rings (three six-membered rings and one five-membered ring) classified by a specific carbon structure conventionally numbered and denoted by the letters *A*, *B*, *C*, and *D* [1] (Fig. 27.1.).

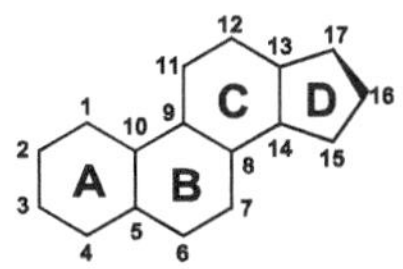

FIG. 27.1 Steroids skeleton.

Steroids vary from one another in the nature of attached groups, their position, and the configuration of the steroid nucleus. The smallest modifications in structures of steroids produce significant differences in their biological activities. Many, but not all, steroids are hormones.

Steroid hormones are essential for the proper function of most organ systems in vertebrates. These hormones affect virtually every tissue and organ in the human body and play major roles in the development, differentiation, and homeostasis of normal individuals. They mediate a wide variety of vital physiological functions ranging from antiinflammatory agents to regulating events during pregnancy.

Naturally occurring steroid hormones in humans include glucocorticoids, mineralocorticoids, androgens, estrogens, and progestins.

Molting hormones of insects; bile acids; sterols, phytosterols and cardiac glycosides found in plants; D vitamins precursors, which occur naturally in milk, in the skin of animals, and in yeasts; and some mushrooms and phytoplankton also belong to this group of steroids.

Interestingly, many processes in plants are influenced by steroids. Practically all members of all steroid groups, except the bile acids, are found in plants. These include cholesterol and the sex hormones estrone, progesterone, and testosterone [2].

The adrenal cortex is a generator of steroid hormones. The adrenal cortex is responsible for production of three major classes of steroid hormones: glucocorticoids, which regulate carbohydrate metabolism; mineralocorticoids, which regulate the body levels of electrolytes, particularly sodium and potassium; and sex

Synthesis of Best-Seller Drugs. http://dx.doi.org/10.1016/B978-0-12-411492-0.00027-4

steroids—androgens, estrogens, and progestins. (Androgens control the development and maintenance of male characteristics and are precursor of all estrogens. Estrogens are naturally occurring hormones in women. Estrogens, together with progestins, promote and maintain uterus function during the gestation period.)

The human organism produces at least 20 to 30 different steroids, whose actions are similar to those of the above-mentioned steroids. Major classes of steroid hormones are usually distinguished from each other by their relationship to their biological source.

The period of the 1930s through the 1950s has been called the golden age of steroid chemistry.

The science surrounding steroids has probably led to the most Nobel Prize's awarded in the drug-creation sphere.

The Nobel Prize in Chemistry 1927 was awarded to Heinrich Wieland "for his investigations of the constitution of the bile acids and related substances."

The Nobel Prize in Chemistry 1928 was awarded to Adolf Windaus "for the services rendered through his research into the constitution of the sterols and their connection with the vitamins."

The Nobel Prize in Chemistry 1939 was divided equally between Adolf Friedrich Johann Butenandt, "for his work on sex hormones," and Leopold Ruzicka, "for his work on polymethylenes and higher terpenes."

Endogenous steroid hormones in vertebrates are secreted in the adrenal glands of both sexes, ovaries, and testes, and regulate a wide range of physiologic functions. Cholesterol is the central precursor of all five major classes of steroid hormones.

All steroid hormones are synthesized from cholesterol through a series of enzyme-mediated transformations.

A major common pathway of cholesterol transformations to steroid hormones starts from its conversion to progesterone and pregnenolone-precursors for all other steroids (Fig. 27.2.).

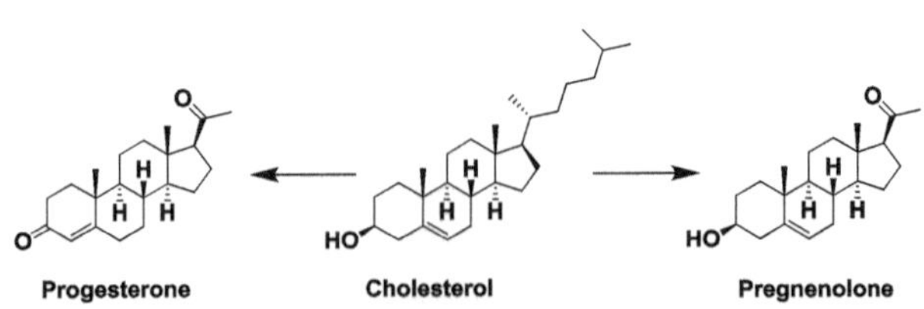

FIG. 27.2 Cholesterol transformations to steroid hormones.

Being the precursor for steroid hormones and bile acids, cholesterol is also a major constituent of the membranes in most eukaryotic cells. It regulates the physical state of the phospholipid bilayer and affects the activity of membrane proteins. Cholesterol is also is the precursor of D vitamins, which play an essential role in the control of calcium and phosphorus metabolism. They are derived from cholesterol by the ring-splitting photoreaction.

Considering the essential functions of cholesterol briefly enumerated above, the enthusiasm around cholesterol lowering seems misguided and may not represent the best treatment of hyperlipidemia.

Hundreds of tons of steroid drugs are produced annually worldwide. Mainly from androstendione, which is, in turn, available in bulk scale by microbiological transformations of inexpensive sterols, such as sitosterol (Fig. 27.3.).

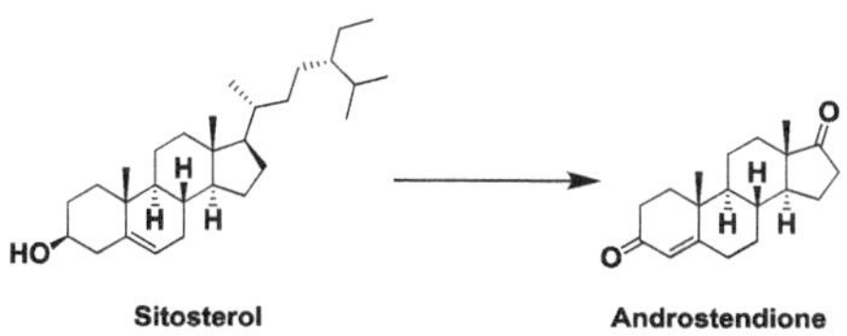

FIG. 27.3 Transformations of sitosterol to androstendione.

As was mentioned above, steroid hormones in humans are divided into five categories: glucocorticoids, mineralocorticoids, estrogens, progestins, and androgens.

27.1 GLUCOCORTICOIDS

Adrenal glands produce adrenal steroids or corticosteroids, which are glucocorticoids (the name of which is composed from abbreviating glucose + cortex + steroid) when defines their role in the regulation of the metabolism of glucose. Glucocorticoids originate in the adrenal cortex and affect mainly metabolism in many different ways.

The Nobel Prize in Physiology or Medicine 1950 was awarded jointly to Edward Calvin Kendall, Tadeus Reichstein, and Philip Showalter Hench "for their discoveries relating to the hormones of the adrenal cortex, their structure and biological effects" [3].

Glucocorticoids decrease inflammation. They are produced in reaction to stress, increasing resistance to stress, help in the metabolism of fats, carbohydrates, and proteins, and have multiple implementations in medicine as antiinflammatories and immunosuppressives, as drugs affecting energy balance, glucose utilization, and fat metabolism; as drugs influencing bone and cartilage formation; and as drugs influencing cognitive function and psychological stress [4-12].

Glucocorticoid receptors are present in all cell types and maintain many metabolic disturbances, infection, regulation of blood pressure, hemorrhage, and anxiety [13].

Synthetic glucocorticoids are extremely effective and one of the most commonly, frequently, and successfully used drugs to treat asthma, rheumatoid arthritis, inflammatory bowel disease, and the lymphoproliferative disorders of leukemia and Hodgkin disease. They are also used for treatment of diseases such as atopic disorders and autoimmune diseases, and as cotreatment remedies in several cancer cases and chemotherapy regimens [14-17].

However, glucocorticoids have side effects that cause several adverse reactions that limit their clinical use, especially at higher doses and for long periods.

Side effects are frequent and long-term use of glucocorticoids can lead to irreversible processes, the most critical and debilitating of which are fat redistribution, obesity, and osteoporosis.

Synthetic and natural glucocorticoids are classified as systemic and as topical corticosteroids. Among the systemic drugs are cortisone (**27.1.1**), which was immediately hailed as a "wonder drug" effective in inflammation-associated conditions, most notably rheumatoid arthritis, and hydrocortisone or cortisol in pharmaceutical preparations (**27.1.2**), both of which, for more than 50 years, have represented the mainstay of second-line therapy for different inflammatory disorders, when the first-line drugs, such as the nonsteroidal antiinflammatory drugs (NSAIDs), proved ineffective. Many other very effective systemic corticosteroids are present and circulating on the pharmaceutical market such as prednisone (**27.1.3**), prednisolone (**27.1.4**), triamcinolone (**27.1.5**), dexamethasone (**27.1.6**), paramethasone (**27.1.7**), flumethasone (**27.1.8**), fludrocortisone (**27.1.9**), triamcinolone acetonide (**27.1.10**), and flurandrenolone acetonide (**27.1.10**) (Fig. 27.4.).

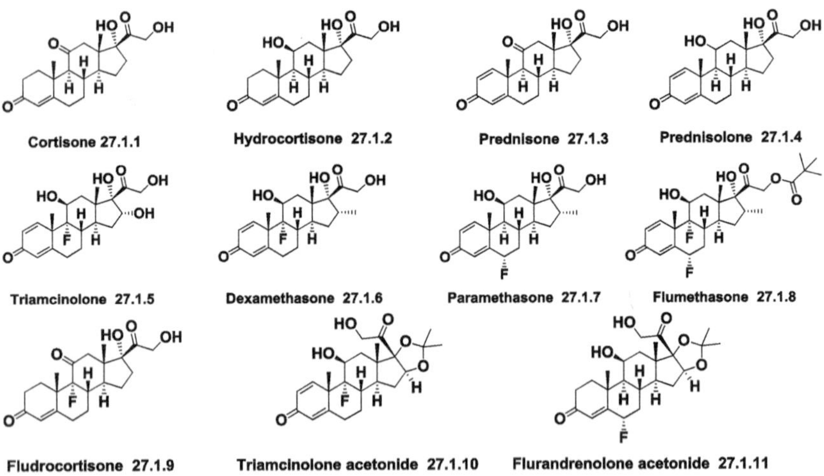

FIG. 27.4 **Systemic corticosteroids available on the pharmaceutical market.**

Topical preparations [18,19] for skin inflammations are prepared on the base of cortisone (**27.1.1**), hydrocortisone (**27.1.2**), fludrocortisone (**27.1.9**), triamcinolone acetonide (**27.1.10**), flurandrenolone acetonide (**27.1.11**), mometasone furoate (**27.1.12**), clobetasol (**27.1.13**), fluticasone (**27.1.14**), betamethasone valerate (**27.1.15**), desonide (**27.1.16**), halcinonide (**27.1.17**), fluocinonide (**27.1.18**), and fluocinolone acetonide (**27.1.19**) (Fig. 27.5.).

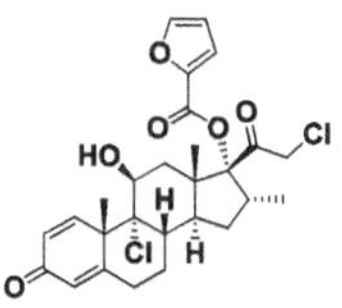

Mometasone furoate
27.1.12

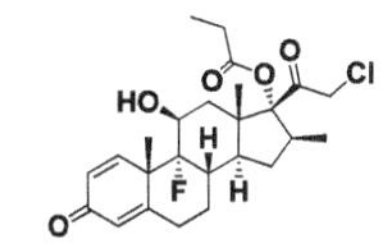

Clobetasol propionate
27.1.13

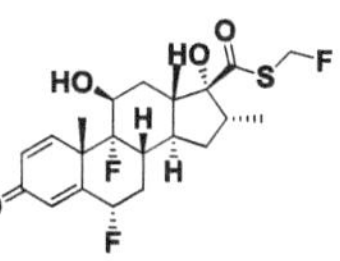

Fluticasone
27.1.14

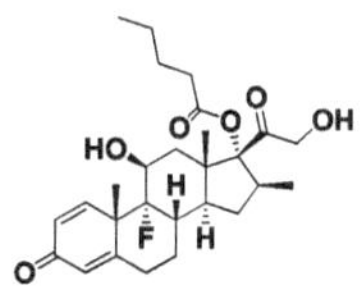

Betamethasone valerate
27.1.15

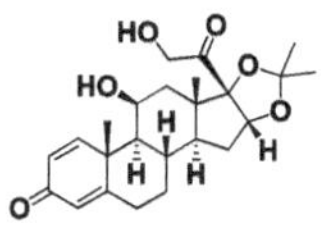

Desonide
27.1.16

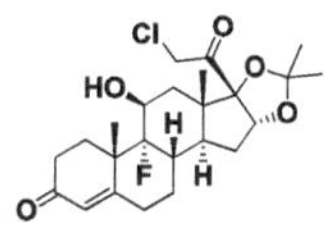

Halcinonide
27.1.17

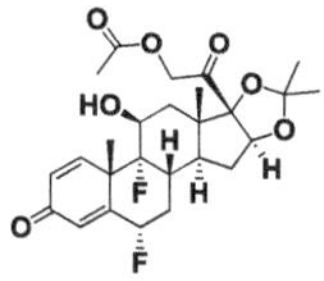

Fluocinonide
27.1.18

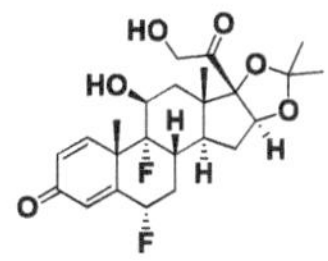

Fluocinolone acetonide
27.1.19

FIG. 27.5 **Topical corticosteroids available on the pharmaceutical market.**

It is interesting to mention, and necessary to consider, that glucocorticoids affect carbohydrate, protein, and lipid metabolism in a manner that is nearly opposite that of insulin, and influence a wide variety of other vital functions, including inflammatory reactions and the capacity to cope with stress.

Receptors for steroid hormones (nuclear receptors) are located inside target cells. A unique property of nuclear receptors that differentiates them from other classes of receptors is their ability to directly interact with and control the expression of genomic DNA. Being lipids, steroid hormones easily enter the cell by simple diffusion across the plasma membrane. These intracellular receptors exist either in the cytoplasm or nucleus. When hormone binds to their receptor, a characteristic series of events occurs.

Glucocorticoids mometasone (**27.1.12**), clobetasol (**27.1.13**), and fluticasone (**27.1.14**) are included in the list of Top 200 Drugs by sales for the 2010s.

Mometasone Furoate–Nasonex

Mometasone furoate (**27.1.12**) is a medium-potency synthetic glucocorticosteroid frequently used topically to reduce inflammation of the skin or in the airways [20-25]. It has a balanced efficacy and safety profile and antipruritic, and vasoconstrictive properties. Mometasone furoate is a prodrug of the free form mometasone (**27.1.20**) (Fig. 27.6.).

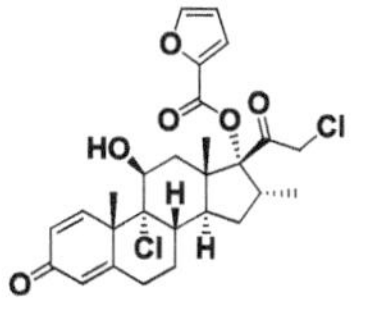

Mometasone furoate 27.1.12

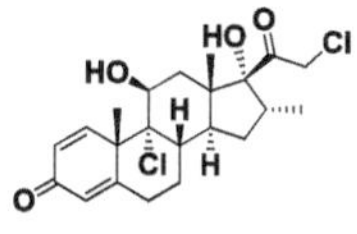

Mometasone 27.1.20

FIG. 27.6 **Structure of mometasone and mometasone furoate.**

Two general approaches for the synthesis of mometasone furoate are described [26-28].

Preparation of mometasone furoate (**27.1.12**) begins with acylation of 16α-methyl-11,4,9(11)-pregnatriene-17α,21-diol-3,20-dione 21-acetate (**27.1.21**) with 2-furoyl chloride in the presence of 4-dimethylamino)pyridine in dichloromethane to produce 17α-furoate. The 17α-furoate was dissolved in THF and the acetyl group was removed by adding 70% $HClO_4$ in water, which produced hydroxyl ketone (**27.1.22**). The hydroxy group of the hydroxyl ketone (**27.1.22**) was replaced with chlorine via transformation to mesylate, followed by nucleophilic substitution of mesyl group for chlorine using LiCl in DMF. The obtained chloride (**27.1.23**) was worked up with 1,3-dichloro-5,5-dimethylhydantoin in water, which, in principle, is equivalent to hypochlorous acid (HOCl), to produce the desired mometasone furoate (**27.1.12**) (Scheme 27.1.).

An alternative route exemplified in the preparation of the desired mometasone (**27.1.12**) starts from the epoxide (**27.1.24**), which, after acidic hydrolysis in water, was transferred to diol (**27.1.25**) which processed with dry HCl in acetic acid to produce a high yield of mometasone furoate (**27.1.12**). Some minor changes are proposed for these approaches [29,30] (Scheme 27.1.).

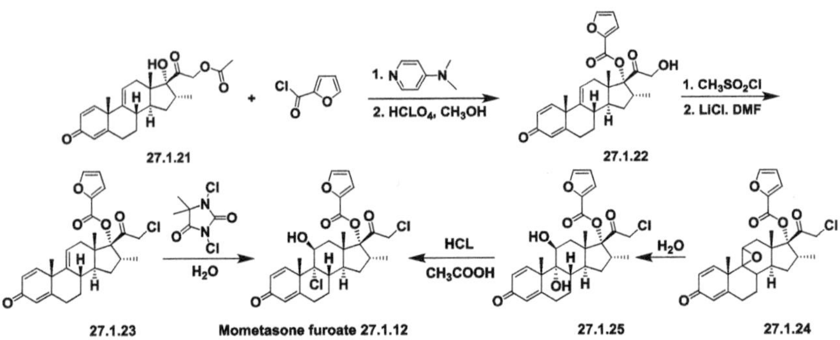

SCHEME 27.1 **Synthesis of mometasone furoate.**

Clobetasol–Clobex

Clobetasol propionate (**27.1.13**) is the most potent of all topical steroids and is used topically to treat various skin disorders such as itching, redness, dryness, crusting, scaling, inflammation, and discomfort of various scalp and skin conditions, including eczema and psoriasis [31-33].

Synthesis of clobetasol propionate (**27.1.13**) starts from the known betamethasone 17-propionate (**27.1.26**), a potent glucocorticoid steroid with antiinflammatory and immunosuppressive properties, which was mesylated with methanesulfonyl chloride in pyridine to produce 9α-fluoro-11β-hydroxy-21-methylsulfonyloxy-16β-methyl 17-propionyloxypregna-1,4-diene-3,20-dione (**27.1.27**). The obtained product was refluxed in acetone, DMF, and dry LiCl mixture to produce the desired clobetasol propionate (**27.1.13**) [34] (Scheme 27.2.).

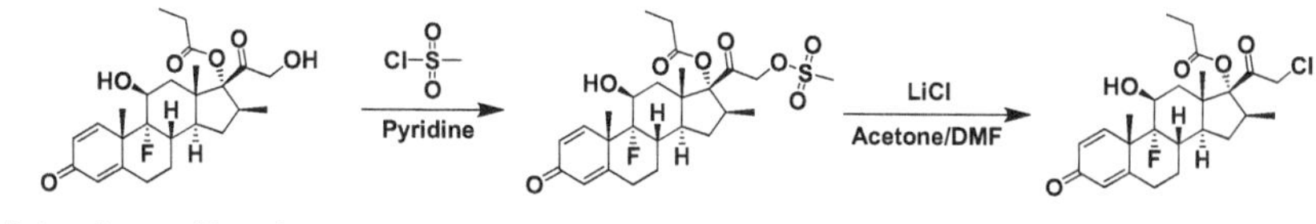

Betamethasone 17-propionate
27.1.26

27.1.27

Clobetasol propionate
27.1.13

SCHEME 27.2 **Synthesis of clobetasol propionate.**

Fluticasone–Flovent

Fluticasone propionate (**27.1.14**) is used for the prophylaxis of asthma and to prevent, ease, or treat allergic rhinitis and various skin disorders [35-41].

Synthesis of fluticasone (**27.1.14**) started with oxidation of 21-hydroxy-pregnan-20 one derivative (**27.1.28**), which with periodic acid in aqueous THF produced intermediate carboxylic acid (**27.1.29**). It was readily 17α-acylated, without concomitant acylation, by reaction with excess propionyl chloride and triethylamine to produce propionate (**27.1.30**), which reacted rapidly with N,N′-carbonyldiimidazole in dimethylformamide to produce species, which, with H_2S in DMF and further acidic hydrolysis, produced the unstable carbothioic acid derivative (**27.1.31**). Reaction of the obtained carbothioic acid with fluoroiodo-methane is a key synthetic stage of flumethasone synthesis. Chloromethyl carbo-thioate was prepared from carbothioate salt by alkylation with chloroiodomethane in DMF to produce the desired fluticasone (**27.1.14**) [42,43] (Scheme 27.3.).

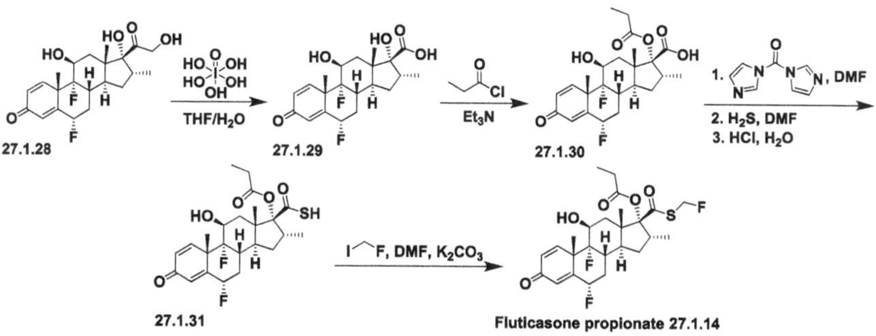

27.1.28 27.1.29 27.1.30 27.1.31

Fluticasone propionate 27.1.14

SCHEME 27.3 **Synthesis of fluticasone propionate.**

Glucocorticoids are one the most potent and widely prescribed antiinflammatory agents worldwide and the synthesis of essential drugs of this series is described in our previous book [44].

Glucocorticoid Receptor Antagonists

Glucocorticoid hormones, are crucial for the physiological regulation of metabolism of carbohydrates, lipid and proteins. Glucocorticoids have strong antiinflammatory and immunosuppressive properties and various actions in the central nervous system.

Excess of glucocorticoid hormones cause a number of clinical diseases. Hypersecretion of glucocorticoids has been suggested to play role in a number of metabolic disorders, expressed, for example, as a cluster of symptoms called Cushing syndrome, hypertension, diabetes mellitus, obesity and anorexia, osteoporosis and immune disorders, several psychiatric disorders, Alzheimer disease and cognitive dysfunctions, and glaucoma [45-50].

Over the years many ligands for steroid receptors have been developed and have found extensive clinical use, but there are relatively few literature reports of selective, glucocorticoid, nonsteroidal antagonists.

The utility of glucocorticoid antagonists has been demonstrated by the use of the only glucocorticoid antagonist available in the clinic, mifepristone (**27.1.32**), a synthetic compound with both antiprogesterone and antiglucocorticoid properties used to treat Cushing syndrome, diabetes, glaucoma, and depression. But mifepristone's primary potency is antigestagenic, making its usefulness as a glucocorticoid antagonist limited.

Other experimental steroid receptor antagonists, such as direct analogues of mifepristone–compounds RU-40555 (**27.1.33**), RU-43044 (**27.1.34**), indole derivative LLY-2707 (**27.3.35**), and hexahydro-pyrazolo-isoquinoline derivative CORT-18279 (**27.3.36**), as well as other compounds, are known and are in different stages of trials (Fig. 27.7.).

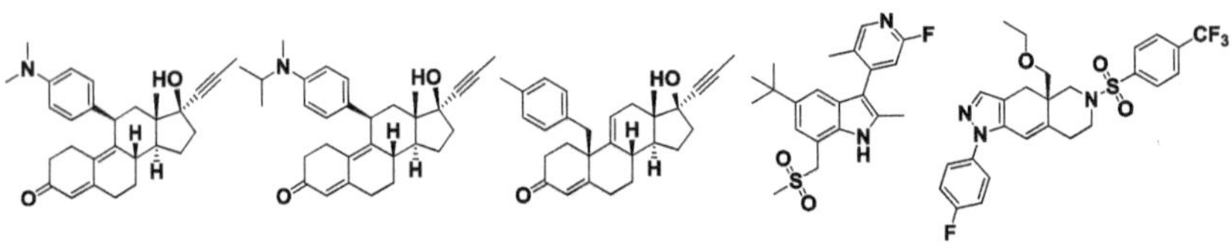

Mifepristone 27.1.32 RU-40555 27.1.33 RU-43044 27.1.34 LLY-2707 27.1.35 CORT- 18279 27.1.36

FIG. 27.7 **Experimental steroid receptor antagonists.**

27.2 MINERALOCORTICOIDS

The mineralocorticoids [51-57] are a group of hormones that originate in the adrenal cortex. They maintain electrolyte and water balance and, in particular, control the retention of sodium and potassium in the kidneys within the body. Normally, aldosterone (**27.2.1**) is the only functioning natural mineralocorticoid in humans.

Aldosterone is synthesized from cholesterol. Its synthesis is stimulated principally by angiotensin II. Secreted aldosterone increases sodium retention by the kidneys. Water follows the osmotic gradient created by the sodium influx, and more water is retained. Aldosterone's actions are mediated by binding to the mineralocorticoid receptor in target tissues, particularly in the kidney.

Other steroid hormones, such as deoxycorticosterone (**27.2.2**), an intermediary in the synthetic pathway of aldosterone, and cortisol, the major glucocorticoid hormone, can also bind with high affinity to mineralocorticoid receptor. However, under normal conditions, their effects on water and electrolyte balance are negligible.

Synthetic fludrocortisone (**27.2.3**) is the only drug available for the treatment of mineralocorticoid deficiency (Fig. 27.8.).

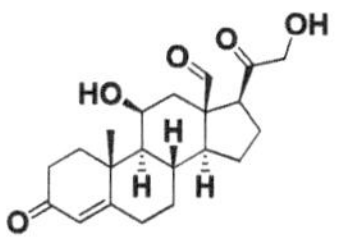

Aldosterone 27.2.1

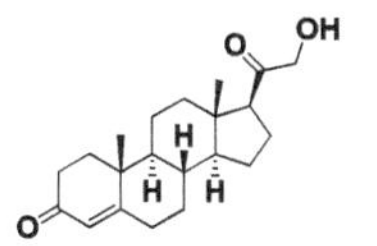

Deoxycorticosterone 27.2.2

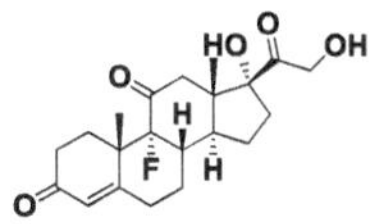

Fludrocortisone 27.2.3

FIG. 27.8 **Structure of mineralocorticoids.**

The discovery of aldosterone in the 1950s had a profound influence on medicine. It helped to explain the mechanism of edematous states, and led to the description of the renin–angiotensin system, which regulates and controls blood pressure.

Aldosterone is now recognized as a key factor in several diseases, including hypertension, heart failure, arrhythmia, and metabolic and kidney diseases, to name only a few.

The response to aldosterone is mediated by the mineralocorticoid receptor. The classical mode of action for this receptor involves the regulation of gene transcription. Several genes have been shown to be regulated by aldosterone in epithelial tissues.

Mineralocorticoids are acutely critical for maintenance of life.

Their primary effects are increasing the reabsorption of sodium and the secretion of potassium. Secondary effects include the reabsorption of water, anion reabsorption, and secretion of hydrogen ions. The net result is maintenance of fluid and electrolyte balance and, therefore, adequate cardiac output.

A deficiency in aldosterone can occur by itself or, more commonly, in conjunction with a glucocorticoid deficiency, and is known as hypoadrenocorticism or Addison disease. A lack of aldosterone is lethal because of electrolyte imbalances and the resulting hypotension and cardiac failure. Treatment is mineralocorticoid replacement therapy, and in all cases of low aldosterone production, includes a mineralocorticoid receptor agonist, primarily fludrocortisone (**27.2.3**).

Aldosterone excess is most commonly observed in cases of elevated plasma potassium (hyperkalemia) and low vascular volume. Plasma potassium and angiotensin II are the major factors that regulate aldosterone secretion. Importantly, it is now recognized that approximately 10% cases of primary hypertension are associated with hyperaldosteronism, most commonly as a result of aldosterone-secreting adrenal tumors or mutations in potassium channels.

Drugs that interfere with the secretion or action of aldosterone are in use as antihypertensives.

Primary aldosteronism, which is the term used for overproduction of aldosterone by the adrenal, is treated with a glucocorticoid dexamethasone (**27.1.6**), which has no mineralocorticoid receptor binding capacity, but suppresses synthesis of aldosterone and endogenous production of glucocorticoids.

Treatment of increased mineralocorticoid activity usually starts with spironolactone (**27.2.4**), which inhibits binding of the excess cortisol to the mineralocorticoid receptor.

Another important mineralocorticoid antagonist is eplerenone (**27.2.5**) (Fig. 27.9.).

Mineralocorticoids are not included in the list of Top 200 Drugs by sales for the 2010s.

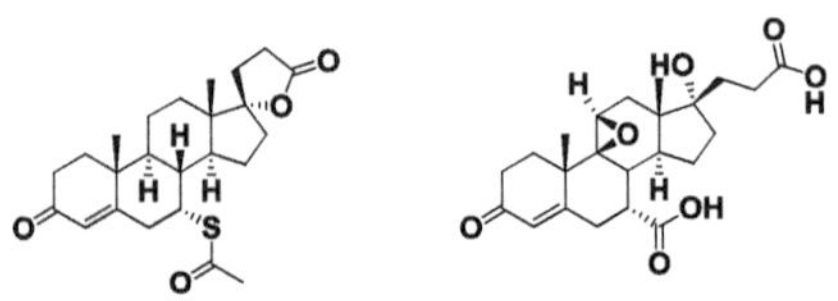

Spironolactone 27.2.4 Eplerenone 27.2.5

FIG. 27.9 **Structure of spironolactone and eplerenone.**

27.3 ESTROGENS

The term *sex steroid* is nearly always synonymous with sex hormone.

Sex steroids (estrogens, androgens, and progestins) play an essential role in the reproductive system. They function at the genomic level through classic receptors that belong to the superfamily of nuclear receptors and can also have extranuclear actions through other receptors.

The two main classes of sex steroids are estrogens and androgens, and their most important representatives in humans are estradiol and testosterone, respectively. Both estrogens and androgens affect sexual development and function. They regulate sexual differentiation, the secondary sex characteristics, and sexual behavior patterns.

Progestins belong to the third class of sex steroids, which maintain the pregnancy process and are distinct from estrogens and androgens.

There are also other sex-related hormones, such as the luteinizing hormone, gonadotropin-releasing hormone, and follicle-stimulating hormone, which are nonsteroids and traditionally not regarded as sex hormones.

Estrogens are one of the two types of female sex hormones. They are compounds regulating menstrual and reproductive cycles. In girls, estrogens originate in the adrenal cortex and gonads, and primarily promote and affect maturation and function of secondary sex organs, and development of the primary and secondary female sex characteristics; they also stimulate linear growth and skeletal maturation and maintain the female sexual determination.

The primary source of the female sex hormones estrogen and progesterone in mature individuals is the ovaries, the female gonads. Limited amounts of these hormones, are produced in the adrenal cortex and adipose tissue, as well as in the placenta during pregnancy.

The three major naturally occurring estrogens in women are estrone (predominant) (**27.3.1**), estradiol (**27.3.2**), and estriol (**27.3.3**) (Fig. 27.10.). Estriol is almost exclusively an estrogen of pregnancy. But there are at least two dozen other identified estrogens produced in a woman's body.

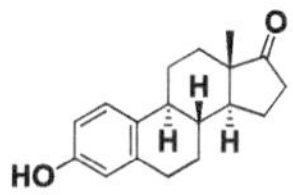

Estrone 27.3.1

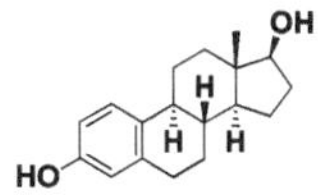

Estradiol 27.3.2

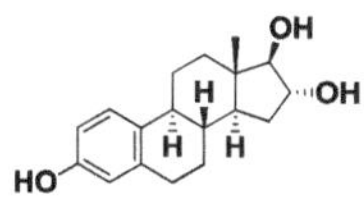

Estriol 27.3.3

FIG. 27.10 **Structure of naturally occurring estrogens.**

The role of estrogens is not limited to reproduction. Their role extends to the skeletal, cardiovascular, and central nervous systems. Various pathologies, such as cancer, inflammatory, neurodegenerative, and metabolic diseases, are often associated with dysfunctions of the estrogenic system. Therapeutic interventions that affect the estrogenic signaling pathway are useful in the treatment of many diseases.

Estrogen is a fundamental regulator of the metabolic system of the female brain and body. Within the brain, estrogens regulate glucose transport, aerobic glycolysis, and mitochondrial function to generate adenosine triphosphate (ATP). In the body, estrogen protects against adiposity, insulin resistance, and type 2 diabetes, and regulates energy intake and expenditure.

Estrogen mediates its diverse effects in the estrogen target tissues through estrogen receptors (α) and (β), that belong to the nuclear receptor family. Estrogen receptor α is mainly expressed in reproductive tissues, kidney, bone, white adipose tissue, and liver, whereas estrogen receptor β is expressed in the ovary, prostate, lung, gastrointestinal tract, bladder, hematopoietic cells, and the central nervous system. Besides the several beneficial effects of estrogen in female health, estrogen also fuels the proliferation of breast cancers and uterine cancers in women.

The most common uses for estrogen in medicine are in birth control pills and contraceptives. Low doses of estrogen combined with progestin are used in birth control pills. The reason why estrogen is used in such a manner is because it prevents ovulation by reducing secretion of the follicle-stimulating hormone and the luteinizing hormone. They are widely used to prevent or treat osteoporosis and as a part of some hormone replacement therapies, as well as for treatment of some disorders of the endocrine system.

Estrogens are the only treatment proven to address the myriad symptoms associated with menopause, including hot flushes and sweating, and for treating vaginal dryness, itching, or burning.

In principle, estrogens, which are recognized to have at least a couple hundred different functions, are misunderstood hormones.

Synthetic estrogens, named parabens, are used also in cosmetics—in body and skin creams, shampoos, conditioners, moisturizers etc.—

The natural estrogens (estradiol, estrone, and estriol) were isolated in the late 1920s and 1930s. The first orally active steroidal estrogen, ethinyl estradiol (**27.3.4**), was synthesized in 1938. Later mestranol (**27.3.5**) was synthesized. These two compounds remain the only orally active synthetic estrogens used in birth control formulations today. The first orally active pharmacological nonsteroidal estrogen was diethylstilbestrol (**27.3.6**) (Fig. 27.11.).

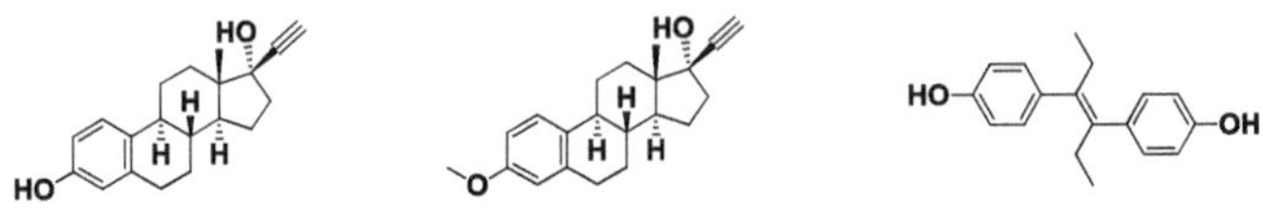

Ethinyl estradiol 27.3.4 Mestranol 27.3.5 Diethylstilbestrol 27.3.6

FIG. 27.11 Structure of synthetic estrogens.

Esterified estrogens, conjugated synthetic A estrogens (composition containing a blend of nine synthetic estrogenic substances), conjugated synthetic B estrogens (composition containing a blend of ten synthetic estrogenic substances), estradiol, estropipate (a salt of estrone sulfate and piperazine) are all widely used in medicinal practice.

Nonsteroidal estrogens are either synthetic (xenoestrogens) or naturally occurring substances that possess estrogenic activity—phytoestrogens and mycoestrogens.

Estrogen receptor agonists are used mainly to alleviate the symptoms associated with postmenopausal syndrome and as contraceptives; however, the risk-to-benefit profile of estrogen substitution therapy is under discussion.

For a long time the idea of antiestrogen compounds had no significant impact on medicinal chemists because of a lack of therapeutic application. That changed when some correlations between breast cancer, one of the most frequently diagnosed cancers among women, and the estrogen content of the tumors was found. It is generally believed that breast tumors depend, at least initially, on the stimulatory effects of estrogens. However, many breast tumors eventually progress to an estrogen-independent growth phenotype. Estrogens and androgens also play important roles in the growth and development of the prostate.

Tamoxifen (**27.3.7**) and the series of its analogues are the first-generation estrogen antagonists that compete with natural estrogens for the receptors in these tumors. Initially, they were designed as antifertility compounds but did the opposite, inducing ovulation in subfertile women. Tamoxifen currently is used to treat all stages of breast cancer and for chemoprevention in women who are at high risk for breast cancer. Tamoxifen also affects bone mineral density in postmenopausal women. It remains the standard of care for the treatment of breast cancer while functioning as an agonist in uterus and skeletal tissue, preserving bone mineral density.

The second-generation antiestrogen compounds, with improved tissue selectivity, include raloxifene (**27.3.8**), which is approved for the prevention of postmenopausal osteoporosis. Raloxifene is an antiestrogen with reduced uterotropic activity compared with tamoxifen. Raloxifene is used to prevent and treat osteoporosis and to decrease the risk of developing invasive breast cancer.

Toremifene (**27.3.9**) is approved for use in advanced metastatic breast cancer. Clomifene (**27.3.10**) is used to induce ovulation and has become popular in bodybuilding.

The newer antiestrogen compounds that are in late development stages include ospemifene (**27.3.11**), which is used to treat women experiencing moderate to severe dyspareunia (pain during sexual intercourse), and lasofoxifene (**27.3.12**), which is under development for the prevention and treatment of osteoporosis and for the treatment of vaginal atrophy. Bazedoxifene (**27.3.13**) and arzoxifene (**27.3.14**) are third-generation antiestrogen compounds. Bazedoxifene is intended to prevent (not treat) postmenopausal osteoporosis; arzoxifene also intended to prevent (not treat) mammary cancer (Fig. 27.12.).

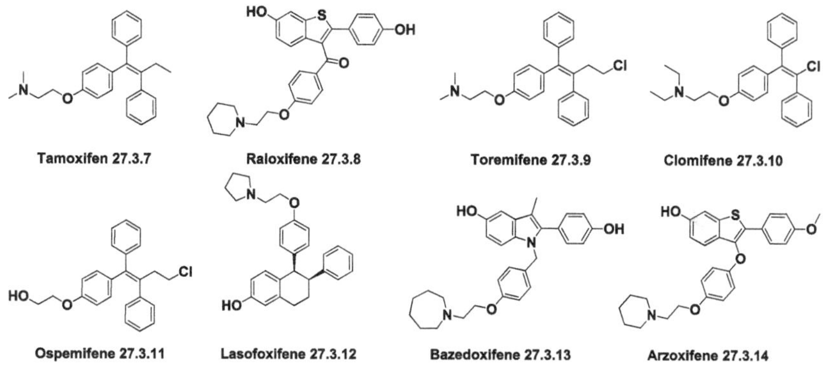

FIG. 27.12 Estrogen antagonists.

Estrogen agonists and antagonists are widely used in clinical practice. The total world market for this class of drugs is worth billions of dollars. But no estrogen agonist or antagonists is included in the list of Top 200 Drugs by sales for the 2010s.

The search for the "ideal" antiestrogen compound, with estrogenic effects on bone and serum lipids, neutral effects on the uterus, and antiestrogenic effects on breast tissue, as well as being devoid of the adverse effects of the drugs used today, is currently under way [58-67].

27.4 PROGESTINS

Progestins are the other class of female steroid hormones that have progesterone-like activity. They play an important role in maintaining pregnancy *(progestation)*. Progestins originate from the ovaries and placenta, and mediate menstrual cycle, maintain pregnancy, and exert their functions via nuclear receptors. The most important progestin is progesterone (**27.4.1**).

Progesterone is essential for the regulation of reproductive function. It plays a significant role in maintaining pregnancy, as well as in stimulating and regulating various other functions in other tissues, particularly in the nervous system and the vessels. The progesterone's ability to suppress ovulation during pregnancy generated a search for analogue compounds that could be administered orally to control ovulation. Progesterone and progestins are approved for the treatment of irregular and anovulatory menstrual cycles and, when combined with estrogen, for contraception and the prevention of endometrial hyperplasia in postmenopausal hormonal replacement therapy.

In hormonal contraceptives, progestins represent the major agent designed for suppressing ovulation and are used in combination with estrogen and ethynyl-estradiol.

Progesterone (**27.4.1**) along with its effective synthetic versions—ethisterone (**27.4.2**), norethisterone (**27.4.3**), medroxyprogesterone acetate (**27.3.2.4**), desogestrel (**27.4.5**), danazol (**27.3.2.6**), and 17α-hydroxyprogesterone caproate (**27.4.7**)—have been in medicinal use since the 1950s (Fig. 27.13.).

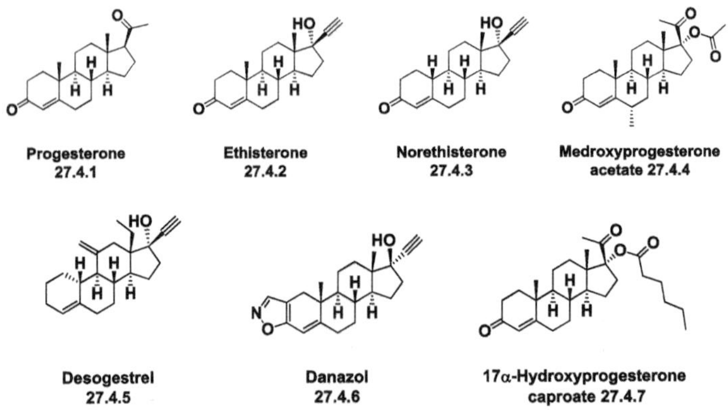

FIG. 27.13 **Progestins.**

Several new progestins have been synthesized. These include dienogest (**27.4.8**), nestorone (**27.4.9**), nomegestrol acetate (**27.4.10**), trimegestone (**27.4.11**), and drospirenone (**27.4.12**) (Fig. 27.14.).

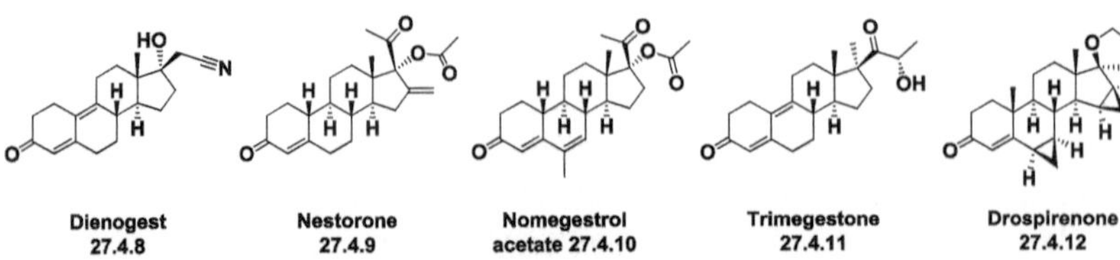

FIG. 27.14 **New progestins.**

In addition to their use as oral contraceptives, they are used in menopausal hormone therapy and for treatment of a variety of other conditions, including abnormal uterine bleeding and amenorrhea (absence of periods), endometriosis, and breast, kidney, and uterine cancer.

Almost all synthetic progestogens are devoid of an antimineralocorticoid effect. They are unable to antagonize the salt-retaining effect of estrogens. This could be one cause of weight gain and increased blood pressure.

The female steroid hormones, estrogen and progesterone, are prescribed widely by physicians, and their risks and benefits have been studied extensively.

The development of new generations of progestins with improved selectivity profiles has been a great challenge [68-78].

One representative of progestins, drospirenone (**27.4.12**), is included in the list of Top 200 Drugs by sales for the 2010s.

Drospirenone–Yaz

The synthesis of drospirenone (**27.4.12**) is believed to have been described for the first time in Wiechert et al [79], with a total yield of approximately 2 to 3% via the pathway presented in Scheme 27.4.

Each compound produced after each reaction step was purified by column chromatography.

Androsta-5,15-diene-3-ol-17-one was methylenated at the 15,16-position (**27.4.13**) and reacted with organometallic reagent (3,3-dimethoxypropyl) lithium prepared from 3-bromo-1,1-dimethoxypropane (**27.4.14**) and lithium in THF to produce the tertiary alcohol (**27.4.15**), which on short-term reflux with toluenesulfonic acid in acetone transformed to cyclic 21,17-hemiacetal (**27.4.16**). Oppenanuer oxidation with aluminium isopropoxide in excess of cyclohexanone in toluene was brought to mild oxidation of both secondary alcohol groups, and the simultaneous isomerization of the 5,6 double bond to the 4,5 position produced the compound (**27.4.17**). The last was oxidized with Jones reagent—chromic trioxide in diluted sulfuric acid—producing conjugated diene-one (**27.4.18**). Corey methylenation of the obtained product with dimethyloxosulfonium methylide in DMSO containing sodium hydride produced the final compound, the desired drospirenone (**27.4.12**).

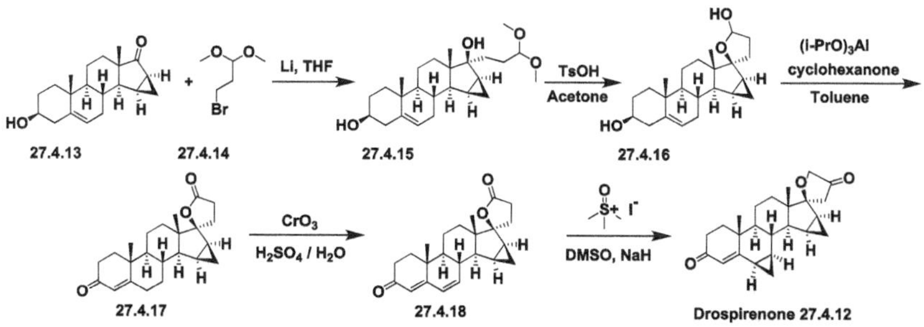

SCHEME 27.4 **Synthesis of drospirenone.**

The following patents and publications [80-83], which differ slightly from one another, disclose similar processes for preparing drospirenone and are presented in Scheme 27.5.

In Scheme 27.5, drospirenone (**27.4.12**) is prepared by converting the key starting compound (**27.4.19**) into the corresponding chloride (**27.4.20**) via reaction with triphenylphosphine and tetrachloromethane under mild conditions. Reductive dechlorination with Zn in acetic acid in THF tetrahydrofuran produced 5-hydroxy-15β,16β-methylene-3β-pivaloyloxy-5β-androst-6-en-17-one (**27.4.21**). The pivaloyl protecting group of the last was removed with the mixture of potassium hydroxide and sodium perchlorate in THF/methanol mixture to produce the diol (**27.4.22**). Simmons–Smith cyclopropanation reaction was applied to this compound. For that purpose, solution of (**27.4.22**) in dimethyl Cellosolve was stirred at 80°C with zinc-copper couple and methylene iodide, which produced the desired compound (**27.4.23**). The compound (**27.4.23**) underwent ethinylation with propargyl alcohol using potassium methylate in THF as a base to produce the 1,4-butindiol derivative (**27.4.24**). The triple bond of the 1,4-butindiol derivative (**27.4.24**) was hydrogenated in aTHF/methanol/pyridine mixture in the presence of palladium on carbon to produce the 1,4-butanediol derivative (**27.4.25**). The obtained compound underwent oxidation–lactonization at 50°C using a solution of CrO3 in water and pyridine to produce the desired drospirenone (**27.4.12**).

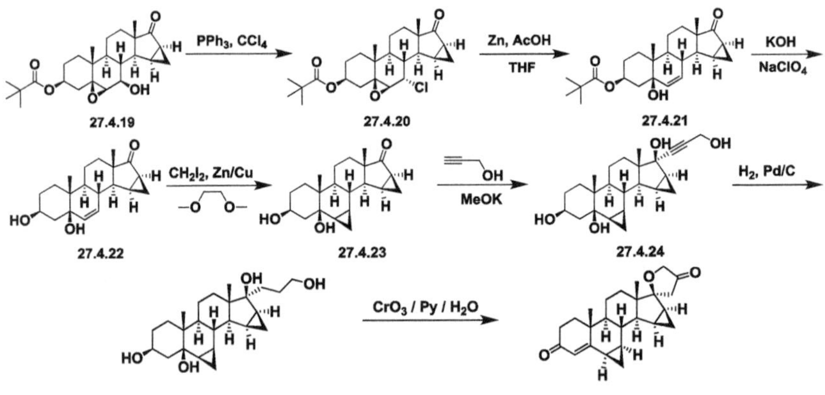

SCHEME 27.5 An alternative synthesis of drospirenone.

Several other synthetic routes for the production of drospirenone have been proposed [84-96], one of which [96] is presented in Scheme 27.6.

According to Scheme 27.6, a mixture of the key starting ketodiol (**27.4.26**), synthesis of which was described previously [84], with ethyl propiolate in THF was added to a solution of lithium hexamethyldisilylamide to produce, after quenching with acetic acid and saturated ammonium chloride solution, ethinyl alcohol (**27.4.27**). This product was hydrogenated on H_2-Pd/C catalyst to produce ethyl 4-hydroxybutanoate (**27.4.28**). The 3-hydroxy group in the obtained product was oxidized to the keto group with (2,2,6,6-tetramethyl-piperidin-1-yl)oxyl, resulting in the compound (**27.4.29**). Treatment of the last with potassium hydroxide in a methanol–water mixture affects both hydrolysis of the ester group and dehydration of 5-hydroxy substituent. Acidification of the resulting intermediate results in drospirenone (**27.4.12**).

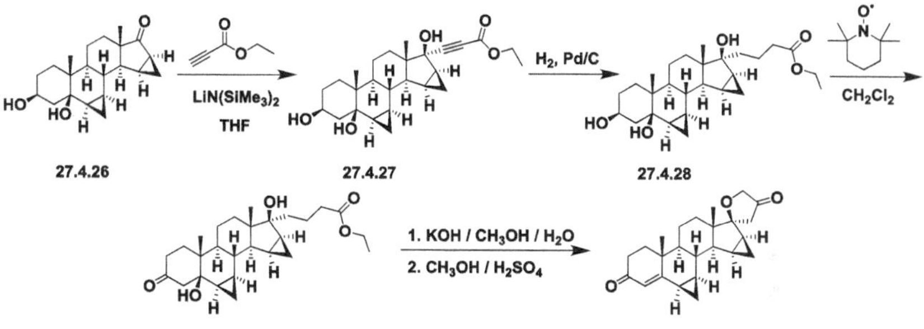

SCHEME 27.6 Another method for synthesizing drospirenone.

Drospirenone is a unique synthetic progestogen derived from 17α-spirolactone; it has a pharmacological profile very similar to that of endogenous progesterone. Drospirenone prevents ovulation and is used in contraceptive pills; it is also used as a postmenopausal hormone replacement. Drospirenone provides reliable and well-tolerated contraception and effective treatment of menopause. It has progestational, antialdosterone, and antiandrogenic properties, but is devoid of any estrogenic, androgenic, glucocorticoid, antiglucocorticoid, and mineralocorticoid activities. The affinity of drospirenone for the mineralocorticoid receptor makes it an antagonist of aldosterone, which is not only important in the renin–angiotensin–aldosterone system, but also means it acts directly on the cardiovascular system. It is progestin with antimineralocorticoid property that acts to suppress gonadotropins. It is thus able to prevent excessive sodium loss and regulate blood pressure. Drospirenone slightly decreases body weight and blood pressure and shares many pharmacodynamic properties with progesterone [97-110].

Progestin Antagonists

The progesterone receptors play a crucial role in the establishment and maintenance of pregnancy, as well as in mediating multiple aspects of the female reproductive system. The receptors also are targets for compounds that can modulate progesterone-dependent events. Many selective progesterone receptor modulators, characterized as full progesterone receptor antagonists or mixed agonist–antagonists have been designed [111-124].

Progesterone receptor antagonists suppress estrogen-dependent mitotic activity in the endometrial glands and block progestational development of the endometrium. Antiprogesterone compounds play an important role in fertility control and in the treatment of hormone-dependent diseases, inhibiting the synthesis of progesterone and antagonizing progesterone action. They could be used not only for pregnancy termination and abortion induction, but also for labor induction, contraception, and cervical ripening. They also may be used in the treatment of endometriosis, fibromyomata, meningiomas, Cushing syndrome, and glaucoma. Areas of further research include the treatment of uterine fibromyomas and the treatment and prevention of hormone-dependent tumors such as breast cancer and meningiomas. Another potential clinical indication could be treatment of corticoid-related disorders. History and perspectives of antiprogestins has been carefully reviewed [125].

Mifepristone (**27.4.30**) is a drug with both antiprogesterone and antiglucocorticoid properties, and is the first and only glucocorticoid receptor antagonist available in the pharmaceutical market. It has been widely and effectively used throughout the world for medical abortion, and to a lesser extent for emergency contraception. It has been used in the treatment of Cushing syndrome, to treat endometriosis, to women with uterine leiomyomas, associated with pain and bleeding with and decrease in fibroid size, for early termination of pregnancy, cervical dilation before surgical termination of pregnancy, and management of early embryonic loss or fetal death [126-129] (Fig. 27.15.).

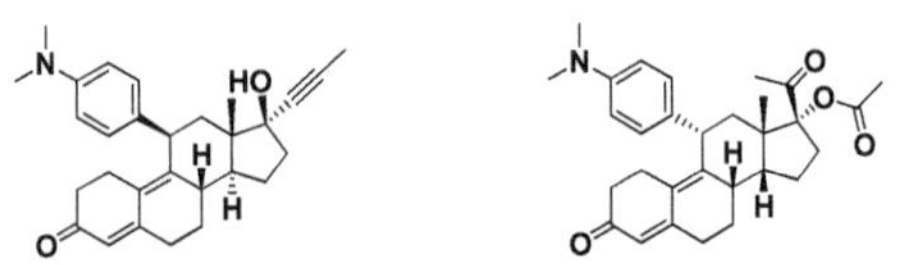

Mifepristone 27.4.30 Ulipristal acetate 27.4.31

FIG. 27.15 Progestin antagonists.

Ulipristal acetate (**27.4.31**) is the first of a new class of selective progesterone receptor modulators. It has been in use since 2010, and is an effective alternative emergency contraception regimen to prevent unintended pregnancies. It acts by inhibiting ovulation and delaying implantation. Its effectiveness is active up to 120 hours after sexual intercourse. Ulipristal acetate is used for contraception and as preoperation treatment of moderate to severe symptoms of uterine fibroids in adult women of reproductive age [130-133].

Among several hundred compounds with antiprogestin activity that are currently under investigation, biological characterization is most advanced for the derivatives ZK-98734 (**27.4.32**), onapristone (**27.4.33**), Org-31710 (**27.4.34**), and Org-33628 (**27.4.35**) (Fig. 27.16.).

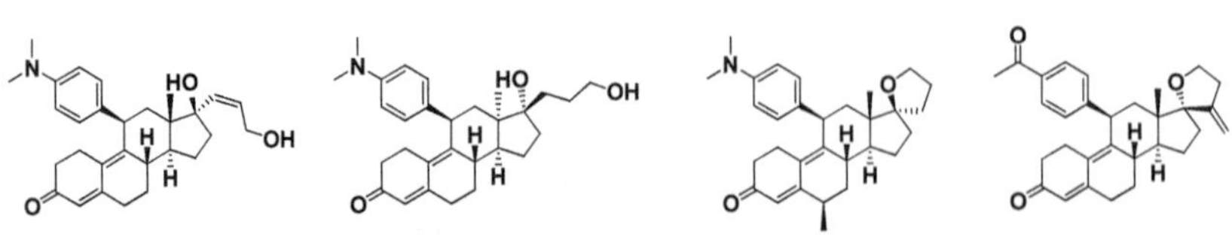

ZK-98734 27.4.32 Onapristone 27.4.33 Org-31710 27.4.34 Org-33628 27.4.35

FIG. 27.16 Experimental progestin antagonists.

A variety of antiprogestins have been synthesized by partially fluorinating the steroid skeleton in positions relevant for receptor binding to obtain partial agonists (mesoprogestins) that have significant applications for antiproliferative and antiovulatory treatment strategies in gynecology therapy, such as uterine fibroids, endometriosis, and heavy menstrual bleeding.

27.5 ANDROGENS

Androgens are a group of male sex hormone steroids that stimulate the development of male sex organs and secondary sexual characteristics; control the development and maintenance of masculine characteristics, including male sexual behavior and reproductive functions; maintain spermatogenesis, muscular development, the growth of facial and body hair; and deepening of the voice. Androgens are secreted by the testes and, in smaller quantities, by the adrenal gland.

Females produce trace quantities of androgens, mostly in the adrenal glands, but also in the ovaries.

In women, androgens play a key role in the hormonal cascade that kick starts puberty, regulate body function before, during, and after menopause, and regulate the function of many other organs, including, bone, kidneys, liver, and muscle. In adult women, androgens are necessary for estrogen synthesis and have been shown to play a key role in the prevention of bone loss, as well as in sexual desire and satisfaction [134-144].

Testosterone (**27.5.1**), dihydrotestosterone (**27.5.2**), and androstenedione (**27.5.3**) are the principal androgens of the testes. Dihydrotestosterone is the more potent androgen in vivo and in vitro (Fig. 27.17.).

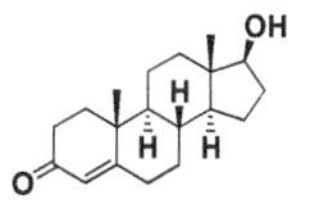

Testosterone 27.5.1

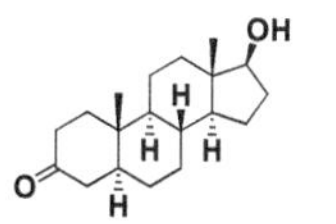

Dihydrotestosterone 27.5.2

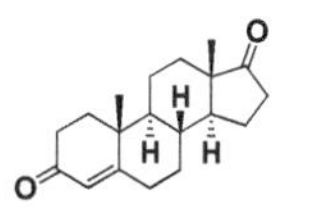

Androstenedione 27.5.3

FIG. 27.17 **Principal androgens.**

Androgen receptor is a member of the steroid and nuclear receptor superfamily. It regulates the gene expression responsible for male sexual differentiation. Androgens can also have extranuclear actions through other receptors.

The actions of androgens in the reproductive tissues, are known as the *androgenic effects*, while their nitrogen-retaining effects in muscles and bones are known as the *anabolic effects*.

The known androgen receptor ligands can be classified based on their structure as steroidal, all of which possess mixed agonist and antagonist activities, or nonsteroidal.

Androgens are widely used in a variety of clinical applications.

The most common indication for androgen therapy is hypogonadism in men. Testosterone replacement in men with hypogonadism induces greater interest in sexual activity and improvement in other aspects of sexual behavior and as hormone replacement therapy in aging males. Androgens are used to treat delayed puberty in adolescent boys, Klinefelter syndrome, anemia secondary to chronic renal failure, protein wasting diseases associated with cancer, burns, traumas, breast cancer, and hereditary angioedema, and to treat breast cancer in women. They are also used in conditions with hormonal imbalance, such as partial androgen deficiency of the aging male, which is referred to as "andropause."

Structural modification of the naturally occurring testosterone include its esterification, which enhances the lipid solubility of the steroid and permits the formation of a local depot after intramuscular injection. Because the esters are hydrolyzed in the body, testosterone becomes active specie in vivo. Testosterone propionate, enanthate, undecanoate, and cypionate (**27.5.4**) are widely represented in the pharmaceutical market under different trade names.

Other testosterone molecule modifications include insertion of an alkyl group into the testosterone structure. In particular, insertion of a methyl group into the 17α position of steroid skeleton allowed to block the metabolism of newly synthesized testosterone analogues and greatly improved the oral bioavailability of compounds.

Further structural modification of 17α-methyltestosterone (**27.5.5**) led to more potent and orally active steroids, like methandrostenolone (**27.5.6**), fluoxymesterone (**27.5.7**), and stanozolol (**27.5.8**) (Fig. 27.18.).

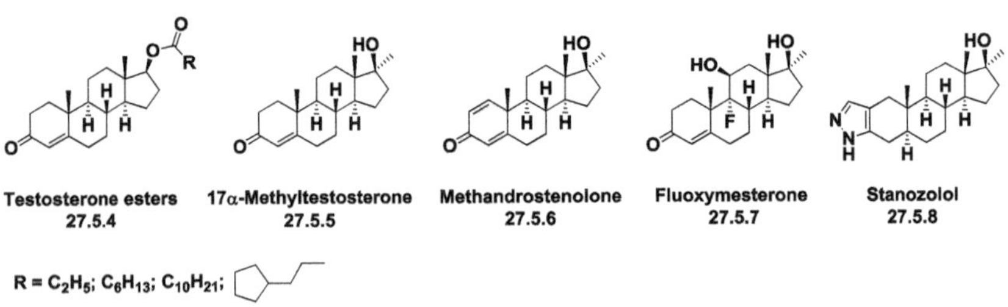

FIG. 27.18 Structural modifications of testosterone.

Although these compounds show improved pharmacokinetic profiles of testosterone, their long-term use is associated with hepatotoxicity, causing masculinizing actions in some women and children, and causing salt and water retention that results in edema.

Consequently, a huge amount of work has been performed to create nonsteroidal ligands to achieve high specificity and better oral bioavailability [145].

Nonsteroidal androgen receptor agonists or modulators include a diversity of compound derivatives of quinolone, including LG-121071 (**27.5.9**) and hydantoin–BMS-564929 (**27.5.10**), and toluidides such as S-1 (**27.5.11**), which are now in different stages of trials (Fig. 27.19.).

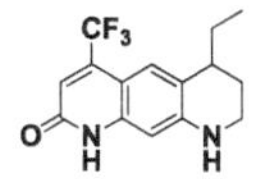
LG-121071 27.5.9

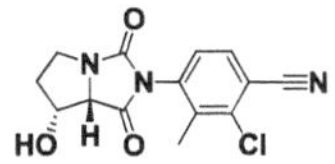
BMS-564929 27.5.10

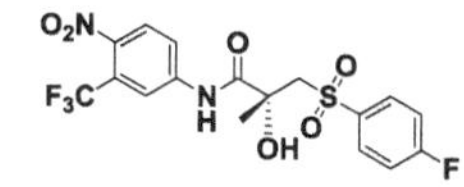
S-1 27.5.11

FIG. 27.19 **Nonsteroidal androgen receptor agonists.**

Androgen Antagonists

Androgen antagonists, by definition, antagonize the actions of testosterone and its active metabolite, 5α-dehydrotestosterone (DHT) by competing for androgen receptor binding sites.

Such compounds prevent androgens from expressing their biological effects on responsive tissues.

Androgen antagonist drugs may be given for any of several conditions or disorders, ranging from skin problems to mental disorders, including to treat early stage and advanced prostate cancer; to treat amenorrhea (the absence of menstrual periods in females); to treat hirsutism (excessive facial and body hair in women); to treat virilization (the development of male pattern baldness, voice changes, and overdevelopment of the skeletal muscles in females); to clear acne and seborrhea; to treat androgenic alopecia; and as a male contraceptive [146-148].

A few compounds with steroidal skeletons are in use as androgen antagonists. These include cyproterone acetate (**27.5.12**), oxendolone (**27.5.13**), and spironolactone (**27.5.14**), and have limited applications because of severe side effects (Fig. 27.20.).

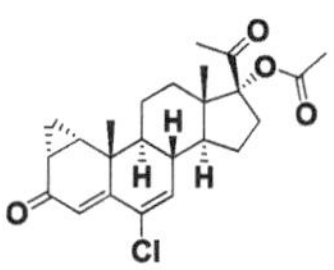
Cyproterone acetate 27.5.12

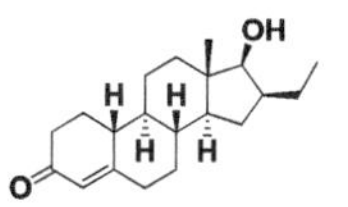
Oxendolone 27.5.13

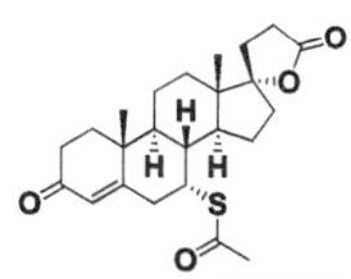
Spironolactone 27.5.14

FIG. 27.20 **Androgen antagonists.**

Many compounds with modified steroidal skeletons have been synthesized and checked as possible antiandrogens (**27.5.15** to **27.5.22**) (Fig. 27.21.).

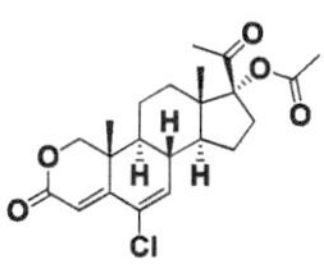
27.5.15

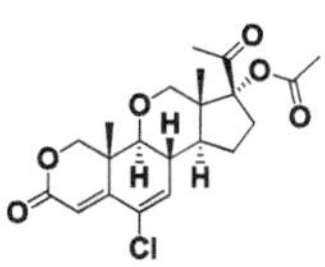
27.5.16

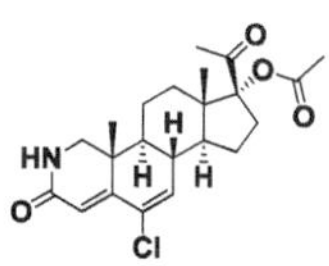
27.5.17

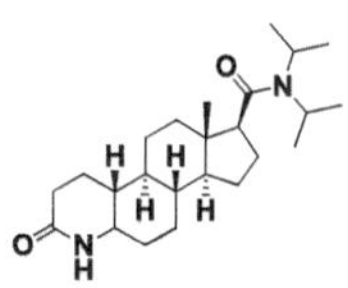
27.5.18

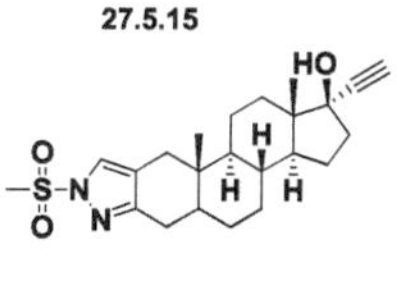
27.5.19

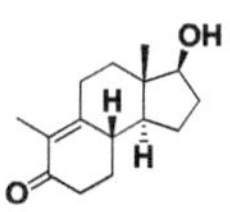
27.5.20

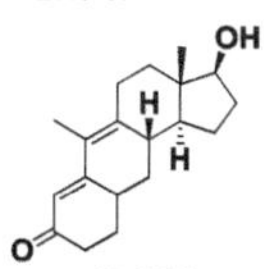
27.5.21

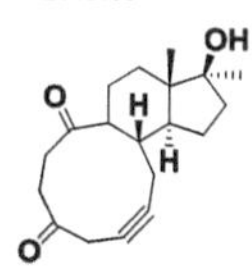
27.5.22

FIG. 27.21 **Androgen antagonists under investigation.**

The first developed nonsteroidal antiandrogens compounds were substituted toluidides, such as bicalutamide (**27.5.23**), flutamide (**27.5.24**), and nilutamide (**27.5.25**) (Fig. 27.22.).

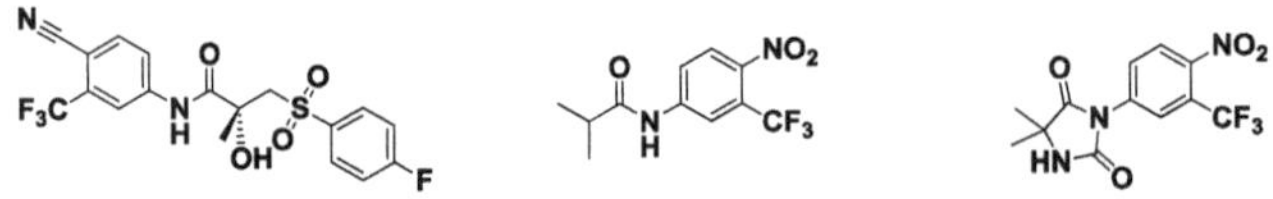

Bicalutamide 27.5.23 Flutamide 27.5.24 Nilutamide 27.5.25

FIG. 27.22 Nonsteroidal antiandrogens compounds.

They are used to treat androgen-sensitive prostate cancer and hyperplasia.

A variety of investigational antiandrogens, such as the steroidal benzimidazole derivative galeterone (**27.5.26**), the hydantoin derivative enzalutamide (**27.5.27**), and the hydrogenated phtalimide derivative BMS-641988 (**27.5.28**), are in clinical development. Pyridoquinoline derivatives of the general type (**27.5.29**) and of substituted 4-methyl-3-methylenecyclohex-1-ene (**27.5.30**) are in preclinical development (Fig. 27.23.).

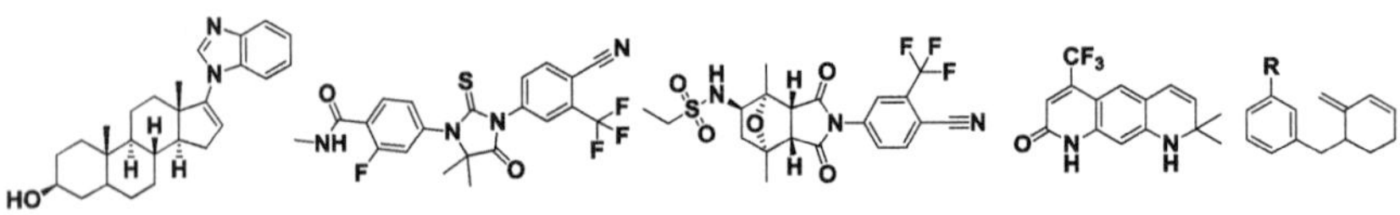

Galeterone 27.5.26 Enzalutamide 27.5.27 BMS-641988 27.5.28 27.5.29 27.5.30

FIG. 27.23 Structures of some investigational antiandrogens.

5α-Reductase Inhibitors

Design of 4-aza steroid analogues such as (**27.5.18**) created a class of antiandrogen drugs that inhibit 5α-reductase. These compounds block the action of the 5α-reductase enzyme, which if not blocked converts testosterone into dihydrotestosterone. Some 5α-reductase inhibitors are approved for the treatment of symptomatic benign prostatic hyperplasia (enlarged prostate gland), prostate cancer, and male pattern hair loss—androgenic disorders believed to be mediated by dihydrotestosterone [149-165].

In general, 5α-reductase inhibitors are able to mitigate the effects that androgens have on the prostate by proliferating prostate cells, decreasing prostate cell apoptosis, and elevating the rate of angiogenesis within the prostate. The increasing of levels of testosterone and decreasing levels of dihydrotestosterone are considered to be a logical treatment for the above-mentioned diseases. That is why inhibition of this enzyme has become a pharmacological strategy for the creation of new antiandrogenic drugs.

Two isoenzymes of 5α-reductase have been discovered. Type 1 is present in most tissues in the body where 5α-reductase is expressed, and is the dominant form in sebaceous glands. Type 2 5α-reductase is the dominant isoenzyme in genital tissues, including the prostate.

Two drugs currently available within this class are finasteride (**27.5.31**), marketed as Proscar and Propecia, and dutasteride (**27.5.32**), marketed as Avodart. Both are included in the list of Top 200 Drugs by sales for the 2010s (Fig. 27.24.).

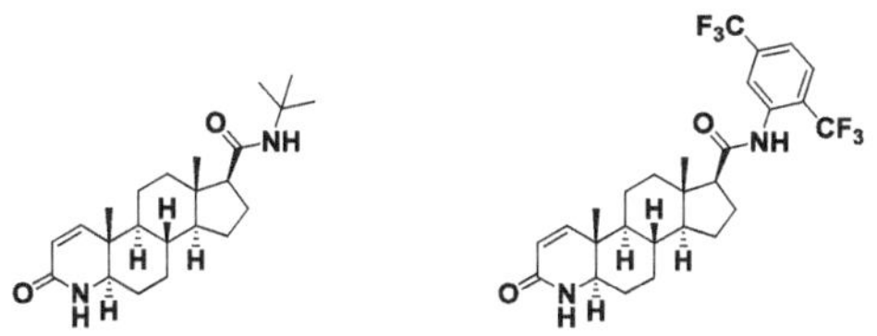

FIG. 27.24 Structure of the 5α-reductase inhibitors finasteride and dutasteride.

Finasteride was found to selectively block the 5α-reductase type 2 enzyme, inhibiting approximately 70% of the conversion of testosterone to DHT within the body, while dutasteride has been shown to inhibit both 5α-reductase type 1 and type 2, preventing 95% of the conversion of testosterone to DHT.

These drugs differ in their length of activity. Finasteride's half-life is estimated to be approximately 6 to 8 hours and dutasteride's half-life is estimated to be 5 weeks [166].

Adverse effects in this class are decreased libido, decreased semen quantity during ejaculation, and impotence. These adverse effects diminish after the first year of treatment [166].

Finasteride–Proscar, Propecia

The synthesis of finasteride, as well as the chemistry of 4-azasteroids in general, are described in the first patent and paper [167,168].

Finasteride (**27.5.31**) was prepared starting from 3-oxo-4-androstene-17β-carboxylic acid (**27.5.33**), which on oxidative cleavage in a mixture of t-butyl alcohol and aqueous Na_2CO_3 with $NaIO_4$ and $KMnO_4$ produced the diacid (**27.5.34**). Ring closure of the diacid (**27.5.34**) with t-butyl amine was carried out in cold ethylene glycol treated with liquid ammonia. The solution was then gradually heated to 180°C, producing the intermediate 3-oxo-4-aza-5-androstene-17β-carboxylic acid (**27.5.35**), which was hydrogenated over a platinum catalyst to produce 4-azasteroid (**27.5.36**). The obtained acid was mixed with dicyclohexylcarbodiimide and N-hydroxybenzotriazole in CH_2Cl_2 and t-butyl amine to produce saturated azasteroid (**27.5.37**), which was oxidized with benzeneseleninic anhydride in chlorobenzene to produce the desired finasteride (**27.5.31**) (Scheme 27.7.).

Some improved processes for the preparation of pure finasteride have been disclosed [169,170].

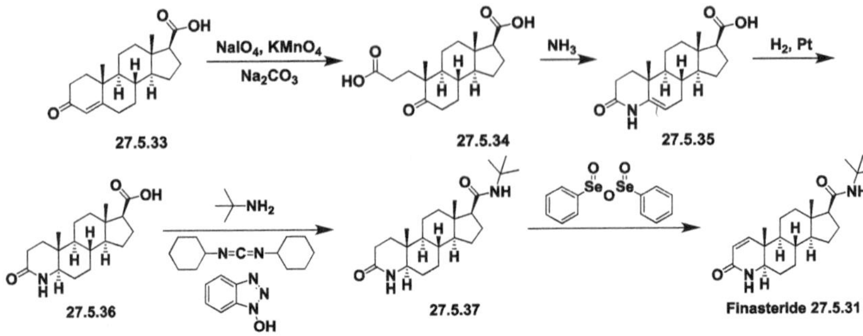

SCHEME 27.7 **Synthesis of finasteride.**

Finasteride, first potent inhibitor of steroid 5-α-reductase, is used alone or in combination with doxazosin to treat benign prostatic hypertrophy and to treat its symptoms such as frequent and difficult urination. It is also used to treat male pattern hair loss [171-180].

Dutasteride–Avodart

The first method reported for the synthesis of dutasteride involves the same first four steps (**27.5.33**→**27.5.36**) implemented for the synthesis of finasteride, which include oxidation of 3-oxo-4-androstene-17β-carboxylic acid (**27.5.33**), its transformation to 5-oxo-a-nor-3,5-secoandrostan-3-oic acid derivative (**27.5.34**), then to 4-aza-androst-5-en-3-one (**27.5.35**), which, after hydrogenation, furnished 4-aza-5α-androstan-3-one (**27.5.36**). Esterification of the 4-aza-5α-androstan-3-one (**27.5.36**) with methanol using sulfuric acid produced the ester (**27.5.38**). Dehydrogenation of the obtained ester using bis(trimethylsilyl) trifluoroacetamide for protection of the amide group and 2,3-dichloro-5,6-dicyano-benzoquinone for suitable dehydrogenation, produced the steroidal ester (**27.5.39**). Hydrolysis of the steroidal ester (**27.5.39**) with sodium hydroxide in a water–methanol mixture yielded dehydrocarboxylic acid (**27.5.40**), which was converted to the corresponding acid halide (**27.5.41**) using thionyl chloride in a toluene/CH_2CL_2/THF mixture, and further reacted with 2,5-bis(trifluoromethyl) aniline to produce dutasteride (**27.5.32**) in an overall yield of 5.13% [181,182] (Scheme 27.8.).

A shorter synthetic sequence, with an overall yield of approximately 40%, starting from 3-oxo-4-aza-5α-androstane-17β-carboxylic acid (**27.5.40**) and using the Ullmann–Goldberg–type condensation, was proposed recently [183]. The mentioned acid (**27.5.40**) was converted to amide (**27.5.42**) by treatment with thionyl chloride in the presence of a catalytic amount of pyridine and further reaction with ammonia to produce the amide (**27.5.42**). Condensation of the (**27.5.42**) with 2-iodo-1,4-bis(trifluoromethyl) benzene using copper powder and potassium carbonate in o-xylene at 140 to 150°C produced the desired dutasteride (**27.5.32**) in 63% crude yield (Scheme 27.8.).

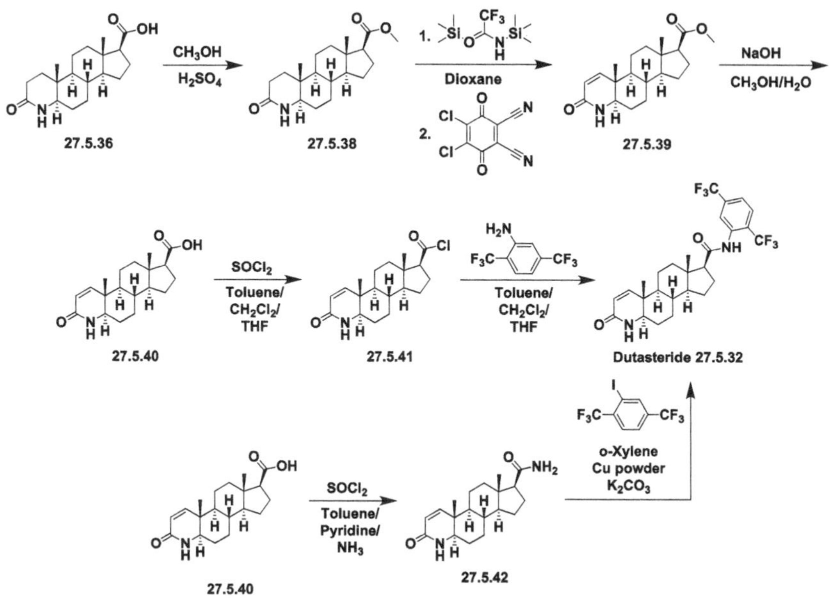

SCHEME 27.8 **Synthesis of dutasteride.**

Further minor improvements have been proposed for the synthesis of dutasteride [184-186].

Dutasteride is a new dual 5α-reductase inhibitor for the treatment of benign prostatic hyperplasia. It differs from finasteride as it inhibits both isoenzymes of 5α-reductase and results in near-complete suppression of serum dihydrotestosterone.

It is an effective treatment option in patients with moderate to severe symptomatic benign prostatic hyperplasia and demonstrable prostatic enlargement. Dutasteride may have the potential to reduce the risk of developing biopsy-detectable prostate cancer. It also reduces the chance of developing acute urinary retention (sudden inability to urinate) [187-198].

Anabolic Steroids

The observation that androgens promote nitrogen retention and muscle mass led to their use to improve physical performance as early as the 1940s. Androgens are now widely used by professional and recreational athletes, weight lifters and bodybuilders, and nonathletes wishing to enhance their performance and appearance.

Anabolism is defined as any state in which nitrogen is differentially retained in lean body mass, either through stimulation of protein synthesis and/or decreased breakdown of protein anywhere in the body.

The term *anabolic steroids* refers to testosterone derivatives that cause nitrogen retention and positive protein metabolism, thereby leading to increased protein synthesis and muscle mass, and that are used either clinically or by athletes for their anabolic properties. Testosterone itself has marked anabolic effects in addition to its effects on reproduction.

No drugs are currently available that are purely anabolic; all possess androgenic properties as well [199-203].

Regular anabolic steroid hormone reception disrupts the normal production of hormones in the body, generating several negative health consequences, including infertility, hair loss, breast development in males, heart attacks, and liver tumors.

Anabolic steroids form one of the classes of doping agents. In spite of the fact that they also produce adverse effects and damage several organs and systems, consumption of anabolic steroids by participants in competitive games is increasing. It is increasingly recognized that androgen use may lead to a dependence syndrome that has both psychological and physiological origins. Androgen dependence likely affects some millions of individuals worldwide.

The World Anti-Doping Agency publishes a list of drugs whose use is prohibited either in or out of competition.

The chemical structures of some commonly used steroids for doping, including testosterone (**27.5.1**), nandrolone (**27.5.43**), prednisone (**27.1.3**), prednisolone (**27.1.4**), methylprednisolone (**27.5.44**), and triamcinolone acetonide (**27.1.10**), are presented on Fig. 27.25.

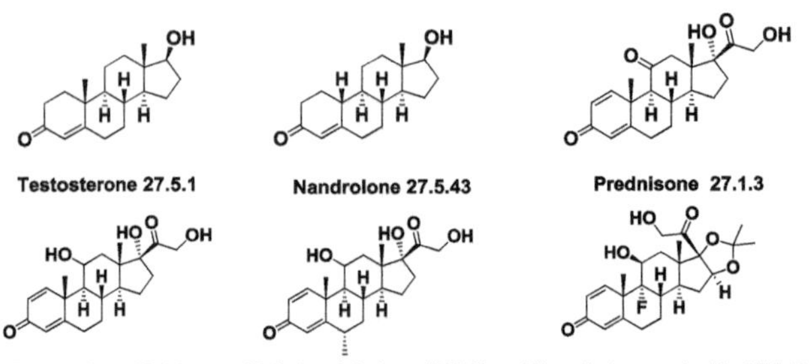

FIG. 27.25 Commonly used steroids for doping.

The development of novel ligands for steroid hormone receptors is an interminable field of research with tremendous potential. New patents describing new ligands for hormone receptors present a wide variety of new structures having strong potential to become compounds of therapeutic utility.

REFERENCES

1. Moss, G. P. Nomenclature of steroids (recommendations 1989). *Pure Appl. Chem.* **1989,** *61* (10), 1783–1822.
2. Saden-Krehula, M.; Kustrak, D. Steroid hormones in the plant kingdom. *Farm. Glas.* **1995,** *51* (2), 23–32.
3. Hench, P. S.; Kendall, E. C.; Slocumb, C. H.; Polley, H. F. Effects of cortisone acetate and pituitary ACTH on rheumatoid arthritis, rheumatic fever and certain other conditions. *Arch. Intern. Med.* **1950,** *85* (4), 545–666.

4. Hillier, S. G. Diamonds are forever: the cortisone legacy. *J. Endocrinol.* **2007,** *195* (1), 1–6.
5. Barnes, P. J. Glucocorticoids: pharmacology and mechanisms. In *Advances in Combination Therapy for Asthma and COPD;* Lotvall, J., Ed.; Wiley-Blackwell, 2012; pp 16–37.
6. Spies, C. M.; Strehl, C.; van der Goes, M. C.; Bijlsma, J. W. J.; Buttgereit, F. Glucocorticoids. *Best Pract. Res., Clin. Rheumatol.* **2011,** *25* (6), 891–900.
7. Corrigan, C. J. Glucocorticoids (corticosteroids). In *Allergy and Allergic Diseases: The New Mechanisms and Therapeutics;* Denburg, J. A., Ed.; Humana, 1998; pp 523–539.
8. Munc, A. Historical introduction. In *Glucocorticoids (Milestones in Drug Therapy);* Goulding, N. J., Flower, R., Eds.; Birkhauser, 2001; pp 17–34.
9. Sneader, W. Adrenal cortex hormones. In *Drug Discovery: A History;* Wiley, 2005; pp 179–184.
10. Whitehouse, M. W. Anti-inflammatory glucocorticoid drugs: reflections after 60 years. *Inflammopharmacology* **2011,** *19* (1), 1–19.
11. Gessi, S.; Merighi, S.; Borea, P. A. Glucocorticoid' pharmacology: past, present and future. *Curr. Pharm. Des.* **2010,** *16* (32), 3540–3553.
12. He, Y.; Yi, W.; Suino-Powell, K.; Zhou, X. E.; Tolbert, W. D.; Tang, X.; Yang, J.; Yang, H.; Shi, J.; Hou, L.; Jiang, H.; Melcher, K.; Xu, H. E. Structures and mechanism for the design of highly potent glucocorticoids. *Cell Res.* **2014,** *24* (6), 713–726.
13. Miner, J. N.; Hong, M. H.; Negro-Vilar, A. New and improved glucocorticoid receptor ligands. *Expert Opin. Invest. Drugs* **2005,** *14* (12), 1527–1545.
14. Miner, J. N. Designer glucocorticoids. *Biochem. Pharmacol. (Amsterdam, Neth.)* **2002,** *64* (3), 355–361.
15. Buttgereit, F.; Burmester, G.-R.; Lipworth, B. J. Optimised glucocorticoid therapy: the sharpening of an old spear. *Lancet* **2005,** *365* (9461), 801–803.
16. Caporali, R.; Todoerti, M.; Sakellariou, G.; Montecucco, C. Glucocorticoids in rheumatoid arthritis. *Drugs* **2013,** *73* (1), 31–43.
17. Vaidya, J. S.; Baldassarre, G.; Thorat, M. A.; Massarut, S. Role of glucocorticoids in breast cancer. *Curr. Pharm. Des.* **2010,** *16* (32), 3593–3600.
18. Katz, M.; Gans, E. H. Topical corticosteroids, structure-activity and the glucocorticoid receptor: discovery and development-a process of "planned serendipity". *J. Pharm. Sci.* **2008,** *97* (8), 2936–2947.
19. Wiedersberg, S.; Leopold, C. S.; Guy, R. H. Bioavailability and bioequivalence of topical glucocorticoids. *Eur. J. Pharm. Biopharm.* **2008,** *68* (3), 453–466.
20. Onrust, S. V.; Lamb, H. M. Mometasone furoate: a review of its intranasal use in allergic rhinitis. *Drugs* **1998,** *56* (4), 725–745.
21. Westergaard, C. G.; Porsbjerg, C.; Backer, V. A review of mometasone furoate/formoterol in the treatment of asthma. *Expert Opin. Pharmacother.* **2013,** *14* (3), 339–346.
22. Prakash, A.; Benfield, P. Topical mometasone: a review of its pharmacological properties and therapeutic use in the treatment of dermatological disorders. *Drugs* **1998,** *55* (1), 145–163.
23. Bousquet, J. Mometasone furoate: an effective anti-inflammatory with a well-defined safety and tolerability profile in the treatment of asthma. *Int. J. Clin. Pract.* **2009,** *63* (5), 806–819.
24. Tan, R. A.; Corren, J. Mometasone furoate in the management of asthma: a review. *Ther. Clin. Risk Manage.* **2008,** *4* (6), 1201–1208.
25. McCormack, P. L.; Plosker, G. L. Inhaled mometasone furoate: a review of its use in persistent asthma in adults and adolescents. *Drugs* **2006,** *66* (8), 1151–1168.
26. Shapiro, E.L. 3,20-Dioxo-1,4-pregnadien-17α-ol 17-aromatic heterocycle carboxylates, US 4472393 (1984).
27. Shapiro, E. L. Aromatic heterocyclic esters of steroids and pharmaceutical compositions containing them, EP 57401 (1982).

28. Shapiro, E. L.; Gentles, M. J.; Tiberi, R. L.; Popper, T. L.; Berkenkopf, J.; Lutsky, B.; Watnick, A. S. 17-Heteroaroyl esters of corticosteroids. 2. 11β-Hydroxy series. *J. Med. Chem.* **1987,** *30* (9), 1581–1588.
29. Heggie, W.; Bandarra, J. Process for the preparation of mometasone furoate, US 6177560 (2001).
30. Hesk, D.; Delduca, P.; Koharski, D.; McNamara, P.; Magatti, C.; Saluja, S.; Thomas, L. Synthesis of tritium labeled mometasone furoate. *J. Labelled Compd. Radiopharm.* **1993,** *33* (5), 439–442.
31. Olsen, E. A.; Cornell, R. C. Topical clobetasol-17-propionate: review of its clinical efficacy and safety. *J. Am. Acad. Dermatol.* **1986,** *15* (2 Pt 1), 246–255.
32. Pels, R.; Sterry, W.; Lademann, J. Clobetasol propionate—where, when, why? *Drugs Today* **2008,** *44* (7), 547–557.
33. Reid, D. C.; Kimball, A. B. Clobetasol propionate foam in the treatment of psoriasis. *Expert Opin. Pharmacother.* **2005,** *6* (10), 1735–1740.
34. Elks, J.; Phillipps, G. H. Halopregnenones, DE 1902340 (1969).
35. Wiseman, L. R.; Benfield, P. Intranasal fluticasone propionate: a reappraisal of its pharmacology and clinical efficacy in the treatment of rhinitis. *Drugs* **1997,** *53* (5), 885–907.
36. Korting, H. C.; Schoellmann, C. Topical fluticasone propionate: intervention and maintenance treatment options of atopic dermatitis based on a high therapeutic index. *J. Eur. Acad. Dermatol. Venereol.* **2012,** *26* (2), 133–140.
37. Staresinic, A. G.; Sorkness, C. A. Fluticasone propionate: a potent inhaled corticosteroid for the treatment of asthma. *Expert Opin. Pharmacother.* **2000,** *1* (6), 1227–1244.
38. Spencer, C. M.; Wiseman, L. R. Topical fluticasone propionate: a review of its pharmacological properties and therapeutic use in the treatment of dermatological disorders. *BioDrugs* **1997,** *7* (4), 318–334.
39. Johnson, M. Development of fluticasone propionate and comparison with other inhaled corticosteroids. *J. Allergy Clin. Immunol.* **1998,** *101* (4 Pt 2), S434–S439.
40. Sastre, J. Pharmacology of fluticasone propionate. *J. Invest. Allergol. Clin. Immunol.* **1997,** *7* (5), 382–384.
41. Harding, S. M. Fluticasone propionate: pharmacology and implications for clinical practice. *Adv. Ther.* **1997,** *14* (4), 153–159.
42. Phillipps, G. H.; Bain, B. M.; Williamson, C.; Steeples, I. P. Androstane carbothioates, CA 1205464 (1986).
43. Phillipps, G. H.; Bailey, E. J.; Bain, B. M.; Borella, R. A.; Buckton, J. B.; Clark, J. C.; Doherty, A. E.; English, A. F.; Fazakerley, H.; Laing, S. B.; Lane-Allman, E.; Robinson, J. D.; Sandford, P. E.; Sharratt, P. J.; Steeples, I. P.; Stonehouse, R. D.; Williamson, C. Synthesis and structure-activity relationships in a series of antiinflammatory corticosteroid analogs, halomethyl androstane-17β-carbothioates and -17β-carboselenoates. *J. Med. Chem.* **1994,** *37* (22), 3717–3729.
44. Vardanyan, R. S.; Hruby, V. J. *Synthesis of essential drugs;* Elsevier, 2006.
45. Clark, R. D. Glucocorticoid receptor antagonists. *Curr. Top. Med. Chem.* **2008,** *8* (9), 813–838.
46. McMaster, A.; Ray, D. W. Drug insight: selective agonists and antagonists of the glucocorticoid receptor. *Nat. Clin. Pract. Endocrinol. Metab.* **2008,** *4* (2), 91–101.
47. Nihalani, N. D.; Schwartz, T. L. Drug evaluation: mifepristone, a glucocorticoid antagonist for the potential treatment of psychotic major depression. *Curr. Opin. Invest. Drugs (BioMed Cent.)* **2007,** *8* (7), 563–569.
48. Peeters, B. W. M. M.; Tonnaer, J. A. D. M.; Groen, M. B.; Broekkamp, C. L. E.; van der Voort, H. A. A.; Schoonen, W. G. F. J.; Smets, R. J. M.; Vanderheyden, P. M. L.; Gebhard, R.; Ruigt, G. S. F. Short review: glucocorticoid receptor antagonists: new tools to investigate disorders characterized by cortisol hypersecretion. *Stress* **2004,** *7* (4), 233–241.

49. Mohler, M. L.; He, Y.; Wu, Z.; Hong, S.-S.; Miller, D. Non-steroidal glucocorticoid receptor antagonists: the race to replace RU-486 for anti-glucocorticoid therapy. *Expert Opin. Ther. Pat.* **2007,** *17* (1), 59–81.
50. Morgan, B. P.; Swick, A. G.; Hargrove, D. M.; LaFlamme, J. A.; Moynihan, M. S.; Carroll, R. S.; Martin, K. A.; Lee, E.; Decosta, D.; Bordner, J. Discovery of potent, non-steroidal, and highly selective glucocorticoid receptor antagonists. *J. Med. Chem.* **2002,** *45* (12), 2417–2424.
51. Collin, M.; Niemann, F.; Jaisser, F. Mineralocorticoid receptor modulators: a patent review (2007-2012). *Expert Opin. Ther. Pat.* **2014,** *24* (2), 177–183.
52. Sica, D. A. The risks and benefits of aldosterone antagonists. *Curr. Heart Failure Rep.* **2005,** *2* (2), 65–71.
53. Horton, R. Aldosterone and aldosteronism. *Steroids* **2003,** *68* (14), 1135–1138.
54. Ghulam, A.; Vantyghem, M. C.; Wemeau, J. L.; Boersma, A. Adrenal mineralocorticoids pathway and its clinical applications. *Clin. Chim. Acta* **2003,** *330* (1–2), 99–110.
55. Gordon, R. D. Mineralocorticoid excess syndromes. In *Adrenal Disorders;* Margioris, A. N., Chrousos, G. P., Eds.; Humana, 2001; pp 355–377.
56. Rogerson, F. M.; Fuller, P. J. Mineralocorticoid action. *Steroids* **2000,** *65* (2), 61–73.
57. Espiner, E. A. Mineralocorticoidsm. *Basic Clin. Endocrinol.* **1984,** *4,* 371–401.
58. Dahlman-Wright, K.; Cavailles, V.; Fuqua, S. A.; Jordan, V. C.; Katzenellenbogen, J. A.; Korach, K. S.; Maggi, A.; Muramatsu, M.; Parker, M. G.; Gustafsson, J.-A. International union of pharmacology. LXIV. Estrogen receptors. *Pharmacol. Rev.* **2006,** *58* (4), 773–781.
59. Rettberg, J. R.; Yao, J.; Brinton, R. D. Estrogen: A master regulator of bioenergetic systems in the brain and body. *Front. Neuroendocrinol.* **2014,** *35* (1), 8–30.
60. Matsumoto, A. M. Reproductive endocrinology: estrogens—not just female hormones. *Nat. Rev. Endocrinol.* **2013,** *9* (12), 693–694.
61. Ferguson, R. D.; Gallagher, E. J.; Scheinman, E. J.; Damouni, R.; LeRoith, D. The epidemiology and molecular mechanisms linking obesity, diabetes, and cancer. *Vitam. Horm. (London, U. K.)* **2013,** *93,* 51–98.
62. Falkenstein, E.; Tillmann, H. C.; Christ, M.; Feuring, M.; Wehling, M. Multiple actions of steroid hormones—-a focus on rapid, nongenomic effects. *Pharmacol. Rev.* **2000,** *52* (4), 513–556.
63. Jensen, E. V. The contribution of "alternative approaches" to understanding steroid hormone action. *Mol. Endocrinol.* **2005,** *19* (6), 1439–1442.
64. Macgregor, J. I.; Jordan, V. C. Basic guide to the mechanisms of antiestrogen action. *Pharmacol. Rev.* **1998,** *50* (2), 151–196.
65. Maximov, P. Y.; Lee, T. M.; Jordan, V. C. The discovery and development of selective estrogen receptor modulators (SERMs) for clinical practice. *Curr. Clin. Pharmacol.* **2013,** *8* (2), 135–155.
66. Dhingra, K. Antiestrogens—tamoxifen, SERMs and beyond. *Invest. New Drugs* **1999,** *17* (3), 285–311.
67. Frank, E.; Schneider, G. Synthesis of sex hormone-derived modified steroids possessing antiproliferative activity. *J. Steroid Biochem. Mol. Biol.* **2013,** *137,* 301–315.
68. Sitruk-Ware, R.; El-Etr, M. Progesterone and related progestins: potential new health benefits. *Climacteric* **2013,** *16* (S1), 69–78.
69. Sitruk-Ware, R. New progestogens: a review of their effects in perimenopausal and postmenopausal women. *Drugs Aging* **2004,** *21* (13), 865–883.
70. Benagiano, G.; Primiero, F. M.; Farris, M. Clinical profile of contraceptive progestins, Eur. *J. Contracept. Reprod. Health Care* **2004,** *9* (3), 182–193.

71. Endrikat, J.; Gerlinger, C.; Richard, S.; Rosenbaum, P.; Duesterberg, B. Ovulation inhibition doses of progestins: a systematic review of the available literature and of marketed preparations worldwide. *Contraception* **2011,** *84* (6), 549–557.
72. Sitruk-Ware, R.; Nath, A. The use of newer progestins for contraception. *Contraception* **2010,** *82* (5), 410–417.
73. Belaisch, J. Progestins and medical treatment of endometriosis—physiology, history and society. *Gynecol. Endocrinol.* **2009,** *25* (11), 751–756.
74. Pasqualini, J. R. Progestins in the menopause in healthy women and breast cancer patients. *Maturitas* **2009,** *62* (4), 343–348.
75. Schoellkopf, K.; Schmees, N. Progesterone receptor: overview of modern steroidal progestins and developments in the field of nonsteroidal selective progesterone receptor modulators. *Methods Princ. Med. Chem.* **2008,** *39,* 201–222.
76. McCarty, K. S., Jr.; Nichols, M. D.; Hershberger, P. A.; McCarty, J. C.; McCarty, K. S. Sr., Progestins. In *Holland-Frei Cancer Medicine 7;* Kufe, D. W., Ed.; BC Decker, 2006; pp 843–849.
77. Hapgood, J. P.; Koubovec, D.; Louw, A.; Africander, D. Not all progestins are the same: implications for usage. *Trends Pharmacol. Sci.* **2004,** *25* (11), 554–557.
78. Sitruk-Ware, R. Pharmacological profile of progestins. *Maturitas* **2004,** *47* (4), 277–283.
79. Wiechert, R.; Bittler, D.; Kerb, U.; Vasals-Stenzel, J.; Losert, W. Spirolactones, DE 2652761 (1978).
80. Petzoldt, K.; Laurent, H.; Wiechert, R. 3β,7β-Dihydroxy-Δ5-steroids, EP 75189 (1983).
81. Mohr, J-T.; Nickisch, K. Preparation method for drospirenone, DE 19633685 (1997).
82. Bittler, D.; Hofmeister, H.; Laurent, H.; Nickisch, K.; Nickolson, R.; Petzoldt, K.; Wiechert, R. Synthesis of a new highly effective aldosterone antagonist (spirorenone). *Angew. Chem.* **1982,** *94* (9), 718–719.
83. Laurent, H.; Bittler, D.; Hofmeister, H.; Nickisch, K.; Nickolson, R.; Petzoldt, K.; Wiechert, R. Synthesis and activities of antialdosterones. *J. Steroid Biochem.* **1983,** *19* (1C), 771–776.
84. Galik, G.; Horvath, J.; Soeroes, B.; Maho, S.; Tuba, Z.; Balogh, G. A process for the preparation of 17-hydroxy-6β,7β,15β,16β-bismethylene-17α-pregn-4-ene-3-one-21-carboxylic acid γ-lactone and key-intermediates for this process, WO 2006059168 (2006).
85. Costantino, F.; Lenna, R.; Piuri, S. Process for the preparation of drospirenone, WO 2006061309 (2006).
86. Costantino, F.; Lenna, R.; Piuri, S. Process for the preparation of drospirenone, US 20050192450 (2005).
87. Seilz, C.; Seba, H. Process for the production of 3-oxo-pregn-4-ene-21,17-carbolactones by the metal-free oxidation of 17-(3-hydroxypropyl)-3,17-dihydroxyandrostanes, WO 2007009821 (2005).
88. Mohr, J-T.; Nickisch, K. Preparation method for drospirenone, DE 19633685 (1997).
89. Montorsi, M.; Mariani, E.; Gambarin, L.; Orru, G.; Scalaprice, R.; Merlo, M.; Andriolo, E. Methods for the preparation of drospirenone and intermediates thereof, WO 2012016860 (2012).
90. Pontiroli, A.; Diulgheroff, N.; Scarpitta, F.; Arosio, R.; Pogialli, A.; Villa, M. A process for preparing of drospirenone and intermediates thereof, WO 2008137050 (2008).
91. Cabri, W.; Benedetti, F.; Alpegiani, M.; Rodriguez, M.; Botta, C. Epoxidation of 17-oxo-15,16-methylene steroids with sulfoxonium ylides, EP 1903051 (2008).
92. Soeroes, B; Horvath, J.; Galik, G.; Bodi, J.; Tuba, Z.; Maho, S.; Balogh, G.; Aranyi, A. Industrial process for the preparation of 17-hydroxy-6β,7β;15β,16β-bismethylene-3-oxo-17α-pregn-4-ene-21-carboxylic acid γ-lactone and key intermediates for this process, WO 2006059167 (2006).

93. Seilz, C.; Seba, H. Process for the production of 3-oxopregn-4-ene-21,17-carbolactones by metal free oxidation of 17-(3-hydroxypropyl)-3,17-dihydroxyandrostanes, EP 1746101 (2007).
94. Bandini, M.; Contento, M.; Garelli, A.; Monari, M.; Tolomelli, A.; Umani-Ronchi, A.; Andriolo, E.; Montorsi, M. A nonclassical stereoselective semi-synthesis of drospirenone via cross-metathesis reaction. *Synthesis* **2008,** *23,* 3801–3804.
95. Deng, G.; Huang, Z.; Zhao, X.; Li, Z.; Li, Y.; Jiang, B. Stereospecific synthesis of drospirenone. *Chin. J. Chem.* **2013,** *31* (1), 15–17.
96. Nickisch, K.; Acosta, K.; Santhamma, B. Methods for the preparation of drospirenone, US 20100261896 (2010).
97. Wiechert, R. The history of drospirenone. In *Analogue-based Drug Discovery;* Fischer, J., Ganellin, C. R., Eds.; Wiley-VCH, 2006; pp 395–400.
98. Muhn, P.; Fuhrmann, U.; Fritzemeier, K.; Krattenmacher, R.; Schillinger, E. Drospirenone: A novel progestogen with antimineralocorticoid and antiandrogenic activity. *Ann. N. Y. Acad. Sci.* **1955,** *761,* 311–335.
99. Muhn, P.; Krattenmacher, R.; Beier, S.; Elger, W.; Schillinger, E. Drospirenone: a novel progestogen with antimineralocorticoid and antiandrogenic activity. Pharmacological characterization in animal models. *Contraception* **1995,** *51* (2), 99–110.
100. Sitruk-Ware, R. New progestagens for contraceptive use. *Hum. Reprod. Update* **2006,** *12* (2), 169–178.
101. Rapkin, A. J.; Winer, S. A. Drospirenone: a novel progestin. *Expert Opin. Pharmacother.* **2007,** *8* (7), 989–999.
102. Sitruk-Ware, R. Pharmacology of different progestogens: the special case of drospirenone. *Climacteric* **2005,** *8* (Suppl. 3), 4–12.
103. Krattenmacher, R. Drospirenone: pharmacology and pharmacokinetics of a unique progestogen. *Contraception* **2000,** *62* (1), 29–38.
104. Oelkers, W. Drospirenone, a progestogen with antimineralocorticoid properties: a short review. *Mol. Cell. Endocrinol.* **2004,** *217* (1–2), 255–261.
105. Ylikorkala, O. Drospirenone, a progestin with a unique cardiovascular profile, for safe contraception and treatment of menopausal symptoms. *Climacteric* **2005,** *8* (Suppl. 3), 1–3.
106. Rubig, A. Drospirenone: a new cardiovascular-active progestin with antialdosterone and antiandrogenic properties. *Climacteric* **2003,** *6* (Suppl. 3), 49–54.
107. Norman, P.; Castaner, J.; Castaner, R. M. Drospirenone: contraceptive, hormone replacement therapy, aldosterone antagonist, progestogen. *Drugs Future* **2000,** *25* (12), 1247–1256.
108. Mishell, D. R., Jr. YAZ and the novel progestin drospirenone. *J. Reprod. Med. (St. Louis, MO, U. S.)* **2008,** *53* (9), 721–728.
109. Mallareddy, M.; Hanes, V.; White, W. B. Drospirenone, a new progestogen, for postmenopausal women with hypertension. *Drugs Aging* **2007,** *24* (6), 453–466.
110. Motivala, A.; Pitt, B. Drospirenone for oral contraception and hormone replacement therapy. Are its cardiovascular risks and benefits the same as other progestogens? *Drugs* **2007,** *67* (5), 647–655.
111. Benagiano, G.; Bastianelli, C.; Farris, M.; Brosens, I. Selective progesterone receptor modulators: an update, Expert Opin. *Pharmacother.* **2014,** *15* (10), 1403–1415.
112. Allan, G.; Macielag, M. Progesterone receptor agonists and antagonists. *Expert Opin. Ther. Pat.* **1999,** *9* (7), 955–962.
113. Lanari, C.; Wargon, V.; Rojas, P.; Molinolo, A. A. Antiprogestins in breast cancer treatment: are we ready? *Endocr.-Relat. Cancer* **2012,** *19* (3), R35–R50.
114. Chakraborty, A.; Chatterjee, S.; Roy, P. Progesterone receptor agonists and antagonists as anticancer agents. *Mini-Rev. Med. Chem.* **2010,** *10* (6), 506–517.

115. Schwenkhagen, A.; Schaudig, K. Antiprogestins: new therapeutic options in gynecology Gynaekol. *Endokrinol.* **2008,** *6* (4), 216–220.
116. Spitz, I. M. Progesterone receptor antagonists and selective progesterone receptor modulators: proven and potential clinical applications. *Expert Rev. Obstet. Gynecol.* **2007,** *2* (2), 227–242.
117. Spitz, I. M. Progesterone receptor antagonists. *Curr. Opin. Invest. Drugs (BioMed Cent.), Drugs* **2006,** *7* (10), 882–889.
118. Spitz, I. M. Progesterone antagonists and progesterone receptor modulators: an overview. *Steroids* **2003,** *68* (10-13), 981–993.
119. Spitz, I. M. Clinical applications of progesterone receptor antagonists and selective progesterone receptor modulators. *Endocrinologist (Hagerstown, MD, U. S.)* **2005,** *15* (6), 391–400.
120. von Hertzen, H.; Van Look, P. F. A. Antiprogestins for contraception? *Semin. Reprod. Med.* **2005,** *23* (1), 92–100.
121. Hess-Stumpp, H.; Hoffmann, J.; Fuhrmann, U. Progesterone receptor antagonists (antiprogestins). *Drugs Future* **2002,** *27* (11), 1113–1123.
122. Klijn, J. G. M.; Setyono-Han, B.; Foekens, J. A. Progesterone antagonists and progesterone receptor modulators in the treatment of breast cancer. *Steroids* **2000,** *65* (10–11), 825–830.
123. Baird, D. T. Clinical uses of antiprogestogens. *J. Soc. Gynecol. Invest.* **2000,** *7* (Suppl. 1), S49–S52.
124. Michna, H.; Nishino, Y.; Parczyk, K.; Schneider, M. R. Antiprogestins: past, present, and future. In *Estrogens, Progestins, and Their Antagonists,* 1st ed.; Pavlik, E. J., Ed; Birkhauser, 1997; pp 297–319.
125. Teutsch, G.; Philibert, D. History and perspectives of antiprogestins from the chemist's point of view. *Hum. Reprod.* **1994,** *9* (Suppl. 1), 12–31.
126. Ashok, P. W.; Wagaarachchi, P. T.; Templeton, A. The antiprogestogen mifepristone: a review. *Med. Chem.: Immunol., Endocr. Metab. Agents* **2002,** *2* (2), 71–90.
127. Permezel, M. The antiprogesterone steroid, RU 486 (mifepristone). *Aust. N. Z. J. Obstet. Gynaecol.* **1990,** *30* (1), 77–80.
128. Brogden, R. N.; Goa, K. L.; Faulds, D. Mifepristone: a review of its pharmacodynamic and pharmacokinetic properties, and therapeutic potential. *Drugs* **1993,** *45* (3), 384–409.
129. Sun, Y.; Fang, M.; Davies, H.; Hu, Z. Mifepristone: a potential clinical agent based on its anti-progesterone and anti-glucocorticoid properties. *Gynecol. Endocrinol.* **2014,** *30* (3), 169–173.
130. McKeage, K.; Croxtall, J. D. Ulipristal acetate: a review of its use in emergency contraception. *Drugs* **2011,** *71* (7), 935–945.
131. Jadav, S. P.; Parmar, D. M. Ulipristal acetate, a progesterone receptor modulator for emergency contraception. *J. Pharmacol. Pharmacother.* **2012,** *3* (2), 109–111.
132. Richardson, A. R.; Maltz, F. N. Ulipristal acetate: review of the efficacy and safety of a newly approved agent for emergency contraception. *Clin. Ther.* **2012,** *34* (1), 24–36.
133. Orihuela, P. A. Ulipristal: a progesterone receptor antagonist as an emergency contraceptive. *Expert Rev. Obstet. Gynecol.* **2010,** *5* (1), 13–17.
134. Jordan, C. L.; Don Carlos, L. Androgens in health and disease: an overview. *Horm. Behav.* **2008,** *53* (5), 589–595.
135. Gooren, L. J. G.; Bunck, M. C. M. Androgen replacement therapy: present and future. *Drugs* **2004,** *64* (17), 1861–1891.
136. Chung, P. H.; Gayed, B. A.; Thoreson, G. R.; Raj, G. V. Emerging drugs for prostate cancer. *Expert Opin. Emerg. Drugs* **2013,** *18* (4), 533–550.
137. Schmidt, L. J.; Tindall, D. J. Androgen receptor: past, present and future. *Curr. Drug Targets* **2013,** *14* (4), 401–407.
138. Cadilla, R.; Turnbull, P. Selective androgen receptor modulators in drug discovery: medicinal chemistry and therapeutic potential. *Curr. Top. Med. Chem.* **2006,** *6* (3), 245–270.

139. Levine, P. M.; Garabedian, M. J.; Kirshenbaum, K. Targeting the androgen receptor with steroid conjugates. *J. Med. Chem.* **2014,** *57* (20), 8224–8237.
140. Myers, J. M.; Meacham, R. B. Androgen replacement therapy in the aging male. *Rev. Urol.* **2003,** *5* (4), 216–226.
141. Chawnshang, C., Ed. *Androgens and Androgen Receptor: Mechanisms, Functions, and Clinical Applications;* Springer, 2002.
142. Gao, W.; Bohl, C. E.; Dalton, J. T. Chemistry and structural biology of androgen receptor. *Chem. Rev. (Washington, DC, U. S.)* **2005,** *105* (9), 3352–3370.
143. Bagatell, C. J.; Bremner, W. J. Androgen in men—uses and abuses. *N. Engl. J. Med.* **1966,** *334* (11), 707–714.
144. Chen, Y.; Sawyers, C. L.; Scher, H. I. Targeting the androgen receptor pathway in prostate cancer. *Curr. Opin. Pharmacol.* **2008,** *8* (4), 440–448.
145. Lin, X.; Huebner, V. Non-steroidal ligands for steroid hormone receptors. *Curr. Opin. Drug Discovery Dev.* **2000,** *3* (4), 383–398.
146. Vasaitis, T. S.; Njar, V. C. O. Novel, potent anti-androgens of therapeutic potential: recent advances and promising developments. *Future Med. Chem.* **2010,** *2* (4), 667–680.
147. Singh, S. M.; Gauthier, S.; Labrie, F. Androgen receptor antagonists (antiandrogens): structure-activity relationships. *Curr. Med. Chem.* **2000,** *7* (2), 211–247.
148. Trendel, J. A. The hurdle of antiandrogen drug resistance: drug design strategies. *Expert Opin. Drug Discovery* **2013,** *8* (12), 1491–1501.
149. Aggarwal, S.; Thareja, S.; Verma, A.; Bhardwaj, T. R.; Kumar, M. An overview on 5α-reductase inhibitors. *Steroids* **2010,** *75* (2), 109–153.
150. Bratoeff, E.; Cabeza, M.; Ramirez, E.; Heuze, Y.; Flores, E. Recent advances in the chemistry and pharmacological activity of new steroidal antiandrogens and 5α-reductase inhibitors. *Curr. Med. Chem.* **2005,** *12* (8), 927–943.
151. Foley, C. L.; Bott, S. R. J.; Shergill, I. S.; Kirby, R. S. An update on the use of 5α-reductase inhibitors. *Drugs Today* **2004,** *40* (3), 213–223.
152. Flores, E.; Bratoeff, E.; Cabeza, M.; Ramirez, E.; Quiroz, A.; Heuze, I. Steroid 5α-reductase inhibitors. *Mini-Rev. Med. Chem.* **2003,** *3* (3), 225–237.
153. Machetti, F.; Guarna, A. Novel inhibitors of 5α-reductase. *Expert Opin. Ther. Pat.* **2002,** *12* (2), 201–215.
154. Chen, Y.; Sawyers, C. L.; Scher, H. I. Targeting the androgen receptor pathway in prostate cancer. *Curr. Opin. Pharmacol.* **2008,** *8* (4), 440–448.
155. McConnell, J. D.; Stoner, E. 5α-Reductase inhibitors. *Adv. Protein Chem.* **2001,** *56,* 143–180.
156. Marberger, M. Drug insight: 5α-Reductase inhibitors for the treatment of benign prostatic hyperplasia. *Nat. Clin. Pract. Urol.* **2006,** *3* (9), 495–503.
157. Gomella, L. G. Prostate cancer: alcohol, cancer and 5α-reductase inhibitors-is there a link? *Nat. Rev. Urol.* **2014,** *11* (5), 253–254.
158. Hamilton, R. J.; Freedland, S. J. 5α-Reductase inhibitors and prostate cancer prevention: where do we turn now? *BMC Med.* **2011,** *9,* 105.
159. Tarter, T. H.; Vaughan, E. D., Jr. Inhibitors of 5α-reductase in the treatment of benign prostatic hyperplasia. *Curr. Pharm. Des.* **2006,** *12* (7), 775–783.
160. Sun, J.; Xiang, H.; Yang, L.-L.; Chen, J.-B. A review on steroidal 5α-reductase inhibitors for treatment of benign prostatic hyperplasia. *Curr. Med. Chem.* **2011,** *18* (23), 3576–3589.
161. Schmidt, L. J.; Tindall, D. J. Steroid 5α-reductase inhibitors targeting BPH and prostate cancer. *J. Steroid Biochem. Mol. Biol.* **2011,** *125* (1–2), 32–38.
162. Bratoeff, E.; Ramirez, E.; Murillo, E.; Flores, G.; Cabeza, M. Steroidal antiandrogens and inhibitors of 5α-reductase. *Curr. Med. Chem.* **1999,** *6* (12), 1107–1123.

163. Rittmaster, R. S. 5α-Reductase inhibitors. *J. Androl.* **1997,** *18* (6), 582–587.
164. Harris, G. S.; Kozarich, J. W. Steroid inhibitors 5α-reductase in androgen-dependent disorders. *Curr. Opin. Chem. Biol.* **1997,** *1* (2), 254–259.
165. Frye, S. V. Inhibitors of 5α-reductase. *Curr. Pharm. Des.* **1996,** *2* (1), 59–84.
166. Kumar, V. L.; Wahane, V. D. Current status of 5-alpha reductase inhibitors in the treatment of benign hyperplasia of prostate. *Indian J. Med. Sci.* **2008,** *62,* 167–175.
167. Rasmusson, G. H.; Reynolds, G. F. 17β-Substituted 4-aza-5α-androstenones and their use as testosterone 5α-reductase inhibitors, EP 155096 (1985).
168. Rasmusson, G. H.; Reynolds, G. F.; Steinberg, N. G.; Walton, E.; Patel, G. F.; Liang, T.; Cascieri, M. A.; Cheung, A. H.; Brooks, J. R.; Berman, C. Azasteroids: structure-activity relationships for inhibition of 5α-reductase and of androgen receptor binding. *J. Med. Chem.* **1986,** *29* (11), 2298–2315.
169. Wang, Z-X.; Ceccarelli, A. P.; Raheem, M. A.; Guntoori, B. R. Process for the preparation of 17-N-substituted-carbamoyl-4-azaandrost-1-en-3-ones, WO 2008101308 (2008).
170. Dandala, R.; Rao, Divvela V.; Prabahar, K. J.; Rao, G. V.; Sivakumaran, M. Process for the preparation of pure finasteride, US 20060084671 (2006).
171. Gormley, G. J. Finasteride: a clinical review. *Biomed. Pharmacother.* **1995,** *49* (7/8), 319–324.
172. De Nunzio, C.; Miano, R.; Trucchi, A.; Agro, E. F.; Tubaro, A. Finasteride for prostatic disease: an updated and comprehensive review. *Expert Opin. Drug Metab. Toxicol.* **2008,** *4* (12), 1561–1568.
173. Finn, D. A.; Beadles-Bohling, A. S.; Beckley, E. H.; Ford, M. M.; Gililland, K. R.; Gorin-Meyer, R. E.; Wiren, K. M. A new look at the 5α-reductase inhibitor finasteride. *CNS Drug Rev.* **2006,** *12* (1), 53–76.
174. de Vere White, R. W. Finasteride for chemoprevention of prostate cancer: why has it not been embraced? *J. Clin. Oncol.* **2007,** *25* (21), 2999–3000.
175. Wilde, M. I.; Goa, K. L. Finasteride: an update of its use in the management of symptomatic benign prostatic hyperplasia. *Drugs* **1999,** *57* (4), 557–581.
176. McClellan, K. J.; Markham, A. Finasteride: a review of its use in male pattern hair loss. *Drugs* **1999,** *57* (1), 111–126.
177. Williams, J. M. R. α-Reductase inhibitors—the finasteride story. In *The Art of Process Chemistry;* Yasuda, N., Ed.; Wiley-VCH, 2011; pp 77–115.
178. Thompson, I. M.; Klein, E. A.; Lippman, S. M.; Coltman, C. A.; Djavan, B. Prevention of prostate cancer with finasteride: US/European perspective. *Eur. Urol.* **2003,** *44* (6), 650–655.
179. Bostwick, D. G.; Qian, J.; Civantos, F.; Roehrborn, C. G.; Montironi, R. Does finasteride alter the pathology of the prostate and cancer grading? *Clin. Prostate Cancer* **2004,** *2* (4), 228–235.
180. Bartsch, G.; Rittmaster, R. S.; Klocker, H. Dihydrotestosterone and the concept of 5α-reductase inhibition in human benign prostatic hyperplasia. *Eur. Urol.* **2000,** *37* (4), 367–380.
181. Batchelor, K. W.; Frye, S. V.; Dorsey, G. F., Jr.; Mook, R. A., Jr. Preparation and formulation of an androstenone derivative for treatment of androgen-related diseases, US 5565467 (1996).
182. Batchelor, K. W.; Frye, S. V. Androstenone derivative, WO 9507927 (1995).
183. Satyanarayana, K.; Srinivas, K.; Himabindu, V.; Reddy, G. M. A scalable synthesis of dutasteride: a selective 5α-reductase inhibitor. *Org. Process Res. Dev.* **2007,** *11* (5), 842–845.
184. Mulla, S.; Mukhopadhyay, R. N.; Kulkarni, S. R.; Bhure, S.; Iyer, K. Synthesis solation and characterization of isomeric impurity of dutasteride. *IOSR J. Appl. Chem.* **2013,** *3* (4), 37–44.
185. Sarma, M. S. P.; Sukumar, N.; Reddy, G. B.; Naresh, A.; Rani, A.; Islam, A.; Sivakumaran, M. An improved process for the preparation of dutasteride, IN 2009CH01634 (2013).
186. Zhang, K.-P.; Lei, X.-P. Synthesis of dutasteride. *J. Chin. Pharm. Sci.* **2007,** *16* (3), 233–235.
187. Evans, H. C.; Goa, K. L. Dutasteride. *Drugs Aging* **2003,** *20* (12), 905–916.

188. Keam, S. J.; Scott, L. J. Dutasteride: a review of its use in the management of prostate disorders. *Drugs* **2008,** *68* (4), 463–485.
189. Graul, A.; Silvestre, J.; Castaner, J. Dutasteride: steroid 5α-reductase inhibitor treatment of BPH. *Drugs Future* **1999,** *24* (3), 246–253.
190. Marihart, S.; Harik, M.; Djavan, B. Dutasteride: A Review of Current Data on a Novel Dual Inhibitor of 5α Reductase. *Rev. Urol.* **2005,** *7* (4), 203–210.
191. Frye, S. V. Discovery and clinical development of dutasteride, a potent dual 5α-reductase inhibitor. *Curr. Top. Med. Chem.* **2006,** *6* (5), 405–421.
192. Dolder, C. R. Dutasteride: a dual 5-α reductase inhibitor for the treatment of symptomatic benign prostatic hyperplasia. *Ann. Pharmacother.* **2006,** *40* (4), 658–665.
193. Djavan, B.; Milani, S.; Fong, Y. K. Dutasteride: a novel dual inhibitor of 5α-reductase for benign prostatic hyperplasia. *Expert Opin. Pharmacother.* **2005,** *6* (2), 311–317.
194. Pohlman, G. D.; Pohlman, E. A.; David, C. E. Dutasteride: a review of its use in the management of prostate disorders. *Clin. Med. Insights: Ther.* **2011,** *3,* 171–177.
195. Wu, C.; Kapoor, A. Dutasteride for the treatment of benign prostatic hyperplasia. *Expert Opin. Pharmacother.* **2013,** *14* (10), 1399–1408.
196. Brown, C. T.; Nuttall, M. C. Dutasteride: a new 5α-reductase inhibitor for men with lower urinary tract symptoms secondary to benign prostatic hyperplasia. *Int. J. Clin. Pract.* **2003,** *57* (8), 705–709.
197. Andriole, G. L.; Kirby, R. Safety and tolerability of the dual 5α-reductase inhibitor dutasteride in the treatment of benign prostatic hyperplasia. *Eur. Urol.* **2003,** *44* (1), 82–88.
198. Pohlman, G. D.; Pohlman, E. A.; David, C. E. Dutasteride: a review of its use in the management of prostate disorders. *Clin. Med. Insights: Ther.* **2011,** *3,* 171–177.
199. Kuhn, C. M. Anabolic steroids. *Recent Prog. Horm. Res.* **2002,** *57,* 411–434.
200. Kicman, A. T. Pharmacology of anabolic steroids. *Br. J. Pharmacol.* **2008,** *154* (3), 502–521.
201. Handelsman, D. J.; Heather, A. Androgen abuse in sports. *Asian J. Androl.* **2008,** *10* (3), 403–415.
202. Handelsman, D. J. Commentary: androgens and "anabolic steroids": the one-headed janus. *Endocrinology* **2011,** *152* (5), 1752–1754.
203. Kanayama, G.; Pope, H. G., Jr. Illicit use of androgens and other hormones: recent advances. *Curr. Opin. Endocrinol., Diabetes Obes.* **2012,** *19* (3), 211–219.

Chapter 28

Antineoplastic Agents

Cancer is a term used for group of more than 120 distinct diseases that are characterized by the uncontrolled growth of abnormal cells in the body, and which are able to invade other tissues, eventually causing death.

Cancer has been known as a disease for approximately 3500 years, and herbal management of cancer is described in ancient manuscripts. Some five centuries ago, mercury, zinc, and silver preparations were used. The use of drugs in cancer treatment has been considered since the 1940s.

Cancer types are usually classified by body system: blood, bone, breast, endocrine, genitourinary, digestive/gastrointestinal cancers, respiratory cancers, skin cancers

There are several strategies to treat cancer, including surgery, radiation therapy, immunologic treatment, and chemical-based approaches. Generally, a combination of all these methods is used, and most of the therapeutic approaches to the treatment of cancer include a chemical component.

Up to the present, thousands of synthetic and natural compounds have been evaluated as anticancer agents and thousands of books and papers have been published, among the most recent of which are listed in the references [1-16].

The most common types of cancer based on frequency of diagnosis are: lung cancer, prostate cancer, breast cancer, colorectal cancer, kidney (renal) cancer, bladder cancer, non-Hodgkin lymphoma, thyroid cancer, endometrial cancer, and skin cancer.

The main categories of cancer include: leukemia (cancer of blood-forming tissues); lymphoma and myeloma (cancers of the immune system cell); carcinoma (cancer of the skin or tissues covering internal organs); sarcoma (cancer of bones or other connective tissue); and central nervous system cancers of the brain and spinal cord tissues.

Chemotherapy is one of the three approaches of cancer treatment along with surgical treatment and radiation therapy.

Chemotherapy is the treatment of cancer with cytotoxic antineoplastic drugs, and now is one of the traditional methods. Unfortunately, antitumor drugs commonly used in clinics have a variety of defects, including poor efficacy, side effects, and other problems like taking a long time and so on, as well as having a big input for developing new tumors.

Interestingly, newer studies have found that a large number of safe drugs that have been widely applied in clinics for nontumor diseases may have an

Synthesis of Best-Seller Drugs. http://dx.doi.org/10.1016/B978-0-12-411492-0.00028-6

antitumor effect. Among them are aspirin, metformin, some dietary supplements, such as extracts from *Astragalus, Tripterygium wilfordii,* and so on.

Generally, anticancer drugs are divided into three categories: cytotoxic or chemotherapeutic, immunotherapeutic, and hormonal drugs [17,18].

A new classification based on therapeutic targets has been proposed [19]. Anticancer drugs were grouped according to their targets—cancer cells, endothelium, extracellular matrix, immune system, or host cells. In turn, drugs targeting tumor cells were divided into groups according to their targets—DNA, RNA, protein, etc.

Another classification system is based on cellular biological mechanisms [20]. Anticancer drugs are grouped as cytotoxic drug and modifiers, which could regulate the interaction of tumor, host, and drugs. The modifiers in turn, are subdivided into groups—cell biological modifiers, biological response modifiers, and biochemical modulators—which affect the host's metabolic pathway of the cytotoxic drug.

Traditional chemotherapy drugs generally act against all actively dividing cells.

A relatively new approach is to target only cancer cells, creating drugs that interfere only with targets in tumor cells blocking their growth. This approach is called "targeted cancer therapies" or "molecularly targeted drugs" and nowadays generates new classes of anticancer drugs.

The diversity of potential targets include a number of new possibilities, some of which are:

- Apoptosis inducers that cause cancer cells to undergo a process of controlled death.
- Angiogenesis inhibitors that block the growth of new blood vessels around cancer cells.
- Some immunotherapy drugs, represented mainly by monoclonal antibodies, that recognize specific molecules on the surface of cancer cells and destroy them.
- Signal transduction inhibitors that block the signaling elements that have key roles in cancer cell survival and proliferation.

28.1 CHEMOTHERAPEUTIC AGENTS

Chemotherapeutic antineoplastic agents–cytotoxic drugs are the mainstay of cancer therapy.

Cytotoxic agents are the traditional therapies that damage cancer cells by interfering with DNA or its precursor, inhibiting cellular division. These kinds of agents have the great drawback of killing healthy cells along with cancer cells.

There is no standard classification for chemotherapeutic agents. It varies among national and international agencies. Chemotherapeutic antineoplastic agents include alkylating agents, antimetabolites, microtubule damaging agents, topoisomerase inhibitors, antibiotics, and miscellaneous.

Alkylating Agents

The era of chemotherapy with major types of cytotoxic agents started in the late 1940s with the clinical introduction of the classical alkylating agents, (nitrogen mustards, nitrosoureas, alkylsulfonates, ethyleneimines, methylhydrazines, platinum complexes, triazenes) [21-30].

Many alkylating agents are used as chemotherapeutic drugs and have a long history of clinical application. These agents mechanisms of action have not been clearly established, seems to act directly on DNA, during all phases of the cell cycle, which is a key structure that determines development and reproduction of the leaving organisms, causing DNA strand breaks, leading to abnormal base pairing and eventually resulting in cell death.

Alkylating agents are the oldest class of anticancer agents. The era of alkylating agents started with the approval of mechlorethamine (**28.1.1**) in 1949 for the treatment of hematological malignancies. Even though their clinical use is far beyond the use of new targeted therapies, they still occupy a major place in specific indications and sometimes represent the unique option for the treatment of refractory diseases. One of them, temozolomide (**28.1.29**), is included in the list of Top 200 Drugs by sales for the 2010s.

The representatives of the major classes of alkylating agents are presented in Figs. 28.1. to 28.7.

Nitrogen Mustards

Mechlorethamine (**28.1.1**), chlorambucil (**28.1.2**), melphalan (**28.1.3**), bendamustine (**28.1.4**), cyclophosphamide (**28.1.5**), and ifosfamide (**28.1.6**) are bifunctional alkylating agents which form interstrand and intrastrand crosslinks in DNA strands causing cell cycle blockage and death [31,32] (Fig. 28.1.).

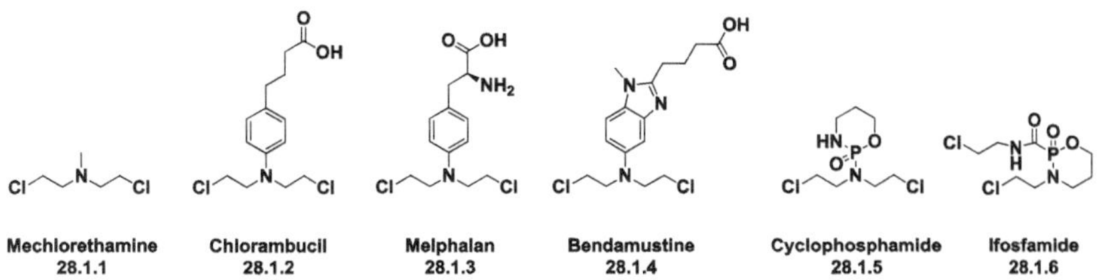

FIG. 28.1 Nitrogen mustards.

Nitrosoureas

Ethylnitrosourea (**28.1.7**), carmustine (**28.1.8**), lomustine (**28.1.9**), semustine (**28.1.10**), streptozotocin (**28.1.11**), fotemustine (**28.1.12**), and nimustine (**28.1.13**) are relatively new nitrosourea derivatives, and tauromustine (TCNU) (**28.1.14**) is a nitrosourea derivative which is still in Phase III clinical trials. Alkylating and crosslinking of DNA causes DNA fragmentation, inhibition of protein synthesis, and cell death [33,34] (Fig. 28.2.).

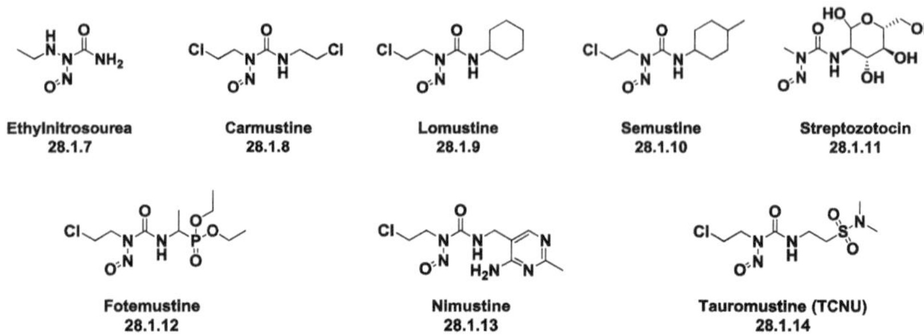

FIG. 28.2 Nitrosoureas.

Alkylsulfonates

Busulfan (**28.1.15**) was approved in 1999 and treosulfan (**28.1.16**) is under clinical development and has not yet received approval from the FDA. However, treosulfan (**28.1.16**) is currently being used in Europe, and is considered a bifunctional alkylating agent because it has a selective immunosuppressive effect on bone marrow. Treosulfan, which is a hydrophilic analogue of busulfan, was the first dimethanesulfonate registered for the treatment of ovarian cancer. Although both drugs are alkylating agents, their mechanisms of action, pharmacokinetics, and toxicity profiles are different, but their proposed mechanism of action remains the same—the intrastrand DNA crosslinks, resulting in the cell undergoing apoptosis [35] (Fig. 28.3.).

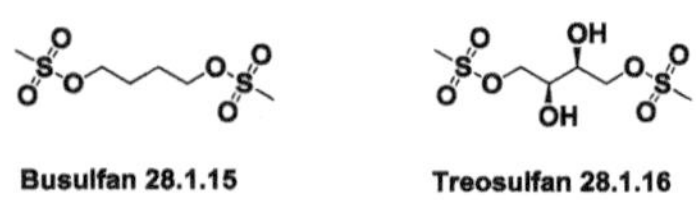

FIG. 28.3 Alkylsulfonates.

Ethyleneimines or Aziridines

Thiotepa (**28.1.17**), a tri(aziridin-1-yl)phosphine sulfide derivative, and altretamine (**28.1.18**), a 1,3,5-triazine-2,4,6-triamine derivative, are polyfunctional alkylating agents that are similar to nitrogen mustards. The precise mechanism by which they exert an anticancer effect is unknown, but they are classified as alkylating agents and it is believed that they stop tumor growth by crosslinking DNA double-helix strands. This makes the strands unable to uncoil and separate, which is necessary in DNA replication. As a result, cancer cells can no longer divide and die. Thiotepa has been in use for more than 50 years; clinical trials of altretamine began during the 1960s. A family of aziridine-containing natural compounds—mitomycins—have antitumor and antibiotic activity. Mitomycin C (**28.1.19**) is in use as a potent bis-alkylating DNA strands crosslinker. There is another, synthetic quinone derivative, diaziquone (**28.1.20**), which is still in clinical trials, that contains two ethyleneimine groups with potential antineoplastic activity [36] (Fig. 28.4.).

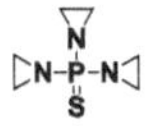
Thiotepa 28.1.17

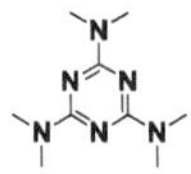
Altretamine 28.1.18

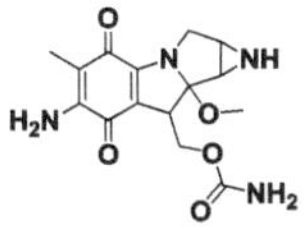
Mitomycin 28.1.19

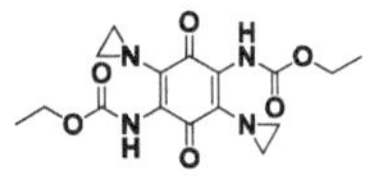
Diaziquone 28.1.20

FIG. 28.4 Ethyleneimines or aziridines.

Methylhydrazines

Methylhydrazines display a tumor inhibitory effect. Several hundred compounds revealed some 40 compounds to be effective tumor inhibitors among which was procarbazine.

Procarbazine (**28.1.21**) was developed in the 1960s as an active agent in lymphoid malignancies. During the last few decades, attempts to clarify the cellular pathways involved in the modes of action of this drug have shown evidence that hydrogen peroxide, which is formed during in vivo autooxidation of the drug to azo-procarbazine, may attack protein sulfhydryl groups contained in residual protein, which is tightly bound to DNA. The drug may also act by inhibition of transmethylation of methyl groups of methionine into t-RNA which causes the cessation of protein synthesis and, consequently, DNA and RNA synthesis. In addition, procarbazine may directly damage DNA [37, 38] (Fig. 28.5.).

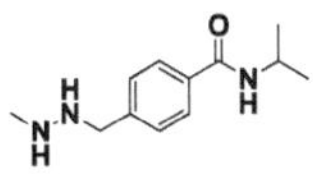
Procarbazine 28.1.21

FIG. 28.5 Methylhydrazines.

Platinum Complexes

Cisplatin (**28.1.22**), carboplatin (**28.1.23**), nedaplatin (**28.1.24**), and oxaliplatin (**28.1.25**) are essential anticancer agents with proven effects against a variety of tumors.

Since the discovery of cisplatin in 1965, cisplatin and its derivatives have appeared as the most important of the chemotherapeutic agents. These agents likely share common mechanisms of action, which consists of inducing DNA adducts, which are responsible for a cellular stress generated by inhibition of DNA synthesis, and the suppressing of RNA transcription, which results in the elimination of the proliferating cells by apoptosis. There are wide gaps in fully understanding the process that translates cisplatin-induced DNA damage into its characteristic drug-mediated cellular effects that are beneficial for the process of apoptosis. Satraplatin (**28.1.26**) and triplatin tetranitrate (**28.1.27**) are drugs under investigation [39] (Fig. 28.6.).

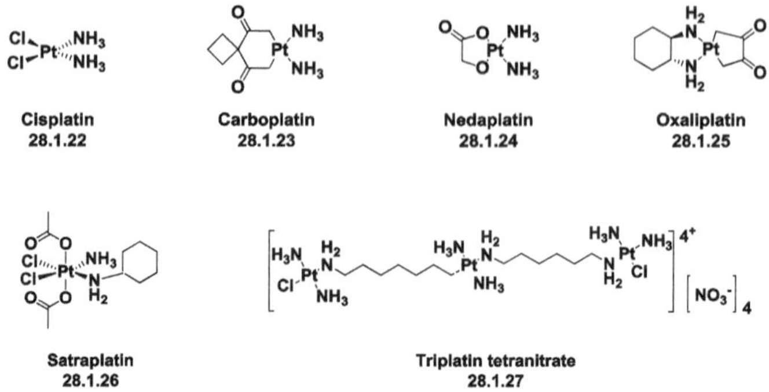

FIG. 28.6 Platinum complexes.

Triazenes

As a potential anticancer class of compounds, triazenes were widely investigated in the 1970s for their potential to treat cancer diseases. However, because of their considerable toxicity and number of side effects, they never became widely used anticancer drugs. One exception, until 2005, was dacarbazine (**28.1.28**). Temozolomide (**28.1.29**) was approved by FDA in 2005. It is an imidazotetrazine, a cyclic analogue of dacarbazine [40,41] (Fig. 28.7.).

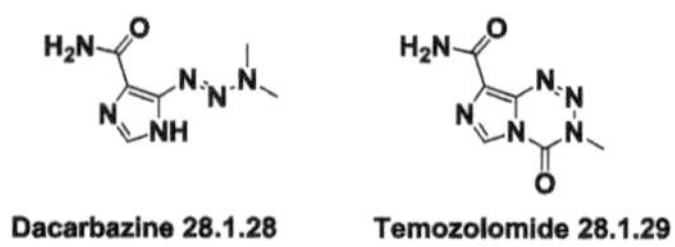

FIG. 28.7 Triazenes.

Small molecules bearing NNN linkages in either cyclic (1,2,3-triazines) or acyclic (triazenes) arrangements possess versatile chemical reactivity, which was used to achieve selective antitumor effect.

Dacarbazine (DTIC) (**28.1.28**) is believed to be demethylated in liver to methyltriazenoimidazole carboxamide (MTIC), which, in turn, converts to 5-amino-1H-imidazole-4-carboxamide (IC), yielding the powerful methylating agent, diazomethane, that attacks nucleophilic groups in DNA (Fig. 28.8.).

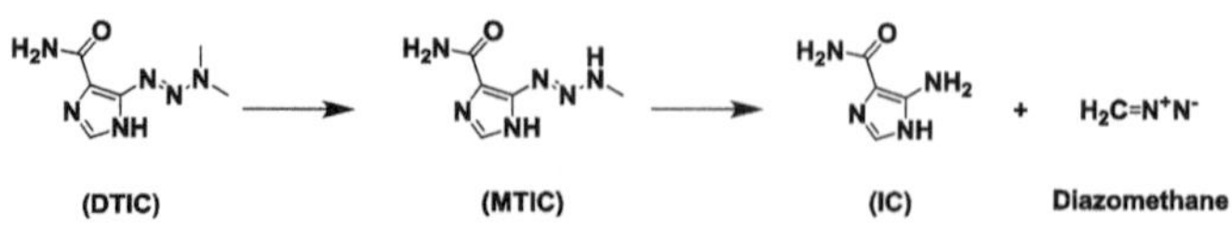

FIG. 28.8 Hypothetical mechanism of action of dacarbazine as a methylating agent.

Temozolomide–Temodar

Temozolomide (**28.1.29**) is included in the list of Top 200 Drugs by sales for the 2010s.

The first synthesis of temozolomide (**28.1.29**), was accomplished by the reaction of 5-diazoimidazole-4-carboxamide (**28.1.31**) obtained by diazotization of 5-amino-4-carboxamide (**28.1.30**) with methyl isocyanate in dichloromethane for 20 days, which probably proceeds via an initial [3+2] cycloaddition to form unstable spirobicycle (version A), which spontaneously rearrange to desired product by a [1,5] sigmatropic shift. This reaction can be understood also as a [7+2] cycloaddition of the diazo-pyrazole to the electron-deficient hetero-double bond of the isocyanate (version B) finally producing the desired imidazotetrazinones [41,42].

The alternative synthesis of temozolomide does not employ volatile methyl isocyanate and propose interaction of ethylisocyanatoacetate and 5-diazoimidazole-4-carboxamide (**28.1.31**), which produced the ester (**28.1.32**) in high yield. Hydrolysis of the ester to the acid (**28.1.33**) was easily accomplished in hydrochloric acid. All efforts to decarboxylate the acid (**28.1.33**) directly to temozolomide failed. For this purpose a radical reaction known as the Barton decarboxylation discovered by Nobel laureate Sir Derek Barton was implemented. The acid (**28.1.33**) was first converted to the reactive ester (**28.1.34**) using isobutylchloroformate/methylmorpholine in dry DMF and then to thiohydroxamate ester (**28.1.35**), which is commonly referred to as a Barton ester, with 2-mercaptopyridine N-oxide in triethylamine.

The product was then heated in the presence of a radical initiator 2,2′-azobisisobutyronitrile (ALBN) and tributyltin hydride as hydrogen donor in DMF to complete the reductive decarboxylation to produce the desired compound temozolomide (**28.1.29**) [43] (Scheme 28.1.).

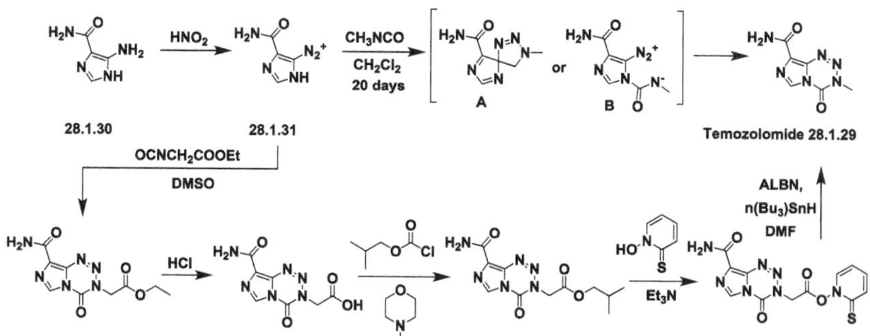

SCHEME 28.1 **Synthesis of temozolomide.**

Three modified alternative pathways for the synthesis of the antitumor drug temozolomide were proposed using intermediate imidazolecarboxamides such as (**28.1.36**), (**28.1.37**), and (**28.1.38**) [44,45] (Fig. 28.9.).

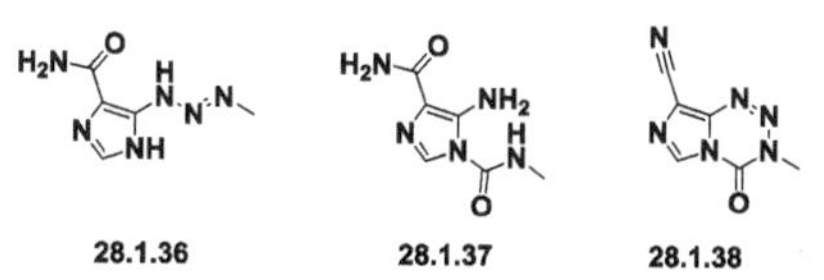

FIG. 28.9 **Alternative starting compounds for the synthesis of temozolomide.**

Temozolomide is a tolerable and effective oral anticancer agent, that inhibits DNA replication by methylating nucleotide bases. It is approved for the treatment of glioblastoma in combination with radiotherapy.

Malignant gliomas (glioblastoma multiform and anaplastic astrocytoma) occur more frequently than other types of primary central nervous system tumors and are the most common primary intracerebral tumors.

This novel oral alkylating agent has demonstrated promising activity not only in brain tumors, but in a variety of solid tumors, including malignant melanoma [46-60].

Alkylating agents have been used for the treatment of cancer for more than 6 decades, yet their repertoire continues to grow. These agents mechanisms which have not been clearly established seems to act directly on DNA, causing DNA strand breaks, leading to abnormal base pairing, inhibition of cell division and eventually resulting in cell death.

Antimetabolites

Antimetabolites have been in use for the treatment of cancer for approximately 60 years, since the discovery [61] that aminopterin (**28.1.39**) could cause remission of leukemia.

Antimetabolites are a class of anticancer drugs defined as compounds, structurally similar to natural purine or pyrimidine base, nucleoside or nucleotides, molecules needed to carry out primary metabolic reactions that by virtue of their similarity act as analogues of a normal metabolites, interfere with the normal metabolic processes within cells and, thus, preventing the synthesis of DNA, RNA, and cell division.

Their first mechanism of action is to induce depletion in nucleotide, inducing, in turn, an inhibition of DNA replication. Another mechanism of action is that some of them are inserted fraudulently into nucleic acids, inducing structural abnormalities that lead to cell death.

Considering that folic acid derivatives play an important role in RNA and DNA synthesis, folate analogues are also considered as antimetabolites. Folate vitamin B is important for cells and tissues that rapidly divide, and compounds that interfere with folate metabolism are used to treat cancer.

Antimetabolites are the most widely used and most efficacious group of anticancer drugs.

Although some were designed in the 1950s, these drugs are still used today in treatment of leukemia, breast cancer, and many other cancers.

Five types of antimetabolites are represented in this group. They include folate antagonists (methotrexate, pemetrexed, and pralatrexate); pyrimidine analogues (fluoropyrimidines-5-fluorouracil, and various nucleoside derivatives such as cytarabine, floxuridine, gemcitabine, and capecitabine, and drugs that formally belonged to the same series of 1,3,5-triazines, which were considered as 5-azapyrimidines, namely 5-azacytidine and decitabine); purine analogues (mercaptopurine, thioguanine, fludarabine, deoxycoformycin, clofarabine, and cladribine) and three new compounds that are currently being evaluated clinically (immucillin-H, nelarabine, and 8-chloroadenosine), thiopurines (6-mercaptopurine, 6-thioguanine, and azathioprine); sugar-modified analogues of purines (cytarabine, fludarabine) and ribonucleotide reductase inhibitors (hydroxyureas) [62-72].

Antimetabolites widely used in the United States for treatment of cancer diseases are listed here in order of their approval date: methotrexate (1953), 6-mercaptopurine (1953), 5-fluorouracil (1962), 6-thioguanine (1966), hydroxyurea (1967), cytarabine (1969), floxuridine (1970), leucovorin (1991), pentostatin (1991), fludarabine (1991), cladribine (1992), gemcitabine (1996), capecitabine (1998), pemetrexed (2004), 5-azacytidine (2004), clofarabine (2004), nelarabine (2005), decitabine (2006), and pralatrexate (2009).

Folate Antagonists

Folic acid, or vitamin B, is an essential and necessary element for the synthesis of nucleotides and other biomolecules after reduction by dihydrofolate reductase.

It was empirically observed in 1947 that folic acid worsened leukemia, and that a diet deficient in folic acid could, conversely, produce improvement in leukemia patients. This observation led to a search for and creation of dihydrofolate reductase inhibitors. Found compounds, particularly methotrexate (**28.1.40**), still occupy a very significant place in cancer chemotherapy.

The metabolic processes of the cell are very complex and involve many reactions and enzymes.

One of them in humans is conversion of folic acid to dihydrofolic acid, tetrahydrofolic acid, and other derivatives, which have various biological meanings, one of which is considered a very important metabolic process—synthesis of building blocks of nucleic acids, purines, and pyrimidines nucleotide precursors, including thymidylate and several amino acids. One of the most important players responsible for these transformations is an enzyme–dihydrofolate reductase, which is responsible for the conversion of dihydrofolic acid (vitamin B_9) to tetrahydrofolic acid (Fig. 28.10.).

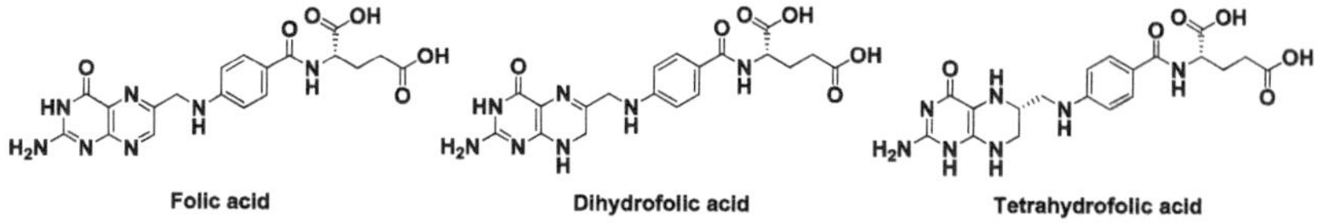

FIG. 28.10 Folic, dihydrofolic, and tetrahydrofolic acids.

The inhibition of dihydrofolate reductase results in depletion of intracellular pools of reduced folates. The lack of reduced folates impairs synthesis of purine nucleotides, thymidylate, and certain amino acids, which can lead to cell death.

First dihydrofolate reductase inhibitor, aminopterin (**28.1.39**), was discovered during development of folic acid analogues. Later, methotrexate (**28.1.40**), the first clinically accepted dihydrofolate reductase inhibitor was synthesized (Fig. 28.11.).

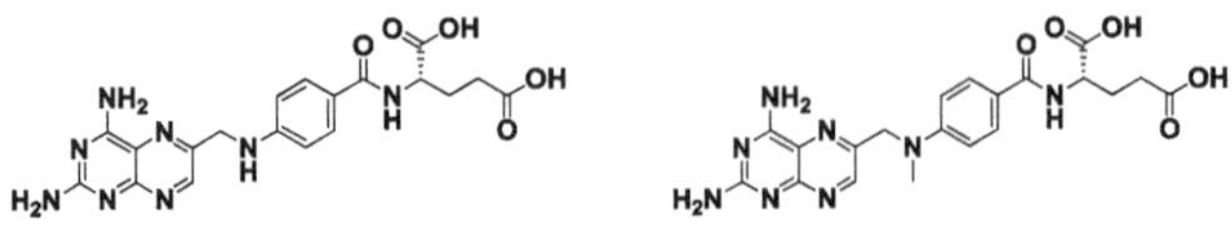

Aminopterin 28.1.39

Methotrexate 28.1.40

FIG. 28.11 **Structures of aminopterin and methotrexate.**

Methotrexate, the most commonly used and the most widely studied dihydrofolate reductase inhibitor is an effective therapeutics agent available to treat many solid tumors and autoimmune diseases such as rheumatoid arthritis.

Other approved folate antagonists are pemetrexed (**28.1.41**) and pralatrexate (**28.1.42**). Pemetrexed (**28.1.41**) is a multitargeted antifolate and works by simultaneous inhibition of all four enzymes used in purine and pyrimidine synthesis: thymidylate synthase, dihydrofolate reductase, aminoimidazole carboxamide ribonucleotide formyl transferase, and glycinamide ribonucleotide formyltransferase. Pemetrexed is effective in the treatment of mesothelioma and non–small cell lung cancer. Pralatrexate (**28.1.42**) is approved for the treatment of patients with relapsed or refractory peripheral T-cell lymphoma [73-79] (Fig. 28.12.).

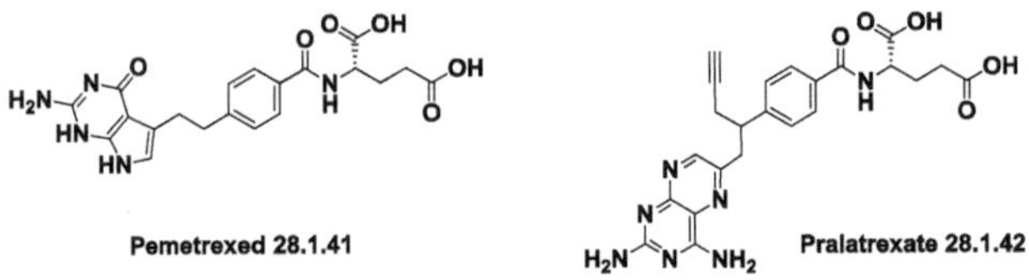

Pemetrexed 28.1.41

Pralatrexate 28.1.42

FIG. 28.12 **Structure of pemetrexed and pralatrexate.**

Some new folate antagonists, such as edatrexate (**28.1.43**), lometrexol (**28.1.44**), LY-309887 (**28.1.45**), trimetrexate (**28.1.46**), and BW-301U (**28.1.47**), are in clinical trials (Fig. 28.13.).

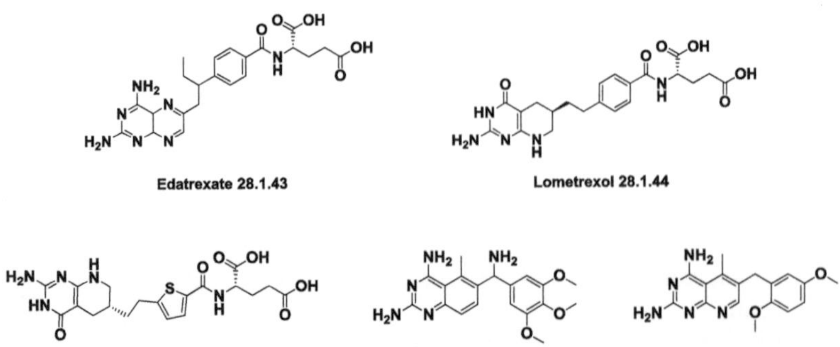

Edatrexate 28.1.43

Lometrexol 28.1.44

LY-309887 28.1.45

Trimetrexate 28.1.46

BW-301U 28.1.47

FIG. 28.13 **Structure of new folate antagonists.**

Another enzyme, one of three used in cell purine and pyrimidine synthesis, is thymidylate synthase, the only enzyme in folate metabolism in which the 5,10-methylenetetrahydrofolate is oxidized during one-carbon transfer, which plays a crucial role in DNA biosynthesis. Thymidylate synthase inhibitors also have become an important target for cancer treatment. Compounds structurally closely related to dihydrofolate reductase inhibitors, such as raltitrexed (**28.1.48**), plevitrexed (**28.1.49**), GW 1843 (**28.1.50**), and nolatrexed (**28.1.51**), are in clinical trials (Fig. 28.14.).

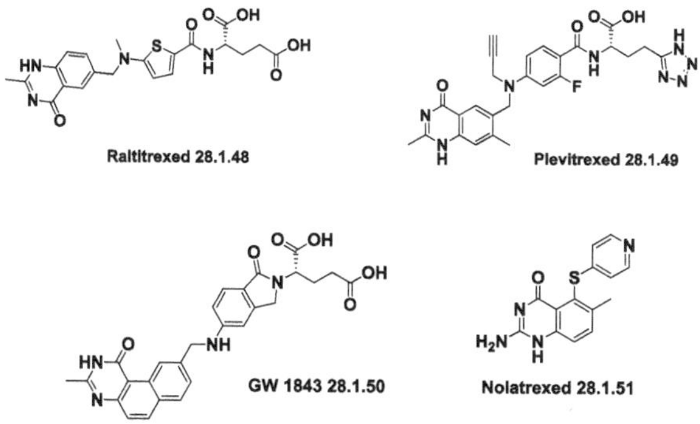

Raltitrexed 28.1.48

Plevitrexed 28.1.49

GW 1843 28.1.50

Nolatrexed 28.1.51

FIG. 28.14 **Thymidylate synthase inhibitors in clinical trials.**

None of the folate antagonists is included in the list of Top 200 Drugs by sales for the 2010s.

Purines, Pyrimidines, and Their Nucleoside Analogues

Pyrimidines

Pyrimidines and their nucleoside analogues are essential drugs for cancer therapy. These compounds act as antimetabolites interfering DNA replication in cells via converting to analogues of cellular nucleotides by the metabolic regular pathway thus causing DNA damage and induction of apoptosis [80-84].

5-Fluorouracil (**28.1.53**), a fluorinated analogue of uracil (**28.1.52**), which is one of the four bases coding genetic information in the polynucleotide chain of RNA, is the first and single example of an anticancer drug used to treat various types of cancer (Fig. 28.15.). In general, fluoropyrimidine therapy has been a mainstay in the treatment of cancers for nearly half of a century.

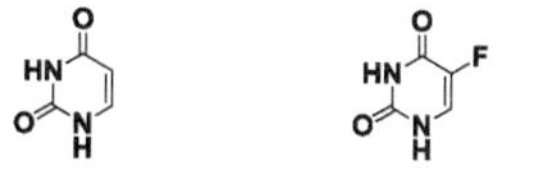

Uracil 28.1.52

5-Fluorouracil 28.1.53

FIG. 28.15 **Structures of uracil and 5-fluorouracil.**

Other pyrimidine anticancer drugs—cytarabine (**28.1.55**), floxuridine (**28.1.54**), gemcitabine (**28.1.56**), capecitabine (**28.1.57**), and drugs that formally belonged to the same 1,3,5-triazine series, which are considered as 5-azapyrimidines, namely, 5-azacytidine (**28.1.58**) and decitabine (**28.1.59**), are various nucleoside derivatives, including deoxynucleoside analogues, which are compounds with one hydroxyl group removed from the 2′ position for example (**28.1.54**) and (**28.1.59**) (Fig. 28.16.).

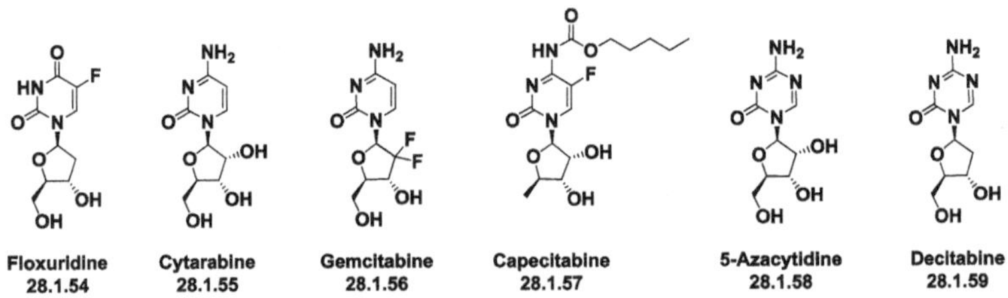

FIG. 28.16 **Pyrimidine anticancer drugs.**

Some new compounds of this series, such as troxacitabine (**28.1.60**), tezacitabine (**28.1.61**), thiarabine (**28.1.62**), and sapacitabine (**28.1.63**), are in clinical trials (Fig. 28.17.).

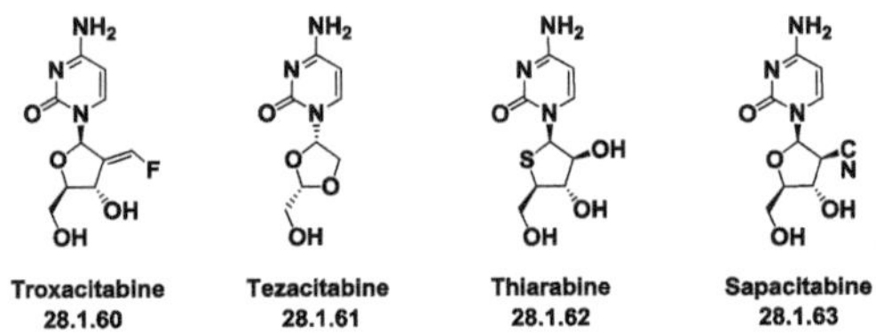

FIG. 28.17 **New pyrimidine anticancer drugs**

Capecitabine–Xeloda

Capecitabine (**28.1.57**), which is included in the list of Top 200 Drugs by sales for the 2010s, is an orally administered chemotherapeutic agent, a prodrug of 5-fluorouracil, which is converted to 5-fluorouracil in the cancer cell by enzymatic degradation. It generates 5-fluorouracil preferentially within tumors through exploitation of the high intratumoral activity of thymidine phosphorylase. It is believed that capecitabine is the only cytotoxic agent without cumulative toxicity. It exhibits synergistic activity when used in combination with a wide range of other cytotoxic agents.

Information on capecitabine concerning clinical pharmacology and efficacy, mechanism of action, pharmacokinetic and pharmacodynamic properties, for different types of cancer, adverse-effect profile, drug interactions, dosage and administration, and future directions of ongoing research, are presented in multiple reviews [85-106].

Several schemes for the synthesis of capecitabine, which differ only in details, have been proposed [107-115].

One scheme is a three-step synthesis started from the 5-fluorocytosine (**28.1.64**). For that purpose it was sialylated with hexamethyldisilazane in toluene to produce the intermediate (**28.1.65**), which without separation was ribosylated with 1,2,3-tri-O-acetyl-5-deoxyribose using anhydrous stannic chloride as a Lewis acid in dichloromethane. After hydrolysis of the trimethylsilyl- protecting groups with sodium bicarbonate water solution, 5′-deoxy-2′,3′-di-O-acetyl-5-fluorocytidne (**28.1.66**) was obtained. It was acylated with n-pentyl chloroformate in CH_2Cl_2 using pyridine as a base. The acetyl groups obtained crude 2′,3′-di-O-acetyl-5′-deoxy-5-fluoro-N-4-(pentyloxycarbonyl)cytidine (**28.1.67**) was removed in MeOH/NaOH/H_2O mixture on cooling. Then, concentrated HCl was added to the reaction, adjusting the pH of the mixture from 4 to 5 to produce the desired capecitabine (**28.1.57**) [115] (Scheme 28.2.).

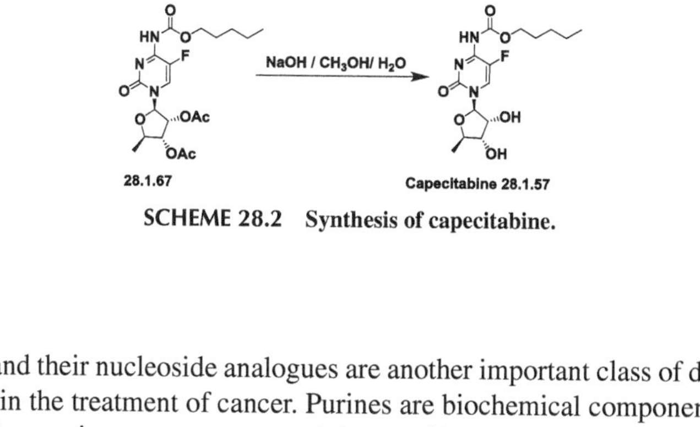

SCHEME 28.2 **Synthesis of capecitabine.**

Purines

Purines and their nucleoside analogues are another important class of drugs that are used in the treatment of cancer. Purines are biochemical components of the tumor microenvironment, potent modulators of immune cell responses, and key players in host–tumor interaction, which directly affect tumor cell growth. Most of purine drugs currently used in oncology kill cancer cells or via inhibiting the synthesis of DNA or interfering with its function in one manner or another.

The first compound of purine series, 6-mercaptopurine (**28.1.68**), was approved by the FDA in 1953 for the treatment of cancer, particularly for childhood leukemia.

Another agent, which was first proposed for treatment of acute lymphoblastic leukemia in children, thioguanine (**28.1.69**), is now widely used for treating some forms of inflammatory bowel diseases, particularly ulcerative colitis. Thioguanine is a drug on the list of Essential Drugs established by the World Health Organization (WHO) (Fig. 28.18.).

FIG. 28.18 **Purines for the treatment of cancer.**

The purine nucleoside analogues—fludarabine (**28.1.70**), nelarabine (**28.1.71**), clofarabine (**28.1.72**), cladribine (**28.1.73**), and pentostatin (**28.1.74**)—have been approved by the FDA as new drugs for different types of cancer. 8-Chloroadenosine (**28.1.75**) and forodesine (**28.1.76**) are still under development [116-118] (Fig. 28.19.).

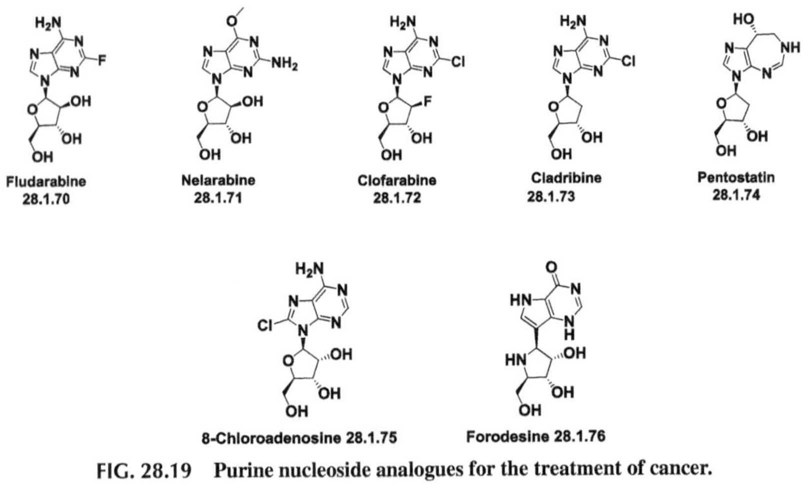

FIG. 28.19 **Purine nucleoside analogues for the treatment of cancer.**

It is evident that compounds in this class are structurally similar and share many mechanistic details, but they differ in the metabolism of these agents which have a profound impact on their antitumor activity. "...[O]ne of the remarkable features of purine and pyrimidine nucleoside analogues that remains unexplained is how drugs with such similar structural features, that share metabolic pathways, and elements of their mechanism of action show such diversity in their clinical activities" [119].

Ribonucleotide Reductase Inhibitors

Ribonucleotide reductase is an enzyme responsible for the reduction of ribonucleotides to deoxyribonucleotides, which are building blocks for DNA replication and repair, and plays a critical for the generation of the cytosine, adenine, and guanine deoxyribonucleotide 5′-triphosphate [120-125].

The role of ribonucleotide reductase in DNA synthesis and cell growth has made it an important target for anticancer therapy. Its inhibition precludes DNA transcription and repair, from which results cell apoptosis.

Hydroxyurea (**28.1.77**) is the only ribonucleotide reductase inhibitor in clinical usage as a first-line treatment of myeloproliferative disorders, such as polycythemia vera, essential thrombocythemia, and primary myelofibrosis, apart from having a vital role in combination therapy for management of malignant melanoma, head and neck cancers and brain tumors [126,127] (Fig. 28.20.).

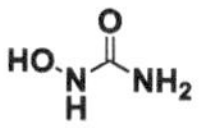

Hydroxyurea 28.1.77

FIG. 28.20 Structure of hydroxyurea.

Some of hydroxamic acids such as didox (**28.1.78**) and trimidox (**28.1.79**) also reduce the activity of ribonucleotide reductase [127]. Thiosemicarbazones such as Triapine (**28.1.80**) [128] attributed to the inhibition of ribonucleotide reductase and represent potential anticancer drugs [128-130] (Fig. 28.21.).

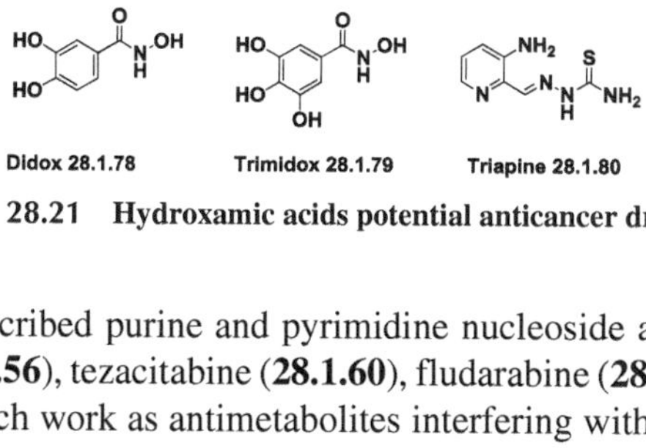

Didox 28.1.78 Trimidox 28.1.79 Triapine 28.1.80

FIG. 28.21 Hydroxamic acids potential anticancer drugs.

The above-described purine and pyrimidine nucleoside analogues, such as gemcitabine (**28.1.56**), tezacitabine (**28.1.60**), fludarabine (**28.1.70**) and cladribine (**28.1.73**), which work as antimetabolites interfering with DNA replication, display also ribonucleotide reductase inhibitors properties.

Natural Products and Their Synthetic Modifications

Natural products isolated from plants, marine flora, and microorganisms have long been and are a rich and fertile source of drugs in medicine, and particularly as potential drugs for treatment of cancer [131-134].

Many anticancer drugs are obtained from natural sources. These include the well-known vincristine, paclitaxel, podophyllotoxin, and paclitaxel. Nature has provided many other effective anticancer drugs, such as the antibiotics doxorubicin, dactinomycin, and bleomycin. Structural modifications have led to more potent compounds than the prototypes. A number of promising agents, such as flavopiridol, roscovitine, combretastatin A-4, betulinic acid, and silvestrol are in development.

"The influence of natural products upon anticancer drug discovery and design cannot be overestimated. Approx. 60% of all drugs now in clinical trials for the multiplicity of cancers are either natural products, compounds derived from natural products, contain pharmacophores derived from active natural products or are 'old drugs in new clothes,' where (modified) natural products are attached to targeting systems" [131].

"Currently, more than 30 compounds of natural origin are in different phases of clinical study for the treatment of different types of cancer." Of importance are ixabepilone (epothilone B from *Sorangium cellulosum*), romidepsin (depsipeptide from *Chromobacterium violaceum*), and the dibenzodiazepine ECO-4601 (from *Micromonospora* species.) [132].

Microtubule Damaging Agents

Cytoskeleton is a network of fibers throughout the cell, constructed from globular proteins called tubulins, intermediate filaments, and microfilaments, that helps the cell maintain its shape and assist in the movement of vesicles, granules, organelles like mitochondria, and chromosomes via special attachment proteins playing a huge role in cell's vital functions.

Tubulins are structural subunits of microtubules that are involved in nucleic and cell division, organization of intracellular structure, cell–cell contacts, and other functions made of microtubules, which are a valuable target for cancer chemotherapy. Inhibition of microtubule functioning induces persistent modification of biological processes and signaling pathways, ultimately leading to apoptosis, the process of programmed cell death.

A large number of chemically diverse compounds, most of which are derived from natural products, are able to interfere with tubulin or microtubules and inhibit proliferation and are widely used in anticancer medicine [135,136]. Among them are the following compounds.

Vinca Alkaloids

Vinca alkaloids—vinblastine (**28.1.81**), vincristine (**28.1.82**), vindesine (**28.1.83**), and vinorelbine (**28.1.84**)—are isolated from the plant *Catharanthus rosea.* A novel drug, which is a fluorinated derivative of vinorelbine (**28.1.84**), is vinflunine (**28.1.85**) (Fig. 28.22.). Interfering with the formation and growth of microtubules vinca alkaloids and their derivatives prevent cancerous cells from undergoing mitosis. The mechanism of action of these cell-cycle–dependent agents is the inhibition of tubulin polymerization into microtubules.

Vinca alkaloids have similar structures but show differences both in their activity spectrum and in their toxicity. Vinca alkaloid drug also affects the division of normal cells, thus producing several undesirable side effects [137-140].

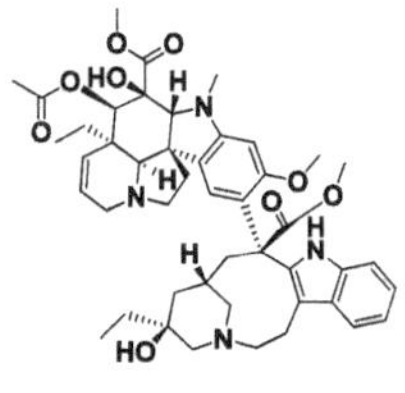

Vinblastine 28.1.81

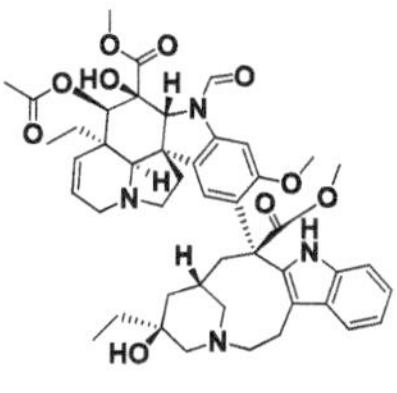

Vincristine 28.1.82

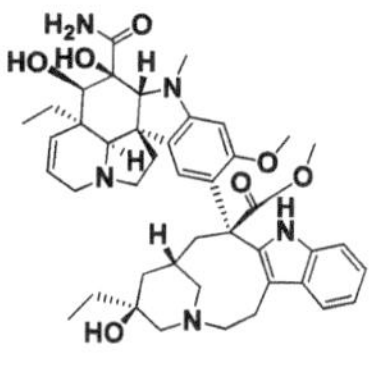

Vindesine 28.1.83

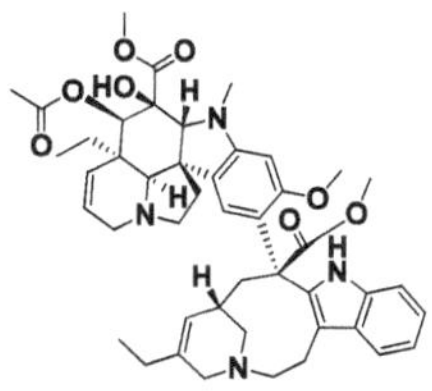

Vinorelbine 28.1.84

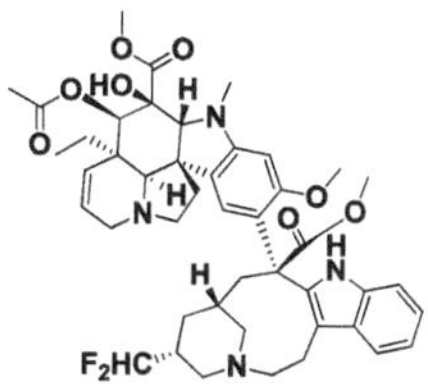

Vinflunine 28.1.85

FIG. 28.22 **Vinca alkaloids.**

Taxanes

Taxanes are novel microtubule-stabilizing agents and have shown efficacy in cancer since the 1990s [141-143]. They are diterpenoids separated from *Taxus brevifolia* species. Representatives of taxanes available in the pharmaceutical market are paclitaxel (**28.1.86**) and its semisynthetic versions docetaxel (**28.1.87**) and cabazitaxel (**28.1.88**) (Fig. 28.23.).

Taxanes are core therapeutic components for several advanced malignancies. Acting as microtubulin-stabilizing agents, they disrupt the normal function of microtubules, whereas vinca alkaloids destabilize this process. Vinca alkaloids affect the rates of tubulin polymerization, whereas the taxanes induce the inhibition of microtubule depolymerization.

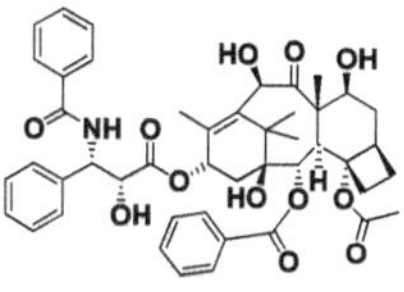

Paclitaxel 28.1.86

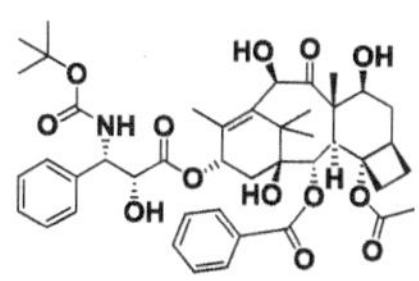

Docetaxel 28.1.87

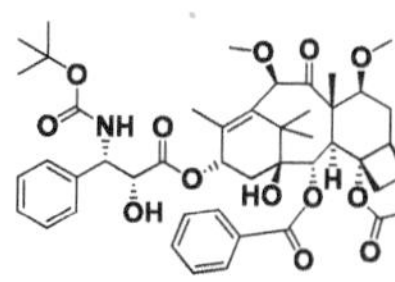

Cabazitaxel 28.1.88

FIG. 28.23 **Structure of taxanes.**

Epothilones

Epothilones are a novel class of natural cytotoxic compounds, inhibitors of microtubule function that were originally isolated from the myxobacterium *S. cellulosum* in the early 1990s. They represent a new class of cancer drugs with excellent Taxol-like antitumor activity, but more effective than Taxol, and are considered as "post-taxanes" with great market potential [144-156].

Epothilones, a new class of cancer drugs, are also inhibitors of microtubule function and are represented by ixabepilone (**28.1.89**) and eribulin (**28.1.90**). Ixabepilone is a semisynthetic analogue of epothilone B that binds to the same site on β-tubulin as paclitaxel and may be a more potent polymerizer of tubulin. It is in Phase II evaluation.

Eribulin (**28.1.90**) is another epothilone representative which was recently approved in patients progressing after being treated with anthracyclines and taxanes (Fig. 28.24.)

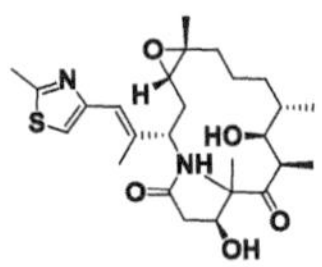

Ixabepilone 28.1.89

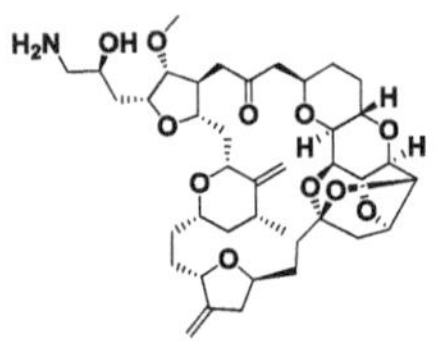

Eribulin 28.1.90

FIG. 28.24 Structure of epothilones.

Topoisomerase Inhibitors

DNA strand separation for transcription and replication, the flawless segregation of two identical copies in cells, is believed to be regulated with a family of enzymes—DNA topoisomerases.

DNA topoisomerases are involved in the topological problems associated with DNA replication, transcription, recombination, and supercoiling, which is why it became a target of anticancer and antibacterial drugs.

There are a number of different types of topoisomerases, each specializing in a different aspect of DNA manipulation [157-159].

Topoisomerase I Inhibitors

Topoisomerase I is required for DNA replication, transcription and relaxation during these critical cellular functions and other processes. Topoisomerase I inhibitors are a relatively new class of anticancer agents which interrupt DNA replication in cancer cells, resulting in their death. Most, if not all, known topoisomerase I inhibitors are derivatives of the plant extract camptothecin.

Inhibition of topoisomerase I activities leads to replication-mediated DNA damage, which leads to cell death. Thus topoisomerase I inhibitors that include camptothecins are quinoline alkaloids of *Camptotheca* and *Nothapodytes* species and their derivatives. Camptothecin (**28.1.91**) and its analogues topotecan (**28.1.92**) and irinotecan (**28.1.93**) are promising natural anticancer molecules that inhibit the DNA enzyme topoisomerase I [160] (Fig. 28.25.).

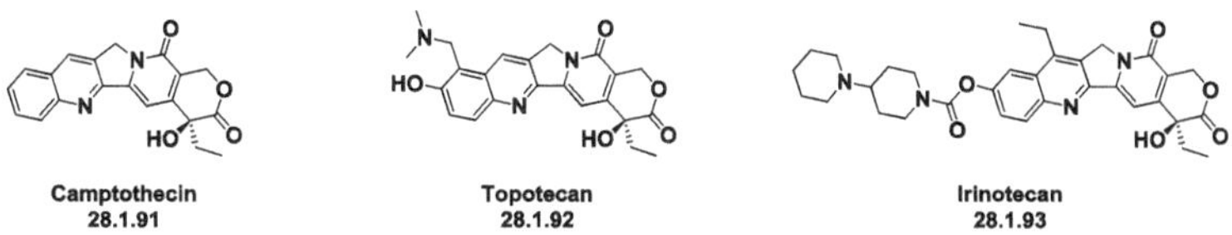

FIG. 28.25 **Topoisomerase I inhibitors.**

Topoisomerase II Inhibitors

DNA topoisomerase II enzymes are involved in DNA replication, transcription, transferring one double helix through a transient break in another DNA double helix, and regulating conformational changes in DNA topology. They act on double-stranded DNA, catalyzing its relaxation and unknotting during replication, transcription, chromosome condensation, and decondensation.

Etoposide (**28.1.95**) and teniposide (**28.1.96**) are semisynthetic derivatives of a lignan-podophyllotoxin (**28.1.94**) derived from *Podophyllum* species and which acts as anticancer drugs by inhibiting the topoisomerase II DNA enzyme (Fig. 28.26.). Podophyllotoxin-like lignans inhibit the polymerization of tubulin, arresting the cell cycle in the metaphase.

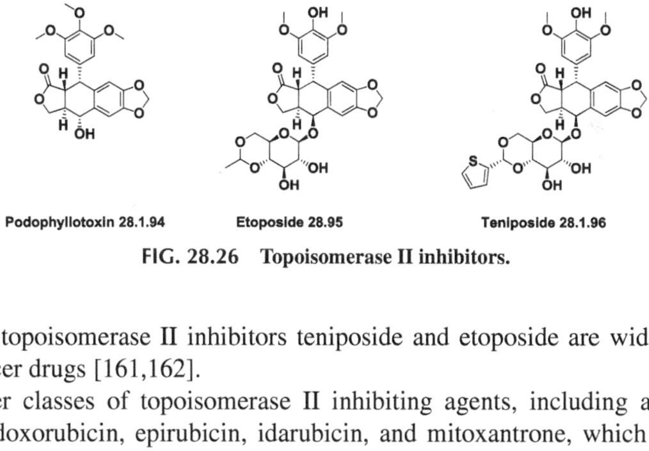

FIG. 28.26 **Topoisomerase II inhibitors.**

The topoisomerase II inhibitors teniposide and etoposide are widely used anticancer drugs [161,162].

Other classes of topoisomerase II inhibiting agents, including anthracyclines, doxorubicin, epirubicin, idarubicin, and mitoxantrone, which are discussed below, are potent inducers of double-strand breaks in DNA, and can cause arrest in the cell cycle.

Cytotoxic Antibiotics

Cytotoxic antibiotics are made from natural products. These drugs act during multiple phases of the cell cycle and are considered cell-cycle specific. There are several types of antitumor antibiotics: anthracyclines, chromomycins, and an antibiotics series called miscellaneous, which includes mitomycin and bleomycin [163-167].

Anthracycline Antibiotics

The most commonly used antineoplastic antibiotics are anthracycline antibiotics, which represent a rich family of natural products isolated from microorganisms with activity against breast cancer, leukemias, lymphomas, and sarcomas. Anthracycline antibiotics inhibit topoisomerase II, but also intercalate into DNA and form metabolites that interact with many intracellular molecules. Thus, the effects of the anthracyclines may not be based solely on topoisomerase II inhibition. Anthracycline antibiotics currently approved for use in the United States are daunorubicin (**28.1.97**), doxorubicin (**28.1.98**), idarubicin (**28.1.99**) and epirubicin (**28.1.100**) (Fig. 28.27.).

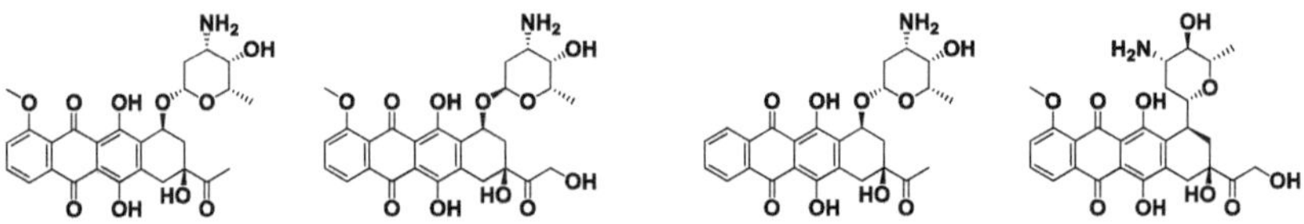

Daunorubicin 28.1.97 Doxorubicin 28.1.98 Idarubicin 28.1.99 Epirubicin 28.1.100

FIG. 28.27 **Anthracycline antibiotics.**

Closely related group of anthraquinone antibiotics are represented by mitoxantrone (**28.1.101**) used for the treatment of advanced prostate cancer and certain forms of leukemia. Another group of antibiotics is represented by acridine derivative called amsacrine (**28.1.102**), which is now a less frequently used compound in general oncology, but essentially still available for the treatment of leukemia (Fig. 28.28.).

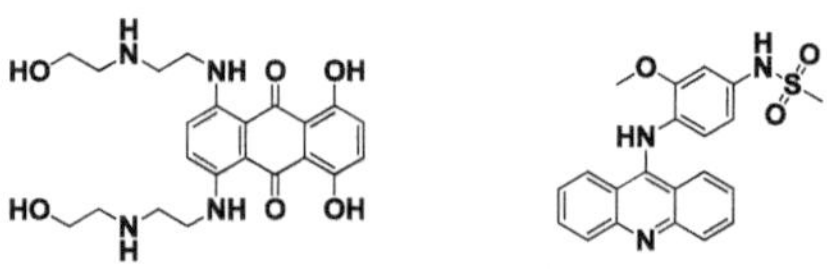

Mitoxantrone 28.1.101 Amsacrine 28.1.102

FIG. 28.28 **Anthraquinone antibiotics.**

Novel anthracyclines are continuously studied and tested in humans, including these: valrubicin (**28.1.103**), pirarubicin (**28.1.104**), sabarubicin (**28.1.105**), berubicin (**28.1.106**), and amrubicin (**28.1.107**) (Fig. 28.29.).

Chromomycin Antibiotics

One of the older anticancer drugs and most commonly used in treatment of a variety of cancers is dactinomycin (**28.1.108**). It is the most significant member of actinomycin antibiotics, which are isolated from *Streptomyces parvulus* soil bacteria. It is a compound composed of two cyclic peptides attached

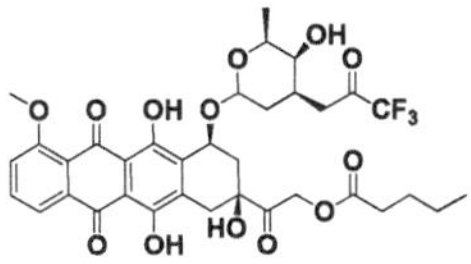
Valrubicin 28.1.103

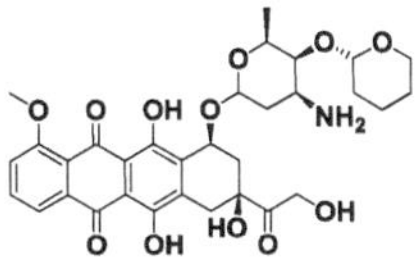
Pirarubicin 28.1.104

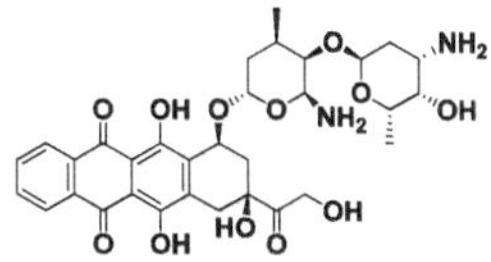
Sabarubicin 28.1.105

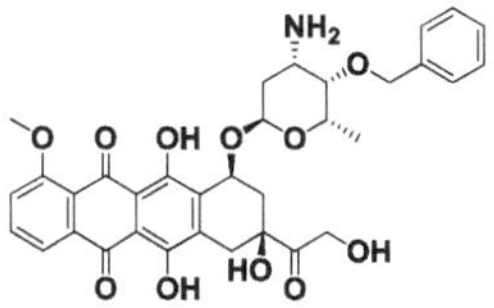
Berubicin 28.1.106

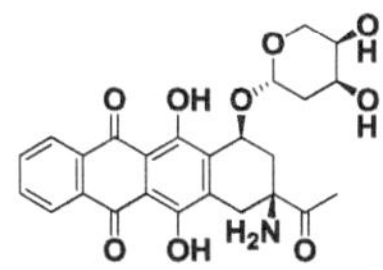
Amrubicin 28.1.107

FIG. 28.29 Novel anthracyclines under investigation.

to a phenoxazine ring system. It binds to DNA and inhibits RNA transcription. This medication is classified as an "alkylating agent." Another anticancer antibiotic–RNA synthesis inhibitor produced by *Streptomyces plicatus* is plicamycin (**28.1.109**). The manufacturer discontinued its production in 2000 (Fig. 28.30.).

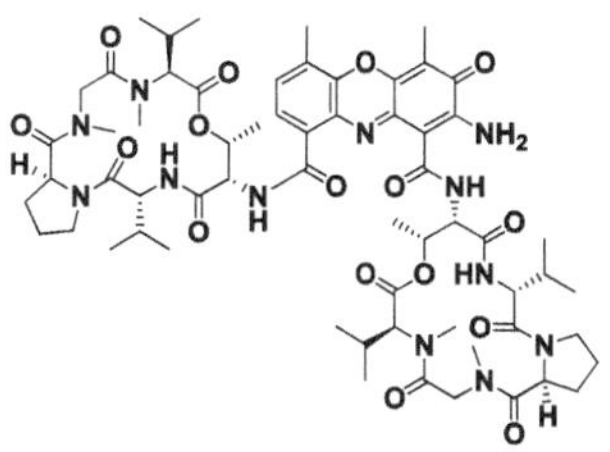
Dactinomycin 28.1.108

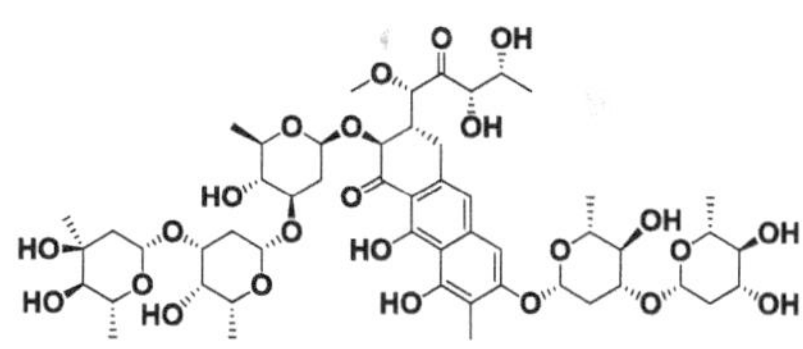
Plicamycin 28.1.109

FIG. 28.30 Chromomycin antibiotics.

Miscellaneous Antibiotics

Miscellaneous antibiotics in cancer chemotherapy include mitomycin and bleomycin (Fig. 28.31.).

Mitomycin (**28.1.110**) is an antineoplastic antibiotic against a wide variety of cancer types produced by *Streptomyces caespitosus*. It is an important antitumor compound, and being an alkylating agent, has an extraordinary ability to crosslink DNA with high efficiency. Mitomycin serve as promising scaffolds for small molecule anticancer drugs.

Bleomycin (**28.1.111**) is cytotoxic glycopeptide antibiotic isolated from a strain of *Streptomyces verticillus.* It acts by induction of DNA strand breaks. Bleomycin is used to treat several types of cancer, including testicular cancer, lymphoma, squamous cell cancer of the head, neck, and cervix.

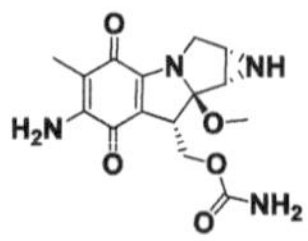

Mitomycin 28.1.110

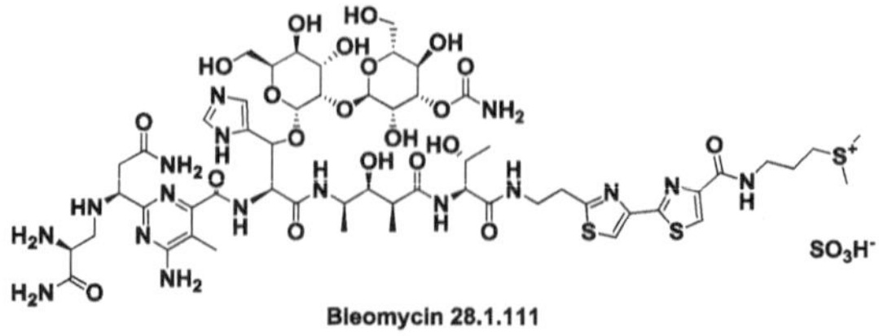

Bleomycin 28.1.111

FIG. 28.31 Miscellaneous antibiotics.

28.2 TARGETED CANCER THERAPY AGENTS

Tyrosine Kinase Inhibitors

Classical chemotherapy agents remain the mainstay of cancer treatment for tumors. Traditional chemotherapy drugs, described above act against all actively dividing cells, but the era of molecular targeted cancer therapy which differentiate cancer and normal cells has clearly arrived. Targeted cancer therapies imply creation of drugs that will be able to interfere with substances involved only in cancer cell growth.

The increasing understanding of cancer biology over the past decade has resulted in an explosive growth of potential targets for drug discovery [168-171].

Many different targeted therapies have been approved for use in cancer treatment, including signal transduction inhibitors, which interfere with inappropriate signaling that stimulates cells to divide continuously; angiogenesis inhibitors, which block the growth of new blood vessels to tumors; apoptosis inducers, which cause cancer cells to undergo a process of controlled cell death; monoclonal antibodies, which deliver toxic molecules that can cause the death of cancer cells specifically; immunotherapies, which trigger the immune system to destroy cancer cells; and hormone therapies, which stop the growth of hormone-sensitive tumors.

Cancer vaccines and gene therapy sometimes also are considered targeted therapies.

Protein kinases are enzymes that catalyze the transfer of a phosphate group from adenosine triphosphate (ATP) to amino acids in the side chain of a protein. Protein kinanses govern many complex cellular processes. Dysregulation of kinase signaling is associated with many human diseases, particularly with cancers.

Tyrosine kinases, a subgroup of protein kinases that selectively phosphorylates protein tyrosine residues, play essential roles in regulating cell proliferation and differentiation. Deregulation of tyrosine kinases can lead to cellular events associated with tumor maintenance and progression.

Tyrosine kinases are key regulators and important mediators of the signaling cascade, determining diverse biological processes like growth, differentiation, metabolism and apoptosis in response to external and internal stimuli. Recent advances implicate the role of tyrosine kinases in the pathophysiology of cancer.

Tyrosine kinases are subdivided into two classes based on their structure, function, and localization: nonreceptor tyrosine kinases (NRTKs) found in the cytoplasm, and receptor tyrosine kinases (RTKs), which are the transmembrane receptor type and transduce signals from both outside and inside the cell.

NRTKs—nine families of which have been identified— include proto-oncogene tyrosine-protein kinase (SRC), Abelson tyrosine kinase (ABL), which caused more than 90% of chronic myelogenous carcinoma, focal adhesion kinase (FAK), and Janus kinase. The NRTKs are a group of proteins that play a significant role in regulation, cell proliferation, differentiation, metabolism, migration and pathogenesis of many types of cancers and have become a target for treatment this disease.

Transmembrane RTKs are also involved in many cellular processes, such as differentiation, metabolism, and pathogenesis, and have oncogenic potential. RTKs are distributed into 20 subfamilies, including epidermal growth factor receptor (EGFR), platelet-derived growth factor receptor (PDGFR), fibroblast growth factor receptor (FGFR), and the insulin receptor (IR).

The RTKs are not only cell-surface transmembrane receptors, but are also enzymes having kinase activity. They are expressed on the surface of tumor and endothelial cells and represent targets for tyrosine kinase inhibitors, which can cause arrest of tumor cell proliferation, impact tumor angiogenesis, and induce apoptosis and tumor migration.

Several excellent recent reviews have described the functions of various tyrosine kinases [172-186].

Tyrosine kinase inhibitors are a new class of drugs with unique mechanism of action.

Tyrosine kinase inhibitors are currently one of the most important classes of cancer drugs and much experimental and clinical data confirm the idea that better results in cancer therapy can be obtained by simultaneous blocking of several biochemical pathways of tumor cells.

These compounds, which are one of the most impressive approaches of targeted cancer therapy, were introduced in the late 1990s and have revolutionized the management of cancer by significantly attenuating cancer cell survival and growth. Tyrosine kinase inhibitors currently are one of the most important classes of anticancer drugs.

It is believed that about one-third of the current effort of the pharmaceutical industry is devoted to the development of protein kinase inhibitors, especially tyrosine kinase inhibitors, and that more than 50% of current cancer drug discovery programs are focused on protein kinase inhibitors.

Two of the tyrosine kinase inhibitors—the breakthrough medicine imatinib (**28.2.1**), one of the first targeted therapy drugs ever used to treat cancer, and erlotinib (**28.2.11**) are included in the list of Top 200 Drugs by sales for the 2010s.

There are several attempts to systematize these drugs, from different points of view, but no one of the proposed classifications seems perfect.

Some authors subdivide tyrosine kinase inhibitors into two or even three categories: compounds that recognize and bind to the active conformation of a kinase, for example, sunitinib (**28.2.21**); compounds like imatinib (**28.2.1**) and sorafenib (**28.2.18**), which recognize the inactive conformation of a kinase; and compounds that covalently bind to cysteines at specific sites of the kinases, known as "covalent" inhibitors, of which vandetanib (**28.2.15**) is representative [187].

Another group of authors separates tyrosine kinase inhibitors into three broad categories: inhibitors of the EGFR tyrosine kinase family (gefitinib, erlotinib), inhibitors of the split kinase domain RTK subgroup (vatalanib, sunitinib), and inhibitors of tyrosine kinases from multiple subgroups such as imatinib (**28.2.1**) [188].

Others "grouped the RTKs into three distinct classes: (i) an EGFR/FGFR1/c-Met class constituting epidermal growth factor receptor, fibroblast growth factor receptor 1, and the hepatocyte growth factor receptor c-Met; (ii) an IGF-1R/NTRK2 class constituting insulin-like growth factor 1 receptor and neurotrophic tyrosine receptor kinase 2; and (iii) a PDGFRβ class constituting platelet-derived growth factor receptor β" [189].

Finally, a kinase inhibitor classification is proposed based on a compound's ability to inhibit selectively NRTKs or RTKs. Nevertheless a huge group of compounds are referred to as multitarget tyrosine kinase inhibitors [190].

Nonreceptor Tyrosine Kinase Inhibitors

NRTK inhibitors are divided into first- and second-generation drugs. A first-generation drug that is considered to be representative of NRTK inhibitors is imatinib (**28.2.1**), which inhibits oncogenic tyrosine kinase BCR-ABL. However, it additionally blocks other RTKs, such as KIT and PDGFRα. Dasatinib (**28.2.2**) is considered a second-generation anticancer drug whose potency against BCR-ABL is 325 times greater than that of imatinib. Nilotinib (**28.2.3**), tofacitinib (**28.2.4**), bosutinib (**28.2.5**), and ponatinib (**28.2.6**), are other representatives of second-generation drugs that inhibit several kinases, such as BCR-ABL, KIT, LCK, and ZAK. Ruxolitinib (**28.2.7**), crizotinib (**28.2.8**), lestaurtinib (**28.2.9**), and pacritinib (**28.2.10**) are experimental tyrosine kinase inhibitor drug candidates of diverse structures (Fig. 28.32.).

Receptor Tyrosine Kinase Inhibitors

Transmembrane RTKs are expressed on the surface of tumor and endothelial cells and represent targets for tyrosine kinase inhibitors, which can cause arrest of tumor cell proliferation, impact tumor angiogenesis, and induce apoptosis and tumor migration.

First-generation RTK inhibitors are compounds that inhibit the HER family of human epidermal growth factors, which consists of four structurally related members: HER1 (EGFR), HER2, HER3, and HER4. Three drugs, which are currently in clinical use, erlotinib (**28.2.11**), gefitinib (**28.2.12**), and lapatinib (**28.2.13**), considered to be first-generation RTK inhibitors (Fig. 28.33.).

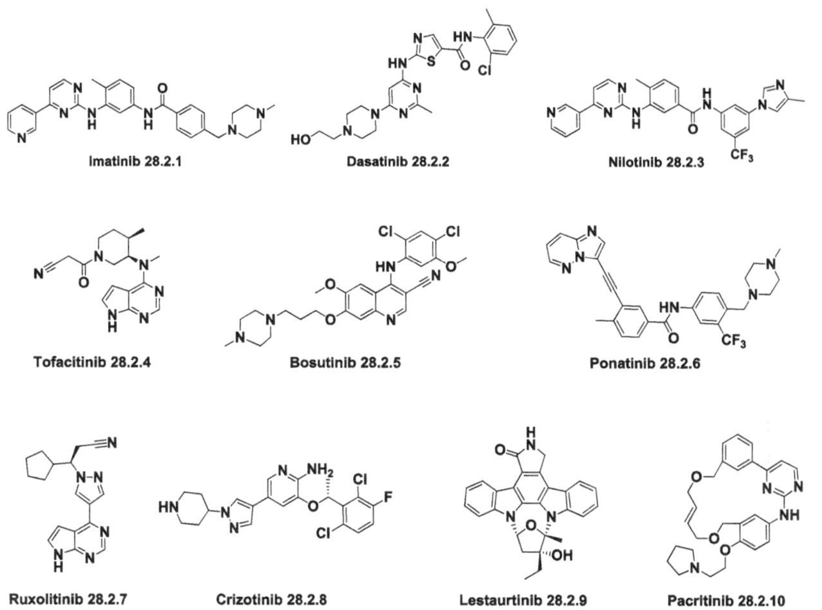

FIG. 28.32 Nonreceptor tyrosine kinase inhibitors.

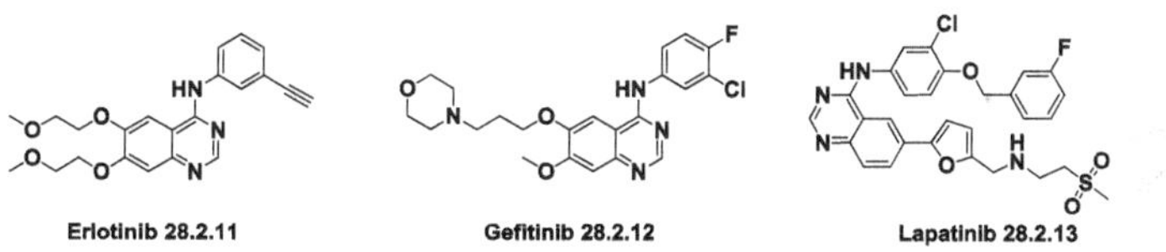

FIG. 28.33 Receptor tyrosine kinase inhibitors.

Other approved drugs of this series are afatinib (**28.2.14**), vandetanib (**28.2.15**), pazopanib (**28.2.16**), tivozanib (**28.2.17**), sorafenib (**28.2.18**), regorafenib (**28.2.19**), toceranib (**28.2.20**), sunitinib (**28.2.21**), and axitinib (**28.2.22**) (Fig. 28.34.).

Cediranib (**28.2.23**), canertinib (**28.2.24**), neratinib (**28.2.25**), pelitinib (**28.2.26**), semaxinib (**28.2.27**), and quizartinib (**28.2.28**) are RTK inhibitor drug candidates that are under investigation (Fig. 28.35.).

Multitarget Tyrosine Kinase Inhibitors

In the field of oncology, lack of specificity has not proved to be a barrier to clinical approval. The simultaneous inhibition of several kinases can even be advantageous. Much data about different tumors show that better cancer therapy can be obtained by blocking several biochemical pathways of tumor cells. Moreover, protein kinases, in general, are components of so-called cascade systems in which several protein kinases activate one another sequentially.

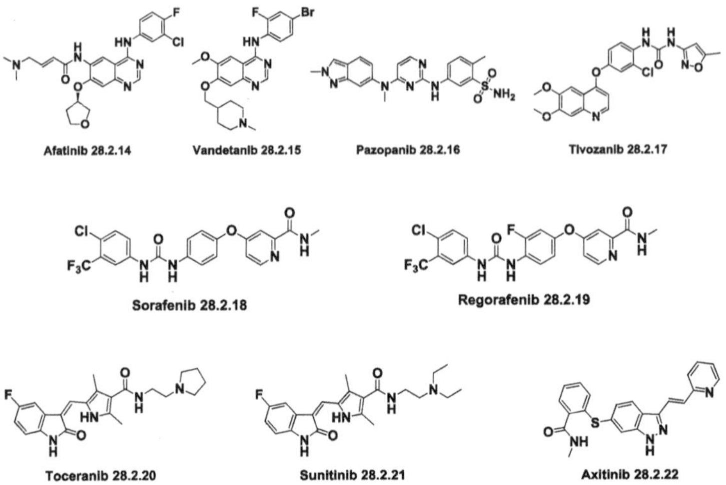

FIG. 28.34 Other approved kinase inhibitor drugs.

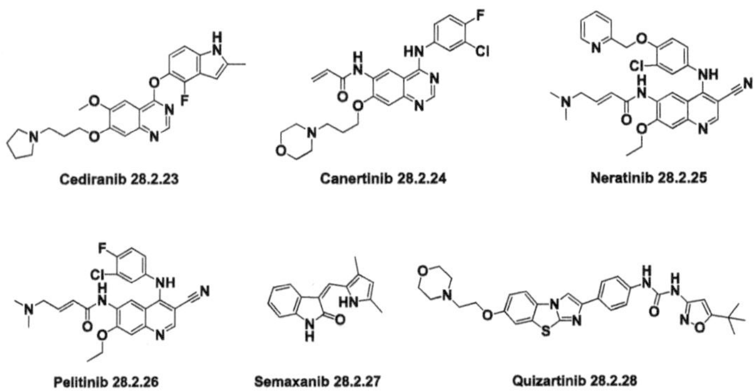

FIG. 28.35 New receptor tyrosine kinase inhibitor drug candidates that are under investigation.

Multikinase inhibitors inhibit multiple tyrosine kinases and related signaling pathways. For example, sunitinib (**28.2.21**), which is considered a first-generation multikinase inhibitor, simultaneously blocks vascular endothelial growth factor receptors (VEGFRs), PDGFRs (PDGFRα and PDGFRβ), stem cell factor receptor (KIT), FMS-like tyrosine kinase-3 (FLT3), glial cell-line–derived neurotrophic factor receptor (RET), and the receptor of macrophage-colony stimulating factor (CSF1R). Sorafenib (**28.2.18**) is another first-generation representative of multikinase inhibitors (see Fig. 28.34.).

Representatives of second-generation multikinase inhibitors include vandetanib (**28.2.15**), pazopanib (**28.2.16**), and axitinib (**28.2.22**).

A little bit out of the implemented classifications are compounds that act on vascular endothelial growth factor (VEGF) and its receptor (VEGFR).

The VEGF–VEGFR Signal Pathway Inhibitors

VEGF and its receptors (VEGFRs) are important regulatory factors in formation of tumor blood vessels. The VEGF–VEGFR signal pathway inhibitors have been successfully used in clinical practice.

Classification of VEGF receptors and ligands, their roles in angiogenesis, and the current status of VEGF-targeted drugs are reviewed in several articles [191-193].

VEGF is often found overexpressed in tumors that require a vascular supply to grow and could achieve this via the expression of angiogenic growth factors, including members of the VEGF family of five structurally related members: VEGF-A, VEGF-B, VEGF-C, VEGF-D and placenta growth factor (PLGF). VEGF pathways are considered the most validated approach for blocking tumor angiogenesis. VEGFR is an important tyrosine kinase as it is involved in angiogenesis formation of the circulatory system and growth of new blood vessels. Three different, but related, RTKs regulate different aspects of vascular function.

Angiogenesis inhibitors, representatives of the tyrosine kinases inhibitor series, are designed to prevent the formation of new blood vessels, thereby stopping or slowing the growth or spread of tumors. That is the philosophy of VEGFR-based antiangiogenic therapy.

Five tyrosine kinase inhibitors targeting VEGF or its receptors—pazopanib (**28.2.16**), sorafenib (**28.2.18**), regorafenib (**28.2.19**), sunitinib (**28.2.21**), and axitinib (**28.2.22**)—have become the cornerstone for treatment of several malignancies and are now in the pharmaceutical market.

VEGF pathway-targeted drugs—afatinib (**28.2.14**) and vandetanib (**28.2.15**)—also were developed in recent years.

Several pharmaceutical companies have developed new small molecule inhibitors of the VEGFRs. Investigational drugs in this series are cediranib (**28.2.29**), lenvatinib (**28.2.30**), motesanib (**28.2.31**), and vatalanib (**28.2.32**) (Fig. 28.36.).

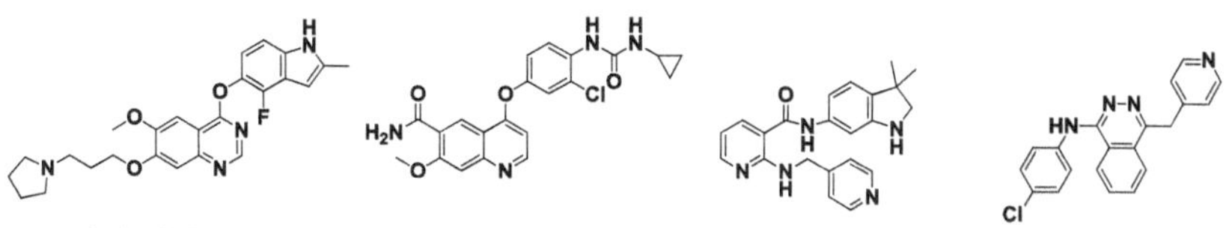

Cediranib 28.2.29 **Lenvatinib 28.2.30** **Motesanib 28.2.31** **Vatalanib 28.2.32**

FIG. 28.36 **Investigational VEGFR inhibitor drugs.**

Some 20 drugs that target tyrosine kinases have been approved for clinical use over the past decade, and hundreds more are undergoing clinical trials.

Unfortunately, these drugs are also associated with an increase in the risk of potentially life-threatening adverse events, such as arterial thrombotic events and bleeding.

Drugs That Target Tyrosine Kinases

Imatinib–Gleevec

The advent of imatinib, the first representative of the tyrosine kinase inhibitors as anticancer drugs, started a new era in the management of patients with BCR-ABL–positive chronic myelogenous leukemia, a kind of bone marrow multiplicative blood disease that is characterized by the amplification of hematopoietic cells caused by an abnormal Philadelphia chromosome, gastrointestinal tumors, myeloproliferative neoplasms, and dermatofibrosarcoma protuberans [194-215].

For more than 10 years, imatinib has been the standard first-line therapy for chronic myeloid and myelomonocytic leukemias, gastrointestinal stromal tumors, and myeloproliferative neoplasms. Treatment with imatinib is generally well tolerated with a low incidence of severe side effects. The most common adverse events include mild to moderate edema, muscle cramps, diarrhea, nausea, skin rashes, and myelosuppression.

The creation of imatinib is an example of modern interdisciplinary rational drug design. The high throughput screening identified the 2-phenylaminopyrimidine scaffold as a lead compound, which was further modified to produce imatinib (**28.2.1**).

The first synthesis of imatinib was realized starting from the reaction of 3-nitroaniline (**28.2.33**) nitrate in ethanol with cyanamide, which yielded 3-nitrophenyl-guanidine nitrate (**28.2.34**). This compound was condensed with 3-dimethylamino-1-(3-pyridyl)-2-propen-1-one (**28.2.36**) prepared from 3-acetylpyridine (**28.2.35**) via formylation with ethyl formate in presence of sodium methoxide and further conversion to the desired enaminone (**28.2.36**) which reacted with dimethylamine in toluene in the presence acetic acid. Condensation of 3-nitrophenyl-guanidine nitrate (**28.2.34**) and enaminone (**28.2.36**) in isopropanol in the presence of sodium hydroxide produced N-(3-nitrophenyl)-4-(3-pyridyl)-2-pyrimidine-amine (**28.2.37**), which was hydrogenated using a palladium on carbon (5%) catalyst under normal pressure of hydrogen, which yielded N-(3-aminophenyl)-4-(3-pyridyl)-2-pyrimidine-amine (**28.2.38**). The last was acylated with the benzoyl chloride derivative (**28.2.39**) in pyridine to produce the desired imatinib (**28.2.1**) [216-219] (Scheme 28.3.).

According to another scheme, methyl 4-formylbenzoate (**28.2.40**) undergoes reductive amination with 1-methylpiperazine implementing Pt/C catalyzed hydrogenation. The obtained ester-methyl 4-((4-methylpiperazin-1-yl)methyl) benzoate (**28.2.41**) is converted to the corresponding amide reacting with 3-bromo-4-methylaniline (**28.2.42**) in toluene in the presence of trimethyl aluminum to produce benzamide (**28.2.43**). The last is involved in a palladium-catalyzed Buchwald–Hartwig cross-coupling reaction with aminopyrimidine (**28.2.44**) in $CHCl_3$ using t-BuONa as a base, and $Pd_2(dba)_3$-(tris(dibenzylideneacetone) dipalladium (0) with rac-BINAP-(2,2′-bis(diphenylphosphino)-1,1′-binaphthyl) as a catalyst to produce imatinib (**28.2.1**) [220] (Scheme 28.3.).

Closely related alternative routes for the synthesis of imatinib are proposed in many patents and papers [221-228], and have been reviewed [229-231].

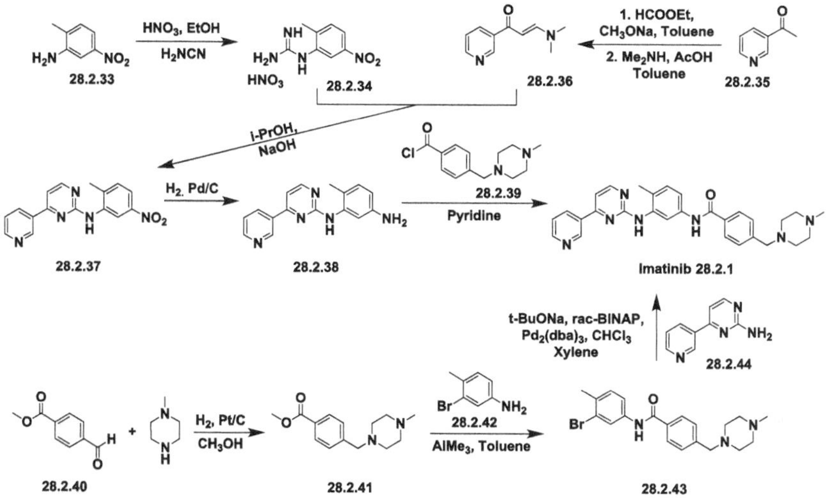

SCHEME 28.3 Synthesis of imatinib.

Erlotinib–Tarceva

Synthesis of erlotinib (**28.2.11**), which comprises reaction of 6,7-bis(2-methoxyethoxy)-4-chloroquinazoline (**28.2.50**) with3-ethynylaniline, was first disclosed in patents [232,233].

The synthesis started from 3,4-dihydroxybenzoic acid (**28.2.44**), which was converted to ethyl 3,4-dihydroxybenzoate (**28.2.45**) in the traditional manner, via heating in ethanol in the presence of sulfuric acid. The obtained product was then etherified with 2-bromoethyl methyl ether in the presence of tetrabutylammonium iodide (TBAI) in acetone potassium carbonate media, to prepare ethyl 3,4-bis(2-methoxy-ethoxy)benzoate (**28.2.46**). This product in acetic acid was treated with concentrated HNO_3 at 5°C for 24 hours to afford ethyl 4,5-bis-(2-methoxy-ethoxy)-2-nitro-benzoate (**28.2.47**). The last was hydrogenated under 45 psi H_2 for 6 hours in ethanol containing 1 equivalent of HCl to produce the hydrochloride salt of ethyl 2-amino-4,5-bis-(2-methoxy-ethoxy)benzoate (**28.2.48**). This material, with an equimolar quantity of ammonium formate, was dissolved in formamide and the stirred mixture was heated to 160 to 165°C under N_2 to produce 6,7-bis(2-methoxy-ethoxy)-quinazolone (**28.2.49**).

The obtained product was dissolved in $CHCl_3$ containing 1 drop of DMF, and oxalylchloride was added in several portions to afford 6,7-bis(2-methoxyethoxy)-4-chloroquinazoline (**28.2.50**). 3-Ethynylaniline was added to prepared 4-chloro-6,7-bis(2-methoxyethoxy)quinazoline dissolved in isopropanol containing pyridine, and after refluxing the mixture for 4 hours under nitrogen, produced the desired erlotinib (**28.2.11**) base (Scheme 28.4.).

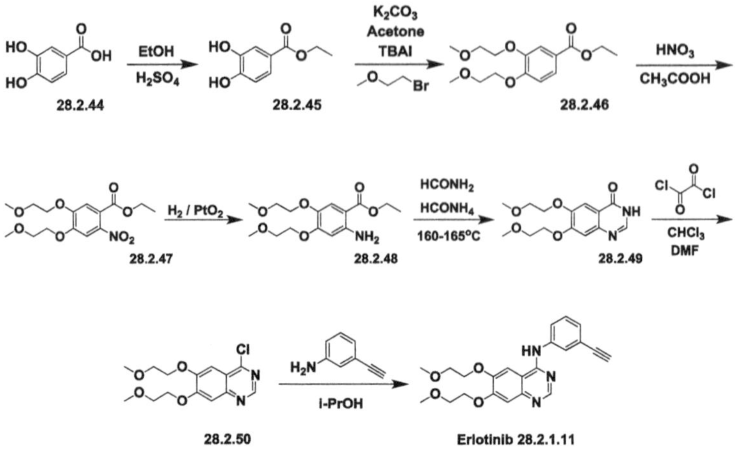

SCHEME 28.4 **Synthesis of erlotinib.**

A slightly different approach—reacting 6,7-bis(2-methoxyethoxy)-4-methoxyquinazoline with 3-ethynylaniline—was demonstrated recently [234].

The starting compound, 6,7-dimethoxy-4-(3H)-quinazolineone (**28.2.51**), was demethylated in 48% HBr via heating for 1 hour at 110°C and then at 140°C for 15 hours to produce dihydroxyquinazolinone (**28.2.52**). The obtained product was dissolved in DMF and after the adding of five-fold excess of $POCl_3$, the mixture was heated at 120°C for 3 hours. After the removal of excess of $POCl_3$, toluene and 1 equivalent of K_2CO_3 were added, and then a large excess of methanol was carefully added, keeping the temperature under 40°C, which produced 4-methoxyquinazoline-6,7-diol (**28.2.53**). The obtained diol was converted to 2-methoxyethyl ether (**28.2.54**) in the regular way. It was dissolved in DMF in the presence of 5 equivalents of K_2CO_3, and bromomethyl ethyl ether (4 equivalents) was added. After stirring at 30°C for 24 hours, methoxyquinazoline (**28.2.54**) was produced.

A n-BuLi solution in toluene was added to the solution of 3-ethynylaniline in THF. After stirring for 30 minutes, a solution of 6,7-bis(2-methoxyethoxy)-4-methoxyquinazoline (**28.2.54**) in THF was added on cooling and the reaction mixture was stirred at room temperature for 4 hours to produce the desired erlotinib (**28.2.11**) (Scheme 28.5.).

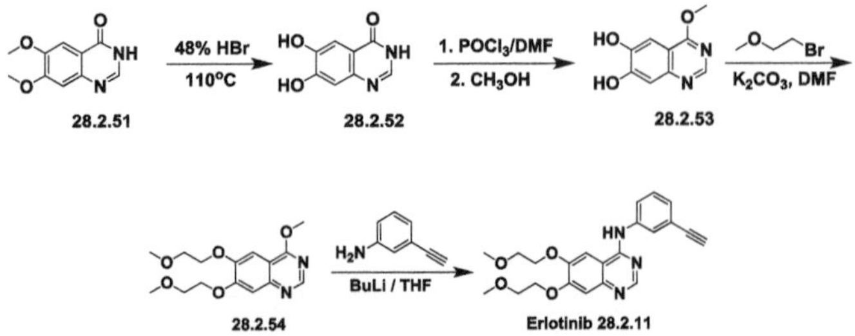

SCHEME 28.5 **Synthesis of erlotinib.**

Other synthetic schemes closely related to those described above have been proposed [235-241].

Erlotinib, a potent inhibitor of EGFR activity, is the only drug of its class approved as a monotherapy for the treatment of advanced non–small cell lung cancer and colon cancer and in combination with gemcitabine for advanced pancreatic cancer [242-256].

28.3 IMMUNOTHERAPEUTIC AGENTS

Some cancer treatments are considered immunotherapy; that is, they stimulate the immune system to fight the disease. The immune system is the built-in host defense mechanism against infectious agents such as cancer.

The main types of immunotherapy now being used to treat cancer include monoclonal antibodies, cancer vaccines, and nonspecific immunotherapies.

Monoclonal Antibodies

Having recognized that one of the significant differences between cancerous and normal cells is that cancerous cells contain specific antigens, it has been proposed to design monoclonal antibodies specifically to target cancer cells and to eliminate them. Emerging evidence supports the use of monoclonal antibodies, which have become a general modality in therapeutic development and now occupy an increasing niche in the arsenal available to treat cancer. Monoclonal antibodies represent the fastest growing type of biopharmaceuticals in the arsenal available to treat cancer.

This is a new therapeutic option with a big potential to generate anticancer immunity, and the mechanisms behind their efficacy are multifaceted. Monoclonal antibodies can act through a number of mechanisms, such as blocking of targeted molecule functions, inducing apoptosis of cells that express the target, or by modulating signaling pathways [257-267].

Monoclonal antibodies are intended to target aberrant oncogenic signaling within tumors and their microenvironment. These targets include ligands such as VEGFs, their receptors, human EGFR (HER), cluster of differentiation 20 (CD20), insulin-like growth factor I receptor (IGF-1R), tumor necrosis factor-related apoptosis-inducing ligand receptor (TRAIL-R), integrin and c-Met, cell-surface counter receptors and their receptor-bound ligands, etc. Monoclonal antibodies are directed at the extracellular domain of the receptor, whereas tyrosine kinase inhibitors act intracellularly.

The basic structural unit of most mammalian antibodies is a glycoprotein (MW ~150,000 daltons) comprising four polypeptide chains—two light chains and two heavy chains—which are connected by disulfide bonds. This area is outside the borders of routine organic synthesis, but the ability to genetically engineer monoclonal antibodies has significantly improved their effectiveness and revealed novel therapeutic targets.

In principle, two types of monoclonal antibodies are used in cancer treatments.

Naked monoclonal antibodies are those without any drug or radioactive material attached to them, and are the most commonly used monoclonal antibodies at this time.

Conjugated monoclonal antibodies are those joined to a chemotherapy drug, radioactive particle, or other cancer cell killing agent.

More than 20 monoclonal antibodies have been approved by the FDA, among them 14 are approved for oncology indications. The ones approved for oncological indications are alemtuzumab, bevacizumab, brentuximab, catumaxomab, cetuximab, edrecolomab, gemtuzumab, ibritumomab, ipilimumab, ofatumumab, panitumumab, rituximab, trastuzumab, and tositumomab.

The only conjugated antibody approved for treating cancer is gemtuzumab. It represents an antibody against the CD33 antigen, which is present on most leukemia cells with attached antibiotic calicheamicin able to cause DNA strand scission.

Monoclonal antibodies are very specific drugs with moderate toxicity.

Cancer Vaccines

Vaccines are substances used to stimulate the production of antibodies and provide immunity against diseases.

Vaccines can confer immunity against a harmful agent by stimulating the immune system. Once stimulated by a vaccine, the antibody-producing cells remain ready to respond to the specific agent.

Exploiting a naturally occurring defense system, the immunotherapeutic approach embodies an ideal nontoxic treatment for cancer.

A huge amount of research has shown that the immune system can be actively polarized against malignant cells by means of a variety of vaccination strategies, and that in some cases, vaccination associated with tumor regression [268-274].

Although cancer immunotherapy seems to be a safe and hopeful method of treatment, only few vaccines have been approved. Probably it is due the fact that although cancer cells express a wide profile of specific compounds related to cancer cells only, the immune system is highly tolerant to them and does not react to these self-antigens. Active immunotherapy aims to reverse this immune tolerance so that the immune system can respond appropriately to self-antigens.

Cancer vaccines in medicinal use are classified as preventive vaccines and treatment (or therapeutic) vaccines. Preventive vaccines target infectious agents that cause the development of cancer. Only a few have been approved and others are still undergoing investigation, although cancer immunotherapy seems to be a safe and hopeful method of treatment.

Two preventive vaccines, Gardasil and Cervarix, and one cancer treatment vaccine, Provenge, are approved in the United States.

More good evidence is needed for vaccines to be considered a globally standard immunotherapy treatment for cancer.

Nonspecific Cancer Immunotherapies

Strong, nonspecific, immunostimulatory therapies sometimes lead to a good immune response against cancer cells.

Cytokines, a large group of proteins, peptides, or glycoproteins that are secreted by specific cells of immune system, are released in response to a diverse range of cellular stresses, including carcinogen-induced injury. Cytokines that are released in response to infection, inflammation, and immunity can function to inhibit tumor development and progression.

These immune-modulating effects allow them to be used as drugs to provoke an immune response against the tumor. Nonspecific immunotherapies don't target cancer cells specifically. Some nonspecific immunotherapies are given by themselves as cancer treatments. Others are used as adjuvants.

Commonly used cytokines are the interferons and interleukins.

Interferons (IFN-α, -β, -γ) are glycoproteins that have antitumor and antiviral activity. Depending on dose, interferons may either enhance or decrease cellular and humoral immune functions. Interferons also inhibit division and certain synthetic processes in a variety of cells.

Interleukins (IL-α, IL-2, IL-7, IL-12, and IL-21) are a class of glycoproteins produced by leukocytes for regulating immune responses. Interleukins can be used as a single-drug treatment or can be combined with chemotherapy or with other cytokines.

Other cytokines used in cancer treatment are granulocyte-macrophage colony-stimulating factor (GM-CSF), which causes the bone marrow to make more of certain types of immune system cells and blood cells [275].

Some nonspecific immunotherapy agents can be considered agents that target immune system checkpoints—compounds on immune cells that need to be modulated to start an immune response. A group of them are agents that target CTLA-4, a protein found on the surface of immune cells called T cells.

PD-1 and PD-L1 are other proteins on the surface of some T cells. Drugs that target either PD-1 or PD-L1 can enhance the immune response against cancer and show great promise in clinical trials.

Some other drugs, like thalidomide, lenalidomide, pomalidomide, and imiquimod, boost the immune system in a nonspecific way, similar to cytokines.

Bacille Calmette-Guérin (BCG) is a germ that activates the immune system. It also is considered to be a nonspecific immunotherapy for some types of cancer.

28.4 HORMONAL ANTICANCER DRUGS

Cancer chemotherapy is based on the belief that cancer cells are more sensitive to cytotoxic or cytostatic drugs than normal cells.

Steroid hormones, like estrogens, progestins, and androgens, act as key inducers and modulators of cell proliferation, differentiation, and reproduction. They are known to influence the development and growth of many human cancers and many of them, such as endocrine-related breast, prostate, and endometrial cancer, can be hormone sensitive. It is observed that changing the hormonal balance of the body by administering steroidal hormones produces useful remission from cancer. Various steroidal molecules are active against different cancer cells.

Some steroid hormones can stop or slow the growth of cancer cells by either changing the level of particular hormones in the body, or by preventing the hormones affecting the cancer cells. Practically all classes of steroid hormones—androgens, estrogens, progestins, glucocorticoids, and mineralocorticoids—are used in cancer chemotherapy [276-284].

Androgens

Androgens mainly testosterone and dihydrotestosterone, are directly related to prostate cancer, which cannot grow or survive without androgens.

Androgens have important physiological effects in women while at the same time they may be implicated in breast cancer pathologies. Androgens are used to treat breast cancer, probably because they can act as estrogen precursors, or by binding to the androgen receptor in breast cancer cells. Study of molecular mechanisms involving androgenic pathways in breast cancer is still in their infancy.

Androgen receptor is a major therapeutic target in prostate cancer pharmacology, and known androgen receptor ligands can be classified as agonists or antagonists depending on their ability to activate or inhibit this receptor.

Antiandrogens (antagonists) include nonsteroidal—flutamide (**28.4.1**), nilutamide (**28.4.2**), and bicalutamide (**28.4.3**)—and steroidal—finasteride (Proscar) (**28.4.4**) and cyproterone acetate (**28.4.5**) (Fig. 28.37.)—drugs that are used to treat benign prostatic hyperplasia.

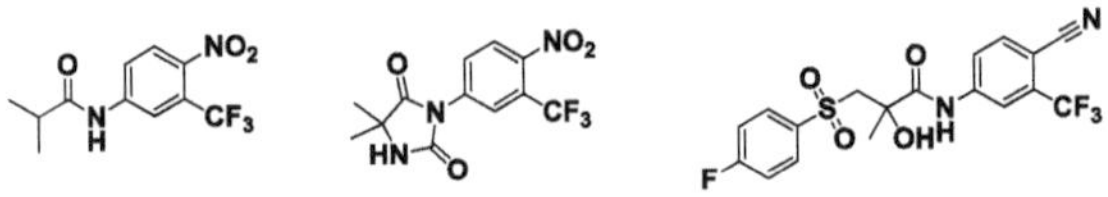

Flutamide 28.4.1 Nilutamide 28.4.2 Bicalutamide 28.4.3

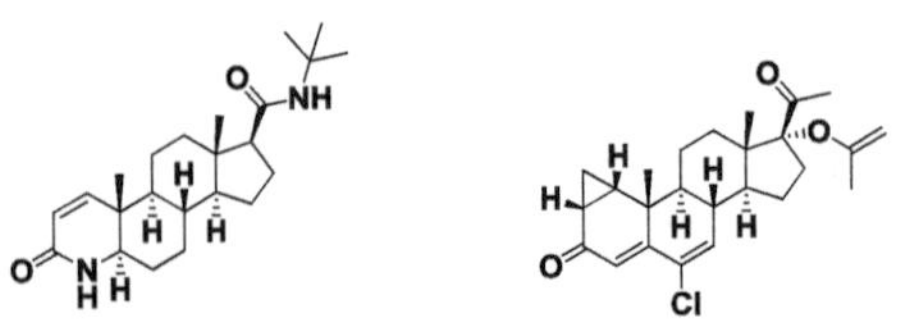

Finasteride 28.4.4 Cyproterone acetate 28.4.5

FIG. 28.37 **Antiandrogens used to treat benign prostatic hyperplasia.**

Androgens (agonists)—testosterone propionate (**28.4.6**), testosterone enanthate (**28.4.7**), testosterone cypionate (**28.4.8**), methyltestosterone (**28.4.9**), fluoxymesterone (**28.4.10**), testolactone (**28.4.11**), and danazol (**28.4.12**) (Fig. 28.38.)—have been used for many years in the treatment of metastatic breast carcinoma.

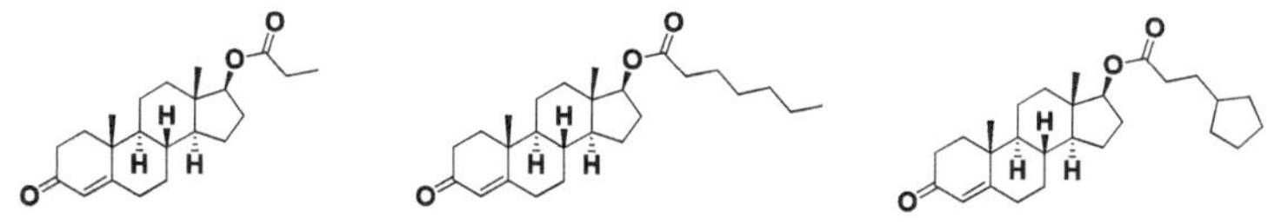

Testosterone propionate 28.4.6 | Testosterone enanthate 28.4.7 | Testosterone cypionate 28.4.8

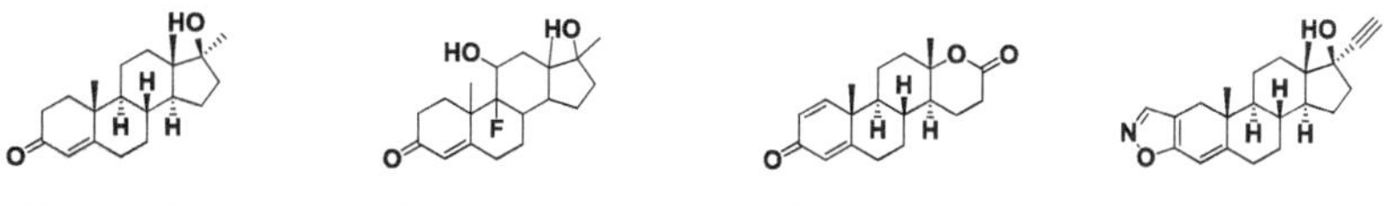

Methyltestosterone 28.4.9 | Fluoxymesterone 28.4.10 | Testolactone 28.4.11 | Danazol 28.4.12

FIG. 28.38 Antiandrogens used for the treatment of metastatic breast carcinoma.

Estrogens

Estrogens—female steroid hormones that exhibit diverse action in multiple physiologic systems—are also implicated in various types of cancer. They promote the growth of breast endometrial and ovarian and prostate cancer cells. They can promote cancer development in one set of circumstances, but assist in preventing cancer development in another.

Diethylstilbestrol (**28.4.13**), chlorotrianisene (**28.4.14**), and ethinyl estradiol (**28.4.15**) are rarely used estrogens in the treatment of prostate and breast carcinoma. But their use is limited to the treatment of postmenopausal women, because of the risk for development of uterine adenocarcinoma, vaginal adenoma, or vaginal adenocarcinoma. Diethylstilbestrol is generally preferred to control inoperable and metastatic neoplasm of the prostate (Fig. 28.39.).

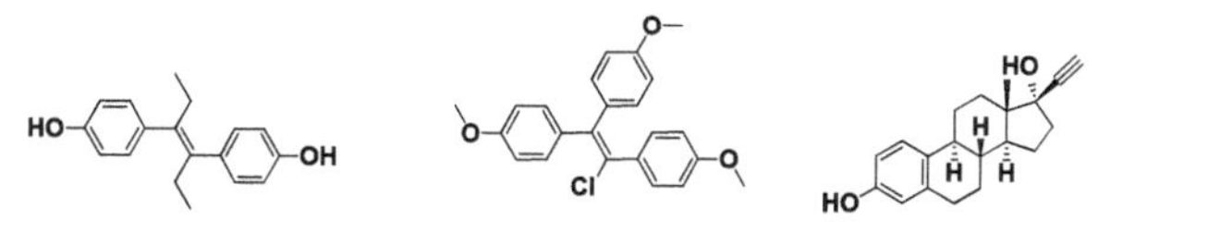

Diethylstilbestrol 28.4.13 | Chlorotrianisene 28.4.14 | Ethinyl estradiol 28.4.15

FIG. 28.39 Estrogens for the treatment of prostate and breast carcinoma.

Antiestrogens

Antiestrogens compete with endogenous estrogens for estrogen receptor binding and also for direct interaction with growth factors which ultimately lead to inhibition of estrogenic action.

Tamoxifen (**28.4.16**), droloxifene (**28.4.17**), toremifene (**28.4.18**), idoxifene (**28.4.19**), trioxifene (**28.4.20**), and raloxifene (**28.4.21**) are triphenylethylene derivatives that are estrogen antagonist or agonist, depending upon the species or organ under study. Tamoxifen has been the classic estrogen receptor antagonist for more than 30 years. It is the first-line hormonal treatment for both pre- and postmenopausal women with breast carcinoma (Fig. 28.40.).

Raloxifene (**28.4.21**) is a second-generation compound in this class that functions as an estrogen antagonist on breast and uterine tissues and an estrogen agonist on bone.

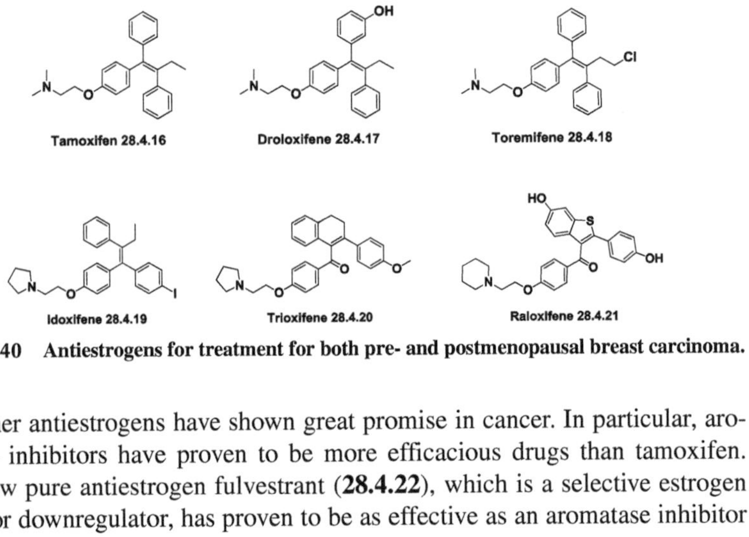

FIG. 28.40 Antiestrogens for treatment for both pre- and postmenopausal breast carcinoma.

Other antiestrogens have shown great promise in cancer. In particular, aromatase inhibitors have proven to be more efficacious drugs than tamoxifen. The new pure antiestrogen fulvestrant (**28.4.22**), which is a selective estrogen receptor downregulator, has proven to be as effective as an aromatase inhibitor (Fig. 28.41.).

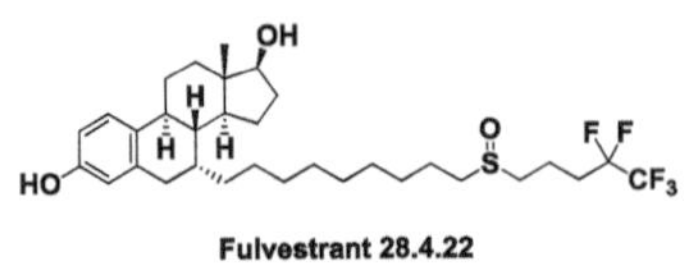

FIG. 28.41 Structure of fulvestrant.

Aromatase Inhibitors

Aromatase is an enzyme that catalyzes the rate-limiting step of testosterone and androstenedione conversion into estradiol and estrone. Decreasing the levels of estrogens in postmenopausal women could be an effective approach to control hormone-sensitive breast cancer. Aromatase inhibitors are very effective in cancer treatment. They were developed starting with aminoglutethimide (**28.4.23**), which was considered a first-generation aromatase inhibitor, but was later withdrawn from the market.

Second-generation aromatase inhibitors include fadrozole (**28.4.24**) and formestane (**28.4.25**), both with improved clinical efficacy.

Third-generation compounds are represented by exemestane (**28.4.26**) and are used in the treatment of postmenopausal breast cancers (Fig. 28.42.).

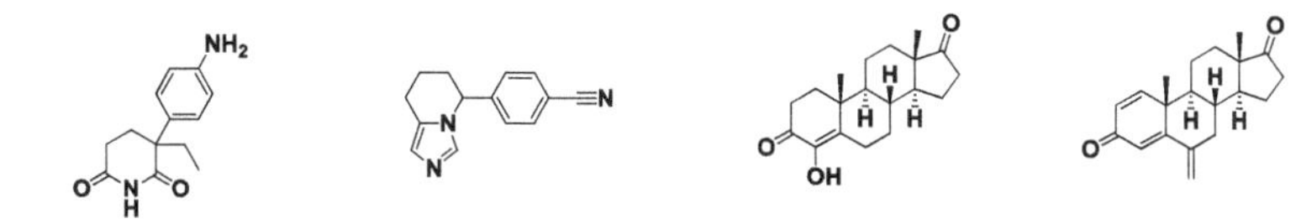

Aminoglutethimide 28.4.23 Fadrozole 28.4.24 Formestane 28.4.25 Exemestane 28.4.26

FIG. 28.42 **Aromatase inhibitors.**

Steroid Sulfatase Inhibitors

Estrogens and androgens are instrumental in the maturation of many hormone-dependent cancers. Consequently, the enzymes involved in their generation have been chosen as cancer therapy targets. One of these enzymes, steroid sulfatase, hydrolyzes estrone, and dehydroepiandrosterone sulfates to estrone and dehydroepiandrosterone, respectively, which are the precursors to the formation of estradiol and androstenediol.

Sulfated steroids are storage reservoirs of steroid hormones. The sulfated steroids themselves are biologically inactive and only become active in vivo, when they are converted into their desulfated form by the enzyme steroid sulfatase. It plays a crucial role in the regulation of tissue concentration of estrogens and androgens in human target organs. Overexpression of steroid sulfatase activity in the tissues leads to a high estrogenic response, which is considered prognostic of various cancers like prostate and breast. Reduction of the estrogen levels by a mechanism of steroid sulfatase inhibition is a promising approach to control the progression of cancer. Inhibitors of steroid sulfatases are considered to be potential therapeutics for the treatment of steroid-dependent cancers.

Estrone-3-O-sulfamate (**28.4.27**) is the prototype inhibitor of a class of potent and irreversible steroid sulfatase inhibitors (Fig. 28.43.).

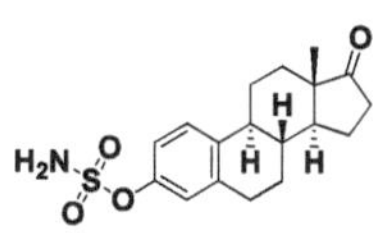

Estrone-3-O-sulfamate 28.4.27

FIG. 28.43 **Structure of estrone-3-O-sulfamate.**

Steroid 5α-Reductase Inhibitors

Benign prostatic hyperplasia and prostatic cancer are age-related proliferative diseases. Their exact etiology is not completely understood, but both appear to be associated with elevated plasma levels of the androgen 5α-dihydrotestosterone. Three types of 5α-reductases—types I, II, and III, which are responsible for the transformation of testosterone into the more potent androgen dihydrotestosterone—are known. Type I is the dominant form.

Several existing 5α-reductase inhibitors can structurally be classified into three main types: azasteroids, 3-carboxylic acids, and pregnane/androstane derivatives.

The most important among them seems to be the azasteroids. The first and best known is the 4-azasteroid finasteride (Proscar) (**28.4.4**) that selectively inhibits the type II isoform. A significant improvement was the development of inhibitors of both type I and type II isoenzymes, which provide more general blockage of dihydrotestosterone synthesis. Dutasteride (Avodart) (**28.4.28**) has emerged as a potent and better-tolerated dual inhibitor. Recently, a novel hybrid, epristeride (**28.4.29**), which combines the structural features of finasteride and dutasteride, was developed (Fig. 28.44.).

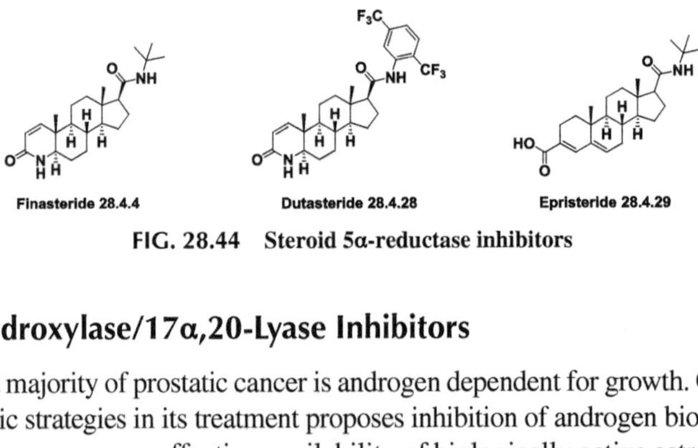

FIG. 28.44 Steroid 5α-reductase inhibitors

17α-Hydroxylase/17α,20-Lyase Inhibitors

The great majority of prostatic cancer is androgen dependent for growth. One of the therapeutic strategies in its treatment proposes inhibition of androgen biosynthesis.

The enzyme group affecting availability of biologically active estrogens and androgens is the family of 17β-hydroxysteroid dehydrogenases, which catalyzes dehydrogenation of 17β-hydroxysteroids.

Inhibition of the key enzyme—17α-hydroxylase/17,20-lyase, which catalyzes the biosynthesis of androgens from pregnane precursors—could prevent androgen biosynthesis and may provide effective treatment of prostate cancer patients.

A great advance in this field was made with the development of abiraterone (**28.4.30**), a D5,16-pregnenolone steroid with a pyridyl group bound to C17 at its 3-position, which was revealed to be a potent CYP17 irreversible inhibitor.

Later, different steroids bearing a heteroaromatic substituent at C17 were developed. Among others, compounds where the pyridyl group is replaced by thiazolyl-, imidazolyl-, pyrazolyl- oxazolyl, oxazolfuranyl, thiophenyl, and other heterocylic residues (Fig. 28.45.).

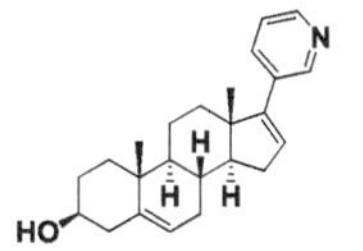

Abiraterone 28.4.30

FIG. 28.45 Structure of abiraterone.

Progestins

Progestins—17-hydroxyprogesterone caproate (**28.4.31**), medroxyprogesterone acetate (**28.4.32**), and megestrol acetate (**28.4.33**)—have emerged as the second choice of hormone therapy, after tamoxifen, for breast carcinoma in postmenopausal women (Fig. 28.46.).

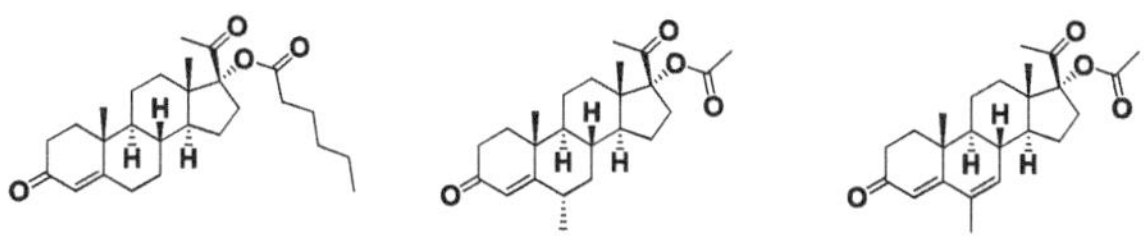

17-Hydroxyprogesterone caproate 28.4.31 · Medroxyprogesterone acetate 28.4.32 · Megestrol acetate 28.4.33

FIG. 28.46 Progestins used as anticancer agents.

Antiprogestins

Antiprogestins are mainly used in reproductive medicines and are less explored as anticancer agents. However, mifepristone (**28.4.34**) has been extensively explored as a cytostatic agent that inhibits growth of neuroblastomas and some breast and ovarian cancers cell lines. Onapristone (**28.4.35**) is a new antiprogestin that is being tested in clinical trials (Fig. 28.47.).

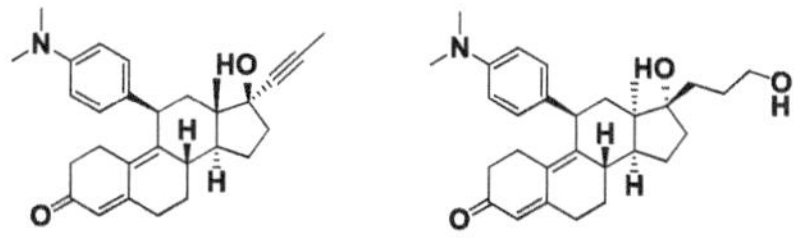

Mifepristone 28.4.34 · Onapristone 28.4.35

FIG. 28.47 Antiprogestins used as anticancer agents.

Corticosteroids

Corticosteroids are commonly used in the treatment of patients with advanced cancer. The use of corticosteroids in advanced cancer revolves around their glucocorticoid effects.

The well-known corticosteroids—prednisone (**28.4.36**), prednisolone (**28.4.37**), cortisol (**28.4.38**), dexamethasone (**28.4.39**), triamcinolone acetonide 21-oic acid methyl ester (**28.4.40**)—are used in cancer treatment, but their therapeutic effects are still disappointing.

Adrenocorticosteroids possess lympholytic effects and are able to suppress mitosis in lymphocytes, whereas glucocorticoids are useful in the treatment of malignant lymphoma and acute leukemias. These are usually used in combination with other antineoplastic agents (Fig. 28.48.).

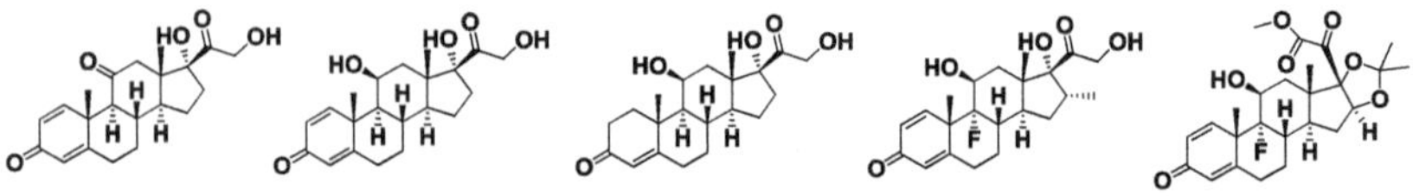

FIG. 28.48 Corticosteroids used as anticancer agents.

Despite the availability of this plethora of chemotherapeutic antineoplastic agents the cancer treatment problem remains unmet and basic approaches for that are constantly changing.

REFERENCES

1. Ambhaikar, N. B. Cancer drugs. In *Drug Discovery: Practices, Processes, and Perspectives;* Li, J. J., Corey, E. J., Eds.; Wiley, 2013; pp 287–336.
2. Roche, V. F. Cancer and chemotherapy. In *Foye's Principles of Medicinal Chemistry,* 6th ed.; Lemke, T. L., Williams, D. A., Eds. Lippincott Williams & Wilkins, 2008; pp 1147–1192.
3. Larsen, I. K.; Kastrup, J. S. Anticancer agents. In *Textbook of Drug Design and Discovery,* 3rd ed.; Krogsgaard-Larsen, P., Liljefors, T., Madsen, U., Eds. CRC Press, 2002; pp 511–558.
4. Weinmann, H.; Ottow, E. Recent development in novel anticancer therapies. In 8th ed.; Taylor, J. B., Triggle, D. J., Eds.; *Comprehensive Medicinal Chemistry II,* Vol. 7; Elsevier, 2006; pp 221–251.
5. Kumar, P. V.; Singh, V.; Suvagiya, V. A review on recent approaches for cancer treatment. *J. Pharm. Res. (Bangalore, India)* **2012,** *5* (1), 274–276.
6. Dixit, S.; Sharma, P. K.; Kumar, N.; Kaushik, N.; Singh, A.; Varshney, J. Recent advances in anticancer drugs. *Pharmacologyonline* **2010,** (3), 944–960.
7. Dieci, M. V.; Guarneri, V.; Conte, P. The future of chemotherapy in the era of personalized medicine. *Curr. Breast Cancer Rep.* **2013,** *5* (1), 57–68.
8. Ali, R.; Mirza, Z.; Ashraf, G. M. D.; Kamal, M. A.; Ansari, S. A.; Damanhouri, G. A.; Abuzenadah, A. M.; Chaudhary, A. G.; Sheikh, I. A. New anticancer agents: recent developments in tumor therapy. *Anticancer Res.* **2012,** *32* (7), 2999–3006.
9. Lu, Y.-H.; Gao, X.-Q.; Wu, M.; Zhang-Negrerie, D.; Gao, Q. Strategies on the development of small molecule anticancer drugs for targeted therapy. *Mini-Rev. Med. Chem.* **2011,** *11* (7), 611–624.
10. Yap, A.; Sandhu, S. K.; Workman, P.; de Bono, J. S. Envisioning the future of early anticancer drug development. *Nat. Rev. Cancer* **2010,** *10* (7), 514–523.

11. Ma, X.; Wang, Z. Anticancer drug discovery in the future: an evolutionary perspective. *Drug Discovery Today* **2009,** *14* (23/24), 1136–1142.
12. Azmi, A. S. Network pharmacology for cancer drug discovery: are we there yet? *Future Med. Chem.* **2012,** *4* (8), 939–941.
13. Los, M.; Burek, C. J.; Stroh, C.; Benedyk, K.; Hug, H.; MacKiewicz, A. Anticancer drugs of tomorrow: apoptotic pathways as targets for drug design. *Drug Discovery Today* **2003,** *8* (2), 67–77.
14. Hoelder, S.; Clarke, P. A.; Workman, P. Discovery of small molecule cancer drugs: successes, challenges and opportunities. *Mol. Oncol.* **2012,** *6* (2), 155–176.
15. Lammers, T.; Kiessling, F.; Hennink, W. E.; Storm, G. Drug targeting to tumors: principles, pitfalls and (pre-) clinical progress. *J. Controlled Release* **2012,** *161* (2), 175–187.
16. Dutt, R.; Madan, A. K. Classification models for anticancer activity. *Curr. Top. Med. Chem.* **2012,** *12* (24), 2705–2726.
17. Burger, A. Classification of drugs. In Hansch, C., Sammes, P. G., Taylor, J. B., Eds.; *Comprehensive Medicinal Chemistry*, Vol. 1; Pergamon Press, 1990; pp 249–260.
18. Binetti, R.; Costamagna, F. M.; Marcello, I. Classifications for carcinogenesis of antitumoral drugs. *J. Exp. Clin. Cancer Res.* **2003,** *22* (4), 651–672.
19. Espinosa, E.; Zamora, P.; Feliu, J.; Gonzalez Baron, M. Classification of anticancer drugs—a new system based on therapeutic targets. *Cancer Treat. Rev.* **2003,** *29* (6), 515–523.
20. Wu, X.-Z. A new classification system of anticancer drugs—based on cell biological mechanisms. *Med. Hypotheses* **2006,** *66* (5), 883–887.
21. Francisco, A. P.; Perry, M. J.; Moreira, R.; Mendes, E. Alkylating agents. In *Anticancer Therapeutics;* Missailidis, S., Ed.; Wiley, 2008; pp 133–158.
22. Celkan, T. Alkylating agents in chemotherapy. In *Alkylating Agents as Environmental Carcinogen and Chemotherapy Agents;* Dincer, Y., Ed.; Nova Science, 2013; pp 69–99.
23. Ben, A. ,F.; Gazzah, A.; Ousbane, A.; Gutierrez, M.; Brain, E. Alkylating agents. *Oncologie* **2007,** *9* (11), 751–757.
24. Pourquier, P. Alkylating agents. *Bull. Cancer* **2011,** *98* (11), 1237–1251.
25. Ralhan, R.; Kaur, J. Alkylating agents and cancer therapy. *Expert Opin. Ther. Pat.* **2007,** *17* (9), 1061–1075.
26. Bignold, L. P. Alkylating agents and DNA polymerases. *Anticancer Res.* **2006,** *26* (2B), 1327–1336.
27. Hubbard, R. D.; Fidanze, S. Alkylating and platinum antitumor compounds. In 8th ed.; Taylor, J. B., Triggle, D. J., Eds.; *Comprehensive Medicinal Chemistry II*, Vol. 7; Elsevier, 2006; pp 129–148.
28. Colvin, M. Alkylating agents and platinum antitumor compounds. In *Holland-Frei Cancer Medicine 7;* Kufe, D. W., Ed.; BC Decker, 2006; pp 675–687.
29. Izbicka, E.; Tolcher, A. W. Development of novel alkylating drugs as anticancer agents. *Curr. Opin. Invest. Drugs (BioMed Cent.)* **2004,** *5* (6), 587–591.
30. Huitema, A. D. R.; Smits, K. D.; Mathot, R. A. A.; Schellens, J. H. M.; Rodenhuis, S.; Beijnen, J. H. The clinical pharmacology of alkylating agents in high-dose chemotherapy. *Anti-Cancer Drugs* **2000,** *11* (7), 515–533.
31. Chabner, B. A.; Roberts, T. G., Jr. Timeline: chemotherapy and the war on cancer. *Nat. Rev. Cancer* **2005,** *5* (1), 65–72.
32. Dy, G. K.; Adjei, A. A. Systemic cancer therapy: evolution over the last 60 years. *Cancer* **2008,** *113* (7 Spec. Iss.), 1857–1887.
33. McCormick, J. E.; McElhinney, R. S. Nitrosoureas rom chemist to physician: classification and recent approaches to drug design. *Eur. J. Cancer* **1990,** *26* (3), 207–221.

34. Schabel, F. M., Jr. Nitrosoureas: a review of experimental antitumor activity. *Cancer Treat. Rep.* **1976,** *60* (6), 665–698.
35. Galaup, A.; Paci, A. Pharmacology of dimethanesulfonate alkylating agents: busulfan and treosulfan. *Expert Opin. Drug Metab. Toxicol.* **2013,** *9* (3), 333–347.
36. Hata, Y.; Watanabe, M. Metabolism of aziridines and the mechanism of their cytotoxicity. *Drug Metab. Rev.* **1994,** *26* (3), 575–604.
37. Maddry, J. A. Procarbazine. In *Cancer Chemotherapeutic Agents;* Foye, W. O., Ed.; American Chemical Society, 1995; pp 197–201.
38. Dhar, S.; Lippard, S. J. Current status and mechanism of action of platinum-based anticancer drugs. In *Bioinorganic Medicinal Chemistry;* Alessio, E., Ed.; Wiley-VCH, 2011; pp 79–95.
39. Marchesi, F.; Turriziani, M.; Tortorelli, G.; Avvisati, G.; Torino, F.; De Vecchis, L. Triazene compounds: mechanism of action and related DNA repair systems. *Pharmacol. Res.* **2007,** *56* (4), 275–287.
40. Stevens, M. F. G.; Newlands, E. S. From triazines and triazenes to temozolomide. *Eur. J. Cancer* **1993,** *29A* (7), 1045–1047.
41. Stevens, M. F. G.; Hickman, J. A.; Stone, R.; Gibson, N. W.; Baig, G. U.; Lunt, E.; Newton, C. G. Antitumour imidazotetrazines. 1. Synthesis and chemistry of 8-carbamoyl-3-(2-chloroethyl) imidazo[5,1-d]-1,2,3,5-tetrazin-4(3H)-one, a novel broad-spectrum antitumor agent. *J. Med. Chem.* **1984,** *27* (2), 196–201.
42. Lunt, E.; Stevens, M. F. G.; Stone, R.; Wooldridge, K. R. H., Tetrazine derivatives and pharmaceutical compositions containing them, DE 3231255 (1983).
43. Wang, Y.; Stevens, M. F. G.; Thomson, W. Alternative syntheses of the antitumor drug temozolomide avoiding the use of methyl isocyanate. *J. Chem. Soc., Chem. Commun.* **1994,** *14,* 1687–1688.
44. Wang, Y.; Stevens, M. F. G.; Chan, T.; DiBenedetto, D.; Ding, Z.; Gala, D.; Hou, D.; Kugelman, M.; Leong, W.; Kuo, S.; Mas, J. L.; Schumacher, D. P.; Shutts, B. P.; Smith, L.; Zhan, Z.-Y. J.; Thomson, W. T. Antitumor imidazotetrazines. 35. New synthetic routes to the antitumor drug temozolomide. *J. Org. Chem.* **1997,** *62* (21), 7288–7294.
45. Kuo, S.-C.; Mas, J. L.; Hou, D., Synthesis of temozolomide and analogs, US 20020095036 (2002).
46. Newlands, E. S.; Stevens, M. F. G.; Wedge, S. R.; Wheelhouse, R. T.; Brock, C. Temozolomide: a review of its discovery, chemical properties, pre-clinical development and clinical trials. *Cancer Treat. Rev.* **1997,** *23* (1), 35–61.
47. Friedman, H. S.; Kerby, T.; Calvert, H. Temozolomide and treatment of malignant glioma. *Clin. Cancer Res.* **2000,** *6* (7), 2585–2597.
48. Agarwala, S. S.; Kirkwood, J. M. Temozolomide, a novel alkylating agent with activity in the central nervous system, may improve the treatment of advanced metastatic melanoma. *Oncologist* **2000,** *5* (2), 144–151.
49. Danson, S.; Middleton, M. R. Temozolomide: a novel oral alkylating agent. *Expert Rev. Anticancer Ther.* **2001,** *1* (1), 13–19.
50. Tatar, Z.; Thivat, E.; Planchat, E.; Gimbergues, P.; Gadea, E.; Abrial, C.; Durando, X. Temozolomide and unusual indications: review of literature. *Cancer Treat. Rev.* **2013,** *39* (2), 125–135.
51. Marucci, G. Treatment of pituitary neoplasms with temozolomide: a review. *Cancer* **2011,** *117* (17), 4101–4102.
52. Bei, R.; Marzocchella, L.; Turriziani, M. The use of temozolomide for the treatment of malignant tumors: clinical evidence and molecular mechanisms of action. *Recent Pat. Anti-Cancer Drug Discovery* **2010,** *5* (3), 172–187.

53. Stevens, M. F. G. Temozolomide: from cytotoxic to molecularly-targeted agent. In *Cancer Drug Design and Discovery;* Neidle, S., Ed.; Academic Press, 2008; pp 157–172.
54. Mutter, N.; Stupp, R. Temozolomide: a milestone in neuro-oncology and beyond? *Expert Rev. Anticancer Ther.* **2006,** *6* (8), 1187–1204.
55. Mason, W. P.; Cairncross, J. G. Drug insight: temozolomide as a treatment for malignant glioma-impact of a recent trial. *Nat. Clin. Pract. Neurol.* **2005,** *1* (2), 88–95.
56. Stupp, R.; van den Bent, M. J.; Hegi, M. E. Optimal role of temozolomide in the treatment of malignant gliomas. *Curr. Neurol. Neurosci. Rep.* **2005,** *5* (3), 198–206.
57. Weller, M.; Steinbach, J. P.; Wick, W. Temozolomide: a milestone in the pharmacotherapy of brain tumors. *Future Oncol.* **2005,** *1* (6), 747–754.
58. Nagasubramanian, R.; Dolan, M. E. Temozolomide: realizing the promise and potential. *Curr. Opin. Oncol.* **2003,** *15* (6), 412–418.
59. Gaya, A.; Rees, J.; Greenstein, A.; Stebbing, J. The use of temozolomide in recurrent malignant gliomas. *Cancer Treat. Rev.* **2002,** *28* (2), 115–120.
60. Darkes, M. J. M.; Plosker, G. L.; Jarvis, B. Temozolomide: a review of its use in the treatment of malignant gliomas, malignant melanoma and other advanced cancers. *Am. J. Cancer (Auckland, N. Z.)* **2002,** *1* (1), 55–80.
61. Farber, S.; Diamond, L. K.; Mercer, R. D.; Sylvester, R. F.; Wolff, J. A. Temporary remissions in acute leukemia in children produced by folic acid antagonist, 4-aminopteroyl-glutamic acid (aminopterin). *N. Engl. J. Med.* **1948,** *238,* 787–793.
62. Tiwari, M. Antimetabolites: established cancer therapy. *J. Cancer Res. Ther. (Mumbai, India)* **2012,** *8* (4), 510–519.
63. Lansiaux, A. Antimetabolites. *Bull. Cancer* **2011,** *98* (11), 1263–1274.
64. Lansiaux, A. What's the news about antimetabolites in oncology? *Oncologie* **2007,** *9* (11), 746–750.
65. Scaife, J.; Kerr, D. Antimetabolites in cancer therapy. In *Anticancer Therapeutics;* Missailidis, S., Ed.; Wiley, 2008; pp 91–110.
66. Mader, M. M.; Henry, J. R. Antimetabolites, Comprehensive Medicinal Chemistry II. In 8th ed.; Taylor, J. B., Triggle, D. J., Eds.; Vol. 7; Elsevier, 2006; pp 55–79.
67. Cole, P. D.; Zebala, J. A.; Kamen, B. A. Antimetabolites: a new perspective. *Drug Discovery Today: Ther. Strategies* **2005,** *2* (4), 337–342.
68. Scagliotti, G. V.; Selvaggi, G. Antimetabolites and cancer: emerging data with a focus on antifolates. *Expert Opin. Ther. Pat.* **2006,** *16* (2), 189–200.
69. Johnston, P. G.; Takimoto, C. H.; Grem, J. L.; Fidias, P.; Grossbard, M. L.; Chabner, B. A.; Allegra, C. J.; Chu, E. Antimetabolites. *Cancer Chemother. Biol. Response Modif.* **1997,** *17,* 1–39.
70. Clarke, S. J.; Jackman, A. L.; Harrap, K. R. Antimetabolites in cancer chemotherapy. *Adv. Exp. Med. Biol.* **1991,** *309A,* 7–13.
71. Grem, J. L.; Takimoto, C. H.; Multani, P.; Chu, E.; Ryan, D.; Chabner, B. A.; Allegra, C. J.; Johnston, P. G. Antimetabolites. *Cancer Chemother. Biol. Response Modif.* **1999,** *18,* 1–38.
72. Kaye, S. B. New antimetabolites in cancer chemotherapy and their clinical impact. *Br. J. Cancer* **1998,** *78* (Suppl. 3), 1–7.
73. Sirotnak, F. M., Ensminger, W. D., Burchall, J. J., Montgomery, J. A., Eds. *Folate Antagonists as Therapeutic Agents*, 2 Vols; Academic Press, 1984.
74. Cole, P. D.; Kamen, B. A.; Bertino, J. R. Folate antagonists. In *Holland-Frei Cancer Medicine 7;* Kufe, D. W., Ed.; BC Decker, 2006; pp 648–660.
75. Kamen, B. Folate and antifolate pharmacology. *Semin. Oncol.* **1997,** *24* (5 Suppl. 18). S18/30–S18/39.

76. Berman, E. M.; Werbel, Leslie M. The renewed potential for folate antagonists in contemporary cancer chemotherapy. *J. Med. Chem.* **1991,** *34* (2), 479–485.
77. Fry, D. W.; Jackson, R. C. Biological and biochemical properties of new anticancer folate antagonists. *Cancer Metastasis Rev.* **1987,** *5* (3), 251–270.
78. Chabner, B. A.; Johns, D. G. Folate antagonists. In Becker, F. F., Ed.; *Cancer: A Comprehensive Treatise*, Vol. 5; Springer, 1977; pp 363–577.
79. Purcell, W. T.; Ettinger, D. S. Novel antifolate drugs. *Curr. Oncol. Rep.* **2003,** *5* (2), 114–125.
80. Pizzorno, G.; Diasio, R. B.; Cheng, Y.-C. Pyrimidine and purine antimetabolites. In *Holland-Frei Cancer Medicine 7;* Kufe, D. W., Ed.; BC Decker, 2006; pp 661–674.
81. Parker, W. B. Enzymology of purine and pyrimidine antimetabolites used in the treatment of cancer. *Chem. Rev. (Washington, DC, U.S.)* **2009,** *109* (7), 2880–2893.
82. Galmarini, C. M.; Jordheim, L.; Dumontet, C. Pyrimidine nucleoside analogs in cancer treatment. *Expert Rev. Anticancer Ther.* **2003,** *3* (5), 717–728.
83. Dudhe, R.; Sharma, P. K.; Verma, P.; Chaudhary, A. Pyrimidine as anticancer agent: a review. *J. Adv. Sci. Res.* **2011,** *2* (3), 10–17.
84. de Bono, J. S.; Twelves, C. J. The oral fluorinated pyrimidines. *Invest. New Drugs* **2001,** *19* (1), 41–59.
85. Ishitsuka, H. Discovery and preclinical pharmacology of capecitabine. In *Fluoropyrimidines in Cancer Therapy;* Rustum, Y. M., Ed.; Humana, 2003; pp 249–259.
86. Walko, C. M.; Lindley, C. Capecitabine: a review. *Clin. Ther.* **2005,** *27* (1), 23–44.
87. Wagstaff, A. J.; Ibbotson, T.; Goa, K. L. Capecitabine. A review of its pharmacology and therapeutic efficacy in the management of advanced breast cancer. *Drugs* **2003,** *63* (2), 217–236.
88. McGavin, J. K.; Goa, K. L. Capecitabine: A review of its use in the treatment of advanced or metastatic colorectal cancer. *Drugs* **2001,** *61* (15), 2309–2326.
89. Johnston, P. G.; Kaye, S. Capecitabine: a novel agent for the treatment of solid tumors. *Anti-Cancer Drugs* **2001,** *12* (8), 639–646.
90. Budman, D. R. Capecitabine. *Invest. New Drugs* **2000,** *18* (4), 355–363.
91. Aprile, G.; Mazzer, M.; Moroso, S.; Puglisi, F. Pharmacology and therapeutic efficacy of capecitabine: focus on breast and colorectal cancer. *Anti-Cancer Drugs* **2009,** *20* (4), 217–229.
92. Aguado, C.; Garcia-Paredes, B.; Sotelo, M. J.; Sastre, J.; Diaz-Rubio, E. Should capecitabine replace 5-fluorouracil in the first-line treatment of metastatic colorectal cancer? *World J. Gastroenterol.* **2014,** *20* (20), 6092–6101.
93. Lyros, E.; Walter, S.; Keller, I.; Papanagiotou, P.; Fassbender, K. Subacute reversible toxic encephalopathy related to treatment with capecitabine: A case report with literature review and discussion of pathophysiology. *NeuroToxicol.* **2014,** *42,* 8–11.
94. Solimando, D. A., Jr.; Waddell, J. A. Capecitabine and gemcitabine (CapGem, CG, GemCap) for advanced pancreatic and biliary tract cancer. *Hosp. Pharm.* **2014,** *49* (2), 127–133.
95. Ang, C.; Kornbluth, M.; Thirlwell, M. P.; Rajan, R. D. Capecitabine-induced cardiotoxicity: case report and review of the literature. *Curr. Oncol.* **2010,** *17* (1), 59–63.
96. Midgley, R.; Kerr, D. J. Capecitabine: have we got the dose right? *Nat. Clin. Pract. Oncol.* **2009,** *6* (1), 17–24.
97. Bang, Y.-J. Capecitabine in gastric cancer. *Expert Rev. Anticancer Ther.* **2011,** *11* (12), 1791–1806.
98. Hameed, H.; Cassidy, J. Use of capecitabine in management of early colon cancer. *Cancer Manage. Res.* **2011,** *3,* 295–299.
99. Hirsch, B. R.; Zafar, S. Y. Capecitabine in the management of colorectal cancer. *Cancer Manage. Res.* **2011,** *3,* 79–89.

100. Okines, A.; Chau, I.; Cunningham, D. Capecitabine in gastric cancer. *Drugs Today* **2008,** *44* (8), 629–640.
101. Koukourakis, G. V.; Kouloulias, V.; Koukourakis, M. J.; Zacharias, G. A.; Zabatis, H.; Kouvaris, J. Efficacy of the oral fluorouracil pro-drug capecitabine in cancer treatment: a review. *Molecules* **2008,** *13* (8), 1897–1922.
102. Schellens, J. H. M. Capecitabine. *Oncologist* **2007,** *12* (2), 152–155.
103. Ershler, W. B. Capecitabine monotherapy: safe and effective treatment for metastatic breast cancer. *Oncologist* **2006,** *11* (4), 325–335.
104. McKendrick, J.; Coutsouvelis, J. Capecitabine: effective oral fluoropyrimidine chemotherapy. *Expert Opin. Pharmacother.* **2005,** *6* (7), 1231–1239.
105. Twelves, C. Vision of the future: Capecitabine. *Oncologist* **2001,** *6* (Suppl. 4), 35–39.
106. Hoshi, A.; Castaner, J. Capecitabine. Antineoplastic. Ro-09-(1978). *Drugs Future* **1996,** *21* (4), 358–360.
107. Fujiu, M.; Ishitsuka, H.; Miwa, M.; Umeda, I.; Yokose, K., Preparation of fluorocytidine derivatives as antitumors and pharmaceutical compositions containing them, EP 316704 (1989).
108. Arasaki, M. N. R.; Ishitsuka, H.; Kuruma, I.; Miwa, M.; Murasaki, C.; Shimma, N.; Umeda, I. I. H., N-Oxycarbonyl-substituted 5′-deoxy-5-fluorocytidines as antitumor agents, EP 602454 (1994).
109. Kamiya, T.; Ishiduka, M.; Nakajima, H., Novel process for producing N4-acyl-5′-deoxy-5-fluorocytidine derivatives, EP 602478 (1994).
110. Brinkman, H. R.; Kalaritis, P.; Morrissey, J. F., Process for producing N4-(alkoxycarbonyl)-5′-deoxy-5-fluorocytidine derivatives by selective dealkoxycarbonylation of N4,2′O,3′O-tris(alkoxycarbonyl) derivatives, US 5476932 (1995).
111. MacDonald, P. L.; Rossetto, P.; Gallina, M., Process for the preparation of capecitabine, WO 2009088989 (2009).
112. Lin, K-C.; Chien, C., Novel synthesis of 5-deoxy-5′-fluorocytidine compounds via condensation reaction, US 20130184451 (2013).
113. Roberts, C. R.; Wong, J-W., Process for producing N4-acyl-5′-deoxy-5-fluorocytidine from 5-fluorocytosine via acid-catalyzed silylation and stereoselective glycosylation reactions, US 20050137392 (2005).
114. Miwa, M.; Ishikawa, T.; Eda, H.; Ryu, M.; Fujimoto, K.; Ninomiya, Y.; Umeda, I.; Yokose, K.; Ishitsuka, H. Comparative studies on the antitumor and immunosuppressive effects of the new fluorouracil derivative N4-trimethoxybenzoyl-5′-deoxy-5-fluorocytidine and its parent drug 5′-deoxy-5-fluorouridine. *Chem. Pharm. Bull.* **1990,** *38* (4), 998–1003.
115. Shimma, N.; Umeda, I.; Arasaki, M.; Murasaki, C.; Masubuchi, K.; Kohchi, Y.; Miwa, M.; Ura, M.; Sawada, N.; Tahara, H.; Kuruma, I.; Horii, I.; Ishitsuka, H. The design and synthesis of a new tumor-selective fluoropyrimidine carbamate, capecitabine. *Bioorg. Med. Chem.* **2000,** *8* (7), 1697–1706.
116. Parker, W. B.; Secrist, J. A., III.; Waud, W. R. Purine nucleoside antimetabolites in development for the treatment of cancer. *Curr. Opin. Invest. Drugs (BioMed Cent.)* **2004,** *5* (6), 592–596.
117. Robak, T.; Korycka, A.; Lech-Maranda, E.; Robak, P. Current status of older and new purine nucleoside analogues in the treatment of lymphoproliferative diseases. *Molecules* **2009,** *14* (3), 1183–1226.
118. Elgemeie, G. H. Thioguanine, mercaptopurine: their analogs and nucleosides as antimetabolites. *Curr. Pharm. Des.* **2003,** *9* (31), 2627–2642.
119. Plunkett, W.; Gandhi, V. Purine and pyrimidine nucleoside analogs. *Cancer Chemother. Biol. Response Modif.* **2001,** *19,* 21–45.

120. Shao, J.; Zhou, B.; Chu, Bernard; Yen, Y. Ribonucleotide reductase inhibitors and future drug design. *Curr. Cancer Drug Targets* **2006,** *6* (5), 409–431.
121. Shao, J.; Liu, X.; Zhu, L.; Yun, Y. Targeting ribonucleotide reductase for cancer therapy. *Expert Opin. Ther. Targets* **2013,** *17* (12), 1423–1437.
122. Cerqueira, N. M. F.S.A.; Fernandes, P. A.; Ramos, M. J. Targeting ribonucleotide reductase for cancer chemotherapy. *Front. Anti-Cancer Drug Discovery* **2010,** *1,* 1–30.
123. Perez, M. A. S.; Cerqueira, N. M. F.S.A.; Fernandes, P. A.; Ramos, M. J. Ribonucleotide reductase: a mechanistic portrait of substrate analogues inhibitors. *Curr. Med. Chem.* **2010,** *17* (26), 2854–2872.
124. Shao, J.; Zhou, B.; Chu, B.; Yen, Y. Ribonucleotide reductase inhibitors and future drug design. *Curr. Cancer Drug Targets* **2006,** *6* (5), 409–431.
125. Cerqueira, N. M. F.S.A.; Fernandes, P. A.; Ramos, M. J. Ribonucleotide reductase: a critical enzyme for cancer chemotherapy and antiviral agents. *Recent Pat. Anti-Cancer Drug Discovery* **2007,** *2* (1), 11–29.
126. Madaan, K.; Kaushik, D.; Verma, T. Hydroxyurea: a key player in cancer chemotherapy. *Expert Rev. Anticancer Ther.* **2012,** *12* (1), 19–29.
127. Saban, N.; Bujak, M. Hydroxyurea and hydroxamic acid derivatives as antitumor drugs. *Cancer Chemother. Pharmacol.* **2009,** *64* (2), 213–221.
128. Miah, A. B.; Harrington, K. J.; Nutting, C. M. Triapine in clinical practice. *Eur. J. Clin. Med. Oncol.* **2010,** *2* (1), 1–6.
129. Kalinowski, D. S.; Quach, P.; Richardson, D. R. Thiosemicarbazones: the new wave in cancer treatment. *Future Med. Chem.* **2009,** *1* (6), 1143–1151.
130. Yu, Y.; Kalinowski, D. S.; Kovacevic, Z.; Siafakas, A. R.; Jansson, P. J.; Stefani, C.; Lovejoy, D. B.; Sharpe, P. C.; Bernhardt, P. V.; Richardson, D. R. Thiosemicarbazones from the old to new: iron chelators that are more than just ribonucleotide reductase inhibitors. *J. Med. Chem.* **2009,** *52* (17), 5271–5294.
131. Cragg, G. M.; Newman, D. J. Antineoplastic agents from natural sources: achievements and future directions. *Expert Opin. Invest. Drugs* **2000,** *9* (12), 2783–2797.
132. Gordaliza, M. Natural products as leads to anticancer drugs. *Clin. Transl. Oncol.* **2007,** *9* (12), 767–776.
133. Mukherjee, A. K.; Basu, S.; Sarkar, N.; Ghosh, A. C. Advances in cancer therapy with plant based natural products. *Curr. Med. Chem.* **2001,** *8* (12), 1467–1486.
134. Das, B.; Das, R. Biotechnological applications in anticancer medicinal plants. In Govil, J. N., Kumar, P. Ananda, Singh, V. K., Eds.; *Recent Progress in Medicinal Plants: Biotechnology and Genetic Engineering*, Vol. 4; Stadium Press, 2004; pp 53–62.
135. Mollinedo, F.; Gajate, C. Microtubules, microtubule -interfering agents and apoptosis. *Apoptosis* **2003,** *8* (5), 413–450.
136. Sarabia, F.; Garcia-Castro, M.; Sanchez-Ruiz, A. Chemistry and biology of novel microtubule-destabilizing agents that bind α-tubulin. *Curr. Bioact. Compd.* **2006,** *2* (3), 269–299.
137. Gueritte, F.; Fahy, J. The vinca alkaloids. In *Anticancer Agents from Natural Products;* Cragg, G. M., Kingston, D. G. I., Newman, D. J., Eds.; CRC Press, 2005; pp 123–135.
138. Roussi, F.; Gueritte, F.; Fahy, J. The vinca alkaloids. In *Anticancer Agents from Natural Products,* 2nd ed.; Cragg, G. M., Kingston, D. G. I., Newman, D. J., Eds. CRC Press, 2012; pp 177–198. (1 plate).
139. Fumoleau, P.; Guiu, S. New vinca alkaloids in clinical development. *Curr. Breast Cancer Rep.* **2013,** *5* (1), 69–72.
140. Duflos, A.; Kruczynski, A.; Barret, J.-M. Novel aspects of natural and modified vinca alkaloids. *Curr. Med. Chem.: Anti-Cancer Agents* **2002,** *2* (1), 55–70.

141. Wang, Y.-F.; Shi, Q.-W.; Dong, M.; Kiyota, H.; Gu, Y.-C.; Cong, B. Natural taxanes: developments since 1828. *Chem. Rev. (Washington, DC, U. S.)* **2011,** *111* (12), 7652–7709.
142. Muggia, F.; Kudlowitz, D. Novel taxanes. *Anti-Cancer Drugs* **2014,** *25* (5), 593–598.
143. Cortes, J.; Vidal, M. Beyond taxanes: the next generation of microtubule-targeting agents. *Breast Cancer Res. Treat.* **2012,** *133* (3), 821–830.
144. Kaiser, S.; Muller, J. J.; Froehlich, P. E.; Cristina, B. G. S.; Bergold, A. M. From bacteria to antineoplastic: epothilones a successful history. *Anti-Cancer Agents Med. Chem.* **2013,** *13* (7), 1057–1068.
145. Pfeiffer, B.; Gaugaz, F. Z.; Schiess, R.; Altmann, K.-H. Epothilones as lead structures for new anticancer drugs. *RSC Drug Discovery Ser.* **2012,** *25,* 339–373.
146. Ferrandina, G.; Mariani, M.; Andreoli, M.; Shahabi, S.; Scambia, G.; Ferlini, C. Novel drugs targeting microtubules: the role of epothilones. *Curr. Pharm. Des.* **2012,** *18* (19), 2793–2803.
147. Michaud, L. B. The epothilones: how pharmacology relates to clinical utility. *Ann. Pharmacother.* **2009,** *43* (7/8), 1294–1309.
148. Cheng, K. L.; Bradley, T.; Budman, D. R. Novel microtubule-targeting agents—the epothilones. *Biol.: Targets Ther.* **2008,** *2* (4), 789–811.
149. Lee, J. J.; Kelly, W. K. Epothilones: tubulin polymerization as a novel target for prostate cancer therapy. *Nat. Clin. Pract. Oncol.* **2009,** *6* (2), 85–92.
150. Mulzer, J.; Altmann, K.-H.; Hoefle, G.; Mueller, R.; Prantz, K. Epothilones—a fascinating family of microtubule stabilizing antitumor agents. *C. R. Chim.* **2008,** *11* (11-12), 1336–1368.
151. Trivedi, M.; Budihardjo, I.; Loureiro, K.; Reid, T. R.; Ma, J. D. Epothilones: a novel class of microtubule-stabilizing drugs for the treatment of cancer. *Future Oncol.* **2008,** *4* (4), 483–500.
152. Lee, J. J.; Swain, S. M. The epothilones: translating from the laboratory to the clinic. *Clin. Cancer Res.* **2008,** *14* (6), 1618–1624.
153. Feyen, F.; Cachoux, F.; Gertsch, J.; Wartmann, M.; Altmann, K.-H. Epothilones as lead structures for the synthesis-based discovery of new chemotypes for microtubule stabilization. *Acc. Chem. Res.* **2008,** *41* (1), 21–31.
154. Cortes, J.; Baselga, J. Targeting the microtubules in breast cancer beyond taxanes: the epothilones. *Oncologist* **2007,** *12* (3), 271–280.
155. Altmann, K.-H.; Pfeiffer, B.; Arseniyadis, S.; Pratt, B. A.; Nicolaou, K. C. The chemistry and biology of epothilones— the wheel keeps turning. *ChemMedChem* **2007,** *2* (4), 396–423.
156. Kolman, A. Activity of epothilones. *Curr. Opin. Invest. Drugs (BioMed Cent.)* **2005,** *6* (6), 616–622.
157. Dezhenkova, L. G.; Tsvetkov, V. B.; Shtil, A. A. Topoisomerase I and II inhibitors: chemical structure, mechanisms of action and role in cancer chemotherapy. *Russ, Chem. Rev.* **2014,** *83* (1), 82–94.
158. Khadka, D. B.; Cho, W.-J. Topoisomerase inhibitors as anticancer agents: a patent update. *Expert Opin. Ther. Pat.* **2013,** *23* (8), 1033–1056.
159. Cortes, F.; Pastor, N.; Mateos, S.; Dominguez, I. Topoisomerase Inhibitors as therapeutic weapons. *Expert Opin. Ther. Pat.* **2007,** *17* (5), 521–532.
160. Pommier, Y. DNA Topoisomerase I inhibitors: chemistry, biology, and interfacial inhibition. *Chem. Rev. (Washington, DC, U. S.)* **2009,** *109* (7), 2894–2902.
161. Bailly, C. Contemporary challenges in the design of topoisomerase II inhibitors for cancer chemotherapy. *Chem. Rev. (Washington, DC, U. S.)* **2012,** *112* (7), 3611–3640.
162. Thakur, D. S. Topoisomerase II inhibitors in cancer treatment. *Int. J. Pharm. Sci. Nanotechnol.* **2011,** *3* (4), 1173–1181.
163. Paz, M. M. Antitumour antibiotics. In *Anticancer Therapeutics;* Missailidis, S., Ed.; Wiley, 2008; pp 111–131.

164. Galm, U.; Hager, M. H.; Van Lanen, S. G.; Ju, J.; Thorson, J. S.; Shen, B. Antitumour antibiotics: bleomycin, enediynes, and mitomycin. *Chem. Rev. (Washington, DC, U. S.)* **2005,** *105* (2), 739–758.
165. Rajski, S. R.; Williams, R. M. DNA crosslinking agents as antitumor drugs. *Chem. Rev. (Washington, DC, U. S.)* **1998,** *98* (8), 2723–2795.
166. Remers, W. A.; Iyengar, B. S. Antitumour antibiotics, Cancer Chemotherapeutic Agents. In Foye, W. O., Ed.; American Chemical Society, 1995; pp 577–679.
167. Hecht, S. M. Bleomycin group antitumor agents, Anticancer Agents from Natural Products. In Cragg, G. M., Kingston, D. G. I., Newman, D. J., Eds., 2nd ed.; CRC Press, 2012; pp 451–478.
168. Sawyers, C. Targeted cancer therapy. *Nature* **2004,** *432* (7015), 294–297.
169. Lu, Y.-H.; Gao, X.-Q.; Wu, M.; Zhang-Negrerie, D.; Gao, Q. Strategies on the development of small molecule anticancer drugs for targeted therapy. *Mini-Rev. Med. Chem.* **2011,** *11* (7), 611–624.
170. Wu, H.-C.; Chang, D.-K.; Huang, C.-T. Targeted therapy for cancer. *J. Cancer Mol.* **2006,** *2* (2), 57–66.
171. Storey, S. Targeting apoptosis: selected anticancer strategies. *Nat. Rev. Drug Discovery* **2008,** *7* (12), 971–972.
172. Arora, A.; Scholar, E. M. Role of tyrosine kinase inhibitors in cancer therapy. *J. Pharmacol. Exp. Ther.* **2005,** *315* (3), 971–979.
173. Traxler, P. M. Protein tyrosine kinase Inhibitors in cancer treatment. *Expert Opin. Ther. Pat.* **1997,** *7* (6), 571–588.
174. Levitzki, A. Protein tyrosine kinase inhibitors as novel therapeutic agents. *Pharmacol. Ther.* **1999,** *82* (2-3), 231–239.
175. Fry, D. W. Recent advances in tyrosine kinase Inhibitors. *Annu. Rep. Med. Chem.* **1996,** *31,* 151–160.
176. Bedada, A. T.; Gayesa, R. T. Tyrosine kinase as target for cancer treatment. *Int. J. Pharm. Sci. Res.* **2014,** *5* (1), 1–15.
177. Cui, J. J. A new challenging and promising era of tyrosine kinase inhibitors. *ACS Med. Chem. Lett.* **2014,** *5* (4), 272–274.
178. Zamecnikova, A. Novel approaches to the development of tyrosine kinase Inhibitors and their role in the fight against cancer. *Expert Opin. Drug Discovery* **2014,** *9* (1), 77–92.
179. Morin, M. J. From oncogene to drug: development of small molecule tyrosine kinase inhibitors as antitumor and antiangiogenic agents. *Oncogene* **2000,** *19* (56), 6574–6583.
180. Hunter, T. The role of tyrosine phosphorylation in cell growth and disease. *Harvey Lect.* **2000,** *94,* 81–119.
181. Gibbs, J. B. Anticancer drug targets: growth factors and growth factor signaling. *J. Clin. Invest.* **2000,** *105* (1), 9–13.
182. Sedlacek, H. H. Kinase inhibitors in cancer therapy: a look ahead. *Drugs* **2000,** *59* (3), 435–476.
183. Lawrence, D. S.; Niu, J. Protein kinase inhibitors: the tyrosine-specific protein kinases. *Pharmacol. Ther.* **1998,** *77* (2), 81–114.
184. Klastersky, J. A. Adverse events of targeted therapies. *Curr. Opin. Oncol.* **2014,** *26* (4), 395–402.
185. Cohen, P.; Alessi, D. R. Kinase drug discovery—what's next in the field? *ACS Chem. Biol.* **2013,** *8* (1), 96–104.
186. Baselga, J. Targeting tyrosine kinases in cancer: the second wave. *Science (Washington, DC, U. S.)* **2006,** *312* (5777), 1175–1178.
187. Liu, Yi; Gray, N. S. Rational design of inhibitors that bind to inactive kinase conformations. *Nat. Chem. Biol.* **2006,** *2* (7), 358–364.

188. Laird, A. D.; Cherrington, J. M. Small molecule tyrosine kinases inhibitors: clinical development of anticancer agents. *Expert Opin. Invest. Drugs* **2003,** *12* (1), 51–64.
189. Wagner, J. P.; Wolf-Yadlin, A.; Sevecka, M.; Grenier, J. K.; Root, D. E.; Lauffenburger, D. A.; MacBeath, G. Receptor tyrosine kinases fall into distinct classes based on their inferred signaling networks. *Sci. Signaling* **2013,** *6* (284). ra58, 18 pp.
190. Natoli, C.; Perrucci, B.; Perrotti, F.; Falchi, L.; Iacobelli, S. Tyrosine kinase inhibitors. *Curr. Cancer Drug Targets* **2010,** *10* (5), 462–483.
191. Bridges, A. J. The rationale and strategy used to develop a series of highly potent, irreversible, inhibitors of the epidermal growth factor receptor family of tyrosine kinases. *Curr. Med. Chem.* **1999,** *6* (9), 825–843.
192. Sia, D.; Alsinet, C.; Newell, P.; Villanueva, A. VEGF signaling in cancer treatment. *Curr. Pharm. Des.* **2014,** *20* (17), 2834–2842.
193. Morabito, A.; De Maio, E.; Di Maio, M.; Normanno, N.; Perrone, F. Tyrosine kinase inhibitors of vascular endothelial growth factor receptors in clinical trials: current status and future directions. *Oncologist* **2006,** *11* (7), 753–764.
194. Capdeville, R.; Buchdunger, E.; Zimmermann, J.; Matter, A. Glivec (STI571, imatinib), a rationally developed, targeted anticancer drug. *Nat. Rev. Drug Discovery* **2002,** *1* (7), 493–502.
195. Savage, D. G.; Antman, K. H. Imatinib mesylate—a new oral targeted therapy. *N. Engl. J. Med.* **2002,** *346* (9), 683–693.
196. Joensuu, H.; Dimitrijevic, S. Tyrosine kinase inhibitor imatinib (STI571) as an anticancer agent for solid tumours. *Ann. Med.* **2001,** *33* (7), 451–455.
197. Lyseng-Williamson, K.; Jarvis, B. Imatinib. *Drugs* **2001,** *61* (12), 1765–1774.
198. Druker, B. J. Imatinib as a paradigm of targeted therapies. *Adv. Cancer Res.* **2004,** *91,* 1–30.
199. de Bree, F.; Sorbera, L. A.; Fernandez, R.; Castaner, J. Imatinib mesylate. *Drugs Future* **2001,** *26* (6), 545–552.
200. Hochhaus, A.; La Rosee, P. Imatinib therapy in chronic myelogenous leukemia: strategies to avoid and overcome resistance. *Leukemia* **2004,** *18* (8), 1321–1331.
201. Cohen, M. H.; Williams, G.; Johnson, J. R.; Duan, J.; Gobburu, J.; Rahman, A.; Benson, K.; Leighton, J.; Kim, S. K.; Wood, R.; Rothmann, M.; Chen, G.; Maung, U. K.; Staten, A. M.; Pazdur, R. Approval summary for imatinib mesylate capsules in the treatment of chronic myelogenous leukemia. *Clin. Cancer Res.* **2002,** *8* (5), 935–942.
202. Dagher, R.; Cohen, M.; Williams, G.; Rothmann, M.; Gobburu, J.; Robbie, G.; Rahman, A.; Chen, G.; Staten, A.; Griebel, D.; Pazdur, R. Approval summary: imatinib mesylate in the treatment of metastatic and/or unresectable malignant gastrointestinal stromal tumors. *Clin. Cancer Res.* **2002,** *8* (10), 3034–3038.
203. Radford, I. R. Imatinib Novartis. *Curr. Opin. Invest. Drugs (BioMed Cent.)* **2002,** *3* (3), 492–499.
204. Goldman, J. M. How I treat chronic myeloid leukemia in the imatinib era. *Blood* **2007,** *110* (8), 2828–2837.
205. Druker, B. J. Translation of the Philadelphia chromosome into therapy for CML. *Blood* **2008,** *112* (13), 4808–4817.
206. Maekawa, T.; Ashihara, E.; Kimura, S. The Bcr-Abl tyrosine kinase inhibitor imatinib and promising new agents against Philadelphia chromosome-positive leukemias. *Int. J. Clin. Oncol.* **2007,** *12* (5), 327–340.
207. Mueller, B. A. Imatinib and its successors-how modern chemistry has changed drug development. *Curr. Pharm. Des.* **2009,** *15* (2), 120–133.
208. Waller, C. F. Imatinib mesylate. *Recent Results Cancer Res.* **2010,** *184,* 3–20.

209. Hochhaus, A.; Reiter, A.; Ernst, T.; La Rosee, P. Imatinib and beyond—targeting activated tyrosine kinases in myeloproliferative disorders. *Onkologie* **2012,** *35* (Suppl. 1), 34–41.
210. Apperley, J. F. Imatinib—should we have more of a good thing? *Nat. Rev. Clin. Oncol.* **2010,** *7* (6), 303–304.
211. Ksienski, D. Imatinib mesylate: past successes and future challenges in the treatment of gastrointestinal stromal tumors. *Clin. Med. Insights: Oncol.* **2011,** *5,* 365–379.
212. Quintas-Cardama, A.; Kantarjian, H.; Cortes, J. Imatinib and beyond—exploring the full potential of targeted therapy for CML. *Nat. Rev. Clin. Oncol.* **2009,** *6* (9), 535–543.
213. Soverini, S.; Martinelli, G.; Iacobucci, I.; Baccarani, M. Imatinib mesylate for the treatment of chronic myeloid leukemia. *Expert Rev. Anticancer Ther.* **2008,** *8* (6), 853–864.
214. Siddiqui, M. A. A.; Scott, L. J. Imatinib. A review of its use in the management of gastrointestinal stromal tumours. *Drugs* **2007,** *67* (5), 805–820.
215. Moen, M. D.; McKeage, K.; Plosker, G. L.; Siddiqui, M. A. A.; Berman, E.; Gambacorti-Passerini, C.; Goldman, J. H.; Hochhaus, A.; Larson, R. A.; Rosti, G.; Simonsson, B. Imatinib: a review of its use in chronic myeloid leukaemia. *Drugs* **2007,** *67* (2), 299–320.
216. Muller, B. A. Imatinib and its successors—how modern chemistry has changed drug development. *Curr. Pharm. Des.* **2009,** *15,* 120–133.
217. Zimmermann, J., Preparation of 2-anilinopyrimidines as antiatherosclerotics and neoplasm inhibitors, EP 564409 (1993).
218. Zimmermann, J., Pyrimidine derivatives and processes for the preparation thereof, US 552(1184), (1996).
219. Zimmermann, J.; Buchdunger, E.; Mett, H.; Meyer, T.; Lydon, N. B. Potent and selective inhibitors of the ABL-kinase: phenylaminopyrimidine (PAP) derivatives. *Bioorg. Med. Chem. Lett.* **1997,** *7* (2), 187–192.
220. Loiseleur, O.; Kaufmann, D.; Abel, S.; Buerger, H. M.; Meisenbach, M.; Schmitz, B.; Sedelmeier, G., Preparation of N-(pyridin-3-ylpyrimidin-2-ylaminophenyl)benzamide derivatives, WO 2003066613 (2003).
221. MacDonald, P.; Rossetto, P., Process for the preparation of imatinib, US 20080103305 (2008).
222. Anli, H.; Xing, L.; Zelikovitch, L.; Kaspi, J., Multi-step process for preparing imatinib, US 20060149061 (2006).
223. Kankan, R. N.; Rao, D. R. Preparation of imatinib and salts by reaction of N-(2-methyl-5-aminophenyl-4-(3-pyridyl)-2-pyrimidinamine with 4-(4-methylpiperazinylmethyl)benzoyl halides. *GB* **2004,** *23* (9856). 5.
224. Kompella, A.; Adibhatla, K. S. B. R.; Venkaiah, C. N.; Srinivas, R., Process for the preparation of the anti-cancer drug imatinib and its analogs via aminolysis of a (chloromethyl)benzamide intermediate, WO 2004108699 (2004).
225. Kumar, A. A.; Kumar, J. A.; Bhaskar, B. S.; Lalit, W., Process for the preparation of imatinib and its pharmaceutically acceptable salts, IN 2009DE02038 (2011).
226. Xing, L.; Xungui, H.; Wang, Y.; Bekhazi, M.; Krivonos, S.; Danon, E., Process for the industrial preparation of Imatinib and it mesylate salt, WO 2008135980 (2008).
227. Kamath, A. A.; Pai, G. G.; Ujagare, A. M.; He, X.; Wu, S.; Shen, X.; Yang, J.; Zhan, H., Process for the preparation of imatinib and salts thereof, IN 2009MU02853 (2012).
228. Szakacs, Z.; Beni, S.; Varga, Z.; Oerfi, L.; Keri, G.; Noszal, B. Acid-base profiling of imatinib (Gleevec) and its fragments. *J. Med. Chem.* **2005,** *48* (1), 249–255.
229. Szczepek, W. J. Imatinib mesylate-synthesis methods and preparation of polymorphs. *Przem. Chem.* **2006,** *85* (5), 306–309.
230. Deadman, B. J.; Hopkin, M. D.; Baxendale, I. R.; Ley, S. V. The synthesis of Bcr-Abl inhibiting anticancer pharmaceutical agents Imatinib, nilotinib and dasatinib. *Org. Biomol. Chem.* **2013,** *11* (11), 1766–1800.

231. Al-Hadiya, B. M. H.; Bakheit, A. H. H.; Abd-Elgalil, A. A. Imatinib mesylate. *Profiles Drug Subst., Excipients, Relat. Methodol.* **2014,** *39,* 265–297.
232. Schnur, R. C.; Arnold, L. D., Preparation of N-phenylquinazoline-4-amines as neoplasm inhibitors, WO 9630347 (1996).
233. Schnur, R. C.; Arnold, L. D., Preparation of alkynyl- and azido-substituted 4-anilinoquinazolines for the treatment of hyperproliferative diseases, US 5747498 (1998).
234. Batra, S. M.; Lupon, R. M. P.; Comely, A. C.; Georges, Y. P., Process for the preparation of erlotinib, EP 2348020 (2011).
235. Lehner, R. S.; Norris, T.; Santafianos, D. P., Method for preparation of anticancer 4-(3-ethynylphenylamino)quinazoline derivatives and intermediates thereof, JP 2000290262 (2000).
236. Cruz-Lopez, O.; Conejo-Garcia, A.; Nunez, M. C.; Kimatrai, M.; Garcia-Rubino, M. E.; Morales, F.; Gomez-Perez, V.; Campos, J. M. Novel substituted quinazolines for potent EGFR tyrosine kinase inhibitors. *Curr. Med. Chem.* **2011,** *18* (7), 943–963.
237. Barghi, L.; Aghanejad, A.; Valizadeh, H.; Barar, J.; Asgari, D. Modified synthesis of erlotonib hydrochloride. *Adv. Pharm. Bull.* **2012,** *2* (1), 119–122. 4 pp.
238. Chandregowda, V.; Rao, G. V.; Reddy, G. C. Convergent Approach for Commercial Synthesis of Gefitinib and Erlotinib Org. *Process Res. Dev.* **2007,** *11* (5), 813–816.
239. Asgari, D.; Aghanejad, A.; Mojarrad, J. S. An improved convergent approach for synthesis of erlotinib, a tyrosine kinase inhibitor, via a ring closure reaction of phenyl benzamidine intermediate. *Bull. Korean Chem. Soc.* **2011,** *32* (3), 909–914.
240. Knesl, P.; Roeseling, D.; Jordis, U. Improved synthesis of substituted 6,7-dihydroxy-4-quinazoline amines: tandutinib, erlotinib and gefitinib. *Molecules* **2006,** *11* (4), 286–297.
241. Chandregowda, V.; Rao, Gudapati, V.; Reddy, G. C. Improved synthesis of gefitinib and erlotinib hydrochloride anticancer agents. *Synth. Commun.* **2007,** *37* (19), 3409–3415.
242. Siegel-Lakhai, W. S.; Beijnen, J. H.; Schellens, J. H. M. Current knowledge and future directions of the selective epidermal growth factor receptor inhibitors erlotinib (Tarceva) and gefitinib (Iressa). *Oncologist* **2005,** *10* (8), 579–589.
243. Tang, P. A.; Tsao, M.-S.; Moore, M. J. A review of erlotinib and its clinical use. *Expert Opin. Pharmacother.* **2006,** *7* (2), 177–193.
244. Smith, J. Erlotinib: small-molecule targeted therapy in the treatment of non-small-cell lung cancer. *Clin. Ther.* **2005,** *27* (10), 1513–1534.
245. Lyseng-Williamson, K. A. Erlotinib. *Mol. Diagn. Ther.* **2013,** *17* (1), 57–62.
246. Perez-Soler, R. The role of erlotinib (Tarceva, OSI 774) in the treatment of non-small cell lung cancer. *Clin. Cancer Res.* **2004,** *10* (12 Pt. 2), 4238s–4240s.
247. Brown, E. R.; Shepherd, F. A. Erlotinib in the treatment of non-small cell lung cancer. *Expert Rev. Anticancer Ther.* **2005,** *5* (5), 767–775.
248. Bonomi, P. Erlotinib: a new therapeutic approach for non-small cell lung cancer. *Expert Opin. Invest. Drugs* **2003,** *12* (8), 1395–1401.
249. Herbst, R. S. Erlotinib (Tarceva): an update on the clinical trial program. *Semin. Oncol.* **2003,** *30* (3, Suppl. 7), 34–46.
250. D'Arcangelo, M.; Cappuzzo, F. A review of erlotinib and its clinical use in the first-line treatment of non-small-cell lung cancer. *Expert Rev. Anticancer Ther.* **2013,** *13* (5), 523–533.
251. Steins, M.; Thomas, M.; Geissler, M. Erlotinib. *Recent Results Cancer Res.* **2010,** *184,* 21–31.
252. Gridelli, C.; Maione, P.; Bareschino, M. A.; Schettino, C.; Sacco, P. C.; Ambrosio, R.; Barbato, V.; Falanga, M.; Rossi, A. A review of erlotinib—an oral, selective epidermal growth factor receptor tyrosine kinase inhibitor in the treatment of non-small cell lung cancer: current status and future developments. *Anticancer Res.* **2010,** *30* (4), 1301–1310.
253. Iyer, R.; Bharthuar, A. A review of erlotinib—an oral, selective epidermal growth factor receptor tyrosine kinase inhibitor. *Expert Opin. Pharmacother.* **2010,** *11* (2), 311–320.

254. Kelley, R. K.; Ko, A. H. Erlotinib in the treatment of advanced pancreatic cancer. *Biol.: Targets Ther.* **2008,** *2* (1), 83–95.
255. Ganjoo, K. N.; Wakelee, H. Review of erlotinib in the treatment of advanced non-small cell lung cancer. *Biol.: Targets Ther.* **2007,** *1* (4), 335–346.
256. Welch, S. A.; Moore, M. J. Erlotinib: success of a molecularly targeted agent for the treatment of advanced pancreatic cancer. *Future Oncol.* **2007,** *3* (3), 247–254.
257. Li, Z.; Chen, L.; Rubinstein, M. P. Cancer immunotherapy: are we there yet? *Exp. Hematol. Oncol.* **2013,** *2,* 33/1–33/6.
258. Ghadage, M. D.; Bedse, A. P. A review on: monoclonal antibodies. *World J. Pharm. Pharm. Sci.* **2013,** *2* (4), 1699–1721.
259. Sliwkowski, M. X.; Mellman, I. Antibody therapeutics in cancer. *Science (Washington, DC, U. S.)* **2013,** *341* (6151), 1192–1198.
260. Glassman, P. M.; Balthasar, J. P. Mechanistic considerations for the use of monoclonal antibodies for cancer therapy. *Cancer Biol. Med.* **2014,** *11* (1), 20–33.
261. Ribatti, D. From the discovery of monoclonal antibodies to their therapeutic application: an historical reappraisal. *Immunol. Lett.* **2014,** *161* (1), 96–99.
262. Rogers, L. M.; Veeramani, S.; Weiner, G. J. Complement in monoclonal antibody therapy of cancer. *Immunol. Res.* **2014,** *59* (1-3), 203–210.
263. Pandey, M.; Mahadevan, D. Monoclonal antibodies as therapeutics in human malignancies. *Future Oncol.* **2014,** *10* (4), 609–636.
264. Hardy, B.; Raiter, A. New era in cancer immunotherapy: twenty years to the discovery of monoclonal antibodies harnessing the immune system to eradicate tumors. *Adv. Biosci. Biotechnol.* **2013,** *4* (4A), 34–37.
265. Hess, C.; Venetz, D.; Neri, D. Emerging classes of armed antibody therapeutics against cancer. *MedChemComm* **2014,** *5* (4), 408–431.
266. Bhutani, D.; Vaishampayan, U. N. Monoclonal antibodies in oncology therapeutics: present and future indications. *Expert Opin. Biol. Ther.* **2013,** *13* (2), 269–282.
267. Modjtahedi, H.; Ali, S.; Essapen, S. Therapeutic application of monoclonal antibodies in cancer: advances and challenges. *Br. Med. Bull.* **2012,** *104* (1), 41–59.
268. Melero, I.; Gaudernack, G.; Gerritsen, W.; Huber, C.; Parmiani, G.; Scholl, S.; Thatcher, N.; Wagstaff, J.; Zielinski, C.; Faulkner, I.; Mellstedt, H. Therapeutic vaccines for cancer: an overview of clinical trials. *Nat. Rev. Clin. Oncol.* **2014,** *11* (9), 509–524.
269. Palucka, K.; Banchereau, J. SnapShot: cancer vaccines. *Cell* **2014,** *157* (2). 516–516.e1.
270. Cuppens, K.; Vansteenkiste, J. Vaccination therapy for non-small-cell lung cancer. *Curr. Opin. Oncol.* **2014,** *26* (2), 165–170.
271. Mocellin, S. Cancer vaccines: the challenge of developing an ideal tumor killing system. *Front. Biosci.* **2005,** *10* (Suppl.), 2285–2305.
272. Mocellin, S.; Rossi, C. R.; Nitti, D. Cancer vaccine development: on the way to break immune tolerance to malignant cells. *Exp. Cell Res.* **2004,** *299* (2), 267–278.
273. Berinstein, N. Overview of therapeutic vaccination approaches for cancer. *Semin. Oncol.* **2003,** *30* (3, Suppl. 8), 1–8.
274. Zoller, M.; Matzku, S. Cancer therapy: new concepts on active immunization. *Immunobiology* **1999,** *201* (1), 1–21.
275. Dranoff, G. Cytokines in cancer pathogenesis and cancer therapy. *Nat. Rev. Cancer* **2004,** *4* (1), 11–22.
276. Kendrick-Parker, C. J.; Jordan, V. C. Drugs that block steroid hormone action for the treatment of breast and prostate cancer. In *Cancer Chemotherapeutic Agents;* Foye, W. O., Ed.; American Chemical Society, 1995; pp 389–428.

277. Dowsett, M. The development of new drugs in the treatment of hormone-dependent cancers. In *Hormones and Cancer;* O'Brien, P. M. S., MacLean, A. B., Eds.; RCOG Press, 1999; pp 254–264.
278. Marsaud, V.; Radanyi, C.; Renoir, J. M. Steroid receptors and immunophilin relationships in hormone-dependent cancers: Use of long-circulating anti-steroid hormone and/or immunosuppressant carriers as a promising therapeutic approach. *S.T.P. Pharma* **1999,** *9* (5), 397–409.
279. Ahmad, N.; Kumar, R. Steroid hormone receptors in cancer development: a target for cancer therapeutics. *Cancer Lett. (N. Y., NY, U. S.)* **2011,** *300* (1), 1–9.
280. Levine, P. M.; Garabedian, M. J.; Kirshenbaum, K. Targeting the androgen receptor with steroid conjugates. *J. Med. Chem.* **2014,** *57* (20), 8224–8237.
281. Salvador, J. A. R.; Carvalho, J. F. S.; Neves, M. A. C.; Silvestre, S. M.; Leitao, A. J.; Silva, M. M. C.; Sa e Melo, M. L. Anticancer steroids: linking natural and semi-synthetic compounds. *Nat. Prod. Rep.* **2013,** *30* (2), 324–374.
282. Bansal, R.; Acharya, P. C. Man-made cytotoxic steroids: Exemplary agents for cancer therapy. *Chem. Rev. (Washington, DC, U. S.)* **2014,** *114* (14), 6986–7005.
283. Gupta, A.; Sathish, K. B.; Negi, A. S. Current status on development of steroids as anticancer agents. *J. Steroid Biochem. Mol. Biol.* **2013,** *137,* 242–270.
284. Lupulescu, A. P. Hormones, vitamins, and growth factors in cancer treatment and prevention. A critical appraisal. *Cancer* **1996,** *78* (11), 2264–2280.

Chapter 29

Immunopharmacological Drugs

The immune system is a complex network of organs, tissues, cells, hormones, other chemical signaling molecules, and cell products, such as antibodies that protect the organism from potential pathogens (bacteria, viruses, parasites, fungi, and others), neutralizing them and eliminating disease.

The immune system is daily exposed to a plethora of antigens (antigens are substances that induce an immune response in the body), especially antigens that stimulate the production of antibodies (antibodies are molecules produced by B cells as a primary immune defense) from the environment and food and as a response to antigenic stimuli-releasing cytokines (cytokines are a diverse group of nonantibody proteins that act as mediators between cells) or the production of antibodies.

The immune system can be categorized in a number of different ways, including the innate and adaptive, the passive and active, the cellular and antibody response. All of these categories can be targeted for treatment.

Multiple endogenous and exogenous factors have an effect on the functions of the immune system, so it follows that immune system activity can be regulated by compounds and factors that enhance or inhibit the immune function [1-18].

There are two general classes of immunotherapeutic drugs: those that stimulate and activate the immune system are called *immunostimulants*, and those that suppress the immune system are called *immunosuppressants*.

29.1 IMMUNOSUPPRESSANTS

Immunosuppressants are used to suppress the immune system in organ transplantations as antirejection drugs to prevent rejection of transplanted organs, and to treat autoimmune diseases such as rheumatoid arthritis, lupus, psoriasis, Crohn disease (a chronic inflammation of the digestive tract), multiple sclerosis, and alopecia, and to treat nonautoimmune diseases such as allergy.

There are more than 80 autoimmune diseases and several common allergic conditions in which immunosuppressants could play a role [19-23].

Immunosuppressants are a structurally and functionally heterogeneous group of drugs, which, in principle, can be classified as xenobiotic and biological immunosuppressants.

Xenobiotic immunosuppressants are endogenic compounds, mainly synthetic chemicals, or derived from microbial or plant products that are alien to the organism.

Synthesis of Best-Seller Drugs. http://dx.doi.org/10.1016/B978-0-12-411492-0.00029-8

Xenobiotics, in turn, can be categorized as corticosteroids, antimetabolites, calcineurin inhibitors, protein kinase inhibitors of the mammalian target of rapamycin (mTOR inhibitors), which is a protein kinase that regulates cell growth, cell proliferation, etc.

The mentioned xenobiotics act via a variety of intracellular molecular mechanisms with cells involved in the immune response and have multiple side effects. Most small molecule immunosuppressive agents are derived from microbial products.

Biological immunosuppressants are prepared from biological (human or animal)-based substances, and mainly target the surface of immunological cells. They are usually used as auxiliary agents of xenobiotics.

Xenobiotic Immunosuppressants

Glucocorticosteroids

Glucocorticosteroids—prednisone (**29.1.1**), prednisolone (**29.1.2**), and methylprednisolone (**29.1.3**) (Fig. 29.1.)—are extremely potent, antiinflammatory/immunosuppressive hormones. Their actions are mediated by a variety of mechanisms that alter cell function.

Steroid molecules interact with an intracytoplasmic glucocorticosteroid receptor, which, in turn, inhibits genes that code for the proinflammatory cytokines—proteins such as interleukin (IL)-1, IL-2, IL-6, and IL-8, interferon, and tumor necrosis factor that are secreted by cells, especially cells of the immune system. Glucocorticoids influence all types of inflammatory events, enhance and stimulate the production of lipocortin, which inhibits phospholipase A_2, interrupting arachidonic acid metabolism. This results in reductions in leukotriene synthesis, which partially mediates the antiasthmatic activity.

The diverse antiinflammatory and immunosuppressive activities of glucocorticosteroids made them potent drugs for the treatment of allergic, dermatologic, and autoimmune diseases when they were first used in clinical transplantation.

Unfortunately, their serious side effects, such as development of glucose intolerance, weight gain, osteoporosis, hypertension, gastritis, cataracts, and, occasionally, aseptic necrosis of the large joints, psychogenic effects, and susceptibility to viral, fungal, and mycobacterial infections limit their use.

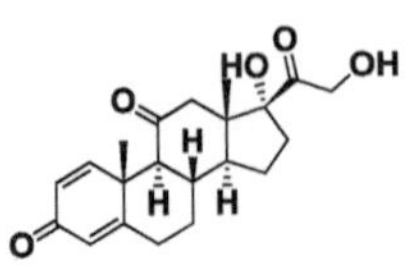

Prednisone 29.1.1

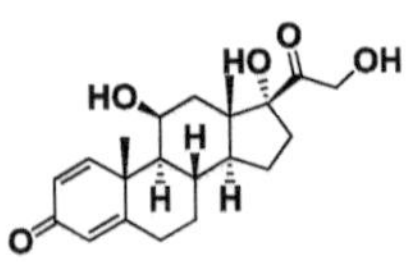

Prednisolone 29.1.2

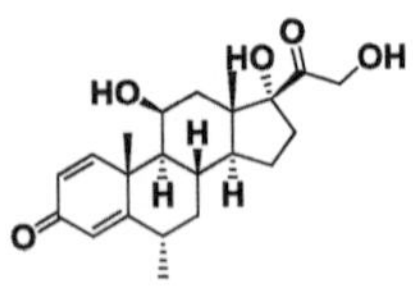

Methylprednisolone 29.1.3

FIG. 29.1 Glucocorticosteroids.

Antimetabolites

Antimetabolites—azathioprine (**29.1.4**), methotrexate (**29.1.5**), cyclophosphamide (**29.1.6**), and the more recently discovered mycophenolate mofetil (**29.1.7**) and mycophenolate acid (**29.1.8**) as sodium salt—interfere with the synthesis of nucleic acids (Fig. 29.2.).

Azathioprine (**29.1.4**) is the oldest pharmacologic agent approved as an immunosuppressive agent and has achieved widespread use.

The drug is a purine analogue, and the accepted mechanism of action is at the level of DNA. It is believed that it is incorporated into replicating DNA and thus can block the pathway of purine synthesis, inhibiting cellular proliferation. But the effects on the blockade of DNA replication have never fully explained azathioprine-induced immunosuppression.

The developers of azathioprine, Gertrude Elion and George Hitchings, were acknowledged by a Nobel Prize in 1988.

Azathioprine is a second-line drug for immunosuppression after the cyclosporine calcineurin inhibitor cyclosporine.

Methotrexate (**29.1.5**) inhibits dihydrofolate reductase, which also inhibits macrophage activation. It is used in immunology for the treatment of rheumatoid arthritis, systemic lupus, and psoriasis.

Cyclophosphamide (**29.1.6**) is an alkylating agent that bonds with DNA, thereby leading to DNA fragmentation, mutations, and cell death. It also suppresses cellular immunity and inhibits antibody and autoantibody production.

Mycophenolate mofetil (**29.1.7**) is a semisynthetic derivative of the fungal antibiotic that is converted into its active metabolite mycophenolic acid (CellCept) (**29.1.8**), which is included in the list of Top 200 Drugs by sales for the 2010s. Mycophenolic acid inhibits the enzyme inosine monophosphate dehydrogenase, thus inhibiting the de novo pathway of guanosine nucleotide synthesis, without incorporating into DNA.

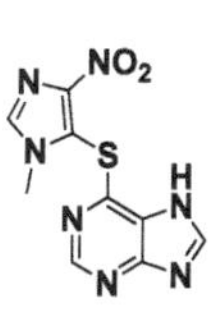

Azathioprine 29.1.4

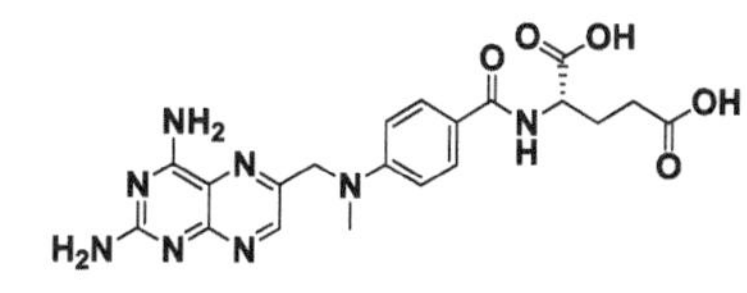

Methotrexate 29.1.5

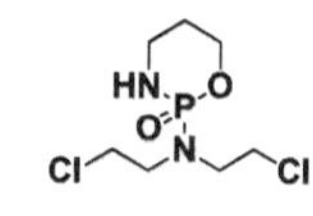

Cyclophosphamide 29.1.6

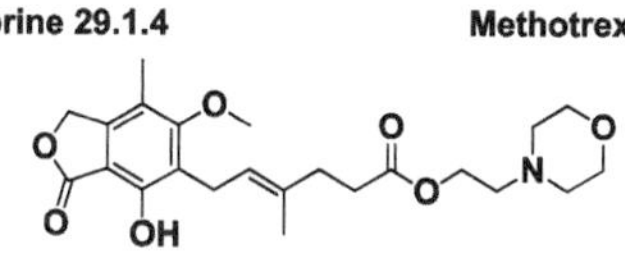

Mycophenolate mofetil 29.1.7

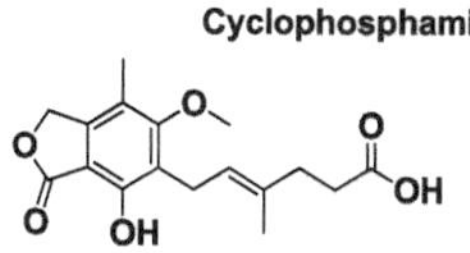

Mycophenolic acid 29.1.8
(sodium salt)

FIG. 29.2 **Antimetabolites.**

Because T and B lymphocytes are critically dependent on de novo purine synthesis for DNA replication, their action results in potent cytostatic effects on T and B lymphocytes. Mycophenolate derivatives inhibit T-cell proliferation after mitogen and allogeneic stimulation, inhibit antibody production.

Mycophenolic Acid–CellCept

Mycophenolic acid is produced by soil fungi *Penicillium brevicompactum*, *P. stoloniferum*, and related species such as *P. scabrosum*, *P. nagemi*, *P. szaferi*, *P. patus-mei*, *P. griseobrunneum*, and *P. viridicatum* [24], and was first isolated nearly a century ago [25,26]. Antibiotic activity was previously reported for this compound [27], but lately it has become of interest as an immunosuppressant.

Several industrial methods for production of mycophenolic acid are disclosed for preparation and isolation of mycophenolic acid, which include fermenting *Penicillium brevicompactum* in an appropriate culture medium containing assimilable nutrient sources of carbon, nitrogen, and inorganic salts such as glucose, glycine, methionine, $MgSO_4$, KH_2PO_4, $FeSO_4$, $CuSO_4$, $ZnSO_4$, $MnSO_4$, K_2MnO_4, and $CaCO_3$ [28-34].

In reported fermentation processes, the yields are very low, in the range of 1.3 mg/L to 363 mg/L. Solid-state fermentation is reported to result in a yield of 3.286 g/kg of biomass [35].

Although several biological fermentation methods are reported for production of mycophenolic, synthetic pathways for its synthesis are proposed also.

Mycophenolic acid was originally synthesized starting from 1,3-dimethoxy-4,6-dimethylcyclohexa-1,3-diene (**29.1.9**), which, in turn, was prepared by rearrangement of 1,5-dimethoxy-2,4-cyclohexa-1,4-diene synthesized by Birch reduction of 4,6-dimethoxy-1,3-dimethylbenzene. The method cannot be considered preparative or industrial.

Thus, 1,3-dimethoxy-2,6-dimethylcyclohexa-1,3-diene (**29.1.9**) was involved in an Alder–Rickert reaction to produce the intermediate bicyclic product (**29.1.10**), which on heating was transformed to dimethyl 4,6-dimethoxy-3-methylphthalate (**29.1.11**).

Selective demethylation of the last with boron trichloride at −10°C produced the ester (**29.1.12**), which on basic hydrolysis with sodium hydroxide at room temperature was converted into the phthalic anhydride derivative (**29.1.13**). Reduction of the phthalic anhydride derivative (**29.1.13**) with zinc dust in a mixture of acetic and hydrochloric acids produced 7-hydroxy-5-methoxyisobenzofuran-1(3H)-one (**29.1.14**). The phenolic hydroxyl group of the last was etherified with allyl bromide in standard potassium carbonate/acetone conditions to produce allyl ether (**29.1.15**), a Claisen rearrangement of which produced 6-allyl-7-hydroxy-5-methoxyisobenzofuran-1(3H)-one (**29.1.16**). Very careful selective ozonolysis of the last produced aldehyde (**29.1.17**). The obtained aldehyde was converted to 2-methylbut-2-enal (**29.1.18**) using a Wittig reaction with appropriate triphenylphosphanylidene reagent.

Reaction of (**29.1.19**) with the carbanion from triethyl phosphonoacetate produced dienoic ester (**29.1.20**), the disubstituted double bond that was selectively hydrogenated by diimide generated in situ through the oxidation of hydrazine. The final hydrolysis of the ester group was carried out with sodium hydroxide in a water–ethanol mixture to produce the desired mycophenolic acid (**29.1.8**) as a sodium salt [26] (Scheme 29.1.).

Other proposed synthetic methods are more or less related to that described above and also proceed to produce in low total yields [36-42] and are reviewed by Cholewinski et al [42].

Mycophenolic acid (MPA) is an immunosuppressive drugs used in clinics against rejection in solid-organ transplantations [43-56]. It inhibits an enzyme, inosine monophosphate dehydrogenase, thereby blocking purine synthesis and inhibiting the proliferation of B and T lymphocytes. Consequently, MPA functions as an effective immunosuppressive agent in transplantation.

Two forms of mycophenolate are available in practice: MPA (**29.1.8**) itself as a sodium salt and mycophenolate mofetil (**29.1.7**). Both products are approved for prophylaxis of organ rejection.

Calcineurin Inhibitors

Calcineurin (phosphatase enzyme involved in T-cell signaling transcription) inhibitors such as cyclosporine (**29.1.21**) and tacrolimus (**29.1.22**) block the activation of lymphocytes and other immune system cells. They are prodrugs that engage cyclophilin, an intracellular protein of the immunophilin family, forming a complex that then engages calcineurin. The introduction of these drugs to transplant management greatly improves organ transplantation protocols and are one of the major breakthroughs in modern medicine. A chemically modified cyclosporine, voclosporin (**29.1.23**), is under development (Fig. 29.3.).

Tacrolimus inhibits calcineurin with greater molar potency than does cyclosporine.

Cyclosporine and tacrolimus are the most powerful immunosuppressants; they are cornerstones of immunosuppression and are included in the list of Top 200 Drugs by sales for the 2010s.

Cyclosporine–Restasis

The cyclosporines are cyclic, 11-membered oligopeptides wherein some of the amino acid moieties are different. At present, approximately 25 different cyclosporine antibiotics, listed as cyclosporines A to Z, are known [57].

Cyclosporine (**29.1.21**) was the first among them [58] to be separated and investigated as an antifungal antibiotic. Significant immunosuppressive effect in human organ transplantations was recognized later [59].

Cyclosporines are metabolites of several fungi, including *Trichoderma polysporum*, *Cylindrocarpon lucidum*, and *Tolypocladium inflatum*.

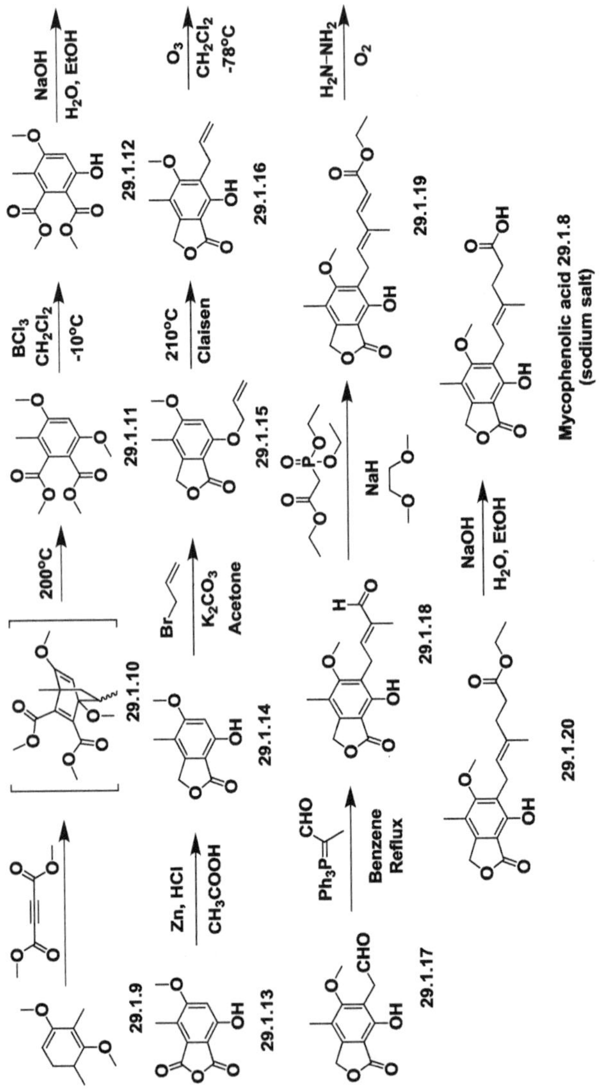

SCHEME 29.1 Synthesis of mycophenolic acid.

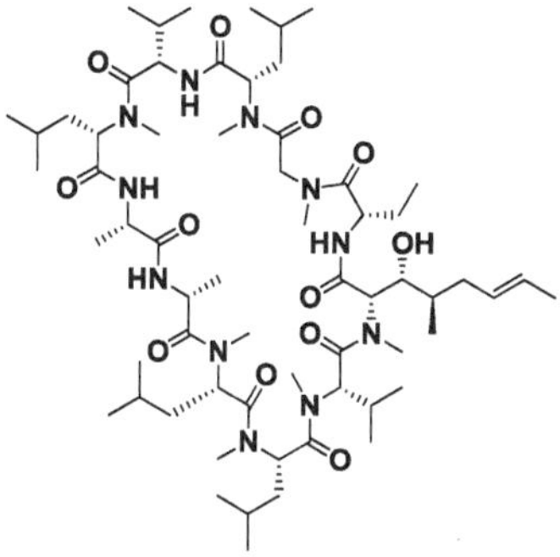

Ciclosporin
29.1.21

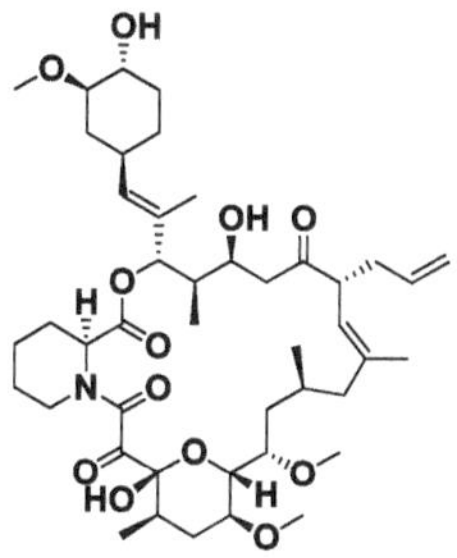

Tacrolimus
29.1.22

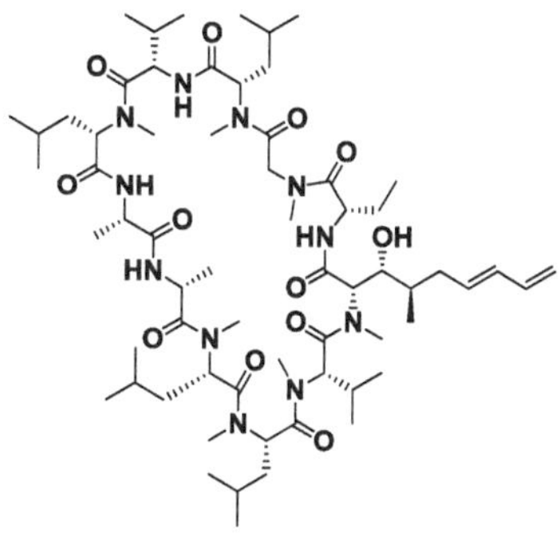

Voclosporin
29.1.23

FIG. 29.3 **Calcineurin inhibitors.**

In the course of the fermentation processes carried out with the above-mentioned microorganisms in different medias, such as media containing glucose, casein acid hydrolysate, malt extract, peptone, DL-α-amino butyric acid, $NaNO_3$, KH_2SO_4, KCl, $MgSO_4$, and $FeSO_4$, production of desired cyclosporines achieved approximately 2.2 g/L medium or 5.85 g/kg biomass [60-65].

Cyclosporine is the most commonly used immunosuppressive agent during organ transplantation. It selectively inhibits T-cell activation and is primarily applied in rejection of organ transplantation. There are different other effects induced by cyclosporine mediating inhibition of DNA repair, synthesis of transforming growth factor (TGF)-β, and others.

Cyclosporine binds to a cytosolic protein called cyclophilin. The cyclosporine–cyclophilin complex, in turn, binds to calcineurin. (Calcineurin dephosphorylates a transcription factor, thereby triggering transcription of numerous genes associated with T-cell proliferation.) When the cyclosporine–cyclophilin complex binds to calcineurin, T-cell proliferation is suppressed. The inhibition of T-cell proliferation results in the suppression of the activation process associated with invasion by foreign bodies. As a consequence, transplant tissues, which are foreign bodies, are not rejected [66-71]. One of the most feared adverse effects of cyclosporine is the appearance of de novo cancers [72].

Tacrolimus–Prograf

Tacrolimus (**29.1.22**) is a 23-membered potent macrolide immunosuppressant, isolated as a metabolite from the whole fermentation broth of *Streptomyces* species [73], which is widely used to prevent rejection.

As an immunosuppressive agent, tacrolimus is 100 times more potent than cyclosporine [74]. Differences between cyclosporine and tacrolimus are widely discussed [75-77].

Tacrolimus is approved for prophylaxis of transplant rejection in liver, kidney, or heart allograft recipients, and for the treatment of allograft rejection resistant to treatment with other immunosuppressive medicinal products [78-88]. Moreover, tacrolimus and its derivatives are effective in treating a number of diseases, such as asthma, inflammatory diseases, and hyperproliferative skin disease. Tacrolimus is a drug of choice in the treatment of atopic dermatitis [89,90]. Unfortunately therapeutic use of tacrolimus is complicated because of its narrow therapeutic index.

At the present time, the large-scale industrial production of tacrolimus is based on the fermentation broth of various *Streptomyces* species cultivated in submerged cultures [91-101]. As a result of high production cost and low yield, the cost of tacrolimus pharmacotherapy is very high.

Mammalian Target of Rapamycin Inhibitors

Mammalian Target of Rapamycin Inhibitor (mTOR) is a widely expressed protein kinase that is centrally involved in control of cell growth, proliferation, angiogenesis, differentiation, and cell-cycle regulation [102-106]. The mTOR inhibitors in clinical use are macrolides sirolimus (rapamycin) (**29.1.24**) [107] and everolimus (**29.1.25**) [108] (Fig. 29.4.).

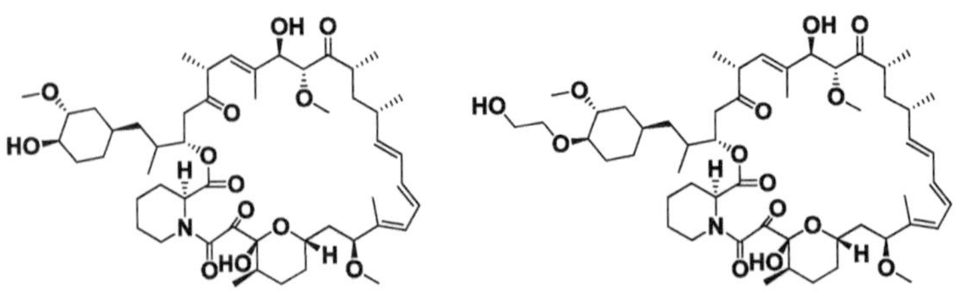

Sirolimus 29.1.24 Everolimus 29.1.25

FIG. 29.4 **Mammalian target of rapamycin inhibitors.**

The mTOR inhibitors work as immunosuppressants blocking cell proliferation by preventing the binding of growth factors to their receptors in the cell, which in the case of lymphocytes is IL-2. They disrupt production of this kinase sharing the same intracellular receptor [109-115].

As inhibitors of mTOR, sirolimus and everolimus suppress T-cell stimulation, promote regulatory T-cell proliferation, inhibit dendritic cell maturation, and play antirejection roles. Because mTOR is a crucial mediator of tumor progression, it may be a promising target in cancer [111]. The immunosuppressants sirolimus and everolimus can serve entirely different purposes [107,108].

Inhibition of mTOR is also suspected to induce metabolic and stress responses that favor longevity, extending maximal life span in both genders in a mammalian species, although exactly how this occurs is not clear.

The discovery that sirolimus increased the life span of mice was recognized by *Science* as one of the top 10 scientific breakthroughs of 2009 [116-118].

Immune suppression by sirolimus is associated with potentially serious side effects, including increased risk of infection and lymphoma. The initial enthusiasm for the advent of a potentially nonnephrotoxic immunosuppressant has been muted by some data on generated nephrotoxicity, anemia, thrombocytopenia, hyperlipidemia, and diabetogenesis, which have limited its use.

Sirolimus and everolimus are produced by the methods of biotechnological fermentation of bacterial species in the genus *Streptomyces*, particularly *Streptomyces hygroscopicus*, generating the antibiotic in medium that contains flour, dextrin, glycerol, L-lysine, KH_2PO_4, K_2HPO_4, $(NH_4)_2SO_4$, and trace elements [119-126]. Synthesis of temsirolimus from rapamycin is patented [127]. Synthetic approaches to sirolimus are also published [128].

The described xenobiotics are widely used to prevent transplant rejection. However, they have multiple side effects, which is why there has been a search for new drugs to minimize these side effects and to improve patients' quality of life.

Biological Immunosuppressants

Biological agents have been proposed as maintenance immunosuppressive agents. The majority of these new immunosuppressive agents are polyclonal or monoclonal antibodies, and the so-called fusion proteins may be the start of a new era of biological immunosuppression for maintenance regimens.

The choice of agent varies from country to country, clinic to clinic, and patient to patient, but generalizing it is possible to conclude that currently, the three most commonly used antibodies are antithymocyte globulin (ATG), alemtuzumab, and basiliximab.

Polyclonal Antibodies

The first biological immunosuppressants used in human therapeutics are currently called polyclonal antibodies. Antibodies are proteins produced by the B lymphocytes of the immune system in response to foreign proteins called antigens. Polyclonal antibodies contain a mixture of antibodies against various lymphocytes. They are usually prepared by injection of lymphocytes or thymocytes into animals, mainly rabbits, triggering an antibody response from which the immunoglobulin fraction containing antibodies of all specificities is extracted and purified to be used in human therapy. These are commonly known as ATG, antilymphocyte globulin (ALG), and muromonab–CD3.

The rationale behind their use is their powerful ability to destroy the lymphocytes that they target.

Rabbit ATG, is an immunosuppressive drug approved for the treatment of renal transplant rejection.

Monoclonal Antibodies

Antibodies produced by a single clone of cells or cell line and consisting of identical antibody molecules are considered monoclonal antibodies.

The main feature of the monoclonal antibody–epitope relationship is key to their use. Monoclonal antibody is specific antibody that binds only to its particular epitope (a short amino acid sequence of the antigen that the antibody binds to).

The two most commonly used antibodies are alemtuzumab and basiliximab.

Alemtuzumab is a humanized monoclonal antibody used for the treatment of chronic lymphocytic leukemia by targeting CD52, a broadly expressed cell surface molecule on immune cells present on the surface of mature lymphocytes.

Initially, it was approved for use in patients with B-cell chronic lymphocytic leukemia for several years, but recently was approved in the European Union and several other countries for use in adult patients with active relapsing–remitting multiple sclerosis and as an induction agent in adult kidney transplant recipients.

Basiliximab is a monoclonal antibody directed at the IL-2 receptor (CD25). It is used in transplantation and is widely employed to lower the risk of acute rejection after organ transplantation, most frequently in kidney and liver transplantation in adults with renal dysfunction, but not yet significantly used in autoimmune diseases.

It is a chimeric monoclonal antibody directed against the α chain of IL-2 receptor (IL-2R). It selectively binds to the α-subunit of these receptors and is widely employed in lowering the risk of acute rejection after organ transplantation.

Development of monoclonal antibodies is a growing segment of pharmaceutical industry. Many new monoclonal antibodies have been approved for a variety of therapeutic applications; approximately 500 derivatives are currently in different stages of development, including:

- Daclizumab is a monoclonal antibody specific for the IL-2Rα chain (CD25), and has multiple mechanisms of action, which may contribute to beneficial effects in immune-related disease, particularly in relapsing and remitting multiple sclerosis (RRMS).

- Rituximab (anti-CD20 monoclonal antibody) is licensed for use against B-cell lymphomas, but there are newly published data of its effectiveness in systemic lupus erythematosus, rheumatoid arthritis, and dermatomyositis.
- Monoclonal antibodies against tumor necrosis factor, infliximab, which also prevents tumor necrosis factor binding to its receptor, and adalimumab are proposed for use in rheumatoid arthritis ankylosing spondylitis, psoriatic arthritis, and inflammatory bowel disease.
- Over the last few years, new generations of monoclonal antibodies have been developed. These include the second-generation monoclonal antibodies, such as atumumab, veltuzumab, ocrelizumab, and others, and the third-generation monoclonal antibodies such as AME-133v, PRO131921, and GA-101.

Among antibodies, it is possible to separate the depleting from the nondepleting ones. Depleting agents are antibodies that destroy T cells, B cells, or both. T-cell depletion is often accompanied by the release of cytokines, which produces severe systemic symptoms.

Nondepleting monoclonal antibodies reduce responsiveness without compromising lymphocyte populations.

29.2 IMMUNOSTIMULANTS

A number of disorders, such as immunodeficiency state, autoimmune disease, cancer, and viral infection, can be a result of immunodeficiencies, that is, disorders of the immune system that manifest as a group of diseases that are associated with an increased susceptibility to infections, malignancy, and allergy. Immunological therapy with immunostimulants drugs for these disorders significantly improves patient management [128-143].

The immune system detects the pathogens and immediately responds by activating immune components of cells—cytokines and chemokines—and by releasing inflammatory mediators, which modulate and potentiate the immune system.

Immunostimulants or immunopotentiators are chemical substances capable of increasing the overall activity of the immune system. They represent an emerging class of drugs, which has shown some promise in the treatment of primary immunodeficiencies, cancers, and viral infections such as HIV and AIDS.

The two main categories of immunostimulants are:

- *Specific immunostimulants*, which are those that stimulate an immune response to specific antigenic types.
- *Nonspecific immunostimulants*, which are those do not have antigenic specificity and are widely used in chronic infections, immunodeficiency, autoimmunity, and neoplastic diseases.

They can act by different pathways. For instance, some of immunostimulants are enhancing the activity of macrophage, others enhancing the immunogenicity and stability of antigen substance and promoting the synthesizing and secreting of antibody to strengthen the specificity and nonspecificity immune response.

Many natural and synthetic chemical entities exert stimulating activities on various functions of the immune system. They include initiators of antibody production, compounds increasing resistance to infections, compounds promoting rejection of malignant cells, etc.

Because no immunostimulant is included in the list of Top 200 Drugs by sales for the 2010s and because most of them are prepared by the newest biotechnological protocols, no attempt to exhaustively cover this field will be done.

Bacteria-Derived Immunostimulants

Whole Bacteria and Bacterial Extracts

Approximately 40 years ago, attenuated, live culture of the bacillus Calmette-Guérin (BCG, the antituberculosis vaccine), the first nonspecific bacterial immunomodulator, was used to enhance resistance of vertebrate organism against several experimental infections. Bacterial vaccines contain killed or attenuated bacteria that activate the immune system. How exactly they work is not known; it is known, however, that they work by stimulating the body to produce corresponding antibodies.

Preparations containing formalinized *Corynebacterium* species and other whole bacteria, such as *Lactobacillus casei*, *Bifidobacterium* species, and *Saccharomyces boulardii,* for stimulation of the body's defense abilities have also been implemented successfully to prevent and treat illnesses caused by bacteria.

A significant number of other bacterial products, of various degrees of purification, could find medical applications to prevent recurrent infections. Several preparations containing bacterial lyophilized lysates of *Haemophilus influenzae, Streptococcus pneumoniae, Streptococcus pyogenes, Neisseria catarrhalis, Klebsiella, Klebsiella ozaenae,* and *Staphylococcus aureus* have been proposed as immunomodulators.

Anthrax vaccine, typhoid vaccine, meningococcal toxoid conjugate vaccine, tetanus toxoid adsorbed vaccine, and haemophilus conjugate vaccine causing the immune system to produce antibodies that help to prevent influenza are successfully implemented in practical medicine to improve immunity to a particular disease [144,145].

Chemically Defined and Semisynthetic Molecules of Bacterial and Nonbacterial Origin

A critical component of the BCG cell wall—muramyl dipeptide (MDP)—is the minimal immunomodulatory structure of bacterial cell wall peptidoglycan, which is a simple dipeptide derivative of muramic acid, consisting of N-acetyl muramic acid attached to a short amino acid chain of L-Ala-D-isoGln, that has retained most of the immunostimulatory properties of the native BCG. Other derivatives of muramyl dipeptide-nor-MDP, des-muramyl peptides and muramyl tripeptide phosphatidyl ethanolamine (L-MTP-PE or mifamurtide), are considered as immunostimulants and adjuvants in the immunotherapy of cancer and infections.

Among immunomodulators are cell wall products of nonbacterial origin—β-glucans, a heterogeneous group of polysaccharides of D-glucose linked by β-glycosidic bonds and found in fungi, plants, some bacteria, and seaweeds—that have been shown experimentally to exert antitumor effects and confer protection against viruses [146,147]. Clinical trials of genetically modified glucan from the cell walls of the yeast *Saccharomyces cerevisiae*—Betafectin (PGG) β-(1,6) branched β-(1,3)-D-glucan—are in progress.

Lentinan, another 1-3-β-D-glucan isolated from mushroom, is licensed as an adjunctive for antitumor therapy in Japan. It can confer protection against influenza virus and *Listeria*, and prevent relapse of *Mycobacterium tuberculosis* infections.

Lipopolysaccharides of Gram-negative bacteria, which are heteropolymers consisting of polysaccharides covalently bound to a nitrogen-containing phospholipid called lipid A, exert a wide spectrum of biological activities stimulating antibody production. For example, lauryl derivative of tetrapeptide L-AIa-D-Glu(L,L-A_2pm(Gly))NH_2, exhibited marked immunopotentiating activities.

Immunostimulants of Mammalian Origin

The thymus is a specialized organ in the immune system. The thymus-derived hormones control T-lymphocyte activities and various other aspects of the immune system. Thymomodulin and thymostimulin are purified extract from bovine thymus and are used clinically in some countries to treat primary immunodeficient states, bone marrow failure, autoimmune disorders, chronic skin diseases, recurrent infections, allergies.

Peptides, such as thymopoietin, thymic humoral factor, thymulin, and thymosin, are separated from thymic extracts.

Some peptide sequences can modulate T-cell proliferation, and helper, suppressor, cytotoxic, and cytokine functions.

Pentapeptide thymopentin (Arg–Lys–Asp–Val–Tyr) is a synthetic peptide representing the active site of the 49-amino-acid thymic hormone, thymopoietin. Thymopentin downregulates a hyperresponsive immune system to normal, has regulatory function on cells of the myeloic line of cells and on T-cell populations. It normalizes immune reactions, reduces cases of recurrent respiratory tract infections, as well as progression of diseases of acquired immune deficiency syndrome.

Immunostimulants of Endogenous Origin

Hormone-like peptide molecules, cytokines—a diverse group of nonantibody proteins that act as mediators between cells—are involved in immune responses.

Cytokines that play a major role in the innate immune system include interferons, tumor necrosis factor, and interleukins.

Interferons

Interferons are a family of structurally related, species-specific proteins that enhance the immune system released by host cells in response to the presence of pathogens such as viruses, bacteria, parasites, and tumor cells. They play an important role in host defense against pathogens and in regulation of immune responses and can confer resistance to cells.

Commercially available interferons are human interferons manufactured using recombinant DNA technologies. Interferons are classified as: α, β, and γ, depending of their origin.

Natural interferon-α is licensed for the treatment of a rare form of cancer: hairy cell leukemia. Recombinant interferon-α is licensed for treatment of chronic hepatitides B and C virus infections and also is approved for treating condyloma acuminata caused by human papilloma virus and for treating cancer (hairy cell leukemia, AIDS-related Kaposi sarcoma, malignant melanoma).

Interferon-β is licensed for use in severe, uncontrolled, virus-mediated diseases such as viral encephalitis, herpes zoster, and varicella in immunosuppressed patients, and to treat or slow down the progression of multiple sclerosis.

Interferon-γ is important for protection against microbial infections, is central to protection against parasitic infections, and as a therapeutic adjunct for use in patients with chronic (septic) granulomatosis

Interleukins

Interleukins (ILs) consist of a group of naturally occurring proteins synthesized by lymphocytes, monocytes, macrophages, and certain other cells, that stimulate immune responses, mediate communication between cells, and regulate cell growth and differentiation. Fifteen different types of interleukins are known.

IL-1 and IL-2 are responsible for activating T and B lymphocytes. IL-4 leads to an increase in antibody secretion. IL-12 stimulates production of leukocytes cytotoxic T cells and natural killer cells.

The IL-1 family is a group of 11 cytokines that produce an inflammatory response of the body against infection and are used to prolong survival against lethal infections.

IL-2, or aldesleukin, increases the growth and activity of other T and B lymphocytes, and affect the development of the immune system. It is used to treat skin melanomas and kidney cancer, and other cancers, as well as some other diseases. It is the only drug approved in the United States for the treatment of metastatic renal cell cancer.

IL-11, or oprelvekin, is recombinant interleukin that stimulates production of platelets to prevent blood clotting caused by some chemotherapy.

IL-12 is currently in clinical trials as a treatment of cancer and HIV-infected patients.

The genes coding for many of interleukins have been identified and sequenced. Several alternative biopharmaceutical protein production platforms based on recombinant DNA technology have been proposed for producing polypeptides having corresponding Interleukin activity.

Colony-Stimulating Factors

Colony-stimulating factors are glycoproteins that promote production of granulocytes, such as neutrophils, in response to infection. They stimulate the stem cells in the bone marrow to produce phagocytic cells, such as neutrophils, monocytes, macrophages, mast, and dendritic cells, to fight the infection.

Colony-stimulating factor drugs—pegfilgrastim, filgrastim, and sargramostim—are produced by recombinant DNA technologies and are used for reduction of the risk of infection in patients treated with myelotoxic chemotherapy.

Small Molecule Immunostimulants

Some small molecules, including even inorganics, exert immunostimulant properties. Among them are sodium selenite (**29.2.1**), ammonium trichloro(dioxoethylene-O,O′) tellurate (AS-101) (**29.2.2**), and propagermanium (**29.2.3**) homopolymer.

The imidazothiazole tetramisole (**29.2.4**) and its levorotatory isomer levamisole (**29.2.5**) are known anthelminthics and have been shown to potentiate an immune response. Immunomodulatory effects (either immunosuppression or immunostimulation) of other anthelmintics, such as fenvalerate, dieldrin, carbofuran, aminocarb, thiabendazole, fenbendazole, oxfendazole, and ivermectin, are also known [148].

Some 7-allyl-8-oxoguanisine derivatives, such as loxoribine (**29.2.6**), enhance antibody production in response to several antigens. Inosine-5′-methylmonophosphate (**29.2.7**) is described as a thyromimetic immunomodulator.

Imiquimod (**29.2.8**), resiquimod (**29.2.9**), and pidotimod (**29.2.10**) are immune-response modifiers. Tucaresol (**29.2.11**) is representative of the so-called small Schiff base-forming molecules, which is a complex of one part of 9-[(2R,3R,4S,5R)-3,4-dihydroxy-5-(hydroxymethyl)oxolan-2-yl]-3H-purin-6-one and three parts of (dimethylamino)propan-2-ol salt of the 4-acetamidobenzoic acid named isoprinosine (**29.2.12**). Some older compounds, such as nosantine (**29.2.13**) and Therafectin (**29.2.14**), also are known as immunostimulants (Fig. 29.5.).

Miscellaneous

Immunostimulants could be considered as alternatives for antibiotics that will boost the immune system.

Antibiotics can interact directly with the immune system. To quantify and to compare immunomodulatory properties of antibiotics, an immune index has

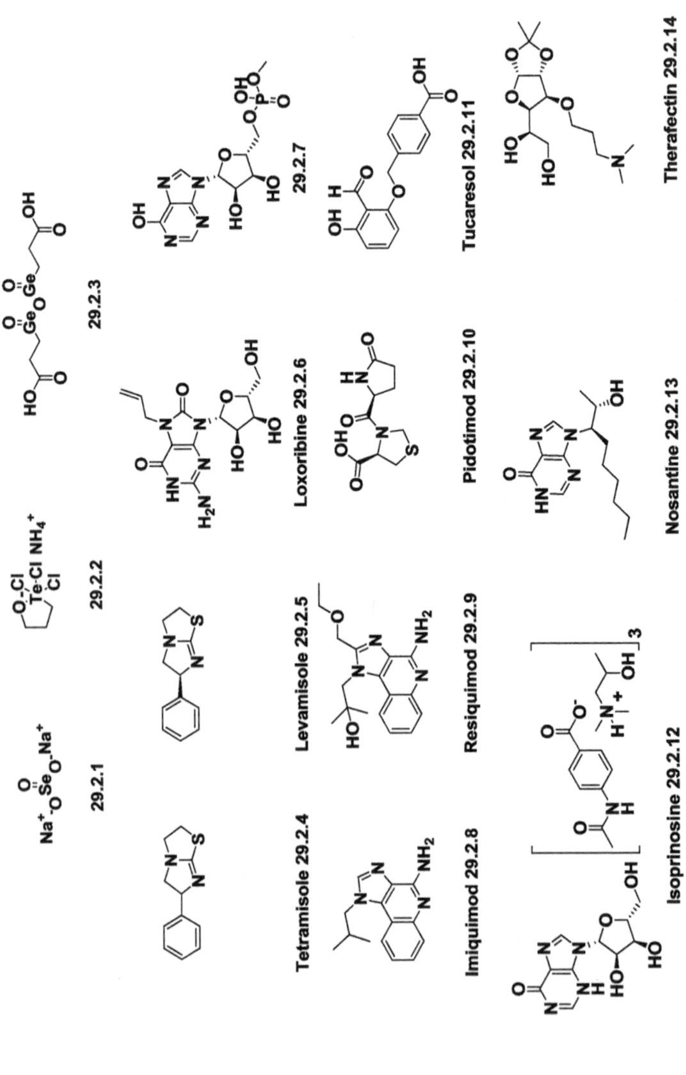

FIG. 29.5 Small molecule immunostimulants.

been calculated. Three markedly immunoenhancing antibiotics (imipenem, cefodizime, and clindamycin) and eight markedly immunodepressing antibiotics (erythromycin, roxithromycin, cefotaxime, tetracycline, rifampicin, gentamicin, teicoplanin, and ampicillin) have been described [149].

In addition to the chemical compounds and biologicals described above, there are many ways to stimulate the immune system to fight infectious diseases and cancer using products that are mainly of plant origin [150-153]. Herbs and mushrooms, such as echinacea, astragalus, garlic, ginseng, green tea, guduchi, and maca; probiotics, vitamins, and minerals (zinc, iron); and animal products, such as propolis, are considered to have immunostimulant properties.

REFERENCES

1. Gilman, S. C., Rogers, T. J., Eds. *Immunopharmacology;* Telford Press, 1989.
2. Nijkamp, F. P., Parnham, M. J., Eds. *Principles of Immunopharmacology;* Birkhauser Verlag, 1999.
3. Rose, N. R., Mackay, I. R., Eds. *Autoimmune Diseases,* 5th ed.; Academic Press, 2014.
4. O'Shea, J. J.; Kanno, Y.; Chan, A. C. In search of magic bullets: the golden age of immunotherapeutics. *Cell* **2014,** *157* (1), 227–240.
5. Parveen, P.; Usha, P.; Sowjanya, Ch.; Sravika, S. V. R. L.; Ragala, B.; Ramya, M. G. An overview of immunology—systemic review. *Am. J. Phytomed. Clin. Ther.* **2013,** *1* (8), 628–644.
6. Enever, C.; Batuwangala, T.; Plummer, C.; Sepp, A. Next generation immunotherapeutic-honing the magic bullet. *Curr. Opin. Biotechnol.* **2009,** *20* (4), 405–411.
7. Miller, M. J.; Foy, K. C.; Kaumaya, P. T. P. Cancer immunotherapy: present status, future perspective, and a new paradigm of peptide immunotherapeutics. *Discov. Med.* **2013,** *15* (82), 166–176.
8. Patil, U. S.; Jaydeokar, A. V.; Bandawane, D. D. Immunomodulators: A pharmacological review. *Int. J. Pharm. Pharm. Sci.* **2012,** *4* (Suppl. 1), 30–36.
9. Periti, P.; Romani, L.; Bonmassar, E.; Puccetti, P. Drugs and the immune system: the emerging era of immunopharmacology. *Trends Immunol.* **2001,** *22* (4), 178–180.
10. Nelson, R. P.; Ballow, M. Immunomodulation and immunotherapy: drugs, cytokines, cytokine receptors, and antibodies. *J. Allergy Clin. Immunol.* **2003,** *111* (2 Suppl. 2), S720–S743.
11. Ballow, M. Biologic immune modifiers: trials and tribulations—are we there yet? *J. Allergy Clin. Immunol.* **2006,** *118* (6), 1209–1215.
12. Ballow, M. Primary immunodeficiency disorders: antibody deficiency. *J. Allergy Clin. Immunol.* **2002,** *109* (4), 581–591.
13. Ballow, M.; Nelson, R. Immunopharmacology: immunomodulation and immunotherapy. *JAMA, J. Am. Med. Assoc.* **1977,** *278* (22), 2008–2017.
14. Cassatella, M. A.; Perretti, M. Immunomodulation: the promises of immunopharmacology for the development of new drugs. *Curr. Opin. Pharmacol.* **2006,** *6,* 376–378.
15. Cohen, A. From pharmacology to immunopharmacology. *Br. J. Pharmacol.* **2006,** *62* (4), 379–382.
16. Zigler, M.; Shir, A.; Levitzki, A. Targeted cancer immunotherapy. *Curr. Opin. Pharmacol.* **2013,** *13* (4), 504–510.
17. Cheever, M. A. Twelve immunotherapy drugs that could cure cancers. *Immunol. Rev.* **2008,** *222,* 357–368.

18. Li, Z.; Chen, L.; Rubinstein, M. P. Cancer immunotherapy: are we there yet? *Exp. Hematol. Oncol.* **2013,** *2,* 33/1–33/6.
19. Allison, A. C. Immunosuppressive drugs: the first 50 years and a glance forward. *Immunopharmacology* **2000,** *47* (2–3), 63–83.
20. Hughes, S. E.; Gruber, S. A. New immunosuppressive drugs in organ transplantation. *J. Clin. Pharmacol.* **1996,** *36* (12), 1081–1092.
21. Grinyo, J. M.; Cruzado, J. M.; Bestard, O.; Castineira, J. R. V.; Torras, J. Immunosuppression in the era of biological agents. *Adv. Exp. Med. Biol.* **2012,** *741,* 60–72.
22. Halloran, P. F. Immunosuppressive drugs for kidney transplantation. *N. Engl. J. Med.* **2004,** *351* (26), 2715–2729.
23. Boraschi, D.; Penton-Rol, G. Perspectives in immunopharmacology: the future of immunosuppression. *Immunol. Lett.* **2014,** *161* (2), 211–215.
24. Clutterbuck, P. W.; Oxford, A. E.; Raistrick, H.; Smith, G. Studies in the biochemistry of micro-organisms: the metabolic products of the *Penicillium brevicompactum* series. *Biochem. J.* **1932,** *26* (5), 1441–1458.
25. Williams, R. H.; Boeck, L. D.; Cline, J. C.; DeLong, D. C.; Gerzon, K.; Gordee, R. S.; Gorman, M.; Holmes, R. E.; Larsen, S. H.; Lively, D. H.; Matthews, T. R.; Nelson, J. D.; Poore, G. A.; Stark, W. M.; Sweeney, M. J. Fermentation, isolation, and biological properties of mycophenolic acid. *Antimicrob. Agents Chemother.* **1968,** *8,* 229–233.
26. Birch, A. J.; Wright, J. H. Total synthesis of mycophenolic acid. *Aust. J. Chem.* **1969,** *22* (12), 2635–2644.
27. Florey, H. W.; Jennings, M. A.; Gilliver, K.; Sanders, A. G. Mycophenolic acid; an antibiotic from *Penicillium brevicompactum* Dierckx. *Lancet* **1946,** *1* (6385), 46–49.
28. Takao, K., Takehiko I.; Hiroshiro S. Method for production of mycophenolic acid by fermentation, US 4452891 (1984).
29. Queener, S. W.; Nash, C. H., III. *Penicillium* species mutants with improved ability to synthesize mycophenolic acid, US 4115197 (1978).
30. Borrow, A.; Jefferys, E. G.; Mills, S. D.; Turner, W. B. Mycophenolic acid from *Penicillium* species, GB 1157099 (1969).
31. Manufacture of mycophenolic acid, JP 57050890 (1982).
32. Pokluda, Z.; Satke, J.; Vala, V.; Valik, J. Regulation of acid metabolite production, WO 200800(2665), (2008).
33. Barta, I.; Boros, S.; Ambrus, G.; Horvath, G.; Szabo, A.; Szabo, I. M.; Jekkel, A.; Konya, A.; Mozes, J.; Salat, J.; Somogyi, G. Process for the preparation of mycophenolic acid and derivatives thereof, WO 2001021607 (2001).
34. Gomes, R.; Jorge C. B.; Clemente, J. J. V. M.; Thomaz, M. M. A. B. F.; Pereira da Cunha, A. E. P. B. Process for the production of mycophenolic acid, EP 1624070 (2006).
35. Sadhukhan, A. K.; Murthy, M. V. Ramana; Kumar, R. Ajaya; Mohan, E. V. S.; Vandana, G.; Bhar, C.; Rao, K. V. Optimization of mycophenolic acid production in solid state fermentation using response surface methodology. *J. Ind. Microbiol. Biotechnol.* **1999,** *22* (1), 33–38.
36. Birch, A. J.; Wright, J. J. Total synthesis of mycophenolic acid. *J. Chem. Soc., Chem. Commun.* **1969,** *14,* 788–789.
37. Canonica, L.; Rindone, B.; Santaniello, E.; Scolastico, C. Total synthesis of mycophenolic acid. *Tetrahedron Lett.* **1971,** *28,* 2691–2692.
38. Canonica, L.; Rindone, B.; Santaniello, E.; Scolastico, C. Total synthesis of mycophenolic acid, some analogs, and some biogenetic intermediates. *Tetrahedron* **1972,** *28* (16), 4395–4404.
39. Patterson, J. W. Synthesis of mycophenolic acid. *Tetrahedron* **1993,** *49* (22), 4789–4798.

40. Danheiser, R. L.; Gee, S. K.; Perez, J. J. Total synthesis of mycophenolic acid. *J. Am. Chem. Soc.* **1986,** *108* (4), 806–810.
41. Malachowska, M.; Cholewinski, G.; Dzierzbicka, K.; Wardowska, A.; Trzonkowski, P. Mycophenolic acid (MPA) and its analog. Synthesis and biological activity. *Wiad. Chem.* **2009,** *63* (3–4), 309–332.
42. Cholewinski, G.; Malachowska-Ugarte, M.; Dzierzbicka, K. The chemistry of mycophenolic acid—synthesis and modifications towards desired biological activity. *Curr. Med. Chem.* **2010,** *17* (18), 1926–1941.
43. Sievers, T. M.; Rossi, S. J.; Ghobrial, R. M.; Arriola, E.; Nishimura, P.; Kawano, M.; Holt, C. D. Mycophenolate mofetil. *Pharmacotherapy* **1997,** *17* (6), 1178–1197.
44. Staatz, C. E.; Tett, S. E. Pharmacology and toxicology of mycophenolate in organ transplant recipients: an update. *Arch. Toxicol.* **2014,** *88* (7), 1351–1389.
45. Ransom, J. T. Mechanism of action of the immunosuppressant prodrug mycophenolate mofetil. *Ther. Drug Monit.* **1995,** *17* (6), 681–684.
46. Sollinger, H. W. Mycophenolate mofetil for the prevention of acute rejection in primary cadaveric renal allograft recipients. U.S. Renal Transplant Mycophenolate Mofetil Study Group. *Transplantation* **1995,** *60* (3), 225–232.
47. Shaw, L. M.; Sollinger, H. W.; Halloran, P.; Morris, R. E.; Yatscoff, R. W.; Ransom, J.; Tsina, I.; Keown, P.; Holt, D. W. Mycophenolate mofetil: a report of the consensus panel. *Ther. Drug Monit.* **1995,** *17* (6), 690–699.
48. Wu, J. C. Mycophenolate mofetil: molecular mechanisms of action. *Perspect. Drug Discovery Des.* **1994,** *2* (1), 185–204.
49. Sollinger, H. W. Mycophenolates in transplantation. *Clin. Transplant.* **2004,** *18* (5), 485–492.
50. Allison, A. C.; Kowalski, W. J.; Muller, C. D.; Eugui, E. M. Mechanisms of action of mycophenolic acid. *Ann. N. Y. Acad. Sci.* **1993,** *696,* 63–87.
51. Marinaro, W. A. Mycophenolate mofetil. *Biotechnol.: Pharm. Aspects* **2007,** *5* (Pt. 2, Prodrugs: Challenges and Rewards), 599–604.
52. Allison, A. C.; Eugui, E. M. Mycophenolate mofetil and its mechanisms of action. *Immunopharmacology* **2000,** *47* (2–3), 85–118.
53. Bardsley-Elliot, A.; Noble, S.; Foster, R. H. Mycophenolate mofetil: a review of its use in the management of solid organ transplantation. *BioDrugs* **1999,** *12* (5), 363–410.
54. Hood, K. A.; Zarembski, D. G. Mycophenolate mofetil: a unique immunosuppressive agent. *Am. J. Health-Syst. Pharm.* **1997,** *54* (3), 285–294.
55. Sievers, T. M.; Rossi, S. J.; Ghobrial, R. M.; Arriola, E.; Nishimura, P.; Kawano, M.; Holt, C. D. Mycophenolate mofetil. *Pharmacotherapy* **1997,** *17* (6), 1178–1197.
56. Morris, R. E.; Hoyt, E. G.; Murphy, M. P.; Eugui, E. M.; Allison, A. C. Mycophenolic acid morpholinoethylester (RS-61443) is a new immunosuppressant that prevents and halts heart allograft rejection by selective inhibition of T- and B-cell purine synthesis. *Transplant. Proc.* **1990,** *22* (4), 1659–1962.
57. Traber, R.; Hofmann, H.; Loosli, H. R.; Ponelle, M.; Von Wartburg, A. Novel cyclosporins from *Tolypocladium inflatum*. Cyclosporins K-Z. *Helv. Chim. Acta* **1987,** *70* (1), 13–36.
58. Traber, R.; Kuhn, M.; Loosli, H. R.; Pache, W.; Von Wartburg, A. New cyclopeptides from *Trichoderma polysporum* (Link ex Pers.) Rifai: cyclosporins B, D and E. *Helv. Chim. Acta* **1977**.
59. Borel, J. F.; Feurer, C.; Magnee, C.; Staehelin, H. Effects of the new anti-lymphocytic peptide cyclosporin A in animals. *Immunolology* **1977,** *32* (6), 1017–1025.
60. Traber, R. P.; Hofmann, H.; Haerri, E. Monocyclic peptide and its use, CH 637123 (1983).

61. Jekkel, A.; Ambrus, G.; Sarudy, E. T.; Szabo, I. M.; Huelber, A.; Andor, A.; Albrecht, K.; Koenczoel, K.; Polya, K.; Erdei, J.; Kiss, L.; Nagy, K.; Patolas, B.; Deli nee Konzski, E.; Buzasi, K.; Molnar nee Antal, A.; Santha, J.; Szaszhegyesi, V.; Tamori nee Joszt, E.; Moravcsik, I. Cyclosporin fermentation with *Tolypocladium varium*, FR 2640641 (1990).
62. Klokkers, K.; Fischer, W. Liquid cyclosporine A preparation for oral or topical administration, DE 19521974 (1996).
63. Balaraman, K.; Mathew, N. Optimization of media composition for the production of cyclosporin A by *Tolypocladium* species. *Indian J. Med. Res.* **2006,** *123* (4), 525–530.
64. Kapturczak, M. H.; Meier-Kriesche, H. U.; Kaplan, B. Pharmacology of calcineurin antagonists. *Transplant. Proc.* **2004,** *36* (2S), 25S–32S.
65. Sharmila, K.; Thillaimaharani, K. A.; Logesh, A. R.; Sathishkumar, A.; Kalaiselvam, M. Production of cyclosporin A by saprophytic filamentous fungus *Fusarium oxysporum. Int. J. Pharm. Pharm. Sci.* **2012,** *4* (4), 149–153.
66. Cohen, D. J.; Loertscher, R.; Rubin, M. F.; Tilney, N. L.; Carpenter, C. B.; Strom, T. B. The mechanism of action of cyclosporine: a continuing puzzle. *Ann. Intern. Med.* **1984,** *101* (5), 667–682.
67. Wenger, R. M.; Payne, T. G.; Schreier, M. H. Cyclosporine: chemistry, structure-activity relationships and mode of action. *Prog. Clin. Biochem. Med.* **1986,** *3,* 157–191.
68. Strijack, B.; Keown, P. A. Cyclosporine: molecular action to clinical therapeutics. In *Immunotherapy in Transplantation: Principles and Practice;* Kaplan, B., Burckart, J., Lakkis, F. G., Eds.; Wiley-Blackwell, 2012; pp 197–223.
69. Tedesco, D.; Haragsim, L. Cyclosporine: a review. *J. Transplant.* **2012,** 631–641.
70. Kahan, B. D. Therapeutic drug monitoring of cyclosporine: 20 years of progress. *Transplant. Proc.* **2004,** *36* (2S), 378S–391S.
71. Dunn, C. J.; Wagstaff, A. J.; Perry, C. M.; Plosker, G. L.; Goa, K. L. Cyclosporin: an updated review of the pharmacokinetic properties, clinical efficacy and tolerability of a microemulsion-based formulation (Neoral) in organ transplantation. *Drugs* **2001,** *61* (13), 1957–2016.
72. Ponticelli, C. Generic cyclosporine: a word of caution. *J. Nephrol.* **2004,** *17* (Suppl. 8), S20–S24.
73. Goto, T.; Kino, T.; Hatanaka, H.; Nishiyama, M.; Okuhara, M.; Kosaka, M.; Aoki, H.; Imanaka, H. Discovery of FK-506, a novel immunosuppressant isolated from *Streptomyces tsukubaensis. Transpl. Proceed.* **1987,** *5* (Suppl. 6), 4–8.
74. Sigal, N. H.; Lin, C. S.; Siekierka, J. J. Inhibition of human T-cell activation by FK 506, rapamycin, and cyclosporine A. *Transplant. Proc.* **1991,** *23* (2 Suppl. 2), 1–5.
75. Maes, B. D.; Vanrenterghem, Y. F. Ch. Cyclosporine: advantages versus disadvantages vis-a-vis tacrolimus. *Transplant. Proc.* **2004,** *36* (2S), 40S–49S.
76. Amaya, T.; Hiroi, J.; Lawrence, I. D. Tacrolimus and other immunosuppressive macrolides in clinical practice. In *Macrolide Antibiotics: Chemistry, Biology, and Practice,* 2nd ed.; Omura, S., Ed.; Academic Press, 2002; pp 421–452.
77. Jiang, H.; Kobayashi, M. Differences between cyclosporin A and tacrolimus in organ transplantation. *Transplant. Proc.* **1999,** *31* (5), 1978–1980.
78. Scott, L. J.; McKeage, K.; Keam, S. J.; Plosker, G. L. Tacrolimus: a further update of its use in the management of organ transplantation. *Drugs* **2003,** *63* (12), 1247–1297.
79. Plosker, G. L.; Foster, R. H. Tacrolimus: a further update of its pharmacology and therapeutic use in the management of organ transplantation. *Drugs* **2000,** *59* (2), 323–389.
80. Ruzicka, T.; Assmann, T.; Homey, B. Tacrolimus: The drug for the turn of the millennium? *Arch. Dermatol.* **1999,** *135* (5), 574–580.

81. Fung, J. J. Tacrolimus and transplantation: a decade in review. *Transplantation* **2004,** *77* (9 Suppl.), S41–S43.
82. Rath, T. Tacrolimus in transplant rejection. *Expert Opin. Pharmacother.* **2013,** *14* (1), 115–122.
83. Fitzsimmons, W. E. Tacrolimus. In *Immunotherapy in Transplantation: Principles and Practice;* Kaplan, B., Burckart., J., Lakkis, F. G., Eds.; Wiley-Blackwell, 2012; pp 224–240.
84. Letko, E.; Bhol, K.; Pinar, V.; Foster, C. S.; Ahmed, A. R. Tacrolimus (FK 506). *Ann. Allergy, Asthma, Immunol.* **1999,** *83* (3), 179–190.
85. Bowman, L. J.; Brennan, D. C. The role of tacrolimus in renal transplantation. *Expert Opin. Pharmacother.* **2008,** *9* (4), 635–643.
86. Kaminska, B.; Gaweda-Walerych, K.; Zawadzka, M. Molecular mechanisms of neuroprotective action of immunosuppressants-facts and hypotheses. *J. Cell. Mol. Med.* **2004,** *8* (1), 45–58.
87. Staatz, C. E.; Tett, S. E. Clinical pharmacokinetics and pharmacodynamics of tacrolimus in solid organ transplantation. *Clin. Pharmacokinet.* **2004,** *43* (10), 623–653.
88. Taube, D.; Jones, G.; O'Beirne, J.; Wennberg, L.; Connor, A.; Rasmussen, A.; Backman, L. Generic tacrolimus in solid organ transplantation. *Clin. Transplant.* **2014,** *28* (5), 623–632.
89. Yamamoto, T. Topical tacrolimus for psoriasis. *Open Allergy J.* **2009,** *2,* 51–55.
90. Bieber, T. Topical tacrolimus (FK 506): a new milestone in the management of atopic dermatitis. *J. Allergy Clin. Immunol.* **1998,** *102* (4 Pt. 1), 555–557.
91. Gajzlerska, W.; Kurkowiak, J.; Turlo, J. Use of three-carbon chain compounds as biosynthesis precursors to enhance tacrolimus production in *Streptomyces tsukubaensis. New Biotechnol.* **2015,** *32* (1), 32–39.
92. Barreiro, C.; Martinez-Castro, M. Trends in the biosynthesis and production of the immunosuppressant tacrolimus (FK506). *Appl. Microbiol. Biotechnol.* **2014,** *98* (2), 497–507.
93. Singh, B. P.; Behera, B. K. Regulation of tacrolimus production by altering primary source of carbons and amino acids. *Lett. Appl. Microbiol.* **2009,** *49* (2), 254–259.
94. Okuhara, M.; Tanaka, H.; Goto, T.; Kino, T.; Hatanaka, H. Preparation of tricyclo compounds as immunosuppressants and medical fungicides, 1990 Preparation of tricyclo compounds as immunosuppressants and medical fungicides, US 4929611 (1990).
95. Fermentative preparation of tacrolimus using genetically modified Streptomyces, EP 2272963 (2011).
96. Salituro, G. M.; Huang, L.; Dumont, F.; Jones, E. T. T.; Garrity, G. M.; Omstead, M. N.; Martin, F. I.; Diez, M. T. Manufacture of FK-506 with Streptomyces, EP 497515 (1992).
97. Edmunds, A. J. F.; Grassberger, M. Preparation of FK506 and FR520 analogs as immunosuppressants and anti-inflammatories, EP 402931 (1990).
98. Kumar, P.; Sharma, S.; Shukla, A.; Kumar, S.; Maurya, R. K.; Katial, V.; Mitra, A.; Gigras, P. Production of tacrolimus (FK-506) using *Streptomyces glaucescens* MTCC (5115), WO 2005038009 (2005).
99. Dumont, F.; Garrity, G. M.; Fernandez, I. M.; Matas, T. D. Manufacture of the immunosuppressant FK-506 with *Streptomyces*, US 5116756 (1992).
100. Vaid, S.; Narula, P. Process for producing tacrolimus (FK-506) using vegetable oil as sole source of carbon, WO 2006011156 (2006).
101. Kumar, P.; Malviya, H. K.; Maurya, R. K.; Shukla, A. Fed-batch fermentation processes for the preparation of tacrolimus, WO 2006011156 (2007).
102. Wu, C.-C.; Chou, P.-C.; Jacinto, E. The target of rapamycin: structure and functions. In *Protein Kinases;* Xavier, G., Da, S., Eds.; , 2012; pp 1–40(InTech).
103. Weichhart, T. Mammalian target of rapamycin: a signaling kinase for every aspect of cellular life. *Methods Mol. Biol. (NY, U. S.)* **2012,** *821* (m-TOR), 1–14.

104. Swiech, L. J.; Urbanska, M.; Macias, M.; Skalecka, A.; Jaworski, J. Mammalian target of rapamycin. *Neuromethods* **2012,** *68,* 291–318.
105. Ryther, R. C. C.; Wong, M. Mammalian target of rapamycin (mTOR) inhibition: potential for antiseizure, antiepileptogenic, and epileptostatic therapy. *Curr. Neurol. Neurosci. Rep.* **2012,** *12* (4), 410–418.
106. Chong, Z. Z.; Shang, Y. C.; Wang, S.; Maiese, K. Shedding new light on neurodegenerative diseases through the mammalian target of rapamycin. *Prog. Neurobiol. (Oxford, U. K.)* **2012,** *99* (2), 128–148.
107. Li, J.; Kim, S. G.; Blenis, J. Rapamycin: one drug, many effects. *Cell Metab.* **2014,** *19* (3), 373–379.
108. Lebwohl, D.; Anak, O.; Sahmoud, T.; Klimovsky, J.; Elmroth, I.; Haas, T.; Posluszny, J.; Saletan, S.; Berg, W. Development of everolimus, a novel oral mTOR inhibitor, across a spectrum of diseases. *Ann. N. Y. Acad. Sci.* **2013,** *1291,* 14–32.
109. Touzot, M.; Soulillou, J. P.; Dantal, J. Mechanistic target of rapamycin inhibitors in solid organ transplantation: from benchside to clinical use. *Curr. Opin. Organ Transplant.* **2012,** *17* (6), 626–633.
110. Kahan, B. D. Inhibitors of mammalian target of rapamycin. In *Immunotherapy in Transplantation: Principles and Practice;* Kaplan, B., Burckart, J., Lakkis, F. G., Eds.; Wiley-Blackwell, 2012; pp 241–272.
111. Malaguti, P.; Vari, S.; Cognetti, F.; Fabi, A. The mammalian target of rapamycin inhibitors in breast cancer: current evidence and future directions. *Anticancer Res.* **2013,** *33* (1), 21–28.
112. Thorpe, M.; Karteris, E. Mammalian target of rapamycin (mTOR) signaling. In *Cancer Cell Signalling;* Harvey, A., Ed.; Wiley, 2013; pp 93–113.
113. Benjamin, O.; Gatien, M.; Berengere, G.; Francoise, R.; Dominique, H. mTOR inhibitors (rapamycin and its derivatives) and nitrogen containing bisphosphonates: bi-functional compounds for the treatment of bone tumours. *Front. Med. Chem.* **2012,** *6,* 189–202.
114. Nelson, V.; Altman, J. K.; Platanias, L. C. Next generation of mammalian target of rapamycin inhibitors for the treatment of cancer. *Expert Opin. Invest. Drugs* **2013,** *22* (6), 715–722.
115. Langone, A. J.; Helderman, J. H. Mammalian target of rapamycin inhibitors in organ transplantation: an unkept promise. *Chest* **2012,** *142* (3), 734–737.
116. Ehninger, D.; Neff, F.; Xie, K. Longevity, aging and rapamycin. *Cell. Mol. Life Sci.* **2014,** *71* (22), 4325–4346.
117. Blagosklonny, M. V. Rapamycin extends life- and health span because it slows aging. *Aging* **2013,** *5* (8), 592–598.
118. Richardson, A. Rapamycin, anti-aging, and avoiding the fate of Tithonus. *J. Clin. Invest.* **2013,** *123* (8), 3204–3206.
119. Sehgal, S. N.; Blazekovic, T. M.; Vezina, C. Rapamycin, a new antibiotic, DE 2347682 (1974).
120. Huang, M.; Li, M.; Feng, Z.; Liu, Y.; Chu, Y.; Tian, Y. Enhanced rapamycin production in *Streptomyces hygroscopicus* by integrative expression of aveR, a LAL family transcriptional regulator. *World J. Microbiol. Biotechnol.* **2011,** *27* (9), 2103–2109.
121. Sinha, R.; Singh, S.; Srivastava, P. Studies on process optimization methods for rapamycin production using *Streptomyces hygroscopicus* ATCC 2(9253). *Bioprocess Biosyst. Eng.* **2014,** *37* (5), 829–840.
122. Rani, P. B.; Kumar, B. S.; Rao, A. K. S. B.; Chowdary, N. V. Fermentation process for the production of rapamycin, IN 2012CH01444 (2013).
123. Patil, N. S.; Hussaini, S. I.; Singh, A. K.; Mendhe, R. B. A pure form of rapamycin and a process for recovery and purification thereof, IN 2006CH02079 (2006).

124. Garg, S.; Melarkode, R.; Gururaja, R.; Suryanarayan, S. Solid state fed-batch fermentation for production of rapamycin by *Streptomyces hygroscopicus*, WO 2004022767 (2004).
125. Kojima, N.; Kojima, Y.; Sakakibara, T.; Yamauchi, Y. Manufacture of rapamycin with a novel *Actinoplanes*, WO 9322446 (1993).
126. Song, H.; Tang, L.; Chen, W.; Li, Z.; Li, J.; Sun, Z.; Feng, J. Process for the preparation of temsirolimus from rapamycin, US 20130296550 (2013).
127. Norley, M. C. Synthetic approaches to rapamycin. *Contemp. Org. Synth.* **1996,** *3* (5), 345–371.
128. Patil, U. S.; Jaydeokar, A. V.; Bandawane, D. D. Immunomodulators: a pharmacological review. *Int. J. Pharm. Pharm. Sci.* **2012,** *4* (Suppl. 1), 30–36.
129. Mostboeck, S. Cytokine/antibody complexes: an emerging class of immunostimulants. *Curr. Pharm. Des.* **2009,** *15* (7), 809–825.
130. Portales, P.; Clot, J. Immunostimulants revisited: focus on the pharmacology of Ribomunyl. *BioDrugs* **2006,** *20* (2), 81–84.
131. Talmadge, J. E. Immunostimulants in cancer therapy. In *Principles of Immunopharmacology,* 2nd ed.; Nijkamp, F. P., Parnham, M. J., Eds. Birkhauser, 2005; pp 345–376.
132. Prescott, S. L.; Thornton, C. A. An overview of immunotherapy for allergic disease: new developments and future strategies. *Med. Chem. Rev.–Online* **2004,** *1* (2), 163–177.
133. Lavelle, E. C.; McGuirk, P.; Mills, K. H. G. Molecules of infectious agents as immunomodulatory drugs. *Curr. Top. Med. Chem.* **2004,** *4* (5), 499–508.
134. Williams, R. J. Stimulation of the innate immune system: a paradigm for the future identification of disease modifying agents to treat asthma and allergic diseases. *Emerging Ther. Targets* **2000,** *4* (3), 313–321.
135. Galeotti, M. Some aspects of the application of immunostimulants and a critical review of methods for their evaluation. *J. Appl. Ichthyol.* **1998,** *14* (3–4), 189–199.
136. Werner, G. H.; Jolles, P. Immunostimulating agents. What next? A review of their present and potential medical applications. *Eur. J. Biochem.* **1996,** *242* (1), 1–19.
137. Hadden, J. W. Immunomodulators. *Immunopharmacol. Rev.* **1996,** *2,* 3–24.
138. Berek, L.; Petri, I. B.; Molnar, J.; Shoyama, Y.; Kawase, M.; Motohashi, N. Structure-activity relationship of the immunomodulatory drugs. *Recent Res. Dev. Chem. Pharm. Sci.* **2002,** *2,* 125–135.
139. Aagaard, L.; Hansen, E. H. Side effects of antineoplastic and immunomodulating medications reported by European consumers. *J. Res. Pharm. Pract.* **2013,** *2* (1), 44–49.
140. Rhodes, J. Discovery of immunopotentiatory drugs: current and future strategies. *Clin. Exp. Immunol.* **2002,** *130* (3), 363–369.
141. Vorob'ev, A. A. Immunomodulators: classification principles and medical application strategy. *Vestn. Ross. Akad. Med. Nauk* **2002,** *4,* 3–6.
142. Masihi, K. N. Fighting infection using immunomodulatory agents. *Expert Opin. Biol. Ther.* **2001,** *1* (4), 641–653.
143. Masihi, K. N. Immunomodulatory agents for prophylaxis and therapy of infections. *Int. J. Antimicrob. Agents* **2000,** *14* (3), 181–191.
144. Schijns, V. E. J.C.; Degen, W. G. J. Vaccine immunopotentiators of the future. *Clin. Pharmacol. Ther. (Hoboken, NJ, U. S.)* **2007,** *82* (6), 750–755.
145. Hodge, J. W.; Schlom, J.; Abrams, S. I. Vaccines and immunostimulants. In *Holland-Frei Cancer Medicine 7;* Kufe, D. W., Ed.; BC Decker, 2006; pp 786–801.
146. Chihara, G. Immunopharmacology of lentinan and the glucans. *EOS–Riv. Immunol. Immunofarmacol.* **1984,** *4* (2), 85–96.
147. Goldman, R. C. Biological response modification by β-D-glucans. *Annu. Rep. Med. Chem.* **1995,** *30,* 129–138.

148. Sajid, M. S.; Iqbal, Z.; Muhammad, G.; Iqbal, M. U. Immunomodulatory effect of various anti-parasitics: a review. *Parasitology* **2006,** *132* (3), 301–313.
149. Van Vlem, B.; Vanholder, R.; De Paepe, P.; Vogelaers, D.; Ringoir, S. Immunomodulating effects of antibiotics: literature review. *Infection* **1996,** *24* (4), 275–291.
150. Wagner, H. Immunostimulants of plant origin. *Croat. Chem. Acta* **1995,** *68* (3), 615–626.
151. Tan, B. K. H.; Vanitha, J. Immunomodulatory and antimicrobial effects of some traditional Chinese medicinal herbs: a review. *Curr. Med. Chem.* **2004,** *11* (11), 1423–1430.
152. Srisilam, K.; Sumalatha, D. V.; Thilagam, E.; Veeresham, C. Immunomodulators from higher plants. *Indian J. Nat. Prod.* **2004,** *20* (1), 3–15.
153. Agarwal, S. S.; Singh, V. K. Immunomodulators: a review of studies on Indian medicinal plants and synthetic peptides. Part II: synthetic peptides. *PINSA-B: Proc. Indian Natl. Sci. Acad., Part B* **1999,** *65* (6), 377–392.

Chapter 30

Antibiotics

Antibacterial agents indicated for clinical use are agents that selectively destroy bacteria by interfering with bacterial growth or survival. Growth inhibition and complete loss of viability of bacteria is often the result of a cascade of events elicited upon treatment with an antibacterial agent whose mode of action usually involves more than one single target.

Antibacterial drugs currently used target distinct cellular constituents, such as membranes, RNA, DNA, enzymes, and enzyme substrates.

Antibacterial-induced cell death or stopping of the growth involve many biochemical and genetic pathways and is not very well understood and proven. In general, mechanisms of action are associated with double-stranded DNA breaks, with the arrest of DNA-dependent RNA synthesis, with cellular energetics in bacteria's, and with alterations of their central metabolism, among others.

There is growing evidence that certain antibiotics exert their beneficial effects not only by killing or inhibiting the growth of bacterial pathogens, but also indirectly by immunomodulation.

Antibacterial agents are used for treatment or prevention of bacterial infection.

Among existing antibacterial agents, antibiotics may be informally defined as compounds that are produced by a living organisms;, are derived from bacterial, fungal, mold, plant, and animal sources; are used to treat bacterial infections; and are one of the largest-selling classes of drugs worldwide.

Although there are several schemes of classification for antibiotics, based on their origin, type of activity, or mode of action—bactericidal versus bacteriostatic (antibiotics that kill bacteria are called "bactericidal," whereas antibiotics that stop the growth of bacteria are called "bacteriostatic"); bacterial spectrum or range of action—broad versus narrow (ie, do they effect many pathogens or only some select strains); the general microbial group they act against (antibacterial, antifungal, or antiprotozoan); route of administration (injectable versus oral versus topical), are the most suitable approaches for this book is a classification based on their chemical structure.

Different classifications of antibiotics by families based on chemical structure, such as macrocyclic lactones, carbohydrates, N- or O-containing heterocycles, compounds having an aromatic skeleton, and those having an aliphatic chain, have been proposed.

Synthesis of Best-Seller Drugs. http://dx.doi.org/10.1016/B978-0-12-411492-0.00030-4

Probably, the mostly accepted main classification of antibiotics based on chemical structure is the following:

1. β-Lactam antibiotics-penicillins, cephalosporins, carbapenems, monobactams;
2. Tetracyclines;
3. Macrolide antibiotics;
4. Aminoglycosides;
5. Peptide antibiotics;.
6. Lincosamides;
7. Streptogramins.

Most antibiotics are either natural products and their semisynthetic derivative, or synthetic compounds mimicking the structure of the natural products. Even after the discovery of penicillin approximately 80 years ago, the search for new antibiotics still continues. Today, approximately 16,500 effective antibiotics have been described [1-3]. Approximately 100 of them have approvals and medical applications.

More than 20 novel classes of antibiotics were produced between the 1930s and 1960s, but only four of them were marketed. These "Big Four" classes of antibiotics are β-lactams, tetracyclines, macrolides, and aminoglycosides. It is necessary to mention here that antibiotics within a structural class as a rule show similar patterns of effectiveness and toxicity.

Antibiotics, in general, are derived biotechnologically from natural sources, and often are further chemically modified to confer better drug properties.

The main producing microorganisms of antibiotics are *Actinobacteria*, especially, *Streptomyces* species, fungi and other filamentous bacteria, which still represent inexhaustible sources for new antibiotics in future.

To our knowledge more than 500 books and 45,000 reviews on the historical development, classification, nomenclature, and chemotherapeutic use of antibiotics have been published. The most recent are quoted here [4-30].

Synthesis of antibiotics is an immense theme that is strongly related to biotechnology and gene engineering, which does not belong to the subject of organic synthesis and extends beyond the margins of this book.

Here we show briefly representatives of classes of antibiotics and general schematic approaches to their production.

30.1 β-LACTAM ANTIBIOTICS—PENICILLINS, CEPHALOSPORINS, CARBAPENEMS, MONOBACTAMS

Antibiotics, which revolutionized medical care in the 20th century, are one of the most successful achievements in the history of medicinal chemistry and in the history of medicine.

Science associates the beginning of the modern "antibiotic era" with the name of Alexander Fleming, who discovered the first antibiotic—penicillin—in

1929. In late 1930s–early 1940s, E. B. Chain and H. W. Florey developed large-scale fermentation processes for the production of antibiotics.

In 1945, Dorothy Hodgkins, using X-ray crystallography, finally solved the structure of penicillin (benzylpenicillin or penicillin G). (Hodgkins also published data on the three-dimensional structure of vitamin B_{12} in 1948, for which she was later awarded the Nobel Prize (1964). Later, in 1969, she published also data on the structure of insulin.)

The Nobel Prize in Physiology or Medicine 1945 was awarded jointly to Sir Alexander Fleming, Ernst Boris Chain, and Sir Howard Walter Florey "for the discovery of penicillin and its curative effect in various infectious diseases."

It is interesting to note that some ancient civilizations, such as those of the Romans, Greeks, and Egyptians, used specially selected mold and other natural source extracts to treat a variety of infections.

Penicillin G was and is used to treat infections generated with Gram-positive bacteria and some Gram-negative cocci. The role of penicillin G in World War II gave it a fantastic reputation that led to its ubiquitous use.

Penicillin G is a narrow-spectrum antibiotic that acts by inhibiting biosynthesis of cell-wall mucopeptide; it is effective mainly against Gram-positive bacteria. Penicillin G is highly active against staphylococci, streptococci (groups A, B, C, G, H, L, and M), pneumococci, and *Neisseria meningitidis*. Other organisms susceptible to penicillin G are *Neisseria gonorrhoeae, Corynebacterium diphtheriae, Bacillus anthracis, Clostridia, Actinomyces* species, *Spirillum minus, Streptobacillus moniliformis, Listeria monocytogenes, Leptospira leptospira,* and *Treponema pallidum.*

Some species of Gram-negative bacteria were previously considered susceptible to penicillin G, including some strains of *Escherichia coli, Proteus mirabilis, Salmonella, Shigella, Enterobacter aerogenes* (formerly *Aerobacter aerogenes*) and *Alcaligenes faecalis*. Penicillin G is no longer considered a drug of choice for infections caused by these organisms.

Penicillin G was the first discovered antibiotic and the first to be industrially produced. About a decade after its discovery, a group of other active penicillins (F, V, G, K, O, X) were isolated. Their mechanism of action is the same as that of penicillin G and consists of inhibition of the synthesis of peptidoglycan, the main component of the bacterial cell wall, thus causing irreversible damage to and death of the bacteria.

β-Lactam antibiotics are bactericidal antibiotics. They are not effective against bacteria and fungi that lack a cell wall.

The β-lactam antibiotics group [31-54] consists of two large "classic" families of antibiotics: the penicillins ("penams"), and the little-bit-later discovered cephalosporins ("cephems"). Another small group of drugs, collectively referred to as "nonclassic" β-lactams, includes the carbapenems and monobactams [55-62].

All of these are constructed on a base of a four-membered β-lactam ring (azetidin-2-one), which is a necessary element for exhibiting antibacterial activity. In penicillins, the β-lactam ring is fused to a five-membered thiazolidine ring and in cephalosporins to a six-membered dihydrothiazine ring. In carbapenemes, 1-azabicyclo[3.2.0]heptanes, the four-membered β-lactam ring, is fused to a five-membered ring sharing the same nitrogen heteroatom. Monobactams have a monocyclic β-lactam structure in which a side sulfo group is attached to a nitrogen atom of azetidin-2-one structure (Fig. 30.1.).

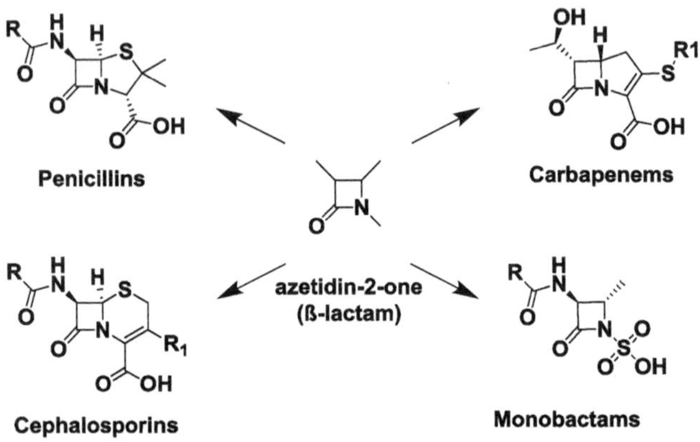

FIG. 30.1 The β-lactam antibiotics groups.

Penicillins

The existence of several natural penicillins (F, V, G, K, O, X) (**30.1.1** to **30.1.6**) (Fig. 30.2.), which differed only in the structures of their side chains, suggested further variations of the chain, producing a plethora of semisynthetic antibiotics.

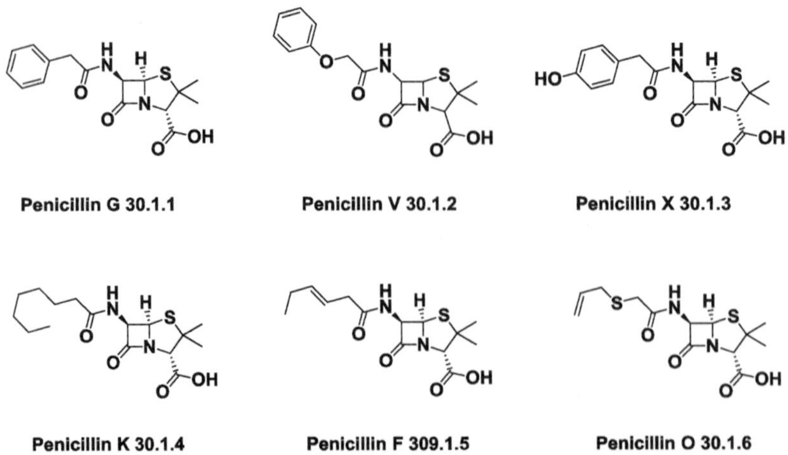

FIG. 30.2 Natural penicillins (F, V, G, K, O, X).

To produce them, different carboxylic acids could be added to the fermentation medium. For example, enzymatic synthesis of β-lactam antibiotics by adding phenylacetic acid into the reaction medium produces penicillin G; adding phenoxyacetic acid produces penicillin V.

But to prepare wide variety of semisynthetic penicillins or cephalosporins, diverse acyl donors-acid derivatives are usually coupled with easily available aminopenicillanic acid (6-APA), 7-aminocephalosporanic acid (7-ACA), or aminodeacetoxycephalosporanic acid (7-ADCA).

6-APA and 7-ACA are produced in industrial scale by enzymatic or chemical deacylation of easily available penicillins (penicillin G or V) or cephalosporins (cephalosporin C) using a variety of microbials produced by hydrolasing enzymes. The most used enzyme is penicillin G deacylase from *E. coli*.

In 2007, the 50th anniversary of the discovery of 6-APA, the precursor of all semisynthetic penicillins and cephalosporins, was chronicled with a special review [63].

Industrial production of semisynthetic antibiotics are based on the key step reaction of condensation of 6-APA or 7-ACA and 7-ADCA with a newly designed side-chain acyl donor (Fig. 30.3.). Synthesis techniques for coupling the acyl side chain with the β-lactam nucleus, such as acid chlorides, mixed anhydrides, acyl azides or acids implementing various coupling protocols, were and are a special focus of researchers [64-74].

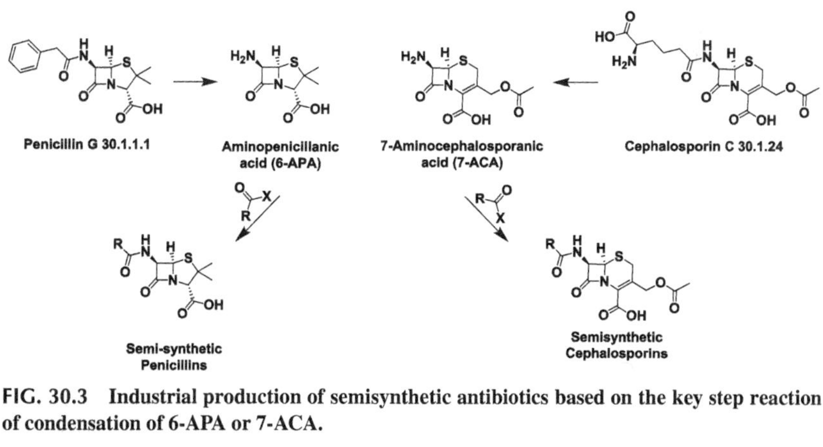

FIG. 30.3 Industrial production of semisynthetic antibiotics based on the key step reaction of condensation of 6-APA or 7-ACA.

7-ADCA is usually prepared by chemical or enzymatic ring expansion reaction (catalytic rearrangement) of 6-APA, in which the five-membered thiazolidine ring is expanded to the six-membered dihydrothiazine ring of 7-ADCA [75]. The enzyme performing the β-lactam ring expansion is named deacetoxycephalosporin C synthase or expandase (Fig. 30.4.).

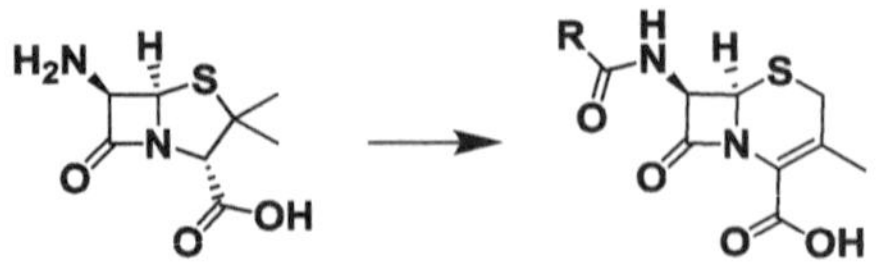

FIG. 30.4 Catalytic rearrangement of 6-APA to 7-ADCA.

The total pure chemical synthesis of β-lactam antibiotics is also carried out. The synthesis of 6-APA penicillins was the result of more than 15 years of efforts, which finally succeeded in 1959 [76,77].

The total synthesis of cephalosporin C was announced in 1965 in Woodward's Nobel lecture [78,79].

The first semisynthetic new β-lactam antibiotic successful in the management of a wide range of infections, was ampicillin. Ampicillin is effective against many bacteria, including *Haemophilus influenzae*, *N. gonorrhoeae*, *E. coli*, *Salmonella*, *Shigella*, streptococci, and certain strains of staphylococci. This was followed by methicillin, cloxacillin, and flucloxacillin. In 1967, carbenicillin was introduced. It was the first penicillin to be clinically active against *Pseudomonas aeruginosa*. It then was synthesized to produce amoxicillin. The discovery of carbenicillin was followed by ticarcillin, then piperacillin, mezlocillin, and azlocillin, all of which are active against *P. aeruginosa*.

The penicillins on the market are ampicillin (**30.1.7**), amoxicillin (**30.1.8**), methicillin (**30.1.9**), nafcillin (**30.1.10**), oxacillin (**30.1.11**), cloxacillin (**30.1.12**), dicloxacillin (**30.1.13**), flucloxacillin (**30.1.14**), carbenicillin (**30.1.15**), ticarcillin (**30.1.16**), temocillin (**30.1.17**), azlocillin (**30.1.18**), mezlocillin (**30.1.19**), piperacillin (**30.1.20**), pivampicillin (**30.1.21**), bacampicillin (**30.1.22**), and pivmecillinam (**30.1.23**) (Fig. 30.5.).

Production of antibiotics is one of the most outstanding achievements of industrial biotechnology. Penicillins and cephalosporins, which represent the world's major biotechnological products and share approximately 70% of the antibiotic market, are produced industrially by a process of fermentation, where the source microorganism (producer organisms—*Penicillium chrysogenum* or *Cephalosporium acremonium*) are grown in reactors in a medium containing a carbon source, nitrogen, phosphates, and trace elements with adjusted oxygen concentration, temperature, and pH. Various carbon sources have been adopted for the fermentation, including glucose, sucrose, and other crude sugars. Ammonia and ammonium sulfate represent major nitrogen sources. The essential precursor substances are phenyl acetic acid (for penicillin G) and phenoxyacetic acid (for penicillin V). Once the process is complete, the antibiotic is separated and purified to a crystalline product.

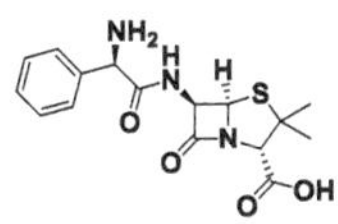

Ampicillin 30.1.7

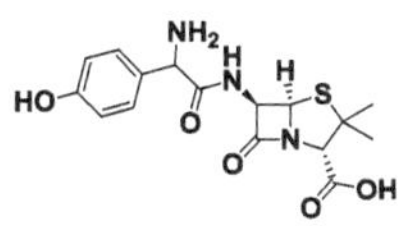

Amoxicillin 30.1.8

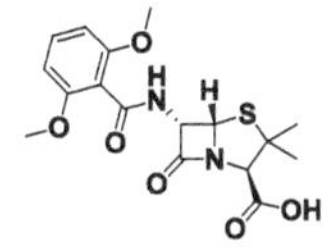

Meticillin 30.1.9

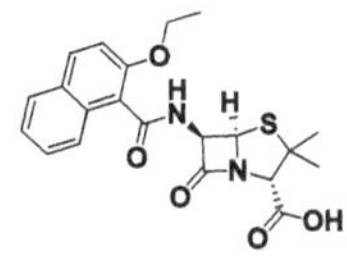

Nafcillin 30.1.10

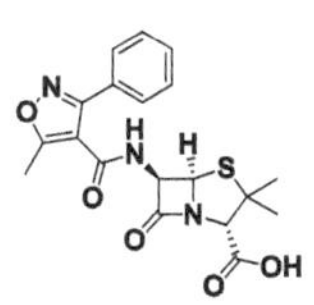

Oxacillin 30.1.11

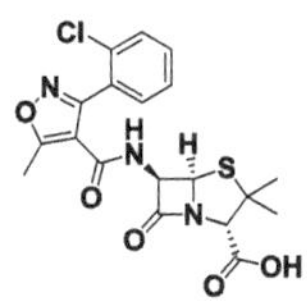

Cloxacillin 30.1.12

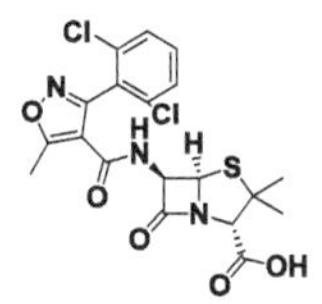

Dicloxacillin 30.1.13

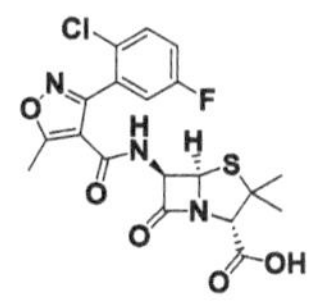

Flucloxacillin 30.1.14

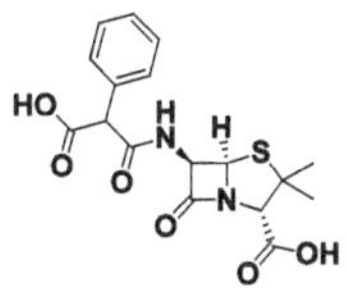

Carbenicillin 30.1.15

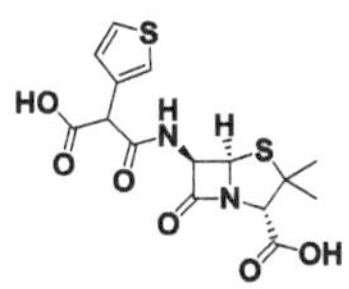

Ticarcillin 30.1.16

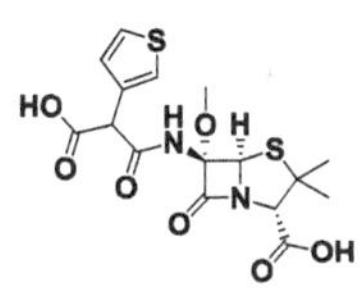

Temocillin 30.1.17

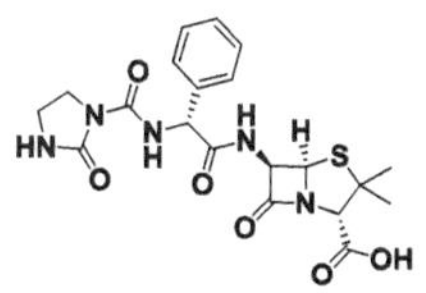

Azlocillin 30.1.18

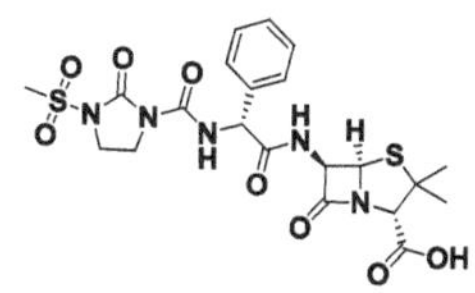

Mezlocillin 30.1.19

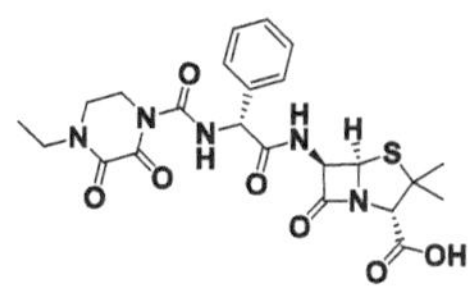

Piperacillin 30.1.20

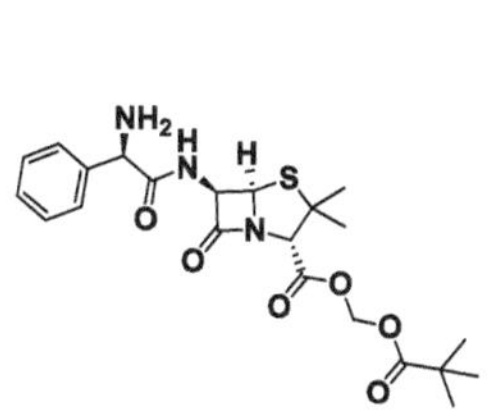

Pivampicillin 30.1.21

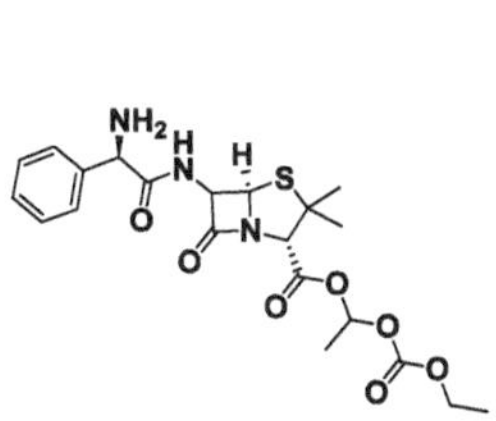

Bacampicillin 30.1.22

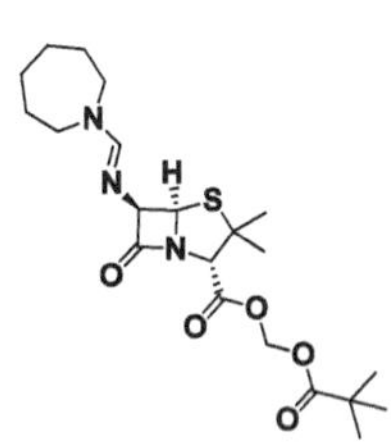

Pivmecillinam 30.1.23

FIG. 30.5 The penicillins available in the market.

Major β-lactam antibiotic manufacturers are producing now 40 to 50 g/L for penicillin and 20 to 25 g/L for cephalosporin. Approximately 75% of the total bulk penicillin volume manufactured is used for the production of semisynthetic penicillins and cephalosporins.

The penicillin nucleus 6-APA, which world market capacity in the 2000s is around 35,000 tons per year, served as an excellent base for development of many semisynthetic penicillins [17,67,71,73,80-86].

Although antibiotics are very effective therapeutics, prolonged use of β-lactams has caused development of resistance in target organisms.

Resistance to β-lactams in particular, and to antibiotics in general, is currently a major health concern in treating infectious diseases. Antibiotic resistance can occur via three general mechanisms: prevention of interaction of the drug with target, efflux of the antibiotic from the cell, and direct destruction or modification of the drug.

The most common mechanism of resistance to β-lactam antibiotics is adaptive bacterial resistance, that is, production of β-lactamases, which destroy β-lactam antibiotics before they reach the bacterial target.

Approaches for overcoming these resistance mechanisms include the development of novel β-lactamase–stable β-lactams, β-lactamase inhibitors to be employed with existing β-lactams, and agents that potentiate the activity of existing β-lactams [87-94].

The synthesis of essential semisynthetic penicillin is described in our previous book [95].

The accelerated emergence of bacteria that are resistant to multiple antibiotic types now appears to be a serious threat to the continuing success of antibiotics in the treatment of infectious diseases.

Bacterial resistance to antibiotics is growing up day by day. Faced with scientific, regulatory and mainly economic challenges, the pharmaceutical companies seems unable to respond to antibiotic discovery crisis. Even Pfizer, the historical champion of antibiotic development, closed its antibiotic research facility for financial reasons in 2011, ending research into Gram-negative bacteria, which are the bacteria that are resistant to numerous drugs and antibiotics. Such giants of pharma as Sanofi, Eli Lilly, and Bristol-Myers Squibb haven't researched and developed antibiotics since 1990.

Cephalosporins

Cephalosporins are the second group of β-lactam antibiotic family to be discovered after penicillin [96].

Antibacterial substance cephalosporin C (**30.1.24**) (Fig. 30.6.), which showed broad spectrum of action against *Staphylococcus aureus, Vibrio cholerae,* and *B. anthracis*, but moderate antibacterial activity was isolated from the fermentation broth of a strain of *A. chrysogenum*. Later a process by which the D-α-amino adipoyl side chain of cephalosporin C could be removed, generating 7-ACA, was proposed [97,98].

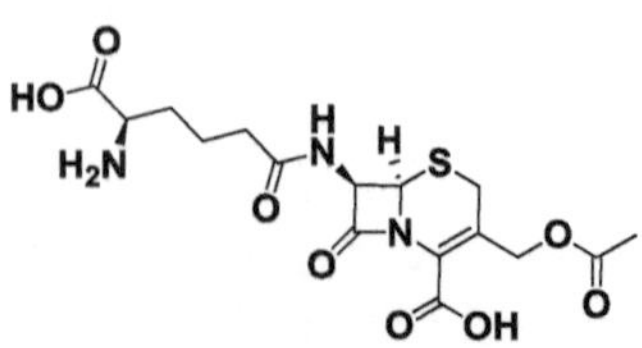

Cephalosporin C 30.1.24

FIG. 30.6 Cephalosporin C.

The 7-ACA nucleus is not sufficiently potent for clinical use, but this invention paved the way for wide modifications and creation of a plethora of semisynthetic cephalosporins coupling 7-ACA with different carbonic acid derivatives chemically or enzymatically. The chemical methods are gradually replaced by the enzymatic methods [99].

Now the 7-ACA is generally produced by hydrolysis of cephalosporin C via chemical or enzymatic methods, but both industrial production protocols have serious limitations [100,101].

Cephalosporins are a group of broad-spectrum, semisynthetic β-lactam antibiotics that currently constitute the most widely prescribed class of antibiotics and are used to treat diseases caused by both Gram-positive and Gram-negative bacteria with wide safety margins compared to penicillins.

Favorable attributes of the cephalosporins include low toxicity, a relatively broad spectrum of activity, and ease of administration. Various cephalosporins are effective for treatment of pneumonia, skin and soft-tissue infections, bacteremia, and meningitis. Differences among the numerous cephalosporin antimicrobial agents are sometimes subtle.

Cephalosporins are bactericidal antibiotics, chemically closely related to penicillins, and have the same mode of action, disrupting the synthesis of the peptidoglycan layer of bacterial cell walls of bacteria, causing their death.

Cephalosporins are grouped into "generations" (although controversies exist) based on their spectrum of antimicrobial activity, which is consistent with the chronological order in which they were produced. The cephalosporins have been classified into five groups according to their antimicrobial spectrum. In general, progression from first to fourth generation is associated with a broadening of the Gram-negative antibacterial spectrum.

Cephalosporins constitute approximately 50% of the all β-lactam antibiotics available in the market.

First-Generation Cephalosporins

First-generation cephalosporins are broad-spectrum antibiotics that are effective against Gram-positive *Streptococcus pyogenes*, Streptococcus agalactiae, viridans streptococci, and Gram-negative bacteria, such as *Staphylococcus*, *Streptococcus*, some strains of *E. coli*, and pulmonary infections caused by *Klebsiella pneumoniae*. They are used in cases of uncomplicated, community-acquired infections of the skin and soft tissue, and urinary and respiratory tract infections caused by penicillin-sensitive *Streptococcus*. Parenteral first-generation cephalosporins are used for surgical wound prophylaxis.

Prototype drugs of the series became cefazolin (for intravenous use) and cefalexin (oral use).

To our knowledge cefacetrile (**30.1.25**), cefaloglycin (**30.1.26**), cefalotin (**30.1.27**), cefapirin (**30.1.28**), cefalexin (**30.1.29**), cefadroxil (**30.1.30**), cefradine (**30.1.31**) (Fig. 30.7.), cefatrizine (**30.1.32**), cefazolin (**30.1.33**), cefazedone (**30.1.34**), cefazaflur (**30.1.35**), ceftezole (**30.1.36**) (Fig. 30.8.), cefroxadine (**30.1.37**), cefaloridine (**30.1.38**), and cefalonium (**30.1.39**) are considered first-generation cephalosporins (Fig. 30.9.).

Here we have attempted to arrange them in accordance to their belonging to: aminocephalosporanic acid (7-ACA) derivatives (**30.1.25** to **30.1.28**); 7-ADCA (7-ADCA) derivatives (**30.1.29** to **30.1.31**) (Fig. 30.7.); (6S,7S)-7-amino-8-oxo-5-thia-1-azabicyclo[4.2.0]oct-2-ene-2-carboxylic acid derivatives (**30.1.32** to **30.1.36**) (Fig. 30.8.); and (6S,7S)-7-amino-3-(methoxy)- or (3-(pyridin-1-ium-1-ylmethyl)-8-oxo-5-thia-1-azabicyclo[4.2.0]oct-2-ene-2-carboxylic acid derivatives (**30.1.37** to **30.1.39**) (Fig. 30.9.).

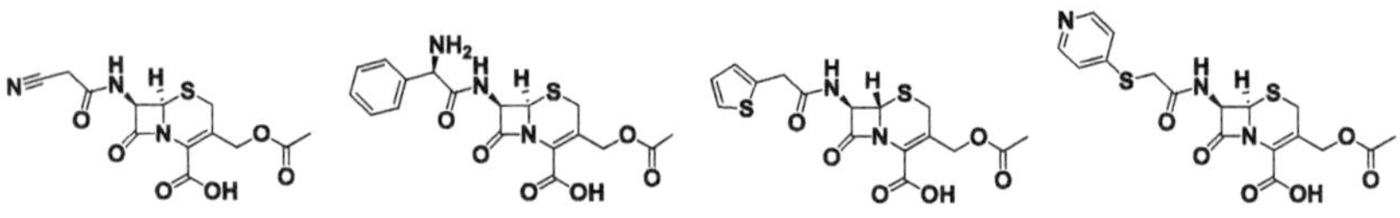

Cefacetrile 30.1.25 Cefaloglycin 30.1.26 Cefalotin 30.1.27 Cefapirin 30.1.28

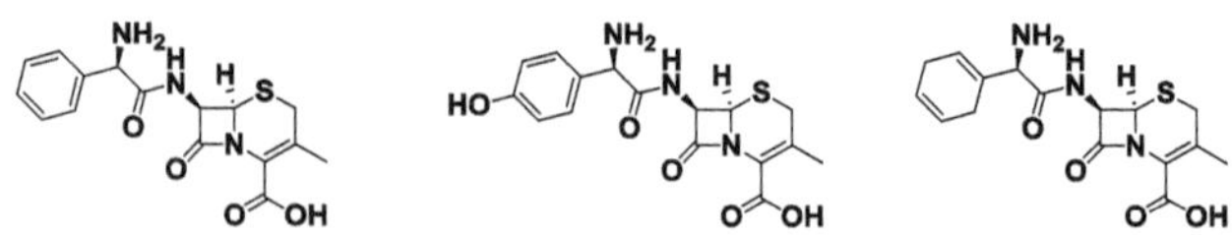

Cefalexin 30.1.29 Cefadroxil 30.1.30 Cefradine 30.1.31

FIG. 30.7 Aminocephalosporanic acid (7-ACA) derivatives (30.1.25 to 30.1.28) and amino-deacetoxycephalosporanic acid (7-ADCA) derivatives (30.1.29 to 30.1.31).

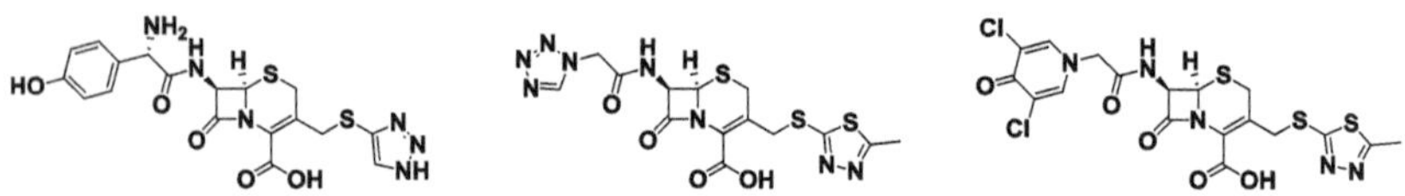

Cefatrizine 30.1.32 Cefazolin 30.1.33 Cefazedone 30.1.34

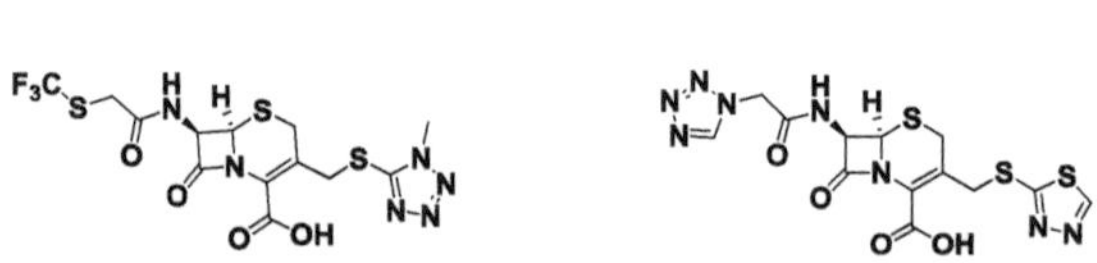

Cefazaflur 30.1.35 Ceftezole 30.1.36

FIG. 30.8 (6S,7S)-7-Amino-8-oxo-5-thia-1-azabicyclo[4.2.0]oct-2-ene-2-carboxylic acid derivatives.

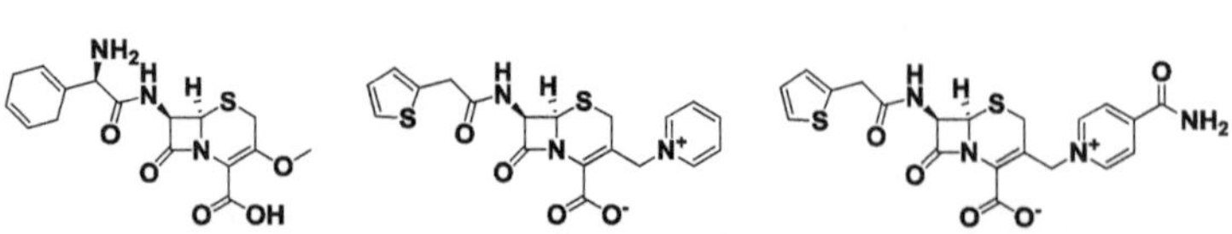

Cefroxadine 30.1.37 Cephaloridine 30.1.38 Cefalonium 30.1.39

FIG. 30.9 Azabicyclo[4.2.0]oct-2-ene-2-carboxylic acid derivatives.

Second-Generation Cephalosporins

Second-generation cephalosporins are more effective against Gram-negative bacterial species that are resistant to the first-generation cephalosporins. They have proven effective against Gram-positive *Staphylococcus* and *Streptococcus* bacteria, and some Gram-negative bacteria, including *Bacteroides fragilis, H. influenzae, Moraxella catarrhalis, N. meningitidis, N. gonorrhoeae,* and some *Enterobacteriaceae.*

Within the second-generation cephalosporins are two groups of compounds that differ structurally, in spectrum of activity, and in side effects. These groups are known as the "true" second-generation cephalosporins—cefamandole (**30.1.40**), cefonicid (**30.1.41**), cefuroxime (**30.1.42**), cefuzonam (**30.1.43**), cefaclor (**30.1.44**), cefprozil (**30.1.45**) (Fig. 30.10.), and cephamycins or 7-methoxycephalosporins.

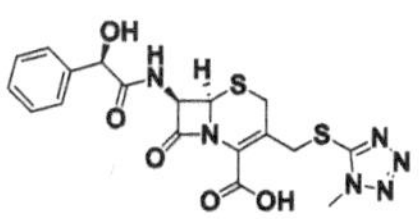

Cefamandole 30.1.40

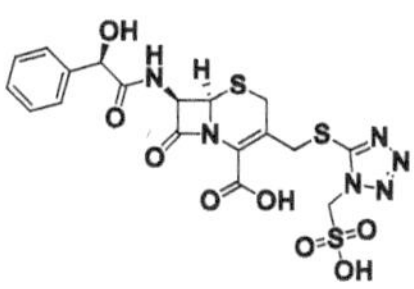

Cefonicid 30.1.41

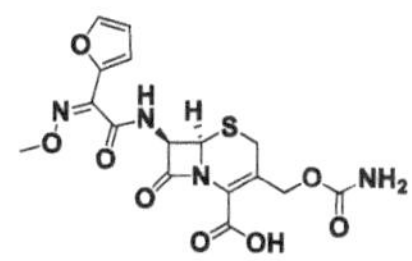

Cefuroxime 30.1.42

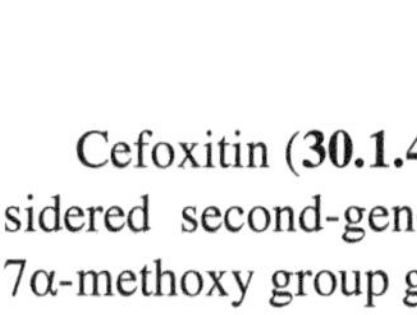

Cefuzonam 30.1.43

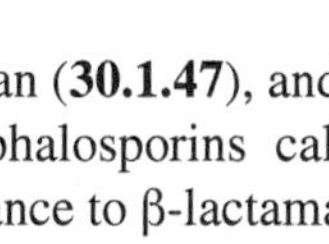

Cefaclor 30.1.44

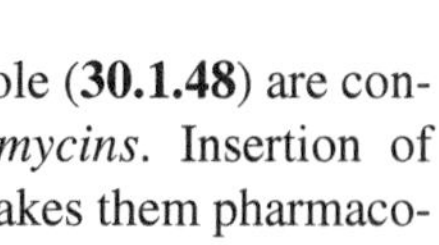

Cefprozil 30.1.45

FIG. 30.10 Second-generation cephalosporins.

Cefoxitin (**30.1.46**), cefotetan (**30.1.47**), and cefmetazole (**30.1.48**) are considered second-generation cephalosporins called *cephamycins*. Insertion of 7α-methoxy group gives resistance to β-lactamases and makes them pharmacologically different from other cephalosporins (Fig. 30.11.).

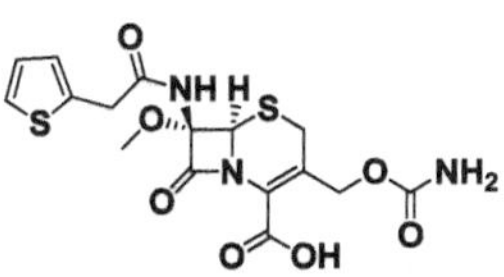

Cefoxitin 30.1.46

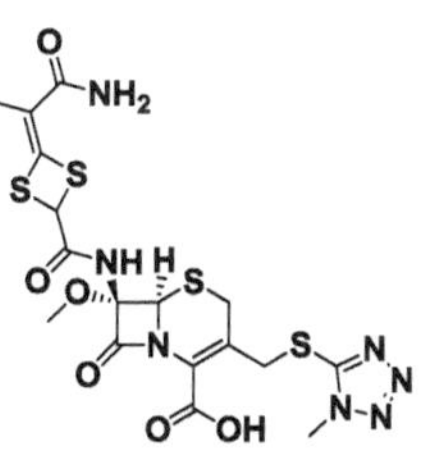

Cefotetan 30.1.47

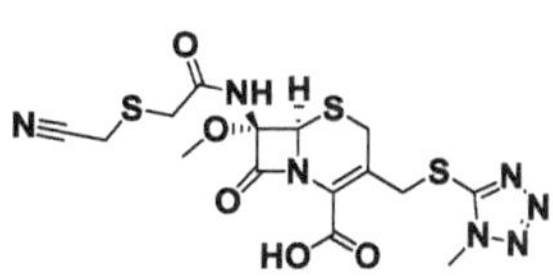

Cefmetazole 30.1.48

FIG. 30.11 Second-generation cephalosporins—the cephamycins.

The "true" second-generation agents are useful for community-acquired infections of the respiratory tract and uncomplicated urinary tract infections. The cephamycin group is useful for mixed aerobic–anaerobic infections of the skin and soft tissues, intraabdominal, and gynecologic infections, and surgical prophylaxis.

Third-Generation Cephalosporins

Third-generation cephalosporins have further expanded activity on the hard-to-treat Gram-negative bacteria spectrum that includes *Enterobacteriaceae*, which are usually associated with hospital-acquired infections, *Serratia*, and *Pseudomonas*. They are active against Gram-positive methicillin-susceptible *S. aureus*, very active against groups A and B streptococci, and viridans streptococci. They are characterized by a broader spectrum of activity and increased stability to β-lactamases compared to the first- and second-generation cephalosporins.

They are used for intraabdominal infections, as well as for infections of the respiratory tract, blood, skin and soft tissue, and urinary tract. The clinical situations requiring use of third-generation cephalosporins are likely to be encountered in patients who have recently received antibiotics or who are immunocompromised.

Third-generation cephalosporins include cefotaxime (**30.1.49**), cefpodoxime (**30.1.50**), cefcapene (**30.1.51**), ceftibuten (**30.1.52**), ceftizoxime (**30.1.53**), cefetamet (**30.1.54**), cefdinir (**30.1.55**), cefditoren (**30.1.56**), cefixime (**30.1.57**), cefmenoxime (**30.1.58**), ceftriaxone (**30.1.59**), cefoperazone (**30.1.60**), cefsulodin (**30.1.61**), and ceftazidime (**30.1.62**). Three third-generation cephalosporins—flomoxef (**30.1.63**) and latamoxef or moxalactam (**30.1.64**)—are not derivatives of 7-ACA, but its derivative where the sulfur atom in the 1,3-thiazinane ring of 7-ACA is replaced by oxygen and a methoxy group is attached to 7-aminoazetidin-one part.

The three most frequently used third-generation cephalosporins are cefotaxime (**30.1.49**), ceftriaxone (**30.1.59**), and ceftazidime (**30.1.62**) (Fig. 30.12.).

Fourth-Generation Cephalosporins

Fourth-generation cephalosporins are cephalosporins with greater activity against both organisms than third-generation agents. Cefepime (**30.1.65**), cefquinome (**30.1.66**), cefpirome (**30.1.67**), cefoselis (**30.1.68**), cefclidine (**30.1.69**), cefluprenam (**30.1.70**), cefozopran (**30.1.71**) (Fig. 30.13.) have broadest spectrum of activity against Gram-positive *Streptococcus pneumoniae* and groups A and B streptococci, and against a broad array of Gram-negative bacteria such as *Enterobacter, E. coli, Klebsiella, Proteus, Pseudomonas, H. influenzae,* and *Neisseria meningitides,* and offer an alternative for the treatment of infections caused by some drug-resistant microorganisms.

They also have a greater resistance to β-lactamases than the third-generation cephalosporins.

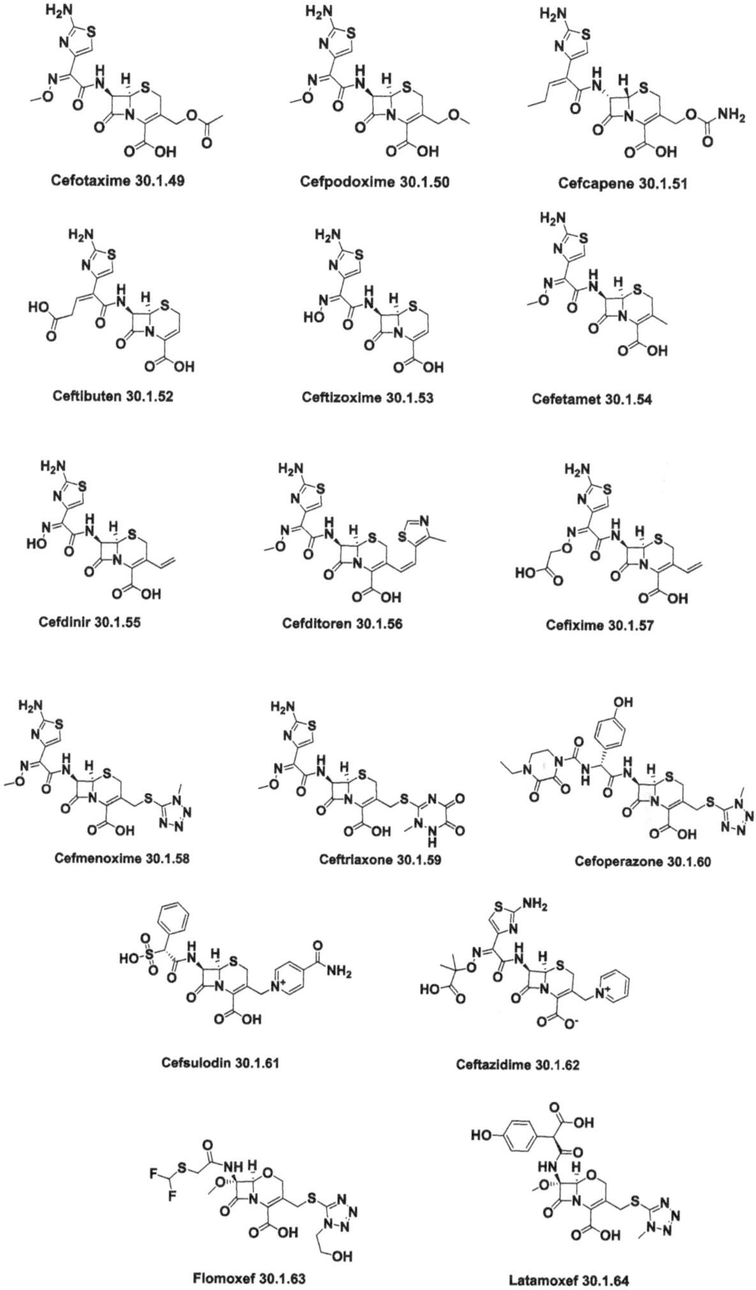

FIG. 30.12 The third-generation cephalosporins.

All of the fourth-generation cephalosporins are zwitterions that can penetrate the outer membrane of Gram-negative bacteria and can be used in the treatment of meningitis.

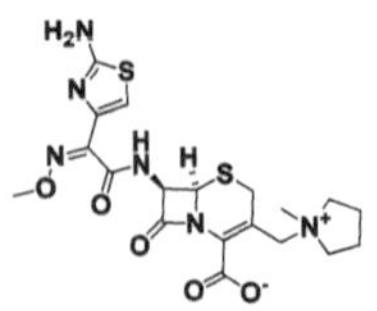
Cefepime 30.1.65

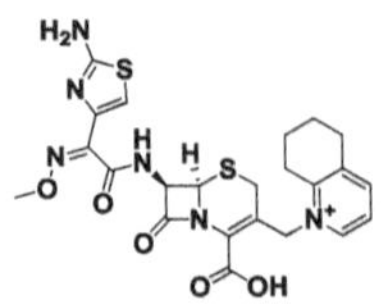
Cefquinome 30.1.66

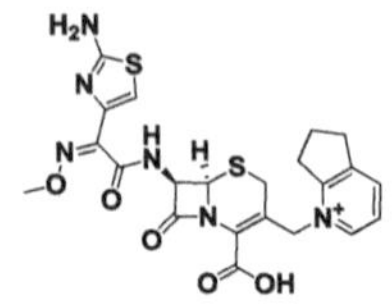
Cefpirome 30.1.67

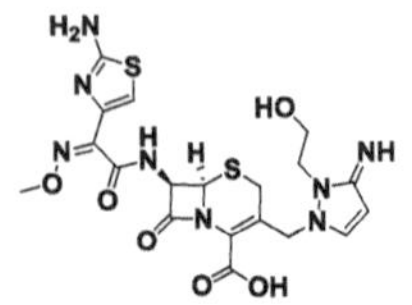
Cefoselis 30.1.68

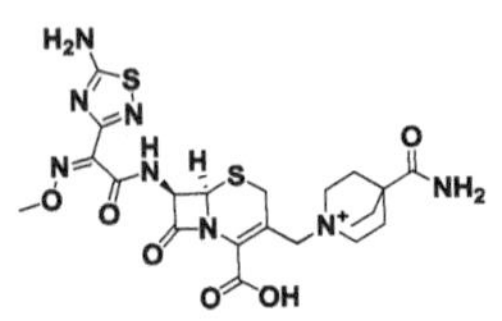
Cefclidin 30.1.69

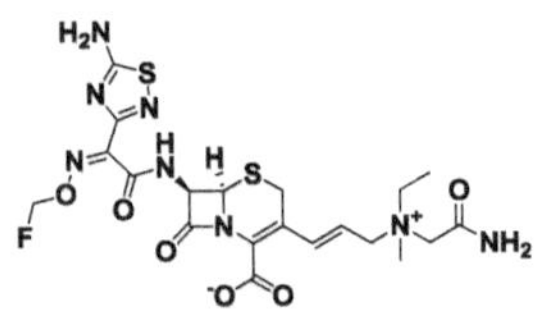
Cefluprenam 30.1.70

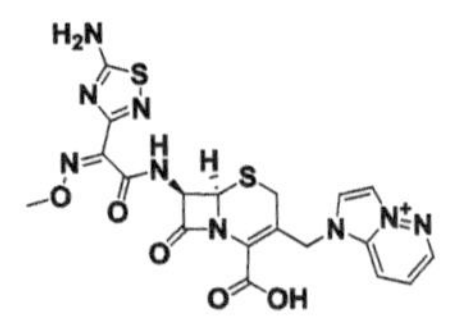
Cefozopran 30.1.71

FIG. 30.13 **The fourth-generation cephalosporins.**

Fifth-Generation Cephalosporins

The antibacterial spectra of fifth-generation cephalosporins ceftobiprole (**30.1.72**) and ceftaroline (**30.1.73**) (Fig. 30.14.) is wider than those of the other cephalosporins, with improved Gram-positive activity and higher activity against methicillin-resistant *S. aureus*. They have higher activity and stability to β-lactamases, and less-serious adverse reactions. Ceftaroline is not active for enterococci. These drugs are used mainly for treating complicated infections of skin and soft tissue.

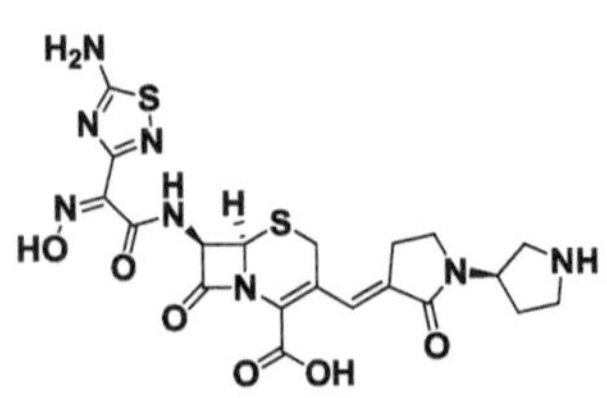
Ceftobiprole 30.1.72

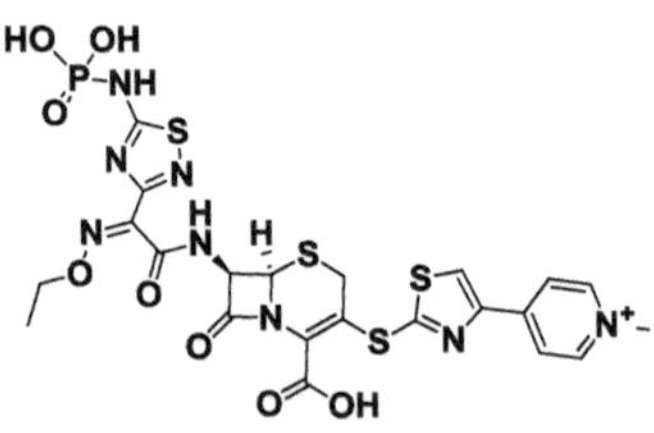
Ceftaroline 30.1.73

FIG. 30.14 **The fifth-generation cephalosporins.**

Not Classified Cephalosporins

There exists several not classified cephalosporins, including cefaloram (**30.1.74**), cefuracetime (**30.1.75**), cefoxazole (**30.1.76**), ceftioxide (**30.1.774**), cefedrolor (**30.1.78**), cefsumide (**30.1.79**), cefrotil (**30.1.80**), cefovecin (**30.1.81**), cefempidone (**30.1.82**), cefetrizole (**30.1.83**), cefivitril (**30.1.84**), cefcanel (**30.1.85**), cefaparole (**30.1.86**), cefmatilen (**30.1.87**), and cefmepidium (**30.1.88**) (Fig. 30.15.).

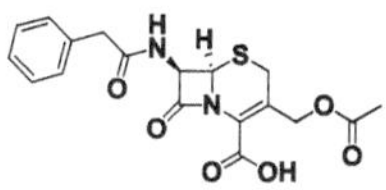

Cefaloram 30.1.74

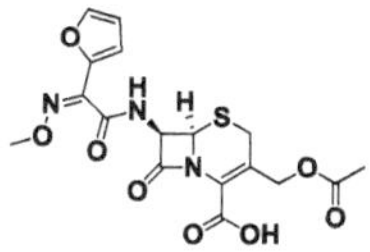

Cefuracetime 30.1.75

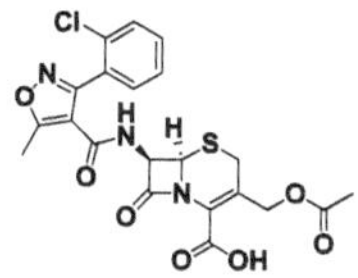

Cefoxazole 30.1.76

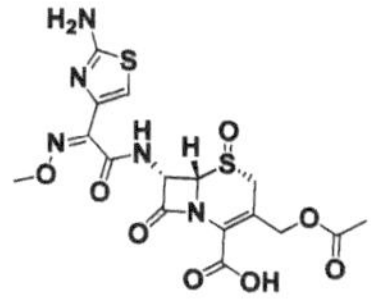

Ceftioxide 30.1.77

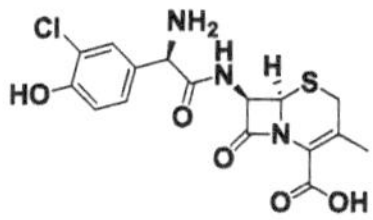

Cefedrolor 50.1.78

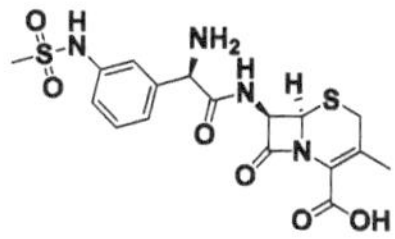

Cefsumide 30.1.79

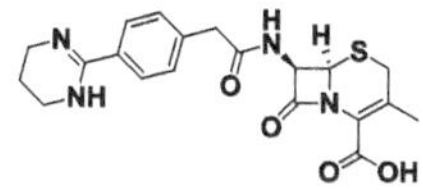

Cefrotil 30.1.80

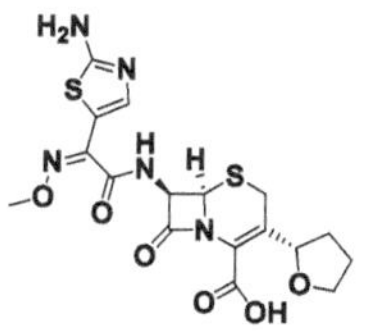

Cefovecin 30.1.81

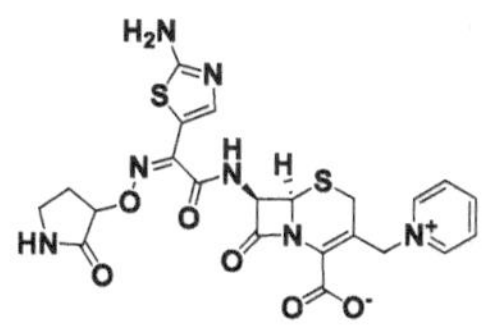

Cefempidone 30.1.82

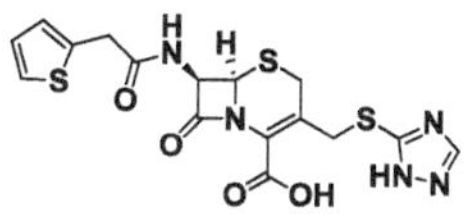

Cefetrizole 30.1.83

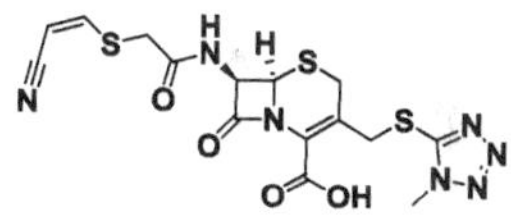

Cefivitril 30.1.84

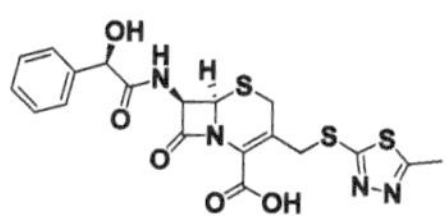

Cefcanel 30.1.85

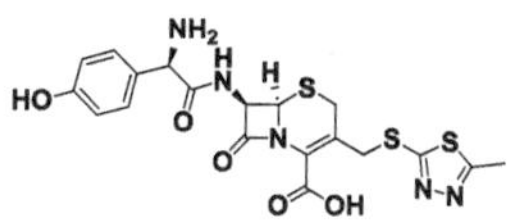

Cefaparole 30.1.86

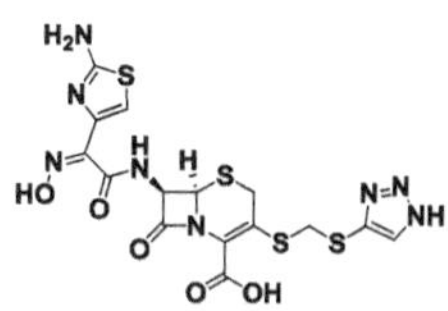

Cefmatilen 30.1.87

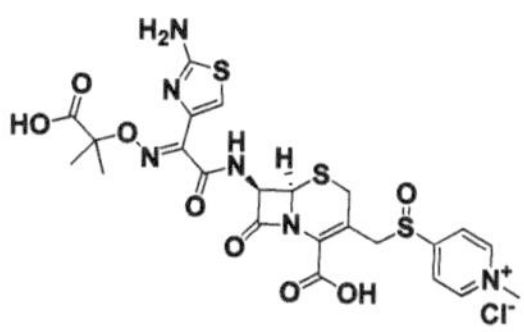

Cefmepidium 30.1.88

FIG. 30.15 Not classified cephalosporins.

Synthesis of many of the listed cephalosporins are described in our previous book [95].

Carbapenems

The first representative of carbapenem class of β-lactam antibiotics—thienamycin (**30.1.89**)—and several other carbapenems were isolated from actinomycete strains.

While sharing general β-lactam features, carbapenems are regarded as the class that is most potent and as having the widest spectrum of antimicrobial activity. Carbapenems are characterized by a β-lactam ring condensed with a pyrroline ring and thus differ from penicillins in the substitution of a methylene for the sulfur atom. In addition, instead of the amide chain in position 6 they have a 6-((R)-1-hydroxyethyl) short chain.

Carbapenems are a class of atypical β-lactam antibiotics with broad-spectrum high antibacterial activity, excellent coverage of many Gram-positive and Gram-negative aerobic and anaerobic bacteria, and stable to most β-lactamases.

The first thienamycin analogue, imipenem (**30.1.90**), was introduced in 1984. The new carbapenem drugs include meropenem (**30.1.91**) doripenem (**30.1.92**), and ertapenem (**30.1.93**) (Fig. 30.16.). Biapenem (**30.1.94**) is one of the new carbapenem antibiotics in clinical trials.

Carbapenems play a significant role in the current antibiotic armamentarium exerting the same mode of action as other β-lactams—inhibiting synthesis of the cell wall of bacteria—and are used in severe infections and in complicated infections after failure of initial antibiotic treatment [102-111].

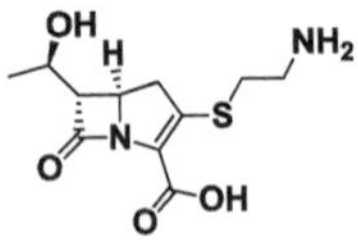

Thienamycin 30.1.89

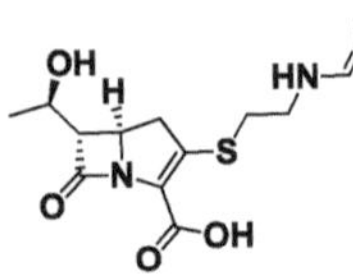

Imipenem 30.1.90

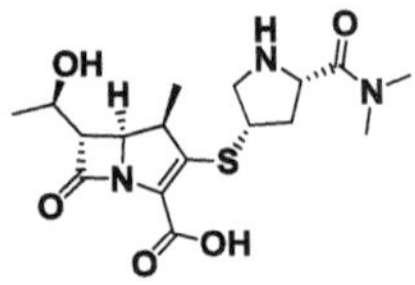

Meropenem 30.1.91

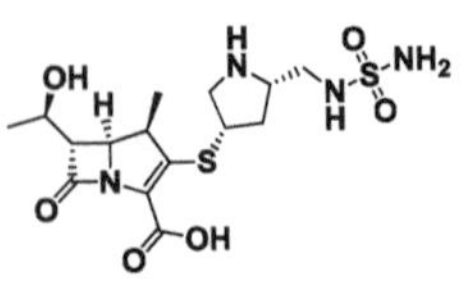

Doripenem 30.1.92

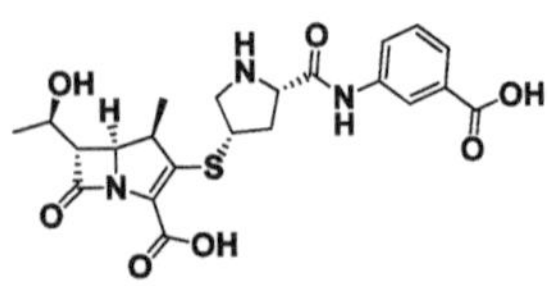

Ertapenem 30.1.93

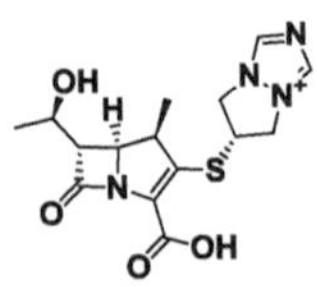

Biapenern 30.1.94

FIG. 30.16 **Carbapenems.**

Monobactams

Carbapenems and monobactams are two classes of new β-lactam antibiotics that were introduced in the 1980s.

Monobactams history concerns attempts of isolation of new β-lactam antibiotics produced by less-common microorganisms, which succeeded with the separation of sulfazecin (**30.1.95**) and isosulfazecin (**30.1.96**), monobactam—β-lactam compounds wherein the β-lactam ring is not fused to another ring, and is produced by new species of *Pseudomonas, Pseudomonas acidophila* and *Pseudomonas mesoacidophila*.

Several alternative compounds were synthesized to prepare derivatives of the mentioned naturally occurring monobactams, which produced new compounds with potent activity against aerobic Gram-negative bacteria. Two monocyclic β-lactamase–resistant antibiotics include aztreonam (**30.1.97**) and carumonam (**30.1.98**). Aztreonam is the single monobactam approved in the United States. Carumonam is a medicine available in a number of countries worldwide. Compounds SQ 83360 (**30.1.99**) and tigemonam (**30.1.100**) are very close in their antibacterial activity to aztreonam, but for unknown reasons did not enter the clinic (Fig. 30.17.).

Aztreonam is active against most strains of Gram-negative microorganisms—aerobic and anaerobic—and indicated for the treatment of the infections caused by susceptible these microorganisms. It is indicated for lower respiratory tract infections, septicemia, skin, intraabdominal, and gynecological infections.

Aztreonam is also indicated for adjunctive therapy to surgery in the management of infections caused by susceptible organisms, including abscesses, infections complicating hollow viscus perforations, cutaneous infections, and infections of serous surfaces. Aztreonam should be used only to treat or prevent infections that are proven or strongly suspected to be caused by susceptible bacteria [112-116].

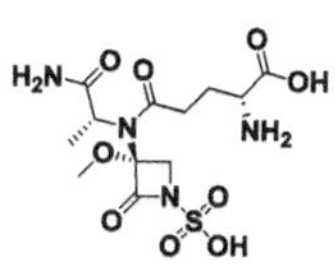

Sulfazecin 30.1.95

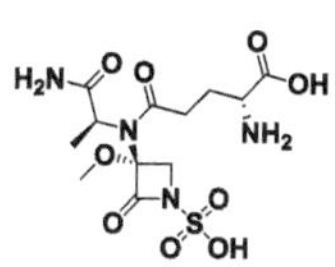

Isosulfazecin 30.1.96

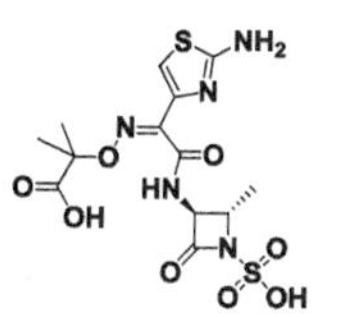

Azthreonam 30.1.97

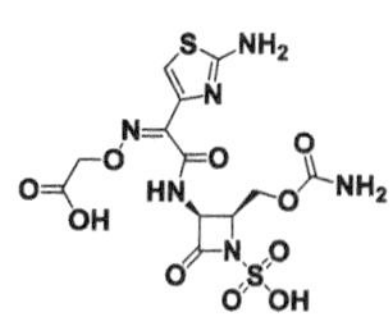

Carumonam 30.1.98

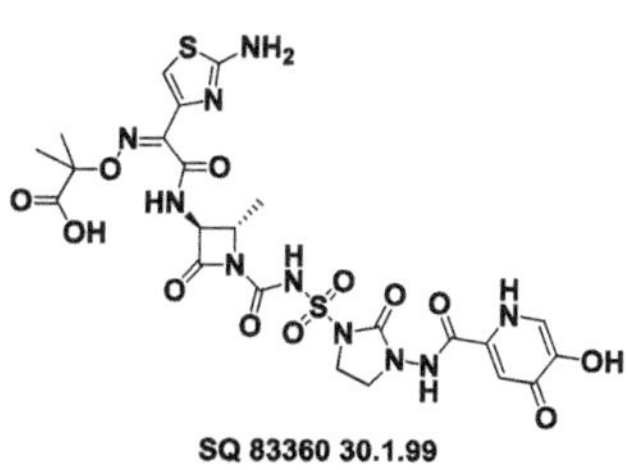

SQ 83360 30.1.99

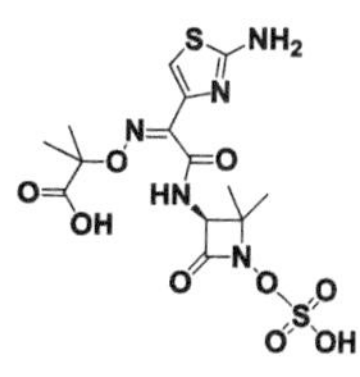

Tigemonam 30.1.100

FIG. 30.17 Monobactams.

β-Lactamase Inhibitors

Since the 1940s β-lactam antibiotics, the largest class of antibiotics, have been used to treat bacterial infections. However, dissemination of β-lactam resistance has reached the point where many marketed β-lactams are no longer effective.

The most common mechanism of resistance to β-lactam antibiotics is the production of β-lactamases, which disrupt the four-member ring of β-lactam antibiotics before they reach the bacterial target.

β-Lactamases constitute a heterogenous group of enzymes. Many (more than 300) distinct β-lactamases have been isolated and identified.

Two classification schemes for β-lactamases are currently in use. One is based on the amino acid sequence and divides β-lactamases into class A, class C, and class D enzymes, which utilize serine for β-lactam hydrolysis, and class B Zn-dependent metalloenzymes, which require Zn^{2+} ions for substrate hydrolysis [117].

β-Lactamase inhibitors and compounds that interfere with the ability of the bacteria to deactivate implemented β-lactam antibiotic and the use of β-lactamase inhibitors in combination with β-lactam antibiotics just by coadministration are currently the most successful strategies to combat a specific resistance mechanism [118-132].

The three classical β-lactamase inhibitors are clavulanic acid (**30.1.101**), sulbactam (**30.1.102**), and tazobactam (**30.1.103**). Avibactam (**30.1.104**), a non–β-lactam, β-lactamase inhibitor, is the most recent one approved (Fig. 30.18.).

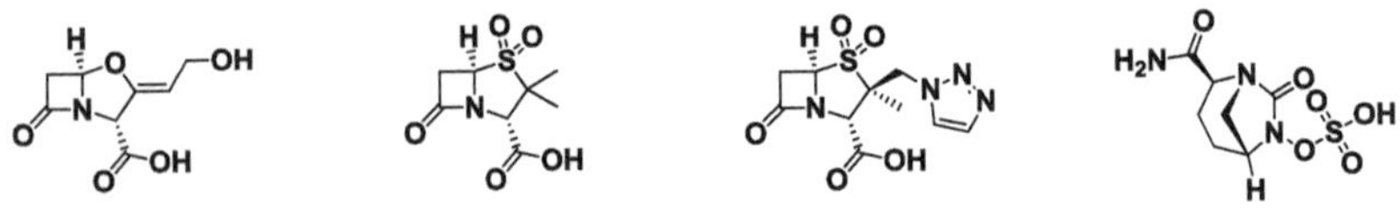

Clavulanic acid 30.1.101 **Sulbactam 30.1.102** **Tazobactam 30.1.103** **Avibactam 30.1.104**

FIG. 30.18 **β-Lactamase inhibitors.**

Several new β-lactamase inhibitors belonging to penicillanic and clavulanic acid derivatives, penems, cephems, and monobactams, are currently under development.

30.2 TETRACYCLINES

Tetracyclines were first reported in the late 1940s as naturally occurring antibiotics that were discovered as products of vital activity of *Actinomycetes* soil bacteria.

Their introduction in therapy began in 1948 with chlortetracycline, produced by *Streptomyces aureofaciens*. Later oxytetracycline, isolated from the cultural broth of *Streptomyces rimosus* was discovered, and then tetracycline, was obtained as a product of vital activity of *S. aureofaciens* in the cultural medium that excluded the presence of chlorine. Tetracycline later was prepared also semisynthetically by catalytic hydrogenation of chlortetracycline.

As one of the earliest antibiotics to be marketed after penicillin, and because of their convenient oral dosing, tetracyclines have achieved wide clinical usage.

The effect of the tetracyclines is principally bacteriostatic.

All three compounds mentioned above show a very similar broad spectrum of activity, encompassing both Gram-positive and Gram-negative organisms. The spectrum of their activity also includes some anaerobic bacteria, *Chlamydia*, mycoplasmas, *Rickettsia*, and protozoan parasites. They lack activity against fungi and viruses [133-146].

Chemically, tetracyclines are a subclass of polyketides, natural or semisynthetic, based on a octahydrotetracene-2-carboxamide skeleton, and only differ from each other by substituent variations.

Tetracyclines comprise a linear fused tetracycline nucleus, with rings designated A, B, C, and D, with naturally occurring stereochemical configurations at the 4a, 12a (A-B ring junction), and 4 dimethylamino group positions to which different functional groups can be attached. The simplest tetracycline to exhibit antibacterial activity is 6-deoxy-6-demethyltetracycline and it is considered the minimum pharmacophore (Fig. 30.19.).

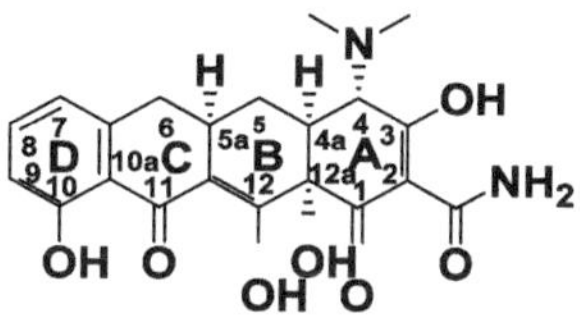

FIG. 30.19 Linear fused tetracycline nucleus.

Another type of tetracyclic antibiotic, daunomycin and adriamycin (Fig. 30.20.) was obtained later from species of *Streptomyces* and *Actinomadura* and showed antibacterial and antitumor activities. Their linear tetracyclic ring system is slightly different from the traditional tetracyclines, and they are named anthracyclines.

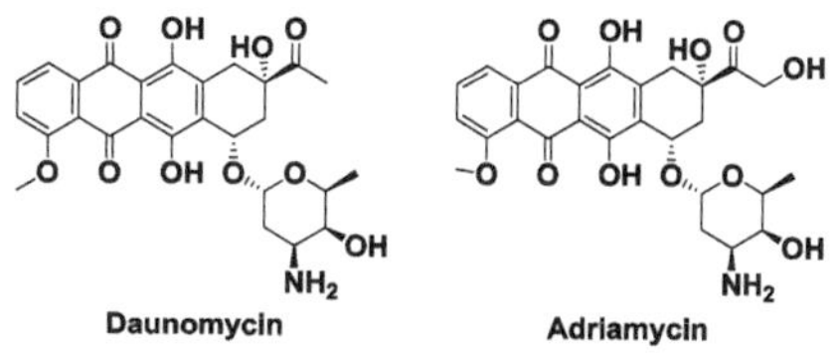

FIG. 30.20 Daunomycin and Adriamycin.

Although approximately 1000 tetracycline derivatives exist, only seven have seen extensive clinical use as antibacterials: tetracycline, chlortetracycline, oxytetracycline, methacycline, demeclocycline, doxycycline, and minocycline.

Tetracyclines inhibit bacterial protein synthesis by binding to the 30S ribosomal subunit at a position that blocks the binding of the aminoacyl-tRNA to the acceptor site on the mRNA-ribosome complex, preventing formation of the protein building initiation, which leads to bacteriostatic effect.

Tetracyclines enter the bacterial cell in two ways: (a) passive diffusion through hydrophilic pores in the outer cell membrane and (b) by active transport system in the cytoplasmic membrane which pumps the drug into the bacterial cell and is energy dependent. This system results in concentrations of tetracycline in bacterial cells only because an analogous mechanism is not present in mammalian cells.

In contrast to β-lactam antibiotics, tetracyclines are infrequently inactivated by resistant bacteria. Resistance to these agents develops primarily by preventing accumulation of the drug inside the cell, either by decreasing influx or increasing efflux. Once resistance develops to one of the drugs in this class, it is typically conferred on all tetracyclines.

Bacteria can use three strategies to become resistant to tetracyclines: (a) limiting the access of a tetracycline to the ribosomes by pumping the antibiotic out of the cell; (b) altering the ribosome making it insensitive to tetracycline, thus preventing effective binding of tetracyclines, and (c) producing distinct tetracycline-inactivating enzymes [147,148].

Tetracyclines are formally divided into three generations.

The first generation consists of the older natural or semisynthetic compounds, agents obtained in the years 1948 to 1963, and include tetracycline (**30.2.1**), oxytetracycline (**30.2.2**), chlortetracycline (**30.2.3**), demeclocycline (**30.2.4**), methacycline (**30.2.5**), rolitetracycline (**30.2.6**), and lymecycline (**30.2.7**). All of them are, characterized by low lipophilicity and poor absorption after oral administration. They are usually available in peroral form only, except rolitetracycline (**30.2.6**) (Fig. 30.21.).

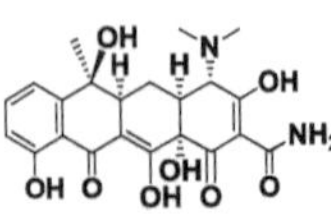

Tetracycline
30.2.1

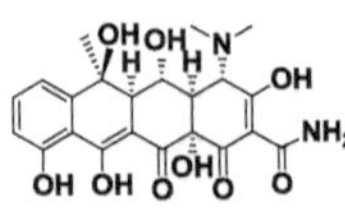

Oxytetracycline
30.2.2

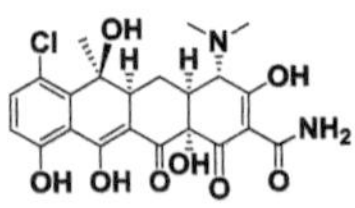

Chlortetracycline
30.2.3

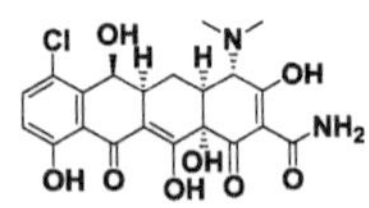

Demeclocycline
30.2.4

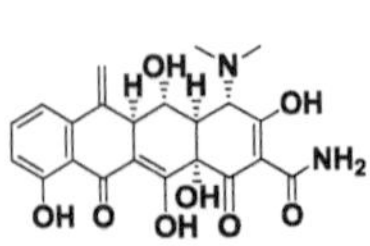

Methacycline
30.2.5

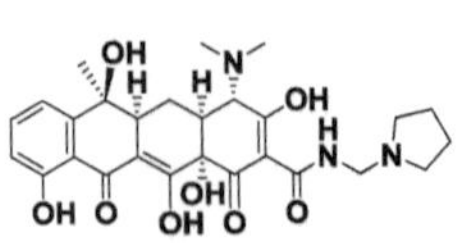

Rolitetracycline
30.2.6

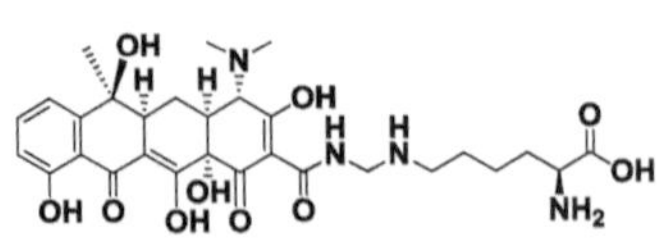

Lymecycline
30.2.7

FIG. 30.21 **The first-generation tetracyclines.**

The second generation of tetracycline compounds, represented by doxycycline (**30.2.8**) and minocycline (**30.2.9**), appeared in the years 1965 to 1972. These two tetracyclines are included in the list of Top 200 Drugs by sales for the 2010s. They have a better oral absorption profile, higher lipophilicity (3 to 5 times more than drugs of the first generation), and a longer half-life of elimination. Formulations for intravenous administration of these drugs have been developed (Fig. 30.22.).

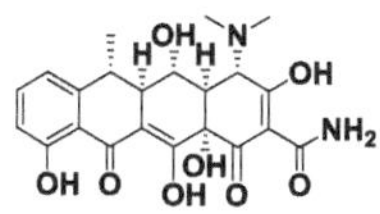

Doxycycline 30.2.8

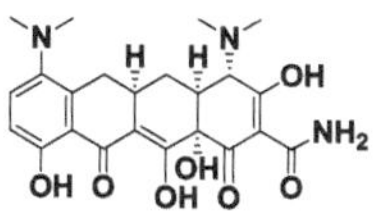

Minocycline 30.2.9

FIG. 30.22 The second-generation tetracyclines.

The third generation of tetracyclines, the glycylcyclines, is represented by a single, in medicinal practice, compound named tigecycline (**30.2.10**), which is 9-tert-butyl glycylamido derivative of minocycline. Tigecycline exhibits broader antibacterial spectrum than compounds of the first and second generations. Amadacyclin (**30.2.11**) is another new glycylcycline under development.

The glycylcyclines exhibit more potent antibacterial activities typical for the whole tetracycline series including tetracycline-resistant organisms. The glycylcyclines are active against other resistant pathogens including methicillin-resistant staphylococci, penicillin-resistant *S. pneumoniae*, and vancomycin-resistant enterococci. Tigecycline is also available in injectable formulation [149-153] (Fig. 30.23.).

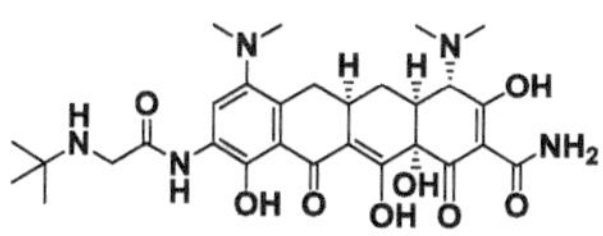

Tigecycline 30.2.10

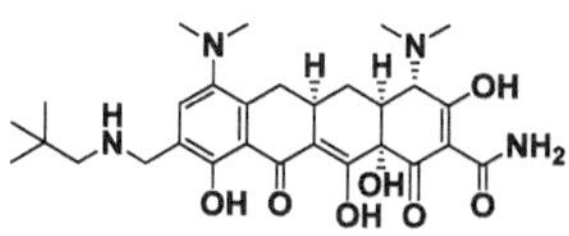

Amadacyclin 30.2.11

FIG. 30.23 The third generation of tetracyclines.

Tetracyclines also may be divided into "typical" tetracyclines—tetracycline, chlortetracycline, minocycline, and doxycycline—which act by inhibiting protein synthesis of bacteria, and "atypical" tetracyclines—chelocardin (**30.2.12**), anhydrotetracycline (**30.2.13**), 4-epi-anhydrotetracycline (**30.2.14**), anhydrochlortetracycline (**30.2.15**), and 6-thiatetracycline (**30.2.16**), which act by interfering with the electrochemical gradient of the bacterial membrane, promoting cell lysis and death via stimulation of autolytic enzyme activity (Fig. 30.24.).

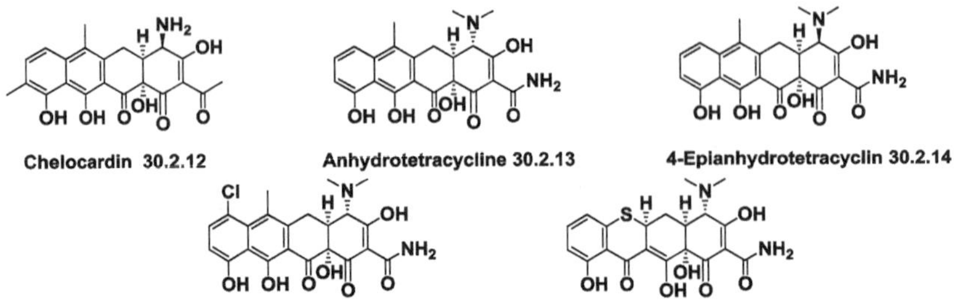

Chelocardin 30.2.12 Anhydrotetracycline 30.2.13 4-Epianhydrotetracyclin 30.2.14

Anhydrochlortetracycline 30.2.15 6-Thiatetracycline 30.2.16

FIG. 30.24 "Atypical" tetracyclines.

In recent years, several studies have reported diverse new pharmacological properties of tetracyclines, such as antiinflammatory, immune-modulating, and neuroprotective effects. They are proposed, as remedies for treatment of some dermatological disorders, as pain relievers. These phenomenons are just beginning to be understood [154-157].

Although many antibiotics had been synthesized chemically, they are not generally available in commercial quantities.

Doxycycline, one of the Top 200 Drugs by sales for the 2010s, is produced semisynthetically from oxytetracycline or methacycline.

Doxycycline–Doryx

Doxycycline is currently the most commonly used tetracycline and is considered an essential drug by the World Health Organization. Although some tetracyclines have been synthesized chemically, they are not generally available in commercial quantities.

Industrial approach for preparation of doxycycline is based on chemical modifications of fermentation tetracyclines—oxytetracycline (**30.2.2**) or methacycline (**30.2.5**). Oxytetracycline (**30.2.2**) is a major starting material for doxycycline production.

One of the first approaches is the hydrogenation of oxytetracycline (**30.2.2**) over a 5% Rh/C catalyst to produce the desired doxycycline (**30.2.8**) [158,159] (Scheme 30.1.).

Another approach came with the observation that interaction of oxytetracycline with N-chlorosuccinimide in a solvent such as 1,2-dimethoxyethane results in the formation of the 11α-chlorotetracycline-6,12-hemiketal (**30.2.9**), which on dehydration with liquid hydrogen fluoride produced 6-methylene-11α-chloro tetracycline (**30.2.11**). It is readily dehalogenated under mild conditions using sodium hydrosulfite to methacycline (**30.2.5**) [160,161].

Another way for preparing methacycline consists in adding pyridine-SO_3 complex to oxytetracycline (**30.2.2**), which produces 5-oxytetracycline 6,12-hemiketal-12-sulfuric acid ester (**30.2.10**), which on workup with liquid HF produced the desired methacycline (**30.2.5**) [162]. The HF could be replaced by other dehydrating acids. The obtained methacycline reacts with mercaptans, particularly with benzene thiol to produce adduct (**30.2.12**), which on reduction on Raney-Ni catalyst gave desired doxycycline (**30.2.8**) [160,161] (see Scheme 30.1.).

For the preparation of doxycycline and other α-6-deoxytetracyclines, the direct catalytic hydrogenation of corresponding 6-methylene intermediates (**30.2.9**) and (**30.2.10**) was described, where it was shown that reduction of 6-methylene-11a-chloro derivatives of tetracyclines with Rh catalysts gave high yields and did not attack the 7-halo group, while Pd dehalogenated the 7-position [161] (see Scheme 30.1.).

Subsequent efforts have been directed to the development of syntheses for producing the 6-deoxytetracyclines in greater yields and with greater stereoselectivity of formation of the desired α epimers. The use of other noble metal or noble metal salt compositions as heterogeneous hydrogenation catalysts in the production of doxycycline has also been disclosed [162-166]. The use of rhodium chloride/triphenylphosphine (Wilkinson's catalyst) and similar complexes as homogeneous, stereospecific hydrogenation catalysts in the production of doxycycline and other α-6-deoxy-5-oxytetracyclines has been discussed [167-173].

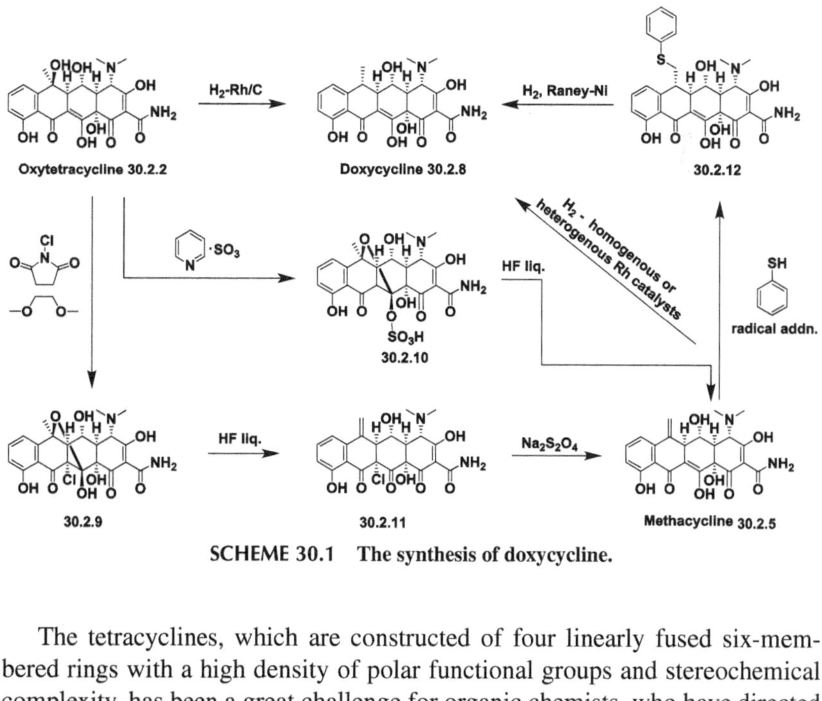

SCHEME 30.1 **The synthesis of doxycycline.**

The tetracyclines, which are constructed of four linearly fused six-membered rings with a high density of polar functional groups and stereochemical complexity, has been a great challenge for organic chemists, who have directed much research emphasis toward the total synthesis of tetracyclines.

The task is becoming extremely hard to do, taking into account the chemical sensitivity of these molecules, their lability in acidic and basic media.

The first total 22-step synthesis of racemic 6-des-methyl-6-deoxytetracycline with an overall yield of 0.003% was reported in 1962 by the legendary Woodward and colleagues [174]. The first enantioselective synthesis of (-)-tetracycline in 34 steps was reported by Tatsuda and colleagues with a 0.002% overall yield [175]. Other approaches to the synthesis of tetracycline were demonstrated by Shemyakin [176], Muxfeldt (22 steps, 0.06% yield) [177], Stork [178], and Myers [179-182]. The early chemistry of tetracyclines has been reviewed [183,184].

The Myers synthesis of doxycycline (**30.2.8**) started by whole-cell microbial hydroxylation of benzoic acid (**30.2.13**) using a mutant strain of *A. eutrophus* B9, which produced the diol (**30.2.14**) with greater than 95% enantiomeric excess and 79% yield. Epoxidation of the obtained product with m-chloroperbenzoic acid provided the epoxide (**30.2.15**), esterification of which with trimethylsilyldiazomethane, followed by bis-silylation and concomitant epoxide isomerization in the presence of tert-butyldimethylsilyl trifluoromethanesulfonate afforded the epoxy ester (**30.2.16**).

The organolithium reagent obtained from 3-benzyloxy-5-dimethylaminomethylisoxazole (**30.2.17**) and n-butyllithium, was reacted with obtained epoxy ester (**30.2.16**), forming the ketone (**30.2.18**).

One of the a key steps in this synthesis-is cyclization achieved by heating of the ketone (**30.2.18**) with lithium triflate at 60°C, followed by selective removal of the allylic silyl ether group with trifluoroacetic acid, which produced the compound (**30.2.19**). It is believed that these transformations involve initial SN-prime opening of the allylic epoxide by the N,N-dimethyl amino group, with corresponding ylide formation followed by [2,3]-sigmatropic rearrangement, a process that is reminiscent of the Sommelet-Hauser rearrangement.

The next step included replacement of the secondary allylic hydroxyl group of (**30.2.19**) with a thiophenyl group with net stereochemical retention. For that purpose, triphenylphosphine and carbon tetrabromide were added to a solution of the allylic alcohol in acetonitrile. The resulting allylic bromide was reacted with benzenethiol in presence of triethylamine to produce the desired allylic thiol ether (**30.2.20**). Diastereoselective sulfoxidation with a chiral oxidant (-)-[(8,8)-(dichlorocamphoryl)sunfonyl]oxaziridine (**30.2.21**) gave allylic sulfoxide (**30.2.22**), which underwent a 2,3-sigmatropic Mislow-Evans rearrangement using the standard reagent-trimethylphosphite in methanol. This reaction produced a new allylic alcohol (**30.2.23**).

After protection of the hydroxyl group in obtained alcohol (**30.2.23**) with benzyl chloroformate, the tert-butyldimethylsilyl group in synthesized (**30.2.24**) was removed using tetrabutylammonium fluoride in acetic acid to produce diol (**30.2.25**). The product was oxidized with 2-iodoxybenzoic acid in dimethylsulfoxide to produce (**30.2.26**). The remaining hydroxyl group was again protected with tert-butyldimethylsilyl trifluoromethanesulfonate and the obtained compound (**30.2.27**) was coupled with Boc protected lithium phenyl 2-ethyl-6-hydroxybenzoate (**30.2.28**) in the presence of N,N,N′N′-tetramethylethylenediamine.

The tetracyclic coupling product of Michael-Claisen cyclization (**30.2.29**) was isolated chromatographically as a single diastereomer. After deprotection of the hydroxyl groups with HF–acetonitrile solution, the compound (**30.2.30**) underwent a subsequent final reductive deprotection using hydrogen on palladium catalyst, which produced the desired doxycycline (**30.2.8**) [179,182] (Scheme 30.2.).

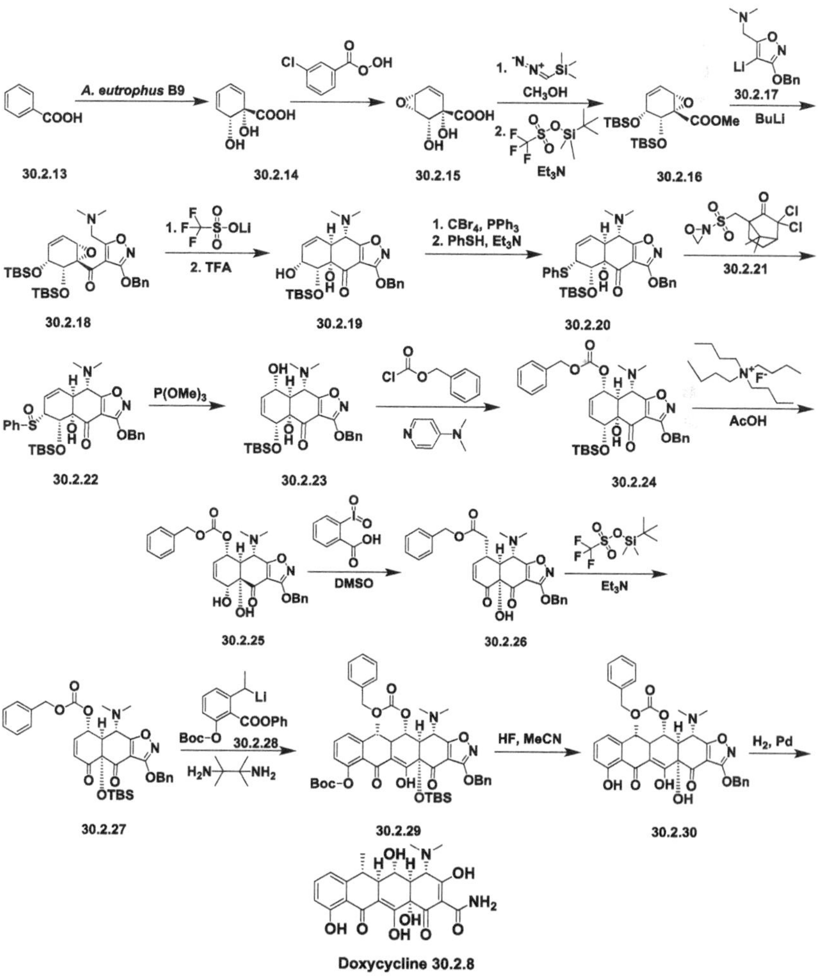

SCHEME 30.2 **The Myers synthesis of doxycycline.**

These synthetic sequences allow preparation of 5 to 20 mg quantities of different tetracycline analogues.

Although many tetracycline antibiotics had been synthesized chemically, they are not generally available in commercial quantities.

Doxycycline is currently the most commonly used tetracycline and is considered an essential drug by the World Health Organization and has been used in medicine for more than 40 years. It has a variety of applications to common respiratory and genitourinary tract infections, but also amongst atypical infections, such as malaria, rickettsial infections, leptospirosis, brucellosis and some of the bioterrorist agents, including anthrax Cytostatic and cytotoxic activity as shown against cell lines of various tumor origins. It is a well-tolerated drug that is bacteriostatic and which is generally given at a dose of 100 mg daily or twice daily. It is well absorbed and has generally good tissue penetration. The serum half-life is 18 to 22 hours depending on dosage. Major side effects are gastrointestinal and dermatological and it is generally contraindicated in pregnancy or childhood [185-191].

Minocycline–Solodyn

Minocycline is another antibiotic included in the list of Top 200 Drugs by sales for the 2010s.

Original synthesis of minocycline (**30.2.9**) started from 6-demethyltetracycline (**30.2.31**), which is available by cultivating *S. aureofaciens* NCIB 9502. Its hydrogenation over Pd-C catalyst gives 6-demethyl-6-deoxytetracycline (**30.2.32**). After nitration of the obtained product with nitronium tetrafluoroborate in an acetonitrile 9-nitro compound (**30.2.33**) was prepared. The nitro group of the last was hydrogenated using PtO_2 catalyst to give the 9-amino derivative (**30.2.34**).

Nitration of the obtained compound with a standard HNO_3/H_2SO_4 mixture gave a product (**30.2.35**) with a nitro group in position 7 of the tetracycline skeleton. After the diazotation of the obtained product with butyl nitrite/sulfuric acid mixture, the obtained intermediate 9-diazonium salt (**30.2.36**) was hydrogenated on the Pd-C catalyst in methylcellosolve and 40% aqueous CH_2O (reductive methylation) with the loose of nitrogen to produce the desired minocycline (**30.2.9**) [192-196] (Scheme 30.3.).

Another approach consists of the reaction of 7-iodosancycline (**30.2.37**) with (N,N-dimethylamino)trimethyltin in the presence of $PdCl_2(P(o\text{-tolyl})_3)_2$ catalyst (Buchwald–Hartwig amination), which affords minocycline (**30.2.8**) in good yield [197] (Scheme 30.4.).

Michael-Claisen cyclization approach [179] for the synthesis of minocycline is described in [198], where treatment of lithiated phenyl ester (**30.2.39**) in the presence of tetramethylethylenediamine, followed by addition of enone (**30.2.38**) provided cyclization product (**30.2.40**) which after two-step deprotection and purification afforded minocycline (**30.2.9**) (Scheme 30.5.).

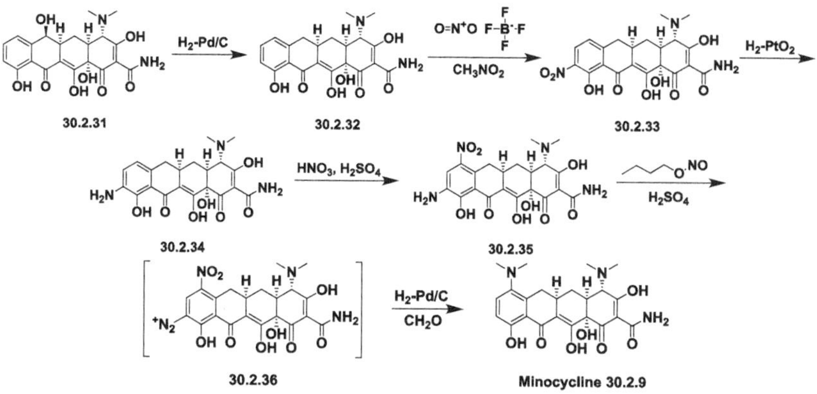

SCHEME 30.3 The synthesis of minocycline.

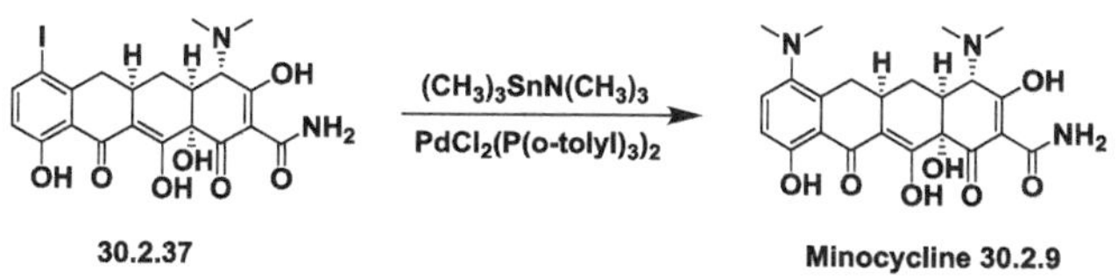

SCHEME 30.4 The alternative synthesis of minocycline from 7-iodosancycline.

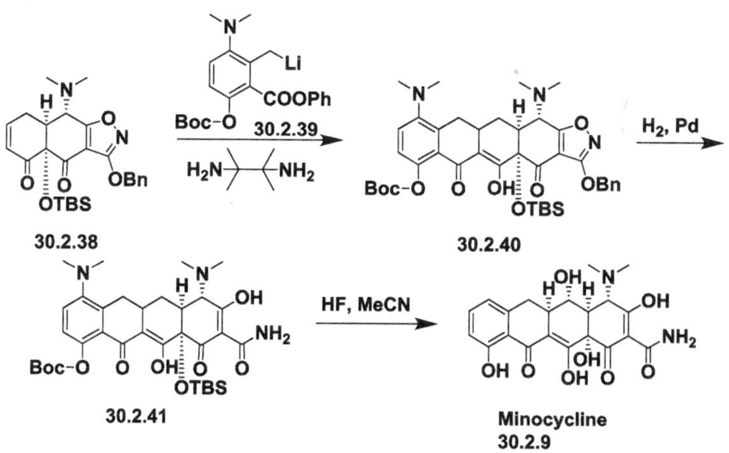

SCHEME 30.5 Michael-Claisen cyclization approach for the alternative synthesis of minocycline.

Minocycline belongs to the second-generation class of tetracyclines. It was synthesized in 1967 and marketed in 1972. The pharmacokinetics of minocycline is characterized by an excellent absorption, a long half-life and an important lipophilic property. It has been in therapeutic use for four decades as a result of its antibiotic properties against both Gram-positive and

Gram-negative bacteria. Its antiinfectious spectrum is similar to that of other tetracyclines, notably against *Chlamydia*, *Treponema*, and *Propionibacterium acnes*. It is mainly used in the treatment of acne vulgaris and some sexually transmitted diseases.

There is a huge volume of convincing data that tetracyclines, and minocycline particularly, in addition to their antimicrobial activity, can exert antiinflammatory activity and be a useful remedy for periodontitis, atherosclerosis, rheumatoid arthritis, and dermatitis, and could be used to inhibit proteolysis, angiogenesis, and tumor metastasis. Tetracyclines provide neuroprotection. This has been confirmed in experimental models of neuropathic pain, ischemia, traumatic brain injury and several neurodegenerative conditions, including Parkinson disease, Huntington disease, amyotrophic lateral sclerosis, Alzheimer disease, multiple sclerosis, and spinal cord injury.

Moreover, preclinical studies show the ability of minocycline to inhibit malignant cell growth and activation, as well as replication of human immunodeficiency virus.

Growing evidence suggests that oxidative stress, inflammation, changes in glutamatergic pathways and neurotrophins play important roles in many psychiatric illnesses, including mood disorders, schizophrenia, and addiction, and that minocycline has serious therapeutic potential in psychiatry [199-210].

30.3 MACROLIDE ANTIBIOTICS

Macrolides are a group of antibiotics produced by various *Streptomyces* strains and make up a large group of compounds with a distinctive 12- to 16-member macrocyclic lactone ring decorated with aminosugars (cladinose, desosamine) [211-237]. Macrolides are generally bacteriostatic, although some of these drugs, at very high concentrations, may be bactericide. The mechanism of action of macrolides is a matter of controversy, but it is accepted that the antibacterial action involves inhibition of protein synthesis in the bacterial cell during translocation.

The prototypical macrolide is erythromycin A, which came into clinical practice in the 1950s. Erythromycin B, a cometabolite of erythromycin A, differs from erythromycin A only in the absence of a hydroxyl group at C12, yet it has never been licensed for clinical use.

There are many criteria for the classification of macrolide. Natural macrolides generally consist of a macrocyclic lactone ring with glycosidic linkages. Depending on the number of atoms in the lactone ring, macrolides are defined according to the size of the ring, the most important of which are 14-, 15-, and 16-membered.

Among 14-membered macrolides are erythromycin A (**30.3.1**), the first macrolide antibiotic isolated in the early 1950s, clarithromycin (**30.3.2**), roxithromycin (**30.3.3**), and dirithromycin (**30.3.4**) (Fig. 30.25.).

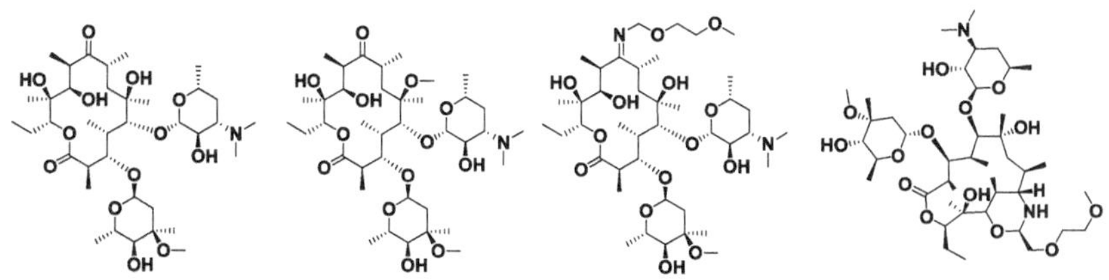

Erythromycin A 30.3.1 Clarithromycin 30.3.2 Roxithromycin 30.3.3 Dirithromycin 30.3.4

FIG. 30.25 14-Membered macrolide antibiotics.

15-Membered macrolides are represented by azithromycin (**30.3.5**), which is 9-deoxo-9a-methyl-9a-aza-9a-homoetrythromycin (Fig. 30.26.).

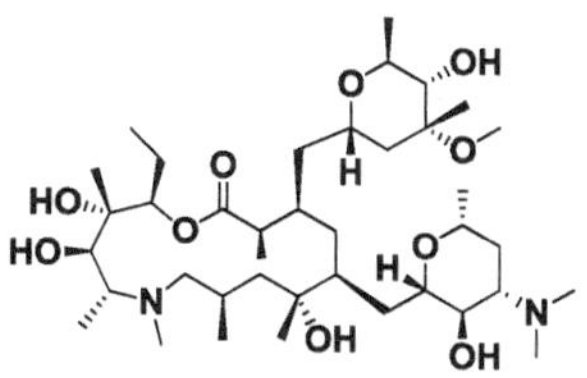

Azithromycin 30.3.5

FIG. 30.26 15-Membered macrolide antibiotics.

16-Membered macrolides include spiramycin (**30.3.6**), josamycin (**30.3.7**), and midecamycin (**30.3.8**), which are available in the pharmaceutical market [234-237] (Fig. 30.27.).

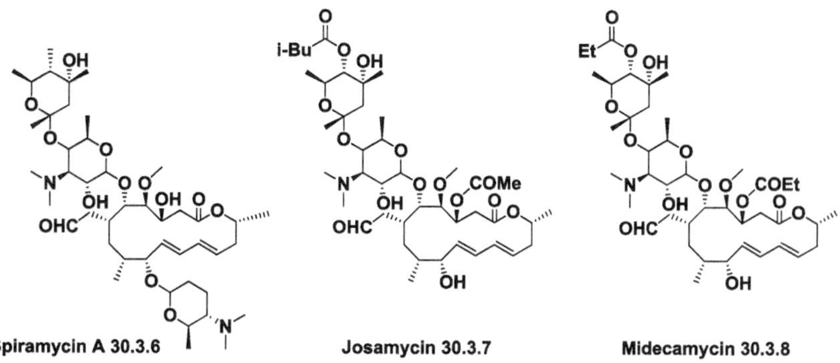

Spiramycin A 30.3.6 Josamycin 30.3.7 Midecamycin 30.3.8

FIG. 30.27 16-Membered macrolide antibiotics.

Spiramycin, which is still considered an experimental drug in the United States, but can sometimes be obtained by special permission from the FDA for toxoplasmosis, consists of a mixture of three major components, spiramycin

A, B, and C, which differ only by substituent in position C3 of macrocycle (spiramycin A (OH), B (O-acetyl), C (O-propionyl)) and 3 minor components, neospiramycin I, II, and III.

Macrolides are also subdivided also to subclasses as acylides, azalides, and ketolides.

Acylides are obtained by introduction of an acyl group to the 3-O position of erythromycin A derivatives instead of to L-cladinose.

Azalides are semisynthetic macrolides, in which a nitrogen atom is introduced into a macrolactone ring. Azithromycin (**30.3.5**), for example, belongs to the family of azalides. It is derived from erythromycin and differs from erythromycin in having a 15-membered ring and a methyl substituted nitrogen inserted at the 9a position in the aglycon ring. This structural modification confers favorable microbiological and pharmacokinetic characteristics, as well as greater acid stability, compared with erythromycin. Azithromycin is also one of the very-good-selling antibiotics.

Ketolides, a group of 14-membered macrolides such as telithromycin (**30.3.9**) and cethromycin (**30.3.10**), are derived by substituting a keto function for the α-L-cladinose sugar moiety at position 3 of the 14-membered erythronolide. These modifications give ketolides much broader spectrum than other macrolides (Fig. 30.28.).

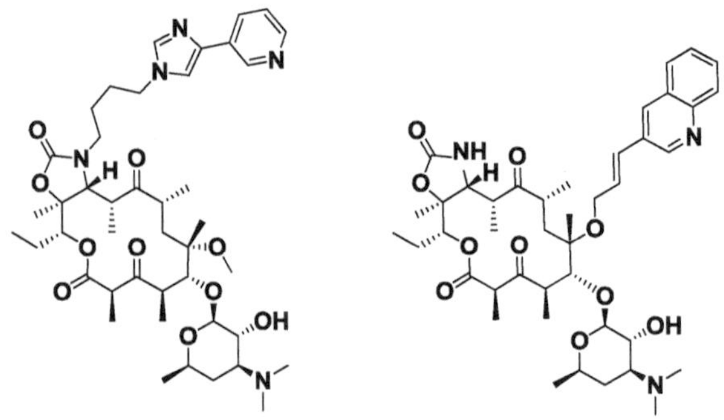

Telithromycin 30.3.9 **Cethromycin 30.3.10**

FIG. 30.28 Ketolides are 14-membered macrolides.

Macrolide drugs also are classified as of the first generation (erythromycin), second generation (14-membered ring clarithromycin and 15-membered ring azithromycin), third generation, which includes ketolides, acylides, and azalides and is represented by telithromycin (**30.3.9**). It is the only ketolide, approved by FDA. Another ketolide, cethromycin (**30.3.10**), is still undergoing clinical trials.

Macrolides are considered the safest antibiotics. Macrolides mainly affect Gram-positive cocci and intracellular pathogens such as *Mycoplasma, Chlamydia,* and *Legionella.*

They have been considered the drug of choice for group A streptococcal and pneumococcal infections when penicillin cannot be used. They are used in the treatment of susceptible Gram-positive cocci and bacilli, as well as for infections caused by Gram-negative *Bordetella pertussis, Campylobacter* species, *Chlamydia* species, and *Mycoplasma pneumoniae.*

Some macrolides are active against *Legionella pneumophila, Rickettsia* strains, and *Spirochaeta*. Most anaerobic bacteria, especially oral anaerobic flora, are susceptible to macrolides.

Macrolides are indicated mainly for lower and upper respiratory tract infections: pneumonia, attack of chronic bronchitis, streptococcal tonsillopharyngitis, acute sinusitis, and acute otitis; infections of the oral cavity, such as periodontitis and periostitis; diphtheria and skin and soft-tissue infections; chlamydia, syphilis, and venereal lymphogranuloma.

About six decades after the discovery of macrolides and their introduction into medical practice there is only a sketchy understanding of how these drugs work. It has been firmly established that they reversibly inhibit bacterial protein biosynthetic machinery interacting with the ribosomes with bacteriostatic actions by binding to the subunit of 50S ribosome of bacteria.

Macrolide resistance has become increasingly serious [238].

Antifungal Polyene Macrolides

Polyene macrolide antibiotics are one of the most effective antifungal agents, which are widely used in clinics and food industries [239,240]. Polyene antifungal antibiotics have been isolated from various strains of *Actinomyces*. The complete biosynthetic processes of polyene macrolide antibiotics are not fully explored.

Polyene macrolides are structurally related cyclic lactones with rings of 25 to 38 carbon atoms, where are present series of conjugated double bonds, such as trienes, tetraenes, and a series of hydroxyl groups in positions opposite to the double bonds. Other common substituents are a carboxyl group and an aminosugar—mycosamine.

The polyene macrolides bind to sterol components in the phospholipid-sterol membranes of fungal cells to form complexes that induce physical changes in the membrane. They are typically active against fungi, sometimes against protozoa, and only exceptionally against bacteria.

Polyene antifungal antibiotics are mainly represented by natamycin (**30.3.11**) and nystatin (**30.3.12**). Nystatin is by far the most widely used. There is also a heptaene antibiotic with a 38-atom ring—amphotericin B (**30.3.13**) (Fig. 30.29.).

Amphotericin B is an antifungal antibiotic very active, against several pathogenic fungi, such as *Candida, Cryptococcus*, and *Histoplasma*, and less active against filamentous fungi, such as *Trichophyton*.

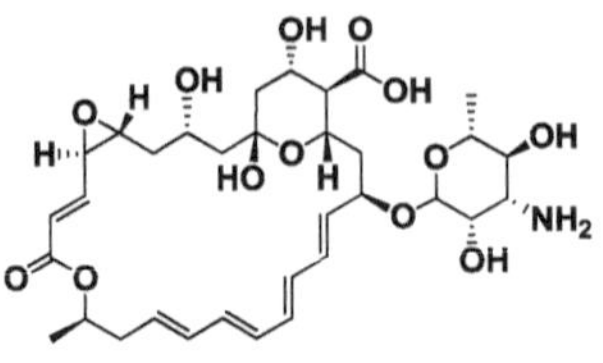

Natamycin 30.3.11

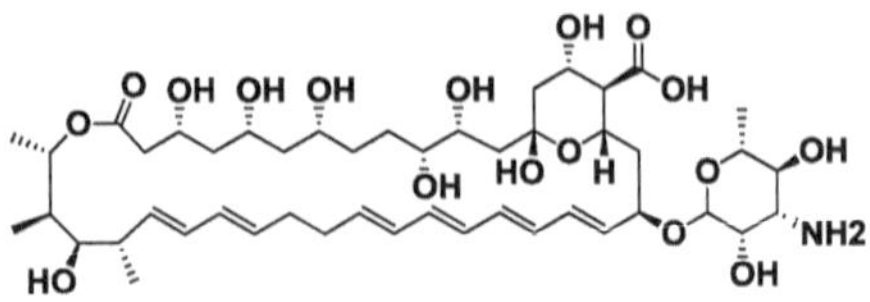

Nystatin 30.3.12

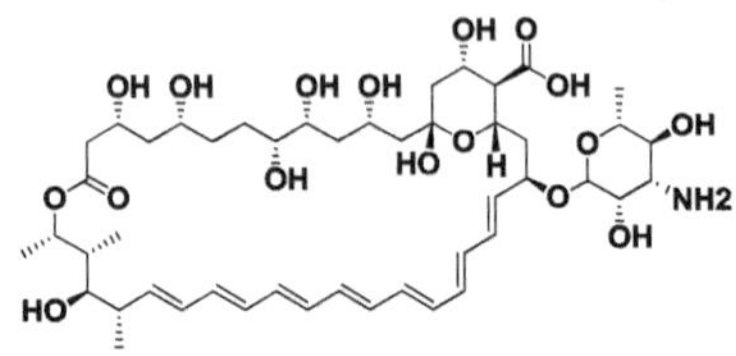

Amphotericin B 30.3.13

FIG. 30.29 **Polyene macrolide antibiotics.**

30.4 AMINOGLYCOSIDES

Aminoglycoside antibiotics, which are some of the oldest antibiotics and are members of the "Big Four" classes of antibiotics (β-lactams, tetracyclines macrolides, and aminoglycosides), have proved an invaluable part of our antimicrobial armamentarium since their introduction into practice more than 60 years ago.

Despite their inherent toxicity, aminoglycosides remain valuable constituent of the antibiotic armamentarium. Nephrotoxicity, ototoxicity, and neurotoxicity are the main adverse effects of aminoglycosides.

All aminoglycosides cause nephrotoxicity, neurotoxicity, and ototoxicity, prolonging the effects of neuromuscular blockers. Aminoglycosides can cause fetal harm when administered to a pregnant woman.

In general, aminoglycosides are active across a broad spectrum of aerobic Gram-negative and Gram-positive organisms, as well as mycobacteria. Of note, anaerobic bacteria are inherently resistant to aminoglycosides.

Compared with other classes of antibiotics, the aminoglycosides demonstrate relative stability against the development of resistance.

Despite the introduction of newer, less-toxic antimicrobial agents, the aminoglycosides continue to serve a useful role in the treatment of serious enterococcal, mycobacterial, and Gram-negative bacillary infections [241-255].

All of these drugs have similar chemical structure, representing a series of glycosidically linked aminocyclitols—streptamine, 2-deoxystreptamine, or streptidine—and amino sugars such as D-glucosamine and kanosamine (Fig. 30.30.).

This group of antibiotics includes at least eight drugs: streptomycin (**30.4.1**), neomycin (**30.4.2**), and paromomycin (**30.4.3**) (Fig. 30.31.), and structurally very related gentamicin (**30.4.4**), kanamycin (**30.4.5**), amikacin (**30.4.6**), netilmicin (**30.4.7**), and tobramycin (**30.4.8**) (Fig. 30.32.).

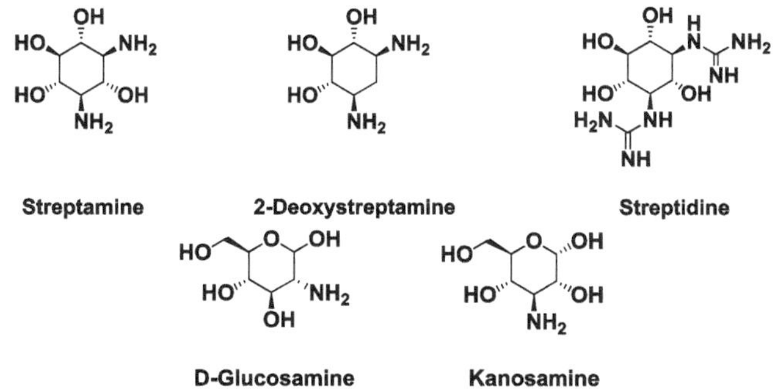

FIG. 30.30 Aminocyclitols–streptamine, 2-deoxystreptamine, streptidine, and the amino sugars D-glucosamine and kanosamine.

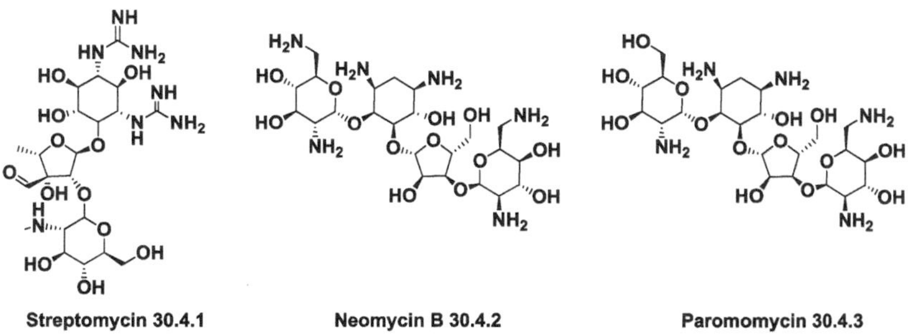

FIG. 30.31 Aminoglycoside antibiotics streptomycin, neomycin, and paromomycin.

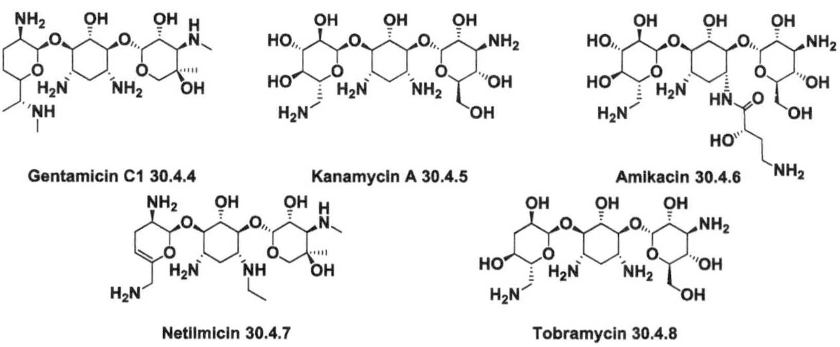

FIG. 30.32 Aminoglycoside antibiotics gentamicin, kanamycin, amikacin, netilmicin, and tobramycin.

Of these, gentamicin and tobramycin are the most frequently prescribed.

Streptomycin (**30.4.1**) was the first of the class of aminoglycoside antibiotics to be discovered, and was the first antibiotic remedy for tuberculosis. Streptomycin is a broad-spectrum drug that is effective against both Gram-positive and Gram-negative bacteria proved to be highly effective in the treatment of a large number of infectious diseases. However, its major use remains as an antituberculosis drug.

Streptomycin is derived from the actinobacterium *Streptomyces griseus* in a medium that contains glucose as a carbon source, soybean meal as nitrogen source, and number of inorganic salts (NaCl, K_2HPO_4, $CaCl_2$, $FeSO_4$).

Closely related aminoglycoside antibiotics are neomycin (**30.4.2**), which is prepared by fermentation by actinomycetes *Streptomyces marinensis*, and paromomycin (**30.4.3**), which is prepared by fermentation of various *Streptomyces* strains such *as S. rimosus*.

Neomycin consists of a mixture of two isomers, neomycin B (**30.4.2**) and neomycin C, along with small amounts of a degradation product-neomycin A. Neomycin is effective against most Gram-negative organisms except *P. aeruginosa* and anaerobic bacteria. It is mainly used in topical formulations for the prevention and treatment of superficial skin, ear, and eye infections, as well as in solutions for urinary instillations (catheters) and in peritoneal irrigation solutions in abdominal surgery.

Paromomycin (**30.4.3**) is a is a relatively new broad-spectrum aminoglycoside antibiotic that is active against Gram-negative and many Gram-positive bacteria, as well as some protozoa and cestodes. Oral paromomycin is used to treat giardiasis, amebiasis, and cryptosporidiosis. Topical paromomycin is currently used to treat visceral leishmaniasis.

Fermentation by *Micromonospora purpurea* or *Micromonospora echinospora* in a broth containing starch as a carbon source, soybean meal as a nitrogen source, and inorganic salts, such as K_2HPO_4, $CaCO_3$, $FeSO_4$, and $CoCl_2$, produces a complex mixture—a family of antibiotics called *gentamicin*—which is composed of a mixture of related components and fractions. Three major components are gentamicin C1 (**30.4.4**) (see Fig. 30.32.), gentamicin C1a, and gentamicin C2, which make up approximately 80% of gentamicin and have been found to have the highest antibacterial activity.

Gentamicin was discovered in 1963 and widely used in medicinal applications to treat the infectious disorders generated by *K. pneumoniae*, *E. coli*, *Serratia marcescens*, *Citrobacter*, *Enterobacteriaceae*, *Pseudomonas* species, and *Staphylococcus* infectious diseases. According to FDA-labeled indications, gentamicin is intended for treatment of pyelonephritis; appendicitis; cystic fibrosis; diverticulitis; adjunct regimen for febrile neutropenia; female genital infection; uterine infection; postnatal infection; necrotizing enterocolitis in fetus or newborn; osteomyelitis; pelvic inflammatory disease; plague; gonorrhea; tularemia; prophylaxis of postcholecystectomy infection, transrectal prostate biopsy, and posttympanostomy-related infection; malignant otitis externa; and intratympanically or transtympanically for Meniere disease.

Kanamycin is a mixture of three main components: kanamycin A (**30.4.5**) (see Fig. 30.32.), B, and C. Kanamycin A is the major component in kanamycin.

Kanamycin is produced by fermentation of *Streptomyces kanamyceticus* and has the same mode of action as streptomycin. Kanamycin is indicated in the short-term treatment of serious infections caused by *E. coli, Proteus* species, *E. aerogenes, K. pneumoniae, S. marcescens,* and *Acinetobacter* species. Kanamycin also has the potential to induce auditory, and sometimes vestibular, toxicity, renal toxicity, and neuromuscular blockade.

Amikacin (**30.4.6**)-1-N-[L-(-)-γ-Amino-α-hydroxybutyryl]kanamycin A (see Fig. 30.32.) is the first semisynthetic aminoglycoside derived from kanamycin A. It shows similar rates of clinical efficacy as gentamicin and it's potential for nephrotoxicity and ototoxicity is close to that of other aminoglycosides. But the potential for cochlear toxicity warrants attention. Moreover, combination of drug costs and toxicity are the reasons that amikacin has not been widely used clinically.

Netilmicin (**30.4.7**) (see Fig. 30.32.) is also a semisynthetic aminoglycoside sisomicin derived from sisomicin via 1-N-ethylation. It is active against most Gram-negative and some Gram-positive bacteria, including many gentamicin-resistant strains. Netilmicin has proved to be effective in Gram-negative infections of the urinary tract, skin and skin structure, and lower respiratory tract, as well as in intraabdominal infections, septicemia, and other miscellaneous infections. It is proposed to be used in blood infection caused by different bacteria, but netilmicin has not been demonstrated to have significant advantages over other aminoglycosides and it is more expensive. Thus, its potential value is limited. Moreover, this drug has the potential to cause kidney problems, nerve damage, or permanent hearing loss, even at usual doses.

Tobramycin (**30.4.8**) (see Fig. 30.32.) is another aminoglycoside, which, when compared with gentamicin, has a similar spectrum of activity against Gram-negative bacteria, but higher activity against *Pseudomonas* species and is less likely to cause nephrotoxicity. Tobramycin is a product of an actinomycete *Streptomyces tenebrarius* fermentation. The fermentation broths typically consist of cell culture media, of a very complex sample matrix that includes number of inorganic salts.

Tobramycin is the aminoglycoside of choice for the treatment of confirmed *Pseudomonas* species infections and mixed infections, including *P. aeruginosa* with other organisms. Tobramycin is similarly preferred for the initial, emergency treatment of patients who have, or are at risk of developing, severe *P. aeruginosa*–related infections. At present, tobramycin is the only FDA-approved antibiotic for treatment of cystic fibrosis patients with formulation that is tobramycin for aerosolized delivery (solution for inhalation or inhalation powder).

Aminoglycosides are highly potent antibiotics that exert their effects on bacterial cells by interfering with the translation process, leading to aberrant protein synthesis that usually results in cell death.

The bactericidal ability of aminoglycosides has not been fully explained. But it is known that the drug attaches to a bacterial cell wall and is drawn into the cell via cellular membrane porins. Once inside the cell, the aminoglycosides bind to the aminoacyl site of 16S ribosomal RNA within the 30S ribosomal subunit, leading to misreading of the genetic code and inhibition of translocation.

30.5 PEPTIDE ANTIBIOTICS

Antimicrobial substances have been discovered throughout nature in both plants and animals and even in some single-cell organism defense systems. Hundreds of antibiotics, predominantly small peptides with astonishing diversity in structure, have been found and isolated from species of both invertebrates and vertebrates, including mammals.

Antimicrobial peptides, the majority of which have a net cationic charge, have been found within the granules of neutrophils and leukocytes, and in secretions of epithelial cells. Despite continuous attacks by a variety of microbes, mammals remain relatively free from infectious diseases.

It is interesting to note that urine, saliva, blood, milk, and skin secretions that have been used in ancient or traditional medicine, contain a range of substances exerting bacteriostatic, bactericidal, virus-neutralizing, anticarcinogenic, and even cytokine-regulating activities [256].

A very heterogeneous naturally occurring group of peptide antibiotics are substances with broad-spectrum activity against Gram-negative and Gram-positive bacteria, mycobacteria, viruses, and fungi, and even transformed or cancerous cells. Besides antimicrobial action, they play multiple roles as mediators of inflammation, influencing diverse processes such as cell proliferation, immune induction, wound healing, cytokine release, chemotaxis, and protease–antiprotease balance [257-265].

Classification of antibiotic peptide families is very complicated and controversial, and different authors divide these heterogeneous molecules according to their origin, type of expression, and chemical composition. No standard and unified, generally accepted classification system is as yet available.

From the chemist's point of view peptide antibiotics can be divided into seven groups: linear peptides; cyclic peptides; glycopeptides; lipoglycopeptides; lipopeptides; thiazolopeptides and thiopeptides; and chromopeptides. In turn, each of these groups can be divided into subgroups according to their chemical structures or antibacterial activities [266].

Another classification system groups the peptide antibiotics according to their size, conformation, or predominant amino acid composition. This approach divides peptide antibiotics to four main classes: Group I includes linear peptides with an α-helical structure such as bombinins, cecropins, magainins, styelins, clavanin, melittin, and LL-37; Group II includes peptides with β-sheet structures stabilized by disulphide bridges and involves protegrin, protegrin, tachyplesins, Θ-defensins, and drosomycin; Group III peptides are defined as peptides with a predominance of one or more amino acids—PR-39, Bac5, Bac7, drosocin, and metchnikowin; Group IV includes peptides with loop structures and consists of bactenecin and ranalexin [267,268].

From the biologist's point of view, peptide antibiotics are classifiable based on whether they are ribosomally or nonribosomally synthesized [269-274].

Nonribosomally Synthesized Peptide Antibiotics

The nonribosomally synthesized peptides group is a very diverse group of natural products that consists of compounds such as the gramicidins, polymyxins, bacitracins, glycopeptides usually and largely produced by bacteria and fungi.

They often have a cyclic or branched structures, sometimes contain nonproteinogenic amino acids, including D-amino acids, can be N-methylated or N-formylated, or can be glycosylated, acylated, halogenated, or hydroxylated. Cyclization of amino acids against the peptide "backbone" is often performed, resulting in oxazolines and thiazolines. Among them are also peptaibiotics [275], nonribosomally synthesized peptides from various ascomycetes, are uniquely characterized by di-alkylated α-amino acids, a rigid helical conformation, and membrane permeation properties.

The nonribosomal peptides group, produced mostly by bacteria, is a diverse group of natural products with an extremely broad range of biological activities and pharmacological properties, and are already used for clinical applications. Among them are gramicidin, polymyxin B, and colistin (polymyxin E, colimycin), bacitracin, vancomycin, and anticancer antibiotics actinomycin D and dactinomycin.

Gramicidin

Gramicidin, a naturally occurring heterogeneous mixture of six linear pentadecapeptides that form cation-selective channels in lipid bilayers, produced by *Bacillus brevis* species, was isolated in 1939. Gramicidin exhibits significant antibacterial activity against all Gram-positive bacteria.

The peptides are modified at both ends and the six forms differ at 1st (Val or Ile) and 11th (Trp, Phe, or Tyr) positions. The most populated, gramicidin A (**30.5.1**), has a sequence CHO-Val$_1$-Gly-Ala-D-Leu-Ala-D-Val-Val-D-VaL-Trp-D-Leu-Trp$_{11}$-D-Leu-Trp-D-Leu-Trp-NHCH$_2$CH$_2$OH [276] (Fig. 30.33.).

Gramicidin is used as local application to skin, wounds, burns and nose and mouth infections.

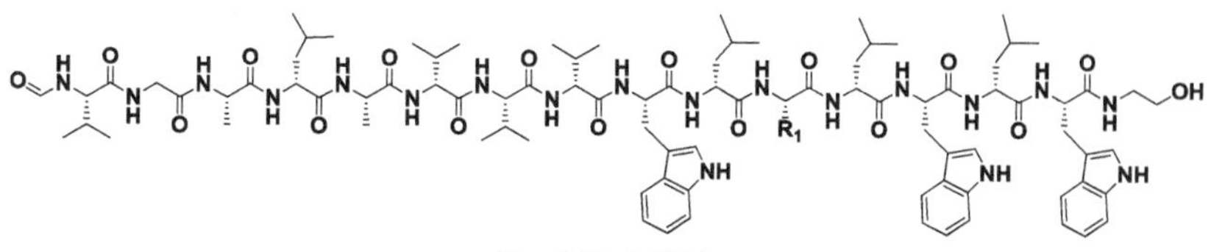

Gramicidin A 30.5.1

FIG. 30.33 Structure of gramicidin.

Gramicidin S (**30.5.2**) is a cyclic decapeptide consisting of two identical pentapeptides [D-Phe-L-Pro-L-Val-LOra-L-Leu]$_2$, which are also produced by certain strains of *B. brevis* [277] (Fig. 30.34.).

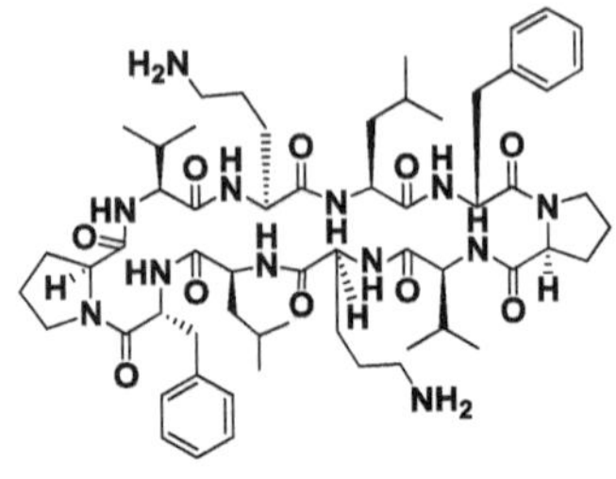

FIG. 30.34 **Structure of gramicidin S.**

Gramicidin S has historically been employed as a topical antibiotic for the treatment of infections from superficial wounds. It exhibits strong antibiotic activity against a broad spectrum of Gram-negative and Gram-positive bacteria, and against several pathogenic fungi.

Polymyxin

Polymyxin, a mixture consisting of five polypeptide antibiotics derived from various species of soil bacteria that are active against Gram-negative and Gram-positive bacteria, was discovered more than six decades ago. Polymyxin is defined as the general name for polypeptides—polymyxins A, B, C, D, and E. All of them contain α,γ-diaminobutyric acid, threonine, and a branched C9 fatty acid. All except C contain leucine. Only polymyxins B (**30.5.3**) and E (colistin) (Fig. 30.35.) are used clinically. Their therapeutic use is in the treatment of infections involving Gram-negative bacteria resistant to other broad-spectrum antibiotics. Polymyxin B (**30.5.3**) is applied topically to treat eye, ear skin and urinary bladder infections, for various bacterial infection threatening, in particular to sepsis. Polymyxin E (colistin) (**30.5.4**) is used frequently to treat diarrhea in children. The polymyxins are often used in combination with other antibiotics. The availability of better antibiotics limits the use of polymyxins [278,279] (Fig. 30.35.).

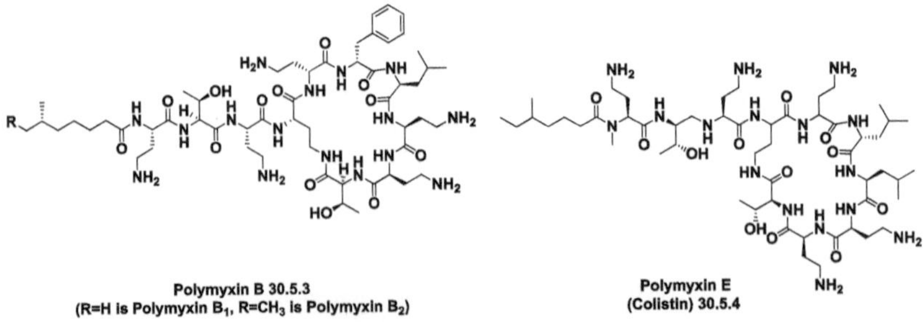

FIG. 30.35 **Structures of polymyxins.**

Bacitracin

Bacitracin is composed of related compounds with varying degrees of antibacterial activity, which include bacitracin A, A1, B, B1, B2, C, D, E, F, G, and X. The most active compound is bacitracin A (**30.5.5**) (Fig. 30.36.).

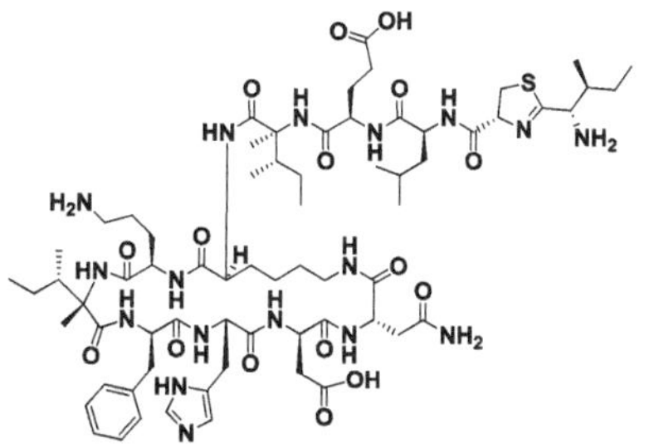

FIG. 30.36 Structure of bacitracin A.

The bacitracin peptides were first discovered in 1945 as a product of a *Bacillus licheniformis* strain isolated from a wound sustained by a patient named Margaret Tracy.

Bacitracin is an antibiotic that is produced by *Bacillus subtilis*, with a potent bactericidal activity directed primarily against Gram-positive organisms used to prevent minor skin infections caused by small cuts and scrapes. Bacitracin is used in several types of consumer products, including cosmetics and ophthalmic and cutaneous ointment. Among systemic diseases, only staphylococcal infections qualify for consideration of bacitracin therapy. Intramuscular bacitracin is indicated in the treatment of infants with pneumonia and empyema caused by staphylococci shown to be susceptible to the drug [280,281].

Vancomycin

Vancomycin (**30.5.6**) is the first glycopeptide antibiotic discovered from the fermentation of *Amycolatopsis orientalis* isolated from the soil of Borneo in 1956 (Fig. 30.37.).

Vancomycin is enlisted as a drug of the last resort for the treatment of resistant Gram-positive bacterial infections and is the antibiotic of first choice for the treatment of methicillin-resistant *S. aureus* infection. Early use of vancomycin was associated with nephrotoxicity and ototoxicity [282,283].

Actinomycin D or Dactinomycin

Actinomycin D or dactinomycin (**30.5.7**), which is isolated from *Streptomyces* species, is the most significant member of the actinomycins, a family of structurally related chromopeptide antibiotics with a common phenoxazine chromophore attached to two pentapeptide lactone moieties, vary in their amino acid content.

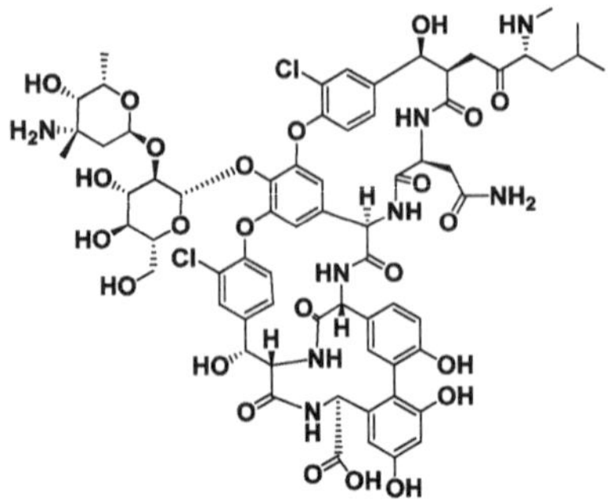

Vancomycin 30.5.6

FIG. 30.37 Structure of vancomycin.

Actinomycin D is a well-known antibiotic that exhibits high antibacterial and antitumor activity. It has been widely used in clinical practice since 1954 as an anticancer drug to treat many tumors, such as Wilms and Ewing tumors, testicular cancer, sarcomas, and choriocarcinoma. It was the first antibiotic shown to have anticancer activity [284,285] (Fig. 30.38.).

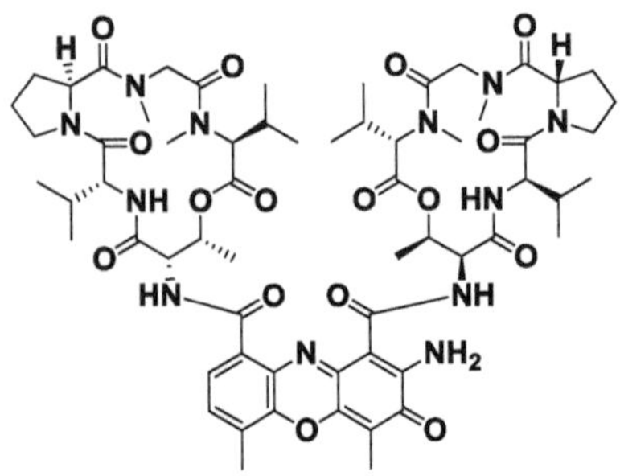

Dactinomycin 30.5.7

FIG. 30.38 Structure of dactinomycin.

Ribosomally Synthesized Peptide Antibiotics

The ribosomally synthesized peptides are produced by all species of life—from prokaryotes to humans—and comprise a host's natural defense against the daily exposure to millions of potential pathogens. They are key elements of innate immunity, which can directly kill multiple bacterial, viral, and fungal pathogens. Typically, ribosomally synthesized peptides are cationic peptides with at least half of the amino acid residues being hydrophobic and a smaller number of neutral or negatively charged residues. Probably their amphipathic structure with opposing hydrophobic and lipophilic faces aids in the perturbation of the bacterial cell wall.

They are very widespread, ranging from mammals, amphibians, insects, plants, bacteria, and even viruses to synthetic peptides. In mammals, antimicrobial peptides have been found within the granules of neutrophils and in mucosal or skin secretions of epithelial cells.

Neutrophils are the major type of polymorphonuclear leukocytes cells of bone marrow origin that ingest and kill invading microorganisms. They contain a range of antimicrobial proteins and peptides, including defensins, cathelicidins, lactoferrin, and others listed below. They progressively upregulate corresponding receptors and release various effector and regulatory molecules that respond to infection and attack bacteria and other foreign invaders.

Series of peptide antibiotics of different families have been identified in humans [286,287].

Defensins, a major family of antimicrobial peptides expressed predominantly in neutrophils and epithelial cells, play important roles in innate immune defense against infectious pathogens. The term *defensin* relates to a set of host defense peptides in vertebrates, invertebrates, plants, and molds that display structural similarities based on a cystine stabilized antiparallel β-sheet core, with an α-helical stretch also present in some members. This is a highly complex group of compounds defined as α-, β-, and θ-defensins. They have a range of activities against bacteria, fungi, and viruses [288-297].

Bactericidal/permeability-increasing proteins present on the neutrophil cell surface and in the specific granules of eosinophils, providing the first line of defense against different pathogens, including bacteria, fungi, viruses, and parasites [298].

Calprotectin is a heterodimer (MRPB/14, S100A8/S100A9, 27E10 antigen) of two calcium-binding proteins present in the cytoplasm of neutrophils and expressed on the membrane of monocytes that inhibits the growth of pathogenic microorganisms by sequestering essential metal nutrients in the extracellular space [299].

Cathelicidin family of antimicrobial polypeptides is expressed in epithelial cells as well as in leukocytes characterized by a highly conserved region (cathelin domain) and a variable linear peptides cathelicidin peptide domain. LL-37 is the only member of the cathelicidin family of host defense peptides expressed in humans. It mediates a wide range of biological responses: direct killing of microorganisms, chemotaxis and chemokine induction, and regulation of inflammatory responses, as well as adjuvant, angiogenic, and wound healing effects [300,301].

Histatins are salivary cationic peptides that provide the first line of defense against pathogens and play a role in wound closure [302-304].

Lactoferrin is an iron-binding glycoprotein of the transferrin family; it is expressed in most biological fluids and with particularly high levels in mammalian milk with a primary function as an iron carrier. In the intestinal tract, it produces antiinfective functions, and has antitumor value in occurrence and development of tumor.

At the same time, lactoferrin occupies the food market because of its important versatile properties having a wide spectra of antiviral, antimicrobial, antiprotozoal, and antioxidant activities, and because it modulates the cell growth rate [305-307].

Lysozyme, discovered by Alexander Fleming in 1922, is present in many biological fluids and tissues of mammals. The hen's egg is the richest source of it. Lysozyme acts against Gram-positive bacteria by hydrolyzing the peptidoglycan layer between the N-acetylglucosamine and N-acetylmuramic acid.

The enzyme lysozyme can be used as a preservative in cheese to prevent late gas blowing caused by *Clostridia* and provides an alternative to nitrate or the bacteriocin nisin [308,309].

Phospholipases A_2, a group of enzymes found in secondary granules of neutrophils, in macrophages, in murine Paneth and liver cells, and secreted into the tears, constitute the major components from Bothrops snake venoms. There is a considerable body of evidence that they have potent microbicidal properties and display a range of many relevant biological effects, including: bactericidal; anti-HIV, antitumoral, antimalarial, and antiparasitic; myotoxic; cytotoxic; induction of edema; anticoagulant; neuromuscular; inhibition of platelet aggregation; and hypotension [310-312].

Serprocidins are antimicrobial serine proteinases (elastase, cathepsin G, proteinase 3 and proteolytic inactive homolog-azurocidin) that are localized in azurophil granules of human polymorphonuclear–leukocyte broad-spectrum antimicrobial activity and manifest antifungal and cytotoxic activity against mammalian cells [313].

Thrombocidins are another group of antibacterial proteins found in human blood platelets, and are an important human defense mechanism. They protect the skin and epithelia against invading microorganisms and assisting neutrophils and platelets. Thrombocidins are components of the innate immune system found in many organisms and produced by a variety of cell types [314,315].

Granulysin. Recent publications have highlighted that eosinophil granules, macrophages, and monocytes also contain antimicrobial proteins. Human natural killer cells and cytolytic T lymphocytes contain granulysin, a cytolytic granule protein with a broad range of antimicrobial and tumoricidal activities [316,317].

The α-melanocyte–stimulating hormone, an endogenous peripheral neural regulatory peptide, represents a very ancient molecule. It is at least 300 million years old because it is formed by both mammals and by invertebrates not endowed with an adaptive type of immunity. α-Melanocyte–stimulating hormone has a wide range of biological activities, including strongly effective antimicrobial activity, anti-AIDS, as well as antimelanoma, antiinflammation, cardiovascular protection [318].

Antimicrobial peptides also have been discovered in:

- *Amphibians*, the first representatives of which was bombinin and later magainins, dermaseptins, buforin II, and ranalexin.

- *Insects*, antimicrobial peptides of which are known as cecropins or venoms such as bee melittin.
- *Plants,* where thionins were the first antimicrobial peptides to be isolated. Other antimicrobial peptides were found in plants to be structurally related to insect and mammal defensins and named "plant defensins."
- *Bacteria,* among which peptides are secreted from both Gram-positive and Gram-negative bacteria and have been classified as bacteriocins—defined as ribosomally synthesized antimicrobial peptides—grouped by different classifications, the most accepted of which organizes them into four groups: antibiotics, cystibiotics, thiolbiotics, no cysteine containing bacteriocins, although there are suggestions for revised classifications.
- *Viruses*, because viral-produced antimicrobial peptides were also identified.

Synthetic Peptides

A wide range of synthetic peptides have been designed that have also been shown to have a variety of interesting activities, including antifungal, antiviral, antiparasitic, and anticancer activities.

Despite thousands of articles about hundreds of these polypeptide antibiotics over several decades and several preclinical studies by small biotechnology companies, no clinical applications of ribosomally synthesized peptides have been reported and there are no published preclinical studies.

Because of problems in application, such as toxicity, susceptibility to proteolysis, manufacturing, and cost, it is very uncertain whether peptide antibiotics of vertebrate origin will enter the Big Pharma armamentarium in the near future. For whatever reasons and despite their promise no one (pharma, biotech, government, entrepreneur, etc.) has been willing to invest in their development.

Although their therapeutic activity is known, the precise description of the mechanism of action of different peptide antibiotics remains controversial.

Generally, it is believed that their strong antibacterial properties are a result of selective bacterial membrane disruption.

Different publications showed that peptide antibiotics selectively disrupt bacterial DNA synthesis, protein biosynthesis, cell wall biosynthesis, and membrane integrity, and can act by triggering the bacterial toxin-antitoxin chromosomal module, thereby causing the bacteria to commit suicide [319-321].

Different peptides display significantly different activities with various target cells.

In words of one syllable, it is necessary to follow a framework—one antibiotic, one inhibitory mechanism.

Finally, although there is a general consensus that highly cationic peptides kill bacteria primarily by injuring their membranes, an additional hypothesis has been proposed, which suggests that a large variety of cationic peptides might also activate their autolytic wall enzymes–muramidases (a "Trojan Horse" phenomenon), which results in bacteriolysis [322].

Vancomycin–Vancocin

Vancomycin (**30.5.6**) (see Fig. 30.37.) is included in the list of Top 200 Drugs by sales for the 2010s.

It belongs to a class of glycopeptide antibiotics that consists of a glycosylated cyclic or polycyclic nonribosomal peptide and that also includes teicoplanin, ramoplanin, and decaplanin.

Vancomycin, the leading member of this class, has been used extensively since the late 1950s and remains the gold standard in the antibiotic armamentarium [323-328]. It is a narrow-spectrum antibiotic that is primarily active against Gram-positive organisms. Bacterial resistance develops rarely because of its numerous modes of action, which results in the inhibition of peptidoglycan.

It became the drug of choice to treat infections caused by methicillin-resistant staphylococci. Vancomycin is also used for the treatment of patients who are undergoing cancer chemotherapy or who are on dialysis.

Despite nearly five decades of clinical use, vancomycin has retained a significant niche in the antibiotics arsenal. Nevertheless, major vancomycin toxicities, in particular, nephrotoxicity and ototoxicity, have been reported in the literature.

Total synthesis of vancomycin has been proposed, with substantial input in synthetic chemistry [329-336], but the complexity and the cost of the synthetic way makes fermentative routes the only viable route to bulk production.

Methods for producing vancomycin with this strain includes fermentation under aerobic conditions in a medium, composed of, carbon and nitrogen sources and some inorganics. For example, it could be glucose, soybean meal, corn steep solids or liquor, NaCl, $CaCO_3$, K_2HPO_4 [337], or a fermentative broth consisting of dextrin, bean powder, potato protein, NaCl and a small amount of other inorganics [338]. The influencing factors of the process are the temperature and pH value, each of which can affect the product synthesized by regulating enzyme activity.

30.6 MISCELLANEOUS ANTIBIOTICS

Lincosamides

Lincosamides are a class of antibiotics derived from the lincomycin, first isolated from *Streptomyces lincolnensis* in a soil bacterium sample from Lincoln, Nebraska (hence the bacterial name).

Lincosamides are bacteriostatics, mainly active against Gram-positive bacteria inhibiting bacterial protein synthesis by binding the 23S portion of the 50S ribosomal subunit and interfering with tRNA activity during translation. Lincosamides do not interfere with protein synthesis in eukaryotic cells [339-341].

In principle there are only two lincosamides in medicinal practice: lincomycin (**30.6.1**) itself and its semisynthetic chlorinated derivative, clindamycin (**30.6.2**) [342,343] (Fig. 30.39.).

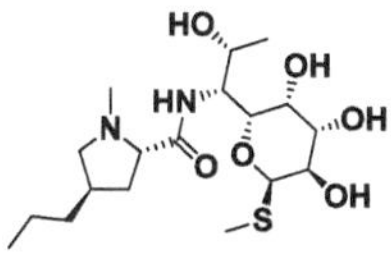

Lincomycin 30.6.1

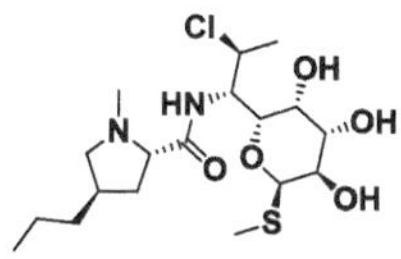

Clindamycin 30.6.2

FIG. 30.39 The lincosamides lincomycin and clindamycin.

They are approved for the treatment of anaerobic, streptococcal, and staphylococcal infections. They do not affect Gram-negative bacteria, but exhibit a significant antibiotic activity against some anaerobic bacteria. Lincomycin and clindamycin are used therapeutically, especially in cases where synergistic effects of a mixed anaerobic and aerobic microflora are anticipated. Clindamycin is usually more active than lincomycin, in particular in cases caused by anaerobic species. It can also be used for the treatment of important protozoal diseases, for example, malaria, most effectively in combination with primaquine. As compared with lincomycin, clindamycin is highly effective in the treatment of toxoplasmosis and pneumocystosis in AIDS patients.

Lincomycin consists of a carbohydrate portion represented by methyl 6-amino-6,8-dideoxy-1-thio D-erythro-α-D-galacto-octopyranoside, commonly referred to as methylthiolincosamide, bound to an unusual aminoacid (L-trans-4-n-propylhygric acid) by an amide linkage.

Several lincomycin syntheses have been described [344,345]. But the main source of lincomycin is cultivation of *Streptomyces espinosus* in a standard biological media medium containing glucose and Pharmamedia (cottonseed flour) [346,347].

Replacement of the hydroxyl in position 7 of lincomycin with a chlorine atom with inversion of the configuration causes a marked increase in activity yielding a new lincosamide antibiotic clindamycin. It is interesting to note that if the replacement takes place without inversion of configuration, the activity is not changed [348-351].

Because of the simplicity of the molecule, several other analogues of lincomycin have been synthesized leading to conclusions that the acyl portion can be modified without significant decrease of activity and that methyl on the sulfur could be replaced by other alkyl groups, producing products that are more active in vitro but not in vivo [351-353].

Ansamycins

Ansamycins are another important class of natural macrolide products. It is a group of antibiotics produced by strains of several *Actinomycetes*. The ansamycins have proved to be very potent molecules displaying anticancer, antibacterial, and antiviral activities.

They are named ansamycins—ansa from the Latin for handle or grip, because of their unique structure which comprises of an aromatic moiety bridged by an aliphatic chain. Ansamycins belong to the macrolides, but they are lactams, not lactones [354-357].

The ansamycins are typically divided into two groups—the benzenoid and naphthalenoid ansamycins—which are further identified according to the difference in the length of their "ansa" chains.

The benzenoid ansamycins are divided into two group: C15 (geldanamycin, herbimycin, macbecin, ansamitocin, maytansine, and TAN-420) and C17 ansa chains (ansatrienin, cytotrienin, hydroxymycotrienin, mycotrienin, thiazinotrienomycin, and trienomycin). The C17 ansa chains are considered potential antitumor antibiotics and have been extensively studied by cancer researchers for nearly four decades [358,359].

Geldanamycin (**30.6.3**) and herbimycin A (**30.6.4**) (Fig. 30.40.) were isolated many years ago from *Streptomyces,* and their synthetic and semisynthetic analogues were originally identified as potent inhibitors of certain kinases and were later shown to act by stimulating kinase degradation, specifically by targeting "molecular chaperones," for example, heat shock protein 90 (HSP90), which are involved in folding, activation and assembly of a wide range of proteins, including key proteins involved in signal transduction, cell cycle control, and transcriptional regulation.

The benzenoid ansamycins are among the most promising compounds currently being studied as anticancer drugs. Geldanamycin, is a lead compound for which more than 20 clinical trials have been initiated [360-364].

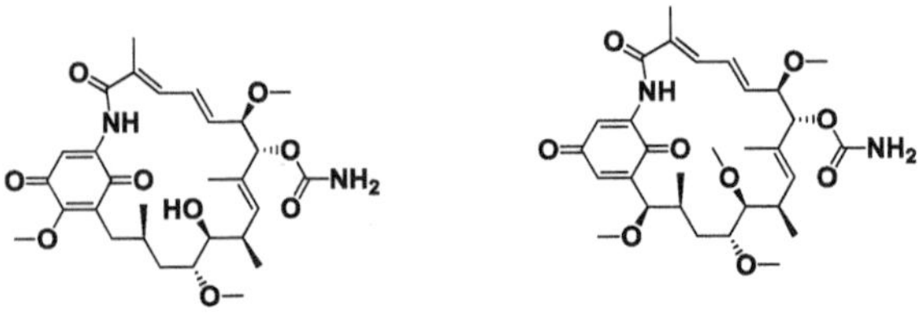

Geldanamycin 30.6.3 Herbimycin A 30.6.4

FIG. 30.40 The ansamycins geldanamycin and herbimycin.

Naphthalenoid ansamycins can be divided into three groups: naphthalenoids with C17 ansa chains (rifamycin, halomicin, streptovaricin, ansathiazin, awamycin, CP-50833, damavaricin, kanglemycin, proansamycin, protorifamycin, protostreptovaricin, and tolypomycin), naphthalenoids, naphthalenoids with C23 ansa chains (actamycin, naphthomycin, naphthomycinol, and naphthoquinomycin), and, at last, naphthalenoids with C9 ansa chains (rubradirin, protorubradirin).

Naphthalenoid ansamycins with C17 ansa chains can be further divided into three groups: rifamycin group, protostreptovaricin, and streptovaricin.

Ansamycins in general are a very specific class of macrocyclic antibiotics of which the rifamycins are among the better known members.

Rifamycins are antibiotics that are active against a large variety of organisms, including bacteria, eukaryotes, and viruses, and are sometimes called "wonder drugs." Clinically, they are particularly useful for the treatment of tuberculosis where they work selectively to inhibit RNA polymerase. Rifampicin SV is the first and the last antituberculosis drug in use. It is now used to treat leprosy and AIDS-related mycobacterial infections.

In general, the rifamycins (A, B, C, D, E), are produced by the *Streptomyces mediterranei* species. Rifamycin B was the only stable isolate component of this mixture, but it was poorly active. Although the natural rifamycins are not used in therapy, some semisynthetic rifamycins have therapeutical roles. It has been observed that its oxidation, when followed by hydrolysis, generated a quinone molecule, rifamycin S, a compound with potent antimicrobial activity. Subsequent reduction to the hydroquinone form produced rifamycin SV (**30.6.5**) with high potency against mycobacteria.

Many other rifamycin derivatives have been synthesized. The antibacterial activity of some of these derivatives is considerably better than that of rifamycin S and rifamycin SV.

Further synthesis of new rifamycin SV derivatives produced rifampicin (**30.6.6**), and other new ansamycins [365-368] (Fig. 30.41.).

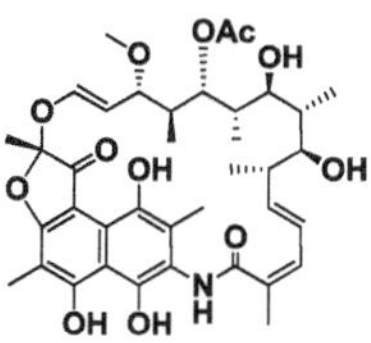

Rifamycin SV 30.6.5

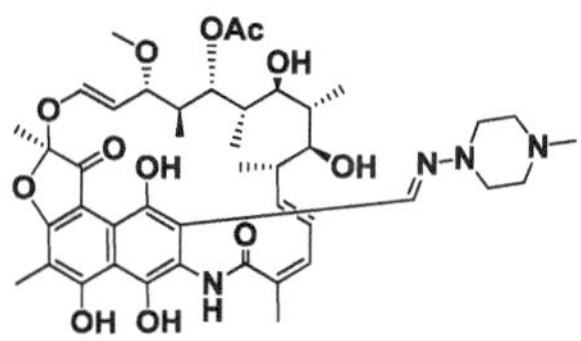

Rifampicin 30.6.6

FIG. 30.41 Structures of rifamycin SV and rifampicin.

In addition to their potent antimicrobial activity, ansamycins are hypolipidemic compounds, which dramatically lower high-density lipoprotein (HDL) cholesterol levels, in addition to reducing the levels of other lipoprotein classes. Some rifamycin derivatives exhibit antiinflammatory, antiallergic, and immunomodulatory activities.

Streptogramins

The streptogramins are a unique class of complex antibiotics, each member of which consists of at least two structurally unrelated molecules belonging to group A, which are polyunsaturated macrolactones, and to group B, representing cyclic hexadepsipeptides. Whereas the type A or type B compound alone has a moderate bacteriostatic activity, their combination has strong bacteriostatic, and often bactericidal, activity. Both group A and group B constituents of streptogramins separately inhibit protein synthesis at the ribosomal level, but they act in combination with a synergistic effect 100-fold higher than that of the individual components.

It is believed that both the groups (A and B) of compounds bind to the peptidyltransferase domain of the bacterial ribosome. The group A compounds prevent the formation of peptide bonds, while the B compounds stimulate the dissociation of the peptidyl-tRNA of the completed polypeptide.

Streptogramins are produced naturally by the corresponding the *Streptomyces*, the largest genus of *Actinobacteria*. Virginiamycin, the first streptogramin, was isolated from *Streptomyces virginiae*. Many other streptogramins, such as mikamycin, vernamycin, and griseoviridin/viridogrisein, which were produced by different strains of *Streptomyces*, have been reported.

Virginiamycin has been used in animal production as a feed additive. Its close analogue pristinamycin is widely used in human therapy for the treatment of staphylococcal and streptococcal infections [369-378].

Virginiamycin is the name of an antibiotic composed of combination of pristinamycin IIA (**30.6.7**) and virginiamycin S1 (**30.6.8**) (Fig. 30.42.). Both of them are of industrial interest and are produced in a scale of several tens of tons a year.

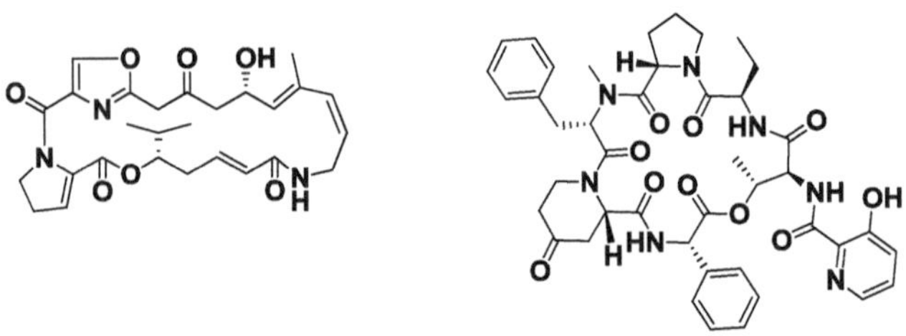

Pristinamycin IIA 30.6.7 **Virginiamycin S1 30.6.8**

FIG. 30.42 Structures of pristinamycin and virginiamycin.

The therapeutic use of the natural streptogramins is limited because of their limited dissolubility in water. New semisynthetic derivatives, in particular the injectable Synercid, which is a mixture of the semisynthetic A group–type component dalfopristin (**30.6.9**) and a B group component quinupristin (**30.6.10**), is successfully used for treating a number of infections that are caused by multiply resistant bacteria (Fig. 30.43.).

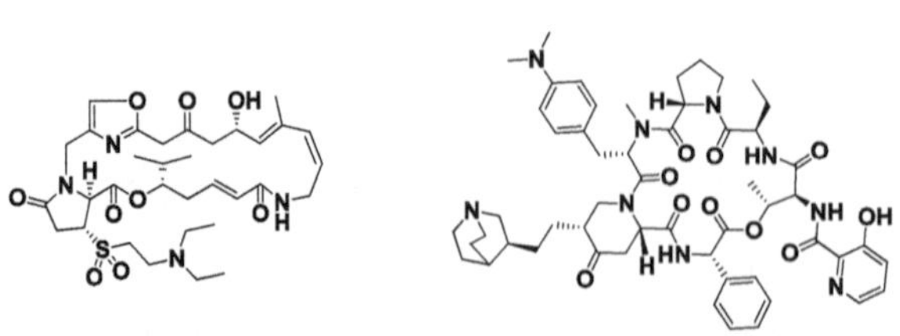

Dalfopristin 30.6.9 **Quinupristin 30.6.10**

FIG. 30.43 Structures of dalfopristin and quinupristin.

This unique class of antibacterials seems to have a significant clinical impact on increasing multidrug resistance affecting the Gram-positive cocci, especially staphylococci and pneumococci.

Anthracyclines

Anthracyclines are anticancer antibiotics were originally derived from certain species of *Streptomyces* by submerged aerobic fermentation.

They are a highly efficacious anticancer compounds for the treatment of solid and hematological malignancies, including acute myeloid leukemia, acute lymphoblastic leukemia, multiple myeloma, Hodgkin disease, and non-Hodgkin lymphoma, demonstrate significant disease activity in breast cancer, ovarian cancer. The anthracyclines are the most effective compounds among the antitumor antibiotics, but their clinical use is limited because of the development of dose-dependent cardiotoxicity, which can result in premature death.

The main products of this family are daunorubicin (daunomycin) (**30.6.11**), the first structurally characterized anthracycline; doxorubicin (**30.6.12**), a hydroxy derivative of daunorubicin; epirubicin (**30.6.13**), an epimer of doxorubicin that differs only in the orientation of the C4 hydroxy group on the sugar part of molecule; idarubicin (**30.6.14**), an analogue of daunorubicin that lacks the C-4 methoxy group; and valrubicin (**30.6.15**) an N-trifluoroacetyl, 14-valerate derivative of doxorubicin [379-389] (Fig. 30.44.).

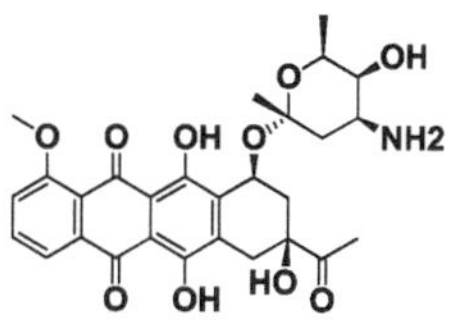

Daunorubicin 30.6.11

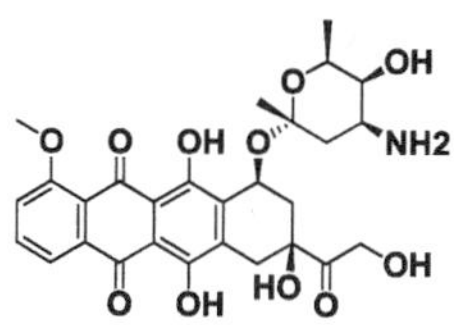

Doxorubicin 30.6.12

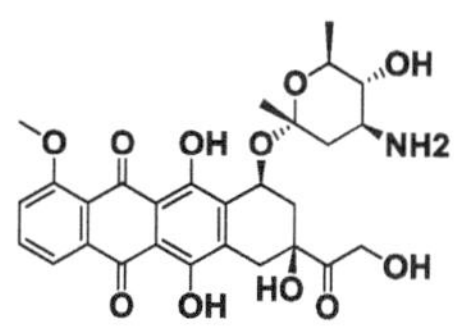

Epirubicin 30.6.13

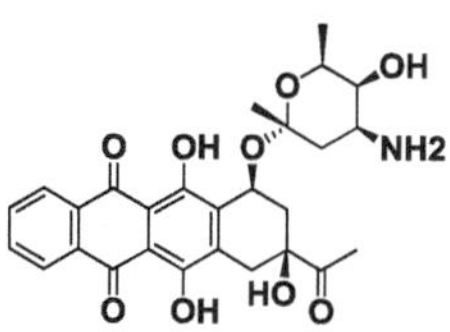

Idarubicin 30.6.14

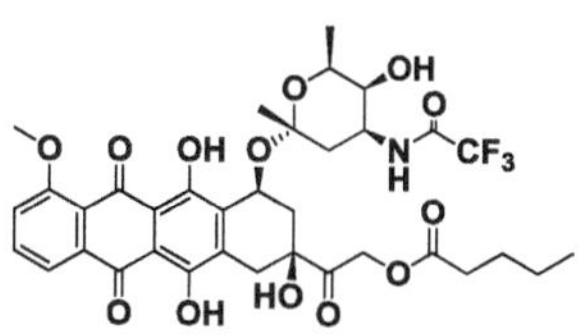

Valrubicin 30.6.15

FIG. 30.44 The family of anthracyclines.

Multiple mechanisms have been proposed to explain the cytostatic and cytotoxic actions of anthracyclines including free radical formation, lipid peroxidation, and direct membrane effects. The most accepted, however, is interactions with DNA or the DNA–topoisomerase II complex, which are responsible for disturbances in DNA replication and transcription, and then the induction of DNA repair or apoptotic cell death.

Phenicols

The phenicols are antibiotics with the simplest chemical structure of all antibiotics. The phenicol antibiotics used in humans are chloramphenicol (**30.6.16**) and thiamphenicol (**30.6.17**). Florfenicol (**30.6.18**) was authorized only as a new veterinary drug [390-400] (Fig. 30.45.).

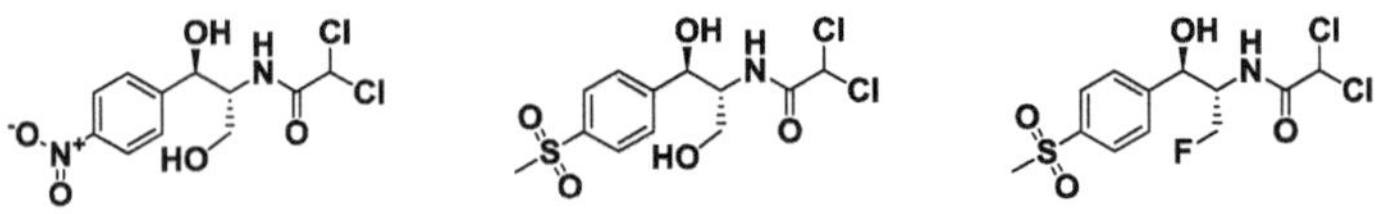

Chloramphenicol 30.6.16 Thiamphenicol 30.6.17 Florfenicol 30.6.18

FIG. 30.45 **Phenicols.**

Phenicols are antibiotics with broad antibacterial activity, including activity against Gram-positive and Gram-negative bacteria, anaerobes, spirochetes, rickettsiae, chlamydiae, and mycoplasma.

Chloramphenicol was originally derived from the bacterium *Streptomyces venezuelae* in the early 1950s, but its use was largely restricted because of concerns regarding its safety profile and the possible development of aplastic anemia, even leukemia, because of bone marrow depression. Less-serious side effects include some gastrointestinal disturbances, and, because of the broad spectrum of the antibiotic's action, secondary infections resulting from elimination of the normal bacterial flora.

Nevertheless, chloramphenicol remains a potential valuable resort in the fight against many multidrug-resistant pathogens. It is the drug of choice for treatment of typhus and salmonellosis, and it is still widely used in topical preparations.

All three phenicols, which are close structural analogues, have a similar spectrum of activity.

Phenicols are primarily bacteriostatic. They bind to the 50S subunit of the ribosome, thereby inhibiting bacterial protein synthesis.

All three structurally related phenicols contain two asymmetric centers. Four stereoisomers are possible, but only the D-threo isomers are active.

Chloramphenicol is among the very few antibiotics produced industrially by synthesis rather than by fermentation.

Fusidic Acid

Fusidic acid (**30.6.19**) is an antibiotic that belongs to a group of its own—the fusidanes [401]. It was obtained from cultures of a fungus, *Fusidium coccineum*, which was originally isolated from monkey feces. It has been available in the market since 1962.

Fusidic acid has asteroid-type structure, as do some other antibiotics produced by fungi, but does not possess any steroid activity. The structure is thought to be responsible for the steroid-like high penetration. The antimicrobial activity of fusidic acid is specifically aimed at the most common skin pathogens, including *S. aureus*, toward which it is one of the most potent antibiotics [402-404] (Fig. 30.46.).

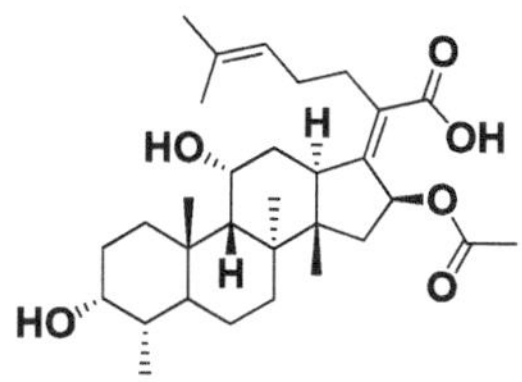

Fusidic acid 30.6.19

FIG. 30.46 Structure of fusidic acid.

Fusidic acid inhibits bacterial protein synthesis by interference with elongation factor G, which promotes translocation on the ribosome after peptide bond formation, preventing further elongation by inhibiting the guanosine triphosphatase (GTPase) function of the elongation factor G.

Fusidic acid, a narrow-spectrum antibiotic, is used both systemically and topically to treat primary skin infections, including impetigo. Moreover, it has been used for a wide variety of less-common infections [405], in the treatment of *Clostridium difficile* colitis, and in staphylococcal infections in patients with cystic fibrosis. It is also effective in bacterial conjunctivitis and other minor external eye infections, and may be effective in reducing bacterial flora in the conjunctival sac prior to eye surgery. Fusidic acid has a potential role as a prophylaxis during neurosurgical interventions.

Reports have been published indicating that fusidic acid has an immunomodulatory effect. It has been investigated in a number of immunologically mediated diseases, HIV infection, Behçet disease, Crohn disease, uveitis, and scleroderma [406].

Fosfomycin

Fosfomycin (**30.6.20**) is a naturally occurring broad-spectrum antibiotic with activity against both Gram-positive and Gram-negative bacteria. It was isolated in 1969 in Spain from *Streptomyces fradiae*.

Fosfomycin irreversibly inhibits phosphoenolpyruvate UDP-N-acetylglucosamine-3-O-enolpyruvyltransferase (enolpyruvyltransferase), the enzyme that catalyzes the first step of peptidoglycan biosynthesis, and shows almost no toxicity to humans. Multidrug-resistant Gram-negative bacteria such as

P. aeruginosa and *Stenotrophomonas maltophilia,* as well as *Enterobacteriaceae* strains, are frequently sensitive to fosfomycin [407-413].

Fosfomycin belongs to the class of phosphonic antibiotics and is characterized by unique structural functions such as a carbon-phosphorus (C-P) bond and an epoxide (Fig. 30.47.).

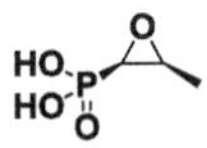

Fosfomycin 30.6.20

FIG. 30.47 Structure of fosfomycin.

Fosfomycin is an antibiotic that has varying application indications across the globe. Its European, Asian, African, and South American usage and implementations are much broader, than its currently limited application in the United States, where uncomplicated urinary tract infection represents the only indication for its use.

Fosfomycin was shown to have immunomodulatory effects, although it is still difficult to precisely identify the cellular targets or the cascade of events induced by the antibiotic [414].

The main limitation for its use in clinics is the high frequency of resistant mutants; consequently, it is mainly used in combination with other antibiotics. Its toxicity is low except for some adverse gastric effects and irritation at the site of injection.

Fosfomycin is now produced synthetically because of the simplicity of its structure.

Mupirocin

Mupirocin (**30.6.21**) is a novel, unique, natural polyketide antibiotic produced by *Pseudomonas fluorescens,* and is totally unrelated in chemical structure and mode of action to any other clinically useful class of antibiotics. Its greatest antibacterial activity is against a wide range of aerobic Gram-positive cocci, namely, *S. aureus*, *Staphylococcus epidermidis*, *S. pyogenes*, and other β-hemolytic streptococci (Fig. 30.48.).

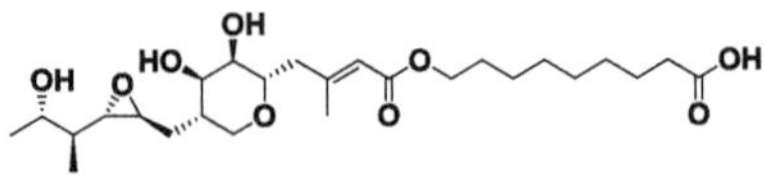

Mupirocin 30.6.21

FIG. 30.48 Structure of mupirocin.

Mupirocin was introduced into clinical practice in 1985, and has proved to be an extremely effective treatment for skin infections and one of the most successful topical antibiotics.

The drug has become the agent of choice for topical treatment of the nasal vestibulum in patients carrying methicillin-resistant *S. aureus* [415-420].

Mupirocin is not active systemically as it is rapidly metabolized to an inactive product, monic acid.

The multidrug-resistant bacteria that is growing day-by-day in both community and hospital settings are a major public health concern, making development of new antibiotics one of the most challenging areas in drug design. Unfortunately, great efforts in developing new antibacterial agents, which include modification of existing ones, exploring new antibacterial targets, and designing new compounds, has resulted only in the invention of two first-in-class antibiotics, fidaxomicin and bedaquiline, which have been proposed for clinical trials [421-425].

The mode of action of the reviewed antibiotics can be summarized as follows:

- Penicillins, cephalosporin, vancomycin, bacitracin, and cycloserine act via inhibition of bacterial cell wall synthesis;
- Tetracyclines and aminoglycosides inhibit protein synthesis in bacteria;
- Polymyxin and amphotericin B act via alteration of cell membrane function in bacteria;
- Erythromycin, clindamycin, and chloramphenicol inhibit protein synthesis in bacteria;
- The rifamycin group inhibits nucleic acid synthesis in bacteria.

REFERENCES

1. Maffioli, S. I. A chemist's survey of different antibiotic classes. In *Antibiotics: Targets, Mechanisms and Resistance;* Gualerzi, C. O., Brandi, L., Fabbretti, A., Pon, C. L., Eds.; Wiley-VCH, 2014; pp 1–22.
2. Bryskier, A. Antibiotics and antibacterial agents: classifications and structure-activity relationship. In *Antimicrobial Agents: Antibacterials and Antifungals;* Bryskier, A., Ed.; ASM Press, 2005; pp 13–38.
3. Felton, L. A., Ed. *Remington: Essentials of Pharmaceutics;* Pharmaceutical Press, 2012; p 86.
4. Gualerzi, C. O., Brandi, L., Fabbretti, A., Pon, C. L., Eds. *Antibiotics: Targets, Mechanisms and Resistance;* Wiley-VCH, 2014.
5. Franklin, T. J., Ed. *Biochemistry and Molecular Biology of Antimicrobial Drug Action,* 6th ed.; Springer, 2004.
6. Greenwood, D., Ed. *Antimicrobial Drugs: Chronicle of a Twentieth Century Medical Triumph;* Oxford University Press, 2008.
7. Greenwood, D.; Finch, R.; Davey, P.; Wilcox, M. *Antimicrobial Chemotherapy*, 5th ed.; Oxford University Press, 2007.
8. Greenwood, D. *Medical Microbiology: A Guide to Microbial Infections: Pathogenesis, Immunity, Laboratory Diagnosis and Control*, 17th ed.; Churchill Livingstone, 2007.
9. O'Grady, F., Finch, R. G., Lambert, H. P., Greenwood, D., Eds. *Antibiotic and Chemotherapy: Anti-Infective Agents and Their Use in Therapy,* 7th ed.; Churchill Livingstone, 1997.
10. Finch, R. G., Greenwood, D., Norrby, S. R., Whitly, R. J., Eds. *Antibiotics and Chemotherapy: Anti-Infective Agents and Their Use in Therapy,* 9th ed.; Saunders, 2010.

11. Walsh, C. *Antibiotics: Actions, Origins, Resistance;* ASM Press, 2003.
12. Hodges, N. Antibiotics and synthetic antimicrobial agents: their properties and uses. In *Hugo and Russell's Pharmaceutical Microbiology,* 8th ed.; Denyer, S. P., Hodges, N., Gorman, S. P., Gilmore, B., Eds. Wiley-Blackwell, 2011; pp 169–199.
13. Axelsen, P. H., Ed. *Essentials of Antimicrobial Pharmacology: A Guide to Fundamentals for Practice;* Humana Press, 2001.
14. Cunha, B. *Antibiotics Essentials*, 12th ed.; Jones & Bartlett Learning, 2013.
15. Park, H.; Thomas, M. Antibiotics in the pipeline. In *Antibiotics: Current Innovations and Future Trends;* Sanchez, S., Demain, A. L., Eds.; Caister Academic Press, 2015; pp 395–416.
16 Meinert, S.; John, E. 80 Years of antibiotics application in medicine. Indispensable against bacteria: antibiotics. *Chem. Unserer Zeit* **2009,** *43* (5), 296–306.
17. Demain, A. L.; Spizek, J. The antibiotic crisis. *Adv. Mol. Cell. Microbiol.* **2012,** *22,* 26–43.
18. Zotchev, S. B. Antibiotics. *Nat. Prod. Chem. Biol.* **2012,** 269–285.
19. Wright, G. D. Antibiotics: a new hope. *Chem. Biol.* **2012,** *19* (1), 3–10.
20. Aminov, R. A brief history of the antibiotic era: lessons learned and challenges for the future. *Front. Microbiol.* **2010,** *1* (134), 1–7.
21. Butler, M. S.; Cooper, M. A. Antibiotics in the clinical pipeline in 2011. *J. Antibiot.* **2011,** *64* (6), 413–425.
22. Butler, M. S.; Blaskovich, M. A.; Cooper, M. A. Antibiotics in the clinical pipeline in 2013. *J. Antibiot.* **2013,** *66* (10), 571–591.
23. Kumbhar, C.; Watve, M. Why antibiotics: a comparative evaluation of different hypotheses for the natural role of antibiotics and an evolutionary synthesis. *Nat. Sci.* **2013,** *5* (4A), 26–40.
24. Lewis, K. Platforms for antibiotic discovery. *Nat. Rev. Drug Discovery* **2013,** *12* (5), 371–387.
25. Weber, T. Antibiotics: biosynthesis, generation of novel compounds. In Flickinger, M. C., Ed.; *Encyclopedia of Industrial Biotechnology*, Vol. 1; Wiley, 2010; pp 311–323.
26. Kohanski, M. A.; Dwyer, D. J.; Collins, J. J. How antibiotics kill bacteria: from targets to networks. *Nat. Rev. Microbiol.* **2010,** *8* (6), 423–435.
27. Herdewijn, P. Antibiotics. In *Textbook of Drug Design and Discovery,* 4th ed.; Krogsgaard-Larsen, P., Stroemgaard, K., Madsen, U., Eds. CRC Press, 2010; pp 420–447.
28. Demain, A. L. Antibiotics: natural products essential to human health. *Med. Res. Rev.* **2009,** *29* (6), 821–842.
29. Behal, V. Antibiotics. *Biotechnol. Annu. Rev.* **2002,** *8,* 227–265.
30. Ferrier, R. J.; Blattner, R.; Field, R. A.; Furneaux, R. H.; Gardiner, J. M.; Hoberg, J. O.; Kartha, K. P. R.; Tilbrook, D. M. G.; Tyler, P. C.; Wightman, R. H. Antibiotics. *Carbohydr. Chem.* **2002,** *33,* 257–274.
31. Hubschwerlen, C. β-Lactam antibiotics. In *Comprehensive Medicinal Chemistry II*, Vol. 7; 8th ed.; Taylor, J. B., Triggle, D. J., Eds.; Elsevier, 2006; pp 479–518.
32. Demain, A. L.; Elander, R. P. β-Lactam antibiotics: past, present, and future. *Antonie van Leeuwenhoek* **1999,** *75* (1–2), 5–19.
33. Sammes, P. G. *Topics in Antibiotic Chemistry*, Vol. 3,4; Ellis Horwood, 1980.
34. Mitsuhashi, S., Ed. *Beta-Lactam Antibiotics;* Springer, 1981.
35. Singh, G. S.; Sudheesh, S. β-Lactams. In *Natural Lactones and Lactams: Synthesis, Occurrence and Biological Activity;* Janecki, T., Ed.; Wiley-VCH, 2014; pp 101–145.
36. Flynn, E. H., Ed. *Cephalosporins and Penicillins: Chemistry and Biology;* Academic Press, 1972.
37. Newall, C. E.; Hallam, P. D. β-Lactam antibiotics: penicillins and cephalosporins. In *Comprehensive Medicinal Chemistry;* Hansch, C., Sammes, P. G., Taylor, J. B., Eds.; Vol. 2, Pergamon Press, 1990; pp 609–653.

38. Hoemann, M. Z. Penicillin and cephalosporin antibiotics. In *Bioactive Heterocyclic Compound Classes: Pharmaceuticals;* Dinges, J., Lamberth, C., Eds.; Wiley-VCH, 2012; pp 237–253.
39. Morin, R. B., Gorman, M., Eds. *Chemistry and Biology of β-Lactam Antibiotics, Vol. 1: Penicillins and Cephalosporins; Vol. 2: Nontraditional β-Lactam Antibiotics; Vol. 3: The Biology of β-Lactam Antibiotics;* Academic Press, 1982.
40. Elks, J., Ed. *Chemical Society Special Publication, No. 28: Recent Advances in the Chemistry of β-Lactam Antibiotics;* Burlington House, London, 1977.
41. Gregory, G. I., Ed. *Royal Society of Chemistry, Special Publication No. 38: Recent Advances in the Chemistry of β-Lactam Antibiotics;* Burlington House, London, 1980.
42. Bush, K.; Macielag, M. J. New β-Lactam antibiotics and β-lactamase inhibitors. *Expert Opin. Ther. Pat.* **2010,** *20* (10), 1277–1293.
43. Walsh, T. F. β-Lactam antibiotics. *Annu. Rep. Med. Chem.* **1988,** *23*. 121–31.
44. Llarrull, L. I.; Testero, S. A.; Fisher, J. F.; Mobashery, S. The future of the β-lactams. *Curr. Opin. Microbiol.* **2010,** *13* (5), 551–557.
45. Bazan, J. A.; Martin, S. I.; Kaye, K. M. Newer β-lactam antibiotics: doripenem, ceftobiprole, ceftaroline, and cefepime. *Med. Clin. North Am.* **2011,** *95* (4), 743–760.
46. Xing, B.; Rao, J.; Liu, R. Novel beta-lactam antibiotics derivatives: their new applications as gene reporters, antitumor prodrugs and enzyme inhibitors. *Mini-Rev. Med. Chem.* **2008,** *8* (5), 455–471.
47. Jovetic, S.; Zhu, Y.; Marcone, G. L.; Marinelli, F.; Tramper, J. β-Lactam and glycopeptide antibiotics: first and last line of defense? *Trends Biotechnol.* **2010,** *28* (12), 596–604.
48. Demain, A. L.; Spizek, J. The β-Lactam and glycopeptide antibiotics: first and last line of defense? Crisis. *Adv. Mol. Cell. Microbiol.* **2012,** *22,* 26–43.
49. Brown, A. G.; Pearson, M. J.; Southgate, R. Other β-lactam agents. In *Comprehensive Medicinal Chemistry;* Hansch, C., Sammes, P. G., Taylor, J. B., Eds.; Vol. 2, Pergamon Press, 1990; pp 655–702.
50. Neuhauser, M. M.; Danziger, L. H. β-Lactam antibiotics. In *Drug Interactions in Infectious Diseases,* 2nd ed.; Piscitelli, S. C., Rodvold, K. A., Eds. GSK, 2005; pp 255–287.
51. Neuhauser, M. M.; Danziger, L. H. β-Lactam antibiotics. In *Drug Interactions in Infectious Diseases;* Piscitelli, S. C., Rodvold, K. A., Eds.; Humana Press, 2001; pp 151–184.
52. Kong, K.-F.; Schneper, L.; Mathee, K. Beta-lactam antibiotics: from antibiosis to resistance and bacteriology. *APMIS* **2010,** *118* (1), 1–36.
53. Qin, W.; Panunzio, M.; Biondi, S. Beta-lactam antibiotics: from antibiosis to resistance and bacteriology renaissance. *Antibiotics* **2014,** *3* (2), 193–215.
54. Prescott, J. F. β-lactam antibiotics: cephalosporins. In *Antimicrobial Therapy in Veterinary Medicine,* 5th ed.; Giguere, S., Prescott, J. F., Dowling, P. M., Eds. Wiley-Blackwell, 2013; pp 153–173.
55. Gales, A. C.; Sader, H. S. Novel β-lactams. *Braz. J. Infect. Dis.* **2008,** *12* (Suppl. 2), 46–58.
56. Sedelmeier, G. A new generation of β-lactam antibiotics. In *Organic Synthesis Highlights II;* Waldmann, H., Ed.; VCH, 1995; pp 277–288.
57. Ortega, E.; Escobar, M. A.; Gaforio, J. J.; Alvarez de Cienfuegos, G. In *Monobactams and carbapenems: only antimicrobial agents? in New Approaches in the Use of Antibiotics;* Ramos-Cormenzana, A., Ed.; Research Signpost, 2003; pp 123–147.
58. Singh, G. S. β-Lactams in the new millennium. Part I: monobactams and carbapenems. *Mini-Rev. Med. Chem.* **2004,** *4* (1), 69–92.
59. Hellinger, W. C.; Brewer, N. S. Monobactams and carbapenems for treatment of intraabdominal infections. *Mayo Clin. Proc.* **1999,** *74* (4), 420–434.

60. Galal, A. M.; Gul, W.; Noreddin, A. M.; Slade, D. An update on the synthesis and antibacterial effects of carbapenems. *Recent Pat. Anti-Infect. Drug Discovery* **2010,** *5* (1), 23–43.
61. Kawamoto, I.; Ohya, S. New aspects in carbapenem antibiotics. *Annu. Rep. Sankyo Res. Lab.* **1998,** *50,* 1–14.
62. El-Gamal, M. I.; Oh, C.-H. Current status of carbapenem antibiotics. *Curr. Top. Med. Chem.* **2010,** *10* (18), 1882–1897.
63. Rolinson, G. N.; Geddes, A. M. The 50th anniversary of the discovery of 6-aminopenicillanic acid (6-APA). *Int. J. Antimicrob. Agents* **2007,** *29* (1), 3–8.
64. Thirkettle, J. E.; Schofield, C. J.; Walter, M. W. β-Lactam chemistry. *Amino Acids, Pept., Proteins* **1997,** *28,* 281–333.
65. Martin, J. F.; Gutierrez, S.; Demain, A. L. Antibiotics β-Lactam. In *Fungal Biotechnology;* Anke, T., Ed.; Chapman & Hall, 1977; pp 91–127.
66. Deaguero, A. L.; Blum, J. K.; Bommarius, A. S. Biocatalytic synthesis of β-lactam antibiotics. In *Encyclopedia of Industrial Biotechnology;* Flickinger, M. C., Ed.; Vol. 1, Wiley, 2010; pp 535–566.
67. Mateo, C.; Abian, O.; Grazu, V.; Fernandez-Lorente, G.; Palomo, J. M.; Fuentes, M.; Segura, R. L.; Montes, T.; Lopez-Gallego, F.; Wilson, L.; Torres, R.; Guisan, J. M.; Fernandez-Lafuente, R. Recent advances in the industrial enzymatic synthesis of semi-synthetic β-lactam antibiotics. *Med. Chem. Rev.–Online* **2005,** *2* (3), 207–218.
68. Alkema, W. B. L.; De Vries, E. J.; Hensgens, C. M. H.; Polderman-Tijmes, J. J.; Dijkstra, B. W.; Janssen, D. B. Engineering enzymes for the synthesis of semi-synthetic antibiotics. In *Synthesis of β-Lactam Antibiotics;* Bruggink, A., Ed.; Springer, 2001; pp 250–279.
69. Sheldon, R. A.; Van Rantwijk, F.; Van Langen, L. M.; Wegman, M. A.; Cao, L.; Janssen, M. H. A. Biocatalysts and biocatalysis in the synthesis of β-lactam antibiotics. In *Synthesis of β-Lactam Antibiotics;* Bruggink, A., Ed.; Springer, 2001; pp 102–148.
70. Bruggink, A.; Roy, P. D. Industrial synthesis of semisynthetic antibiotics. In *Synthesis of β-Lactam Antibiotics;* Bruggink, A., Ed.; Springer, 2001; pp 12–54.
71. Fernandez-Lafuente, R.; Mateo, C.; Abian, O.; Fernandez-Lorente, G.; Palomo, J. M.; Fuentes, M.; Guisan, J. M. Industrial synthesis of semi-synthetic β-lactam antibiotics: recent developments in enzyme biocatalysis for improved and MORE sustainable processes. *Curr. Med. Chem.: Anti-Infect. Agents* **2002,** *1* (4), 375–387.
72. Andersson, I.; Van Scheltinga, A. C. T.; Valegard, K. Towards new β-lactam antibiotics. *Cell. Mol. Life Sci.* **2001,** *58* (12/13), 1897–1906.
73. Wegman, M. A.; Janssen, M. H. A.; Van Rantwijk, F.; Sheldon, R. A. Towards biocatalytic synthesis of β-lactam antibiotics. *Adv. Synth. Catal.* **2001,** *343* (6+7), 559–576.
74. Bellgardt, K.-H. Process models for production of β-lactam antibiotics. *Adv. Biochem. Eng./Biotechnol.* **1998,** *60,* 153–194.
75. Velasco, J.; Adrio, J. L.; Moreno, M. A.; Diez, B.; Soler, G.; Barredo, J. L. Environmentally safe production of 7-aminodeacetoxycephalosporanic acid (7-ADCA) using recombinant strains of *Acremonium chrysogenum. Nat. Biotechnol.* **2000,** *18* (8), 857–861.
76. Sheehan, J. C.; Henery-Logan, K. R. General synthesis of the penicillins. *J. Am. Chem. Soc.* **1959,** *81,* 5838–5839.
77. Sheehan, J. C.; Henry-Logan, K. R. The total and partial general syntheses of the penicillins. *J. Am. Chem. Soc.* **1962,** *84,* 2983–2990.
78. Woodward, R. B. Recent advances in the chemistry of natural products. *Science (Washington, DC, U. S.)* **1966,** *153* (3735), 487–493.
79. Woodward, R. B.; Heusler, K.; Gosteli, J.; Naegeli, P.; Oppolzer, W.; Ramage, R.; Ranganathan, S.; Vorbrueggen, H. Total synthesis of cephalosporin C. *J. Am. Chem. Soc.* **1966,** *88* (4), 852–853.

80. Elander, R. P. Industrial production of β-lactam antibiotics. *Appl. Microbiol. Biotechnol.* **2003,** *61* (5–6), 385–392.
81. Bellgardt, K.-H. β-Lactam antibiotics production with *Penicillium chrysogenum* and *Acremonium chrysogenum.* In *Bioreaction Engineering;* Schuegerl, K., Bellgardt, K.-H., Eds.; Springer, 2000; pp 391–432.
82. Burgdorf, K. Industrial production of β-lactam antibiotics. Part 2. *Schweiz. Lab.-Z.* **2004,** *61* (1–2), 9–14.
83. Weber, S. S.; Bovenberg, R. A. L.; Driessen, A. J. M. Biosynthetic concepts for the production of β-lactam antibiotics in Penicillium chrysogenum. *Biotechnol. J.* **2012,** *7* (2), 225–236.
84. Kim, O. K.; Fung-Tomc, J. Patents on β-lactam antibacterials: January 1999 to March 2001. *Expert Opin. Ther. Pat.* **2001,** *11* (8), 1267–1276.
85. Jensen, S. E.; Demain, A. L. β-lactam. *Biotechnol. Ser.* **1995,** *28,* 239–268.
86. Sklyarenko, A. V.; Kurochkina, V. B.; Egorov, A. M. Enzymatic transformation and synthesis of beta-lactam antibiotics. In *New Research on Biotechnology and Medicine;* Egorov, A. M., Zaikov, G., Eds.; Nova Science Publishers Inc, 2006; pp 73–86.
87. Fisher, J. F.; Meroueh, S. O.; Mobashery, S. Bacterial resistance to β-lactam antibiotics: Compelling opportunism, compelling opportunity. *Chem. Rev. (Washington, DC, U. S.)* **2005,** *105* (2), 395–424.
88. Bradford, P. A. Extended-spectrum β-lactamases in the 21st century: characterization, epidemiology, and detection of this important resistance threat. *Clin. Microbiol. Rev.* **2001,** *14* (4), 933–951.
89. Kernodle, D. S. Mechanisms of resistance to β-lactam antibiotics. In *Gram-Positive Pathogens;* Fischetti, V. A., Novick, R. P., Ferreti, J. J., Portnoy, A. A., Rood, J. I., Eds.; American Society for Microbiology, 2000; pp 609–620.
90. Spratt, B. G. Resistance to β-lactam antibiotics. *New Compr. Biochem.* **1994,** *27,* 517–534.
91. Wright, G. D. Bacterial resistance to antibiotics: enzymatic degradation and modification. *Adv. Drug Delivery Rev.* **2005,** *57* (10), 1451–1470.
92. Poole, K. Resistance to β-lactam antibiotics. *Cell. Mol. Life Sci.* **2004,** *61* (17), 2200–2223.
93. Sandanayaka, V. P.; Prashad, A. S. Resistance to β-lactam antibiotics: structure and mechanism based design of β-lactamase inhibitors. *Curr. Med. Chem.* **2002,** *9* (12), 1145–1165.
94. Wright, G. D. Something old, something new: revisiting natural products in antibiotic drug discovery. *Can. J. Microbiol.* **2014,** *60* (3), 147–154.
95. Vardanyan, R. S.; Hruby, V. J. *Synthesis of essential drugs;* Elsevier, 2006.
96. Wise, R. β-Lactams: cephalosporins. In *Antibiotic and Chemotherapy,* 7th ed.; O'Grady, F., Ed.; Churchill Livingstone, 1997; pp 202–255.
97. Morin, R. B.; Jackson, B. G.; Flynn, E. H.; Roeske, R. W.; Andrews, S. 7-Aminocephalosporanic acid. *BE* **1962,** *61* (5955).
98. Morin, R. B.; Jackson, B. G.; Flynn, E. H.; Roeske, R. W.; Andrews, S. Chemistry of cephalosporin antibiotics. XIV. The reaction of cephalosporin C with nitrosyl chloride. *J. Am. Chem. Soc.* **1969,** *91* (6), 1396–1400.
99. Pollegioni, L.; Rosini, E.; Molla, G. Cephalosporin C acylase: dream and(/or) reality. *Appl. Microbiol. Biotechnol.* **2013,** *97* (6), 2341–2355.
100. Blaser, H. U., Schmidt, E., Eds. *Asymmetric Catalysis on Industrial Scale: Challenges, Approaches and Solutions;* Wiley-VCH, 2004.
101. Blaser, H. U., Federsel, H. J., Eds. *Asymmetric Catalysis on Industrial Scale: Challenges, Approaches, and Solutions,* 2nd ed.; Wiley-VCH, 2010.
102. Bryskier, A. Carbapenems. In *Antimicrobial Agents: Antibacterials and Antifungals;* Bryskier, A., Ed.; ASM Press, 2005; pp 269–318.

103. Breilh, D.; Texier-Maugein, J.; Allaouchiche, B.; Saux, M.-C.; Boselli, E.; Carbapenems. *J. Chemother. (London, U. K.)* **2013,** *25* (1), 1–17.
104. Papp-Wallace, K. M.; Endimiani, A.; Taracila, M. A.; Bonomo, R. A. Carbapenems: past, present, and future. *Antimicrob. Agents Chemother.* **2011,** *55* (11), 4943–4960.
105. Kattan, J. N.; Villegas, M. V.; Quinn, J. P. New developments in carbapenems. *Clin. Microbiol. Infect.* **2008,** *14* (12), 1102–1111.
106. Shah, P. M. Carbapenems—an overview. *Chemother. J.* **2008,** *17* (4), 114–119.
107. Lo, T. S.; Welch, J. M.; Alonto, A. M.; Vicaldo-Alonto, E. A. R. A review of the carbapenems in clinical use and clinical trials. *Recent Pat. Anti-Infect. Drug Discovery* **2008,** *3* (2), 123–131.
108. Nicolau, D. P. Carbapenems: a potent class of antibiotics. *Expert Opin. Pharmacother.* **2008,** *9* (1), 23–37.
109. Zhanel, G. G.; Wiebe, R.; Dilay, L.; Thomson, K.; Rubinstein, E.; Hoban, D. J.; Noreddin, A. M.; Karlowsky, J. A. Comparative review of the carbapenems. *Drugs* **2007,** *67* (7), 1027–1052.
110. Ozcengiz, G.; Demain, A. L. Recent advances in the biosynthesis of penicillins, cephalosporins and clavams and its regulation. *Biotechnol. Adv.* **2013,** *31* (2), 287–311.
111. Birnbaum, J.; Kahan, F. M.; Kropp, H.; Macdonald, J. S. Carbapenems, a new class of beta-lactam antibiotics. Discovery and development of imipenem/cilastatin. *Am. J. Med.* **1985,** *78* (6A), 3–21.
112. Parker, W. L.; O'Sullivan, J.; Sykes, R. B. Naturally occurring monobactams. *Adv. Appl. Microbiol.* **1986,** *31,* 181–205.
113. Bonner, D. P.; Sykes, R. B. The monobactams. *Med. Microbiol.* **1984,** *4,* 171–197.
114. Czachor, J. S.; Gleckman, R. A. Monobactams. *Infect. Dis. Ther.* **1994,** *9,* 125–133.
115. Brewer, N. S.; Hellinger, W. C. The Monobactams. *Mayo Clin. Proc.* **1991,** *66* (11), 1152–1157.
116. Sykes, R. B.; Koster, W. H.; Bonner, D. P. The new monobactams: chemistry and biology. *J. Clin. Pharmacol.* **1988,** *28* (2), 113–119.
117. Bush, K.; Jacoby, G. A. Updated functional classification of β-lactamases. *Antimicrob. Agents Chemother.* **2010,** *54* (3), 969–976.
118. Perez-Llarena, F. J.; Bou, G. β-Lactamase inhibitors: the story so far. *Curr. Med. Chem.* **2009,** *16* (28), 3740–3765.
119. Maiti, S. N.; Phillips, O. A.; Micetich, R. G.; Livermore, D. M. β-Lactamase inhibitors: agents to overcome bacterial resistance. *Cur. Med. Chem.* **1998,** *5* (6), 441–456.
120. Biondi, S.; Long, S.; Panunzio, M.; Qin, W. L. Current trends in β-lactam based β-lactamases inhibitors. *Curr. Med. Chem.* **2011,** *18* (27), 4223–4236.
121. Qin, W.; Panunzio, M.; Biondi, S. β-Llactam antibiotics renaissance. *Antibiotics* **2014,** *3* (2), 193–215.
122. Drawz, S. M.; Papp-Wallace, K. M.; Bonomo, R. A. New β-lactamases inhibitors: a therapeutic renaissance in an MDR world. *Antimicrob. Agents Chemother.* **2014,** *58* (4), 1835–1846.
123. Maiti, S. N.; Phillips, O. A. β-Lactamase inhibitors and β-lactam antibiotics: patent highlights June 1998 to November (1998). *Curr. Opin. Anti-Infect. Invest. Drugs* **1999,** *1* (1), 40–44.
124. Phillips, O. A. β-Lactamase inhibitors: a survey of the patent literature 2000-(2004). *Expert Opin. Ther. Pat.* **2006,** *16* (3), 319–331.
125. Buynak, J. D. β-Lactamase inhibitors: a review of the patent literature 2010-2013. *Expert Opin. Ther. Pat.* **2013,** *23* (11), 1469–1481.
126. Chen, J.; Shang, X.; Hu, F.; Lao, X.; Gao, X.; Zheng, H.; Yao, W. β-Lactamase: an update. *Mini-Rev. Med. Chem.* **2013,** *13* (13), 1846–1861.

127. Worthington, R. J.; Melander, C. Overcoming resistance to β-lactam antibiotics. *J. Org. Chem.* **2013,** *78* (9), 4207–4213.
128. Coleman, K. Diazabicyclooctanes (DBOs): a potent new class of non-β-lactam β-lactamase inhibitors. *Curr. Opin. Microbiol.* **2011,** *14* (5), 550–555.
129. Maiti, S. N.; Babu, R. P. K.; Shan, R. Overcoming bacterial resistance: role of β-lactamase inhibitors. *Top. Heterocycl. Chem.* **2006,** *2* (Heterocyclic Antitumor Antibiotics), 207–246.
130. Kazmierczak, A. β-Llactamase inhibitors. In *Antimicrobial Agents: Antibacterials and Antifungals;* Bryskier, A., Ed.; ASM Press, 2005; pp 401–409.
131. Micetich, R. G.; Salama, S. M.; Maiti, S. N.; Reddy, A. V. N.; Singh, R. β-Lactamases and their inhibitors: an update. *Curr. Med. Chem.: Anti-Infect. Agents* **2002,** *1* (3), 193–213.
132. Page, M. G. P. β-Lactamase inhibitors. *Drug Resist. Updates* **2000,** *3* (2), 109–125.
133. Nelson, M. L.; Levy, S. B. The history of the tetracyclines. *Ann. N. Y. Acad. Sci.* **2011,** *1241,* 17–32.
134. Chopra, I.; Roberts, M. Tetracycline antibiotics: mode of action, applications, molecular biology, and epidemiology of bacterial resistance. *Microbiol. Mol. Biol. Rev.* **2001,** *65* (2), 232–260.
135. Chopra, I.; Hawkey, P. M.; Hinton, M. Tetracyclines, molecular and clinical aspects. *J. Antimicrob. Chemother.* **1992,** *29* (3), 245–277.
136. Klein, N. C.; Cunha, B. A. Tetracyclines. *Med. Clin. North Am.* **1995,** *79* (4), 789–801.
137. Behal, V.; Bucko, M.; Hostalek, Z. Tetracyclines. *Biotechnol. Ser.* **1983,** *2,* 255–276.
138. Bahrami, F.; Morris, D. L.; Pourgholami, M. H. Tetracyclines: drugs with huge therapeutic potential. *Mini-Rev. Med. Chem.* **2012,** *12* (1), 44–52.
139. Nelson, M. L.; Ismail, M. Y. The antibiotic and nonantibiotic tetracyclines. In *Comprehensive Medicinal Chemistry II,* Vol. 7; 8th ed.; Taylor, J. B., Triggle, D. J., Eds.; Elsevier, 2006; pp 597–628.
140. Shlaes, D. M. An update on tetracyclines. *Curr. Opin. Invest. Drugs (BioMed Cent.)* **2006,** *7* (2), 167–171.
141. Hostalek, Z.; Vanek, Z. Tetracyclines. In *Biotechnology. A comprehensive treatise in 8 volumes;* Vol. 4; Pape, H., Rehm, H.-J., Eds.; Verlagsgesellschaft, Weinheim, 1986; pp 393–429.
142. Finch, R. G. Tetracyclines. In *Antibiotic and Chemotherapy: Anti-Infective Agents and Their Use in Therapy,* 7th ed.; O'Grady, F., Lambert, H. P., Eds. Churchill Livingstone, 1997; pp 469–484.
143. Nelson, M. L. The chemistry and biology of the tetracyclines. *Annu. Rep. Med. Chem.* **2002,** *37,* 105–114.
144. Gupta, S.; Dodwad, V. Chemically modified tetracyclines: an emerging host modulatory therapy. *J. Pharm. Biomed. Sci.* **2012,** (21), 13.
145. Sum, P.-E.; Sum, F.-W.; Projan, S. J. Recent developments in tetracycline antibiotics. *Curr. Pharm. Des.* **1998,** *4* (2), 119–132.
146. Holmes, N. E.; Charles, P. G. P. Safety and efficacy review of doxycycline. *Clin. Med.: Ther.* **2009,** *1,* 471–482.
147. Speer, B. S.; Shoemaker, N. B.; Salyers, A. A. Bacterial resistance to tetracycline: mechanisms, transfer, and clinical significance. *Clin. Microbiol. Rev.* **1992,** *5* (4), 387–399.
148. Schnappinger, D.; Hillen, W. Tetracyclines. Antibiotic action, uptake, and resistance mechanisms. *Arch. Microbiol.* **1996,** *165* (6), 359–369.
149. Townsend, M. L.; Pound, M. W.; Drew, R. H. Tigecycline: a new glycylcycline antimicrobial. *Int. J. Clin. Pract.* **2006,** *60* (12), 1662–1672.
150. Pankey, G. A.; Tigecycline. *J. Antimicrob. Chemother.* **2005,** *56* (3), 470–480.

151. Zhanel, G. G.; Homenuik, K.; Nichol, K.; Noreddin, A.; Vercaigne, L.; Embil, J.; Gin, A.; Karlowsky, J. A.; Hoban, D. J. The glycylcyclines: a comparative review with the tetracyclines. *Drugs* **2004,** *64* (1), 63–88.
152. Chopra, I. Glycylcyclines: third-generation tetracycline antibiotics. *Curr. Opin. Pharmacol.* **2001,** *1* (5), 464–469.
153. Tally, F. T.; Ellestad, G. A.; Testa, R. T. Glycylcyclines: a new generation of tetracyclines. *J. Antimicrob. Chemother.* **1995,** *35* (4), 449–452.
154. Bahrami, F.; Morris, D. L.; Pourgholami, M. H. Tetracyclines: drugs with huge therapeutic potential. *Mini-Rev. Med. Chem.* **2012,** *12* (1), 44–52.
155. Garrido-Mesa, N.; Zarzuelo, A.; Galvez, J. Minocycline: far beyond an antibiotic. *Br. J. Pharmacol.* **2013,** *169* (2), 337–352.
156. Bryskier, A. Tetracyclines under investigation. In *Antimicrobial Agents: Antibacterials and Antifungals;* Bryskier, A., Ed.; ASM Press, 2005; pp 652–667.
157. Bryskier, A. In *Tetracyclines, Antimicrobial Agents: Antibacterials and Antifungals;* Bryskier, A., Ed.; ASM Press, 2005; pp 642–651.
158. McCormick, J. R. D.; Jensen, E. R. Catalytic reduction of 6-hydroxyhydronaphthacenes, US 3019260 (1963).
159. McCormick, J. R. D.; Jensen, E. R. 6-Deoxyhydronaphthacenes, DE 1082905 (1958).
160. Blackwood, R. K.; Beereboom, J. J.; Rennhard, H. H.; von Wittenau, M. S.; Stephens, C. R., Jr. 6-Methylenetetracyclines. III. Preparation and properties. *J. Am. Chem. Soc.* **1963,** *85* (24), 3943–3953.
161. Assignee: Chas. Pfizer & Co., Inc. 6-Epi-6-deoxytetracyclines. *GB* **1965,** *99* (5032).
162. Blackwood, R. K. 6-Methylene-5-oxytetracycline, US 3026354 (1962).
163. McCormick, J. R. D.; Jensen, E. R. Catalytic reduction of 6-hydroxyhydronaphthacenes, US 3019260 (1962).
164. Korst, J. J. α-6-Deoxytetracyclines, ZA 6800905 (1968).
165. Morris, T. A. 6-Deoxy-5-hydroxytetracycline, DE 2418499 (1974).
166. Faubl, H.; Belton, A. M. Rhodium-containing catalyst and use thereof in preparation of α-6-deoxy-5-oxytetracycline, US 3962131 (1976).
167. Scanio, C. J. V. α-6-Deoxy-5-hydroxytetracycline, US 3907890 (1975).
168. Cotti, G. α-6-Deoxytetracyclines, DE 2446587 (1975).
169. Faubl, H. α-6-Deoxy-5-hydroxytetracycline, US 4001321 (1977).
170. Broggi, R.; Cotti, G. α-6-Deoxytetracyclines, DE 2308227 (1973).
171. Page, P. R.; Villax, I. Homogeneous catalytic system for hydrogenation of methylenetetracyclines and a process for the preparation of same, US 4743699 (1988).
172. Brennan, T. M.; Faubl, H. Hydrogenation of the exocyclic methylene groups of a 6- methylenetetracycline, DE 2403714 (1974).
173. Krueger, W.; Rudolf, G.; Krause, H. W.; Kuhn, P. Process for the preparation of α-6-deoxy-5-hydroxytetracycline (α-doxycycline, Vibramycin) by hydrogenation of 6-deoxy-6-demethyl-6-methylene-5-hydroxytetracycline (methacycline) over a catalyst containing rhodium and a complex diphosphane ligand, DD 297809 (1992).
174. Conover, L. H.; Butler, K.; Johnston, J. D.; Korst, J. J.; Woodward, R. B. Total synthesis of 6-demethyl-6-deoxytetracycline. *J. Am. Chem. Soc.* **1962,** *84,* 3222–3224.
175. Tatsuta, K.; Yoshimoto, T.; Gunji, H.; Okado, Y.; Takahashi, M. The first total synthesis of natural (-)-tetracycline. *Chem. Lett.* **2000,** *6,* 646–647.
176. Gurevich, A. I.; Karapetyan, M. G.; Kolosov, M. N.; Korobko, V. G.; Onoprienko, V. V.; Popravko, S. A.; Shemyakin, M. M. Tetracycline series. XLIV. Synthesis of 12a-deoxy-5a,6-anhydrotetracycline. *The first synthesis of the naturally occurring tetracycline, Tetrahedron Lett.* **1967,** *2,* 131–134.

177. Muxfeldt, H.; Haas, G.; Hardtmann, G.; Kathawala, F.; Mooberry, J. B.; Vedejs, E. Tetracyclines. 9. Total synthesis of dl-Terramycin. *J. Am. Chem. Soc.* **1979,** *101* (3), 689–701.
178. Stork, G.; La Clair, J. J.; Spargo, P.; Nargund, R. P.; Totah, N. Stereocontrolled synthesis of (±)-12a-deoxytetracycline. *J. Am. Chem. Soc.* **1996,** *118* (22), 5304–5305.
179. Charest, M. G.; Lerner, C. D.; Brubaker, J. D.; Siegel, D. R.; Myers, A. G. A convergent enantioselective route to structurally diverse 6-deoxytetracycline antibiotics. *Science (Washington, DC, U. S.)* **2005,** *308* (5720), 395–398.
180. Brubaker, J. D.; Myers, A. G. A practical, enantioselective synthetic route to a key precursor to the tetracycline antibiotics. *Org. Lett.* **2007,** *9* (18), 3523–3525.
181. Kummer, D. A.; Li, D.; Dion, A.; Myers, A. G. A practical, convergent route to the key precursor to the tetracycline antibiotics. *Chem. Sci.* **2011,** *2* (9), 1710–1718.
182. Myers, A. G.; Charest, M. G.; Lerner, C. D.; Brubaker, J. D.; Siegel, D. R. Synthesis of tetracyclines and analogues thereof, US 7807842 (2010).
183. Kametani, T.; Fukumoto, K. Syntheses of linear tetracyclic antibiotics and anthracyclines. *Med. Res. Rev.* **1981,** *1* (1), 23–72.
184. Clive, D. L. J. Chemistry of tetracyclines. *Q. Rev., Chem. Soc.* **1968,** *22* (4). 435–56.
185. Kogawa, A. C.; Salgado, H. R. N. Doxycycline hyclate: a review of properties, applications and analytical methods. *Int. J. Life Sci. Pharma Res.* **2012,** *2* (4), 11–25.
186. Sagar, J. Doxycycline in clinical medicine. *Clin. Med. Insights: Ther.* **2010,** *2,* 133–136.
187. Joshi, N.; Miller, D. Q. Doxycycline revisited. *Arch. Intern. Med.* **1997,** *157* (13), 1421–1428.
188. Cunha, B. A.; Sibley, C. M.; Ristuccia, A. M. Doxycycline. *Ther. Drug Monit.* **1982,** *4* (2), 115–135.
189. Shehwaro, N.; Langlois, A.-L.; Gueutin, V.; Gauthier, M.; Casenave, M.; Izzedine, H. Doxycycline or how to create new with the old? *Therapie* **2014,** *69* (2), 129–141.
190. Dodd, B. R.; Spence, R. A. Doxycycline inhibition of abdominal aortic aneurysm growth a systematic review of the literature. *Curr. Vasc. Pharmacol.* **2011,** *9* (4), 471–478.
191. Holmes, N. E.; Charles, P. G. P. Safety and efficacy review of doxycycline. *Clin. Med.: Ther.* **2009,** *1,* 471–482.
192. Boothe, J. H.; Petisi, J. Reductive alkylation of tetracycline amines, US 3148212 (1964).
193. Petisi, J.; Boothe, J. H., 7- and 9-alkylamino-6-deoxytetracycline, US 3226436 (1965).
194. Martell, M. J., Jr.; Boothe, J. H. 6-Deoxytetracyclines. VII. Alkylated aminotetracyclines possessing unique antibacterial activity. *J. Med. Chem.* **1967,** *10* (1), 44–46.
195. Winterbottom, R.; Kissman, H. M. Substituted 7- and 9-amino tetracyclines, US 3345410 (1967).
196. Church, R. F. R.; Schaub, R. E.; Weiss, M. J. Synthesis of 7-dimethylamino-6-demethyl-6-deoxytetracycline (minocycline) via 9-nitro-6-demethyl-6-deoxytetracycline. *J. Org. Chem.* **1971,** *36* (5), 723–725.
197. Koza, D. J.; Nsiah, Y. A. Palladium catalyzed C-N bond formation in the synthesis of 7-amino-substituted tetracyclines. *J. Org. Chem.* **2002,** *67* (14), 5025–5027.
198. Sun, C.; Wang, Q.; Brubaker, J. D.; Wright, P. M.; Lerner, C. D.; Noson, K.; Charest, M.; Siegel, D. R.; Wang, Y.-M.; Myers, A. G. A robust platform for the synthesis of new tetracycline antibiotics. *J. Am. Chem. Soc.* **2008,** *130* (52), 17913–17927.
199. Jonas, M.; Cunha, B. A. Minocycline. *Ther. Drug Monit.* **1982,** *4* (2), 137–145.
200. Zbinovsky, V.; Chrekian, G. P. Minocycline. *Anal. Profiles Drug Subst.* **1977,** *6,* 323–339.
201. Allen, J. C. Minocycline. *Ann. Intern. Med.* **1976,** *85* (4), 482–487.
202. Bernier, C.; Dreno, B. Minocycline. *Ann. Dermatol. Venereol.* **2001,** *128* (5), 627–637.
203. Ochsendorf, F. Minocycline in acne vulgaris: benefits and risks. *Am. J. Clin. Dermatol.* **2010,** *11* (5), 327–341.
204. Garrido-Mesa, N.; Zarzuelo, A.; Galvez, J. Minocycline: far beyond an antibiotic. *Br. J. Pharmacol.* **2013,** *169* (2), 337–352.

205. Garrido-Mesa, N.; Zarzuelo, A.; Galvez, J. What is behind the non-antibiotic properties of minocycline. *Pharmacol. Res.* **2013,** *67* (1), 18–30.
206. Dean, O. M.; Data-Franco, J.; Giorlando, F.; Berk, M. Minocycline: therapeutic potential in psychiatry. *CNS Drugs* **2012,** *26* (5), 391–401.
207. Kim, H.-S.; Suh, Y.-H. Minocycline and neurodegenerative diseases. *Behav. Brain Res.* **2009,** *196* (2), 168–179.
208. Elewa, H. F.; Hilali, H.; Hess, D. C.; Machado, L. S.; Fagan, S. C. Minocycline for short-term neuroprotection. *Pharmacotherapy* **2006,** *26* (4), 515–521.
209. Stirling, D. P.; Koochesfahani, K. M.; Steeves, J. D.; Tetzlaff, W. Minocycline as a neuroprotective agent. *Neuroscientist* **2005,** *11* (4), 308–322.
210. Blum, D.; Chtarto, A.; Tenenbaum, L.; Brotchi, J.; Levivier, M. Clinical potential of minocycline for neurodegenerative disorders. *Neurobiol. Dis.* **2004,** *17* (3), 359–366.
211. Sharma, G. V. M.; Doddi, V. R. Macrolactones. In *Natural Lactones and Lactams: Synthesis, Occurrence and Biological Activity;* Janecki, T., Ed.; Wiley-VCH, 2014; pp 229–272.
212. Kirst, H. A. Introduction to the macrolide antibiotics. In *Macrolide Antibiotics;* Schoenfeld, W., Kirst, H. A., Eds.; Springer-Verlag, 2002; pp 1–13.
213. Kirst, H. A. Recent developments with macrolide antibiotics. *Expert Opin. Ther. Pat.* **1988,** *8* (2), 111–120.
214. Williams, J. D.; Sefton, A. M. Comparison of macrolide antibiotics. *J. Antimicrob. Chemother.* **1993,** *31* (Suppl. C), 11–26.
215. Labro, M.-T. Macrolide antibiotics: current and future uses. *Expert Opin. Pharmacother.* **2004,** *5* (3), 541–550.
216. Omura, S., Ed. *Macrolide Antibiotics: Chemistry, Biology, and Practice,* 2nd ed.; Academic Press, 2002.
217. Iacoviello, V. R.; Zinner, S. H. Macrolides: a clinical overview. In *Macrolide Antibiotics;* Schoenfeld, W., Kirst, H. A., Eds.; Springer-Verlag, 2002; pp 15–24.
218. Katz, L.; Donadio, S. Macrolides. *Biotechnol. Ser.* **1995,** *28* (Genetics and Biochemistry of Antibiotic Production), 385–420.
219. Tatsuta, K. Total synthesis of macrolide antibiotics. In *Recent Progress in the Chemical Synthesis of Antibiotics;* Lukacs, G., Ohno, M., Eds.; Springer, 1990; pp 1–38.
220. Kirst, H. A. Structural modification of macrolide antibiotics. In *Recent Progress in the Chemical Synthesis of Antibiotics;* Lukacs, G., Ohno, M., Eds.; Springer, 1990; pp 39–63.
221. Kirst, H. A. Macrolide antibiotics. *Annu. Rep. Med. Chem.* **1990,** *25,* 119–128.
222. Omura, S.; Tanaka, Y. Macrolide antibiotics. In *Biotechnology;* Pape, H., Rehm, H.-J., Eds.; Vol. 4, Academic Press Inc., 1986; pp 359–391.
223. Omura, S., Ed. *Macrolide Antibiotics: Chemistry, Biology, and Practice;* Academic Press, 1984.
224. Kobayashi, Y. Macrolide antibiotics in Japan. *Drugs Today* **1987,** *23* (3), 159–162.
225. Henninger, T. C. Recent progress in the field of macrolide antibiotics. *Expert Opin. Ther. Pat.* **2003,** *1396,* 787–805.
226. Kaneko, T.; McArthur, H.; Sutcliffe, J. Recent developments in the area of macrolide antibiotics. *Expert Opin. Ther. Pat.* **2000,** *10* (4), 403–425.
227. Fernandes, P. Use of antibiotic core structures to generate new and useful macrolide antibiotics. In *Antibiotics: Current Innovations and Future Trends;* Sanchez, S., Demain, A. L., Eds.; Caister Academic Press, 2015; pp 375–393.
228. Pechere, J.-C. New perspectives on macrolide antibiotics. *Int. J. Antimicrob. Agents* **2001,** *18* (Suppl. 1), S93–S97.
229. Bryskier, A.; Agouridas, C. Macrolide antibiotics. Structure-activity relationship and new medical opportunities. *Antiinfect. Drugs Chemother.* **1996,** *14* (2), 147–154.

230. Bahal, N.; Nahata, M. C. The new macrolide antibiotics: azithromycin, clarithromycin, dirithromycin, and roxithromycin. *Ann. Pharmacother.* **1992,** *26* (1), 46–55.
231. Kirst, H. A.; Sides, G. D. New directions for macrolide antibiotics: structural modifications and in vitro activity. *Antimicrob. Agents Chemother.* **1989,** *33* (9), 1413–1418.
232. Masamune, S.; Bates, G. S.; Corcoran, J. W. Macrolides. Recent advances in their chemistry and biochemistry. *Angew. Chem.* **1977,** *89* (9), 602–624.
233. Mankin, A. S. Macrolide myths. *Curr. Opin. Microbiol.* **2008,** *11* (5), 414–421.
234. Kirst, H. A. Semi-synthetic derivatives of 16-membered macrolide antibiotics. *Prog. Med. Chem.* **1994,** *31,* 265–295.
235. Ajito, K.; Miura, T.; Furuuchi, T.; Tamura, A. Sixteen-membered macrolides: chemical modifications and future applications. *Heterocycles* **2014,** *89* (2), 281–352.
236. Cui, W.; Ma, S. Recent advances in the field of 16-membered macrolide antibiotics. *Mini-Rev. Med. Chem.* **2011,** *11* (12), 1009–1018.
237. Przybylski, P. Modifications and biological activity of natural and semisynthetic 16-membered macrolide antibiotics. *Curr. Org. Chem.* **2011,** *15* (3), 328–374.
238. Nakajima, Y. Mechanisms of bacterial resistance to macrolide antibiotics. *J. Infect. Chemother.* **1999,** *5* (2), 61–74.
239. Hamilton-Miller, J. M. T. Chemistry and biology of the polyene macrolide antibiotics. *Bacteriol. Rev.* **1973,** *37* (2), 166–196.
240. Solovieva, S. E.; Olsufyeva, E. N.; Preobrazhenskaya, M. N. Chemical modification of antifungal polyene macrolide antibiotics. *Russ. Chem. Rev.* **2011,** *80* (2), 103–126.
241. Kirst, H. A.; Allen, N. E. Aminoglycosides antibiotics. In *Comprehensive Medicinal Chemistry II*, Vol. 1; 8th ed.; Mandell, G. L., Bennett, J., Dolin, R., Eds.; Churchill Livingstone, 2009; pp 359–385.
242. Umezawa, S.; Kondo, S.; Ito, Y. Aminoglycoside antibiotics. In Pape, H., Rehm, H.-J., Eds.; *Biotechnology*, Vol. 4; Academic Press Inc., 1986. (309–307.
243. Davies, J. E. Aminoglycosides: ancient and modern. *J. Antibiot.* **2006,** *59* (9), 529–532.
244. Jackson, J.; Chen, C.; Buising, K. Aminoglycosides: how should we use them in the 21st century? *Curr. Opin. Infect. Dis.* **2013,** *26* (6), 516–525.
245. Houghton, J. L.; Green, K. D.; Chen, W.; Garneau-Tsodikova, S. The future of aminoglycosides: the end or renaissance? *ChemBioChem* **2010,** *11* (7), 880–902.
246. Durante-Mangoni, E.; Grammatikos, A.; Utili, R.; Falagas, M. E. Do we still need the aminoglycosides? *Int. J. Antimicrob. Agents* **2009,** *33* (3), 201–205.
247. Gilbert, D. N. Aminoglycosides. In *Mandell, Douglas, and Bennett's Principles and Practice of Infectious Diseases*, Vol. 1; 7th ed.; Mandell, G. L., Bennett, J., Dolin, R., Eds.; Churchill Livingstone, 2009; pp 359–385.
248. Guo, L.; Wan, Y.; Wang, X.; Wang, P. G.; Zhao, W. Development of aminoglycoside antibiotics by carbohydrate chemistry. *Mini-Rev. Med. Chem.* **2012,** *12* (14), 1533–1541.
249. Becker, B.; Cooper, M. A. Aminoglycoside antibiotics in the 21st Century. *ACS Chem. Biol.* **2013,** *8* (1), 105–115.
250. Pagkalis, S.; Mantadakis, E.; Mavros, M. N.; Ammari, C.; Falagas, M. E. Pharmacological considerations for the proper clinical use of aminoglycosides. *Drugs* **2011,** *71* (17), 2277–2294.
251. Kim, M.-K.; Nicolau, D. P. Aminoglycosides. In *Antimicrobial Pharmacodynamics in Theory and Clinical Practice, Second Edition (Infectious Disease and Therapy) Nightingale*, Vol. 4; C. H., Ambrose, P. G., Drusano, G. L., Murakawa, T., Eds.; CRC Press, 2007; pp 147–175.
252. Hanberger, H.; Edlund, C.; Furebring, M.; Giske, C. G.; Melhus, A.; Nilsson, L. E.; Petersson, J.; Sjoelin, J.; Ternhag, A.; Werner, M.; Eliasson, E. Rational use of aminoglycosides—review and recommendations by the Swedish Reference Group for Antibiotics (SRGA). *Scand. J. Infect. Dis.* **2013,** *45* (3), 161–175.

253. Avent, M. L.; Rogers, B. A.; Cheng, A. C.; Paterson, D. L. Current use of aminoglycosides: indications, pharmacokinetics and monitoring for toxicity. *Intern. Med. J.* **2011,** *41* (6), 441–449.
254. Arya, D. P.; Shaw, N.; Xi, H. Novel targets for aminoglycosides. In *Aminoglicoside Antibiotics: From Chemical Biology to Drug Discovery;* Arya, D. P., Ed.; John Wiley & Sons, Inc., 2007; pp 289–314 (plate 1).
255. Berkov-Zrihen, Y.; Fridman, M. Synthesis of aminoglycosides. In *Modern Synthetic Methods in Carbohydrate Chemistry: From Monosaccharides to Complex Glycoconjugates;* Werz, D. B., Vidal, S., Eds.; Wiley-VCH, 2014; pp 161–190.
256. Ganz, T.; Lehrer, R. I. Antibiotic peptides from higher eukaryotes: biology and applications. *Mol. Med. Today* **1999,** *5* (7), 292–297.
257. Boman, H. G. Peptide antibiotics and their role in innate immunity. *Annu. Rev. Immunol.* **1955,** *13,* 61–92.
258. Dutton, C. J., Haxell, M. A., McArthur, H. A. I., Wax, R. G., Eds. *Peptide Antibiotics: Discovery, Modes of Action and Applications;* Marcel Dekker, 2002.
259. Conde, R.; Arguello, M.; Izquierdo, J.; Noguez, R.; Moreno, M.; Lanz, H. Natural antimicrobial peptides from eukaryotic organisms. In *Antimicrobial Agents;* Bobbarala, V., Ed.; InTech, 2012; pp 51–72.
260. Rossi, L. M.; Rangasamy, P.; Zhang, J.; Qiu, X.-Q.; Wu, G. Y. Research advances in the development of peptide antibiotics. *J. Pharm. Sci.* **2008,** *97* (3), 1060–1070.
261. Dubin, A.; Mak, P.; Dubin, G.; Rzychon, M.; Stec-Niemczyk, J.; Wladyka, B.; Maziarka, K.; Chmiel, D. New generation of peptide antibiotics. *Acta Biochim. Pol.* **2005,** *52* (3), 633–638.
262. Jenssen, H. Clinical development of peptide antibiotics. *PharmaChem* **2009,** *8* (10), 22–26.
263. Stachelhaus, T. The road to new peptide antibiotics. *Bioforum* **2003,** *26* (10), 626–628.
264. Kamysz, W.; Okroj, M.; Lukasiak, J. Novel properties of antimicrobial peptides. *Acta Biochim. Pol.* **2003,** *50* (2), 461–469.
265. Vaara, M. New approaches in peptide antibiotics. *Curr. Opin. Pharmacol.* **2009,** *9* (5), 571–576.
266. Bryskier, A. Peptide antibiotics. In *Antimicrobial Agents: Antibacterials and Antifungals;* Bryskier, A., Ed.; ASM Press, 2005; pp 826–879.
267. Hancock, R. E. W. Peptide antibiotics. *Lancet* **1997,** *349* (9049), 418–422.
268. Koczulla, A. R.; Bals, R. Antimicrobial peptides: current status and therapeutic potential. *Drugs* **2003,** *63* (4), 389–406.
269. Hancock, R. E. W.; Chapple, D. S. Peptide antibiotics. *Antimicrob. Agents Chemother.* **1999,** *43* (6), 1317–1323.
270. Zhang, L.; Hancock, R. E. W. Peptide antibiotics. In *Antibiotic Development and Resistance;* Hughes, D., Andersson, D. I., Eds.; Harwood Academic, 2001; pp 209–232.
271. Nicolas, P.; Mor, A. Peptides as weapons against microorganisms in the chemical defense system of vertebrates. *Ann. Rev. Microbiol.* **1995,** *49,* 227–304.
272. Sima, P.; Trebichavsky, I.; Sigler, K. Mammalian antibiotic peptides. *Folia Microbiol. (Dordrecht, Neth.)* **2003,** *48* (2), 123–137.
273. Mankelow, D. P.; Neilan, B. A. Non-ribosomal peptide antibiotics. *Expert Opin. Ther. Pat.* **2000,** *10* (10), 1583–1591.
274. Singh, N.; Abraham, J. Ribosomally synthesized peptides from natural sources. *J. Antibiot.* **2014,** *67* (4), 277–289.
275. Tavano, R.; Malachin, G.; De Zotti, M.; Peggion, C.; Biondi, B.; Formaggio, F.; Papini, E. The peculiar N- and C-termini of trichogin GA IV are needed for membrane interaction and human cell death induction at doses lacking antibiotic activity. *Biochim. Biophys. Acta, Biomembr.* **2015,** *1848* (Part 1), 134–144.

276. Ramachandran, L. K. Gramicidins. *J. Sci. Ind. Res.* **1975,** *34* (5), 249–265.
277. Mogi, T.; Kita, K. Gramicidin S and polymyxins: the revival of cationic cyclic peptide antibiotics. *Cell. Mol. Life Sci.* **2009,** *66* (23), 3821–3826.
278. Velkov, T.; Thompson, P. E.; Nation, R. L.; Li, J. Structure-activity relationships of polymyxin antibiotics. *J. Med. Chem.* **2010,** *53* (5), 1898–1916.
279. Kassamali, Z.; Jain, R.; Danziger, L. H. An update on the arsenal for multidrug-resistant *Acinetobacter* infections: polymyxin antibiotics. *Int. J. Infect. Dis.* **2014,** *30C,* 125–132.
280. Toscano, W. A., Jr.; Storm, D. R. Bacitracin. *Pharmacol. Ther.* **1982,** *16* (2), 199–210.
281. Weinberg, E. D. Bacitracin. *Antibiotics (USSR)* **1967,** *1,* 90–101.
282. Gorman, J. J. Vancomycin group antibiotics: from biosynthesis to improving on nature's design. *Chemtracts* **2008,** *21* (11), 415–455.
283. Usach, I.; Melis, V.; Peris, J.-E. Vancomycin: use, dosing and therapeutic drug monitoring. In *Vancomycin: Biosynthesis, Clinical Uses and Adverse Effects;* Hossion, A. G., Ed.; Nova Science, 2014; pp 55–79.
284. Mauger, A. B.; Lackner, H. The actinomycins. In *Anticancer Agents from Natural Products;* Cragg, G. M., Kingston, D. G. I., Newman, D. J., Eds.; CRC Press, 2005; pp 281–297.
285. Graves, D. E. Actinomycin D: sixty years of progress in characterizing a sequence-selective DNA-binding agent. In *Sequence-Specific DNA Binding Agents;* Waring, M., Ed.; RSC Publishing, 2006; pp 109–129.
286. De Smet, K.; Contreras, R. Human antimicrobial peptides: defensins, cathelicidins and histatins. *Biotechnol. Lett.* **2005,** *27* (18), 1337–1347.
287. Lehrer, R. I.; Ganz, T. Antimicrobial peptides in mammalian and insect host defense. *Curr. Opin. Immunol.* **1999,** *11* (1), 23–27.
288. Hazlett, L.; Wu, M. Defensins in innate immunity. *Cell Tissue Res.* **2011,** *343* (1), 175–188.
289. Antcheva, N.; Guida, F.; Tossi, A. Defensins. In *Handbook of Biologically Active Peptides,* 2nd ed.; Kastin, A. J., Ed.; Academic Press, 2013; pp 101–118.
290. Jarczak, J.; Kosciuczuk, E. M.; Lisowski, P.; Strzalkowska, N.; Jozwik, A.; Horbanczuk, J.; Krzyzewski, J.; Zwierzchowski, L.; Bagnicka, E. Defensins: natural component of human innate immunity. *Hum. Immunol.* **2013,** *74* (9), 1069–1079.
291. Olvera, D. P. R.; Gutierrez, C. C.; Ramirez, J. I. P.; Zavala, M. E. M. Defensins: characteristics, mechanisms of action and viral infection. *Curr. Top. Virol.* **2012,** *10,* 39–51.
292. Wilson, S. S.; Wiens, M. E.; Smith, J. G. Antiviral mechanisms of human defensins. *J. Mol. Biol.* **2013,** *425* (24), 4965–4980.
293. Lehrer, R. I.; Lu, W. α-Defensins in human innate immunity. *Immunol. Rev.* **2012,** *245* (1), 84–112.
294. Ouellette, A. J. Paneth cell α-defensins in enteric innate immunity. *Cell. Mol. Life Sci.* **2011,** *68* (13), 2215–2229.
295. Semple, F.; Dorin, J. R. β-Defensins: multifunctional modulators of infection, inflammation and more? *J. Innate Immun.* **2012,** *4* (4), 337–348.
296. Conibear, A. C.; Craik, D. J. The chemistry and biology of theta defensins. *Angew. Chem., Int. Ed.* **2014,** *53* (40), 10612–10623.
297. Lehrer, R. I.; Cole, A. M.; Selsted, M. E. θ-Defensins: cyclic peptides with endless potential. *J. Biol. Chem.* **2012,** *287* (32), 27014–27019.
298. Mannion, B. A.; Weiss, J.; Elsbach, P. Separation of sublethal and lethal effects of the bactericidal/permeability increasing) proteins on *Escherichia coli. J. Clin. Invest.* **1990,** *85* (3), 853–860.
299. Ehrchen, J. M.; Sunderkoetter, C.; Foell, D.; Vogl, T.; Roth, J. The endogenous toll-like receptor 4 agonist S100A8/S100A9 (calprotectin) as innate amplifier of infection, autoimmunity, and cancer. *J. Leukocyte Biol.* **2009,** *86* (3), 557–566.

300. van der Does, A. M.; Bergman, P.; Agerberth, B.; Lindbom, L. Induction of the human cathelicidin LL-37 as a novel treatment against bacterial infections. *J. Leukocyte Biol.* **2012,** *92* (4), 735–742.
301. Mendez-Samperio, P. The human cathelicidin hCAP18/LL-37: a multifunctional peptide involved in mycobacterial infections. *Peptides* **2010,** *31* (9), 1791–1798.
302. Melino, S.; Santone, C.; Di Nardo, P.; Sarkar, B. Histatins: salivary peptides with copper(II)- and zinc(II)-binding motifs. Perspectives for biomedical applications. *FEBS J.* **2014,** *281* (3), 657–672.
303. van't Hof, W.; Oudhoff, M. J.; Veerman, E. C. I. Histatins: multifunctional salivary antimicrobial peptides. In *Antimicrobial Peptides and Innate Immunity;* Hiemstra, P. S., Zaat, S. A. J., Eds.; Springer, 2013; pp 167–181.
304. Kavanagh, K.; Dowd, S. Histatins: antimicrobial peptides with therapeutic potential. *J. Pharm. Pharmacol.* **2004,** *56* (3), 285–289.
305. Morgenthau, A.; Pogoutse, A.; Adamiak, P.; Moraes, T. F.; Schryvers, A. B. Bacterial receptors for host transferrin and lactoferrin: molecular mechanisms and role in host-microbe interactions. *Future Microbiol.* **2013,** *8* (12), 1575–1585.
306. Legrand, D. Lactoferrin, a key molecule in immune and inflammatory processes. *Biochem. Cell Biol.* **2012,** *90* (3), 252–268.
307. Brock, J. H. Lactoferrin-50 years on. *Biochem. Cell Biol.* **2012,** *90* (3), 245–251.
308. Dumoulin, M.; Johnson, R. J. K.; Bellotti, V.; Dobson, C. M. Human lysozyme. *Protein Rev.* **2007,** 285–308 (3 plates).
309. Biziulevicius, G. A.; Biziuleviciene, G.; Kazlauskaite, J. Lysozyme and similar lytic enzyme preparations should be considered antibiotics. *Med. Hypotheses* **2007,** *68* (6), 1420.
310. Samy, R. P.; Gopalakrishnakone, P.; Stiles, B. G.; Girish, K. S.; Swamy, S. N.; Hemshekhar, M.; Tan, K. S.; Rowan, E. G.; Sethi, G.; Chow, V. T. K. Snake venom phospholipases A2: a novel tool against bacterial diseases. *Curr. Med. Chem.* **2012,** *19* (36), 6150–6162.
311. Nevalainen, T. J.; Graham, G. G.; Scott, K. F. Antibacterial actions of secreted phospholipases A2. Review. *Biochim. Biophys. Acta, Mol. Cell Biol. Lipids* **2008,** *1781* (1–2), 1–9.
312. Buckland, A. G.; Wilton, D. C. The antibacterial properties of secreted phospholipases A2. *Biochim. Biophys. Acta, Mol. Cell Biol. Lipids* **2000,** *1488* (1–2), 71–82.
313. Levy, O. Antibiotic proteins of polymorphonuclear leukocytes. *Eur. J. Haematol.* **1996,** *56* (5), 263–277.
314. Kwakman, P. H. S.; Krijgsveld, J.; de Boer, L.; Nguyen, L. T.; Boszhard, L.; Vreede, J.; Dekker, H. L.; Speijer, D.; Drijfhout, J. W.; te Velde, A. A.; Crielaard, W.; Vogel, H. J.; Vandenbroucke-Grauls, C. M. J.E.; Zaat, S. A. J. Native thrombocidin-1 and unfolded thrombocidin-1 exert antimicrobial activity via distinct structural elements. *J. Biol. Chem.* **2011,** *286* (50), 43506–43514.
315. Krijgsveld, J.; Zaat, S. A. J.; Meeldijk, J.; Van Veelen, P. A.; Fang, G.; Poolman, B.; Brandt, E.; Ehlert, J. E.; Kuijpers, A. J.; Engbers, G. H. M.; Feijen, J.; Dankert, J. Thrombocidins, microbicidal proteins from human blood platelets, are C-terminal deletion products of CXC chemokines. *J. Biol. Chem.* **2000,** *275* (27), 20374–20381.
316. Kumar, J.; Okada, S.; Clayberger, C.; Krensky, A. M. Granulysin. A novel antimicrobial. *Expert Opin. Invest. Drugs* **2001,** *10* (2), 321–329.
317. Krensky, A. M.; Clayberger, C. Biology and clinical relevance of granulysin. *Tissue Antigens* **2009,** *73* (3), 193–198.
318. Singh, M.; Mukhopadhyay, K. C-terminal amino acids of alpha-melanocyte stimulating hormone are requisite for its antibacterial activity against Staphylococcus aureus. *Antimicrob. Agents Chemother.* **2011,** *55* (5), 1920–1929.

319. Lohner, K.; Blondelle, S. E. Molecular mechanisms of membrane perturbation by antimicrobial peptides and the use of biophysical studies in the design of novel peptide antibiotics. *Comb. Chem. High Throughput Screening* **2005,** *8* (3), 241–256.
320. McCafferty, D. G.; Cudic, P.; Yu, M. K.; Behenna, D. C.; Kruger, R. Synergy and duality in peptide antibiotic mechanisms. *Curr. Opin. Chem. Biol.* **1999,** *3* (6), 672–680.
321. Andreu, D.; Rivas, L. Animal antimicrobial peptides: an overview. *Biopolymers* **1999,** *47* (6), 415–433.
322. Ginsburg, I. Bactericidal cationic peptides can also function as bacteriolysis-inducing agents mimicking beta-lactam antibiotics? it is enigmatic why this concept is consistently disregarded. *Med. Hypotheses* **2004,** *62* (3), 367–374.
323. McCormick, M. H.; Stark, W. M.; Pittenger, G. E.; Pittenger, R. C.; McGuire, J. M. Vancomycin, a new antibiotic. I. Chemical and biologic properties. *Antibiot. Annu.* **1955-1956,** 606–611.
324. Levine, J. F. Vancomycin: a review. *Med. Clin. North Am.* **1987,** *71* (6), 1135–1145.
325. Kahne, D.; Leimkuhler, C.; Lu, W.; Walsh, C. Glycopeptide and lipoglycopeptide antibiotics. *Chem. Rev. (Washington, DC, U. S.)* **2005,** *105* (2), 425–448.
326. James, R. C.; Pierce, J. G.; Okano, A.; Xie, J.; Boger, D. L. Redesign of glycopeptide antibiotics: back to the future. *ACS Chem. Biol.* **2012,** *7* (5), 797–804.
327. Butler, M. S.; Hansford, K. A.; Blaskovich, M. A. T.; Halai, R.; Cooper, M. A. Glycopeptide antibiotics: back to the future. *J. Antibiot.* **2014,** *67* (9), 631–644.
328. Tan, H.; Guo, H.; Wang, S.; Kong, Q.; Li, W.; Zeng, W. Chemistry and Biology of Glycopeptides with Antibiotic Activity. *Protein & Peptide Letters* **2014,** *21* (10), 1031–1047.
329. Ashford, P.-A.; Bew, S. P. Recent advances in the synthesis of new glycopeptide antibiotics. *Chem. Soc. Rev.* **2012,** *41* (3), 957–978.
330. Evans, D. A.; Wood, M. R.; Trotter, B. W.; Richardson, T. I.; Barrow, J. C.; Katz, J. L. Total syntheses of vancomycin and eremomycin aglycons. *Angew. Chem., Int. Ed.* **1998,** *37* (19), 2700–2704.
331. Thoresen, L. H.; Burgess, K. Total synthesis of vancomycin. In *Organic Synthesis Highlights V;* Schmalz, H.-G., Wirth, T., Eds.; Wiley-VCH, 2003; pp 297–306.
332. Nicolaou, K. C.; Boddy, C. N. C.; Brase, S.; Winssinger, N. Chemistry, biology, and medicine of the glycopeptide antibiotics. *Angew. Chem., Int. Ed.* **1999,** *38* (15), 2097–2152.
333. Zhang, A. J.; Burgess, K. Total synthesis of vancomycin. *Angew. Chem., Int. Ed.* **1999,** *38* (5), 634–636.
334. Rao, A. V. R. Studies directed on the synthesis of vancomycin and related cyclic peptides. *Pure Appl. Chem.* **1998,** *70* (2), 391–396.
335. Hubbard, B. K.; Walsh, C. T. Vancomycin assembly: nature's way. *Angew. Chem., Int. Ed.* **2003,** *42* (7), 730–765.
336. Boger, D. L. Vancomycin, teicoplanin, and ramoplanin: synthetic and mechanistic studies. *Med. Res. Rev.* **2001,** *21* (5), 356–381.
337. An antibiotic vancomycin and its production by fermentation, GB 795289 (1958).
338. Kim, S. Y.; Kim, D. S.; Jung, H. M.; Lee, J. K. Mutant strain of *Amycolatopsis orientalis* and process for preparing vancomycin hydrochloride, US 20080193986 (2008).
339. Bryskier, A. Lincosamines. In *Antimicrobial Agents: Antibacterials and Antifungals;* Bryskier, A., Ed.; ASM Press, 2005; pp 592–603.
340. Rezanka, T.; Spizek, J.; Sigler, K. Medicinal use of lincosamides and microbial resistance to them. *Anti-Infect. Agents Med. Chem.* **2007,** *6* (2), 133–144.
341. Greenwood, D. In *Lincosamides, Antibiotic and Chemotherapy: Anti-Infective Agents and Their Use in Therapy,* 7th ed.; O'Grady, F., Lambert, H. P., Eds. Churchill Livingstone, 1997; pp 371–376.

342. Spizek, J.; Rezanka, T. Lincomycin, clindamycin and their applications. *Appl. Microbiol. Biotechnol.* **2004,** *64* (4), 455–464.
343. Rimmer, D. M. D.; Sales, J. E. L. Lincomycin and clindamycin. *Antibiot. Chemother. (Basel)* **1978,** *25,* 204–216.
344. Howarth, G. B.; Szarek, W. A.; Jones, J. K. N. Synthesis of lincomycin. *J. Chem. Soc. C* **1970,** *16,* 2218–2224.
345. Knapp, S.; Kukkola, P. J. Stereocontrolled lincomycin synthesis. *J. Org. Chem.* **1990,** *55* (5), 1632–1636.
346. Spizek, J.; Rezanka, T. Lincomycin, cultivation of producing strains and biosynthesis. *Appl. Microbiol. Biotechnol.* **2004,** *63* (5), 510–519.
347. Spizek, J.; Novotna, J.; Rezanka, T. Lincosamides: chemical structure, biosynthesis, mechanism of action, resistance, and applications. *Adv. Appl. Microbiol.* **2004,** *56,* 121–154.
348. Birkenmeyer, R. D.; Kagan, F. Lincomycin. XI. Synthesis and structure of clindamycin, a potent antibacterial agent. *J. Med. Chem.* **1970,** *13* (4), 616–619.
349. Bowden, K.; Stevens, G. P. An alternative synthesis of clindamycin. *J. Serb. Chem. Soc.* **2000,** *65* (10), 691–694.
350. Li, Z.; Zhang, Y.; Lin, M.; Ouyang, P.; Ge, J.; Liu, Z. Lipase-catalyzed one-step and regioselective synthesis of clindamycin palmitate. *Org. Process Res. Dev.* **2013,** *17* (9), 1179–1182.
351. Sinkula, A. A.; Morozowich, W.; Rowe, E. L. Chemical modification of clindamycin: synthesis and evaluation of selected esters. *J. Pharm. Sci.* **1973,** *62* (7), 1106–1111.
352. Achmatowicz, O.; Szechner, B. Lincomycin: an organic chemistry perspective. In *Carbohydrates in Drug Design;* Witczak, Z. J., Nieforth, K. A., Eds.; Marcel Dekker, 1997; pp 579–653.
353. Golebiowski, A.; Jurczak, J. Total synthesis of lincomycin and related chemistry. In *Recent Progress in the Chemical Synthesis of Antibiotics;* Lukacs, G., Ohno, M., Eds.; Springer, 1990; pp 365–385.
354. Prelog, V.; Oppoolzer, W. Rifamycins. 4. Ansamycins, a novel class of microbial metabolism products. *Helv. Chim. Acta* **1973,** *56* (7), 2279–2287.
355. Isobe, M. Ansamacrolides. In *Recent Progress in the Chemical Synthesis of Antibiotics;* Lukacs, G., Ohno, M., Eds.; Springer, 1990; pp 103–134.
356. Lancini, G., Ed. *Ansamycins, Biotechnology;* Vol. 4, Pape, H., Rehm, H.-J., Eds. Academic Press Inc., 1986; pp 431–463.
357. Rinehart, K. L., Jr.; Shield, L. S. Chemistry of the ansamycin antibiotics. *Fortschr. Chem. Org. Naturst.* **1976,** *33,* 231–307.
358. Funayama, S.; Cordell, G. A. Ansamycin antibiotics discovery, classification, biosynthesis and biological activities. *Stud. Nat. Prod. Chem.* **2000,** *23* (Part D), 51–106.
359. Wrona, I. E.; Agouridas, V.; Panek, J. S. Design and synthesis of ansamycin antibiotics. *C. R. Chim.* **2008,** *11* (11–12), 1483–1522.
360. Neckers, L.; Schulte, T. W.; Mimnaugh, E. Geldanamycin as a potential anticancer agent: Its molecular target and biochemical activity. *Invest. New Drugs* **1999,** *17* (4), 361–373.
361. Ochel, H.-J.; Eichhorn, K.; Gademann, G. Geldanamycin: the prototype of a class of antitumor drugs targeting the heat shock protein 90 family of molecular chaperones. *Cell Stress Chaperones* **2001,** *6* (2), 105–112.
362. Franke, J.; Eichner, S.; Zeilinger, C.; Kirschning, A. Targeting heat-shock-protein 90 (Hsp90) by natural products: geldanamycin, a show case in cancer therapy. *Nat. Prod. Rep.* **2013,** *30* (10), 1299–1323.
363. Kabakov, A. E. Geldanamycin derivatives as promising anticancer drugs: therapy via Hsp90 inhibition. In *Anticancer Drugs: Design, Delivery and Pharmacology;* Spencer, P., Holt, W., Eds.; Nova Science, 2009; pp 87–113.

364. Fukuyo, Y.; Hunt, C. R.; Horikoshi, N. Geldanamycin and its anti-cancer activities. *Cancer Lett. (N. Y., NY, U. S.)* **2010,** *290* (1), 24–35.
365. Floss, H. G.; Yu, T.-W. Rifamycin—mode of action, resistance, and biosynthesis. *Chem. Rev. (Washington, DC, U. S.)* **2005,** *105* (2), 621–632.
366. Lancini, G.; Cavalleri, B.; Rifamycins In *Biotechnology of antibiotics,* 2nd ed.; Strohl, W. R., Ed. Series, Drugs and the pharmaceutical sciences, *82,* M. Dekker: New York, 1997; pp 521–549.
367. Parenti, F.; Lancini, G. Rifamycins. In *Antibiotic and Chemotherapy: Anti-Infective Agents and Their Use in Therapy,* 7th ed.; O'Grady, F., Lambert, H. P., Eds. Churchill Livingstone, 1997; pp 453–459.
368. Chiao, J. S.; Xia, T. H.; Mei, B. G.; Jin, Z. K.; Gu, W. L. Rifamycin SV and related ansamycins. *Biotechnol. Ser.* **1995,** *28* (Genetics and Biochemistry of Antibiotic Production), 477–489.
369. Pechere, J. C. Streptogramins. In *Antibiotic and Chemotherapy: Anti-Infective Agents and Their Use in Therapy,* 7th ed.; O'Grady, F., Lambert, H. P., Eds. Churchill Livingstone, 1997; pp 416–418.
370. Pechere, J. C. Streptogramins: a unique class of antibiotics. *Drugs* **1996,** *51* (Suppl. 1), 13–19.
371. Mast, Y.; Wohlleben, W. Streptogramins—two are better than one!. *Int. J. Med. Microbiol.* **2014,** *304* (1), 44–50.
372. Kreter, B.; Dowzicky, M. Research in streptogramins. *Infect. Dis. Ther.* **2000,** *23,* 109–116.
373. Creixell, C. ,E.; Juarez, G. J. C. Streptogramins: current status. *Aten. Farm. (1999-2012)* **1999,** *1* (1), 28–30, 32–34, 36–37.
374. Barriere, J. C.; Bouanchaud, D. H.; Desnottes, J. F.; Paris, J. M. Streptogramin analogs. *Expert Opin. Invest. Drugs* **1994,** *3* (2), 115–131.
375. Khosla, R.; Verma, D. D.; Kapur, A.; Aruna, R. V.; Khanna, N. Streptogramins: a new class of antibiotics. *Indian J. Med. Sci.* **1999,** *53* (3), 111–119.
376. Bonfiglio, G.; Furneri, P. M. Novel streptogramin antibiotics. *Expert Opin. Invest. Drugs* **2001,** *10* (2), 185–198.
377. Barriere, J. C.; Berthaud, N.; Beyer, D.; Dutka-Malen, S.; Paris, J. M.; Desnottes, J. F. Recent developments in streptogramin research. *Curr. Pharm. Des.* **1998,** *4* (2), 155–180.
378. Allington, D. R.; Rivey, M. P. Quinopristin/dalfopristin: a therapeutic review. *Clin. Ther.* **2001,** *23* (1), 24–44.
379. Minotti, G.; Menna, P.; Salvatorelli, E.; Cairo, G.; Gianni, L. Anthracyclines: molecular advances and pharmacologic developments in antitumor activity and cardiotoxicity. *Pharmacol. Rev.* **2004,** *56* (2), 185–229.
380. Arcamone, F.-M. Anthracyclines. In *Anticancer Agents from Natural Products,* 2nd ed.; Cragg, G. M., Kingston, D. G. I., Newman, D. J., Eds. CRC Press, 2012; pp 383–405 (1 plate).
381. Hamilton, D.; Batist, G. Anthracyclines. *Cancer Chemother. Biol. Response Modif.* **2005,** *22,* 19–33.
382. Gianni, L.; Grasselli, G.; Cresta, S.; Locatelli, A.; Vigano, L.; Minotti, G. Anthracyclines Cancer Chemother. *Biol. Response Modif.* **2003,** *21,* 29–40.
383. Nadas, J.; Sun, D. Anthracyclines as effective anticancer drugs. *Expert Opin. Drug Discovery* **2006,** *1* (6), 549–568.
384. Laatsch, H.; Fotso, S. Naturally occurring anthracyclines. *Top. Curr. Chem.* **2008,** *282,* 3–74.
385. Piekarski, M.; Jelinska, A. Anthracyclines still prove effective in anticancer therapy. *Mini-Rev. Med. Chem.* **2013,** *13* (5), 627–634.
386. Davidson, A.; Gelmon, K. Do anthracyclines still have a role in adjuvant chemotherapy of breast cancer? *Future Oncol.* **2011,** *7* (1), 37–55.

387. Robson, D.; Verma, S. Anthracyclines in early-stage breast cancer: is it the end of an era? *Oncologist* **2009,** *14* (10), 950–958.
388. Gianni, L.; Valagussa, P. Anthracyclines and early breast cancer: the end of an era? *J. Clin. Oncol.* **2009,** *27* (8), 1155–1157.
389. Cortes-Funes, H.; Coronado, C. Role of anthracyclines in the era of targeted therapy. *Cardiovasc. Toxicol.* **2007,** *7* (2), 56–60.
390. Fisch, A.; Bryskier, A. Phenicols. In *Antimicrobial Agents: Antibacterials and Antifungals;* Bryskier, A., Ed.; ASM Press, 2005; pp 925–929.
391. Wilcox, M. H. Chloramphenicol and thiamphenicol. In *Antibiotic and Chemotherapy: Anti-Infective Agents and Their Use in Therapy,* 7th ed.; O'Grady, F., Lambert, H. P., Eds. Churchill Livingstone, 1997; pp 328–332.
392. Fuglesang, J.; Bergan, T. Chloramphenicol and thiamphenicol. *Antibiot. Chemother. (Basel)* **1982,** *31,* 1–21.
393. Vining, L. C.; Stuttard, C. Chloramphenicol. *Biotechnol. Ser.* **1995,** *28,* 505–530.
394. Pestka, S. Chloramphenicol. In *Antibiotics: Mechanism of Action of Antimicrobial and Antitumor Agents*, Vol. 3; Gottlieb, D., Shaw, P. D., Corcoran, J. W., Eds.; Springer-Verlag, 1975; pp 370–395.
395. Ingall, D.; Sherman, J. D. Chloramphenicol. In *Antimicrobial Therapy;* Kagan, B. M., Ed.; Saunders, 1970; pp 61–77.
396. Thadepalli, H.; Hancz, D. Chloramphenicol. *Infect. Dis. Ther.* **1994,** *9,* 379–390.
397. Al-Badr, A. A.; El-Obeid, H. A. Chloramphenicol. *Anal. Profiles Drug Subst.* **1986,** *15.* 701–760.
398. Hahn, F. E. Chloramphenicol. *Antibiotics* **1983,** *6,* 34–45.
399. Rosenkranz, H. S. Chloramphenicol: magic bullet or double-edge sword? *Mutat. Res., Rev. Genet. Toxicol.* **1988,** *196* (1), 1–16.
400. Vining, L. C.; Westlake, D. W. S. Chloramphenicol: properties, biosynthesis, and fermentation. *Drugs Pharm. Sci.* **1984,** *22,* 387–411.
401. Greenwood, D. Fusidanes. In *Antibiotic and Chemotherapy: Anti-Infective Agents and Their Use in Therapy,* 7th ed.; O'Grady, F., Lambert, H. P., Eds. Churchill Livingstone, 1997; pp 360–362.
402. Musmade, P. B.; Tumkur, A.; Trilok, M.; Bairy, K. L. Fusidic acid-topical antimicrobial in the management of *Staphylococcus aureus. Int. J. Pharm. Pharm. Sci.* **2013,** *5* (Suppl. 4), 381–390.
403. Schoefer, H.; Simonsen, L. Fusidic acid in dermatology: an updated review. *Eur. J. Dermatol.* **2010,** *20* (1), 6–15.
404. Long, B. H. Fusidic acid in skin and soft-tissue infections. *Acta Derm.-Venereol., Suppl.* **2009,** *216,* 14–20.
405. Golledge, C. Fusidic acid in other infections. *Int. J. Antimicrob. Agents* **1999,** *12* (Suppl. 2), S11–S15.
406. Christiansen, K. Fusidic acid non-antibacterial activity. *Int. J. Antimicrob. Agents* **1999,** *12* (Suppl. 2), S73–S78.
407. Falagas, M. E.; Kastoris, A. C.; Kapaskelis, A. M.; Karageorgopoulos, D. E. Fosfomycin for the treatment of multidrug-resistant, including extended-spectrum β-lactamase producing, Enterobacteriaceae infections: a systematic review. *Lancet Infect. Dis.* **2010,** *10* (1), 43–50.
408. Michalopoulos, A. S.; Livaditis, I. G.; Gougoutas, V. The revival of fosfomycin. *Int. J. Infect. Dis.* **2011,** *15* (11), e732–e739.
409. Popovic, M.; Steinort, D.; Pillai, S.; Joukhadar, C. Fosfomycin: an old, new friend? *Eur. J. Clin. Microbiol. Infect. Dis.* **2010,** *29* (2), 127–142.

410. Keating, G. M. Fosfomycin trometamol: a review of its use as a single-dose oral treatment for patients with acute lower urinary tract infections and pregnant women with asymptomatic bacteriuria. *Drugs* **2013,** *73* (17), 1951–1966.
411. Reffert, J. L.; Smith, W. J. Fosfomycin for the treatment of resistant Gram-negative bacterial infections insights from the Society of Infectious Diseases Pharmacists. *Pharmacotherapy* **2014,** *34* (8), 845–857.
412. Raz, R. Fosfomycin: an old-new antibiotic. *Clin. Microbiol. Infect.* **2012,** *18* (1), 4–7.
413. Fromtling, R. A. Fosfomycin (MK-0955): an overview. *Drugs Today* **1987,** *23* (3), 151–158.
414. Morikawa, K.; Torii, I.; Morikawa, S. Immunomodulatory activity of fosfomycin. *Recent Res. Dev. Antimicrob. Agents Chemother.* **1999,** *3* (Pt. 2), 371–381.
415. Bryskier, A. Mupirocin. In *Antimicrobial Agents: Antibacterials and Antifungals;* Bryskier, A., Ed.; ASM Press, 2005; pp 964–971.
416. Casewell, M. W. Mupirocin. In *Antibiotic and Chemotherapy: Anti-Infective Agents and Their Use in Therapy,* 7th ed.; O'Grady, F., Lambert, H. P., Eds. Churchill Livingstone, 1997; pp 394–395.
417. Ward, A.; Campoli-Richards, D. M. Mupirocin. A review of its antibacterial activity, pharmacokinetic properties and therapeutic use. *Drugs* **1986,** *32* (5), 425–444.
418. Gurney, R.; Thomas, C. M. Mupirocin: biosynthesis, special features and applications of an antibiotic from a Gram-negative bacterium. *Appl. Microbiol. Biotechnol.* **2011,** *90* (1), 11–21.
419. Thomas, C. M.; Hothersall, J.; Willis, C. L.; Simpson, T. J. Resistance to and synthesis of the antibiotic mupirocin. *Nat. Rev. Microbiol.* **2010,** *8* (4), 281–289.
420. Cookson, B. D. The emergence of mupirocin resistance: a challenge to infection control and antibiotic prescribing practice. *J. Antimicrob. Chemother.* **1998,** *41* (1), 11–18.
421. Patel, J. B.; Gorwitz, R. J.; Jernigan, J. A. Mupirocin resistance. *Clin. Infect. Diseases* **2009,** *49* (6), 935–941.
422. Lange, R. P.; Locher, H. H.; Wyss, P. C.; Then, R. L. The targets of currently used antibacterial agents: lessons for drug discovery. *Front. Med. Chem.* **2012,** *6,* 62–96.
423. Meinert, S.; John, E. 80 years of antibiotics application in medicine. Indispensible against bacteria: antibiotics. *Chem. Unserer Zeit* **2009,** *43* (5), 296–306.
424. Tegos, G.; Mylonakis, E. Antimicrobial drug discovery: emerging strategies. In *Advances in Molecular and Cellular Microbiology*, Vol. 22; Tegos, G., Ed.; CABI Publishing, 2012; pp 26–43.
425. Gualerzi, C. O., Brandi, L., Fabbretti, A., Pon, C. L., Eds. *Antibiotics: Targets, Mechanisms and Resistance;* Wiley-VCH, 2014.

Chapter 31

Antibacterial Drugs

31.1 SULFONAMIDES

Antibacterial drugs is a general term that refers to a group of drugs that includes antibiotics, antifungals, antivirals, and antiprotozoals.

The term *antibacterials* also includes antibiotics, but antibiotics are more often referred to as substances produced by microorganisms. The story of antimicrobials begins with the observations of Louis Pasteur and Jules Francois Joubert, who discovered that one type of bacteria could prevent the growth of another.

Another significant milestone event in the field of antimicrobials, was the discovery of sulfonamides, synthetic compounds that have activity against both Gram-positive and Gram-negative bacteria [1-5].

Early in the 1930s, Bayer chemists Josef Klarer and Fritz Mietzsch synthesized some azo dyes. The potential antiinfectious properties of the azo dyes was studied by the bacteriologist and pathologist Gerhard Domagk, the appointed director of Bayer's Institute of Pathology and Bacteriology. Domagk began testing the effect of each newly synthesized dye on streptococci in vitro.

A red azo dye, prontosil (**31.1.1**), was effective in controlling *Streptococcus* and *Staphylococcus* bacterial infections. It was introduced into medicine in the 1930s for the treatment of general bacterial infections in humans. Later it was found that the reduction of prontosil in an organism, a reaction catalyzed by aldo-keto reductases present in the tissues, disrupted it to form p-aminobenzenesulfonamide–sulfanilamide (**31.1.2**), which became the first representative of sulfanilamide class of antibacterials, replacing prontosil (**31.1.1**) in clinical use (Fig. 31.1.).

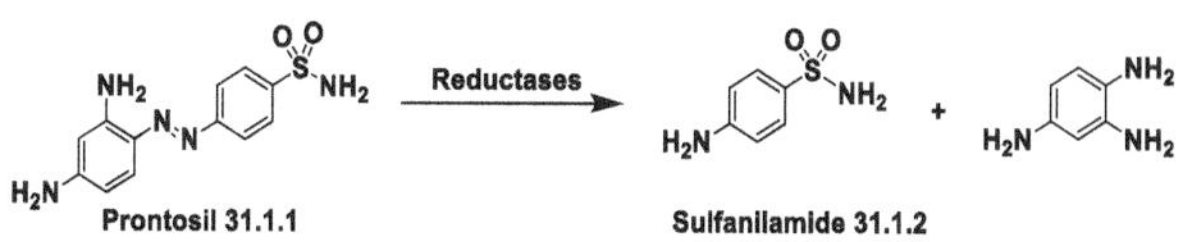

FIG. 31.1 **Discovery of p-aminobenzenesulfonamide, the first sulfanilamide.**

Gerhard Domagk was awarded the 1939 Nobel Prize for Physiology or Medicine for his "Discovery of the antibacterial effects of Prontosil, the first of the sulfonamide drugs."

Synthesis of Best-Seller Drugs. http://dx.doi.org/10.1016/B978-0-12-411492-0.00031-6

Sulfonamides, the first class of antimicrobial agents for cure of bacterial infections, are the result of a combination of intelligent research and "serendipity."

Thousands of new compounds have been synthesized since its discovery. All of them are structural analogues of p-aminobenzoic acid, and all of them are derivatives of sulfanilamide (**31.1.2**). All have the same nucleus to which various functional groups have been added to the amido group, or in which various substitutions on the amino group are made. These changes produce compounds with varying physical, chemical, pharmacologic, and antibacterial properties. The general conclusion of this huge work was that the p-amino group is essential for activity and should be unsubstituted for except in the case of prodrugs in which it could be used as an amide linkage that will be hydrolyzed to produce the active free form; the sulfonamide and the amino group must be directly attached to the benzene ring in para position to each other (any extra substitution will reduce activity); the type of aromatic (heteroaromatic) ring and which sulfonamide group are important for the nuances of antibacterial activity; and sulfonamide nitrogen must be either primary or secondary.

Certain bacteria require p-aminobenzoic acid in order to synthesize dihydrofolic acid, which is required to produce nucleic acids. p-Aminobenzoic acid, one of the two substrates of the dihydropteroate synthase, is a key enzyme in folic acid synthesis.

Sulfonamides are antimetabolites that substitute for p-aminobenzoic acid, resulting in blockade of several enzymes needed for the biogenesis of purine bases; for the transfer of desoxy-uridine to thymidine; and for the biosynthesis of methionine, glycine, and formylmethionyl-tRNA.

The sulfonamides competitively inhibit dihydropteroate synthetase, a vital enzyme that facilitates p-aminobenzoic acid as a substrate for the synthesis of dihydrofolic acid. Dihydrofolate is a precursor for formation of tetrahydrofolate, an essential methyl group shuttle required for the de novo synthesis of purines, thymidylic acid, and certain amino acids (Fig. 31.2.) When the folate synthesis is inhibited in bacteria cells, the inhibition process selectively kills bacteria.

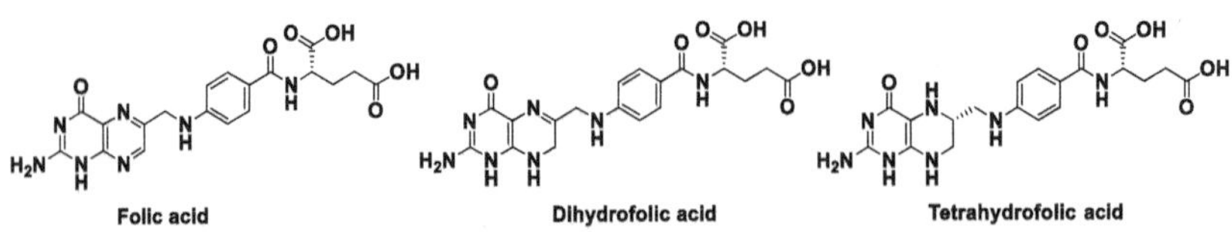

FIG. 31.2 Structures of folic, dihydrofolic, and tetrahydrofolic acids.

Mammals do not synthesize folic acid and take it as vitamin B from food. That's why they are not affected by p-aminobenzoic acid inhibitors. This difference in folate requirement between bacteria and mammals results in selective toxicity.

Sulfonamides are the oldest and remain among the widely used antibacterials. Although all of the sulfonamides have the same mechanism of action and even antimicrobial spectrum, they differ because of the variety of their minor physiochemical characteristics.

The spectrum of all sulfonamides is generally the same. Sulfonamides have a wide range of antimicrobial activity against both Gram-positive and Gram-negative organisms: *Nocardia, Actinomyces* species, and some protozoa such as *Coccidia* and *Toxoplasma* species. More active sulfonamides may include several species of *Streptococcus, Staphylococcus, Salmonella, Pasteurella,* and even *Escherichia coli.* Strains of *Pseudomonas, Klebsiella, Proteus, Clostridium, Leptospira, Rickettsiae, Mycoplasmas,* and most of *Chlamydia* species are highly resistant to sulfonamides.

The synergistic action of sulfonamides with specific diaminopyrimidines (trimethoprim or pyrimethamine) renders these drugs much more effective compared to sulfonamides alone (co-trimoxazole, co-trimazine).

Unfortunately, bacterial resistance to sulfonamides is now common, and their use has decreased with the introduction of antibiotics. However, sulfonamides are still widely used, especially for urinary tract infections in combination with trimethoprim, and for treatment or prevention of parasitic (toxoplasmosis, *Pneumocystis jiroveci*) and malarial infections, usually combined with trimethoprim or pyrimethamine.

Sulfonamides are categorized into several types, based mainly on their duration of action and indications.

Short-acting sulfonamides are sulfanilamide (**31.1.2**), sulfathiourea (**31.1.3**), sulfapyridine (**31.1.4**), sulfathiazole (**31.1.5**), sulfamethizole (**31.1.6**), sulfisomidine (**31.1.7**), and sulfadimidine (sulfamethazine) (**31.1.8**). (Fig. 31.3.).

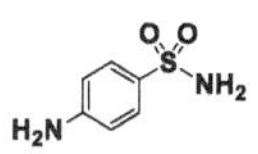
Sulfanilamide 31.1.2

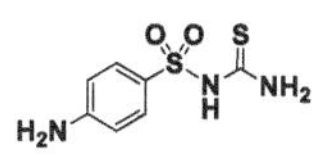
Sulfathiourea 31.1.3

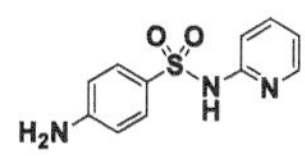
Sulfapyridine 31.1.4

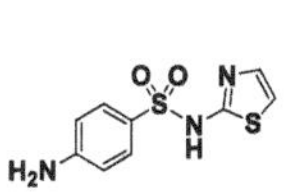
Sulfathiazole 31.1.5

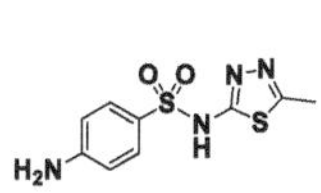
Sulfamethizole 31.1.6

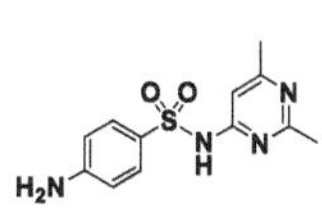
Sulfisomidine 31.1.7

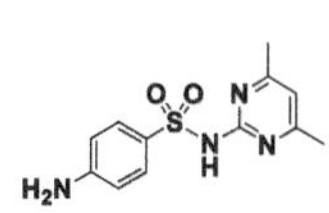
Sulfadimidine 31.1.8 (sulfamethazine)

FIG. 31.3 **Short-acting sulfonamides.**

Intermediate-acting sulfonamides are sulfamethoxazole (**31.1.8**), sulfamoxole (**31.1.9**), and sulfadiazine (**31.1.10**) (Fig. 31.4.).

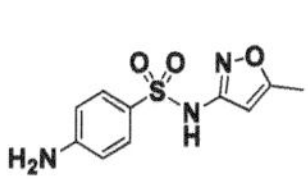
Sulfamethoxazole 31.1.8

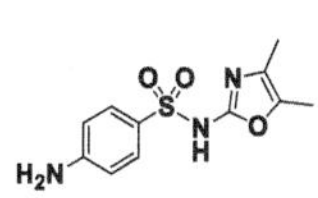
Sulfamoxole 31.1.9

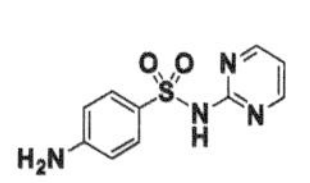
Sulfadiazine 31.1.10

FIG. 31.4 **Intermediate-acting sulfonamides.**

Long-acting sulfonamides are sulfadimethoxine (**31.1.11**), sulfadoxine (**31.1.12**), sulfametomidine (**31.1.13**), sulfaperin (**31.1.14**), sulfamerazine (**31.1.15**), sulfametoxydiazine (**31.1.16**), sulfamethoxypyridazine (**31.1.17**), sulfamazone (**31.1.18**), sulfalene (**31.1.19**), and sulfaphenazole (**31.1.20**) (Fig. 31.5.).

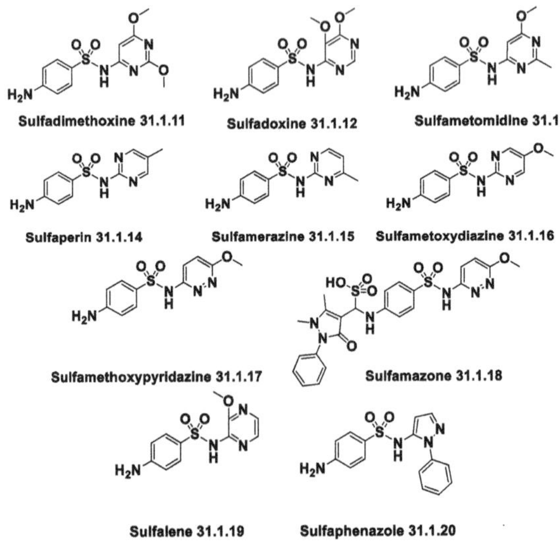

FIG. 31.5 Long-acting sulfonamides.

Ungrouped sulfonamides, such as sulfadicramide (**31.1.21**) and sulfametrole (**31.1.22**), are antiinfectives; sulfacetamide (**31.1.23**) is used as a topical antibacterial in the eye because of low irritation; phthalylsulfathiazole (**31.1.24**) is used to sterilize the gut prior to bowel surgery; dapsone (**31.1.25**), which is not "true" sulfonamide, is used as an antileprosy drug. Another sulfonamide-type medication is mafenide (**31.1.26**), which is used topically as an antibacterial to treat severe burns (Fig. 31.6.).

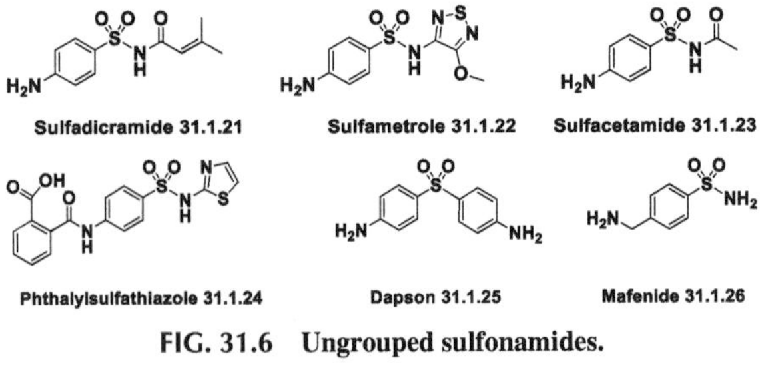

FIG. 31.6 Ungrouped sulfonamides.

Synthesis of many of the listed sulfonamides are described in our previous book [6]. Traditional and general methods for the synthesis of sulfonamides have been reviewed [7]. The general method usually is to couple sulfonyl chloride with an appropriate amine. For example, sulfamethoxazole (**31.1.8**) was prepared by treating p-acetamidobenzenesulfonyl chloride (**31.1.27**) with 3-amino-5-methyloxazole (**31.1.28**) in pyridine to produce 3-acetylsulfanilamido-5-methylisoxazole (**31.1.29**), which was hydrolyzed to by 10% NaOH on heating, producing the desired product [8] (Scheme 31.1.).

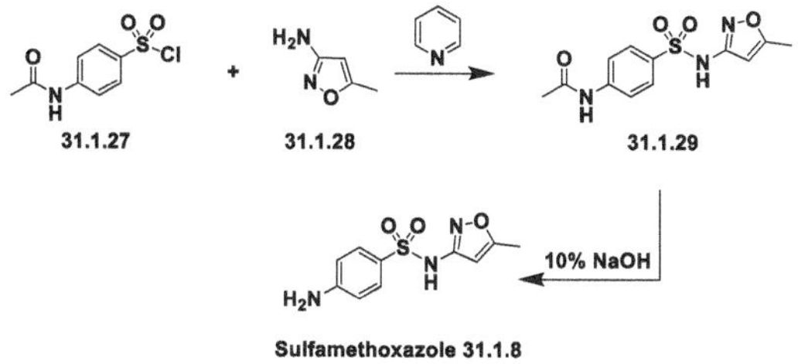

SCHEME 31.1 **General method for the synthesis of sulfonamides.**

None of the sulfonamides is included in the list of Top 200 Drugs by sales for the 2010s.

Sulfamethoxazole-Trimethoprim (Co-Trimoxazole, Bactrim)

Sulfamethoxazole (**31.1.8**) plus trimethoprim (**31.1.30**) (a synthetic antibacterial drug acting as a dihydrofolate reductase inhibitor) in a 5:1 fixed ratio named co-trimoxazole (Bactrim) is considered the first-line drug of choice for the treatment and prophylaxis of patients with urinary tract infections. Co-trimoxazole is a second- or third-line treatment, for patients who have respiratory tract infections. It is used for treatment of sexually transmitted diseases, Gram-negative sepsis, enteric infections, and typhoid fever. It is reported to also be effective against numerous bacterial, fungal, and protozoal pathogens [9-12].

There exists another combination drug, *co-trimazine*, which consists of 5 parts sulfadiazine (**31.1.10**) to 1 part trimethoprim (**31.1.30**) (Fig. 31.7.).

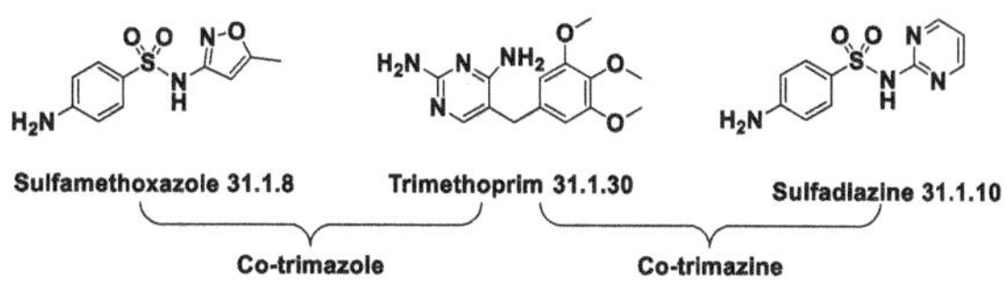

FIG. 31.7 **Combination drugs of sulfonamides and trimethoprim.**

Sulfonamides are not only antibacterials, but exhibit a broad range of biological activities [13-17].

Other groups of sulfonamide drugs have been developed by exploiting observations made during clinical evaluation of sulfonamide derivatives. Nowadays, the sulfonamide group is present in many drugs as antibiotics, antimalarics, diuretics, hypoglycemics, antiinflammatories, antihypertensives, antitumor compounds, antithyroid compounds, etc., showing the broad spectra of therapeutic implementation. Several of these drugs are widely used in therapy. Among them are glipizide, glyburide, glymidine, zonisamide, tolazamide, chlorpropamide, acetohexamide (antihyperglycemic agents); celecoxib, nimesulide (nonsteroidal antiinflammatory drugs); acetazolamide, methazolamide, ethoxzolamide, dichlorphenamide, dorzolamide, brinzolamide (carbonic anhydrase inhibitors); furosemide (loop diuretic), delavirdine (nonnucleoside reverse transcriptase inhibitor); thiothixene sulpiride (antipsychotic drug), sotalol (nonselective β-blocker), tolbutamide (potassium channel blocker); and famotidine (H2-receptor antagonist) are drugs in medicinal practice that contain a sulfonamide group.

31.2 ANTIMICROBIAL QUINOLONES

Quinolones are synthetic, bactericidal antibacterial agents with broad-spectrum activity. They inhibit the enzyme topoisomerase II (named DNA gyrase) that is necessary for the replication of the bacteria. Four main classes (I through IV) of bacterial topoisomerases have been identified. Bacterial topoisomerases I and III interact with single-stranded DNA, whereas bacterial topoisomerases II and IV interact with double-stranded DNA. Moreover, there are distinct structural differences from the mammalian enzyme counterparts that allows the creation of ligand inhibitors, especially for bacterial enzymes. So, by inhibiting topoisomerase II in bacteria, its DNA replication and transcription are blocked, stopping bacterial multiplication. These targets and processes are universal to all bacteria. The main classes of antibacterial topoisomerase inhibitors have been reviewed [18-20].

In 1975 it was shown [21,22] that nalidixic acid (**31.2.1**), a substance with antibacterial activity and known since 1962, which was isolated from the mother liquor during the synthesis of chloroquine, inhibits an enzyme important for bacterial multiplication. In 1976 the enzyme itself, named DNA gyrase (topoisomerase II), was isolated, purified, and identified [23]. This data gave a strong impetus for the search for new antibacterials.

The antibacterial properties of nalidixic acid were limited to the treatment of Gram-negative urinary tract infections. The 4-quinolone nucleus motif became an important multivalent scaffold in many other areas of medicinal chemistry, but remains a milestone for creation of new agents for the treatment of bacterial infection [24].

The unique mechanism of action of this class of antibacterial agents with its economically and clinically proven track record generated considerable scientific efforts and resulted in creation of a new family of synthetic broad-spectrum antibacterial drugs - quinolones [25-35].

Nalidixic acid was effective against Gram-negative pathogens, but more recent 4-quinolone derivatives have improved Gram-positive activity. Resistance is a result of mutations in the gyrase and topoisomerase enzyme targets or efflux of drug out of bacteria. Currently marketed quinolones are most active against aerobic Gram-negative bacteria. These include *Enterobacteriaceae, Haemophilus,* and *Neisseria* species, *Moraxella catarrhalis, Pseudomonas aeruginosa,* and *Acinetobacter baumannii.* Except for norfloxacin, quinolones are active against staphylococci, but methicillin-resistant strains are typically resistant to quinolones. The newer quinolones have increased Gram-positive activity and are effective against streptococci, but not enterococci. These class of compounds also active against mycobacteria (*Mycobacterium tuberculosis, Mycobacterium kansasii,* and *Mycobacterium fortuitum*).

Numerous derivatives synthesized varying structure of nalidixic acid have become important pharmaceutical products.

The first 4-quinolone introduced for clinical use nalidixic acid (**31.2.1**), became the ancestor of a series of other important 4-quinolones of historical interest. Variations in the structure of nalidixic acid brought to creation rosoxacin (**31.2.2**) and oxolinic acid (**31.2.3**), followed by cinoxacin (**31.2.2**) where the quinolin-4-one core is replaced with cinnolin-4-one, and then piromidic (**31.2.5**) and pipemidic (**31.2.6**) acids in which a quinolin-4-one nucleus was replaced with a pyridopyrimidin. These compounds represent the first generation of the quinolone series of antibacterial drugs (Fig. 31.8.).

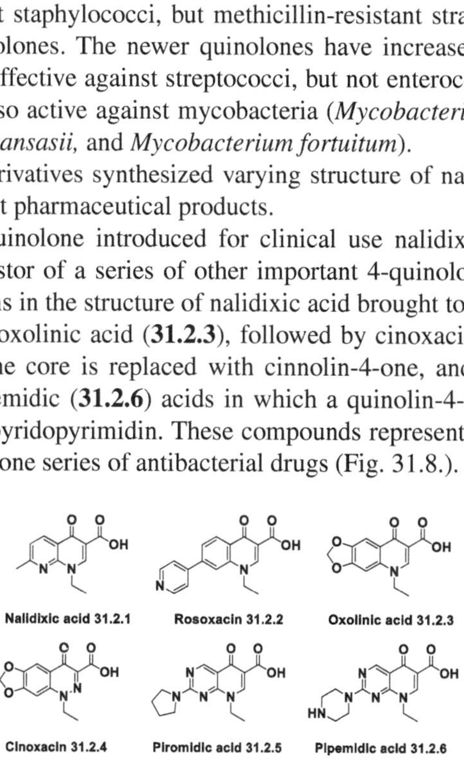

FIG. 31.8 Variations in the structure of nalidixic acid.

The structures of the quinolones have been developed further [36,37]. Perhaps the most important finding was the insertion of fluorine at position C6, which gave the impetus for development of fluoroquinolones, leading to a significant increase in antibacterial activity.

The first fluoroquinolone was flumequine (**31.2.7**). It was in use until ocular toxicity was reported.

Subsequently, newer agents with increased antimicrobial activity—grepafloxacin (**31.2.8**) sparfloxacin (**31.2.9**), and sitafloxacin (**31.2.10**) (Fig. 31.9.)—were developed, but as a result of significant phototoxicity, they were removed from the pharmaceutical market. Only sitafloxacin remains in very limited use.

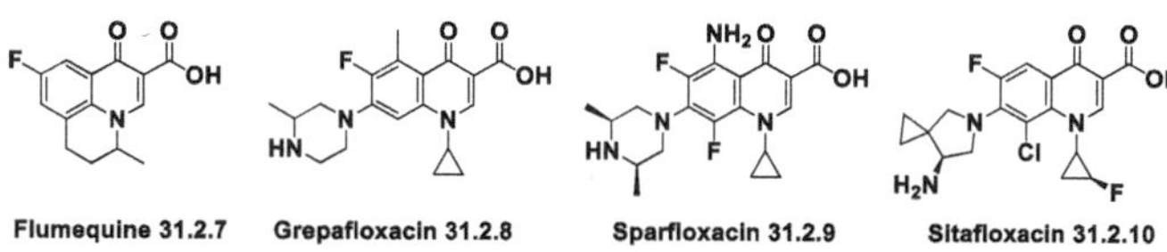

FIG. 31.9 Newer quinolones removed from the pharmaceutical market.

The start for second-generation quinolone agents was launched, epitomized by ciprofloxacin (**31.2.11**), which remains one the most potent and popular quinolone antibacterials. Ciprofloxacin is marketed worldwide, with well over 300 different brand names. It has a wide spectrum of in vitro antibacterial activity, in particular against Gram-negative bacteria, and is effective in many respiratory infections [38,39].

Ciprofloxacin, was patented in 1983 by Bayer A.G. [40], and was approved in the United States in 1987. Since the introduction of ciprofloxacin into medicinal practice, fluoroquinolones have gone beyond the treatment of urinary tract infections and are used for the treatment of upper and lower respiratory infections, having good activity against *Streptococcus pneumoniae*, gastrointestinal infections, gynecologic infections, sexually transmitted diseases, and some skin and soft-tissue infections [41-47].

Current second-generation quinolones are represented by ciprofloxacin (**31.2.11**), norfloxacin (**31.2.12**), pefloxacin (**31.2.13**), ofloxacin (**31.2.14**), rufloxacin (**31.2.15**), lomefloxacin (**31.2.16**), nadifloxacin (**31.2.17**), and enoxacin (**31.2.18**) (Fig. 31.10.).

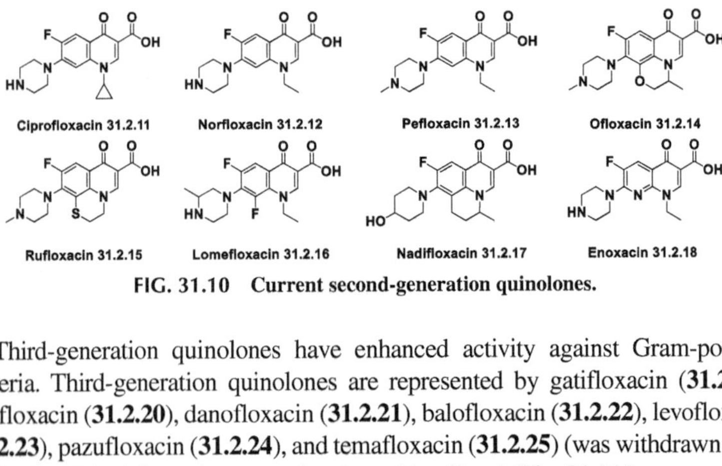

FIG. 31.10 Current second-generation quinolones.

Third-generation quinolones have enhanced activity against Gram-positive bacteria. Third-generation quinolones are represented by gatifloxacin (**31.2.19**), tosufloxacin (**31.2.20**), danofloxacin (**31.2.21**), balofloxacin (**31.2.22**), levofloxacin (**31.2.23**), pazufloxacin (**31.2.24**), and temafloxacin (**31.2.25**) (was withdrawn from sale in the United States because of serious side effects) (Fig. 31.11.).

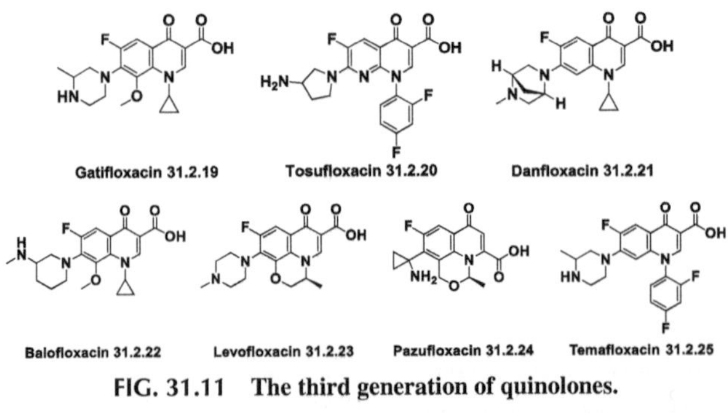

FIG. 31.11 The third generation of quinolones.

Of this series of drugs, levofloxacin and moxifloxacin are active against *S. pneumoniae* and *Staphylococcus aureus*. Moreover, moxifloxacin is active against *M. tuberculosis,* which lacks topoisomerase IV.

The fourth generation of quinolones has enhanced potency and a broader spectrum that includes anaerobic bacteria. The fourth-generation quinolones are clinafloxacin (**31.2.26**), garenoxacin (**31.2.27**), gemifloxacin (**31.2.28**), besifloxacin (**31.2.29**), trovafloxacin (**31.2.30**), moxifloxacin (**31.2.31**), and prulifloxacin (**31.2.32**) (Fig. 31.12.).

Gemifloxacin (**31.2.28**) and moxifloxacin (**31.2.31**) are currently the most potent fluoroquinolones. They are used against community-acquired pneumonia and acute bronchitis. Trovafloxacin (**31.2.30**) was used to treat intraabdominal and pelvic infections, but was withdrawn from the market because of the risk of hepatotoxicity.

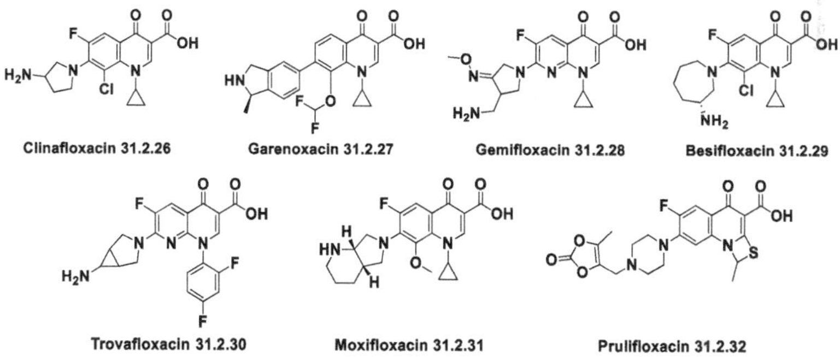

FIG. 31.12 The fourth-generation of quinolones.

This pattern of antibacterial quinolones shows the possibility of transformation of quinoline core to pyridopyrimidine, cinnoline, and naphthyridine scaffolds, and the possibilities of wide varying of side-chain structures, with associated improvements in activity, pharmacokinetics, and tolerability, through the selection of molecules that remain useful and well tolerated [28,31,43,44,48].

Antibacterial quinolones are divided into generations based mainly on their antibacterial spectrum [31,49]. The first-generation drugs are the oldest and least-often used quinolones. Use of these drugs is restricted to the treatment of uncomplicated urinary tract infections.

The second-generation quinolones have increased Gram-negative activity, as well as some Gram-positive and atypical pathogen coverage, and are widely used in the treatment of complicated urinary tract infections and pyelonephritis, sexually transmitted diseases, selected pneumonias, and skin infections. Ciprofloxacin and ofloxacin are the most widely used second-generation quinolones because of their availability in oral and intravenous formulations.

The third-generation quinolones have expanded activity against Gram-positive organisms, particularly penicillin-sensitive and penicillin-resistant *S. pneumoniae*, and atypical pathogens such as *Mycoplasma pneumoniae* and

Chlamydia pneumoniae; they are less active than ciprofloxacin against *Pseudomonas* species. Because of their expanded antimicrobial spectrum, they are useful in the treatment of community-acquired pneumonia, acute sinusitis, and acute exacerbations of chronic bronchitis.

The fourth-generation compounds, have significant antimicrobial activity against anaerobes while maintaining the Gram-positive and Gram-negative activity of the third-generation quinolones. They retain activity against *Pseudomonas* species comparable to that of ciprofloxacin.

Because of concerns about hepatotoxicity, they should be reserved for life-threatening infections requiring inpatient treatment and should be taken for a short time.

Quinolone antibacterials gave a new impetus in the treatment of infectious diseases. Despite these favorable properties, the development of resistance to these group of compounds is becoming a serious problem.

The antibacterial quinolones levofloxacin and moxifloxacin are included in the list of Top 200 Drugs by sales for the 2010s.

Levofloxacin–Levaquin

Levofloxacin (**31.2.23**) is a chiral version L-isomer of the earlier drug–racemic fluoroquinolone ofloxacin (**31.2.14**). Levofloxacin departs from the typical quinolones by having a more complex fused ring connected to the oxazinoquinoline core. Levofloxacin is twice as potent as ofloxacin.

It interferes with critical processes in the bacterial cell, such as DNA replication, transcription, repair, and recombination by inhibiting bacterial topoisomerases.

Levofloxacin has broad-spectrum activity against several causative bacterial pathogens and is approved for use in the treatment of community-acquired pneumonia, acute bacterial sinusitis, complicated urinary tract infections and acute pyelonephritis. It is considered a reserve antibiotic to prevent the occurrence of drug resistance and should be the second antibiotic choice when resistant microorganisms or hypersensitivity to other drugs are found [50-59].

Several approaches for the synthesis of levofloxacin are present in literature, which are based on Gould-Jacobs and Grohe-Heitzer reactions and Chu-Mitscher and Gerster-Hayakawa synthetic approaches.

The first data [60-62] to describe synthesis of levofloxacin are based on implementation of the Gould-Jacobs cyclization reaction for the synthesis of ofloxacin (**31.2.14**) and further resolution of the obtained mixture of [(S)-(-)] and [(R)-(+)] enantiomers via optical, enzymatic, or crystallization methods [63].

According to these methods, the starting material for the synthesis—2,3-difluoro-6-nitrophenol (**31.2.33**)—was prepared from 2,3,4-trifluoro-1-nitrobenzen by displacement of the ortho to the nitro group fluorine atom to the hydroxyl group on reaction with potassium hydroxide in DMSO.

The prepared 2,3-difluoro-6-nitrophenol (**31.2.33**) was reacted with chloroacetone (**31.2.34**) in the presence of potassium carbonate and potassium iodide to produce 1-(2,3-difluoro-6-nitrophenoxy)propan-2-one (**31.2.35**). The obtained product was hydrogenated in the presence of Raney nickel in ethanol to give a cyclic product, 7,8-difluoro-3-methyl-3,4-dihydro-2H-benzo[b][1,4] oxazine, which resulted from sequential formation of amine, cyclic imine, and its consequent (**31.2.36**).

Diethyl ethoxymethylenemalonate (**31.2.37**) was reacted with the benzoxazine derivative (**31.2.36**) using conditions of the well-precedented Gould-Jacobs reaction at 130-140°C to produce the required benzoxazin-yl methylenemalonate (**31.2.38**), which on heating with polyphosphoric ester at 140 to 145°C yielded 9,10-difluoro-3-methyl-7-oxo-2,3-dihydro-7H-[1,4]oxazino[2,3,4-ij]quinoline-6-carboxylic acid ethyl ester (**31.2.39**). It was hydrolyzed in acetic acid and concentrated hydrochloric acid mixture to produce benzoxazine-6-carboxylic acid (**31.2.40**). The obtained product was coupled with N-methylpiperazine in dimethyl sulfoxide. Displacement reaction with a secondary amine smoothly introduces an amino substituent at the C7 position because of the activation by the C4 carbonyl substituent. After appropriate workup, the desired title compound—ofloxacin (**31.2.14**)—was obtained (Scheme 31.2.).

Resolution of enantiomers via optical, enzymatic, or crystallization methods showed that one of the isomers, (-)-ofloxacin, was eightfold to 128-fold more potent at inhibiting the multiplication of Gram-positive and Gram-negative bacteria than (+)-ofloxacin, and approximately twofold more active than the racemate, (±)-ofloxacin [63].

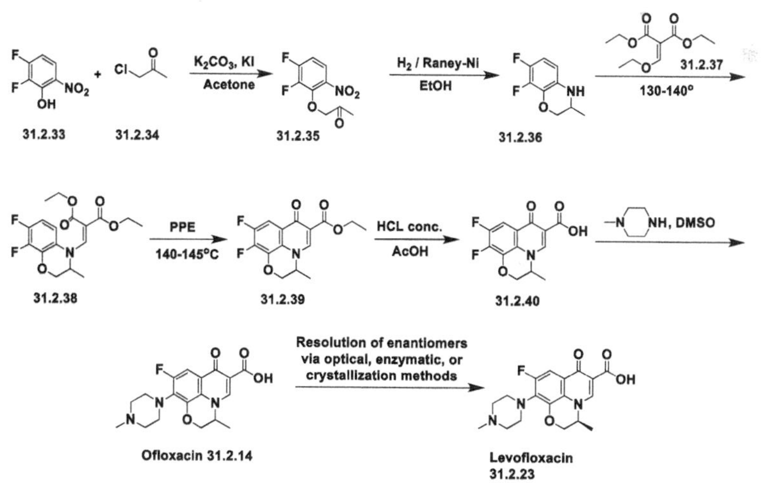

SCHEME 31.2 Synthesis of levofloxacin.

In a slightly modified method [64], coupling of 2,3-difluoro-6-nitrophenol (**31.2.33**) with 1-acetoxy-3-chloro-2-propane (**31.2.41**) in standard conditions (acetone/K_2CO_3, KI) to produce (±)-3-acetoxymethyl-7,8-difluoro-2,3-dihydro-4H-[1,4]benzoxazine (**31.2.42**). The obtained product was hydrogenated in the presence of Raney nickel in methanol to produce a cyclic product, which resulted from the sequence of reactions described above (amine, imine, benzoxazine) to produce (±)-3-acetoxymethyl-7,8-difluoro-2,3-dihydro-4H-[1,4]benzoxazine (**31.2.43**). The last was benzoylated in tetrahydrofuran/pyridine media with 3,5-dinitrobenzoyl chloride (**31.2.44**) to obtain a dinitrobenzoyl derivative of (±)-3-acetoxymethyl-7,8-difluoro-2,3-dihydro-4H-[1,4] benzoxazine (**31.2.45**). When the racemate (**31.2.45**) is treated with lipoprotein lipase "Amano" 3, the (-)-compound is preferentially hydrolyzed, and the product of hydrolysis rich in the (-)-isomer that was separated. The last also was possible to produce by appropriate crystallization or chromatographic methods. After additional benzoylation and separation, the product was hydrolyzed with 1.0 N potassium hydroxide water solution in ethanol to produce (-)-7,8-difluoro-2,3-dihydro-3-hydroxymethyl-4H-[1,4]benzoxazine (**31.2.46**).

The obtained compound was chlorinated with thionyl chloride in pyridine and then the intermediate chloro derivative was reduced with sodium borohydride in dimethyl sulfoxide producing the desired (-)-7,8-difluoro-2,3-dihydro-3-methyl-4H-[1,4]benzoxazine (**31.2.47**). Other approaches for transformation (**31.2.46**) → (**31.2.47**) are also proposed in the same patent.

Diethyl ethoxymethylenemalonate (**31.2.37**) was then reacted with the obtained benzoxazine at 130 to 140°C, and the reaction product—methylenemalonate (**31.2.48**)—was subjected to cyclization in the mixture of acetic anhydride and concentrated sulfuric acid at 50 to 60°C to yield (-)-ethyl 9,10-difluoro-3-methyl-7-oxo-2,3-dihydro-7H-pyrido[1,2,3-de][1,4]benzoxazine-6-carboxylate (**31.2.49**). It was hydrolyzed in acetic acid and concentrated hydrochloric acid to produce crystals of (-)-9,10-difluoro-3-methyl-7-oxo-2,3-dihydro-7H-pyrido[1,2,3-de][1,4] benzoxazine-6-carboxylic acid (**31.2.50**).

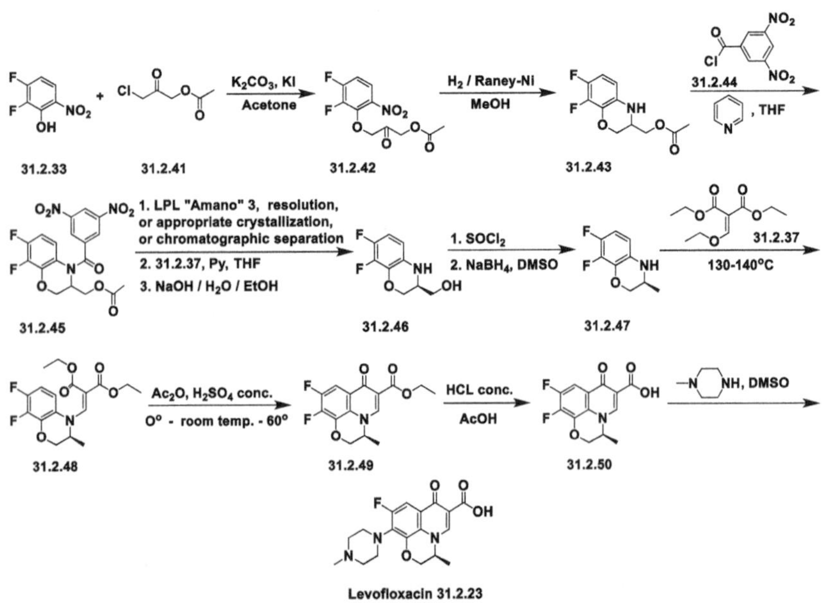

SCHEME 31.3 Synthesis of levofloxacin.

The obtained product was coupled with N-methylpiperazine in dimethyl sulfoxide to produce the desired compound-levofloxacin (**31.2.23**) (Scheme 31.3.).

Another strategy characterized by the construction of benzoxazine-6-carboxylic acid (**31.2.50**) from the appropriate benzoylacrylates (Grohe-Heitzer reaction) has been proposed [65]. For this purpose, (S)-2-aminopropan-1-ol (**31.2.51**) was involved in reaction with ethyl propiolate (**31.2.52**) (Michael addition) in acetonitrile at room temperature to produce in 99% yield chiral 3-((1-hydroxypropan-2-yl)amino)acrylate (**31.2.53**). The obtained compound was acylated with acetyl chloride to produce acrylate (**31.2.54**) and then bezoylated with 3,4,5-trifluoro-2-nitrobenzoyl chloride (**31.2.55**) 2,3,4,5-tetrafluorobenzoyl chloride in improved Grohe method conditions to prepare benzoylacrylate (**31.2.56**). After treatment of the obtained benzoylacrylate with potassium hydroxide in THF at 0°C to room temperature for 1 hour resulted in intramolecular nucleophilic aromatic substitution, the intermediate ethyl 4-oxo-1,4-dihydroquinoline-3-carboxylate (**31.2.57**) to which 10% potassium hydroxide aqueous solution was added and the reaction mixture was refluxed to produce pure benzoxazine-6-carboxylic acid (**31.2.50**) in 92% yield starting from (**31.2.56**) (Scheme 31.4.).

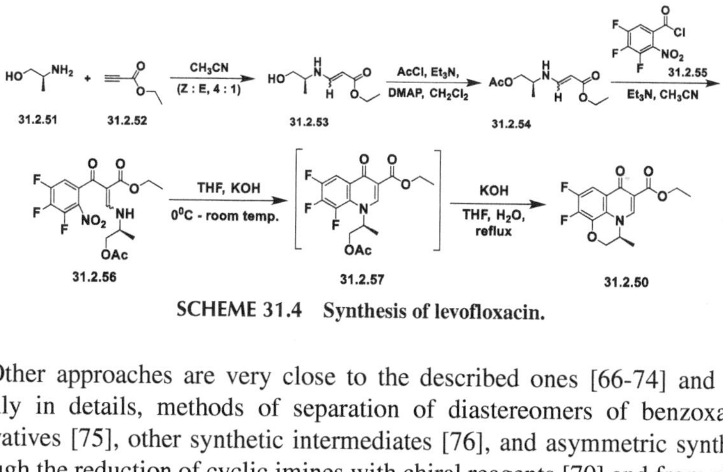

31.2.51 31.2.52 31.2.53 31.2.54

31.2.56 31.2.57 31.2.50

SCHEME 31.4 Synthesis of levofloxacin.

Other approaches are very close to the described ones [66-74] and vary mainly in details, methods of separation of diastereomers of benzoxazine derivatives [75], other synthetic intermediates [76], and asymmetric synthesis through the reduction of cyclic imines with chiral reagents [70] and from chiral materials [77,78], which are comprehensively reviewed in detail [79].

Moxifloxacin–Avelox

Moxifloxacin is a novel fourth-generation fluoroquinolone with a broad spectrum of antibacterial activity against Gram-positive and Gram-negative bacteria, anaerobes, and atypical organisms.

Its lower minimal inhibitory concentration value and high antibacterial effects make it a suitable antibiotic for treating various infectious diseases.

Moxifloxacin is a highly active with superb anti–*S. pneumoniae, Haemophilus influenzae,* and *M. catarrhalis* activity. Its activity against Gram-negative bacteria is comparable to ciprofloxacin; only *P. aeruginosa* is significantly less susceptible to moxifloxacin.

In its clinical applications, such as in the treatment of community-acquired pneumonia, multidrug resistant *S. pneumoniae*, acute exacerbation of chronic bronchitis, urinary and reproductive tract infection, acute bacterial sinusitis, and uncomplicated skin and skin structure infections, it is generally well tolerated. This makes moxifloxacin an important option in the treatment of bacterial infections [80-96].

Common side effects of moxifloxacin include abdominal discomfort, diarrhea, vomiting, mouth sores, headache, and vaginal discomfort (itching or burning sensation).

In general, 5- and 6-membered nitrogen heterocycles at the C7 position as side-chain substituents have been proven to be the optimal for antibacterial properties of 4-quinolone-3-carboxylic acid series and its analogues [97].

The introduction of the pyrrolidine derivatives into this position resulted in a dramatic improvement of their Gram-positive activity and many new quinolones used currently in the clinic, such as tosufloxacin, sitafloxacin, trovafloxacin, gemifloxacin, and moxifloxacin, contain an aminopyrrolidine residue at the mentioned C7 position.

Two general approaches are used for the production of moxifloxacin. One is carried out by the reaction of 1-cyclopropyl-6,7-difluoro-1,4-dihydro-8-methoxy-4-oxo-3-quinolinecarboxylic acid (**31.2.51**) with S,S-2,8-diazabicyclo[4.3.0]nonane (**31.2.52**)

The second approach is based on (**31.2.53**), which on treatment with potassium tert-butoxide in tetrahydrofuran/methanol mixture, undergoes replacement of fluorine by methoxy group in the C8 position of the 4-oxo-3-quinolinecarboxylic acid scaffold to produce the desired moxifloxacin (**31.2.31**) (Fig. 31.13.)

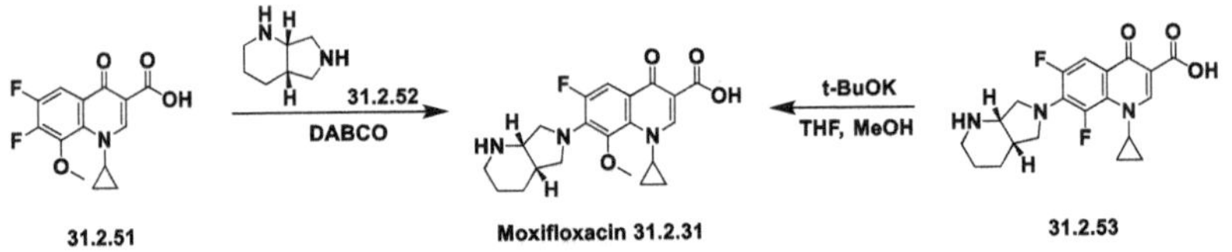

31.2.51 Moxifloxacin 31.2.31 31.2.53

FIG. 31.13 Synthesis of moxifloxacin.

The first prior art synthesis process of moxifloxacin comprises the coupling reaction between 1-cyclopropril-6,7-difluoro-1,4-dihydro-8-methoxy-4-oxo-3-quinolinic acid with S,S-2,8-diazabicyclo[4.3.0]nonane [98].

The methoxyquinolone (**31.2.51**) was prepared starting from 3-hydroxy-2.4,5-trifluorobenzoic acid (**31.2.54**) using the synthetic sequence outlined in Scheme 31.5.

Alkylation of the phenol and acid moieties of the (**31.2.54**) with iodomethane, in DMF using sodium hydride produced methyl 3-methoxy-2,4,5-trifluorobenzoate (**31.2.55**), which on hydrolysis with 1.0 N NaOH in methanol at room temperature produced 3-methoxy-2,4,5-trifluorobenzoic acid (**31.2.56**).

The last was transformed into acid chloride (**31.2.57**) with oxalyl chloride in CH_2Cl_2, and a catalytic amount of DMF.

Ethyl (3-methoxy-2,4,5-trifluorobenzoyl)acetate (**31.2.58**) was prepared by standard literature procedures via acylation of the dianion formed from monoethyl malonate and n-butyllithium with prepared benzoyl chloride (**31.2.57**), followed with decarboxylation of obtained 2-benzoyl-3-methoxy-3-oxopropanoic acid derivative with concentrated HCl.

A solution of β-ketoester (**31.2.58**), triethyl orthoformate, and acetic anhydride was heated at reflux to give ethyl 2-benzoyl-3-ethoxyacrylate derivative (**31.2.59**), which on treatment with cyclopropylamine in ethanol smoothly gave ethyl (cyclopropylmethyl)aminoacrylate (**31.2.60**). The last was cyclized to ethyl 1-cyclopropyl-6,7-difluoro-1,4-dihydro-8-methoxy-4-oxo-3-quinolinecarboxylate (**31.2.61**) in THF using n-butyllithium or sodium hydride, and the ester group was hydrolyzed with concentrated HCl to give desired 1-cyclopropyl-6,7-difluoro-1,4-dihydro-8-methoxy-4-oxo-3-quinolinecarboxylic acid (**31.2.51**) (see Scheme 31.5.).

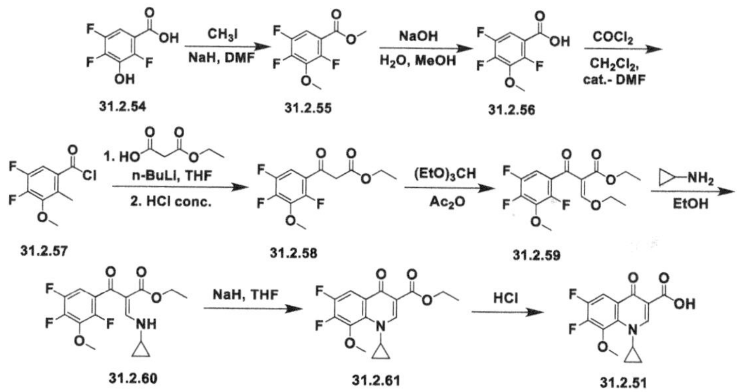

SCHEME 31.5 Synthesis of 1-cyclopropyl-6,7-difluoro-1,4-dihydro-8-methoxy-4-oxo-3-quinolinecarboxylic acid.

The enantiomerically pure amine (**31.2.52**) was prepared starting from pyridine-2,3-dicarboxylic acid (**31.2.62**), which was converted to anhydride (**31.2.63**) on reflux for 30 minutes with acetic anhydride and then was heated with benzylamine in acetic acid for 1 hour to give pyridine-2,3-dicarboxylic acid N-benzylimide (**31.2.64**). The pyridine ring was hydrogenated in the presence of Pd-C to produce succinimide (**31.2.65**), both rings of which were cis-bonded. The subsequent reduction of the succinimide with $LiAlH_4$ in THF produced 8-benzyl-2,8-diazabicyclo[4.3.0]-nonane (**31.2.66**).

The desired amine (**31.2.52**) was prepared by resolution of the racemic (**31.2.66**) using L-R,R(+)-tartaric acid (natural isomer), whereupon the diastereomerically pure R,R-tartrate of the R,R-enantiomer (**31.2.67b**) was crystallized from dimethylformamide; the target S,S-enantiomer (**31.2.67a**) was contained in the mother liquor. It was converted into the free base with sodium hydroxide, and then, for the purpose of further purification, precipitated with D-S,S(−)-tartaric acid (an unnatural isomer) to produce the diastereomerically pure S,S-tartrate (**31.2.67a**), which was then liberated with sodium hydroxide solution (Scheme 31.6.).

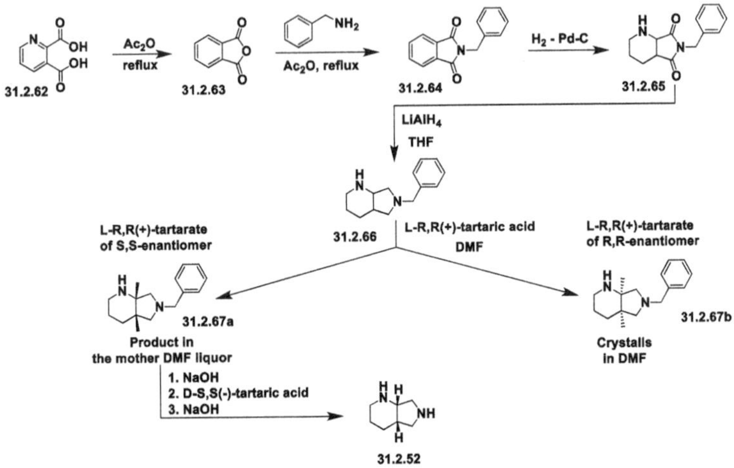

SCHEME 31.6 **Synthesis of S,S-enantiomer of 8-benzyl-2,8-diazabicyclo[4.3.0]-nonane.**

Separation of the enantiomers can also be carried out with high optical yields in an aqueous/alcoholic solution [99]. The debenzylation was achieved on hydrogenation on Pd-C catalyst.

A closely related method for the synthesis of moxifloxacin comprises the reaction between quinolinecarboxylate (**31.2.61**) with boric acid and acetic anhydride to form an intermediate borate complex, which is reacted with amine (**31.2.52**), and prepared complex being subsequently hydrolyzed to produce moxifloxacin [100]. A similar process is implemented using boric acid and propionic anhydride [101]. Another variation describes the reaction of quinoline carboxylic acid (**31.2.51**) with trifluoride etherate to produce a difluoroborate intermediate, which is reacted with amine (**31.2.52**) in the presence of a base, producing a moxifloxacin difluoroborate complex that is hydrolyzed to moxifloxacin [102]. Several other modifications of this approach are proposed [103-110].

The second, much-less-popular method for the synthesis of moxifloxacin is based on replacement of fluorine by methoxy group in the C8 position of the 4-oxo-3-quinoline carboxylic acid scaffold to produce moxifloxacin, which took place on treatment of 1-cyclopropyl-6,8-difluoro-7-((4aS,7aS)-octahydro-6H-pyrrolo[3,4-b]pyridin-6-yl)-4-oxo-1,4-dihydroquinoline-3-carboxylic acid (**31.2.53**) with potassium tert-butoxide in tetrahydrofuran/methanol mixture [111].

31.3 ANTIMICROBIAL NITROFURANS

The nitrofurans comprise a class of synthetic antibacterials characterized by the 5-nitro-2-furanyl scaffold.

Nitrofurans are broad-spectrum agents, characterized by effectiveness against both Gram-positive and Gram-negative bacteria, including *Salmonella, Giardia*, and *Salmonella* species, trichomonads, amebae, and some coccidial species. Despite the broad spectrum of activity, other antibacterial drugs are usually more effective against susceptible organisms.

Nitrofurans were represented in pharmaceutical market by nifuroxime (**31.3.1**), nihydrazone (furacilin) (**31.3.2**), nifuraldezone (**31.3.3**), nitrofurazone (**31.3.4**), furazolidone (**31.3.5**), nifuratel (**31.3.6**), nitrofurantoin (**31.3.7**), nifurtoinol (**31.3.8**), furylfuramide (**31.3.9**), furium (**31.3.10**), furalazine (**31.3.11**), acetylfuratrizine (**31.3.12**), nitrofurathiazide (**31.3.13**), and nitrovin (**31.3.14**) (Fig. 31.14.).

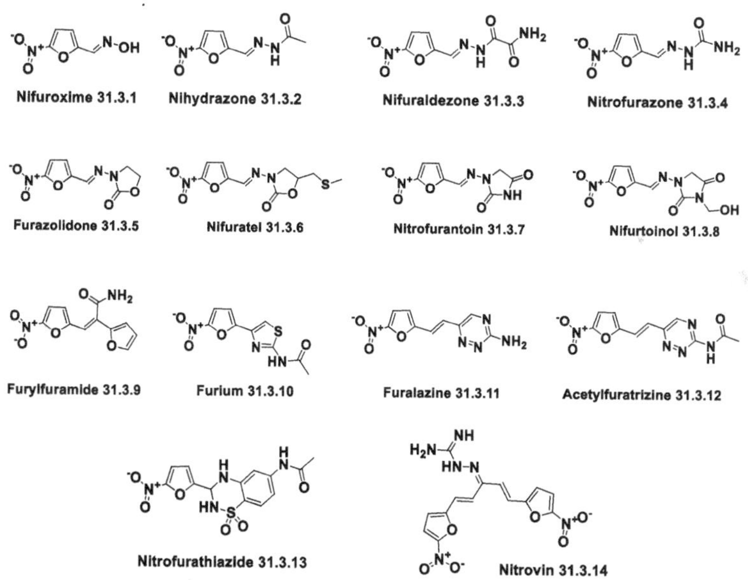

FIG. 31.14 **Antimicrobial nitrofurans.**

For the most part, nitrofurans are bacteriostatic, although at high doses they can be bactericidal.

The nitrofurans have been an important component of the antimicrobial armamentarium for approximately 50 years. Throughout the years, the nitrofurans have remained clinically useful drugs against a wide spectrum of common urinary tract pathogens. They are used to treat urinary infections caused by *E. coli, S. aureus, Streptococcus pyogenes,* and *Aerobacter aerogenes,* and can be administered orally or parenterally. *Proteus* species,

P. aeruginosa, and *Streptococcus faecalis* are usually resistant to this class of antibacterials.

Nitrofurazone (**31.3.4**) and nitrofurantoin (**31.3.7**) are the only nitrofurans that have become established in clinical use in the 20th century. These two nitrofurans have remained clinically useful.

The basic mechanism of action of nitrofuran antibacterials remains unclear. Nitrofurans inhibit many microbial enzyme systems, including those involved in carbohydrate metabolism.

Drugs of this class are usually metabolized to nitroreductases that lead to formation of nitro anion radicals that inhibit genetic translation, the ribosomal process whereby mRNA specifies the amino acid sequence in a polypeptide chain.

Interestingly, nitrofurans have continued to be active against organisms that have developed resistance to other antibacterials. Some nitrofurans are carcinogenic. However, nitrofurans are still used presently in medicine [112-116].

None of the nitrofurans is included in the list of Top 200 Drugs by sales for the 2010s.

REFERENCES

1. Scholar, E. M.; Pratt, W. B. *The Antimicrobial Drugs*, 2nd ed.; Oxford University Press, 2000; pp 207–233.
2. Shukla, N. P. Sulphonamides: a novel approach for antimicrobial chemotherapy. *Biosci., Biotechnol. Res. Asia* **2003,** *1* (1), 57–62.
3. Sammes, P. G. Sulphonamides and sulfones. In Hansch, C., Sammes, P. G., Taylor, J. B., Eds.; *Comprehensive Medicinal Chemistry*; Vol. 2; Pergamon Press, 1990; pp 255–270.
4. Scozzafava, A.; Carta, F.; Supuran, C. T. Secondary and tertiary sulphonamides: a patent review (2008-2012). *Expert Opin. Ther. Pat.* **2013,** *23* (2), 203–213.
5. Pareek, A.; Rani, P.; Kishore, D. A short review on: sulphonamides. *Int. J. Pharma Bio Sci.* **2013,** *4* (4), 812–820.
6. Vardanyan, R. S.; Hruby, V. J. *Synthesis of essential drugs;* Elsevier, 2006.
7. Ashfaq, M.; Shah, S. S. A.; Saheen, T. N. S.; Rivera, G. Synthetic routes of sulfonamide derivatives: a brief review. *Mini-Rev. Med. Chem.* **2013,** *13,* 70–86.
8. Kano, H.; Nishimura, H.; Nakajima, K.; Ogata, K. Sulfonamides, US 2888455 (1959).
9. Wormser, G. P.; Keusch, G. T.; Heel, R. C. Co-trimoxazole (trimethoprim-sulfamethoxazole). An updated review of its antibacterial activity and clinical efficacy. *Drugs* **1982,** *24* (6), 459–518.
10. Masters, P. A.; O'Bryan, T. A.; Zurlo, J.; Miller, D. Q.; Joshi, N. Trimethoprim-sulfamethoxazole revisited. *Arch. Intern. Med.* **2003,** *163* (4), 402–410.
11. Church, J. A.; Fitzgerald, F.; Walker, A. S.; Gibb, D. M.; Prendergast, A. J. The expanding role of co-trimoxazole in developing countries. *Lancet Infect. Dis.* **2015,** *15* (3), 327–339.
12. Huovinen, P. Resistance to trimethoprim-sulfamethoxazole. *Clin. Infect. Dis.* **2001,** *32* (11), 1608–1614.
13. Bhat, M. A.; Imran, M.; Khan, S. A.; Siddiqui, N. Biological activities of sulphonamides Indian. *J. Pharm. Sci.* **2005,** *67* (2), 151–159.
14. Brana, M. F.; Cacho, M.; Guisado, C. Sulphonamides: the magic group. *An. R. Acad. Farm.* **2006,** *72* (2), 317–341.

15. Shah, S. S. A.; Rivera, G.; Ashfaq, M. Recent advances in medicinal chemistry of sulphonamides. Rational design as anti-tumoral, anti-bacterial and anti-inflammatory agents. *Mini-Rev. Med. Chem.* **2013,** *13* (1), 70–86.
16. Tiwari, M.; Kishore, D. Introduction to the chemistry of sulphonamides and sulphones. *Int. J. Chem. Sci.* **2007,** *5* (5), 2454–2462.
17. Scozzafava, A.; Owa, T.; Mastrolorenzo, A.; Supuran, C. T. Anticancer and antiviral sulphonamides. *Curr. Med. Chem.* **2003,** *10* (11), 925–953.
18. Emami, S.; Shafiee, A.; Foroumadi, A. Structural features of new quinolones relationship to antibacterial activity against Gram-positive bacteria. *Mini-Rev. Med. Chem.* **2006,** *6* (4), 375–386.
19. Tse-Dinh, Y.-C. Exploring DNA topoisomerases as targets of novel therapeutic agents in the treatment of infectious diseases. *Infect. Disord.: Drug Targets* **2007,** *7* (1), 3–9.
20. Bradbury, B. J.; Pucci, M. J. Recent advances in bacterial topoisomerase inhibitors. *Curr. Opin. Pharmacol.* **2008,** *8* (5), 574–581.
21. Crumplin, G. C.; Smith, J. T. Nalidixic acid: an antibacterial paradox. *Antimicrob. Agents Chemother.* **1975,** *8* (3), 251–261.
22. Smith, J. T. Awakening the slumbering potential of the 4-quinolone antibacterials. *Pharm. J.* **1984,** *233* (6295), 299–305.
23. Mizuuchi, K.; Mizuuchi, M.; O'Dea, M. H.; Gellert, M. Cloning and simplified purification of *Escherichia coli* DNA gyrase A and B proteins. *J. Biol. Chem.* **1984,** *259* (14), 9199–9201.
24. Mugnaini, C.; Pasquini, S.; Corelli, F. The 4-quinolone-3-carboxylic acid motif as a multivalent scaffold in medicinal chemistry. *Curr. Med. Chem.* **2009,** *16* (14), 1746–1767.
25. Wagman, A. S.; Wentland, M. P. Quinolone antimicrobial agents. In *Comprehensive Medicinal Chemistry II*, Vol. 7; 8th ed.; Taylor, J. B., Triggle, D. J., Eds.; Elsevier, 2006; pp 567–596.
26. David, C.; Hooper, D. C.; Rubinstein, E. *Quinolone Antimicrobial Agents*, 3rd ed.; ASM Press, 2003.
27. Sissi, C.; Palumbo, M. The quinolone family: from antibacterial to anticancer agents. *Curr. Med. Chem.: Anti-Cancer Agents* **2003,** *3* (6), 439–450.
28. Wiles, J. A.; Bradbury, B. J.; Pucci, M. J. New quinolone antibiotics: a survey of the literature from 2005 to 2010. *Expert Opin. Ther. Pat.* **2010,** *20* (10), 1295–1319.
29. Heeb, S.; Fletcher, M. P.; Chhabra, S. R.; Diggle, S. P.; Williams, P.; Camara, M. Quinolones: from antibiotics to autoinducers. *FEMS Microbiol. Rev.* **2011,** *35* (2), 247–274.
30. Appelbaum, P. C. Quinolone activity against most anaerobes. *Drugs* **1999,** *58* (Suppl. 2), 60–64.
31. Ball, P. Quinolone generations: natural history or natural selection? *J. Antimicrob. Chemother.* **2000,** *46* (Topic 1), 17–24.
32. Stein, G. E. The 4-quinolone antibiotics: past, present, and future. *Pharmacotherapy* **1988,** *8* (6), 301–314.
33. Edwards, D. J.; Bowles, S. K.; Svensson, C. K.; Rybak, M. J. Inhibition of drug metabolism by quinolone antibiotics. *Clin. Pharmacokinet.* **1988,** *15* (3), 194–204.
34. Saravana, K. N.; Dhivya, D.; Vijayakumar, B. A focus on quinolones and its medicinal importance. *Int. J. Novel Trends Pharm. Sci.* **2011,** *1* (1), 28–36.
35. Cheng, G.; Hao, H.; Dai, M.; Liu, Z.; Yuan, Z. Antibacterial action of quinolones: from target to network. *Eur. J. Med. Chem.* **2013,** *66,* 555–562.
36. Boteva, A. A.; Krasnykh, O. P. The methods of synthesis, modification, and biological activity of 4-quinolones (review). *Chem. Heterocycl. Compd.* **2009,** *45,* 757–785.
37. Mitscher, L. A. Bacterial topoisomerase inhibitors: quinolone and pyridone antibacterial agents. *Chem. Rev. (Washington, DC, U. S.)* **2005,** *105* (2), 559–592.

38. Castro, W.; Navarro, M.; Biot, C. Medicinal potential of ciprofloxacin and its derivatives. *Future Med. Chem.* **2013,** *5* (1), 81–96.
39. Sharma, P. C.; Jain, A.; Jain, S.; Pahwa, R.; Yar, M. S. Ciprofloxacin: review on developments in synthetic, analytical, and medicinal aspects. *J. Enzyme Inhib. Med. Chem.* **2010,** *25* (4), 577–589.
40. Grohe, K.; Zeiler, H. J.; Metzger, K. 1-Cyclopropyl-6-fluoro-1,4-dihydro-4-oxo-7-piperazinoquinoline-3-carboxylic acids and an antibacterial agent containing them. *DE* **1983,** *314* (2854).
41. Ronald, A. R., Low, D. E., Eds. *Fluoroquinolone Antibiotics (Milestones in Drug Therapy.);* Springer Basel AG, 2003.
42. Bosso, J. A. An overview of the new fluoroquinolone antibiotics. *J. Infect. Dis. Pharmacother.* **1998,** *3* (3), 1–8.
43. De Souza, M.; Vinicius, N. New fluoroquinolones: a class of potent antibiotics. *Mini-Rev. Med. Chem.* **2005,** *5* (11), 1009–1017.
44. Da Silva, A. D.; De Almeida, M. V.; De Souza, M. V. N.; Couri, M. R. C. Biological activity and synthetic methodologies for the preparation of fluoroquinolones, a class of potent antibacterial agents. *Curr. Med. Chem.* **2003,** *10* (1), 21–39.
45. Ball, P.; Tillotson, G. Tolerability of fluoroquinolone antibiotics: past, present and future. *Drug Saf.* **1995,** *13* (6), 343–358.
46. Walker, R. C.; Wright, A. J. The fluoroquinolones. *Mayo Clin. Proc.* **1991,** *66* (12), 1249–1259.
47. Paton, J. H.; Reeves, D. S. The fluoroquinolone antibiotics. Microbiology, pharmacokinetics and clinical use. *Drugs* **1988,** *36* (2), 193–228.
48. Bradbury, B. J.; Pucci, M. J. Recent advances in bacterial topoisomerase inhibitors. *Curr. Opin. Pharmacol.* **2008,** *8* (5), 574–581.
49. King, D. E.; Malone, R.; Lilley, S. H. New classification and update on the quinolone antibiotics. *Am. Fam. Physician* **2000,** *61* (9), 2741–2748.
50. Davis, R.; Bryson, H. M. Levofloxacin: a review of its antibacterial activity, pharmacokinetics and therapeutic efficacy. *Drugs* **1994,** *47* (4), 677–700.
51. Wimer, S. M.; Schoonover, L.; Garrison, M. W. Levofloxacin: a therapeutic review. *Clin. Ther.* **1998,** *20* (6), 1049–1070.
52. North, D. S.; Fish, D. N.; Redington, J. J. Levofloxacin, a second-generation fluoroquinolone. *Pharmacotherapy* **1998,** *18* (5), 915–935.
53. Langtry, H. D.; Lamb, H. M. Levofloxacin. Its use in infections of the respiratory tract, skin, soft tissues and urinary tract. *Drugs* **1998,** *56* (3), 487–515.
54. Norrby, S. R. Levofloxacin. *Expert Opin. Pharmacother* **1999,** *1* (1), 109–119.
55. Anderson, V. R.; Perry, C. M. Levofloxacin: a review of its use as a high-dose, short-course treatment for bacterial infection. *Drugs* **2008,** *68* (4), 535–565.
56. Martin, S. J.; Meyer, J. M.; Chuck, S. K.; Jung, R.; Messick, C. R.; Pendland, S. L. Levofloxacin and sparfloxacin: new quinolone antibiotics. *Ann. Pharmacother.* **1998,** *32* (3), 320–336.
57. Noreddin, A. M.; Elkhatib, W. F. Levofloxacin in the treatment of community-acquired pneumonia. *Expert Rev. Anti-Infect. Ther.* **2010,** *8* (5), 505–514.
58. Croom, K. F.; Goa, K. L. Levofloxacin: a review of its use in the treatment of bacterial infections in the United States. *Drugs* **2003,** *63* (24), 2769–2802.
59. Hurst, M.; Lamb, H. M.; Scott, L. J.; Figgitt, D. P. Levofloxacin: An updated review of its use in the treatment of bacterial infections. *Drugs* **2002,** *62* (14), 2127–2167.
60. Hayakawa, I.; Tanaka, Y.; Hiramitsu, T. Benzoxazine derivatives, EP 47005 (1982).

61. Hayakawa, I.; Hiramitsu, T.; Tanaka, Y. Synthesis and antibacterial activities of substituted 7-oxo-2,3-dihydro-7H-pyrido[1,2,3-de][1,4]benzoxazine-6-carboxylic acids. *Chem. Pharm. Bull.* **1984,** *32* (12), 4907–4913.
62. Mitscher, L. A.; Chu, D. T. Process for preparation of racemic and optically active ofloxacin and related derivatives, US 4777253 (1988).
63. Hayakawa, I.; Atarashi, S.; Yokohama, S.; Imamura, M.; Sakano, K.; Furukawa, M. Synthesis and antibacterial activities of optically active ofloxacin. *Antimicrob. Agents Chemother.* **1986,** *29* (1), 163–164.
64. Hayakawa, I.; Atarashi, S.; Yokohama, S.; Imamura, M.; Sakano, K.; Higashihashi, N.; Ohshima, M. Optically active (S)-(-)-pyridobenzoxazinecarboxylate derivatives, their intermediates, use as antimicrobials, EP 206283 (1986).
65. Kang, S. B.; Park, S.; Kim, Y. H.; Kim, Y. An improved synthesis of levofloxacin. *Heterocycles* **1977,** *45* (1), 137–145.
66. Atarashi, S.; Yokohama, S.; Yamazaki, K.; Sakano, K.; Imamura, M.; Hayakawa, I. Synthesis and antibacterial activities of optically active ofloxacin and its fluoromethyl derivative. *Chem. Pharm. Bull.* **1987,** *35* (5), 1896–1902.
67. Bower, J. F.; Szeto, P.; Gallagher, T. Enantiopure 1,4-benzoxazines via 1,2-cyclic sulfamidates, synthesis of levofloxacin. *Org. Lett.* **2007,** *9* (17), 3283–3286.
68. Berridge, M. S.; Burnazi, E. M. Synthesis of [11C] levofloxacin. *J. Labelled Compd. Radiopharm.* **2001,** *44* (12), 859–864.
69. Sato, K.; Hoshino, K.; Tanaka, M.; Hayakawa, I.; Osada, Y. Antimicrobial activity of DU-(6859), a new potent fluoroquinolone, against clinical isolates. *Antimicrob. Agents Chemother.* **1992,** *36* (7), 1491–1498.
70. Atarashi, S.; Tsurumi, H.; Fujiwara, T.; Hayakawa, I. Asymmetric reduction of 7,8-difluoro-3-methyl-2H-1,4-benzoxazine. Synthesis of a key intermediate of (S)-(-)-ofloxacin (DR-3355). *J. Heterocycl. Chem.* **1991,** *28* (2), 329–331.
71. Hayakawa, I.; Furuhama, K.; Takayama, S.; Osada, Y. Levofloxacin, a new quinolone antibacterial agent: an introductory overview. *Arzneim. Forsch.* **1992,** *42* (3A), 363–364.
72. Schriewer, M.; Grohe, K.; Zeiler, H. J.; Metzger, K. G. Preparation of chiral-bridged quinolone bactericides, including S-ofloxacin, DE 3543513 (1987).
73. Van Zoest, W. J.; Marx, A. F.; Koger, H. S.; Booy, J. Optically active benzoxazines and benzothiazines and a process for their stereospecific preparation, EP 368410 (1990).
74. Egawa, H.; Miyamoto, T.; Matsumoto, J. Pyridonecarboxylic acids as antibacterial agents. Part 6. A new synthesis of 7H-pyrido[1,2,3-de][1,4]benzoxazine derivatives including an antibacterial agent, ofloxacin.. *Chem. Pharm. Bull.* **1986,** *34* (10), 4098–4102.
75. Sakano, K.; Yokohama, S.; Hayakawa, I.; Atarashi, S.; Kadoya, S. Optical resolution of (R,S)-3-(acetoxymethyl)-7,8-difluoro-2,3-dihydro-4H-[1,4]benzoxazine. *Agric. Biol. Chem.* **1987,** *51* (5), 1265–1270.
76. Al-Trawneh, S. A.; Zahra, J. A.; Kamal, M. R.; El-Abadelah, M. M.; Zani, F.; Incerti, M.; Cavazzoni, A.; Alfieri, R. R.; Petronini, P. G.; Vicini, P. Synthesis and biological evaluation of tetracyclic fluoroquinolones as antibacterial and anticancer agents. *Bioorg. Med. Chem.* **2010,** *18* (16), 5873–5884.
77. Van Zoest, W. J.; Marx, A. F.; Koger, H. S.; Booy, J. V. Optically active benzoxazines and benzothiazines and a process for their stereospecific preparation, Eur. Pat. Appl., EP 368410 (1990).
78. Fujiwara, T.; Ebata, T. Propoxybenzene derivatives, their preparation and use in the preparation of benzoxazine derivatives, especially antibacterials such as ofloxacin, Eur. Pat. Appl., EP 322815 (1989).

79. Limberakis, C. Quinolone antibiotics: levofloxacin (Levaquin), moxifloxacin, gemifloxacin (Factive), and garenoxacin (T-3811). In *Art of Drug Synthesis;* Johnson, D. S., Li, J. J., Eds.; Wiley, 2007; pp 39–69.
80. Barrett, J. F. Moxifloxacin (Bayer). *Curr. Opin. Invest. Drugs (BioMed Cent.)* **2000,** *1* (1), 45–51.
81. Balfour, J. A. B.; Wiseman, L. R. Moxifloxacin. *Drugs* **1999,** *57* (3), 363–373.
82. MacGowan, A. P. Moxifloxacin (Bay 12-8039): a new methoxy quinolone antibacterial. *Expert Opin. Invest. Drugs* **1999,** *8* (2), 181–199.
83. Keating, G. M.; Scott, L. J. Moxifloxacin: a review of its use in the management of bacterial infections. *Drugs* **2004,** *64* (20), 2347–2377.
84. Ball, P. Moxifloxacin (Avelox): an 8-methoxyquinolone antibacterial with enhanced potency. *Int. J. Clin. Pract.* **2000,** *54* (5), 329–332.
85. Ball, P.; Stahlmann, R.; Kubin, R.; Choudhri, S.; Owens, R. Safety profile of oral and intravenous moxifloxacin: cumulative data from clinical trials and postmarketing studies. *Clin. Ther.* **2004,** *26* (7), 940–950.
86. Petersen, U. Quinolone antibiotics: the development of moxifloxacin. In *Analogue-Based Drug Discovery;* Fischer, J., Ganellin, C. R., Eds.; Wiley-VCH, 2006; pp 315–370.
87. Caeiro, J.-P.; Iannini, P. B. Moxifloxacin (Avelox): a novel fluoroquinolone with a broad spectrum of activity. *Expert Rev. Anti-Infect. Ther.* **2003,** *1* (3), 363–370.
88. Wiederhold, N. P.; Ritchie, D. J. Moxifloxacin: a review of its in vitro activity, clinical efficacy, and adverse effects. *J. Infect. Dis. Pharmacother.* **2002,** *6* (1), 1–26.
89. Balfour, J. A. B.; Lamb, H. M. Moxifloxacin: a review of its clinical potential in the management of community-acquired respiratory tract infections. *Drugs* **2000,** *59* (1), 115–139.
90. Nightingale, C. H. Moxifloxacin, a new antibiotic designed to treat community-acquired respiratory tract infections: a review of microbiologic and pharmacokinetic-pharmacodynamic characteristics. *Pharmacotherapy* **2000,** *20* (3), 245–256.
91. Blondeau, J. ,M.; Hansen, G. T. Moxifloxacin: a review of the microbiological, pharmacological, clinical and safety features. *Expert Opin. Pharmacother.* **2001,** *2* (2), 317–335.
92. Zhanel, G. G.; Noreddin, A. M. Pharmacokinetics and pharmacodynamics of the new fluoroquinilones: focus on respiratory infections. *Curr. Opin. Pharmacol.* **2001,** *1* (5), 459–463.
93. Burkhardt, O.; Welte, T. 10 years' experience with the pneumococcal moxifloxacin. *Expert Rev. Anti-Infect. Ther.* **2009,** *7* (6), 645–668.
94. Van Bambeke, F.; Tulkens, P. M. Safety profile of the respiratory fluoroquinolone moxifloxacin: comparison with other fluoroquinolones and other antibacterial classes. *Drug Saf.* **2009,** *32* (5), 359–378.
95. Lode, H. M.; Schmidt-Ioanas, M. Moxifloxacin: update and perspectives after 8 years of usage. *Expert Rev. Respir. Med.* **2008,** *2* (4), 443–453.
96. Miravitlles, M.; Anzueto, A. Moxifloxacin: a respiratory fluoroquinolone. *Expert Opin. Pharmacother.* **2008,** *9* (10), 1755–1772.
97. Domagala, J. M. Structure-activity and structure-side-effect relationships for the quinolone antibacterials. *J. Antimicrob. Chemother.* **1994,** *33* (4), 685–706.
98. Petersen, U.; Krebs, A.; Schenke, T.; Philipps, T.; Grohe, K.; Bremm, K.; Endermann, R.; Metzger, K. G.; Haller, I. Preparation of (diazabicyclononyl)quinolones and related compounds as antibacterials, EP 550903 (1993).
99. Fey, P. Preparation of (S,S)-8-benzyl-2,8-diazabicyclo[4.3.0]nonane by resolution using L-tartaric acid, WO 9958532 (1999).

100. Chava, S.; Gorantla, S. R.; Vasireddy, U. R.; Dammalapati, V. L. N. Process for preparation of moxifloxacin hydrochloride monohydrate from Et 1-cyclopropyl-6,7-difluoro-8-methoxy-4-oxo-1,4-dihydro-3-quinolinecarboxylate via (4aS-cis)-1-cyclopropyl-7-(2,8-diazabicyclo[4.3.0]non-8-yl)-6-fluoro-8-methoxy-4-oxo-1,4-dihydro-3-quinoline carboxylic acid (O3,O4)-bis(acyloxy) borate, WO 2005012285 (2005).
101. Rao, D. R.; Kankan, R. N.; Srinivas, P. L.; Ravikumar, P.; Gangrade, M.; Kanathala, S. Process for the preparation of moxifloxacin hydrochloride Form CWO 2008059223 (2008).
102. Dandala, R.; Mitra, J.; Gupta, A. K.; Meenakshisunderam, S. Crystalline form of moxifloxacin hydrochloride and process for its preparation, WO 2006134491 (2006).
103. Al Omari, M. M. H.; Jaafari, D. S.; Al-Sou'od, K. A.; Badwan, A. A. Moxifloxacin hydrochloride, Profiles Drug Subst. *Excipients, Relat. Methodol.* **2014,** *39,* 299–431.
104. Ludescher, J.; Pise, A. C.; Holkar, A. G.; Metkar, S. Process for the preparation of moxifloxacin hydrochloride, WO 2008138759 (2008).
105. Somberg, J. C.; Ranade, V. V. Synthesis, characterization and biological action of optically active isomers of floxacins, WO 2006052264 (2006).
106. Palomo, N. F.; Cosme G., A.; Villasante P. J.; Fernandez, L. S. P.; Molina, P. A. Process for preparation of moxifloxacin (hydrochloride) using a one pot method in which a smaller amount of bicyclic amine is used, EP 1832587 (2007).
107. Castellin, A.; Padovan, P.; Liu, J.; Zhou, Y.; Lin, F. Regioselective process for preparing moxifloxacin and salts thereof, US 20110230661 (2011).
108. Iwata, M.; Kimura, T.; Fujiwara, Y.; Katsube, T. Preparation of alkoxyfluoroquinolonecarboxylic acid derivatives as medical bactericides, EP 241206 (1987).
109. Seidel, D.; Conrad, M.; Brehmer, P.; Mohrs, K.; Petersen, U. Synthesis of carbon-14 labeled moxifloxacin hydrochloride. *J. Labelled Compd. Radiopharm.* **2000,** *43* (8), 795–805.
110. Ramesh, P.; Harini, T.; Fadnavis, N. W. Efficient resolution of cis-(±)-dimethyl 1-acetylpiperidine-2,3-dicarboxylate with soluble *Candida antarctica* lipase B (CAL B). *Org. Process Res. Dev.* **2015,** *19* (1), 296–301.
111. Gehring, R.; Mohrs, K.; Heilmann, W.; Diehl, H. Preparation of 8-methoxyquinolone-carboxylates, DE 19751948 (1999).
112. Beck, K.; Julius, H. C. Nitrofurans. *Infect. Dis. Ther.* **1994,** *9,* 391–401.
113. Hamilton-Miller, J. M. T. Nitrofurans. In *Antibiotic and Chemotherapy: Anti-Infective Agents and Their Use in Therapy,* 7th ed.; O'Grady, F., Lambert, H. P., Eds. Churchill Livingstone, 1997; pp 396–403.
114. Ebetino, F. F. Antibacterial agents, synthetic—nitrofurans. In *Kirk-Othmer Encyclopedia of Chemical Technology*, Vol. 2; 3rd ed.; Grayson, M., Eckroth, D., Eds.; Wiley-Interscience, 1978; pp 790–794.
115. Miura, K.; Reckendorf, H. K. Nitrofurans. *Prog. Med. Chem.* **1967,** *5,* 320–381.
116. Guay, D. R. An update on the role of nitrofurans in the management of urinary tract infections. *Drugs* **2001,** *61* (3), 353–564.

Chapter 32

Antimicobacterial Drugs

Mycobacterium is aerobic bacteria of the genus of *Actinobacteria*, a group that includes the causative agents of tuberculosis and leprosy [1-12].

Tuberculosis is an ancient disease of mankind caused by various strains of mycobacteria, usually *Mycobacterium tuberculosis*.

Leprosy, also known as Hansen disease, is a chronic disease caused by the bacteria *Mycobacterium leprae* and *Mycobacterium lepromatosis*.

Antimycobacterials are used in the treatment of diseases caused by members of the *Mycobacterium* genus, including tuberculosis and leprosy, which have affected humans since antiquity.

32.1 ANTITUBERCULOSIS DRUGS

Tuberculosis is one of the deadliest infectious diseases for humans, and lethal in many cases. It is caused by various strains of mycobacteria, usually *M. tuberculosis*, a kind of intracellular bacterium. *M. tuberculosis* can attack many of the human systems, such as the respiratory system, central nervous system, and urinary system.

Tuberculosis remains a major global health problem as it is the second leading cause of death (HIV being the leading cause) from an infectious disease worldwide.

M. tuberculosis has a very puzzling pathogenesis. It is believed that entering into the human body *M. tuberculosis* initiates an immune macrophage reaction. In the early stage of infection, *M. tuberculosis* growth was inhibited by of macrophages. With the increase of the breeding rate of virulent strains human macrophages themselves underwent significant apoptosis after they came into contact with macrophages infected with *M.* Therefore, the inhibition of *M. tuberculosis* by regulation of the macrophage apoptosis process was the key to the prevention and treatment of tuberculosis. The fatality rate and disability rate caused by tuberculosis is high and is rising in many areas of the world. Because of drug resistance, diagnosis and therapy have become harder. The treatment of tuberculosis is still a major world health problem and new drugs are in demand [13-27].

The first-line antituberculosis drugs—isoniazid (**32.1.1**), pyrazinamide (**32.1.2**), ethambutol (**32.1.3**), and rifampicin (**32.1.4**)—form the core of the

Synthesis of Best-Seller Drugs. http://dx.doi.org/10.1016/B978-0-12-411492-0.00032-8

initial treatment regimen. Rifapentine (**32.1.5**) and rifabutin (**32.1.6**) may be used as a substitute for rifampicin in the treatment of tuberculosis caused by organisms that are presumed to be susceptible to this agent (Fig. 32.1.).

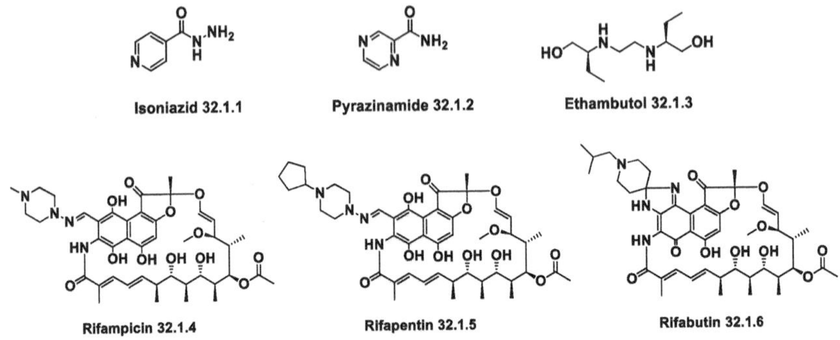

FIG. 32.1 **The first-line antituberculosis drugs.**

The mode of action of isoniazid is complex.

For decades after its introduction, the mechanisms of action of isoniazid remaines unclear. Isoniazid inhibits the synthesis of mycoloic acids, an essential component of the bacterial cell wall. It activate the mycobacterial enzyme KatG, generates reactive species that form adducts with NAD(+) and NADP(+) that are potent inhibitors of lipid and nucleic acid biosynthesis [28].

The mechanism of action of pyrazinamide, despite its discovery more than 50 years ago, is the least understood of all antituberculosis drugs. It is clear that pyrazinamide is a prodrug that is converted into the active form, pyrazinoic acid, by bacterial nicotinamidase/pyrazinamidase, which may inhibit the fatty acid synthetase enzyme of *M. tuberculosis* [29-31].

The mechanism of action of ethambutol also is not completely known. It diffuses into actively growing mycobacteria cells inhibiting the synthesis of one or more metabolites, thus impairing cell metabolism, arresting multiplication, and resulting in cell death [32].

The mechanism of action of rifampicin is an inhibition of the β subunit of the RNA polymerase of prokaryotes, including *M. tuberculosis* [33,34].

There are second-line or reserve drugs that are used for the treatment of tuberculosis. These drugs are reserved for special situations, such as drug intolerance or resistance.

Second-line drugs include ethionamide (**32.1.7**), cycloserine (**32.1.8**), ρ-aminosalicylic acid (**32.1.9**), levofloxacin (**32.1.10**), gatifloxacin (**32.1.11**), moxifloxacin (**32.1.12**), amikacin/kanamycin (**32.1.13/32.1.14**), capreomycin (**32.1.15**), and streptomycin (**32.1.16**). Streptomycin (**32.1.16**) was formerly considered to be a first-line drug and in some instances, is still used in initial treatment, but the increasing prevalence of resistance to streptomycin decreased its overall usefulness (Fig. 32.2.).

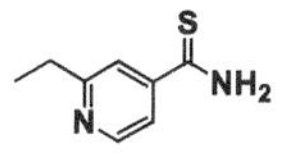

Ethionamide 32.1.7

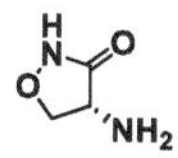

Cycloserine 32.1.8

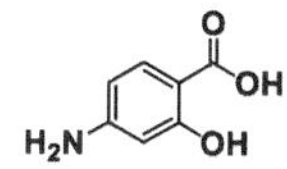

p-Aminosalicylic acid 32.1.9

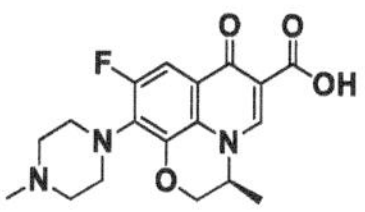

Levofloxacin 32.1.10

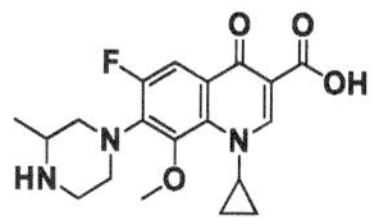

Gatifloxacin 32.1.11

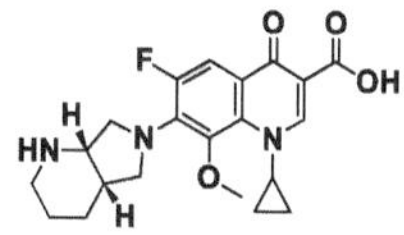

Moxifloxacin 32.1.12

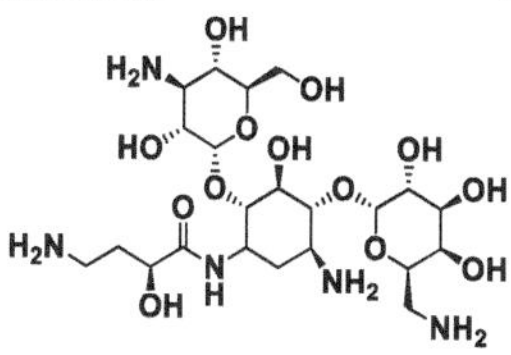

Amikacin 32.1.13

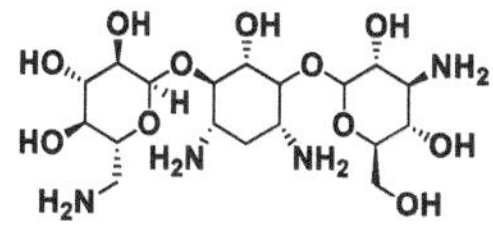

Kanamycin 32.1.14

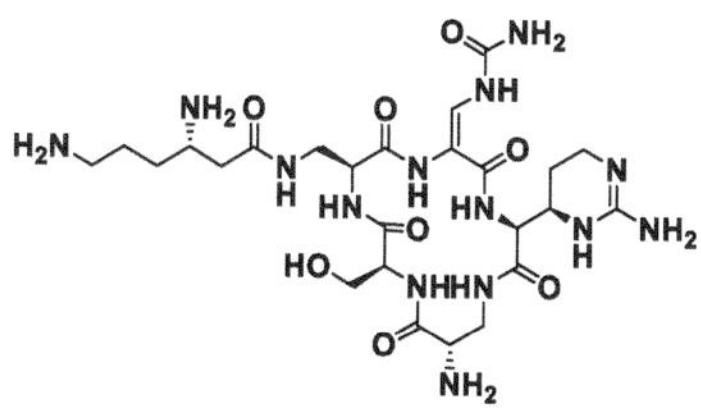

Capreomycin 32.1.15

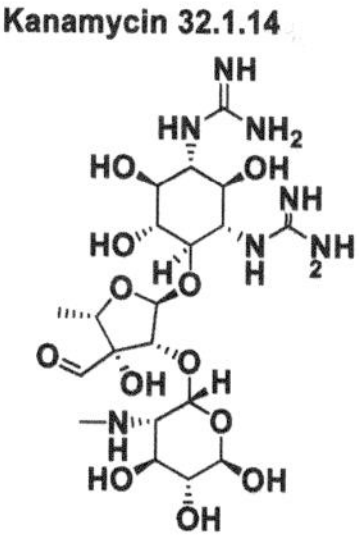

Streptomycin 32.1.16

FIG. 32.2 The second-line antituberculosis drugs.

There have been no new antituberculosis drugs since rifampin was introduced in the 1960s. New drugs designed to attack novel targets would be of enormous global benefit.

The emergence of multidrug-resistant and extensively drug-resistant strains of *M. tuberculosis* require newly developed antituberculosis agents. There is an urgent need for new drugs. Approximately 250 receptors are known and thousands of compounds are available in the literature that, in principle, can be used for the design of new antituberculosis compounds acting by a specific manner as new antituberculosis drugs [12,16].

32.2 ANTILEPROSY DRUGS

Leprosy (Hansen disease) is a chronic infectious disease caused by *M. leprae* that mainly affects the skin and peripheral nerves.

Leprosy has been around since the time of the ancient civilizations of China, Egypt, and India. The first known written mention of leprosy is dated 600 BC. Throughout history, people afflicted were often ostracized by their communities and families. Leprosy cases are found in the relatively highly endemic countries of

Nepal, Madagascar, Myanmar, Indonesia, and especially India and Brazil, where leprosy has affected humans for millennia and remains an important health problem.

The disease is often surrounded by terrifying, negative stigmas and tales, with leprosy patients being shunned as outcasts.

Feared as a highly contagious and devastating disease, it is now well established that leprosy is not highly transmissible, is treatable, and, with early diagnosis and treatment, is not disabling.

Although transmission of *M. leprae* is not entirely understood, it is thought that the respiratory system's long-term exposure to airborne droplets is the main route of infection.

The disease develops slowly (taking from 6 months to 40 years) and results in skin lesions and deformities, most often affecting the cooler places on the body (e.g., eyes, nose, earlobes, hands, feet, and testicles).

Yet leprosy remains the most misunderstood human infectious disease.

It could be best explained as two conjoined diseases. The first is a chronic mycobacterial infection that elicits an extraordinary range of cellular immune responses in humans. The second is a peripheral neuropathy that is initiated by the infection and the accompanying immunological events.

The infection is curable but not preventable, and leprosy remains one of global health problem, especially in the developing world.

Modern antibacterial therapy typically consists of combinations of dapsone (**32.2.1**) and rifampicin (**32.2.2**) with or without clofazimine (**32.2.3**) (Fig. 32.3.).

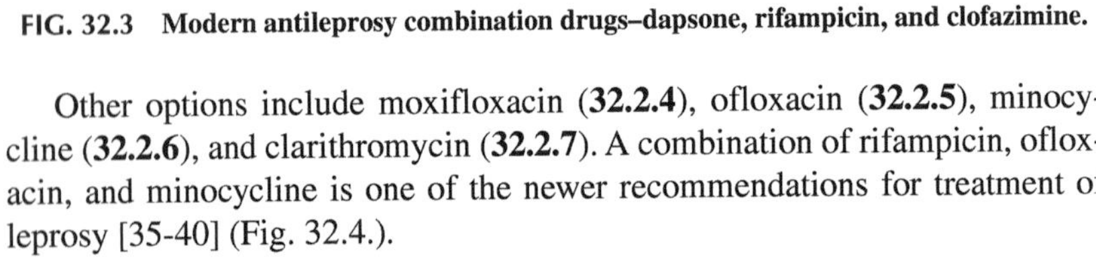

FIG. 32.3 Modern antileprosy combination drugs–dapsone, rifampicin, and clofazimine.

Other options include moxifloxacin (**32.2.4**), ofloxacin (**32.2.5**), minocycline (**32.2.6**), and clarithromycin (**32.2.7**). A combination of rifampicin, ofloxacin, and minocycline is one of the newer recommendations for treatment of leprosy [35-40] (Fig. 32.4.).

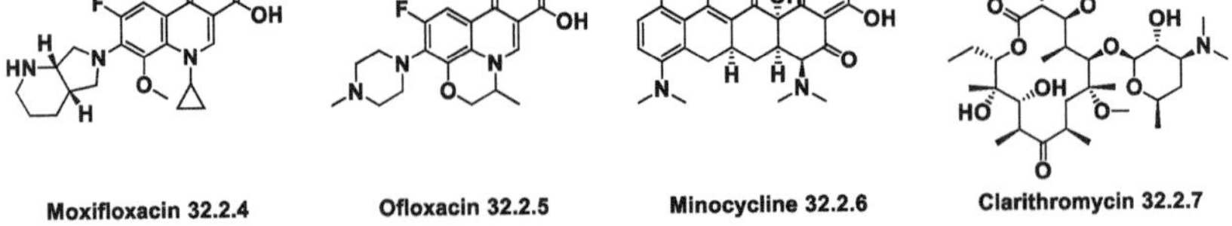

FIG. 32.4 Other drugs for the treatment of leprosy.

The synthesis of essential antimycobacterial drugs is described in our previous book [41].

The tendencies of newly developing compounds are perfectly presented in the recent review by Asif [2].

None of the antimycobacterial drugs is included in the list of Top 200 Drugs by sales for the 2010s.

REFERENCES

1. Biava, M.; Porretta, G. C.; Deidda, D.; Pompei, R. New trends in development of antimycobacterial compounds. *Infect. Disord.: Drug Targets* **2006,** *6* (2), 159–172.
2. Asif, M. A review of antimycobacterial drugs in development. *Mini-Rev. Med. Chem.* **2012,** *12* (13), 1404–1418.
3. Primm, T. P.; Franzblau, S. G. Recent advances in methodologies for the discovery of antimycobacterial drugs. *Curr. Bioact. Compd.* **2007,** *3* (3), 201–208.
4. Ma, Z.; Ginsberg, A. M.; Spigelman, M. Antimycobacterial agents. In 8th ed.; Taylor, J. B., Triggle, D. J., Eds.; *Comprehensive Medicinal Chemistry II, vol. 7*; Elsevier, 2006; pp 699–730.
5. Ballell, L.; Field, R. A.; Duncan, K.; Young, R. J. New small-molecule synthetic antimycobacterials. *Antimicrob. Agents Chemother.* **2005,** *49* (6), 2153–2163.
6. van Daele, I.; van Calenbergh, S. Patent developments in antimycobacterials small-molecule therapeutics. *Expert Opin. Ther. Pat.* **2005,** *15* (2), 131–140.
7. Waisser, K.; Hladuvkova, J.; Hrabalek, A.; Klimesova, V.; Kubicova, L.; Kunes, J.; Palat, K.; Machacek, M.; Vinsova, J.; Buchta, V.; Jilek, P.; Odlerova, Z. New pharmacophores of antimycobacterial and antimycotical activity. *Folia Pharm. Univ. Carol.* **1998,** *21–22,* 69–81.
8. Tomioka, H. Prospects for development of new antimycobacterial drugs. *J. Infect. Chemother.* **2000,** *6* (1), 8–20.
9. Newton, S. M.; Lau, C.; Wright, C. W. A review of antimycobacterial natural products. *Phytother. Res.* **2000,** *14* (5), 303–322.
10. Santhosh, R. S.; Suriyanarayanan, B. Plants: a source for new. *Planta Med.* **2014,** *80* (1), 9–21.
11. Garcia, A.; Bocanegra-Garcia, V.; Palma-Nicolas, J. P.; Rivera, G. Recent advances in antitubercular natural products. *Eur. J. Med. Chem.* **2012,** *49,* 1–23.
12. Speck-Planche, A.; Scotti, M. T.; de Paulo-Emerenciano, V. Current pharmaceutical design of antimycobacterial drugs: future perspectives. *Curr. Pharm. Des.* **2010,** *16* (24), 2656–2665.
13. Tomioka, H. Current status of some antituberculosis drugs and the development of new antituberculous agents with special reference to their in vitro and in vivo antimicrobial activities. *Curr. Pharm. Des.* **2006,** *12* (31), 4047–4070.
14. Tomioka, H. Development of new antituberculous drugs: strategies for new drug targets and drugs delivery. *Drug Des. Rev.–Online* **2005,** *2* (6), 427–434.
15. Sapna, P.; Mathur, A. G.; Amol, P. Advances in anti-tuberculosis drugs. *J. Drug Delivery Ther.* **2014,** *4* (5), 69–73.
16. Janssen, S.; Jayachandran, R.; Khathi, L.; Zinsstag, J.; Grobusch, M. P.; Pieters, J. Exploring prospects of novel drugs for tuberculosis. *Drug Des., Dev. Ther.* **2012,** *6,* 217–224.
17. Dover, L. G.; Coxon, G. D. Current status and research strategies in tuberculosis drug development. *J. Med. Chem.* **2011,** *54* (18), 6157–6165.

18. Janin, Y. L. Antituberculosis drugs: ten years of research. *Bioorg. Med. Chem.* **2007,** *15* (7), 2479–2513.
19. Handbook of anti-tuberculosis agents. Introduction. *Tuberculosis* **2008,** *88* (2), 85–86.
20. Mitchison, D. A. Role of individual drugs in the chemotherapy of tuberculosis. *Int. J. Tuberc. Lung Dis.* **2000,** *4* (9), 796–806.
21. Dartois, V.; Leong, F. J.; Dick, T. Tuberculosis drug discovery: issues, gaps and the way forward. *Drug Discovery Infect. Dis.* **2009,** 415–440.
22. Bhowruth, V.; Dover, L. G.; Besra, G. S. Tuberculosis chemotherapy: recent developments and future perspectives. *Prog. Med. Chem.* **2007,** *45,* 169–203.
23. Duncan, K. Identification and validation of novel drug targets in tuberculosis. *Curr. Pharm. Des.* **2004,** *10* (26), 3185–3194.
24. Grassi, C.; Peona, V. New drugs for tuberculosis. *Eur. Respir. J.* **1995,** *8* (Suppl. 20), 714s–718s.
25. Verma, A.; Kumar, K.; Om, P. Discovery of new drugs against tuberculosis: history guides. *Iran. J. Clin. Infect. Dis.* **2012,** *7* (4), 109–112.
26. Crick, D. C.; Brennan, P. J. Antituberculosis drug research. *Curr. Opin. Anti-Infect. Invest. Drugs* **2000,** *2* (2), 154–163.
27. Speck-Planche, A.; Scotti, M. T.; de Paulo-Emerenciano, V. Current pharmaceutical design of antimycobacterial drugs: future perspectives. *Curr. Pharm. Des.* **2010,** *16* (24), 2656–2665.
28. Timmins, G. S.; Deretic, V. Mechanisms of action of isoniazid. *Mol. Microbiol.* **2006,** *62* (5), 1220–1227.
29. Zimhony, O.; Cox, J. S.; Welch, J. T.; Vilcheze, C.; Jacobs, W. R., Jr. Pyrazinamide inhibits the eukaryotic-like fatty acid synthetase I (FASI) of *Mycobacterium tuberculosis. Nat. Med. (N. Y., NY, U. S.)* **2000,** *6* (9), 1043–1047.
30. Ngo, S. C.; Zimhony, O.; Chung, W. J.; Sayahi, H.; Jacobs, W. R., Jr.; Welch, J. T. Inhibition of isolated *Mycobacterium tuberculosis* fatty acid synthase I by pyrazinamide analogs. *Antimicrob. Agents Chemother.* **2007,** *51* (7), 2430–2435.
31. Zhang, Y.; Wade, M. M.; Scorpio, A.; Zhang, H.; Sun, Z. Mode of action of pyrazinamide: disruption of *Mycobacterium tuberculosis* membrane transport and energetics by pyrazinoic acid. *J. Antimicrob. Chemother.* **2003,** *52* (5), 790–795.
32. Belanger, A. E.; Besra, G. S.; Ford, M. E.; Mikusova, K.; Belisle, J. T.; Brennan, P. J.; Inamine, J. M. The embAB genes of *Mycobacterium avium* encode an arabinosyl transferase involved in cell wall arabinan biosynthesis that is the target for the antimycobacterial drug ethambutol. *Proc. Natl. Acad. Sci. U. S. A.* **1996,** *93* (21), 11919–11924.
33. Campbell, E. A.; Korzheva, N.; Mustaev, A.; Murakami, K.; Nair, S.; Goldfarb, A.; Darst, S. A. Structural mechanism for rifampicin inhibition of bacterial RNA polymerase. *Cell* **2001,** *104* (5), 901–912.
34. De La Iglesia, A. I.; Morbidoni, Y. H. R. Mechanisms of action of and resistance to rifampicin and isoniazid in Mycobacterium tuberculosis: new information on old friends. *Rev. Argent. Microbiol.* **2006,** *38* (2), 97–109.
35. Moschella, S. L. Leprosy: epidemiology and present and possible future therapeutic approaches. *Drugs Future* **2006,** *31* (11), 961–967.
36. Scollard, D. M.; Adams, L. B.; Gillis, T. P.; Krahenbuhl, J. L.; Truman, R. W.; Williams, D. L. The continuing challenges of leprosy. *Clin. Microbiol. Rev.* **2006,** *19* (2), 338–381.
37. Suzuki, K.; Akama, T.; Kawashima, A.; Yoshihara, A.; Yotsu, R. R.; Ishii, N. Current status of leprosy: epidemiology, basic science and clinical perspectives. *J. Dermatol.* **2012,** *39* (2), 121–129.

38. Legendre, D. P.; Muzny, C. A.; Swiatlo, E. Hansen's disease (leprosy): current and future pharmacotherapy and treatment of disease-related immunologic reactions. *Pharmacotherapy* **2012,** *32* (1), 27–37.
39. Hooper, M. The medicinal chemistry of anti-leprosy drugs. *Chem. Soc. Rev.* **1987,** *16* (4), 437–465.
40. Matsuoka, M. Drug resistance in leprosy. *Jpn. J. Infect. Dis.* **2010,** *63* (1), 1–7.
41. Vardanyan, R. S.; Hruby, V. J. *Synthesis of Essential Drugs;* Elsevier, 2006.

Chapter 33

Antifungal Drugs

Human fungal infections have increased dramatically in recent years, becoming a societal burden.

Fungi are a diversified group of microorganisms that are present in the environment, being a part of the normal flora of human and animals, and have the ability to cause injury ranging from mild superficial infections to severe life-threatening invasive infections.

The major human fungal pathogens are the *Candida* species, primarily *Candida albicans* and *Candida glabrata.* Although *C. albicans* is still the primary cause of candidiasis, the incidence of infections associated with *C. glabrata* has grown to the point that this species is the second most common pathogen linked to this class of fungal disease.

Fungi are eukaryotes that absorb nutrients directly through the cell walls and, like animals, obtain their carbon and energy from other organisms. Fungi are considered to be slightly more advanced than bacteria. Fungi have the ability to grow on and in both invertebrate and vertebrate animals. There are relatively few fungal pathogens of vertebrates (200 to 300 species), but some of them can have a devastating impact.

Several different types of fungal infections in humans can be classified as opportunistic (those that develop mainly in immunocompromised hosts, such as AIDS, azotemia, diabetes mellitus, lymphoma, leukemia, other hematologic cancers) or primary (infections that develop in immunocompetent hosts), systemic or local.

The incidence of invasive fungal infections is increasing, primarily because of the rising number of both immunocompromised and critically ill patients. In addition, the spectrum of invasive fungal infections continues to evolve with the emergence of rare and resistant fungal pathogens.

Local fungal infections typically involve the skin (i.e., foot fungus) and mucosal infections (mouth and/or vagina). In systemic fungal infections (fungus in the blood and tissues), the outcome can result in severe inflammatory reactions, which result in morbidity and mortality.

The treatment of fungal diseases depends of location of the infection.

Antifungal drugs are agents that selectively eliminate fungal pathogens from a host with minimal toxicity to the host.

The common fungal infections are superficial and are treated with one or several topical drugs. The systemic mycoses are very difficult to treat.

Synthesis of Best-Seller Drugs. http://dx.doi.org/10.1016/B978-0-12-411492-0.00033-X

When compared to antibacterials, the armamentarium of antifungal drugs armamentarium is very limited and includes five major classes of antifungal medications: polyenes; azoles; echinocandins, allylamines, antimetabolites and some miscellaneous compounds [1-21].

33.1 POLYENES

The polyene compounds are so named because of the alternating conjugated double bonds that constitute a part of their macrolide ring structure (Fig. 33.1.). The polyene antibiotics are produced by *Streptomyces* species. Polyenes target the ergosterols the main component of fungal cell by destabilizing the plasma membrane and initiate cell lysis. Polyenes are also thought to cause oxidative damage. The polyene antifungal agents—amphotericin-B (**33.1.1**) [22-26], nystatin A1 (usually referred to as nystatin) (**33.1.2**) [27,28], and natamycin (**33.1.3**) [29,30]—are the only three members of the polyene/macrolide group of antifungal drugs that are used to treat human infections.

Amphotericin B, discovered in 1956, represents a gold standard, the mainstay antifungal agent for treatment of life-threatening mycoses and for most other mycoses, with the possible exception of the dermatophytoses.

The drug must be administered intravenously and is associated with numerous side effects, ranging from phlebitis at the infusion site to renal toxicity.

Nystatin was the first successful antifungal antibiotic to be developed, and it is still in general use against *Candida* species.

Pimaricin (natamycin) is another polyene; it is used topically to treat superficial mycotic infections of the eye. It is active against both yeasts and molds.

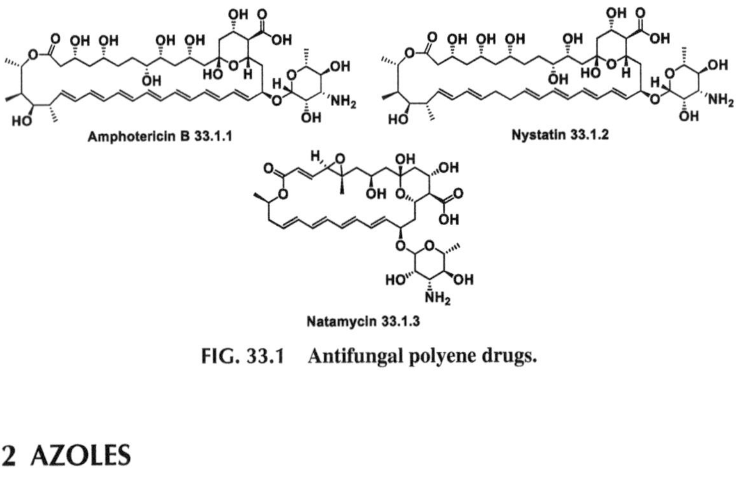

FIG. 33.1 Antifungal polyene drugs.

33.2 AZOLES

The azole group of antifungal drugs is divided into two subgroups: imidazoles and triazoles. A single representative of thiazole antifungals—abafungin—is also known. In general, the azole antifungal agents are thought to inhibit cytochrome P450–dependent enzymes involved in the biosynthesis of cell membrane sterols and at the same time trigger formation of toxic byproducts that are lethal to fungi. Imidazole drugs on the market are represented by clotrimazole (**33.2.1**), bifonazole (**33.2.2**), econazole (**33.2.3**), miconazole (**33.2.4**), fenticonazole (**33.2.5**), isoconazole (**33.2.6**), tioconazole (**33.2.7**), sertaconazole (**33.2.8**), butoconazole (**33.2.9**), sulconazole (**33.2.10**), luliconazole (**33.2.11**), oxiconazole (**33.2.12**), omoconazole (**33.2.13**), and ketoconazole (**33.2.14**), which are imidazole derivatives, the most popular of which are clotrimazole (**33.2.1**), econazole (**33.2.3**), miconazole (**33.2.4**), and ketoconazole (**33.2.14**) (Fig. 33.2.). Ketoconazole set the stage for the orally administered antifungal azoles. It can be administered both orally and topically, and has a range of activity that includes infections caused by *Histoplasma capsulatum* and *Blastomyces dermatitidis*, for which it is often used in nonimmunocompromised patients. It is also active against mucosal candidiasis and a variety of cutaneous mycoses, including dermatophyte infections, pityriasis versicolor, and cutaneous candidiasis. It is not indicated for treatment of aspergillosis or of systemic infections caused by yeasts.

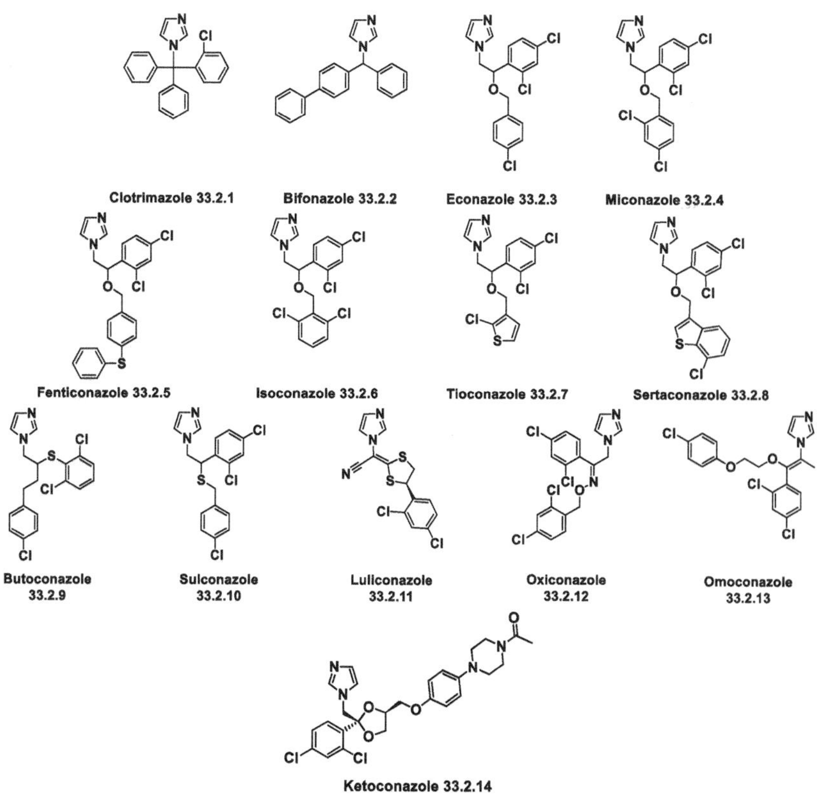

FIG. 33.2 The imidazole group of antifungal drugs.

The triazole group of antifungal drugs are fluconazole (**33.2.15**), voriconazole (**33.2.16**), ravuconazole (**33.2.17**) (under trial), posaconazole (**33.2.18**), and itraconazole (**33.2.19**). Albaconazole (**33.2.20**) and isavuconazole (**33.2.21**) are new antifungals compounds under investigation (Fig. 33.3.).

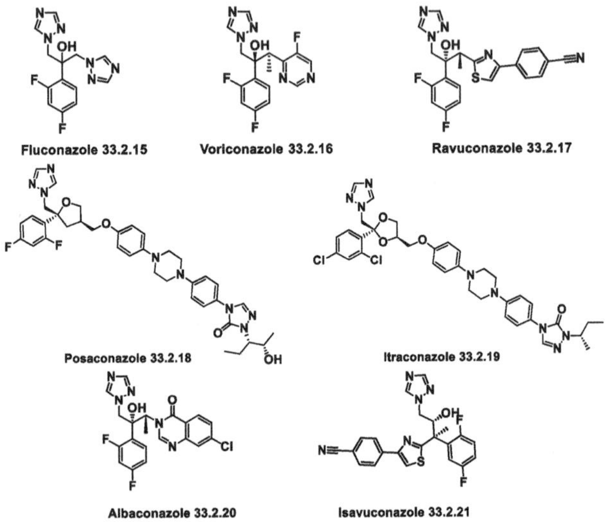

FIG. 33.3 The triazole group of antifungal drugs.

The imidazoles are used mostly as topical agents because of their severe toxicity when taken orally. Triazoles have been regularly modified and are now used to treat fungal infections caused by both yeasts and molds.

The most clinically useful imidazoles are clotrimazole, miconazole, and ketoconazole. Two important triazoles are itraconazole and fluconazole. In general, the azole antifungal agents are thought to inhibit cytochrome P450–dependent enzymes involved in the biosynthesis of cell membrane sterols.

The triazoles (fluconazole, itraconazole) have become the standard for the azoles, and have replaced amphotericin B for managing certain forms of the systemic mycoses. Fluconazole is now routinely used to treat candidemia in nonneutropenic hosts, and is gaining acceptance for use in cryptococcosis and selected forms of coccidioidomycosis. Itraconazole has proven to be effective for histoplasmosis, blastomycosis, sporotrichosis, coccidioidomycosis, consolidation treatment for cryptococcosis, and certain forms of aspergillosis. Fluconazole can be administered either orally, or intravenously. The licensed formulation for itraconazole is oral, but an intravenous formulation is under study, and could be a significant addition directed at bioavailability problems relating to absorption of the oral formulation.

Thiazole antifungal abafungin (**33.2.22**) (Fig. 33.4.) is the first member of a novel class of synthetic antifungal compounds, the arylguanidines, and is a highly effective drug with improved potency and less toxicity. It is believed that abafungin blocks the synthesis of ergosterol, an important component of the fungal cell membrane. It is proposed mainly for treatments of onychomycosis, a fungal infection of the fingernails and toenails that results in thickening, discoloration, splitting of the nails, and lifting of the nail from the nail bed.

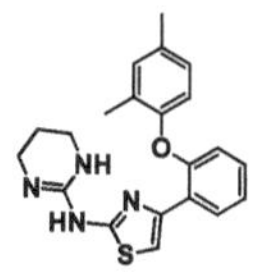

Abafungin 33.2.22

FIG. 33.4 **The thiazole antifungal abafungin.**

Side effects are not as common with the azoles as with amphotericin B, but life-threatening liver toxicity can arise with long-term use. Liver toxicity noted with ketoconazole has been less problematic with the triazoles. Other side effects include nausea and vomiting. Drug interactions are a potential problem between azoles and other drug classes and include cyclosporine, certain antihistamines, anticoagulants, and antiseizure, oral hypoglycemic, and other medications that are metabolized via similar pathways in the liver [31-36].

33.3 ECHINOCANDINS

Echinocandins are the new antifungal agents that are synthetic derivatives of lipopeptides, which are produced by various fungi (*Aspergillus rugulovalvus, Zalerion arboricola,* and *Papularia sphaerosperma*). The echinocandins class is the most recently developed class of antifungals. The three approved echinocandins are caspofungin (**33.3.1**), anidulafungin (**33.3.2**), and micafungin (**33.3.3**) (Fig. 33.5.). Caspofungin is the first and prototype of the echinocandins. Echinocandins inhibit the synthesis of 1,3-β-D-glucan, an essential component of the fungal cell wall, and represent a valuable treatment option for fungal infections. The echinocandins exhibit potent in vitro and in vivo fungicidal activity against *Candida* species, including azole-resistant pathogens. The advantages of the echinocandins include less toxicity, broad-spectrum activity, and synergistic effect in combination therapy. Echinocandins are now often recommended as first-line drugs for many invasive fungal infections. These drugs have markedly changed the approach to antifungal therapy, sometimes even allowing oral treatment of chronic mycoses. Adverse effects such as hepatitis and rash could be developed [37-41].

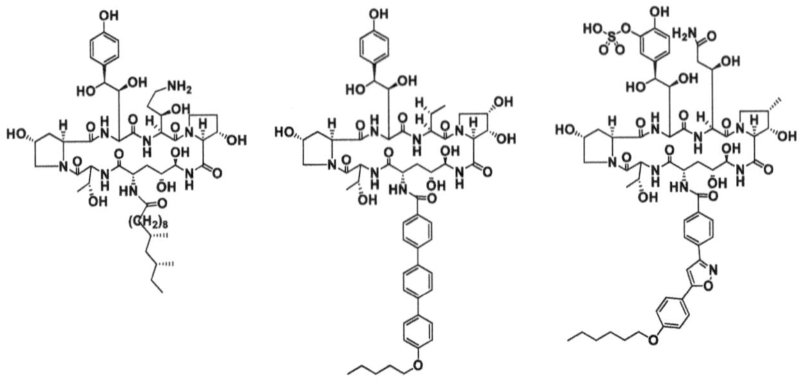

Caspofungin B 33.3.1 Anidulafungin 33.3.2 Micafungin 33.3.3

FIG. 33.5 **Echinocandins, a recently developed class of antifungals.**

33.4 ALLYLAMINES

The allylamines terbinafine (**33.4.1**) [42,43], butenafine (**33.4.2**) [44], and naftifine (**33.4.3**) [45,46] (Fig. 33.6.) reveal fungicidal action against many fungi as a result of their specific inhibition of membrane-bound enzyme-squalene epoxidase, which inhibits the synthesis of ergosterol. The morpholine drug, amorolfine (**33.4.4**) [47-49], is formally considered an allylamine because it inhibits the same biochemical pathway, but at a later step.

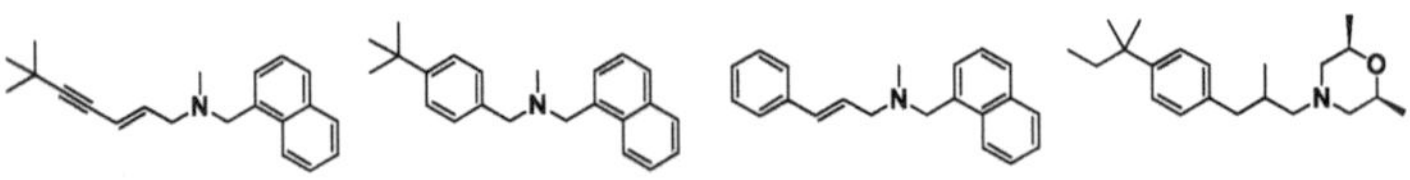

Terbinafine 33.4.1 Butenafine 33.4.2 Naftifine 33.4.3 Amorolfine 33.4.4

FIG. 33.6 **Allylamine antifungals.**

These drugs are all used as topical agents and are effective in treatment of superficial fungal infections and used clinically for topical therapy of superficial mycoses, including dermatophytic fungi. The primary mode of action of the allylamines is the inhibition of fungal ergosterol biosynthesis at the point of squalene epoxidation. Selective toxicity of allylamines is determined by the fact that squalene epoxidase from mammalian liver is orders of magnitude less sensitive to the allylamines than that of fungi.

Treated fungi accumulate squalene while becoming deficient in ergosterol, an essential component of fungi cell membranes. High intracellular squalene concentrations is believed to interfere with fungal membrane function and cell wall synthesis. Thousands of structural analogues of the allylamine type have been synthesized, many of them having antifungal activity.

33.5 ANTIMETABOLITES

This class has only one example—flucytosine or 5-fluorocytosine (**33.5.1**) [50] (Fig. 33.7.). This old antifungal agent was synthesized in 1957, as a potential antitumor agent, but it was not sufficiently effective against tumors. 5-Fluorocytosine was implemented as an effective remedy in treating infections caused by *Cryptococcus neoformans*, *Candida* species, and *Torulopsis glabrata*, and in chromoblastomycosis and phaeohyphomycosis

5-Fluorocytosine exerts its antifungal effects by interfering with both DNA and protein synthesis. It is transported into susceptible fungi by cytosine permease, then deaminated to 5-fluorouracil (**33.5.2**) by cytosine deaminase. The fact of absence of cytosine deaminase in mammalian cells allows selective effects on fungal cells.

The most significant toxicities of 5-fluorocytosine are hematologic, hepatic, and gastrointestinal. Leukopenia and thrombocytopenia, in particular, can limit its use.

5-Fluorocytosin 33.5.1 5-Fluorouracil 33.5.2

FIG. 33.7 Antifungal antimetabolites.

33.6 MISCELLANEOUS ANTIFUNGAL AGENTS

Ciclopirox (**33.6.1**), a broad-spectrum antifungal agent available since 1996, belongs to the chemical class of hydroxypyridones. It also exhibits antiinflammatory and antibacterial activity. This antifungal agent is currently used to treat mild to moderate cutaneous fungal infection.

The mechanism of action of ciclopirox seems to be different from that of other topical antifungal drugs. The high affinity of ciclopirox for trivalent metal cations, resulting in inhibition of the metal-dependent enzymes that are responsible for the degradation of peroxides within the fungal cell, appears to be the major determinant of its antimicrobial activity. This unique and multilevel mechanism of action provides a very low potential for the development of resistance in pathogenic fungi, with cases of resistance rarely reported [51].

Tolnaftate (**33.6.2**) is an antifungal drug that is applied topically as a cream, powder, or solution to treat various fungal infections of the skin, including ringworm. Tolnaftate is also effective against ascomycetes and fungi imperfecti. It is a thiocarbamate derivative whose exact mechanism of action is not entirely known, but it is believed to inhibit squalene epoxidase [52].

Griseofulvin (**33.6.3**) is a chlorine-containing metabolite obtained from culture liquid of *Penicillium griseofulvum*. It has been in clinical use for the treatment of ringworm, and only recently has attracted renewed attention as an antifungal, antiviral, and anticancer drug.

It is an antifungal antibiotic active in vitro against most dermatophytes and has been the drug of choice for chronic infections caused by these fungi (e.g., nail infections with *Trichophyton rubrum*) because it is orally administered and presumably incorporated into actively growing tissue. It is still used in such instances, but is being challenged by some of the newer azole antifungal agents. The drug inhibits mitosis in fungi [53] (Fig. 33.8.).

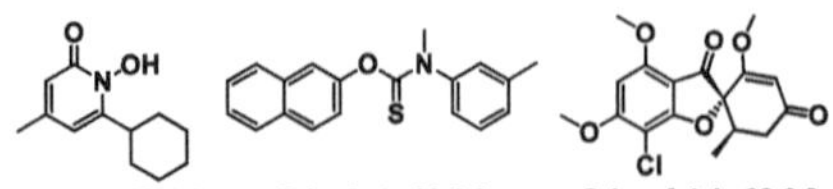

Ciclopirox 33.6.1 Tolnaftate 33.6.2 Griseofulvin 33.6.3

FIG. 33.8 Miscellaneous antifungals.

None of the antifungal drugs is included in the list of Top 200 Drugs by sales for the 2010s.

REFERENCES

1. Nett, J. E.; Andes, D. R. Antifungals: drug class, mechanisms of action, pharmacokinetics/pharmacodynamics, drug-drug interactions, toxicity, and clinical use. In *Candida and Candidiasis,* 2nd ed.; Calderone, R. A., Clancy, C. J., Eds. ASM Press, 2012; pp 345–371.
2. Warnock, D. W. Antifungal agents. In 10th ed.; Versalovic, J., Carroll, K. C., Funke, G., Jorgensen, J. H., Landry, M. L., Warnock, D. W., Eds.; *Manual of Clinical Microbiology, vol.* 2; ASM Press, 2011; pp 1995–2007.
3. Bell, A. S. Major antifungal drugs. In 8th ed.; Taylor, J. B., Triggle, D. J., Eds.; *Comprehensive Medicinal Chemistry II, vol. 7*; Elsevier, 2006; pp 445–468.
4. Gupta, A. K.; Tomas, E. New antifungal agents. *Dermatol. Clin.* **2003,** *21* (3), 565–576.
5. Li, D.; Calderone, R. A. Antifungal drugs, targets and target discovery. In *Pathogenic Fungi: Host interactions and Emerging Strategies for Control;* San-Blas, G., Calderone, R. A., Eds.; Caister Academic Press, 2004; pp 335–355.
6. Alex, Deepu; Li, D. Antifungal drug discovery: the process and outcomes. *Future Microbiol.* **2014,** *9* (6), 791–805.
7. Dixon; G. K., Copping, L. G.; Hollomon, D. W., Eds. *Antifungal Agents: Discovery and Mode of Action;* Bios Scientific, 1995.
8. Fostel, J. M.; Lartey, P. A. Emerging novel antifungal agents. *Drug Discovery Today* **2000,** *5* (1), 25–32.
9. Kauffman, C. A.; Carver, P. L. Antifungal agents in the 1990s. Current status and future developments. *Drugs* **1997,** *53* (4), 539–549.
10. Mohr, J.; Johnson, M.; Cooper, T.; Lewis, J. S., II.; Ostrosky-Zeichner, L. Current options in antifungal pharmacotherapy. *Pharmacotherapy* **2008,** *28* (5), 614–645.
11. Georgopapadakou, N. H. Antifungals: mechanism of action and resistance, established and novel drugs. *Curr. Opin. Microbiol.* **1998,** *1* (5), 547–557.
12. Andes, D. In vivo pharmacodynamics of antifungal drugs in treatment of Candidiasis. *Antimicrob. Agents Chemother.* **2003,** *47* (4), 1179–1186.
13. Calderone, R.; Sun, N.; Gay-Andrieu, F.; Groutas, W.; Weerawarna, P.; Prasad, S.; Alex, D.; Li, D. Antifungal drug discovery: the process and outcomes. *Future Microbiol.* **2014,** *9* (6), 791–805.
14. Tada, R.; Latge, J.-P.; Aimanianda, V. Undressing the fungal cell wall/cell membrane—the antifungal drug targets. *Curr. Pharm. Des.* **2013,** *19* (20), 3738–3747.
15. Paiva, J. A.; Pereira, J. M. New antifungal antibiotics. *Curr. Opin. Infect. Dis.* **2013,** *26* (2), 168–174.

16. Talaviya, S.; Majmudar, F. Recent developments in antifungal agents. *Int. J. Pharm. Pharm. Sci.* **2012,** *4* (Suppl. 4), 4–10.
17. Turel, O. Newer antifungal agents. *Expert Rev. Anti-Infect. Ther.* **2011,** *9* (3), 325–338.
18. Sheng, C.; Zhang, W. New lead structures in antifungal drug discovery. *Curr. Med. Chem.* **2011,** *18* (5), 733–766.
19. Castelli, M. V.; Butassi, E.; Monteiro, M. C.; Svetaz, L. A.; Vicente, F.; Zacchino, S. A. Novel antifungal agents: a patent review (2011-present). *Expert Opin. Ther. Pat.* **2014,** *24* (3), 323–338.
20. Kontoyiannis, D. P.; Lewis, R. E. Antifungal drug resistance of pathogenic fungi. *Lancet* **2002,** *359* (9312), 1135–1144.
21. Neely, M. N.; Ghannoum, M. A. The exciting future of antifungal therapy. *Eur. J. Clin. Microbiol. Infect. Dis.* **2000,** *19* (12), 897–914.
22. Patel, P. A.; Fernandes, C. B.; Pol, A. S.; Patravale, V. B. Oral amphotericin B: challenges and avenues. *Int. J. Pharm. Biosci. Technol.* (IJPBST). **2013,** *1* (1), 1–9; ISSN: 2321-2969. http://www.ijpbst.com/.
23. Volmer, A. A.; Szpilman, A. M.; Carreira, E. M. Synthesis and biological evaluation of amphotericin B derivatives. *Nat. Prod. Rep.* **2010,** *27* (9), 1329–1349.
24. Sedlak, M. Amphotericin B: from derivatives to covalent targeted conjugates. *Mini-Rev. Med. Chem.* **2009,** *9* (11), 1306–1316.
25. Baginski, M.; Czub, J. Amphotericin B and its new derivatives: mode of action. *Curr. Drug Metab.* **2009,** *10* (5), 459–469.
26. Moen, M. D.; Lyseng-Williamson, K. A.; Scott, L. J. Liposomal amphotericin B: a review of its use as empirical therapy in febrile neutropenia and in the treatment of invasive fungal infections. *Drugs* **2009,** *69* (3), 361–392.
27. Arikan, S.; Rex, J. H. Nystatin LF (Aronex/Abbott). *Curr. Opin. Invest. Drugs (BioMed Cent.)* **2001,** *2* (4), 488–495.
28. Witten, V. H.; Katz, S. I. Nystatin. *Med. Clin. North Am.* **1970,** *54* (5), 1329–1337.
29. Brik, H. Natamycin. *Anal. Profiles Drug Subst. Excipients* **1994,** *23,* 399–419.
30. Bogan, J. A. Natamycin. *Drugs Today* **1978,** *14* (6), 254–256.
31. Sandhu, S. S.; Shukla, H.; Aharwal, R. P.; Kumar, S.; Shukla, S. Antifungal azole derivatives and their pharmacological potential: prospects & retrospect. *Nat. Prod. J.* **2014,** *4* (2), 140–152.
32. Musiol, R.; Kowalczyk, W. Azole antimycotics—a highway to new drugs or a dead end? *Curr. Med. Chem.* **2012,** *19* (9), 1378–1388.
33. Ashley, E. S. Dodds Pharmacology of azole antifungal agents. In *Antifungal Therapy;* Ghannoum, M. A., Perfect, J. R., Eds. Informa Healthcare, 2010; pp 199–218.
34. Peng, X.-M.; Cai, G.-X.; Zhou, C.-H. Recent developments in azole compounds as antibacterial and antifungal agent. *Curr. Top. Med. Chem.* **2013,** *13* (16), 1963–2010.
35. Girmenia, C.; Finolezzi, E. New-generation triazole antifungal drugs: review of the phase II and III trials. *Clin. Invest. (London, U. K.)* **2011,** *1* (11), 1577–1594.
36. Saag, M. S.; Dismukes, W. E. Azole antifungal agents: emphasis on new triazoles. *Antimicrob. Agents Chemother.* **1988,** *32* (1), 1–8.
37. Hector, R. F.; Bierer, D. E. New β-glucan inhibitors as antifungal drugs. *Expert Opin. Ther. Pat.* **2011,** *21* (10), 1597–1610.
38. Chen, S. C.-A.; Slavin, M. A.; Sorrell, T. C. Echinocandin antifungal drugs in fungal infections: a comparison. *Drugs* **2011,** *71* (1), 11–41.
39. Grover, N. D. Echinocandins: a ray of hope in antifungal drug therapy. *Indian J. Pharmacol.* **2010,** *42* (1), 9–11.
40. Denning, D. W. Echinocandin antifungal drugs. *Lancet* **2003,** *362* (9390), 1142–1151.

41. Georgopapadakou, N. H. Update on antifungals targeted to the cell wall: focus on β-1,3-glucan synthase inhibitors. *Expert Opin. Invest. Drugs* **2001,** *10* (2), 269–280.
42. Krishnan-Natesan, S. Terbinafine: a pharmacological and clinical review. *Expert Opin. Pharmacother.* **2009,** *10* (16), 2723–2733.
43. Humphreys, F. Terbinafine. *J. Drug Eval.* **2004,** *2* (5), 133–155.
44. McNeely, W.; Spencer, C. M. Butenafine. *Drugs* **1998,** *55* (3), 405–412.
45. Gupta, A. K.; Ryder, J. E.; Cooper, E. A. Naftifine: a review. *J. Cutaneous Med. Surg.* **2008,** *12* (2), 51–58.
46. Monk, J. P.; Brogden, R. N. Naftifine. A review of its antimicrobial activity and therapeutic use in superficial dermatomycoses. *Drugs* **1991,** *42* (4), 659–672.
47. Gupta, A. K. Amorolfine: an overview. *Drugs Today* **1995,** *31* (3), 145–153.
48. Fromtling, R. A. Amorolfine (Loceryl). *Drugs Today* **1992,** *28* (6), 399–403.
49. Haria, M.; Bryson, H. M. Amorolfine. A review of its pharmacological properties and therapeutic potential in the treatment of onychomycosis and other superficial fungal infections. *Drugs* **1995,** *49* (1), 103–120.
50. Vermes, A.; Guchelaar, H.-J.; Dankert, J. Flucytosine: a review of its pharmacology, clinical indications, pharmacokinetics, toxicity and drug interactions. *J. Antimicrob. Chemother.* **2000,** *46* (2), 171–179.
51. Gupta, A. K. Ciclopirox: an overview. *Int. J. Dermatol.* **2001,** *40* (5), 305–310.
52. Dash, A. K. Tolnaftate. *Anal. Profiles Drug Subst. Excipients* **1994,** *23,* 543–570.
53. Petersen, A. B.; Roennest, M. H.; Larsen, T. O.; Clausen, M. H. The chemistry of griseofulvin. *Chem. Rev. (Washington, DC, U. S.)* **2014,** *114* (24), 12088–12107.

Chapter 34

Antiviral Drugs

Viruses are major pathogenic agents causing a variety of serious diseases in humans, other animals, and plants.

Viruses are one of the most widespread of all organisms and are capable of infecting every species of animal from mammals down to insects, plants, and even bacteria. It seems there are more species of viruses in the world than of all other creatures put together.

The most common viral infections are respiratory (infections of the nose, throat, upper airways, and lungs); gastrointestinal (gastroenteritis); liver (hepatitis); and skin (warts or other blemishes, rashes).

Viral diseases include influenza (causing fever, severe aching, and catarrh, often occurring in epidemics); severe acute respiratory syndrome (a form of pneumonia); chickenpox (disease caused by the herpes zoster virus, which manifests in a mild fever and a rash of itchy inflamed blisters); herpes (herpes simplex or herpes zoster, causing the eruption of small blister-like vesicles on the skin or mucous membranes); hepatitis (a disease characterized by inflammation of the liver); cold sores (diseases affecting mouth or genitals); measles (disease causing fever and a red rash on the skin, typically occurring in childhood); shingles (painful inflammation of the nerve ganglia, with a skin eruption often forming a girdle around the middle of the body); poliomyelitis (disease that affects the central nervous system that can cause temporary or permanent paralysis); acquired immunodeficiency syndrome (AIDS) caused by human immunodeficiency virus (HIV) (disease includes dry cough or shortness of breath, difficult or painful swallowing, diarrhea, white spots or unusual blemishes in and around the mouth, pneumonia-like symptoms, fever, vision loss, nausea, abdominal cramps, and vomiting); smallpox (disease started from fever, overall discomfort, headache, severe fatigue, severe back pain, vomiting. A few days later, flat, red spots appear on whole trunk, which become lesions. Occur first in the mouth and spread to the face, then to entire body); rabies (a contagious and fatal viral disease that causes madness and convulsions); dengue (jungle fever) causing sudden fever and acute pains in the joints); Ebola (fatal disease marked by fever and severe internal bleeding); and Lassa (fever with headaches, mouth ulcers, muscle aches, hemorrhages under the skin, heart and kidney failure, and a high mortality rate).

Viruses are transmitted in various ways. Some are swallowed, some are inhaled, some are spread sexually, and some are spread by the bites of insects,

Synthesis of Best-Seller Drugs. http://dx.doi.org/10.1016/B978-0-12-411492-0.00034-1

such as mosquitoes, certain biting flies, ticks or animals, or during transfusion with contaminated blood.

The deadliest, most horrifying diseases the world has ever seen have been initiated by viruses. A smallpox epidemic disaster is described in ancient Sanskrit medical texts, dating to about 1500 BC. In Europe, smallpox likely appeared by about 300 AD.

Rabies has a history dating back to 2300 BC, as described by Babylonians, who went mad and died after being bitten by dogs.

The bubonic plague arrived in Europe in 1347 when 12 Genoese trading ships docked at the Sicilian port. The ships carried goods that brought bubonic plague, which caused the Black Death, killing more than 20 million people in Europe—almost one-third of the continent's population.

It is very believable that plague virus is very similar to the Ebola virus, the recent deadly virus named after the Ebola River in the Democratic Republic of the Congo, which is classified as one of the most dangerous pathogens on the planet.

There are still more than 100 million cases of dengue fever each year. Marburg and Ebola hemorrhagic fever have 90% fatality rates.

In 1847-1848, influenza swept through the Mediterranean to Western Europe.

In 1878, a disease causing high mortality in poultry became known as the "fowl plague." Fowl plague is now called *highly pathogenic avian influenza A*.

In 1918-1919, the most famous disaster generated by viruses occurred: Spanish flu (influenza), which infected 20 to 40% of the world's population and killed from 20 million to 100 million people. This pandemic happened in 1918 and is considered to be one of the worst in human history. Probably, no virus can claim credit for more worldwide pandemics and scares than the constantly mutating thousands of strains of influenza.

In 1957-1958, the "Asian flu" caused the second pandemic of the 20th century. It began in China and killed one million people worldwide.

In 1968-1969, the "Hong Kong flu" caused the last flu pandemic.

In 2009, a swine flu outbreak hit, originating in Mexico City and quickly spreading to more than 10 nations.

The first case of HIV infection in a human was identified in 1959. The first cases of HIV in the United States date back to 1981.

In 1984, the HIV, which causes HIV infection and AIDS, was discovered. The HIV infection is distinct from AIDS, the full-blown syndrome that, along with the consequences of a damaged immune system, is most often fatal. More than 78 million people have been infected with HIV since the start of the epidemic in the 1980s. AIDS-related illness are the sixth leading cause of death.

Viruses are small particles, much smaller than fungi or bacteria. These strange substances are something straddling between living and nonliving particles. Viruses are basically a pack of genetic material—either DNA or RNA—carried in a shell made up of protein. Some viruses have an additional layer called an envelope. Viruses vary in size and complexity, but common features for species of viruses are: nucleic acid, protein coat, lipid membrane (envelope). (Viruses can be divided into two major categories: enveloped and non-enveloped.) A virus particle, consisting of an outer protein shell called a capsid and an inner core of nucleic acid (either ribonucleic or deoxyribonucleic), is called a virion. Viruses are classified as DNA viruses or RNA viruses, depending on whether they use DNA or RNA to replicate. Further classification can be based on the nature of the genetic material that is packaged by the virus (single-stranded and double-stranded RNA viruses, analogue DNA viruses, and others). Viruses can't even be considered cells. They can't metabolize nutrients, produce and excrete wastes, move around on their own, or even reproduce unless they are inside another organism's cells.

Coming into contact with host cells, they fuse themselves to the host cell and insert their genetic material into that cell, taking over its machinery to replicate themselves. Consequently, there is controversy regarding whether viruses should really be considered as living organisms.

The virus has to enter the cell to start it's activity. As a first step, they must attach to a receptor on the cell surface. Each virus has its specific receptor and it is a vital component of the cell surface. The selectivity of the viruses determines the cell preference.

For instance, rhinoviruses have a preference for airways cells; HIV infects mainly T lymphocytes and macrophages with the clusters such as CD4, CCR5, and CXCR4 on the surface of immune cells; Epstein–Barr virus and rabies virus infect B lymphocytes carrying complement receptor type CR2. It is believed that influenza viruses infect cells in a multi-step process. The main targets of the influenza virus are the columnar epithelial cells of the trachea, bronchi and bronchioles. On the first step viruses are internalized via receptor-mediated endocytosis and then are trafficked along the endocytic pathway to endosomes. After that HA protein – hemagglutinin catalyzed fusion between the viral and endosomal membranes takes place, releasing viral ribonucleoproteins which are imported into the nucleus for viral gene expression and replication.

When a virus enters the body, it triggers the body's immune defenses, such as lymphocytes and monocytes, which destroy the virus or the cells it has infected. If the body survives the virus attack, it becomes able to respond to a subsequent infection by the same virus (immunity, which can also be produced by vaccination). Vaccination is possible to prevent infections with some viruses, such as hepatitis B, varicella-zoster, influenzas A and B, Yellow fever, and poliovirus viruses; but not infections caused by HIV, hepatitis C, herpes simplex, cytomegalovirus, and most hemorrhagic fever viruses (except for Yellow fever virus).

One classification of viruses is based on pathogenic property and mode of transmission.

Respiratory viruses—influenza, rhinovirus, adenovirus—are usually acquired by inhalation of droplets and replicate in the respiratory tract. The polioviruses, rotaviruses, reoviruses, and some adenoviruses are enteric viruses. These are viruses that replicate in the gut and cause gastric infections. The most serious complications include meningitis, encephalitis, poliomyelitis, and myocarditis. Arboviruses infect insects that ingest vertebrate blood, replicate in tissue of the insect, and then are transmitted to the vertebrate host. Such viruses include flaviviruses, bunyaviruses, and some rhabdoviruses. Sexually transmitted viruses include HIV, herpes simplex, and papilloma viruses. Hepatitis viruses cause disease of the liver.

More than 30,000 different viruses have been isolated and studied to date. Historically, their classification was disease related. Now they are grouped at different hierarchical levels of order, family, subfamily, genus, and species on the basis of their properties.

Viral infections that cause persistent infections are viruses that belong to the Retroviridae family (HIV, which causes AIDS, and human T-cell lymphotrophic virus [HTLV], which causes leukemia); to the Flaviviridae family (hepatitis C [HCV] and to the Hepadnaviridae family (hepatitis B [HBV]), which cause chronic hepatitis and hepatocellular carcinoma; and to the Herpesviridae family (herpes simplex viruses 1 and 2 [HSV-1 and HSV-2]), which cause recurrent mucocutaneous infections and encephalitis. In addition, varicella zoster virus (VZV) causes recurrent neurological lesions; cytomegalovirus (CMV) causes retinitis, pneumonia, encephalitis; Epstein-Barr virus (EBV) causes lymphoproliferative disorders; and viruses that belong to the Papovaviridae family cause cervical carcinoma and warts.

Human viral infections associated with loss of work hours include rhinoviruses (influenza A virus causing influenza) and Caliciviruses, Norwalk viruses, and Astra diarrhea-causing viruses.

Viruses are also classified on the basis of morphology, chemical composition, and mode of replication, which reflects only a small part of the spectrum of the multitude of different viruses.

Viral diseases are not treatable with antibiotics, which can only cure bacterial diseases and infections.

Drugs that combat viral infections are called antiviral drugs. There are no effective antiviral drugs for many viral infections. However, there are several drugs for influenza, a couple of drugs for herpesviruses, and some new antiviral drugs for treatment of HIV and hepatitis C infections.

The arsenal of antivirals is complex. As of March 2014, it consists of approximately 50 drugs approved by the FDA, approximately half of which are directed against HIV. Antiviral drug creation strategies is focused on two different approaches: targeting the viruses themselves or targeting host cell factors. These approaches are widely reviewed [1-60].

34.1 DIRECT VIRUS-TARGETING ANTIVIRAL DRUGS

Attachment Inhibitors

The first event in viral infection of the host cell is binding of the virus to the cell surface, which involves numerous interactions between the virion surface and the receptor. Some viruses have specific attachment sites widely distributed all over the host cell membrane that recognize molecules on the surface of virus particle. Many viruses use as an attachment site heparan sulphate proteoglycans found at the cell surface.

To block initial aspecific binding of virus to cells, polyanionic compounds (polysulfates, i.e., dextran sulfate, polysulfonates, polycarboxylates, polyoxometalates, and negatively charged albumins) have been suggested. A series of them were potent and selective inhibitors of respiratory viruses under experimental conditions.

Some pyrrolopyridine (6-azaindole) compounds, such as BMS-488043 (**34.1.1**) [61,62] and BMS-663068 (**34.1.2**) [63], which block interactions between a virus and its receptor, thereby significantly reducing HIV-1 proliferation, have been proposed as anti-HIV therapeutics (Fig. 34.1.).

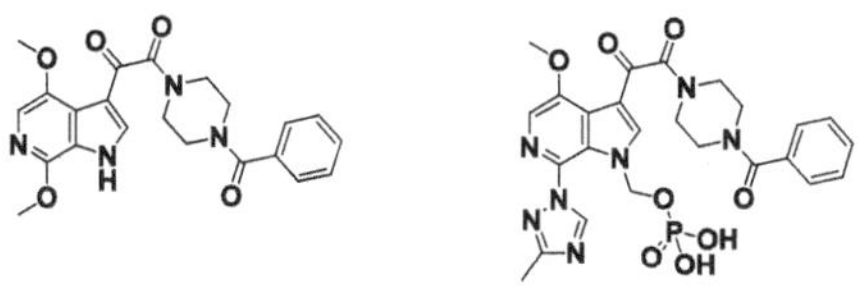

BMS-488043 34.1.1 BMS-663068 34.1.2

FIG. 34.1 Compounds that block the interactions between a virus and its receptor.

Another example of compounds blocking initial binding of virus to cells are hydrolytic enzyme neuraminidase inhibitors.

The influenza neuraminidase is a surface glycoprotein that cleaves the cell-receptor sialic acid residues, thereby allowing the release of the virus to infect new cells [64]. Neuraminidase inhibitors are commonly used in both the prevention and the treatment of influenza.

Several neuraminidase inhibitors have been developed, including oseltamivir (Tamiflu) (**34.1.3**) and zanamivir (**34.1.4**), which were first used clinically as antiflu therapies [65]. Laninamivir (**34.1.5**) and peramivir (**34.1.6**) were also approved in northeast Asia (China, Japan and South Korea) recently. These agents are proven to be safe and effective alone or in combination for the treatment of uncomplicated influenzas [66] (Fig. 34.2.).

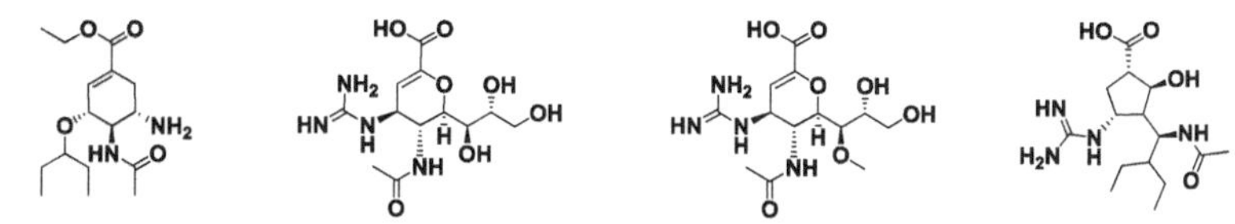

Oseltamivir 34.1.3 Zanamivir 34.1.4 Laninamivir 34.1.5 Peramivir 34.1.6

FIG. 34.2 Neuraminidase inhibitors.

Another way of preventing binding of the virus to the host cell is by vaccination immunization using antibodies (specific immunoglobulins) against the infecting agent.

Entry Inhibitors

After attaching to host cells, the second step in the viral replication cycle is penetration. Enveloped viruses penetrate by fusion of the viral membrane with the cell membrane, while "naked" viruses penetrate the cell by phagocytosis of the virion and releasing viral genome into the host cytoplasm. For this reason entry inhibitors became one of the major concepts in the development of new antiviral drugs and play a very big role for antiviral therapies.

Fusion inhibitors are designed to block the conformational changes that are required for membrane fusion. Enfuvirtide (**34.1.7**), is the first and the only clinically approved fusion inhibitor that can inhibit a broad range of enveloped virus, particularly HIV strains. Enfuvirtide (Ac-Tyr-Thr-Ser-Leu-Ile-His-Ser-Leu-Ile-Glu-Glu-Ser-Gln-Asn-Gln-Gln-Glu-Lys-Asn-Glu-Gln-Glu-Leu-Leu-Glu-Leu-Asp-Lys-Trp-Ala-Ser-Leu-Trp-Asn-Trp-Phe-NH_2) is the first of a novel class of peptide fusion inhibitors considered to be active against HIV-1 by disrupting the HIV-1 molecular machinery at the final stage of fusion with the target cell [67]. A series of modified peptides, including peptide fusion inhibitors possessing potent anti-HIV activity, namely, CP32M [68], sifuvirtide [69], and T2635 [70], have been synthesized and modified by pegylation, glycosylation, and other processes.

A plethora of small molecule fusion inhibitors have been synthesized, but because of the difficulty of their production and delivery, none has been approved for clinical use. Pleconaril (**34.1.8**), an entry inhibitor for nonenveloped viruses, prevents the virus from attaching itself to the host cell. It is the prime example of a human rhinovirus inhibitor [71,72] to have been proposed. Other capsid-targeting molecules, namely, BTA-798 (**34.1.9**) [73,74] and V-073 (**34.1.10**) [74,75], are in clinical development.

Ac-Tyr-Thr-Ser-Leu-Ile-His-Ser-Leu-Ile-Glu-Glu-Ser-Gln-Asn-Gln-Gln-Glu-Lys-Asn-Glu-Gln-Glu-Leu-Leu-Glu-Leu-Asp-Lys-Trp-Ala-Ser-Leu-Trp-Asn-Trp-Phe-NH_2

Enfuvirtide 34.1.7

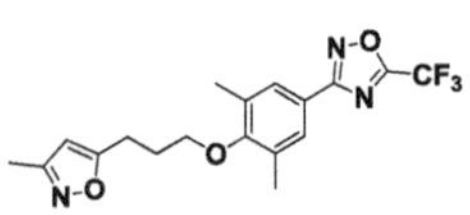

Pleconaril 34.1.8

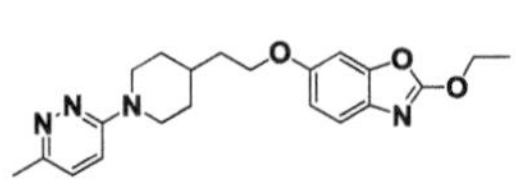

BTA-798 34.1.9

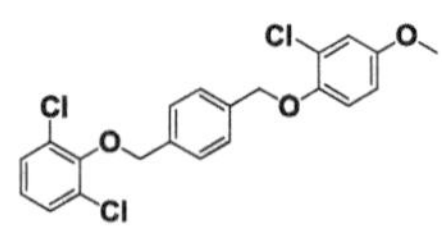

V-073 34.1.10

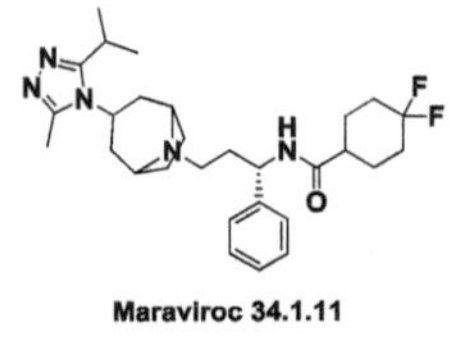

Maraviroc 34.1.11

FIG. 34.3 Entry inhibitors.

The entry inhibitors primarily target the HIV-1 envelope glycoproteins or the cellular receptors, CD4, and the chemokine receptors involved in a number of biological processes. HIV-1 enters CD4-expressing cells via one or both of the chemokine receptors CCR5 and CXCR4. The CC-chemokine receptor 5 (CCR5) antagonist maraviroc (**34.1.11**) is the only approved drug of the novel class of "antiretroviral agents" that prevents the entry of HIV-1 into host cells by blocking the CCR5 coreceptor [74,75] (Fig. 34.3.).

Uncoating Inhibitors

Another antiviral drug target is the uncoating step during viral infection, which is the process of capsid disintegration, retaining the virus in the encapsulated state, and not allowing the virus to release its genomic material into the host cell to interrupt the virus replicative cycle before it proceeds to the reverse transcriptase step. It is believed that such drugs work by blocking ion channels in the virus. Amantadine (**34.1.12**) [76-79] and rimantadine (**34.1.13**) [78-80] are representatives of uncoating inhibitors specifically prevent release of influenza A virus in the cells, but to date no compounds have been shown to act at the uncoating of HIV or retroviruses (Fig. 34.4.).

Amantadine 34.1.12 Rimantadine 34.1.13

FIG. 34.4 Uncoating inhibitors.

Protease Inhibitors

Proteases are essential enzymes that regulate number of processes such as infection, fertilization, allergic reactions, inflammation, blood clotting, cell growth and death, and bone remodeling. They have become a remarkable target for promising therapeutic agents.

Protease inhibitors are enzymes that are critical for diverse biological processes involved in the life cycle of viruses and their replication processes. Protease inhibitors have became very important constituents of the antiretroviral drugs armada. In particular, they prevent viral replication by blocking proteolytic cleavage of protein precursors that are necessary for the production of infectious particles. It should be noted that most viruses also encode proteases, which protect viral proteins by modulating host cell. Protease inhibitors were a major therapeutic breakthrough in the mid-1990s for the treatment of HIV infection, ushering in the era of highly active antiretroviral therapy. Among the anti-HIV drugs developed over the past two decades, inhibitors of HIV-1 protease and reverse transcriptase have found a very prominent clinical use; creation of a series of protease inhibitors is one of the great successes of antiviral drug design. Early research involved in a quest among suitable peptides. But finally it was concluded to decrease peptide-like features, minimize the size of molecules.

Currently approved protease inhibitors, which have been mainly created on principles of structure- and fragment-based drug design, include: nelfinavir (**34.1.14**), amprenavir (**34.1.15**), fosamprenavir (**34.1.16**), darunavir (**34.1.17**), saquinavir (**34.1.18**), atazanavir (**34.1.19**), indinavir (**34.1.20**), lopinavir (**34.1.21**), ritonavir (**34.1.22**), tipranavir (**34.1.23**), simeprevir (**34.1.24**). They share relative similarity in chemical structures and are recognized by suffix–navir and have become the most potent types of antiviral drugs (Fig. 34.5.).

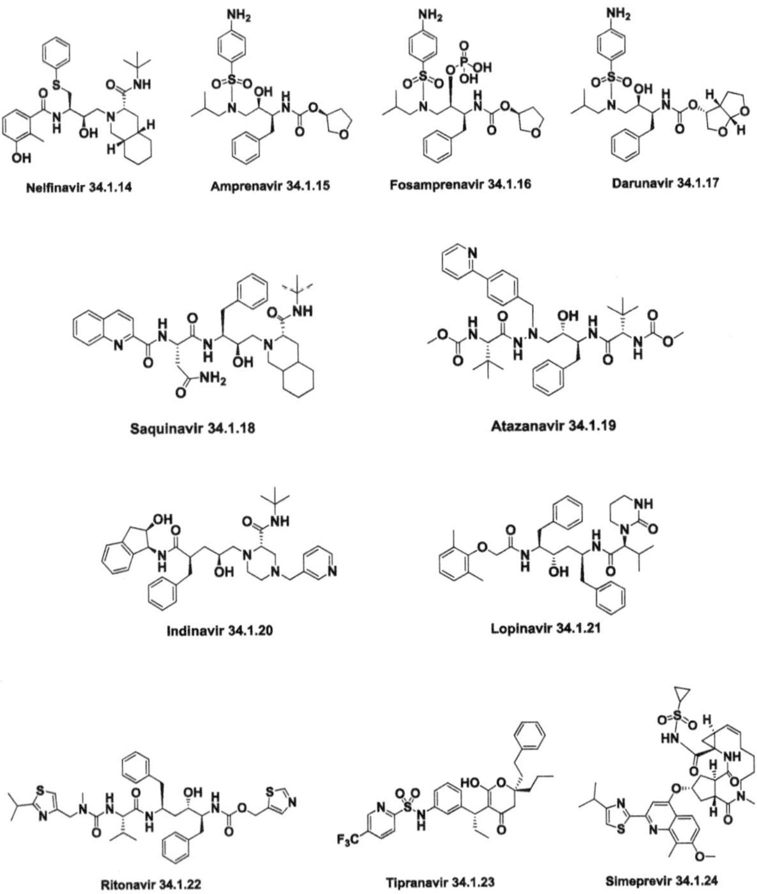

FIG. 34.5 **Protease inhibitors that act as antivirals.**

Darunavir (**34.1.17**), atazanavir (**34.1.19**), and ritonavir (**34.1.22**) are drugs included in the list of Top 200 Drugs by sales for the 2010s.

Darunavir–Prezista

The synthesis of darunavir (**34.1.17**) [81,82] was carried out via coupling mixed carbonate (**34.1.27**) with a benzene sulfonamide derivative of 1,3-diamino-4-phenylbutan-2-ol (**34.1.34**).

For this purpose, optically active bis-THF-ol (**34.1.25**) [83] was converted to mixed carbonate (**34.1.27**) by reaction with N,N′-disuccinimidyl carbonate (**34.1.25**) in the presence of triethylamine in methylene chloride.

For the synthesis of sulfonamide (**34.1.34**), commercially available (S)-1-((S)-oxiran-2-yl)-2-phenylethan-1-amine (**34.1.29**) was reacted with isobutylamine (**34.1.28**) in refluxing iso-propanol to give amino alcohol (**34.1.30**). Reaction of the resulting amino alcohol with p-nitrobenzenesulfonyl chloride (**34.1.31**) in the presence of aqueous $NaHCO_3$ furnished the sulfonamide derivative (**34.1.32**). Catalytic hydrogenation of obtained product over 10% Pd-C in ethyl acetate effected reduction of the nitro group to the corresponding amine (**34.1.33**). The BOC group of the amine (**34.1.33**) was removed with the use of trifluoroacetic acid to produce diamine (**34.1.34**). Reaction of the diamine with the mixed carbonate (**34.1.27**) in the presence of triethylamine provided darunavir (**34.1.17**) in high yield (Scheme 34.1.). Slight modifications of this Scheme have been proposed [84-86].

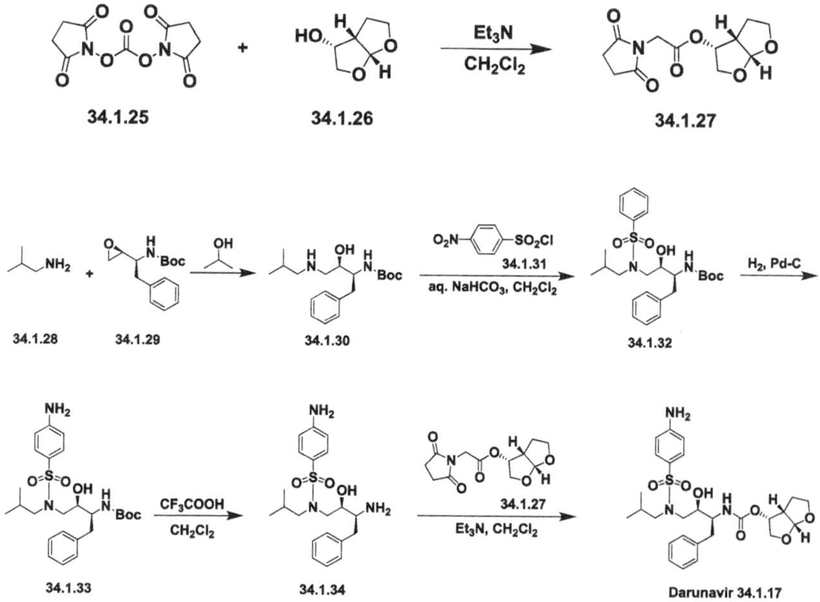

SCHEME 34.1 The synthesis of darunavir.

A series of darunavir analogues have displayed impressive enzymatic and antiviral properties and represent promising lead compounds for further optimization [87].

Darunavir is a unique and the most recent extremely potent protease inhibitor drug used to treat HIV infection. Darunavir usually is used in combination with ritonavir and other medicines. It maintains antiretroviral activity against a variety of multidrug-resistant HIV strains [88-97].

Darunavir can cause serious, life-threatening side effects, including liver problems and severe skin reactions or rash.

Atazanavir–Reyataz

The original synthesis of atazanavir (**34.1.19**) [98,99] was carried out through the reaction of the Boc-(pyridin-2-yl)benzyl)hydrazine (**34.1.39**) with the epoxide (**34.1.40**) prepared according to known procedures [100,101] followed after deprotection by coupling with N-methoxycarbonyl-L-tert-leucine (**34.1.42**) to produce the desired atazanavir (**34.1.19**) (Scheme 34.2.).

The required benzylhydrazine (**34.1.39**) was prepared by the coupling of 4-bromobenzaldehyde dimethyl acetal (**34.1.36**) to 2-bromopyridine catalyzed by the presence of 1,3-bis(diphenylphosphino)propane nickel (II) chloride and diisobutylaluminium hydride (current understanding of the coupling mechanism is limited) followed by acidic hydrolysis.

Obtained 4-(pyridin-2-yl)benzaldehyde (**34.1.38**) easily produced hydrazone with Boc-hydrazine in ethanol, which on hydrogenation on Pd-C catalyst or reduction with sodium cyanoborohydride provided the hydrazine building block (**34.1.39**).

Opening of the N-Boc–protected epoxide (**34.1.40**) with the Boc-protected benzylhydrazine (**34.1.38**) led to the symmetrically N-Boc–protected aza-dipeptide mimetic. Both of the Boc-protected groups were simultaneously cleaved off by acidic treatment (HCl in water/THF) and the obtained product (**34.1.41**) was coupled with N-methoxycarbonyl-L-tert-leucine (**34.1.42**) according to standard peptide synthesis procedures (carboxylic acid activator-carbodiimide-EDCl, racemization suppressor hydroxybenzotriazole [HOBt] base-triethylamine) to produce the target compound, atazanavir (**34.1.19**) (Scheme 34.2.) The improved approach of the same strategy was published [102].

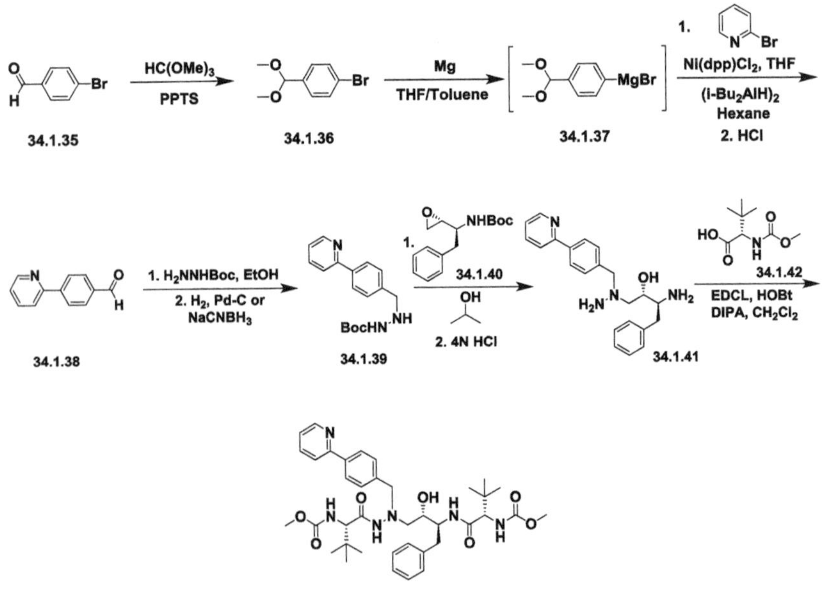

SCHEME 34.2 The original synthesis of atazanavir.

Another efficient and practical synthesis of atazanavir was developed by employing the diastereoselective reduction of ketomethylene aza-dipeptide (**34.1.47**) as the key and final step [103]. It was assembled from two key intermediates: hydrazide (**34.1.44**) and 1-amino-3-chloropropan-2-one derivative (**34.1.46**).

Hydrazide (**34.1.44**) was synthesized from the already known 4-(pyridin-2-yl)benzaldehyde (**34.1.38**), which in this special case was prepared by Kumada coupling of the Grignard reagent to 2-bromopyridine. The condensation of the aldehyde (**34.1.38**) with N-(methoxycarbonyl)-L-tert-leucinyl hydrazine (**34.1.43**) was carried out in ethanol. The resulting mixture was subjected to the next reaction without further purification. The quantitative hydrogenation with the use of $Pd(OH)_2/C$ in ethanol proceeded in high yield and high purity to produce the desired hydrazide (**34.1.44**).

The second key reagent, chloromethyl ketone (**34.1.46**), is prepared by reaction of (S)-3-amino-1-chloro-4-phenylbutan-2-one (**34.1.45**) with mixed anhydride, which in turn, was prepared from N-(methoxycarbonyl)-L-tert-leucine and isobutyl chloroformate. The coupling of the obtained chloromethyl ketone (**34.1.46**) with hydrazide (**34.1.44**) smoothly takes place in acetonitrile in presence of NaI and $NaHCO_3$ to produce the ketone (**34.1.47**).

For the key and final step—reduction of obtained amino ketone (**34.1.47**)—a new reagent–solvent combination $LiAlH(O\text{-}tBu)_3$ in Et_2O was employed, which allowed synthesis of the desired syn-1,2-amino alcohol atazanavir (**34.1.19**) exclusively (Scheme 34.3.) Other minor modifications of the presented two approaches have been proposed [104-106].

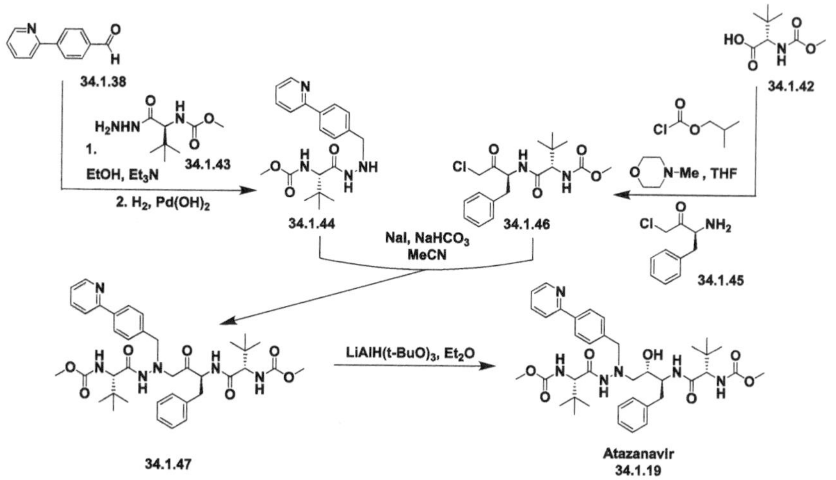

SCHEME 34.3 The synthesis of atazanavir.

Atazanavir is a novel, potent, safe, and generally well-tolerated inhibitor of the HIV protease. It was the first, and to date the only, protease inhibitor designed to be applied once daily. Atazanavir is expected to overcome the problems of earlier agents of this class of drugs, such as unfavorable adverse events like hyperlipidemia, diarrhea, and lipodystrophy. In combination with nucleoside reverse transcriptase inhibitors and boosted with ritonavir, it has been established as the preferred initial regimen in published guidelines [107-119].

Ritonavir–Norvir

The first publications [120,121] describing the synthesis of ritonavir (**34.1.22**) started from methyl N-(benzyloxycarbonyl)-L-phenylalanine methyl ester (**34.1.48**), which on reduction with $LiAlH_4$ was transformed to N-(benzyloxycarbonyl)-L-phenylalaninol (**34.1.49**). Oxidation of the N-(benzyloxycarbonyl)-L-phenylalaninol (**34.1.49**) with oxalyl chloride in DMSO at −60°C produced the corresponding aldehyde (**34.1.50**). The obtained aldehyde was dimerized with Zn dust in dichloromethane, in presence of vanadium trichloride, which catalyzes the pinacol coupling reaction, yielding (2S,3R,4R,5S)-2,5-bis(benzyloxy-carbonylamino)-1,6-diphenylhexane-3,4-diol (**34.1.51**) [along with the (2S,3S,4S,5S)-isomer]. The reaction of the diol (**34.1.51**) with α-acetoxyisobutyryl bromide (**34.1.52**) in hexane/dichloromethane produces (2S,3R,4R,5S)-2,5-bis(benzyloxycarbonylamino)-4-bromo-1,6-diphenylhexan-3-ol acetate ester (**34.1.53**), which was converted into the corresponding epoxide (**34.1.54**) with NaOH in dioxane/water. The reduction of the epoxide (**34.1.54**) with $NaBH_4$ in THF followed by deprotection with CF_3COOH produces (2S,3S,5S)-2,5-diamino-1,6-diphenylhexan-3-ol (**34.1.55**).

The obtained product was condensed with (5-thiazolylmethyl)(4-nitrophenyl) carbonate (**34.1.56**) in THF to produce (2S,3S,5S)-5-amino-1,6-diphenyl-2-(5-thiazolylmethoxycarbonylamino) hexan-3-ol (**34.1.57**). Finally, this compound was condensed with N-[N-(2-isopropylthiazol-4-ylmethyl)-N-methylaminocarbonyl]-L-valine (**34.1.58**) (EDCl, HOBt) in THF to produce the desired ritonavir (**34.1.22**) (Scheme 34.4.).

(5-Thiazolylmethyl)(4-nitrophenyl)carbonate (**34.1.56**), used in Scheme 34.4., was synthesized from 5-thiazolylmethanol (**34.1.65**), which, in turn, was prepared via cyclization on heating of thioformamide (**34.1.60**) with 2-chloro-3-oxopropionic acid ethyl ester (**34.1.61**). The synthesis of thioformamide (**34.1.60**) was accomplished with P_2S_5 in ethyl ether from formamide (**34.1.59**). The starting β-aldehydoester (**34.1.63**) was obtained by condensation of ethyl chloroacetate (**34.1.61**) with ethyl formate (**34.1.62**) in the presence of t-BuOK in THF. The reduction of (**34.1.64**) with $LiAlH_4$ in THF affords the 5-thiazolylmethanol (**34.1.65**), which was then esterified with 4-nitrophenyl chloroformate (**34.1.66**) in dichloromethane in the presence of 4-methylmorpholine to produce the desired product (**34.1.56**) (Scheme 34.5.).

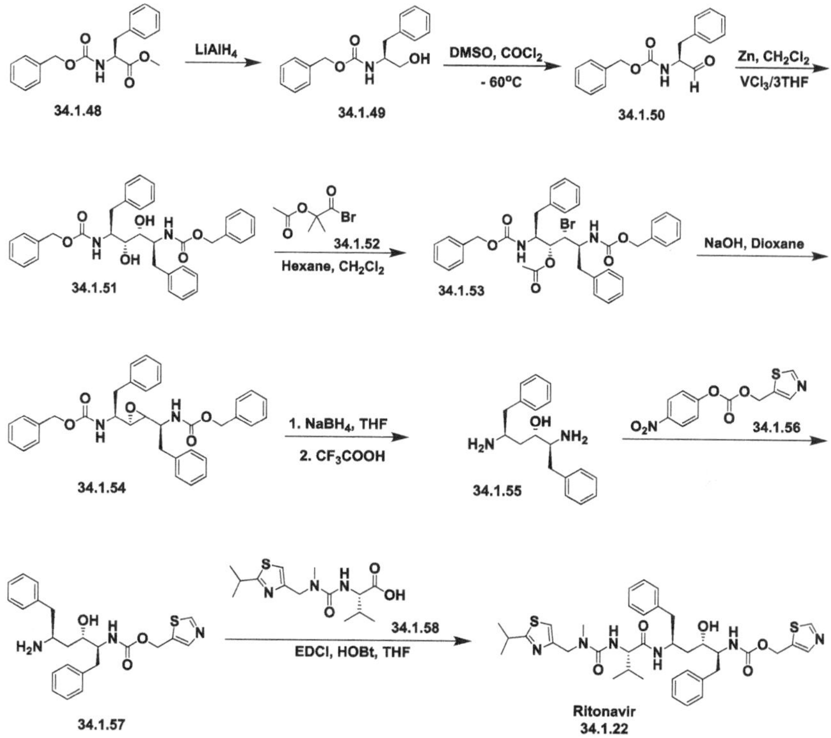

SCHEME 34.4 The synthesis of ritonavir.

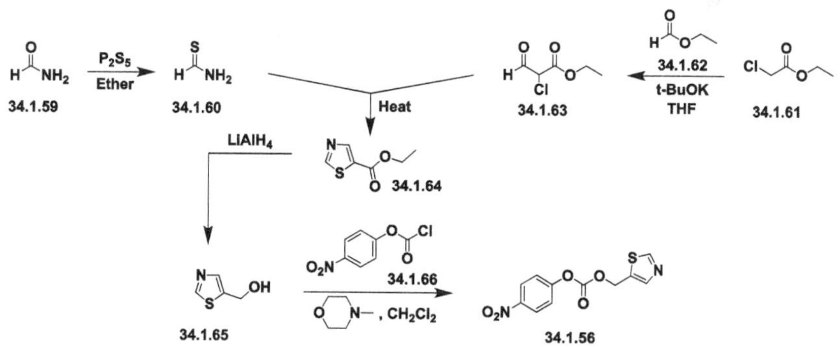

SCHEME 34.5 The synthesis of ritonavir (5-thiazolylmethyl)(4-nitrophenyl)carbonate.

The second thiazoly moiety-N-[N-(2-isopropylthiazol-4-ylmethyl)-N-methylaminocarbonyl]-L-valine (**34.1.58**) was synthesized by an analogues scheme.

The reaction of isobutyramide (**34.1.67**) with P_2S_5 in ethyl ether produced the corresponding thioamide (**34.1.68**), which was cyclized with 1,3-dichloroacetone (**34.1.69**) by means of $MgSO_4$ in refluxing acetone, yielding 4-(chloromethyl)-2-isopropylthiazole (**34.1.70**). The reaction of the 4-(chloromethyl)-2-isopropylthiazole (**34.1.70**) with methylamine in water produced N-(2-isopropylthiazol-4-ylmethyl)-N-methylamine (**34.1.71**), which was condensed with N-(4-nitrophenoxycarbonyl)-L-valine methyl ester (**34.1.73**), which was synthesized by reaction of chloroformate (**34.1.66**) with L-valine methyl ester (**34.1.72**) in the presence of 4-methylmorpholine in dichloromethane.

Finally, this compound was converted into the corresponding free acid (**34.1.58**) with LiOH in dioxane/water mixture (Scheme 34.6.).

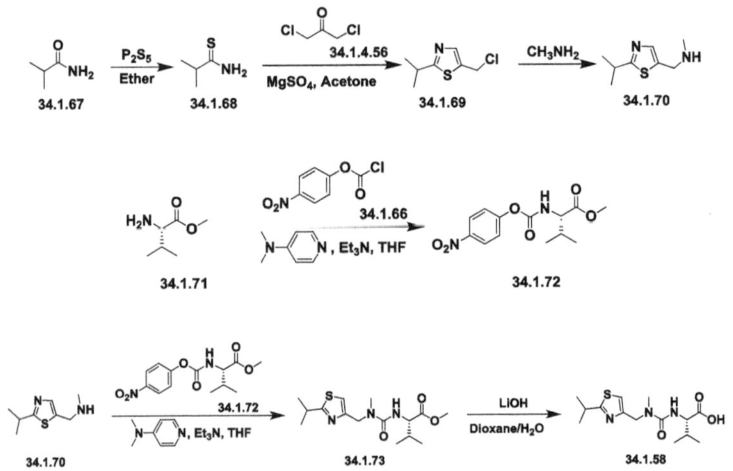

SCHEME 34.6 **The synthesis of N-[N-(2-isopropylthiazol-4-ylmethyl)-N-methylaminocarbonyl]-L-valine.**

A slightly different approach was demonstrated in patents [122,123], which describe a new method for the preparation of a key diaminoalcohol intermediate (**34.1.55**) in the synthesis of ritonavir, has been described: The reaction of L-phenylalanine (**34.1.74**) with benzyl chloride using K_2CO_3 and KOH in hot water produced L-phenylalanine benzyl ester (**34.1.75**), which was condensed first with acetonitrile in presence of $NaNH_2$ and then reacted with benzylmagnesium chloride, in methyl tert-butyl ether, yielding 5-amino-2-(dibenzylamino)-1,6-diphenylhex-4-en-3-one (**34.1.76**). The reduction of the obtained product with $NaBH_4$ resulted in an enhanced enantioselectivity toward the desired (S,S,S)-enantiomer of 5-amino-2-(dibenzylamino)-1,6-diphenyl-3-hexanol (**34.1.77**). Finally, this compound was debenzylated by hydrogenation with ammonium formate over Pd/C in methanol/water to provide the target chiral diaminoalcohol (**34.1.55**) transformed to ritonavir (**34.1.22**) (Scheme 34.7.).

Different variations of these approaches have been published [124-127].

Ritonavir is a potent inhibitor of the protease encoded by the HIV-1 and is clinically applied to suppress HIV-1 replication in AIDS patients. Ritonavir

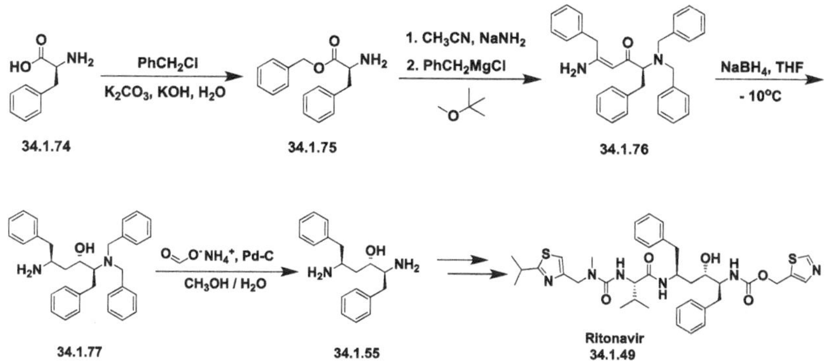

SCHEME 34.7 **The synthesis of ritonavir.**

does not cure HIV or AIDS, but it may slow the progress of the disease. It reduces the amount of virus in the body, has good oral bioavailability, and may increase the bioavailability of other protease inhibitors, including saquinavir, nelfinavir, and indinavir [128-133]. Ritonavir, can cause clinically significant increases in serum levels of other protease inhibitors such as darunavir, fosamprenavir, lopinavir, and saquinavir, and is often used in combination with them.

Lopinavir–Kaletra

The first report of a new protease inhibitor candidate lopinavir (**34.1.21**) seems to be Abbott's patent [134].

The core of lopinavir is identical to that of ritonavir. The 5-thiazolyl end group in ritonavir was replaced by the phenoxyacetyl group, and the 2-isopropylthiazolyl group in ritonavir was replaced by a modified valine in which the amino terminus had a six-membered cyclic urea attached.

Synthetic strategy employed for the synthesis of multikilogram quantities of lopinavir is very similar to that implemented for ritonavir (**34.1.22**), but using the protected version (**34.1.79**) of the "core" diamino alcohol (**34.1.55**), which was sequentially acylated with the acid chlorides of (S)-3-methyl-2-(2-oxotetrahydropyrimidin-1(2H)-yl)butanoic acid (**34.1.89**) and 2-(2,6-dimethylphenoxy) acetic acid (**34.1.94**) [135,136].

The bulk synthesis of protected diamino alcohol (**34.1.79**) was proposed [137,138] by a method closely related to the method [122,123] for ritonavir. For that purpose L-phenylalanine (**34.1.74**) was sequentially trialkylated with benzyl chloride using a K_2CO_3/water system at reflux to produce a tribenzylated product (**34.1.75**). A solution with generated acetonitrile anion in THF was added to the benzylated product (**34.1.75**) at less than −40°C to yield the cyanomethylketone (**34.1.76**), which was exposed to Grignard reagent–benzyl magnesium chloride to produce an enaminone (**34.1.77**). No racemization was observed in these two steps. The addition of the obtained enaminone (**34.1.77**) in THF/PrOH to a solution of $NaBH_4$ and MsOH in THF at 5°C produced an intermediate aminoketone (**34.1.78**). No further reduction of the keto group occurs under these conditions. Reduction of the keto group could proceed by addition of a preformed solution of sodium tris(trifluoroacetoxy)borohydride in tetrahydrofuran [$NaBH_3$(OTFA)], which produces a mixture of amino alcohols composed of 93% of the desired (2S)-5-amino-2-(dibenzylamino)-1,6-diphenylhexan-3-ol (**34.1.79**) along with 7% of the three undesired diastereomers. The crude mixture was debenzylated (Pd-C, $HCONH_4$), and the product was purified by precipitation from iPrOH/HCl (aq) to produce (**34.1.55**) in greater than 99% purity and in high yield (Scheme 34.8.).

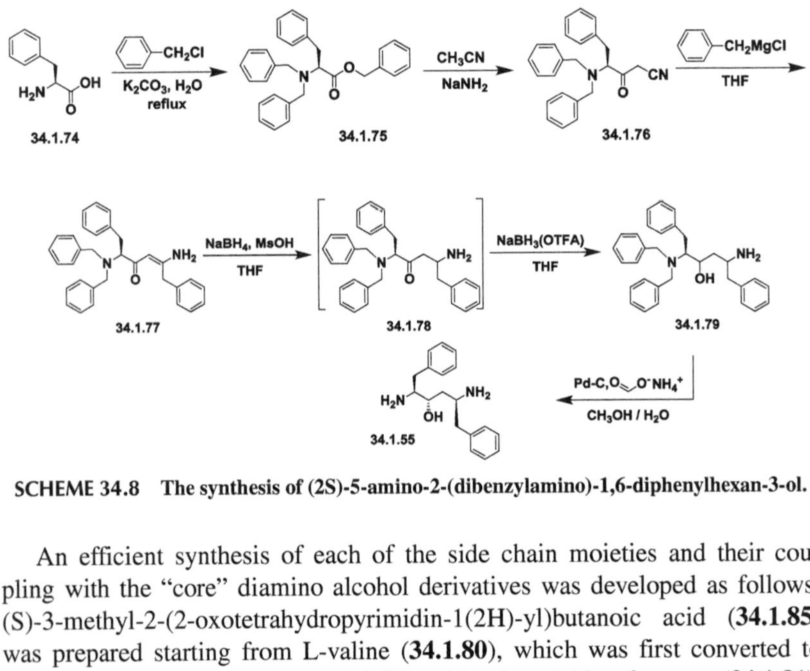

SCHEME 34.8 The synthesis of (2S)-5-amino-2-(dibenzylamino)-1,6-diphenylhexan-3-ol.

An efficient synthesis of each of the side chain moieties and their coupling with the "core" diamino alcohol derivatives was developed as follows: (S)-3-methyl-2-(2-oxotetrahydropyrimidin-1(2H)-yl)butanoic acid (**34.1.85**) was prepared starting from L-valine (**34.1.80**), which was first converted to N-phenoxycarbonyl-L-valine (**34.1.82**) with phenylchloroformate (**34.1.81**). Accurate pH monitoring (pH 9.5 to 10.2) was necessary and LiOH was found to be a superior base. Control of pH was essential as the valine dimer and its derivatives were formed as reaction byproducts outside of this pH margin. LiCl was added to provide a lower freezing point to the aqueous solution and neutral Al_2O_3 was added to prevent gumming and emulsion formation during the course of the reaction.

Treatment of N-phenoxycarbonyl-L-valine (**34.1.82**) with 3-chloropropylamine hydrochloride (**34.1.3**) and solid NaOH in THF produced the unisolated salt of chloropropylurea (**34.1.84**), which was then treated with t-BuOK, effecting cyclization to produce the desired acid (**34.1.85**) in 75 to 85% yield and in greater than 99% enantiomeric excess. Acylation of (**34.1.79**) with synthesized acid (**34.1.85**) was initially achieved by well-known peptide coupling methods. Optimization of this transformation allowed the discovery of a more cost-effective method for implementing acyl chloride (**34.1.86**), which was easily prepared using thionyl chloride in THF at room temperature.

The reaction of dibenzylamino alcohols (**34.1.79**) with acyl chloride (**34.1.86**) in the presence of 3.0 equivalents of imidazole in EtOAc and DMF produced acylated intermediate (**34.1.87**) as a mixture of diastereomers, which, without any further purification, was subjected to debenzylation with Pd/C and HCO_2NH_4 in MeOH at 50°C, which proceeded without significant complications to produce (**34.1.88**).

Exposure of crude (**34.1.88**) to L-pyroglutamic acid (**34.1.89**) in dioxane at 50°C followed by cooling, allowed for the isolation of (S)-N-((2S,4S,5S)-5-amino-4-hydroxy-1,6-diphenylhexan-2-yl)-3-methyl-2-(2-oxotetrahydropyrimidin-1(2H)-yl)butanamide (**34.1.90**) pyroglutamic salt as virtually a single diastereomer in high yield (Scheme 34.9.).

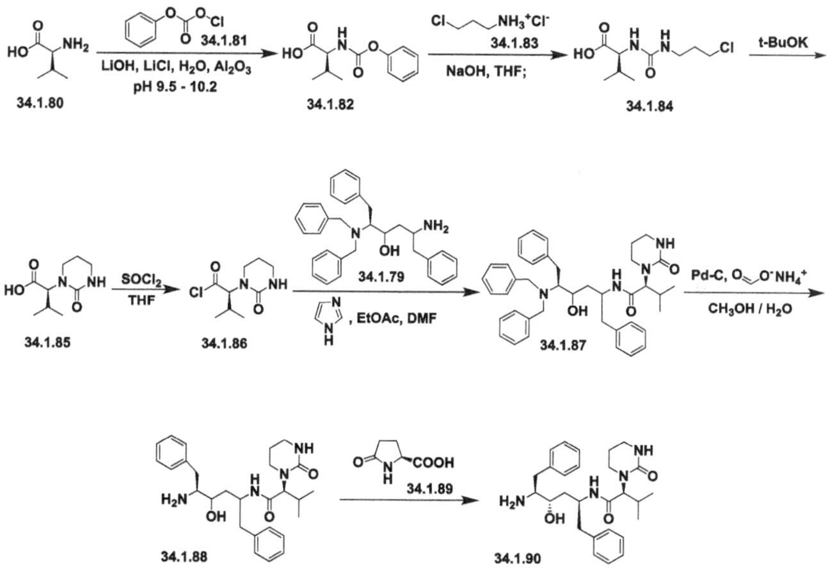

SCHEME 34.9 **The synthesis of (S)-N-((2S,4S,5S)-5-amino-4-hydroxy-1,6-diphenylhexan-2-yl)-3-methyl-2-(2-oxotetrahydropyrimidin-1(2H)-yl)butanamide.**

Acyl chloride (**34.1.92**) was prepared by the reaction of 2-(2,6-dimethylphenoxy)acetic acid (**34.1.91**) with thionyl chloride in EtOAc, at room temperature adding a single drop of DMF, and warming the slurry to 50°C, which produced a clear solution of (**34.1.92**) that was used in the subsequent acylation of amine (**34.1.90**). Reaction of pyroglutamate salt (**34.1.90**) with acyl chloride (**34.1.95**) in ethyl acetate under Schotten-Baumann reaction conditions (use of a two-phase solvent system) in the presence of a water solution of $NaHCO_3$ for liberation of free amine, produced the desired lopinavir (**34.1.21**) in high yield and purity (Scheme 34.10.).

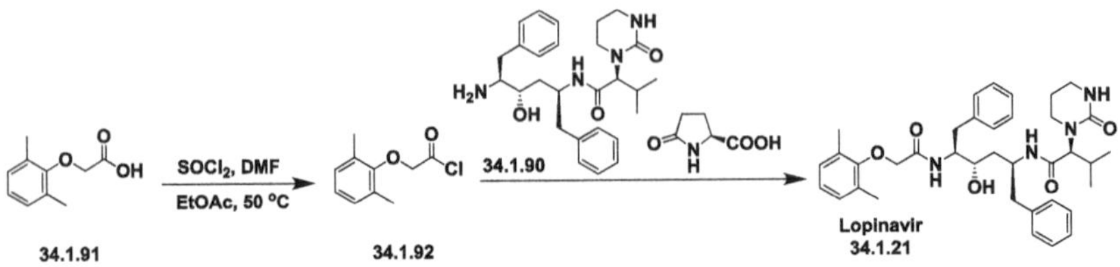

34.1.91 34.1.92

SCHEME 34.10 **The synthesis of lopinavir.**

Lopinavir is a novel and strong protease inhibitor developed from ritonavir with high specificity for HIV-1 protease [139-141]. It is indicated, in combination with other antiretroviral agents, for the treatment of HIV-1 infection. Numerous clinical trials have shown that lopinavir/ritonavir (Kaletra) is highly effective as a component of highly active antiretroviral therapy [142,143].

Polymerase Inhibitors

The polymerases are enzymes essentially required for the replication of viruses. Viral DNA and RNA polymerases are responsible for copying the genetic materials of viruses, their transcription and replication, and therefore are central components in the life cycles of viruses.

Polymerase inhibitors block enzymatic function, thus preventing virus from multiplying. This group of antiviral medications consist of two classes: nucleoside inhibitors and nonnucleoside inhibitors. In contrast to the nucleoside inhibitors that bind to the active site of the polymerase, the nonnucleoside inhibitors bind to allosteric binding sites within the polymerase, thus blocking its action.

Along with nucleoside and nonnucleoside polymerase inhibitors there exist a class of nucleotide and nonnucleotide polymerase inhibitors.

Polymerase inhibitors are classified also as DNA or RNA polymerase inhibitors.

Nucleoside viral DNA polymerases are the specific target of a number of antiviral drugs currently used to inhibit viral replication. Most antiviral approved drugs which inhibit a DNA polymerase are nucleoside analogues. They represent the most productive source of antiviral agents. These agents need to be phosphorylated to their active form. Active forms inhibit polymerases by competing with natural substrates incorporation into the growing DNA chain, and in this way terminating viral DNA elongation.

The DNA polymerase inhibitors are acyclovir (**34.1.93**), valacyclovir (**34.1.94**), penciclovir (**34.1.95**), famciclovir (**34.1.96**), ganciclovir (**34.1.97**), and valganciclovir (**34.1.98**) (Fig. 34.6.).

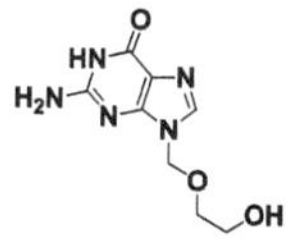

Aciclovir 34.1.93

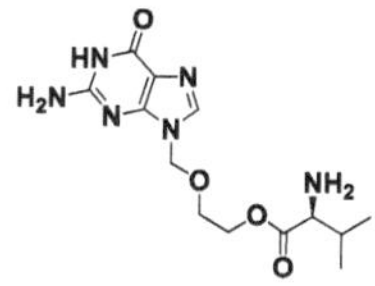

Valaciclovir 34.1.94

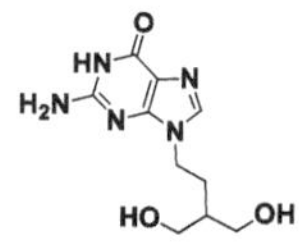

Penciclovir 34.1.95

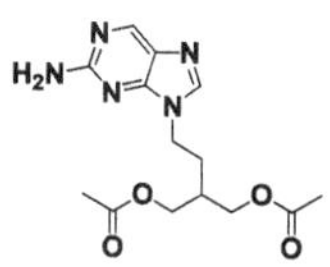

Famciclovir 34.1.96

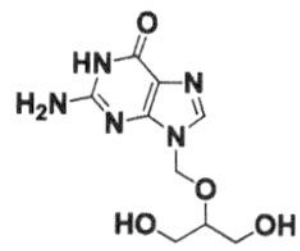

Ganciclovir 34.1.97

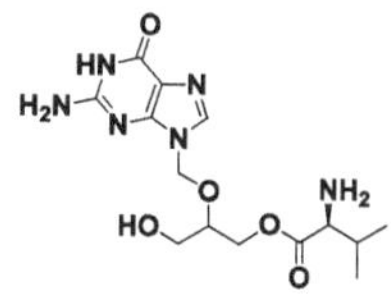

Valganciclovir 34.1.98

FIG. 34.6 Structure of DNA polymerase inhibitors.

A special place among viral DNA polymerases inhibitors is occupied by foscarnet (Foscavir) (**34.1.99**), which selectively inhibits the pyrophosphate binding site on viral DNA polymerases, blocking the release of pyrophosphate from the terminal nucleoside, and does not affect human DNA polymerases (Fig. 34.7.).

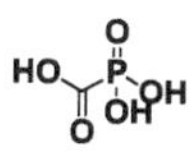

Foscarnet 34.1.99

FIG. 34.7 Structure of Foscavir.

Nucleoside and nucleotide DNA polymerase inhibitors are drugs identified by the suffix -ovir.

Viral RNA polymerase inhibitors are represented by the single drug ribavirin (**34.1.100**) (Fig. 34.8.).

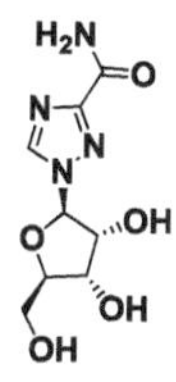

Ribavirin 34.1.100

FIG. 34.8 Structure of ribavirin.

Ribavirin (**34.1.100**) inhibits guanosine triphosphate formation, prevents capping of viral mRNA, and blocks derivative that resembles viral RNA-dependent RNA polymerase activity.

In recent years, antiviral drug discovery platforms utilizing high-throughput screening technology (HTS) have enabled discovery of many initial lead series of nonnucleoside viral RNA polymerase inhibitors.

The viral DNA polymerase inhibitors acyclovir, valacyclovir, valganciclovir, and tenofovir are included in the list of Top 200 Drugs by sales for the 2010s.

Acyclovir–Zovirax

The synthesis of the forerunner to polymerase inhibitors—acyclovir (**34.1.93**), as an antiherpes drug—was first described [144] and then disclosed in detail later [145].

The synthesis started with benzonitrile (**34.1.101**), which was heated at reflux in ethylene glycol for 72 hours to produce ethylene glycol monobenzoate (**34.1.102**). A cold mixture of the obtained ethylene glycol monobenzoate and paraformaldehyde in dry dichloroethane was saturated with dry HCl, producing 1-benzoyloxy-2-chloromethoxyethane (**34.1.103**). The last was added to a solution of 2,6-chloropurine (**34.1.104**) and triethylamine in dimethylformamide, and after the exothermic reaction the product—2,6-chloro-9-(2-benzoyloxyethoxymethyl)purine (**34.1.105**)—was separated. A solution of the 2,6-chloro-9-(2-benzoyloxyethoxymethyl)purine (**34.1.105**) in ammonia methanol solution was heated in autoclave at 95°C. As a result of the differences in the chemical reactivity in the 2- and 6- positions in the pyrimidine ring, selective substitution of the 6-chloro group takes place with simultaneous deprotection of the side chain to produce 2-chloro-9-(hydroxyethoxymethyl)adenine (**34.1.106**). Treatment of the last with nitrous acid, followed by reaction of deaminated intermediate with methanolic ammonia to displace the 2-chloro group, produces a moderate yield of acyclovir (**34.1.93**) (Scheme 34.11.).

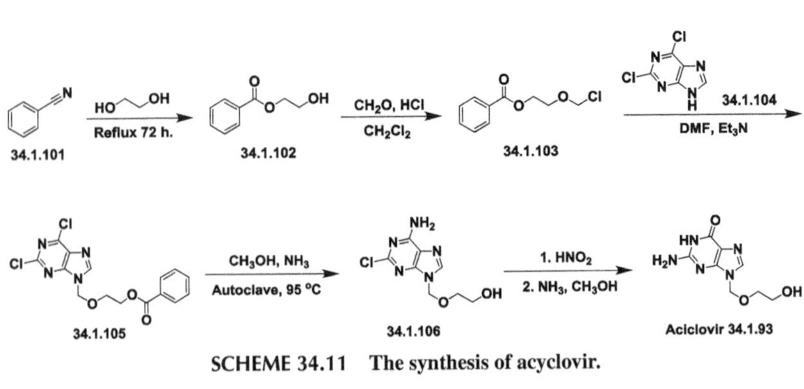

SCHEME 34.11 **The synthesis of acyclovir.**

A more efficient method to prepare acyclovir, which consists of alkylation of 2-chloro-6-iodopurine (**34.1.109**) with iodomethyl[(trimethylsilyl)oxy]ethyl ether (**34.1.108**), has been performed. The synthesis of (**34.1.108**) involved the reaction of 1,3-dioxolane (**34.1.107**) with trimethylsilyl iodide in cyclohexene to produce the desired side-chain moiety (**34.1.108**). Treatment of the anion of 2-chloro-6-iodo-purine (**34.1.109**) generated with NaH in dry DMF with prepared (**34.1.108**) at −63°C followed by hydrolysis and ammonolysis reactions, yielded the desired acyclovir (**34.1.93**). Hydrolysis was accomplished by adding aqueous solution of K_2CO_3 at room temperature to the solution of synthesized 2-chloro-9-[(2-hydroxyethoxy)methyl]-6-iodopurine (**34.1.110**) in dioxane. Ammonolysis occurred by heating of product of hydrolysis with NH_3 in methanol in a sealed tube that was heated to 110°C to produce acyclovir (**34.1.93**) [146,147] (Scheme 34.12.).

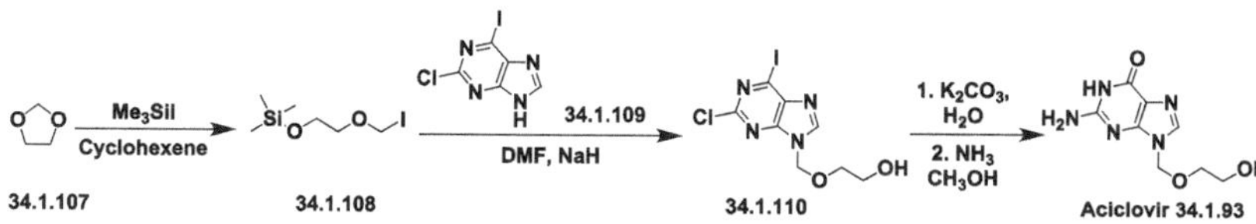

SCHEME 34.12 **The synthesis of acyclovir.**

A more convenient and economical synthesis of acyclovir has been reported [148]. The synthesis started from the easily available mixture of N,N′-diacetylguanines (**34.1.112**), which was prepared by acylation of guanine (**34.1.111**) with acetic anhydride/acetic acid mixture [149]. The obtained commixture of compounds was condensed with 2-oxabutane-1,4-diol diacetate (**34.1.113**) in the presence of p-toluenesulfonic acid or without solvent [149] or in DMSO [148], which produced a satisfactory yield of the desired separable crystalline intermediate (**34.1.114**), without the necessity for column purification. Deprotection of the compound (**34.1.114**) with 40% aqueous $MeNH_2$ for 20 minutes at 100°C produced the acyclovir (**34.1.96**) for an overall yield of 33%. Effective hydrolysis of both acetyl protective groups is possible also by using ammonia in MeOH, 5% aqueous NaOH at room temperature, pyrrolidine at room temperature, via refluxing in glycol, whatever etc. Variations of this method have been reported [148-150] and reviewed [151,152] (Scheme 34.13.).

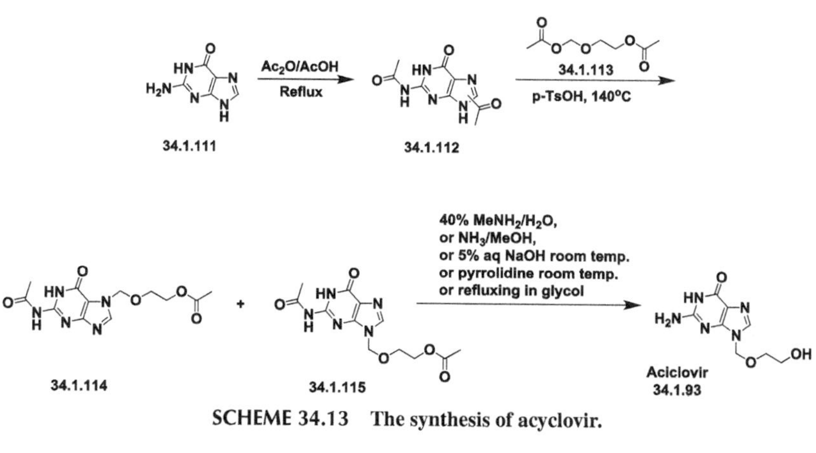

SCHEME 34.13 **The synthesis of acyclovir.**

Another approach [153] for the synthesis of acyclovir (**34.1.93**) was demonstrated; it started from the available 7-formamidofurazanopyrimidine (**34.1.115**) [154], which was alkylated with 2-(benzoyloxy)ethoxymethyl chloride (**34.1.103**) in dimethylformamide in the presence of triethylamine to produce a mixture of the desired compound (**34.1.116**) and some deformylated product. The mixture was reformylated with acetic-formic anhydride to produce crude (**34.1.116**). Reductive cleavage of the furazan ring in the obtained product with zinc dust in acetic acid followed by heating facilitated cyclization to produce 2-(methy1thio)adenine (**34.1.117**), the protective benzoyl group of which was cleaved with aqueous methylamine by heating on a steam bath to produce (**34.1.118**). The 2-methylthio group of (**34.1.118**) failed to react in liquid ammonia at 90°C. For this reason, the 6-amino group of the mentioned compound (**34.1.118**) was transformed to a hydroxyl group with sodium nitrite in acetic acid, which yielded (**34.1.119**). Subsequent amination with ammonia saturated ethanol at 140°C in an autoclave produced the desired acyclovir (**34.1.93**) (Scheme 34.14.).

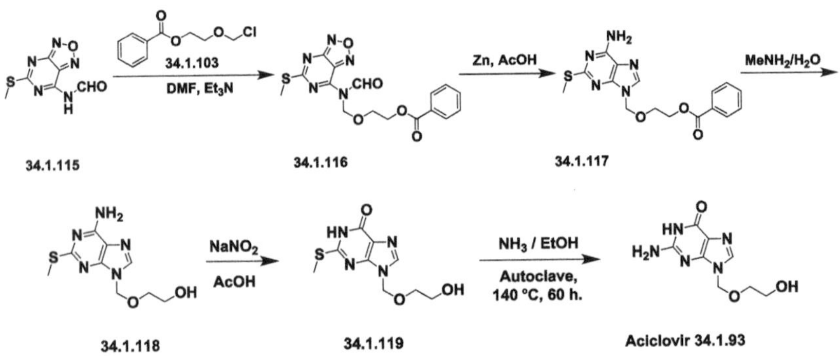

SCHEME 34.14 **The synthesis of acyclovir.**

Other approaches, which differ in details from described above, have been reviewed [151,152].

Acyclovir is closely related to the natural component of DNA, guanine deoxyriboside, and acts to prevent the DNA replication of a DNA virus at concentrations far below those that affect cellular DNA.

The discovery of acyclovir, a nucleoside analogue, more than 30 years ago, represents a milestone. Acyclovir is the drug of choice for the prophylactic and curative treatment of herpes simplex (genital herpes) virus and varicella-zoster (shingles) virus infection, which remain a challenge in the 21st century. Acyclovir does not cure herpes, but may prevent a breakout of herpes sores or blisters. The viruses continue to live in the body even between outbreaks. But acyclovir decreases the severity and length of these outbreaks. It helps the sores heal faster, keeps new sores from forming, and decreases pain and itching. Acyclovir formulations include injection, oral and topical forms [153-156].

Valacyclovir–Valtrex

Valacyclovir is a prodrug derived by esterifying acyclovir with L-valine. It is quickly absorbed and well tolerated. Upon administration, valacyclovir is rapidly and completely converted to acyclovir by enzymatic hydrolysis, which increases the oral bioavailability three- to fivefold.

In the first synthesis of valacyclovir (**34.1.94**) [157,158], acyclovir (**34.1.93**) was condensed with N-carbobenzyloxy-L-valine (**34.1.120**) in DMF and in the presence of dicyclohexylcarbodiimide, providing N-carbobenzyloxy–protected valacyclovir (**34.1.121**), which was subjected to palladium catalyzed deprotection (Pd/Al_2O_3 in DMF) to furnish valacyclovir (**34.1.94**). An efficient and scaleable process according to this Scheme 34.15 was developed later [159].

The same approach was implemented with another protecting group on an amino acid moiety and, valacyclovir was prepared by reaction of N-(Boc)-L-valine with acyclovir using 1-(3-dimethyl-aminopropyl)-3-ethylcarbodiimide hydrochloride (EDC) as coupling agent and HCl in the deprotection step [160].

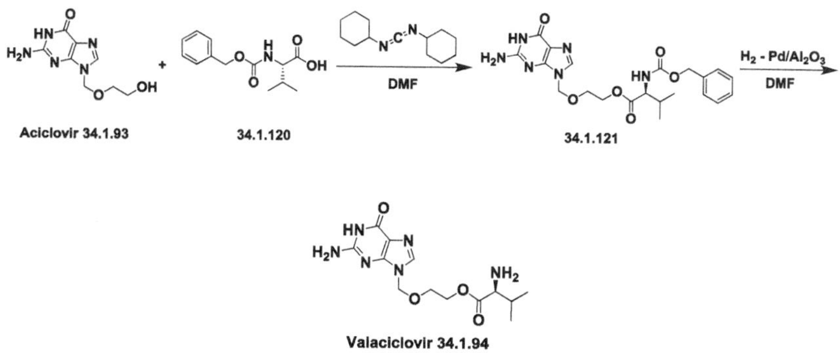

SCHEME 34.15 **The synthesis of valacyclovir.**

Valacyclovir is used for the treatment of herpes, varicella zoster, and cytomegaloviruses. A discussion on the clinical pharmacology, antiviral activity, clinical efficacy, and other therapeutic issues is presented in reviews [161-166].

Valganciclovir–Valcyte

Valganciclovir (**34.1.98**) is a mono-L-valyl ester prodrug of the antiviral compound ganciclovir. Most of the known literature on the synthesis of valganciclovir involves the coupling of the protected form of ganciclovir (**34.1.97**) with N-protected L-valine derivatives followed by deprotection.

The synthesis of ganciclovir for the synthesis of valganciclovir started from epichlorohydrin (**34.1.122**), which on reaction with benzyl alcohol in presence of 50% aqueous NaOH (at room temperature) produced 1,3-di-O-benzylglycerol (**34.1.123**) in good yield. Chloromethylation of (**34.1.123**) with HCl and paraformaldehyde in methylene chloride gave the chloromethyl ether (**34.1.124**), reaction of which with potassium acetate in acetone yielded

2-O-(acetoxymethyl)-1,3-di-O-benzylglycerol (**34.1.125**). Condensation of the obtained product with a mixture of N,N′-diacetylguanines (**34.1.112**) in the presence of a catalytic amount of p-TsOH in sulfolane or DMF, produced a 3:2 mixture of N^2-acetyl-9-[[1,3-bis(benzyloxy)-2-propoxy]methyl]guanine (**34.1.126**) and its corresponding N^7 isomer. The desired isomer (**34.1.126**) was separated by crystallization from toluene. Debenzylation of the obtained product (**34.1.126**) using palladium hydroxide on carbon (cyclohexene, ethanol) produced compound (**34.1.127**), which was deacetylated with a concentrated NH_4OH/methanol solution to produce ganciclovir (**34.1.97**) [167,168] (Scheme 34.16.).

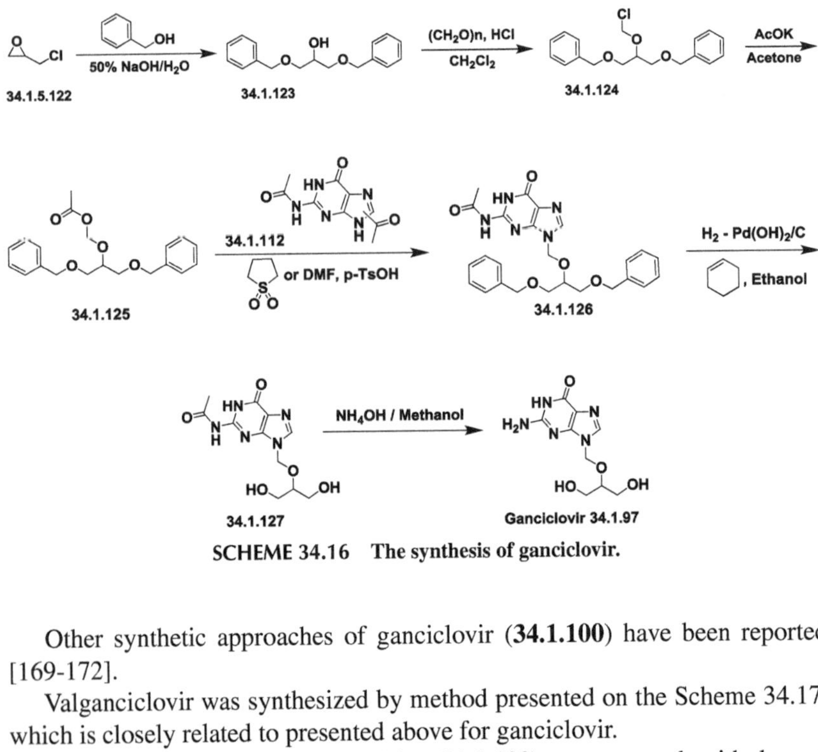

SCHEME 34.16 **The synthesis of ganciclovir.**

Other synthetic approaches of ganciclovir (**34.1.100**) have been reported [169-172].

Valganciclovir was synthesized by method presented on the Scheme 34.17, which is closely related to presented above for ganciclovir.

For this purpose, epichlorohydrin (**34.1.122**) was reacted with benzyl alcohol in the presence of powdered KOH in toluene at room temperature to produce benzyloxymethyloxirane (**34.1.128**). Gaseous HCl was bubbled into a stirred mixture of the obtained oxirane (**34.1.132**) and paraformaldehyde in dichloromethane to produce (1-chloro-2-chloromethoxy-3-benzyloxy)propane (**34.1.129**). This chloromethyl ether was reacted with potassium acetate in acetone to produce (1-chloro-2-acetoxymethoxy-3-benzyloxy)propane (**34.1.130**). A solution of diacetylguanine (**34.1.112**), (1-chloro-2-chloromethoxy-3-benzyloxy)propane (**34.1.130**), and p-TsOH in DMF was heated at 100°C for 6 hours to produce, after flash chromatography purification, a chloro derivative (**34.1.131**). The product from the previous step, and a large excess of potassium acetate in DMF, were heated to reflux for 5 hours to produce 2-(2-amino-1,6-dihydro-6-oxo-purin-9-yl)methoxy-1-acetoxy-3-benzyloxy-propane (**34.1.132**). Stirring of the last in 30% ammonia/methanol at ambient temperature produced the deprotected compound (**34.1.133**). N-Benzyloxycarbonyl-L-valine (**34.1.120**)/dicyclohexylcarbodiimide complex prepared in dichloromethane was added to the suspension of (**34.1.133**) in DMF followed by 4-dimethylaminopyridine. The workup of the mixture after 18 hours produced valinate (**34.1.134**). The obtained product was deprotected by hydrogenation over palladium hydroxide in refluxing ethanol to produce the desired valganciclovir (**34.1.98**) [173,174] (Scheme 34.17.).

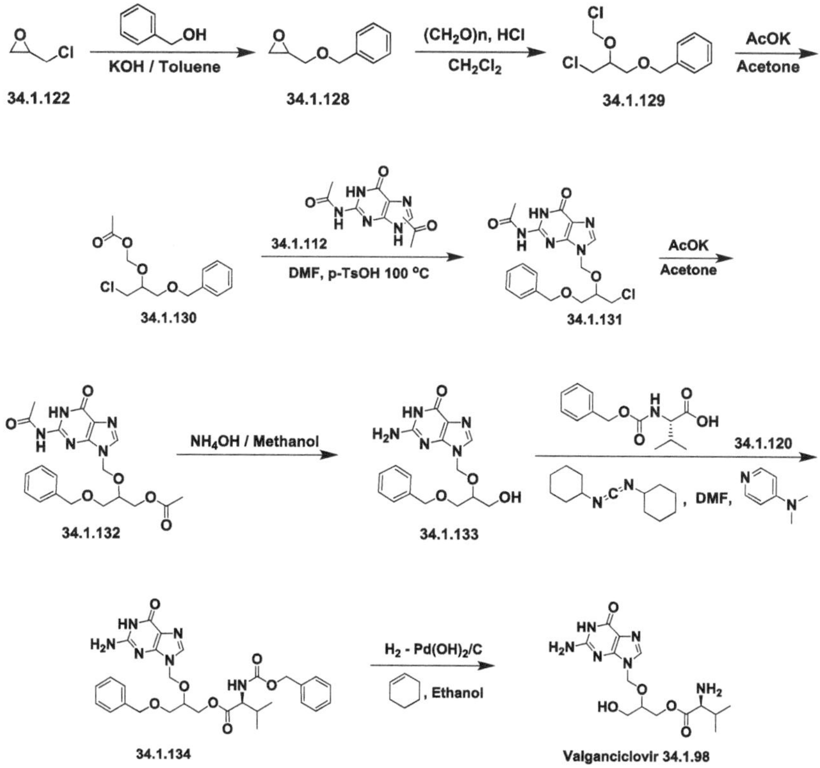

SCHEME 34.17 **The synthesis of valganciclovir.**

Valganciclovir (**34.1. 98**) has traditionally been synthesized by employing protection–deprotection strategies of either of the two hydroxy groups of optionally amino-protected ganciclovir and treating the monohydroxy-protected ganciclovir with protected L-valine (**34.1.120**). Most of the syntheses involve the use of key starting material ganciclovir (**34.1.97**).

In one approach, ganciclovir reacted with trimethyl orthoformate (or any other orthoformate) to produce cyclic orthoester (**34.1.135**), which was treated with formic acid in DMF/H_2O to yield ganciclovir O-monoformate (**34.1.136**). The condensation of the ganciclovir O-monoformate (**34.1.136**) with N-benzyloxycarbonyl-L-valine- (**34.1.120**) produced the monovalinate (**34.1.137**), which upon deformylation with HCl in dichloromethane/methanol mixture of solvents gave rise to compound (**34.1.138**), which was deprotected by hydrogenation with Pd/C catalyst [175] to produce the desired valganciclovir (**34.1.98**) (Scheme 34.18.).

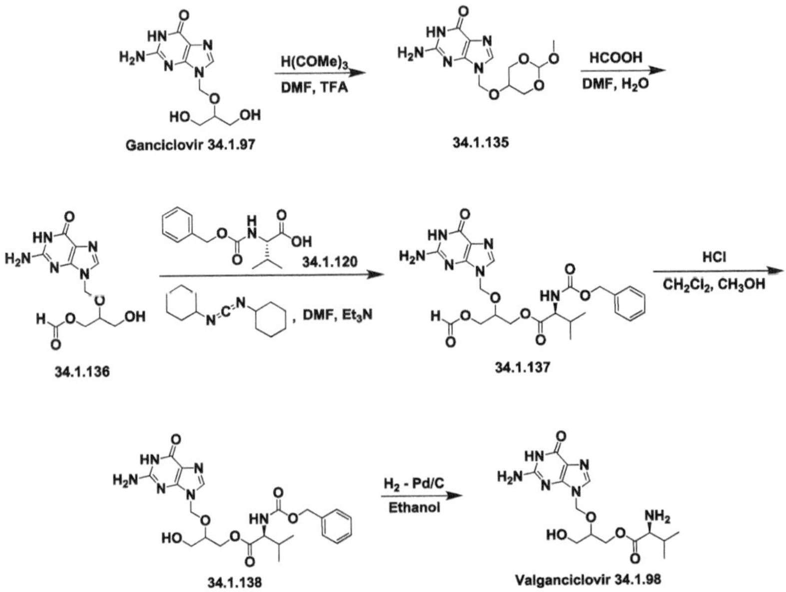

SCHEME 34.18 The synthesis of valganciclovir.

In another strategy, ganciclovir (**34.1.97**) was protected with tritylchloride or its derivatives in DMF in the presence of triethylamine and DMAP to produce the corresponding N,O-bis(trityl)–protected compound (**34.1.139**), which was condensed with N-(benzyloxycarbonyl)-L-valine (**34.1.120**) or N-(tert-butoxycarbonyl)-L-valine to yield the corresponding valine ester (**34.1.140**). Thereafter, the trityl-protecting group was removed using hot aqueous hydrochloric, acetic acid or trifluoroacetic acid, producing (**34.1.141**) followed by benzyloxycarbonyl removal by hydrogenation with Pd/C to obtain valganciclovir (**34.1.98**) [173,176-178] (Scheme 34.19.).

Many other closely related approaches also have been published [179-184].

Valganciclovir is effective for the treatment of AIDS-related CMV retinitis, and for the prophylaxis of cytomegalovirus infection and disease in high-risk solid organ transplant recipients. The drug is generally well tolerated and has a similar tolerability profile. It is devoid of adverse events related to intravenous or indwelling catheter access associated with the use of intravenous ganciclovir [185-192].

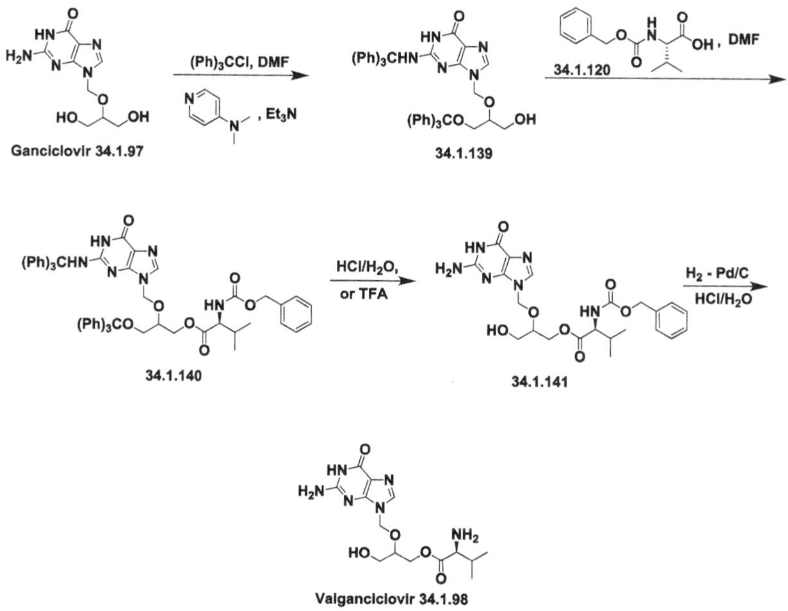

SCHEME 34.19 The synthesis of valganciclovir.

Nucleoside and Nucleotide Reverse Transcriptase Inhibitors

This group includes antiviral agents that are mainly recognized for the treatment of HIV, usually in combination with other retroviral drugs.

By 1980, HIV had spread to all continents. Identification of HIV as the causative agent of AIDS started a novel era in medicinal chemistry. As a result, the first antiretroviral agent, azidothymidine (Zidovudine, AZT) (**34.1.146**), was developed, which targeted the very nature of this class of viruses, reverse transcriptase, an enzyme that controls the replication of the genetic material of HIV. This enzyme is essential for HIV, but is not present in eukaryotic cells. These discoveries established the first class of antiretroviral agents: nucleoside and nucleotide reverse transcriptase inhibitors.

Nucleoside and nucleotide reverse transcriptase inhibitors are analogues of endogenous nucleosides and nucleotides. From the chemical point they can be represented as cytidine, guanosine, thymidine, and adenosine derivatives. They all are inactive in their parent forms and require phosphorylation steps by host cell kinases and phosphotransferases to form triphosphate derivatives capable of viral inhibition. In triphosphate forms, they compete with their corresponding endogenous deoxynucleoside triphosphates for incorporation by HIV reverse transcriptase. Once incorporated, they serve as chain-terminators of viral reverse transcripts, thus, acting on the viral replication cycle by inhibiting a critical step of proviral DNA synthesis prior to integration into the host cell genome [193-199].

The use of nucleoside and nucleotide reverse transcriptase inhibitors has revolutionized the treatment of infection by HIV and hepatitis-B virus. Nucleoside and nucleotide reverse transcriptase inhibitors became essential components in first-line therapy for HIV infection.

A variety of nucleoside reverse transcriptase inhibitors was synthesized, starting with azidothymidine (**34.1.142**), zalcitabine (**34.1.143**), stavudine (**34.1.144**), lamivudine (**34.1.145**), emtricitabine (**34.1.146**), didanosine (**34.1.147**), and abacavir (**34.1.148**), and, later, some nucleotide reverse transcriptase inhibitors, namely, cidofovir (**34.1.149**), adefovir (**34.1.150**), and tenofovir (**34.1.151**) appeared on the pharmaceutical market. Tenofovir (**34.1.151**) is used as tenofovir disoproxil fumarate, a prodrug for oral delivery, and is included in the list of Top 200 Drugs by sales for the 2010s (Fig. 34.9.).

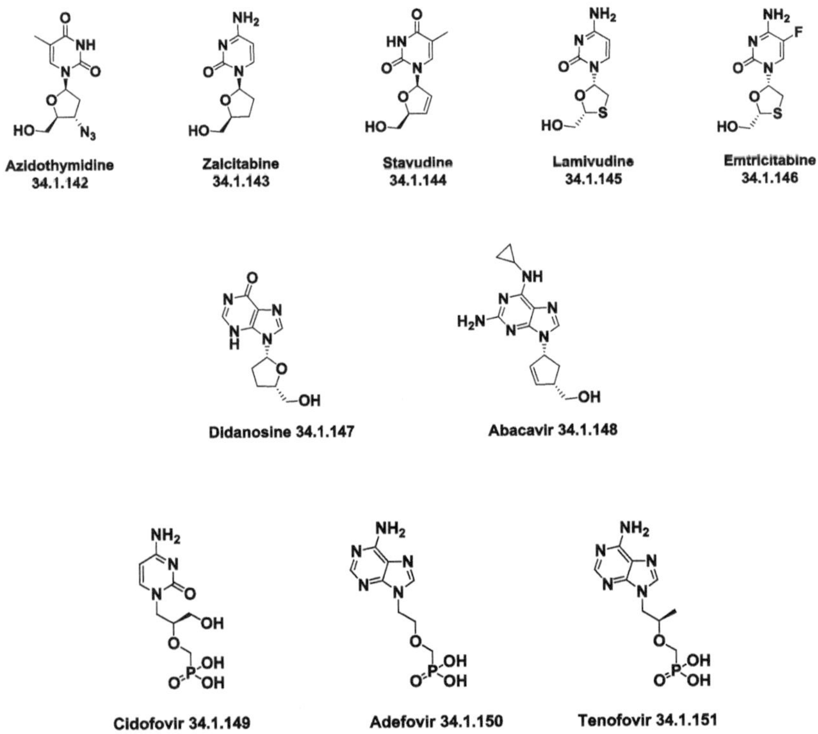

FIG. 34.9 **Nucleoside reverse transcriptase inhibitors.**

Tenofovir–Viread

Two approaches for the synthesis of tenofovir (**34.1.151**) have been reported, both of which employ readily available reagents.

One is based on alkylation of adenine (**34.1.152**) in DMF in the presence of $CsCO_3$ with chiral p-toluenesulfonyloxymethanephosphonate (**34.1.153**), which, in turn, was prepared in seven steps from D-(+)-isobutyl lactate [200]. Deprotection of the side chain of the obtained (**34.1.154**) using a standard deprotection/cleavage procedure with trimethylsilyl bromide to reflux acetonitrile produces tenofovir (**34.1.151**) [201] (Scheme 34.20.).

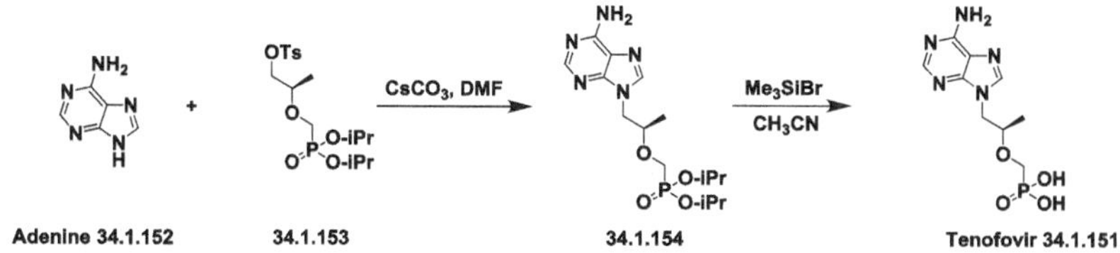

SCHEME 34.20 **The synthesis of tenofovir.**

In the second approach [202], adenine (**34.1.152**) was transformed to (R)-9-(2-hydroxypropyl)-adenine (**34.1.156**) by reaction with (R)-2-O-tetrahydropyranyl-1-O-p-toluenesulfonylpropane-1,2-diol (**34.1.155**) followed by deprotection with 0.25M sulfuric acid to produce the desired compound (**34.1.156**). After selective protection of N6 in (**34.1.156**) with benzoyl group, which was achieved by selective silylation with chlorotrimethylsilane in pyridine, followed by reaction with benzoyl chloride, the obtained 9-(R)-(2-hydroxypropyl)-N6-benzoyladenine (**34.1.157**) was alkylated with diisopropyl p-toluenesulfonyloxymethanephosphonate in DMF in the presence of sodium hydride to produce the compound (**34.1.158**). Debenzoylation of the N6 amino group was achieved using an ammonium hydroxide solution that produced (**34.1.158**), whose side chain was deprotected using trimethylsilyl bromide in reflux acetonitrile, which produced tenofovir (**34.1.151**) (Scheme 34.21.)

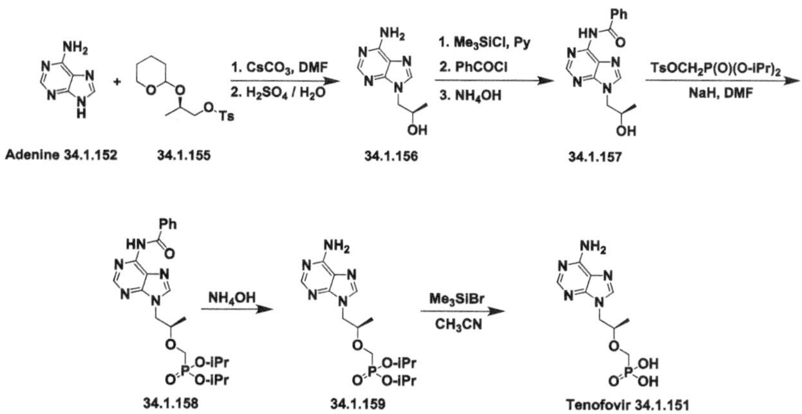

SCHEME 34.21 **The synthesis of tenofovir.**

The third approach was based on condensation of adenine (**34.1.152**) with propylene carbonate (**34.1.160**) to produce the (R)-9-(2-hydroxypropyl)-adenine (**34.1.161**). Coupling of the prepared adenine (**34.1.161**) with tosylated hydroxymethylphosphonate diester (**34.1.162**) using lithium tert-butoxide in THF, and after quenching obtained product with acetic acid and in dichloromethane/water mixture, the diethylphosphonate ester (**34.1.163**), which was subjected to excess bromotrimethylsilane in refluxing acetonitrile produced the desired tenofovir (**34.1.151**) [203] (Scheme 34.22.).

This method was improved by converting it to a manufacturing process for the large-scale synthesis of tenofovir disoproxil fumarate [204,205].

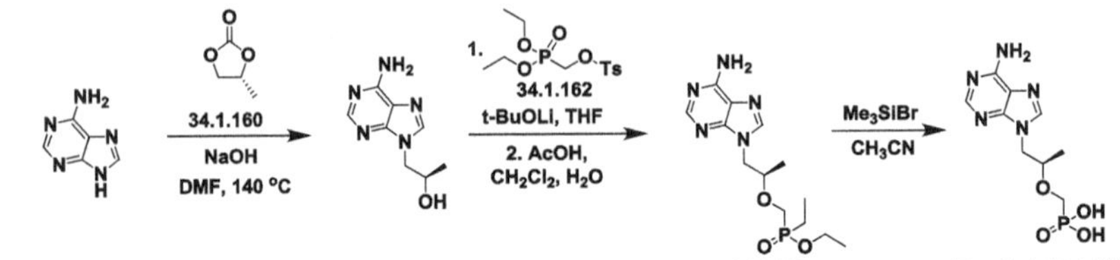

SCHEME 34.22 The synthesis of tenofovir.

Tenofovir became available in 2001 and today it is an effective and widely used treatment for both HIV and hepatitis B virus infection.

Tenofovir is a component of the preferred first-line combination antiretroviral therapy. The efficacy, tolerability, prolonged half-life allowing for once-daily administration, and availability as a component of several fixed-dose formulations make tenofovir an attractive choice for treatment-naïve and treatment-experienced HIV-infected patients. It can be used in combination with other anti-HIV drugs [206-219]. Since its approval in 2001, tenofovir has become one of the most frequently prescribed agents against HIV infection [206-219].

Nonnucleoside Reverse-Transcriptase Inhibitors

Nonnucleoside reverse-transcriptase inhibitors represent one of the most significant classes of drugs for the treatment of AIDS/HIV infection and are a crucial component of current antiretroviral therapy [220-224].

These drugs are not competitive with nucleoside reverse transcriptase inhibitors. They are not incorporated into the viral DNA and work at a different site of the enzyme, preventing the enzyme's action.

After the initial discovery of 1-(2-hydroxyethoxymethyl)-6-(phenylthio) thymine (HEPT) (**34.1.164**) followed by the synthesis of emivirine (**34.1.165**) and compounds of the IQP-0410 (**34.1.166**) series [225], several other chemical classes of compounds with nonnucleoside reverse transcriptase inhibitors properties were discovered. These include derivatives of α-anilinophenyl-acetamides (α-APA) such as loviride (**34.1.167**) and iminothioureas such as ITP (**34.1.168**) with activity against a wide variety of HIV-1 mutant strains [226], diaryltriazines such as R-106168 (**34.1.169**) [227], tetrahydroimidazo[4,5,1-jkj] [1,4] benzodiazepin-2(1H)-one and -thione derivatives (TIBO) such as tivirapine (**34.1.170**) [228] and various other compounds. The hallmark of nonnucleoside reverse-transcriptase inhibitors has been their ability to interact with a specific site ("pocket") of HIV-1 reverse-transcriptase (Fig. 34.10.).

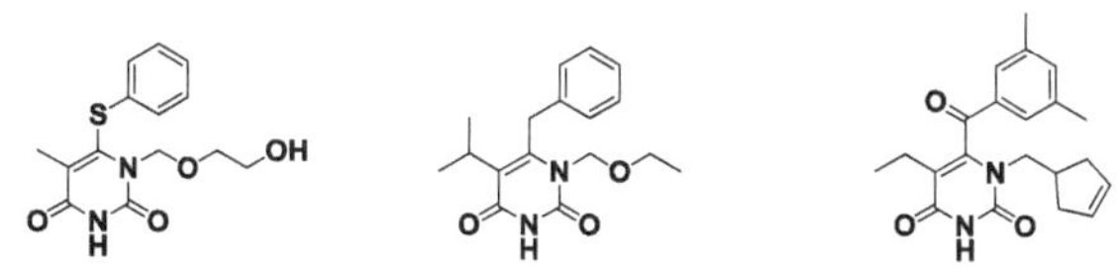

HEPT 34.1.164 Emivirine 34.1.165 IQP-0410 34.1.166

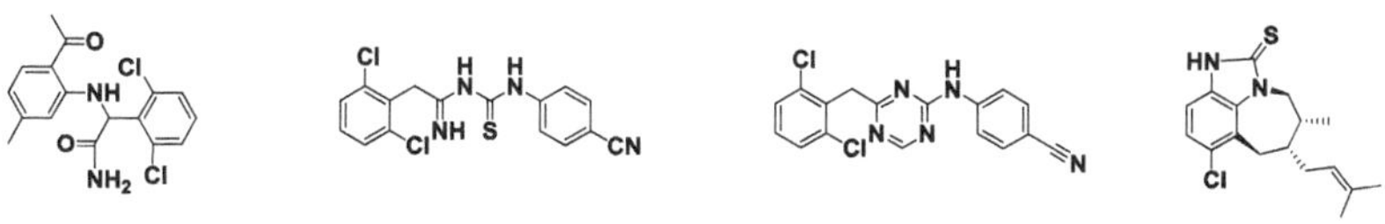

Loviride 34.1.167 ITU 34.1.168 R-106168 34.1.169 Tivirapine 34.1.170

FIG. 34.10 **Nonnucleoside reverse-transcriptase inhibitors.**

These findings followed with the discovery of the first drugs of this series, known as the first generation of nonnucleoside reverse-transcriptase inhibitors, which include efavirenz (**34.1.171**), the first nonnucleoside reverse transcriptase inhibitor (approved in 1998) [229-232], delavirdine (**34.1.172**) [233-235], which is rarely used nowadays, and nevirapine (**34.1.173**) [236-239]; all were commercialized as anti-HIV drugs (Fig. 34.11.).

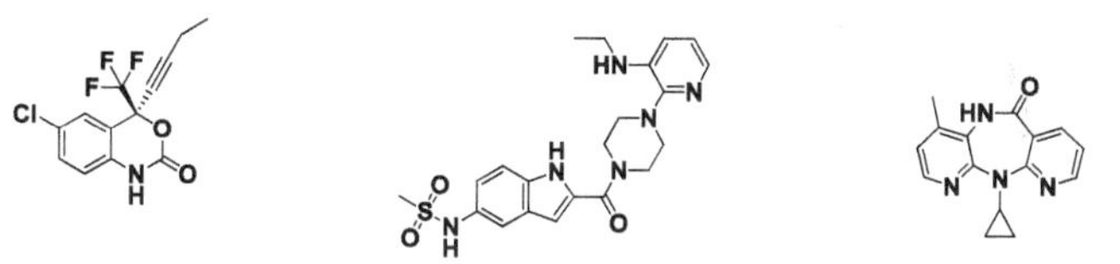

Efavirenz 34.1.171 Delavirdine 34.1.172 Nevirapine 34.1.173

FIG. 34.11 **Commercialized nonnucleoside reverse-transcriptase inhibitors.**

Sustained efforts in this area that were based on molecular modeling studies led to the identification of many promising hits, leads, and candidates, yielding the second-generation of nonnucleoside reverse-transcriptase inhibitors: etravirine (**34.1.174**), which was approved in 2008 [240,241], and rilpivirine (**34.1.175**) [242-249], which was approved in 2011 (Fig. 34.12.).

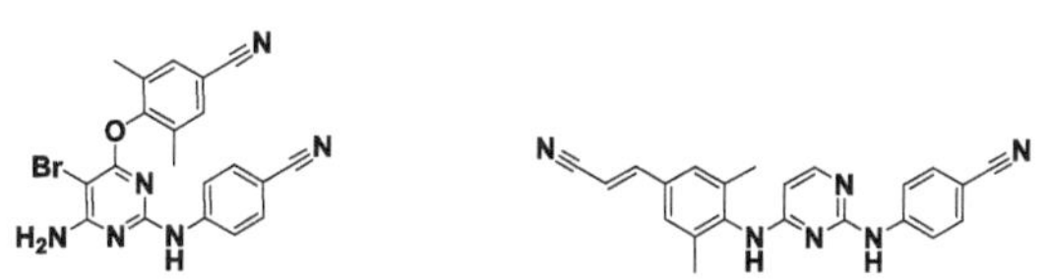

Etravirine 34.1.174 Rilpivirine 34.1.175

FIG. 34.12 **Second generation of nonnucleoside reverse-transcriptase inhibitors.**

The nonnucleoside reverse-transcriptase inhibitors in clinical use today are efavirenz (**34.1.171**), delavirdine (**34.1.172**), nevirapine (**34.1.173**), etravirine (**34.1.174**), and rilpivirine (**34.1.175**).

Lersivirine (**34.1.176**) is a novel second-generation nonnucleoside reverse transcriptase inhibitor. Its development was recently stopped in Phase IIb clinical trials [250].

Capravirine (**34.1.177**) [251], is another second-generation nonnucleoside reverse-transcriptase inhibitor. Studies showed that it had no specific advantages over currently used drugs and, consequently, clinical trials were discontinued after Phase IIb trials (Fig. 34.13.).

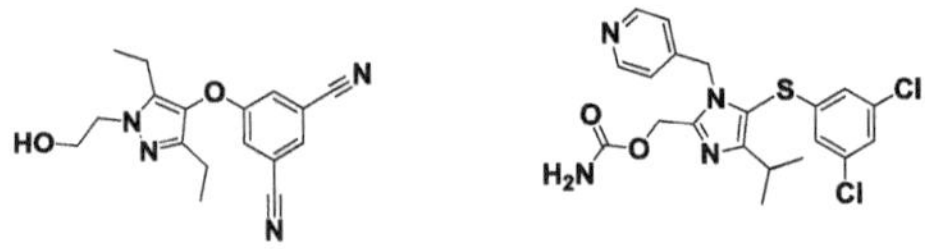

Lersivirine 34.1.176 Capravirine 34.1.177

FIG. 34.13 Novel second-generation nonnucleoside reverse-transcriptase inhibitors.

Human immunodeficiency virus infections are typically treated with drug combinations consisting of at least three different antiretroviral drugs; nonnucleoside reverse-transcriptase inhibitors are an essential component of antiretroviral therapy [252].

Integrase Inhibitors

Integrase inhibitors are a promising group of novel antiretroviral drugs that suppress the integrase-enzyme that facilitates the incorporation of HIV's proviral DNA into the host cell genome and catalyzes a function vital to viral replication, via inhibiting the "integration" of the viral DNA into the hosts' DNA genome [253-257].

Inhibitors of this enzyme represent the newest class of antiretroviral drugs in our armamentarium to treat viral infection.

Early integrase inhibitors included polyhydroxylated aromatic compounds, peptides, nucleotides, and DNA complexes, none of which were able to be developed into an effective drug.

The first major breakthrough was the discovery of the pyrrolo-diketo acids [258] as integrase inhibitors flowed by indole-diketo acids [259], naphthyridines [260-262], and dihydroxypyrimidine carboxamides [263], which finally led to discovery of raltegravir (**34.1.178**) [264,265].

Raltegravir (**34.1.178**) was the first integrase inhibitor (it was approved in 2007) for the treatment of HIV infections [266-281]. Raltegravir is included in the list of Top 200 Drugs by sales for the 2010s.

Elvitegravir (**34.1.179**), approved in late 2012, is another potent inhibitor of viral integrase [282-284]. Both raltegravir and elvitegravir are considered first-generation integrase inhibitors and are highly efficacious as first-line antiretroviral therapy.

Dolutegravir (**34.1.180**) is a second-generation integrase inhibitor that is currently under review by the FDA for marketing approval [285-289] (Fig. 34.14.).

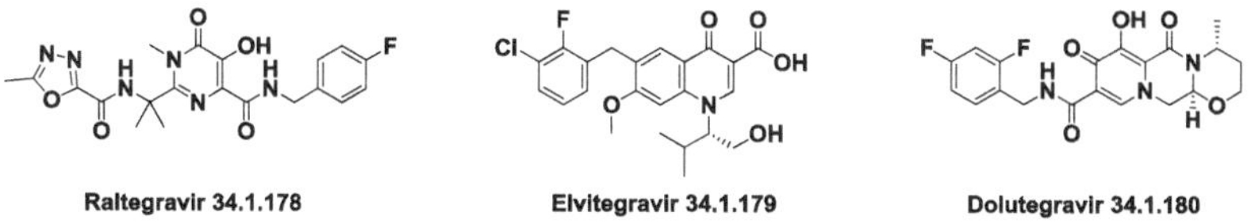

FIG. 34.14 **Integrase inhibitors.**

Raltegravir–Isentress

The general synthetic route starts from the transformation of acetone cyanohydrin (**34.1.181**) to aminonitrile (**34.1.182**) under Strecker reaction conditions. Obtained aminonitrile (**34.1.182**) was converted to the N-Cbz–protected intermediate (**34.1.183**) using benzyl chloroformate in sodium carbonate water solution. For the preparation of amidoxime (**34.1.184**), hydroxylamine hydrochloride was added to a solution of (**34.1.183**) and potassium hydroxide in methanol. The amidoxime (**34.1.184**) was treated with dimethylacetylenedicarboxylate (**34.1.185**) in chloroform to produce (**34.1.186**), which was taken into xylene and heated at 145°C for 48 hours to be cyclized to pyrimidine-4-carboxylate (**34.1.187**). The obtained compound was benzoylated with benzoic acid chloride in pyridine to produce (**34.1.188**), which was purified by flash column chromatography and then N-methylated with dimethyl sulfate in dioxane using lithium hydride as a base, which provided the compound (**34.1.189**). The prepared (**34.1.189**) was hydrogenated in the presence of 10% Pd/C to produce the N-deprotected product (**34.1.190**), which was acylated with freshly prepared 5-methyl-1,3,4- oxadiazole-2-carbonyl chloride (**34.1.191**) in the presence of triethylamine in dichloromethane. Finally, refluxing prepared (**34.1.192**) overnight with p-fluorobenzylamine in methanol produced the desired raltegravir (**34.1.178**) [265,267,290] (Scheme 34.23.). Further developments for efficient manufacturing of raltegravir are published [291-295].

Raltegravir is used along with other medications to treat HIV infection. It can cause serious, life-threatening side effects such as allergic reactions, skin reactions, and liver problems.

A number of other classes of compounds as potential antiviral drugs have attracted researcher's attention in the last few years. Among them are methyltransferase inhibitors [296] (methyltransferase catalyzes the transfer of a methyl group from S-adenosyl-methionine to viral RNA, and is essential for the life cycle of many significant human pathogen viruses); helicase inhibitors [297-299] (helicases catalytically unwind duplex DNA or RNA and are required to displace the single-stranded genome after replication); neuraminidase inhibitors [300,301] (neuraminidase inhibitors interfere with the release of virus from infected host cells); replication and transcription

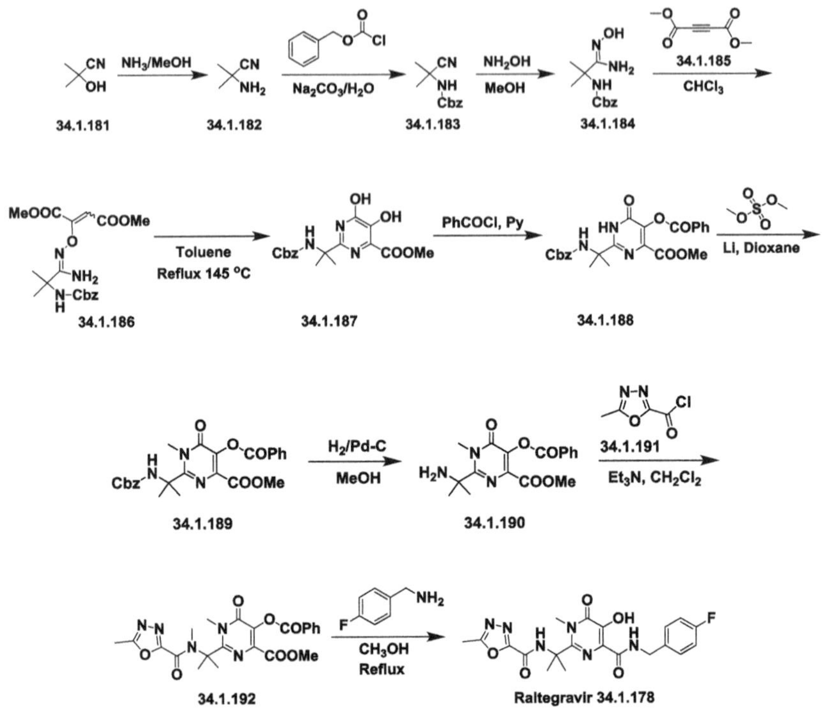

SCHEME 34.23 **The synthesis of raltegravir.**

complex blockers [302,303] (reagents that can efficiently block assembling of viral replication and transcription complex responsible for the production of the viral genome); and ribonucleoprotein complex inhibitors [304] (these inhibitors are thought to act as "molecule staples" that stabilize interactions between viral nucleoprotein monomers, promoting the formation of nonfunctional aggregates).

34.2 INDIRECT VIRUS-TARGETING ANTIVIRALS

While the antivirals currently in use exclusively target viral factors, several approaches now focus on cellular factors or pathways that indirectly interact with virus replication [305-307].

Among them are inhibitors of intracellular signaling cascades that are essential for virus replication.

Replication and transcription complex blockers block the formation of the viral replication and transcription complex, responsible for the production of the viral genome or other nucleic acids and ribonucleoprotein complex inhibitors which triggers the aggregation of and inhibits the nuclear accumulation of virally encoded nucleoprotein thereby inhibiting the replication of virus.

Among them a new compound – nucleozin (**34.2.1**) which inhibits the replication of various influenza A virus strains [308] (Fig. 34.15.).

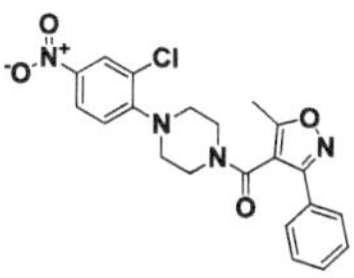

Nucleozin 34.2.1

FIG. 34.15 **Structure of nucleozin.**

REFERENCES

1. De Clercq, E. Antiviral drugs in current clinical use. *J. Clin. Virol.* **2004,** *30* (2), 115–133.
2. De Clercq, E. Antiviral drug discovery and development: where chemistry meets with biomedicine. *Antiviral Res.* **2005,** *67* (2), 56–75.
3. Balfour, H. H., Jr. Antiviral drugs. *N. Engl. J. Med.* **1999,** *340* (16), 1255–1268.
4. Bonhoeffer, S.; May, R. M.; Shaw, G. M.; Nowak, M. A. Virus dynamics and drug therapy. *Proc. Natl. Acad. Sci. U. S. A.* **1997,** *94* (13), 6971–6976.
5. De Clercq, E. Antiviral agents active against influenza A viruses. *Nat. Rev. Drug Discovery* **2006,** *5* (12), 1015–1025.
6. Boltz, D. A.; Aldridge, J. R., Jr.; Webster, R. G.; Govorkova, E. A. Drugs in development for influenza. *Drugs* **2010,** *70* (11), 1349–1362.
7. Biron, K. K. Antiviral drugs for cytomegalovirus diseases. *Antiviral Res.* **2006,** *71* (2-3), 154–163.
8. Coen, D. M.; Schaffer, P. A. Antiherpesvirus drugs: a promising spectrum of new drugs and drug targets. *Nat. Rev. Drug Discovery* **2003,** *2* (4), 278–288.
9. Arora, A.; Mendoza, N.; Tyring, S. K. Antiviral market overview. In *Development of Therapeutic Agents Handbook;* Gad, S. C., Ed.; Wiley, 2012; pp 127–143.
10. De Clercq, E. Antivirals: past, present and future. *Biochem. Pharmacol. (Amsterdam, Neth.)* **2013,** *85* (6), 727–744.
11. De Clercq, E. Antiviral drugs. In *Textbook of Drug Design and Discovery,* 4th ed.; Krogsgaard-Larsen, P., Stroemgaard, K., Madsen, U., Eds.; CRC Press, 2010; pp 393–417.
12. De Clercq, E. Highlights in the discovery of antiviral drugs: a personal retrospective. *J. Med. Chem.* **2010,** *53* (4), 1438–1450.
13. De Clercq, E. The discovery of antiviral agents: ten different compounds, ten different stories. *Med. Res. Rev.* **2008,** *28* (6), 929–953.
14. De Clercq, E. Emerging antiviral drugs. *Expert Opin. Emerg. Drugs* **2008,** *13* (3), 393–416.
15. De Clercq, E. Antivirals: current state of the art. *Future Virol.* **2008,** *3* (4), 393–405.
16. De Clercq, E. Status presens of antiviral drugs and strategies: part I: RNA viruses and retroviruses. *Adv. Antiviral Drug Des.* **2007,** *5,* 1–58.
17. De Clercq, E. Status presens of antiviral drugs and strategies: part II: RNA viruses (except retroviruses). *Adv. Antiviral Drug Des.* **2007,** *5,* 59–112.
18. De Clercq, E.; Holy, A. Case history: acyclic nucleoside phosphonates: a key class of antiviral drugs. *Nat. Rev. Drug Discovery* **2005,** *4* (11), 928–940.
19. De Clercq, E. Recent highlights in the development of new antiviral drugs. *Curr. Opin. Microbiol.* **2005,** *8* (5), 552–560.

20. De Clercq, E. Strategies in the design of antiviral drugs. *Nat. Rev. Drug Discovery* **2002,** *1* (1), 13–25.
21. De Clercq, E. Antiviral drugs: current state of the art. *J. Clin. Virol.* **2001,** *22* (1), 73–89.
22. De Clercq, E. Dancing with chemical formulae of antivirals: a personal account. *Biochem. Pharmacol. (Amsterdam, Neth.)* **2013,** *86* (6), 711–725.
23. De Clercq, E. Dancing with chemical formulae of antivirals: a panoramic view (Part 2). *Biochem. Pharmacol. (Amsterdam, Neth.)* **2013,** *86* (10), 1397–1410.
24. De Clercq, E. The design of drugs for HIV and HCV. *Nat. Rev. Drug Discovery* **2007,** *6* (12), 1001–1018.
25. Burke, J. D.; Fish, E. N. Antiviral strategies: the present and beyond. *Curr. Mol. Pharmacol.* **2009,** *2* (1), 32–39.
26. Meanwell, N. A.; Kadow, J. F.; Scola, P. M. Antiviral agents. *Annu. Rep. Med. Chem.* **2002,** *37,* 133–147.
27. Eigen, M. Error catastrophe and antiviral strategy. *Proc. Natl. Acad. Sci. U. S. A.* **2002,** *99* (21), 13374–13376.
28. Zhang, Z.; Wang, H.; Du, L.-Y.; Chen, K.-B.; Xiao, S.-L.; Yu, F.; Zhang, L.-H.; Zhou, D.-M. Discovery and development of antiviral drugs. *J. Chin. Pharm. Sci.* **2010,** *19* (6), 409–422.
29. Lou, Z.; Sun, Y.; Rao, Z. Current progress in antiviral strategies. *Trends Pharmacol. Sci.* **2014,** *35* (2), 86 102.
30. Jones, P. S. Strategies for antiviral drug discovery. *Antiviral Chem. Chemother.* **1998,** *9* (4), 283–302.
31. Tyring, S. K. General (non-antiretroviral) antiviral drugs. *Infect. Dis. Ther.* **2005,** *37,* 123–289.
32. He, H. Vaccines and antiviral agents. In *Current Issues in Molecular Virology: Viral Genetics and Biotechnological Applications;* Romanowski, V., Ed.; InTech, 2013; pp 239–250.
33. Arbuthnot, P., Ed. *Antiviral Drugs: Aspects of Clinical Use and Recent Advances;* InTech, 2012.
34. Antonelli, G.; Turriziani, O. Antiviral therapy: old and current issues. *Int. J. Antimicrob. Agents* **2012,** *40* (2), 95–102.
35. Kazmierski, W. M., Ed. *Antiviral Drugs: From Basic Discovery Through Clinical Trials;* Wiley, 2011.
36. Chen, T.-C.; Weng, K.-F.; Chang, S.-C.; Lin, J.-Y.; Huang, P.-N.; Shih, S.-R. Development of antiviral agents for enteroviruses. *J. Antimicrob. Chemother.* **2008,** *62* (6), 1169–1173.
37. Magri, A.; Bocchetta, S.; Burlone, M. E.; Minisini, R.; Pirisi, M. Recent advances in HCV entry. *Future Virol.* **2014,** *9* (2), 189–205.
38. Hazuda, D. J.; Burroughs, M.; Howe, A. Y. M.; Wahl, J.; Venkatraman, S. Development of boceprevir: a first-in-class direct antiviral treatment for chronic hepatitis C infection. *Ann. N. Y. Acad. Sci.* **2013,** *1291,* 69–76.
39. Masgala, A.; Nikolopoulos, G.; Tsiodras, S.; Bonovas, S.; Sitaras, N. M. Antiviral drugs in the prophylaxis of HBV infection. *Curr. Med. Chem.* **2012,** *19* (35), 5940–5946.
40. Mlynarczyk-Bonikowska, B.; Majewska, A.; Malejczyk, M.; Mlynarczyk, G.; Majewski, S. Antiviral medication in sexually transmitted diseases. Part I: HSV, HPV. *Mini-Rev. Med. Chem.* **2013,** *13* (13), 1837–1845.
41. Wegzyn, C. M.; Wyles, D. L. Antiviral drug advances in the treatment of human immunodeficiency virus (HIV) and chronic hepatitis C virus (HCV). *Curr. Opin. Pharmacol.* **2012,** *12* (5), 556–561.
42. Gallay, P. A. Cyclophilin inhibitors: a novel class of promising host-targeting ant-HCV agents. *Immunol. Res.* **2012,** *52* (3), 200–210.
43. Buti, M.; Esteban, R. Drugs in development for hepatitis B. *Drugs* **2005,** *65* (11), 1451–1460.

44. Jordan, R. Discovery and development of antiviral drugs for treatment of pathogenic human orthopoxvirus infections. *RSC Drug Discovery Ser.* **2013,** *32,* 81–110.
45. Arts, E. J.; Hazuda, D. J. HIV-1 Targeting the host or the virus: current and novel concepts for antiviral approaches against influenza virus infection drug therapy. *Perspect. Med.* **2012,** *2* (4). a007161/1-a007161/23.
46. Stellbrink, H.-J. Antiviral drugs in the treatment of AIDS: what is in the pipeline? *Eur. J. Med. Res.* **2007,** *12* (9), 483–495.
47. Karmon, S. L.; Markowitz, M. Next-generation integrase inhibitors. *Drugs* **2013,** *73* (3), 213–228.
48. Laver, G. Antiviral drugs for influenza: Tamiflu past, present and future. *Future Virol.* **2006,** *1* (5), 577–586.
49. Hayden, F. G. Antivirals for influenza: historical perspectives and lessons learned. *Antiviral Res.* **2006,** *71* (2-3), 372–378.
50. Wee, T.; Jenssen, H. Influenza drugs—current standards and novel alternatives. *J. Antivirals Antiretrovirals* **2009,** *1* (1), 001–010.
51. Lee, S. M.-Y.; Yen, H.-L. Targeting the host or the virus: current and novel concepts for antiviral approaches against influenza virus infection. *Antiviral Res.* **2012,** *96* (3), 391–404.
52. Saravolac, E. G.; Wong, J. P. Recent patents on development of nucleic acid-based antiviral drugs against seasonal and pandemic influenza virus infections. *Recent Pat. Anti-Infect. Drug Discovery* **2007,** *2* (2), 140–147.
53. Saravolac, E. G.; Wong, J. P. Recent patents on development of nucleic acid-based antiviral drugs against seasonal and pandemic influenza virus infections. *Front. Anti-Infect. Drug Discovery* **2010,** *1,* 409–425.
54. Loregian, A.; Mercorelli, B.; Nannetti, G.; Compagnin, C.; Palu, G. Antiviral strategies against influenza virus: towards new therapeutic approaches. *Cell. Mol. Life Sci.* **2014,** *71* (19), 3659–3683.
55. Park, S.; Kim, Jin I.; Park, M.-S. Antiviral agents against influenza viruses. *J. Bacteriol. Virol.* **2012,** *42* (4), 284–293.
56. Driscoll, J. S., Ed. *Antiviral drugs;* (Wiley), 2002.
57. Gilbert, C.; Boivin, G. Human cytomegalovirus resistance to antiviral drugs. *Antimicrob. Agents Chemother.* **2005,** *49* (3), 873–883.
58. Schang, L. M. Herpes simplex viruses in antiviral drug discovery. *Curr. Pharm. Des.* **2006,** *12* (11), 1357–1370.
59. Schafer, J. J.; Squires, K. E. Integrase inhibitors: a novel class of antiretroviral agents. *Ann. Pharmacother.* **2010,** *44* (1), 145–156.
60. Van Westreenen, M.; Boucher, C. A. B. Classes of antiviral drugs. In *Practical Guidelines in Antiviral Therapy;* Boucher, C. A. B., Galasso, G. A., Katzenstein, D. A., Cooper, D. A., Eds.; Elsevier, 2002; pp 1–12.
61. Da, L.-T.; Quan, J.-M.; Wu, Y.-D. Understanding the binding mode and function of BMS-488043 against HIV-1 viral entry. *Proteins: Struct., Funct., Genet.* **2011,** *79* (6), 1810–1819.
62. Yang, Z.; Zadjura, L. M.; Marino, A. M.; D'Arienzo, C. J.; Malinowski, J.; Gesenberg, C.; Lin, P.-F.; Colonno, R. J.; Wang, T.; Kadow, J. F.; Meanwell, N. A.; Hansel, S. B. Utilization of in vitro Caco-2 permeability and liver microsomal half-life screens in discovering BMS-48(8043), a novel HIV-1 attachment inhibitor with improved pharmacokinetic properties. *J. Pharm. Sci.* **2010,** *99* (4), 2135–2152.
63. Chen, K.; Risatti, C.; Bultman, M.; Soumeillant, M.; Simpson, J.; Zheng, B.; Fanfair, D.; Mahoney, M.; Mudryk, B.; Fox, R. J.; Hsaio, Y.; Murugesan, S.; Conlon, D. A.; Buono, F. G.; Eastgate, M. D. Synthesis of the 6-azaindole containing HIV-1 attachment inhibitor pro-drug, BMS-66(3068). *J. Org. Chem.* **2014,** *79* (18), 8757–8767.

64. Colman, P. M.; Varghese, J. N.; Laver, W. G. Structure of the catalytic and antigenic sites in influenza virus neuraminidase. *Nature (London, U. K.)* **1983,** *303* (5912), 41–44.
65. Ison, M. G. Antivirals and resistance: influenza virus. *Curr. Opin. Virol.* **2011,** *1* (6), 563–573.
66. Ison, M. G. Clinical use of approved influenza antivirals: therapy and prophylaxis. *Influenza Other Respir. Viruses* **2013,** *7* (Suppl. 1), 7–13.
67. Joly, V.; Jidar, K.; Tatay, M.; Yeni, P. Enfuvirtide: from basic investigations to current clinical use. *Expert Opin. Pharmacother.* **2010,** *11* (16), 2701–2713.
68. Yao, X.; Chong, H.; Zhang, C.; Qiu, Z.; Qin, B.; Han, R.; Waltersperger, S.; Wang, M.; He, Y.; Cui, S. Structural basis of potent and broad HIV-1 fusion inhibitor CP32M. *J. Biol. Chem.* **2012,** *287* (32), 26618–26629.
69. Zhang, X.; Wu, H.; Wang, F. Sifuvirtide, a novel HIV-1 fusion inhibitor. In *Peptide Drug Discovery and Development: Translational Research in Academia and Industry;* Castanho, M., Santos, N. C., Eds.; Wiley-VCH, 2011; pp 231–243.
70. Eggink, D.; Langedijk, J. P. M.; Bonvin, A. M. J.J.; Deng, Y.; Lu, M.; Berkhout, B.; Sanders, R. W. Detailed mechanistic insights into HIV-1 sensitivity to three generations of fusion inhibitors. *J. Biol. Chem.* **2009,** *284* (39), 26941–26950.
71. Xiao, C.; McKinlay, M. A.; Rossmann, M. G. Design of capsid-binding antiviral agents against human rhinoviruses. *RSC Biomol. Sci.* **2011,** *21* (Structural Virology), 319–337.
72. Zhang, G.; Zhou, F.; Gu, B.; Ding, C.; Feng, D.; Xie, F.; Wang, J.; Zhang, C.; Cao, Q.; Deng, Y.; Hu, W.; Yao, K. In vitro and in vivo evaluation of ribavirin and pleconaril antiviral activity against enterovirus 71 infection. *Arch. Virol.* **2012,** *157* (4), 669–679.
73. Thibaut, H. J.; De Palma, A. M.; Neyts, J. Combating enterovirus replication: state-of-the-art on antiviral research. *Biochem. Pharmacol. (Amsterdam, Neth.)* **2012,** *83* (2), 185–192.
74. Dorr, P.; Stammen, B.; van der Ryst, E. Discovery and development of maraviroc, a CCR5 antagonist for the treatment of HIV infection. In *Case Studies in Modern Drug Discovery and Development;* Huang, X., Aslanian, R. G., Eds.; Wiley, 2012; pp 196–226.
75. Lieberman-Blum, S. S.; Fung, H. B.; Bandres, J. C. Maraviroc: a CCR5-receptor antagonist for the treatment of HIV-1 infection. *Clin. Ther.* **2008,** *30* (7), 1228–1250.
76. Hubsher, G.; Haider, M.; Okun, M. S. Amantadine: the journey from fighting flu to treating Parkinson disease. *Neurology* **2012,** *78* (14), 1096–1099.
77. Saito, R.; Li, D.; Sato, M.; Suzuki, H. Amantadine. *Virus Rep.* **2006,** *3* (1), 40–47.
78. Fleming, D. M. Managing influenza: amantadine, rimantadine and beyond. *Int. J. Clin. Pract.* **2001,** *55* (3), 189–195.
79. Hayden, F. G. Amantadine and rimantadine—clinical aspects. In *Antiviral Drug Resistance;* Richman, D. D., Ed.; Wiley, 1996; pp 59–77.
80. Hay, A. J. Amantadine and rimantadine-mechanisms. In *Antiviral Drug Resistance;* Richman, D. D., Ed.; (Wiley), 1996; pp 43–58.
81. Ghosh, A. K.; Leshchenko, S.; Noetzel, M. Stereoselective photochemical 1,3-dioxolane addition to 5-alkoxymethyl-2(5H)-furanone: synthesis of bis-tetrahydrofuranyl ligand for HIV protease inhibitor UIC-94017 (TMC-114). *J. Org. Chem.* **2004,** *69* (23), 7822–7829.
82. Ghosh, A. K.; Kincaid, J. F.; Cho, W.; Walters, D. E.; Krishnan, K.; Hussain, K. A.; Koo, Y.; Cho, H.; Rudall, C.; Holland, L.; Buthod, J. Potent HIV protease inhibitors incorporating high-affinity P2-ligands and (R)-[(hydroxyethyl)amino]sulfonamide isostere. *Bioorg. Med. Chem. Lett.* **1998,** *8* (6), 687–690.
83. Ghosh, A. K.; Leshchenko, S.; Noetzel, M. W. Method of preparing (3R,3aS,6aR)-3-hydroxyhexahydrofuro[2,3-b]furan and related compounds, WO 2004033462 (2004).
84. Erickson, J. W.; Gulnik, S. V., Fitness assay and associated methods, and applications to drug resistance and HIV protease inhibitors and other drugs with reduced resistance, WO 9967417 (1999).

85. Ghosh, A. K.; Sridhar, P. R.; Kumaragurubaran, N.; Koh, Y.; Weber, I. T.; Mitsuya, H. Bis-tetrahydrofuran: a privileged ligand for darunavir and a new generation of HIV protease inhibitors that combat drug resistance. *ChemMedChem* **2006,** *1* (9), 939–950.
86. Ghosh, A. K. Capturing the essence of organic synthesis: from bioactive natural products to designed molecules in today's medicine. *J. Org. Chem.* **2010,** *75* (23), 7967–7989.
87. Ghosh, A. K.; Gemma, S.; Simoni, E.; Baldridge, A.; Walters, D. E.; Ide, K.; Tojo, Y.; Koh, Y.; Amano, M.; Mitsuya, H. Synthesis and biological evaluation of novel allophenylnorstatine-based HIV-1 protease inhibitors incorporating high affinity P2-ligands. *Bioorg. Med. Chem. Lett.* **2010,** *20* (3), 1241–1246.
88. Ruela Correa, J. C.; D'Arcy, D. M.; dos Reis Serra, C. H.; Nunes Salgado, H. R. Darunavir: a critical review of its properties, use and drug interactions. *Pharmacology* **2012,** *90* (1-2), 102–109.
89. de Bethune, M.-P.; Peeters, A.; Wigerinck, P. From saquinavir to darunavir: the impact of 10 years of medicinal chemistry on a lethal disease. *Methods Princ. Med. Chem.* **2011,** *50,* 73–90.
90. Deeks, E. D. Darunavir: a review of its use in the management of HIV-1 infection. *Drugs* **2014,** *74* (1), 99–125.
91. de Bethune, M.-P.; Sekar, V.; Spinosa-Guzman, S.; Vanstockem, M.; De Meyer, S.; Wigerinck, P.; Lefebvre, E. Darunavir (prezista, TMC114): from bench to clinic, improving treatment options for HIV-infected patients. In *Antiviral Drugs: From Basic Discovery Through Clinical Trials;* Kazmierski, W. M., Ed.; Wiley, 2011; pp 31–45.
92. Kogawa, A. C.; Salgado, H. R. N. Characteristics, complexation and analytical methods of Darunavir. *Br. J. Pharm. Res.* **2014,** *4* (11), 1276–1286.
93. Phung, B.-C.; Yeni, P. Darunavir: an effective protease inhibitor for HIV-infected patients. *Expert Rev. Anti-Infect. Ther.* **2011,** *9* (6), 631–643.
94. El-Atrouni, W. I.; Temesgen, Z. Darunavir. *Drugs Today* **2007,** *43* (10), 671–679.
95. Molina, J.-M.; Hill, A. Darunavir (TMC114): a new HIV-1 protease inhibitor. *Expert Opin. Pharmacother.* **2007,** *8* (12), 1951–1964.
96. Sorbera, L. A.; Castaner, J.; Bayes, M. Darunavir. *Drugs Future* **2005,** *30* (5), 441–449.
97. McKeage, K.; Perry, C. M.; Keam, S. J. Darunavir: a review of its use in the management of HIV infection in adults. *Drugs* **2009,** *69* (4), 477–503.
98. Bold, G.; Faessler, A.; Capraro, H.-G.; Cozens, R.; Klimkait, T.; Lazdins, J.; Mestan, J.; Poncioni, B.; Roesel, J.; Stover, D.; Tintelnot-Blomley, M.; Acemoglu, F.; Beck, W.; Boss, E.; Eschbach, M.; Huerlimann, T.; Masso, E.; Roussel, S.; Ucci-Stoll, K.; Wyss, D.; Lang, M. New aza-dipeptide analogs as potent and orally absorbed HIV-1 protease inhibitors: candidates for clinical development. *J. Med. Chem.* **1998,** *41* (18), 3387–3401.
99. Fassler, A.; Bold, G.; Capraro, H-G.; Steiner, H. Process for the preparation of hydrazine derivatives useful as intermediates for the preparation of peptide analogs, PCT Int. Appl. (1997), WO 9746514 A1 19971211.
100. Thompson, W. J.; Fitzgerald, P. M. D.; Holloway, M. K.; Emini, E. A.; Darke, P. L.; McKeever, B. M.; Schleif, W. A.; Quintero, J. C.; Zugay, J. A.; Tucker, T. J.; Schwering, J. E.; Homnick, C. F.; Nunberg, J.; Springer, J. P.; Huff, J. R. Synthesis and antiviral activity of a series of HIV-1 protease inhibitors with functionality tethered to the P1 or P1' phenyl substituents: x-ray crystal structure assisted design. *J. Med. Chem.* **1992,** *35,* 1685–1701.
101. Luly, J. R.; Dellaria, J. F.; Plattner, J. J.; Soderquist, J. L.; Yi, N. A. Synthesis of protected aminoalkyl epoxides from R-amino acids. *J. Org. Chem.* **1987,** *52,* 1487–1492.
102. Xu, Z.; Singh, J.; Schwinden, M. D.; Zheng, B.; Kissick, T. P.; Patel, B.; Humora, M. J.; Quiroz, F.; Dong, L.; Hsieh, D.-M.; Heikes, J. E.; Pudipeddi, M.; Lindrud, M. D.; Srivastava, S. K.; Kronenthal, D. R.; Mueller, R. H. Process research and development for an efficient synthesis of the HIV protease inhibitor BMS-23(2632). *Org. Process Res. Dev.* **2002,** *6* (3), 323–328.

103. Fan, X.; Song, Y.-L.; Long, Y.-Q. An efficient and practical synthesis of the HIV protease inhibitor atazanavir via a highly diastereoselective reduction approach. *Org. Process Res. Dev.* **2008,** *12* (1), 69–75.
104. Giordano, C.; Pozzoli, C.; Benedetti, F., Process for the preparation of aryl-pyridinyl compounds, WO 2001027083 (2001).
105. Chen, W. Process for synthesizing atazanavir, WO 2009130534 (2009).
106. Simhadri, S.; Mohammad, Y.; Indukuri, V. S. K.; Gorantla, S. R., Preparation of atazanavir bisulfate, WO 2014030173 (2014).
107. Farajallah, A.; Bunch, R. T.; Meanwell, N. A. Discovery and development of atazanavir. In *Antiviral Drugs: From Basic Discovery Through Clinical Trials;* Kazmierski, W. M., Ed.; Wiley, 2011; pp 3–17.
108. Goldsmith, D. R.; Perry, C. M. Atazanavir. *Drugs* **2003,** *63* (16), 1679–1693.
109. Busti, A. J.; Hall, R. G., II.; Margolis, D. M. Atazanavir for the treatment of human immunodeficiency virus infection. *Pharmacotherapy* **2004,** *24* (12), 1732–1747.
110. Orrick, J. J.; Steinhart, C. R. Atazanavir. *Ann. Pharmacother.* **2004,** *38* (10), 1664–1674.
111. Havlir, D. V.; O'Marro, S. D. Atazanavir: new option for treatment of HIV infection. *Clin. Infect. Dis.* **2004,** *38* (11), 1599–1604.
112. Bentue-Ferrer, D.; Arvieux, C.; Tribut, O.; Ruffault, A.; Bellissant, E. Clinical pharmacology, efficacy and safety of atazanavir: a review. *Expert Opin. Drug Metab. Toxicol.* **2009,** *5* (11), 1455–1468.
113. Croom, K. F.; Dhillon, S.; Keam, S. J. Atazanavir: a review of its use in the management of HIV-1 infection. *Drugs* **2009,** *69* (8), 1107–1140.
114. Wood, R. Atazanavir: its role in HIV treatment. *Expert Rev. Anti-Infect. Ther.* **2008,** *6* (6), 785–796.
115. von Hentig, N. Atazanavir/ritonavir: a review of its use in HIV therapy. *Drugs Today* **2008,** *44* (2), 103–132.
116. Gianotti, N.; Soria, A.; Lazzarin, A. Antiviral activity and clinical efficacy of atazanavir in HIV-1-infected patients: a review. *New Microbiol.* **2007,** *30* (2), 79–88.
117. Harrison, T. S.; Scott, L. J. Atazanavir: a review of its use in the management of HIV infection. *Drugs* **2005,** *65* (16), 2309–2336.
118. Piliero, P. J. Atazanavir: A novel once-daily protease inhibitor. *Drugs Today* **2004,** *40* (11), 901–912.
119. Piliero, P. J. Atazanavir: a novel HIV-1 protease inhibitor. *Expert Opin. Invest. Drugs* **2002,** *11* (9), 1295–1301.
120. Kempf, D. J.; Norbeck, D. W.; Sham, H. L.; Zhao, C.; Sowin, T. J.; Reno, D. S.; Haight, A. R.; Cooper, A. J. Preparation of peptide analogs as retroviral protease inhibitors, WO 9414436 (1994).
121. Kempf, D. J.; Sham, H. L.; Marsh, K. C.; Flentge, C. A.; Betebenner, D.; Green, B. E.; McDonald, E.; Vasavanonda, S.; Saldivar, A.; Wideburg, N. E.; Kati, W. M.; Ruiz, L.; Zhao, C.; Fino, L. M.; Patterson, J.; Molla, A.; Plattner, J. J.; Norbeck, D. W. Discovery of ritonavir, a potent inhibitor of HIV protease with high oral bioavailability and clinical efficacy. *J. Med. Chem.* **1998,** *41* (4), 602–617.
122. Stuk, T. L.; Allen, M. S.; Haight, A. R.; Kerdesky, F. A.; Langridge, D. C.; Leanna, M. R.; Lijewski, L. M.; Melcher, L.; Morton, H. E.; Robbins, T. A.; Sowin, T. J. Process for the stereoselective preparation of a substituted 2,5-diamino-3-hydroxyhexane as an intermediate for HIV protease inhibitors, US 5491253 (1996).
123. Haight, A. R.; Goodmonson, O. J.; Parekh, S. I.; Robbins, T. A.; Seif, L. S. Process for the preparation of a phenyl-disubstituted 2,5-diamino-3-hydroxyhexane, WO 9604232 (1996).

124. Tien, J J.; Menzia, J. A.; Cooper, A. J. Process for the preparation of HIV protease inhibiting peptide analogs, US 5567823 (1996).
125. Adamo, I.; Benedetti, F.; Berti, F.; Campaner, P. Stereoselective hydroazidation of amino enones: synthesis of the ritonavir/lopinavir core. *Org. Lett.* **2006,** *8* (1), 51–54.
126. Cheng, Y.; Tanaka, H.; Baba, M. Preparation of 2′,3′-dideoxy and 2′,3′-didehydro nucleoside analogs as prodrugs for treating viral infections, most notably HIV, US 20040167096 (2004).
127. Bellani, P.; Frigerio, M.; Castoldi, P. A process for the synthesis of ritonavir, WO 2001021603 (2001).
128. Kempf, D. J.; Marsh, K. C.; Denissen, J. F.; McDonald, E.; Vasavanonda, S.; Flentge, C. A.; Green, B. E.; Fino, L.; Park, C. H.; Kong, X. P. ABT-538 is a potent inhibitor of human immunodeficiency virus protease and has high oral bioavailability in humans. *Proc. Natl. Acad. Sci. U. S. A.* **1995,** *92* (7), 2484–2488.
129. Lea, A. P.; Faulds, D. Ritonavir. *Drugs* **1996,** *52* (4), 541–546. discussion 547–548.
130. Hoetelmans, R. M. W.; Meenhorst, P. L.; Mulder, J. W.; Burger, D. M.; Koks, C. H. W.; Beijnen, J. H. Clinical pharmacology of HIV protease inhibitors: focus on saquinavir, indinavir, and ritonavir. *Pharm. World Sci.* **1997,** *19* (4), 159–175.
131. Hull, M. W.; Montaner, J. S. G. Ritonavir-boosted protease inhibitors in HIV therapy. *Ann. Med.* **2011,** *43* (5), 375–388.
132. Sahali, S.; Chaix, M.-L.; Delfraissy, J.-F.; Ghosn, J. Ritonavir-boosted protease inhibitor monotherapy for the treatment of HIV-1 infection. *AIDS Rev.* **2008,** *10* (1), 4–14.
133. Cooper, C. L.; van Heeswijk, R. P. G.; Gallicano, K.; Cameron, D. W. A review of low-dose ritonavir in protease inhibitor combination therapy. *Clin. Infect. Dis.* **2003,** *36* (12), 1585–1592.
134. Sham, H. L.; Norbeck, D. W.; Chen, X.; Betebenner, D. A. Preparation of peptide analogs as retroviral protease inhibitors, US 5914332 (1999).
135. Stoner, E. J.; Cooper, A. J.; Dickman, D. A.; Kolaczkowski, L.; Lallaman, J. E.; Liu, J.-H.; Oliver-Shaffer, P. A.; Patel, K. M.; Paterson, J. B., Jr.; Plata, D. J.; Riley, D. A.; Sham, H. L.; Stengel, P. J.; Tien, J.-H. J. Synthesis of HIV protease inhibitor ABT-378 (Lopinavir). *Org. Process Res. Dev.* **2000,** *4* (4), 264–269.
136. Stoner, E. J.; Stengel, P. J.; Cooper, A. J. Synthesis of ABT-378, an HIV protease inhibitor candidate: avoiding the use of carbodiimides in a difficult peptide coupling. *Org. Process Res. Dev.* **1999,** *3* (2), 145–148.
137. Stuk, T. L.; Haight, A. R.; Scarpetti, D.; Allen, M. S.; Menzia, J. A.; Robbins, T. A.; Parekh, S. I.; Langridge, D. C.; Tien, J.-H. J.; Pariza, R. J.; Kerdesky, F. A. An efficient stereocontrolled strategy for the synthesis of hydroxyethylene dipeptide isosteres. *J. Org. Chem.* **1994,** *59* (15), 4040–4041.
138. Haight, A. R.; Stuk, T. L.; Menzia, J. A.; Robbins, T. A. A convenient synthesis of enaminones using tandem acetonitrile condensation. *Grignard addition, Tetrahedron Lett.* **1997,** *38* (24), 4191–4194.
139. Nunes, E. P.; Santini de Oliveira, M.; Grinsztejn, B. Lopinavir: the old champion. *Future Virol.* **2011,** *6* (5), 561–570.
140. Hurst, M.; Faulds, D. Lopinavir. *Drugs* **2000,** *60* (6), 1371–1379.
141. Cvetkovic, R. S.; Goa, K. L. Lopinavir/ritonavir: a review of its use in the management of HIV infection. *Drugs* **2003,** *63* (8), 769–802.
142. Chandwani, A.; Shuter, J. Lopinavir/ritonavir in the treatment of HIV-I infection: a review. *Ther. Clin. Risk Manage.* **2008,** *4* (5), 1023–1033.
143. Oldfield, V.; Plosker, G. L. Lopinavir/ritonavir: a review of its use in the management of HIV infection. *Drugs* **2006,** *66* (9), 1275–1299.

144. Schaeffer, H. J.; Beauchamp, L.; De Miranda, P.; Elion, G. B.; Bauer, D. J.; Collins, P. 9-(2-Hydroxyethoxymethyl)guanine activity against viruses of the herpes group. *Nature (London, U. K.)* **1978,** *272* (5654), 583–585.
145. Schaeffer, H. J. Compositions for treating viral infections and guanine acyclic nucleosides, US 4199574 (1980).
146. Keyser, G. E.; Bryant, J. D.; Barrio, J. R. Iodomethyl ethers from 1,3-dioxolane and 1,3-oxathiolane: preparation of acyclic nucleoside analogs. *Tetrahedron Lett.* **1979,** *35,* 3263–3264.
147. Barrio, J. R.; Bryant, J. D.; Keyser, G. E. A direct method for the preparation of 2-hydroxyethoxymethyl derivatives of guanine, adenine, and cytosine. *J. Med. Chem.* **1980,** *23* (5), 572–574.
148. Matsumoto, H.; Kaneko, C.; Yamada, K.; Takeuchi, T.; Mori, T.; Mizuno, Y. A convenient synthesis of 9-[(2-hydroxyethoxy)methyl]guanine (acyclovir) and related compounds. *Chem. Pharm. Bull.* **1988,** *36* (3), 1153–1155.
149. Beauchamp, L. M.; Dolmatch, B. L.; Schaeffer, H. J.; Collins, P.; Bauer, D. J.; Keller, P. M.; Fyfe, J. A. Modifications on the heterocyclic base of acyclovir: syntheses and antiviral properties. *J. Med. Chem.* **1985,** *28* (8), 982–987.
150. Stimac, A.; Kobe, J. A new synthesis of acyclovir prodrugs. N2-acetylacyclovir and 6-deoxyacyclovir. *Synthesis* **1990,** *6,* 461–464.
151. Gao, H.; Mitra, A. K. Synthesis of acyclovir ganciclovir and their prodrugs: a review. *Synthesis* **2000,** *3,* 329 351.
152. Chu, C. K.; Cutler, S. J. Chemistry and antiviral activities of acyclonucleosides. *J. Het. Chem.* **1986,** *23* (2), 289–319.
153. Kelley, J. L.; Schaeffer, H. J. Purine acyclic nucleosides. Unambiguous synthesis of acyclovir via a furazano[3,4-d]pyrimidine. *J. Heterocycl. Chem.* **1986,** *23* (1), 271–273.
154. Taylor, E. C.; Beardsley, G. P.; Maki, Y. New, general synthesis of 2-, 8-, and 9-substituted adenines. *J. Org. Chem.* **1971,** *36* (21), 3211–3217.
155. Laskin, O. L. Acyclovir. Pharmacology and clinical experience. *Arch. Intern. Med.* **1984,** *144* (6), 1241–1246.
156. Richards, D. M.; Carmine, A. A.; Brogden, R. N.; Heel, R. C.; Speight, T. M.; Avery, G. S. Acyclovir. A review of its pharmacodynamic properties and therapeutic efficacy. *Drugs* **1983,** *26* (5), 378–438.
157. Krenitsky, T. A.; Beauchamp, L. M. Acycloxic esters with valine and isoleucine as herpes virucides, EP 308065 (1989).
158. Beauchamp, L. M.; Krenitsky, T. A. Acyclovir prodrugs: the road to valaciclovir. *Drugs Future* **1993,** *18,* 619–628.
159. Prasada Raju, V. V. N.K.V.; Vedantham, R.; Khunt, M. D.; Mathad, V. T.; Dubey, P. K.; Chakravarthy, A. K. An efficient and large scale process for synthesis of valacyclovir. *Asian J. Chem.* **2010,** *22* (5), 4092–4098.
160. Etinger, M. Y.; Yudovich, L. M.; Yuzefovich, M.; Nisnevich, G. A.; Dolitzki, B. Z.; Pertsikov, B.; Tishin, B.; Blasberger, D. Synthesis and purification of valaciclovir, WO 2003041647 (2003).
161. Acosta, E. P.; Fletcher, C. V. Valaciclovir. *Ann. Pharmacother.* **1997,** *31* (2), 185–191.
162. Beutner, K. R. Valaciclovir: a review of its antiviral activity, pharmacokinetic properties, and clinical efficacy. *Antiviral Res.* **1995,** *28* (4), 281–290.
163. Smiley, M. L.; Murray, A.; De Miranda, P. Valaciclovir HCl (Valtrex): an acyclovir prodrug with improved pharmacokinetics and better efficacy for treatment of zoster. *Adv. Exp. Med. Biol.* **1996,** *394* (Antiviral Chemotherapy 4), 33–39.
164. Antman, M. D.; Gudmundsson, O. S. Valaciclovir: a prodrug of acyclovir. *Biotechnol.: Pharm. Aspects* **2007,** *5* (Pt. 2, Prodrugs: Challenges and Rewards), 669–676.

165. Tyring, S. K.; Baker, D.; Snowden, W. Valaciclovir for herpes simplex virus infection: long-term safety and sustained efficacy after 20 years' experience with acyclovir. *J. Infect. Dis.* **2002,** *186* (Suppl. 1), S40–S46.
166. Perry, C. M.; Faulds, D. Valaciclovir: a review of its antiviral activity, pharmacokinetic properties and therapeutic efficacy in herpesvirus infections. *Drugs* **1996,** *52* (5), 754–772.
167. Verheyden, J. P. H.; Martin, J. C. 9-(1,3-Dihydroxy-2-propoxymethyl)guanine as antiviral agent, US 4423050 (1983).
168. Martin, J. C.; Dvorak, C. A.; Smee, D. F.; Matthews, T. R.; Verheyden, J. P. H. 9-(1,3-Dihydroxy-2-propoxymethyl)guanine: a new potent and selective antiherpes agent. *J. Med. Chem.* **1983,** *26* (5), 759–761.
169. Ogilvie, K. K.; Cheriyan, U. O.; Radatus, B. K.; Smith, K. O.; Galloway, K. S.; Kennell, W. L. Biologically active acyclonucleoside analogs. II. The synthesis of 9-[[2-hydroxy-1-(hydroxymethyl) ethoxy]methyl]guanine (BIOLF-62). *Can. J. Chem.* **1982,** *60* (24), 3005–3010.
170. Field, A. K.; Davies, M. E.; DeWitt, C.; Perry, H. C.; Liou, R.; Germershausen, J.; Karkas, J. D.; Ashton, W. T.; Johnston, D. B. R.; Tolman, R. L. 9-{[2-Hydroxy-1-(hydroxymethyl)ethoxy] methyl}guanine: a selective inhibitor of herpes group virus replication. *Proc. Natl. Acad. Sci. U. S. A.* **1983,** *80* (13), 4139–4143.
171. Ashton, W. T.; Karkas, J. D.; Field, A. K.; Tolman, R. L. Activation by thymidine kinase and potent antiherpetic activity of 2′-nor-2′-deoxyguanosine (2′NDG). *Biochem. Biophys. Res. Commun.* **1982,** *108* (4), 1716–1721.
172. McGee, D. P. C.; Martin, J. C.; Verheyden, J. P. H. Synthesis of 9-[(1,3-dihydroxy-2-propoxy) methyl]guanine (DHPG) via condensation of N2,9-diacetylguanine with a sulfinylmethyl ether. *Synth. Commun.* **1988,** *18* (14), 1651–1660.
173. Nestor, J. J.; Womble, S. W.; Maag, H. 2-(2-Amino-1,6-dihydro-6-oxo-purin-9-yl)methoxy-1,3-propanediol L-monovaline ester [ganciclovir valine ester] as an antiviral with improved oral absorption, EP 694547 (1996).
174. Ramchandra, R. D.; Narayanrao, K. R.; Purushottam, P. V. Preparation of valganciclovir, EP 1837336 (2007).
175. Arzeno, H. B.; Humphreys, E. R.; Wong, J.; Roberts, C. R. Process for preparing 2-(2-amino-1,6-dihydro-6-oxo-purin-9-yl)methoxy-1,3-propanediol (ganciclovir) mono-L-valinate ester, US 5840890 (1988).
176. Arzeno, H. B.; Humphreys, E. R. Preparation of L-monovaline ester of purine acyclic nucleosides as virucides, WO 9727196 (1997).
177. Dvorak, C. A.; Wren, D. L.; Fisher, L. E.; Axt, S. D.; Humphreys, E. R.; A., Humberto B.; Beard, C. C.; Nguyen, S. L.; Han, Y-K.; Roberts, C. R.; Lund, J. P.; Fatheree, P. R. Process for preparing 2-(2-amino-1,6-dihydro-6-oxo-purin-9-yl)methoxy-1,3-propanediol (ganciclovir) mono-L-valinate ester, US 6040446 (2000).
178. Sharma, M. K.; Raina, S.; Panda, A. K.; Kumar, Y.; Khanduri, C. H. Preparation of ganciclovir mono-N-benzyloxycarbonyl-L-valinate ester, WO 2005092891 (2005).
179. Sorbera, L. A.; Castaner, R.; Castaner, J. Valganciclovir hydrochloride. *Drugs Future* **2000,** *25* (5), 474–480.
180. Arzeno, H. B.; Beard, C. C.; Fisher, L. E.; Prince, A. Preparation of L-monovaline ester of ganciclovir purine acyclic nucleosides as virucides, WO 9727194 (1970).
181. Arzeno, H. B. Preparation of L-monovaline ester of ganciclovir purine acyclic nucleosides as virucides, WO 9727195 (1997).
182. Arzeno, H. B.; Humphreys, E. R.; Wong, J-W.; Roberts, C. R. Preparation of L-monovaline ester of ganciclovir purine acyclic nucleosides as virucides, WO 9727197 (1997).

183. Arzeno, H. B.; Axt, S. D.; Beard, C. C.; Dvorak, C. A.; Fatheree, P. R.; Fisher, L. E.; Han, Y-K.; Humphreys, E. R.; Lund, J. P.; Nguyen, S. L.; Wren, D. L. Preparation of L-monovaline ester of purine acyclic nucleosides as virucides, WO 9727198 (1997).
184. Martin, J. C.; Tippe, M. A.; McGee, D. P. C.; Verheyden, J. P. H. Synthesis and antiviral activity of various esters of 9-[(1,3-dihydroxy-2-propoxy)methyl]guanine. *J. Pharm. Sci.* **1987,** *76* (2), 180–184.
185. Curran, M.; Noble, S. Valganciclovir. *Drugs* **2001,** *61* (8), 1145–1150.
186. Cvetkovic, R. S.; Wellington, K. Valganciclovir: a review of its use in the management of CMV infection and disease in immunocompromised patients. *Drugs* **2005,** *65* (6), 859–878.
187. Razonable, R. R.; Paya, C. V. Valganciclovir for the prevention and treatment of cytomegalovirus disease in immunocompromised hosts. *Expert Rev. Anti-Infect. Ther.* **2004,** *2* (1), 27–42.
188. Cocohoba, J. M.; McNicholl, I. R. Valganciclovir: an advance in cytomegalovirus therapeutics. *Ann. Pharmacother.* **2002,** *36* (6), 1075–1079.
189. Reusser, P. Oral valganciclovir: a new option for treatment of cytomegalovirus infection and disease in immunocompromised hosts. *Expert Opin. Invest. Drugs* **2001,** *10* (9), 1745–1753.
190. Pescovitz, M. D. Valganciclovir: recent progress. *Am. J. Transplant.* **2010,** *10* (6), 1359–1364.
191. Perrottet, N.; Decosterd, L. A.; Meylan, P.; Pascual, M.; Biollaz, J.; Buclin, T. Valganciclovir in adult solid organ transplant recipients. *Clin. Pharmacokinet.* **2009,** *48* (6), 399–418.
192. Maag, H. Valganciclovir: a prodrug of ganciclovir. *Biotechnol.: Pharm. Aspects* **2007,** *5* (Pt. 2, Prodrugs: Challenges and Rewards, Part 2), 677–686.
193. Balzarini, J.; De Clercq, E. Nucleoside and nucleotide reverse transcriptase inhibitors. In *Antiretroviral Therapy;* De Clercq, E. D. A., Ed.; American Society Microbiolgy, 2001; pp 31–62.
194. De Clercq, E. New developments in anti-HIV chemotherapy. *Biochim. Biophys. Acta, Mol. Basis Dis.* **2002,** *1587* (2-3), 258–275.
195. Cihlar, T.; Ray, A. S. Nucleoside and nucleotide HIV reverse transcriptase inhibitors: 25 years after zidovudine. *Antiviral Res.* **2010,** *85* (1), 39–58.
196. Waters, L.; Boffito, M. Pharmacology of current and investigational human immunodeficiency virus (HIV) Nucleoside/nucleotide reverse transcriptase inhibitors in adults. *Anti-Infect. Agents Med. Chem.* **2007,** *6* (3), 213–221.
197. Back, D. J.; Burger, D. M.; Flexner, C. W.; Gerber, J. G. The pharmacology of antiretroviral nucleoside and nucleotide reverse transcriptase inhibitors: implications for once-daily dosing. *JAIDS, J. Acquired Immune Defic. Syndr.* **2005,** *39* (Suppl. 1), S1–S23.
198. Das, K.; Arnold, E. HIV-1 reverse transcriptase and antiviral drug resistance. Part 1. *Curr. Opin. Virol.* **2013,** *3* (2), 111–118.
199. Das, K.; Arnold, E. HIV-1 reverse transcriptase and antiviral drug resistance. Part 2. *Curr. Opin. Virol.* **2013,** *3* (2), 119–128.
200. Holy, A.; Rosenberg, I. Preparation of 5′-O-phosphonylmethyl analogs of nucleoside-5′-phosphates, 5′-diphosphates and 5′-triphosphates. *Coll. Czech. Chem. Comm.* **1982,** *47* (12), 3447–3463.
201. Holy, A.; Dvorakova, H.; Masojidkova, M. Synthesis of enantiomeric N-(2-phosphonomethoxypropyl) derivatives of purine and pyrimidine bases. II. The synthon approach. *Coll. Czech. Chem. Comm.* **1995,** *60* (8), 1390–1409.
202. Holy, A.; Masojidkova, M. Synthesis of enantiomeric N-(2-phosphonomethoxypropyl) derivatives of purine and pyrimidine bases. I. The stepwise approach. *Coll. Czech. Chem. Comm.* **1995,** *60* (7), 1196–1212.
203. Schultze, L. M.; Chapman, H. H.; Dubree, N. J. P.; Jones, R. J.; Kent, K. M.; Lee, T. T.; Louie, M. S.; Postich, M. J.; Prisbe, E. J.; Rohloff, J. C.; Yu, R. H. Practical synthesis of the anti-HIV drug. *PMPA, Tetrahedron Lett.* **1998,** *39* (14), 1853–1856.

204. Barral, K.; Priet, S.; Sire, J.; Neyts, J.; Balzarini, J.; Canard, B.; Alvarez, K. Synthesis, in vitro antiviral evaluation, and stability studies of novel α-borano-nucleotide analogues of 9-[2-(phosphonomethoxy)ethyl]adenine and (r)-9-[2-(phosphonomethoxy)propyl]adenine. *J. Med. Chem.* **2006,** *49* (26), 7799–7806.
205. Ripin, D. H. B.; Teager, D. S.; Fortunak, J.; Basha, S. M.; Bivins, N.; Boddy, C. N.; Byrn, S.; Catlin, K. K.; Houghton, S. R.; Jagadeesh, S. T.; Kumar, K. A.; Melton, J.; Muneer, S.; Rao, L. N.; Rao, R. V.; Ray, P. C.; Reddy, N. G.; Reddy, R. M.; Shekar, K. C.; Silverton, T.; Smith, D. T.; Stringham, R. W.; Subbaraju, G. V.; Talley, F.; Williams, A. Process improvements for the manufacture of tenofovir disoproxil fumarate at commercial scale. *Org. Process Res. Dev.* **2010,** *14* (5), 1194–1201.
206. De Clercq, E. Discovery and development of tenofovir disoproxil fumarate. In *Antiviral Drugs: From Basic Discovery Through Clinical Trials;* Kazmierski, W. M., Ed.; Wiley, 2011; pp 85–101.
207. Kearney, B. P.; Flaherty, J. F.; Shah, J. Tenofovir disoproxil fumarate: clinical pharmacology and pharmacokinetics. *Clin. Pharmacokinet.* **2004,** *43* (9), 595–612.
208. Gallant, J. E.; Deresinski, S. Tenofovir disoproxil fumarate. *Clin. Infect. Dis.* **2003,** *37* (7), 944–950.
209. Chapman, T. M.; McGavin, J. K.; Noble, S. Tenofovir disoproxil fumarate. *Drugs* **2003,** *63* (15), 1597–1608.
210. Grim, S. A.; Romanelli, F. Tenofovir disoproxil fumarate. *Ann. Pharmacother.* **2003,** *37* (6), 849–859.
211. Fung, H. B.; Stone, E. A.; Piacenti, F. J. Tenofovir disoproxil fumarate: a nucleotide reverse transcriptase inhibitor for the treatment of HIV infection. *Clin. Ther.* **2002,** *24* (10), 1515–1548.
212. De Clercq, E. Tenofovir: quo vadis anno 2012 (where is it going in the year 2012)? *Med. Res. Rev.* **2012,** *32* (4), 765–785.
213. Celum, C.; Baeten, J. M. Tenofovir-based pre-exposure prophylaxis for HIV prevention: evolving evidence. *Curr. Opin. Infect. Dis.* **2012,** *25* (1), 51–57.
214. Perry, C. M.; Simpson, D. Tenofovir disoproxil fumarate: in chronic hepatitis B. *Drugs* **2009,** *69* (16), 2245–2256.
215. Buti, M.; Homs, M. Tenofovir disoproxil fumarate in the treatment of chronic hepatitis B. *Expert Rev. Gastroenterol. Hepatol.* **2012,** *6* (4), 413–421.
216. Jenh, A. M.; Thio, C. L.; Pham, P. A. Tenofovir for the treatment of hepatitis B virus. *Pharmacotherapy* **2009,** *29* (10), 1212–1227.
217. Pham, P. A.; Gallant, J. E. Tenofovir disoproxil fumarate for the treatment of HIV infection. *Expert Opin. Drug Metab. Toxicol.* **2006,** *2* (3), 459–469.
218. Lyseng-Williamson, K. A.; Reynolds, N. A.; Plosker, G. L. Tenofovir disoproxil fumarate: a review of its use in the management of HIV infection. *Drugs* **2005,** *65* (3), 413–432.
219. Foggia, M.; Nappa, S.; Bonadies, G.; Cotugno, M.; Di Filippo, G.; Borrelli, F.; Orlando, R.; Borgia, G. Tenofovir disoproxil fumarate in the clinical practice: an overview. *Anti-Infect. Agents Med. Chem.* **2008,** *7* (4), 285–295.
220. James, C.; Preininger, L.; Sweet, M. Rilpivirine: a second-generation nonnucleoside reverse transcriptase inhibitor. *Am. J. Health-Syst. Pharm.* **2012,** *69* (10), 857–861.
221. Sahlberg, C.; Zhou, X.-X. Development of non-nucleoside reverse transcriptase inhibitor inhibitors for anti-HIV therapy. *Anti-Infect. Agents Med. Chem.* **2008,** *7* (2), 101–117.
222. Pedersen, O. S.; Pedersen, E. B. Non-nucleoside reverse transcriptase inhibitors: the NNRTI boom. *Antiviral Chem. Chemother.* **1999,** *10* (6), 285–314.
223. Zhan, P.; Liu, X. Novel HIV-1 non-nucleoside reverse transcriptase inhibitors: a patent review (2005 - 2010). *Expert Opin. Ther. Pat.* **2011,** *21* (5), 717–796.

224. Pauwels, R.; Andries, K.; Debyser, Z.; Van Daele, P.; Schols, D.; Stoffels, P.; De Vreese, K.; Woestenborghs, R.; Vandamme, A. M.; Janssen, C. G. M.; Cauwenbergh, J. A. G.; Desmyter, J.; Heykants, J.; Janssen, M. A. C.; De Clercq, E.; Janssen, P. A. J. Potent and highly selective human immunodeficiency virus type 1 (HIV-1) inhibition by a series of α-anilinophenylacetamide derivatives targeted at HIV-1 reverse transcriptase. *Proc. Natl. Acad. Sci. U. S. A.* **1993,** *90* (5), 1711–1715.
225. Buckheit, K. W.; Yang, L.; Buckheit, R. W., Jr. Development of dual-acting pyrimidinediones as novel and highly potent topical anti-HIV microbicides. *Antimicrob. Agents Chemother.* **2011,** *55* (11), 5243–5254.
226. Ludovici, D. W.; Kukla, M. J.; Grous, P. G.; Krishnan, S.; Andries, K.; de Bethune, M.-P.; Azijn, H.; Pauwels, R.; De Clercq, E.; Arnold, E.; Janssen, P. A. J. Evolution of anti-HIV drug candidates. Part 1: From α-Anilinophenylacetamide (α-APA) to imidoyl thiourea (ITU). *Bioorg. Med. Chem. Lett.* **2001,** *11* (17), 2225–2228.
227. Ludovici, D. W.; Kavash, R. W.; Kukla, M. J.; Ho, C. Y.; Ye, H.; De Corte, B. L.; Andries, K.; de Bethune, M.-P.; Azijn, H.; Pauwels, R.; Moereels, H. E. L.; Heeres, J.; Koymans, L. M. H.; de Jonge, M. R.; Van Aken, K. J. A.; Daeyaert, F. F. D.; Lewi, P. J.; Das, K.; Arnold, E.; Janssen, P. A. J.; et al. Evolution of anti-HIV drug candidates. Part 2: Diaryltriazine (DATA) analogues. *Bioorg. Med. Chem. Lett.* **2001,** *11* (17), 2229–2234.
228. De Clercq, E. Perspectives of non-nucleoside reverse transcriptase inhibitors (NNRTIs) in the therapy of HIV-1 infection. *Farmaco* **1999,** *54* (1-2), 26–45.
229. Yasuda, N.; Tan, L. Efavirenz, a non-nucleoside reverse transcriptase inhibitor (NNRTI), and a previous structurally related development candidate. In *The Art of Process Chemistry;* Yasuda, N., Ed.; , 2011; pp 1–43.
230. Rakhmanina, N. Y.; van den Anker, J. N. Efavirenz in the therapy of HIV infection. *Expert Opin. Drug Metab. Toxicol.* **2010,** *6* (1), 95–103.
231. Best, B. M.; Goicoechea, M. Efavirenz—still first-line king? *Expert Opin. Drug Metab. Toxicol.* **2008,** *4* (7), 965–972.
232. Adkins, J. C.; Noble, S. Efavirenz. *Drugs* **1998,** *56* (6), 1055–1064.
233. Adams, W. J.; Aristoff, P. A.; Jensen, R. K.; Morozowich, W.; Romero, D. L.; Schinzer, W. C.; Tarpley, W. G.; Thomas, R. C. Discovery and development of the BHAP nonnucleoside reverse transcriptase inhibitor delavirdine mesylate. *Pharm. Biotechnol.* **1998,** *11,* 285–312.
234. Freimuth, W. W. Delavirdine mesylate, a potent non-nucleoside HIV-1 reverse transcriptase inhibitor. *Adv. Exp. Med. Biol.* **1996,** *394* (4), 279–289.
235. Scott, L. J.; Perry, C. M. Delavirdine: a review of its use in HIV infection. *Drugs* **2000,** *60* (6), 1411–1444.
236. Murphy, R. L.; Montaner, J. Nevirapine: a review of its development, pharmacological profile and potential for clinical use. *Expert Opin. Invest. Drugs* **1996,** *5* (9), 1183–1199.
237. Grozinger, K.; Proudfoot, J.; Hargrave, K. Discovery and development of nevirapine. In Chorghade, M. S., Ed.; Drug Discovery and Development, Vol. 1, Wiley-Interscience, 2006; pp 353–363.
238 Milinkovic, A.; Martinez, E. Nevirapine in the treatment of HIV. *Expert Rev. Anti-Infect. Ther.* **2004,** *2* (3), 367–373.
239. Podzamczer, D.; Fumero, E. The role of nevirapine in the treatment of HIV-1 disease. *Expert Opin. Pharmacother.* **2001,** *2* (12), 2065–2078.
240. Schiller, D. S.; Youssef-Bessler, M. Etravirine: a second-generation nonnucleoside reverse transcriptase inhibitor (NNRTI) active against NNRTI-resistant strains of HIV. *Clin. Ther.* **2009,** *31* (4), 692–704.

241. Kakuda, T. N.; Scholler-Gyure, M.; Hoetelmans, R. M. W. Clinical perspective on antiretroviral drug-drug interactions with the non-nucleoside reverse transcriptase inhibitor etravirine. *Antiviral Ther.* **2010,** *15* (6), 817–829.
242. Janssen, P. A. J.; Lewi, P. J.; Arnold, E.; Daeyaert, F.; de Jonge, M.; Heeres, J.; Koymans, L.; Vinkers, M.; Guillemont, J.; Pasquier, E.; Kukla, M.; Ludovici, D.; Andries, K.; de Bethune, M.-P.; Pauwels, R.; Das, K.; Clark, A. D., Jr.; Frenkel, Y. V.; Hughes, S. H.; Medaer, B.; De Knaep, F.; Bohets, H.; De Clerck, F.; Lampo, A.; Williams, P.; Stoffels, P. In search of a novel anti-HIV drug: multidisciplinary coordination in the discovery of 4-[[4-[[4-[(1e)-2-cyanoethenyl]-2,6-dimethylphenyl]amino]-2- pyrimidinyl]amino]benzonitrile (R27(8474), rilpivirine). *J. Med. Chem.* **2005,** *48* (6), 1901–1909.
243. Sanford, M. Rilpivirine. *Drugs* **2012,** *72* (4), 525–541.
244. Fernandez-Montero, J. V.; Vispo, E.; Anta, L.; de Mendoza, C.; Soriano, V. Rilpivirine: a next-generation non-nucleoside analogue for the treatment of HIV infection. *Expert Opin. Pharmacother.* **2012,** *13* (7), 1007–1014.
245. Garvey, L.; Winston, A. Rilpivirine: a novel non-nucleoside reverse transcriptase inhibitor. *Expert Opin. Invest. Drugs* **2009,** *18* (7), 1035–1041.
246. Ripamonti, D.; Bombana, E.; Rizzi, M. Rilpivirine: drug profile of a second-generation non-nucleoside reverse transcriptase HIV-inhibitor. *Expert Rev. Anti-Infect. Ther.* **2014,** *12* (1), 13–29.
247. Zaharatos, G. J.; Wainberg, M. A. Update on rilpivirine: A new potent non-nucleoside reverse transcriptase inhibitor (NNRTI) of HIV replication. *Ann. Med.* **2013,** *45* (3), 236–241.
248. Schafer, J. J.; Short, W. R. Rilpivirine, a novel non-nucleoside reverse transcriptase inhibitor for the management of HIV-1 infection: a systematic review. *Antiviral Ther.* **2012,** *17* (8), 1495–1502.
249. James, C.; Preininger, L.; Sweet, M. Rilpivirine: a second-generation nonnucleoside reverse transcriptase inhibitor. *Am. J. Health-Syst. Pharm.* **2012,** *69* (10), 857–861.
250. Platten, M.; Faetkenheuer, G. Lersivirine-a new drug for HIV infection therapy. *Expert Opin. Invest. Drugs* **2013,** *22* (12), 1687–1694.
251. Sorbera, L. A.; Castaner, J.; Bayes, M. Capravirine. *Drugs Future* **2003,** *28* (12), 1149–1158.
252. Deeks, E. D. Emtricitabine/rilpivirine/tenofovir disoproxil fumarate single-tablet regimen: a review of its use in HIV infection. *Drugs* **2014,** *74* (17), 2079–2095.
253. Hazuda, D.; Iwamoto, M.; Wenning, L. Emerging pharmacology: inhibitors of human immunodeficiency virus integration. *Annu. Rev. Pharmacol. Toxicol.* **2009,** *49,* 377–394.
254. Di Santo, R. Inhibiting the HIV integration process: past, present, and the future. *J. Med. Chem.* **2014,** *57* (3), 539–566.
255. Johns, B. A.; Kawasuji, T.; Velthuisen, E. J. HIV integrase inhibitors. *RSC Drug Discovery Ser.* **2013,** *32,* 149–188.
256. Metifiot, M.; Marchand, C.; Pommier, Y. HIV integrase inhibitors: 20-year landmark and challenges. *Adv. Pharmacol. (San Diego, CA, U. S.)* **2013,** *67* (Antiviral Agents), 75–105.
257. Dayam, R.; Gundla, R.; Al-Mawsawi, L. Q.; Neamati, N. HIV-1 integrase inhibitors: 2005-2006 update. *Med. Res. Rev.* **2008,** *28* (1), 118–154.
258. Hazuda, D. J.; Felock, P.; Witmer, M.; Wolfe, A.; Stillmock, K.; Grobler, J. A.; Espeseth, A.; Gabryelski, L.; Schleif, W.; Blau, C.; Miller, M. D. Inhibitors of strand transfer that prevent integration and inhibit HIV-1 replication in cells. *Science (Washington, DC, U. S.)* **2000,** *287,* 646–650.
259. Fujishita, T.; Yoshinaga, T.; Sato, A., Aromatic heterocycle compounds having HIV integrase inhibiting activities, WO2000039086 (2000).

260. Zhuang, L.; Wai, J. S.; Embrey, M. W.; Fisher, T. E.; Egbertson, M. S.; Payne, L. S.; Guare, J. P., Jr.; Vacca, J. P.; Hazuda, D. J.; Felock, P. J.; Wolfe, A. L.; Stillmock, K. A.; Witmer, M. V.; Moyer, G.; Schleif, W. A.; Gabryelski, L. J.; Leonard, Y. M.; Lynch, J. J., Jr.; Michelson, S. R.; Young, S. D. Design and synthesis of 8-hydroxy-[1,6]-naphthyridines as novel inhibitors of HIV-1 integrase in vitro and in infected cells. *J. Med. Chem.* **2003,** *46,* 453–456.
261. Guare, J. P.; Wai, J. S.; Gomez, R. P.; Anthony, N. J.; Jolly, S. M.; Cortes, A. R.; Vacca, J. P.; Felock, P. J.; Stillmock, K. A.; Schleif, W. A.; Moyer, G.; Gabryelski, L. J.; Jin, L.; Chen, I. W.; Hazuda, D. J.; Young, S. D. A series of 5-aminosubstituted 4-fluorobenzyl-8-hydroxy-[1,6]naphthyridine-7-carboxamide HIV-1 integrase inhibitors. *Bioorg. Med. Chem. Lett.* **2006,** *16,* 2900–2904.
262. Hazuda, D. J.; Young, S. D.; Guare, J. P.; Anthony, N. J.; Gomez, R. P.; Wai, J. S.; Vacca, J. P.; Handt, L.; Motzel, S. L.; Klein, H. G.; Dornadula, G.; Danovich, R. M.; Witmer, M. V.; Wilson, K. A.; Tussey, L.; Schleif, W. A.; Gabryelski, L. S.; Jin, L.; Miller, M. D.; Casimiro, D. R.; Emini, E. A.; Shiver, J. W. Integrase inhibitors and cellular immunity suppress retroviral replication in rhesus macaques. *Science (Washington, DC, U. S.)* **2004,** *305,* 528–532.
263. Summa, V.; Petrocchi, A.; Matassa, V. G.; Gardelli, C.; Muraglia, E.; Rowley, M.; Paz, O. G.; Laufer, R.; Monteagudo, E.; Pace, P. 4,5-Dihydroxypyrimidine carboxamides and N-alkyl-5-hydroxypyrimidinone carboxamides are potent, selective HIV integrase inhibitors with good pharmacokinetic profiles in preclinical species. *J. Med. Chem.* **2006,** *49,* 6646–6649.
264. Anker, M.; Coralcs, R. B. Raltegravir (MK-0518): a novel integrase inhibitor for the treatment of HIV infection. *Expert Opin. Invest. Drugs* **2008,** *17,* 97–103.
265. Summa, V.; Petrocchi, A.; Bonelli, F.; Crescenzi, B.; Donghi, M.; Ferrara, M.; Fiore, F.; Gardelli, C.; Paz, O. G.; Hazuda, D. J.; Jones, P.; Kinzel, O.; Laufer, R.; Monteagudo, E.; Muraglia, E.; Nizi, E.; Orvieto, F.; Pace, P.; Pescatore, G.; Scarpelli, R.; Stillmock, K.; Witmer, M. V.; Rowley, M. Discovery of raltegravir, a potent, selective orally bioavailable HIV-integrase inhibitor for the treatment of HIV-AIDS infection. *J. Med. Chem.* **2008,** *51,* 5843–5855.
266. Beare, K. D.; Coster, M. J.; Rutledge, P. J. Diketoacid inhibitors of HIV-1 integrase: from L-708,906 to raltegravir and beyond. *Curr. Med. Chem.* **2012,** *19* (8), 1177–1192.
267. Summa, V.; Pace, P. Discovery and development of HIV integrase inhibitor raltegravir. In *Antiviral Drugs: From Basic Discovery Through Clinical Trials;* Kazmierski, W. M., Ed.; Wiley, 2011; pp 181–195.
268. Rowley, M. The discovery of raltegravir, an integrase inhibitor for the treatment of HIV infection. *Prog. Med. Chem.* **2008,** *46,* 1–28.
269. Evering, T. H.; Markowitz, M. Raltegravir: an integrase inhibitor for HIV-1. *Expert Opin. Invest. Drugs* **2008,** *17* (3), 413–422.
270. Hicks, C.; Gulick, R. M. Raltegravir: the first HIV type 1 integrase inhibitor. *Clin. Infect. Dis.* **2009,** *48* (7), 931–939.
271. Cocohoba, J.; Dong, B. J. Raltegravir: the first HIV integrase inhibitor. *Clin. Ther.* **2008,** *30* (10). 2747–1765.
272. Sayana, S.; Khanlou, H. Raltegravir: the first in a new class of integrase inhibitors for the treatment of HIV. *Expert Rev. Anti-Infect. Ther.* **2008,** *6* (4), 419–426.
273. Temesgen, Z.; Siraj, D. S. Raltegravir: first in class HIV integrase inhibitor. *Ther. Clin. Risk Manage* **2008,** *4* (2), 493–500.
274. Croxtall, J. D.; Lyseng-Williamson, K. A.; Perry, C. M. Raltegravir. *Drugs* **2008,** *68* (1), 131–138.
275. Calin, R.; Katlama, C. The place of raltegravir in the clinical management of HIV-1 infection. *Clin. Pract. (London, U. K.)* **2013,** *10* (4), 427–438.
276. Rokas, K. E. E.; Brandon, B. P.; Shamroe, C. L.; Scott, S. S.; Millisor, V. E.; Bryant, J. E.; Weissman, S. B. Role of raltegravir in HIV-1 management. *Ann. Pharmacother.* **2012,** *46* (4), 578–589.

277. Brainard, D. M.; Wenning, L. A.; Stone, J. A.; Wagner, J. A.; Iwamoto, M. Clinical pharmacology profile of raltegravir, an HIV-1 integrase strand transfer inhibitor. *J. Clin. Pharmacol.* **2011,** *51* (10), 1375–1402.
278. Okeke, N. L.; Hicks, C. Role of raltegravir in the management of HIV-1 infection. *HIV/AIDS* **2011,** *3,* 81–92.
279. Nunes, E. P.; Santini de Oliveira, M.; Grinsztejn, B. Clinical use of raltegravir: a review. *HIV Ther.* **2010,** *4* (5), 531–542.
280. Burger, D. M. Raltegravir: a review of its pharmacokinetics, pharmacology and clinical studies. *Expert Opin. Drug Metab. Toxicol.* **2010,** *6* (9), 1151–1160.
281. Evering, T. H.; Markowitz, M. Raltegravir (MK-0518): an integrase inhibitor for the treatment of HIV-1. *Drugs Today* **2007,** *43* (12), 865–877.
282. Shimura, K.; Kodama, E. N. Elvitegravir: a new HIV integrase inhibitor. *Antiviral Chem. Chemother.* **2009,** *20* (2), 79–85.
283. Deeks, E. D. Elvitegravir: a review of its use in adults with HIV-1 infection. *Drugs* **2014,** *74* (6), 687–697.
284. Ramanathan, S.; Mathias, A. A.; German, P.; Kearney, B. P. Clinical pharmacokinetic and pharmacodynamic profile of the HIV integrase inhibitor elvitegravir. *Clin. Pharmacokinet.* **2011,** *50* (4), 229–244.
285. Karmon, S. L.; Markowitz, M. Next-generation integrase inhibitors. *Drugs* **2013,** *73* (3), 213–228.
286. Wainberg, M. A.; Mesplede, T.; Quashie, P. K. The development of novel HIV integrase inhibitors and the problem of drug resistance. *Curr. Opin. Virol.* **2012,** *2* (5), 656–662.
287. Shah, B. M.; Schafer, J. J.; De Simone, J. A., Jr. Dolutegravir: a new integrase strand transfer inhibitor for the treatment of HIV. *Pharmacotherapy* **2014,** *34* (5), 506–520.
288. Max, B.; Vibhakar, S. Dolutegravir: a new HIV integrase inhibitor for the treatment of HIV infection. *Future Virol.* **2014,** *9* (11), 967–978.
289. McCormack, P. L. Dolutegravir: a review of its use in the management of HIV-1 infection in adolescents and adults. *Drugs* **2014,** *74* (11), 1241–1252.
290. Belyk, K., M.; Morrison, H., G.; Jones, P.; Summa, V., Preparation of N-(4-fluorobenzyl)-5-hydroxy-1-methyl-2-(1-methyl-1-{[(5-methyl-1,3,4-oxadiazol-2-yl)carbonyl]amino}ethyl)-6-oxo-1,6-dihydropyrimidine-4-carboxamide potassium salts as HIV integrase inhibitors, WO 2006060712 (2006).
291. Humphrey, G. R.; Pye, P. J.; Zhong, Y.-L.; Angelaud, R.; Askin, D.; Belyk, K. M.; Maligres, P. E.; Mancheno, D. E.; Miller, R. A.; Reamer, R. A.; Weissman, S. A. Development of a second-generation, highly efficient manufacturing route for the HIV integrase inhibitor raltegravir potassium. *Org. Process Res. Dev.* **2011,** *15* (1), 73–83.
292. Hunt, J. A. Raltegravir (Isentress): the first-in-class HIV-1 integrase inhibitor. In *Modern Drug Synthesis;* Li, J. J., Johnson, D. S., Eds.; Wiley, 2010; pp 3–15.
293. Humphrey, G. R.; Zhong, Y.-L. HIV integrase inhibitor: raltegravir. In *The Art of Process Chemistry;* Yasuda, N., Ed.; Wiley-VCH, 2011; pp 165–190.
294. Patil, G. D.; Kshirsagar, S. W.; Shinde, S. B.; Patil, P. S.; Deshpande, M. S.; Chaudhari, A. T.; Sonawane, S. P.; Maikap, G. C.; Gurjar, M. K. Identification, synthesis, and strategy for minimization of potential impurities observed in raltegravir potassium drug substance. *Org. Process Res. Dev.* **2012,** *16* (8), 1422–1429.
295. Gurjar, M. K.; Sonawane, S. P.; Maikap, G. S.; Patil, G. D.; Shinde, S. B.; Patil, P. S.; Mehta, S. S., Process for the preparation of raltegravir, WO 2013098854 (2013).
296. Sippl, W.; Jung, M. DNA methyltransferase inhibitors. *Methods Princ. Med. Chem.* **2009,** *42* (Epigenetic Targets in Drug Discovery), 163–183.

297. Frick, D. N.; Lam, A. M. I. Understanding helicases as a means of virus control. *Curr. Pharm. Des.* **2006,** *12* (11), 1315–1338.
298. Xi, X. G. Helicases as antiviral and anticancer drug targets. *Curr. Med. Chem.* **2007,** *14* (8), 883–915.
299. Belon, C. A.; Frick, D. N. Helicase inhibitors as specifically targeted antiviral therapy for hepatitis C. *Future Virol.* **2009,** *4* (3), 277–293.
300. Amorim, M. J.; Kao, R. Y.; Digard, P. Nucleozin targets cytoplasmic trafficking of viral ribonucleoprotein-Rab11 complexes in influenza A virus infection. *J. Virol.* **2013,** *87* (8), 4694–4703.
301. Gerritz, S. W.; Cianci, C.; Kim, S.; Pearce, B. C.; Deminie, C.; Discotto, L.; McAuliffe, B.; Minassian, B. F.; Shi, S.; Zhu, S.; Zhai, W.; Pendri, A.; Li, G.; Poss, M. A.; Edavettal, S.; McDonnell, P. A.; Lewis, H. A.; Maskos, K.; Mortl, M.; Kiefersauer, R.; Steinbacher, S.; Baldwin, E. T.; Metzler, W.; Bryson, J.; Healy, M. D.; Philip, T.; Zoeckler, M.; Schartman, R.; Sinz, M.; Leyva-Grado, V. H.; Hoffmann, H.-H.; Langley, D. R.; Meanwell, N. A.; Krystal, M. Inhibition of influenza virus replication via small molecules that induce the formation of higher-order nucleoprotein oligomers. *Proc. Natl. Acad. Sci. U. S. A.* **2011,** *108* (37), 15366–15371.
302. Belema, M.; Lopez, O. D.; Bender, J. A.; Romine, J. L.; St. Laurent, D. R.; Langley, D. R.; Lemm, J. A.; O'Boyle, D. R., II.; Sun, J.-H.; Wang, C.; Fridell, R. A.; Meanwell, N. A. Discovery and development of hepatitis C virus NS5A replication complex inhibitors. *J. Med. Chem.* **2014,** *57* (5), 1643–1672.
303. Gentile, I.; Borgia, F.; Coppola, N.; Buonomo, A. R.; Castaldo, G.; Borgia, G. Daclatasvir: the first of a new class of drugs targeted against hepatitis C virus NS5A. *Curr. Med. Chem.* **2014,** *21* (12), 1391–1404.
304. Muratore, G.; Goracci, L.; Mercorelli, B.; Foeglein, A.; Digard, P.; Cruciani, G.; Palu, G.; Loregian, A. Small molecule inhibitors of influenza A and B viruses that act by disrupting subunit interactions of the viral polymerase. *Proc. Natl. Acad. Sci. U. S. A.* **2012,** *109* (16), 6247–6252.
305. Lou, Z.; Sun, Y.; Rao, Z. Current progress in antiviral strategies. *Trends Pharmacol. Sci.* **2014,** *35* (2), 86–102.
306. Ludwig, S. Targeting cell signaling pathways to fight the flu: towards a paradigm change in anti-influenza therapy. *J. Antimicrob. Chemother.* **2009,** *64* (1), 1–4.
307. Mercorelli, B.; Sinigalia, E.; Loregian, A.; Palu, G. Human cytomegalovirus DNA replication: antiviral targets and drugs. *Rev. Med. Virol.* **2008,** *18* (3), 177–210.
308. Cheng, H.; Wan, J.; Lin, M.-I.; Liu, Y.; Lu, X.; Liu, J.; Xu, Y.; Chen, J.; Tu, Z.; Cheng, Y.-S. E.; Ding, K. Design, synthesis, and in vitro biological evaluation of 1H-1,2,3-Triazole-4-carboxamide derivatives as new anti-influenza A agents targeting virus nucleoprotein. *J. Med. Chem.* **2012,** *55* (5), 2144–2153.

Chapter 35

Drugs for Treating Protozoan Infections

Protozoan infections remain a major unsolved medical problem in many parts of our world. A major obstacle is the blatant lack of affordable, effective, and safe medications. Protozoan parasites, unicellular eukaryotic organisms, cause serious infections. Pathogenic protozoa include *Plasmodium* species, the cause of human malaria; *Entamoeba histolytica*, the cause of amebic dysentery; more than 20 *Leishmania* species the cause of leishmaniasis; *Trypanosoma gambiense*, the cause of trypanosomiasis; and *Balantidium coli* and *Isospora belli*, both of which cause diarrhea in humans [1-3].

35.1 ANTIMALARIALS

Malaria is a life-threatening, complex, blood disease prevalent in tropical and subtropical regions caused by four species of the protozoan *Plasmodium*. Malaria begins with a bite from an infected female *Anopheles* mosquito, the carrier for the parasite. Among the various parasitic diseases, malaria is the deadliest. Malaria symptoms include headache, chills, tremors, aches, and shaking.

Once inside the human host, *Plasmodium* multiplies in the liver and transforms into progeny called merozoites (a stage in the life cycle of the *Plasmodium*). After two to three weeks of the initial infection, the merozoites are released into the blood and infect red blood cells, causing severe fever and chills. Relapses may occur months, or even years, after the initial infection.

For the prevention of the disease, chloroquine (**35.1.1**) is the drug of choice.

Pharmacotherapy of malaria attempts to interrupt the life cycle of *Plasmodium*.

Treatment of acute attacks are used to eliminate the merozoites from red blood cells. Chloroquine is the classic antimalarial for that purpose. Other medications are primaquine (**35.1.2**) and the use of a fixed-dose combination of artemether (**35.1.3**)-lumefantrine (**35.1.4**), which is called Coartem. Artemisinin-based combination therapy is considered the best available treatment, particularly for *Plasmodium falciparum* malaria [4-17] (Fig. 35.1.).

Artemether (**35.1.3**), the methyl ether of artemisinin, is obtained from the Chinese herb *Artemisia annua*, which has been known to have antimalarial properties since antiquity. It is believed that artemisinin derivatives are targeting

Synthesis of Best-Seller Drugs. http://dx.doi.org/10.1016/B978-0-12-411492-0.00035-3

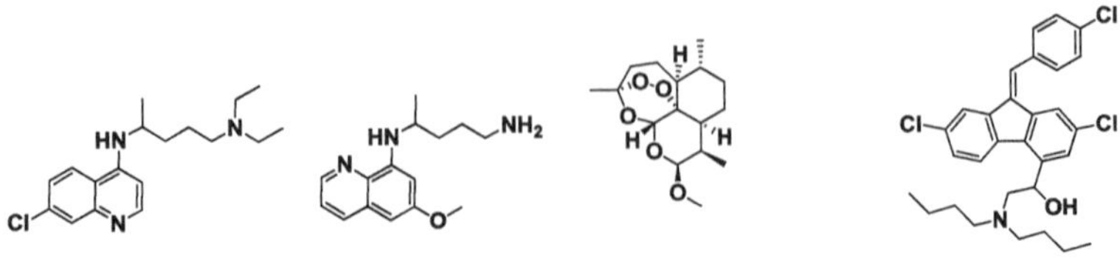

Chloroquine 35.1.1 Primaquine 35.1.2 Artemether 35.1.3 Lumefantrine 35.1.4

Coartem

FIG. 35.1 **Drugs of choice for treatment of malaria.**

erythrocytic schizogony (a stage when the parasite divides several times to produce new merozoites). None of the antimalarial drugs is included in the list of Top 200 Drugs by sales for the 2010s.

As a result of the spread of multidrug resistance to the marketed antimalarial drugs, including to the "magic bullet" artemisinin, discovery and development of new antimalarial drugs is one of the greatest and most important challenges of current medicinal chemistry [18,19].

Other drugs used to treat different forms or stages of malaria include quinine (**35.1.5**), and its congener compounds mefloquine (**35.1.6**), pamaquine (**35.1.7**), and piperaquine (**35.1.8**). Another group is represented by proguanil (**35.1.9**), atovaquone (**35.1.10**), artemotil (**35.1.11**), the ethyl ether of artemisinin, halofantrine (**35.1.12**), the congener of lumefantrine, and a known antibacterial combination sulfadoxine-pyrimethamine-(**35.1.13**)/(**35.1.14**) (Fig. 35.2.) Syntheses of all of these drugs is presented in our previous book [20].

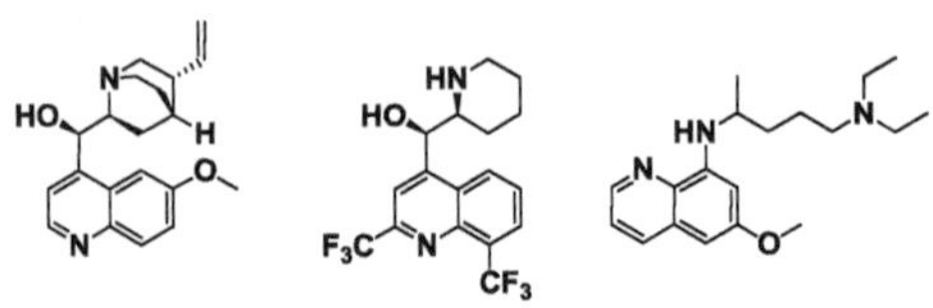

Quinine 35.1.5 Mefloquine 35.1.6 Pamaquine 35.1.7

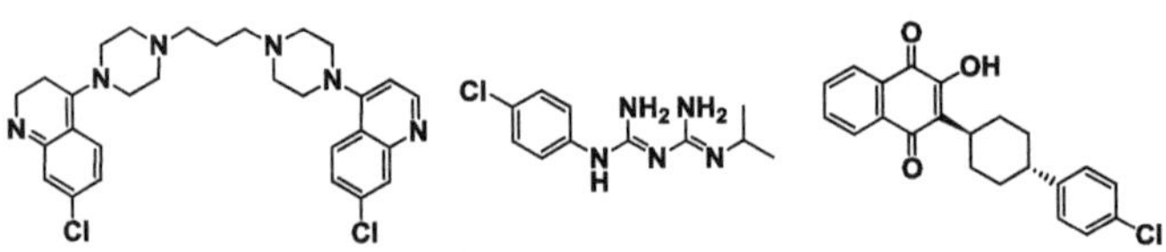

Piperaquine 35.1.8 Proguanil 35.1.9 Atovaquone 35.1.10

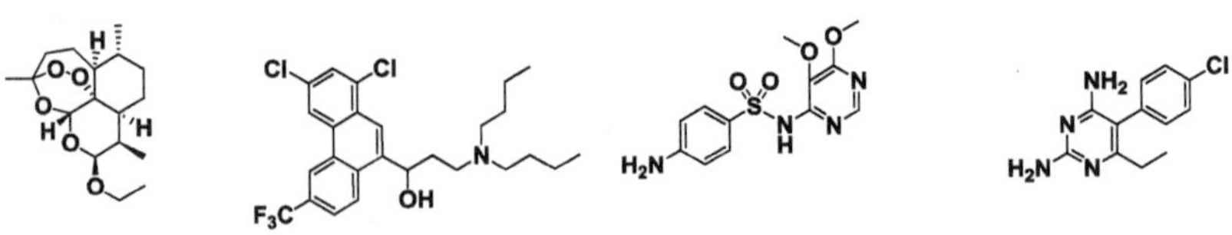

Artemotil 35.1.11 Halofantrine 35.1.12 Sulfadoxine 35.1.13 Pyrimethamine 35.1.14

Sulfadoxine/
Pyrimethamine

FIG. 35.2 **Other drugs used to treat different forms or stages of malaria.**

Unfortunately, drugs such as mefloquine, halofantrine, artether, and artemotil are known to cause clinical manifestations such as neuropsychiatric toxicity, neurotoxicity, cardiorespiratory collapse, and even death.

There are quite a number of antimalarial compounds in different states of preclinical and clinical development. Among them are new compounds with novel or even unknown mechanisms of action and new fixed-dose artemisinin combinations [21-23].

35.2 AMEBIASIS

Amebiasis, a disease caused by a one-celled parasite called *E. histolytica,* is an infection of the intestine, liver, or other tissues.

Some strains of *E. histolytica* are harmless and most cases of amebiasis have very mild symptoms or none. Other strains invade the intestinal wall, causing mucus secretion into the intestines and diarrhea, which is called amebic dysentery. Additional symptoms may include fever, abdominal cramping, and pain. Amebiasis is among the three top causes of death from parasitic infections worldwide, as a result of amebic colitis (dysentery) and liver or brain abscess [24-32].

In cases of amebic colitis (dysentery) *E. histolytica* invades the colonic mucosa, generating bloody diarrhea and abdominal pain and tenderness. Amebic dysentery manifests with episodes of frequent semiliquid stools that often contain blood, mucus, and live trophozoites.

The most common extraintestinal manifestation of disease is amoebic liver-abscess causing fever, right upper quadrant pain, and substantial hepatic tenderness.

Patients with pleuropulmonary amebiasis develop cough, pleuritic chest pain, and respiratory distress.

Amebic brain abscesses are very rare, and are described as appearance of sudden onset of headache, vomiting, seizures, and mental status changes followed in half of the patients with rapid progression to death.

Although many drugs destroy *E. histolytica*, the number of tissue amebicides used to treat invasive amebiasis is relatively limited [33].

The cornerstone of treatment for amebiasis are the nitroimidazole derivatives: metronidazole (**35.2.1**), tinidazole (**35.2.2**), ornidazole (**35.2.3**), and secnidazole (**35.2.4**) [34,35].

Amebic colitis usually is treated by nitroimidazole, followed by a luminal agent—paromomycin (**35.2.5**), iodoquinol (**35.2.6**), or diloxanide furoate (**35.2.7**)—to eradicate colonization.

The most potent amebicide in current use is emetine (**35.2.8**). Despite its potential toxicity (muscle contractions, sometimes leading to cardiac failure in some cases), it continues to be a lifesaving drug when properly used in appropriate situations in hospitals. Dehydroemetine (**35.2.9**) produces fewer side effects (Fig. 35.3.).

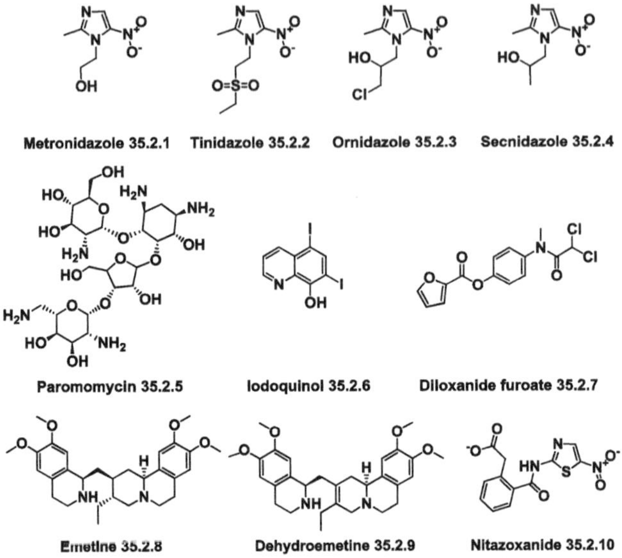

FIG. 35.3 **Drugs for treatment of amebiasis.**

Syntheses of all these drugs are described in our previous book [20].

In cases of hepatic amebiasis, the effectiveness of chloroquine (**35.1.1**) was demonstrated in 1948, soon after its introduction for malaria therapy.

A relatively recent drug that has a broad spectrum of antiparasitic activity, against *E. histolytica* is nitazoxanide (**35.2.10**) (see Fig. 35.3.), which has broad spectrum of antiparasitic activity, against *E. histolytica.* This compound could be the key in the therapy of amebiasis. Nitazoxanide is a first-line choice for the treatment of illness caused by *Cryptosporidium parvum* or *Giardia lamblia* infection.

35.3 LEISHMANIASIS

Leishmaniasis is a deadly disease caused by parasitic protozoa of the genus *Leishmania*; it affects millions of people. Leishmaniasis is found in tropical and subtropical regions and in areas close to the Mediterranean. Humans usually are infected via the bite of phlebotomine sandflies. External leishmaniasis affects the skin and internal leishmaniasis affects the inner organs, such as the spleen and liver. Leishmaniasis broadly manifests as visceral, cutaneous, diffuse, and mucocutaneous.

Visceral leishmaniasis, also known as kala azar, is usually fatal if untreated. It is characterized by high fever, substantial weight loss, swelling of the spleen and liver, and anaemia. It is considered one of the oldest diseases of humanity.

Cutaneous leishmaniasis, the most common form of leishmaniasis, causes skin sores, which lead to disfiguring lesions. In cutaneous forms, skin ulcers usually form on exposed areas, such as the face, arms, and legs. These usually heal within a few months, leaving scars.

Diffuse cutaneous leishmaniasis produces disseminated and chronic skin lesions resembling those of lepromatous leprosy. It is difficult to treat.

In mucocutaneous forms, the lesions can partially or totally destroy the mucous membranes of the nose, mouth, and throat cavities and surrounding tissues.

The treatment strategy of *Leishmania* infection at this time is based on pentavalent antimonials, which are first-line drugs for the treatment of the cutaneous form of leishmaniasis. Second-line drugs include amphotericin B and pentamidine [36-49].

Available antileishmanial drugs are Glucantime (meglumine antimonate) (**35.3.1**) and Pentostam (sodium stibogluconate) (**35.3.2**) (Fig. 35.4.), which require long-time treatment and display severe side effects. Pentavalent antimonials are rapidly absorbed and are converted into trivalent antimonite, which is believed to be the active form of the drug. The reduction of pentavalent to trivalent antimony takes place either in the macrophages or in the parasite.

Myalgia, nausea, liver and cardiac disorders, abdominal pain, headache, and asthenia are side effects often associated with such drugs.

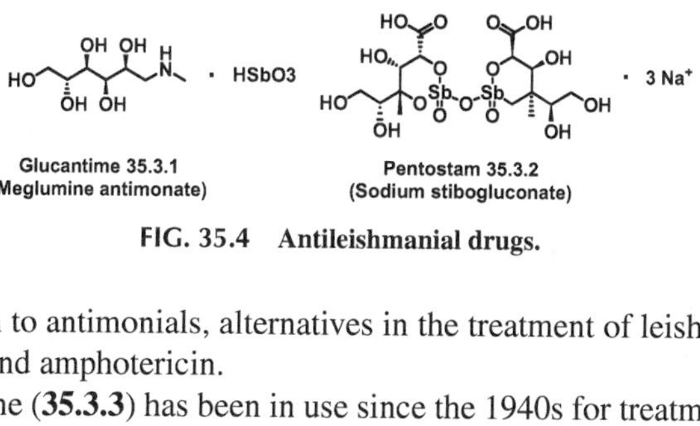

Glucantime 35.3.1
(Meglumine antimonate)

Pentostam 35.3.2
(Sodium stibogluconate)

FIG. 35.4 Antileishmanial drugs.

In addition to antimonials, alternatives in the treatment of leishmaniasis are pentamidine and amphotericin.

Pentamidine (**35.3.3**) has been in use since the 1940s for treatment of sleeping sickness, but became a molecule of great interest in the treatment of antimony refractory leishmaniasis. Its mechanism of action is not well defined. It is an effective drug, but its side effects such as hypoglycemia, arrhythmia, renal failure, pancreatitis, and diabetes mellitus, are severe and prolonged.

Amphotericin B (**35.3.4**), a macrolide antibiotic, was first identified as an antifungal drug. Now it has been proposed as a therapeutic agent of choice and is considered a second-generation leishmanicidal drug. It is extensively used in cases of failure in the treatment with antimony compounds (Fig. 35.5.).

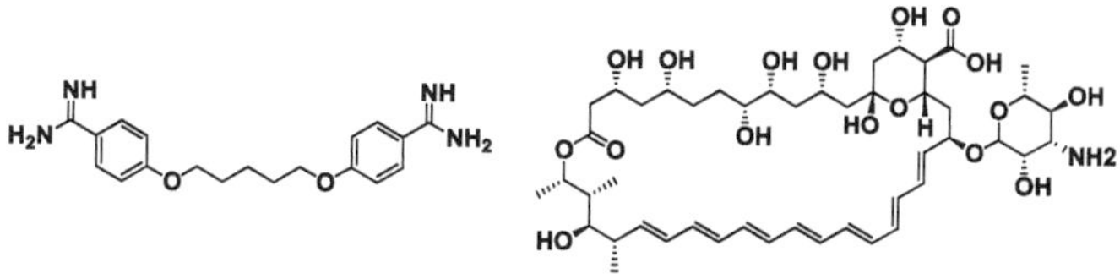

Pentamidine 35.3.3

Amphotericin B 35.3.4

FIG. 35.5 Antileishmanial drugs.

Recently, anticancer alkylphosphocholines have been found to be the most effective oral compounds. Most promising of these are miltefosine (**35.3.5**), edelfosine (**35.3.6**), and ilmofosine (**35.3.7**) (Fig. 35.6.).

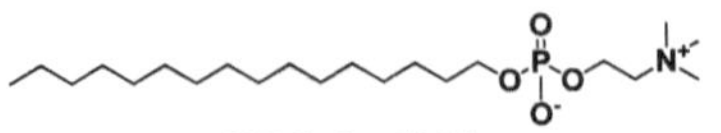

Miltefosine 35.3.5

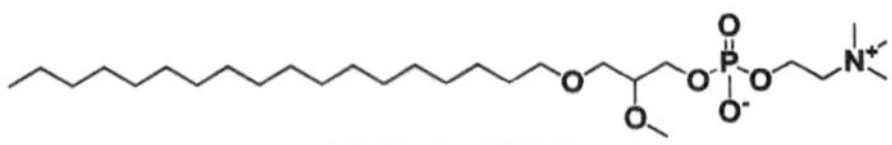

Edelfosine 35.3.6

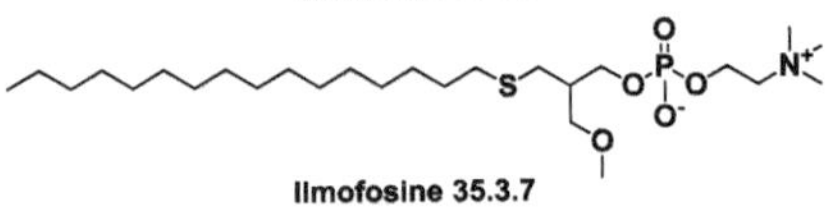

Ilmofosine 35.3.7

FIG. 35.6 Alkylphosphocholines that are potential antileishmanial drugs.

35.4 TRYPANOSOMIASIS

Trypanosomiasis is the name of several diseases in vertebrates caused by parasitic protozoan trypanosomes of the genus *Trypanosoma* [37,50-52].

There are two kinds of African trypanosomiasis: East and West African. Both are also known as sleeping sickness. American trypanosomiasis is also called Chagas disease named in honor of its discoverer, Carlos Chagas. It occurs only on the American continents, from Mexico to Argentina.

The African forms of trypanosomiasis, a potentially fatal disease, is caused by the bite of the tsetse fly. Tsetse flies infected with the protozoan *Trypanosoma brucei rhodesiense* spread East African trypanosomiasis, the most severe form of the disease. The West African variety comes from a fly infected with *T. brucei gambiense*.

Human African trypanosomiasis is commonly known as sleeping sickness.

Patients exposed to tsetse fly bites in the first, hemolymphatic phase, when parasites proliferate in the blood and lymphatic systems, have symptoms such as headache and general malaise. In the second phase, when the parasites have invaded the central nervous system, progressive neurological breakdown starts, including psychiatric disorders, depression, and altered sleep–wake patterns, ensues.

Atoxyl (4-aminophenylarsonic acid) (**35.4.1**), which is of historical interest, was the first medicinal drug against human African trypanosomiasis. It was in use more than 100 years ago. Until recently, only four drugs for human African trypanosomiasis treatments are on the market. Suramin (**35.4.2**), which was first used in 1922, melarsoprol (**35.4.3**), which emerged in the 1940s, pentamidine (**35.3.3**). which was approved for use in the United States in 1984, and eflornithine (**35.4.4**), which was approved in 2000 (Fig. 35.7.).

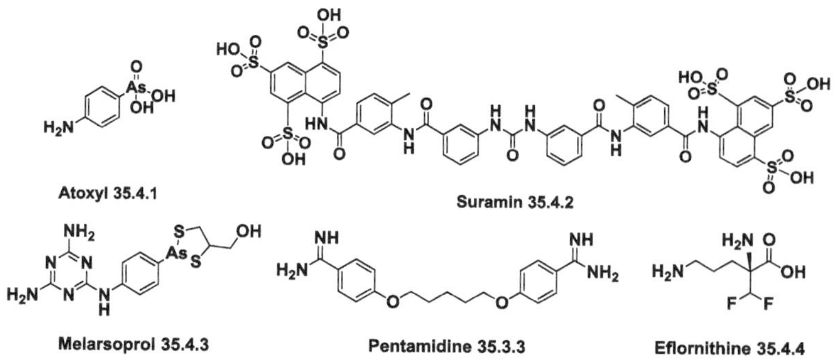

FIG. 35.7 **Drugs for treatment human African trypanosomiasis.**

The use of these drugs depends on the disease phase and causative pathogen. Nifurtimox (**35.4.5**) (Fig. 35.8.), usually used against "Chagas diseases," has been utilized as off-label for treatment of infected with Western African trypanosomiasis. The combination therapy involving the oral drug nifurtimox and eflornithine has been the most recent breakthrough in antitrypanosomiasis drug research.

Unfortunately, none of these medicines fulfill modern pharmaceutical requirements and may be considered as therapeutic "ultima ratio" because of the many, often severe side effects [53-66].

The American variety of trypanosomiasis, is caused by blood-sucking bugs-triatomines (also called assassin, cone-nose, or kissing bugs), which carry the *Trypanosoma cruzi* protozoa. The disease is transmitted to humans usually by the bites or feces of triatomine bugs or, occasionally, by nonvectorial mechanisms, such as blood transfusion and mother to fetus.

American trypanosomiasis remains a serious health problem in Latin America.

Chagas disease develops in two phases. The initial, acute phase, which lasts approximately eight weeks, is considered the time after infection when parasites circulate in the blood practically without symptoms. During the second, chronic phase, patients suffer from cardiac disorders and some patients suffer from digestive, neurological, or mixed alterations. In later years, the infection can lead to sudden death or heart failure caused by progressive destruction of the heart muscle and its nervous system.

Chagas disease can be treated with nifurtimox (**35.4.5**) and benznidazole (**35.4.6**) [67-77] (Fig. 35.8.). Both medicines are nearly 100% effective at curing the disease if given soon after infection at the onset of the acute phase. However, the efficacy of both diminishes the longer a person has been infected.

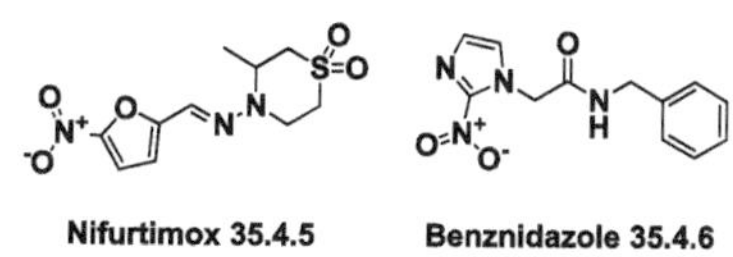

Nifurtimox 35.4.5 Benznidazole 35.4.6

FIG. 35.8 **Drugs for treatment human American trypanosomiasis.**

Finding new ways in the treatment of fatal parasitic diseases is a real challenge of biomedical research. There are some compounds in development for the treatment of the human African trypanosomiasis such as new nitroindole derivative fexinidazole (**35.4.7**), the benzoxaborole derivative SCYX-7158 (**35.4.8**), and a new diamidine derivative CPD-0802 (**35.4.9**), in advanced preclinical development (Fig. 35.9.).

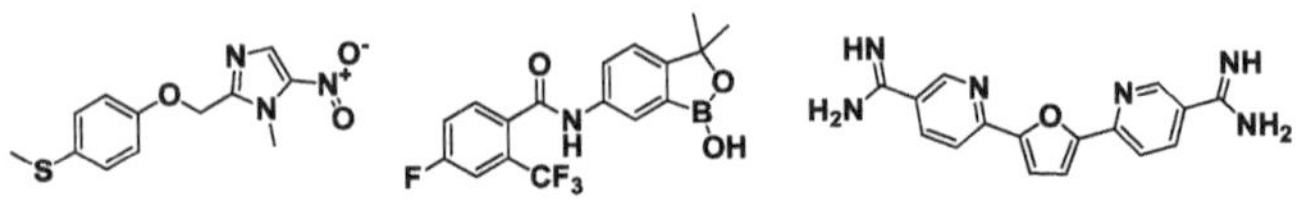

Fexinidazole 35.4.7 **SCYX-7158 35.4.8** **CPD-0802 35.4.9**

FIG. 35.9 **Compounds in development for the treatment of trypanosomiasis.**

35.5 BALANTIDIASIS AND CYSTOISOSPORIASIS

Balantidiasis, the protozoan infection caused by *B. coli*, poses a problem mostly in developing countries, where water sources may be contaminated with swine or human feces. The symptoms of balantidiasis include chronic diarrhea, occasional dysentery, nausea, colitis, and abdominal pain, and can be treated with antibiotics of the tetracycline series, metronidazole, and iodoquinol.

Cystoisosporiasis is caused by *Isospora belli* is also transmitted by ingesting food or water that has been contaminated with feces from someone who is infected. The most common symptom of disease is watery diarrhea. Other symptoms can include abdominal pain, cramps, loss of appetite, nausea, vomiting, and fever. The most common antibiotic that prescribed for cystoisosporiasis treatment is trimethoprim-sulfamethoxazole.

REFERENCES

1. Taylor, J. B.; Triggle, D. J. Antiprotozoal agents (African trypanosomiasis, Chagas disease, and leishmaniasis). *Comprehensive Medicinal Chemistry II*, Vol. 7; 8th ed.; Taylor, J. B., Triggle, D. J., Eds.; Elsevier, 2006; pp 815–843.
2. Njoroge, M.; Njuguna, N. M.; Mutai, P.; Ongarora, D. S. B.; Smith, P. W.; Chibale, K. Recent approaches to chemical discovery and development against malaria and the neglected tropical diseases human African trypanosomiasis and schistosomiasis. *Chem. Rev. (Washington, DC, U. S.)* **2014,** *114* (22), 11138–11163.
3. Smithson, D. C.; Guiguemde, W. A.; Guy, R. K. *Burger's Medicinal Chemistry, Drug Discovery and Development*, Vol. 7; 7th ed.; Abraham, D. J., Rotella, D. P., Eds.; Wiley, 2010; pp 603–712.
4. Schlitzer, M. Malaria chemotherapeutics. Part I: history of antimalarial drug development, currently used therapeutics, and drugs in clinical development. *ChemMedChem* **2007,** *2* (7), 944–986.
5. Ana Maria Madeira M. Faísca Phillips. Anti-malaria chemotherapy: state-of-the-art in prevention and treatment and novel leads for drug development. *Frontiers in Anti-Infect. Drug Discovery* **2014,** *2,* 269–397. http://dx.doi.org/10.2174/9781608058600114020l.
6. Jatakiya, V. P.; Chhipa, N. M. R.; Sen, D. J. Chemical and biochemical history of antimalarials: a podium of plasmodium unicellular unit. *Br. Biomed. Bull.* **2014,** *2* (1), 192–208.

7. Neha, T.; Shukla, S. K. Small bite-big threats: assessment of pernicious repercussion of antimalarial drugs. *Res. J. Recent Sci.* **2014,** *3* (Spec.Iss.), 1–4.
8. Jain, P.; Bele, D. S. Novel antimalarial agents and targets: an optimism over resistance. *Int. J. Pharm., Chem. Biol. Sci.* **2015,** *5* (1), 163–170.
9. Biamonte, M. A.; Wanner, J.; Le Roch, K. G. Recent advances in malaria drug discovery. *Bioorg. Med. Chem. Lett.* **2013,** *23* (10), 2829–2843.
10. Barnett, D. S.; Guy, R. K. Antimalarials in development in 2014. *Chem. Rev. (Washington, DC, U. S.)* **2014,** *114* (22), 11221–11241.
11. Leroy, D.; Campo, B.; Ding, X. C.; Burrows, J. N.; Cherbuin, S. Defining the biology component of the drug discovery strategy for malaria eradication. *Trends Parasitol.* **2014,** *30* (10), 478–490.
12. Barmade, M. A.; Murumkar, P. R.; Sharma, M. K.; Shingala, K. P.; Giridhar, R. R.; Yadav, M. R. Discovery of anti-malarial agents through application of in silico studies. *Comb. Chem. High Throughput Screening* **2015,** *18* (2), 151–187.
13. Huizing, A. P.; Mondal, M.; Hirsch, A. K. H. Fighting malaria: structure-guided discovery of nonpeptidomimetic plasmepsin inhibitors. *J. Med. Chem.* **2015,** *58* (13), 5151–5163.
14. Teixeira, C.; Vale, N.; Perez, B.; Gomes, A.; Gomes, J. R. B.; Gomes, P. "Recycling" classical drugs for malaria. *Chem. Rev. (Washington, DC, U. S.)* **2014,** *114* (22), 11164–11220.
15. Mishra, P. S.; Sharma, H.; Mishra, R.; Gupta, S. K. A review on anti-malarial drug discovery and its screening method. *World J. Pharm. Pharm. Sci.* **2014,** *3* (8), 1288–1304.
16. Kumar, A.; Paliwal, D.; Saini, D.; Thakur, A.; Aggarwal, S.; Kaushik, D. A comprehensive review on synthetic approach for antimalarial agents. *Eur. J. Med. Chem.* **2014,** *85,* 147–178.
17. Marella, A.; Verma, G.; Shaquiquzzaman, M.; Akhter, M.; Alam, M. Malaria: hitches and hopes. *Mini-Rev. Med. Chem.* **2014,** *14* (5), 453–470.
18. Haynes, R. K.; Cheu, K.-W.; N'Da, D.; Coghi, P.; Monti, D. Considerations on the mechanism of action of artemisinin antimalarials: part 1—the "carbon radical" and "heme" hypotheses. *Infect. Disord.: Drug Targets* **2013,** *13* (4), 217–277.
19. Jefford, C. W. Synthetic peroxides as potent antimalarials. News and views. *Curr. Top. Med. Chem.* **2012,** *12* (5), 373–399.
20. Vardanyan, R. S.; Hruby, V. J. *Synthesis of Essential Drugs;* Elsevier, 2006.
21. Anthony, M. P.; Burrows, J. N.; Duparc, S.; Moehrle, J. J.; Wells, T. N. C. The global pipeline of new medicines for the control and elimination of malaria. *Malar. J.* **2012,** *11,* 316.
22. Schrader, F. C.; Barho, M.; Steiner, I.; Ortmann, R.; Schlitzer, M. The antimalarial pipeline—an update. *Int. J. Med. Microbiol.* **2012,** *302* (4-5), 165–171.
23. Wells, T. N. C.; van Huijsduijnen, R. H.; Van Voorhis, W. C. Malaria medicines: a glass half full? *Nat. Rev. Drug Discovery* **2015,** *14,* 424–442.
24. Bruckner, D. A. Amebiasis. *Clin. Microbiol. Rev.* **1992,** *5* (4), 356–369.
25. Stanley, S. L. Amebiasis. *Lancet* **2003,** *361* (9362), 1025–1034.
26. Chacin-Bonilla, L. An update on amebiasis. *Rev. Med. Chile* **2013,** *141* (5), 609–615.
27. Chacin-Bonilla, L. Current pharmacotherapy of amebiasis, advances in new drugs, and design of a vaccine. *Invest. Clin. (Maracaibo, Venez.)* **2012,** *53* (3), 301–314.
28. Mortimer, L.; Chadee, K. *Entamoeba histolytica*: host defense and immune responses. In *Immune Response to Parasitic Infections*, Vol. 1; Jirillo, E., Brandonisio, O., Eds.; Bentham Science, 2010; pp 55–77.
29. Mortimer, L.; Chadee, K. The immunopathogenesis of *Entamoeba histolytica*. *Exp. Parasitol.* **2010,** *126* (3), 366–380.
30. Lejeune, M.; Rybicka, J. M.; Chadee, K. Recent discoveries in the pathogenesis and immune response toward *Entamoeba histolytica*. *Future Microbiol.* **2009,** *4* (1), 105–118.

31. Ali, I. K. M.; Clark, C. G.; Petri, W. A. Molecular epidemiology of amebiasis. *Infect., Genet. Evol.* **2008,** *8* (5), 698–707.
32. Petri, W. A. Pathogenesis of amebiasis. *Curr. Opin. Microbiol.* **2002,** *5* (4), 443–447.
33. Azam, A.; Agarwal, S. M. Targeting amoebiasis: status and developments. *Curr. Bioact. Compd.* **2007,** *3* (2), 121–133.
34. Bendesky, A.; Menendez, D. Metronidazole: a comprehensive view. *Rev. Fac. Med. U.N.A.M.* **2001,** *44* (6), 255–259.
35. Fung, H. B.; Doan, T.-L. Tinidazole: a nitroimidazole antiprotozoal agent. *Clin. Ther.* **2005,** *27* (12), 1859–1884.
36. Silva-Jardim, I.; Thiemann, O. H.; Anibal, F. F. Leishmaniasis and Chagas disease chemotherapy: a critical review. *J. Braz. Chem. Soc.* **2014,** *25* (10), 1810–1823.
37. Nagle, A. S. K.; Shilpi, K.; Arun, B.; Supek, F.; Buchynskyy, A.; Mathison, C. J. N.; Chennamaneni, N. K.; Pendem, N.; Buckner, F. S.; Gelb, M. H.; Molteni, V. Recent developments in drug discovery for leishmaniasis and human African trypanosomiasis. *Chem. Rev. (Washington, DC, U. S.)* **2014,** *114* (22), 11305–11347.
38. Mishra, J.; Saxena, A.; Singh, S. Chemotherapy of leishmaniasis: past, present and future. *Front. Med. Chem.* **2012,** *6,* 97–130.
39. Salotra, P.; Singh, R.; Seifert, K. Visceral leishmaniasis—current treatments and needs. *Drug Discovery Infect. Dis.* **2013,** *4,* 3–15.
40. Grogl, M.; Hickman, M.; Ellis, W.; Hudson, T.; Lazo, J. S.; Sharlow, E. R.; Johnson, J.; Berman, J.; Sciotti, R. J. Review: drug discovery algorithm for cutaneous leishmaniasis. *Am. J. Trop. Med. Hyg.* **2013,** *88* (2), 216–222.
41. Sundar, S.; Chakravarty, J. Leishmaniasis: an update of current pharmacotherapy. *Expert Opin. Pharmacother.* **2013,** *14* (1), 53–63.
42. Hussain, H.; Al-Harrasi, A.; Al-Rawahi, A.; Green, I. R.; Gibbons, S. Fruitful decade for antileishmanial compounds from 2002 to late 2011. *Chem. Rev. (Washington, DC, U. S.)* **2014,** *114* (20), 10369–10428.
43. Croft, S. L.; Olliaro, P. Leishmaniasis chemotherapy—challenges and opportunities. *Clin. Microbiol. Infect.* **2011,** *17* (10), 1478–1483.
44. Tiuman, T. S.; Santos, A. O.; Ueda-Nakamura, T.; Dias Filho, B. P.; Nakamura, C. V. Recent advances in leishmaniasis treatment. *Int. J. Infect. Dis.* **2011,** *15* (8), e525–e532.
45. Pereira, B. A. S.; Souza-Silva, F.; Silva-Almeida, M.; Santos-de-Souza, R.; Goncalves de Oliveira, L. F.; Ribeiro-Guimaraes, M. L.; Alves, C. R. Proteinase inhibitors: a promising drug class for treating leishmaniasis. *Curr. Drug Targets* **2014,** *15* (12), 1121–1131.
46. Dorlo, T. P. C.; Balasegaram, M.; Beijnen, J. H.; de Vries, P. J. Miltefosine: a review of its pharmacology and therapeutic efficacy in the treatment of leishmaniasis. *J. Antimicrob. Chemother.* **2012,** *67* (11), 2576–2597.
47. Singh, N.; Kumar, M.; Singh, R. K. Leishmaniasis: current status of available drugs and new potential drug targets. *Asian Pac. J. Trop. Med.* **2012,** *5* (6), 485–497.
48. Murray, H. W. Review: leishmaniasis in the United States: treatment in 2012. *Am. J. Trop. Med. Hyg.* **2012,** *86* (3), 434–440.
49. Malik, S.; Kumar, S.; Choudhary, A.; Kumar, A.; Singh, A.; Avasthi, G. Leishmaniasis: current treatment strategies and future opportunities. *J. Chem. Pharm. Res.* **2010,** *2* (3), 70–91.
50. Chimelli, L.; Scaravilli, F. Trypanosomiasis. *Brain Pathol.* **1997,** *7* (1), 599–611.
51. Barrett, M. P.; Croft, S. L. Management of trypanosomiasis and leishmaniasis. *Br. Med. Bull.* **2012,** *104* (1), 175–196.
52. Jacobs, R. T.; Nare, B.; Phillips, M. A. State of the art in African trypanosome drug discovery. *Curr. Top. Med. Chem.* **2011,** *11* (10), 1255–1274.

53. Barrett, M. P.; Boykin, D. W.; Brun, R.; Tidwell, R. R. Human African trypanosomiasis: pharmacological re-engagement with a neglected disease. *Br. J. Pharmacol.* **2007,** *152* (8), 1155–1171.
54. Burri, C. Chemotherapy against human African trypanosomiasis: is there a road to success? *Parasitology* **2010,** *137* (14), 1987–1994.
55. Barrett, M. P. Potential new drugs for human African trypanosomiasis: some progress at last. *Curr. Opin. Infect. Dis.* **2010,** *23* (6), 603–608.
56. Schlitzer, M. Drugs for the treatment of African sleeping sickness. *Pharmazie* **2009,** *38* (6), 552–558.
57. Giordani, F.; Mwenechanya, R.; Barrett, M. P. Advances in understanding and treatment of human African trypanosomiasis: divergent diseases caused by distinct parasites. In *Handbook of Pharmacogenomics and Stratified Medicine;* Padmanabhan, S., Ed.; Academic Press, 2014; pp 901–917.
58. Stein, J.; Mogk, S.; Mudogo, C. N.; Sommer, B. P.; Scholze, M.; Meiwes, A.; Huber, M.; Gray, A.; Duszenko, M. Drug development against: sleeping sickness: old wine in new bottles? *Curr. Med. Chem.* **2014,** *21* (15), 1713–1727.
59. Kinoshita, T. Designing sleeping sickness control. *ACS Chem. Biol.* **2008,** *3* (10), 601–603.
60. Ferrins, L.; Rahmani, R.; Baell, J. B. Drug discovery and human African trypanosomiasis: a disease less neglected? *Future Med. Chem., J* **2013,** *5* (15), 1801–1841.
61. Simarro, P. P.; Franco, J.; Diarra, A.; Postigo, J. A. R.; Jannin, J. Update on field use of the available drugs for the chemotherapy of human African trypanosomiasis. *Parasitology* **2012,** *139* (7), 842–846.
62. Babokhov, P.; Sanyaolu, A. O.; Oyibo, W. A.; Fagbenro-Beyioku, A. F.; Iriemenam, N. C. A current analysis of chemotherapy strategies for the treatment of human African trypanosomiasis. *Pathog. Global Health* **2013,** *107* (5), 242–252.
63. Phillips, M. A. Stoking the drug target pipeline for human African trypanosomiasis. *Mol. Microbiol.* **2012,** *86* (1), 10–14.
64. Stich, A.; Schlitzer, M. African sleeping sickness and the drugs for its therapy. *Chemother. J.* **2011,** *20* (4), 115–125.
65. Brun, R.; Don, R.; Jacobs, R. T.; Wang, M. Z.; Barrett, M. P. Development of novel drugs for human African trypanosomiasis. *Future Microbiol.* **2011,** *6* (6), 677–691.
66. Delespaux, V.; de Koning, H. P. Drugs and drug resistance in African trypanosomiasis. *Drug Resist. Updates* **2007,** *10* (1-2), 30–50.
67. Telleria, J., Tibayrenc, M., Eds. *American Trypanosomiasis: Chagas Disease One Hundred Years of Research;* Elsevier, 2010.
68. Kirchhoff, L. V. American trypanosomiasis (Chagas' disease)—-a tropical disease now in the United States. *N. Engl. J. Med.* **1993,** *329* (9), 639–644.
69. Rassi, A., Jr.; Rassi, A. Marcondes de Rezende, J., American trypanosomiasis (Chagas disease). *Infect. Dis. Clin. North Am.* **2012,** *26* (2), 275–291.
70. Gonzalez, C. I.; Mantilla, J. C. Chagas heart disease. In *Cardiomyopathies: From Basic Research to Clinical Management;* Veselka, J., Ed.; InTech, 2012; pp 749–774.
71. Moncayo, A.; Ortiz, Y. M. An update on Chagas disease (human American trypanosomiasis). *Ann. Trop. Med. Parasitol.* **2006,** *100* (8), 663–677.
72. Chatelain, E. Chagas disease drug discovery: toward a new era. *J. Biomol. Screening* **2015,** *20* (1), 22–35.
73. Urbina, J. A. Chemotherapy of Chagas disease. *Curr. Pharm. Des.* **2002,** *8* (4), 287–295.
74. Ilies, M. A. New synthetic strategies for the management of Chagas disease (American trypanosomiasis). *J. Med. Chem.* **2014,** *57* (2), 296–297.

75. Urbina, J. A. New chemotherapeutic approaches for the treatment of Chagas disease (American trypanosomiasis). *Expert Opin. Ther. Pat.* **2003,** *13* (5), 661–669.
76. Urbina, J. A. Specific treatment of Chagas disease: current status and new developments. *Curr. Opin. Infect. Dis.* **2001,** *14* (6), 733–741.
77. Castro, J. A.; Montalto de Mecca, M.; Bartel, L. C. Toxic side effects of drugs used to treat Chagas' disease (American trypanosomiasis). *Hum. Exp. Toxicol.* **2006,** *25* (8), 471–479.

Chapter 36

Anthelmintics

Anthelmintics are drugs used to eradicate helminth infestations of the intestine or tissues of other organs.

The helminths differ from other infectious organisms in that they have a complex body structure and have partial or complete muscular, nervous, digestive, and reproductive systems.

There are three groups of medically important helminths: cestodes (tapeworms), nematodes (roundworms), and trematodes (flukes).

These parasites live in both the body spaces (gut lumen, bile ducts, lungs, oral cavity, etc.) and in tissues (blood, muscles, and skin).

Cestodes (tapeworms) are ribbon-shaped flatworms that as adults dwell entirely in the human small intestine.

There is a large variety of cestodes. Those that are pathogenic to humans include *Taenia solium* (pork tapeworm), *Taenia saginata* (beef tapeworm), *Diphyllobothrium latum* (fish or broad tapeworm), *Hymenolepis nana* (dwarf tapeworm), and *Hymenolepis diminuta* (rat tapeworm).

Taenia species are the most common cestode parasites found in humans. The length of the adult *T. saginata* ranges from 4 to 8 meters and that of *T. solium* ranges in length from 3 to 5 meters. The presence of the adult worm rarely causes symptoms apart from slight abdominal irritation with diarrhea, constipation, or indigestion. On the other side, *H. nana*, the dwarf tapeworm, is the smallest tapeworm to infect humans. Its length ranges from approximately 15 to 40 mm and is 1 mm wide. *H. nana* is thought to be the most common tapeworm worldwide.

Humans become infected by tapeworms via ingesting inadequately cooked beef or pork. The protoscoleces evaginate and pass into the small intestine where they attach themselves to the mucosa and develop into adult worms.

Echinococcus granulosus and *Echinococcus multilocularis* are tapeworms that cause human echinococcosis (hydatid disease) symptoms; the symptoms depend on the site where the cysts have located and are similar to a growing tumor. It could be large abdominal cysts that lead to discomfort, or liver cysts resulting in jaundice. Lung cysts can lead to abscess formation. Brain cysts can cause focal seizures and raised intracranial pressure.

People who accidentally swallow the eggs of the *E. granulosus* tapeworm are at risk for infection. Dogs and other canids are the definitive hosts. Direct contact with infected dogs may lead to human infection. Ingestion of soil, water, and vegetables contaminated with infected dog feces may also lead to infections.

Synthesis of Best-Seller Drugs. http://dx.doi.org/10.1016/B978-0-12-411492-0.00036-5

Sparganosis is infection with larvae of *Spirometra* species or *Sparganum proliferum*. *Spirometra* species have been parasitic intestinal tapeworms of cats since antiquity. Commonly known as the "zipper worm," these tapeworms can be found in the stomach and small intestine of the cat. Humans can become infected by accidental ingestion of copepods from water contaminated by cat or dog feces, ingestion of inadequately cooked flesh from another intermediate host, or contact with poultices containing flesh from these sources.

In humans, larvae typically migrate to subcutaneous tissue or muscle and form slowly growing masses. Other sites, including the central nervous system (CNS), may be involved, but are much less common. Symptoms are caused by mass effect (drugs recommended for the treatment of Cestodes infections are praziquantel, niclosamide, albendazole, and other benzimidazoles).

Nematodes (roundworms) cause various clinical infections in humans. Depending on the habitat in the infected host, they can be classified as blood, liver, lung, or intestinal nematodes.

Gastrointestinal nematode (soil-transmitted helminth) infections are amongst the most prevalent, with the roundworm, pinworm, hookworms, and threadworm. Millions of people worldwide are infected with the intestinal roundworm, *Ascaris lumbricoides,* which can grow to a length of 35 cm. People generally get infected with this parasite by ingestion of the worm's eggs through contaminated food, water, or soil. Like *A. lumbricoides*, the whipworm, *Trichuris trichiura*, is another very common soil-transmitted nematode. The adult is characterized by a "whip-like" appearance. Another common nematode infection is the pinworm, *Enterobius vermicularis*. The adult pinworm is thread-like and grows to a length of up to 13 mm. Hookworms are another intestinal nematode found worldwide. People become infected through skin penetration by the hookworm larvae. Finally, *Trichinella spiralis* is the cause of trichinosis, which could appear after eating infected raw or undercooked meat, as a result of the larvae being released in the stomach, where it migrates to encapsulate in tissue, usually muscle.

For treatment of intestinal round worms are used piperazine, benzimidazoles, morantel, pyrantel, levamisole, avermectins and milbemycins, closantel and other halogenated salicylamides, emodepside, and diethylcarbamazine. Trematode worms (flukes) are the most common species of helminth affecting humans. Depending on the habitat in the infected host, flukes can be classified as blood, liver, lung, or intestinal flukes. People become infected through the consumption of raw or poorly cooked food: fish, crabs, or crayfish and vegetables. Trematodes, can invade the nervous system secondarily after blood, liver, intestinal, or pulmonary infection. Trematodes cause various clinical infections in humans. The most common human intestinal trematode is *Fasciolopsis buski*. The other important intestinal trematodes flukes are *Heterophyes heterophyes*, *Metagonimus yokogawai,* and *Echinostoma* species.

Most trematode-infected people are asymptomatic. Moderate infection presents with occasional loose stools, some weight loss, malaise, and abdominal pain.

Severe infections are manifested with diarrhea alternating with constipation and hunger. As the infection progresses and the worm burden increases, edema of the face, abdominal wall, and lower limbs occurs, as well as ascites and generalized abdominal pain. In people infected with *H. heterophyes*, embolization of the eggs can lead to myocarditis, chronic heart failure, and/or cerebral emboli (praziquantel remains the drug of choice for all trematode infections).

Schistosoma is the only trematode that invades through the skin. Schistosomiasis, also known as bilharzia or "snail fever," is a parasitic disease carried by fresh water snails infected with one of the five varieties of the parasite *Schistosoma*.

Schistosomiasis is transmitted by contact with contaminated fresh water, such as lakes and rivers, inhabited by snails carrying the parasite. Larvae emerge from the snails on contact with an individual and penetrate the skin. Inside the body, the larvae develop into worms, causing urinary, bladder, kidney, intestinal, liver, lung, and spleen schistosomiasis, causing symptoms that vary depending on the type of worm involved and the location of the parasite. The symptoms include: itching and rash at infection site ("swimmer's itch"); frequent, painful or bloody urine, abdominal pain and bloody diarrhea, anemia, fever, liver, spleen, lymph node enlargement; and others. Schistosomiasis has emerged as an important tropical infection and is most common in Asia, Africa, Latin and South America, and the Middle East (the drug of choice for treating all species of schistosomes is praziquantel).

Anthelmintic drugs are substances that expel or destros worms and other internal parasites from the body by either stunning or killing them and without causing significant damage to the host.

Before 1940, the only compounds used to deal with parasitism were natural substances that had some effect on parasites. The modern age of deworming began with the introduction of phenothiazine, which was sometimes combined with lead arsenate to control tapeworms.

In the 1960s and 1970s, the organophosphate anthelmintic Holoxan was introduced. It was eventually removed from the market as a result of toxicity issues.

Although the precise mode of action of many anthelmintics is not fully understood, the sites of action and mechanisms of many of them are generally known.

Anthelmintics can be classified by various criteria in different ways, such as type of parasites controlled; developmental stage affected; mode of action; and chemical structure.

- Type of parasites controlled
  - Nematicides: effective against nematodes;
  - Taeniacides or cestodicides: effective against cestodes;
  - Flukicides or trematodicides: effective against trematodes.

- Type of developmental stage affected by the antihelmintic ingredient:
  - Adulticides kill the adult worms;
  - Larvicides kill the immature worms, that is, the larvae;
  - Ovicides kill the eggs.
- Type of pharmacological action, which is based on interference of drugs with the integrity of parasites, their neuromuscular coordination, or protective mechanisms against host immunity, and directed starvation, paralysis, and expulsion of the parasite.

From this point of view the pharmacological agents are separated into two groups:

  - Compounds that act to infringe the integrity of helminthes, which are commonly referred to as biochemical action and include: β-tubulin ligands (thiabendazole, mebendazole, flubendazole, oxibendazole, and albendazole); lipoxygenase inhibitors (diethylcarbamazine); chitinase inhibitor/ionophore (closantel); SH ligand (melarsomine); pyruvate/ferredoxin oxidoreductase inhibitor (nitazoxanide); isothiocyanate-ATP and cholinesterase inhibition (nitroscanate and amoscanate).
  - Compounds that act on helminth membrane ion channels: γ-aminobutyric acid agonist (piperazine); cholinergic agonists (levamisole), tetrahydropyrimidines (pyrantel, morantel, and oxantel), quaternary/tertiary amines (bephenium and tribendimidine), pyridines (methyridine), and amino-acetonitrile derivatives (monepantel); cholinergic antagonists (phenothiazine and derquantel); allosteric modulators of glutamate-gated chloride channels: avermectins (ivermectin, doramectin, eprinomectin, and abamectin) and milbemycins (moxidectin and milbemycin); and BK (Big Potassium) potassium channel activator (emodepside). This introductory data is perfectly presented in reviews [1-8].
- Types of chemical classes of anthelmintics:
  - The most relevant of them are piperazines, benzimidazoles, imidazothiazoles, tetrahydropyrimidines, isoquinolines, salicylanilides, and few sporadic compounds that do not belong to these chemical classes—clorsulon, monepantel, and nitroxinil.

36.1 PIPERAZINES

Piperazine (**36.1.1**) (Fig. 36.1.) was first used as an anthelmintic in the 1950s and it is still the active constituent of nonprescription remedies for thread worm infection in children. It acts as a weak γ-aminobutyric acid (GABA)-mimetic and causes a flaccid, reversible paralysis of worm body wall muscle. The parasites, paralyzed and depleted of energy, are expelled by peristalsis [9,10].

Diethylcarbamazine (**36.1.2**) (Fig. 36.1.) is a drug that has been used for many years to treat lymphatic filariasis, a significant public health problem across the developing world, in humans and animals; it also affects intestinal nematodes, but its mechanism of action remains unclear. Its mode of action is not well understood, but it is known that it is an inhibitor of arachidonic acid metabolism in helminths, targeting the cyclooxygenase and lipoxygenase pathways, reducing the production of thromboxane, prostacyclin, prostaglandin, and leukotrienes. Diethylcarbamazine blocks host, and possibly parasite, enzymes involved in arachidonic acid metabolism, and enhances the innate, nonspecific immune system [11,12].

36.2 PHENOTHIAZINES

Phenothiazine (**36.2.1**) (Fig. 36.1.) was introduced as an anthelmintic in 1940 and has modest acetylcholine inhibitory effects in nematodes. It is not used today [13,14].

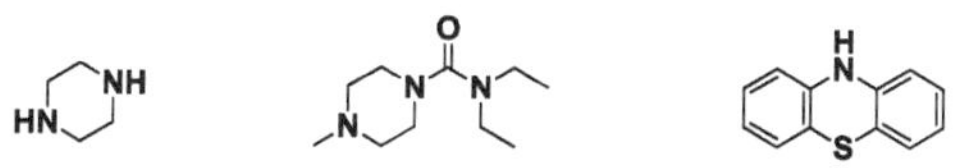

Piperazine 36.1.1 Diethylcarbamazine 36.1.2 Phenothiazine 36.2.1

FIG. 36.1 Piperazine and phenothiazine anthelmintics.

36.3 BENZIMIDAZOLES

The first of this class, thiabendazole (**36.3.1**) [15-19], was discovered in 1961; subsequent discoveries were albendazole (**36.3.2**) [20-27], fenbendazole (**36.3.3**) [28,29], mebendazole (**36.3.4**) [30-32], flubendazole (**36.3.5**) [33], and triclabendazole (**36.3.6**) [34-36], anthelmintics with broad spectrum (Fig. 36.2.). Their anthelmintic efficacy is based on their ability to interact with a parasite's β-tubulin, thereby inhibiting polymerization and its assembly into microtubules. Benzimidazoles produce degenerative changes in the epidermis and cells of the intestine of nematode parasites. That further damages the parasite, immobilizing it and causing it to die.

Twenty years after the first cases of echinococcal disease were treated with albendazole, albendazole has become an important component in the overall management of both cystic and alveolar echinococcosis. Albendazole is the current medication of choice for the treatment of neurocysticercosis and is recommended for symptomatic patients with multiple viable cysts in the brain parenchyma. It is also used in extraparenchymal cysticercosis, when complete surgical resection of the cysts is usually impracticable.

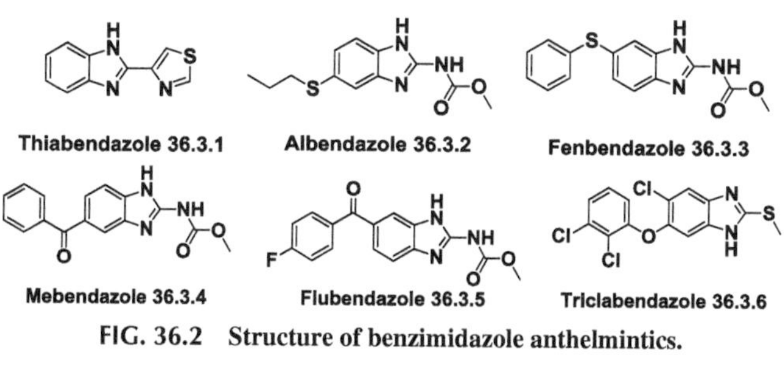

Thiabendazole 36.3.1 Albendazole 36.3.2 Fenbendazole 36.3.3

Mebendazole 36.3.4 Flubendazole 36.3.5 Triclabendazole 36.3.6

FIG. 36.2 Structure of benzimidazole anthelmintics.

36.4 IMIDAZOTHIAZOLES

Levamisole (**36.4.1**) (Fig. 36.3.) is a nicotinic receptor agonist that elicits spastic muscle paralysis as a result of prolonged activation of the excitatory nicotinic acetylcholine receptors on helmints' body wall muscle, producing depolarization, contraction, and spastic paralysis of the parasites, so they are swept away from their location. Levamisole [37] is listed as an anthelmintic for use in humans, but is rarely used in practice because of its low therapeutic index.

36.5 TETRAHYDROPYRIMIDINES

Tetrahydropyrimidines—morantel (**36.5.1**) [38], pyrantel (**36.5.2**) [39,40], and oxantel (**36.5.3**) [41,42] (Fig. 36.3.)—are nicotinic receptor agonists anthelmintics, a chemical family of antiparasitic compounds moderately used on pets (Fig. 36.3.). Pyrantel is also approved for use on humans. Tetrahydropyrimidines act on the nervous system of the worms as inhibitors of acetylcholinesterase. Terminating the transmission of nervous signals where acetylcholine is the neurotransmitter the parasites are paralyzed and cannot keep themselves attached to the intestinal wall.

36.6 ISOQUINOLINES

Isoquinolines are represented by praziquantel (**36.6.1**) (Fig. 36.3.). The molecular mode of action of praziquantel is not precisely known at present. It is believed that it affects the permeability of the calcium ions in the parasites muscular membrane violates carbohydrate metabolism of the parasite. As consequence the helmints are paralyzed and die. Praziquantel is currently the only available antischistosomal drug [43-55].

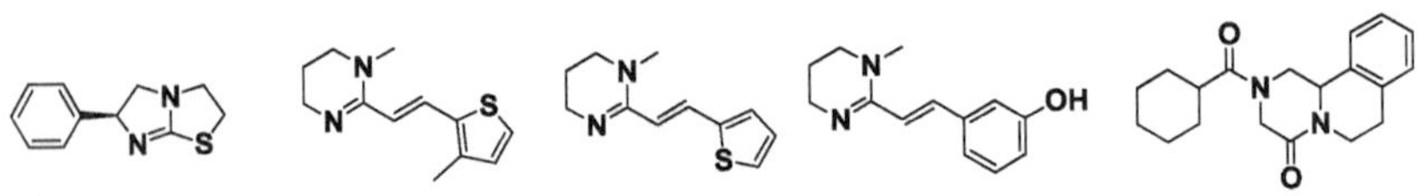

Levamisole 36.4.1 Morantel 36.5.1 Pyrantel 36.5.2 Oxantel 36.5.3 Praziquantel 36.6.1

FIG. 36.3 Structure of imidazothiazole, tetrahydropyrimidine, and isoquinoline anthelmintics.

36.7 SALICYLANILIDES

The salicylanilides—niclosamide (**36.7.1**), closantel (**36.7.2**), rafoxanide (**36.7.3**), and oxyclozanide (**36.7.4**) (Fig. 36.4.)—are anthelmintics that interfere with energy metabolism of parasites by uncoupling oxidative phosphorylation in the fluke. They uncouple the oxidative phosphorylation in the cell mitochondria, which disturbs the production of adenosine triphosphate. This impairs the parasites motility and probably other processes as well.

Niclosamide is an oral antihelminthic drug, which had been used to treat tapeworm infection for approximately 50 years. It playes a big role in the control of schistosomiasis. In addition, it is an efficient molluscicide and is used against aquatic snails. Niclosamide is approved in the United States as a human medicine [56]. Closantel is indicated for use in cattle and sheep to treat and control adult and immature flukes and nematodes; [57]. Rafoxanid and oxyclozanide are also approved as veterinary use [58].

Nitazoxanide (**36.7.5**) (Fig. 36.4.) is a nitrothiazolyl-salicylamide derivative that acts against a broad spectrum of protozoa and helminths in the intestinal tract. The activity of nitazoxanide is believed to be based on interference with the pyruvate:ferredoxin oxidoreductase enzyme-dependent electron transfer reaction, which is essential to anaerobic energy metabolism [59-64].

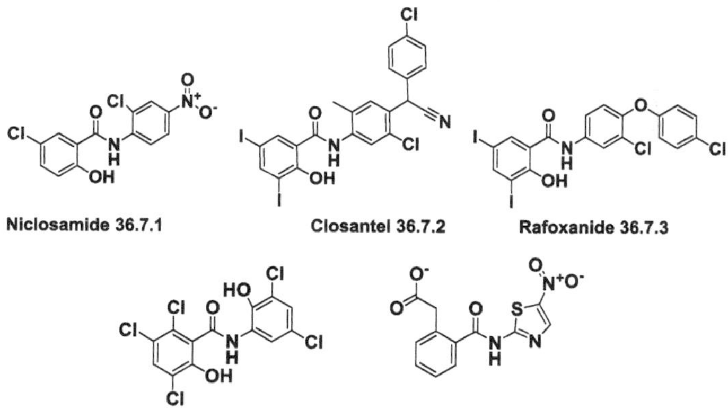
Niclosamide 36.7.1 Closantel 36.7.2 Rafoxanide 36.7.3
Oxyclozanide 36.7.4 Nitazoxanide 36.7.5

FIG. 36.4 Anthelmintic salicylanilides.

36.8 ARYLISOTHIOCYANATES

Nitroscanate (**36.8.1**) [65] and amoscanate (**36.8.2**) [66] (Fig. 36.5.) are effective against the major gastrointestinal nematodes of dogs and cats. They are experimental anthelmintic agents of the arylisothiocyanate class, which was found to be highly effective in treatment of the four major species of schistosomas infecting humans.

The molecular mode of action of nitroscanate has not been elucidated. It is assumed that it acts as an uncoupler of the oxidative phosphorylation in the cell mitochondria, which disturbs the production of ATP, the cellular "fuel." This impairs the parasites motility and probably other processes as well.

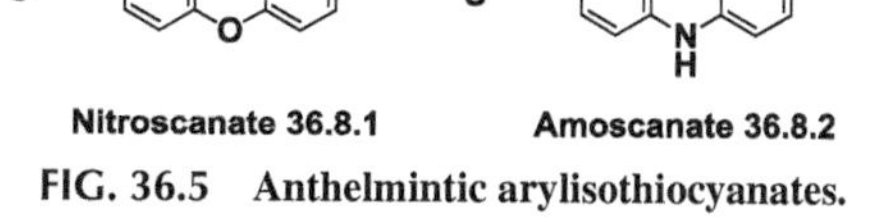
Nitroscanate 36.8.1 Amoscanate 36.8.2

FIG. 36.5 Anthelmintic arylisothiocyanates.

36.9 AMINO ACETONITRILE DERIVATIVES

Monepantel (**36.9.1**) (Fig. 36.6.) is the first drug of a new family of anthelmintics, representative of amino acetonitrile derivatives, recently developed anthelmintic with a novel mode of action. It acts on a particular nicotinic acetylcholine receptor subunit that occurs only in nematodes and not in mammals. Monepantel blocks these receptors and paralyzes the affected worms, which are expelled.

Monepantel presently is used to treat ruminants infected with gastrointestinal nematodes such as *Haemonchus contortus*. It is effective against nematodes that show resistance to levamisole, ivermectin, or albendazole and derquantel [67-69].

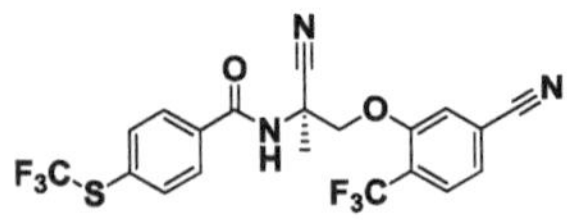

Monepantel 36.9.1

FIG. 36.6 Structure of monepantel.

36.10 SPIROINDOLES

Derquantel (**36.10.1**) (Fig. 36.7.) began to be marketed in 2010 for animal use. It belongs to the spiroindole class of anthelmintics and is prepared semisynthetically by chemical reduction of paraherquamide, which, in turn, is obtained from fermentation of a soil fungi *Penicillium simplicissimum*. Paraherquamide was also found to be an effective anthelmintic but it has not been developed as a commercial product. Derquantel is a nicotinic cholinergic antagonist. The effect on the worms is that they are paralyzed and expelled [70,71].

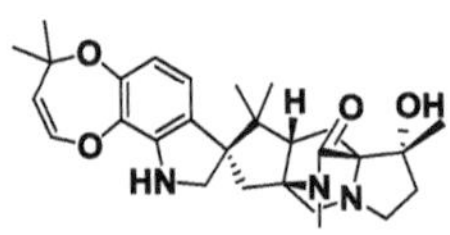

Derquantel 36.10.1

FIG. 36.7 Structure of derquantel.

36.11 MACROCYCLIC LACTONES

The prototype macrocyclic lactones is ivermectin. There are two different groups of macrocyclic lactones: avermectins, a series 16-membered macrocyclic lactones (ivermectin [72-82] [commercial name of a mixture of 22,23-dihydroavermectin B1a (**36.11.1a**) and 22,23-dihydroavermectin B1b (**36.11.1b**)], doramectin (**36.11.2**) [83,84], eprinomectin (**36.11.3**) [82], and abamectin, a mixture of (**36.11.4a**) and (**36.11.4b**) [85,86]), and milbemycins (moxidectin (**36.11.5**) [87,88] and milbemycin (**36.11.6**) [89,90]) (Fig. 36.8.), which are chemically related to avermectins and have a similar mechanism of action, but with a longer half-life. Since their introduction 30 years ago, the avermectins and milbemycins are the largest selling anthelmintics in the world, widely used in veterinary medicine.

Macrocyclic lactones act as GABA agonists and also bind to glutamate-gated chloride channels in nerve and muscle cells of invertebrates. In both cases, these actions block the transmission of neuronal signals of the parasites, which

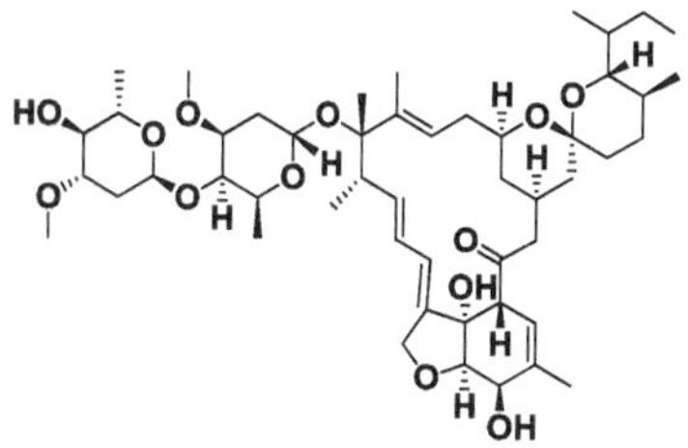

Ivermectin B1a 36.11.1a

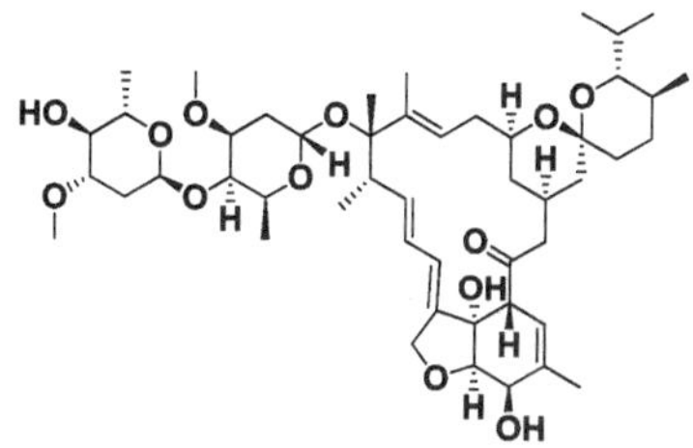

Ivermectin B1b 36.11.1b

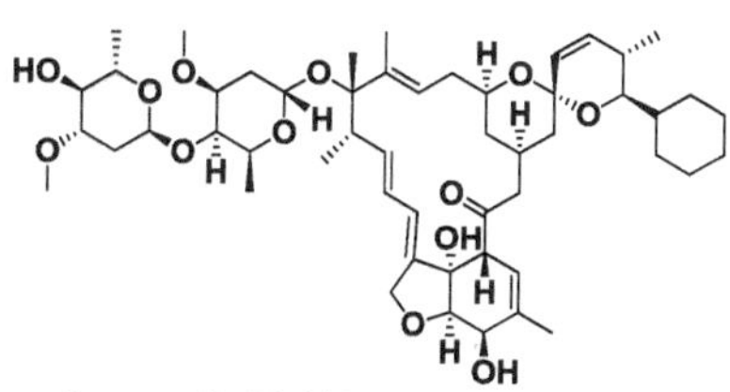

Doramectin 36.11.2

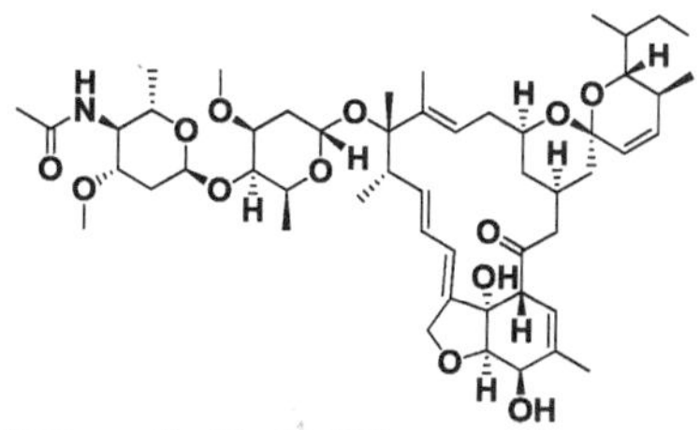

Eprinomectin B1a 36.11.3

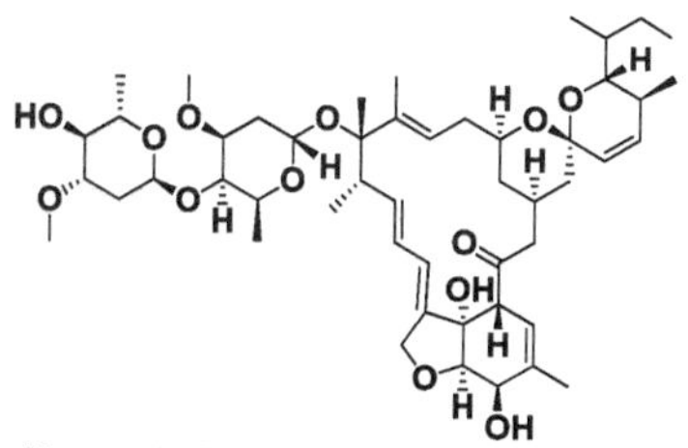

Abamectin B1a 36.11.4a

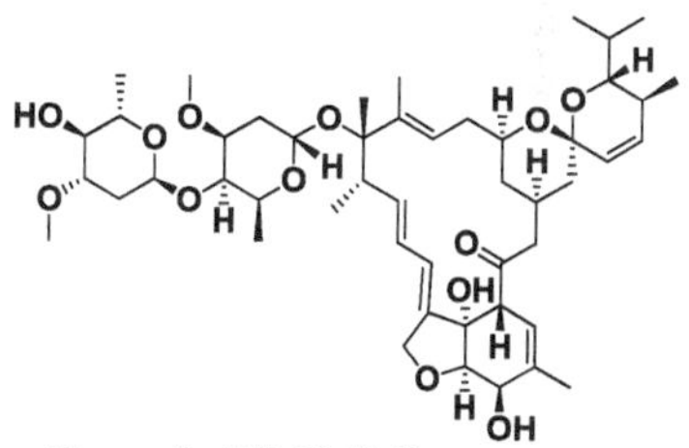

Abamectin B1b 36.11.4b

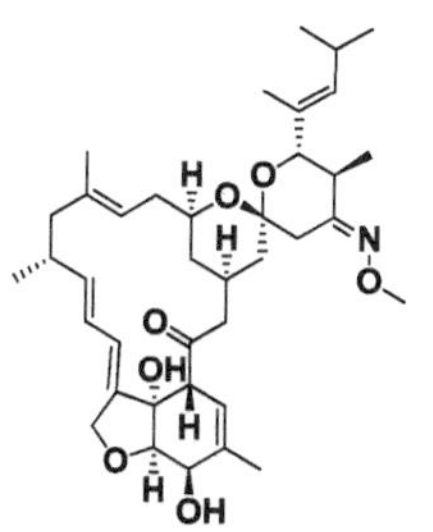

Moxidectin 36.11.5

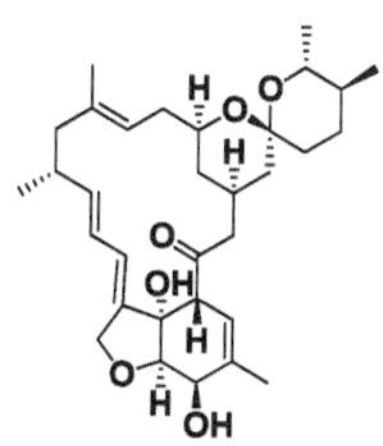

Milbemycin A1 36.11.6

FIG. 36.8 **Structure of macrocyclic lactone anthelmintics.**

are either paralyzed and expelled out of the body, or starved helmints. They also affect the reproduction of some parasites by diminishing oviposition or inducing an abnormal oogenesis [91,92].

Since 1987, ivermectin (both (**36.11.1a**) and (**36.11.1b**)) has a widespread use in veterinary medicine and its use has been extended in humans. It is mainly used in in humans for the treatment of onchocerciasis (river blindness), but it is also effective against other worm infestations. Ivermectin is also of proven benefit in scabies infestation and pediculosis [93].

36.12 MELAMINYLTHIOARSENATES

Melarsomine (**36.12.1**) (Fig. 36.9.) is an arsenic-based drug that treats heartworm infections in animals by killing the adult heartworms that are living in the arteries of the lungs. Melarsomine does not have any effect on heartworm microfilaria [94]. Melarsomine is for use in dogs only and should never be used in cats.

36.13 PEPTIDES (CYCLOOCTADEPSIPEPTIDE)

Emodepside (**36.13.1**) (Fig. 36.9.), which was introduced in 2001, is a semi-synthetic derivative of fermentation products of the fungus *Mycelia sterilia*. Emodepside has a molecular mode of action that is different from the other anthelmintics. It binds to so-called latrophilin receptors in neuromuscular junctions of the worm muscle cells and also interferes with the potassium channels in the neuronal membranes. As a result of these dual actions, parasites are paralyzed and expelled. The drug is licensed for use in cats, but it is in development for use in humans [95,96].

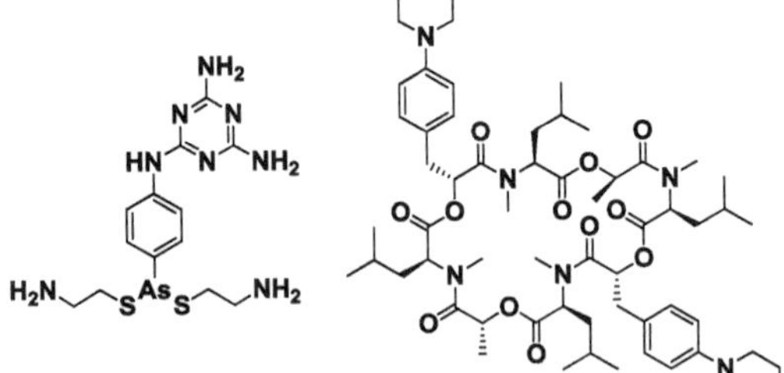

Melarsomine 36.12.1 **Emodepside 36.13.1**

FIG. 36.9 Structure of melarsomine and emodepside.

36.14 QUARTERNARY/TERTIARY AMINES

Bephenium (**36.14.1**) [1] and tribendimidine (**36.14.2**) [97] (Fig. 36.10.) are cholinergic agonists. The mechanism of their action against nematodes is the same as levamisole and pyrantel, producing helmints body wall muscle depolarization, contraction, and spastic paralysis of helmints (bephenium is not FDA-approved and is not available in the United States).

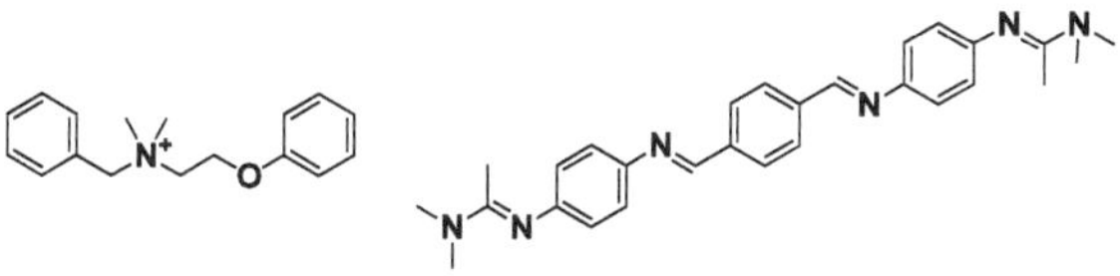

Bephenium 36.14.1 Tribendimidine 36.14.2

FIG. 36.10 Structure of bephenium and tribendimidine.

36.15 PYRIDINES

Methyridine (**36.15.1**) (Fig. 36.11.) acts as a cholinergic agonist on nematode muscle receptors. It is efficient against most helmints found in the small intestine, but less effective against those in the large intestine. The drug appears to show activity against lungworm and to reduce larval development in the feces. It is effective against *Trichuris vulsis* infestation and *Strongyloides* eggs in dogs, completely eliminating the worms. Methyridine is useful in low-grade infection, but severe infections do not respond to a single injection [98].

36.16 BENZENE-SULPHONAMIDES

Clorsulon (**36.16.1**) (Fig. 36.11.) inhibits various enzymes involved in the glycolytic process of flukes, making it impossible to obtain energy from glucose. As a consequence the levels of ATP, the cellular fuel, are depressed, causing death for helminths. Clorsulon is not used for humans, cats, or dogs. Cattle, sheep, and goats tolerate clorsulon very well [99,100].

36.17 HALOGENATED PHENOLS

Nitroxinil (**36.17.1**) (Fig. 36.11.) [101-103] works by uncoupling of the oxidative phosphorylation in the cell mitochondria, which disturbs the production of the cellular "fuel" ATP. This impairs the parasites motility and probably other processes as well. Nitroxinil is not used for humans. It can be used on dairy animals, but is not used in dogs or pets.

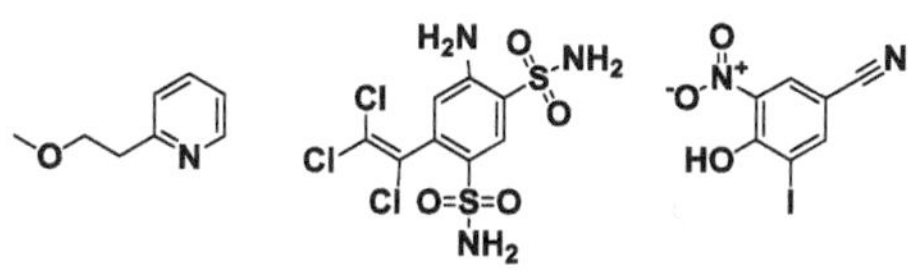

Methyridine 36.15.1 Clorsulon 36.16.1 Nitroxinil 36.17.1

FIG. 36.11 Structure of methyridine and nitroxinil.

Generally speaking anthelmintic therapy is limited to three major chemical classes of anthelmintics: benzimidazoles, imidazothiazoles, and macrocyclic lactones.

REFERENCES

1. Denham, D. A. Anthelmints. In *Antibiotic and Chemotherapy: Anti-Infective Agents and Their Use in Therapy,* 7th ed.; O'Grady, F., Lambert, H. P., Eds. Churchill Livingstone, 1997; pp 513–521.
2. MacKenzie, C. D.; Geary, T. G. Addressing the current challenges to finding new anthelmintic drugs. *Expert Rev. Anti-Infect. Ther.* **2013,** *11* (6), 539–541.
3. de Hostos, E. L.; Nguyen, T. Anthelmintic drugs: tools and shortcuts for the long road from discovery to product. *Drug Discovery Infect. Dis.* **2012,** *3* (Parasitic Helminths), 219–232.
4. Geary, T. G. Are new anthelmintics needed to eliminate human helminthiases? *Curr. Opin. Infect. Dis.* **2012,** *25* (6), 709–717.
5. Robertson, A. P.; Buxton, S. K.; Puttachary, S.; Williamson, S. M.; Wolstenholme, A. J.; Neveu, C.; Cabaret, J.; Charvet, C. L.; Martin, R. J. Antinematodal drugs-modes of action and resistance: and worms will not come to thee (Shakespeare: Cymbeline. IV, ii). *Drug Discovery Infect. Dis.* **2012,** *3* (Parasitic Helminths), 233–249.
6. Geary, T. G.; Gauvry, N. Anthelmintic discovery for human infections. *RSC Drug Discovery Ser.* **2012,** *14* (Neglected Diseases and Drug Discovery), 290–321.
7. Geary, T. G.; Woo, K.; McCarthy, J. S.; MacKenzie, C. D.; Horton, J.; Prichard, R. K.; de Silva, N. R.; Olliaro, P. L.; Lazdins-Helds, J. K.; Engels, D. A.; Bundy, D. A. Unresolved issues in anthelmintic pharmacology for helminthiases of humans. *Int. J. Parasitol.* **2010,** *40* (1), 1–13.
8. Stepek, G.; Buttle, D. J.; Duce, I. R.; Behnke, J. M. Human gastrointestinal nematode infections: are new control methods required? *Int. J. Exp. Pathol.* **2006,** *87* (5), 325–341.
9. Nanivadekar, A. S.; Gadgil, S. D.; Apte, V. V. Efficacy of levamisole, mebendazole, piperazine and pyrantel in roundworm infection. By National Anthelmintic Study Group. *J. Postgrad. Med. (Bombay)* **1984,** *30* (3), 144–152.
10. Mjos, K. Piperazine. In 2nd ed.; Standen, A., Ed.; *Kirk-Othmer Encyclopedia of Chemical Technology*, Vol. 15; Interscience, 1964; pp 638–646.
11. Cesbron, J. Y.; Capron, A.; Vargaftig, B. B.; Lagarde, M.; Pincemail, J.; Braquet, P.; Taelman, H.; Joseph, M. Platelets mediate the action of diethylcarbamazine on microfilariae. *Nature (London, U. K.)* **1987,** *325* (6104), 533–536.
12. MacKenzie, C. D.; Kron, M. A. Diethylcarbamazine: review of pharmacology, mechanisms of action and clinical uses. *Trop. Dis. Bull.* **1985,** *82* (10), R1–R37.
13. Mitchell, S. C. Phenothiazine: the parent molecule. *Curr. Drug Targets* **2006,** *7* (9), 1181–1189.
14. Ohlow, M. J.; Moosmann, B. Phenothiazine: the seven lives of pharmacology's first lead structure. *Drug Discovery Today* **2011,** *16* (3/4), 119–131.
15. Cuckler, A. C.; Mezey, K. C. The therapeutic efficacy of thiabendazole for helminthic infections in man; a literature review. *Arzneim. Forsch.* **1966,** *16* (3), 411–428.
16. Pronk, M. E. J.; Schefferlie, G. J. Tiabendazole (thiabendazole) (addendum). *WHO Food Addit. Ser.* **2002,** *49,* 11–25.
17. Kapoor, V. K. Thiabendazole. *Anal. Profiles Drug Subst.* **1987,** *16,* 611–639.
18. Campbell, W. C.; Cuckler, A. C. Thiabendazole in the treatment and control of parasitic infections in man. *Tex. Rep. Biol. Med.* **1969,** *27* (Suppl. 2), 665–692.

19. Robinson, H. J.; Silber, R. H.; Graessle, O. E. Thiabendazole: toxicological, pharmacological and antifungal properties. *Tex. Rep. Biol. Med.* **1969,** *27* (Suppl. 2), 537–560.
20. Venkatesan, P.; Albendazole, J. Antimicrob. *Chemother.* **1998,** *41* (2), 145–147.
21. Bogan, J. A. Albendazole. *Drugs Today* **1979,** *15* (3), 87–91.
22. Horton, R. J. Albendazole in treatment of human cystic echinococcosis: 12 years of experience. *Acta Trop.* **1997,** *64(1* (2)), 79–93.
23. El Harti, J.; Ansar, M.; Taoufik, J. Albendazole and its analogues. *Int. J. Pharm. Sci. Res.* **2014,** *5* (3), 102–107.
24. Horton, J. The development of albendazole for lymphatic filariasis. *Ann. Trop. Med. Parasitol.* **2009,** *103* (Suppl. 1), S33–S40.
25. Critchley, J.; Addiss, D.; Ejere, H.; Gamble, C.; Garner, P.; Gelband, H. The International Filariasis Review Group, Albendazole for the control and elimination of lymphatic filariasis: systematic review. *Trop. Med. Int. Health* **2005,** *10* (9), 818–825.
26. Albanese, G.; Venturi, C. Albendazole: a new drug for human parasitoses. *Dermatol. Clin.* **2003,** *21* (2), 283–290.
27. Horton, J. Albendazole for the treatment of echinococcosis. *Fundam. Clin. Pharmacol.* **2003,** *17* (2), 205–212.
28. Villar, D.; Cray, C.; Zaias, J.; Altman, N. H. Biologic effects of fenbendazole in rats and mice: a review. *J. Am. Assoc. Lab. Anim. Sci.* **2007,** *46* (6), 8–15.
29. Booze, T. F.; Oehme, F. W. A literature review of the anthelmintic, fenbendazole. *Vet. Hum. Toxicol.* **1982,** *24* (1), 49–52.
30. Bennett, A.; Guyatt, H. Reducing intestinal nematode infection: efficacy of albendazole and mebendazole. *Parasitol. Today* **2000,** *16* (2), 71–77.
31. Al-Badr, A. A.; Tariq, M. Mebendazole. *Anal. Profiles Drug Subst.* **1987,** *16,* 291–326.
32. Van den Bossche, H.; Rochette, F.; Hoerig, C. Mebendazole and related anthelmintics. *Adv. Pharmacol. Chemother.* **1982,** *19,* 67–128.
33. Bogan, J. A. Flubendazole. *Drugs Today* **1980,** *16* (9), 306–310.
34. Fairweather, I. Triclabendazole progress report, 2005-2009: an advancement of learning? *J. Helminthol.* **2009,** *83* (2), 139–150.
35. Keiser, J.; Engels, D.; Buescher, G.; Utzinger, J. Triclabendazole for the treatment of fascioliasis and paragonimiasis. *Expert Opin. Invest. Drugs* **2005,** *14* (12), 1513–1526.
36. Fairweather, I. Triclabendazole: new skills to unravel an old(ish) enigma. *J. Helminthol.* **2005,** *79* (3), 227–234.
37. Thienpont, D.; Vanparijs, O. F. J.; Raeymaekers, A. H. M.; Vandenberk, J.; Demoen, P. J. A.; Allewijn, F. T. N.; Marsboom, R. P. H.; Niemegeers, C. J. E.; Schellekens, K. H. L.; Janssen, Paul, A. J. Tetramisole (R 8299), a new, potent broad spectrum anthelmintic. *Nature (London, U. K.)* **1966,** *209* (5028), 1084–1086.
38. Bogan, J. A. Morantel tartrate. *Drugs Today* **1977,** *13* (11), 507–509.
39. Kopp, S. R.; Kotze, A. C.; McCarthy, J. S.; Traub, R. J.; Coleman, G. T. Pyrantel in small animal medicine: 30 years on. *Vet. J.* **2008,** *178* (2), 177–184.
40. Hatchuel, W. New anthelminthic-Combantrin (Pyrantel Pamoate). *Cent. Afr. J. Med.* **1973,** *19* (5), 102–104.
41. Speich, B.; Ame, S. M.; Ali, S. M.; Alles, R.; Huwyler, J.; Hattendorf, J.; Utzinger, J.; Albonico, M.; Keiser, J. Oxantel pamoate-albendazole for *Trichuris trichiura* infection. *N. Engl. J. Med.* **2014,** *370* (7), 610–620.
42. Keiser, J.; Tritten, L.; Silbereisen, A.; Speich, B.; Adelfio, R.; Vargas, M. Activity of oxantel pamoate monotherapy and combination chemotherapy against Trichuris muris and hookworms: revival of an old drug. *PLoS Neglected Trop. Dis.* **2013,** *7* (3), e2119.

43. Doenhoff, M. J.; Cioli, D.; Utzinger, J. Praziquantel: mechanisms of action, resistance and new derivatives for schistosomiasis. *Curr. Opin. Infect. Dis.* **2008,** *21* (6), 659–667.
44. Doenhoff, M. J.; Pica-Mattoccia, L. Praziquantel for the treatment of schistosomiasis: its use for control in areas with endemic disease and prospects for drug resistance. *Expert Rev. Anti-Infect. Ther.* **2006,** *4* (2), 199–210.
45. Redman, C. A.; Robertson, A.; Fallon, P. G.; Modha, J.; Kusel, J. R.; Doenhoff, M. J.; Martin, R. J. Praziquantel: an urgent and exciting challenge. *Parasitol. Today* **1996,** *12* (1), 14–20.
46. Andrews, P. Praziquantel: mechanisms of antischistosomal activity. *Pharmacol. Ther.* **1985,** *29* (1), 129–156.
47. Cioli, D.; Valle, C.; Angelucci, F.; Miele, A. E. Will new antischistosomal drugs finally emerge? *Trends Parasitol.* **2008,** *24* (9), 379–382.
48. Cioli, D.; Pica-Mattoccia, L.; Basso, A.; Guidi, A. Schistosomiasis control: praziquantel forever? *Mol. Biochem. Parasitol.* **2014,** *195* (1), 23–29.
49. Greenberg, R. M. Praziquantel: mechanism of action. In *Parasitic Flatworms: Molecular Biology, Biochemistry, Immunology and Physiology;* Maule, A. G., Marks, N. J., Eds.; Cabi, 2006; pp 269–281.
50. El-Subbagh, H. I.; Al-Badr, A. A. Praziquantel. *Anal. Profiles Drug Subst. Excipients* **1988,** *25,* 463–500.
51. Olliaro, P.; Delgado-Romero, P.; Keiser, J. The little we know about the pharmacokinetics and pharmacodynamics of praziquantel (racemate and R-enantiomer). *J. Antimicrob. Chemother.* **2014,** *69* (4), 863–870.
52. Chai, J.-Y. Praziquantel treatment in trematode and cestode infections: an update. *Infect. Chemother.* **2013,** *45* (1), 32–43.
53. Pica-Mattoccia, L.; Cioli, D. Praziquantel: too good to be replaced? *Drug Discovery Infect. Dis.* **2012,** *3,* 309–321.
54. Doemling, A.; Khoury, K. Praziquantel and Schistosomiasis. *ChemMedChem* **2010,** *5* (9), 1420–1434.
55. Bygott, J. M.; Chiodini, P. L. Praziquantel: neglected drug? Ineffective treatment? Or therapeutic choice in cystic hydatid disease? *Acta Trop.* **2009,** *111* (2), 95–101.
56. Al-Hadiya, B. M. H. Niclosamide: comprehensive profile. *Profiles Drug Subst., Excipients, Relat. Methodol.* **2005,** *32,* 67–96.
57. Michiels, M.; Meuldermans, W.; Heykants, J. The metabolism and fate of closantel (Flukiver) in sheep and cattle. *Drug Metab. Rev.* **1987,** *18* (2-3), 235–251.
58. Swan, G. E. The pharmacology of halogenated salicylanilides and their anthelmintic use in animals. *J. S. Afr. Vet. Assoc.* **1999,** *70* (2), 61–70.
59. Fox, L. M.; Saravolatz, L. D. Nitazoxanide: a new thiazolide antiparasitic agent. *Clin. Infect. Dis.* **2005,** *40* (8), 1173–1180.
60. White, A. C., Jr. Nitazoxanide: a new broad spectrum antiparasitic agent. *Expert Rev. Anti-Infect. Ther.* **2004,** *2* (1), 43–50.
61. Anderson, V. R.; Curran, M. P. Nitazoxanide: a review of its use in the treatment of gastrointestinal infections. *Drugs* **2007,** *67* (13), 1947–1967.
62. Hemphill, A.; Mueller, J.; Esposito, M. Nitazoxanide, a broad-spectrum thiazolide anti-infective agent for the treatment of gastrointestinal infections. *Expert Opin. Pharmacother.* **2006,** *7* (7), 953–964.
63. Halsey, J. L. Current approaches to the treatment of gastrointestinal infections: focus on nitazoxanide. *Clin. Med.: Ther.* **2009,** *1,* 263–275.
64. Aslam, S.; Musher, D. M. Nitazoxanide: clinical studies of a broad-spectrum anti-infective agent. *Future Microbiol.* **2007,** *2* (6), 583–590.

65. Boray, J. C.; Strong, M. B.; Allison, J. R.; Von Orelli, M.; Sarasin, G.; Gfeller, W. Nitroscante: a new broad spectrum anthelmintic against nematodes and cestodes of dogs and cats. *Aust. Vet. J.* **1979,** *55* (2), 45–53.
66. Greene, L. K.; Grenan, M. M.; Davidson, D. E., Jr.; Jones, D. H.; Shedd, T. R. Amoscante as a topically applied chemical for prophylaxis against *Schistosoma mansoni* infections in mice. *Am. J. Trop. Med. Hyg.* **1983,** *32* (6), 1356–1363.
67. Lecova, L.; Stuchlikova, L.; Prchal, L.; Skalova, L. Monepantel: the most studied new anthelmintic drug of recent years. *Parasitology* **2014,** *141* (13), 1686–1698.
68. Kaminsky, R.; Ducray, P.; Jung, M.; Clover, R.; Rufener, L.; Bouvier, J.; Weber, S. S.; Wenger, A.; Wieland-Berghausen, S.; Goebel, T.; Gauvry, N.; Pautrat, F.; Skripsky, T.; Froelich, O.; Komoin-Oka, C.; Westlund, B.; Sluder, A.; Mäser, P. A new class of anthelmintics effective against drug-resistant nematodes. *Nature (London, U. K.)* **2008,** *452* (7184), 176–180.
69. Lecova, L.; Stuchlikova, L.; Prchal, L.; Skalova, L. Monepantel: the most studied new anthelmintic drug of recent years. *Parasitology* **2014,** *141* (13), 1686–1698.
70. Chamberlain, P. L.; Jeong, S.-H.; Cerniglia, C.; Greenlees, K. Derquantel. *WHO Food Addit. Ser.* **2012,** *66* (Toxicological Evaluation of Certain Veterinary Drug Residues in Food), 65–118.
71. Woods, D. J.; Maeder, S. J.; Robertson, A. P.; Martin, R. J.; Geary, T. G.; Thompson, D. P.; Johnson, S. S.; Conder, G. A. Discovery, mode of action, and commercialization of derquantel. *Drug Discovery Infect. Dis.* **2012,** *3* (Parasitic Helminths), 297–307.
72. Omura, S.; Crump, A. Timeline: tropical infectious diseases: the life and times of ivermectin—a success story. *Nat. Rev. Microbiol.* **2004,** *2* (12), 984–989.
73. Richard-Lenoble, D.; Chandenier, J.; Gaxotte, P. Ivermectin and filariasis. *Fundam. Clin. Pharmacol.* **2003,** *17* (2), 199–203.
74. Omura, S. Ivermectin: 25 years and still going strong. *Int. J. Antimicrob. Agents* **2008,** *31* (2), 91–98.
75. Geary, T. G. Ivermectin 20 years on: maturation of a wonder drug. *Trends Parasitol.* **2005,** *21* (11), 530–532.
76. Campbell, W. C. Ivermectin, an antiparasitic agent. *Med. Res. Rev.* **1993,** *13* (1), 61–79.
77. Gonzalez, P.; Gonzalez, F. A.; Ueno, K. Ivermectin in human medicine, an overview of the current status of its clinical applications. *Curr. Pharm. Biotechnol.* **2012,** *13* (6), 1103–1109.
78. Burkhart, C. N. Ivermectin: an assessment of its pharmacology, microbiology and safety. *Vet. Hum. Toxicol.* **2000,** *42* (1), 30–35.
79. Cupp, E. W.; Sauerbrey, M.; Richards, F. Elimination of human onchocerciasis: history of progress and current feasibility using ivermectin (Mectizan) monotherapy. *Acta Trop.* **2011,** *120* (Suppl. 1), S100–S108.
80. Crump, A.; Omura, S. Ivermectin "wonder drug" from Japan: the human use perspective. *Proc. Jpn. Acad., Ser. B* **2011,** *87* (2), 13–28.
81. Dourmishev, A. L.; Dourmishev, L. A.; Schwartz, R. A. Ivermectin: pharmacology and application in dermatology. *Int. J. Dermatol.* **2005,** *44* (12), 981–988.
82. Shoop, W.; Soll, M. Ivermectin, abamectin and eprinomectin. In *Macrocyclic Lactones in Antiparasitic Therapy;* Vercruysse, J., Rew, R. S., Eds.; CABI, 2002; pp 1–29.
83. Chamberlain, P. L. Doramectin. *WHO Food Addit. Ser.* **2002,** *49,* 3–10.
84. Conder, G. A.; Baker, W. J. Chemistry, pharmacology and safety: doramectin and selamectin. In *Macrocyclic Lactones in Antiparasitic Therapy;* Vercruysse, J., Rew, R. S., Eds.; CABI, 2002; pp 30–50.
85. Dybas, R. A. Abamectin use in crop protection. In *Ivermectin and Abamectin;* Campbell, W. C., Ed.; Springer, 1989; pp 287–310.

86. Tusseau, E.; Gillham, M.; Fougeroux, A.; Bernard, J.-L. News on the mode of action of abamectin. *Phytoma* **2005,** *579,* 24–26.
87. Rock, D. W.; DeLay, R. L.; Gliddon, M. J. Chemistry, pharmacology and safety: moxidectin. In *Macrocyclic Lactones in Antiparasitic Therapy;* Vercruysse, J., Rew, R. S., Eds.; CABI, 2002; pp 75–96.
88. Kieran, P. J. Moxidectin against ivermectin-resistant nematodes—a global view. *Aust. Vet. J.* **1994,** *71* (1), 18–20.
89. Wolstenholme, A. J.; Rogers, A. T. Glutamate-gated chloride channels and the mode of action of the avermectin/milbemycin anthelmintics. *Parasitology* **2005,** *131* (Suppl.), S85–S95.
90. Ide, J.; Okazaki, T.; Ono, M.; Saito, A.; Nakagawa, K.; Naito, S.; Sato, K.; Tanaka, K.; Yoshikawa, H.; Ando, M. Milbemycin: discovery and development. *Sankyo Kenkyusho Nenpo* **1993,** *45,* 1–98.
91. Davies, H. G.; Green, R. H. Avermectins and milbemycins. Part I. *Chem. Soc. Rev.* **1991,** *20* (2), 211–269.
92. Davies, H. G.; Green, R. H. Avermectins and milbemycins. Part II. *Chem. Soc. Rev.* **1991,** *20* (2), 271–339.
93. Canga, A. G.; Prieto, A. M. S.; Liebana, M. J. D.; Martinez, N. F.; Vega, M. S.; Vieitez, J. J. G. The pharmacokinetics and interactions of ivermectin in humans—a mini-review. *AAPS J.* **2008,** *10* (1), 42–46.
94. McTier, T. L.; McCall, J. W.; Dzimianski, M. T.; Raynaud, J. P.; Strickland, J. E. Use of melarsomine dihydrochloride (RM 340) for adulticidal treatment of dogs with naturally acquired infections of *Dirofilaria immitis* and for clinical prophylaxis during reexposure for 1 year. *Vet. Parasitol.* **1994,** *55* (3), 221–233.
95. Martin, R. J.; Buxton, S. K.; Neveu, C.; Charvet, C. L.; Robertson, A. P. Emodepside and SL0-1 potassium channels: a review. *Exp. Parasitol.* **2012,** *132* (1), 40–46.
96. Amliwala, K.; Bull, K.; Willson, J.; Harder, A.; Holden-Dye, L.; Walker, R. J. Emodepside, a cyclo-octadepsipeptide anthelmintic with a novel mode of action. *Drugs Future* **2004,** *29* (10), 1015–1024.
97. Xiao, S.-H.; Wu, H.-M.; Tanner, M.; Utzinger, J.; Wang, C. Tribendimidine: a promising, safe and broad-spectrum anthelmintic agent. *Acta Trop.* **2005,** *94* (1), 1–14.
98. Girardi, C.; Pancaldi, P.; Valle, V. C.; Graziano, E. Methyridine in the therapy of trichocephaliasis in dogs. *Ann. Fac. Med. Vet. Torino* **1979,** *26,* 398–405.
99. Foreyt, W. J. Evaluation of clorsulon against immature *Fascioloides magna* in cattle and sheep. *Am. J. Vet. Res.* **1988,** *49* (7), 1004–1006.
100. Mossallam, S. F.; Ali Safia, M.; El Zawawy, L. A.; Said, D. E. The efficacy of antihelminthic compound; clorsulon against experimental *Schistosoma mansoni* infection. *J. Egypt. Soc. Parasitol.* **2007,** *37* (1), 171–188.
101. Telvekar, V. N.; Herlekar, O. P., Process for the preparation of 4-hydroxy-3-iodo-5-nitrobenzonitrile as a fasciolicide, IN 2014MU01443 (2014).
102. Makkar, M. S. Some pharmacological studies on nitroxynil (4-cyano-2-iodo-6-nitrophenol). *Indian J. Anim. Res.* **1974,** *8* (2), 81–83.
103. Davis, M.; Rosenbaum, J.; Wright, D. E. Chemotherapy of fascioliasis. II. 4-Cyano-2-iodo-6-nitrophenol (nitroxynil) and related compounds. *J. Sci. FoodAgric.* **1969,** *20* (12), 748–754.

Chapter 37

Proton Pump Inhibitors

Stomach acid is natural contributor to orderly digestion. But its excess is a menace, inflaming and irritating the esophagus, causing heartburn with possible development of ulcers in the stomach and the duodenum.

In the mid-1970s cimetidine (Tagamet), known as an H_2 blocker, appeared on the market for the treatment of overproduction of stomach acid. It is believed that it was the first blockbuster drug. Other H_2 blockers, such as famotidine (Pepcid) and ranitidine (Zantac) quickly followed this drug, but soon were substituted with a new class of drugs—proton-pump inhibitors, which proved to be superior to the H_2 blockers.

Proton pump inhibitors relieve various problems that occur when stomach acid exceeds the normal level, including a condition called gastroesophageal reflux disease (GERD), a condition when stomach content occasionally flows back into the food pipe (esophagus), which is among the most common disorders of the gastrointestinal tract, and Zollinger-Ellison syndrome, a condition in which tumors, called gastrinomas, form in pancreas or the upper part of small intestine (duodenum), causing overproduction of acid by the stomach.

Proton pump inhibitors represent the most effective treatment option for heartburn, GERD, and Zollinger-Ellison syndrome, relieving symptoms, healing erosions, and maintaining a healed mucosa.

Proton pump inhibitors suppress the secretion of gastric acid by blocking the enzyme hydrogen/potassium adenosine triphosphatase (H+,K+ ATPase), referred as the proton pump. This enzyme causes parietal cells of stomach lining to produce acid.

Proton pump inhibitors are potent inhibitors of gastric acid secretion, and are widely regarded as the agents of choice for the treatment of peptic acid disorders. For patients with upper gastrointestinal symptoms of uncertain etiology, improvement with proton pump inhibitors therapy is considered prima facie evidence. In addition to antisecretory effects, proton pump inhibitors have antioxidant properties and direct effects on neutrophils, monocytes, endothelial, and epithelial cells that might prevent inflammation. Currently, they are the mainstay of medical treatment of acid-related disorders [1-25].

Omeprazole (1979) (**37.1**) was the first proton pump inhibitor, having been marketed in 1988. The S-enantiomer of omeprazole, esomeprazole (2001) (**37.2**), was marketed much later. Omeprazole discovery was followed by discovery of lansoprazole (1995) (**37.3**), pantoprazole (1997) (**37.5**), and rabeprazole (1999)

Synthesis of Best-Seller Drugs. http://dx.doi.org/10.1016/B978-0-12-411492-0.00037-7

(**37.6**). The R-enantiomer, dexlansoprazole (**37.4**), appeared on the market in 2009. These proton pump inhibitors are irreversible proton pump inhibitors and share the core structure of pyridinylmethylsulfinyl benzimidazoles. Among them, their pharmacokinetics and pharmacodynamics, on both speed and degree of gastric acid suppression, differ slightly [26].

Proton pump inhibitors have demonstrated an excellent safety profile after approximately two decades of clinical use. There are some unspecific side effects, such as headache, nausea, and diarrhea [27-30] (Fig. 37.1.).

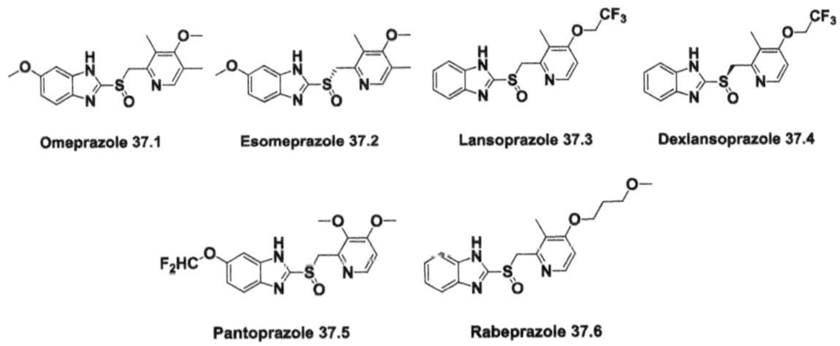

FIG. 37.1 Proton pump inhibitors.

New pump inhibitors drug candidates, ilaprazole (**37.7**) and tenatoprazole (**37.8**), are still in clinical trials.

A promising experimental compound is pyridylmethylsulfinyl thienoimidazole (saviprazole (**37.9**)) is currently undergoing laboratory investigations (Fig. 37.2.).

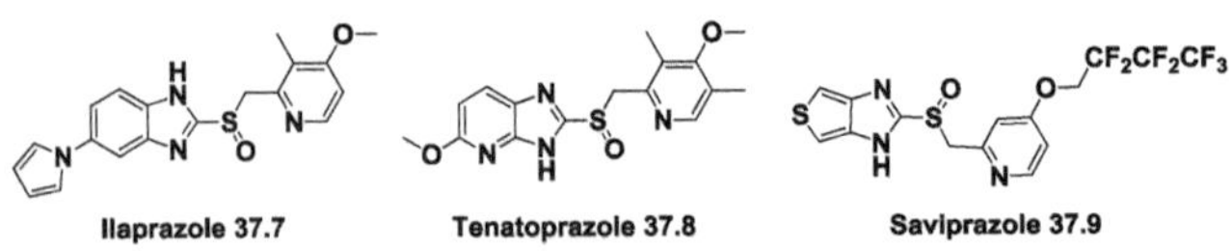

FIG. 37.2 New proton pump inhibitors.

Compound of principally other, structure 2-(pyrimidin-4-yl)-1,2,3,4-tetrahydroisoquinoline, revaprazan (**37.10**), is approved for use in Korea, but not in Europe or the United States (Fig. 37.3.).

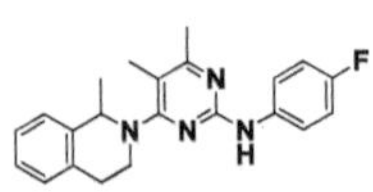

Revaprazan 37.10

FIG. 37.3 Structure of revaprazan.

To overcome some drawbacks associated with the use of irreversible proton pump inhibitors, compounds that are considered reversible inhibitors were synthesized and include the imidazopyridine derivative SCH-28080 (**37.11**), the imidazopyrazin derivative SCH-32651 (**37.12**), and the quinoline derivatives SK&F-97574 (**37.13**) and SK&F-96067 (**37.14**) (Fig. 37.4.).

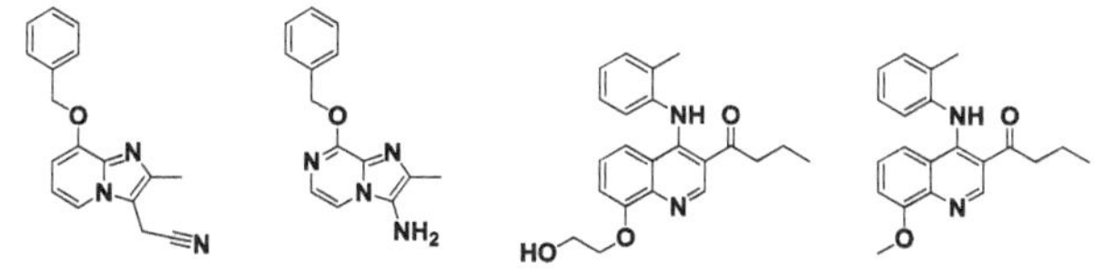

SCH-28080 37.11 SCH-32651 37.12 SK&F-97574 37.13 SK&F-96067 37.14

FIG. 37.4 **New, experimental proton pump inhibitors are imidazopyridine and imidazopyrazin derivatives.**

Proton pump inhibitors are the second most commonly prescribed drug class in the United States and esomeprazole (Nexium) (**37.2**), lansoprazole (Prevacid) (**37.3**), dexlansoprazole (Kapidex, Dexilant) (**37.4**), pantoprazole (Protonix) (**37.5**), and rabeprazole (AcipHex) (**37.6**) are included in the list of Top 200 Drugs by sales for the 2010s.

ESOMEPRAZOLE–NEXIUM

Esomeprazole (S-omeprazole) (**37.2**) is the single optical S-enantiomer of omeprazole (**37.1**), the first drug of the class of proton pump inhibitors.

Proposed syntheses of esomeprazole are based mainly on the reaction sequence for omeprazole followed by separating obtained enantiomers by chromatography in analytical [31], or preparative scale [32,33], or implementing methods of asymmetric oxidation [34].

The methods of synthesizing omeprazole (**37.1**) are very close to each other and mainly based on the first patent approach where its synthesis was disclosed [35] (Scheme 37.1.).

2,3,5-Collidine (**37.15**) was oxidized by hydrogen peroxide in acetic acid to produce the N-oxide (**37.16**). The obtained N-oxide was nitrated in a mixture of nitric acid and sulfuric acids to produce the 4-nitro derivative (**37.17**). The nitro group in (**37.17**) was displaced by methoxy group on heating in MeOH in the presence of NaOH to yield (**37.18**) [36]. The product was heated with Ac_2O, which simultaneously reduced the ring and formed an acetyl derivative (Boekelheide rearrangement) to produce the hydroxymethyl-pyridine acetyl derivative (**37.19**). The corresponding alcohol (**37.20**) was formed by the treatment with NaOH, followed by transformation to chloride-2-chloromethyl-4-methoxy-2,3,5-trimethylpyridine (**37.21**) using thionyl chloride.

Separately, condensing 4-methoxy-o-phenylendiamine (**37.22**) with potassium ethylxantogenate (**37.23**) in the traditional manner synthesized 2-mercapto-5-methoxy benzimidazole (**37.24**).

Reaction of obtained mercapto benzimidazole (**37.24**) with 2-chloromethylpyridine derivative (**37.21**) using NaOH in refluxing water–ethanol mixture, or in phase transfer catalysis conditions (benzene 40% NaOH, tetrabutyl ammonium bromide) produced thioether-5-methoxy-2-[((4-methoxy-3,5-dimethyl-2-pyridinyl)methyl)thio]-1H-benzimidzole (**37.25**) (pyrmetazole), which on oxidation by 3-chloroperbenzoic acid in chloroform or hydrogen peroxide formed the corresponding sulfoxide, omeprazole (**37.1**). Alternative and improved routes for the synthesis of omeprazole [37-45] differ slightly from the "mother" patent [35]. A major drawback of this approach is the problem of incomplete oxidation or overoxidation to sulfone and formation of sulfone N-oxide.

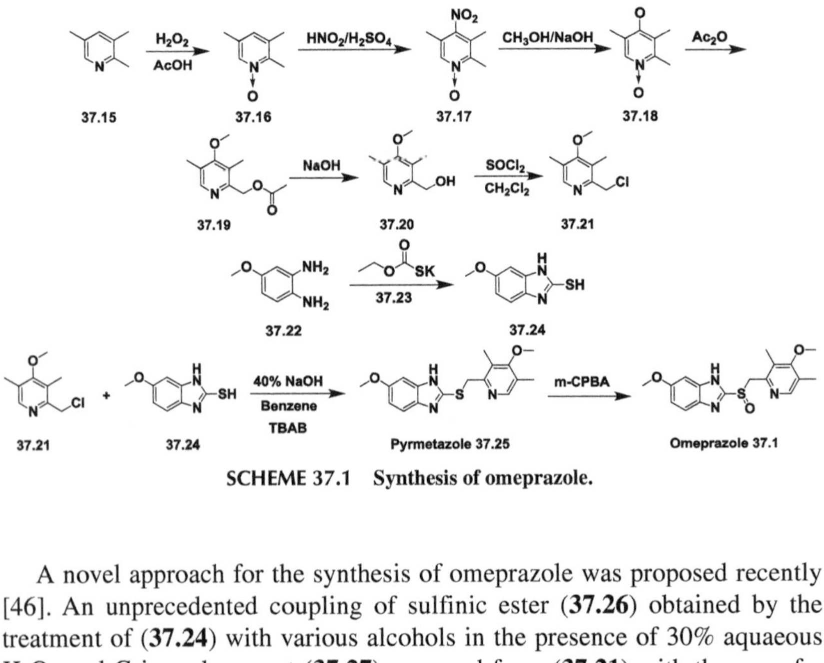

SCHEME 37.1 **Synthesis of omeprazole.**

A novel approach for the synthesis of omeprazole was proposed recently [46]. An unprecedented coupling of sulfinic ester (**37.26**) obtained by the treatment of (**37.24**) with various alcohols in the presence of 30% aqueous H_2O_2 and Grignard reagent (**37.27**) prepared from (**37.21**) with the use of a magnesium–antharacene complex [Mg(antracene)(THF)$_3$] was successfully demonstrated to synthesize omeprazole. The proposed method probably can be utilized for the synthesis of the other members of the "prazole" class of molecules (Scheme 37.2.).

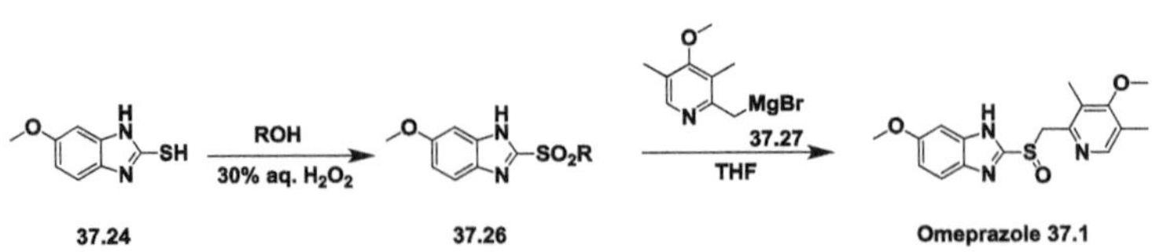

SCHEME 37.2 **Synthesis of omeprazole.**

Resolution of racemates of omeprazole in preparative scale requires either the preparation of fenchyloxymethyl chloride and then reaction with the racemic compound, or introduction of chloromethyl group at the 1-position of benzimidazole ring, followed by reaction with the chiral auxiliary to obtain a diastereomeric mixture. Diastereomers are then separated and desired isomer is liberated from a separated diastereomer [32,33].

A highly efficient synthesis of esomeprazole has been developed [47] based on the Kagan modification [48] of the Sharpless reagent. The synthesis, which is based on a titanium and chiral tartrate diester-mediated asymmetric oxidation of the corresponding prochiral sulphide, is suitable for large-scale production.

Large scale (kilogram) asymmetric synthesis of esomeprazole is described in Scheme 37.3.

Suspension of pyrmeprazole (**37.25**) in toluene was added to a water solution of (S,S)-diethyl tartrate and titanium tetraisopropoxide. After stirring for 50 minutes, N,N-diisopropylethylamine and cumene hydroperoxide were added. After one hour the conversion of omeprazole was 92%, and the enantiomeric excess of crude sulphoxide was greater than 94%. After transformation of the obtained product to sodium salt using concentric sodium hydroxide solution and acetonitrile, esomeprazole sodium as a solid with an enantiomeric excess of 99.5% was obtained. The reaction mechanism is not understood in detail. The role of N,N-diisopropylethylamine in the oxidizing system is also unclear.

A variety of modifications to this asymmetric synthesis of esomeprazole were proposed in literature [49-54].

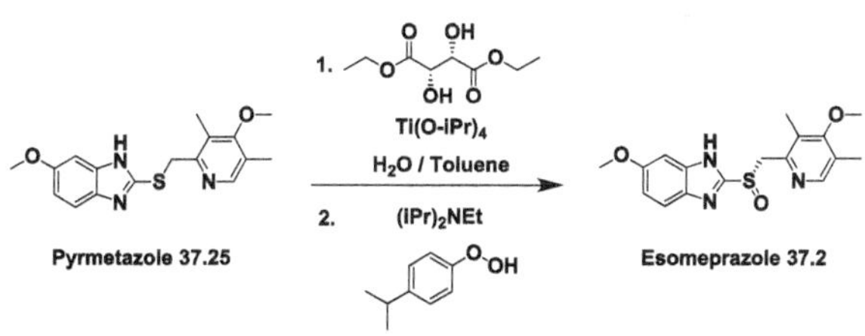

SCHEME 37.3 Asymmetric synthesis of esomeprazole.

Esomeprazole was approved in 2001 for the management of gastroesophageal reflux disease, the prevention and treatment of nonsteroidal antiinflammatory drug–associated gastric ulcer disease, and treatment of duodenal ulcer disease associated with *Helicobacter pylori* infection, as well as for the treatment of Zollinger-Ellison syndrome [55-72].

LANSOPRAZOLE–PREVACID

Lansoprazole (**37.3**) is the second approved gastric acid pump inhibitor. The common approach for the synthesis of lansoprazole involves coupling of mercapto-benzimidazole (**37.24**) with a new 2-chloromethylpyridine derivative

(**37.32**) followed by oxidation of the prochiral sulfide group with m-chloroperbenzoic acid or hydrogen peroxide was first disclosed by Nohara and Maki [73], with followed improvements in patents [74-78] and briefly summed up in papers [79-80].

Lansoprazole synthesis is represented on the Scheme 37.4.

In principle it repeats the synthesis Scheme of omeprazole, differing in details and characteristics, for example, in place of 2,3,5-collidine (**37.15**) as a starting material, 2,3-lutidine (**37.27**) was selected, and the methoxy group in the fourth position of pyridine ring was replaced by the 2,2,2-trifluoroethoxy group.

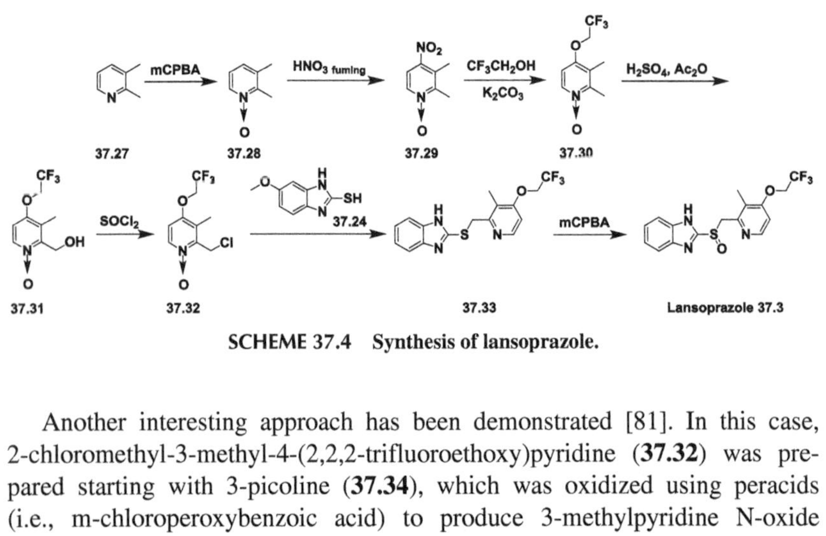

SCHEME 37.4 Synthesis of lansoprazole.

Another interesting approach has been demonstrated [81]. In this case, 2-chloromethyl-3-methyl-4-(2,2,2-trifluoroethoxy)pyridine (**37.32**) was prepared starting with 3-picoline (**37.34**), which was oxidized using peracids (i.e., m-chloroperoxybenzoic acid) to produce 3-methylpyridine N-oxide (**37.35**). The obtained product was nitrated with fuming nitric acid to produce 3-methyl-4-nitropyridine N-oxide (**37.36**). The prepared N-oxide was treated with dimethylsulfate at 65 to 70°C to form N-methoxypyridinium salt (**37.37**), the aqueous solution of which on cooling was treated with sodium cyanide to produce an after formation of intermediate (**37.38**) and elimination of methanol 2-cyano-3-methyl-4-nitropyridine (**37.39**). This method for the synthesis of 2-cyanopyridines via addition of cyanide ion to N-alkoxy-quaternary salts of pyridines, supplements the plethora of Reissert-Kaufmann reactions in the quinoline and isoquinoline series previously described [82]. The nitro group in (**37.39**) was replaced by the 2,2,2-trifluoroethoxy group by a direct reaction with sodium trifluoroethoxide in trifluoroethanol that produced ether (**37.40**). The next step—transformation of nitrile group in prepared 2-cyanopyridine (**37.40**) to 2-carboxypyridine (**37.41**)—was carried out in a one-pot procedure by heating the 2-cyano compound in the presence of concentrated sulfuric acid followed by reaction of the intermediate amide with sodium nitrite under aqueous acidic conditions [83,84]. The obtained acid was esterified in methanol with a catalytic amount of sulfuric acid to produce ester (**37.42**). The ester (**37.42**) was reduced by $NaBH_4$, producing the above-described 2-hydroxymethyl- pyridine derivative (**37.31**) followed by a reaction with thionyl chloride in dioxane that produced the required 2-chloromethylpyridine compound (**37.32**). Direct reaction of the last with 2-mercaptobenzimidazole (**37.34**) in methanol, even without use of any base, produced a sulfide (**37.33**) in high yield. The oxidation of the last to lansoprazole (**37.3**) has been carried out by various oxidants and catalysts, which, together with the desired sulfoxide, produced a certain amount of overoxidized product. Oxidizing sulfide (**37.33**) with a new oxidation method made up of the use of the composite metal oxide catalyst, $LiNbMoO_6$, in methanol and 35% H_2O_2 as an oxidant sulfide (**37.33**) was successfully oxidized to desired lansoprazole (**37.3**) (Scheme 37.5.).

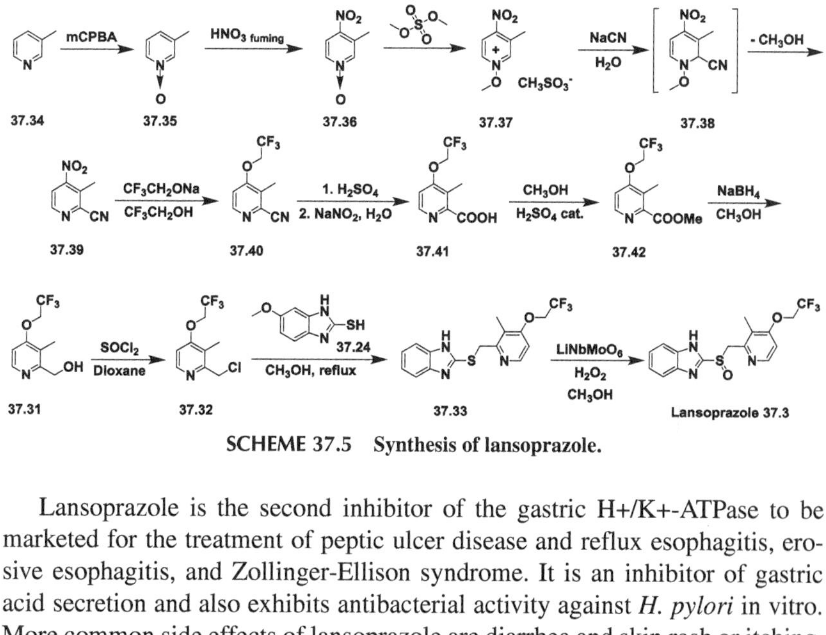

SCHEME 37.5 **Synthesis of lansoprazole.**

Lansoprazole is the second inhibitor of the gastric H+/K+-ATPase to be marketed for the treatment of peptic ulcer disease and reflux esophagitis, erosive esophagitis, and Zollinger-Ellison syndrome. It is an inhibitor of gastric acid secretion and also exhibits antibacterial activity against *H. pylori* in vitro. More common side effects of lansoprazole are diarrhea and skin rash or itching. Less-common side effects are abdominal pain, joint pain, nausea, vomiting, and increased or decreased appetite [85-91].

DEXLANSOPRAZOLE–KAPIDEX, DEXILANT

Dexlansoprazole (**37.4**) is the R-enantiomer, the dextrorotatory enantiomer, of lansoprazole with better pharmacokinetic characteristics, including a longer mean residence time and much smaller elimination rate.

It is indicated for healing all grades of esophagitis, maintaining the healing of erosive esophagitis and treating heartburn associated with nonerosive gastroesophageal reflux disease [92,93].

Patents [94-96] and an article [97] disclose the synthesis of dexlansoprazole (**37.4**) by stereoselective oxidation, according to the Kagan-Modena method [48], with an oxidizing agent (hydrogen peroxide, tert-butyl hydroperoxide, cumene hydroperoxide, etc.) in the presence of a catalyst for asymmetry induction (titanium isopropoxide).

It is a complex reaction mainly when carried out on industrial scale and, generally, brings to the formation of a mixture of products comprising the sulfoxide of formula having the desired stereochemistry, the enantiomer thereof, variable quantities of the staring material, and the superoxidation product. But the method remains the most straight and convenient approach for the synthesis of chiral sulfoxides.

Still there is a need for a process that can be used in large-scale production for the preparation of the single enantiomers of prazoles.

All known methods of optical resolution (fractional recrystallization, chiral column method, diastereomer method, etc.) were implemented for separation of R-enantiomer of lansoprazole–dexlansoprazole [98-100].

Surprisingly, the racemates of prazoles and, in particular, of dexlansoprazole are very selectively precipitated from a solvent yielding the single enantiomers with an enhanced optical purity. Many patents describe methods for separation of (+)- and (-)-enantiomers of dexlansoprazole using different solvents and solvent mixtures [101-104] (Fig. 37.5.).

(It is curious and controversial that in the case of omeprazole, the S-enantiomer esomeprazole is considered more effective than the mixture, and in the case of lansoprazole, the R-enantiomer is considered more effective.)

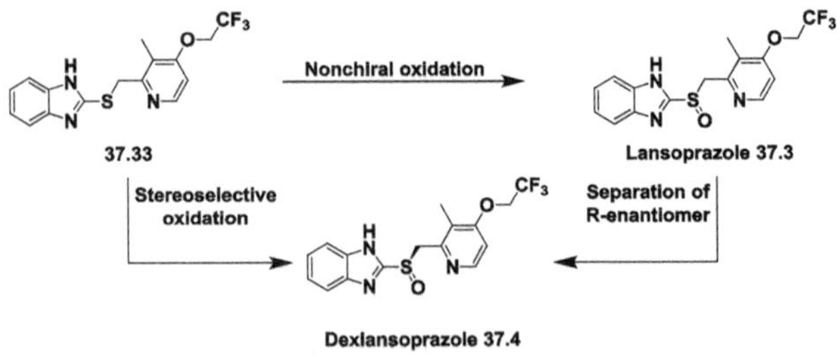

FIG. 37.5 **The methods that can be used for the preparation of dexlansoprazole.**

PANTOPRAZOLE–PROTONIX

Pantoprazole is the third proton pump inhibitor to be launched for the treatment of peptic acid diseases. Like other drugs in this class, pantoprazole binds irreversibly and specifically to the proton pump, thereby reducing gastric acid secretion and causing long-lasting inhibition of acid secretion by inactivating the parietal cell H+/K+-ATPase. Pantoprazole more rapidly heals ulcer, and has far greater efficacy than omeprazole and ranitidinein esophageal reflux disease.

Pantoprazole appears to have a lower potential for interactions with other medications, and has been used in more than 100 different countries worldwide [105-113].

Pantoprazole (**37.5**) was synthesized according to the same methods used for the synthesis of omeprazole and pantoprazole, that is, by condensation of 2-(chloromethyl)-3,4-dimethoxypyridine (**37.40**) and 5-(difluoromethoxy)-2-mercapto-1H-benzimidazole (**37.41**) [114-116].

The known sequence of reactions is presented in Fig. 37.6.

3-Methoxy-2-methylpyridine (**37.34**) was consequentially oxidized to N-oxide (**37.35**), nitrated in the fourth position (**37.36**), the nitro group was replaced by the methoxy group (**37.37**), isomerized (Boekelheide rearrangement) (**37.38**) to produce after hydrolysis 2-(hydroxy)-3,4-dimethoxypyridine (**37.39**), which was transformed to one of the key starting materials, 2-(chloromethyl)-pyridine (**37.40**). The last was condensed with (difluoromethoxy)-2-mercapto-lH-benzimidazole (**37.41**) to form a 2-((pyridin-2-ylmethyl)thio)-1H-benzo[d] imidazole derivative (**37.42**), which, after oxidation, produced the desired pantoprazole [**37.5**] (Fig. 37.6.).

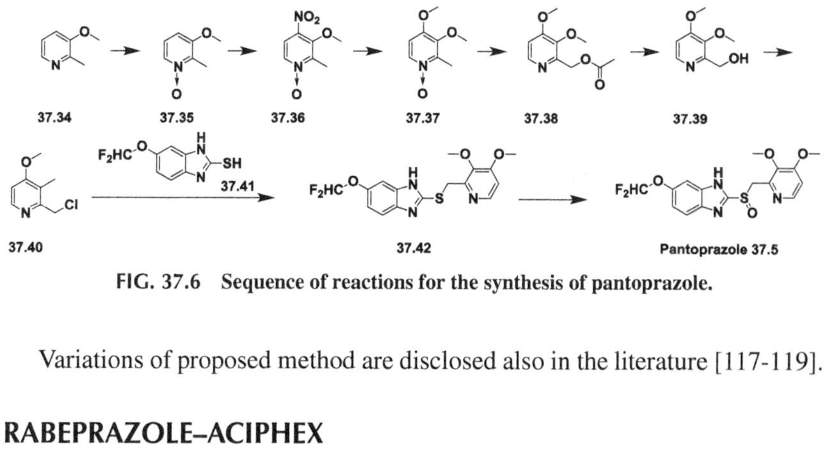

FIG. 37.6 Sequence of reactions for the synthesis of pantoprazole.

Variations of proposed method are disclosed also in the literature [117-119].

RABEPRAZOLE–ACIPHEX

Rabeprazole (**37.6**) is another proton pump inhibitor that is used to treat conditions requiring a reduction of gastric acid secretion, such as erosive or ulcerative gastroesophageal reflux disease, nonerosive reflux disease, duodenal and gastric ulcers, and other pathological hypersecretory conditions, including Zollinger-Ellison syndrome. It is also used as a part of combination therapy for the eradication of *H. pylori*, a pathogen frequently implicated in the development of gastric and duodenal ulcers. Rabeprazole is a more potent inhibitor of H+,K+-ATPase and acid secretion, and is a more rapid inhibitor of proton pumps, than omeprazole, lansoprazole, or pantoprazole [120-133].

Synthesis of rabeprazole (**37.6**) differs a little bit from that described above pyridinylmethylsulfinyl benzimidazoles. It starts from

4-nitro-2,3-dimethylpyridine-N-oxide (**37.29**), which on heating with concentrated hydrochloric acid gives 4-chloro-2,3-dimethylpyridine-N-oxide (**37.43**). Reaction of the last with sodium 3-methoxypropan-1-olate in DMSO yields the corresponding 4-alkoxy intermediate (**37.44**). The 4-alkoxy intermediate (**37.44**), on further reaction with acetic anhydride, underwent Boekelheide rearrangement to produce O-acetyl intermediate (**37.45**), which, after alkaline hydrolysis, produced pyridin-2-ylmethanol (**37.46**), which was chlorinated using thionyl choride to furnish a chloro intermediate (**37.46**). Condensation of the obtained compound with 2-mercapto benzimidazole (**37.48**) in the presence of base produced a 2-((pyridin-2-ylmethyl)thio)-1H-benzo[d]imidazole derivative (**37.49**), which oxidized with meta-chloroperbenzoic acid to produce the final rabeprazole (**37.6**) [132-134] (Scheme 37.6.).

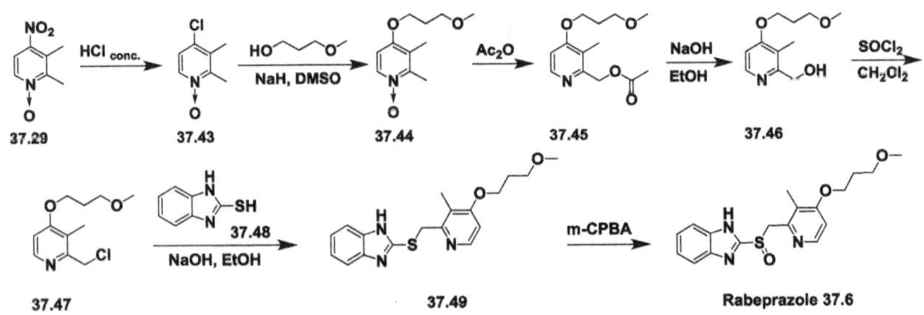

SCHEME 37.6 Synthesis of rabeprazole.

Comparisons of different proton pump inhibitors [135-142] show that they all have similar potency and efficacy. Rabeprazole, however, displays a slightly more rapid onset of acid inhibition than the others. Esomeprazole, the S-isomer of omeprazole, exhibits a somewhat higher potency than the other proton pump inhibitors.

REFERENCES

1. Sachs, G.; Shin, J. M.; Briving, C.; Wallmark, B.; Hersey, S. The pharmacology of the gastric acid pump: the H+,K+ ATPase. *Annu. Rev. Pharmacol. Toxicol.* **1995,** *35,* 277–305.
2. Shi, S.; Klotz, U. Proton pump inhibitors: an update of their clinical use and pharmacokinetics. *Eur. J. Clin. Pharmacol.* **2008,** *64* (10), 935–951.
3. Schubert, M. L.; Peura, D. A. Control of gastric acid secretion in health and disease. *Gastroenterology* **2008,** *134* (7), 1842–1860.
4. Jain, K. S.; Shah, A. K.; Bariwal, J.; Shelke, S. M.; Kale, A. P.; Jagtap, J. R.; Bhosale, A. V. Recent advances in proton pump inhibitors and management of acid-peptic disorders. *Bioorg. Med. Chem.* **2007,** *15* (3), 1181–1205.
5. Mullin, J. M.; Gabello, M.; Murray, L. J.; Farrell, C. P.; Bellows, J.; Wolov, K. R.; Kearney, K. R.; Rudolph, D.; Thornton, J. J. Proton pump inhibitors: actions and reactions. *Drug Discovery Today* **2009,** *14* (13/14), 647–660.

6. Andersson, T.; Weidolf, L. Stereoselective disposition of proton pump inhibitors. *Clin. Drug Invest.* **2008,** *28* (5), 263–279.
7. Campbell, D. R.; Hurwitz, A.; Parkinson, A. Proton pump inhibitors. In *Current Clinical Topics in Gastrointestinal Pharmacology;* Lewis, J. H., Dubois, A., Eds.; Wiley-Blackwell, 1997; pp 303–324.
8. Richardson, P.; Hawkey, C. J.; Stack, W. A. Proton pump inhibitors: pharmacology and rationale for use in gastrointestinal disorders. *Drugs* **1998,** *56* (3), 307–335.
9. Sachs, G. Proton pump inhibitors and acid-related diseases. *Pharmacotherapy* **1997,** *17* (1), 22–37.
10. Huber, R.; Kohl, B.; Sachs, G.; Senn-Bilfinger, J.; Simon, W. A.; Sturm, E. Review article: The continuing development of proton-pump inhibitors with particular reference to pantoprazole. *Aliment. Pharmacol. Ther.* **1995,** *9* (4), 363–378.
11. Kromer, W.; Horbach, S.; Luhmann, R. Relative efficacies of gastric proton pump inhibitors. Their clinical and pharmacological basis. *Pharmacology* **1999,** *59* (2), 57–77.
12. Scarpignato, C.; Hunt, R. H. Proton pump inhibitors: the beginning of the end or the end of the beginning? *Curr. Opin. Pharmacol.* **2008,** *8* (6), 677–684.
13. Spector, R.; Vesell, E. S. The power of pharmacological sciences: the example of proton pump inhibitors. *Pharmacology* **2006,** *76* (3), 148–156.
14. Vakil, N. Review article: new pharmacological agents for the treatment of gastro-esophageal reflux disease. *Aliment. Pharmacol. Ther.* **2004,** *19* (10), 1041–1049.
15. Robinson, M.; Horn, J. Clinical pharmacology of proton pump inhibitors: what the practicing physician needs to know. *Drugs* **2003,** *63* (24), 2739–2754.
16. Furuta, T.; Shirai, N.; Sugimoto, M.; Ohashi, K.; Ishizaki, T. Pharmacogenomics of proton pump inhibitors. *Pharmacogenomics* **2004,** *5* (2), 181–202.
17. Chong, E.; Ensom, M. H. H. Pharmacogenetics of the proton pump inhibitors: a systematic review. *Pharmacotherapy* **2003,** *23* (4), 460–471.
18. McColl, K. E. L.; Kennerley, P. Proton pump inhibitors—differences emerge in hepatic metabolism. *Dig. Liver Dis.* **2002,** *34* (7), 461–467.
19. Robinson, M. Proton pump inhibitors: update on their role in acid-related gastrointestinal diseases. *Int. J. Clin. Pract.* **2005,** *59* (6), 709–715.
20. Edwards, S. J.; Lind, T.; Lundell, L. Systematic review: proton pump inhibitors (PPIs) for the healing of reflux esophagitis-a comparison of esomeprazole with other PPIs. *Aliment. Pharmacol. Ther.* **2006,** *24* (5), 743–750.
21. Sachs, G.; Shin, J. M.; Howden, C. W. The clinical pharmacology of proton pump inhibitors. *Aliment. Pharmacol. Ther.* **2006,** *23* (Suppl. 2), 2–8.
22. Fass, R.; Shapiro, M.; Dekel, R.; Sewell, J. Systematic review: proton-pump inhibitor failure in gastro-oesophageal reflux disease-where next? *Aliment. Pharmacol. Ther.* **2005,** *22* (2), 79–94.
23. Kahrilas, P. J.; Boeckxstaens, G. Failure of reflux inhibitors in clinical trials: bad drugs or wrong patients? *Postgrad. Med. J.* **2013,** *89* (1048), 111–119.
24. Laine, L.; Hennekens, C. Proton pump inhibitor and clopidogrel interaction: fact or fiction? *Am. J. Gastroenterol.* **2010,** *105* (1), 34–41.
25. Vakil, N. Prescribing proton pump inhibitors: is it time to pause and rethink? *Drugs* **2012,** *72* (4), 437–445.
26. Horn, J. The proton-pump inhibitors: similarities and differences. *Clin. Ther.* **2000,** *22* (3), 266–280.
27. Abraham, N. S. Proton pump inhibitors: potential adverse effects. *Curr. Opin. Gastroenterol.* **2012,** *28* (6), 615–620.

28. Lodato, F.; Azzaroli, F.; Turco, L.; Mazzella, N.; Buonfiglioli, F.; Zoli, M.; Mazzella, G. Adverse effects of proton pump inhibitors. *Best Pract. Res., Clin. Gastroenterol.* **2010,** *24* (2), 193–201.
29. Targownik, L. E. Another bad break for proton-pump inhibitors? *Nat. Rev. Rheumatol.* **2009,** *5* (9), 478–480.
30. Kuipers, E. J. Proton pump inhibitors and gastric neoplasia. *Gut* **2006,** *55* (9), 1217–1221.
31. Erlandsson, P.; Isaksson, R.; Lorentzon, P.; Lindberg, P. Resolution of the enantiomers of omeprazole and some of its analogs by liquid chromatography on a trisphenylcarbamoyl cellulose-based stationary phase. The effect of the enantiomers of omeprazole on gastric glands. *J. Chromatogr.* **1990,** *532* (2), 305–319.
32. Kohl, B.; Senn-Bilfinger, J., Enantiomerically pure (pyridylmethylsulfinyl)benzimidazoles useful as drugs, and their preparation from racemates, DE 4035455 (1992).
33. Lindberg, P. L.; Von Unge, S., Process for the preparation of optically pure crystalline salts of omeprazole, WO 9427988 (1994).
34. Stepankova, H.; Zezula, J.; Hajicek, J.; Kral, V., Process for preparation of esomeprazole via asymmetric oxidation using hydroperoxide and a chiral metal complex of lactic acid, WO 2010091652 (2010).
35. Junggren, U. K.; Sjostrand, S. E., Substituted pyridylsulfinylbenzimidazoles having gastric acid secretion properties, EP 5129 (1979).
36. Braendstroem, A. E.; Lamm, B. R., 3,5-Dimethyl-4-methoxypyridine 1-oxides, EP 103553 (1984).
37. Braendstroem, A. E., Method for synthesis of omeprazole, WO 9118895 (1991).
38. Smahovsky, V.; Oremus, V.; Heleyova, K.; Zlatoidsky, P.; Gattnar, O.; Varga, I.; Stalmach, V.; Jezek, L., Method of omeprazole preparation, WO 9809962 (1998).
39. Bekhazi, M.; Zoghbi, M., Synthesis of omeprazole-type pyridine derivatives via 1,4-dihydropyridine intermediates, WO 9729103 (1997).
40. Zoghbi, M.; Chen, L., Synthesis of pyridine derivatives useful as pharmaceutical intermediates under free radical conditions, WO 9850361 (1998).
41. Gustavsson, A.; Kallstrom, A., Method for the synthesis of a benzimidazole compound, WO 9722603 (1997).
42. Clausen, F. P.; Mccluskey, K. K.; Preikschat, H. F.; Pedersen, S. B., Process for the preparation of 2-[[(2-pyridinyl)methyl]sulfinyl]-1H-benzimidazoles, WO 9840378 (1998).
43. Kato, M.; Toyoshima, Y.; Iwano, N., Production of 2-(2-pyridylmethylsulfinyl)benzimidazole as ulcer inhibitors via S-oxidation using hydrogen peroxide and vanadium catalysts, EP 302720 (1989).
44. McManus, J. W.; Anousis, N.; Banks, B. N.; Liu, H.; Zhou, L., Preparation of omeprazole from pyrmetazole, US 6191148 (2001).
45. Prasad, K., Durga Intermediates and an improved process for the preparation of omeprazole, US 6303787 (2001).
46. Bhalerao, D. S.; Kondaiah, G. C. M.; Dwivedi, N.; Mylavarappu, R. K.; Reddy, L. A.; Roy, A.; Nagaraju, G.; Reddy, P. P.; Bhattacharya, A.; Bandichhor, R. Novel approach to the synthesis of omeprazole: an antipeptic ulcer agent. *Synth. Commun.* **2010,** *40* (20), 2983–2987.
47. Cotton, H.; Elebring, T.; Larsson, M.; Li, L.; Sorensen, H.; von Unge, S. Asymmetric synthesis of esomeprazole. *Tetrahedron: Asymmetry* **2000,** *11* (18), 3819–3825.
48. Pitchen, P.; Duiiach, E.; Deshmukh, M. N.; Kagan, H. B. An efficient asymmetric oxidation of sulfides to sulfoxides. *J. Am. Chem. Soc.* **1984,** *106,* 8188–8193.
49. Julien Legros, J.; Dehli, J. R.; Bolm, C. Applications of catalytic asymmetric sulfide oxidations to the syntheses of biologically active sulfoxides. *Adv. Synth. Catal.* **2005,** *347,* 19–31.

50. Khomenko, T. M.; Volcho, K. P.; Komarova, N. I.; Salakhutdinov, N. F. An efficient procedure for the synthesis of esomeprazole using a titanium complex with two chiral ligands. *Russ. J. Org. Chem.* **2008,** *44* (1), 124–127.
51. Song, J. J.; Frutos, R. P.; Tampone, T.; Senanayake, C. H.; Krishnamurthy, D. Industrial applications of asymmetric synthesis: asymmetric synthesis as an enabler of green chemistry. In Carreira, E. M., Yamamoto, H., Eds.; *Comprehensive Chirality*, Vol. 9; Elsevier Science, 2012; pp 46–72.
52. Li, H.-Y.; Liu, R.; Behrens, C.; Ni, C.-Y. Industrial application of chiral technologies. In *Chiral Drugs: Chemistry and Biological Action;* Lin, G.-Q., You, Q.-D., Cheng, J.-F., Eds.; Wiley, 2011; pp 253–296.
53. Federsel, H.-J.; Larsson, M. An innovative asymmetric sulfide oxidation: the process development history behind the new antiulcer agent esomeprazole. In *Asymmetric Catalysis on Industrial Scale: Challenges, Approaches and Solutions;* Blaser, H.-U., Schmidt, E., Eds.; Wiley-VCH, 2004; pp 413–436.
54. Li, Z.; Kong, X.; Mai, W.; Sun, G.; Zhao, S. In *Synthesis of esomeprazole through asymmetric oxidation*, Vol. 881-883; Advanced Materials Research: Durnten-Zurich, Switzerland, 2014; pp 351–355. Issue (Chemical, Material and Metallurgical Engineering III).
55. Spencer, C. M.; Faulds, D. Esomeprazole. *Drugs* **2000,** *60* (2), 321–329.
56. Graul, A.; Castaner, R.; Castaner, J. Esomeprazole magnesium. *Drugs Future* **1999,** *24* (11), 1178–1183.
57. Scott, L. J.; Dunn, C. J.; Mallarkey, G.; Sharpe, M. Esomeprazole: a review of its use in the management of acid-related disorders. *Drugs* **2002,** *62* (10), 1503–1538.
58. Edwards, S. J.; Lind, T.; Lundell, L. Systematic review: proton pump inhibitors (PPIs) for the healing of reflux esophagitis—a comparison of esomeprazole with other PPIs. *Aliment. Pharmacol. Ther.* **2006,** *24* (5), 743–750.
59. Lindberg, P.; Keeling, D.; Fryklund, J.; Andersson, T.; Lundborg, P.; Carlsson, E. Review article: esomeprazole-enhanced bio-availability, specificity for the proton pump and inhibition of acid secretion. *Aliment. Pharmacol. Ther.* **2003,** *17* (4), 481–488.
60. McKeage, K.; Blick, S. K. A.; Croxtall, J. D.; Lyseng-Williamson, K. A.; Keating, G. M. Esomeprazole: a review of its use in the management of gastric acid-related diseases in adults. *Drugs* **2008,** *68* (11), 1571–1607.
61. Johnson, D. A. Review of esomeprazole in the treatment of acid disorders. *Expert Opin. Pharmacother.* **2003,** *4* (2), 253–264.
62. Kale-Pradhan, P. B.; Landry, H. K.; Sypula, W. T. Esomeprazole for acid peptic disorders. *Ann. Pharmacother.* **2002,** *36* (4), 655–663.
63. Johnson, T. J.; Hedge, D. D. Esomeprazole: a clinical review. *Am. J. Health-Syst. Pharm.* **2002,** *59* (14), 655–663.
64. Vachhani, R.; Olds, G.; Velanovich, V. Esomeprazole: a proton pump inhibitor. *Expert Rev. Gastroenterol. Hepatol.* **2009,** *3* (1), 15–27.
65. Kulkarni, S.; Tripathi, S.; Mehta, P. D.; Lodhi, N. S.; Sengar, N. P. S. Esomeprazole in the treatment of acidic disorder: an overview. *Asian J. Biochem. Pharm. Res.* **2011,** *1* (2), 562–566.
66. Al-Judaibi, B.; Chande, N.; Dresser, G. K.; Sultan, N.; Gregor, J. C. Gastric acid-related diseases: focus on esomeprazole. *Clin. Med. Insights: Ther.* **2010,** *2,* 439–452.
67. Dent, J. Review article: pharmacology of esomeprazole and comparisons with omeprazole. *Aliment. Pharmacol. Ther.* **2003,** *17* (Suppl. 1), 5–9.
68. Saccar, C. L. The pharmacology of esomeprazole and its role in gastric acid related diseases. *Expert Opin. Drug Metab. Toxicol.* **2009,** *5* (9), 1113–1124.

69. Kalaitzakis, E.; Bjoernsson, E. A review of esomeprazole in the treatment of gastroesophageal reflux disease (GERD). *Ther. Clin. Risk Manage.* **2007,** *3* (4), 653–663.
70. Lindberg, P.; Carlsson, E. Esomeprazole in the framework of proton-pump inhibitor development. In *Analogue-Based Drug Discovery;* Fischer, J., Ganellin, C. R., Eds.; Wiley-VCH, 2006; pp 81–113.
71. Andersson, T.; Hassan-Alin, M.; Hasselgren, G.; Rohss, K.; Weidolf, L. Pharmacokinetic studies with esomeprazole, the (S)-isomer of omeprazole. *Clin. Pharmacokinet.* **2001,** *40* (6), 411–426.
72. Olbe, L.; Carlsson, E.; Lindberg, P. A proton-pump inhibitor expedition: the case histories of omeprazole and esomeprazole. *Nat. Rev. Drug Discovery* **2003,** *2* (2), 132–139.
73. Nohara, A.; Maki, Y., (Pyridylmethylthio)benzimidazoles and their sulfoxides, EP 174726 (1986).
74. Nohara, A.; Maki, Y., Preparation of 2-[(2-pyridylmethyl)thio or -sulfinyl]benzimidazoles as antiulcer agents, US 4689333 (1987).
75. Bosch R. A.; Dalmases B., P.; Marquillas O. F.; Caldero Ges, J. M., Process for preparation of the antiulcer agent 2-[[[3-methyl-4-(2,2,2-trifluoroethoxy)-2-pyridinyl]methyl]sulfinyl]-1H-benzimidazole [lansoprazole], ES 2023609 (1992).
76. Buxade, V. A., New process for the synthesis of a 2-(2-pyridylmethylsulfinyl)benzimidazole derivative [lansoprazole], and new intermediates prepared in the process, ES 2060541 (1994).
77. Kato, M.; Toyoshima, Y.; Iwano, N., Production of 2-(2-pyridylmethylsulfinyl)benzimidazole as ulcer inhibitors via S-oxidation using hydrogen peroxide and vanadium catalysts, EP 302720 (1989).
78. Kohl, B.; Sturm, E.; Klemm, K.; Riedel, R.; Figala, V.; Rainer, G.; Schaefer, H.; Senn-Bilfinger, J., [[(Dialkoxypyridyl)methyl]thio]benzimidazoles, EP 166287 (1986).
79. Kubo, K.; Oda, K.; Kaneko, T.; Satoh, H.; Nohara, A. Synthesis of 2-[[(4-fluoroalkoxy-2-pyridyl)methyl]sulfinyl]-1H-benzimidazoles as antiulcer agents. *Chem. Pharm. Bull.* **1990,** *38* (10), 2853–2858.
80. Gangula, S.; Elati, C. R.; Neredla, A.; Baddam, S. R.; Neelam, U. K.; Bandichhor, R.; Dongamanti, A. An improved process for the production of lansoprazole: investigation of key parameters that influence the water content in final API. *Org. Process Res. Dev.* **2010,** *14* (1), 229–233.
81. Ahn, K.-H.; Kim, H.; Kim, J. R.; Jeong, S. C.; Kang, T. S.; Shin, H. T.; Lim, G. J. A new synthetic process of lansoprazole. *Bull. Korean Chem. Soc.* **2002,** *23* (4), 626–628.
82. Matsumura, E.; Ariga, M.; Ohfuji, T. Reissert-Kaufmann-type reaction of 4-nitropyridine N-oxide and its homologs. *Bull. Chem. Soc. Jpn.* **1970,** *43* (10), 3210–3214.
83. Sarel, S.; Newman, M. S. Synthesis of branched primary and secondary alkyl acetates. *J. Am. Chem. Soc.* **1956,** *78,* 5416–5420.
84. Tsia, L.; Miwa, T.; Newman, M. S. Steric effects in hydrolysis of hindered amides and nitriles. *J. Am. Chem. Soc.* **1957,** *79,* 2530–2533.
85. Spencer, C. M.; Faulds, D. Lansoprazole. A reappraisal of its pharmacodynamic and pharmacokinetic properties, and its therapeutic efficacy in acid-related. *Drugs* **1994,** *48* (3), 404–430.
86. Barradell, L. B.; Faulds, D.; McTavish, D. Lansoprazole: a review of its pharmacodynamic and pharmacokinetic properties and its therapeutic efficacy in acid-related. *Drugs* **1992,** *44* (2), 225–250.
87. Matheson, A. J.; Jarvis, B. Lansoprazole: an update of its place in the management of acid-related disorders. *Drugs* **2001,** *61* (12), 1801–1833.
88. Langtry, H. D.; Wilde, M. I. Lansoprazole: an update of its pharmacological properties and clinical efficacy in the management of acid-related disorders. *Drugs* **1997,** *54* (3), 473–500.

89. Garnett, W. R. Lansoprazole: a proton pump inhibitor. *Ann. Pharmacother.* **1996,** *30* (12), 1425–1436.
90. Satoh, H. Discovery of lansoprazole and its unique pharmacological properties independent from anti-secretory activity. *Curr. Pharm. Des.* **2013,** *19* (1), 67–75.
91. Naito, Y.; Takagi, T.; Yoshikawa, T. Lansoprazole, a proton pump inhibitor, to reduce gastrointestinal inflammation via heme oxygenase-1 induction. *Mol. Cell. Pharmacol.* **2010,** *2* (2), 53–60.
92. Fock, K. M.; Ang, T. L. Dexlansoprazole, a modified release formulation of an enantiomer of lansoprazole, for the treatment of reflux esophagitis. *Curr. Opin. Invest. Drugs (BioMed Cent.)* **2008,** *9* (10), 1109–1115.
93. Hershcovici, T.; Jha, L. K.; Fass, R. Dexlansoprazole MR: a review. *Ann. Med.* **2011,** *43* (5), 366–374.
94. Larsson, M. E.; Stenhede, U. J.; Sorensen, H.; Von Unge, S. P. O.; Cotton, H. K., Preparation of unsymmetry heterocyclylsulfoxide derivatives for treating gastrointestinal disorders, US 5948789 (1999).
95. Hashimoto, H.; Urai, T., Process for producing optically active pyridylmethylsulfinylbenzimidazole derivatives, WO 2001083473 (2001).
96. Attolino, E.; Lucchini, V., Process for the preparation of dexlansoprazole, US 20100125143 (2010).
97. Raju, M. N.; Kumar, N. U.; Reddy, B. S.; Anitha, N.; Srinivas, G.; Bhattacharya, A.; Mukkanti, K.; Kolla, N.; Bandichhor, R. An efficient synthesis of dexlansoprazole employing asymmetric oxidation strategy. *Tetrahedron Lett.* **2011,** *52* (42), 5464–5466.
98. Von Unge, S., Optical purification of enantiomerically enriched 2-[(arylmethyl)sulfinyl]benzimidazole derivatives, WO 9702261 (1977).
99. Kohl, B.; Senn-Bilfinger, J., Enantiomerically pure (pyridylmethylsulfinyl)benzimidazoles useful as drugs, and their preparation from racemates, DE 4035455 (1992).
100. Satyanarayana, R. M.; Eswaraiah, S.; Venkatesh, M., Process for the preparation of dexlansoprazole via S-oxidation of N-camphorsulfonyl lansoprazole sulfide precursor, WO 2010095144 (2012).
101. Hashimoto, H.; Maruyama, H., Process for producing crystal of optically active 2-[[[3-methyl-4-(2,2,2-trifluoroethoxy)-2-pyridyl]methyl]sulfinyl]benzimidazole, WO 2001087874 (2001).
102. Hashimoto, H.; Urai, T., Process for the crystallization of (R)- or (S)-lansoprazole, WO 2002044167 (2002).
103. Fujishima, A.; Aoki, I.; Kamiyama, K., Purification and crystallization of (R)-lansoprazole as antiulcer agent, WO 2000078745 (2000).
104. Vladiskovic, C.; Restelli, A.; Razzetti, G., Process for the preparation of crystalline form of dexlansoprazole, US 20110028728 (2011).
105. Senn-Bilfinger, J.; Sturm, E. The development of a new proton-pump inhibitor: the case history of pantoprazole. In *Analogue-Based Drug Discovery;* Fischer, J., Ganellin, C. R., Eds.; Wiley-VCH, 2006; pp 115–136.
106. Fitton, A.; Wiseman, L. Pantoprazole: a review of its pharmacological properties and therapeutic use in acid-related disorders. *Drugs* **1996,** *51* (3), 460–482.
107. Cheer, S. M.; Prakash, A.; Faulds, D.; Lamb, H. M. Pantoprazole: an update of its pharmacological properties and therapeutic use in the management of acid-related disorders. *Drugs* **2003,** *63* (1), 101–132.
108. Bardhan, K. D. Pantoprazole: a new proton pump inhibitor in the management of upper gastrointestinal disease. *Drugs Today* **1999,** *35* (10), 773–808.
109. Poole, P. Pantoprazole. *Am. J. Health-Syst. Pharm.* **2001,** *58* (11), 999–1008.

110. Mathews, S.; Reid, A.; Tian, C.; Cai, Q. An update on the use of pantoprazole as a treatment for gastroesophageal reflux disease. *Clin. Exp. Gastroenterol.* **2010,** *3,* 11–16.
111. Dias, L. M. Pantoprazole a proton pump inhibitor. *Clin. Drug Invest.* **2009,** *29* (Suppl. 2), 3–12.
112. Beil, W.; Sewing, K.-F.; Kromer, W. Basic aspects of selectivity of pantoprazole and its pharmacological actions. *Drugs Today* **1999,** *35* (10), 753–764.
113. Garner, A.; Fadlallah, H. Pantoprazole: a new and more specific proton pump inhibitor. *Expert Opin. Invest. Drugs* **1997,** *6* (7), 885–893.
114. Kohl, B.; Sturm, E.; Klemm, K.; Riedel, R.; Figala, V.; Rainer, G.; Schaefer, H.; Senn-Bilfinger, J., [[(Dialkoxypyridyl)methyl]thio]benzimidazoles, EP 166287 (1986).
115. Senn-Bilfinger, J.; Schaeffer, H.; Figala, V.; Kurt, K.; Klemm, K.; Hartmann, S.; Figala, V., Fluoroalkoxy compounds, ZA 8403288 (1984).
116. Kohl, B.; Sturm, E.; Senn-Bilfinger, J.; Simon, W. A.; Krueger, U.; Schaefer, H.; Rainer, G.; Figala, V.; Klemm, K. (H+, K+)-ATPase inhibiting 2-[(2-pyridylmethyl)sulfinyl]benzimidazoles. 4. A novel series of dimethoxypyridyl-substituted inhibitors with enhanced selectivity. The selection of pantoprazole as a clinical candidate. *J. Med. Chem.* **1992,** *35* (6), 1049–1057.
117. Zoghbi, M.; Chen, L., Synthesis of pyridine derivatives useful as pharmaceutical intermediates under free radical conditions, WO 9850361 (1998).
118. Bekhazi, M.; Zoghbi, M., Synthesis of omeprazole-type pyridine derivatives via 1,4-dihydropyridine intermediates, WO 9729103 (1997).
119. Palomo, N., Francisco, E.; Pastor Del Castillo, A.; Molina, P. A., Process for preparing 2-(2-pyridylmethyl)-sulfinyl-1H-benzimidazoles and the intermediate compounds used therein, EP 1992619 (2008).
120. Dadabhai, A.; Friedenberg, F. K. Rabeprazole: a pharmacologic and clinical review for acid-related disorders. *Expert Opin. Drug Saf.* **2009,** *8* (1), 119–126.
121. Pace, F.; Pallotta, S.; Casalini, S.; Porro, G. B. A review of rabeprazole in the treatment of acid-related diseases. *Ther. Clin. Risk Manage.* **2007,** *3* (3), 363–379.
122. Lim, P. W. Y.; Goh, K. L. Review article: efficacy and safety of rabeprazole in treating gastro-esophageal reflux disease. *J. Gastroenterol. Hepatol.* **2004,** *19* (Suppl. 3), S61–S68.
123. Fuhr, U.; Jetter, A. Rabeprazole: pharmacokinetics and pharmacokinetic drug interactions. *Pharmazie* **2002,** *57* (9), 595–601.
124. Williams, M. P.; Pounder, R. E. The pharmacology of rabeprazole. *Aliment. Pharmacol. Ther* **1999,** *13* (Suppl), 3–10.
125. Thjodleifsson, B. Review of rabeprazole in the treatment of gastro-oesophageal reflux disease. *Expert Opin. Pharmacother.* **2004,** *5* (1), 137–149.
126. Prakash, A.; Faulds, D. Rabeprazole. *Drugs* **1998,** *55* (2), 261–267.
127. Langtry, H. D.; Markham, A. Rabeprazole. A review of its use in acid-related gastrointestinal disorders. *Drugs* **1999,** *58* (4), 725–742.
128. Carswell, C. I.; Goa, K. L. Rabeprazole: an update of its use in acid-related disorders. *Drugs* **2001,** *61* (15), 2327–2356.
129. Madan, M. R. M.; Mahidhar, R. D.; Jyothirmai, K.; Rambabu, P.; Anka, R. E.; Revathi, S.; Dayasagar, B. A novel review on anti-intestinal proton pump inhibitor: rabeprazole. *PHARMANEST* **2013,** *4* (1), 66–75.
130. Marelli, S.; Pace, F. Rabeprazole for the treatment of acid-related disorders. *Expert Rev. Gastroenterol. Hepatol.* **2012,** *6* (4), 423–435.
131. Baldwin, C. M.; Keam, S. J. Rabeprazole: a review of its use in the management of gastric acid-related diseases in adults. *Drugs* **2009,** *69* (10), 1373–1401.

132. Souda, S.; Ueda, N.; Miyazawa, S.; Tagami, K.; Nomoto, S.; Okita, M.; Shimomura, N.; Kaneko, T.; Fujimoto, M.; Murakami, M.; Oketani, K.; Fijisaki, H.; Shibata, H.; Wakabayashi, T., Preparation of 2-[(4-alkoxypyrid-2-yl)methylthio]benzimidazoles, -benzothiazoles, and -benzoxazoles as ulcer inhibitors, EP295603, (1988).
133. Souda, S.; Miyazawa, S.; Ueda, N.; Tagami, K.; Nomoto, S.; Okita, M.; Shimomura, N.; Kaneko, T.; Fujimoto, M.; Murakami, M., Preparation of pyridiniobenzimidazoles and analogs as antiulcer agents, WO 8910927 (1989).
134. Reddy, P. R.; Himabindu, V.; Jaydeepkumar, L.; Reddy, G. M.; Kumar, J. V.; Reddy, G. M. An improved process for the production of rabeprazole sodium substantially free from the impurities. *Org. Process Res. Dev.* **2009,** *13* (5), 896–899.
135. Sachs, G.; Shin, J. M.; Briving, C.; Wallmark, B.; Hersey, S. The pharmacology of the gastric acid pump: the H+,K+ ATPase. *Annu. Rev. Pharmacol. Toxicol.* **1995,** *35,* 277–305.
136. Edwards, S. J.; Lind, T.; Lundell, L. Systematic review: proton pump inhibitors (PPIs) for the healing of reflux esophagitis—a comparison of esomeprazole with other PPIs. *Aliment. Pharmacol. Ther.* **2006,** *24* (5), 743–750.
137. Blume, H.; Donath, F.; Warnke, A.; Schug, B. S. Pharmacokinetic drug interaction profiles of proton pump inhibitors. *Drug Saf.* **2006,** *29* (9), 769–784.
138. Hellstroem, P. M.; Vitols, S. The choice of proton pump inhibitor: does it matter? *Basic Clin. Pharmacol. Toxicol.* **2004,** *94* (3), 106–111.
139. Vakil, N.; Fennerty, M. B. Systematic review: direct comparative trials of the efficacy of proton pump inhibitors in the management of gastro-oesophageal reflux disease and peptic ulcer disease. *Aliment. Pharmacol. Ther.* **2003,** *18* (6), 559–568.
140. Sachs, G. Proton pump inhibitors and acid-related diseases. *Pharmacotherapy* **1997,** *17* (1), 22–37.
141. Robinson, M.; Horn, J. Clinical pharmacology of proton pump inhibitors: what the practicing physician needs to know. *Drugs* **2003,** *63* (24), 2739–2754.
142. Stedman, C. A. M.; Barclay, M. L. Comparison of the pharmacokinetics, acid suppression and efficacy of proton pump inhibitors. *Aliment. Pharmacol. Ther.* **2000,** *14* (8), 963–978.

Chapter 38

Drugs for Treatment of Erectile Dysfunction

Erectile dysfunction is defined as the persistent inability to achieve or maintain penile erection sufficient for satisfactory sexual performance [1]. In other words, erectile dysfunction, commonly known as impotence, is the inability to get and keep an erection firm enough for sex.

Impotence, in the course of the history, has been attributed to the mental pathology of anxiety. A myriad of herbal medications, natural aphrodisiacs, have been used since antiquity to enhance "sexual strength" in many cultures. However, in most cases, the real boundary between myth and reality is yet to be established by scientific methods.

There are several etiologies for erectile dysfunction, including vasculogenic, neurogenic, psychogenic, and hormonal. Arterial insufficiency, alcoholism, venosinusoidal, neurological, and endocrine- and drug-related disorders are known to cause erectile dysfunction.

Erectile dysfunction is a disorder that afflicts as many as 30 million men between the ages of 40 to 70 years in the United States [2].

Penile erection occurs in response to cavernous smooth muscle relaxation, increased blood flow to the penis, and restriction of venous outflow [3].

These events are regulated by extremely complex processes, which include interactions between visual, psychological, hormonal, vascular, neural, hormonal, and other factors. Dopamine, acetylcholine, nitric oxide (NO), peptides, such as oxytocin, and adrenocorticotropic/α-melanocyte–stimulating hormones, have a facilitatory role, whereas serotonin may be either facilitatory or inhibitory, and enkephalins are inhibitory. The balance between them controls the state of the penis.

The etiology of erectile dysfunction is predominantly vascular, and is currently explained by NO metabolism disturbances being in the background [4-7].

At least three neurotransmitters are capable of relaxing the cavernous smooth muscle—NO, acetylcholine, and vasoactive intestinal polypeptide, of which NO is believed to be the most important [8-10].

There are two main intracellular mechanisms for relaxing the cavernosal smooth muscle: the guanylate cyclase and adenylate cyclase pathways.

The cavernous nerves and endothelial cells contain neuronal NO synthase. They directly release NO in the penis, where it stimulates guanylyl cyclase to

Synthesis of Best-Seller Drugs. http://dx.doi.org/10.1016/B978-0-12-411492-0.00038-9

produce cyclic guanidine monophosphate (cGMP) and lowers intracellular Ca^{2+} levels. This triggers relaxation of arterial and trabecular smooth muscle, leading to arterial dilation, venous constriction, and erection.

Phosphodiesterase enzymes are widely distributed throughout the body and are found in corpus cavernosum, heart, lung, platelets, prostate, urethra, bladder, liver, brain, and stomach, having numerous effects and functions. In the penis, the actions of cGMP are curtailed primarily by phosphodiesterase type 5 (PDE-5). PDE-5 inhibitors prevent the hydrolysis of cGMP, leading to prolongation of the action of cGMP, thereby amplifying the NO signal. This enhances events leading to penile erection. In addition, NO may reduce norepinephrine release from noradrenergic nerves [8-10].

Efferent fibers within the cavernous nerves also contain acetylcholine and vasoactive intestinal polypeptide (VIP). Acetylcholine activates endothelium via binding to M3 muscarinic receptors, which leads to production of NO [11].

VIP acts through adenylate cyclase to trigger a rise in cyclic adenosine monophosphate (cAMP) [8,9]. A rise in cAMP, like a rise in cGMP, results in a fall in cytosolic Ca^{2+} in cavernous smooth muscle, which triggers relaxation of cavernous smooth muscle.

Another mechanism is involved in maintenance of the erectile process is the phosphatidylinositol 3-kinase (PI3-kinase) pathway that activates the serine/threonine protein kinase. It causes direct phosphorylation of endothelial NO synthase, reducing the enzyme's Ca^{2+} requirement and causing increased production of NO.

Penile erection also occurs through inhibition of contractile mechanisms mediated by norepinephrine (NE) and endothelins (ETs) [8], which activate α_1-adrenergic receptor and inositol triphosphate (IP3) receptors, resulting in a rise in cytosolic Ca^{2+} [12].

The pharmacology of drugs for the treatment of erectile dysfunction has been reviewed [13-23].

In summary, general strategies for treatment of erectile dysfunction could be focused on agents that raise cGMP, agents that raise cAMP, and agents that prevent IP3 formation, which are the basis of current pharmacological therapies. Moreover, drugs that raise cytosolic calcium either prevent or abort erection. Conversely, drugs that lower cytosolic calcium relax smooth muscle and can initiate penile erection.

Until recently, the system used to classify drugs used to treat erectile dysfunction was based on the route of administration—intracavernosal injections and nonintracavernosal administration. Currently, oral PDE-5 inhibitors are the drugs of choice worldwide.

38.1 DRUGS FOR INTRACAVERNOSAL ADMINISTRATION

Intracavernous agents [24,25] used to treat erectile dysfunction are peripheral initiators. They are also the first approved drugs for the treatment of erectile dysfunction.

Papaverine

Papaverine (**38.1.1**) is an alkaloid found in opium. It is a direct-acting smooth-muscle relaxant. It acts as a vasodilator, and is the first clinically effective drug for the treatment of erectile dysfunction. Papaverine is administered by intracavernosal injections and has a very complex mode of action. It is regarded as "multilevel acting drug" [24-27] (Fig. 38.1.).

Phentolamine

Phentolamine (**38.1.2**) is a competitive α-adrenergic receptor antagonist with similar affinity for α_1- and α_2-adrenergic receptors; this is considered its main mechanism of action [10]. This drug can also block serotonin (5-HT) and inhibit release of histamine from mast cells [8]. Because intracavernosal phentolamine injection sometimes does not result in a satisfactory erection, the drug is often used in combination with other drugs. Studies on its use for the treatment of erectile dysfunction have been reviewed [28] (Fig. 38.1.).

Thymoxamine

Thymoxamine (**38.1.3**) produces erection when injected intracavernosally. It is a relatively selective α_1-adrenergic receptor antagonist with some antihistamine activity. Thymoxamine produces erections of moderate quality, but it seems to be no longer used [29] (Fig. 38.1.).

Prostaglandin E_1

Prostaglandin E_1 (**38.1.4**) administered intracavernously, either alone or in combination with other drugs, is the most common drug used by this route; today it is a second-line treatment for erectile dysfunction. The actions of prostaglandin E_1 are believed to be mediated by adenylate cyclase, which causes a rise in cAMP [30,31] (Fig. 38.1.).

Linsidomine

Linsidomine (**38.1.5**), an active metabolite of the antianginal drug molsidomine, is a NO donor and believed to act by nonenzymatic liberation of NO, making it an interesting alternative [32]. However, it is no longer used therapeutically [8] (Fig. 38.1.).

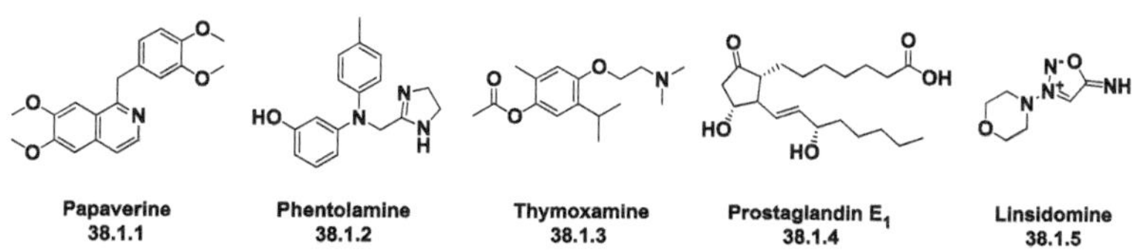

FIG. 38.1 Intracavernous agents used to treat erectile dysfunction.

38.2 DRUGS FOR NONINTRACAVERNOSAL ADMINISTRATION

Oral agents have been used for treatment of erectile dysfunction for decades.

Yohimbine

Yohimbine (**38.2.1**) is a natural peripherally and centrally acting α_2-adrenergic receptor antagonist; it is not currently recommended in most guidelines on the management of erectile dysfunction. It has been used in traditional medicine since antiquity as an aphrodisiac, but it possesses little efficacy [33] (Fig. 38.2.).

Phentolamine

Phentolamine (**38.1.2**) has had mixed results when given orally [34-37] (Fig. 38.1.).

Prostaglandin E_1

Prostaglandin E_1 (**38.1.4**) may be a reasonable treatment option for erectile dysfunction on nonintracavernosal administration [38,39] (Fig. 38.1.).

Naloxone and Naltrexone

It is well known that chronic use of narcotics leads to decreased libido and erectile dysfunction. Some sporadic reports suggest that the opiate receptor antagonists naloxone (**38.2.2**) and naltrexone (**38.2.3**) demonstrate usefulness for erectile dysfunction [40,41] (Fig. 38.2.).

Trazodone

Contradictory results have been obtained with the use of the "atypical" antidepressant trazodone (**38.2.4**). This drug selectively inhibits central 5-HT uptake and increases the turnover of brain dopamine, but does not prevent the peripheral reuptake of noradrenaline [42-44] (Fig. 38.2.).

Minoxidil

Minoxidil (**38.2.5**) is an antihypertensive agent. It is a K-channel opener that has been tried as a drug for treatment of erectile dysfunction [8,45] (Fig. 38.2.). Topically applied 2% minoxidil solution has been reported to increase diameter, rigidity and arterial flow to the penis [45].

Apomorphine

Apomorphine (**38.2.6**) is a centrally acting, nonselective dopamine agonist with more potent D_2-like effects. It acts by binding to dopaminergic receptors of hypothalamic neurons. Currently, apomorphine is under investigation in Phase II trials [46].

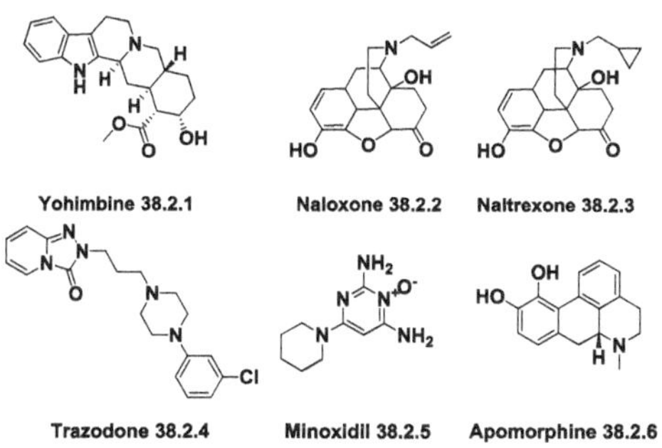

FIG. 38.2 Drugs for nonintracavernosal administration.

None of the described drugs is included in the list of Top 200 Drugs by sales for the 2010s.

38.3 PDE-5 INHIBITORS

The most important advance in the treatment of erectile dysfunction has been the development of oral PDE-5 inhibitors, which hydrolyze cGMP, resulting in smooth muscle contraction. Because of their high efficacy, they have revolutionized treatment of erectile dysfunction. Current erectile dysfunction treatment guidelines recommend them as the first-line therapy [47-59].

The first commercially available PDE-5 inhibitor was sildenafil (Viagra) (**38.3.1**) in 1998, which was followed by the worldwide launch of vardenafil (Levitra) (**38.3.2**) and tadalafil (Cialis) (**38.3.3**). More recently, udenafil (Zydena) (**38.3.4**) and avanafil (Stendra) (**38.3.5**) were approved in the United States. Mirodenafil (**38.3.6**) has been launched in South Korea. Another PDE-5 inhibitor, lodenafil (**38.3.7**), is currently under development (Fig. 38.3.).

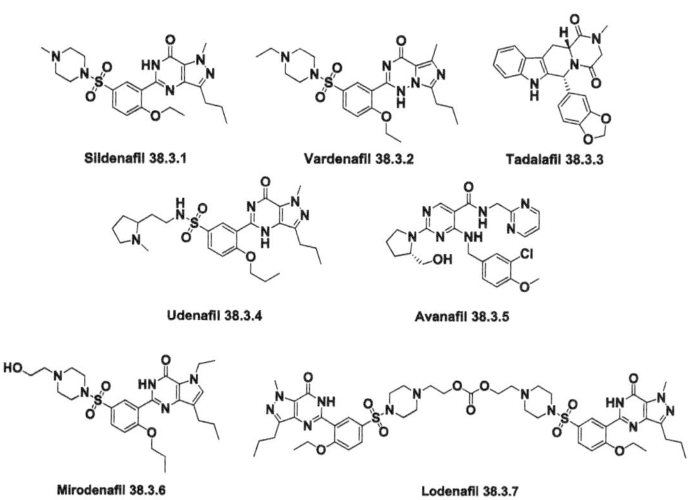

FIG. 38.3 Oral PDE-5 inhibitors.

There are subtle differences in how long each drug works and how quickly each starts to work. Vardenafil works a little longer (5 hours) than sildenafil (4 hours). They both take effect in approximately 30 minutes. Tadalafil works within approximately 15 minutes, and its effects last up to 36 hours in some cases. Avanafil can start working in as little as 15 minutes and last up to 6 hours [60,61].

Sildenafil (Viagra), vardenafil (Levitra), and tadalafil (Cialis) are included in the list of Top 200 Drugs by sales for the 2010s.

Sildenafil–Viagra

Sildenafil (Viagra) is first in class, oral drug. It's discovery started from a task to find a molecule that would bind to PDE, so that it could convert cGMP to the inactive GMP form as a potential drug for treatment hypotension and angina.

In Phase I studies, volunteers reported about some side effects, among which was the unexpected side effect of a change in erectile function. This observation changed the medical management of erectile dysfunction. Phase II trials of sildenafil started in 1994. In 1998, the FDA approved use of Viagra to treat impotence.

Sildenafil, a selective inhibitor of PDE-5, the main phosphodiesterase isoform responsible for hydrolysis of intracellular cyclic guanosine monophosphate, has been used by millions of men for the treatment of erectile dysfunction with significant improvements in erectile function [62-76].

Accounts of sildenafil use in clinical practice and postmarketing data reflect findings of effectiveness in a broad spectrum of other possible implementation areas for sildenafil [77]. Sildenafil provides a prolonged benefit in various other diseases. Promising results have been reported for patients with treated prostate cancer, end-stage renal failure, Parkinson disease, spina bifida, and in multiple organ transplant recipients. It has been shown that sildenafil has a potential therapeutic efficacy for disorders related to the central nervous system, exerting neuroprotective effects in multiple sclerosis and significant memory-enhancing action [77], plays a big role in the treatment of pulmonary hypertension [78,79], has shown great promise as a possible drug for cardioprotection [80], and increases muscle protein synthesis and reduces muscle fatigue [81]. Recently, literature regarding the use of sildenafil in treatment of female sexual dysfunction has appeared [82-84]. Side effects are rare. Most common side effects could be represented as diarrhea, dizziness, flushing, headache, mild pain, redness, or swelling.

Sildenafil also exerts a minor inhibitory action against PDE-6, which is present exclusively in rod and cone photoreceptors. At higher doses it can cause mild and transient visual symptoms in a minority of patients [85].

The first synthesis of sildenafil [86,87] have been commenced with the preparation of the pyrazole (**38.3.9**) from the ethyl 3-butyrylpyruvate (**38.3.8**) and hydrazine hydrate in ethanol followed by the selective N-methylation of the pyrazole ring with dimethyl sulfate in sodium hydroxide solution to produce (**38.3.10**).

Nitration followed by hydrolysis produced the carboxylic acid, which was transformed to carboxamide (**38.3.11**) through formation of acid chloride and then with reaction of this intermediate with concentrated ammonium hydroxide solution. The nitro group in the obtained compound was reduced to an amino group by stannous chloride/hydrochloric acid in ethanol, which produced the key pyrazole intermediate (**38.3.12**). Alkylating the prepared aminopyrazole with 2-ethoxy benzoyl chloride (**38.3.13**) in dichloromethane and cyclization of the obtained amide (**38.3.14**) under basic conditions produced pyrimidinone (**38.3.15**). The original patent proposed to cyclize the compound (**38.3.14**) with an aqueous alcoholic solution of sodium hydroxide in presence of hydrogen peroxide. Later alternative conditions for conducting this reaction (t-BuOK/t-BuOH) were proposed; the alternatives allowed avoidance of any side product and to have 100% of the product (**38.3.15**) [88]. The obtained pyrimidinone (**38.3.15**) derivative was reacted with chlorosulphonic acid at 0°C in a nitrogen atmosphere. Chlorosulphonylation proceeded selectively on the 5′ position of the phenyl ring, producing a sulfonyl chloride derivative (**38.3.16**) that was readily coupled with N-methylpiperazine (**38.3.17**), producing the sulfonamide product, sildenafil (**38.3.1**) (Scheme 38.1.).

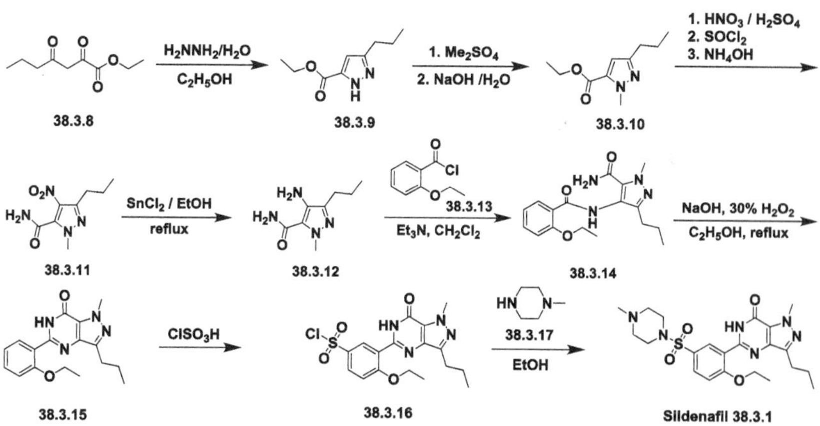

SCHEME 38.1 Synthesis of sildenafil.

An alternative, "commercial" route for synthesizing of sildenafil was proposed later [88,89]. The synthesis started from 2-ethoxy benzoyl chloride (**38.3.13**), which was chlorosulphonated with chlorosulphonic acid, although it was essential to add thionyl chloride to ensure the intermediate sulphonic acid was converted to the sulphonyl chloride. The intermediate product acid chloride was chilled with water to produce 5-(chlorosulfonyl)-2-ethoxybenzoic acid, which reacted with N-methylpiperazine (**38.3.17**) in water solution of NaOH, producing, after pH adjustment, 2-ethoxy-5-((4-methylpiperazin-1-yl) sulfonyl)benzoic acid sulfonamide (**38.3.19**). Coupling of the obtained product with key pyrazole intermediate (**38.3.12**), during the synthesis of which the $SnCl_2$ reduction of the nitro group (**38.3.11**) [(**38.3.11** → **38.3.12**)] was replaced with palladium, catalyzed hydrogenation was realized using as a coupling reagent 1,1′-carbonyldiimidazole (**38.3.20**). The final bond forming the cyclization reaction was performed using potassium t-butoxide at reflux in t-butanol, producing the desired high-quality sildenafil (**38.3.19**) (Scheme 38.2.).

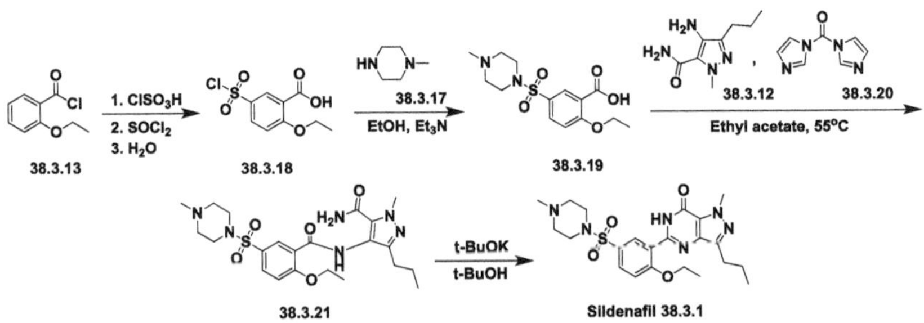

SCHEME 38.2 **Synthesis of sildenafil.**

Other minor modifications of this general approach are proposed in the literature [90-95].

Vardenafil-Levitra

Vardenafil (**38.3.2**) is another selective PDE-5 inhibitor that is approved for the treatment of erectile dysfunction. It is effective in a broad spectrum of underlying conditions of patients. Vardenafil potentiates the increase in intracellular cGMP in the corpora cavernosa in response to sexual stimuli, resulting in enhanced and prolonged erections. The overall tolerability and safety profile is acceptable. The major side effects observed by some patients are headache, flushing, rhinitis, and dyspepsia [96-107]. Vardenafil is considered five to 10 times more potent than sildenafil.

Structurally, vardenafil (**38.3.2**) is very similar to sildenafil (**38.3.1**) and differs in that the pyrazolo-pyrimidine heterocycle is replaced by an isosteric imidazo-triazine core.

The first synthesis reported is based on 2-hydroxybenzonitrile (**38.3.22**), the hydroxyl group of which was alkylated with ethyl bromide in standard conditions (K_2CO_3, acetone, reflux) to produce the intermediate-2-ethoxybenzonitrile, which was converted to the amidine (**38.3.23**) by reaction with $AlMeClNH_2$, which was generated in situ by reacting NH_4Cl with $AlMe_3$. The obtained amidine (**38.3.23**) was converted to amidrazone (**38.3.24**) by reaction with hydrazine hydrate.

In parallel, D,L-alanine (**38.3.25**) was N-acylated with butyryl chloride using in situ trimethyl silyl protection of acid group with trimethyl silyl chloride (TMSCl) to prepare butyrylalanine (**38.3.26**), which was subjected to a Dakin-West reaction with ethoxy oxalylchloride in THF using pyridine as a base and 4-dimethylaminopyridine as a catalyst, which produced the second intermediate-mixed acid anhydride (**38.3.27**). On reflux in ethanol, the mixture of amidrazone (**38.3.24**) and prepared mixed acid anhydride (**38.3.27**) produced a condensation product (**38.3.28**). Dehydration–cyclization reaction of the last with $POCl_3$ produced imidazolotriazinone (**38.3.29**), which was subjected to chlorosulphonation to produce (**38.3.30**) and then followed by sulfonamide formation in a sequence analogous to that implemented in the synthesis of sildenafil, to produce the desired vardenafil (**38.3.2**). [108,109] (Scheme 38.3.).

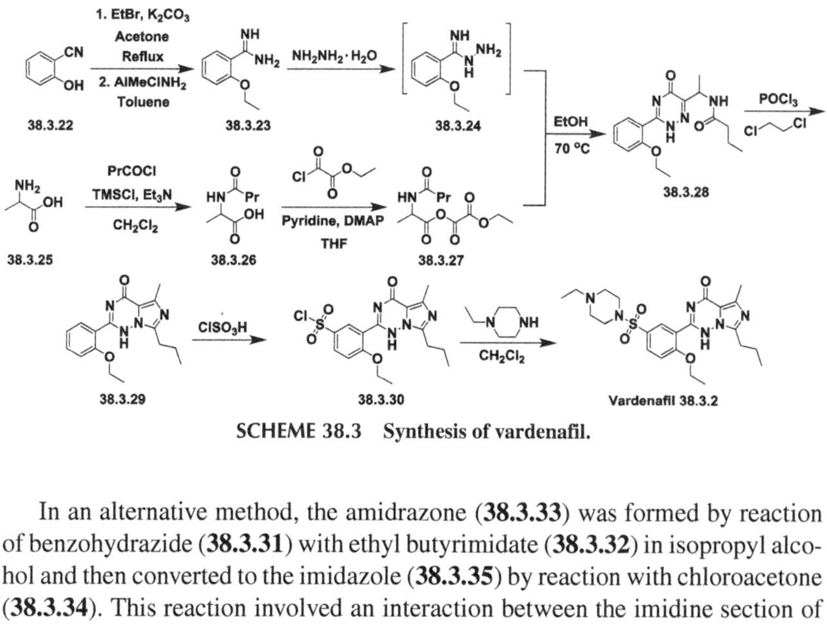

SCHEME 38.3 Synthesis of vardenafil.

In an alternative method, the amidrazone (**38.3.33**) was formed by reaction of benzohydrazide (**38.3.31**) with ethyl butyrimidate (**38.3.32**) in isopropyl alcohol and then converted to the imidazole (**38.3.35**) by reaction with chloroacetone (**38.3.34**). This reaction involved an interaction between the imidine section of amidrazone (**38.3.33**) with α-chloroketone (**38.3.34**). The obtained imidazole was then brominated with Br_2 in acetic acid to produce the bromo derivative (**38.3.36**), which was converted to the nitrile (**38.3.37**) by reaction with copper (I) cyanide. The obtained nitrile (**38.3.37**) was cyclized to the requested imidazolotriazinone (**38.3.29**) during hydrolysis of the nitrile group on heating in 48% sulfuric acid at 70°C. The imidazolotriazinone (**38.3.29**) underwent sulfochlorination, and sulfonamide formation with 1-ethylpiperazine in sequence by the same route which had been used in the previous approach (Scheme 38.3), giving an opportunity for multikilogram industrial scale production of vardenafil (**38.3.2**) [110] (Scheme 38.4.).

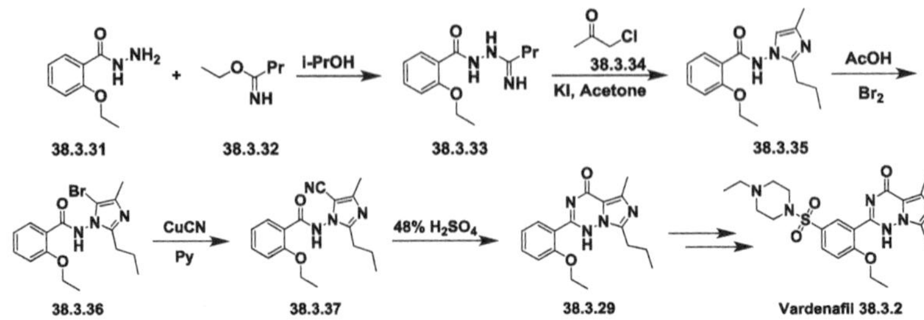

SCHEME 38.4 Synthesis of vardenafil.

Further improvements of described approaches have been proposed [111-113] and reviewed [91].

Tadalafil–Cialis

In 1994, while conducting cardiovascular testing of an experimental heart medication, a new PDE-5 enzyme inhibitor IC351, researchers observed an unusual symptom: the male test subjects got unexplained erections when on the drug. This "accident" led to the discovery of tadalafil (**38.3.3**), a new compound for the treatment of male erectile dysfunction, which differs structurally from sildenafil and vardenafil.

Tadalafil is a new potent and selective PDE-5 inhibitor–a derivative of β-carboline scaffold with fused pyrazino-1,4-dione moiety—that is very different from sildenafil and vardenafil, and provides effective treatment for erectile dysfunction [114-125].

Tadalafil was approved by FDA in 2003 to not only treat erectile dysfunction, but to also treat the symptoms of benign prostatic hyperplasia, which include difficulty urinating, weak stream, incomplete bladder emptying, painful urination, and urinary frequency and urgency in adult men [126-128]. Moreover, in 2009, it was approved for the treatment of pulmonary arterial hypertension [129,130].

Tadalafil (**38.3.3**) is the newest and most versatile PDE-5 inhibitor in the clinical armamentarium. Its most unique characteristic for the treatment of erectile dysfunction is its long half-life. It became nicknamed "the weekend pill" because its effects last up to 36 hours, whereas Viagra and Levitra are effective for up to 8 hours.

The general synthesis of tadalafil (**38.3.3**) is based on the Pictet-Spengler reaction between racemic tryptophan ester (**38.3.38**) and piperonyl (**38.3.39**) in modified Pictet-Spengler reaction conditions (CH_2Cl_2, CF_3COOH, room temperature) to prepare a mixture of *cis*- and *trans*-isomers of β-carbolines (**38.3.40 a,b**) which can be separated by flash chromatography or crystallization. Acylation of the *cis*-carboline (**38.3.40a**) in chloroform with chloroacetyl chloride in the presence of triethylamine or sodium hydrogen carbonate as base provided the *cis*-chloroacetyl derivative (**38.3.41**). Further treatment of (**38.3.41**) with an excess of 33% solution of methylamine in methanol under reflux produced the desired tadalafil (**38.3.3**) [131-134] (Scheme 38.5.).

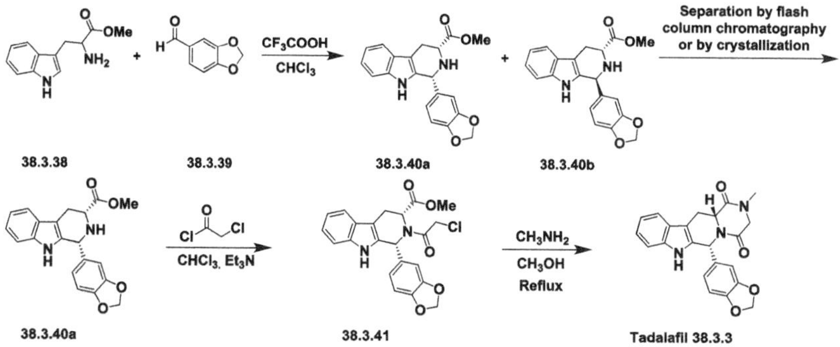

SCHEME 38.5 **Synthesis of tadalafil.**

In an alternative way, D-tryptophan methyl ester (**38.3.38-D**) was acylated with piperonyloyl chloride (**38.3.42**) in dichloromethane and in presence of trimethylamine to produce an amide (**38.3.43**). The amide (**38.3.43**) was heated in dimethoxyethane with Lawesson's reagent (2,4-bis(4-methoxyphenyl)-1,3,2,4-dithiadiphosphetane-2,4-disulfide) to produce thioamide (**38.3.44**). A solution of obtained thioamide (**38.3.44**) and methyl iodide in dichloromethane was heated to form intermediate methyl benzimidothioate (**38.3.45**), which spontaneously underwent cyclocondensation, with the loss of methanethiol producing a cyclic imine-4,9-dihydro-3H-pyrido[3,4-b]indole derivative (**38.3.46**). The imine double bond in the obtained compound was reduced with sodium borohydride, producing *cis*-β-carboline (**38.3.40a**), which was described in Scheme 38.5. Applying the same sequence of reactions with chloroacetyl chloride and then with methylamine, produced the desired tadalafil (**38.3.3**) [132] (Scheme 38.6.).

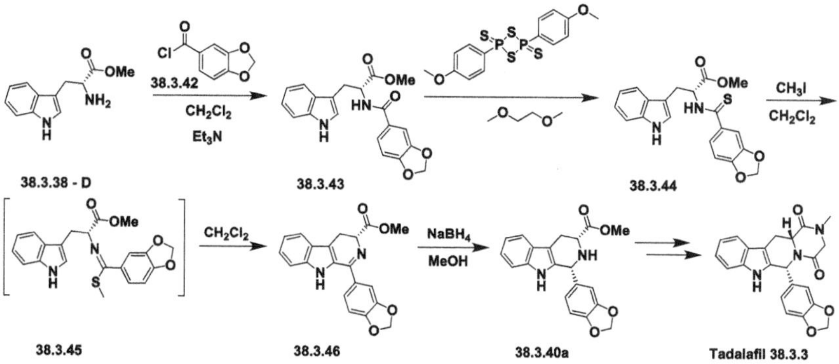

SCHEME 38.6 **Synthesis of tadalafil.**

Most of other reported syntheses [135-144] are all variations of the above two routes, which are well-reviewed by Dunn [91].

Currently, at least 46 different experimental analogues are reported and more are expected [145].

REFERENCES

1. NIH Consensus Conference. Impotence. NIH Consensus Development Panel on Impotence. *JAMA, J. Am. Med. Assoc.* **1993,** *270* (1), 83–90.
2. Lewis, R. W. Epidemiology of erectile dysfunction. *Urol. Clin. North Am.* **2001,** *28,* 209–216.
3. Steers, W. D. Pharmacologic treatment of erectile dysfunction. *Rev. Urol.* **2002,** *4* (Suppl. 3), S17–S25.
4. Ignarro, L. J.; Bush, P. A.; Buga, G. M.; Wood, K. S.; Fukuto, J. M.; Rajfer, J. Nitric oxide and cyclic GMP formation upon electrical field stimulation cause relaxation of corpus cavernosum smooth muscle. *Biochem. Biophys. Res. Commun.* **1990,** *170* (2), 843–850.
5. Leite, R.; Giachini, F. R. C.; Carneiro, F. S.; Nunes, K. P.; Tostes, R. C.; Webb, R. C. Targets for the treatment of erectile dysfunction: is NO/cGMP still the answer? *Recent Pat. Cardiovasc. Drug Discovery* **2007,** *2* (2), 119–132.
6. Toda, N.; Ayajiki, K.; Okamura, T. Nitric oxide and penile erectile function. *Pharmacol. Ther.* **2005,** *106* (2), 233–266.
7. Burnett, A. L. Role of nitric oxide in the physiology of erection. *Biol. Reprod.* **1995,** *52* (3), 485–489.
8. Andersson, K. E. Pharmacology of penile erection. *Pharmacol. Rev.* **2001,** *53,* 417–450.
9. Andersson, K. E. Mechanisms of penile erection and basis for pharmacological treatment of erectile dysfunction. *Pharmacol. Rev.* **2011,** *63* (4), 811–859.
10. Andersson, K. E.; Stief, C. Oral alpha adrenoceptor blockade as a treatment of erectile dysfunction. *World J. Urol.* **2001,** *19,* 9–13.
11. Moreland, R. B.; Hsieh, G.; Nakane, M.; Brioni, J. The biochemical and neurologic basis for the treatment of male erectile dysfunction. *J. Pharmacol. Exp. Ther.* **2001,** *296,* 225–234.
12. Khan, M. A.; Calvert, R. C.; Sullivan, M. E.; Thompson, C. S.; Mumtaz, F. H.; Morgan, R. J.; Mikhailidis, D. P. Normal and pathological erectile function: the potential clinical role of endothelin-1 antagonists. *Curr. Drug Targets* **2000,** *1,* 247–260.
13. Lue, T. F. Erectile dysfunction. *N. Engl. J. Med.* **2000,** *342* (24), 1802–1813.
14. Shamloul, R.; Ghanem, H. Erectile dysfunction. *Lancet* **2013,** *381* (9861), 153–165.
15. Decaluwe, K.; Pauwels, B.; Boydens, C.; Van de Voorde, J. Treatment of erectile dysfunction: new targets and strategies from recent research. *Pharmacol. Biochem. Behav.* **2014,** *121,* 146–157.
16. Andersson, K.-E. Mechanisms of penile erection and basis for pharmacological treatment of erectile dysfunction. *Pharmacol. Rev.* **2011,** *63* (4), 811–859.
17. Hatzimouratidis, K.; Hatzichristou, D. G. A comparative review of the options for treatment of erectile dysfunction: which treatment for which patient? *Drugs* **2005,** *65* (12), 1621–1650.
18. Feifer, A.; Carrier, S. Pharmacotherapy for erectile dysfunction. *Expert Opin. Invest. Drugs* **2008,** *17* (5), 679–690.
19. Hatzimouratidis, K.; Hatzichristou, D. G. Looking to the future for erectile dysfunction therapies. *Drugs* **2008,** *68* (2), 231–250.
20. Albersen, M.; Shindel, A. W.; Mwamukonda, K. B.; Lue, T. F. The future is today: emerging drugs for the treatment of erectile dysfunction. *Expert Opin. Emerg. Drugs* **2010,** *15* (3), 467–480.

21. Alwaal, A.; Al-Mannie, R.; Carrier, S. Future prospects in the treatment of erectile dysfunction: focus on avanafil. *Drug Des., Dev. Ther.* **2011,** *5,* 435–443.
22. Peak, Taylor C.; Yafi, Faysal A.; Sangkum, Premsant; Hellstrom, Wayne J. G. Emerging drugs for the treatment of erectile disfunction. *Exp. Opin. Emerging Drugs* **2015,** *20* (2), 263–275.
23. Benet, A. E.; Rehman, J.; Melman, A. The medical treatment of erectile dysfunction. *Drugs Today* **1996,** *32* (6), 483–499.
24. Bella, A. J.; Brock, G. B. Intracavernous pharmacotherapy for erectile dysfunction. *Endocrine* **2004,** *23* (2–3), 149–155.
25. Ledda, A. Erectile dysfunction: intracavernous treatment. *Curr. Med. Res. Opin.* **2001,** *16* (Suppl. 1), S59–S62.
26. Kuchler, J. Erectile impotence, drug therapy with papaverine—a review of the literature. *Fortschr. Med.* **1988,** *106* (23), 481–484.
27. Andersson, K. E. Pharmacology of erection: agents which initiate and terminate erection. *Sex. Disabil.* **1994,** *12,* 53–79.
28. Graul, A.; Castaner, J. Phentolamine mesylate (Regitine, Vasomax, Z-Max): treatment of erectile dysfunction. *Drugs Future* **1998,** *23* (7), 725–728.
29. Brindley, G. S. Pilot experiments on the actions of drugs injected into the human corpus cavernosum penis. *Br. J. Pharmacol.* **1986,** *87,* 495–500.
30. Albersen, M.; Mwamukonda, K. B.; Shindel, A. W.; Lue, T. F. Evaluation and treatment of erectile dysfunction. *Med. Clin. North Am.* **2011,** *95,* 201–212.
31. Alexandre, B.; Lemaire, A.; Desvaux, P.; Amar, E. Intracavernous injections of, prostaglandin E1 for erectile dysfunction: patient satisfaction and quality of sex life on long-term treatment. *J. Sex. Med.* **2007,** *4,* 426–431.
32. Truss, M. C.; Becker, A. J.; Djamilian, M. H.; Stief, C. G.; Jonas, U. Role of the nitric oxide donor linsidomine chlorhydrate (SIN-1) in the diagnosis and treatment of erectile dysfunction. *Urology* **1994,** *44* (4), 553–556.
33. Morales, A. Yohimbine in erectile dysfunction: the facts. *Int. J. Impotence Res.* **2000,** *12* (Suppl. 1), S70–S74.
34. Goldstein, I.; Carson, C.; Rosen, R.; Islam, A. Vasomax for the treatment of male erectile dysfunction. *World J. Urol.* **2001,** *19* (1), 51–56.
35. Brock, G. Oral phentolamine (Vasomax). *Drugs Today* **2000,** *36* (2–3), 121–124.
36. Zorgniotti, A. W. "On demand" erection with oral preparations for impotence: 3-(N-(2-imidazoline-2-ylmethyl)-p-toluidinol) phenol mesylate. *Int. J. Impotence Res.* **1992,** *4* (Suppl. 2), A99.
37. Zorgniotti, A. W. Experience with buccal phentolamine mesylate for impotence. *Int. J. Impotence Res.* **1994,** *6,* 37–41.
38. Jaffe, J. S.; Antell, M. R.; Greenstein, M.; Ginsberg, P. C.; Mydlo, J. H.; Harkaway, R. C. Use of intraurethral alprostadil in patients not responding to sildenafil citrate. *Urology* **2004,** *63,* 951–954.
39. Urciuoli, R.; Cantisani, T. A.; Carlini, M.; Giuglietti, M.; Botti, F. M. Prostaglandin E1 for treatment of erectile dysfunction. *Cochrane Database Syst. Rev.* **2004,** (2), CD 001784.
40. Sathe, R. S.; Komisaruk, B. R.; Ladas, A. K.; Godbole, S. V. Naltrexone-induced augmentation of sexual response in men. *Arch. Med. Res.* **2001,** *32* (3), 221–226.
41. Brennemann, W.; Stitz, B.; Van Ahlen, H.; Brensing, K. A.; Klingmuller, D. Treatment of idiopathic erectile dysfunction in men with the opiate antagonist naltrexone—a double-blind study. *J. Androl.* **1993,** *14* (6), 407–410.
42. Lance, R.; Albo, M.; Costabile, R. A.; Steers, W. D. Oral trazodone as empirical therapy for erectile dysfunction: a retrospective review. *Urology* **1995,** *46,* 117–120.

43. Georgotas, A.; Forsell, T. L.; Mann, J. J.; Kim, M.; Gershon, S. Trazodone hydrochloride: a wide spectrum antidepressant with a unique pharmacological profile. *A review of its neurochemical effects, pharmacology, clinical efficacy, and toxicology, Pharmacotherapy* **1982,** *2,* 255–265.
44. Fink, H. A.; Macdonald, R.; Rutks, I. R.; Wilt, T. J. Trazodone for erectile dysfunction: a systematic review and meta-analysis. *BJU Int.* **2003,** *92* (4), 441–446.
45. Radomski, S. B.; Herschorn, S.; Rangaswamy, S., Topical minoxidil in the treatment of male erectile dysfunction. *J. Urol.* **1994,** *151* (5), 1225–1226.
46. Albersen, M.; Shindel, A. W.; Lue, T. F. Sexual dysfunction in the older man. *Rev. Clin. Gerontol.* **2009,** *19* (4), 237–248.
47. Corbin, J. D.; Francis, S. H. Pharmacology of phosphodiesterase-5 inhibitors. *Int. J. Clin. Pract.* **2002,** *56* (6), 453–459.
48. Rotella, D. P. Phosphodiesterase 5 inhibitors: current status and potential applications. *Nat. Rev. Drug Discovery* **2002,** *1* (9), 674–682.
49. Gresser, U.; Gleiter, C. H. Erectile dysfunction: comparison of efficacy and side effects of the PDE-5 inhibitors sildenafil, vardenafil and tadalafil review of the literature. *Eur. J. Med. Res.* **2002,** *7* (10), 435–446.
50. Stamford, A. W. Phosphodiesterase 5 inhibitors. *Annu. Rep. Med. Chem.* **2002,** *37,* 53–64.
51. Gibson, A. Phosphodiesterase 5 inhibitors and nitrergic transmission-from zaprinast to sildenafil. *Eur. J. Pharmacol.* **2001,** *411* (1/2), 1–10.
52. Gupta, M.; Kovar, A.; Meibohm, B. The clinical pharmacokinetics of phosphodiesterase-5 inhibitors for erectile dysfunction. *J. Clin. Pharmacol.* **2005,** *45* (9), 987–1003.
53. Sandner, P.; Huetter, J.; Tinel, H.; Ziegelbauer, K.; Bischoff, E. PDE5 inhibitors beyond erectile dysfunction. *Int. J. Impotence Res.* **2007,** *19* (6), 533–543.
54. Bruzziches, R.; Francomano, D.; Gareri, P.; Lenzi, A.; Aversa, A. An update on pharmacological treatment of erectile dysfunction with phosphodiesterase type 5 inhibitors. *Expert Opin. Pharmacother.* **2013,** *14* (10), 1333–1344.
55. Palit, V.; Eardley, I. An update on new oral PDE5 inhibitors for the treatment of erectile dysfunction. *Nat. Rev. Urol.* **2010,** *7* (11), 603–609.
56. Dorsey, P.; Keel, C.; Klavens, M.; Hellstrom, W. J. G. Phosphodiesterase type 5 (PDE5) inhibitors for the treatment of erectile dysfunction. *Expert Opin. Pharmacother.* **2010,** *11* (7), 1109–1122.
57. Aversa, A.; Bruzziches, R.; Pili, M.; Spera, G. Phosphodiesterase 5 inhibitors in the treatment of erectile dysfunction. *Curr. Pharm. Des.* **2006,** *12* (27), 3467–3484.
58. Rosen, R. C.; Kostis, J. B. Overview of phosphodiesterase 5 inhibition in erectile dysfunction. *Am. J. Cardiol.* **2003,** *92* (9A), 9M–18M.
59. Eros, D.; Szantai-Kis, C.; Kiss, R.; Keri, Gy.; Hegymegi-Barakonyi, B.; Kovesdi, I.; Orfi, L. Structure–activity relationships of PDE5 inhibitors. *Curr. Med. Chem.* **2008,** *15* (16), 1570–1585.
60. Wright, P. J. Comparison of phosphodiesterase type 5 (PDE5) inhibitors. *Int. J. Clin. Pract.* **2006,** *60* (8), 967–975.
61. Gresser, U.; Gleiter, C. H. Erectile dysfunction: comparison of efficacy and side effects of the PDE-5 inhibitors sildenafil, vardenafil and tadalafil review of the literature. *Eur. J. Med. Res.* **2002,** *7* (10), 435–446.
62. Ghofrani, H. A.; Osterloh, I. H.; Grimminger, F. Sildenafil: from angina to erectile dysfunction to pulmonary hypertension and beyond. *Nat. Rev. Drug Discovery* **2006,** *5* (8), 689–702.
63. Kukreja, R. C.; Salloum, F.; Das, A.; Ockaili, R.; Yin, C.; Bremer, Y. A.; Fisher, P. W.; Wittkamp, M.; Hawkins, J.; Chou, E.; Kukreja, A. K.; Wang, X.; Marwaha, V. R.; Xi, L. Pharmacological preconditioning with sildenafil: basic mechanisms and clinical implications. *Vasc. Pharmacol.* **2005,** *42* (5-6), 219–232.

64. Langtry, H. D.; Markham, A. Sildenafil: a review of its use in erectile dysfunction. *Drugs* **1999,** *57* (6), 967–989.
65. Moreland, R. B.; Goldstein, I.; Kim, N. N.; Traish, A. Sildenafil citrate, a selective phosphodiesterase type 5 inhibitor: research and clinical implications in erectile dysfunction. *Trends Endocrinol. Metab.* **1999,** *10* (3), 97–104.
66. Salonia, A.; Rigatti, P.; Montorsi, F. Sildenafil in erectile dysfunction: a critical review. *Curr. Med. Res. Opin.* **2003,** *19* (4), 241–262.
67. Morgentaler, A.; Goldstein, I. Oral sildenafil in the treatment of erectile dysfunction. *J. Sex. Med.* **2014,** *11* (suppl. 2), 117–119.
68. Leoni, L. A. B.; Leite, G. S.; Wichi, R. B.; Rodrigues, B. Sildenafil: two decades of benefits or risks? *Aging Male* **2013,** *16* (3), 85–91.
69. Fenig, D. M.; McCullough, A. Sildenafil in the treatment of erectile dysfunction. *Aging Health* **2007,** *3* (3), 295–303.
70. Padma-Nathan, H. Sildenafil citrate (Viagra) treatment for erectile dysfunction: an updated profile of response and effectiveness. *Int. J. Impotence Res.* **2006,** *18* (5), 423–431.
71. Francis, S. H.; Corbin, J. D. Sildenafil: efficacy, safety, tolerability and mechanism of action in treating erectile dysfunction. *Expert Opin. Drug Metab. Toxicol.* **2005,** *1* (2), 283–293.
72. Jackson, G.; Gillies, H.; Osterloh, I. Past, present, and future: a 7-year update of Viagra (sildenafil citrate). *Int. J. Clin. Pract.* **2005,** *59* (6), 680–691.
73. Sadovsky, R.; Miller, T.; Moskowitz, M.; Hackett, G. Three-year update of sildenafil citrate (Viagra) efficacy and safety. *Int. J. Clin. Pract.* **2001,** *55* (2), 115–128.
74. Boyce, E. G.; Umland, E. M. Sildenafil citrate: a therapeutic update. *Clin. Ther.* **2001,** *23* (1), 2–23.
75. Brock, G. Sildenafil citrate (Viagra). *Drugs Today* **2000,** *36* (2–3), 125–134.
76. Cartledge, J.; Eardley, I. Sildenafil. *Expert Opin. Pharmacother.* **1999,** *1* (1), 137–147.
77. Uthayathas, S.; Karuppagounder, S. S.; Thrash, B. M.; Parameshwaran, K.; Suppiramaniam, V.; Dhanasekaran, M. Versatile effects of sildenafil: recent pharmacological applications. *Pharmacol. Rep.* **2007,** *59* (2), 150–163.
78. Barnett, C. F.; Machado, R. F. Sildenafil in the treatment of pulmonary hypertension. *Vasc. Health Risk Manage.* **2006,** *2* (4), 411–422.
79. Hemnes, A. R.; Champion, H. C. Sildenafil, a PDE5 inhibitor, in the treatment of pulmonary hypertension. *Expert Rev. Cardiovasc. Ther.* **2006,** *4* (3), 293–300.
80. Raja, S. G. Cardioprotection with sildenafil: implications for clinical practice. *Curr. Med. Chem.* **2006,** *13* (26), 3155–3164.
81. Sheffield-Moore, M.; Wiktorowicz, J. E.; Soman, K. V.; Danesi, C. P.; Kinsky, M. P.; Dillon, E. L.; Randolph, K. M.; Casperson, S. L.; Gore, D. C.; Horstman, A. M.; Lynch, J. P.; Doucet, B. M.; Mettler, J. A.; Ryder, J. W.; Ploutz-Snyder, L. L.; Hsu, J. W.; Jahoor, F.; Jennings, K.; White, G. R.; McCammon, S. D.; Durham, W. J. Sildenafil increases muscle protein synthesis and reduces muscle fatigue. *Clin. Translat. Sci.* **2013,** *6* (6), 463–468.
82. Brown, D. A.; Kyle, J. A.; Ferrill, M. J. Assessing the clinical efficacy of sildenafil for the treatment of female sexual dysfunction. *Ann. Pharmacother.* **2009,** *43* (7), 1275–1285.
83. Schoen, C.; Bachmann, G. Sildenafil citrate for female sexual arousal disorder: a future possibility? *Nat. Rev. Urol.* **2009,** *6* (4), 216–222.
84. Shields, K. M.; Hrometz, S. L. Use of sildenafil for female sexual dysfunction. *Ann. Pharmacother.* **2006,** *40* (5), 931–934.
85. Laties, A. M.; Zrenner, E. Viagra (sildenafil citrate) and ophthalmology. *Prog. Retinal Eye Res.* **2002,** *21* (5), 485–506.
86. Bell, A. S.; Brown, D.; Terrett, N. K. Preparation of pyrazolo[4,3-d]pyrimidin-7-ones as cardiovascular agents, EP 463756 (1992).

87. Terrett, N. K.; Bell, A. S.; Brown, D.; Ellis, P. Sildenafil (Viagra), a potent and selective inhibitor of type 5 cGMP phosphodiesterase with utility for the treatment of male erectile dysfunction. *Bioorg. Med. Chem. Lett.* **1996,** *6* (15), 1819–1824.
88. Dale, D. J.; Dunn, P. J.; Golightly, C.; Hughes, M. L.; Levett, P. C.; Pearce, A. K.; Searle, P. M.; Ward, G.; Wood, A. S. The chemical development of the commercial route to sildenafil: a case history. *Org. Process Res. Dev.* **2000,** *4* (1), 17–22.
89. Dunn, P. J.; Wood, A. S. Process for preparation of sildenafil by cyclization, EP 812845 (1997).
90. Dunn, P. J. The chemical development of the commercial route to sildenafil citrate. In *Process Chemistry in the Pharmaceutical Industry;* Gadamasetti, K., Braish, T., Eds.; Challenges in an Ever Changing Climate, Vol. 2, CRC Press, 2008; pp 267–277.
91. Dunn, P. J. Synthesis of commercial phosphodiesterase (V) inhibitors. *Org. Process Res. Dev.* **2005,** *9* (1), 88–97.
92. Dunn, P. J.; Galvin, S.; Hettenbach, K. The development of an environmentally benign synthesis of sildenafil citrate (Viagra) and its assessment by Green Chemistry metrics. *Green Chem.* **2004,** *6* (1), 43–48.
93. Patel, P. V.; Joshi, N.; Panchal, D. P. Process for preparation of 5-(2-ethoxy-5-((4-methylpiperazin-1-yl)sulfonyl)phenyl)-3-isobutyl-1-methyl-1H-pyrazolo[4,3-d]pyrimidin-7(6H)-one (sildenafil citrate impurity). *Heterocycl. Lett.* **2012,** *2* (4), 467–478.
94. Kumar, I. V. S.; Ramanjaneyulu, G. S.; Bindu, V. H. Synthesis of sildenafil citrate and process related impurities. *Lett. Org. Chem.* **2011,** *8* (9), 668–673.
95. Hariharakrishnan, V.; Rao, B. V.; Singh, A. N.; Prasad, K. H.; Kumar, A. P. Synthesis of sildenafil using a new intermediate. *Indian J. Heterocycl. Chem.* **2006,** *16* (2), 201–202.
96. Keating, G. M.; Scott, L. J. Vardenafil: a review of its use in erectile dysfunction. *Drugs* **2003,** *63* (23), 2673–2703.
97. Reffelmann, T.; Kloner, R. A. Vardenafil: a selective inhibitor of phosphodiesterase-5 for the treatment of erectile dysfunction. *Expert Opin. Pharmacother.* **2007,** *8* (7), 965–974.
98. Martin-Morales, A.; Rosen, R. C. Effective treatment of erectile dysfunction with vardenafil. *Drugs Today* **2003,** *39* (1), 51–59.
99. Young, J. M. Vardenafil. *Expert Opin. Invest. Drugs* **2002,** *11* (10), 1487–1496.
100. Ormrod, D.; Easthope, S. E.; Figgitt, D. P. Vardenafil. *Drugs Aging* **2002,** *19* (3), 217–227.
101. Hatzimouratidis, K.; Hatzichristou, D. G. Vardenafil in the treatment of erectile dysfunction: a review of clinical data. *Aging Health* **2005,** *1* (3), 367–377.
102. Montorsi, F.; Salonia, A.; Briganti, A.; Barbieri, L.; Zanni, G.; Suardi, N.; Cestari, A.; Montorsi, P.; Rigatti, P. Vardenafil for the treatment of erectile dysfunction: a critical review of the literature based on personal clinical experience. *Eur. Urol.* **2005,** *47* (5), 612–621.
103. Kendirci, M.; Bivalacqua, T. J.; Hellstrom, W. J. G. Vardenafil: a novel type 5 phosphodiesterase inhibitor for the treatment of erectile dysfunction. *Expert Opin. Pharmacother.* **2004,** *5* (4), 923–932.
104. Markou, S.; Perimenis, P.; Gyftopoulos, K.; Athanasopoulos, A.; Barbalias, G. Vardenafil (Levitra) for erectile dysfunction: a systematic review and meta-analysis of clinical trial reports. *Int. J. Impotence Res.* **2004,** *16* (6), 470–478.
105. Arcaniolo, D.; Imbimbo, C.; Grillo, M.; Fusco, F. Vardenafil for the treatment of erectile dysfunction. *Clin. Pract.* **2012,** *9* (1), 39–48.
106. Rice, K. R.; Dean, R. C. Vardenafil in erectile dysfunction: the evidence of its therapeutic value. *Clin. Med. Insights: Ther.* **2010,** *2,* 965–975.
107. Morales, A. M.; Mirone, V.; Dean, J.; Costa, P. Vardenafil for the treatment of erectile dysfunction: an overview of the clinical evidence. *Clin. Interventions Aging* **2009,** *4,* 463–472.

108. Niewohner, U.; Es-Sayed, M.; Haning, H.; Schenke, T.; Schlemmer, K-H.; Keldenich, J.; Bischoff, E.; Perzborn, E.; Dembowsky, K.; Serno, P.; Nowakowski, M. Preparation of 2-phenylimidazotriazinones as phosphodiesterase inhibitors, WO 9924433 (1999).
109. Haning, H.; Niewohner, U.; Schenke, T.; Es-Sayed, M.; Schmidt, G.; Lampe, T.; Bischoff, E. Imidazo[5,1-f]triazin-4(3H)-ones, a new class of potent PDE 5 inhibitors. *Bioorg. Med. Chem. Lett.* **2002,** *12* (6), 865–868.
110. Nowakowski, M.; Vetter, A. Preparation of 2-(2-ethoxyphenyl)imidazotriazinones, DE 10063106 (2002).
111. Heim-Riether, A.; Healy, J. A novel method for the synthesis of imidazo[5,1-f][1,2,4]triazin-4(3H)-ones. *J. Org. Chem.* **2005,** *70* (18), 7331–7337.
112. Mao, Y.; Tian, G.; Liu, Z.; Shen, J.; Shen, J. An improved synthetic route for preparative process of vardenafil. *Org. Process Res. Dev.* **2009,** *13* (6), 1206–1208.
113. Sorbera, L. A.; Martin, L.; Rabasseda, X.; Castaner, J. Vardenafil. *Drugs Future* **2001,** *26* (2), 141–144.
114. Eardley, I.; Cartledge, J. Tadalafil (Cialis) for men with erectile dysfunction. *Int. J. Clin. Pract.* **2002,** *56* (4), 300–304.
115. Meuleman, E. J. H. Review of tadalafil in the treatment of erectile dysfunction. *Expert Opin. Pharmacother.* **2003,** *4* (11), 2049–2056.
116. Curran, M. P.; Keating, G. M. Tadalafil. *Drugs* **2003,** *63* (20), 2203–2212.
117. Porst, H. Tadalafil—a new phosphodiesterase-5 inhibitor. *Arzneimitteltherapie* **2004,** *22* (3), 65–70.
118. Pomerol, J. M.; Rabasseda, X. Tadalafil, a further innovation in the treatment of sexual dysfunction. *Drugs Today* **2003,** *39* (2), 103–113.
119. Coward, R. M.; Carson, C. C. Tadalafil in the treatment of erectile dysfunction. *Ther. Clin. Risk Manage.* **2008,** *4* (6), 1315–1329.
120. Washington, S. L., III.; Shindel, A. W. A once-daily dose of tadalafil for erectile dysfunction: compliance and efficacy. *Drug Des., Dev. Ther.* **2010,** *4,* 159–171.
121. Kuan, J.; Brock, G. Selective phosphodiesterase type 5 inhibition using tadalafil for the treatment of erectile dysfunction. *Expert Opin. Invest. Drugs* **2002,** *11* (11), 1605–1613.
122. Frajese, G. V.; Pozzi, F.; Frajese, G. Tadalafil in the treatment of erectile dysfunction; an overview of the clinical evidence. *Clin. Interventions Aging* **2006,** *1* (4), 439–449.
123. Bella, A. J.; Brock, G. B. Tadalafil: a clinical update. *Aging Health* **2005,** *1* (2), 203–214.
124. Bella, A. J.; Brock, G. B. Tadalafil: a comprehensive update. *J. Drug Eval.* **2004,** *2* (8), 225–246.
125. Wrishko, R.; Sorsaburu, S.; Wong, D.; Strawbridge, A.; McGill, J. Safety, efficacy, and pharmacokinetic overview of low-dose daily administration of tadalafil. *J. Sex. Med.* **2009,** *6* (7), 2039–2048.
126. Andersson, K.-E.; de Groat, W. C.; McVary, K. T.; Lue, T. F.; Maggi, M.; Roehrborn, C. G.; Wyndaele, J. J.; Melby, T.; Viktrup, L. Tadalafil for the treatment of lower urinary tract symptoms secondary to benign prostatic hyperplasia: pathophysiology and mechanism(s) of action. *Neurourol. Urodyn.* **2011,** *30* (3), 292–301.
127. Carson, C. C.; Rosenberg, M.; Kissel, J.; Wong, D. G. Tadalafil-a therapeutic option in the management of BPH-LUTS. *Int. J. Clin. Pract.* **2014,** *68* (1), 94–103.
128. Cantrell, M. A.; Baye, J.; Vouri, S. M. Tadalafil: a phosphodiesterase-5 inhibitor for benign prostatic hyperplasia. *Pharmacotherapy* **2013,** *33* (6), 639–649.
129. Frey, M. K.; Lang, I. Tadalafil for the treatment of pulmonary arterial hypertension. *Expert Opin. Pharmacother.* **2012,** *13* (5), 747–755.
130. Arif, S. A.; Poon, H. Tadalafil: a long-acting phosphodiesterase-5 inhibitor for the treatment of pulmonary arterial hypertension. *Clin. Ther.* **2011,** *33* (8), 993–1004.

131. Daugan, A. C. M.; Gellibert, F. Tetracyclic cGMP-specific phosphodiesterase inhibitors and their use in disease treatment, US 6143746 (2000).
132. Daugan, A. C. M. Preparation of pyrazinopyridoindolediones as inhibitors of cyclic guanosine 3′,5′-monophosphate specific phosphodiesterase, WO 9519978 (1995).
133. Daugan, A.; Grondin, P.; Ruault, C.; Le Monnier de Gouville, A.-C.; Coste, H.; Kirilovsky, J.; Hyafil, F.; Labaudiniere, R. The discovery of tadalafil: a novel and highly selective PDE5 inhibitor. 1: 5,6,11,11a-tetrahydro-1H-imidazo[1′,5′:1,6]pyrido[3,4-b]indole-1,3(2H)-dione analogues. *J. Med. Chem.* **2003,** *46* (21), 4525–4532.
134. Daugan, A.; Grondin, P.; Ruault, C.; Le Monnier de Gouville, A.-C.; Coste, H.; Linget, J. M.; Kirilovsky, J.; Hyafil, F.; Labaudiniere, R. The discovery of tadalafil: a novel and highly selective PDE5 inhibitor. 2: 2,3,6,7,12,12a-hexahydropyrazino[1′,2′:1,6]pyrido[3,4-b]indole-1,4-dione analogue. *J. Med. Chem.* **2003,** *46* (21), 4533–4542.
135. Maw, G. N.; Allerton, C. M. N.; Gbekor, E.; Million, W. A. Design, synthesis and biological activity of β-carboline-based type-5 phosphodiesterase inhibitors. *Bioorg. Med. Chem. Lett.* **2003,** *13* (8), 1425–1428.
136. Beghyn, T.; Hounsou, C.; Deprez, B. P. PDE5 inhibitors: an original access to novel potent arylated analogues of tadalafil. *Bioorg. Med. Chem. Lett.* **2004,** *17* (3), 789–792.
137. Revell, J. D.; Srinivasan, N.; Ganesan, A. Two concise syntheses of Cialis via the N-acyliminium Pictet-Spengler reaction. *Synlett* **2004,** (8), 1428–1430.
138. Xiao, S.; Shi, X.-X.; Xing, J.; Yan, J.-J.; Liu, S.-L.; Lu, W.-D. Synthesis of tadalafil (Cialis) from L-tryptophan. *Tetrahedron: Asymmetry* **2009,** *20* (18), 2090–2096.
139. Sajja, E.; Vetukuri, V. N. K. V. P. R.; Ningam, S. R.; Vedantham, R.; Bodepudi, R. Process for producing tadalafil, WO 2007100387 (2007).
140. Dolitzky, B-Z.; Diller, D. Process of synthesizing tadalafil, WO 2006091975 (2006).
141. Orme, W.; Sawyer, J. S.; Schultze, L. M. Preparation of pyrazino[1′,2′:1,6]pyrido[3,4-b]indole derivatives as phosphoesterase inhibitors for use as therapeutic agents, WO 2002036593 (2002).
142. Orme, M. W.; Martinelli, M. J.; Doecke, C. W.; Pawlak, J. M.; Chelius, E. C. Preparation of tetrahydro-β-carboline diastereomers by modified Pictet-Spengler reaction, WO 2004011463 (2004).
143. Crasto, A. M.; Joshi, N. S.; Pradhan, N. S. C. Process for preparation of tadalafil, WO 2007110734 (2007).
144. Lohray, B. B.; Lohray, V. B.; Patel, S. I. Process for preparing tadalafil and its intermediates, WO 2005068464 (2005).
145. Venhuis, B. J.; de Kaste, D. Towards a decade of detecting new analogues of sildenafil, tadalafil and vardeanafil in food supplements: A history, analytical aspects and health risks. *J. Pharm. Biomed. Anal.* **2012,** *69,* 196–208.

Drugs Index

Note: Page numbers with "f" indicate figures.

A

B

C

D

E

F

G

N

P

Q

R

S

T

U

V

W

X

Y

Z

Subject Index

Note: Page numbers followed by "f" indicate figures and "t" indicate tables.

A

C

D

E

F

G

H

I

J

K

L

M

N

Q

R

T

Z

FSC
www.fsc.org
MIX
Papier | Fördert
gute Waldnutzung
FSC® C083411

Kolibri 360 GmbH
Ferdinand-Jühlke-Straße 7
99095 Erfurt, Deutschland
produktsicherheit@kolibri360.de